ENERGY MANAGEMENT
HANDBOOK

EDITORIAL BOARD

ENERGY MANAGEMENT HANDBOOK

BY

WAYNE C. TURNER

SCHOOL OF INDUSTRIAL ENGINEERING AND MANAGEMENT

OKLAHOMA STATE UNIVERSITY

Published by
THE FAIRMONT PRESS, INC.
700 Indian Trail
Lilburn, GA 30247

Library of Congress Cataloging-in-Publication Data

Turner, Wayne C., 1942-
    Energy management handbook / by Wayne C. Turner. --3rd ed.
    p. cm.
    Includes bibliographical references and index.
    ISBN 0-88173-228-1
  1. Power resources--Handbooks, manuals, etc. 2. Energy
conservation--Handbooks, manuals, etc. I. Title.
TJ163.2.T87   1996   658.2--dc20   96-42134
                                                            CIP

Energy Management Handbook / By Wayne C. Turner, Third Edition.

Published by The Fairmont Press, Inc.
700 Indian Trail
Lilburn, GA 30247

Printed in the United States of America

10 9 8 7 6 5 4 3 2 1

ISBN 0-88173-228-1   FP

ISBN 0-13-728098-X   PH

While every effort is made to provide dependable information, the publisher, authors, and
editors cannot be held responsible for any errors or omissions.

Distributed by Prentice Hall PTR
Prentice-Hall, Inc.
A Simon & Schuster Company
Upper Saddle River, NJ 07458

Prentice-Hall International (UK) Limited, London
Prentice-Hall of Australia Pty. Limited, Sydney
Prentice-Hall Canada Inc., Toronto
Prentice-Hall Hispanoamericana, S.A., Mexico
Prentice-Hall of India Private Limited, New Delhi
Prentice-Hall of Japan, Inc., Tokyo
Simon & Schuster Asia Pte. Ltd., Singapore
Editora Prentice-Hall do Brasil, Ltda., Rio de Janeiro

CONTENTS

FOREWORD

The energy "roller coaster" never ceases with new turns and spirals which make for a challenging ride. Those who started on this ride in the 1970's have witnessed every event possible which certainly would make a good novel rather than an historical docudrama. In 1995 the winds of change are once again blowing and wiping away our preconceived notions about electric power.

Who would have envisioned that the stable utility infrastructure would be turned upside down? Utilities have downsized and reduced staff in the mid 1990's, utility stock prices tumbled by 30% or more and now there is the reality of retail competition. Global economic competition is creating pressures for lower electricity prices. Retail wheeling has already become a reality. The United Kingdom is implementing a common carrier electricity distribution system that allows retail customers to select from competing suppliers. Daily developments in California, Michigan, Wisconsin and other states indicate that companies must reevaluate the way they purchase power. New opportunities will certainly open up, but with every new scenario there are hidden risks. Reliability of power and how to structure a power marketing deal are just some of the new factors energy managers must evaluate. At the start of 1995 there were over 100 companies which have been granted the status of power marketers. Their role in helping customers find power at the lowest cost will become a new factor in buying power.

Needless to stay the deregulation of the electric utility marketplace is one of the milestones in the "energy roller coast ride" of the 1990's. Energy managers need to have all the tools to evaluate both supply side and demand side options. They need to know how to put into perspective new technologies as they impact energy use. They need to see the whole picture and understand how the nontechnical issues impact their decision making. Probably the most important reference source to help energy managers cope with these challenges is the *Energy Management Handbook* by Wayne C. Turner. I am pleased to have played a part in contributing to this impressive work which has guided energy managers for over a decade. This book has helped students learn the basic principles of energy management as well as shown seasoned professionals advanced energy technologies. This newly revised edition in sure to play a key role in helping energy managers meet the new challenges ahead.

Albert Thumann, P.E., C.E.M.
Executive Director
Association of Energy Engineers
Atlanta, GA

PREFACE TO THE THIRD EDITION

The First Edition lasted 10 years before the Second Edition was published; but the Second Edition lasted only 4 years. Does this mean the Second Edition was not as noteworthy as the First or does it say the Energy Management is marching on at an accelerated pace. I believe it's the latter.

Look at the change in lighting technology over the last few years (especially fluorescent lights). Fluorescent technology exists today that can provide equal lumens for slightly less than half the energy input for conventional bulbs. Other technologies are accelerating almost as rapidly. Look at the use of Thermal Energy Storage today compared to 10 years ago. Who would have thought that now we can build a system with chiller and ice storage for about the same cost as a conventional chiller and get better control with reduced humidity? This discussion could go on but the point is the TECHNOLOGY IS MATURING RAPIDLY, OPENING NEW DOORS FOR US AS ENERGY MANAGERS.

In a similar fashion, legislation, codes, and standards are changing rapidly. Just when we start to understand ASHRAE 62, the experts feel another dramatic change is needed and is being finalized as I print these words. Then, OSHA says IAQ is such a big problem that we need formal Indoor Air Quality Programs. This has been proposed and is being worked on now. EPACT 92 may very well lead to the most significant changes we energy managers have had to face. Some of these changes may be beneficial to us and some will likely not be so beneficial.

Thus, a Third Edition was needed. Some new authors, some new chapters, some dramatic revisions to old chapters, and few with little to no revisions make up the changes that go into the Third Edition. We all hope you enjoy and profit from the Third Edition. Some very talented people donated lots of time and effort to the Third Edition. With all my heart, I thank them.

Finally, you should feel great about what you do. What other discipline can claim to save money for our companies or clients, protect the environment, and save resources for future generations? Keep up the fight.

Wayne C. Turner
Stillwater, OK
March 17, 1996
(Happy St. Pat's Day)

PREFACE TO THE SECOND EDITION

Few books last ten years without a revision; but this one did. Sales have been brisk but most importantly the profession has been extremely active. For example, the Association of Energy Engineers is now an international organization with members in several countries and they have never experienced a year of declining membership. Energy consumed per dollar GNP continues to drop and energy engineers are still in high demand. Is the bloom off the energy rose? Definitely not!!!!

The profession is changing but the basics remain the same. That's why a second edition was not needed for so long. Now, however, we feel a second edition is required to bring in some new subjects and revise some old material. We have added several new chapters and have chosen to rewrite several of the older ones. New material has been added on cogeneration, thermal energy storage systems, fuels procurement, energy economics, energy management control systems, and a host of other fast changing and developing areas. We are proud of the book after this second edition and sincerely hope you enjoy and profit from using it.

I said some things about the energy crisis and about the professionals working on this crisis in the first edition. We have purposely left that Preface intact so you can see some of the changes that have taken place and yet how the basics still hold. "The more things change, the more they stay the same."

Special thanks go to Mr. Mike Gooch of South Carolina Gas & Electric who provided significant editorial comments.

Wayne C. Turner
Stillwater, Oklahoma
April 1992

PREFACE TO THE FIRST EDITION

"There is no such thing as a problem without a gift for you in its hands. You seek problems because you need their gifts." (Richard Bach, *Illusions*, Dell, New York, 1979, p. 71.)

The energy crisis is here! It is also real, substantial, and will likely be long lasting. Energy costs are rising rapidly, conventional energy supplies are dwindling, and previously secure energy sources are highly questionable. With this myriad of energy-related problems, prudent management of any organization is or soon will be initiating and conducting energy management programs.

This book is a handbook for the practicing engineer or highly qualified technician working in the area of energy management. The *Energy Management Handbook is* designed to be a practical and "stand-alone" reference. Attempts have been made to include all data and information necessary for the successful conducting and management of energy management programs. It does not, of course, contain sufficient technical or theoretical development to answer all questions on any subject; but it does provide you, the reader, with enough information to successfully accomplish most energy management activities.

Industry is responding. This is demonstrated vividly by the fact that energy to GNP ratio declined an average of 2.8% per year in the period 1973-1977. We have to ask ourselves, however, if this much has been done thus far, *how much more can we do with more effort?*

Large savings are possible. Most companies find that 5 to 15% comes easily. A dedicated program often yields 30%, and some companies have reached 40, 50, and even 60% savings. The potential for substantial cost reduction is real.

Energy management may be defined as *the judicious use of energy to accomplish prescribed objectives.* For private enterprise, these objectives are normally to ensure survival, maximize profits, and enhance competitive positions. For nonprofit organizations, survival and cost reduction are normally the objectives. This handbook is designed to help you accomplish these objectives.

Several highly dedicated professionals worked many long, arduous hours pulling this handbook together. The list is too long to repeat here, but the Associate Editors, all the authors, and many graduate students at Oklahoma State University who helped review the material all spent many hours. As Senior Editor, I am grateful and can only hope that seeing this in print justifies their efforts. With professionals like this group, we will solve all our future energy problems.

Wayne C. Turner
Stillwater, Oklahoma January 1982

Chapter 1

Introduction

DR. WAYNE C. TURNER, PROFESSOR
Director, Oklahoma Industrial Assessment Center
Oklahoma State University
Stillwater, Ok.

DR. BARNEY L. CAPEHART, PROFESSOR
Director, Florida Industrial Assessment Center
University of Florida
Gainesville, Fla.

1.1 BACKGROUND

Mr. Al Thumann, Executive Director of the Association of Energy Engineers, said it well in the Foreword. "The energy 'roller coaster' never ceases with new turns and spirals which make for a challenging ride." Those professionals who boarded the ride in the late 70's and stayed on board have experienced several ups and downs. First, being an energy manager was like being a mother, John Wayne, and a slice of apple pie all in one. Everyone supported the concept and success was around every bend. Then, the mid-80's plunge in energy prices caused some to wonder "Do we really need to continue energy management?"

Sometime in the late 80's, the decision was made. Energy management is good business but it needs to be run by professionals. The Certified Energy Manager Program of the Association of Energy Engineers became popular and started a very steep growth curve that is continuing today (January, 1996). AEE continued to grow in membership and stature.

About the same time (late 80's), the impact of the Natural Gas Policy Act began to be felt. Now, energy managers found they could sometimes save significant amounts of money by buying "spot market" natural gas and arranging transportation. About the only thing that could be done in purchasing electricity was to choose the appropriate rate schedule and optimize parameters (power factor, demand, ratchet clauses, time of use, etc.— see the Appendix on energy rate schedules). Then, the Energy Policy Act of 1992 burst upon the scene. Soon, energy managers should be able to purchase electricity from wherever the best deal can be found and wheel the electric energy through the grid. At the time of this writing, FERC is issuing orders and Notice of Proposed Regulations (NOPR's) to make all this happen. Energy managers throughout the country and even the world are watching this with great anticipation and a bit of

apprehension as a new skill must be learned.

However, EPACT's impact is further reaching. If utilities must compete with other producers of electricity, then they must be "lean and mean." As Mr. Thumann mentions in the Foreword, this means many of the Demand Side Management (DSM) and other conservation activities of the utilities are being cut or eliminated. The roller coaster ride goes on.

The Presidential Executive Orders mentioned in Chapter 20 created the Federal Energy Management Program (FEMP) to aid the federal sector in meeting federal energy management goals. The potential FEMP savings are mammoth and new professionals affiliated with Federal, as well as State and Local Governments have joined the energy manager ranks. However, as Congress changes complexion, the FEMP and even DOE itself face at best uncertain futures. The roller coaster ride continues.

As utility DSM programs shrink, while private sector businesses and the Federal Government expand their needs for energy management programs, the door is opening for the ESCOs (Energy Service Companies), Shared Savings Providers, Performance Contractors, and other similar organizations. These groups are providing the auditing, energy and economic analyses, capital and monitoring to help other organizations reduce their energy consumption and reduce their expenditures for energy services. By guaranteeing and sharing the savings from improved energy efficiency and improved productivity, both groups benefit and prosper.

Throughout it all, energy managers have proven time and time again, that energy management is cost effective. Furthermore, energy management is vital to our national security, environmental welfare, and economic productivity. This will be discussed in the next section.

1.2 THE VALUE OF ENERGY MANAGEMENT

Business, industry and government organizations have all been under tremendous economic and environmental pressures in the last few years. Being economically competitive in the global marketplace and meeting increasing environmental standards to reduce air and water pollution have been the major driving factors in most of the recent operational cost and capital cost investment decisions for all organizations. Energy management has been an important tool to help organizations meet these critical objectives for their short term survival and long-term success.

The problems that organizations face from both their individual and national perspectives include:

•   Meeting more stringent environmental quality standards, primarily related to reducing global warming and reducing acid rain.

Energy management helps improv environmental quality. For example, the primary culprit in global warming is carbon dioxide, CO_2. Equation 1.1, a balanced chemistry equation involving the combustion of methane (natural gas is mostly methane), shows that 2.75 pounds of carbon dioxide is produced for every pound of methane combusted. Thus, energy management, by reducing the combustion of methane can dramatically reduce the amount of carbon dioxide in the atmosphere and help reduce global warming. Commercial and industrial energy use accounts for about 45 percent of the carbon dioxide released from the burning of fossil fuels, and about 70 percent of the sulfur dioxide emissions from stationary sources.

$$CH_4 + 2 O_2 = CO_2 + 2 H_2O$$
$$(12 + 4*1) + 2(2*16) =$$
$$(12 + 2*16) + 2(2*1 + 16) \qquad (1.1)$$

Thus, 16 pounds of methane produces 44 pounds of carbon dioxide; or 2.75 pounds of carbon dioxide is produced for each pound of methane burned.

Energy management reduces the load on power plants as fewer kilowatt hours of electricity are needed. If a plant burns coal or fuel oil, then a significant amount of acid rain is produced from the sulphur dioxide emitted by the power plant. Acid rain problems then are reduced through energy management.

Less energy consumption means less petroleum field development and subsequent on-site pollution. Less energy consumption means less thermal pollution at power plants and less cooling water discharge. Reduced cooling requirements or more efficient satisfaction of those needs means less CFC usage and reduced ozone depletion in the stratosphere. The list could go on almost indefinitely, but the bottom line is that energy management helps improve environmental quality.

•   Becoming—or continuing to be—economically competitive in the global marketplace, which requires reducing the cost of production or services, reducing industrial energy intensiveness, and meeting customer service needs for quality and delivery times.

Significant energy and dollar savings are available through energy management. Most facilities (manufacturing plants, schools, hospitals, office buildings, etc) can save according to the profile shown in Figure 1.1. Even more savings have been accomplished by some programs.

√	Low cost activities first year or two: 5 to 15%
√	Moderate cost, significant effort, three to five years: 15 to 30%
√	Long-term potential, higher cost, more engineering: 30 to 50%

**Figure 1.1   Typical Savings
Through Energy Management**

Thus, large savings can be accomplished often with high returns on investments and rapid paybacks. Energy management can make the difference between profit and loss and can establish real competitive enhancements for most companies.

Energy management in the form of implementing new energy efficiency technologies, new materials and new manufacturing processes and the use of new technologies in equipment and materials for business and industry is also helping companies improve their productivity and increase their product or service quality. Often, the energy savings is not the main driving factor when companies decide to purchase new equipment, use new processes, and use new high-tech materials. However, the combination of increased productivity, increased quality, reduced environmental emissions, and reduced energy costs provides a powerful incentive for companies and organizations to implement these new technologies.

Total Quality Management (TQM) is another emphasis that many businesses and other organizations have developed over the last decade. TQM is an integrated approach to operating a facility, and energy cost control should be included in the overall TQM program. TQM is based on the principle that front-line employees should have the authority to make changes and other decisions at the lowest operating levels of a facility. If employees have energy management training, they can make informed decisions and recommendations about energy operating costs.

•   Maintaining energy supplies that are:
    — Available without significant

interruption, and

— Available at costs that do not fluctuate too rapidly.

Once again, the country is becoming dependent on imported oil. During the time of the 1979 oil price crisis, the U.S. was importing almost 50% of our total oil consumption. By 1995, the U.S. was again importing 50% of our consumption. Thus, the U.S. is once again vulnerable to an oil embargo or other disruption of supply. The major difference is that there is a better balance of oil supply among countries friendly to the U.S. Nonetheless, much of the oil used in this country is not produced in this country. The trade balance would be much more favorable if we imported less oil.

• Helping solve other national concerns which include:
— Need to create new jobs
— Need to improve the balance of payments by reducing costs of imported energy
— Need to minimize the effects of a potential limited energy supply interruption

None of these concerns can be satisfactorily met without having an energy efficient economy. Energy management plays a key role in helping move toward this energy efficient economy.

1.3 THE ENERGY MANAGEMENT PROFESSION

Energy management skills are important to people in many organizations, and certainly to people who perform duties such as energy auditing, facility or building management, energy and economic analysis, and maintenance. The number of companies employing professionally trained energy managers is large and growing. A partial list of job titles is given in Figure 1.2. Even though this is only a partial list, the breadth shows the robustness of the profession.

For some of these people, energy management will be their primary duty, and they will need to acquire in-

depth skills in energy analysis as well as knowledge about existing and new energy using equipment and technologies. For others—such as maintenance managers—energy management skills are simply one more area to cover in an already full plate of duties and expectations. The authors are writing this *Energy Management Handbook* for both of these groups of readers and users.

Fifteen years ago, few university faculty members would have stated their primary interest was energy management, yet today there are numerous faculty who prominently list energy management as their principal specialty. In 1995, there were at least 30 Universities throughout the country listed by DOE as Industrial Assessment Centers or Energy Analysis and Diagnostic Centers. Other Universities offer coursework and/or do research in energy management but do not have one of the above centers. Finally, several professional Journals and Magazines now publish exclusively for energy managers while we know of none that existed 15 years ago.

The Federal Energy Management Program (FEMP) started during the Bush Administration but it received a significant boost on March 8, 1994 when President Clinton issued Executive Order 12902. A brief summary of the requirements of that order is given in Figure 1.3. This program should dramatically reduce government expenditures for energy and water.

Like energy management itself, utility DSM programs have had their ups and downs. DSM efforts peaked in the late 80s and early 90s, and have since retrenched significantly in the mid-90s as utility deregulation and the movement to retail wheeling have caused utilities to reduce staff and cut costs as much as possible. This short-term cost cutting is seen by many utilities as their only way to become a competitive low-cost supplier of electric power. Once their large customers have the choice of their power supplier, they want to be able to hold on to these customers by offering rates that are competitive with other producers around the country. In the meantime, the other energy services provided by the utility are being reduced or eliminated in this corporate downsizing effort.

This reduction in electric utility incentive and re-

• Plant Energy Manager   • Building/Facility Energy Manager
• Utility Energy Auditor   • Utility Energy Analyst
• State Agency Energy Analyst   • Federal Energy Analyst
• Consulting Energy Manager   • Consulting Energy Engineer
• DSM Auditor/Manager

Figure 1.2 Typical Energy Management Job Titles

√ Reduce Energy Consumption per square foot in Federal Buildings by 10% between 1985 and 1995.

√ Reduce Energy Consumption per square foot in Federal Buildings by 20% between 1985 and 2000.

√ Reduce Energy Consumption per square foot in Federal Buildings by 30% between 1990 and 2005.

√ Reduce energy consumption in Federal Agency Industrial Facilities by 20% between 1990 and 2005.

√ Provide for federal Agency Participation in DSM services offered by utilities.

Figure 1.3 FEMP Program Objectives Summary

bate programs, as well as the reduction in customer support, has produced a gap in energy service assistance that is being met by a growing sector of equipment supply companies and energy service consulting firms that are willing and able to provide the technical and financial assistance that many organizations previously got from their local electric utility. New business opportunities and many new jobs are being created in this shift away from utility support to energy service company support. Energy management skills are extremely important in this rapidly expanding field, and even critical to those companies that are in the business of identifying energy savings and providing a guarantee of the savings results.

Thus, the future for energy management is extremely promising. It is cost effective, it improves environmental quality, it helps reduce the trade deficit, and it helps reduce dependence on foreign fuel supplies. Energy Management will continue to grow in size and importance.

1.4 SOME SUGGESTED PRINCIPLES OF ENERGY MANAGEMENT

(The material in this section is repeated verbatim from the First and Second Editions of this Handbook. Mr. Roger Sant who was then Director of the Energy Productivity Center of the Carnegie-Mellon Institute of Research in Arlington, Va., wrote this section for the First

Edition. It was unchanged for the Second Edition. Now, the Third Edition is being printed. The Principles developed in this section are still sound. Some of the number quoted may now be a little old; but the principles are still sound. Amazing, but what was right 14 years ago for energy management is still right today. The game has changed, the playing field has moved; but the Principles stay the same).

If energy productivity is an important opportunity for the nation as a whole, it is a necessity for the individual company. It represents a real chance for creative management to reduce that component of product cost that has risen the most since 1973.

Those who have taken advantage of these opportunities have done so because of the clear intent and commitment of the top executive. Once that commitment is understood, managers at all levels of the organization can and do respond seriously to the opportunities at hand. Without that leadership, the best designed energy management programs produce few results. In addition, we would like to suggest four basic principles which, if adopted, may expand the effectiveness of existing energy management programs or provide the starting point of new efforts.

The first of these is to *control the costs of the energy function or service provided, but not the Btu of energy.* As most operating people have noticed, energy is just a means of providing some service or benefit. With the possible exception of feedstocks for petrochemical production, energy is not consumed directly. It is always converted into some useful function. The existing data are not as complete as one would like, but they do indicate some surprises. In 1978, for instance, the aggregate industrial expenditure for energy was $55 billion. Thirty-five percent of that was spent for machine drive from electric motors, 29% for feedstocks, 27% for process heat, 7% for electrolytic functions, and 2% for space conditioning and light. As shown in Table 1.3, this is in blunt contrast to measuring these functions in Btu. Machine drive, for example, instead of 35% of the dollars, required only 12% of the Btu.

In most organizations it will pay to be even more specific about the function provided. For instance, evaporation, distillation, drying, and reheat are all typical of the uses to which process heat is put. In some cases it has also been useful to break down the heat in terms of temperature so that the opportunities for matching the heat source to the work requirement can be utilized.

In addition to energy costs, it is useful to measure the depreciation, maintenance, labor, and other operating costs involved in providing the conversion equipment necessary to deliver required services. These costs add as much as 50% to the fuel cost.

It is the total cost of these functions that must be

managed and controlled, not the Btu of energy. The large difference in cost of the various Btu of energy can make the commonly used Btu measure extremely misleading. In November 1979, as shown in Table 1.4, the cost of 1 Btu of electricity was nine times that of 1 Btu of steam coal.

Availabilities also differ and the cost of maintaining fuel flexibility can affect the cost of the product. And as shown before, the average annual price increase of natural gas has been almost three times that of electricity. Therefore, an energy management system that controls Btu per unit of product may completely miss the effect of the changing economics and availabilities of energy alternatives and the major differences in usability of each fuel. Controlling the total cost of energy functions is much more closely attuned to one of the principal interests of the executives of an organization—controlling costs.

A second principle of energy management is to *control energy functions as a product cost, not as a part of manufacturing or general overhead.* It is surprising how many companies still lump all energy costs into one general or manufacturing overhead account without identifying those products with the highest energy function cost. In most cases, energy functions must become part of the standard cost system so that each function can be assessed as to its specific impact on the product cost.

The minimum theoretical energy expenditure to produce a given product can usually be determined en route to establishing a standard energy cost for that product. The seconds of 25-hp motor drive, the minutes necessary in a 2200°F furnace to heat a steel part for fabrication, or the minutes of 5-V electricity needed to make an electrolytic separation, for example, can be determined as theoretical minimums and compared with the actual figures. As in all production cost functions, the minimum standard is often difficult to meet, but it can serve as an indicator of the size of the opportunity.

In comparing actual values with minimum values, four possible approaches can be taken to reduce the variance, usually in this order:

1. An hourly or daily control system can be installed to keep the function cost at the desired level.
2. Fuel requirements can be switched to a cheaper and more available form.
3. A change can be made to the process methodology to reduce the need for the function.
4. New equipment can be installed to reduce the cost of the function.

The starting point for reducing costs should be in achieving the minimum cost possible with the present equipment and processes. Installing management control

Table 1.3 Industrial Energy Functions by Expenditure and Btu, 1978

Function	Dollar Expenditure (billions)	Percent of Expenditure	Percent of Total Btu
Machine drive	19	35	12
Feedstocks	16	29	35
Process steam	7	13	23
Direct heat	4	7	13
Indirect heat	4	7	13
Electrolysis	4	7	3
Space conditioning and lighting	1	1	1
Total	55	100	100

Source: Technical Appendix, *The Least-Cost Energy Strategy,* Carnegie-Mellon University Press, Pittsburgh, Pa., 1979, Tables 1.2.1 and 11.3.2.

Table 1.4 Cost of Industrial Energy per Million Btu, 1979

Fuel	Cost
Steam coal	$1.11
Natural gas	2.75
Residual oil	2.95
Distillate oil	4.51
Electricity	10.31

Source: Monthly Comparative Fuel Supplement, November 1979.

systems can indicate what the lowest possible energy use is in a well-controlled situation. It is only at that point when a change in process or equipment configuration should be considered. An equipment change prior to actually minimizing the expenditure under the present system may lead to oversizing new equipment or replacing equipment for unnecessary functions.

The third principle is to *control and meter only the main energy functions*—the roughly 20% that make up 80% of the costs. As Peter Drucker pointed out some time ago, a few functions usually account for a majority of the costs. It is important to focus controls on those that represent the meaningful costs and aggregate the remaining items in a general category. Many manufacturing plants in the United States have only one meter, that leading from the gas main or electric main into the plant from the outside source. Regardless of the reasonableness of the standard cost established, the inability to measure actual consumption against that standard will render such a system useless. Submetering the main functions can provide the information not only to measure but to control costs in a short time interval. The cost of metering and submetering is usually incidental to the potential for realizing significant cost improvements in the main energy functions of a production system.

The fourth principle is to put *the major effort of an energy management program into installing controls and achieving results*. It is common to find general knowledge about how large amounts of energy could be saved in a plant. The missing ingredient is the discipline necessary to achieve these potential savings. Each step in saving energy needs to be monitored frequently enough by the manager or first-line supervisor to see noticeable changes. Logging of important fuel usage or behavioral observations are almost always necessary before any particular savings results can be realized. Therefore, it is critical that an energy director or committee have the authority from the chief executive to install controls, not just advise line management. Those energy managers who have achieved the largest cost reductions actually install systems and controls; they do not just provide good advice.

As suggested earlier, the overall potential for increasing energy productivity and reducing the cost of energy services is substantial. The 20% or so improvement in industrial energy productivity since 1972 is just the beginning. To quote the energy director of a large chemical company: "Longterm results will be much greater."

Although no one knows exactly how much we can improve productivity in practice, the American Physical Society indicated in their 1974 energy conservation study that it is theoretically possible to achieve an eightfold improvement of the 1972 energy/production ratio.[9] Most certainly, we are a long way from an economic saturation of the opportunities (see, e.g., Ref. 10). The common argument that not much can be done after a 15 or 20% improvement has been realized ought to be dismissed as baseless. Energy productivity provides an expanding opportunity, not a last resort. The chapters in this book provide the information that is necessary to make the most of that opportunity in each organization.

REFERENCES

1. *Statistical Abstract of the United States*, U.S. Government Printing Office, Washington, D.C., 1979, Table 785.
2. *Energy User News*, Jan. 14, 1980.
3. JOHN G. WINGER et al., *Outlook for Energy in the United States to* 1985, The Chase Manhattan Bank, New York, 1972, p 52.
4. DONELLA H. MEADOWS et al., *The Limits to Growth*, Universe Books, New York, 1972, pp. 153-154.
5. JIMMY E. CARTER, July 15, 1979, "Address to the Nation," *Washington Post*, July 16, 1979, p. A14.
6. *Monthly Energy Review*, Jan. 1980, U.S. Department of Energy, Washington, D.C., p. 16.
7. *Monthly Energy Review*, Jan. 1980, U.S. Department of Energy, Washington D.C., p. 8; *Statistical Abstract of the United States*, U.S. Government Printing Office, Washington, D.C., 1979, Table 1409; *Energy User News*, Jan. 20, 1980, p. 14.
8. American Association for the Advancement of Science, "U.S. Energy Demand: Some Low Energy Futures," *Science*, Apr. 14, 1978, p. 143.
9. American Physical Society Summer Study on Technical Aspects of Efficient Energy Utilization, 1974. Available as W.H. CARNAHAN et al., *Efficient Use of Energy, a Physics Perspective*, from NTIS PB-242-773, or in *Efficient Energy Use, Vol.* 25 of the American Institute of Physics Conference Proceedings.
10. R.W. SANT, *The Least-Cost Energy Strategy*, Carnegie-Mellon University Press, Pittsburgh, Pa., 1979

Chapter 2

Effective Energy Management

WILLIAM H. MASHBURN, P.E., CEM
Professor & Director
Energy Management Institute
Mechanical Engineering Department
Virginia Polytechnic Institute & State University
Blacksburg, Virginia

2.1 INTRODUCTION

Most efforts to manage energy are in a stratified layer within the organization. Its tentacles do not extend upward into management nor downward into the worker level. Consequently, maximum benefits are not obtained. Energy management is a combination of technical skills and business management. Techniques for managing energy have not been incorporated into the curriculum of business schools, so future managers of organizations have no concept of the benefits to be obtained. Engineering schools deal mostly with the technical and not the management aspects of energy. So, it is a new concept that must integrate both engineering and business principles.

With all the alternative energy sources now available, this country could easily reduce its energy consumption by 50%—if there were no barriers to the implementation. But of course, there are barriers, mostly economic. Therefore, we might conclude that **managing energy is not a technical problem, but one of how to best implement those technical changes within economic limits, and with a minimum of disruption.**

The movement from a national to a global economy has caused organizations to downsize with fewer people working longer hours. As a result, patchworking the system to keep the product flowing is the norm, leaving little time and resources for optimizing the system. Consequently, while profits may increase, efficiency of operation may decrease.

Companies that have extracted all possible savings from downsizing, are now looking for other ways to become more competitive. Better management of energy is a viable way, so there is an upward trend in the number of companies that are establishing an energy management program.

Unlike other management fads that have come and gone, such as value analysis and quality circles, the need to manage energy will be permanent within our society. There are several reasons for this:

- There is a direct economic return. Most opportunities found in an energy survey have less than a two year payback. Some are immediate, such as load shifting or going to a new electric rate schedule.

- Most manufacturing companies are looking for a competitive edge. A reduction in energy costs to manufacture the product can be immediate and permanent. In addition, products that use energy, such as motor driven machinery, are being evaluated to make them more energy efficient, and therefore more marketable. Many foreign countries where energy is more critical, now want to know the maximum power required to operate a piece of equipment.

- Energy technology is changing so rapidly that state-of-the-art techniques have a half life of ten years at the most. Someone in the organization must be in a position to constantly evaluate and update this technology.

- Energy security is a part of energy management. Without a contingency plan for temporary shortages or outages, and a strategic plan for long range plans, organizations run a risk of major problems without immediate solutions. Future price shocks will occur. When world energy markets swing wildly with only a five percent decrease in supply, as they did in 1979, it is reasonable to expect that such occurrences will happen again.

Those people then who choose—or in many cases are drafted—to manage energy will do well to recognize this continuing need, and exert the extra effort to become skilled in this emerging and dynamic profession.

The purpose of this chapter is to provide the fundamentals of an energy management program that can be, and have been, adapted to organizations large and small. **Developing a working organizational structure may be the most important thing an energy manager can do.**

2.2 ENERGY MANAGEMENT PROGRAM

All the components of a comprehensive energy management program are depicted in figure 2.1. These components are the organizational structure, a policy, and plans for audits, education, reporting, and strategy. In addition to the organizational structure shown in Figure 2.1, if the organization is large, it may be beneficial to have a committee for overall direction. Some have a technical staff devoted specifically to energy that supplements that of other parts of the organization. It is hoped that by understanding the fundamentals of managing energy, the energy manager can then adapt a good working program to the existing organizational structure. Each component is discussed in detail below.

2.3. ORGANIZATIONAL STRUCTURE

The organizational chart for energy management shown in Figure 2.1 is generic. It must be adapted to fit into an existing structure for each organization. For example, the presidential block may be the general manager, and VP blocks may be division managers, but the fundamental principles are the same. The main feature of the chart is the location of the energy manager. This position should be high enough in the organizational structure to have access to key players in management, and to have a knowledge of current events within the company. For example, the timing for presenting energy projects can be critical. Funding availability and other management priorities should be known and understood. The organizational level of the energy manager is also

indicative of the support management is willing to give to the position.

2.3.1 Energy Manager

The person selected for this position should be one with the vision of what managing energy can do for the company. Every successful program has had this one thing in common—one person who is a shaker and mover that makes things happen. The program is then built around this person.

There is a great tendency for the energy manager to become an energy engineer, or a prima donna, and attempt to conduct the whole effort alone. Much has been accomplished in the past with such individuals working alone, but for the long haul, managing the program by involving everyone at the facility is much more productive and permanent.

In addition to enthusiasm, the following qualifications are desirable in an energy manager:

- Varied technical background
- For manufacturing plants, experience in production process or plant engineering
- Business and management skills and experience
- Communication skills—both oral and written
- Planning skills
- Training in the energy management discipline

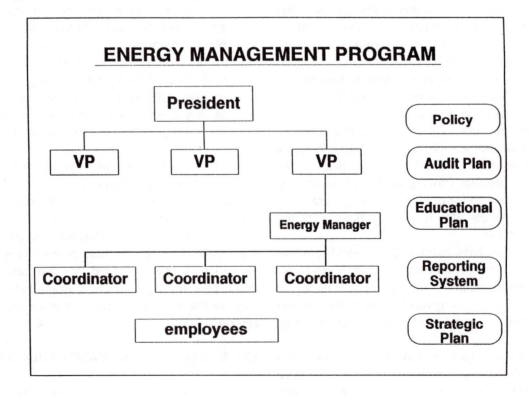

Figure 2.1

Very few people will be found that have all these qualifications, so being trainable in those where they are weak should be a consideration.

Most selections for energy managers are made from persons in-house. The advantage is their familiarity with the technical aspects of the operation as well as the organizational structure and personalities involved.

2.3.2 Coordinators

The coordinators shown in Figure 2.1 represent the energy management team within one given organizational structure, such as one company within a corporation. This group is the core of the program. The main criteria for membership should be an indication of interest. There should be a representative from the administrative group such as accounting or purchasing, someone from facilities and/or maintenance, and a representative from each major department.

This energy team of coordinators should be appointed for a specific time period, such as one year. Rotation can then bring new people with new ideas, can provide a mechanism for tactfully removing non-performers, and involve greater numbers of people in the program in a meaningful way.

Coordinators should be selected to supplement skills lacking in the energy manager since, as pointed out above, it is unrealistic to think one energy manager can have all the qualifications outlined. So, total skills needed for the team, including the energy manager may be defined as follows:

- Have enough technical knowledge within the group to either understand the technology used by the organization, or be trainable in that technology.

- Have a knowledge of potential new technology that may be applicable to the program.

- Have planning skills that will help establish the organizational structure, plan energy surveys, determine educational needs, and develop a strategic energy management plan.

- Understand the economic evaluation system used by the organization, particularly payback and life cycle cost analysis.

- Have good communication and motivational skills since energy management involves everyone within the organization.

The strengths of each team member should be evaluated in light of the above desired skills, and their assignments made accordingly.

Energy management programs can, and have, originated within one division of a large corporation. The division, by example and savings, motivates people at corporate level to pick up on the program and make energy management corporate wide. Many also originate at corporate level with people who have facilities responsibility, and have implemented a good corporate facilities program. They then see the importance and potential of an energy management program, and take a leadership role in implementing one. In every case observed by the author, **good programs have been instigated by one individual who has recognized the potential, is willing to put forth the effort in addition to regular duties—will take the risk of pushing new concepts, and is motivated by a seemingly higher calling to implement an energy management program.**

If initiated at corporate level, there are some advantages and some precautions. Some advantages are:

- More resources are available to implement the program, such as budget, staff, and facilities.

- If top management is secured at corporate level, getting management support at division level is easier.

- Total personnel expertise throughout the corporation is better known and can be identified and made known to division energy managers.

- Expensive test equipment can be purchased and maintained at corporate level for use by divisions as needed.

- A unified reporting system can be put in place.

- Creative financing may be the most needed and the most important assistance to be provided from corporate level. One major company the author is working with is considering contracting, at corporate level, a lighting retrofit program that includes a creative financing package which each division then has the option of using.

- Impacts of energy and environmental legislation can best be determined at corporate level.

- Electrical utility rates and structures, as well as effects of unbundling of electric utilities, can be evaluated at corporate level.

Some precautions are:

- Many people at division level may have already done a good job of saving energy, and are cautious about corporate level staff coming in and taking credit for their work.

- All divisions don't progress at the same speed. Work with those who are most interested first, then through the reporting system to top management give them credit. Others will then request assistance.

- Individual assistance is needed by division energy managers. This would include such things as training technical support, instrumentation, and help with financing.

2.3.3 Employees

Employees are shown as a part of the organizational structure, and are perhaps the greatest untapped resource in an energy management program. A structured method of soliciting their ideas for more efficient use of energy will prove to be the most productive effort of the energy management program. A good energy manager will devote 20% of total time working with employees. Too many times' employee involvement is limited to posters that say "Save Energy."

Employees in manufacturing plants generally know more about the equipment than anyone else in the facility because they operate it. They know how to make it run more efficiently, but because there is no mechanism in place for them to have an input, their ideas go unsolicited.

An understanding of the psychology of motivation is necessary before an employee involvement program can be successfully conducted. Motivation may be defined as the amount of physical and mental energy that a worker is willing to invest in his or her job. Three key factors of motivation are listed below:

- Motivation is already within people. The task of the supervisor is not to provide motivation, but to know how to release it.

- The amount of energy and enthusiasm people are willing to invest in their work varies with the individual. Not all are over-achievers, but not all are lazy either.

- The amount of personal satisfaction to be derived determines the amount of energy an employee will invest in the job.

Achieving personal satisfaction has been the subject of much research by industrial psychologists, and they have emerged with some revealing facts. For example, they have learned that most actions taken by people are done to satisfy a physical need—such as the need for food—or an emotional need—such as the need for acceptance, recognition, or achievement.

Research has also shown that many efforts to motivate employees deal almost exclusively with trying to satisfy physical needs, such as raises, bonuses, or fringe benefits. These methods are effective only for the short term, so we must look beyond these to other needs that may be sources of releasing motivation,

A study done by Heresy and Blanchard[1] in 1977 asked workers to rank job related factors listed below. The results were as follows:

1. Full appreciation for work done
2. Feeling "in" on things
3. Understanding of personal problems
4. Job security
5. Good wages
6. Interesting work
7. Promoting and growth in the company
8. Management loyalty to workers
9. Good working conditions
10. Tactful discipline of workers

This priority list would no doubt change with time and with individual companies, but the rankings of what supervisors thought employees wanted were almost diametrically opposed. They ranked good wages as first.

It becomes obvious from this that **job enrichment is a key to motivation.** Knowing this, the energy manager can plan a program involving employees that can provide job enrichment by some simple and inexpensive recognitions.

Some things to consider in employee motivation are as follows:

- There appears to be a positive relationship between fear arousal and persuasion if the fear appeals deal with topics primarily of significance to the individual, e.g., personal well being.

- The success of persuasive communication is directly related to the credibility of the source of communication and may be reduced if recommended changes deviate too far from existing beliefs and practices.

- When directing attention to conservation, display the reminder at the point of action at the appropriate time for action, and specify who is responsible for taking the action and when it should occur. Generic posters located in the work area are not effective.

- Studies have shown that pro-conservation attitudes and actions will be enhanced through associations with others with similar attitudes, such as being part of an energy committee.

- Positive effects are achieved with financial incentives if the reward is in proportion to the savings, and represents respectable increments of spendable income.

- Consumers place considerable importance on the potential discomfort in reducing their consumption of energy. Changing thermostat settings from the comfort zone should be the last desperate act for an energy manager.

- Social recognition and approval is important, and can occur through such things as the award of medals, designation of employee of the month, and selection to membership in elite sub-groups. Note that the dollar cost of such recognitions is minimal.

- The potentially most powerful source of social incentives for conservation behavior—but the least used—is the commitment to others that occurs in the course of group decisions.

Before entering seriously into a program involving employees, be prepared to give a heavy commitment of time and resources. In particular, have the resources to respond quickly to their suggestions.

2.4 ENERGY POLICY

A well written energy policy that has been authorized by management is as good as the proverbial license to steal. It provides the energy manager with the authority to be involved in business planning, new facility location and planning, the selection of production equipment, purchase of measuring equipment, energy reporting, and training things that are sometimes difficult to do.

If you already have an energy policy, chances are that it is too long and cumbersome. To be effective, the policy should be short—two pages at most. Many people confuse the policy with a procedures manual. It should be bare bones, but contain the following items as a minimum:

- Objectives—this can contain the standard motherhood and flag statements about energy, but the most important is that the organization will incorporate energy efficiency into facilities and new equipment, with emphasis on life cycle cost analysis rather than lowest initial cost.

- Accountability—This should establish the organizational structure and the authority for the energy manager, coordinators, and any committees or task groups.

- Reporting—Without authority from top management, it is often difficult for the energy manager to require others within the organization to comply with reporting requirements necessary to properly manage energy. The policy is the place to establish this. It also provides a legitimate reason for requesting funds for instrumentation to measure energy usage.

- Training—If training requirements are established in the policy, it is again easier to include this in budgets. It should include training at all levels within the organization.

Many companies, rather that a comprehensive policy encompassing all the features described above, choose to go with a simpler policy statement.

Appendices A and B give two sample energy policies. Appendix A is generic and covers the items discussed above. Appendix B is a policy statement of a multinational corporation.

2.5 PLANNING

Planning is one of the most important parts of the energy management program, and for most technical people is the least desirable. It has two major functions in the program. First, a good plan can be a shield from disruptions. Second, by scheduling events throughout the year, continuous emphasis can be applied to the energy management program, and will play a major role in keeping the program active.

Almost everyone from top management to the custodial level will be happy to give an opinion on what can be done to save energy. Most suggestions are worthless. It is not always wise from a job security standpoint to say this to top management. However, if you inform people—especially top management—that you will evaluate their suggestion, and assign a priority to it in your plan, not only will you not be disrupted, but may be considered effective because you do have a plan.

Many programs were started when the fear of energy shortages was greater, but they have declined into oblivion. By planning to have events periodically through the year, a continued emphasis will be placed on energy management. Such events can be training programs, audits, planning sessions, demonstrations, research projects, lectures, etc.

The secret to a workable plan is to have people who are required to implement the plan involved in the planning process. People feel a commitment to making things work if they have been a part of the design. This is fundamental to any management planning, but more often that not is overlooked. However, in order to prevent the most outspoken members of a committee from dominating with their ideas, and rejecting ideas from less outspoken members, a technique for managing committees must be used. A favorite of the author is the Nominal Group Technique developed at the University of Wisconsin in the late 1980's by Andre Delbecq and Andrea Van de Ven[2]. This technique consists of the following basic steps:

1. Problem definition—The problem is clearly defined to members of the group.

2. Grouping—Divide large groups into smaller groups of seven to ten, then have the group elect a recording secretary.

3. Silent generation of ideas—Each person silently and independently writes as many answers to the problem as can be generated within a specified time.

4. Round-robin listing—Secretary lists each idea individually on an easel until all have been recorded.

5. Discussion—Ideas are discussed for clarification, elaboration, evaluation and combining.

6. Ranking—Each person ranks the five most important items. The total number of points received for each idea will determine the first choice of the group.

2.6   AUDIT PLANNING

The details of conducting audits are discussed in a comprehensive manner in Chapter 4, but planning should be conducted prior to the actual audits. The planning should include types of audits to be performed, team makeup, and dates.

By making the audits specific rather than general in nature, much more energy can be saved. Examples of some types of audits that might be considered are:

- Tuning-Operation-Maintenance (TOM)
- Compressed air
- Motors
- Lighting
- Steam system
- Water
- Controls
- HVAC
- Employee suggestions

By defining individual audits in this manner, it is easy to identify the proper team for the audit. Don't neglect to bring in outside people such as electric utility and natural gas representatives to be team members. Scheduling the audits, then, can contribute to the events that will keep the program active.

2.7   EDUCATIONAL PLANNING

A major part of the energy manager's job is to provide some energy education to persons within the organization. In spite of the fact that we have been concerned with it for the past two decades, **there is still a sea of ignorance concerning energy.**

Raising the energy education level throughout the organization can have big dividends. The program will operate much more effectively if management understands the complexities of energy, and particularly the potential for economic benefit; the coordinators will be more effective is they are able to prioritize energy conservation measures, and are aware of the latest technology; the quality and quantity of employee suggestions will improve significantly with training.

Educational training should be considered for three distinct groups—management, the energy team, and employees.

2.7.1 Management Training

It is difficult to gain much of management's time, so subtle ways must be developed to get them up to speed. Scheduling time during a regular meeting to provide updates on the program is one way. When the momentum of the program gets going, it may be advantageous to have a half or one day presentation for management.

A good concise report periodically can be a tool to educate management. Short articles that are pertinent to your educational goals, taken from magazines and newspapers can be attached to reports and sent selectively. Having management be a part of a training program for either the energy team and/or employees, can be an educational experience since we learn best when we have to make a presentation.

Ultimately, the energy manager should aspire to be a part of business planning for the organization. A strategic plan for energy should be a part of every business plan. This puts the energy manager into a position for more contact with management people, and thus the opportunity to inform and teach.

2.7.2 Energy Team Training

Because the energy team is the core group of the energy management program, proper and thorough training for them should have the highest priority. Training is available from many sources and in many forms.

- Self study—this necessitates having a good library of energy related materials from which coordinators can select.

- In-house training—may be done by a qualified member of the team—usually the energy manager, or someone from outside.

- Short courses offered by associations such as the Association of Energy Engineers[3], by individual consultants, by corporations, and by colleges and universities.

- Comprehensive courses of one to four weeks duration offered by universities, such as the one at the University of Wisconsin, and the one being run cooperatively by Virginia Tech and N.C. State University.

For large decentralized organizations with perhaps ten or more regional energy managers, an annual two- or three-day seminar can be the base for the educational program. Such a program should be planned carefully. The following suggestions should be incorporated into such a program:

- Select quality speakers from both inside and outside the organization.

- This is an opportunity to get top management support. Invite a top level executive from the organization to give opening remarks. It may be wise to offer to write the remarks, or at least to provide some material for inclusion.

- Involve the participants in workshop activities so they have an opportunity to have an input into the program. Also, provide some practical tips on energy savings that they might go back and implement immediately. One or two good ideas can sometimes pay for their time in the seminar.

- Make the seminar first class with professional speakers; a banquet with an entertaining—not technical—after dinner speaker; a manual that includes a schedule of events, biosketches of speakers, list of attendees, information on each topic presented, and other things that will help pull the whole seminar together. Vendors will contribute things for door prizes.

- You may wish to develop a logo for the program, and include it on small favors such as cups, carrying cases, etc.

2.7.3 Employee Training

A systematic approach for involving employees should start with some basic training in energy. This will produce a much higher quality of ideas from them. Employees place a high value on training, so a side benefit is that morale goes up. Simply teaching the difference between electrical demand and kilowatt hours of energy, and that compressed air is very expensive is a start. Short training sessions on energy can be injected into other ongoing training for employees, such as safety. A more comprehensive training program should include:

- Energy conservation in the home
- Fundamentals of electric energy
- Fundamentals of energy systems
- How energy surveys are conducted and what to look for

2.8   Strategic Planning

Developing an objective, strategies, programs, and action items constitutes strategic planning for the energy management program. It is the last but perhaps the most important step in the process of developing the program, and unfortunately is where many stop. The very name "Strategic Planning" has an ominous sound for those who are more technically inclined. However, by using a simplified approach and involving the energy management team in the process, a plan can be developed using a flow chart that will define the program for the next five years.

If the team is involved in developing each of the components of objective, strategies, programs, and action items—using the Nominal Group Technique—the result will be a simplified flow chart that can be used for many purposes. First, it is a protective plan that discourages intrusion into the program, once it is established and approved. It provides the basis for resources such as funding and personnel for implementation. It projects strategic planning into overall planning by the organization, and hence legitimizes the program at top management level. By involving the implementors in the planning process, there is a strong commitment to make it work.

Appendix C, Figures 2.2 through 2.7, contains flow charts depicting a strategic plan developed in a workshop conducted by the author by a large defense organization. It is a model plan in that it deals not only with the technical aspects of energy management but also the funding, communications, education, and behavior modification.

2.9 REPORTING

There is no generic form that can be used for reporting. There are too many variables such as organization size, product, project requirements, and procedures already in existence. The ultimate reporting system is one used by a chemical company making a textile product. The Btu/lb of product is calculated on a computer system that gives an instantaneous reading. This is not only a reporting system, but one that detects maintenance problems. Very few companies are set up to do this, but many do have some type of energy index for monthly reporting.

In previous years when energy prices were fluctuating wildly, the best energy index was one based on Btu's. Now that prices have stabilized somewhat, the best index is dollars. However, there are still many factors that will influence any index, such as weather, production, expansion or contraction of facilities, new technologies, etc.

The bottom line is that any reporting system has to be customized to suit individual circumstances. And, while reporting is not always the most glamorous part of managing energy, it can make a contribution to the program by providing the bottom line on its effectiveness. It is also a straight pipeline into management, and can be a tool for promoting the program.

The report is probably of most value to the one who prepares it. It is a forcing function that requires all information to be pulled together in a coherent manner. This requires much thought and analysis that might not otherwise take place.

By making reporting a requirement of the energy policy, getting the necessary support can be easier. In many cases, the data may already be collected on a periodic basis and put into a computer. It may simply require combining production data and energy data to develop an energy index.

Keep the reporting requirements as simple as possible. The monthly report could be something as simple as adding to an ongoing graph that compares present usage to some baseline year. Any narrative should be short, with data kept in a file that can be provided for any supporting in-depth information.

2.10 TIPS FOR SUCCESS

In observing successful energy management programs, the following tips for success have been compiled:

- Have a plan. A plan dealing with organization, surveys, training, and strategic planning—with events scheduled—has two advantages. It prevents disruptions by non-productive ideas, and it sets up scheduled events that keeps the program active.

- Give away—or at least share—ideas for saving energy. The surest way to kill a project is to be possessive. If others have a vested interest they will help make it work.

- Be aggressive. The energy team—after some training—will be the most energy knowledgeable group within the company. Too many management decisions are made with a meager knowledge of the effects on energy.

- Use proven technology. Many programs get bogged down trying to make a new technology work, and lose sight of the easy projects with good payback. Don't buy serial number one. In spite of price breaks and promise of vendor support, it can be all consuming to make the system work.

- Go with the winners. Not every department within a company will be enthused about the energy program. Report the better departments to top management, and all will follow.

- A final major tip—ask the machine operator what should be done to reduce energy. Then make sure they get proper recognition for ideas.

2.11 SUMMARY

Let's now summarize by assuming you have just been appointed energy manager of a fairly large company. What are the steps you might consider in setting up an energy management program? Here is a suggested procedure.

2.11.1 Situation analysis

Determine what has been done before. Was there a previous attempt to establish an energy management program? What were the results of this effort? Next, plot the energy usage for all fuels for the past two—or more—years, then project the usage, and cost, for the next five years at the present rate. This will not only help you sell your program, but will identify areas of concentration for reducing energy.

2.11.2 Policy

Develop some kind of acceptable policy that gives authority to the program. This will help later on with such things as reporting requirements, and need for measurement instrumentation.

2.11.3 Organization

Set up the energy committee and/or coordinators.

2.11.4 Training

With the committee involvement, develop a training plan for the first year.

2.11.5 Audits

Again with the committee involvement, develop an auditing plan for the first year.

2.11.6 Reporting

Develop a simple reporting system.

2.11.7 Schedule

From the above information develop a schedule of events for the next year, timing them so as to give periodic actions from the program, which will help keep the program active and visible.

2.11.8 Implement the program.

2.12 REFERENCES

1 Hersey, Paul and Kenneth H. Blanchard, *Management of Organizational Behavior: Utilizing Human Resources*, Harper and Row, 1970
2 Delbecq, Andre L., Andrew H. Van de Ven, and David H. Gustafson, *Group Techniques for Program Planning*, Green Briar Press, 1986.
3 Association of Energy Engineers, 4025 Pleasantdale Road, Suite 420, Atlanta, GA.
4 Mashburn, William H., *Managing Energy Resources in Times of Dynamic Change*, Fairmont Press, 1992
5 Turner, Wayne, *Energy Management Handbook*, 2nd edition, chapter 2, Fairmont Press, 1993.

APPENDIX A

ENERGY POLICY

Acme Manufacturing Company
Policy and Procedures Manual
Subject: Energy Management Program

I. Policy

Energy Management shall be practiced in all areas of the Company's operation.

II. Energy Management Program Objectives

It is the Company's objective to use energy efficiently and provide energy security for the organization for both immediate and long range by:

- Utilizing energy efficiently throughout the Company's operations.

- Incorporating energy efficiency into existing equipment and facilities, and in the selection and purchase of new equipment.

- Complying with government regulations—Federal, state, and local.

- Putting in place an Energy Management Program to accomplish the above objectives.

III. Implementation

A. Organization
The Company's Energy Management Program shall be administered through the Facilities Department.

1. Energy Manager
The Energy Manager shall report directly to the Vice President of Facilities, and shall have overall responsibility for carrying out the Energy Management Program.

2. Energy Committee

The Energy Manager may appoint an Energy Committee to be comprised of representatives from various departments. Members will serve for a specified period of time. The purpose of the Energy Committee is to advise the Energy Manager on the operation of the Energy Management Program, and to provide assistance on specific tasks when needed.

3. Energy Coordinators

Energy Coordinators shall be appointed to represent a specific department or division. The Energy Manager shall establish minimum qualification standards for Coordinators, and shall have joint approval authority for each Coordinator appointed.

Coordinators shall be responsible for maintaining an ongoing awareness of energy consumption and expenditures in their assigned areas. They shall recommend and implement energy conservation projects and energy management practices.

Coordinators shall provide necessary information for reporting from their specific areas.

They may be assigned on a full-time or part-time basis, as required to implement programs in their areas.

B. Reporting

The energy Coordinator shall keep the Energy Office advised of all efforts to increase energy efficiency in their areas. A summary of energy cost savings shall be submitted each quarter to the Energy Office.

The Energy Manager shall be responsible for consolidating these reports for top management.

C. Training

The Energy Manager shall provide energy training at all levels of the Company.

IV. Policy Updating

The Energy Manager and the Energy Advisory Committee shall review this policy annually and make recommendations for updating or changes.

APPENDIX B

POLICY STATEMENT

Acme International Corporation is committed to the efficient, cost effective, and environmentally responsible use of energy throughout its worldwide operations. Acme will promote energy efficiency by implementing cost-effective programs that will maintain or improve the quality of the work environment, optimize service reliability, increase productivity, and enhance the safety of our workplace.

APPENDIX C

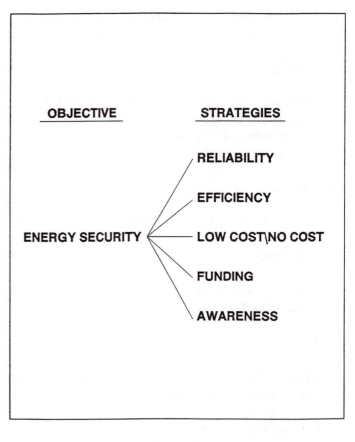

Figure 2.2

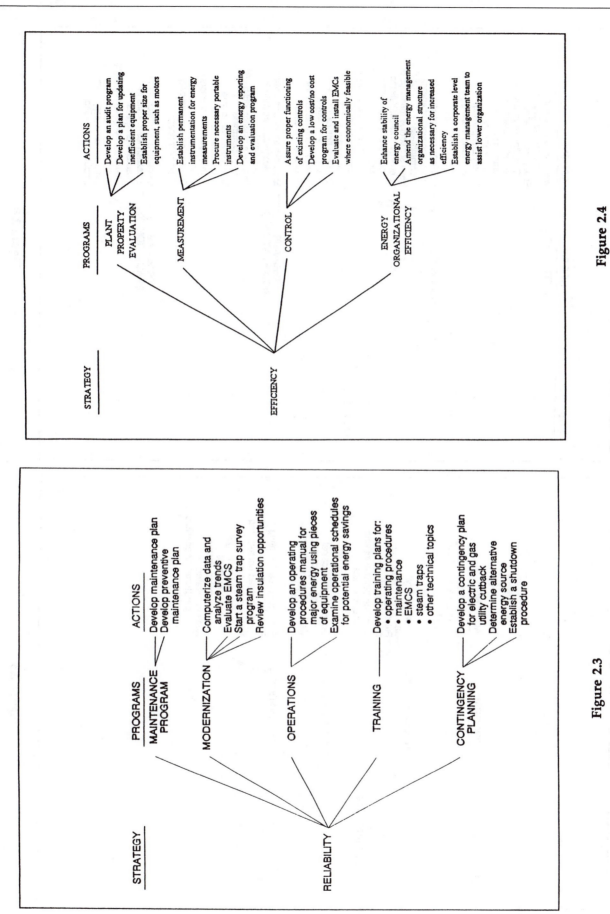

Figure 2.4

Figure 2.3

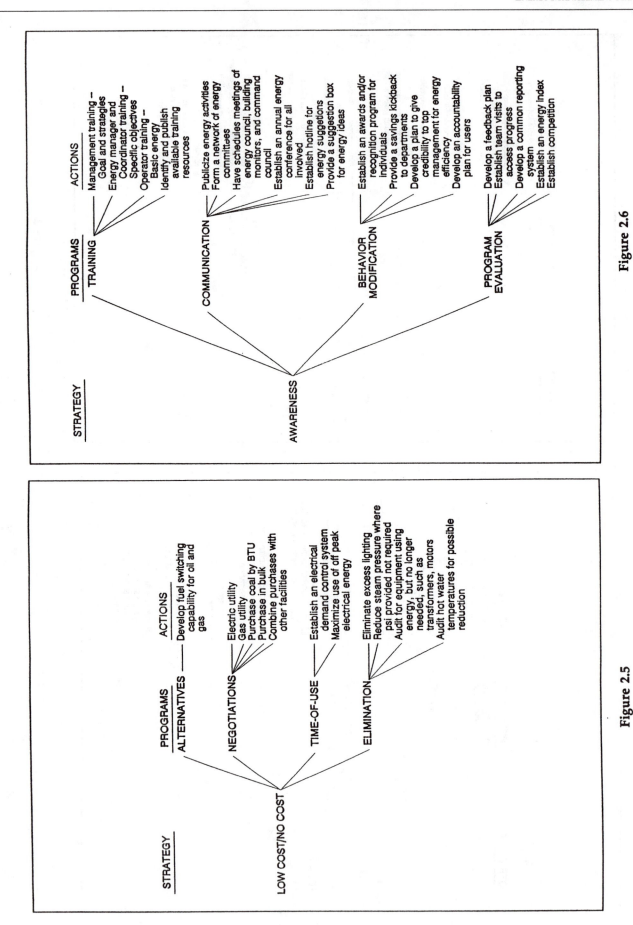

Figure 2.6

Figure 2.5

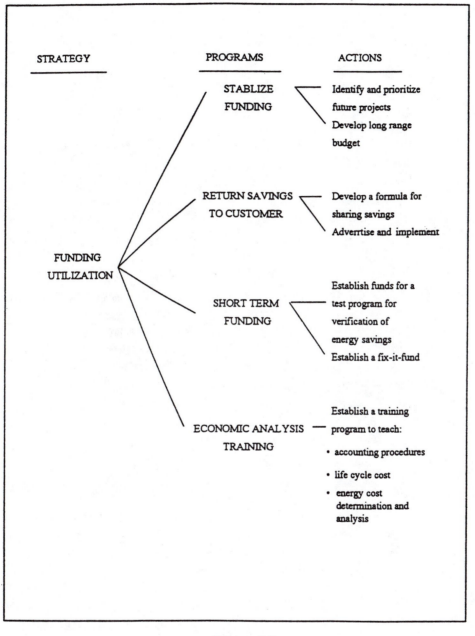

STRATEGY   PROGRAMS   ACTIONS

                       STABLIZE   Identify and prioritize
                       FUNDING   future projects
                                              Develop long range
                                              budget

                       RETURN SAVINGS   Develop a formula for
                       TO CUSTOMER   sharing savings
                                              Adverrtise and implement

FUNDING
UTILIZATION
                                              Establish funds for a
                       SHORT TERM   test program for
                       FUNDING   verification of
                                              energy savings
                                              Establish a fix-it-fund

                                              Establish a training
                       ECONOMIC ANALYSIS   program to teach:
                       TRAINING
                                              • accounting procedures

                                              • life cycle cost

                                              • energy cost
                                                determination and
                                                analysis

Figure 2.7

CHAPTER 3

ENERGY AUDITING

BARNEY L. CAPEHART AND MARK B. SPILLER
University of Florida — Gainesville Regional Utilities
Gainesville, FL

3.1 INTRODUCTION

Saving money on energy bills is attractive to businesses, industries, and individuals alike. Customers whose energy bills use up a large part of their income, and especially those customers whose energy bills represent a substantial fraction of their company's operating costs, have a strong motivation to initiate and continue an ongoing energy cost-control program. No-cost or very low-cost operational changes can often save a customer or an industry 10-20% on utility bills; capital cost programs with payback times of two years or less can often save an additional 20-30%. In many cases these energy cost control programs will also result in both reduced energy consumption and reduced emissions of environmental pollutants.

The energy audit is one of the first tasks to be performed in the accomplishment of an effective energy cost control program. An energy audit consists of a detailed examination of how a facility uses energy, what the facility pays for that energy, and finally, a recommended program for changes in operating practices or energy-consuming equipment that will cost-effectively save dollars on energy bills. The energy audit is sometimes called an energy survey or an energy analysis, so that it is not hampered with the negative connotation of an audit in the sense of an IRS audit. The energy audit is a positive experience with significant benefits to the business or individual, and the term "audit" should be avoided if it clearly produces a negative image in the mind of a particular business or individual.

3.2 ENERGY AUDITING SERVICES

Energy audits are performed by several different groups. Electric and gas utilities throughout the country offer free residential energy audits. A utility's residential energy auditors analyze the monthly bills, inspect the construction of the dwelling unit, and inspect all of the energy-consuming appliances in a house or an apartment. Ceiling and wall insulation is measured, ducts are inspected, appliances such as heaters, air conditioners, water heaters, refrigerators, and freezers are examined, and the lighting system is checked.

Some utilities also perform audits for their industrial and commercial customers. They have professional engineers on their staff to perform the detailed audits needed by companies with complex process equipment and operations. When utilities offer free or low-cost energy audits for commercial customers, they usually only provide walk-through audits rather than detailed audits. Even so, they generally consider lighting, HVAC systems, water heating, insulation and some motors.

Large commercial or industrial customers may hire an engineering consulting firm to perform a complete energy audit. Other companies may elect to hire an energy manager or set up an energy management team whose job is to conduct periodic audits and to keep up with the available energy efficiency technology.

The U.S. Department of Energy (U.S.DOE) funds a program where universities around the country operate Energy Analysis and Diagnostic Centers which perform free energy audits for small and medium sized manufacturing companies. There are currently 18 EADC's funded by the Industrial Division of the U.S. DOE.

The Institutional Conservation Program (ICP) is another energy audit service funded by the U.S. Department of Energy. It is usually administered through state energy offices. This program pays for audits of schools, hospitals, and other institutions, and has some funding assistance for energy conservation improvements.

3.3 BASIC COMPONENTS OF AN ENERGY AUDIT

An initial summary of the basic steps involved in conducting a successful energy audit is provided here, and these steps are explained more fully in the sections that follow. This audit description primarily addresses the steps in an industrial or large-scale commercial audit, and not all of the procedures described in this section are required for every type of audit.

The audit process starts by collecting information about a facility's operation and about its past record of utility bills. This data is then analyzed to get a picture of how the facility uses—and possibly wastes—energy, as well as to help the auditor learn what areas to examine to reduce energy costs. Specific changes—called Energy Conservation Opportunities (ECO's)—are identified and evaluated to determine their benefits and their cost-effectiveness. These ECO's are assessed in terms of their costs and benefits, and an economic comparison is made to rank the various ECO's. Finally, an Action Plan is created where certain ECO's are selected for implementation, and the actual process of saving energy and saving money begins.

3.3.1 The Auditor's Toolbox

To obtain the best information for a successful energy cost control program, the auditor must make some measurements during the audit visit. The amount of equipment needed depends on the type of energy-consuming equipment used at the facility, and on the range of potential ECO's that might be considered. For example, if waste heat recovery is being considered, then the auditor must take substantial temperature measurement data from potential heat sources. Tools commonly needed for energy audits are listed below:

Tape Measures

The most basic measuring device needed is the tape measure. A 25-foot tape measure l" wide and a 100-foot tape measure are used to check the dimensions of walls, ceilings, windows and distances between pieces of equipment for purposes such as determining the length of a pipe for transferring waste heat from one piece of equipment to the other.

Lightmeter

One simple and useful instrument is the lightmeter which is used to measure illumination levels in facilities. A lightmeter that reads in footcandles allows direct analysis of lighting systems and comparison with recommended light levels specified by the Illuminating Engineering Society. A small lightmeter that is portable and can fit into a pocket is the most useful. Many areas in buildings and plants are still significantly over-lighted, and measuring this excess illumination then allows the auditor to recommend a reduction in lighting levels through lamp removal programs or by replacing inefficient lamps with high efficiency lamps that may not supply the same amount of illumination as the old inefficient lamps.

Thermometers

Several thermometers are generally needed to measure temperatures in offices and other worker areas, and to measure the temperature of operating equipment. Knowing process temperatures allows the auditor to determine process equipment efficiencies, and also to identify waste heat sources for potential heat recovery programs. Inexpensive electronic thermometers with interchangeable probes are now available to measure temperatures in both these areas. Some common types include an immersion probe, a surface temperature probe, and a radiation shielded probe for measuring true air temperature. Other types of infra-red thermometers and thermographic equipment are also available. An infra-red "gun" is valuable for measuring temperatures of steam lines that are not readily reached without a ladder.

Voltmeter

An inexpensive voltmeter is useful for determining operating voltages on electrical equipment, and especially useful when the nameplate has worn off of a piece of equipment or is otherwise unreadable or missing. The most versatile instrument is a combined volt-ohm-ammeter with a clamp-on feature for measuring currents in conductors that are easily accessible. This type of multimeter is convenient and relatively inexpensive.

Wattmeter/Power Factor Meter

A portable hand-held wattmeter and power factor meter is very handy for determining the power consumption and power factor of individual motors and other inductive devices. This meter typically has a clamp-on feature which allows an easy connection to the current-carrying conductor, and has probes for voltage connections.

Combustion Analyzer

Combustion analyzers are portable devices capable of estimating the combustion efficiency of furnaces, boilers, or other fossil fuel burning machines. Two types are available: digital analyzers and manual combustion analysis kits. Digital combustion analysis equipment performs the measurements and reads out in percent combustion efficiency. These instruments are fairly complex and expensive.

The manual combustion analysis kits typically require multiple measurements including exhaust stack: temperature, oxygen content, and carbon dioxide content. The efficiency of the combustion process can be calculated after determining these parameters. The manual process is lengthy and is frequently subject to human error.

Airflow Measurement Devices

Measuring air flow from heating, air conditioning or ventilating ducts, or from other sources of air flow is one of the energy auditor's tasks. Airflow measurement devices can be used to identify problems with air flows, such as whether the combustion air flow into a gas heater is correct. Typical airflow measuring devices include a velometer, an anemometer, or an airflow hood. See section 3.4.3 for more detail on airflow measurement devices.

Blower Door Attachment

Building or structure tightness can be measured with a blower door attachment. This device is frequently used in residences and in office buildings to determine the air leakage rate or the number of air changes per hour in the facility. This is often helps determine whether the facility has substantial structural or duct leaks that need to be found and sealed. See section 3.4.2 for additional information on blower doors.

Smoke Generator

A simple smoke generator can also be used in residences, offices and other buildings to find air infiltration and leakage around doors, windows, ducts and other structural features. Care must be taken in using this device, since the chemical "smoke" produced may be hazardous, and breathing protection masks may be needed. See section 3.4.1 for additional information on the smoke generation process, and use of smoke generators.

Safety Equipment

The use of safety equipment is a vital precaution for any energy auditor. A good pair of safety glasses is an absolute necessity for almost any audit visit. Hearing protectors may also be required on audit visits to noisy plants or areas with high horsepower motors driving fans and pumps. Electrical insulated gloves should be used if electrical measurements will be taken, and asbestos gloves should be used for working around boilers and heaters. Breathing masks may also be needed when hazardous fumes are present from processes or materials used. Steel-toe and steel-shank safety shoes may be needed on audits of plants where heavy materials, hot or sharp materials or hazardous materials are being used. (See section 3.3.3 for an additional discussion of safety procedures.)

3.3.2 Preparing for the Audit Visit

Some preliminary work must be done before the auditor makes the actual energy audit visit to a facility. Data should be collected on the facility's use of energy through examination of utility bills, and some preliminary information should be compiled on the physical description and operation of the facility. This data should then be analyzed so that the auditor can do the most complete job of identifying Energy Conservation Opportunities during the actual site visit to the facility.

Energy Use Data

The energy auditor should start by collecting data on energy use, power demand and cost for at least the previous 12 months. Twenty-four months of data might be necessary to adequately understand some types of billing methods. Bills for gas, oil, coal, electricity, etc. should be compiled and examined to determine both the amount of energy used and the cost of that energy. This data should then be put into tabular and graphic form to see what kind of patterns or problems appear from the tables or graphs. Any anomaly in the pattern of energy use raises the possibility for some significant energy or cost savings by identifying and controlling that anomalous behavior. Sometimes an anomaly on the graph or in the table reflects an error in billing, but generally the deviation shows that some activity is going on that has not been noticed, or is not completely understood by the customer.

Rate Structures

To fully understand the cost of energy, the auditor must determine the rate structure under which that energy use is billed. Energy rate structures may go from the extremely simple ones—for example, $1.25 per gallon of Number 2 fuel oil, to very complex ones—for example, electricity consumption which may have a customer charge, energy charge, demand charge, power factor charge, and other miscellaneous charges that vary from month to month. Few customers or businesses really understand the various rate structures that control the cost of the energy they consume. The auditor can help here because the customer must know the basis for the costs in order to control them successfully.

• Electrical Demand Charges: The demand charge is based on a reading of the maximum power in kW that a customer demands in one month. Power is the rate at which energy is used, and it varies quite rapidly for many facilities. Electric utilities average the power reading over intervals from fifteen minutes to one hour, so that very short fluctuations do not adversely affect customers. Thus, a customer might be billed for demand for a month based on a maximum value of a fifteen minute integrated average of their power use.

- Ratchet Clauses: Some utilities have a rachet clause in their rate structure which stipulates that the minimum power demand charge will be the highest demand recorded in the last billing period or some percentage (i.e., typically 70%) of the highest power demand recorded in the last year. The rachet clause can increase utility charges for facilities during periods of low activity or where power demand is tied to extreme weather.

- Discounts/Penalties: Utilities generally provide discounts on their energy and power rates for customers who accept power at high voltage and provide transformers on site. They also commonly assess penalties when a customer has a power factor less than 0.9. Inductive loads (e.g., lightly loaded electric motors, old fluorescent lighting ballasts, etc.) reduce the power factor. Improvement can be made by adding capacitance to correct for lagging power factor, and variable capacitor banks are most useful for improving the power factor at the service drop. Capacitance added near the loads can effectively increase the electrical system capacity. Turning off idling or lightly loaded motors can also help.

- Wastewater charges: The energy auditor also frequently looks at water and wastewater use and costs as part of the audit visit. These costs are often related to the energy costs at a facility. Wastewater charges are usually based on some proportion of the metered water use since the solids are difficult to meter. This can needlessly result in substantial increases in the utility bill for processes which do not contribute to the wastewater stream (e.g., makeup water for cooling towers and other evaporative devices, irrigation, etc.). A water meter can be installed at the service main to supply the loads not returning water to the sewer system. This can reduce the charges by up to two-thirds.

Energy bills should be broken down into the components that can be controlled by the facility. These cost components can be listed individually in tables and then plotted. For example, electricity bills should be broken

Summary of Energy Usage and Costs

Month	kWh Used (kWh)	kWh Cost ($)	Demand (kW)	Demand Cost ($)	Total Cost ($)
Mar	44960	1581.35	213	1495.26	3076.61
Apr	47920	1859.68	213	1495.26	3354.94
May	56000	2318.11	231	1621.62	3939.73
Jun	56320	2423.28	222	1558.44	3981.72
Jul	45120	1908.16	222	1558.44	3466.60
Aug	54240	2410.49	231	1621.62	4032.11
Sept	50720	2260.88	222	1558.44	3819.32
Oct	52080	2312.19	231	1621.62	3933.81
Nov	44480	1954.01	213	1495.26	3449.27
Dec	38640	1715.60	213	1495.26	3210.86
Jan	36000	1591.01	204	1432.08	3023.09
Feb	42880	1908.37	204	1432.08	3340.45
Totals	569,360	24,243.13	2,619	18,385.38	42,628.51
Monthly Averages	47,447	2,020.26	218	1,532.12	3,552.38

down into power demand costs per kW per month, and energy costs per kWh. The following example illustrates the parts of a rate structure for an industry in Florida.

Example: A company that fabricates metal products gets electricity from its electric utility at the following general service demand rate structure.

Rate structure:

Customer cost	=	$21.00 per month
Energy cost	=	$0.051 per kWh
Demand cost	=	$6.50 per kW per month
Taxes	=	Total of 8%
Fuel adjustment	=	A variable amount per kWh each month

The energy use and costs for that company for a year are summarized below:

The auditor must be sure to account for all the taxes, the fuel adjustment costs, the fixed charges, and any other costs so that the true cost of the controllable energy cost components can be determined. In the electric rate structure described above, the quoted costs for a kW of demand and a kWh of energy are not complete until all these additional costs are added. Although the rate structure says that there is a basic charge of $6.50 per kW per month, the actual cost including all taxes is $7.02 per kW per month. The average cost per kWh is most easily obtained by taking the data for the twelve month period and calculating the cost over this period of time. Using the numbers from the table, one can see that this company has an average energy cost of $0.075 per kWh.

These data are used initially to analyze potential ECO's and will ultimately influence which ECO's are recommended. For example, an ECO that reduces peak demand during a month would save $7.02 per kW per month. Therefore, the auditor should consider ECO's that would involve using certain equipment during the night shift when the peak load is significantly less than the first shift peak load. ECO's that save both energy and demand on the first shift would save costs at a rate of $0.075 per kWh. Finally, ECO's that save electrical energy during the off-peak shift should be examined too, but they may not be as advantageous; they would only save at the rate of $0.043 per kWh because they are already used off-peak and there would not be any additional demand cost savings.

Physical and Operational Data for the Facility

The auditor must gather information on factors likely to affect the energy use in the facility. Geographic location, weather data, facility layout and construction, operating hours, and equipment can all influence energy use.

- **Geographic Location/Weather Data:** The geographic location of the facility should be noted, together with the weather data for that location. Contact the local weather station, the local utility or the state energy office to obtain the average degree days for heating and cooling for that location for the past twelve months. This degree-day data will be very useful in analyzing the need for energy for heating or cooling the facility. Bin weather data would also be useful if a thermal envelope simulation of the facility were going to be performed as part of the audit.

- **Facility Layout:** Next the facility layout or plan should be obtained, and reviewed to determine the facility size, floor plan, and construction features such as wall and roof material and insulation levels, as well as door and window sizes and construction. A set of building plans could supply this information in sufficient detail. It is important to make sure the plans reflect the "as-built" features of the facility, since many original building plans do not get used without alterations.

- **Operating Hours:** Operating hours for the facility should also be obtained. Is there only a single shift? Are there two shifts? Three? Knowing the operating hours in advance allows some determination as to whether some loads could be shifted to off-peak times. Adding a second shift can often be cost effective from an energy cost view, since the demand charge can then be spread over a greater amount of kWh.

- **Equipment List:** Finally, the auditor should get an equipment list for the facility and review it before conducting the audit. All large pieces of energy-consuming equipment such as heaters, air conditioners, water heaters, and specific process-related equipment should be identified. This list, together with data on operational uses of the equipment allows a good understanding of the major energy-consuming tasks or equipment at the facility. As a general rule, the largest energy and cost activities should be examined first to see what savings could be achieved. The greatest effort should be devoted to the ECO's which show the greatest savings, and the least effort to those with the smallest savings potential.

The equipment found at an audit location will depend greatly on the type of facility involved. Residential audits for single-family dwellings generally involve smaller-sized lighting, heating, air conditioning and refrigeration systems. Commercial operations such as grocery stores, office buildings and shopping centers usually have equipment similar to residences, but much larger in size and in energy use. However, large residential structures such as apartment buildings have heating, air conditioning and lighting that is very similar to many commercial facilities. Business operations is the area where commercial audits begin to involve equipment substantially different from that found in residences.

Industrial auditors encounter the most complex equipment. Commercial-scale lighting, heating, air conditioning and refrigeration, as well as office business equipment, is generally used at most industrial facilities. The major difference is in the highly specialized equipment used for the industrial production processes. This can include equipment for chemical mixing and blending, metal plating and treatment, welding, plastic injection molding, paper making and printing, metal refining, electronic assembly, and making glass, for example.

3.3.3 Safety Considerations

Safety is a critical part of any energy audit. The audit person orteam should be thoroughly briefed on safety equipment and procedures, and should never place themselves in a position where they could injure themselves or other people at the facility. Adequate safety equipment should be worn at all appropriate times. Auditors should be extremely careful making any measurements on electrical systems, or on high temperature devices such as boilers, heaters, cookers, etc. Electrical gloves or asbestos gloves should be worn as appropriate.

The auditor should be careful when examining any operating piece of equipment, especially those with open drive shafts, belts or gears, or any form of rotating machinery. The equipment operator or supervisor should be notified that the auditor is going to look at that piece of equipment and might need to get information from some part of the device. If necessary, the auditor may need to come back when the machine or device is idle in order to safely get the data. The auditor should never approach a piece of equipment and inspect it without the operator or supervisor being notified first.

Safety Checklist

1. Electrical:
    a. Avoid working on live circuits, if possible.
    b. Securely lock off circuits and switches before working on a piece of equipment.
    c. Always keep one hand in your pocket while making measurements on live circuits to help prevent cardiac arrest.
2. Respiratory:
    a. When necessary, wear a full face respirator mask with adequate filtration particle size.
    b. Use activated carbon cartridges in the mask when working around low concentrations of noxious gases. Change the cartridges on a regular basis.
    c. Use a self-contained breathing apparatus for work in toxic environments.
3. Hearing:
    a. Use foam insert plugs while working around loud machinery to reduce sound levels up to 30 decibels.

3.3.4 Conducting the Audit Visit

Once the information on energy bills, facility equipment and facility operation has been obtained, the audit equipment can be gathered up, and the actual visit to the facility can be made.

Introductory Meeting

The audit person—or team—should meet with the facility manager and the maintenance supervisor and briefly discuss the purpose of the audit and indicate the kind of information that is to be obtained during the visit to the facility. If possible, a facility employee who is in a position to authorize expenditures or make operating policy decisions should also be at this initial meeting.

Audit Interviews

Getting the correct information on facility equipment and operation is important if the audit is going to be most successful in identifying ways to save money on energy bills. The company philosophy towards investments, the impetus behind requesting the audit, and the expectations from the audit can be determined by interviewing the general manager, chief operating officer, or other executives. The facility manager or plant manager is one person that should have access to much of the operational data on the facility, and a file of data on facility equipment. The finance officer can provide any necessary financial records (e.g.; utility bills for electric, gas, oil, other fuels, water and wastewater, expenditures for maintenance and repair, etc.).

The auditor must also interview the floor supervisors and equipment operators to understand the building and process problems. Line or area supervisors usually have the best information on times their equipment is used. The maintenance supervisor is often the primary person to talk to about types of lighting and lamps, sizes of motors, sizes of air conditioners and space heaters, and electrical loads of specialized process equipment. Finally, the maintenance staff must be interviewed to find the equipment and performance problems.

The auditor should write down these people's names, job functions and telephone numbers, since it is frequently necessary to get additional information after the initial audit visit.

Walk-through Tour

A walk-through tour of the facility or plant tour should be conducted by the facility/plant manager, and should be arranged so the auditor or audit team can see the major operational and equipment features of the facility. The main purpose of the walkthrough tour is to obtain general information. More specific information should be obtained from the maintenance and operational people after the tour. .

Getting Detailed Data

Following the facility or plant tour, the auditor or audit team should acquire the detailed data on facility equipment and operation that will lead to identifying the significant Energy Conservation Opportunities (ECO's) that may be appropriate for this facility. This includes data on lighting, HVAC equipment, motors, water heating, and specialized equipment such as refrigerators, ovens, mixers, boilers, heaters, etc. This data is most easily recorded on individualized data sheets that have been prepared in advance.

What to Look for

- **Lighting:** Making a detailed inventory of all lighting is important. Data should be recorded on numbers of each type of light fixtures and lamps, wattages of lamps, and hours of operation of groups of lights. A lighting inventory data sheet should be used to record this data. Using a lightmeter, the auditor should also record light intensity readings for each area. Taking notes on types of tasks performed in each area will help the auditor select alternative lighting technologies that might be

more energy efficient. Other items to note are the areas that may be infrequently used and may be candidates for occupancy sensor controls of lighting, or areas where daylighting may be feasible.

- **HVAC Equipment:** All heating, air conditioning and ventilating equipment should be inventoried. Prepared data sheets can be used to record type, size, model numbers, age, electrical specifications or fuel use specifications, and estimated hours of operation. The equipment should be inspected to determine the condition of the evaporator and condenser coils, the air filters, and the insulation on the refrigerant lines. Air velocity measurement may also be made and recorded to assess operating efficiencies or to discover conditioned air leaks. This data will allow later analysis to examine alternative equipment and operations that would reduce energy costs for heating, ventilating, and air conditioning.

- **Electric Motors:** An inventory of all electric motors over 1 horsepower should also be taken. Prepared data sheets can be used to record motor size, use, age, model number, estimated hours of operation, other electrical characteristics, and possibly the operating power factor. Measurement of voltages, currents, and power factors may be appropriate for some motors. Notes should be taken on the use of motors, particularly recording those that are infrequently used and might be candidates for peak load control or shifting use to off-peak times. All motors over 1 hp and with times of use of 2000 hours per year or greater, are likely candidates for replacement by high efficiency motors—at least when they fail and must be replaced.

- **Water Heaters:** All water heaters should be examined, and data recorded on their type, size, age, model number, electrical characteristics or fuel use. What the hot water is used for, how much is used, and what time it is used should all be noted. Temperature of the hot water should be measured.

- **Waste Heat Sources:** Most facilities have many sources of waste heat, providing possible opportunities for waste heat recovery to be used as the substantial or total source of needed hot water. Waste heat sources are air conditioners, air compressors, heaters and boilers, process cooling systems, ovens, furnaces, cookers, and many others. Temperature measurements for these waste heat sources are necessary to analyze them for replacing the operation of the existing water heaters.

- **Peak Equipment Loads:** The auditor should particularly look for any piece of electrically powered equipment that is used infrequently or whose use could be controlled and shifted to offpeak times. Examples of infrequently used equipment include trash compactors, fire sprinkler system pumps (testing), certain types of welders, drying ovens, or any type of back-up machine. Some production machines might be able to be scheduled for off-peak. Water heating could be done off-peak if a storage system is available, and off-peak thermal storage can be accomplished for onpeak heating or cooling of buildings. Electrical measurements of voltages, currents, and wattages may be helpful. Any information which leads to a piece of equipment being used off-peak is valuable, and could result in substantial savings on electric bills. The auditor should be especially alert for those infrequent on-peak uses that might help explain anomalies on the energy demand bills.

- **Other Energy-Consuming Equipment:** Finally, an inventory of all other equipment that consumes a substantial amount of energy should be taken. Commercial facilities may have extensive computer and copying equipment, refrigeration and cooling equipment, cooking devices, printing equipment, water heaters, etc. Industrial facilities will have many highly specialized process and production operations and machines. Data on types, sizes, capacities, fuel use, electrical characteristics, age, and operating hours should be recorded for all of this equipment.

Preliminary Identification of ECO's: As the audit is being conducted, the auditor should take notes on potential ECO's that are evident. Identifying ECO's requires a good knowledge of the available energy efficiency technologies that can accomplish the same job with less energy and less cost. For example, overlighting indicates a potential lamp removal or lamp change ECO, and inefficient lamps indicates a potential lamp technology change. Motors with high use times are potential ECO's for high efficiency replacements. Notes on waste heat sources should indicate what other heating sources they might replace, and how far away they are from the end use point. Identifying any potential ECO's during the walk-through will make it easier later on to analyze the data and to determine the final ECO recommendations.

3.3.5 Post-Audit Analysis

Following the audit visit to the facility, the data collected should be examined, organized and reviewed for completeness. Any missing data items should be obtained from the facility personnel or from a re-visit to the facility. The preliminary ECO's identified during the audit visit should now be reviewed, and the actual analysis of the equipment or operational change should be conducted. This involves determining the costs and the benefits of the potential ECO, and making a judgment on the cost-effectiveness of that potential ECO.

Cost-effectiveness involves a judgment decision that is viewed differently by different people and different companies. Often, Simple Payback Period (SPP) is used to measure cost-effectiveness, and most facilities want a SPP of two years or less. The SPP for an ECO is found by taking the initial cost and dividing it by the annual savings. This results in finding a period of time for the savings to repay the initial investment, without using the time value of money.One other common measure of cost-effectiveness is the discounted benefit-cost ratio. In this method, the annual savings are discounted when they occur in future years, and are added together to find the present value of the annual savings over a specified period of time. The benefit-cost ratio is then calculated by dividing the present value of the savings by the initial cost. A ratio greater than one means that the investment will more than repay itself, even when the discounted future savings are taken into account.

Several ECO examples are given here in order to illustrate the relationship between the audit information obtained and the technology and operational changes recommended to save on energy bills.

Lighting ECO

First, an ECO technology is selected—such as replacing an existing 400 watt mercury vapor lamp with a 325 watt multi-vapor lamp when it burns out. The cost of the replacement lamp must be determined. Product catalogs can be used to get typical prices for the new lamp—about $10 more than the 400 watt mercury vapor lamp. The new lamp is a direct screw-in replacement, and no change is needed in the fixture or ballast. Labor cost is assumed to be the same to install either lamp. The benefits—or cost savings—must be calculated next. The power savings is 400-325 = 75 watts. If the lamp operates for 4000 hours per year and electric energy costs $0.075/ kWh, then the savings is (.075 kW)(4000 hr/ year)($0.075/kWh) = $22.50/year. This gives a SPP = $10/$22.50/yr =.4 years, or about 5 months. This would be considered an extremely cost-effective ECO. (For illustration purposes, ballast wattage has been ignored

Motor ECO

A ventilating fan at a fiberglass boat manufacturing company has a standard efficiency 5 hp motor that runs two shifts a day, or 4160 hours per year. When this motor wears out, the company will have an ECO of using a high efficiency motor. A high efficiency 5 hp motor costs around $80 more to purchase than the standard efficiency motor. The standard motor is 83% efficient and the high efficiency model is 88.5% efficient. The cost savings is found by calculating (5 hp)(4160 hr/yr)(.746 kW/hp)[(1/.83) −(1/.885)]($.075/kWh) = (1162 kWh)*($0.075) = $87.15/year. The SPP = $80/$87.15/yr =.9 years, or about 11 months. This is also a very attractive ECO when evaluated by this economic measure.

The discounted benefit-cost ratio can be found once a motor life is determined, and a discount rate is selected. Companies generally have a corporate standard for the discount rate used in determining their measures used to make investment decisions. For a 10 year assumed life, and a 10% discount rate, the present worth factor is found as 6.144 (see Appendix IV). The benefit-cost ratio is found as B/C = ($87.15)(6.144)/$80 = 6.7. This is an extremely attractive benefit-cost ratio.

Peak Load Control ECO

A metals fabrication plant has a large shot-blast cleaner that is used to remove the rust from heavy steel blocks before they are machined and welded. The cleaner shoots out a stream of small metal balls—like shotgun pellets—to clean the metal blocks. A 150 hp motor provides the primary motive force for this cleaner. If turned on during the first shift, this machine requires a total electrical load of about 180 kW which adds directly to the peak load billed by the electric utility. At $7.02/kW/month, this costs (180 kW)*($7.02/kW/month) = $1263.60/month. Discussions with line operating people resulted in the information that the need for the metal blocks was known well in advance, and that the cleaning could easily be done on the evening shift before the blocks were needed. Based on this information, the recommended ECO is to restrict the shot-blast cleaner use to the evening shift, saving the company $15,163.20 per year. Since there is no cost to implement this ECO, the SPP = O; that is, the payback is immediate.

3.3.6 The Energy Audit Report

The next step in the energy audit process is to prepare a report which details the final results and recommendations. The length and detail of this report will vary depending on the type of facility audited. A residential audit may result in a computer printout from the utility. An industrial audit is more likely to have a detailed explanation of the ECO's and benefit-cost analyses. The following discussion covers the more detailed audit reports.

The report should begin with an executive summary that provides the owners/managers of the audited facility with a brief synopsis of the total savings available and the highlights of each ECO. The report should then describe the facility that has been audited, and provide information on the operation of the facility that relates to its energy costs. The energy bills should be presented, with tables and plots showing the costs and consumption. Following the energy cost analysis, the recommended ECO's should be presented, along with the calculations for the costs and benefits, and the cost-effectiveness criterion.

Regardless of the audience for the audit report, it should be written in a clear, concise and easy-to understand format and style. The executive summary should be tailored to non-technical personnel, and technical jargon should be minimized. A client who understands the report is more likely to implement the recommended ECO's. An outline for a complete energy audit report is shown below.

Energy Audit Report Format

Executive Summary
  A brief summary of the recommendations and cost savings
Table of Contents
Introduction
  Purpose of the energy audit
  Need for a continuing energy cost control program
Facility Description
  Product or service, and materials flow
  Size, construction, facility layout, and hours of operation
  Equipment list, with specifications
Energy Bill Analysis
  Utility rate structures
  Tables and graphs of energy consumptions and costs
  Discussion of energy costs and energy bills
Energy Conservation Opportunities
  Listing of potential ECO's

Cost and savings analysis
Economic evaluation
Action Plan
Recommended ECO's and an implementation schedule
Designation of an energy monitor and ongoing program
Conclusion
Additional comments not otherwise covered

3.3.7 The Energy Action Plan

The last step in the energy audit process is to recommend an action plan for the facility. Some companies will have an energy audit conducted by their electric utility or by an independent consulting firm, and will then make changes to reduce their energy bills. They may not spend any further effort in the energy cost control area until several years in the future when another energy audit is conducted. In contrast to this is the company which establishes a permanent energy cost control program, and assigns one person—or a team of people—to continually monitor and improve the energy efficiency and energy productivity of the company. Similar to a Total Quality Management program where a company seeks to continually improve the quality of its products, services and operation, an energy cost control program seeks continual improvement in the amount of product produced for a given expenditure for energy.

The energy action plan lists the ECO's which should be implemented first, and suggests an overall implementation schedule. Often, one or more of the recommended ECO's provides an immediate or very short payback period, so savings from that ECO—or those ECO's can be used to generate capital to pay for implementing the other ECO's. In addition, the action plan also suggests that a company designate one person as the energy monitor for the facility. This person can look at the monthly energy bills and see whether any unusual costs are occurring, and can verify that the energy savings from ECO's is really being seen. Finally, this person can continue to look for other ways the company can save on energy costs, and can be seen as evidence that the company is interested in a future program of energy cost control.

3.4 SPECIALIZED AUDIT TOOLS

3.4.1 Smoke Sources

Smoke is useful in determining airflow characteristics in buildings, air distribution systems, exhaust hoods and systems, cooling towers, and air intakes. There are several ways to produce smoke. Ideally, the smoke should be neutrally buoyant with the air mass around it so that no motion will be detected unless a force is applied. Cigarette and incense stick smoke, although inexpensive, do not meet this requirement.

Smoke generators using titanium tetrachloride ($TiCl_4$) provide an inexpensive and convenient way to produce and apply smoke. The smoke is a combination of hydrochloric acid (HCl) fumes and titanium oxides produced by the reaction of $TiCl_4$ and atmospheric water vapor. This smoke is both corrosive and toxic so the use of a respirator mask utilizing activated carbon is strongly recommended. Commercial units typically use either glass or plastic cases. Glass has excellent longevity but is subject to breakage since smoke generators are often used in difficult-to-reach areas. Most types of plastic containers will quickly degrade from the action of hydrochloric acid.

Small Teflon* squeeze bottles (i.e., 30 ml) with attached caps designed for laboratory reagent use resist degradation and are easy to use. The bottle should be stuffed with 2-3 real cotton balls then filled with about 0.15 fluid ounces of liquid $TiCl_4$. Synthetic cotton balls typically disintegrate if used with titanium tetrachloride. This bottle should yield over a year of service with regular use. The neck will clog with debris but can be cleaned with a paper clip.

Some smoke generators are designed for short time use. These bottles are inexpensive and useful for a day of smoke generation, but will quickly degrade. Smoke bombs are incendiary devices designed to emit a large volume of smoke over a short period of time. The smoke is available in various colors to provide good visibility. These are useful in determining airflow capabilities of exhaust air systems and large-scale ventilation systems. A crude smoke bomb can be constructed by placing a stick of elemental phosphorus in a metal pan and igniting it. A large volume of white smoke will be released. This is an inexpensive way of testing laboratory exhaust hoods since many labs have phosphorus in stock.

More accurate results can be obtained by measuring the chemical composition of the airstream after injecting a known quantity of tracer gas such as sulphur hexafluoride into an area. The efficiency of an exhaust system can be determined by measuring the rate of tracer gas removal. Building infiltration/exfiltration rates can also be estimated with tracer gas.

3.4.2 Blower Door

The blower door is a device containing a fan, controller, several pressure gauges, and a frame which fits in the doorway of a building. It is used to study the pressurization and leakage rates of a building and its air distribution system under varying pressure conditions. The units currently available are designed for use in residences although they can be used in small commercial buildings as well. The large quantities of ventilation air limit blower door use in large commercial and industrial buildings.

An air leakage/pressure curve can be developed for the building by measuring the fan flow rate necessary to achieve a pressure differential between the building interior and the ambient atmospheric pressure over a range of values. The natural air infiltration rate of the building under the prevailing pressure conditions can be estimated from the leakage/pressure curve and local air pressure data. Measurements made before and after sealing identified leaks can indicate the effectiveness of the work.

The blower door can help to locate the source of air leaks in the building by depressurizing to 30 Pascals and searching potential leakage areas with a smoke source. The air distribution system typically leaks on both the supply and return air sides. If the duct system is located outside the conditioned space (e.g., attic, under floor, etc.), supply leaks will depressurize the building and increase the air infiltration rate; return air leaks will pressurize the building, causing air to exfiltrate. A combination of supply and return air leaks is difficult to detect without sealing off the duct system at the registers and measuring the leakage rate of the building compared to that of the unsealed duct system. The difference between the two conditions is a measure of the leakage attributable to the air distribution system.

3.4.3 Airflow Measurement Devices

Two types of anemometers are available for measuring airflow: vane and hot-wire. The volume of air moving through an orifice can be determined by estimating the free area of the opening (e.g., supply air register, exhaust hood face, etc.) and multiplying by the air speed. This result is approximate due to the difficulty in determining the average air speed and the free vent area. Regular calibrations are necessary to assure the accuracy of the instrument. The anemometer can also be used to optimize the face velocity of exhaust hoods by adjusting the door opening until the anemometer indicates the desired airspeed.

Airflow hoods also measure airflow. They contain an airspeed integrating manifold which averages the velocity across the opening and reads out the airflow volume. The hoods are typically made of nylon fabric

supported by an aluminum frame. The instrument is lightweight and easy to hold up against an air vent. The lip of the hood must fit snugly around the opening to assure that all the air volume is measured. Both supply and exhaust airflow can be measured. The result must be adjusted if test conditions fall outside the design range.

3.5 INDUSTRIAL AUDITS

3.5.1 Introduction

Industrial audits are some of the most complex and most interesting audits because of the tremendous variety of equipment found in these facilities. Much of the industrial equipment can be found during commercial audits too. Large chillers, boilers, ventilating fans, water heaters, coolers and freezers, and extensive lighting systems are often the same in most industrial operations as those found in large office buildings or shopping centers. Small cogeneration systems are often found in both commercial and industrial facilities.

The highly specialized equipment that is used in industrial processes is what differentiates these facilities from large commercial operations. The challenge for the auditor and energy management specialist is to learn how this complex—and often unique—industrial equipment operates, and to come up with improvements to the processes and the equipment that can save energy and money. The sheer scope of the problem is so great that industrial firms often hire specialized consulting engineers to examine their processes and recommend operational and equipment changes that result in greater energy productivity.

3.5.2 Audit Services

A few electric and gas utilities are large enough, and well-enough staffed, that they can offer industrial audits to their customers. These utilities have a trained staff of engineers and process specialists with extensive experience who can recommend operational changes or new equipment to reduce the energy costs in a particular production environment. Many gas and electric utilities, even if they do not offer audits, do offer financial incentives for facilities to install high efficiency lighting, motors, chillers, and other equipment. These incentives can make many ECO's very attractive.

Small and medium-sized industries that fall into the Manufacturing Sector—SIC 2000 to 3999, and are in the service area of one of the Energy Analysis and Diagnostic Centers funded by the U.S. Department of Energy, can receive free energy audits throughout this program. There are presently eighteen EADC's operating primarily in the eastern and mid-western areas of the U.S. These EADC's are administered by the University City Science Center in Philadelphia, PA. Companies that are interested in knowing if an EADC is located near them, and if they qualify for an EADC audit can call 215 387-2255 and ask for information on the Energy Analysis and Diagnostic Center program.

3.5.3 Industrial Energy Rate Structures

Except for the smallest industries, facilities will be billed for energy services through a large commercial or industrial rate category. It is important to get this rate structure information for all sources of energy—electricity, gas, oil, coal, steam, etc. Gas, oil and coal are usually billed on a straight cost per unit basis—e.g. $0.90 per gallon of #2 fuel oil. Electricity and steam most often have complex rate structures with components for a fixed customer charge, a demand charge, and an energy charge. Gas, steam, and electric energy are often available with a time of day rate, or an interruptible rate that provides much cheaper energy service with the understanding that the customer may have his supply interrupted (stopped) for periods of several hours at a time. Advance notice of the interruption is almost always given, and the number of times a customer can be interrupted in a given period of time is limited.

3.5.4 Process and Technology Data Sources

For the industrial audit, it is critical to get in advance as much information as possible on the specialized process equipment so that study and research can be performed to understand the particular processes being used, and what improvements in operation or technology are available. Data sources are extremely valuable here; auditors should maintain a library of information on processes and technology and should know where to find additional information from research organizations, government facilities, equipment suppliers and other organizations.

EPRI/GRI

The Electric Power Research Institute (EPRI) and the Gas Research Institute (GRI) are both excellent sources of information on the latest technologies of using electric energy or gas. EPRI has a large number of on-going projects to show the cost-effectiveness of electro-technologies using new processes for heating,

drying, cooling, etc. GRI also has a large number of projects underway to help promote the use of new cost-effective gas technologies for heating, drying, cooling, etc. Both of these organizations provide extensive documentation of their processes and technologies; they also have computer data bases to aid customer inquiries.

U.S. DOE Industrial Division

The U.S. Department of Energy has an Industrial Division that provides a rich source of information on new technologies and new processes. This division funds research into new processes and technologies, and also funds many demonstration projects to help insure that promising improvements get implemented in appropriate industries. The Industrial Division of USDOE also maintains a wide network of contacts with government-related research laboratories such as Oak Ridge National Laboratory, Brookhaven National Laboratory, Lawrence Berkeley National Laboratory, Sandia National Laboratory, and Battelle National Laboratory. These laboratories have many of their own research, development and demonstration programs for improved industrial and commercial technologies.

State Energy Offices

State energy offices are also good sources of information, as well as good contacts to see what kind of incentive programs might be available in the state. Many states offer programs of free boiler tune-ups, free air conditioning system checks, seminars on energy efficiency for various facilities, and other services. Most state energy offices have well-stocked energy libraries, and are also tied into other state energy research organizations, and to national laboratories and the USDOE.

Equipment Suppliers

Equipment suppliers provide additional sources for data on energy efficiency improvements to processes. Marketing new, cost-effective processes and technologies provides sales for the companies as well as helping industries to be more productive and more economically competitive. The energy auditor should compare the information from all of the sources described above.

3.5.5 Conducting the Audit

Safety Considerations

Safety is the primary consideration in any industrial audit. The possibility of injury from hot objects, hazardous materials, slippery surfaces, drive belts, and electric shocks is far greater than when conducting residential and commercial audits. Safety glasses, safety shoes, durable clothing and possibly a safety hat and breathing mask might be needed during some audits. Gloves should be worn while making any electrical measurements, and also while making any measurements around boilers, heaters, furnaces, steam lines, or other very hot pieces of equipment.In all cases, adequate attention to personal safety is a significant feature of any industrial audit.

Lighting

Lighting is not as great a percent of total industrial use as it is in the commercial sector on the average, but lighting is still a big energy use and cost area for many industrial facilities. A complete inventory of all lighting should be taken during the audit visit. Hours of operation of lights are also necessary, since lights are commonly left on when they are not needed. Timers, Energy Management Systems, and occupancy sensors are all valuable approaches to insuring that lights that are not needed are not turned on. It is also important to look at the facility's outside lighting for parking and for storage areas.

During the lighting inventory, types of tasks being performed should also be noted, since light replacement with more efficient lamps often involves changing the color of the resultant light. For example, high pressure sodium lamps are much more efficient than mercury vapor lamps or even metal halide lamps, but they produce a yellowish light that makes fine color distinction difficult. However, many assembly tasks can still be performed adequately under high pressure sodium lighting. These typically include metal fabrication, wood product fabrication, plastic extrusion, and many others.

Electric Motors

A common characteristic of many industries is their extensive use of electric motors. A complete inventory of all motors over 1 HP should be taken, as well as recording data on how long each motor operates during a day. For motors with substantial usage times, replacement with high-efficiency models is almost always cost effective. In addition, consideration should be given to replacement of standard drive belts with synchronous belts which transmit the motor energy more efficiently. For motors which are used infrequently, it may be possible to shift the use to off-peak times, and to achieve a kW demand reduction which would reduce energy cost.

HVAC Systems

An inventory of all space heaters and air conditioners should be taken. Btu per hour ratings and efficiencies of all units should be recorded, as well as usage patterns. Although many industries do not heat or air condition the production floor area, they almost always have office areas, cafeterias, and other areas that are normally heated and air conditioned. For these conditioned areas, the construction of the facility should be noted—how much insulation, what are the walls and ceilings made of, how high are the ceilings. Adding additional insulation might be a cost effective ECO.

Production floors that are not air conditioned often have large numbers of ventilating fans that operate anywhere from one shift per day to 24 hours a day. Plants with high heat loads and plants in the mild climate areas often leave these ventilating fans running all year long. These are good candidates for high efficiency motor replacements. Timers or an Energy Management System might be used to turn off these ventilating fans when the plant is shut down.

Boilers

All boilers should be checked for efficient operation using a stack gas combustion analyzer. Boiler specifications on Btu per hour ratings, pressures and temperatures should be recorded. The boiler should be varied between low-fire, normal-fire, and high-fire, with combustion gas and temperature readings taken at each level. Boiler tune-up is one of the most common, and most energy-saving operations available to many facilities. The auditor should check to see whether any waste heat from the boiler is being recovered for use in a heat recuperator or for some other use such as water heating. If not, this should be noted as a potential ECO.

Specialized Equipment

Most of the remaining equipment encountered during the industrial audit will be the highly specialized process production equipment and machines. This equipment should all be examined and operational data taken, as well as noting hours and periods of use. All heat sources should be considered carefully as to whether they could be replaced with sources using waste heat, or whether a particular heat source could serve as a provider of waste heat to another application. Operations where both heating and cooling occur periodically—such as a plastic extrusion machine—are good candidates for reclaiming waste heat, or in sharing heat from a machine needing cooling with another machine needing heat.

Air Compressors

Air compressors should be examined for size, operating pressures, and type (reciprocating or screw), and whether they use outside cool air for intake. Large air compressors are typically operated at night when much smaller units are sufficient. Also, screw-type air compressors use a large fraction of their rated power when they are idling, so control valves should be installed to prevent this loss. Efficiency is improved with intake air that is cool, so outside air should be used in most cases—except in extremely cold temperature areas.

The auditor should determine whether there are significant air leaks in air hoses, fittings, and in machines. Air leaks are a major source of energy loss in many facilities, and should be corrected by maintenance action. Finally, air compressors are a good source of waste heat. Nearly 90% of the energy used by an air compressor shows up as waste heat, so this is a large source of low temperature waste heat for heating input air to a heater or boiler, or for heating hot water for process use.

3.6 COMMERCIAL AUDITS

3.6.1 Introduction

Commercial audits span the range from very simple audits for small offices to very complex audits for multi-story office buildings or large shopping centers. Complex commercial audits are performed in substantially the same manner as industrial audits. The following discussion highlights those areas where commercial audits are likely to differ from industrial audits.

Commercial audits generally involve substantial consideration of the structural envelope features of the facility, as well as significant amounts of large or specialized equipment at the facility. Office buildings, shopping centers and malls all have complex building envelopes that should be examined and evaluated. Building materials, insulation levels, door and window construction, skylights, and many other envelope features must be considered in order to identify candidate ECO's.

Commercial facilities also have large capacity equipment, such as chillers, space heaters, water heaters, refrigerators, heaters, cookers, and office equipment such as computers and copy machines. Small cogeneration systems are also commonly found in commercial facilities and institutions such as schools and hospitals.

Much of the equipment in commercial facilities is the same type and size as that found in manufacturing or industrial facilities. Potential ECO's would look at more efficient equipment, use of waste heat, or operational changes to use less expensive energy.

3.6.2 Commercial Audit Services

Electric and gas utilities, as well as many engineering consulting firms, perform audits for commercial facilities. Some utilities offer free walk-through audits for commercial customers, and also offer financial incentives for customers who change to more energy efficient equipment. Schools, hospitals and some other government institutions can qualify for free audits under the ICP program described in the first part of this chapter. Whoever conducts the commercial audit must initiate the ICP process by collecting information on the rate energy rate structures, the equipment in use at the facility, and the operational procedures used there.

3.6.3 Commercial Energy Rate Structures

Small commercial customers are usually billed for energy on a per energy unit basis, while large commercial customers are billed under complex rate structures containing components related to energy, rate of energy use (power), time of day or season of year, power factor, and numerous other elements. One of the first steps in a commercial audit is to obtain the rate structures for all sources of energy, and to analyze at least one to two year's worth of energy bills. This information should be put into a table and also plotted.

3.6.4 Conducting the Audit

A significant difference in industrial and commercial audits arises in the area of lighting. Lighting in commercial facilities is one of the largest energy costs—sometimes accounting for half or more of the entire electric bill. Lighting levels and lighting quality are extremely important to many commercial operations. Retail sales operations, in particular, want light levels that are far in excess of standard office values. Quality of light in terms of color is also a big concern in retail sales, so finding acceptable ECO's for reducing lighting costs is much more difficult for retail facilities than for office buildings. The challenge is to find new lighting technologies that allow high light levels and warm color while reducing the wattage required. New T8 and T10 fluorescent lamps, and metal halide lamp replacements for mercury vapor lamps offer these features, and usually represent cost-effective ECO's for retail sales and other facilities.

3.7 RESIDENTIAL AUDITS

Audits for large, multi-story apartment buildings can be very similar to commercial audits. (See section 3.6.) Audits of single-family residences, however, are generally fairly simple. For single-family structures, the energy audit focuses on the thermal envelope and the appliances such as the heater, air conditioner, water heater, and "plug loads."

The residential auditor should start by obtaining past energy bills and analyzing them to determine any patterns or anomalies. During the audit visit, the structure is examined to determine the levels of insulation, the conditions of and seals for windows and doors, and the integrity of the ducts. The space heater and/or air conditioner is inspected, along with the water heater. Equipment model numbers, age, size, and efficiencies are recorded. The post-audit analysis then evaluates potential ECO's such as adding insulation, adding double-pane windows, window shading or insulated doors, and changing to higher efficiency heaters, air conditioners, and water heaters. The auditor calculates costs, benefits, and Simple Payback Periods and presents them to the owner or occupant. A simple audit report—often in the form of a computer printout is given to the owner or occupant.

3.8 INDOOR AIR QUALITY

3.8.1 Introduction

Implementation of new energy-related standards and practices has contributed to a degradation of indoor air quality. In fact, the quality of indoor air has been found to exceed the Environmental Protection Agency (EPA) standards for outdoor air in many homes, businesses, and factories. Thus, testing for air quality problems is done in some energy audits both to prevent exacerbating any existing problems and to recommend ECO's that might improve air quality. Air quality standards for the industrial environment have been published by the American Council of Governmental Industrial Hygienists (ACGIH) in their booklet "Threshold Limit Values." No such standards currently exist for the residential and commercial environments although the ACGIH standards are typically and perhaps inappropriately used. The EPA has been working to develop residential and commercial standards for quite some time.

3.8.2 Symptoms of Air Quality Problems

Symptoms of poor indoor air quality include, but are not limited to: headaches; irritation of mucous membranes such as the nose, mouth, throat, lungs; tearing, redness and irritation of the eyes; numbness of the lips, mouth, throat; mood swings; fatigue; allergies; coughing; nasal and throat discharge; and irritability. Chronic exposure to some compounds can lead to damage to internal organs such as the liver, kidney, lungs, and brain; cancer; and death.

3.8.3 Testing

Testing is required to determine if the air quality is acceptable. Many dangerous compounds, like carbon monoxide and methane without odorant added, are odorless and colorless. Some dangerous particulates such as asbestos fibers do not give any indication of a problem for up to twenty years after inhalation. Testing must be conducted in conjunction with pollution-producing processes to ensure capture of the contaminants. Testing is usually performed by a Certified Industrial Hygienist (CIH).

3.8.4 Types of Pollutants

Airstreams have three types of contaminants: particulates like dust and asbestos; gases like carbon monoxide, ozone, carbon dioxide, volatile organic compounds, anhydrous ammonia, Radon, outgassing from urea-formaldehyde insulation, low oxygen levels; and biologicals like mold, mildew, fungus, bacteria, and viruses.

3.8.5 Pollutant Control Measures

Particulates

Particulates are controlled with adequate filtration near the source and in the air handling system. Mechanical filters are frequently used in return air streams, and baghouses are used for particulate capture. The coarse filters used in most residential air conditioners typically have filtration efficiencies below twenty percent. Mechanical filters called high efficiency particulate apparatus (HEPA) are capable of filtering particles as small as 0.3 microns at up to 99% efficiency. Electrostatic precipitators remove particulates by placing a positive charge on the walls of collection plates and allowing negatively charged particulates to attach to the surface. Periodic cleaning of the plates is necessary to maintain high filtration efficiency. Loose or friable asbestos fibers should be removed from the building or permanently encapsulated to prevent entry into the respirable airstream. While conducting an audit, it is important to determine exactly what type of insulation is in use before disturbing an area to make temperature measurements.

Problem Gases

Problem gases are typically removed by ventilating with outside air. Dilution with outside air is effective, but tempering the temperature and relative humidity of the outdoor air mass can be expensive in extreme conditions. Heat exchangers such as heat wheels, heat pipes, or other devices can accomplish this task with reduced energy use. Many gases can be removed from the airstream by using absorbent/adsorbent media such as activated carbon or zeolite. This strategy works well for spaces with limited ventilation or where contaminants are present in low concentrations. The media must be checked and periodically replaced to maintain effectiveness.

Radon gas—Ra 222—cannot be effectively filtered due to its short half life and the tendency for its Polonium daughters to plate out on surfaces. Low oxygen levels are a sign of inadequate outside ventilation air. A high level of carbon dioxide (e.g., 1000-10,000 ppm) is not a problem in itself but levels above 1000 ppm indicate concentrated human or combustion activity or a lack of ventilation air. Carbon dioxide is useful as an indicator compound because it is easy and inexpensive to measure.

Microbiological Contaminants

Microbiological contaminants generally require particular conditions of temperature and relative humidity on a suitable substrate to grow. Mold and mildew are inhibited by relative humidity levels less than 50%. Air distribution systems often harbor colonies of microbial growth. Many people are allergic to microscopic dust mites. Cooling towers without properly adjusted automated chemical feed systems are an excellent breeding ground for all types of microbial growth.

Ventilation Rates

Recommended ventilation quantities are published by the American Society of Heating, Refrigerating, and Air-Conditioning Engineers (ASHRAE) in standard 62-

1989, "Ventilation for Acceptable Air Quality." These ventilation rates are for effective systems. Many existing systems fail in entraining the air mass efficiently. The density of the contaminants relative to air must be considered in locating the exhaust air intakes and ventilation supply air registers.

Liability

Liability related to indoor air problems appears to be a growing but uncertain issue because few cases have made it through the court system. However, in retrospect, the asbestos and ureaformaldehyde pollution problems discovered in the last two decades suggest proceeding with caution and a proactive approach.

3.9 CONCLUSION

Energy audits are an important first step in the overall pro-cess of reducing energy costs for any building, company, or industry. A thorough audit identifies and analyzes the changes in equipment and operations that will result in cost-effective energy cost reduction. The energy auditor plays a key role in the successful conduct of an audit, and also in the implementation of the audit recommendations.

BIBLIOGRAPHY

Instructions For Energy Auditors, Volumes I and II, U.S. Department of Energy, DOE/CS-0041/12&13, September, 1978. Available through National Technical Information Service, Springfield, VA.

Energy Conservation Guide for Industry and Commerce, National Bureau of Standards Handbook 115 and Supplement, 1978. Available through U.S. Government Printing Office, Washington, DC.

Energy Management, Kennedy, William J., and Turner, Wayne C., Prentice-Hall Publishers, Englewood Cliffs, NJ, 1984.

Illuminating Engineering Society, *IES Lighting Handbook*, New York, NY, 1984.

Total Energy Management, A Handbook prepared by the National Electrical Contractors Association and the National Electrical Manufacturers Association, Washington, DC, Third Edition, 1986.

Handbook of Energy Audits, Thumann, Albert, Third Edition, Fairmont Press, Atlanta, GA, 1987.

Industrial Energy Management and Utilization, Witte, Larry C., Schmidt, Philip S., and Brown, David R., Hemisphere Publishing Corporation, Washington, DC, 1988.

Threshold Limit Values for Chemical Substances and Physical Agents and Biological Exposure Indices, 1990-91 American Conference of Governmental Industrial Hygienists.

Ventilation for Acceptable Indoor Air Quality, ASHRAE 62-1989, American Society of Heating, Refrigerating and Air-Conditioning Engineers, Inc., 1989.

Facility Design and Planning Engineering Weather Data, Departments of the Air Force, the Army, and the Navy, 1978.

CHAPTER 4

ECONOMIC ANALYSIS

DR. DAVID PRATT
Industrial Engineering and Management
Oklahoma State University
Stillwater, OK

4.1 OBJECTIVE

The objective of this chapter is to present a coherent, consistent approach to economic analysis of capital investments (energy related or other). Adherence to the concepts and methods presented will lead to sound investment decisions with respect to time value of money principles. The chapter opens with material designed to motivate the importance of life cycle cost concepts in the economic analysis of projects. The next three sections provide foundational material necessary to fully develop time value of money concepts and techniques. These sections present general characteristics of capital investments, sources of funds for capital investment, and a brief summary of tax considerations which are important for economic analysis. The next two sections introduce time value of money calculations and several approaches for calculating project measures of worth based on time value of money concepts. Next the measures of worth are applied to the process of making decisions when a set of potential projects are to be evaluated. The final concept and technique section of the chapter presents material to address several special problems that may be encountered in economic analysis. This material includes, among other things, discussions of inflation, non-annual compounding of interest, and sensitivity analysis. The chapter closes with a brief summary and a list of references which can provide additional depth in many of the areas covered in the chapter.

4.2 INTRODUCTION

Capital investment decisions arise in many circumstances. The circumstances range from evaluating business opportunities to personal retirement planning. Regardless of circumstances, the basic criterion for evaluating any investment decision is that the revenues (savings) generated by the investment must be greater than the costs incurred. The number of years over which the

revenues accumulate and the comparative importance of future dollars (revenues or costs) relative to present dollars are important factors in making sound investment decisions. This consideration of costs over the entire life cycle of the investments gives rise to the name *life cycle cost* analysis which is commonly used to refer to the economic analysis approach presented in this chapter. An example of the importance of life cycle costs is shown in Figure 4.1 which depicts the estimated costs of owning and operating an oil-fired furnace to heat a 2,000-square-foot house in the northeast United States. Of particular note is that the initial costs represent only 23% of the total costs incurred over the life of the furnace. The life cycle cost approach provides a significantly better evaluation of long term implications of an investment than methods which focus on first cost or near term results.

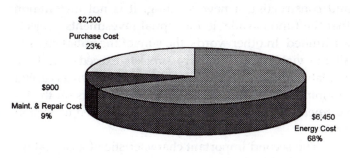

Figure 4.1 15-Year life cycle costs of a heating system

Life cycle cost analysis methods can be applied to virtually any public or private business sector investment decision as well as to personal financial planning decisions. Energy related decisions provide excellent examples for the application of this approach. Such decisions include: evaluation of alternative building designs which have different initial costs, operating and maintenance costs, and perhaps different lives; evaluation of investments to improve the thermal performance of an existing building (wall or roof insulation, window glazing); or evaluation of alternative heating, ventilating, or air conditioning systems. For federal buildings, Congress and the President have mandated, through legislation and executive order, energy conservation goals that must be met using cost-effective measures. The life cycle cost approach is mandated as the means of evaluating cost effectiveness.

4.3 GENERAL CHARACTERISTICS OF CAPITAL INVESTMENTS

4.3.1 Capital Investment Characteristics

When companies spend money, the outlay of cash can be broadly categorized into one of two classifications; expenses or capital investments. Expenses are generally those cash expenditures that are routine, on-going, and necessary for the ordinary operation of the business. Capital investments, on the other hand, are generally more strategic and have long term effects. Decisions made regarding capital investments are usually made at higher levels within the organizational hierarchy and carry with them additional tax consequences as compared to expenses.

Three characteristics of capital investments are of concern when performing life cycle cost analysis. First, capital investments usually require a relatively large initial cost. "Relatively large" may mean several hundred dollars to a small company or many millions of dollars to a large company. The initial cost may occur as a single expenditure such as purchasing a new heating system or occur over a period of several years such as designing and constructing a new building. It is not uncommon that the funds available for capital investments projects are limited. In other words, the sum of the initial costs of all the viable and attractive projects exceeds the total available funds. This creates a situation known as capital rationing which imposes special requirements on the investment analysis. This topic will be discussed in Section 4.8.3.

The second important characteristic of a capital investment is that the benefits (revenues or savings) resulting from the initial cost occur in the future, normally over a period of years. The period between the initial cost and the last future cash flow is the life cycle or life of the investment. It is the fact that cash flows occur over the investment's life that requires the introduction of time value of money concepts to properly evaluate investments. If multiple investments are being evaluated and if the lives of the investments are not equal, special consideration must be given to the issue of selecting an appropriate planning horizon for the analysis. Planning horizon issues are introduced in Section 4.8.5.

The last important characteristic of capital investments is that they are relatively irreversible. Frequently, after the initial investment has been made, terminating or significantly altering the nature of a capital investment has substantial (usually negative) cost consequences. This is one of the reasons that capital investment decisions are usually evaluated at higher levels of the organizational hierarchy than operating expense decisions.

4.3.2 Capital Investment Cost Categories

In almost every case, the costs which occur over the life of a capital investment can be classified into one of the following categories:

- Initial Cost,
- Annual Expenses and Revenues,
- Periodic Replacement and Maintenance, or
- Salvage Value.

As a simplifying assumption, the cash flows which occur during a year are generally summed and regarded as a single end-of-year cash flow. While this approach does introduce some inaccuracy in the evaluation, it is generally not regarded as significant relative to the level of estimation associated with projecting future cash flows.

Initial costs include all costs associated with preparing the investment for service. This includes purchase cost as well as installation and preparation costs. Initial costs are usually nonrecurring during the life of an investment. Annual expenses and revenues are the recurring costs and benefits generated throughout the life of the investment. Periodic replacement and maintenance costs are similar to annual expenses and revenues except that they do not (or are not expected to) occur annually. The salvage (or residual) value of an investment is the revenue (or expense) attributed to disposing of the investment at the end of its useful life.

4.3.3 Cash Flow Diagrams

A convenient way to display the revenues (savings) and costs associated with an investment is a *cash flow diagram*. By using a cash flow diagram, the timing of the cash flows are more apparent and the chances of properly applying time value of money concepts are increased. With practice, different cash flow patterns can be recognized and they, in turn, may suggest the most direct approach for analysis.

It is usually advantageous to determine the time frame over which the cash flows occur first. This establishes the horizontal scale of the cash flow diagram. This scale is divided into time periods which are frequently, but not always, years. Receipts and disbursements are then located on the time scale in accordance with the problem specifications. Individual outlays or receipts are indicated by drawing vertical lines appropriately placed along the time scale. The relative magnitudes can be suggested by the heights, but exact scaling generally does not enhance the meaningfulness of the diagram. Upward directed lines indicate cash inflow (revenues or savings) while downward directed lines indicate cash outflow (costs).

Figure 4.2 illustrates a cash flow diagram. The cash flows depicted represent an economic evaluation of whether to choose a baseboard heating and window air conditioning system or a heat pump for a ranger's house in a national park [Fuller and Petersen, 1994]. The differential costs associated with the decision are:

- The heat pump costs (cash outflow) $1500 more than the baseboard system,

- The heat pump saves (cash inflow) $380 annually in electricity costs,

- The heat pump has a $50 higher annual maintenance costs (cash outflow),

- The heat pump has a $150 higher salvage value (cash inflow) at the end of 15 years,

- The heat pump requires.$200 more in replacement maintenance (cash outflow) at the end of year 8.

Although cash flow diagrams are simply graphical representations of income and outlay, they should exhibit as much information as possible. During the analysis phase, it is useful to show the Minimum Attractive Rate of Return (an interest rate used to account for the time value of money within the problem) on the cash flow diagram, although this has been omitted in Figure 4.2. The requirements for a good cash flow diagram are completeness, accuracy, and legibility. The measure of a successful diagram is that someone else can understand the problem fully from it

4.4 SOURCES OF FUNDS

Capital investing requires a source of funds. For large companies multiple sources may be employed. The process of obtaining funds for capital investment is called financing. There are two broad sources of financial funding; debt financing and equity financing. Debt financing involves borrowing and utilizing money which is to be repaid at a later point in time. Interest is paid to the lending party for the privilege of using the money. Debt financing does not create an ownership position for the lender within the borrowing organization. The borrower is simply obligated to repay the borrowed funds plus accrued interest according to a repayment schedule. Car loans and mortgage loans are two examples of this type of financing. The two primary sources of debt capital are loans and bonds. The cost of capital associated with debt financing is relatively easy to calculate since interest rates and repayment schedules are usually clearly documented in the legal instruments controlling the financing arrangements. An added benefit to debt financing under current U.S. tax law (October 1995) is that the interest payments made by corporations on debt capital are tax deductible. This effectively lowers the cost of debt financing since for debt financing with deductible interest payments, the after-tax cost of capital is given by:

$$\text{Cost of Capital}_{AFTERTAX} =$$
$$\text{Cost of Capital}_{BEFORETAX} * (1 - \text{TaxRate})$$
where the tax rate is determine by applicable tax law.

The second broad source of funding is equity financing. Under equity financing the lender acquires an ownership (or equity) position within the borrower's organization. As a result of this ownership position, the

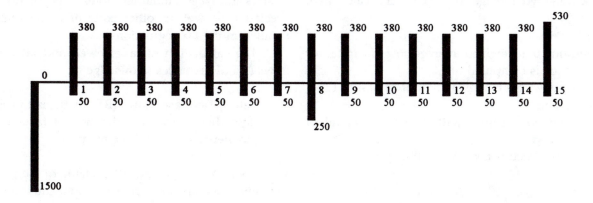

Figure 4.2. Heat pump and baseboard system differential life cycle costs

lender has the right to participate in the financial success of the organization as a whole. The two primary sources of equity financing are stocks and retained earnings. The cost of capital associated with shares of stock is much debated within the financial community. A detailed presentation of the issues and approaches is beyond the scope of this chapter. Additional reference material can be found in Park and Sharp-Bette [1990]. One issue over which there is general agreement is that the cost of capital for stocks is higher than the cost of capital for debt financing. This is at least partially attributable to the fact that interest payments are tax deductible while stock dividend payments are not.

If any subject is more widely debated in the financial community than the cost of capital for stocks, it is the cost of capital for retained earnings. Retained earnings are the accumulation of annual earnings surpluses that a company retains within the company's coffers rather than pays out to the stockholders as dividends. Although these earnings are held by the company, they truly belong to the stockholders. In essence the company is establishing the position that by retaining the earnings and investing them in capital projects, stockholders will achieve at least as high a return through future financial successes as they would have earned if the earnings had been paid out as dividends. Hence, one common approach to valuing the cost of capital for retained earnings is to apply the same cost of capital as for stock. This, therefore, leads to the same generally agreed result. The cost of capital for financing through retained earnings generally exceeds the cost of capital for debt financing.

In many cases the financing for a set of capital investments is obtained by packaging a combination of the above sources to achieve a desired level of available funds. When this approach is taken, the overall cost of capital is generally taken to be the weighted average cost of capital across all sources. The cost of each individual source's funds is weighted by the source's fraction of the total dollar amount available. By summing across all sources, a weighted average cost of capital is calculated.

Example 1

Determine the weighted average cost of capital for financing which is composed of:

25% loans with a before tax cost of capital of 12%/yr and

75% retained earnings with a cost of capital of 10%/yr.

The company's effective tax rate is 34%.

Cost of Capital$_{LOANS}$ = 12% * (1 - 0.34) = 7.92%

Cost of Capital$_{RETAINEDEARNINGS}$ = 10%

Weighted Average Cost of Capital = (0.25)*7.92% + (0.75)*10.00% = 9.48%

4.5 TAX CONSIDERATIONS

4.5.1 After Tax Cash Flows

Taxes are a fact of life in both personal and business decision making. Taxes occur in many forms and are primarily designed to generate revenues for governmental entities ranging from local authorities to the Federal government. A few of the most common forms of taxes are income taxes, ad valorem taxes, sales taxes, and excise taxes. Cash flows used for economic analysis should always be adjusted for the combined impact of all relevant taxes. To do otherwise, ignores the significant impact that taxes have on economic decision making. Tax laws and regulations are complex and intricate. A detailed treatment of tax considerations as they apply to economic analysis is beyond the scope of this chapter and generally requires the assistance of a professional with specialized training in the subject. A high level summary of concepts and techniques that concentrate on Federal income taxes are presented in the material which follows. The focus is on Federal income taxes since they impact most decisions and have relatively wide and general application.

The amount of Federal taxes due are determined based on a tax rate multiplied by a taxable income. The rates (as of October 1995) are determined based on tables of rates published under the Omnibus Reconciliation Act of 1993 as shown in Table 4.1. Depending on income range, the marginal tax rates vary from 15% of taxable income to 39% of taxable income. *Taxable income* is calculated by subtracting *allowable deductions* from *gross income*. Gross income is generated when a company sells its product or service. Allowable deductions include salaries and wages, materials, interest payments, and depreciation as well as other costs of doing business as detailed in the tax regulations.

The calculation of taxes owed and after tax cash flows (ATCF) requires knowledge of:

• Before Tax Cash Flows (BTCF), the net project cash flows before the consideration of taxes due, loan payments, and bond payments;

• Total loan payments attributable to the project, including a breakdown of principal and interest components of the payments;

Table 4.1 Federal tax rates based on the Omnibus Reconciliation Act of 1993

Taxable Income (TI)	Taxes Due	Marginal Tax Rate
$0 < TI ≤ $50,000	0.15*TI	0.15
$50,000 < TI ≤ $75,000	$7,500+0.25(TI-$50,000)	0.25
$75,000 < TI ≤ $100,000	$13,750+0.34(TI-$75,000)	0.34
$100,000 < TI ≤ $335,000	$22,250+0.39(TI-$100,000)	0.39
$335,000 < TI ≤ $10,000,000	$113,900+0.34(TI-$335,000)	0.34
$10,000,000 < TI ≤ $15,000,000	$3,400,000+0.35(TI-$10,000,000)	0.35
$15,000,000 < TI ≤ $18,333,333	$5,150,000+0.38(TI-$15,000,000)	0.38
$18,333,333 < TI	$6,416,667+0.35(TI-$18,333,333)	0.35

- Total bond payments attributable to the project, including a breakdown of the redemption and interest components of the payments; and

- Depreciation allowances attributable to the project.

Given the availability of the above information, the procedure to determine the ATCF on a year-by-year basis proceeds using the following calculation for each year:

- Taxable Income = BTCF - Loan Interest - Bond Interest - Deprecation

- Taxes = Taxable Income * Tax Rate

- ATCF = BTCF - Total Loan Payments - Total Bond Payments - Taxes

An important observation is that Depreciation reduces Taxable Income (hence, taxes) but does not directly enter into the calculation of ATCF since it is not a true cash flow. It is not a true cash flow because no cash changes hands, depreciation is an accounting concept design to stimulate business by reducing taxes over the life of an asset. The next section provides additional information about depreciation.

4.5.2 Depreciation

Most assets used in the course of a business decrease in value over time. U.S. Federal income tax law permits reasonable deductions from taxable income to allow for this. These deductions are called depreciation allowances. To be depreciable, an asset must meet three primary conditions: (1) it must be held by the business for the purpose of producing income, (2) it must wear out or be consumed in the course of its use, and (3) it must have a life longer than a year.

Many methods of depreciation have been allowed under U.S. tax law over the years. Among these methods are straight line, sum-of-the-years digits, declining balance, and the accelerated cost recovery system. Descriptions of these methods can be found in many references including economic analysis text books [White, Agee, Case, 1989]. The method currently used for depreciation of assets placed in service after 1986 is the Modified Accelerated Cost Recovery System (MACRS). Determination of the allowable MACRS depreciation deduction for an asset is a function of (1) the asset's property class, (2) the asset's basis, and (3) the year within the asset's recovery period for which the deduction is calculated.

Eight property classes are defined for assets which are depreciable under MACRS. The property classes and several examples of property that fall into each class are shown in Table 4.2. Professional tax guidance is recommended to determine the MACRS property class for a specific asset.

The basis of an asset is the cost of placing the asset in service. In most cases, the basis includes the purchase cost of the asset plus the costs necessary to place the asset in service (e.g., installation charges).

Given an asset's property class and its depreciable basis the depreciation allowance for each year of the asset's life can be determined from tabled values of MACRS percentages. The MACRS percentages specify the percentage of an asset's basis that are allowable as deductions during each year of an asset's recovery period. The MACRS percentages by recovery year (age of the asset) and property class are shown in Table 4.3.

Table 4.2 MACRS property classes

Property Class	Example Assets
3-Year Property	special handling devices for food special tools for motor vehicle manufacturing
5-Year Property	computers and office machines general purpose trucks
7-Year Property	office furniture most manufacturing machine tools
10-Year Property	tugs & water transport equipment petroleum refining assets
15-Year Property	fencing and landscaping cement manufacturing assets
20-Year Property	farm buildings utility transmission lines and poles
27.5-Year Residential Rental Property	rental houses and apartments
31.5-Year Nonresidential Real Property	business buildings

Example 2

Determine depreciation allowances during each recovery year for a MACRS 5-year property with a basis of $10,000.

Year 1 deduction: $10,000 * 20.00% = $2,000
Year 2 deduction: $10,000 * 32.00% = $3,200
Year 3 deduction: $10,000 * 19.20% = $1,920
Year 4 deduction: $10,000 * 11.52% = $1,152
Year 5 deduction: $10,000 * 11.52% = $1,152
Year 6 deduction: $10,000 * 5.76% = $576

The sum of the deductions calculated in Example 2 is $10,000 which means that the asset is "fully depreciated" after six years. Though not shown here, tables similar to Table 4.3 are available for the 27.5-Year and 31.5-Year property classes. There usage is similar to that outlined above except that depreciation is calculated monthly rather than annually.

4.6 TIME VALUE OF MONEY CONCEPTS

4.6.1 Introduction

Most people have an intuitive sense of the time value of money. Given a choice between $100 today and $100 one year from today, almost everyone would prefer the $100 today. Why is this the case? Two primary factors lead to this time preference associated with money; interest and inflation. Interest is the ability to earn a return on money which is loaned rather than consumed. By taking the $100 today and placing it in an interest bearing bank account (i.e., loaning it to the bank), one year from today an amount greater than $100 would be available for withdrawal. Thus, taking the $100 today and loaning it to earn interest, generates a sum greater than $100 one year from today and thus is preferred. The amount in excess of $100 that would be available depends upon the interest rate being paid by the bank. The next section develops the mathematics of the relationship between interest rates and the timing of cash flows.

The second factor which leads to the time preference associated with money is inflation. Inflation is a complex subject but in general can be described as a decrease in the purchasing power of money. The impact of inflation is that the "basket of goods" a consumer can buy today with $100 contains more than the "basket" the consumer could buy one year from today. This decrease in purchasing power is the result of inflation. The subject of inflation is addressed in Section 4.9.4.

4.6.2 The Mathematics of Interest

The mathematics of interest must account for the amount and timing of cash flows. The basic formula for studying and understanding interest calculations is:

Table 4.3 MACRS percentages by recovery year and property class

Recovery Year	3-Year Property	5-Year Property	7-Year Property	10-Year Property	15-Year Property	20-Year Property
1	33.33%	20.00%	14.29%	10.00%	5.00%	3.750%
2	44.45%	32.00%	24.49%	18.00%	9.50%	7.219%
3	14.81%	19.20%	17.49%	14.40%	8.55%	6.677%
4	7.41%	11.52%	12.49%	11.52%	7.70%	6.177%
5		11.52%	8.93%	9.22%	6.93%	5.713%
6		5.76%	8.92%	7.37%	6.23%	5.285%
7			8.93%	6.55%	5.90%	4.888%
8			4.46%	6.55%	5.90%	4.522%
9				6.56%	5.91%	4.462%
10				6.55%	5.90%	4.461%
11				3.28%	5.91%	4.462%
12					5.90%	4.461%
13					5.91%	4.462%
14					5.90%	4.461%
15					5.91%	4.462%
16					2.95%	4.461%
17						4.462%
18						4.461%
19						4.462%
20						4.461%
21						2.231%

$$F_n = P + I_n$$

where: F_n = a future amount of money at the *end* of the nth year,

P = a present amount of money at the beginning of the year which is n years prior to F_n,

I_n = the amount of accumulated interest over n years, and

n = the number of years between P and F

The goal of studying the mathematics of interest is to develop a formula for F_n which is expressed only in terms of the present amount P, the annual interest rate i, and the number of years n. There are two major approaches for determining the value of I_n; simple interest and compound interest. Under simple interest, interest is earned (charged) only on the original amount loaned (borrowed). Under compound interest, interest is earned (charged) on the original amount loaned (borrowed) plus any interest accumulated from previous periods.

4.6.3 Simple Interest

For simple interest, interest is earned (charged) only on the original principal amount at the rate of i% per year (expressed as i%/yr). Table 4.4 illustrates the annual calculation of simple interest. In Table 4.4 and the formulas which follow, the interest rate i is to be expressed as a decimal amount (e.g., 8% interest is expressed as 0.08).

At the beginning of year 1 (end of year 0), P dollars (e.g., $100) are deposited in an account earning i%/yr (e.g., 8%/yr or 0.08) simple interest. Under simple compounding, during year 1 the P dollars ($100) earn P*i

Table 4.4 The mathematics of simple interest

Year (t)	Amount At Beginning Of Year	Interest Earned During Year	Amount At End Of Year (F_t)
0	-	-	P
1	P	Pi	P + Pi = P (1 + i)
2	P (1 + i)	Pi	P (1+ i) + Pi = P (1 + 2i)
3	P (1 + 2i)	Pi	P (1+ 2i) + Pi = P (1 + 3i)
n	P (1 + (n-1)i)	Pi	P (1+ (n-1)i) + Pi = P (1 + ni)

dollars ($100*0.08 = $8) of interest. At the end of the year 1 the balance in the account is obtained by adding P dollars (the original principal, $100) plus P*i (the interest earned during year 1, $8) to obtain P+P*i ($100+$8=$108). Through algebraic manipulation, the end of year 1 balance can be expressed mathematically as P*(1+i) dollars ($100*1.08=$108).

The beginning of year 2 is the same point in time as the end of year 1 so the balance in the account is P*(1+i) dollars ($108). During year 2 the account again earns P*i dollars ($8) of interest since under simple compounding, interest is paid only on the <u>original</u> principal amount P ($100). Thus at the end of year 2, the balance in the account is obtained by adding P dollars (the original principal) plus P*i (the interest from year 1) plus P*i (the interest from year 2) to obtain P+P*i+P*i ($100+$8+$8=$116). After some algebraic manipulation, this can be written conveniently mathematically as P*(1+2*i) dollars ($100*1.16=$116).

Table 4.4 extends the above logic to year 3 and then generalizes the approach for year n. If we return our attention to our original goal of developing a formula for F_n which is expressed only in terms of the present amount P, the annual interest rate i, and the number of years n, the above development and Table 4.4 results can be summarized as follows:

For Simple Interest
$$F_n = P (1+n*i)$$

Example 3

Determine the balance which will accumulate at the end of year 4 in an account which pays 10%/yr simple interest if a deposit of $500 is made today.

$$F_n = P * (1 + n*i)$$
$$F_4 = 500 * (1 + 4*0.10)$$
$$F_4 = 500 * (1 + 0.40)$$
$$F_4 = 500 * (1.40)$$
$$F_4 = $700$$

4.6.4 Compound Interest

For compound interest, interest is earned (charged) on the original principal amount <u>plus any accumulated interest from previous years</u> at the rate of i% per year (i%/yr). Table 4.5 illustrates the annual calculation of compound interest. In the Table 4.5 and the formulas which follow, i is expressed as a decimal amount (i.e., 8% interest is expressed as 0.08).

At the beginning of year 1 (end of year 0), P dollars (e.g., $100) are deposited in an account earning i%/yr (e.g., 8%/yr or 0.08) compound interest. Under compound interest, during year 1 the P dollars ($100) earn P*i dollars ($100*0.08 = $8) of interest. Notice that this the same as the amount earned under simple compounding. This result is expected since the interest earned in previous years is zero for year 1. At the end of the year 1 the balance in the account is obtain by adding P dollars (the original principal, $100) plus P*i (the interest earned during year 1, $8) to obtain P+P*i ($100+$8=$108). Through algebraic manipulation, the end of year 1 balance can be expressed mathematically as P*(1+i) dollars ($100*1.08=$108).

During year 2 and subsequent years, we begin to see the power (if you are a lender) or penalty (if you are a borrower) of compound interest over simple interest.

Table 4.5 The Mathematics of Compound Interest

Year (t)	Amount At Beginning Of Year	Interest Earned During Year	Amount At End Of Year (F_t)
0	-	-	P
1	P	Pi	$P + Pi$ $= P(1 + i)$
2	$P(1 + i)$	$P(1 + i)i$	$P(1 + i) + P(1 + i)i$ $= P(1 + i)(1 + i)$ $= P(1+i)^2$
3	$P(1+i)^2$	$P(1+i)^2 i$	$P(1+i)^2 + P(1 + i)^2 i$ $= P(1 + i)^2 (1 + i)$ $= P(1+i)^3$
n	$P(1+i)^{n-1}$	$P(1+i)^{n-1} i$	$P(1+i)^{n-1} + P(1 + i)^{n-1} i$ $= P(1 + i)^{n-1}(1 + i)$ $= P(1+i)^n$

The beginning of year 2 is the same point in time as the end of year 1 so the balance in the account is $P*(1+i)$ dollars ($108). During year 2 the account earns i% interest on the original principal, P dollars ($100), <u>and</u> it earns i% interest on the accumulated interest from year 1, $P*i$ dollars ($8). Thus the interest earned in year 2 is $[P+P*i]*i$ dollars ($[$100+$8]*0.08=$8.64). The balance at the end of year 2 is obtained by adding P dollars (the original principal) plus $P*i$ (the interest from year 1) plus $[P+P*i]*i$ (the interest from year 2) to obtain $P+P*i+[P+P*i]*i$ dollars ($100+$8+$8.64=$116.64). After some algebraic manipulation, this can be written conveniently mathematically as $P*(1+i)^n$ dollars ($100*1.082 =$116.64).

Table 4.5 extends the above logic to year 3 and then generalizes the approach for year n. If we return our attention to our original goal of developing a formula for F_n which is expressed only in terms of the present amount P, the annual interest rate i, and the number of years n, the above development and Table 4.5 results can be summarized as follows:

For Compound Interest
$$F_n = P(1+i)^n$$

Example 4

Repeat Example 3 using compound interest rather than simple interest.

$F_n = P * (1 + i)^n$

$F_4 = 500 * (1 + 0.10)^4$

$F_4 = 500 * (1.10)^4$

$F_4 = 500 * (1.4641)$

$F_4 = \$732.05$

Notice that the balance available for withdrawal is higher under compound interest ($732.05 > $700.00). This is due to earning interest on principal plus interest rather than earning interest on just original principal. Since compound interest is by far more common in practice than simple interest, the remainder of this chapter is based on <u>compound interest</u> unless explicitly stated otherwise.

4.6.5 Single Sum Cash Flows

Time value of money problems involving compound interest are common. Because of this frequent need, tables of compound interest time value of money factors can be found in most books and reference manuals that deal with economic analysis. The factor $(1+i)^n$ is known as the <u>single sum, future worth factor</u> or the <u>single payment, compound amount factor</u>. This factor is denoted $(F|P,i,n)$ where F denotes a future amount, P denotes a present amount, i is an interest rate (expressed as a percentage amount), and n denotes a number of years. The factor (F|P,i,n) is read "to find F given P at i% for n years." Tables of values of (F|P,i,n) for selected values of i and n are provided in Appendix 4A. The tables of values in Appendix 4A are organized such that the annual interest rate (i) determines the appropriate page, the time value of money factor (F|P) determines the appropriate column, and the number of years (n) determines the appropriate row.

Example 5
Repeat Example 4 using the single sum, future worth factor.

$F_n = P * (1 + i)^n$

$F_n = P * (F|P,i,n)$

$F_4 = 500 * (F|P,10\%,4)$

$F_4 = 500 * (1.4641)$

$F_4 = 732.05$

The above formulas for compound interest allow us to solve for an unknown F given P, i, and n. What if we want to determine P with known values of F, i, and n? We can derive this relationship from the compound interest formula above:

$$F_n = P (1+i)^n$$
dividing both sides by $(1+i)^n$ yields

$$P = \frac{F_n}{(1+i)^n}$$

which can be rewritten as
$$P = F_n (1+i)^{-n}$$

The factor $(1+i)^{-n}$ is known as the <u>single sum, present worth factor</u> or the <u>single payment, present worth factor</u>. This factor is denoted (P|F,i,n) and is read "to find P given F at i% for n years." Tables of (P|F,i,n) are provided in Appendix 4A.

Example 6
To accumulate $1000 five years from today in an account earning 8%/yr compound interest, how much must be deposited today?

$P = F_n * (1 + i)^{-n}$

$P = F_5 * (P|F,i,n)$

$P = 1000 * (P|F,8\%,5)$

$P = 1000 * (0.6806)$

$P = 680.60$

To verify your solution, try multiplying 680.60 * (F|P,8%,5). What would expect for a result? (Answer: $1000) If your still not convinced, try building a table like Table 4.5 to calculate the year end balances each year for five years.

4.6.6 Series Cash Flows

Having considered the transformation of a single sum to a future worth when given a present amount and vice versa, let us generalize to a series of cash flows. The future worth of a series of cash flows is simply the sum of the future worths of each individual cash flow. Similarly, the present worth of a series of cash flows is the sum of the present worths of the individual cash flows.

Example 7
Determine the future worth (accumulated total) at the end of seven years in an account that earns 5%/yr if a $600 deposit is made today and a $1000 deposit is made at the end of year two?

for the $600 deposit, n=7 (years between today and end of year 7)

for the $1000 deposit, n=5 (years between end of year 2 and end of year 7)

$F7 = 600 * (F|P,5\%,7) + 1000 * (F|P,5\%,5)$

$F7 = 600 * (1.4071) + 1000 * (1.2763)$

$F7 = 844.26 + 1276.30 = \$2120.56$

Example 8
Determine the amount that would have to be deposited today (present worth) in an account paying 6%/yr interest if you want to withdraw $500 four years from today and $600 eight years from today (leaving zero in the account after the $600 withdrawal).

for the $500 deposit n=4, for the $600 deposit n=8

$P = 500 * (P|F,6\%,4) + 600 * (P|F,6\%,8)$

$P = 500 * (0.7921) + 600 * (0.6274)$

$P = 396.05 + 376.44 = \$772.49$

4.6.7 Uniform Series Cash Flows

A uniform series of cash flows exists when the cash flows in a series occur every year and are all equal in value. Figure 4.3 shows the cash flow diagram of a uniform series of withdrawals. The uniform series has length 4 and amount 2000. If we want to determine the amount of money that would have to be deposited today to support this series of withdrawals starting one year from today, we could use the approach illustrated in Example 8 above to determine a present worth component for each individual cash flow. This approach would require us to sum the following series of factors (assuming the interest rate is 9%/yr):

P =   2000*(P|F,9%,1) + 2000*(P|F,9%,2) +
      2000*(P|F,9%,3) + 2000*(P|F,9%,4)

After some algebraic manipulation, this expression can be restated as:

P =   2000*[(P|F,9%,1) + (P|F,9%,2) +
      (P|F,9%,3) + (P|F,9%,4)]
P =   2000*[(0.9174) + (0.8417) + (0.7722) + (0.7084)]
P =   2000*[3.2397] = $6479.40

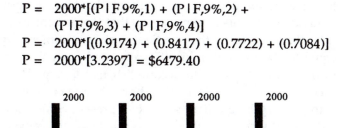

Figure 4.3. Uniform series cash flow

Fortunately, uniform series occur frequently enough in practice to justify tabulating values to eliminate the need to repeatedly sum a series of (P|F,i,n) factors. To accommodate uniform series factors, we need to add a new symbol to our time value of money terminology in addition to the single sum symbols P and F. The symbol "A" is used to designate a uniform series of cash flows. When dealing with uniform series cash flows, the symbol A represents the amount of each annual cash flow and the n represents the number of cash flows in the series. The factor (P|A,i,n) is known as the <u>uniform series, present worth factor</u> and is read "to find P given A at i% for n years." Tables of (P|A,i,n) are provided in Appendix 4A. An algebraic expression can also be derived for the (P|A,i,n) factor which expresses P in terms of A, i, and n. The derivation of this formula is omitted here, but the resulting expression is shown in the summary table (Table 4.6) at the end of this section.

An important observation when using a (P|A,i,n) factor is that the "P" resulting from the calculation occurs one period prior to the first "A" cash flow. In our example the first withdrawal (the first "A") occurred one year after the deposit (the "P"). Restating the example problem above using a (P|A,i,n) factor, it becomes:

P = A * (P|A,i,n)

P = 2000 * (P|A,9%,4)

P = 2000 * (3.2397) = $6479.40

This result is identical (as expected) to the result using the (P|F,i,n) factors. In both cases the interpretation of the result is as follows: if we deposit $6479.40 in an account paying 9%/yr interest, we could make withdrawals of $2000 per year for four years starting one year after the initial deposit to deplete the account at the end of 4 years.

The reciprocal relationship between P and A is symbolized by the factor (A|P,i,n) and is called the <u>uniform series, capital recovery factor</u>. Tables of (A|P,i,n) are provided in Appendix 4A and the algebraic expression for (A|P,i,n) is shown in Table 4.6 at the end of this section. This factor enables us to determine the amount of the equal annual withdrawals "A" (starting one year after the deposit) that can be made from an initial deposit of "P."

Example 9

Determine the equal annual withdrawals that can be made for 8 years from an initial deposit of $9000 in an account that pays 12%/yr. The first withdrawal is to be made one year after the initial deposit.

A = P * (A|P,12%,8)

A = 9000 * (0.2013)

A = $1811.70

Factors are also available for the relationships between a future worth (accumulated amount) and a uniform series. The factor (F|A,i,n) is known as the <u>uniform series future worth</u> factor and is read "to find F given A at i% for n years." The reciprocal factor, (A|F,i,n), is known as the <u>uniform series sinking fund</u> factor and is read "to find A given F at i% for n years." An important observation when using an (F|A,i,n) factor or an (A|F,i,n) factor is that the "F" resulting from the calculation occurs at the same point in time as the last "A" cash flow. The algebraic expressions for (A|F,i,n) and (F|A,i,n) are shown in Table 6 at the end of this section.

Example 10

If you deposit $2000 per year into an individual retirement account starting on your 24th birthday, how much will have accumulated in the account at the time of your deposit on your 65th birthday? The account pays 6%/yr.

n = 42 (birthdays between 24th and 65th, inclusive)

F = A * (F|A,6%,42)

F = 2000 * (175.9505) = \$351,901

Example 11

If you want to be a millionaire on your 65th birthday, what equal annual deposits must be made in an account starting on your 24th birthday? The account pays 10%/yr.

n = 42 (birthdays between 24th and 65th, inclusive)

A = F * (A|F,10%,42)

A = 1000000 * (0.001860) = \$1860

4.6.8 Gradient Series

A gradient series of cash flows occurs when the value of a given cash flow is greater than the value of the previous period's cash flow by a constant amount. The symbol used to represent the constant increment is G. The factor (P|G,i,n) is known as the <u>gradient series, present worth factor</u>. Tables of (P|G,i,n) are provided in Appendix 4A. An algebraic expression can also be derived for the (P|G,i,n) factor which expresses P in terms of G, i, and n. The derivation of this formula is omitted here, but the resulting expression is shown in the summary table (Table 4.6) at the end of this section.

It is not uncommon to encounter a cash flow series that is the sum of a uniform series and a gradient series. Figure 4.4 illustrates such a series. The uniform component of this series has a value of 1000 and the gradient series has a value of 500. By convention the first element of a gradient series has a zero value. Therefore, in Figure 4.4, both the uniform series and the gradient series have length four (n=4). Like the uniform series factor, the "P" calculated by a (P|G,i,n) factor is located one period before the first element of the series (which is the zero element for a gradient series).

Example 12

Assume you wish to make the series of withdrawals illustrated in Figure 4.4 from an account which pays 15%/yr. How much money would you have to deposit today such that the account is depleted at the time of the last withdrawal?

This problem is best solved by recognizing that the cash flows are a combination of a uniform series of value 1000 and length 4 (starting at time=1) plus a gradient series of size 500 and length 4 (starting at time=1).

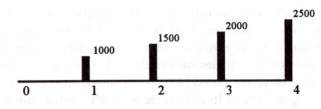

Figure 4.4. Combined uniform series and gradient series cash flow

P = A * (P|A,15%,4) + G * (P|G,15%,4)

P = 1000 * (2.8550) + 500 * (3.7864)

P = 2855.00 + 1893.20 = \$4748.20

Occasionally it is useful to convert a gradient series to an equivalent uniform series of the same length. Equivalence in this context means that the present value (P) calculated from the gradient series is numerically equal to the present value (P) calculated from the uniform series. One way to accomplish this task with the time value of money factors we have already considered is to convert the gradient series to a present value using a (P|G,i,n) factor and then convert this present value to a uniform series using an (A|P,i,n) factor. In other words:

A = [G * (P|G,i,n)] * (A|P,i,n)

An alternative approach is to use a factor known as the <u>gradient-to-uniform series conversion factor</u>, symbolized by (A|G,i,n). Tables of (A|G,i,n) are provided in Appendix 4A. An algebraic expression can also be derived for the (A|G,i,n) factor which expresses A in terms of G, i, and n. The derivation of this formula is omitted here, but the resulting expression is shown in the summary table (Table 4.6) at the end of this section.

4.6.9 Summary of Time Value of Money Factors

Table 4.6 summarizes the time value of money factors introduced in this section. Time value of money factors are useful in economic analysis because they provide a mechanism to accomplish two primary functions: (1) they allow us to replace a cash flow at one point in time with an equivalent cash flow (in a time value of money sense) at a different point in time and (2) they allow us to convert one cash flow pattern to another (e.g., convert a single sum of money to an equivalent cash flow series or convert a cash flow series to an equivalent single sum). The usefulness of these two functions when performing economic analysis of alternatives will become apparent in Sections 4.7 and 4.8 which follow.

Table 4.6 Summary of discrete compounding time value of money factors

To Find	Given	Factor	Symbol	Name
P	F	$(1+i)^{-n}$	$(P\mid F,i,n)$	Single Payment, Present Worth Factor
F	P	$(1+i)^{n}$	$(F\mid P,i,n)$	Single Payment, Compound Amount Factor
P	A	$\dfrac{(1+i)^{n}-1}{i(1+i)^{n}}$	$(P\mid A,i,n)$	Uniform Series, Present Worth Factor
A	P	$\dfrac{i(1+i)^{n}}{(1+i)^{n}-1}$	$(A\mid P,i,n)$	Uniform Series, Capital Recovery Factor
F	A	$\dfrac{(1+i)^{n}-1}{i}$	$(F\mid A,i,n)$	Uniform Series, Compound Amount Factor
A	F	$\dfrac{i}{(1+i)^{n}-1}$	$(A\mid F,i,n)$	Uniform Series, Sinking Fund Factor
P	G	$\dfrac{1-(1+ni)(1+i)^{-n}}{i^{2}}$	$(P\mid G,i,n)$	Gradient Series, Present Worth Factor
A	G	$\dfrac{(1+i)^{n}-(1+ni)}{i\left[(1+i)^{n}-1\right]}$	$(A\mid G,i,n)$	Gradient Series, Uniform Series Factor

4.6.10 The Concepts of Equivalence and Indifference

Up to this point the term "equivalence" has been used several times but never fully defined. It is appropriate at this point to formally define equivalence as well as a related term, indifference.

In economic analysis, "equivalence" means "the state of being equal in value." The concept is primarily applied to the comparison of two or more cash flow profiles. Specifically, two (or more) cash flow profiles are equivalent if their time value of money worths at a common point in time are equal.

Question: Are the following two cash flows equivalent at 15%/yr?
Cash Flow 1: Receive $1,322.50 two years from today
Cash Flow 2: Receive $1,000.00 today

Analysis Approach 1: Compare worths at t=0 (present worth)
PW(1) = 1,322.50*(P|F,15,2) = 1322.50*0.756147 = 1,000
PW(2) = 1,000

Answer: Cash Flow 1 and Cash Flow 2 are equivalent

Analysis Approach 2: Compare worths at t=2 (future worth)
FW(1) = 1,322.50
FW(2) = 1,000*(F|P,15,2) = 1,000*1.3225 = 1,322.50
Answer: Cash Flow 1 and Cash Flow 2 are equivalent

Generally the comparison (hence the determination of equivalence) for the two cash flow series in this example would be made as present worths (t=0) or future worths (t=2), but the equivalence definition holds regardless of the point in time chosen. For example:

Analysis Approach 3: Compare worths at t=1
W1(1) = 1,322.50*(P|F,15,1)
       = 1,322.50*0.869565 = 1,150.00
W1(2) = 1,000*(F|P,15,1) = 1,000*1.15 = 1,150.00
Answer: Cash Flow 1 and Cash Flow 2 are equivalent

Thus, the selection of the point in time, t, at which to make the comparison is completely arbitrary. Clearly however, some choices are more intuitively appealing than others (t= 0 and t=2 in the above example).

In economic analysis, "indifference" means "to have no preference" The concept is primarily applied in the comparison of two or more cash flow profiles. Specifically, a potential investor is indifferent between two (or more) cash flow profiles if they are equivalent.

Question: Given the following two cash flows at 15%/yr which do you prefer?
Cash Flow 1: Receive $1,322.50 two years from today
Cash Flow 2: Receive $1,000.00 today
Answer: Based on the equivalence calculations above, given these two choices, an investor is indifferent.

The concept of equivalence can be used to break a large, complex problem into a series of smaller more manageable ones. This is done by taking advantage of the fact that, in calculating the economic worth of a cash flow profile, any part of the profile can be replaced by an equivalent representation without altering the worth of the profile at an arbitrary point in time.

Question: You are given a choice between (1) receiving P dollars today or (2) receiving the cash flow series illustrated in Figure 4.5. What must the value of P be for you to be indifferent between the two choices if i=10%/yr?

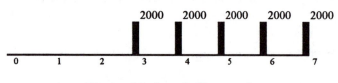

Figure 4.5 A cash flow series

Analysis Approach: To be indifferent between the choices, P must have a value such that the two alternatives are equivalent at 10%/yr. If we select t=0 as the common point in time upon which to base the analysis (present worth approach), then the analysis proceeds as follows.
PW(Alt 1) = P
Because P is already at t=0 (today), no time value of money factors are involved.
PW(Alt 2)
Step 1 - Replace the uniform series (t=3 to 7) with an equivalent single sum, V_2, at t=2 (one period before the first element of the series).
$V_2 = 2,000 * (P|A,10\%,5) = 2,000 * 3.6048 = 7,209.60$
Step 2 - Replace the single sum V_2 with an equivalent value V_0 at t=0:
PW(Alt 2) = $V_0 = V_2 * (P|F,10,2) =$
7,209.60 * 0.7972 = 5,747.49

Answer: To be indifferent between the two alternatives, they must be equivalent at t=0. To be equivalent, P must have a value of $5,747.49

4.7 PROJECT MEASURES OF WORTH

4.7.1 Introduction

In this section measures of worth for investment projects are introduced. The measures are used to evaluate the attractiveness of a single investment opportunity. The measures to be presented are (1) present worth, (2) annual worth, (3) internal rate of return, (4) savings investment ratio, and (5) payback period. All but one of these measures of worth require an interest rate to calculate the worth of an investment. This interest rate is commonly referred to as the Minimum Attractive Rate of Return (MARR). There are many ways to determine a value of MARR for investment analysis and no one way is proper for all applications. One principle is, however, generally accepted. MARR should always exceed the cost of capital as described in Section 4.4, Sources of Funds, presented earlier in this chapter.

In all of the measures of worth below, the following conventions are used for defining cash flows. At any given point in time (t = 0, 1, 2,..., n), there may exist both revenue (positive) cash flows, R_t, and cost (negative) cash flows, C_t. The net cash flow at t, A_t, is defined as $R_t - C_t$.

4.7.2 Present Worth

Consider again the cash flow series illustrated in Figure 4.5. If you were given the opportunity to "buy" that cash flow series for $5,747.49, would you be interested in purchasing it. If you expected to earn a 10%/yr return on your money (MARR=10%), based on the analysis in the previous section, your conclusion would be (should be) that you are indifferent between (1) retaining your $5,747.49 and (2) giving up your $5,747.49

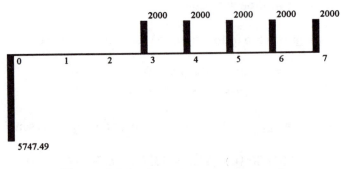

Figure 4.6 An investment opportunity

in favor of the cash flow series. Figure 4.6 illustrates the net cash flows of this second investment opportunity.

What value would you expect if we calculated the present worth (equivalent value of all cash flows at t=0) of Figure 4.6? We must be careful with the signs (directions) of the cash flows in this analysis since some represent cash outflows (downward) and some represent cash inflows (upward).

PW = -5747.49 + 2000*(P|A,10%,5)*(P|F,10%,2)

PW = -5747.49 + 2000*(3.6048)*(0.7972)

PW = -5747.49 + 5747.49 = $0.00

The value of zero for present worth indicates indifference regarding the investment opportunity. We would just as soon do nothing (i.e., retain our $5747.49) as invest in the opportunity.

What if the same returns (future cash inflows) where offered for a $5000 investment (t=0 outflow), would this be more or less attractive? Hopefully, after a little reflection, it is apparent that this would be a more attractive investment because you are getting the same returns but paying less than the indifference amount for them. What happens if calculate the present worth of this new opportunity?

PW = -5000 + 2000*(P|A,10%,5)*(P|F,10%,2)

PW = -5000 + 2000*(3.6048)*(0.7972)

PW = -5000.00 + 5747.49 = $747.49

The positive value of present worth indicates an attractive investment. If we repeat the process with an initial cost greater than $5747.49, it should come as no surprise that the present worth will be negative indicating an unattractive investment.

The concept of present worth as a measure of investment worth can be generalized as follows:

Measure of Worth: Present Worth

Description: All cash flows are converted to a single sum equivalent at time zero using i=MARR.

Calculation Approach: $PW = \sum_{t=0}^{n} A_t (P|F,i,t)$

Decision Rule: If PW ≥0, then the investment is attractive.

Example 13

Installing thermal windows on a small office building is estimated to cost $10,000. The windows are expected to last six years and have no salvage value at that time. The energy savings from the windows are expected to be $2525 each year for the first three years and $3840 for each of the remaining three years. If MARR is 15%/yr and the present worth measure of worth is to be used, is this an attractive investment?

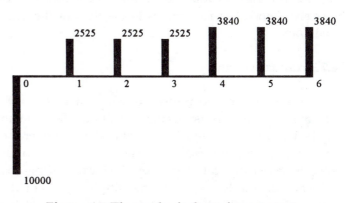

Figure 4.7 Thermal windows investment

The cash flow diagram for the thermal windows is shown in Figure 4.7.

PW = −10000+2525*(P|F,15%,1)+2525*(P|F,15%,2) +2525*(P|F,15%,3)+3840*(P|F,15%,4)+ 3840*(P|F,15%,5)+ 3840*(P|F,15%,6)

PW = −10000+2525*(0.8696)+2525*(0.7561) +2525*(0.6575)+ 3840*(0.5718)+3840*(0.4972)+ 3840*(0.4323)

PW = −10000+2195.74+1909.15+1660.19+2195.71 +1909.25+1660.03

PW = $1530.07

Decision: PW≥0 ($1530.07≥0.0), therefore the window investment is attractive.

An alternative (and simpler) approach to calculating PW is obtained by recognizing that the savings cash flows are two uniform series; one of value $2525 and length 3 starting at t=1 and one of value $3840 and length 3 starting at t=4.

PW =   -10000+2525*(P I A,15%,3)+3840*
        (P I A,15%,3)*(P I F,15%,3)

PW =   -10000+2525*(2.2832)+3840*(2.2832)*
        (0.6575) = $1529.70

Decision: PW≥0 ($1529.70>0.0), therefore the window investment is attractive.

The slight difference in the PW values is caused by the accumulation of round off errors as the various factors are rounded to four places to the right of the decimal point.

4.7.3 Annual Worth

An alternative to present worth is annual worth. The annual worth measure converts all cash flows to an equivalent uniform annual series of cash flows over the investment life using i=MARR. The annual worth measure is generally calculated by first calculating the present worth measure and then multiplying this by the appropriate (A I P,i,n) factor. A thorough review of the tables in Appendix 4A or the equations in Table 4.6 leads to the conclusion that for all values of i (i>0) and n (n>0), the value of (A I P,i,n) is greater than zero. Hence,

if PW>0, then AW>0;

if PW<0, then AW<0; and

if PW=0, then AW=0

because the only difference between PW and AW is multiplication by a positive, non-zero value, namely (A I P,i,n). The decision rule for investment attractiveness for PW and AW are identical; positive values indicate an attractive investment; negative values indicate an unattractive investment; zero indicates indifference. Frequently the only reason for choosing between AW and PW as a measure of worth in an analysis is the preference of the decision maker.

The concept of annual worth as a measure of investment worth can be generalized as follows:

Measure of Worth: Annual Worth

Description: All cash flows are converted to an equivalent uniform annual series of cash flows over the planning horizon using i=MARR.

Calculation Approach: AW = PW (A I P,i,n)

Decision Rule: If AW ≥0, then the investment is attractive.

Example 14

Reconsider the thermal window data of Example 13. If the annual worth measure of worth is to be used, is this an attractive investment?

AW = PW (A I P,15%,6)
AW = 1529.70 (0.2642) = $404.15/yr
Decision: AW ≥0 ($404.15>0.0), therefore the window investment is attractive.

4.7.4 Internal Rate of Return

One of the problems associated with using the present worth or the annual worth measures of worth is that they depend upon knowing a value for MARR. As mentioned in the introduction to this section, the "proper" value for MARR is a much debated topic and tends to vary from company to company and decision maker to decision maker. If the value of MARR changes, the value of PW or AW must be recalculated to determine whether the attractiveness/unattractiveness of an investment has changed.

The internal rate of return (IRR) approach is designed to calculate a rate of return that is "internal" to the project. That is,

if IRR > MARR, the project is attractive,
if IRR < MARR, the project is unattractive,
if IRR = MARR, indifferent.

Thus, if MARR changes, no new calculations are required. We simply compare the calculated IRR for the project to the new value of MARR and we have our decision.

The value of IRR is typically determined through a trial and error process. An expression for the present worth of an investment is written without specifying a value for i in the time value of money factors. Then, various values of i are substituted until a value is found that sets the present worth (PW) equal to zero. The value of i found in this way is the IRR.

As appealing as the flexibility of this approach is, their are two major drawbacks. First, the iterations required to solve using the trial and error approach to solution can be time consuming. This factor is mitigated by the fact that most spreadsheets and financial calculators are pre-programmed to solve for an IRR value given a cash flow series. The second, and more serious, drawback to the IRR approach is that some cash flow series have more than one value of IRR (i.e., more than one value of i sets the PW expression to zero). A detailed discussion of this multiple solution issue is beyond the scope of this chapter, but can be found in White, Agee, and Case [1989] as well as most other economic analysis references. However, it can be shown that, if a cash flow series consists of an initial investment (negative cash flow at t=0) followed by a series of future returns (positive or zero cash flows for all t>0) then a unique IRR exists. If these conditions are not satisfied a unique IRR is not guaranteed and caution should be exercised in making decisions based on IRR.

The concept of internal rate of return as a measure of investment worth can be generalized as follows:

Measure of Worth: Internal Rate of Return

Description: An interest rate, IRR, is determined which yields a present of zero. IRR implicitly assumes the reinvestment of recovered funds at IRR.

Calculation Approach:

$$\text{find IRR such that PW} = \sum_{t=0}^{n} A_t \ (P \mid F, IRR, t) = 0$$

Important Note: Depending upon the cash flow series, multiple IRRs may exist! If the cash flow series consists of an initial investment (net negative cash flow) followed by a series of future returns (net non-negative cash flows), then a unique IRR exists.

Decision Rule: If IRR is unique and IRR ≥MARR, then the investment is attractive.

Example 15

Reconsider the thermal window data of Example 13. If the internal rate of return measure of worth is to be used, is this an attractive investment?

First we note that the cash flow series has a single negative investment followed by all positive returns, therefore, it has a unique value for IRR. For such a cash flow series it can also be shown that as i increases PW decreases.

From example 11, we know that for i=15%:

PW =   -10000+2525*(P | A,15%,3)+3840*(P | A,15%,3)* (P | F,15%,3)

PW =   -10000+2525*(2.2832)+3840*(2.2832)* (0.6575) = $1529.70

Because PW>0, we must increase i to decrease PW toward zero
for i=18%:

PW =   -10000+2525*(P | A,18%,3)+3840* (P | A,18%,3)*(P | F,18%,3)

PW =   -10000+2525*(2.1743)+3840*(2.1743)* (0.6086) = $571.50

Since PW>0, we must increase i to decrease PW toward zero
for i=20%:

PW =   -10000+2525*(P | A,20%,3)+3840* (P | A,20%,3)*(P | F,20%,3)

PW =   -10000+2525*(2.1065)+3840*(2.1065)* (0.5787) = -$0.01

Although we could interpolate for a value of i for which PW=0 (rather than -0.01), for practical purposes PW=0 at i=20%, therefore IRR=20%.

Decision: IRR≥MARR (20%>15%), therefore the window investment is attractive.

4.7.5 Saving Investment Ratio

Many companies are accustomed to working with benefit cost ratios. An investment measure of worth which is consistent with the present worth measure and has the form of a benefit cost ratio is the savings investment ratio (SIR). The SIR decision rule can be derived from the present worth decision rule as follows:

Starting with the PW decision rule

$$PW \geq 0$$

replacing PW with its calculation expression

$$\sum_{t=0}^{n} A_t \ (P \mid F, i, t) \geq 0$$

which, using the relationship $A_t = R_t - C_t$, can be restated

$$\sum_{t=0}^{n} \left(R_t - C_t \right) (P \mid F, i, t) \geq 0$$

which can be algebraically separated into

$$\sum_{t=0}^{n} R_t \ (P \mid F, i, t) - \sum_{t=0}^{n} C_t \ (P \mid F, i, t) \geq 0$$

adding the second term to both sides of the inequality

$$\sum_{t=0}^{n} R_t (P|F,i,t) \geq \sum_{t=0}^{n} C_t (P|F,i,t)$$

dividing both sides of the inequality by the right side term

$$\frac{\sum_{t=0}^{n} R_t (P|F,i,t)}{\sum_{t=0}^{n} C_t (P|F,i,t)} \geq 1$$

which is the decision rule for SIR.

The SIR represents the ratio of the present worth of the revenues to the present worth of the costs. If this ratio exceeds one, the investment is attractive.

The concept of savings investment ratio as a measure of investment worth can be generalized as follows:

Measure of Worth: Savings Investment Ratio

Description: The ratio of the present worth of positive cash flows to the present worth of (the absolute value of) negative cash flows is formed using i=MARR.

$$\text{Calculation Approach: } SIR = \frac{\sum_{t=0}^{n} R_t (P|F,i,t)}{\sum_{t=0}^{n} C_t (P|F,i,t)}$$

Decision Rule: If SIR ≥1, then the investment is attractive.

Example 16

Reconsider the thermal window data of Example 13. If the savings investment ratio measure of worth is to be used, is this an attractive investment?

From example 13, we know that for i=15%:

$$SIR = \frac{\sum_{t=0}^{n} R_t (P|F,i,t)}{\sum_{t=0}^{n} C_t (P|F,i,t)}$$

$$SIR = \frac{2525*(P|A, 15\%,3) + 3840*(P|A, 15\%,3)*(P|F, 15\%,3)}{10000}$$

$$SIR = \frac{11529.70}{10000.00} = 1.15297$$

Decision: SIR≥1.0 (1.15297>1.0), therefore the window investment is attractive.

An important observation regarding the four measures of worth presented to this point (PW, AW, IRR, and SIR) is that they are all consistent and equivalent. In other words, an investment that is attractive under one measure of worth will be attractive under each of the other measures of worth. A review of the decisions determined in Examples 13 through 16 will confirm the observation. Because of their consistency, it is not necessary to calculate more than one measure of investment worth to determine the attractiveness of a project. The rationale for presenting multiple measures which are essentially identical for decision making is that various individuals and companies may have a preference for one approach over another.

4.7.6 Payback Period

The payback period of an investment is generally taken to mean the number of years required to recover the initial investment through net project returns. The payback period is a popular measure of investment worth and appears in many forms in economic analysis literature and company procedure manuals. Unfortunately, all too frequently, payback period is used inappropriately and leads to decisions which focus exclusively on short term results and ignore time value of money concepts. After presenting a common form of payback period these shortcomings will be discussed.

Measure of Worth: Payback Period

Description: The number of years required to recover the initial investment by accumulating net project returns is determined.

Calculation Approach:

$$PBP = \text{the smallest } m \text{ such that } \sum_{t=1}^{m} A_t \geq C_0$$

Decision Rule: If PBP is less than or equal to a predetermined limit (often called a hurdle rate), then the investment is attractive.

<u>Important Note</u>: This form of payback period ignores the time value of money and ignores returns beyond the predetermined limit.

The fact that this approach ignores time value of money concepts is apparent by the fact that no time value of money factors are included in the determination of m. This implicitly assumes that the applicable interest rate to convert future amounts to present amounts is zero. This implies that people are indifferent between $100 today and $100 one year from today, which is an implication that is highly inconsistent with observable behavior.

The short-term focus of the payback period measure of worth can be illustrated using the cash flow diagrams of Figure 4.8. Applying the PBP approach above yields a payback period for investment (a) of PBP=2 (1200>1000 @ t=2) and a payback period for investment (b) of PBP=4 (1000300>1000) @ t=4). If the decision hurdle rate is 3 years (a very common rate), then investment (a) is attractive but investment (b) is not. Hopefully, it is obvious that judging (b) unattractive is not good decision making since a $1,000,000 return four years after a $1,000 investment is attractive under almost any value of MARR. In point of fact, the IRR for (b) is 465% so for any value of MARR less than 465%, investment (b) is attractive.

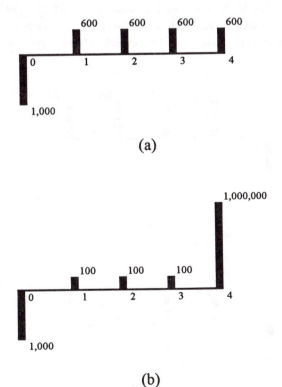

(a)

(b)

Figure 4.8 Two investments evaluated using payback period

4.8 ECONOMIC ANALYSIS

4.8.1 Introduction

The general scenario for economic analysis is that a set of investment alternatives are available and a decision must be made regarding which ones (if any) to accept and which ones (if any) to reject. If the analysis is deterministic, then an assumption is made that cash flow amounts, cash flow timing, and MARR are known with certainty. Frequently, although this assumption does not hold exactly, it is not considered restrictive in terms of potential investment decisions. If however the lack of certainty is a significant issue then the analysis is stochastic and the assumptions of certainty are relaxed using probability distributions and statistical techniques to conduct the analysis. The remainder of this section deals with deterministic economic analysis so the assumption of certainty will be assumed to hold. Stochastic techniques are introduced in Section 4.9.5.

4.8.2 Deterministic Unconstrained Analysis

Deterministic economic analysis can be further classified into unconstrained deterministic analysis and constrained deterministic analysis. Under unconstrained analysis, all projects within the set available are assumed to be independent. The practical implication of this independence assumption is that an accept/reject decision can be made on each project without regard to the decisions made on other projects. In general this requires that (1) there are sufficient funds available to undertake all proposed projects, (2) there are no mutually exclusive projects, and (3) there are no contingent projects.

A funds restriction creates dependency since, before deciding on a project being evaluated, the evaluator would have to know what decisions had been made on other projects to determine whether sufficient funds were available to undertake the current project. Mutual exclusion creates dependency since acceptance of one of the mutually exclusive projects precludes acceptance of the others. Contingency creates dependence since prior to accepting a project, all projects on which it is contingent must be accepted.

If none of the above dependency situations are present and the projects are otherwise independent, then the evaluation of the set of projects is done by evaluating each individual project in turn and accepting the set of projects which were individually judged acceptable. This accept or reject judgment can be made using either the PW, AW, IRR, or SIR measure of worth. The unconstrained decision rules for each or these measures of worth are restated below for convenience.

Unconstrained PW Decision Rule: If PW ≥0, then the project is attractive.

Unconstrained AW Decision Rule: If AW ≥0, then the project is attractive.

Unconstrained IRR Decision Rule: If IRR is unique and IRR ≥MARR, then the project is attractive.

Unconstrained SIR Decision Rule: If SIR ≥1, then the project is attractive.

Example 17

Consider the set of four investment projects whose cash flow diagrams are illustrated in Figure 4.9. If MARR is 12%/yr and the analysis is unconstrained, which projects should be accepted?

Using present worth as the measure of worth:

PW_A = -1000+600*(P|A,12%,4) = -1000+600(3.0373) = $822.38 ⟹ Accept A

PW_B = -1300+800*(P|A,12%,4) = -1300+800(3.0373) = $1129.88 ⟹ Accept B

PW_C = -400+120*(P|A,12%,4) = -400+120(3.0373) = -$35.52 ⟹ Reject C

PW_D = -500+290*(P|A,12%,4) = -500+290(3.0373) = $380.83 ⟹ Accept D

Therefore,

Accept Projects A, B, and D and Reject Project C

4.8.3 Deterministic Constrained Analysis

Constrained analysis is required any time a dependency relationship exists between any of the projects within the set to be analyzed. In general dependency exists any time (1) there are insufficient funds available to undertake all proposed projects (this is commonly referred to as capital rationing), (2) there are mutually exclusive projects, or (3) there are contingent projects.

Several approaches have been proposed for selecting the best set of projects from a set of potential projects under constraints. Many of these approaches will select the optimal set of acceptable projects under some conditions or will select a set that is near optimal. However, only a few approaches are guaranteed to select the optimal set of projects under all conditions. One of these approaches is presented below by way of a continuation of Example 17.

The first steps in the selection process are to specify the cash flow amounts and cash flow timings for each project in the potential project set. Additionally, a value of MARR to be used in the analysis must be specified. These issues have been addressed in previous sections so further discussion will be omitted here. The next step

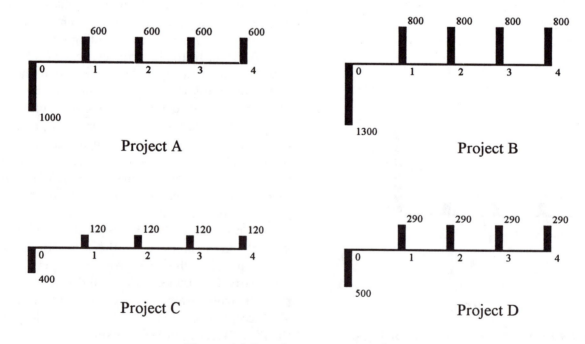

Project A

Project B

Project C

Project D

Figure 4.9 Four investments projects

is to form the set of all possible decision alternatives from the projects. A single decision alternative is a collection of zero, one, or more projects which could be accepted (all others not specified are to be rejected). As an illustration, the possible decision alternatives for the set of projects illustrated in Figure 4.9 are listed in Table 4.7. As a general rule, there will be 2^n possible decision alternatives generated from a set of n projects. Thus, for the projects of Figure 4.9, there are $2^4 = 16$ possible decision alternatives. Since this set represents all possible decisions that could be made, one, and only one, will be selected as the best (optimal) decision. The set of decision alternatives developed in this way has the properties of being collectively exhaustive (all possible choices are listed) and mutually exclusive (only one will be selected).

The next step in the process is to eliminate decisions from the collectively exhaustive, mutually exclusive set that represent choices which would violate one (or more) of the constraints on the projects. For the projects of Figure 4.9, assume the following two constraints exist:

Project B is contingent on Project C, and

A budget limit of $1500 exists on capital expenditures at t=0.

Table 4.7 The decision alternatives from four projects

Accept A only Accept B only Accept C only Accept D only
Accept A and B only Accept A and C only Accept A and D only Accept B and C only Accept B and D only Accept C and D only
Accept A, B, and C only Accept A, B, and D only Accept A, C, and D only Accept B, C, and D only
Accept A, B, C, and D (frequently called the do everything alternative)
Accept none (frequently called the do nothing or null alternative)

Table 4.8 The decision alternatives with constraints imposed

Accept A only	OK
Accept B only	infeasible, B contingent on C
Accept C only	OK
Accept D only	OK
Accept A and B only	infeasible, B contingent on C
Accept A and C only	OK
Accept A and D only	OK
Accept B and C only	infeasible, capital rationing
Accept B and D only	infeasible, B contingent on C
Accept C and D only	OK
Accept A, B, and C only	infeasible, capital rationing
Accept A, B, and D only	infeasible, B contingent on C
Accept A, C, and D only	infeasible, capital rationing
Accept B, C, and D only	infeasible, capital rationing
Do Everything	infeasible, capital rationing
null	OK

Based on these constraints the following decision alternatives must be removed from the collectively exhaustive, mutually exclusive set: any combination that includes B but not C (B only; A&B; B&D; A&B&D), any combination not already eliminated whose t=0 costs exceed $1500 (B&C, A&B&C, A&C&D, B&C&D, A&B&C&D). Thus, from the original set of 16 possible decision alternatives, 9 have been eliminated and need not be evaluated. These results are illustrated in Table 4.8. It is frequently the case in practice that a significant percentage of the original collectively exhaustive, mutually exclusive set will be eliminated before measures of worth are calculated.

The next step is to create the cash flow series for the remaining (feasible) decision alternatives. This is a straight forward process and is accomplished by setting a decision alternative's annual cash flow equal to the sum of the annual cash flows (on a year by year basis) of all projects contained in the decision alternative. Table 4.9 illustrates the results of this process for the feasible decision alternatives from Table 4.8.

The next step is to calculate a measure of worth for each decision alternative. Any of the four consistent measures of worth presented above (PW, AW, IRR, or SIR but NOT PBP) can be used. The measures are entirely consistent and will lead to the same decision alternative being selected. For illustrative purposes, PW will be calculated for the decision alternatives of Table 4.9 assuming MARR=12%.

PW_A = $-1000 + 600*(P|A,12\%,4)$ = $-1000 + 600 (3.0373)$ = $822.38

PW_C = $-400 + 120*(P|A,12\%,4)$ = $-400 + 120 (3.0373)$ = $-35.52

PW_D = $-500 + 290*(P|A,12\%,4)$ = $-500 + 290 (3.0373)$ = $380.83

$PW_{A\&C}$ = $-1400 + 720*(P|A,12\%,4)$ = $-1400 + 720 (3.0373)$ = $786.86

$PW_{A\&D}$ = $-1500 + 890*(P|A,12\%,4)$ = $-1500 + 890 (3.0373)$ = $1203.21

$PW_{C\&D}$ = $-900 + 410*(P|A,12\%,4)$ = $-900 + 410 (3.0373)$ = $345.31

PW_{null} = $-0 + 0*(P|A,12\%,4)$ = $-0 + 0 (3.0373)$ = $0.00

The decision rules for the various measures of worth under constrained analysis are list below.

Constrained PW Decision Rule: Accept the decision alternative with the highest PW.

Constrained AW Decision Rule: Accept the decision alternative with the highest AW.

Constrained IRR Decision Rule: Accept the decision alternative with the highest IRR.

Constrained SIR Decision Rule: Accept the decision alternative with the highest SIR.

For the example problem, the highest present worth ($1203.21) is associated with accepting projects A and D (rejecting all others). This decision is guaranteed to be optimal (i.e., no feasible combination of projects has a higher PW, AW, IRR, or SIR).

Table 4.9 The decision alternatives cash flows

yr \ Alt	A only	C only	D only	A&C	A&D	C&D	null
0	-1000	-400	-500	-1400	-1500	-900	0
1	600	120	290	720	890	410	0
2	600	120	290	720	890	410	0
3	600	120	290	720	890	410	0
4	600	120	290	720	890	410	0

4.8.4   Some Interesting Observations Regarding Constrained Analysis

Several interesting observations can be made regarding the approach, measures of worth, and decisions associated with constrained analysis. Detailed development of these observations is omitted here but may be found in many engineering economic analysis texts [White, Agee, and Case, 1989].

- The present worth of a decision alternative is the sum of the present worths of the projects contained within the alternative. (From above $PW_{A\&D} = PW_A + PW_D$).

- The annual worth of a decision alternative is the sum of the annual worths of the projects contained within the alternative.

- The internal rate of return of a decision alternative is NOT the sum of internal rates of returns of the projects contained within the alternative. The IRR for the decision alternative must be calculated by the trial and error process of finding the value of i that sets the PW of the decision alternative to zero.

- The savings investment ratio of a decision alternative is NOT the sum of the savings investment ratios of the projects contained within the alternative. The SIR for the decision alternative must be calculated from the cash flows of the decision alternative.

- A common, but flawed, procedure for selecting the projects to accept from the set of potential projects involves ranking the projects (not decision alternatives) in preferred order based on a measure of worth calculated for the project (e.g., decreasing project PW) and then accepting projects as far down the list as funds allow. While this procedure will select the optimal set under some conditions (e.g., it works well if the initial investments of all projects are small relative to the capital budget limit), it is not guaranteed to select the optimal set under all conditions. The procedure outlined above will select the optimal set under all conditions.

- Table 4.10 illustrates that the number of decision alternatives in the collectively exhaustive, mutually exclusive set can grow prohibitively large as the number of potential projects increases. The mitigating factor in this combinatorial growth problem is that in most practical situations a high percentage of the possible decisions alternatives are infeasible and do not require evaluation.

Table 4.10 The number of decision alternatives as a function of the number of projects

Number of Projects	Number of Decision Alternatives
1	2
2	4
3	8
4	16
5	32
6	64
7	128
8	256
9	512
10	1,024
15	32,768
20	1,048,576
25	33,554,432

4.8.5 The Planning Horizon Issue

When comparing projects, it is important to compare the costs and benefits over a common period of time. The intuitive sense of fairness here is based upon the recognition that most consumers expect an investment that generates savings over a longer period of time to cost more than an investment that generates savings over a shorter period of time. To facilitate a fair, comparable evaluation a common period of time over which to conduct the evaluation is required. This period of time is referred to as the planning horizon. The planning horizon issue arises when at least one project has cash flows defined over a life which is greater than or less than the life of at least one other project. This situation did not occur in Example 17 of the previous section since all projects had 4 year lives.

There are four common approaches to establishing a planning horizon for evaluating decision alternatives. These are (1) shortest life, (2) longest life, (3) least common multiple of lives, and (4) standard. The shortest life planning horizon is established by selecting the project with the shortest life and setting this life as the planning horizon. A significant issue in this approach is how to value the remaining cash flows for projects whose lives are truncated. The typical approach to this valuation is to estimate the value of the remaining cash flows as the salvage value (market value) of the investment at that point in its life.

Example 18

Determine the shortest life planning horizon for projects A, B, C with lives 3, 5, and 6 years, respectively.

The shortest life planning horizon is 3 years based on Project A. A salvage value must be established at t=3 for B's cash flows in years 4 and 5. A salvage value must be established at t=3 for C's cash flows in years 4, 5, and 6.

The longest life planning horizon is established by selecting the project with the longest life and setting this life as the planning horizon. The significant issue in this approach is how to handle projects whose cash flows don't extend this long. The typical resolution for this problem is to assume that shorter projects are repeated consecutively (end-to-end) until one of the repetitions extends at least as far as the planning horizon. The assumption of project repeatability deserves careful consideration since in some cases it is reasonable and in others it may be quite unreasonable. The reasonableness of the assumption is largely a function of the type of investment and the rate of innovation occurring within the investment's field (e.g., assuming repeatability of investments in high technology equipment is frequently ill advised since the field is advancing rapidly). If in repeating a project's cash flows, the last repetition's cash flows extend beyond the planning horizon, then the truncated cash flows (those that extend beyond the planning horizon) must be assigned a salvage value as above.

Example 19

Determine the longest life planning horizon for projects A, B, C with lives 3, 5, and 6 years, respectively.

The longest life planning horizon is 6 years based on Project C. Project A must be repeated twice, the second repetition ends at year 6, so no termination of cash flows is required. Project B's second repetition extends to year 10, therefore, a salvage value at t=6 must be established for B's repeated cash flows in years 7, 8, 9, and 10.

An approach that eliminates the truncation salvage value issue from the planning horizon question is the least common multiple approach. The least common multiple planning horizon is set by determining the smallest number of years at which repetitions of all projects would terminate simultaneously. The least common multiple for a set of numbers (lives) can be deter-

mined mathematically using algebra. Discussion of this approach is beyond the scope of this chapter. For a small number of projects, the value can be determined by trial and error by examining multiples of the longest life project.

Example 20

Determine the least common multiple planning horizon for projects A, B, C with lives 3, 5, and 6 years, respectively.

The least common multiple of 3, 5, and 6 is 30. This can be obtained by trial and error starting with the longest project life (6) as follows:

1st trial: 6*1=6; 6 is a multiple of 3 but not 5; reject 6 and proceed

2nd trial: 6*2=12; 12 is a multiple of 3 but not 5; reject 12 and proceed

3rd trial: 6*3=18; 18 is a multiple of 3 but not 5; reject 18 and proceed

4th trial: 6*4=24; 24 is a multiple of 3 but not 5; reject 24 and proceed

5th trial: 6*5=30; 30 is a multiple of 3 and 5; accept 30 and stop

Under a 30-year planning horizon, A's cash flows are repeated 10 times, B's 6 times, and C's 5 times. No truncation is required.

The standard planning horizon approach uses a planning horizon which is independent of the projects being evaluated. Typically, this type of planning horizon is based on company policies or practices. The standard horizon may require repetition and/or truncation depending upon the set of projects being evaluated.

Example 21

Determine the impact of a 5 year standard planning horizon on projects A, B, C with lives 3, 5, and 6 years, respectively.

With a 5-year planning horizon:

Project A must be repeated one time with the second repetition truncated by one year.

Project B is a 5 year project and does not require repetition or truncation.

Project C must be truncated by one year.

There is no single accepted approach to resolving the planning horizon issue. Companies and individuals generally use one of the approaches outlined above. The decision of which to use in a particular analysis is generally a function of company practice and consideration of the reasonableness of the project repeatability assumption and the availability of salvage value estimates at truncation points.

4.9 SPECIAL PROBLEMS

4.9.1 Introduction

The preceding sections of this chapter outline an approach for conducting deterministic economic analysis of investment opportunities. Adherence to the concepts and methods presented will lead to sound investment decisions with respect to time value of money principles. This section addresses several topics that are of special interest in some analysis situations.

4.9.2 Interpolating Interest Tables

All of the examples previously presented in this chapter conveniently used interest rates whose time value of money factors were tabulated in Appendix 4A. How does one proceed if non-tabulated time value of money factors are needed? There are two viable approaches; calculation of the exact values and interpolation. The best and theoretically correct approach is to calculate the exact values of needed factors based on the formulas in Table 4.6.

Example 22
Determine the exact value for $(F|P,13\%,7)$.

From Table 4.6,

$(F|P,i,n) = (1+i)^n = (1+.13)^7 = 2.3526$

Interpolation is often used instead of calculation of exact values because, with practice, interpolated values can be calculated quickly. Interpolated values are not "exact" but for most practical problems they are "close enough," particularly if the range of interpolation is kept as narrow as possible. Interpolation of some factors, for instance $(P|A,i,n)$, also tends to be less error prone than the exact calculation due simpler mathematical operations.

Interpolation involves determining an unknown time value of money factor using two known values which bracket the value of interest. An assumption is made that the values of the time value of money factor vary linearly between the known values. Ratios are then used to estimate the unknown value. The example below illustrates the process.

Example 23
Determine an interpolated value for $(F|P,13\%,7)$.

The narrowest range of interest rates which bracket 13% and for which time value of money factor tables are provided in Appendix 4A is 12% to 15%.

The values necessary for this interpolation are

i values	$(F|P,i\%,7)$
12%	2.2107
13%	$(F|P,13\%,7)$
15%	2.6600

The interpolation proceeds by setting up ratios and solving for the unknown value, $(F|P,13\%,7)$, as follows:

$$\frac{\text{change between rows 2 \& 1 of left column}}{\text{change between rows 3 \& 1 of left column}} =$$

$$\frac{\text{change between rows 2 \& 1 of right column}}{\text{change between rows 3 \& 1 of right column}}$$

$$\frac{0.13-0.12}{0.15-0.12} = \frac{(F|P,13\%,7)-2.2107}{2.6600-2,2107}$$

$$\frac{0.01}{0.03} = \frac{(F|P,13\%,7)-2.2107}{0.4493}$$

$$0.1498 = (F|P,13\%,7) - 2.2107$$

$$(F|P,13\%,7) = 2.3605$$

The interpolated value for $(F|P,13\%,7)$, 2.3605, differs from the exact value, 2.3526, by 0.0079. This would imply a $7.90 difference in present worth for every thousand dollars of return at t=7. The relative importance of this interpolation error can be judged only in the context of a specific problem.

4.9.3 Non-Annual Interest Compounding

Many practical economic analysis problems involve interest that is not compounded annually. It is common practice to express a non-annually compounded interest rate as follows:

12% per year compounded monthly or 12%/yr/mo.

When expressed in this form, 12%/yr/mo is known as the *nominal* annual interest rate The techniques covered in this chapter up to this point can not be used directly to solve an economic analysis problem of this type because the interest period (per year) and compounding period (monthly) are not the same. Two approaches can be used to solve problems of this type. One approach involves determining a *period* interest rate, the other involves determining an *effective* interest rate.

To solve this type of problem using a period interest rate approach, we must define the period interest rate:

$$\text{Period Interest Rate} = \frac{\text{Nominal Annual Interest Rate}}{\text{Number of Interest Periods per Year}}$$

In our example,

$$\text{Period Interest Rate} = \frac{12\%/\text{yr}/\text{mo}}{12\,\text{mo}/\text{yr}} = 1\%/\text{mo}/\text{mo}$$

Because the interest period and the compounding period are now the same, the time value of money factors in Appendix 4A can be applied directly. Note however, that the number of interest periods (n) must be adjusted to match the new frequency.

Example 24

$2,000 is invested in an account which pays 12% per year compounded monthly. What is the balance in the account after 3 years?

Nominal Annual Interest Rate = 12%/yr/mo

$$\text{Period Interest Rate} = \frac{12\%/\text{yr}/\text{mo}}{12\,\text{mo}/\text{yr}} = 1\%/\text{mo}/\text{mo}$$

Number of Interest Periods = 3 years × 12 mo/yr = 36 interest periods (months)

F = P (F|P,i,n) = $2,000 (F|P,1,36) = $2,000 (1.4308) = $2,861.60

Example 25

What are the monthly payments on a 5-year car loan of $12,500 at 6% per year compounded monthly?

Nominal Annual Interest Rate = 6%/yr/mo

$$\text{Period Interest Rate} = \frac{6\%/\text{yr}/\text{mo}}{12\,\text{mo}/\text{yr}} = 0.5\%/\text{mo}/\text{mo}$$

Number of Interest Periods = 5 years × 12 mo/yr = 60 interest periods

A = P (A|P,i,n) = $12,500 (A|P,0.5,60) = $12,500 (0.0193) = $241.25

To solve this type of problem using an effective interest rate approach, we must define the effective interest rate. The effective annual interest rate is the annualized interest rate that would yield results equivalent to the period interest rate as previously calculated. Note however that the effective annual interest rate approach should not be used if the cash flows are more frequent than annual (e.g., monthly). In general, the interest rate for time value of money factors should match the frequency of the cash flows (e.g., if the cash flows are monthly, use the period interest rate approach with monthly periods).

As an example of the calculation of an effective interest rate, assume that the nominal interest rate is 12%/yr/qtr, therefore the period interest rate is 3%/qtr/qtr. One dollar invested for 1 year at 3%/qtr/qtr would have a future worth of:

F = P (F|P,i,n) = $1 (F|P,3,4) = $1 (1.03)^4
    = $1 (1.1255) = $1.1255

To get this same value in 1 year with an annual rate the annual rate would have to be of 12.55%/yr/yr. This value is called the effective annual interest rate. The effective annual interest rate is given by (1.03)^4 - 1 = 0.1255 or 12.55%.

The general equation for the Effective Annual Interest Rate is:

Effective Annual Interest Rate = $(1 + r/m)^m - 1$
where: r = nominal annual interest rate
          m = number of interest periods per year

Example 26

What is the effective annual interest rate if the nominal rate is 12%/yr compounded monthly?

nominal annual interest rate = 12%/yr/mo

period interest rate = 1%/mo/mo

effective annual interest rate = $(1+0.12/12)^{12} - 1 =$ 0.1268 or 12.68%

4.9.4 Economic Analysis Under Inflation

Inflation is characterized by a decrease in the purchasing power of money caused by an increase in general price levels of goods and services without an accompanying increase the value of the goods and services. Inflationary pressure is created when more dollars are put into an economy without an accompanying increase in goods and services. In other words, printing more money without an increase in economic output generates inflation. A complete treatment of inflation is beyond the scope of this chapter. A good summary can be found in Sullivan and Bontadelli [1980].

When consideration of inflation is introduced into economic analysis, future cash flows can be stated in terms of either constant-worth dollars or then-current dollars. Then-current cash flows are expressed in terms of the face amount of dollars (actual number of dollars) that will change hands when the cash flow occurs. Alternatively, constant-worth cash flows are expressed in terms of the purchasing power of dollars relative to a fixed point in time known as the base period.

Example 27

For the next 4 years, a family anticipates buying $1000 worth of groceries each year. If inflation is expected to be 3%/yr what are the then-current cash flows required to purchase the groceries.

To buy the groceries, the family will need to take the following face amount of dollars to the store. We will somewhat artificially assume that the family only shops once per year, buys the same set of items each year, and that the first trip to the store will be one year from today.

Year 1: dollars required $1000.00*(1.03)=$1030.00
Year 2: dollars required $1030.00*(1.03)=$1060.90
Year 3: dollars required $1060.90*(1.03)=$1092.73
Year 4: dollars required $1092.73*(1.03)=$1125.51

What are the constant-worth cash flows, if today's dollars are used as the base year.

The constant worth dollars are inflation free dollars, therefore the $1000 of groceries costs $1000 each year.

Year 1: $1000.00
Year 2: $1000.00
Year 3: $1000.00
Year 4: $1000.00

The key to proper economic analysis under inflation is to base the value of MARR on the types of cash flows. If the cash flows contain inflation, then the value of MARR should also be adjusted for inflation. Alternatively, if the cash flows do not contain inflation, then the value of MARR should be inflation free. When MARR does not contain an adjustment for inflation, it is referred to as a real value for MARR. If it contains an inflation adjustment, it is referred to as a combined value for MARR. The relationship between inflation rate, the real value of MARR, and the combined value of MARR is given by:

$$1 + MARR_{COMBINED} = (1 + \text{inflation rate}) * (1 + MARR_{REAL})$$

Example 28

If the inflation rate is 3%/yr and the real value of MARR is 15%/yr, what is the combined value of MARR?

$$1 + MARR_{COMBINED} = (1 + \text{inflation rate}) * (1 + MARR_{REAL})$$
$$1 + MARR_{COMBINED} = (1 + 0.03) * (1 + 0.15)$$
$$1 + MARR_{COMBINED} = (1.03) * (1.15)$$
$$1 + MARR_{COMBINED} = 1.1845$$
$$MARR_{COMBINED} = 1.1845 - 1 = 0.1845 = 18.45\%$$

If the cash flows of a project are stated in terms of then-current dollars, the appropriate value of MARR is the combined value of MARR. Analysis done in this way is referred to as *then current analysis*. If the cash flows of a project are stated in terms of constant-worth dollars, the appropriate value of MARR is the real value of MARR. Analysis done in this way is referred to as then *constant worth analysis*.

Example 29

Using the cash flows of Examples 27 and interest rates of Example 28, determine the present worth of the grocery purchases using a constant worth analysis.

Constant worth analysis requires constant worth cash flows and the real value of MARR.

PW = 1000 * (P I A,15%,4)
       = 1000 * (2.8550) = \$2855.00

Example 30

Using the cash flows of Examples 27 and interest rates of Example 28, determine the present worth of the grocery purchases using a then current analysis.

Then current analysis requires then current cash flows and the combined value of MARR.

PW = 1030.00 * (P I F,18.45%,1) + 1060.90 * (P I F,18.45%,2) + 1092.73 * (P I F,18.45%,3) +1125.51 * (P I F,18.45%,4)

PW = 1030.00 * (0.8442) + 1060.90 * (0.7127) + 1092.73 * (0.6017) +1125.51 * (0.5080)

PW = 869.53 + 756.10 + 657.50 + 571.76 = 2854.89

The notable result of Examples 29 and 30 is that the present worths determined by the constant-worth approach (\$2855.00) and the then-current approach (\$2854.89) are equal (the \$0.11 difference is due to rounding). This result is often unexpected but it is mathematically sound. The important conclusion is that if care is taken to appropriately match the cash flows and value of MARR, the level of general price inflation is not a determining factor in the acceptability of projects. To make this important result hold, inflation must either (1) be included in both the cash flows and MARR (the then-current approach) or (2) be included in neither the cash flows nor MARR (the constant-worth approach).

4.9.5 Sensitivity Analysis and Risk Analysis

Often times the certainty assumptions associate with deterministic analysis are questionable. These certainty assumptions include certain knowledge regarding amounts and timing of cash flows as well as certain knowledge of MARR. Relaxing these assumptions requires the use of sensitivity analysis and risk analysis techniques.

Initial sensitivity analyses are usually conducted on the optimal decision alternative (or top two or three) on a single factor basis. Single factor sensitivity analysis involves holding all cost factors except one constant while varying the remaining cost factor through a range of percentage changes. The effect of cost factor changes on the measure of worth is observed to determine whether the alternative remains attractive under the evaluated changes and to determine which cost factor effects the measure of worth the most.

Example 31

Conduct a sensitivity analysis of the optimal decision resulting from the constrained analysis of the data in Example 17. The sensitivity analysis should explore the sensitivity of present worth to changes in annual revenue over the range -10% to +10%.

The PW of the optimal decision (Accept A & D only) was determined in Section 4.8.3 to be:

$PW_{A\&D}$ = -1500 + 890*(P I A,12%,4) = -1500 + 890 (3.0373) = \$1203.21

If annual revenue decreases 10%, it becomes 890 - 0.10*890 = 801 and PW becomes

$PW_{A\&D}$ = -1500 + 801*(P I A,12%,4) = -1500 + 801 (3.0373) = \$932.88

If annual revenue increases 10%, it becomes 890 + 0.10*890 = 979 and PW becomes

$PW_{A\&D}$ = -1500 + 979*(P I A,12%,4) = -1500 + 979 (3.0373) = \$1473.52

The sensitivity of PW to changes in annual revenue over the range -10% to +10% is +\$540.64 from \$932.88 to \$1473.52.

Example 32

Repeat Example 31 exploring of the sensitivity of present worth to changes in initial cost over the range -10% to +10%.

The PW of the optimal decision (Accept A & D only) was determined in Section 4.8.3 to be:

$PW_{A\&D}$ = -1500 + 890*(P I A,12%,4) = -1500 + 890 (3.0373) = \$1203.21

If initial cost decreases 10% it becomes 1500 - 0.10*1500 = 1350 and PW becomes

$PW_{A\&D}$ = -1350 + 890*(P I A,12%,4) = -1350 + 890 (3.0373) = \$1353.20

If initial cost increases 10% it becomes 1500 + 0.10*1500 = 1650 and PW becomes

$PW_{A\&D}$ = -1650 + 890*(P I A,12%,4) = -1500 + 890 (3.0373) = \$1053.20

The sensitivity of PW to changes in initial cost over the range -10% to +10% is -$300.00 from $1353.20 to $1053.20.

Example 33

Repeat Example 31 exploring the sensitivity of the present worth to changes in MARR over the range -10% to +10%.

The PW of the optimal decision (Accept A & D only) was determined in Section 4.8.3 to be:

$$PW_{A\&D} = -1500 + 890*(P|A,12\%,4) = -1500 + 890 (3.0373) = \$1203.21$$

If MARR decreases 10% it becomes 12% - 0.10*12% = 10.8% and PW becomes

$$PW_{A\&D} = -1500 + 890*(P|A,10.8\%,4) = -1500 + 890 (3.1157) = \$1272.97$$

If MARR increases 10% it becomes 12% + 0.10*12% = 13.2% and PW becomes

$$PW_{A\&D} = -1500 + 890*(P|A,13.2\%,4) = -1500 + 890 (2.9622) = \$1136.36$$

The sensitivity of PW to changes in MARR over the range -10% to +10% is -$136.61 from $1272.97 to $1136.36.

The sensitivity data from Examples 31, 32, and 33 are summarized in Table 4.11. A review of the table reveals that the decision alternative A&D remains attractive (PW ≥0) within the range of 10% changes in annual revenues, initial cost, and MARR. An appealing way to summarize single factor sensitivity data is using a "spider" graph. A spider graph plots the PW values determined in the examples and connects them with lines, one line for each factor evaluated. Figure 4.10 illustrates the spider graph for the data of Table 4.11. On this graph, lines with large positive or negative slopes (angle relative to horizontal regardless of whether it is increasing or decreasing) indicate factors to which the present value measure of worth is sensitive. Figure 4.10 shows that PW is least sensitive to changes in MARR (the MARR line is the most nearly horizontal) and most sensitivity to changes in annual revenue (the annual revenue line has the steepest slope). Additional sensitivities could be explored in a similar manner.

Table 4.11 Sensitivity analysis data table

Factor \ Percent Change	- 10%	Base	+ 10%
1st Cost	1353.20	1203.21	1053.20
Annual Revenue	932.88	1203.21	1473.52
MARR	1272.97	1203.21	1136.36

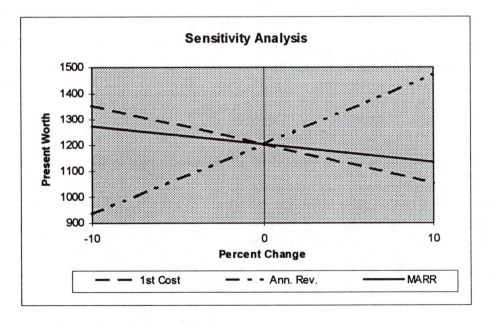

Figure 4.10. Sensitivity analysis "spider" graph

When single factor sensitivity analysis is inadequate to assess the questions which surround the certainty assumptions of a deterministic analysis, risk analysis techniques can be employed. One approach to risk analysis is the application of probabilistic and statistical concepts to economic analysis. These techniques require information regarding the possible values that uncertain quantities may take on as well as estimates of the probability that the various values will occur. A detailed treatment of this topic is beyond the scope of this chapter. A good discussion of this subject can be found in Park and Sharp-Bette [1990].

A second approach to risk analysis in economic analysis is through the use of simulation techniques and simulation software. Simulation involves using a computer simulation program to sample possible values for the uncertain quantities in an economic analysis and calculating the measure of worth. This process is repeated many times using different samples each time. After many samples have been taken, probability statements regarding the measure of worth may be made. A good discussion of this subject can be found in Park and Sharp-Bette [1990].

4.10 SUMMARY AND ADDITIONAL EXAMPLE APPLICATIONS

In this chapter a coherent, consistent approach to economic analysis of capital investments (energy related or other) has been presented. To conclude the chapter, this section provides several additional examples to illustrate the use of time value of money concepts for energy related problems. Additional example applications as well as a more in depth presentation of conceptual details can be found in the references listed at the end of the chapter. These references are by no means exclusive; many other excellent presentations of the subject matter are also available. Adherence to the concepts and methods presented here and in the references will lead to sound investment decisions with respect to time value of money principles.

Example 34

In Section 4.3.3 an example involving the evaluation of a baseboard heating and window air conditioner versus a heat pump was introduced to illustrate cash flow diagramming (Figure 4.2). A summary of the differential costs is repeat here for convenience.

- The heat pump costs $1500 more than the baseboard system,

- The heat pump saves $380 annually in electricity costs,

- The heat pump has a $50 higher annual maintenance costs,

- The heat pump has a $150 higher salvage value at the end of 15 years,

- The heat pump requires $200 more in replacement maintenance at the end of year 8.

If MARR is 18%, is the additional investment in the heat pump attractive?

Using present worth as the measure of worth:

$$PW = -1500 + 380*(P|A,18\%,15) - 50*(P|A,18\%,15) + 150*(P|F,18\%,15) - 200*(P|F,18\%,8)$$

$$PW = -1500 + 380*(5.0916) - 50*(5.0916) + 150*(0.0835) - 200*(0.2660)$$

$$PW = -1500.00 + 1934.81 - 254.58 + 12.53 - 53.20 = \$139.56$$

Decision: PW≥0 ($139.56>0.0), therefore the additional investment for the heat pump is attractive.

Example 35

A homeowner needs to decide whether to install R-11 or R-19 insulation in the attic of her home. The R-19 insulation costs $150 more to install and will save approximately 400 kWh per year. If the planning horizon is 20 years and electricity costs $0.08/kWh is the additional investment attractive at MARR of 10%?

At $0.08/kWh, the annual savings are: 400 kWh * $0.08/kWh = $32.00

Using present worth as the measure of worth:

$$PW = -150 + 32*(P|A,10\%,20)$$

$$PW = -150 + 32*(8.5136) = -150 + 272.44 = \$122.44$$

Decision: PW≥0 ($122.44>0.0), therefore the R-19 insulation is attractive.

Example 36

The homeowner from Example 35 can install R-30 insulation in the attic of her home for $200 more than the

R-19 insulation. The R-30 will save approximately 250 kWh per year over the R-19 insulation. Is the additional investment attractive?

Assuming the same MARR, electricity cost, and planning horizon, the additional annual savings are: 250 kWh * $0.08/kWh = $20.00

Using present worth as the measure of worth:

PW = -200 + 20*(P | A,10%,20)

PW = -200 + 20*(8.5136) = -200 + 170.27 = -$29.73

Decision: PW<0 (-$29.73<0.0), therefore the R-30 insulation is not attractive.

Example 37

An economizer costs $20,000 and will last 10 years. It will generate savings of $3,500 per year with maintenance costs of $500 per year. If M1ARR is 10% is the economizer an attractive investment.

Using present worth as the measure of worth:

PW = -20000 + 3500*(P | A,10%,10)
      - 500*(P | A,10%,10)

PW = -20000 + 3500*(6.1446) - 500*(6.1446)

PW = -20000.00 + 21506.10 - 3072.30 = -$1566.20

Decision: PW<0 (-$1566.20<0.0), therefore the economizer is not attractive.

Example 38

If the economizer from Example 37 has a salvage value of $5000 at the end of 10 years is the investment attractive?

Using present worth as the measure of worth:

PW = -20000 + 3500*(P | A,10%,10)
      - 500*(P | A,10%,10) + 5000*(P | F,10%,10)

PW = -20000 + 3500*(6.1446) - 500*(6.1446)
      + 5000*(0.3855)

PW = -20000.00 + 21506.10 - 3072.30 + 1927.50
      = $361.30

Decision: PW≥0 ($361.30≥0.0), therefore the economizer is now attractive.

4.11 References

Brown, R.J. and R.R. Yanuck, 1980, *Life Cycle Costing: A Practical Guide for Energy Managers*, The Fairmont Press, Inc., Atlanta, GA.

Fuller, S.K. and S.R. Petersen, 1994, NISTIR 5165: *Life-Cycle Costing Workshop for Energy Conservation in Buildings: Student Manual*, U.S. Department of Commerce, Office of Applied Economics, Gaithersburg, MD.

Fuller, S.K. and S.R. Petersen, 1995, *NIST Handbook 135: Life-Cycle Costing Manual for the Federal Energy Management Program*, National Technical Information Service, Springfield, VA.

Park, C.S. and G.P. Sharp-Bette, 1990, Advanced Engineering Economics, John Wiley & Sons, New York, NY.

Sullivan, W.G. and J.A. Bontadelli, 1980, "The Industrial Engineer and Inflation," *Industrial Engineering*, Vol. 12, No. 3, 24-33.

Thuesen, G.J. and W.J. Fabrycky, 1993, *Engineering Economy, 8th Edition*, Prentice Hall, Englewood Cliffs, NJ.

White, J.A., M.H. Agee, and K.E. Case, 1989, *Principles of Engineering Economic Analysis*, 3rd Edition, John Wiley & Sons, New York, NY.

Appendix 4A

Time Value of Money Factors - Discrete Compounding
i = 1%

n	Single Sums		Uniform Series				Gradient Series	
	To Find F Given P (F|P,i%,n)	To Find P Given F (P|F,i%,n)	To Find F Given A (F|A,i%,n)	To Find A Given F (A|F,i%,n)	To Find P Given A (P|A,i%,n)	To Find A Given P (A|P,i%,n)	To Find P Given G (P|G,i%,n)	To Find A Given G (A|G,i%,n)
1	1.0100	0.9901	1.0000	1.0000	0.9901	1.0100	0.0000	0.0000
2	1.0201	0.9803	2.0100	0.4975	1.9704	0.5075	0.9803	0.4975
3	1.0303	0.9706	3.0301	0.3300	2.9410	0.3400	2.9215	0.9934
4	1.0406	0.9610	4.0604	0.2463	3.9020	0.2563	5.8044	1.4876
5	1.0510	0.9515	5.1010	0.1960	4.8534	0.2060	9.6103	1.9801
6	1.0615	0.9420	6.1520	0.1625	5.7955	0.1725	14.3205	2.4710
7	1.0721	0.9327	7.2135	0.1386	6.7282	0.1486	19.9168	2.9602
8	1.0829	0.9235	8.2857	0.1207	7.6517	0.1307	26.3812	3.4478
9	1.0937	0.9143	9.3685	0.1067	8.5660	0.1167	33.6959	3.9337
10	1.1046	0.9053	10.4622	0.0956	9.4713	0.1056	41.8435	4.4179
11	1.1157	0.8963	11.5668	0.0865	10.3676	0.0965	50.8067	4.9005
12	1.1268	0.8874	12.6825	0.0788	11.2551	0.0888	60.5687	5.3815
13	1.1381	0.8787	13.8093	0.0724	12.1337	0.0824	71.1126	5.8607
14	1.1495	0.8700	14.9474	0.0669	13.0037	0.0769	82.4221	6.3384
15	1.1610	0.8613	16.0969	0.0621	13.8651	0.0721	94.4810	6.8143
16	1.1726	0.8528	17.2579	0.0579	14.7179	0.0679	107.2734	7.2886
17	1.1843	0.8444	18.4304	0.0543	15.5623	0.0643	120.7834	7.7613
18	1.1961	0.8360	19.6147	0.0510	16.3983	0.0610	134.9957	8.2323
19	1.2081	0.8277	20.8109	0.0481	17.2260	0.0581	149.8950	8.7017
20	1.2202	0.8195	22.0190	0.0454	18.0456	0.0554	165.4664	9.1694
21	1.2324	0.8114	23.2392	0.0430	18.8570	0.0530	181.6950	9.6354
22	1.2447	0.8034	24.4716	0.0409	19.6604	0.0509	198.5663	10.0998
23	1.2572	0.7954	25.7163	0.0389	20.4558	0.0489	216.0660	10.5626
24	1.2697	0.7876	26.9735	0.0371	21.2434	0.0471	234.1800	11.0237
25	1.2824	0.7798	28.2432	0.0354	22.0232	0.0454	252.8945	11.4831
26	1.2953	0.7720	29.5256	0.0339	22.7952	0.0439	272.1957	11.9409
27	1.3082	0.7644	30.8209	0.0324	23.5596	0.0424	292.0702	12.3971
28	1.3213	0.7568	32.1291	0.0311	24.3164	0.0411	312.5047	12.8516
29	1.3345	0.7493	33.4504	0.0299	25.0658	0.0399	333.4863	13.3044
30	1.3478	0.7419	34.7849	0.0287	25.8077	0.0387	355.0021	13.7557
36	1.4308	0.6989	43.0769	0.0232	30.1075	0.0332	494.6207	16.4285
42	1.5188	0.6584	51.8790	0.0193	34.1581	0.0293	650.4514	19.0424
48	1.6122	0.6203	61.2226	0.0163	37.9740	0.0263	820.1460	21.5976
54	1.7114	0.5843	71.1410	0.0141	41.5687	0.0241	1.002E+03	24.0945
60	1.8167	0.5504	81.6697	0.0122	44.9550	0.0222	1.193E+03	26.5333
66	1.9285	0.5185	92.8460	0.0108	48.1452	0.0208	1.392E+03	28.9146
72	2.0471	0.4885	104.7099	9.550E-03	51.1504	0.0196	1.598E+03	31.2386
120	3.3004	0.3030	230.0387	4.347E-03	69.7005	0.0143	3.334E+03	47.8349
180	5.9958	0.1668	499.5802	2.002E-03	83.3217	0.0120	5.330E+03	63.9697
360	35.9496	0.0278	3.495E+03	2.861E-04	97.2183	0.0103	8.720E+03	89.6995

Time Value of Money Factors - Discrete Compounding
i = 2%

	Single Sums		Uniform Series				Gradient Series	
	To Find F Given P (F|P,i%,n)	To Find P Given F (P|F,i%,n)	To Find F Given A (F|A,i%,n)	To Find A Given F (A|F,i%,n)	To Find P Given A (P|A,i%,n)	To Find A Given P (A|P,i%,n)	To Find P Given G (P|G,i%,n)	To Find A Given G (A|G,i%,n)
n								
1	1.0200	0.9804	1.0000	1.0000	0.9804	1.0200	0.0000	0.0000
2	1.0404	0.9612	2.0200	0.4950	1.9416	0.5150	0.9612	0.4950
3	1.0612	0.9423	3.0604	0.3268	2.8839	0.3468	2.8458	0.9868
4	1.0824	0.9238	4.1216	0.2426	3.8077	0.2626	5.6173	1.4752
5	1.1041	0.9057	5.2040	0.1922	4.7135	0.2122	9.2403	1.9604
6	1.1262	0.8880	6.3081	0.1585	5.6014	0.1785	13.6801	2.4423
7	1.1487	0.8706	7.4343	0.1345	6.4720	0.1545	18.9035	2.9208
8	1.1717	0.8535	8.5830	0.1165	7.3255	0.1365	24.8779	3.3961
9	1.1951	0.8368	9.7546	0.1025	8.1622	0.1225	31.5720	3.8681
10	1.2190	0.8203	10.9497	0.0913	8.9826	0.1113	38.9551	4.3367
11	1.2434	0.8043	12.1687	0.0822	9.7868	0.1022	46.9977	4.8021
12	1.2682	0.7885	13.4121	0.0746	10.5753	0.0946	55.6712	5.2642
13	1.2936	0.7730	14.6803	0.0681	11.3484	0.0881	64.9475	5.7231
14	1.3195	0.7579	15.9739	0.0626	12.1062	0.0826	74.7999	6.1786
15	1.3459	0.7430	17.2934	0.0578	12.8493	0.0778	85.2021	6.6309
16	1.3728	0.7284	18.6393	0.0537	13.5777	0.0737	96.1288	7.0799
17	1.4002	0.7142	20.0121	0.0500	14.2919	0.0700	107.5554	7.5256
18	1.4282	0.7002	21.4123	0.0467	14.9920	0.0667	119.4581	7.9681
19	1.4568	0.6864	22.8406	0.0438	15.6785	0.0638	131.8139	8.4073
20	1.4859	0.6730	24.2974	0.0412	16.3514	0.0612	144.6003	8.8433
21	1.5157	0.6598	25.7833	0.0388	17.0112	0.0588	157.7959	9.2760
22	1.5460	0.6468	27.2990	0.0366	17.6580	0.0566	171.3795	9.7055
23	1.5769	0.6342	28.8450	0.0347	18.2922	0.0547	185.3309	10.1317
24	1.6084	0.6217	30.4219	0.0329	18.9139	0.0529	199.6305	10.5547
25	1.6406	0.6095	32.0303	0.0312	19.5235	0.0512	214.2592	10.9745
26	1.6734	0.5976	33.6709	0.0297	20.1210	0.0497	229.1987	11.3910
27	1.7069	0.5859	35.3443	0.0283	20.7069	0.0483	244.4311	11.8043
28	1.7410	0.5744	37.0512	0.0270	21.2813	0.0470	259.9392	12.2145
29	1.7758	0.5631	38.7922	0.0258	21.8444	0.0458	275.7064	12.6214
30	1.8114	0.5521	40.5681	0.0246	22.3965	0.0446	291.7164	13.0251
36	2.0399	0.4902	51.9944	0.0192	25.4888	0.0392	392.0405	15.3809
42	2.2972	0.4353	64.8622	0.0154	28.2348	0.0354	497.6010	17.6237
48	2.5871	0.3865	79.3535	0.0126	30.6731	0.0326	605.9657	19.7556
54	2.9135	0.3432	95.6731	0.0105	32.8383	0.0305	715.1815	21.7789
60	3.2810	0.3048	114.0515	8.768E-03	34.7609	0.0288	823.6975	23.6961
66	3.6950	0.2706	134.7487	7.421E-03	36.4681	0.0274	930.3000	25.5100
72	4.1611	0.2403	158.0570	6.327E-03	37.9841	0.0263	1.034E+03	27.2234
120	10.7652	0.0929	488.2582	2.048E-03	45.3554	0.0220	1.710E+03	37.7114
180	35.3208	0.0283	1.716E+03	5.827E-04	48.5844	0.0206	2.174E+03	44.7554
360	1.248E+03	8.016E-04	6.233E+04	1.604E-05	49.9599	0.0200	2.484E+03	49.7112

Time Value of Money Factors - Discrete Compounding
i = 3%

	Single Sums		Uniform Series				Gradient Series	
	To Find F Given P (F|P,i%,n)	To Find P Given F (P|F,i%,n)	To Find F Given A (F|A,i%,n)	To Find A Given F (A|F,i%,n)	To Find P Given A (P|A,i%,n)	To Find A Given P (A|P,i%,n)	To Find P Given G (P|G,i%,n)	To Find A Given G (A|G,i%,n)
n								
1	1.0300	0.9709	1.0000	1.0000	0.9709	1.0300	0.0000	0.0000
2	1.0609	0.9426	2.0300	0.4926	1.9135	0.5226	0.9426	0.4926
3	1.0927	0.9151	3.0909	0.3235	2.8286	0.3535	2.7729	0.9803
4	1.1255	0.8885	4.1836	0.2390	3.7171	0.2690	5.4383	1.4631
5	1.1593	0.8626	5.3091	0.1884	4.5797	0.2184	8.8888	1.9409
6	1.1941	0.8375	6.4684	0.1546	5.4172	0.1846	13.0762	2.4138
7	1.2299	0.8131	7.6625	0.1305	6.2303	0.1605	17.9547	2.8819
8	1.2668	0.7894	8.8923	0.1125	7.0197	0.1425	23.4806	3.3450
9	1.3048	0.7664	10.1591	0.0984	7.7861	0.1284	29.6119	3.8032
10	1.3439	0.7441	11.4639	0.0872	8.5302	0.1172	36.3088	4.2565
11	1.3842	0.7224	12.8078	0.0781	9.2526	0.1081	43.5330	4.7049
12	1.4258	0.7014	14.1920	0.0705	9.9540	0.1005	51.2482	5.1485
13	1.4685	0.6810	15.6178	0.0640	10.6350	0.0940	59.4196	5.5872
14	1.5126	0.6611	17.0863	0.0585	11.2961	0.0885	68.0141	6.0210
15	1.5580	0.6419	18.5989	0.0538	11.9379	0.0838	77.0002	6.4500
16	1.6047	0.6232	20.1569	0.0496	12.5611	0.0796	86.3477	6.8742
17	1.6528	0.6050	21.7616	0.0460	13.1661	0.0760	96.0280	7.2936
18	1.7024	0.5874	23.4144	0.0427	13.7535	0.0727	106.0137	7.7081
19	1.7535	0.5703	25.1169	0.0398	14.3238	0.0698	116.2788	8.1179
20	1.8061	0.5537	26.8704	0.0372	14.8775	0.0672	126.7987	8.5229
21	1.8603	0.5375	28.6765	0.0349	15.4150	0.0649	137.5496	8.9231
22	1.9161	0.5219	30.5368	0.0327	15.9369	0.0627	148.5094	9.3186
23	1.9736	0.5067	32.4529	0.0308	16.4436	0.0608	159.6566	9.7093
24	2.0328	0.4919	34.4265	0.0290	16.9355	0.0590	170.9711	10.0954
25	2.0938	0.4776	36.4593	0.0274	17.4131	0.0574	182.4336	10.4768
26	2.1566	0.4637	38.5530	0.0259	17.8768	0.0559	194.0260	10.8535
27	2.2213	0.4502	40.7096	0.0246	18.3270	0.0546	205.7309	11.2255
28	2.2879	0.4371	42.9309	0.0233	18.7641	0.0533	217.5320	11.5930
29	2.3566	0.4243	45.2189	0.0221	19.1885	0.0521	229.4137	11.9558
30	2.4273	0.4120	47.5754	0.0210	19.6004	0.0510	241.3613	12.3141
36	2.8983	0.3450	63.2759	0.0158	21.8323	0.0458	313.7028	14.3688
42	3.4607	0.2890	82.0232	0.0122	23.7014	0.0422	385.5024	16.2650
48	4.1323	0.2420	104.4084	9.578E-03	25.2667	0.0396	455.0255	18.0089
54	4.9341	0.2027	131.1375	7.626E-03	26.5777	0.0376	521.1157	19.6073
60	5.8916	0.1697	163.0534	6.133E-03	27.6756	0.0361	583.0526	21.0674
66	7.0349	0.1421	201.1627	4.971E-03	28.5950	0.0350	640.4407	22.3969
72	8.4000	0.1190	246.6672	4.054E-03	29.3651	0.0341	693.1226	23.6036
120	34.7110	0.0288	1.124E+03	8.899E-04	32.3730	0.0309	963.8635	29.7737
180	204.5034	4.890E-03	6.783E+03	1.474E-04	33.1703	0.0301	1.076E+03	32.4488
360	4.182E+04	2.391E-05	1.394E+06	7.173E-07	33.3325	0.0300	1.111E+03	33.3247

Time Value of Money Factors - Discrete Compounding
  i = 4%

	Single Sums		Uniform Series				Gradient Series	
	To Find F Given P (F|P,i%,n)	To Find P Given F (P|F,i%,n)	To Find F Given A (F|A,i%,n)	To Find A Given F (A|F,i%,n)	To Find P Given A (P|A,i%,n)	To Find A Given P (A|P,i%,n)	To Find P Given G (P|G,i%,n)	To Find A Given G (A|G,i%,n)
n								
1	1.0400	0.9615	1.0000	1.0000	0.9615	1.0400	0.0000	0.0000
2	1.0816	0.9246	2.0400	0.4902	1.8861	0.5302	0.9246	0.4902
3	1.1249	0.8890	3.1216	0.3203	2.7751	0.3603	2.7025	0.9739
4	1.1699	0.8548	4.2465	0.2355	3.6299	0.2755	5.2670	1.4510
5	1.2167	0.8219	5.4163	0.1846	4.4518	0.2246	8.5547	1.9216
6	1.2653	0.7903	6.6330	0.1508	5.2421	0.1908	12.5062	2.3857
7	1.3159	0.7599	7.8983	0.1266	6.0021	0.1666	17.0657	2.8433
8	1.3686	0.7307	9.2142	0.1085	6.7327	0.1485	22.1806	3.2944
9	1.4233	0.7026	10.5828	0.0945	7.4353	0.1345	27.8013	3.7391
10	1.4802	0.6756	12.0061	0.0833	8.1109	0.1233	33.8814	4.1773
11	1.5395	0.6496	13.4864	0.0741	8.7605	0.1141	40.3772	4.6090
12	1.6010	0.6246	15.0258	0.0666	9.3851	0.1066	47.2477	5.0343
13	1.6651	0.6006	16.6268	0.0601	9.9856	0.1001	54.4546	5.4533
14	1.7317	0.5775	18.2919	0.0547	10.5631	0.0947	61.9618	5.8659
15	1.8009	0.5553	20.0236	0.0499	11.1184	0.0899	69.7355	6.2721
16	1.8730	0.5339	21.8245	0.0458	11.6523	0.0858	77.7441	6.6720
17	1.9479	0.5134	23.6975	0.0422	12.1657	0.0822	85.9581	7.0656
18	2.0258	0.4936	25.6454	0.0390	12.6593	0.0790	94.3498	7.4530
19	2.1068	0.4746	27.6712	0.0361	13.1339	0.0761	102.8933	7.8342
20	2.1911	0.4564	29.7781	0.0336	13.5903	0.0736	111.5647	8.2091
21	2.2788	0.4388	31.9692	0.0313	14.0292	0.0713	120.3414	8.5779
22	2.3699	0.4220	34.2480	0.0292	14.4511	0.0692	129.2024	8.9407
23	2.4647	0.4057	36.6179	0.0273	14.8568	0.0673	138.1284	9.2973
24	2.5633	0.3901	39.0826	0.0256	15.2470	0.0656	147.1012	9.6479
25	2.6658	0.3751	41.6459	0.0240	15.6221	0.0640	156.1040	9.9925
26	2.7725	0.3607	44.3117	0.0226	15.9828	0.0626	165.1212	10.3312
27	2.8834	0.3468	47.0842	0.0212	16.3296	0.0612	174.1385	10.6640
28	2.9987	0.3335	49.9676	0.0200	16.6631	0.0600	183.1424	10.9909
29	3.1187	0.3207	52.9663	0.0189	16.9837	0.0589	192.1206	11.3120
30	3.2434	0.3083	56.0849	0.0178	17.2920	0.0578	201.0618	11.6274
36	4.1039	0.2437	77.5983	0.0129	18.9083	0.0529	253.4052	13.4018
42	5.1928	0.1926	104.8196	9.540E-03	20.1856	0.0495	302.4370	14.9828
48	6.5705	0.1522	139.2632	7.181E-03	21.1951	0.0472	347.2446	16.3832
54	8.3138	0.1203	182.8454	5.469E-03	21.9930	0.0455	387.4436	17.6167
60	10.5196	0.0951	237.9907	4.202E-03	22.6235	0.0442	422.9966	18.6972
66	13.3107	0.0751	307.7671	3.249E-03	23.1218	0.0432	454.0847	19.6388
72	16.8423	0.0594	396.0566	2.525E-03	23.5156	0.0425	481.0170	20.4552
120	110.6626	9.036E-03	2.742E+03	3.648E-04	24.7741	0.0404	592.2428	23.9057
180	1.164E+03	8.590E-04	2.908E+04	3.439E-05	24.9785	0.0400	620.5976	24.8452
360	1.355E+06	7.379E-07	3.388E+07	2.952E-08	25.0000	0.0400	624.9929	24.9997

Time Value of Money Factors - Discrete Compounding
i = 5%

n	Single Sums		Uniform Series				Gradient Series	
	To Find F Given P (F|P,i%,n)	To Find P Given F (P|F,i%,n)	To Find F Given A (F|A,i%,n)	To Find A Given F (A|F,i%,n)	To Find P Given A (P|A,i%,n)	To Find A Given P (A|P,i%,n)	To Find P Given G (P|G,i%,n)	To Find A Given G (A|G,i%,n)
1	1.0500	0.9524	1.0000	1.0000	0.9524	1.0500	0.0000	0.0000
2	1.1025	0.9070	2.0500	0.4878	1.8594	0.5378	0.9070	0.4878
3	1.1576	0.8638	3.1525	0.3172	2.7232	0.3672	2.6347	0.9675
4	1.2155	0.8227	4.3101	0.2320	3.5460	0.2820	5.1028	1.4391
5	1.2763	0.7835	5.5256	0.1810	4.3295	0.2310	8.2369	1.9025
6	1.3401	0.7462	6.8019	0.1470	5.0757	0.1970	11.9680	2.3579
7	1.4071	0.7107	8.1420	0.1228	5.7864	0.1728	16.2321	2.8052
8	1.4775	0.6768	9.5491	0.1047	6.4632	0.1547	20.9700	3.2445
9	1.5513	0.6446	11.0266	0.0907	7.1078	0.1407	26.1268	3.6758
10	1.6289	0.6139	12.5779	0.0795	7.7217	0.1295	31.6520	4.0991
11	1.7103	0.5847	14.2068	0.0704	8.3064	0.1204	37.4988	4.5144
12	1.7959	0.5568	15.9171	0.0628	8.8633	0.1128	43.6241	4.9219
13	1.8856	0.5303	17.7130	0.0565	9.3936	0.1065	49.9879	5.3215
14	1.9799	0.5051	19.5986	0.0510	9.8986	0.1010	56.5538	5.7133
15	2.0789	0.4810	21.5786	0.0463	10.3797	0.0963	63.2880	6.0973
16	2.1829	0.4581	23.6575	0.0423	10.8378	0.0923	70.1597	6.4736
17	2.2920	0.4363	25.8404	0.0387	11.2741	0.0887	77.1405	6.8423
18	2.4066	0.4155	28.1324	0.0355	11.6896	0.0855	84.2043	7.2034
19	2.5270	0.3957	30.5390	0.0327	12.0853	0.0827	91.3275	7.5569
20	2.6533	0.3769	33.0660	0.0302	12.4622	0.0802	98.4884	7.9030
21	2.7860	0.3589	35.7193	0.0280	12.8212	0.0780	105.6673	8.2416
22	2.9253	0.3418	38.5052	0.0260	13.1630	0.0760	112.8461	8.5730
23	3.0715	0.3256	41.4305	0.0241	13.4886	0.0741	120.0087	8.8971
24	3.2251	0.3101	44.5020	0.0225	13.7986	0.0725	127.1402	9.2140
25	3.3864	0.2953	47.7271	0.0210	14.0939	0.0710	134.2275	9.5238
26	3.5557	0.2812	51.1135	0.0196	14.3752	0.0696	141.2585	9.8266
27	3.7335	0.2678	54.6691	0.0183	14.6430	0.0683	148.2226	10.1224
28	3.9201	0.2551	58.4026	0.0171	14.8981	0.0671	155.1101	10.4114
29	4.1161	0.2429	62.3227	0.0160	15.1411	0.0660	161.9126	10.6936
30	4.3219	0.2314	66.4388	0.0151	15.3725	0.0651	168.6226	10.9691
36	5.7918	0.1727	95.8363	0.0104	16.5469	0.0604	206.6237	12.4872
42	7.7616	0.1288	135.2318	7.395E-03	17.4232	0.0574	240.2389	13.7884
48	10.4013	0.0961	188.0254	5.318E-03	18.0772	0.0553	269.2467	14.8943
54	13.9387	0.0717	258.7739	3.864E-03	18.5651	0.0539	293.8208	15.8265
60	18.6792	0.0535	353.5837	2.828E-03	18.9293	0.0528	314.3432	16.6062
66	25.0319	0.0399	480.6379	2.081E-03	19.2010	0.0521	331.2877	17.2536
72	33.5451	0.0298	650.9027	1.536E-03	19.4038	0.0515	345.1485	17.7877
120	348.9120	2.866E-03	6.958E+03	1.437E-04	19.9427	0.0501	391.9751	19.6551
180	6.517E+03	1.534E-04	1.303E+05	7.673E-06	19.9969	0.0500	399.3863	19.9724
360	4.248E+07	2.354E-08	8.495E+08	1.177E-09	20.0000	0.0500	399.9998	20.0000

Time Value of Money Factors - Discrete Compounding
i = 6%

	Single Sums		Uniform Series				Gradient Series	
	To Find F Given P (F|P,i%,n)	To Find P Given F (P|F,i%,n)	To Find F Given A (F|A,i%,n)	To Find A Given F (A|F,i%,n)	To Find P Given A (P|A,i%,n)	To Find A Given P (A|P,i%,n)	To Find P Given G (P|G,i%,n)	To Find A Given G (A|G,i%,n)
n								
1	1.0600	0.9434	1.0000	1.0000	0.9434	1.0600	0.0000	0.0000
2	1.1236	0.8900	2.0600	0.4854	1.8334	0.5454	0.8900	0.4854
3	1.1910	0.8396	3.1836	0.3141	2.6730	0.3741	2.5692	0.9612
4	1.2625	0.7921	4.3746	0.2286	3.4651	0.2886	4.9455	1.4272
5	1.3382	0.7473	5.6371	0.1774	4.2124	0.2374	7.9345	1.8836
6	1.4185	0.7050	6.9753	0.1434	4.9173	0.2034	11.4594	2.3304
7	1.5036	0.6651	8.3938	0.1191	5.5824	0.1791	15.4497	2.7676
8	1.5938	0.6274	9.8975	0.1010	6.2098	0.1610	19.8416	3.1952
9	1.6895	0.5919	11.4913	0.0870	6.8017	0.1470	24.5768	3.6133
10	1.7908	0.5584	13.1808	0.0759	7.3601	0.1359	29.6023	4.0220
11	1.8983	0.5268	14.9716	0.0668	7.8869	0.1268	34.8702	4.4213
12	2.0122	0.4970	16.8699	0.0593	8.3838	0.1193	40.3369	4.8113
13	2.1329	0.4688	18.8821	0.0530	8.8527	0.1130	45.9629	5.1920
14	2.2609	0.4423	21.0151	0.0476	9.2950	0.1076	51.7128	5.5635
15	2.3966	0.4173	23.2760	0.0430	9.7122	0.1030	57.5546	5.9260
16	2.5404	0.3936	25.6725	0.0390	10.1059	0.0990	63.4592	6.2794
17	2.6928	0.3714	28.2129	0.0354	10.4773	0.0954	69.4011	6.6240
18	2.8543	0.3503	30.9057	0.0324	10.8276	0.0924	75.3569	6.9597
19	3.0256	0.3305	33.7600	0.0296	11.1581	0.0896	81.3062	7.2867
20	3.2071	0.3118	36.7856	0.0272	11.4699	0.0872	87.2304	7.6051
21	3.3996	0.2942	39.9927	0.0250	11.7641	0.0850	93.1136	7.9151
22	3.6035	0.2775	43.3923	0.0230	12.0416	0.0830	98.9412	8.2166
23	3.8197	0.2618	46.9958	0.0213	12.3034	0.0813	104.7007	8.5099
24	4.0489	0.2470	50.8156	0.0197	12.5504	0.0797	110.3812	8.7951
25	4.2919	0.2330	54.8645	0.0182	12.7834	0.0782	115.9732	9.0722
26	4.5494	0.2198	59.1564	0.0169	13.0032	0.0769	121.4684	9.3414
27	4.8223	0.2074	63.7058	0.0157	13.2105	0.0757	126.8600	9.6029
28	5.1117	0.1956	68.5281	0.0146	13.4062	0.0746	132.1420	9.8568
29	5.4184	0.1846	73.6398	0.0136	13.5907	0.0736	137.3096	10.1032
30	5.7435	0.1741	79.0582	0.0126	13.7648	0.0726	142.3588	10.3422
36	8.1473	0.1227	119.1209	8.395E-03	14.6210	0.0684	170.0387	11.6298
42	11.5570	0.0865	175.9505	5.683E-03	15.2245	0.0657	193.1732	12.6883
48	16.3939	0.0610	256.5645	3.898E-03	15.6500	0.0639	212.0351	13.5485
54	23.2550	0.0430	370.9170	2.696E-03	15.9500	0.0627	227.1316	14.2402
60	32.9877	0.0303	533.1282	1.876E-03	16.1614	0.0619	239.0428	14.7909
66	46.7937	0.0214	763.2278	1.310E-03	16.3105	0.0613	248.3341	15.2254
72	66.3777	0.0151	1.090E+03	9.177E-04	16.4156	0.0609	255.5146	15.5654
120	1.088E+03	9.190E-04	1.812E+04	5.519E-05	16.6514	0.0601	275.6846	16.5563
180	3.590E+04	2.786E-05	5.983E+05	1.672E-06	16.6662	0.0600	277.6865	16.6617
360	1.289E+09	7.760E-10	2.148E+10	4.656E-11	16.6667	0.0600	277.7778	16.6667

Time Value of Money Factors - Discrete Compounding
i = 7%

n	Single Sums		Uniform Series				Gradient Series	
	To Find F Given P $(F|P,i\%,n)$	To Find P Given F $(P|F,i\%,n)$	To Find F Given A $(F|A,i\%,n)$	To Find A Given F $(A|F,i\%,n)$	To Find P Given A $(P|A,i\%,n)$	To Find A Given P $(A|P,i\%,n)$	To Find P Given G $(P|G,i\%,n)$	To Find A Given G $(A|G,i\%,n)$
1	1.0700	0.9346	1.0000	1.0000	0.9346	1.0700	0.0000	0.0000
2	1.1449	0.8734	2.0700	0.4831	1.8080	0.5531	0.8734	0.4831
3	1.2250	0.8163	3.2149	0.3111	2.6243	0.3811	2.5060	0.9549
4	1.3108	0.7629	4.4399	0.2252	3.3872	0.2952	4.7947	1.4155
5	1.4026	0.7130	5.7507	0.1739	4.1002	0.2439	7.6467	1.8650
6	1.5007	0.6663	7.1533	0.1398	4.7665	0.2098	10.9784	2.3032
7	1.6058	0.6227	8.6540	0.1156	5.3893	0.1856	14.7149	2.7304
8	1.7182	0.5820	10.2598	0.0975	5.9713	0.1675	18.7889	3.1465
9	1.8385	0.5439	11.9780	0.0835	6.5152	0.1535	23.1404	3.5517
10	1.9672	0.5083	13.8164	0.0724	7.0236	0.1424	27.7156	3.9461
11	2.1049	0.4751	15.7836	0.0634	7.4987	0.1334	32.4665	4.3296
12	2.2522	0.4440	17.8885	0.0559	7.9427	0.1259	37.3506	4.7025
13	2.4098	0.4150	20.1406	0.0497	8.3577	0.1197	42.3302	5.0648
14	2.5785	0.3878	22.5505	0.0443	8.7455	0.1143	47.3718	5.4167
15	2.7590	0.3624	25.1290	0.0398	9.1079	0.1098	52.4461	5.7583
16	2.9522	0.3387	27.8881	0.0359	9.4466	0.1059	57.5271	6.0897
17	3.1588	0.3166	30.8402	0.0324	9.7632	0.1024	62.5923	6.4110
18	3.3799	0.2959	33.9990	0.0294	10.0591	0.0994	67.6219	6.7225
19	3.6165	0.2765	37.3790	0.0268	10.3356	0.0968	72.5991	7.0242
20	3.8697	0.2584	40.9955	0.0244	10.5940	0.0944	77.5091	7.3163
21	4.1406	0.2415	44.8652	0.0223	10.8355	0.0923	82.3393	7.5990
22	4.4304	0.2257	49.0057	0.0204	11.0612	0.0904	87.0793	7.8725
23	4.7405	0.2109	53.4361	0.0187	11.2722	0.0887	91.7201	8.1369
24	5.0724	0.1971	58.1767	0.0172	11.4693	0.0872	96.2545	8.3923
25	5.4274	0.1842	63.2490	0.0158	11.6536	0.0858	100.6765	8.6391
26	5.8074	0.1722	68.6765	0.0146	11.8258	0.0846	104.9814	8.8773
27	6.2139	0.1609	74.4838	0.0134	11.9867	0.0834	109.1656	9.1072
28	6.6488	0.1504	80.6977	0.0124	12.1371	0.0824	113.2264	9.3289
29	7.1143	0.1406	87.3465	0.0114	12.2777	0.0814	117.1622	9.5427
30	7.6123	0.1314	94.4608	0.0106	12.4090	0.0806	120.9718	9.7487
36	11.4239	0.0875	148.9135	6.715E-03	13.0352	0.0767	141.1990	10.8321
42	17.1443	0.0583	230.6322	4.336E-03	13.4524	0.0743	157.1807	11.6842
48	25.7289	0.0389	353.2701	2.831E-03	13.7305	0.0728	169.4981	12.3447
54	38.6122	0.0259	537.3164	1.861E-03	13.9157	0.0719	178.8173	12.8500
60	57.9464	0.0173	813.5204	1.229E-03	14.0392	0.0712	185.7677	13.2321
66	86.9620	0.0115	1.228E+03	8.143E-04	14.1214	0.0708	190.8927	13.5179
72	130.5065	7.662E-03	1.850E+03	5.405E-04	14.1763	0.0705	194.6365	13.7298
120	3.358E+03	2.978E-04	4.795E+04	2.085E-05	14.2815	0.0700	203.5103	14.2500
180	1.946E+05	5.139E-06	2.780E+06	3.598E-07	14.2856	0.0700	204.0674	14.2848
360	3.786E+10	2.641E-11	5.408E+11	1.849E-12	14.2857	0.0700	204.0816	14.2857

Time Value of Money Factors - Discrete Compounding
i = 8%

	Single Sums		Uniform Series				Gradient Series	
	To Find F Given P (F|P,i%,n)	To Find P Given F (P|F,i%,n)	To Find F Given A (F|A,i%,n)	To Find A Given F (A|F,i%,n)	To Find P Given A (P|A,i%,n)	To Find A Given P (A|P,i%,n)	To Find P Given G (P|G,i%,n)	To Find A Given G (A|G,i%,n)
n								
1	1.0800	0.9259	1.0000	1.0000	0.9259	1.0800	0.0000	0.0000
2	1.1664	0.8573	2.0800	0.4808	1.7833	0.5608	0.8573	0.4808
3	1.2597	0.7938	3.2464	0.3080	2.5771	0.3880	2.4450	0.9487
4	1.3605	0.7350	4.5061	0.2219	3.3121	0.3019	4.6501	1.4040
5	1.4693	0.6806	5.8666	0.1705	3.9927	0.2505	7.3724	1.8465
6	1.5869	0.6302	7.3359	0.1363	4.6229	0.2163	10.5233	2.2763
7	1.7138	0.5835	8.9228	0.1121	5.2064	0.1921	14.0242	2.6937
8	1.8509	0.5403	10.6366	0.0940	5.7466	0.1740	17.8061	3.0985
9	1.9990	0.5002	12.4876	0.0801	6.2469	0.1601	21.8081	3.4910
10	2.1589	0.4632	14.4866	0.0690	6.7101	0.1490	25.9768	3.8713
11	2.3316	0.4289	16.6455	0.0601	7.1390	0.1401	30.2657	4.2395
12	2.5182	0.3971	18.9771	0.0527	7.5361	0.1327	34.6339	4.5957
13	2.7196	0.3677	21.4953	0.0465	7.9038	0.1265	39.0463	4.9402
14	2.9372	0.3405	24.2149	0.0413	8.2442	0.1213	43.4723	5.2731
15	3.1722	0.3152	27.1521	0.0368	8.5595	0.1168	47.8857	5.5945
16	3.4259	0.2919	30.3243	0.0330	8.8514	0.1130	52.2640	5.9046
17	3.7000	0.2703	33.7502	0.0296	9.1216	0.1096	56.5883	6.2037
18	3.9960	0.2502	37.4502	0.0267	9.3719	0.1067	60.8426	6.4920
19	4.3157	0.2317'	41.4463	0.0241	9.6036	0.1041	65.0134	6.7697
20	4.6610	0.2145	45.7620	0.0219	9.8181	0.1019	69.0898	7.0369
21	5.0338	0.1987	50.4229	0.0198	10.0168	0.0998	73.0629	7.2940
22	5.4365	0.1839	55.4568	0.0180	10.2007	0.0980	76.9257	7.5412
23	5.8715	0.1703	60.8933	0.0164	10.3711	0.0964	80.6726	7.7786
24	6.3412	0.1577	66.7648	0.0150	10.5288	0.0950	84.2997	8.0066
25	6.8485	0.1460	73.1059	0.0137	10.6748	0.0937	87.8041	8.2254
26	7.3964	0.1352	79.9544	0.0125	10.8100	0.0925	91.1842	8.4352
27	7.9881	0.1252	87.3508	0.0114	10.9352	0.0914	94.4390	8.6363
28	8.6271	0.1159	95.3388	0.0105	11.0511	0.0905	97.5687	8.8289
29	9.3173	0.1073	103.9659	9.619E-03	11.1584	0.0896	100.5738	9.0133
30	10.0627	0.0994	113.2832	8.827E-03	11.2578	0.0888	103.4558	9.1897
36	15.9682	0.0626	187.1021	5.345E-03	11.7172	0.0853	118.2839	10.0949
42	25.3395	0.0395	304.2435	3.287E-03	12.0067	0.0833	129.3651	10.7744
48	40.2106	0.0249	490.1322	2.040E-03	12.1891	0.0820	137.4428	11.2758
54	63.8091	0.0157	785.1141	1.274E-03	12.3041	0.0813	143.2229	11.6403
60	101.2571	9.876E-03	1.253E+03	7.979E-04	12.3766	0.0808	147.3000	11.9015
66	160.6822	6.223E-03	1.996E+03	5.010E-04	12.4222	0.0805	150.1432	12.0867
72	254.9825	3.922E-03	3.175E+03	3.150E-04	12.4510	0.0803	152.1076	12.2165
120	1.025E+04	9.753E-05	1.281E+05	7.803E-06	12.4988	0.0800	156.0885	12.4883
180	1.038E+06	9.632E-07	1.298E+07	7.706E-08	12.5000	0.0800	156.2477	12.4998
360	1.078E+12	9.278E-13	1.347E+13	7.422E-14	12.5000	0.0800	156.2500	12.5000

Time Value of Money Factors - Discrete Compounding
i = 9%

n	Single Sums		Uniform Series				Gradient Series	
	To Find F Given P (F|P,i%,n)	To Find P Given F (P|F,i%,n)	To Find F Given A (F|A,i%,n)	To Find A Given F (A|F,i%,n)	To Find P Given A (P|A,i%,n)	To Find A Given P (A|P,i%,n)	To Find P Given G (P|G,i%,n)	To Find A Given G (A|G,i%,n)
1	1.0900	0.9174	1.0000	1.0000	0.9174	1.0900	0.0000	0.0000
2	1.1881	0.8417	2.0900	0.4785	1.7591	0.5685	0.8417	0.4785
3	1.2950	0.7722	3.2781	0.3051	2.5313	0.3951	2.3860	0.9426
4	1.4116	0.7084	4.5731	0.2187	3.2397	0.3087	4.5113	1.3925
5	1.5386	0.6499	5.9847	0.1671	3.8897	0.2571	7.1110	1.8282
6	1.6771	0.5963	7.5233	0.1329	4.4859	0.2229	10.0924	2.2498
7	1.8280	0.5470	9.2004	0.1087	5.0330	0.1987	13.3746	2.6574
8	1.9926	0.5019	11.0285	0.0907	5.5348	0.1807	16.8877	3.0512
9	2.1719	0.4604	13.0210	0.0768	5.9952	0.1668	20.5711	3.4312
10	2.3674	0.4224	15.1929	0.0658	6.4177	0.1558	24.3728	3.7978
11	2.5804	0.3875	17.5603	0.0569	6.8052	0.1469	28.2481	4.1510
12	2.8127	0.3555	20.1407	0.0497	7.1607	0.1397	32.1590	4.4910
13	3.0658	0.3262	22.9534	0.0436	7.4869	0.1336	36.0731	4.8182
14	3.3417	0.2992	26.0192	0.0384	7.7862	0.1284	39.9633	5.1326
15	3.6425	0.2745	29.3609	0.0341	8.0607	0.1241	43.8069	5.4346
16	3.9703	0.2519	33.0034	0.0303	8.3126	0.1203	47.5849	5.7245
17	4.3276	0.2311	36.9737	0.0270	8.5436	0.1170	51.2821	6.0024
18	4.7171	0.2120	41.3013	0.0242	8.7556	0.1142	54.8860	6.2687
19	5.1417	0.1945	46.0185	0.0217	8.9501	0.1117	58.3868	6.5236
20	5.6044	0.1784	51.1601	0.0195	9.1285	0.1095	61.7770	6.7674
21	6.1088	0.1637	56.7645	0.0176	9.2922	0.1076	65.0509	7.0006
22	6.6586	0.1502	62.8733	0.0159	9.4424	0.1059	68.2048	7.2232
23	7.2579	0.1378	69.5319	0.0144	9.5802	0.1044	71.2359	7.4357
24	7.9111	0.1264	76.7898	0.0130	9.7066	0.1030	74.1433	7.6384
25	8.6231	0.1160	84.7009	0.0118	9.8226	0.1018	76.9265	7.8316
26	9.3992	0.1064	93.3240	0.0107	9.9290	0.1007	79.5863	8.0156
27	10.2451	0.0976	102.7231	9.735E-03	10.0266	0.0997	82.1241	8.1906
28	11.1671	0.0895	112.9682	8.852E-03	10.1161	0.0989	84.5419	8.3571
29	12.1722	0.0822	124.1354	8.056E-03	10.1983	0.0981	86.8422	8.5154
30	13.2677	0.0754	136.3075	7.336E-03	10.2737	0.0973	89.0280	8.6657
36	22.2512	0.0449	236.1247	4.235E-03	10.6118	0.0942	99.9319	9.4171
42	37.3175	0.0268	403.5281	2.478E-03	10.8134	0.0925	107.6432	9.9546
48	62.5852	0.0160	684.2804	1.461E-03	10.9336	0.0915	112.9625	10.3317
54	104.9617	9.527E-03	1.155E+03	8.657E-04	11.0053	0.0909	116.5642	10.5917
60	176.0313	5.681E-03	1.945E+03	5.142E-04	11.0480	0.0905	118.9683	10.7683
66	295.2221	3.387E-03	3.269E+03	3.059E-04	11.0735	0.0903	120.5546	10.8868
72	495.1170	2.020E-03	5.490E+03	1.821E-04	11.0887	0.0902	121.5917	10.9654
120	3.099E+04	3.227E-05	3.443E+05	2.905E-06	11.1108	0.0900	123.4098	11.1072
180	5.455E+06	1.833E-07	6.061E+07	1.650E-08	11.1111	0.0900	123.4564	11.1111
360	2.975E+13	3.361E-14	3.306E+14	3.025E-15	11.1111	0.0900	123.4568	11.1111

Time Value of Money Factors - Discrete Compounding
i = 10%

	Single Sums		Uniform Series				Gradient Series	
	To Find F Given P (F|P,i%,n)	To Find P Given F (P|F,i%,n)	To Find F Given A (F|A,i%,n)	To Find A Given F (A|F,i%,n)	To Find P Given A (P|A,i%,n)	To Find A Given P (A|P,i%,n)	To Find P Given G (P|G,i%,n)	To Find A Given G (A|G,i%,n)
n								
1	1.1000	0.9091	1.0000	1.0000	0.9091	1.1000	0.0000	0.0000
2	1.2100	0.8264	2.1000	0.4762	1.7355	0.5762	0.8264	0.4762
3	1.3310	0.7513	3.3100	0.3021	2.4869	0.4021	2.3291	0.9366
4	1.4641	0.6830	4.6410	0.2155	3.1699	0.3155	4.3781	1.3812
5	1.6105	0.6209	6.1051	0.1638	3.7908	0.2638	6.8618	1.8101
6	1.7716	0.5645	7.7156	0.1296	4.3553	0.2296	9.6842	2.2236
7	1.9487	0.5132	9.4872	0.1054	4.8684	0.2054	12.7631	2.6216
8	2.1436	0.4665	11.4359	0.0874	5.3349	0.1874	16.0287	3.0045
9	2.3579	0.4241	13.5795	0.0736	5.7590	0.1736	19.4215	3.3724
10	2.5937	0.3855	15.9374	0.0627	6.1446	0.1627	22.8913	3.7255
11	2.8531	0.3505	18.5312	0.0540	6.4951	0.1540	26.3963	4.0641
12	3.1384	0.3186	21.3843	0.0468	6.8137	0.1468	29.9012	4.3884
13	3.4523	0.2897	24.5227	0.0408	7.1034	0.1408	33.3772	4.6988
14	3.7975	0.2633	27.9750	0.0357	7.3667	0.1357	36.8005	4.9955
15	4.1772	0.2394	31.7725	0.0315	7.6061	0.1315	40.1520	5.2789
16	4.5950	0.2176	35.9497	0.0278	7.8237	0.1278	43.4164	5.5493
17	5.0545	0.1978	40.5447	0.0247	8.0216	0.1247	46.5819	5.8071
18	5.5599	0.1799	45.5992	0.0219	8.2014	0.1219	49.6395	6.0526
19	6.1159	0.1635	51.1591	0.0195	8.3649	0.1195	52.5827	6.2861
20	6.7275	0.1486	57.2750	0.0175	8.5136	0.1175	55.4069	6.5081
21	7.4002	0.1351	64.0025	0.0156	8.6487	0.1156	58.1095	6.7189
22	8.1403	0.1228	71.4027	0.0140	8.7715	0.1140	60.6893	6.9189
23	8.9543	0.1117	79.5430	0.0126	8.8832	0.1126	63.1462	7.1085
24	9.8497	0.1015	88.4973	0.0113	8.9847	0.1113	65.4813	7.2881
25	10.8347	0.0923	98.3471	0.0102	9.0770	0.1102	67.6964	7.4580
26	11.9182	0.0839	109.1818	9.159E-03	9.1609	0.1092	69.7940	7.6186
27	13.1100	0.0763	121.0999	8.258E-03	9.2372	0.1083	71.7773	7.7704
28	14.4210	0.0693	134.2099	7.451E-03	9.3066	0.1075	73.6495	7.9137
29	15.8631	0.0630	148.6309	6.728E-03	9.3696	0.1067	75.4146	8.0489
30	17.4494	0.0573	164.4940	6.079E-03	9.4269	0.1061	77.0766	8.1762
36	30.9127	0.0323	299.1268	3.343E-03	9.6765	0.1033	85.1194	8.7965
42	54.7637	0.0183	537.6370	1.860E-03	9.8174	0.1019	90.5047	9.2188
48	97.0172	0.0103	960.1723	1.041E-03	9.8969	0.1010	94.0217	9.5001
54	171.8719	5.818E-03	1.709E+03	5.852E-04	9.9418	0.1006	96.2763	9.6840
60	304.4816	3.284E-03	3.035E+03	3.295E-04	9.9672	0.1003	97.7010	9.8023
66	539.4078	1.854E-03	5.384E+03	1.857E-04	9.9815	0.1002	98.5910	9.8774
72	955.5938	1.046E-03	9.546E+03	1.048E-04	9.9895	0.1001	99.1419	9.9246
120	9.271E+04	1.079E-05	9.271E+05	1.079E-06	9.9999	0.1000	99.9860	9.9987
180	2.823E+07	3.543E-08	2.823E+08	3.543E-09	10.0000	0.1000	99.9999	10.0000
360	7.968E+14	1.255E-15	7.968E+15	1.255E-16	10.0000	0.1000	100.0000	10.0000

Time Value of Money Factors - Discrete Compounding
i = 12%

	Single Sums		Uniform Series				Gradient Series	
	To Find F Given P (F|P,i%,n)	To Find P Given F (P|F,i%,n)	To Find F Given A (F|A,i%,n)	To Find A Given F (A|F,i%,n)	To Find P Given A (P|A,i%,n)	To Find A Given P (A|P,i%,n)	To Find P Given G (P|G,i%,n)	To Find A Given G (A|G,i%,n)
n								
1	1.1200	0.8929	1.0000	1.0000	0.8929	1.1200	0.0000	0.0000
2	1.2544	0.7972	2.1200	0.4717	1.6901	0.5917	0.7972	0.4717
3	1.4049	0.7118	3.3744	0.2963	2.4018	0.4163	2.2208	0.9246
4	1.5735	0.6355	4.7793	0.2092	3.0373	0.3292	4.1273	1.3589
5	1.7623	0.5674	6.3528	0.1574	3.6048	0.2774	6.3970	1.7746
6	1.9738	0.5066	8.1152	0.1232	4.1114	0.2432	8.9302	2.1720
7	2.2107	0.4523	10.0890	0.0991	4.5638	0.2191	11.6443	2.5515
8	2.4760	0.4039	12.2997	0.0813	4.9676	0.2013	14.4714	2.9131
9	2.7731	0.3606	14.7757	0.0677	5.3282	0.1877	17.3563	3.2574
10	3.1058	0.3220	17.5487	0.0570	5.6502	0.1770	20.2541	3.5847
11	3.4785	0.2875	20.6546	0.0484	5.9377	0.1684	23.1288	3.8953
12	3.8960	0.2567	24.1331	0.0414	6.1944	0.1614	25.9523	4.1897
13	4.3635	0.2292	28.0291	0.0357	6.4235	0.1557	28.7024	4.4683
14	4.8871	0.2046	32.3926	0.0309	6.6282	0.1509	31.3624	4.7317
15	5.4736	0.1827	37.2797	0.0268	6.8109	0.1468	33.9202	4.9803
16	6.1304	0.1631	42.7533	0.0234	6.9740	0.1434	36.3670	5.2147
17	6.8660	0.1456	48.8837	0.0205	7.1196	0.1405	38.6973	5.4353
18	7.6900	0.1300	55.7497	0.0179	7.2497	0.1379	40.9080	5.6427
19	8.6128	0.1161	63.4397	0.0158	7.3658	0.1358	42.9979	5.8375
20	9.6463	0.1037	72.0524	0.0139	7.4694	0.1339	44.9676	6.0202
21	10.8038	0.0926	81.6987	0.0122	7.5620	0.1322	46.8188	6.1913
22	12.1003	0.0826	92.5026	0.0108	7.6446	0.1308	48.5543	6.3514
23	13.5523	0.0738	104.6029	9.560E-03	7.7184	0.1296	50.1776	6.5010
24	15.1786	0.0659	118.1552	8.463E-03	7.7843	0.1285	51.6929	6.6406
25	17.0001	0.0588	133.3339	7.500E-03	7.8431	0.1275	53.1046	6.7708
26	19.0401	0.0525	150.3339	6.652E-03	7.8957	0.1267	54.4177	6.8921
27	21.3249	0.0469	169.3740	5.904E-03	7.9426	0.1259	55.6369	7.0049
28	23.8839	0.0419	190.6989	5.244E-03	7.9844	0.1252	56.7674	7.1098
29	26.7499	0.0374	214.5828	4.660E-03	8.0218	0.1247	57.8141	7.2071
30	29.9599	0.0334	241.3327	4.144E-03	8.0552	0.1241	58.7821	7.2974
36	59.1356	0.0169	484.4631	2.064E-03	8.1924	0.1221	63.1970	7.7141
42	116.7231	8.567E-03	964.3595	1.037E-03	8.2619	0.1210	65.8509	7.9704
48	230.3908	4.340E-03	1.912E+03	5.231E-04	8.2972	0.1205	67.4068	8.1241
54	454.7505	2.199E-03	3.781E+03	2.645E-04	8.3150	0.1203	68.3022	8.2143
60	897.5969	1.114E-03	7.472E+03	1.338E-04	8.3240	0.1201	68.8100	8.2664
66	1.772E+03	5.644E-04	1.476E+04	6.777E-05	8.3286	0.1201	69.0948	8.2961
72	3.497E+03	2.860E-04	2.913E+04	3.432E-05	8.3310	0.1200	69.2530	8.3127
120	8.057E+05	1.241E-06	6.714E+06	1.489E-07	8.3333	0.1200	69.4431	8.3332
180	7.232E+08	1.383E-09	6.026E+09	1.659E-10	8.3333	0.1200	69.4444	8.3333
360	5.230E+17	1.912E-18	4.358E+18	2.295E-19	8.3333	0.1200	69.4444	8.3333

Time Value of Money Factors - Discrete Compounding
i = 15%

n	Single Sums		Uniform Series				Gradient Series	
	To Find F Given P (F|P,i%,n)	To Find P Given F (P|F,i%,n)	To Find F Given A (F|A,i%,n)	To Find A Given F (A|F,i%,n)	To Find P Given A (P|A,i%,n)	To Find A Given P (A|P,i%,n)	To Find P Given G (P|G,i%,n)	To Find A Given G (A|G,i%,n)
1	1.1500	0.8696	1.0000	1.0000	0.8696	1.1500	0.0000	0.0000
2	1.3225	0.7561	2.1500	0.4651	1.6257	0.6151	0.7561	0.4651
3	1.5209	0.6575	3.4725	0.2880	2.2832	0.4380	2.0712	0.9071
4	1.7490	0.5718	4.9934	0.2003	2.8550	0.3503	3.7864	1.3263
5	2.0114	0.4972	6.7424	0.1483	3.3522	0.2983	5.7751	1.7228
6	2.3131	0.4323	8.7537	0.1142	3.7845	0.2642	7.9368	2.0972
7	2.6600	0.3759	11.0668	0.0904	4.1604	0.2404	10.1924	2.4498
8	3.0590	0.3269	13.7268	0.0729	4.4873	0.2229	12.4807	2.7813
9	3.5179	0.2843	16.7858	0.0596	4.7716	0.2096	14.7548	3.0922
10	4.0456	0.2472	20.3037	0.0493	5.0188	0.1993	16.9795	3.3832
11	4.6524	0.2149	24.3493	0.0411	5.2337	0.1911	19.1289	3.6549
12	5.3503	0.1869	29.0017	0.0345	5.4206	0.1845	21.1849	3.9082
13	6.1528	0.1625	34.3519	0.0291	5.5831	0.1791	23.1352	4.1438
14	7.0757	0.1413	40.5047	0.0247	5.7245	0.1747	24.9725	4.3624
15	8.1371	0.1229	47.5804	0.0210	5.8474	0.1710	26.6930	4.5650
16	9.3576	0.1069	55.7175	0.0179	5.9542	0.1679	28.2960	4.7522
17	10.7613	0.0929	65.0751	0.0154	6.0472	0.1654	29.7828	4.9251
18	12.3755	0.0808	75.8364	0.0132	6.1280	0.1632	31.1565	5.0843
19	14.2318	0.0703	88.2118	0.0113	6.1982	0.1613	32.4213	5.2307
20	16.3665	0.0611	102.4436	9.761E-03	6.2593	0.1598	33.5822	5.3651
21	18.8215	0.0531	118.8101	8.417E-03	6.3125	0.1584	34.6448	5.4883
22	21.6447	0.0462	137.6316	7.266E-03	6.3587	0.1573	35.6150	5.6010
23	24.8915	0.0402	159.2764	6.278E-03	6.3988	0.1563	36.4988	5.7040
24	28.6252	0.0349	184.1678	5.430E-03	6.4338	0.1554	37.3023	5.7979
25	32.9190	0.0304	212.7930	4.699E-03	6.4641	0.1547	38.0314	5.8834
26	37.8568	0.0264	245.7120	4.070E-03	6.4906	0.1541	38.6918	5.9612
27	43.5353	0.0230	283.5688	3.526E-03	6.5135	0.1535	39.2890	6.0319
28	50.0656	0.0200	327.1041	3.057E-03	6.5335	0.1531	39.8283	6.0960
29	57.5755	0.0174	377.1697	2.651E-03	6.5509	0.1527	40.3146	6.1541
30	66.2118	0.0151	434.7451	2.300E-03	6.5660	0.1523	40.7526	6.2066
36	153.1519	6.529E-03	1.014E+03	9.859E-04	6.6231	0.1510	42.5872	6.4301
42	354.2495	2.823E-03	2.355E+03	4.246E-04	6.6478	0.1504	43.5286	6.5478
48	819.4007	1.220E-03	5.456E+03	1.833E-04	6.6585	0.1502	43.9997	6.6080
54	1.895E+03	5.276E-04	1.263E+04	7.918E-05	6.6631	0.1501	44.2311	6.6382
60	4.384E+03	2.281E-04	2.922E+04	3.422E-05	6.6651	0.1500	44.3431	6.6530
66	1.014E+04	9.861E-05	6.760E+04	1.479E-05	6.6660	0.1500	44.3967	6.6602
72	2.346E+04	4.263E-05	1.564E+05	6.395E-06	6.6664	0.1500	44.4221	6.6636
120	1.922E+07	5.203E-08	1.281E+08	7.805E-09	6.6667	0.1500	44.4444	6.6667
180	8.426E+10	1.187E-11	5.617E+11	1.780E-12	6.6667	0.1500	44.4444	6.6667
360	7.099E+21	1.409E-22	4.733E+22	2.113E-23	6.6667	0.1500	44.4444	6.6667

Time Value of Money Factors - Discrete Compounding
  i = 18%

	Single Sums		Uniform Series				Gradient Series	
	To Find F Given P (F|P,i%,n)	To Find P Given F (P|F,i%,n)	To Find F Given A (F|A,i%,n)	To Find A Given F (A|F,i%,n)	To Find P Given A (P|A,i%,n)	To Find A Given P (A|P,i%,n)	To Find P Given G (P|G,i%,n)	To Find A Given G (A|G,i%,n)
n								
1	1.1800	0.8475	1.0000	1.0000	0.8475	1.1800	0.0000	0.0000
2	1.3924	0.7182	2.1800	0.4587	1.5656	0.6387	0.7182	0.4587
3	1.6430	0.6086	3.5724	0.2799	2.1743	0.4599	1.9354	0.8902
4	1.9388	0.5158	5.2154	0.1917	2.6901	0.3717	3.4828	1.2947
5	2.2878	0.4371	7.1542	0.1398	3.1272	0.3198	5.2312	1.6728
6	2.6996	0.3704	9.4420	0.1059	3.4976	0.2859	7.0834	2.0252
7	3.1855	0.3139	12.1415	0.0824	3.8115	0.2624	8.9670	2.3526
8	3.7589	0.2660	15.3270	0.0652	4.0776	0.2452	10.8292	2.6558
9	4.4355	0.2255	19.0859	0.0524	4.3030	0.2324	12.6329	2.9358
10	5.2338	0.1911	23.5213	0.0425	4.4941	0.2225	14.3525	3.1936
11	6.1759	0.1619	28.7551	0.0348	4.6560	0.2148	15.9716	3.4303
12	7.2876	0.1372	34.9311	0.0286	4.7932	0.2086	17.4811	3.6470
13	8.5994	0.1163	42.2187	0.0237	4.9095	0.2037	18.8765	3.8449
14	10.1472	0.0985	50.8180	0.0197	5.0081	0.1997	20.1576	4.0250
15	11.9737	0.0835	60.9653	0.0164	5.0916	0.1964	21.3269	4.1887
16	14.1290	0.0708	72.9390	0.0137	5.1624	0.1937	22.3885	4.3369
17	16.6722	0.0600	87.0680	0.0115	5.2223	0.1915	23.3482	4.4708
18	19.6733	0.0508	103.7403	9.639E-03	5.2732	0.1896	24.2123	4.5916
19	23.2144	0.0431	123.4135	8.103E-03	5.3162	0.1881	24.9877	4.7003
20	27.3930	0.0365	146.6280	6.820E-03	5.3527	0.1868	25.6813	4.7978
21	32.3238	0.0309	174.0210	5.746E-03	5.3837	0.1857	26.3000	4.8851
22	38.1421	0.0262	206.3448	4.846E-03	5.4099	0.1848	26.8506	4.9632
23	45.0076	0.0222	244.4868	4.090E-03	5.4321	0.1841	27.3394	5.0329
24	53.1090	0.0188	289.4945	3.454E-03	5.4509	0.1835	27.7725	5.0950
25	62.6686	0.0160	342.6035	2.919E-03	5.4669	0.1829	28.1555	5.1502
26	73.9490	0.0135	405.2721	2.467E-03	5.4804	0.1825	28.4935	5.1991
27	87.2598	0.0115	479.2211	2.087E-03	5.4919	0.1821	28.7915	5.2425
28	102.9666	9.712E-03	566.4809	1.765E-03	5.5016	0.1818	29.0537	5.2810
29	121.5005	8.230E-03	669.4475	1.494E-03	5.5098	0.1815	29.2842	5.3149
30	143.3706	6.975E-03	790.9480	1.264E-03	5.5168	0.1813	29.4864	5.3448
36	387.0368	2.584E-03	2.145E+03	4.663E-04	5.5412	0.1805	30.2677	5.4623
42	1.045E+03	9.571E-04	5.799E+03	1.724E-04	5.5502	0.1802	30.6113	5.5153
48	2.821E+03	3.545E-04	1.566E+04	6.384E-05	5.5536	0.1801	30.7587	5.5385
54	7.614E+03	1.313E-04	4.230E+04	2.364E-05	5.5548	0.1800	30.8207	5.5485
60	2.056E+04	4.865E-05	1.142E+05	8.757E-06	5.5553	0.1800	30.8465	5.5526
66	5.549E+04	1.802E-05	3.083E+05	3.244E-06	5.5555	0.1800	30.8570	5.5544
72	1.498E+05	6.676E-06	8.322E+05	1.202E-06	5.5555	0.1800	30.8613	5.5551
120	4.225E+08	2.367E-09	2.347E+09	4.260E-10	5.5556	0.1800	30.8642	5.5556
180	8.685E+12	1.151E-13	4.825E+13	2.073E-14	5.5556	0.1800	30.8642	5.5556
360	7.543E+25	1.326E-26	4.190E+26	2.386E-27	5.5556	0.1800	30.8642	5.5556

Time Value of Money Factors - Discrete Compounding
 i = 20%

n	Single Sums To Find F Given P (F|P,i%,n)	Single Sums To Find P Given F (P|F,i%,n)	Uniform Series To Find F Given A (F|A,i%,n)	Uniform Series To Find A Given F (A|F,i%,n)	Uniform Series To Find P Given A (P|A,i%,n)	Uniform Series To Find A Given P (A|P,i%,n)	Gradient Series To Find P Given G (P|G,i%,n)	Gradient Series To Find A Given G (A|G,i%,n)
1	1.2000	0.8333	1.0000	1.0000	0.8333	1.2000	0.0000	0.0000
2	1.4400	0.6944	2.2000	0.4545	1.5278	0.6545	0.6944	0.4545
3	1.7280	0.5787	3.6400	0.2747	2.1065	0.4747	1.8519	0.8791
4	2.0736	0.4823	5.3680	0.1863	2.5887	0.3863	3.2986	1.2742
5	2.4883	0.4019	7.4416	0.1344	2.9906	0.3344	4.9061	1.6405
6	2.9860	0.3349	9.9299	0.1007	3.3255	0.3007	6.5806	1.9788
7	3.5832	0.2791	12.9159	0.0774	3.6046	0.2774	8.2551	2.2902
8	4.2998	0.2326	16.4991	0.0606	3.8372	0.2606	9.8831	2.5756
9	5.1598	0.1938	20.7989	0.0481	4.0310	0.2481	11.4335	2.8364
10	6.1917	0.1615	25.9587	0.0385	4.1925	0.2385	12.8871	3.0739
11	7.4301	0.1346	32.1504	0.0311	4.3271	0.2311	14.2330	3.2893
12	8.9161	0.1122	39.5805	0.0253	4.4392	0.2253	15.4667	3.4841
13	10.6993	0.0935	48.4966	0.0206	4.5327	0.2206	16.5883	3.6597
14	12.8392	0.0779	59.1959	0.0169	4.6106	0.2169	17.6008	3.8175
15	15.4070	0.0649	72.0351	0.0139	4.6755	0.2139	18.5095	3.9588
16	18.4884	0.0541	87.4421	0.0114	4.7296	0.2114	19.3208	4.0851
17	22.1861	0.0451	105.9306	9.440E-03	4.7746	0.2094	20.0419	4.1976
18	26.6233	0.0376	128.1167	7.805E-03	4.8122	0.2078	20.6805	4.2975
19	31.9480	0.0313	154.7400	6.462E-03	4.8435	0.2065	21.2439	4.3861
20	38.3376	0.0261	186.6880	5.357E-03	4.8696	0.2054	21.7395	4.4643
21	46.0051	0.0217	225.0256	4.444E-03	4.8913	0.2044	22.1742	4.5334
22	55.2061	0.0181	271.0307	3.690E-03	4.9094	0.2037	22.5546	4.5941
23	66.2474	0.0151	326.2369	3.065E-03	4.9245	0.2031	22.8867	4.6475
24	79.4968	0.0126	392.4842	2.548E-03	4.9371	0.2025	23.1760	4.6943
25	95.3962	0.0105	471.9811	2.119E-03	4.9476	0.2021	23.4276	4.7352
26	114.4755	8.735E-03	567.3773	1.762E-03	4.9563	0.2018	23.6460	4.7709
27	137.3706	7.280E-03	681.8528	1.467E-03	4.9636	0.2015	23.8353	4.8020
28	164.8447	6.066E-03	819.2233	1.221E-03	4.9697	0.2012	23.9991	4.8291
29	197.8136	5.055E-03	984.0680	1.016E-03	4.9747	0.2010	24.1406	4.8527
30	237.3763	4.213E-03	1.182E+03	8.461E-04	4.9789	0.2008	24.2628	4.8731
36	708.8019	1.411E-03	3.539E+03	2.826E-04	4.9929	0.2003	24.7108	4.9491
42	2.116E+03	4.725E-04	1.058E+04	9.454E-05	4.9976	0.2001	24.8890	4.9801
48	6.320E+03	1.582E-04	3.159E+04	3.165E-05	4.9992	0.2000	24.9581	4.9924
54	1.887E+04	5.299E-05	9.435E+04	1.060E-05	4.9997	0.2000	24.9844	4.9971
60	5.635E+04	1.775E-05	2.817E+05	3.549E-06	4.9999	0.2000	24.9942	4.9989
66	1.683E+05	5.943E-06	8.413E+05	1.189E-06	5.0000	0.2000	24.9979	4.9996
72	5.024E+05	1.990E-06	2.512E+06	3.981E-07	5.0000	0.2000	24.9992	4.9999
120	3.175E+09	3.150E-10	1.588E+10	6.299E-11	5.0000	0.2000	25.0000	5.0000
180	1.789E+14	5.590E-15	8.945E+14	1.118E-15	5.0000	0.2000	25.0000	5.0000
360	3.201E+28	3.124E-29	1.600E+29	6.249E-30	5.0000	0.2000	25.0000	5.0000

Time Value of Money Factors - Discrete Compounding
i = 25%

n	Single Sums		Uniform Series				Gradient Series	
	To Find F Given P (F|P,i%,n)	To Find P Given F (P|F,i%,n)	To Find F Given A (F|A,i%,n)	To Find A Given F (A|F,i%,n)	To Find P Given A (P|A,i%,n)	To Find A Given P (A|P,i%,n)	To Find P Given G (P|G,i%,n)	To Find A Given G (A|G,i%,n)
1	1.2500	0.8000	1.0000	1.0000	0.8000	1.2500	0.0000	0.0000
2	1.5625	0.6400	2.2500	0.4444	1.4400	0.6944	0.6400	0.4444
3	1.9531	0.5120	3.8125	0.2623	1.9520	0.5123	1.6640	0.8525
4	2.4414	0.4096	5.7656	0.1734	2.3616	0.4234	2.8928	1.2249
5	3.0518	0.3277	8.2070	0.1218	2.6893	0.3718	4.2035	1.5631
6	3.8147	0.2621	11.2588	0.0888	2.9514	0.3388	5.5142	1.8683
7	4.7684	0.2097	15.0735	0.0663	3.1611	0.3163	6.7725	2.1424
8	5.9605	0.1678	19.8419	0.0504	3.3289	0.3004	7.9469	2.3872
9	7.4506	0.1342	25.8023	0.0388	3.4631	0.2888	9.0207	2.6048
10	9.3132	0.1074	33.2529	0.0301	3.5705	0.2801	9.9870	2.7971
11	11.6415	0.0859	42.5661	0.0235	3.6564	0.2735	10.8460	2.9663
12	14.5519	0.0687	54.2077	0.0184	3.7251	0.2684	11.6020	3.1145
13	18.1899	0.0550	68.7596	0.0145	3.7801	0.2645	12.2617	3.2437
14	22.7374	0.0440	86.9495	0.0115	3.8241	0.2615	12.8334	3.3559
15	28.4217	0.0352	109.6868	9.117E-03	3.8593	0.2591	13.3260	3.4530
16	35.5271	0.0281	138.1085	7.241E-03	3.8874	0.2572	13.7482	3.5366
17	44.4089	0.0225	173.6357	5.759E-03	3.9099	0.2558	14.1085	3.6084
18	55.5112	0.0180	218.0446	4.586E-03	3.9279	0.2546	14.4147	3.6698
19	69.3889	0.0144	273.5558	3.656E-03	3.9424	0.2537	14.6741	3.7222
20	86.7362	0.0115	342.9447	2.916E-03	3.9539	0.2529	14.8932	3.7667
21	108.4202	9.223E-03	429.6809	2.327E-03	3.9631	0.2523	15.0777	3.8045
22	135.5253	7.379E-03	538.1011	1.858E-03	3.9705	0.2519	15.2326	3.8365
23	169.4066	5.903E-03	673.6264	1.485E-03	3.9764	0.2515	15.3625	3.8634
24	211.7582	4.722E-03	843.0329	1.186E-03	3.9811	0.2512	15.4711	3.8861
25	264.6978	3.778E-03	1.055E+03	9.481E-04	3.9849	0.2509	15.5618	3.9052
26	330.8722	3.022E-03	1.319E+03	7.579E-04	3.9879	0.2508	15.6373	3.9212
27	413.5903	2.418E-03	1.650E+03	6.059E-04	3.9903	0.2506	15.7002	3.9346
28	516.9879	1.934E-03	2.064E+03	4.845E-04	3.9923	0.2505	15.7524	3.9457
29	646.2349	1.547E-03	2.581E+03	3.875E-04	3.9938	0.2504	15.7957	3.9551
30	807.7936	1.238E-03	3.227E+03	3.099E-04	3.9950	0.2503	15.8316	3.9628
36	3.081E+03	3.245E-04	1.232E+04	8.116E-05	3.9987	0.2501	15.9481	3.9883
42	1.175E+04	8.507E-05	4.702E+04	2.127E-05	3.9997	0.2500	15.9843	3.9964
48	4.484E+04	2.230E-05	1.794E+05	5.575E-06	3.9999	0.2500	15.9954	3.9989
54	1.711E+05	5.846E-06	6.842E+05	1.462E-06	4.0000	0.2500	15.9986	3.9997
60	6.525E+05	1.532E-06	2.610E+06	3.831E-07	4.0000	0.2500	15.9996	3.9999
66	2.489E+06	4.017E-07	9.957E+06	1.004E-07	4.0000	0.2500	15.9999	4.0000
72	9.496E+06	1.053E-07	3.798E+07	2.633E-08	4.0000	0.2500	16.0000	4.0000
120	4.258E+11	2.349E-12	1.703E+12	5.871E-13	4.0000	0.2500	16.0000	4.0000
180	2.778E+17	3.599E-18	1.111E+18	8.998E-19	4.0000	0.2500	16.0000	4.0000
360	7.720E+34	1.295E-35	3.088E+35	3.238E-36	4.0000	0.2500	16.0000	4.0000

Time Value of Money Factors - Discrete Compounding
i = 30%

n	Single Sums		Uniform Series				Gradient Series	
	To Find F Given P (F|P,i%,n)	To Find P Given F (P|F,i%,n)	To Find F Given A (F|A,i%,n)	To Find A Given F (A|F,i%,n)	To Find P Given A (P|A,i%,n)	To Find A Given P (A|P,i%,n)	To Find P Given G (P|G,i%,n)	To Find A Given G (A|G,i%,n)
1	1.3000	0.7692	1.0000	1.0000	0.7692	1.3000	0.0000	0.0000
2	1.6900	0.5917	2.3000	0.4348	1.3609	0.7348	0.5917	0.4348
3	2.1970	0.4552	3.9900	0.2506	1.8161	0.5506	1.5020	0.8271
4	2.8561	0.3501	6.1870	0.1616	2.1662	0.4616	2.5524	1.1783
5	3.7129	0.2693	9.0431	0.1106	2.4356	0.4106	3.6297	1.4903
6	4.8268	0.2072	12.7560	0.0784	2.6427	0.3784	4.6656	1.7654
7	6.2749	0.1594	17.5828	0.0569	2.8021	0.3569	5.6218	2.0063
8	8.1573	0.1226	23.8577	0.0419	2.9247	0.3419	6.4800	2.2156
9	10.6045	0.0943	32.0150	0.0312	3.0190	0.3312	7.2343	2.3963
10	13.7858	0.0725	42.6195	0.0235	3.0915	0.3235	7.8872	2.5512
11	17.9216	0.0558	56.4053	0.0177	3.1473	0.3177	8.4452	2.6833
12	23.2981	0.0429	74.3270	0.0135	3.1903	0.3135	8.9173	2.7952
13	30.2875	0.0330	97.6250	0.0102	3.2233	0.3102	9.3135	2.8895
14	39.3738	0.0254	127.9125	7.818E-03	3.2487	0.3078	9.6437	2.9685
15	51.1859	0.0195	167.2863	5.978E-03	3.2682	0.3060	9.9172	3.0344
16	66.5417	0.0150	218.4722	4.577E-03	3.2832	0.3046	10.1426	3.0892
17	86.5042	0.0116	285.0139	3.509E-03	3.2948	0.3035	10.3276	3.1345
18	112.4554	8.892E-03	371.5180	2.692E-03	3.3037	0.3027	10.4788	3.1718
19	146.1920	6.840E-03	483.9734	2.066E-03	3.3105	0.3021	10.6019	3.2025
20	190.0496	5.262E-03	630.1655	1.587E-03	3.3158	0.3016	10.7019	3.2275
21	247.0645	4.048E-03	820.2151	1.219E-03	3.3198	0.3012	10.7828	3.2480
22	321.1839	3.113E-03	1.067E+03	9.370E-04	3.3230	0.3009	10.8482	3.2646
23	417.5391	2.395E-03	1.388E+03	7.202E-04	3.3254	0.3007	10.9009	3.2781
24	542.8008	1.842E-03	1.806E+03	5.537E-04	3.3272	0.3006	10.9433	3.2890
25	705.6410	1.417E-03	2.349E+03	4.257E-04	3.3286	0.3004	10.9773	3.2979
26	917.3333	1.090E-03	3.054E+03	3.274E-04	3.3297	0.3003	11.0045	3.3050
27	1.193E+03	8.386E-04	3.972E+03	2.518E-04	3.3305	0.3003	11.0263	3.3107
28	1.550E+03	6.450E-04	5.164E+03	1.936E-04	3.3312	0.3002	11.0437	3.3153
29	2.015E+03	4.962E-04	6.715E+03	1.489E-04	3.3317	0.3001	11.0576	3.3189
30	2.620E+03	3.817E-04	8.730E+03	1.145E-04	3.3321	0.3001	11.0687	3.3219
36	1.265E+04	7.908E-05	4.215E+04	2.372E-05	3.3331	0.3000	11.1007	3.3305
42	6.104E+04	1.638E-05	2.035E+05	4.915E-06	3.3333	0.3000	11.1086	3.3326
48	2.946E+05	3.394E-06	9.821E+05	1.018E-06	3.3333	0.3000	11.1105	3.3332
54	1.422E+06	7.032E-07	4.740E+06	2.110E-07	3.3333	0.3000	11.1110	3.3333
60	6.864E+06	1.457E-07	2.288E+07	4.370E-08	3.3333	0.3000	11.1111	3.3333
66	3.313E+07	3.018E-08	1.104E+08	9.054E-09	3.3333	0.3000	11.1111	3.3333
72	1.599E+08	6.253E-09	5.331E+08	1.876E-09	3.3333	0.3000	11.1111	3.3333
120	4.712E+13	2.122E-14	1.571E+14	6.367E-15	3.3333	0.3000	11.1111	3.3333
180	3.234E+20	3.092E-21	1.078E+21	9.275E-22	3.3333	0.3000	11.1111	3.3333
360	1.046E+41	9.559E-42	3.487E+41	2.868E-42	3.3333	0.3000	11.1111	3.3333

CHAPTER 5

BOILERS AND FIRED SYSTEMS

S.A. PARKER

Senior Research Engineer, Energy Division
Pacific Northwest National Laboratory
Richland, Washington

R.B. SCOLLON

Corporate Manager, Energy Conservation

R.D. SMITH

Manager, Energy Generation and Feedstocks
Allied Corporation
Morristown, New Jersey

5.1 INTRODUCTION

Boilers and other fired systems are the most significant energy consumers. Almost two-thirds of the fossil-fuel energy consumed in the United States involves the use of a boiler, furnace, or other fired system. Even most electric energy is produced using fuel-fired boilers. Over 68% of the electricity generated in the United States is produced through the combustion of coal, fuel oil, and natural gas. (The remainder is produced through nuclear, 22%; hydroelectric, 10%; and geothermal and others, <1%.) Unlike many electric systems, boilers and fired systems are not inherently energy efficient.

This chapter and the following chapter on Steam and Condensate Systems examine how energy is consumed, how energy is wasted, and opportunities for reducing energy consumption and costs in the operation of boiler and steam plants. A list of energy and cost reduction measures is presented, categorized as: load reduction, waste heat recovery, efficiency improvement, fuel cost reduction, and other opportunities. Several of the key opportunities for reducing operating costs are presented ranging from changes in operating procedures to capital improvement opportunities. The topics reflect recurring opportunities identified from numerous in-plant audits. Several examples are presented to demonstrate the methodology for estimating the potential energy savings associated with various opportunities. Many of these examples utilize easy to understand nomographs and

charts in the solution techniques.

In addition to energy saving opportunities, this chapter also describes some issues relevant to day-to-day operations, maintenance, and troubleshooting. Considerations relative to fuel comparison and selection are also discussed. Developing technologies relative to alternative fuels and types of combustion equipment are also discussed. Some of the technologies discussed hold the potential for significant cost reductions while alleviating environmental problems.

The chapter concludes with a brief discussion of some of the major regulations impacting the operation of boilers and fired systems. It is important to emphasize the need to carefully assess the potential impact of federal, state, and local regulations.

5.2 ANALYSIS OF BOILERS AND FIRED SYSTEMS

5.2.1 Boiler Energy Consumption

Boiler and other fired systems, such as furnaces and ovens, combust fuel with air for the purpose of releasing the chemical heat energy. The purpose of the heat energy may be to raise the temperature of an industrial product as part of a manufacturing process, it may be to generate high-temperature high-pressure steam in order to power a turbine, or it may simply be to heat a space so the occupants will be comfortable. The energy consumption of boilers, furnaces, and other fire systems can be determined simply as a function of load and efficiency as expressed in the equation:

Energy consumption =
$$\int (load) \times (1/efficiency) \, dt \tag{5.1}$$

Similarly, the cost of operating a boiler or fired system can be determined as:

Energy cost =
$$\int (load) \times (1/efficiency) \times (fuel\ cost) \, dt \tag{5.2}$$

As such, the opportunities for reducing the energy consumption or energy cost of a boiler or fired system can be put into a few categories. In order to reduce boiler energy consumption, one can either reduce the load, increase the operating efficiency, reduce the unit fuel energy cost, or combinations thereof.

Of course equations 5.1 and 5.2 are not always that simple because the variables are not always constant. The *load* varies as a function of the process being supported. The *efficiency* varies as a function of the *load* and other functions, such as time or weather. In addition, the *fuel cost* may also vary as a function of time (such as in seasonal, time-of-use, or spot market rates) or as a function of load (such as declining block or spot market rates.) Therefore, solving the equation for the energy consumption or energy cost may not always be simplistic.

5.2.2 Balance Equations

Balance equations are used in an analysis of a process which determines inputs and outputs to a system. There are several types of balance equations which may prove useful in the analysis of a boiler or fired-system. These include a heat balance and mass balance.

Heat Balance

A heat balance is used to determine where all the heat energy enters and leaves a system. Assuming that energy can neither be created or destroyed, all energy can be accounted for in a system analysis. Energy in equals energy out. Whether through measurement or analysis, all energy entering or leaving a system can be determined. In a simple furnace system, energy enters through the combustion air, fuel, and mixed-air duct. Energy leaves the furnace system through the supply-air duct and the exhaust gases.

In a boiler system, the analysis can become more complex. Energy input comes from the following: condensate return, make-up water, combustion air, fuel, and maybe a few others depending on the complexity of the system. Energy output departs as the following: steam, blowdown, exhaust gases, shell/surface losses, possibly ash, and other discharges depending on the complexity of the system.

Mass Balance

A mass balance is used to determine where all mass enters and leaves a system. There are several methods in which a mass balance can be performed that can be useful in the analysis of a boiler or other fired system. In the case of a steam boiler, a mass balance can be used in the form of a water balance (steam, condensate return, make-up water, blowdown, and feedwater.) A mass balance can also be used for water quality or chemical balance (total dissolved solids, or other impurity.) The mass balance can also be used in the form of a combustion analysis

(fire-side mass balance consisting of air and fuel in and combustion gasses and excess air out.) This type of analysis is the foundation for determining combustion efficiency and determining the optimum air-to-fuel ratio.

For analyzing complex systems, the mass and energy balance equations may be used simultaneously such as in solving multiple equations with multiple unknowns. This type of analysis is particularly useful in determining blowdown losses, waste heat recovery potential, and other interdependent opportunities.

5.2.3 Efficiency

There are several different measures of efficiency used in boilers and fired systems. While this may lead to some confusion, the different measures are used to convey different information. Therefore, it is important to understand what is being implied by a given efficiency measure.

The basis for testing boilers is the American Society of Mechanical Engineers (ASME) Power Test Code 4.1 (PTC-4.1-1964.) This procedure defines and established two primary methods of determining efficiency: the input-output method and the heat-loss method. Both of these methods result in what is commonly referred to as the gross thermal efficiency. The efficiencies determined by these methods are "gross" efficiencies as apposed to "net" efficiencies which would include the additional energy input of auxiliary equipment such as combustion air fans, fuel pumps, stoker drives, etc. For more information on these methods, see the ASME PTC-4.1-1964 or Taplin 1991.

Another efficiency term commonly used for boilers and other fired systems is combustion efficiency. Combustion efficiency is similar to the heat loss method, but only the heat losses due to the exhaust gases are considered. Combustion efficiency can be measured in the field by analyzing the products of combustion the exhaust gases.

Typically measuring either carbon dioxide (CO_2) or oxygen (O_2) in the exhaust gas can be used to determine the combustion efficiency as long as there is excess air. Excess air is defined as air in excess of the amount required for stoichiometric conditions. In other words, excess air is the amount of air above that which is theoretically required for complete combustion. In the real world, however, it is not possible to get perfect mixture of air and fuel to achieve complete combustion without some amount of excess air. As excess air is reduced toward the fuel rich side, incomplete combustion begins to occur resulting in the formation of carbon monoxide, carbon, smoke, and in extreme cases, raw unburned fuel. Incomplete combustion is inefficient, expensive, and frequently unsafe. Therefore, some amount of excess air is required to ensure complete and safe combustion.

However, excess air is also inefficient as it results in the excess air being heated from ambient air temperatures to exhaust gas temperatures resulting in a form of heat loss. Therefore while some excess air is required it is also desirable to minimize the amount of excess air.

As illustrated in Figure 5.1, the amount of carbon dioxide, percent by volume, in the exhaust gas reaches a maximum with no excess air stoichiometric conditions. While carbon dioxide can be used as a measure of complete combustion, it can not be used to optimally control the air-to-fuel ratio in a fired system. A drop in the level of carbon dioxide would not be sufficient to inform the control system if it were operating in a condition of excess air or insufficient air. However, measuring oxygen in the exhaust gases is a direct measure of the amount of excess air. Therefore, measuring oxygen in the exhaust gas is a more common and preferred method of controlling the air-to-fuel ratio in a fired system.

5.2.4 Energy Conservation Measures

As noted above, energy cost reduction opportunities can generally be placed into one of the following categories: reducing load, increasing efficiency, and reducing unit energy cost. As with most energy conservation and cost reducing measures there are also a few additional opportunities which are not so easily categorized. Table 5.1 lists several energy conservation measures that have been found to be very cost effective in various boilers and fired-systems.

5.3　KEY ELEMENTS FOR MAXIMUM EFFICIENCY

There are several opportunities for maximizing efficiency and reducing operating costs in a boiler or other fired-system as noted earlier in Table 5.1. This section examines in more detail several key opportunities for energy and cost reduction, including excess air, stack temperature, load balancing, boiler blowdown, and condensate return.

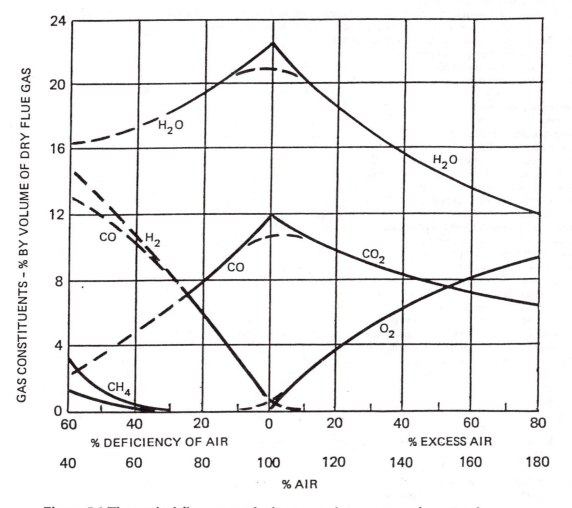

Figure 5.1 Theoretical flue gas analysis versus air percentage for natural gas.

Table 5.1 Energy Conservation measures
for boilers and fired systems(a)

Load Reduction
Insulation
—steam lines and distribution system
—condensate lines and return system
—heat exchangers
—boiler or furnace
Repair steam leaks
Repair failed steam straps
Return condensate to boiler
Reduce boiler blowdown
Improve feedwater treatment
Improve make-up water treatment
Repair condensate leaks
Shut off steam tracers during the summer
Shut off boilers during long periods of no use
Eliminate hot standby
Reduce flash steam loss
Install stack dampers or heat traps in natural draft boilers
Replace continuous pilots with electronic ignition pilots

Waste Heat Recovery (a form of load reduction)
Utilize flash steam
Preheat feedwater with an economizer
Preheat make-up water with an economizer
Preheat combustion air with a recouperator
Recover flue gas heat to supplement other heating system, such as domestic or service hot water, or unit space heater
Recover waste heat from some other system to preheat boiler make-up or feedwater
Install a heat recovery system on incinerator or furnace
Install condensation heat recovery system
—indirect contact heat exchanger
—direct contact heat exchanger

Efficiency Improvement
Reduce excess air
Provide sufficient air for complete combustion
Install combustion efficiency control system
—Constant excess air control
—Minimum excess air control
—Optimum excess air and CO control
Optimize loading of multiple boilers
Shut off unnecessary boilers
Install smaller system for part-load operation
—Install small boiler for summer loads
—Install satellite boiler for remote loads
Install low excess air burners
Repair or replace faulty burners
Replace natural draft burners with forced draft burners
Install turbulators in firetube boilers
Install more efficient boiler or furnace system
—high-efficiency, pulse combustion, or condensing boiler or furnace system
Clean heat transfer surfaces to reduce fouling and scale
Improve feedwater treatment to reduce scaling
Improve make-up water treatment to reduce scaling

Fuel Cost Reduction
Switch to alternate utility rate schedule
—interruptible rate schedule
Purchase natural gas from alternate source, self procurement of natural gas
Fuel switching
—switch between alternate fuel sources
—install multiple fuel burning capability
—replace electric boiler with a fuel-fired boiler

Switch to a heat pump
—use heat pump for supplemental heat requirements
—use heat pump for baseline heat requirements

Other Opportunities
Install variable speed drives on feedwater pumps
Install variable speed drives on combustion air fan
Replace boiler with alternative heating system
Replace furnace with alternative heating system
Install more efficient combustion air fan
Install more efficient combustion air fan motor
Install more efficient feedwater pump
Install more efficient feedwater pump motor
Install more efficient condensate pump
Install more efficient condensate pump motor

[a]Reference: F.W. Payne, Efficient Boiler Operations Sourcebook, 3rd ed., Fairmont Press, Lilburn, GA, 1991.

5.3.1 Excess Air

In combustion processes, excess air is generally defined as air introduced above the stoichiometric or theoretical requirements to effect complete and efficient combustion of the fuel.

There is an optimum level of excess-air operation for each type of burner or furnace design and fuel type. Only enough air should be supplied to ensure complete combustion of the fuel, since more than this amount increases the heat rejected to the stack, resulting in greater fuel consumption for a given process output.

To identify the point of minimum excess-air opera-

tion for a particular fired system, curves of combustibles as a function of excess O_2 should be constructed similar to that illustrated in Figure 5.2. In the case of a gas-fueled system, the combustible monitored would be carbon monoxide (CO), whereas, in the case of a liquid- or solid-fueled system, the combustible monitored would be the Smoke Spot Number (SSN). The curves should be developed for various firing rates as the minimal excess-air operating point will also vary as a function of the firing rate (percent load). Figure 5.2 illustrates two potential curves, one for high-fire and the other for low-fire. The optimal excess-air-control set point should be set at some margin (generally 0.5 to 1%) above the minimum O_2

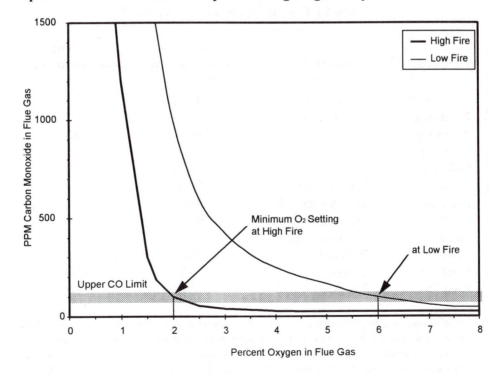

Figure 5.2 Hypothetical CO-O_2 characteristic curve for a gas-fired industrial boiler.

point to allow for response and control variances. It is important to note that some burners may exhibit a gradual or steep CO-O_2 behavior and this behavior may even change with various firing rates. It is also important to note that some burners may experience potentially unstable operation with small changes in O_2 (steep CO-O_2 curve behavior). Upper control limits for carbon monoxide vary depending on the referenced source. Points referenced for gas-fired systems are typically 400 ppm, 200 ppm, or 100 ppm. Today, local environmental regulations may dictate acceptable upper limits. Maximum desirable SSN for liquid fuels is typically SSN=1 for No. 2 fuel oil and SSN=4 for No. 6 fuel oil. Again, local environmental regulations may dictate lower acceptable upper limits.

Typical optimum levels of excess air normally attainable for maximum operating efficiency are indicated in Table 5.2 and classified according to fuel type and firing method.

The amount of excess air (or O_2) in the flue gas, unburned combustibles, and the stack temperature rise above the inlet air temperature are significant in defining the efficiency of the combustion process. Excess oxygen (O_2) measured in the exhaust stack is the most typical method of controlling the air-to-fuel ratio. However, for more precise control, carbon monoxide (CO) measurements may also be used to control air flow rates in combination with O_2 monitoring. Careful attention to furnace operation is required to ensure an optimum level of performance.

Figures 5.3, 5.4, and 5.5 can be used to determine the combustion efficiency of a boiler or other fired system burning natural gas, No. 2 fuel oil, or No. 6 fuel oil respectively so long as the level of unburned combustibles is considered negligible. These figures were derived from H. R. Taplin, Jr., *Combustion Efficiency Tables*, Fairmont Press, Lilburn, GA, 1991. For more information on combustion efficiency including combustion efficiencies using other fuels, see Taplin 1991.

Where to Look for Conservation Opportunities

Fossil-fuel-fired steam generators, process fired heaters/furnaces, duct heaters, and separately fired superheaters may benefit from an excess-air-control program. Specialized process equipment, such as rotary kilns, fired calciners, and so on, can also benefit from an air control program.

How to Test for Relative Efficiency

To determine relative operating efficiency and to establish energy conservation benefits for an excess-air-control program, you must determine: (1) percent oxygen (by volume) in the flue gas (typically dry), (2) stack temperature rise (the difference between the flue gas tem-

Table 5.2 Typical Optimum Excess Air[a]

Fuel Type	Firing Method	Optimum Excess Air (%)	Equivalent O_2 (by Volume)
Natural gas	Natural draft	20-30	4-5
Natural gas	Forced draft	5-10	1-2
Natural gas	Low excess air	0.4-0.2	0.1-0.5
Propane	—	5-10	1-2
Coke oven gas	—	5-10	1-2
No. 2 oil	Rotary cup	15-20	3-4
No. 2 oil	Air-atomized	10-15	2-3
No. 2 oil	Steam-atomized	10-15	2-3
No. 6 oil	Steam-atomized	10-15	2-3
Coal	Pulverized	15-20	3-3.5
Coal	Stoker	20-30	3.5-5
Coal	Cyclone	7-15	1.5-3

[a]To maintain safe unit output conditions, excess-air requirements may be greater than the optimum levels indicated. This condition may arise when operating loads are substantially less than the design rating. Where possible, check vendors' predicted performance curves. If unavailable, reduce excess-air operation to minimum levels consistent with satisfactory output.

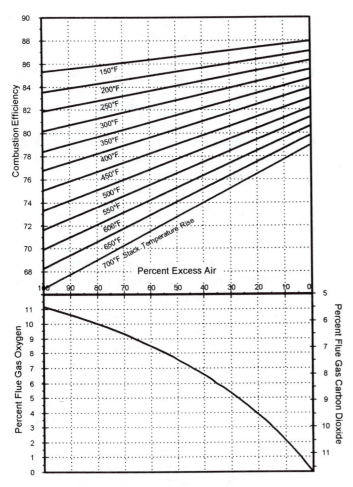

Figure 5.3 Combustion efficiency chart for natural gas.

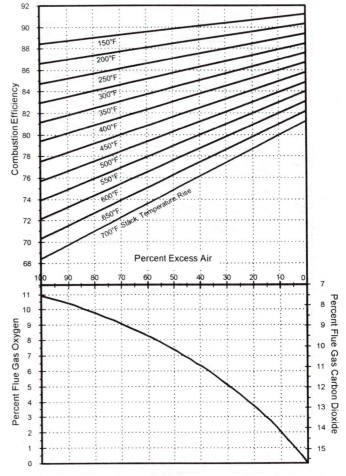

Figure 5.4 Combustion efficiency chart for number 2 fuel oil.

perature and the combustion air inlet temperature), and (3) fuel type.

To accomplish optimal control over avoidable losses, continuous measurement of the excess air is a necessity. There are two types of equipment available to measure flue-gas oxygen and corresponding "excess air": (1) portable equipment an Orsat flue-gas analyzer, heat prover, electronic gas analyzer, or equivalent analyzing device; and (2) permanent-type installations probe-type continuous oxygen analyzers (available from various manufacturers), which do not require external gas sampling systems.

The major advantage of permanently mounted equipment is that the on-line indication or recording allows remedial action to be taken frequently to ensure continuous operation at optimum levels. Computerized systems which allow safe control of excess air over the boiler load range have proven economic for many installations. Even carbon monoxide-based monitoring and control systems, which are notably more expensive than simple oxygen-based monitoring and control systems,

prove to be cost effective for larger industrial-and utility-sized boiler systems.

] Portable equipment only allows performance checking on an intermittent or spot-check basis. Periodic monitoring may be sufficient for smaller boilers or boilers which do not undergo significant change in operating conditions. However, continuous monitoring and control systems have the ability to respond more rapidly to changing conditions, such as load and inlet air conditions.

The stack temperature rise may be obtained with portable thermocouple probes in conjunction with a potentiometer or by installing permanent temperature probes within the exhaust stack and combustion air inlet and providing continuous indication or recording. Each type of equipment provides satisfactory results for the planning and operational results desired.

An analysis to establish performance can be made with the two measurements, percent oxygen and the stack temperature rise, in addition to the particular fuel fired. As an illustration, consider the following example.

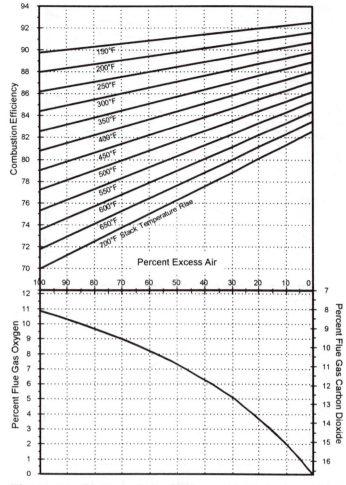

Figure 5.5 Combustion efficiency chart for number 6 fuel oil.

Example: Determine the potential energy savings associated with reducing the amount of excess air to an optimum level for a natural gas-fired steam boiler.

Operating Data.

Current energy consumption	1,100,000 therms/yr
Boiler rated capacity	600 boiler horsepower
Operating hours	8,500 hr/yr
Current stack gas analysis	9% Oxygen (by volume, dry)
	Minimal CO reading
Combustion air inlet temperature	80°F
Exhaust gas stack temperature	580°F
Proposed operating condition	2% Oxygen (by volume, dry)

Calculation and Analysis.

STEP 1: Determine current boiler combustion efficiency using Figure 5.6 for natural gas. Note that this is the same figure as Figure 5.3.

A) Determine the current stack temperature rise.
    STR = (exhaust stack temperature)

– (combustion air temperature)
STR = 580°F - 80°F = 500°F

B) Enter the chart with an oxygen level of 9% and following a line to the curve, read the percent excess air to be approximately 66%.

C) Continue the line to the curve for a stack temperature rise of 500°F and read the current combustion efficiency to be 76.4%.

STEP 2: Determine the proposed boiler combustion efficiency using the same figure.

D) Repeat steps A through C for the proposed combustion efficiency assuming the same stack temperature conditions. Read the proposed combustion efficiency to be 81.4%.

Note that in many cases reducing the amount of excess air will tend to reduce the exhaust stack temperature, resulting in an even more efficient

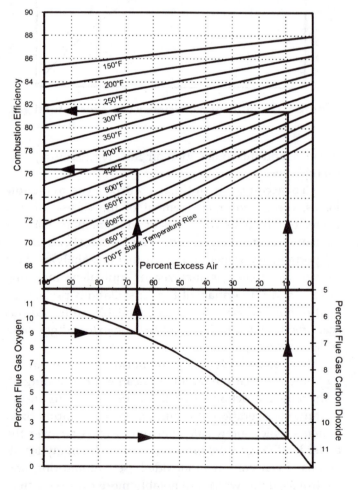

Figure 5.6. Combustion efficiency curve for reducing excess air example.

operating condition. Unfortunately, it is difficult to predict the extent of this added benefit.

STEP 3: Determine the fuel savings.

E) Percent fuel savings = [(new efficiency) – (old efficiency)]/ (new efficiency)
Percent fuel savings = [(81.4%) – (76.4%)] / (81.4%)
Percent fuel savings = 6.14%

F) Fuel savings =(current fuel consumption) × (percent fuel savings)
Fuel savings = (1,100,000 therms/yr) × (6.14%)
Fuel savings = 67,540 therms/yr

Conclusions.

This example assumes that the results of the combustion analysis and boiler load are constant. Obviously this is an oversimplification of the issue. Because the air-to-fuel ratio (excess air level) is different for different boiler loads, a more thorough analysis should take this into account. One method to accomplish this would be to use perform the analysis at various firing rates, such as high-fire and low-fire. For modulating type boilers which can vary between high- and low-firing rates, a modified bin analysis approach or other bin-type methodology could be employed.

Requirements to Effect Maximum Economy

To obtain the maximum benefits of an excess-air-control program, the following modifications, additions, checks, or procedures should be considered:

Key Elements for Maximum Efficiency

1. Ensure that the furnace boundary walls and flue work are airtight and not a source of air infiltration or exfiltration.
   a. Recognized leakage problem areas include (1) test connection for oxygen analyzer or portable Orsat connection; (2) access doors and ash-pit doors; (3) penetration points passing through furnace setting; (4) air seals on soot-blower elements or sight glasses; (5) seals around boiler drums and header expansion joints; (6) cracks or breaks in brick settings or refractory; (7) operation of the furnace at too negative a pressure; (8) burner penetration points; and (9) deterioration of air preheater radial seals or tube-sheet expansion and cracks on tubular air heater applications.
   b. Tests to locate leakage problems: (1) a light test whereby a strong spotlight is placed in the furnace and the unit inspected externally; (2) the use of a pyrometer to obtain a temperature profile on the outer casing. This test generally indicates points where refractory or insulation has deteriorated; (3) a soap-bubble test on suspected penetration points or seal welds; (4) a smoke-bomb test and an external examination for traces of smoke; (5) holding a lighted candle along the casing seams has pinpointed leakage problems on induced- or natural-draft units; (6) operating the forced draft fan on high capacity with the fire out, plus use of liquid chemical smoke producers has helped identify seal leaks; and (7) use of a thermographic device to locate "hot spots" which may indicate faulty insulation or flue-gas leakage.

2. Ensure optimum burner performance.
   a. Table 5.3 lists common burner difficulties that can be rectified through observation and maintenance.
   b. Ascertain integrity of air volume control: (1) the physical condition of fan vanes, dampers, and operators should be in optimum working condition; and (2) positioning air volume controls should be checked for responsiveness and adequacy to maintain optimum air/fuel ratios. Consult operating manual or control manufacturer for test and calibration.
   c. Maintain or purchase high-quality gas analyzing systems: calibrate instrument against a known flue-gas sample.
   d. Purchase or update existing combustion controls to reflect the present state of the art.
   e. Consider adapting "oxygen trim" feature to existing combustion control system.

3. Establish a maintenance program.
   a. Table 5.4 presents a summary of frequent boiler system problems and possible causes.
   b. Perform period maintenance as recommended by the manufacturer.
   c. Keep a boiler operator's log and monitor key parameters.
   d. Perform periodic inspections.

Guidelines for Day-to-Day Operation

The following steps must be taken to assure peak boiler efficiency and minimum permissible excess-air operation.

1. Check the calibration of the combustion gas analyzer frequently and check the zero point daily.
2. If a sampling system is employed, check to assure proper operation of the sampling system.
3. The forced-draft damper should be checked for its physical condition to ensure that it is not broken or damaged.
4. Casing leakage must be detected and stopped.
5. Routinely check control drives and instruments.
6. If the combustion gas analyzer is used for monitoring purposes, the excess air must be checked daily. The control may be manually altered to reduce excess air, without shortcutting the safety of operation.
7. The fuel flow and air flow charts should be carefully checked to ensure that the fuel follows the air on increasing load with proper safety margin and also that the fuel leads the air on decreasing load. This should be compared on a daily shift basis to ensure consistency of safe and efficient operation.
8. Check the burner flame configuration frequently during each shift and note burner register changes in the operator's log.
9. Periodically check flue-gas CO levels to ensure complete combustion. If more than a trace amount of CO is present in the flue gas, investigate burner conditions identified on Table 5.3 or fuel supply quality limits such as fuel-oil viscosity/temperature or coal fineness and temperature.

5.3.2 Exhaust Stack Temperature

Another primary factor affecting unit efficiency and ultimately fuel consumption is the temperature of combustion gases rejected to the stack. Increased operating efficiency with a corresponding reduction in fuel input can be achieved by rejecting stack gases at the lowest practical temperature consistent with basic design principles. In general, the application of additional heat recovery equipment can realize this energy conservation objective when the measured flue-gas temperature exceeds approximately 250°F. For a more extensive coverage of waste-heat recovery, see Chapter 8.

Where to Look

Steam boilers, process fired heaters, and other combustion or heat-transfer furnaces can benefit from a heat-recovery program.

The adaptation of heat-recovery equipment to existing units as discussed in this section will be limited to flue gas/liquid and/or flue gas/air preheat exchangers. Specifically, economizers and air preheaters come under this category. Economizers are used to extract heat energy from the flue gas to heat the incoming liquid process feedstream to the furnace. Flue gas/air preheaters lower the flue-gas temperature by exchanging heat to the incoming combustion air stream.

Table 5.3 Malfunctions in Fired Systems

Malfunction	Fuel			Detection	Action
	Coal	Oil	Gas		
Uneven air distribution to burners	x	x	x	Observe flame patterns	Adjust registers (trial and error)
Uneven fuel distribution to burners	x	x	x	Observe fuel pressure gages, or take coal sample and analyze	Consult manufacturer
Improperly positioned guns or impellers	x	x		Observe flame patterns	Adjust guns (trial and error)
Plugged or worn burners	x	x		Visual inspection	Increase frequency of cleaning; install strainers (oil)
Damaged burner throats	x	x	x	Visual inspection	Repair

Table 5.4 Boiler Performance Troubleshooting

System	Problem	Possible Cause
Heat transfer related	High stack gas temperature	Buildup of gas- or water-side deposits
		Improper water treatment procedure
		Improper suit blower operation
Combustion related	High excess air	Improper control system operation
		Low fuel supply pressure
		Change in fuel heating value
		Change on oil viscosity
		Decrease in inlet air temperature
	Low excess air	Improper control system operation
		Fan limitations
		Increase in inlet air temperature
	High carbon monoxide and combustible emissions	Plugged gas burners
		Unbalanced fuel and air distribution in multiburner furnaces
		Improper air register settings
		Deterioration of burner throat refractory
		Stoker grate condition
		Stoker fuel distribution orientation
		Low fineness on pulverized systems
Miscellaneous	Casing leakage	Damaged casing and insulation
	Air heater leakage	Worn or improper adjusted seals on rotary heaters
		Tube corrosion
	Coal pulverizer power	Pulverizer in poor repair
		Too low classifier setting
	Excessive blowdown	Improper operation
	Steam leaks	Holes in waterwall tube Valve packing
	Missing or loose insulation	Overheating
		Weathering
	Excessive sootblower operation	Arbitrary operation schedule that is in excess of requirements

Planning-quality guidelines will be presented to determine the final sink temperature, as well as comparative economic benefits to be derived by the installation of heat-recovery equipment. Costs to implement this energy conservation opportunity can then be compared against the potential benefits.

How to Test for Heat-Recovery Potential

In assessing overall efficiency and potential for heat recovery, the parameters of significant importance are temperature and fuel type/sulfur content. To obtain a meaningful operating flue-gas temperature measurement and a basis for heat-recovery selection, the unit under consideration should be operating at, or very close to, design and optimum excess-air values as defined on Table 5.2.

Temperature measurements may be made by mercury or bimetallic element thermometers, optical pyrometers, or an appropriate thermocouple probe. The most adaptable device is the thermocouple probe in which an iron or chromel constantan thermocouple is used. Temperature readout is accomplished by connecting the thermocouple leads to a potentiometer. The output of the potentiometer is a voltage reading which may be correlated with the measured temperature for the particular thermocouple element employed.

To obtain a proper and accurate temperature measurement, the following guidelines should be followed:

1. Locate the probe in an unobstructed flow path and sufficient distance, approximately five diameters downstream or upstream, of any major change of direction in the flow path.

2. Ensure that the probe entrance connection is relatively leak free.

3. Take multiple readings by traversing the cross-sectional area of the flue to obtain an average and representative flue-gas temperature.

Modifications or Additions for Maximum Economy

The installation of economizers and/or flue-gas air preheaters on units not presently equipped with heat-recovery devices and those with minimum heat-recovery equipment are practical ways of reducing stack temperature while recouping flue-gas sensible heat normally rejected to the stack.

There are no "firm" exit-temperature guidelines that cover all fuel types and process designs. However, certain guiding principles will provide direction to the lowest practical temperature level of heat rejection. The elements that must be considered to make this judgment include (1) fuel type, (2) flue-gas dew-point considerations, (3) heat-transfer criteria, (4) type of heat-recovery surface, and (5) relative economics of heat-recovery equipment.

Tables 5.5 and 5.6 may be used for selecting the lowest practical exit-gas temperature achievable with installation of economizers and/or flue-gas air preheaters.

As an illustration of the potential and methodology for recouping flue-gas sensible heat by the addition of heat-recovery equipment, consider the following example.

Example: Determine the energy savings associated with installing an economizer or flue-gas air preheater on the boiler from the previous example. Assume that the excess-air control system from the previous example has already been implemented.

Available Data.

Current energy consumption	1,032,460 therms/yr
Boiler rated capacity	600 boiler horsepower
Operating hours	8,500 hr/yr
Exhaust stack gas analysis	2% Oxygen (by volume, dry)
	Minimal CO reading

Current operating conditions:

Combustion air inlet temperature	80°F
Exhaust gas stack temperature	580°F
Feedwater temperature	180°F
Operating steam pressure	110 psia
Operating steam temperature	335°F

Proposed operating condition:

Combustion air inlet temperature	80°F
Exhaust gas stack temperature	380°F

Calculation and Analysis.

STEP 1: Compare proposed stack temperature against minimum desired stack temperature.

A) Heat transfer criteria:
$T_g = T_1 + 100°F$ (minimum)
$T_g = 180 + 100°F$ (minimum)
$T_g = 280°F$ (minimum)

B) Flue-gas dew point:
$T_g = 120°F$ (from Figure 5.8)

C) Proposed stack temperature
$T_g = 380°F$ is acceptable

STEP 2: Determine current boiler combustion efficiency using Figure 5.7 for natural gas. Note that this is the same figure as Figure 5.3.

Table 5.5 Economizers

Fuel Type	Test for Determination of Exit Flue-Gas Temperatures
Gaseous fuel (minimum percent sulphur)	Heat-transfer criteria: $T_g = T_1 + 100°F$ (minimum): typically the higher of (a) or (b) below.
Fuel oils and coal	(a) Heat-transfer criteria: $T_g = T_1 + 100°F$ (min.) (b) Flue-gas dew point (from Figure 5.8 for a particular fuel and percent sulphur by weight

Where: T_g = Final stack flue temperature
         T_1 = Process liquid feed temperature

Table 5.6 Flue-Gas/Air Preheaters

Fuel Type	Test for Determination of Exit Flue-Gas Temperatures
Gaseous fuel	Historic economic breakpoint: T_g (min.) = approximately 250°F
Fuel oils and coal	Average cold-end considerations; see Figure 5.9 for determination of T_{ce}; the exit-gas temperature relationship is $T_g = 2T_{ce} - T_a$

Where: T_g = Final stack flue temperature
         T_{ce} = Flue gas air preheater recommended average cold end temperature
         T_a = Ambient air temperature

A) Determine the stack temperature rise.
STR = (exhaust stack temperature)
– (combustion air temperature)
STR = 580°F - 80°F = 500°F

B) Enter the chart with an oxygen level of 2% and following a line to the curve, read the percent excess air to be approximately 9.3%.

C) Continue the line to the curve for a stack temperature rise of 500°F and read the current combustion efficiency to be 81.4%.

STEP 3: Determine the proposed boiler combustion efficiency using the same figure.

D) Repeat steps A through C for the proposed combustion efficiency assuming the new exhaust stack temperature conditions. Read the proposed combustion efficiency to be 85.0%.

STEP 4: Determine the fuel savings.
E) Percent fuel savings = [(new efficiency) – (old efficiency)]/(new efficiency)
Percent fuel savings = [(85.0%) - (81.4%)]/(85.0%)
Percent fuel savings = 4.24%

F) Fuel savings =(current fuel consumption) × (percent fuel savings)
Fuel savings = (1,032,460 therms/yr) × (4.24%)
Fuel savings = 43,776 therms/yr

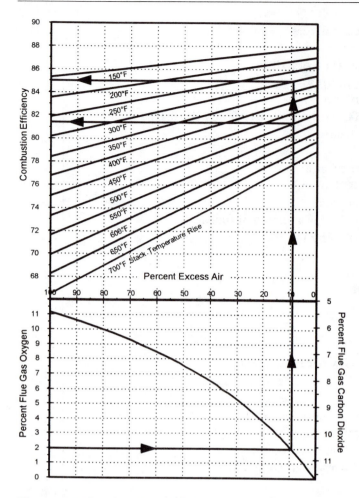

Figure 5.7 Combustion efficiency curve for stack temperature reduction example.

Conclusion.

As with the earlier example, this analysis methodology assumes that the results of the combustion analysis and boiler load are constant. Obviously this is an oversimplification of the issue. Because the air-to-fuel ratio (excess air level) is different for different boiler loads, a more thorough analysis should take this into account.

Additional considerations in flue-gas heat recovery include:

1. Space availability to accommodate additional heating surface within furnace boundary walls or adjacent area to stack.

2. Adequacy of forced-draft and/or induced-draft fan capacity to overcome increased resistance of heat-recovery equipment.

3. Adaptability of sootblowers for maintenance of heat-transfer-surface cleanliness when firing ash- and soot-forming fuels.

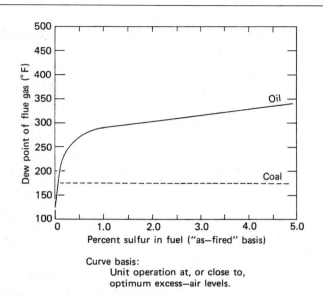

Figure 5.8 Flue-gas dew point. Based on unit operation *at* **or** *close to* **"optimal" excess-air.**

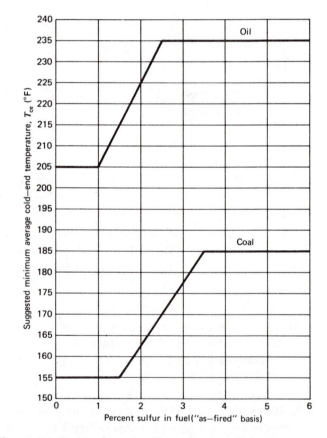

Figure 5.9 Guide for selecting flue-gas air preheaters.

4. Design considerations to maintain average cold-end temperatures for flue gas/air preheater applications in cold ambient surroundings.

5. Modifications required of flue and duct work and additional insulation needs.

6. The addition of structural steel supports.

7. Adequate pumping head to overcome increased fluid pressure drop for economizer applications.

8. The need for bypass arrangements around economizers or air preheaters.

9. Corrosive properties of gas, which would require special materials.

10. Direct flame impingement on recovery equipment.

Guidelines for Day-to-Day Operation

1. Maintain operation at goal excess air levels and stack temperature to obtain maximum efficiency and unit thermal performance.

2. Log percent O_2 or equivalent excess air, inlet air temperature, and stack temperatures, once per shift or more frequent, noting the unit load and fuel fired.

3. Use oxygen analyzers with recorders for units larger than about 35×10^6 Btu/hr output.

4. Maintain surface cleanliness by soot blowing at least once per shift for ash- and soot-forming fuels.

5. Establish a more frequent cleaning schedule when heat-exchange performance deteriorates due to firing particularly troublesome fuels.

6. External fouling can also cause high excess air operation and higher stack temperatures than normal to achieve desired unit outputs. External fouling can be detected by use of draft loss gauges or water manometers and periodically (once a week) logging the results.

7. For flue gas/air preheaters, oxygen checks should be taken once a month before and after the heating surface to assess condition of circumferential and radial seals. If O_2 between the two readings varies in excess of 1% O_2, air heater leakage is excessive to the detriment of operating efficiency and fan horsepower.

8. Check fan damper operation weekly. Adjust fan damper or operator to correspond to desired excess air levels.

9. Institute daily checks on continuous monitoring equipment measuring flue-gas conditions. Check calibration every other week.

10. Establish an experience guideline on optimum time for cleaning and changing oil guns and tips.

11. Receive the "as-fired" fuel analysis on a monthly basis from the supplier. The fuel base may have changed, dictating a different operating regimen.

12. Analyze boiler blowdown every two months for iron. Internal surface cleanliness is as important to maintaining heat-transfer characteristics and performance as external surface cleanliness.

13. When possible, a sample of coal, both raw and pulverized, should be analyzed to determine if operating changes are warranted and if the design coal fineness is being obtained.

5.3.3 Waste-Heat-Steam Generation

Plants that have fired heaters and/or low-residence-time process furnaces of the type designed during the era of cheap energy may have potentially significant energy-saving opportunities. This section explores an approach to maximize energy efficiency and provide an analysis to determine overall project viability.

The major problem on older units is to determine a practical and economical approach to utilize the sensible heat in the exhaust flue gas. Typically, many vintage units have exhaust-flue-gas temperatures in the range 1050 to 1600°F. In this temperature range, a conventional flue-gas air preheater normally is not a practical approach because of materials of construction requirements and significant burner front modifications. Additionally, equipping these units with an air preheater could materially alter the inherent radiant characteristics of the furnace, thus adversely affecting process heat transfer. An alternative approach to utilizing the available flue-gas sensible heat and maximizing overall plant energy efficiency is to consider: (1) waste-heat-steam generation: (2) installing an unfired or supplementary fired recirculating hot-oil loop or ethylene glycol loop to effectively utilize transferred heat to a remote location: and (3) installing a process feed economizer.

Because most industrial process industries have a need for steam, the example is for the application of an unfired waste-heat-steam generator.

The hypothetical plant situation is a reformer furnace installed in the plant in 1963 at a time when it was not considered economical to install a waste-heat-steam generator. As a result, the furnace currently vents hot flue gas (1562°F) to the atmosphere after inspiriting ambient air to reduce the exhaust temperature so that standard materials of construction could be utilized.

The flue-gas temperature of 1562°F is predicated on a measured value by thermocouple and is based on a typical average daily process load on the furnace. This induced-draft furnace fires a No. 2 fuel oil and has been optimized for 20% excess air operation. Flue-gas flow is calculated at 32,800 lb/hr. The plant utilizes approximately 180,000 lb/hr of 300-psig saturated steam from three boilers each having a nameplate capacity of 75,000 lb/hr. The plant steam load is shared equally by the three operating boilers, each supplying 60,000 lb/hr. Feedwater to the units is supplied at 220°F from a common water-treating facility. The boilers are fired with low-sulfur (0.1% sulphur by weight) No. 2 fuel oil. Boiler efficiency averages 85% at load. Present fuel costs are $0.76/gal or $5.48/10^6 Btu basis of No. 2 fuel oil having a heating value of 138,800 Btu/gal. The basic approach to enhancing plant energy efficiency and minimizing cost is to generate maximum quantities of "waste" heat steam by recouping the sensible heat from the furnace exhaust flue gas.

Certain guidelines would provide a "fix" on the amount of steam that could be reasonably generated. The flue-gas temperature drop could practically be reduced to 65 to 100°F above the boiler feedwater temperature of 220°F. Using an approach temperature of 65°F yields an exit-flue gas temperature of 220 + 65 = 285°F. This assumes that an economizer would be furnished integral with the waste-heat-steam generator.

A heat balance on the flue-gas side (basis of flue-gas temperature drop) would provide the total heat duty available for steam generation. The sensible heat content of the flue gas is derived from Figures 5.10a and 5.10b based on the flue-gas temperature and percent moisture in the flue gas.

Percentage moisture (by weight) in the flue gas is a function of the type of fuel fired and percentage excess-air operation. Typical values of percentage moisture are indicated in Table 5.7 for various fuels and excess air. For No. 2 fuel oil firing at 20% excess air, percent moisture by

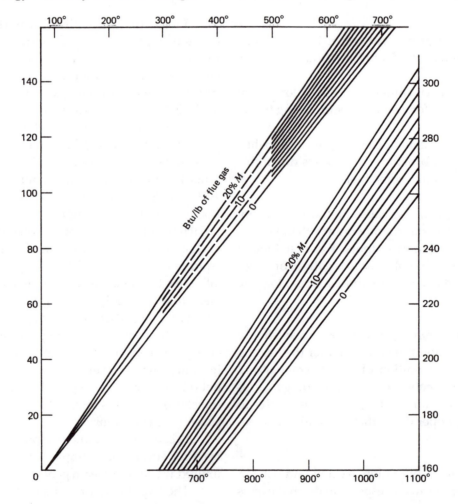

Figure 5.10a Heat in flue gases vs. percent moisture by weight. (Derived from Keenan and Kayes 1948.)

weight in flue gas is approximately 6.8%.

Therefore, a flue-gas heat balance becomes

Flue-Gas Temperature Drop (°F)	Sensible Heat in Flue Gas (Btu/lb W.G.)	
1562	412	(Fig. 5.15)
285	52	(Fig. 5.14)
1277	360	

Table 5.7 Percent Moisture by Weight in Flue Gas

Fuel Type	Percent Excess Air			
	10	15	20	25
Natural gas	12.1	11.7	11.2	10.8
No. 2 fuel oil	7.3	7.0	6.8	6.6
Coal (varies)	6.7-5.1	6.4-4.9	6.3-4.7	6.1-4.6
Propane	10.1	9.7	9.4	9.1

The total heat available from the flue gas for steam generation becomes

$$(32,800 \text{ lb.W.G.}) \times (360 \text{ Btu/lb.W.G.}) = (11.8 \times 10^6 \text{ Btu/h})$$

The amount of steam that may be generated is determined by a thermodynamic heat balance on the steam circuit.

Enthalpy of steam at 300 psig saturated
$$h_3 = 1203 \text{ Btu/lb}$$

Enthalpy of saturated liquid at drum pressure of 300 psig
$$h_f = 400 \text{ Btu/lb}$$

Enthalpy corresponding to feedwater temperature of 200°F
$$h_1 = 188 \text{ Btu/lb}$$

For this example, assume that boiler blowdown is 10% of steam flow. Therefore, feedwater flow through the economizer to the boiler drum will be 1.10 times the steam outflow from the boiler drum. Let the steam outflow be designated as x. Equating heat absorbed by the waste-heat-steam generator to the heat available from reducing the flue-gas temperature from 1562°F to 285°F yields the following steam flow:

$$(1.10)(x)(h_f-h_1) + (x)(h_3-h_f) = 11.8 \times 10^6 \text{ Btu/hr}$$

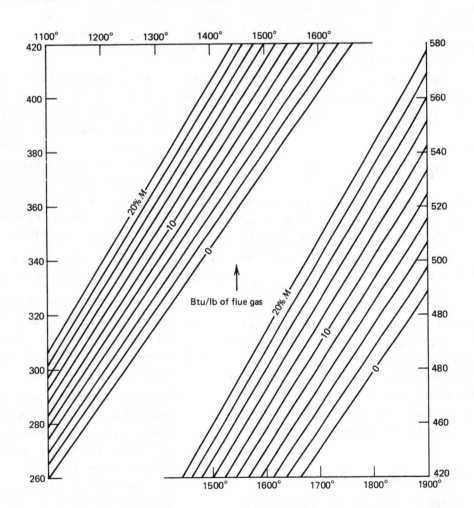

Figure 5.10b Heat in flue gases vs. percent moisture by weight, cont. (Derived from Keenan and Kayes 1948.)

Therefore,
steam flow, x   = 11,388 lb/hr
feedwater flow   = 1.10(x)= 1.10(11,388)= 12,527 lb/hr
boiler blowdown   = 12,527 − 11,388 = 1,139 lb/hr

Determine the equivalent fuel input in conventional fuel-fired boilers corresponding to the waste heat-steam generator capability. This would be defined as follows:

Fuel input to conventional boilers
    = (output) / (boiler efficiency)

Therefore,
Fuel input   = (11.8 × 10^6 Btu/h) / (0.85)
              = 13.88 × 10^6 Btu/h

This suggests that with the installation of the waste-heat-steam generator utilizing the sensible heat of the reformer furnace flue gas, the equivalent of 13.88×10^6 Btu/hr of fossil-fuel input energy could be saved in the firing of the conventional boilers while still satisfying the overall plant steam demand.

As with other capital projects, the waste-heat-steam generator must compete for capital, and to be viable, it must be profitable. Therefore, the decision to proceed becomes an economical one. For a project to be considered life-cycle cost effective it must have a net-present value greater than or equal to zero, or an internal rate of return greater than the company's hurdle rate. For a thorough coverage of economic analysis, see Chapter 4.

5.3.4 Load Balancing

Energy Conservation Opportunities

There is an inherent variation in the energy conversion efficiencies of boilers and their auxiliaries with the operating load imposed on this equipment. It is desirable, therefore, to operate each piece of equipment at the capacity that corresponds to its highest efficiency.

Process plants generally contain multiple boiler units served by common feedwater and condensate return facilities. The constraints imposed by load variations and the requirement of having excess capacity on line to provide reliability seldom permit operation of each piece of equipment at optimum conditions. The energy conservation opportunities therefore lie in the establishment of an operating regimen which comes closest to attaining this goal for the overall system in light of operational constraints.

How to Test for Energy Conservation Potential

Information needed to determine energy conservation opportunities through load-balancing techniques re-

quires a plant survey to determine (1) total steam demand and duration at various process throughputs (profile of steam load versus runtime), and (2) equipment efficiency characteristics (profile of efficiency versus load).

Steam Demand

Chart recorders are the best source for this information. Individual boiler steam flowmeters can be totalized for plant output. Demands causing peaks and valleys should be identified and their frequency estimated.

Equipment Efficiency Characteristics

The efficiency of each boiler should be documented at a minimum of four load points between half and maximum load. A fairly accurate method of obtaining unit efficiencies is by measuring stack temperature rise and percent O_2 (or excess air) in the flue gas or by the input/output method defined in the ASME power test codes. Unit efficiencies can be determined with the aid of Figure 5.3, 5.4, or 5.5 for the particular fuel fired. For pump(s) and fan(s) efficiencies, the reader should consult manufacturers' performance curves.

An example of the technique for optimizing boiler loading follows.

Example: A plant has a total installed steam-generating capacity of 500,000 lb/hr, and is served by three boilers having a maximum continuous rating of 200,000, 200,000, and 100,000 lb/hr, respectively. Each unit can deliver superheated steam at 620 psig and 700°F with feedwater supplied at 250°F. The fuel fired is natural gas priced into the operation at $3.50/10^6 Btu. Total plant steam averages 345,000 lb/hr and is relatively constant.

The boilers are normally operated according to the following loading (top of page 103):

Analysis. Determine the savings obtainable with optimum steam plant load-balancing conditions.

STEP 1. Begin with approach (a) or (b).
a) Establish the characteristics of the boiler(s) over the load range suggested through the use of a consultant and translate the results graphically as in Figures 5.11 and 5.12.

b) The plant determines boiler efficiencies for each unit at four load points by measuring unit stack temperature rise and percent O_2 in the flue gas. With these parameters known, efficiencies are obtained from Figures 5.3, 5.4, or 5.5. Tabulate the results and graphically plot unit efficiencies and unit heat inputs as a function of steam load. The results of such an analysis are shown in the

Boiler No.	Size Boiler (10^3 lb/hr)	Normal Boiler Load (10^3 lb/hr)	——Measured—— Stack Temp. (°F)	O_2 (%)	Unit Eff. (%)
1	200	140	290	5	85.0
2	200	140	540	6	77.4
3	100	<u>65</u>	540	7	76.5
Plant steam demand		345			

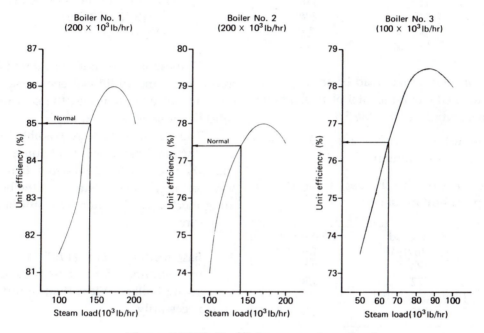

Figure 5.11 Unit efficiency vs. steam load.

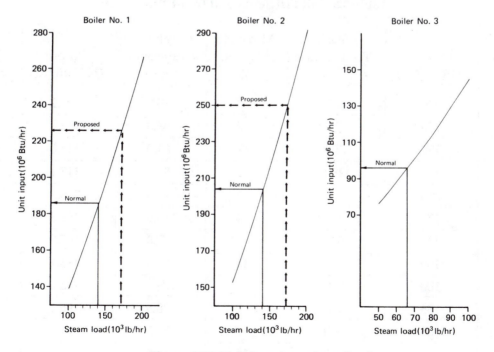

Figure 5.12 Unit input vs. steam load.

tabulation and graphically illustrated in Figures 5.11 and 5.12.

(Unit input) = (unit output) / (efficiency)

STEP 2. Sum up the total unit(s) heat input at the present normal operating steam plant load conditions. From Figure 5.12:

Boiler No.	Steam Load (10^3 lb/hr)	Heat Input (10^6 Btu/hr)
1	140	186
2	140	204
3	65	96
Plant totals	345	486

STEP 3. Optimum steam plant load-balancing conditions are satisfied when the total plant steam demand is met according to Table 5.8.

(Boiler No. 1 input) + (Boiler No. 2 input) + (Boiler No. 3 input) +... = minimum

By trial and error and with the use of Figure 5.12, optimum plant heat input is:

Boiler No.	Steam Load (10^3 lb/hr)	Heat Input (10^6 Btu/hr)
1	173	226
2	172	250
3	(Banked standby)	—
Plant totals	345	476

STEP 4. The annual fuel savings realized from optimum load balancing is the difference between the existing boiler input and the optimum boiler input.

Steam plant energy savings
= (existing input) – (optimum input)
= 486 - 476 × 10^6 Btu/hr
= 10 × 10^6 Btu/hr
or annually:
= (10 × 10^6 Btu/hr) × (8500 hr/yr)
× ($3.50/$10^6$ Btu)
= $297,500 /yr

Costs that were not considered in the preceding example are the additional energy savings due to more efficient fan operation and the cost of maintaining the third boiler in banked standby.

The cost savings were possible in this example because the plant had been maintaining a high ratio of total capacity in service to actual steam demand. This results in low-load inefficient operation of the boilers. Other operating modes which generally result in inefficient energy usage are:

1. Base-loading boilers at full capacity. This can result in operation of the base-loaded boilers and the swing boilers at less than optimum efficiency unnecessarily.

Table 5.8 Unit Efficiency and Input Tabulation

Boiler No.	Steam Load (10^3 lb/hr)	Stack Temperature (°F)	Measured Oxygen (%)	Combustion Efficiency (%)	Output (10^6 Btu/hr)	Fuel Input (10^6 Btu/hr)
1	200	305	2	85.0	226.2	266.1
	170	280	2	86.0	192.3	223.6
	130	300	7	84.0	147.0	175.0
	100	280	12	81.5	113.1	138.8
2	200	625	2	77.5	226.2	291.0
	170	570	4	78.0	192.3	246.5
	130	520	7	77.0	147.0	190.9
	100	490	11	74.0	113.1	152.8
3	100	600	2	78.0	113.1	145.0
	85	570	2	78.5	96.1	122.5
	65	540	7	76.5	73.5	96.1
	50	500	11	73.5	56.6	76.9

2. Operation of high-pressure boilers to supply low-pressure steam demands directly via letdown steam.

3. Operation of an excessive number of auxiliary pumps. This results in throttled, inefficient operation.

Requirements for Maximum Economy

Establish a Boiler Loading Schedule. An optimized loading schedule will allow any plant steam demand to be met with the minimum energy input. Some general points to consider when establishing such a schedule are as follows:

1. Boilers generally operate most efficiently at 65 to 85% full-load rating; centrifugal fans at 80 to 90% design rating. Equipment efficiencies fall off at higher or lower load points, with the decrease most pronounced at low-load conditions.

2. It is usually more efficient to operate a lesser number of boilers at higher loads than a larger number at low loads.

3. Boilers should be put into service in order of decreasing efficiency starting with the most efficient unit.

4. Newer units and units with higher capacity are generally more efficient than are older, smaller units.

5. Generally, steam plant load swings should be taken in the smallest and least efficient unit.

Optimize the Use of High-Pressure Boilers. The boilers in a plant that operate at the highest pressure are usually the most efficient. It is, therefore, desirable to supply as much of the plant demand as possible with these units provided that the high-grade energy in the steam can be effectively used. This is most efficiently done by installation of back-pressure turbines providing useful work output, while providing the exhaust steam for low-pressure consumers.

Degrading high-pressure steam through a pressure reducing and desuperheating station is the least efficient method of supplying low-pressure steam demands. Direct generation at the required pressure is usually more efficient by comparison.

Establish an Auxiliary Loading Schedule. A schedule for cutting plant auxiliaries common to all boilers in and out of service with rising or falling plant load should be established.

Establish Procedures for Maintaining Boilers in Standby Mode. It is generally more economical to run fewer boilers at a higher rating. On the other hand, the integrity of the steam supply must be maintained in the face of forced outage of one of the operating boilers. Both conditions can sometimes be satisfied by maintaining a standby boiler in a "live bank" mode. In this mode the boiler is isolated from the steam system at no load but kept at system operating pressure. The boiler is kept at a pressure by intermittent firing of either the ignitors or a main burner to replace ambient heat losses. Guidelines for live banking of boilers are as follows:

1. Shut all dampers and registers to minimize heat losses from the unit.

2. Establish and follow strict safety procedures for ignitor/burner light-off.

3. For units supplying turbines, take measures to ensure that any condensate which has been formed during banking is not carried through to the turbines. Units with pendant-type superheaters will generally form condensate in these elements.

Operators should familiarize themselves with emergency startup procedures and it should be ascertained that the system pressure decay which will be experienced while bringing the banked boiler(s) up to load can be tolerated.

Guidelines for Day-to-Day Operation

1. Monitor all boiler efficiencies continuously and immediately correct items that detract from performance. Computerized load balancing may prove beneficial.

2. Ensure that load-balancing schedules are followed.

3. Reassess the boiler loading schedule whenever a major change in the system occurs, such as an increase or decrease in steam demand, derating of boilers, addition/decommissioning of boilers, or addition/removal of heat-recovery equipment.

4. Recheck parameters and validity of established operating mode.

5. Measure and record fuel usage and correlate to steam production and flue-gas analysis for determination of the unit heat input relationship.

6.   Keep all monitoring instrumentation calibrated and functioning properly.

7.   Optimize excess air operation and minimize boiler blowdown.

Computerized Systems Available

There are commercially available direct digital control systems and proprietary sensor devices which accomplish optimal steam/power plant operation, including tie-line purchased power control. These systems control individual boilers to minimum excess air, SO_2, NO_x, CO (and opacity if desired), and control boiler and cogeneration complexes to reduce and optimize fuel input.

Boiler plant optimization is realized by boiler controls which ensure that the plant's steam demands are met in the most cost-effective manner, continuously recognizing boiler efficiencies that differ with time, load, and fuel quality. Similarly, computer control of cogeneration equipment can be cost effective in satisfying plant electrical and process steam demands.

As with power boiler systems, the efficiencies for electrical generation and extraction steam generation can be determined continuously and, as demand changes occur, loading for optimum overall efficiency is determined.

Fully integrated computer systems can also provide electric tie-line control, whereby the utility tie-line load is continuously monitored and controlled within the electrical contract's limits. For example, loads above the peak demand can automatically be avoided by increasing in-plant power generation, or in the event that the turbines are at full capacity, shedding loads based on previously established priorities.

5.3.5 Boiler Blowdown

In the generation of steam, most water impurities are not evaporated with the steam and thus concentrate in the boiler water. The concentration of the impurities is usually regulated by the adjustment of the continuous blowdown valve, which controls the amount of water (and concentrated impurities) purged from the steam drum.

When the amount of blowdown is not properly established and/or maintained, either of the following may happen:

1.   If too little blowdown, sludge deposits and carryover will result.

2.   If too much blowdown, excessive hot water is removed, resulting in increased boiler fuel requirements, boiler feedwater requirements, and boiler chemical requirements.

Significant energy savings may be realized by utilizing the guides presented in this section for (1) establishing optimum blowdown levels to maintain acceptable boiler-water quality and to minimize hot-water losses, and (2) the recovery of heat from the hot-water blowdown.

Where to Look For Energy-Saving Opportunities

The continuous blowdown from any steam-generating equipment has the potential for energy savings whether it is a fired boiler or waste-heat-steam generator. The following items should be carefully considered to maximize savings:

1.   Reduce blowdown (BD) by adjustment of the blowdown valve such that the controlling water impurity is held at the maximum allowable level

2.   Maintain blowdown continuously at the minimum acceptable level. This may be achieved by frequent manual adjustments or by the installation of automatic blowdown controls. At current fuel costs, automatic blowdown controls often prove to be economic

3.   Minimize the amount of blowdown required by:
     a.   Recovering more clean condensate, which reduces the concentration of impurities coming into the boiler.
     b.   Establishing a higher allowable drum solids level than is currently recommended by ABMA standards (see below). This must be done only on recommendation from a reputable water treatment consultant and must be followed up with lab tests for steam purity.
     c.   Selecting the raw-water treatment system which has the largest effect on reducing makeup water impurities. This is generally considered applicable only to grass-roots or revamp projects.

4.   Recover heat from the hot blowdown water. This is typically accomplished by flashing the water to a low pressure. This produces low-pressure steam (for utilization in an existing steam header) and hot water which may be used to preheat boiler makeup water.

Tests and Evaluations

STEP 1: *Determine Actual Blowdown.* Obtain the following data:

T = ppm of impurities in the makeup water to the deaerator from the treatment plant; obtain average value through lab tests

B = ppm of concentrated impurities in the boiler drum water (blowdown water); obtain average value through lab tests

lb/hr MU = lb/hr of makeup water to the deaerator from the water treatment plant; obtain from flow indicator

lb/hr BFW = lb/hr of boiler feedwater to each

lb/hr STM = lb/hr of steam output from each boiler; obtain from flow indicator

lb/hr CR = lb/hr of condensate return

Note: percentages for BFW, MU, and CR are determined as a percentage of STM.

Calculate the following:

$$\%MU = lb/hr\ MU \times 100\% / (total\ lb/hr\ BFW)$$
$$= lb/hr\ MU \times 100\% / [(boiler\ no.\ 1\ lb/hr\ BFW) + (boiler\ no.\ 2\ lb/hr\ BFW) + ...]$$

$$\%MU = 100\% - \%CR \qquad (5.3)$$

$$A = ppm\ of\ impurity\ in\ BFW = T \times \%MU \qquad (5.4)$$

Now actual blowdown (BD) may be calculated as a function (percentage) of steam output:

$$\%BD = (A \times 100\%) / (B-A) \qquad (5.5)$$

Converting to lb/hr BD yields

$$lb/hr\ BD = \%\ BD \times lb/hr\ STM \qquad (5.6)$$

Note: In using all curves presented in this section. Blowdown must be based on steam output from the boiler as calculated above. Boiler blowdown based on boiler feedwater rate (percent BD BFW) to the boiler should not be used. If blowdown is reported as a percent of the boiler feedwater rate, it may be converted to a percent of steam output using

$$\%BD = \%BD_{BFW} \times (1)/(1 - \%BD_{BFW}) \qquad (5.7)$$

STEP 2: *Determine Required Blowdown.* The amount of blowdown required for satisfactory boiler operation is normally based on allowable limits for water impurities as established by the American Boiler Manufacturers Association (ABMA).

These limits are presented in Table 5.9. Modifications to these limits are possible as discussed below. The required blowdown may be calculated using the equations presented above by substituting the ABMA limit for B (concentration of impurity in boiler).

$$\%\ BD_{required} = (A) / (B_{required} - A) \times 100\% \qquad (5.8)$$

$$lb/hr\ BD_{required} = \%\ BD_{required} \times lb/hr\ STM \qquad (5.9)$$

STEP 3: *Evaluate the Cost of Excess Blowdown.* The amount of actual boiler blowdown (as calculated in equation 5.4) that is in excess of the amount of required blowdown (as calculated in equation

Table 5.9 Recommended Limits for Boiler-Water Concentrations

Drum Pressure (psig)	Total Solids		Alkalinity		Suspended Solids		Silica
	ABMA	Possible	ABMA	Possible	ABMA	Possible	ABMA
0 to 300	3500	6000	700	1000	300	250	125
301 to 450	3000	5000	600	900	250	200	90
451 to 600	2500	4000	500	500	150	100	50
601 to 750	2000	2500	400	400	100	50	35
751 to 900	1500	—	300	300	60	—	20
901 to 1000	1250	—	250	250	40	—	8
1001 to 1500	1000	—	200	200	20	—	2

5.6) is considered as wasting energy since this water has already been heated to the saturation temperature corresponding to the boiler drum pressure. The curves presented in Figure 5.13 provide an easy method of evaluating the cost of excess blowdown as a function of various fuel costs and boiler efficiencies.

As an illustration of the cost of boiler blowdown, consider the following example.

Example: Determine the potential energy savings associated with reducing boiler blowdown from 12% to 10% using Figure 5.13.

Operating Data.

Average boiler load	75,000 lb/hr
Steam pressure	150 psig
Make up water temperature	60°F
Operating hours	8,200 hr/yr
Boiler efficiency	80%
Average fuel cost	$2.00 /10^6 Btu

Calculation and Analysis.

Using the curves in Figure 5.13, enter Chart A at 10% blowdown to the curve for 150 psig boiler drum pressure. Follow the line over to chart B and the curve for a unit efficiency of 80%. Then follow the line down to Chart C and the curve for a fuel cost of $2,.00 /10^6 Btu. Read the scale for the equivalent fuel value in blowdown. The cost of the blowdown is estimated at $8.00/hr per 100,000 lb/hr of steam generated. Repeat the procedure for the blowdown rate of 12% and find the cost of the blowdown is $10.00/hr per 100,000 lb/hr of steam generated.

Potential energy savings then is estimated to be
= ($10.00 - 8.00 /hr/100,000 lb/hr)
$\times$ (75,000 lb/hr) $\times$ (8,200 hr/yr)
= $12,300 /yr

Energy Conservation Methods

1. **Minimize Blowdown by Manual Adjustment.** This is accomplished by establishing an operating proce-

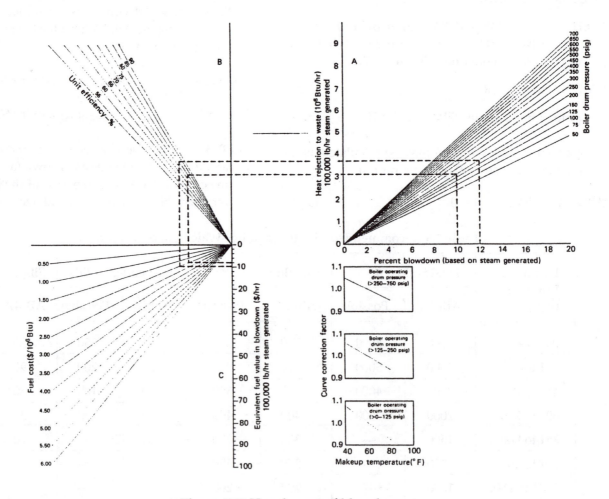

Figure 5.13 Hourly cost of blowdown.

dure requiring frequent water quality testing and re-adjustment of blowdown valves so that water impurities in the boiler are held at the allowable limit. Continuous indicating/recording analyzers may be employed allowing the operator to establish quickly the actual level of water impurity and manually read-just blowdown valves.

2. **Minimize Blowdown by Automatic Adjustment.** The adjustment of blowdown may be automated by the installation of automatic analyzing equipment and the replacement of manual blowdown valves with control valves (see Figure 5.14). The cost of this equipment is frequently justifiable, particularly when there are frequent load changes on the steam-generating equipment since the automation allows continuous maintenance of the highest allowable level of water contaminants. Literature has approximated that the average boiler plant can save about 20% blowdown by changing from manual control to automatic adjustment.

3. **Decrease Blowdown by Recovering More Condensate.** Since clean condensate may be assumed to be essentially free of water impurities, addition of condensate to the makeup water serves to dilute the concentration of impurities. The change in required

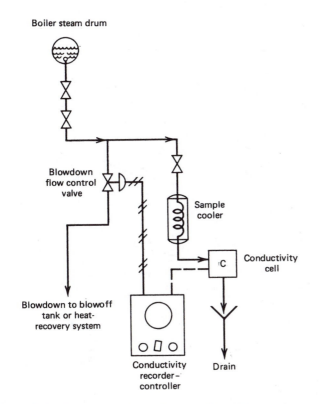

Boiler steam drum

Blowdown flow control valve

Sample cooler

C Conductivity cell

Blowdown to blowoff tank or heat-recovery system

Conductivity recorder-controller

Drain

Figure 5.14 Automated continuous blowdown system.

blowdown may be calculated using equations 5.3 and 5.5.

Example: Determine the effect on boiler blowdown of increasing the rate of condensate return from 50 to 75%:

Operating Data.
MU water impurities = 10 ppm = T
Maximum allowable limit in drum
= 100 ppm = B

Calculate:
For 50% return condensate:
MU = 100% – 50% = 50%
A = 10 × 0.5 = 5 ppm in BFW
% BD = A/(B-A) = 5/(100-5) = 5.3%

For 75% return condensate:
MU = 100 – 75% = 25%
A = 10 × 0.25 = 2.5 ppm
% BD = 2.5/(100-2.5) = 2.6%

Conclusion.
These values may then be used with the curves on Figure 5.13 to approximate the potential energy savings.

4. **Increase Allowable Drum Solids Level.** In some instances it may be possible to increase the maximum allowable impurity limit without adversely affecting the operation of the steam system. However, it must be emphasized that a water treatment consultant should be contacted for recommendation on changes in the limits as given in Table 5.9. The changes must also be followed by lab tests for steam purity to verify that the system is operating as anticipated.

The energy savings may be evaluated by using the foregoing equations for blowdown and the graphs in Figures 5.13 and 5.15. Consider the following example.

Example: Determine the blowdown rates as a percentage of steam flow required to maintain boiler drum water impurity concentrations at an average of 3000 ppm and of 6,000 ppm.

Operating Data.
Average makeup water impurity
(measurement) ....350 ppm
Condensate return (percent of steam flow) 25%
Assume condensate return free from impurities

Calculation and Analysis.

Calculate the impurity concentration in the boiler feedwater (BFW):

A = MU impurity × (1.00 - % CR)
A = 350 ppm × (1.00 - 0.25)
A = 262 ppm

Mathematical solution.
For drum water impurity level of 3000 ppm:
    % BD = A / (B - A)
    % BD = 262 / (3000 - 262)
    % BD = 9.6%

For drum water impurity level of 6000 ppm:
    % BD = A / (B - A)
    % BD = 262 / (6000 - 262)
    % BD = 4.6%

Graphical Solution. Referring to Figure 5.15
Enter the graph at feedwater impurity level of 262 ppm and follow the line to the curves for 3000 ppm and 6000 ppm boiler drum water impurity level. Then read down to the associated boiler blowdown percentage.

Conclusion.
The blowdown percentages may not be used in conjunction with Figure 5.13 to determine the annual cost of blowdown and the potential energy cost savings associated with reducing boiler blowdown.

5. **Select Raw-Water Treatment System for Largest Reduction in Raw-Water Impurities.** Since a large investment would be associated with the installation of new equipment, this energy conservation method is usually applicable to new plants or revamps only. A water treatment consultant should be retained to recommend the type of treatment applicable. An example of how water treatment affects blowdown follows.

Example: Determine the effects on blowdown of using a sodium zeolite softener producing a water quality of 350 ppm solids and of using a demineralization unit producing a water quality of 5 ppm solids. The makeup water rate is 30% and the allowable drum solids level is 3000 ppm.

Solution:
For sodium zeolite:
    % BD = (350 × 0.3 × 100%) / [300- (350 × 0.3)] = 3.6%

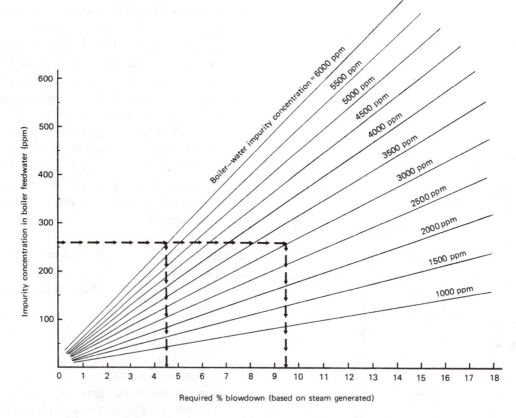

Figure 5.15 Required percent blowdown. Based on equation 5.5.

For demineralization unit:

%BD $= (5 \times 0.3 \times 100\%) / [3000 - (5 \times 0.3)]$

$\quad\quad = 0.6\%$

Therefore, the percentage blowdown would be reduced by 3.0%.

A secondary benefit derived from increasing feedwater quality is the reduced probability of scale formation in the boiler. Internal scale reduces the effectiveness of heat-transfer surfaces and can result in a reduction of as much as 1 to 2% in boiler efficiency in severe cases.

6. **Heat Recovery from Blowdown.** Since a certain amount of continuous blowdown must be maintained for satisfactory boiler performance, a significant quantity of heat is removed from the boiler. A large amount of the heat in the blowdown is recoverable by using a two-stage heat-recovery system as shown in Figure 5.16 before discharging to the sewer. In this system, blowdown lines from each boiler discharge into a common flash tank. The flashed steam may be tied into an existing header, used directly by process, or used in the deaerator. The remaining hot water may be used to preheat makeup water to the deaerator or preheat other process streams.

    The following procedure may be used to calculate the total amount of heat that is recoverable using this system and the associated cost savings.

STEP 1. Determine the annual cost of blowdown using the percent blowdown, steam flow rate (lb/hr), unit efficiency, and fuel cost. This can be accomplished in conjunction with Figure 5.13.

STEP 2. Determine:

Flash % = percent of blowdown that is flashed to steam (using Figure 5.17, curve B, at the flash tank pressure or using equation 5.10a or 5.10b)

COND % = 100% - Flash %

h_{tk} = enthalpy of liquid leaving the flash tank (using Figure 5.17, curve A, at the flash tank pressure)

h_{ex} = enthalpy of liquid leaving the heat exchanger [using Figure 5.17, curve C; for planning purposes, a 30 to 40°F approach temperature (condensate discharge to makeup water temperature) may be used]

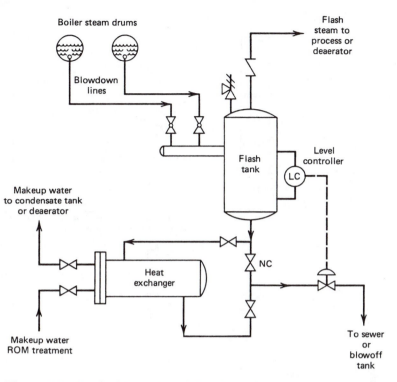

Figure 5.16 Typical two-stage blowdown heat-recovery system.

STEP 3. Calculate the amount of heat recoverable from the condensate (% QC) using

$$\%QC = [(h_{tk} - h_{ex}) / h_{tk}] \times COND\,\%$$

STEP 4. Since all of the heat in the flashed steam is recoverable, the total percent of heat recoverable (% Q) from the flash tank and heat-exchanger system is

$$\% Q = \% QC + Flash\,\%$$

STEP 5. The annual savings from heat recovery may then be determined by using this percent (% Q) with the annual cost of blowdown found in step 1:

$$annual\ savings = (\% QC / 100) \times BD\ cost$$

To further illustrate this technique consider the following example.

Example: Determine the percent of heat recoverable (%Q) from a 150 psig boiler blowdown waste stream, if the stream is sent to a 20 psig flash tank and heat exchanger.

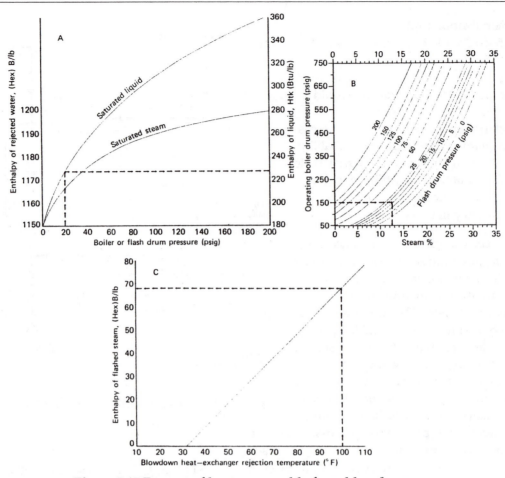

Figure 5.17 Percent of heat recoverable from blowdown.

Available Data.

Boiler drum pressure	150 psig
Flash tank pressure	20 psig
Makeup water temperature	70°F

Assume a 30°F approach temperature between condensate discharge and makeup water temperature

Calculation and Analysis.

Referring to Figure 5.17.

Determine Flash % using Chart B:

    Entering chart B with a boiler drum pressure of 150 psig and following a live to the curve for a flash tank pressure of 20 psig, read the Steam percentage (Flash %) to be 12.5%.

Determine COND %:

    COND % = 100 - Flash %

    COND % = 100 - 12.5 %

    COND % = 87.5%

Determine htk using Chart A:

    Entering chart A with a flash tank pressure of 20 psig and following a line to the curve for saturated liquid, read the enthalpy of the drum water (h_{tk}) to be 226 Btu/lb.

Determine h_{ex} using Chart C:

    Assuming a 30°F approach temperature between condensate discharge and makeup water temperature, the temperature of the blowdown discharge is equal to the makeup water temperature plus the approach temperature which equals 100°F (70°F + 30°F).

    Entering chart C with a blowdown heat exchanger rejection temperature of 100°F and following a line to the curve, read the enthalpy of the blowdown discharge water to be 68 Btu/lb.

Determine the % QC:

    % QC = [(h_{tk} - h_{ex}) / h_{tk}] × COND %

    % QC = [(226 Btu/lb - 68 Btu/lb) / 226 Btu/lb]

        × 87.5%

    % QC = 61.2%

Determine the % Q:

    % Q = % QC + Flash %

    % Q = 61.2% + 12.5%

    % Q = 73.7%

Conclusion.

Therefore, approximately 73.7% of the heat energy

can be recovered using this blowdown heat recovery technique.

More on Flash Steam

To determine the amount of flash steam that is generated by high-pressure, high-temperature condensate being reduced to a lower pressure you can use the following equation:

$$\text{Flash \%} = (h_{HPl} - h_{LPl}) \times 100\% / (h_{LPv} - h_{LPl}) \quad (5.10a)$$
or
$$\text{Flash \%} = (h_{HPl} - h_{LPl}) \times 100\% / (h_{LPevp}) \quad (5.10b)$$

where: Flash % = amount of flash steam as a percent of total mass
h_{HPl} = enthalpy of the high pressure liquid
h_{LPl} = enthalpy of the low pressure liquid
h_{LPv} = enthalpy of the low pressure vapor
h_{LPevap} = evaporation enthalpy of the low pressure liquid = $(h_{LPv} - h_{LPl})$

Guidelines for Day-to-Day Operation

1. Maintain concentration of impurities in the boiler drum at the highest allowable level. Frequent checks should be made on water quality and blowdown valves adjusted accordingly.

2. Continuous records of impurity concentration in makeup water and boiler drum water will indicate trends in deteriorating water quality so that early corrective actions may be taken.

3. Control instruments should be calibrated on a weekly basis.

5.3.6 Condensate Return

In today's environment of ever-increasing fuel costs, the return and utilization of the heat available in clean steam condensate streams can be a practical and economical energy conservation opportunity. Refer to Chapter 6 for a comprehensive discussion of condensate return. The information below is presented to summarize briefly and emphasize the benefits and major considerations pertinent to optimum steam generator operations. Recognized benefits of return condensate include:

Reduction in steam power plant raw-water makeup and associated treatment costs.
Reduction in boiler blowdown requirements resulting in

direct fuel savings. Refer to section on boiler blowdown.
Reduced steam required for boiler feedwater deaeration.
Raw-water and boiler-water chemical cost reduction.
Opportunities for increased useful work output without additional energy input.
Reduces objectionable environmental discharges from contaminated streams.

Where to Look

Examine and survey all steam-consuming units within a plant to determine the present disposition of any condensate produced or where process modifications can be made to produce "clean" condensate. Address the following:

1. Is the condensate clean and being sewered?

2. Is the stream essentially clean but on occasion becomes contaminated?

3. If contaminated, can return to the steam system be justified by polishing the condensate?

4. Can raw makeup or treated water be substituted for condensate presently consumed?

5. Is condensate dumped for operating convenience or lack of chemical purity?

Results from chemical purity tests, establishing battery limit conditions and analysis of these factors, provide the basis of obtaining maximum economy.

Modifications Required for Maximum Economy

Often, the only requirement to gain the benefits of return condensate is to install the necessary piping and/or pumping facilities. Other solutions are more complex and accordingly, require a more in-depth analysis. Chemically "clean" or "contaminated" condensate can be effectively utilized by:

Providing single- or multistage flashing for contaminated streams, and recouping the energy of the flashed steam. Recovering additional heat from the flash drum condensate by indirect heat exchange is also a possibility.

Collecting condensate from an atmospheric flash drum with "automatic" provision to dump on indication of stream contamination. This concept, when conditions warrant, allows "normally clean" condensate to be used within the system.

Installing ion-exchange polishing units for condensate streams which may be contaminated but are significant in quantity and heat value.

Providing a centrally located collection tank and pump to return the condensate to the steam system. This avoids a massive and complex network of individual return lines.

Using raw water in lieu of condensate and returning the condensate to the system. An example is the use of condensate to regenerate water treatment units.

Changing barometric condensers or other direct-contact heat exchangers to surface type or indirect exchangers, respectively, and returning the clean condensate to the system.

Collecting condensate from sources normally overlooked, such as space heating, steam tracing, and steam traps.

Providing flexibility to isolate and sewer individual return streams to maintain system integrity. Providing "knockout" or disengaging drum(s) to ensure clean condensate return to the system.

Recovering the heat content from contaminated condensate by indirect heat exchangers. An example is using an exchanger to heat the boiler makeup water.

Returning the contaminated condensate stream to a clarifier or hot lime unit to cleanup for boiler makeup rather than sewering the stream.

Allowing provision for manual water testing of the condensate stream suspected of becoming contaminated.

Using the contaminated condensate for noncritical applications, such as space heating, tank heating, and so on.

Guidelines for Day-to-Day Operation

1. Maintain the system, including leak detection and insulation repair.

2. Periodically test the return water at its source of entry within the steam system for (a) contamination, (b) corrosion, and (c) acceptable purity.

3. Maintain and calibrate monitoring and analyzing equipment.

4. Ensure that the proper operating regimen is followed; that is, the condensate is returned and not sewered.

5.4 FUEL CONSIDERATIONS

The selection and application of fuels to various combustors are becoming increasingly complex. Most existing units have limited flexibility in their ability to fire alternative fuels and new units must be carefully planned to assure the lowest first costs without jeopardizing the future capability to switch to a different fuel. This section presents an overview of the important considerations in boiler and fuel selection. Also refer to Section 5.5.

5.4.1 Natural Gas

Natural-gas firing in combustors has traditionally been the most attractive fuel type, because:

1. Gas costs have been held artificially low through government control.

2. Only limited fuel-handling equipment typically consisting of pipelines, metering, a liquid knockout drum, and appropriate controls is required.

3. Boiler costs are minimized due to smaller boiler sizes; which result from highly radiant flame characteristics and higher velocities, resulting in enhanced heat transfer and less heating surface.

4. Freedom from capital and operating costs associated with pollution control equipment.

Natural gas, being the cleanest readily available conventional form of fuel, also makes gas-fired units the easiest to operate and maintain.

However, as discussed elsewhere, the continued use of natural gas as fuel to most combustors will probably be limited in the future by government regulations, rising fuel costs, and inadequate supplies. One further disadvantage, which often seems to be overlooked, is the lower boiler efficiency that results from firing gas, particularly when compared to oil or coal.

5.4.2 Fuel Oil

Classifications

Influential in the storage, handling, and combustion efficiency of a liquid fuel are its physical and chemical characteristics.

Fuel oils are graded as No. 1, No. 2, No. 4, No. 5 (light), No. 5 (heavy), and No. 6. Distillates are Nos. 1 and 2 and residual oils are Nos. 4, 5, and 6. Oils are classified according to physical characteristics by the American Society for Testing and Materials (ASTM) according to Standard D-396.

No. 1 oil is used as domestic heating oil and as a

light grade of diesel fuel. Kerosene is generally in a lighter class; however, often both are classified the same. No. 2 oil is suitable for industrial use and home heating. The primary advantage of using a distillate oil rather than a residual oil is that it is easier to handle, requiring no heating to transport and no temperature control to lower the viscosity for proper atomization and combustion. However, there are substantial purchase cost penalties between residual and distillate.

It is worth noting that distillates can be divided into two classes: straight-run and cracked. A straight-run distillate is produced from crude oil by heating it and then condensing the vapors. Refining by cracking involves higher temperatures and pressures or catalysts to produce the required oil from heavier crudes. The difference between these two methods is that cracked oils contain substantially more aromatic and olivinic hydrocarbons which are more difficult to burn than the paraffinic and naphthenic hydrocarbons from the straight-run process. Sometimes a cracked distillate, called industrial No. 2, is used in fuel-burning installations of medium size (small package boiler or ceramic kilns for example) with suitable equipment.

Because of the viscosity range permitted by ASTM, No. 4 and No. 5 oil can be produced in a variety of ways: blending of No. 2 and No. 6, mixture of refinery by-products, through utilization of off-specification products, and so on. Because of the potential variations in characteristics, it is important to monitor combustion performance routinely to obtain optimum results. Burner modifications may be required to switch from, say, a No. 4 that is a blend and a No. 4 that is a distillate.

Light (or cold) No. 5 fuel oil and heavy (or hot) are distinguished primarily by their viscosity ranges: 150 to 300 SUS (Saybolt Universal Seconds) at 100°F and 350 to 750 SUS at 100°F respectively. The classes normally delineate the need for preheating with heavy No. 5 requiring some heating for proper atomization.

No. 6 fuel oil is also referred to as residual, Bunker C, reduced bottoms, or vacuum bottoms. It is a very heavy oil or residue left after most of the light volatiles have been distilled from crude. Because of its high viscosity, 900 to 9000 SUS at 100°F, it can only be used in systems designed with heated storage and sufficient temperature/viscosity at the burner for atomization.

Heating Value

Fuel oil heating content can be expressed as higher (or gross) heating value and low (or net) heating value. The higher heating value (HHV) includes the water content of the fuel, whereas the lower heating value (LHV)

does not. For each gallon of oil burned, approximately 7 to 9 lb of water vapor is produced. This vapor, when condensed to 60°F, releases 1058 Btu. Thus the HHV is about 1000 Btu/lb or 8500 Btu/gal higher than the LHV. While the LHV is representative of the heat produced during combustion, it is seldom used in the United States except for exact combustion calculations.

Viscosity

Viscosity is a measure of the relative flow characteristics of an oil an important factor in the design and operation of oil-handling and -burning equipment, the efficiency of pumps, temperature requirements, and pipe sizing. Distillates typically have low viscosities and can be handled and burned with relative ease. However, No. 5 and No. 6 oils may have a wide range of viscosities, making design and operation more difficult.

Viscosity indicates the time required in seconds for 60 cm³ of oil to flow through a standard-size orifice at a specific temperature. Viscosity in the United States is normally determined with a Saybolt viscosimeter. The Saybolt viscosimeter has two variations (Universal and Furol) with the only difference being the size of orifice and sample temperature. The Universal has the smallest opening and is used for lighter oils. When stating an oil's viscosity, the type of instrument and temperature must also be stated.

Flash Point

Flash point is the temperature at which oil vapors flash when ignited by an external flame. As heating continues above this point, sufficient vapors are driven off to produce continuous combustion. Since flash point is an indication of volatility, it indicates the maximum temperature for safe handling. Distillate oils normally have flash points from 145 to 200°F, whereas the flash point for heavier oils may be up to 250°F. Thus under normal ambient conditions, fuel oils are relatively safe to handle (unless contaminated).

Pour Point

Pour point is the lowest temperature at which an oil flows under standard conditions. It is 5°F above the oil's solidification temperature. The wax content of the oil significantly influences the pour point (the more wax, the higher the pour point). Knowledge of an oil's pour point will help determine the need for heated storage, storage temperature, and the need for pour-point depressant. Also, since the oil may cool while being transferred, burner preheat temperatures will be influenced and should be watched.

Sulfur Content

The sulfur content of an oil is dependent upon the source of crude oil. Typically, 70 to 80% of the sulfur in a crude oil is concentrated in the fuel product, unless expensive desulfurization equipment is added to the refining process. Fuel oils normally have sulfur contents of from 0.3 to 3.0% with distillates at the lower end of the range unless processed from a very high sulfur crude. Often, desulfurized light distillates are blended with high-sulfur residual oil to reduce the residual's sulfur content. Sulfur content is an important consideration primarily in meeting environmental regulations.

Ash

During combustion, impurities in oil produce a metallic oxide ash in the furnace. Over 25 different metals can be found in oil ash, the predominant being nickel, iron, calcium, sodium, aluminum, and vanadium. These impurities are concentrated from the source crude oil during refining and are difficult to remove since they are chemically bound in the oil. Ash contents vary widely: distillates have about 0 to 0.1% ash and heavier oil 0.2 to 1.5%. Although percentages are small, continuous boiler operations can result in considerable accumulations of ash in the firebox.

Problems associated with ash include reduction in heat-transfer rates through boiler tubes, fouling of superheaters, accelerated corrosion of boiler tubes, and deterioration of refractories. Ashes containing sodium, vanadium, and/or nickel are especially troublesome.

Other Contaminants

Other fuel-oil contaminants include water, sediment, and sludge. Water in fuel oil comes from condensation, leaks in storage equipment, and/or leaking heating coils. Small amounts of water should not cause problems. However, if large concentrations (such as at tank bottoms) are picked up, erratic and inefficient combustion may result. Sediment comes from dirt carried through with the crude during processing and impurities picked up in storage and transportation. Sediment can cause line and strainer plugging, control problems, and burned/nozzle plugging. More frequent filter cleaning may be required.

Sludge is a mixture of organic compounds that have precipitated after different heavy oils are blended. These

are normally in the form of waxes or asphaltenese, which can cause plugging problems.

Additives

Fuel-oil additives may be used in boilers to improve combustion efficiency, inhibit high-temperature corrosion, and minimize cold-end corrosion. In addition, additives may be useful in controlling plugging, corrosion, and the formation of deposits in fuel-handling systems. However, caution should be used in establishing the need for and application of any additive program. Before selecting an additive, clearly identify the problem requiring correction and the cause of the problem. In many cases, solutions may be found which would obviate the need and expense of additives. Also, be sure to understand clearly both the benefits and the potential debits of the additive under consideration.

Additives to fuel-handling systems may be warranted if corrosion problems persist due to water which cannot be removed mechanically. Additives are also available which help prevent sludge and/or other deposits from accumulating in equipment, which could result in increased loading due to increased pressure drops on pumps and losses in heat-transfer-equipment efficiencies.

Additive vendors claim that excess air can be controlled at lower values when catalysts are used. Although these claims appear to be verifiable, consideration should be given to mechanically controlling O_2 to the lowest possible levels. Accurate O_2 measurement and control should first be implemented and then modifications to burner assemblies considered. Catalysts, consisting of metallic oxides (typically manganese and barium), have demonstrated the capability of reducing carbon carryover in the flue gas and thus would permit lower O_2 levels without smoking. Under steady load conditions, savings can be achieved. However, savings may be negligible under varying loads, which necessitate prevention of fuel-rich mixtures by maintaining air levels higher than optimal.

Other types of combustion additives are available which may be beneficial to specific boiler operating problems. However, these are not discussed here since they are specific in nature and are not necessarily related to improved boiler efficiency. Generally, additives are used when a specific problem exists and when other conventional solutions have been exhausted.

Atomization

Oil-fired burners, kilns, heat-treating furnaces, ovens, process reactors, and process heaters will realize increased efficiency when fuel oil is effectively atomized. The finer the oil is atomized, the more complete combustion and higher overall combustion efficiency. Obtaining the optimum degree of atomization depends on maintaining a precise differential between the pressure of the oil and pressure of the atomizing agent normally steam or air. The problem usually encountered is that the steam (or air pressure) remains constant while the oil pressure can vary substantially. One solution is the addition of a differential pressure regulator which controls the steam pressure so that the differential pressure to the oil is maintained. Other solutions, including similar arrangements for air-atomized systems, should be reviewed with equipment vendors.

Fuel-Oil Emulsions

In general, fuel-oil emulsifier systems are designed to produce an oil/water emulsion that can be combusted in a furnace or boiler.

The theory of operation is that micro-size droplets of water are injected and evenly dispersed throughout the oil. As combustion takes place, micro explosions of the water droplets take place which produce very fine oil droplets. Thus more surface area of the fuel is exposed which then allows for a reduction in excess air level and improved efficiency. Unburned particles are also reduced.

Several types of systems are available. One uses a resonant chamber in which shock waves are started in the fluid, causing the water to cavitate and breakdown into small bubbles. Another system produces an emulsion by injecting water into oil. The primary technical difference among the various emulsifier systems currently marketed is the water particle-size distribution that is formed. The mean particle size as well as the particle-size distribution are very important to fuel combustion performance, and these parameters are the primary reasons for differing optimal water content for various systems (from 3% to over 200%). Manufacturers typically claim improvements in boiler efficiency of from 1 to 6% and further savings due to reduced particulate emissions which allow burning of higher-sulfur (lower-cost) fuels. However, recent independent testing seems to indicate that while savings are achieved, these are often at the low end of the range, particularly for units that are already operating near optimum conditions. Greater savings are more probable from emulsifiers that are installed on small, inefficient, and poorly instrumented boilers. Tests have confirmed reductions in particulate and NO_x emissions and the ability to keep boiler internals cleaner. Flames exhibit characteristics of more transparency and

higher radiance (similar to gas flames).

Before considering installation of an emulsifier system, existing equipment should be tuned up and put in optimum condition for maximum efficiency. A qualified service technician may be helpful in obtaining best results. After performance tests are completed, meaningful tests with an emulsified oil can be run. It is suggested that case histories of a manufacturer's equipment be reviewed and an agreement be based on satisfaction within a specific time frame.

5.4.3 Coal

The following factors have a negative impact on the selection of coal as a fuel:

1. Environmental limitations which necessitate the installation of expensive equipment to control particulates, SO_2, and NO_x. These requirements, when combined with the artificially low price of oil and gas in the late 1960s and early 1970s, forced many existing industrial and utility coal-fired units to convert to oil or gas. And new units typically were designed with no coal capability at all.

2. Significantly higher capital investments, not only for pollution abatement but also for coal-receiving equipment; raw-coal storage; coal preparation (crushing, conveying, pulverizing, etc.): prepared-coal storage; and ash handling.

3. Space requirements for equipment and coal storage.

4. Higher maintenance costs associated with the installation of more equipment.

5. Concern over uninterrupted availability of coal resulting from strikes.

6. Increasing transportation costs.

The use of coal is likely to escalate even in light of the foregoing factors, owing primarily to its relatively secure availability both on a short-term and long-term basis, and its lower cost. The substantial operating cost savings at current fuel prices of coal over oil or gas can economically justify a great portion (if not all) of the significantly higher capital investments required for coal. In addition, most predictions of future fuel costs indicate that oil and gas costs will escalate much faster than coal.

Enhancing Coal-Firing Efficiency

A potentially significant loss from the combustion of coal fuels is the unburned carbon loss. All coal-fired steam generators and coal-fired vessels inherently suffer an efficiency debit attributable to unburned carbon. At a given process output, quantities in excess of acceptable values for the particular unit design and coal rank detract greatly from the unit efficiency, resulting in increased fuel consumption.

It is probable that pulverized coal-fired installations suffer from an inordinately high unburned carbon loss when any of the following conditions are experienced:

A change in the raw-fuel quality from the original design basis.
Deterioration of the fuel burners, burner throats, or burner swirl plates or impellers.
Increased frequency of soot blowing to maintain heat-transfer-surface cleanliness.
Noted increase in stack gas opacity.
Sluggish operation of combustion controls of antiquated design.
Uneven flame patterns characterized by a particularly bright spot in one sector of the flame and a quite dark spot in another.
CO formation as determined from a flue-gas analysis.
Frequent smoking observed in the combustion zone.
Increases in refuse quantities in collection devices.
Lack of continued maintenance and/or replacement of critical pulverized internals and classifier assembly.
High incidence of coal "hang-up" in the distribution piping to the burners.
Frequent manipulation of the air/coal primary and secondary air registers.

How to Test for Relative Operating Efficiency

The aforementioned items are general symptoms that are suspect in causing a high unburned carbon loss from the combustion of coal. However, the magnitude of the problem often goes undetected and remedial action is never taken because of the difficulty in establishing and quantifying this loss while relating it to the overall unit operating efficiency.

In light of this, it is recommended that the boiler manufacturer be consulted to establish the magnitude of this operating loss as well as to review the system equipment and operating methods.

The general test procedure to determine the unburned carbon loss requires manual sampling of the ash (refuse) in the ash pit, boiler hopper(s), air heater hopper, and dust collector hopper and performing a laboratory analysis of the samples. For reference, detailed methods, test procedures, and results are outlined in the ASME Power Test Codes, publications PTC 3.2 (Solid Fuels) and

PTC 4.1 (Steam Generating Units), respectively. In addition to the sampling of the ash, a laboratory analysis should be performed on a representative raw coal sample and pulverized coal sample.

The results and analysis of such a testing program will:

1. Quantify the unburned carbon loss.

2. Determine if the unburned carbon loss is high for the type of coal fired, unit design, and operating methods.

3. Reveal a reasonable and attainable value for unburned carbon loss.

4. Provide guidance for any corrective action.

5. Allow the plant, with the aid of Figure 5.18, to assess the annual loss in dollars by operating with a high unburned carbon loss.

6. Suggest an operating mode to reduce the excess air required for combustion.

Figure 5.18 can be used to determine the approximate energy savings resulting from reducing unburned coal fuel loss. The unburned fuel is generally collected with the ash in either the boiler ash hopper(s) or in various collection devices. The quantity of ash collected at various locations is dependent and unique to the system design. The boiler manufacturer will generally specify the proportion of ash normally collected in various ash hoppers furnished on the boiler proper. The balance of ash and unburned fuel is either collected in flue-gas cleanup devices or discharged to the atmosphere. A weighted average of total percentage combustibles in the ash must be computed to use Figure 5.18. To further illustrate the potential savings from reducing combustible losses in coal-fired systems using Figure 5.18, consider the following example.

Example: Determine the benefits of reducing the combustible losses for a coal-fired steam generator having a maximum continuous rating of 145,000 lbs/hr at an operating pressure of 210 psig and a temperature of 475°F at the desuperheater outlet, feedwater is supplied at 250°F.

Available Data.

Average boiler load	125,000 lb/hr
Superheater pressure	210 psig
Superheater temperature	475°F

Fuel type	Coal
Measured flue gas oxygen	3.5%
Operating combustibles in ash	40%
Obtainable combustibles in ash	5%
Yearly operating time	8500 hr/yr
Design unit heat output	150×10^6 Btu/hr
Average unit heat output	129×10^6 Btu/hr
Average fuel cost	$1.50 /$10^6$ Btu

Analysis. Referring to Figure 5.18:
Chart A.

Operating combustibles	40%
Obtainable combustibles	5%

Chart B.

Design Unit output	150×10^6 Btu/hr

Chart C.

Average fuel cost	$1.50 /$10^6$ Btu
Annual fuel savings	$210,000

Annual fuel savings corrected to actual operating conditions:

Savings \$ = (curve value) × [(operating heat output) / (design heat output)] × [(actual annual oper-

ating hours) / (8,760 hr/yr)]

= ($210,000 /yr) × [($129 \times 10^6$ Btu/hr) / (150×10^6 Btu/hr)] × [(8,500 hr/yr) / (8,760 hr/yr)]

= $175,200 /yr

Notes: If the unit heat output or average fuel cost exceeds the limit of Figure 5.18, use half the particular value and double the savings obtained from Chart C. The annual fuel savings are based on 8,760 hr/yr. For an operating factor other that the curve basis, apply a correction factor to curve results. The application of the charts assumes operating at, or close to, optimum excess air values.

Modifications for Maximum Economy

It is very difficult to pinpoint and rectify the major problem detracting from efficiency, as there are many interdependent variables and numerous pieces of equipment required for the combustion of coal. Further testing and/or operating manipulations may be required to "zero in" on a solution. Modification(s) that have been instituted with a fair degree of success to reduce high unburned carbon losses and/or high-excess-air operation are:

Modifying or changing the pulverizer internals to increase the coal fineness and thereby enhance combustion characteristics.

Rerouting or modifying air/coal distribution piping to avoid coal hang-up and slug flow going to the burners.

Installing additional or new combustion controls to smooth out and maintain consistent performance.

Purchasing new coal feeders compatible with and responsive to unit demand fluctuations.

Calibrating air flow and metering devices to ensure correct air/coal mixtures and velocities at the burner throats.

Installing new classifiers to ensure that proper coal fines reach the burners for combustion. Optimally setting air register positions for proper air/fuel mixing and combustion. Replacing worn and abraded

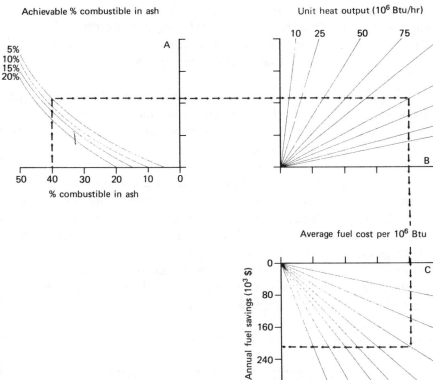

Figure 5.18 Coal annual savings from reducing combustible losses.

burner impeller plates.

Increasing the air/coal mixture temperature exiting the pulverizers to ensure good ignition without coking.

Cleaning the burning throat areas of deposits.

Installing turning vanes or air foils in the secondary air supply duct or air plenum to ensure even distribution and proper air/fuel mixing at each burner.

Purchasing new, or updating existing combustion controls to reflect the present state of the art.

As can be seen, the solutions can be varied, simple or complex, relatively cheap or quite expensive.

The incentives to correct the problem(s) are offered in Figure 5.18. Compare the expected benefits to the boiler manufacturer's solution and cost to implement and judge the merits.

Guidelines for Day-to-Day Operation

To a great extent, the items listed which detract from operating performance should be those checked on a day-to-day basis. A checklist should be initiated and developed by the plant to correct those conditions that do not require a shutdown, such as:

1. Maintaining integrity of distribution dampers and air registers with defined positions for operating load.

2. Ensuring pulverizer exit temperature by maintaining air calibration devices and air-moving equipment.

3. Checking feeder speeds and coal hopper valves, ensuring an even and steady coal feed.

4. Frequent blending of the raw coal pile to provide some measure of uniformity.

Coal Conversion

Converting existing boilers that originally were designed to fire oil and/or natural gas to coal capability is possible for certain types of units provided that significant deratings in steam rate capacities are acceptable to existing operations. Since boiler installations are somewhat custom in design, fuel changes should be addressed on an individual basis. In general, however, conversion of field-erected units to coal firing is technically possible, whereas conversion of shop-assembled boilers is not.

Modifications required to convert shop-assembled units would include installation of ash-handling facilities in the furnace and convection pass, soot-blowing equipment, and modifications to tube spacing each of which is almost physically impossible to do and prohibitive in cost. In addition, even if these modifications could be made, derating of the unit by about two-thirds would be necessary. Alternatives for consideration rather than converting existing shop-assembled units would be to replace with new coal-fired units, or to carefully assess the application of one of an alternative fuel, such as a coal/oil mixture or ultra-fine coal.

Many field-erected boilers can be converted to coal firing without serious compromises being made to good design. The most serious drawback to converting these boilers is the necessity to downrate by 40 to 50%. Although in some instances downrating may be acceptable, most operating locations cannot tolerate losses in steam supply of this magnitude.

Why is downrating necessary? Generally, there are differences between boilers designed for oil and gas and those designed for coal. Burning of coal requires significantly more combustion volume than that needed for oil or gas due to flame characteristics and the need to avoid excessive slagging and fouling of heat-transfer surfaces. This means that a boiler originally designed for oil or gas must be downrated to maintain heat-release rates and firebox exit temperatures within acceptable limits for coal. Other factors influencing downrating are acceptable flue-gas velocities to minimize erosion and tube spacing.

A suggested approach for assuring the viability of coal conversion is:

1. Contact the original boiler manufacturer and determine via a complete inspection the actual modifications that would be required for the boiler.

2. Determine if the site can accommodate coal storage and other associated equipment, including unloading facilities, conveyors, dust abatement, ash disposal, and so on.

3. Select the type of coal to burn, considering cost, availability, and pollution restrictions. Before final selection of a coal, it must be analyzed to determine its acceptability and effect on unit rating. For any installation, final selection must be based on the coal's heating value, moisture content, mineral matter content, ash fusion and chemical characteristics, and/or pulverizers' grindability.

4. Assess over economics.

5.5 DEVELOPING TECHNOLOGIES

As domestic costs of fuels rise, the economics of utilizing alternative fuel types and combustion techniques will become more favorable. Also, advances in technology to reduce equipment and operating costs will help accelerate implementation. A few of these developing technologies warrant description and possible consideration for replacement or new combustor installations once reliability, operability, and overall economics are satisfactorily demonstrated.

It is also likely that there will be continuing pressures placed on companies installing new boilers to fire fuels other than fuel oil or natural gas. Pressures from foreign suppliers can be in the forms of high prices and unreliable supplies, while domestic pressures are in the form of regulations, incentives, and price controls or subsidies. As mentioned in Section 5.6, owners of new boilers above specified capacities will have to file for exemptions with the Department of Energy during preliminary design if the unit under consideration will fire oil or natural gas. This procedure requires the owner to quantify the economics of firing alternative fuels before a permanent or temporary exemption is granted. What this means is that the federal government now has the legal authority to mandate the consideration and utilization of alternative fuels. This regulation may be subject to change or deletion depending on the prevailing administration.

5.5.1 Vaporized Fuel Oil

The vaporization of fuel oil enables the firing of fuel oil in furnaces equipped with natural gas burners. Normally, converting an existing natural-gas unit to liquid-fuel-oil firing requires burner changes, furnace structural modifications, and results in less-than-optimal flame characteristics and minor derating. A recently developed process called Vaporized Fuel Oil (VFO) vaporizes the light components of the oil, which when combined with steam can be handled like natural gas. Thus conversion to fuel-oil capability is possible for a variety of furnaces, kilns, calciners, and ovens. However, economics typically indicate that the VFO process is not competitive with direct-fuel-oil firing in boilers and other combustors which can easily be converted.

Other advantages of VFO are:

Natural-gas headers, controls, and burners require little or no modification, a very important consideration in multi-burner furnaces.

VFO can be installed in a manner that requires very little downtime.

Coking and corrosion problems are reduced because the vaporization step removes nonvolatile contaminants such as vanadium, sulfur, and sodium.

On-line switching from gas to oil is instantaneous, with only minor fluctuations in temperature.

5.5.2 Fluidized-Bed Combustors

In a fluidized-bed combustor, crushed coal is burned in a bed of granulated limestone particles which are maintained in suspension by a stream of air that is directed upward (see Figures 5.19 and 5.20). Energy from combustion in the hot fluid mass converts water passing through tubes located in and above the granular bed to steam. The air velocity is maintained at a rate that overcomes gravitational forces on the particles without carrying the particles out of the furnace.

Fluidized-bed systems have been used for over half a century, primarily in the chemical industry, to enhance reaction rates of materials that are difficult to burn, and in

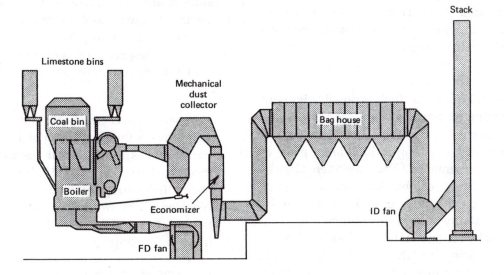

Figure 5.19 General arrangement for fluidized-bed combustor.

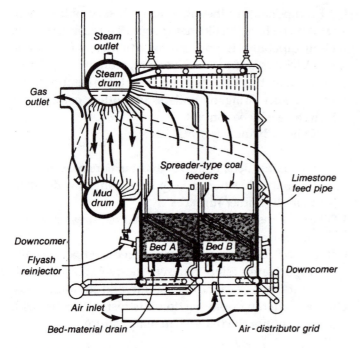

Figure 5.20 Typical fluidized-bed combustor steam generator arrangement.

fluidized drying, oil refining catalytic crackers, and so on. Recent revitalized interest in burning coal has led to the development of fluidized-bed units for steam generation and air heating. Advantages of fluidized-bed units for coal combustion are many; for example:

1. The overall heat-transfer rate is greater than that achievable from conventional units; thus smaller and less expensive units are possible.

2. Combustion temperatures are lower (1500 to 1600°F), which inherently minimizes the formation of NO_x and furnace slag. This also allows considerable flexibility in selecting fuel types since the combustor is relatively insensitive to fuel ash characteristics.

3. Combustion of high-sulfur fuels is possible without flue-gas treatment. Since raw crushed limestone (about 95%) and crushed coal (about 5%) are used as the bed materials, calcining to lime takes place at normal bed temperatures. The lime then absorbs SO_2 released during combustion, producing gypsum ($CaSO_4$) on the surface of the bed material particles. The ability to capture sulfur continuously is maintained by continuously withdrawing the material and replacing it with fresh limestone. Thus pollution control requires only the reduction of particulates by conventional means, such as a baghouse

or electrostatic precipitator, and the disposal of inert calcium sulfate.

4. Grinding costs are less because only crushed coal is required.

5. Higher combustion efficiencies than some conventional units are possible.

Commercial units are currently becoming available. As reliability and operability are demonstrated in industrial and utility applications, designers of combustor installations should increasingly consider the use of a fluidized-bed unit.

5.5.3 Coal-Oil Mixtures

Mixing pulverized coal with oil to provide a composite fuel that can be fired directly in a combustor appears to be one of the most promising near-term methods of reducing fuel costs and reliance on foreign oil. In a coal-oil mixture (COM) process, up to about 50 wt % coal is pulverized and blended with oil to form a fuel that can be stored, pumped, and combusted much as a liquid fuel. Thus units originally designed to fire oil or gas have the potential of being partially converted to coal with only modest modifications. The primary advantages of utilizing COM, other than coal availability, is that the resulting mixture has a lower cost per million Btu (assuming that the delivered cost of coal is less than oil, which is probable) than firing oil only. For example, in a 50% COM, about 35% of the heat content is provided by lower-cost coal.

Many demonstration projects are under way to evaluate technical issues of COM preparation and handling, combustion, and pollution control. Although demonstration results look increasingly promising, to most commercial, industrial, and utility users, operability, reliability, and maintainability have not been proven to the point where widespread use is common.

As COM systems are proven, factors to consider include:

1. Comparison of economics of purchasing a prepared COM versus the installation of a preparation plant. Early studies indicate that installations requiring less than 200,000 to 400,000 lb/hr of steam could not justify their own preparation plants. Purchasing the required COM would be more suitable provided that price and COM characteristics were clearly defined in the contract.

2. Preparation and handling system requirements, such as:

a. Best method of pulverization (dry grinding, wet-grinding ball mill, wet disperser, ultra-fine grinding).

b. Blending and stabilization requirements (coal concentration, homogenization, additives versus no additives, coal and oil characteristics).

c. Storage and pumping equipment requirement and operating procedures (long-range stability, pumpability and remixability, continuous mixing, temperature control, erosion, corrosion, etc.).

d. Application of commercial instrumentation and effect of COM on reliability.

e. Transportation requirements for temperature control (heat source and insulation), pumping, minimizing vibration effects.

3. Combustion related factors such as:
   a. Stability of COM at high temperatures required for atomization.
   b. Burner type and nozzle erosion.
   c. Furnace combustion efficiency, flame characteristics, and potential for slagging.
   d. Ash deposition (and characteristics on heat-transfer surfaces); also, soot-blowing requirements, ash removal, and effect of high temperatures on corrosion/erosion.
   e. Overall boiler efficiency, derating requirements, and maintainability operability.

4. Ability to meet environmental regulations:
   a. What are fly-ash characteristics, and what effect do they have on baghouse and electrostatic precipitators (i.e., particle-size distribution and resistivity)?
   b. Must low-sulfur coal be used or is flue-gas desulfurization economical and practical?
   c. Can NO_x regulations be met?

5.5.4 Coal Gasification

Low-Btu Gases

The high cost of low-Btu gas (typical heating value 100 to 200 Btu/scf) derived from coal is one of the primary reasons for limited application. Lack of current design and operating experience is also a major contributing factor. Low-Btu gasifiers have typically been small, producing less than 2 billion Btu/day of gas. As larger production facilities are developed, the use of low-Btu gas shows promise for both new and existing central stations. However, since transportation is difficult and costs are high, gasifiers must be located close (within 1 mile) to the point of use, thus seriously limiting the number of large central locations. Probably the major potential application is for large units originally designed for coal but converted to oil or gas for environmental reasons. For these applications, gasification and sulfur removal before combustion may have better economics than adding scrubbers.

Several boiler manufacturers indicate that conversion of oil/gas-fired packaged boilers to low-Btu gas is virtually impossible. Furnace volume is too small for complete combustion unless the unit is derated by as much as 50%. Also, new burners may be too large for economic retrofit.

Factors to consider in using low-Btu gas are:

1. Flue-gas volume will increase by about 20% and excess air by about 30%. Note that low-Btu gas requires less air for combustion than does natural gas.

2. Flame temperature is lower for low-Btu gas than for natural gas (about 3200°F and 3560°F respectively) and flame patterns may vary, necessitating changes in heat-transfer surfaces.

3. New fuel-gas piping may be necessary as well as modifications to ignition systems, flame safeguards, and combustion controls.

4. Burners and windbox may need to be replaced and ratings of forced-draft and induced-draft fans increased.

Medium-Btu Gases

The advantage of using medium-Btu gas, typically 200 to 500 Btu/scf, over low-Btu gas is that flame temperatures are within about 100°F of natural gas, combustion air requirements are only about 5% greater, and flue-gas volumes are nearly the same. This means that medium-Btu gas can be fired in many combustors with relatively little equipment modification.

Gasifer systems used to produce low- and medium-Btu gases are virtually the same with one major difference an oxygen plant is required. This addition makes installation in smaller plants (less than 150 billion Btu/day or more) uneconomical. Transportation is practical over much longer distances than low-Btu gas, making large installations a possibility provided that equipment costs can be reduced and reliability demonstrated.

High-Btu Substitute Gases

Substitute natural gas or SNG is produced by upgrading medium-Btu synthesis gas. This is accomplished by reacting carbon monoxide with steam to produce carbon dioxide and hydrogen. The stream is then methanated by purging it of carbon dioxide, hydrogen sulfide, and organics. Methane and by-product steam are then produced by reacting the remaining hydrogen and carbon monoxide in a heated catalyst.

Although small commercial units have been demonstrated, large-scale units have not been built because of very high capital costs.

5.5.5 Coal Liquids

Liquefaction processes that produce clean liquids from coal are many years away from producing commercially significant quantities of oil. Many coal liquid processes are now undergoing intensive development work; however, high capital costs and resulting poor economics make these noncompetitive with other fuel sources in the near future. Even so, political, technical, and economic factors may eventually make the use of coal liquids viable to industry and utilities.

There are three basic processes for liquefaction: direct liquefaction, indirect liquefaction, and pyrolysis. In the direct liquefaction process, coal is slurried in a process-derived oil and then reacted with hydrogen at high pressures and moderate temperatures to form liquid hydrocarbons. Ash and unreacted coal are removed from the product and the unused coal gasified to produce the required hydrogen. This type of process is receiving much emphasis from the Department of Energy because conversion efficiency is high (about 65 to 70% of the energy in the coal feedstock is in the product) and yields are good.

Conversely, indirect liquefaction has a relatively low efficiency (45 to 60%) and low yield, thus little work is being done in this area. In the indirect liquefaction process, coal is first gasified to a synthesis gas and then catalytically converted to methanol or liquid hydrocarbons, which are essentially pollutant free.

The third process, pyrolysis, requires heat to drive off volatile portions of the coal, producing gases, liquids, and solid products.

Initial application of coal liquids in industry could be in new and existing boilers that are prohibited by the government from using oil or natural gas, or for existing combustors that would require expensive flue-gas cleanup systems.

Current limited testing of raw-coal liquid fuels in combustors indicates that burner redesigns and/ or combustion control improvements would be required. Also, because these fuels may not be miscible with oil, separate storage, pumping, and piping systems could be required.

However, if these fuels are upgraded, a relatively expensive process, they are likely to burn in a manner similar to comparable oil products and could be blended without adverse problems.

5.5.6 Solvent-Refined Coal

Clean solid fuels can be produced from coal by utilizing chemical benefitting processes. These processes generally promise sulfur levels low enough to preclude the use of flue-gas scrubbers and remove significant quantities of ash, resulting in lower maintenance.

Solvent-refined coal (SRC) has been proved technically feasible in pilot tests and appears to be the most attractive of the methods of chemical benefitting, which include hydrothermal treating Metha-coal processes, and thermochemical desulfurization. In the SRC process, raw coal is mixed with a solvent derived from the coal and a small quantity of hydrogen, and then heated to 800°F. Sulfur in the fuel is converted to hydrogen sulfide and filterable iron sulfides and other mineral matter. The end product is a liquid as it leaves the reaction process, which can be converted to a solid by cooling.

Although burn tests appear promising, enforcement of rigid Environmental Protection Agency air quality standards may necessitate some form of flue-gas cleanup even with SRC, making economics questionable. Limited testing also indicates that available technologies can readily be applied to handling and combustion in existing and new boilers.

5.5.7 Micron-Size Coal

Micron-size coal is generally a coal that has been ground to particle sizes in the micron range, say 4-micron average with maximum size of 20 to 40 microns. Test results have shown that grinding coal to small particles substantially increases the reactive surface area, and thus the burning rate is greatly increased. The potential benefits are numerous. First, with flame characteristics that closely resemble a gaseous fuel, it may be possible to directly fire micron-size coal in a combustor originally designed for oil or gas with little boiler modification and no derating. On new designs this could mean that smaller, less costly coal-fired units could be fabricated, which may be particularly attractive for the industrial user. Also, the firing of small particles is recognized in theory as one way of increasing combustion efficiency, as in atomizing oil or pulverizing large pieces of coal.

Another possible advantage is that the coal ash apparently moves through the boiler and up the stack without harmful buildup. However, the effect on emission control systems has not been tested sufficiently to quantify the overall ability to meet environmental regulations. Reduced slagging on boiler tubes may also be a potential benefit. Advantages when used in the preparation of coal/oil mixtures are also foreseen, such as more stable mixtures and higher combustion efficiencies.

Several major problems exist. Micron-size particles of coal can be produced in a variety of mills however, none of these has proved reliable over long periods. Fluid energy mills, utilizing high-pressure steam to force collision of the coal particles, have produced micron-size coal for relatively short periods of time and appear to be the best prospect. Other factors associated with the long-term firing of micron-size coal require further demonstration and quantification before commercial application will become a reality. These include combustion control techniques, erosion/corrosion effects, and pollution control affects.

5.5.8 Shale Oil

Cost factors and environmental issues have delayed commercialization of shale oil technology. However, with recent increase in government incentives, commercialization appears more likely.

Oil shale is a sedimentary rock containing kerogen, a solid organic material. Heating the shale to about 900°F in a retort decomposes the kerogen into hydrocarbons and carbonaceous residue. Investigations are also being carried out with an in situ process that potentially reduces fuel costs. Crude shale oil can be hydrotreated and further refined to finished products (gasoline, kerosene, diesel oil, etc.). However, preliminary tests indicate that crude shale oil may be a suitable fuel for use in utility or industrial steam boilers. These tests indicate that handling would be similar to (if not compatible with) firing residue oils.

5.5.9 Methanol

Chemical or fuel-grade methanol can be produced from natural gas, residual oil, coal, tar sands, oil shale, or biomass. Currently, over 1 billion gallons per year are produced from natural gas and residual oil. However, large increases in fuel prices have focused attention on deriving methanol from sources other than oil and gas. Several coal-to-methanol plants are under development, with commercialization likely to appear sooner than for other alternative fuel technologies. In these processes, coal is first converted to synthesis gas (hydrogen and carbon monoxide), from which methane is produced. For the production of methane, a competitive fuel, costs are dependent on the availability of low-cost coal, either low-grade (such as lignite) or high-sulfur.

One major advantage to using methanol as a fuel source is that it is one of the cleanest-burning and most versatile fuels derived from coal. Nitrogen oxide emissions are low because methanol contains no nitrogen, and flame temperatures are lower than for distillates. Although extensive combustion tests have not been made, indications are that sulfur dioxide levels are very low and smoke barely visible. Boiler modifications to burn methanol should be relatively straightforward, requiring primarily burner modifications.

5.5.10 Hydrogen

Hydrogen is a clean fuel, containing approximately 325 Btu/scf, which ignites readily and burns rapidly, producing only water vapor and nitrogen oxides. The use of hydrogen should be considered in any location that generates the gas as a process by-product for which there is no other viable use. Hydrogen can easily be fired in a burner similar to a natural-gas burner.

Large quantities of hydrogen cannot currently be commercially produced, and it is unlikely that large-scale units will be available before the end of the century. Large-scale production of hydrogen currently focuses on splitting water molecules into component elements by electrolysis and/or thermochemical reactions.

5.6 REGULATIONS AND FUEL SELECTION

Energy management entails not only the use of a minimum quantity of Btu to produce a unit of product, but also the manufacture of a product at the lowest possible cost. Recent dramatic changes in both the cost and availability of various fuels have focused much attention on the need to have flexibility in the ability to fire several fuels in a given unit. Until recently, the price of each fuel, and environmental constraints, were the primary selection criteria. However, the selection of appropriate firing capability for existing and new units has become increasingly complex owing to several additional factors, including (1) availability uncertainties for oil and natural gas; (2) legislative and regulatory actions dealing with the environment, such as the Clean Air Act, the Clean Water Act, and the Resource Conservation and Recovery Act; and (3) legislative and regulatory actions dealing with energy, such as the Fuel Use Act and other governmental actions, including natural-gas decontrol and incremental pricing, deregulation of domestic oil, and so on.

Although a definitive review of all legislation and effective and pending regulations that influence the optimum selection of fuels is beyond the scope of this chapter, a synopsis will serve to highlight factors that are critical in the final selection of suitable fuels and as an indication of the need to review carefully all pertinent regulations before selecting a fuel.

5.6.1 Fuel Use Act

The Powerplant and Industrial Fuel Use Act of 1978 (FUA) became effective May 8,1979, replacing the Energy Supply and Environmental Coordination Act of 1974 (ESECA). In the industrial area, the thrust of the FUA is to force conversion of large plants to alternative fuels (primarily coal) by empowering the Department of Energy (DOE) to prohibit the use of petroleum and natural gas, or to require minimum percentages of an alternative fuel, in both new and existing major fuel-burning installations (MFBI). A MFBI is defined as an installation having either (1) a stationary combustion unit with the capacity for burning 100 million Btu/hr or more of any fuel, or (2) any combination of two or more combustion units at the same site with the combined capacity for burning 250 million Btu/hr or more of any fuel (only units with capacities of 50 million Btu/hr or more are included). The implications of these regulations are far-reaching for existing and new combustors:

1. Existing MFBIs having the technical capability to use an alternative primary fuel may be banned from using oil or gas provided that such a switch is deemed financially feasible and substantial physical modification of the unit is not required nor a substantial capacity reduction.

2. Natural gas or oil may not be used as a primary energy source in a new MFBI consisting of a boiler. Prohibitions for new MFBIs other than boilers will be based on relevant factors, such as technical feasibility, facility size, location, and so on.

3. Exemptions for various circumstances are possible; however, the burden of proof in determining who should be exempt from DOE prohibitions now rests with industry, not the government.

4. Prior to issuing exemptions that would allow the burning of gas and oil, the DOE must be satisfied that some part of consumption cannot be fulfilled with alternative fuels. Thus as technology is further developed and demonstrated for fluidized-bed combustion, coal gasification or liquefaction, coal-oil mixture utilization, and so on, technical and economic feasibility will have to be reviewed with the DOE before exemptions are permitted. It is important to realize that the DOE currently has the authority to force conversion to these technologies.

5. Temporary and permanent exemptions can be granted if an alternative fuel supply is not available, if significant site limitations exist, if environmental laws cannot be met, and for certain other specific reasons.

5.6.2 Natural Gas Policy Act (NGPA)

Title II of the NGPA concerns the incremental pricing of natural gas, which is designed to shield residential and commercial customers from the high costs of decontrolled natural gas. Industrial customers on interstate gas pipelines will be charged the bulk of the additional costs of new gas supplies. Currently, the costs that can be passed on to industrial customers are a moving ceiling tied to the price of an alternative fuel in a given geographic area. Therefore, industry's exposure to increased gas costs has risen substantially.

5.6.3 Clean Air Act

The Clean Air Act provides the power to the Environmental Protection Agency (EPA) to adopt and enforce air pollution control regulations. Thus far, standards have been set to limit the maximum allowable ambient concentrations for carbon monoxide, nitrogen oxides, sulfur dioxide, hydrocarbons, photochemical oxidants, lead, and total suspended particulate matter. In addition to issuing air quality standards, the EPA has required states to develop State Implementation Plans (SIPs) indicating how the state will meet and enforce these standards. SIPs include limitations on stationary sources and mobile sources of pollution. They also regulate air quality for new and expanded facilities as well as hazardous air pollutants.

Each state was required to submit documentation to the EPA regarding the attainment status of its air quality control regions for various pollutants. Specifically, regions were designated as a Prevention of Significant Deterioration (PSD) area if air quality was better than National Air Quality Standards (NAQS). PSD areas are further divided into three classes: (1) areas in which very little air quality deterioration would be allowed, (2) areas allowing moderate deterioration, and (3) areas allowing significant deterioration. Areas not meeting these standards were classified as nonattainment areas. States that have submitted SIPs for nonattainment areas have, in many cases, had to revise their plans and may be required to do so in the future.

The EPA has also established New Source Performance Standards (NSPS) which apply to new industrial and municipal plants, as well as modifications to new facilities. These standards require the implementation of the Best Available Pollution Control Technology (BACT).

New Source Pollution Standards have been adopted for electric generating plants for NO_x, SO_2, and particulates which will increase utility costs substantially. Although standards are still in the development stage for industrial boilers, substantial cost increases can be expected to meet Best Available Control Technology requirements. Obtaining permits, particularly in nonattainment areas will probably become increasingly difficult. These factors have been emphasized to indicate the need to consider carefully all aspects critical in the selection of appropriate fuel types, whether conventional or unconventional.

5.6.4 Resource Conservation and Recovery Act (RCRA)

The RCRA was enacted in 1977, covering the management of solid industrial wastes, including residue and sludge by-products from air and water pollution control equipment, as well as ash handling from combustion equipment. Subtitle C of RCRA empowers EPA to identify and regulate hazardous wastes. A waste is considered hazardous if it exhibits certain characteristics, such as ignitability, reactivity, corrosiveness, and toxicity.

Regulations that affect powerhouses have not yet been finalized. However, EPA has created a "special waste" category for flue-gas desulfurization wastes, bottom ash, and fly ash from fossil-fuel-fired utilities. This category currently is exempt from most regulations.

5.6.5 Clean Water Act

The Clean Water Act, while concentrating on toxic pollutants, also considers five-day biochemical oxygen demand, total suspended solids, fecal coliform bacteria, oil and grease, and pH. Of particular concern to industrial powerhouses are wastewater streams associated with coal. These include bottom as sluice water, fly-ash sluice water, wastewater from flue-gas scrubbers, and coal pile runoff. Varying concentrations of toxic heavy metals may be in these streams. Although specific regulations have not been developed, some possibilities include (1) requirements to cover coal piles to prevent pickup of toxic pollutants; (2) dikes around coal piles; and (3) lining of ash ponds, and so on.

Bibliography

AMERICAN SOCIETY OF MECHANICAL ENGINEERS (ASME), ASME Power Test Code (PTC) 4.1-1964; Steam Generating Units, ASME, New York, 1964.

KEENAN, J.H., and J. KAYE, *The Gas Tables*, Wiley, New York, 1948.

PAYNE, F.W., *Efficient Boiler Operations Source Book*, 3rd ed., Fairmont, Lilburn, GA., 1991.

SCOLLON, R.B., and R.D. SMITH, *Energy Conservation Guidebook*, Allied Corporation, Morristown, N.J., 1976

TAPLIN Jr., H.R., *Combustion Efficiency Tables*, Fairmont, Lilburn, GA., 1991.

CHAPTER 6

STEAM AND CONDENSATE SYSTEMS

PHILIP S. SCHMIDT

Department of Mechanical Engineering
University of Texas at Austin
Austin, Texas

6. 1 INTRODUCTION

Nearly half of the energy used by industry goes into the production of process steam, approximately the same total energy usage as that required to heat all the homes and commercial buildings in America. Why is so much of our energy resource expended for the generation of industrial steam?

Steam is one of the most abundant, least expensive, and most effective heat-transfer media obtainable. Water is found everywhere, and requires relatively little modification from its raw state to make it directly usable in process equipment. During boiling and condensation, if the pressure is held constant and both water and steam are present, the temperature also remains constant. Further, the temperature is uniquely fixed by the pressure, and hence by maintaining constant pressure, which is a relatively easy parameter to control, excellent control of process temperature can also be maintained. The conversion of a liquid to a vapor absorbs large quantities of heat in each pound of water. The resulting steam is easy to transport, and because it is so energetic, relatively small quantities of it can move large amounts of heat. This means that relatively inexpensive pumping and piping can be used compared to that needed for other heating media.

Finally, the process of heat transfer by condensation, in the jacket of a steam-heated vessel, for example, is extremely efficient. High rates of heat transfer can be obtained with relatively small equipment, saving both space and capital. For these reasons, steam is widely used as the heating medium in thousands of industries.

Prior to the mid-1970s, many steam systems in industry were relatively energy wasteful. This was not necessarily bad design for the times, because energy was so cheap that it was logical to save first cost, even at the expense of a considerable increase in energy requirements. But things have changed, and today it makes good sense to explore every possibility for improving the energy efficiency of steam systems.

This chapter will introduce some of the language commonly used in dealing with steam systems and will help energy managers define the basic design constraints and the applicability of some of the vast array of manufacturer's data and literature which appears in the marketplace. It will discuss the various factors that produce inefficiency in steam system operations and some of the measures that can be taken to improve this situation. Simple calculation methods will be introduced to estimate the quantities of energy that may be lost and may be partially recoverable by the implementation of energy conservation measures. Some important energy conservation areas pertinent to steam systems are covered in other chapters and will not be repeated here. In particular the reader is referred to Chapters 5 (on boilers), 7 (on cogeneration), and 15 (on industrial insulation).

6.1.1 Components of Steam and Condensate Systems

Figure 6.1 shows a schematic of a typical steam system in an industrial plant. The boiler, or steam generator, produces steam at the highest pressure (and therefore the highest temperature) required by the various processes. This steam is carried from the boiler house through large steam mains to the process equipment area. Here, transfer lines distribute the steam to each piece of equipment. If some processes require lower temperatures, the steam may be throttled to lower pressure through a pressure-regulating valve (designated PRV on the diagram) or through a back-pressure turbine. Steam traps located on the equipment allow condensate to drain back into the condensate return line, where it flows back to the condensate receiver. Steam traps also perform other functions, such as venting air from the system on startup, which is discussed in more detail later.

The system shown in Figure 6.1 is, of course, highly idealized. In addition to the components shown, other elements, such as strainers, check valves, and pumping traps may be utilized. In some plants, condensate may be simply released to a drain and not returned; the potential for energy conservation through recovery of condensate will be discussed later. Also, in the system shown, the flash steam produced in the process of throttling across the steam traps is vented to the atmosphere. Prevention of this loss represents an excellent opportunity to save Btus and dollars.

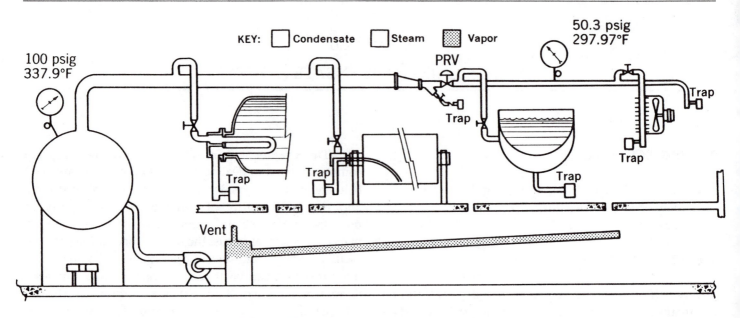

Fig. 6.1 Typical steam system components.

6.1.2 Energy Conservation Opportunities in Steam Systems

Many opportunities for energy savings exist in steam system operations, ranging from simple operating procedure modifications to major retrofits requiring significant capital expenditures. Table 6.1 shows a checklist of energy conservation opportunities applicable to most steam systems. It is helpful for energy conservation managers to maintain such a running list applicable to their own situations. Ideas are frequently presented in the technical and trade literature, and plant operators often make valuable contributions, since after all, they are the people closest to the problem.

To "sell" such improvements to plant management and operating personnel, it is necessary to demonstrate the dollars-and-cents value of a project or operating change. The following sections discuss the thermal properties of steam, how to determine the steam requirements of plant equipment and to estimate the amount of steam required to make up for system losses, how to assign a dollar value to this steam, and various approaches to alleviating these losses.

6.2 THERMAL PROPERTIES OF STEAM

6.2.1 Definitions and Terminology

Before discussing numerical calculations of steam properties for various applications, it is necessary to establish an understanding of some terms commonly used in the operation of steam systems.

British Thermal Unit (Btu). One Btu is the amount of heat required to raise 1 pound of water 1 degree Fahrenheit in temperature. To get a perspective on this quantity, a cubic foot of natural gas at atmospheric pressure will release about 1000 Btu when burned in a boiler with no losses. This same 1000 Btu will produce a little less than 1 pound of steam at atmospheric conditions, starting from tap water.

Boiling Point. The boiling point is the temperature at which water begins to boil at any given pressure. The boiling point of water at sea-level atmospheric pressure is about 212°F. At high altitude, where the atmospheric pressure is lower, the boiling point is also lower. Conversely, the boiling point of water goes up with increasing pressure. In steam systems, we usually refer to the boiling point by another term, "saturation temperature."

Absolute and Gauge Pressure. In steam system literature, we frequently see two different pressures used. The "absolute pressure," designated psia, is the true force per unit of area (e.g., pounds per square inch) exerted by the steam on the wall of the pipe or vessel containing it. We usually measure pressures, however, with sensing devices that are exposed to the atmosphere outside, and which therefore register an indication, not of the true force inside the vessel, but of the difference between that force and the force exerted by the outside atmosphere. We call this difference the "gauge pressure," designated psig. Since atmospheric pressure at sea level is usually around 14.7 psi, we can obtain the absolute pressure by simply adding 14.7 to the gauge pressure reading. In tables of steam

Table 6.1 Checklist of Energy Conservation Opportunities in Steam and Condensate Systems

General Operations

1. Review operation of long steam lines to remote single-service applications. Consider relocation or conversion of remote equipment, such as steam-heated storage tanks.
2. Review operation of steam systems used only for occasional services, such as winter-only steam tracing lines. Consider use of automatic controls, such as temperature-controlled valves, to assure that the systems are used only when needed.
3. Implement a regular steam leak survey and repair program.
4. Publicize to operators and plant maintenance personnel the annual cost of steam leaks and unnecessary equipment operations.
5. Establish a regular steam-use monitoring program, normalized to production rate, to track progress in reduction of steam consumption. Publicize on a monthly basis the results of this monitoring effort.
6. Consider revision of the plant-wide steam balance in multipressure systems to eliminate venting of low-pressure steam. For example, provide electrical backup for currently steam-driven pumps or compressors to permit shutoff of turbines when excess low-pressure steam exists.
7. Check actual steam usage in various operations against theoretical or design requirement. Where significant disparities exist, determine the cause and correct it.
8. Review pressure-level requirements of steam-driven mechanical equipment to evaluate feasibility of using lower pressure levels.
9. Review temperature requirements of heated storage vessels and reduce to minimum acceptable temperatures.
10. Evaluate production scheduling of batch operations and revise if possible to minimize startups and shutdowns.

Steam Trapping

1. Check sizing of all steam traps to assure they are adequately rated to provide proper condensate drainage. Also review types of traps in various services to assure that the most efficient trap is being used for each application.
2. Implement a regular steam trap survey and maintenance program. Train maintenance personnel in techniques for diagnosing trap failure.

Condensate Recovery

1. Survey condensate sources presently being discharged to waste drains for feasibility of condensate recovery.
2. Consider opportunities for flash steam utilization in low-temperature processes presently using first-generation steam.
3. Consider pressurizing atmospheric condensate return systems to minimize flash losses.

Mechanical Drive Turbines

1. Review mechanical drive standby turbines presently left in the idling mode and consider the feasibility of shutting down standby turbines.
2. Implement a steam turbine performance testing program and clean turbines on a regular basis to maximize efficiency.
3. Evaluate the potential for cogeneration in multipressure steam systems presently using large pressure-reducing valves.

Insulation

1. Survey surface temperatures using infrared thermometry or thermography on insulated equipment and piping to locate areas of insulation deterioration. Maintain insulation on a regular basis.
2. Evaluate insulation of all uninsulated lines and fittings previously thought to be uneconomic. Recent rises in energy costs have made insulation of valves, flanges, and small lines desirable in many cases where this was previously unattractive.
3. Survey the economics of retrofitting additional insulation on presently insulated lines, and upgrade insulation if economically feasible.

properties, it is more common to see pressures listed in psia, and hence it is necessary to make the appropriate correction to the pressure indicated on a gauge.

Saturated and Superheated Steam. If we put cold water into a boiler and heat it, its temperature will begin to rise until it reaches the boiling point. If we continue to heat the water, rather than continuing to rise in temperature, it begins to boil and produce steam. As long as the pressure remains constant, the temperature will remain at the saturation temperature for the given pressure, and the more heat we add, the more liquid will be converted to steam. We call this boiling liquid a "saturated liquid" and refer to the steam so generated as "saturated vapor." We can continue to add more and more heat, and we will simply generate more saturated vapor (or simply "saturated steam") until the water is completely boiled off. At this point, if we continue to add heat, the steam temperature will begin to rise once more. We call this "superheated steam." This chapter will concentrate on the behavior of saturated steam, because this is the steam condition most commonly encountered in industrial process heating applications. Superheated steam is common in power generation and is often produced in industrial systems when cogeneration of power and process heat is used.

Sensible and Latent Heat. Heat input that is directly registered as a change in temperature of a substance is called "sensible heat," for the simple reason that we can, in fact, "sense" it with our sense of touch or with a thermometer. For example, the heating of the water mentioned above before it reaches the boiling point would be sensible heating. When the heat goes into the conversion of a liquid to a vapor in boiling, or vice versa in the process of condensation, it is termed "latent heat." Thus when a pound of steam condenses on a heater surface to produce a pound of saturated liquid at the same temperature, we say that it has released its latent heat. If the condensate cools further, it is releasing sensible heat.

Enthalpy. The total energy content of a flowing medium, usually expressed in Btu/lb, is termed its "enthalpy." The enthalpy of steam at any given condition takes into account both latent and sensible heat, and also the "mechanical" energy content reflected in its pressure. Hence steam at 500 psia and 600°F will have a higher enthalpy than steam at the same temperature but at 300 psia. Also, saturated steam at any temperature and pressure has a higher enthalpy than condensate at the same conditions due to the latent heat content of the steam. Enthalpy, as listed in tables of steam properties, does not include the kinetic energy of motion, but this component is insignificant in most energy conservation applications.

Specific Volume. The specific volume of a substance is the amount of space (e.g., cubic feet) occupied by 1 pound of substance. This term will become important in some of our later discussions, because steam normally occupies a much greater volume for a given mass than water (i.e., it has a much greater specific volume), and this must be taken into account when considering the design of condensate return systems.

Condensate. Condensate is the liquid produced when steam condenses on a heater surface. As shown later, this condensate still contains a significant fraction of its energy, and can be returned to the boiler to conserve fuel.

Flash Steam. When hot condensate at its saturation temperature corresponding to the elevated pressure in a heating vessel rapidly drops in pressure, as, for example, when passing through a steam trap or a valve, it suddenly finds itself at a temperature above the saturation temperature for the new pressure. Steam is thus generated which absorbs sufficient energy to drop the temperature of the condensate to the appropriate saturation level. This is called "flash steam," and the pressure-reduction process is called "flashing." In many condensate return systems, flash steam is simply released to the atmosphere, but it may, in fact, have practical applications in energy conservation.

Boiler Efficiency. The boiler efficiency is the percentage of the energy released in the burning of fuel in a boiler which actually goes into the production of steam. The remaining percentage is lost through radiation from the boiler surfaces, blowdown of the boiler water to maintain satisfactory impurity levels, and loss of the hot flue gas up the stack. Although this chapter does go into detail on the subject of boiler efficiency, which is discussed in Chapter 5, it is important to recognize that this parameter relates the energy savings obtainable by conserving steam to the fuel savings obtainable at the boiler, a relation of obvious economic importance. Thus if we save 100 Btu of steam energy and have a boiler with an efficiency of 80%, the actual fuel energy saved would be 100/0.80, or 125 Btu. Because boilers always have an efficiency of less than 100% (and more commonly around 75 to 80%) there is a built-in "amplifier" on any energy savings effected in the steam system.

6.2.2 Properties of Saturated Steam

In calculating the energy savings obtainable through various measures, it is important to understand the quantitative thermal properties of steam and condensate. Table 6.2

Table 6.2 Thermodynamic Properties of Saturated Steam

(1) Gauge Pressure	(2) Absolute Pressure (psia)	(3) Steam Temp. (°F)	(4) Enthalpy of Sat. Liquid (Btu/lb)	(5) Latent Heat (Btu/lb)	(6) Enthalpy of Steam (Btu/lb)	(7) Specific Volume (ft³/lb)
In. vacuum						
29.743	0.08854	32.00	0.00	1075.8	1075.8	3306.00
29.515	0.2	53.14	21.21	1063.8	1085.0	1526.00
27.886	1.0	101.74	69.70	1036.3	1106.0	333.60
19.742	5.0	162.24	130.13	1001.0	1131.1	73.52
9.562	10.0	193.21	161.17	982.1	1143.3	38.42
7.536	11.0	197.75	165.73	979.3	1145.0	35.14
5.490	12.0	201.96	169.96	976.6	1146.6	32.40
3.454	13.0	205.88	173.91	974.2	1148.1	30.06
1.418	14.0	209.56	177.61	971.9	1149.5	28.04
Psig						
0.0	14.696	212.00	180.07	970.3	1150.4	26.80
1.3	16.0	216.32	184.42	967.6	1152.0	24.75
2.3	17.0	219.44	187.56	965.5	1153.1	23.39
5.3	20.0	227.96	196.16	960.1	1156.3	20.09
10.3	25.0	240.07	208.42	952.1	1160.6	16.30
15.3	30.0	250.33	218.82	945.3	1164.1	13.75
20.3	35.0	259.28	227.91	939.2	1167.1	11.90
25.3	40.0	267.25	236.03	933.7	1169.7	10.50
30.3	45.0	274.44	243.36	928.6	1172.0	9.40
40.3	55.0	287.07	256.30	919.6	1175.9	7.79
50.3	65.0	297.97	267.50	911.6	1179.1	6.66
60.3	75.0	307.60	277.43	904.5	1181.9	5.82
70.3	85.0	316.25	286.39	897.8	1184.2	5.17
80.3	95.0	324.12	294.56	891.7	1186.2	4.65
90.3	105.0	331.36	302.10	886.0	1188.1	4.23
100.0	114.7	337.90	308.80	880.0	1188.8	3.88
110.3	125.0	344.33	315.68	875.4	1191.1	3.59
120.3	135.0	350.21	321.85	870.6	1192.4	3.33
125.3	140.0	353.02	324.82	868.2	1193.0	3.22
130.3	145.0	355.76	327.70	865.8	1193.5	3.11
140.3	155.0	360.50	333.24	861.3	1194.6	2.92
150.3	165.0	365.99	338.53	857.1	1195.6	2.75
160.3	175.0	370.75	343.57	852.8	1196.5	2.60
180.3	195.0	379.67	353.10	844.9	1198.0	2.34
200.3	215.0	387.89	361.91	837.4	1199.3	2.13
225.3	240.0	397.37	372.12	828.5	1200.6	1.92
250.3	265.0	406.11	381.60	820.1	1201.7	1.74
	300.0	417.33	393.84	809.0	1202.8	1.54
	400.0	444.59	424.00	780.5	1204.5	1.16
	450.0	456.28	437.20	767.4	1204.6	1.03
	500.0	467.01	449.40	755.0	1204.4	0.93
	600.0	486.21	471.60	731.6	1203.2	0.77
	900.0	531.98	526.60	668.8	1195.4	0.50
	1200.0	567.22	571.70	611.7	1183.4	0.36
	1500.0	596.23	611.60	556.3	1167.9	0.28
	1700.0	613.15	636.30	519.6	1155.9	0.24
	2000.0	635.82	671.70	463.4	1135.1	0.19
	2500.0	668.13	730.60	360.5	1091.1	0.13
	2700.0	679.55	756.20	312.1	1068.3	0.11
	3206.2	705.40	902.70	0.0	902.7	0.05

What is the temperature inside the line? Coming down column (1) we find a pressure of 150, and moving over to column (3), we note that the corresponding steam temperature at this pressure is about 366°F.

Column (4) lists the sensible heat of the saturated liquid in Btu/lb of water. We can see at the head of the column that this sensible heat is designated as 0 at a temperature [column (3)] of 32°F.

shows a typical compilation of the properties of saturated steam.

Columns 1 and 2 list various pressures, either in gauge (psig) or absolute (psia). Note that these two pressures always differ by about 15 psi (14.7 to be more precise). Remember that the former represents the pressure indicated on a normal pressure gauge, while the latter represents the true pressure inside the line. Column 3 shows the saturation temperature corresponding to each of these pressures. Note, for example, that at an absolute pressure of 14.696 (the normal pressure of the atmosphere at sea level) the saturation temperature is 212°F, the figure we are all familiar with. Suppose that we have a pressure of 150 psi indicated on the pressure gauge on a steam line. This is an arbitrary reference point, and therefore the heat indicated at any other temperature tells us the amount of heat added to raise the water from an initial value of 32°F to that temperature. For example, referring back to our 150 psig steam, the water contains about 338.5 Btu/lb: starting from 32°F, 10 lb of water would contain 10 times this number, or about 3385 Btu. We can also subtract one number from another in this column to find the amount of heat necessary to raise the water from one temperature to another. If the water started at 101.74°F, it would contain a heat of 69.7 Btu/lb, and to raise it from this temperature to 366°F would require 338.5—69.7, or about 268.8 Btu for each pound of water. Column (5) shows the latent heat content of a pound of steam for each pressure. For our 150-psig example, we can see that it takes about 857 Btu to convert each pound of saturated water into saturated steam. Note that this is a much larger quantity than the heat content of the water alone, confirming the earlier observation that steam is a very effective carrier of heat; each pound can give up, in this case, 857 Btu when condensed on a surface back to saturated liquid. Column (6), the enthalpy of the saturated steam, represents simply the sum of columns (4) and (5), since each pound of steam contains both the latent heat required to vaporize the water and the sensible heat required to raise the water to the boiling point in the first place.

Column (7) shows the specific volume of the saturated steam at each pressure. Note that as the pressure increases, the steam is compressed; that is, it occupies less space per pound. 150-psig steam occupies only 2.75 ft³/lb; if released to atmospheric pressure (0 psig) it would expand to nearly 10 times this volume. By comparison, saturated liquid at atmospheric pressure has a specific volume of only 0.017 ft³/lb (not shown in the table), and it changes only a few percent over the entire pressure range of interest here. Thus 1 lb of saturated liquid condensate at 212°F will expand more than 1600 times in volume in converting to a vapor. This illustrates that piping systems for the return of condensate from steam-heated equipment must be sized primarily to accommodate the large volume of flashed vapor, and that the volume occupied by the condensate itself is relatively small.

The steam tables can be a valuable tool in estimating energy savings, as illustrated in the following example.

Example: A 100-ft run of 6-in. steam piping carries saturated steam at 95 psig. Tables obtained from an insulation manufacturer indicate that the heat loss from this piping run is presently 110,000 Btu/hr. With proper insulation, the manufacturer's tables indicate that this loss could be reduced to 500 Btu/hr. How many pounds per hour of steam savings does this installation represent, and if the boiler is 80% efficient, what would be the resulting fuel savings?

From the insulation manufacturer's data, we can find the reduction in heat loss:

$$\text{heat-loss reduction} = 110{,}000 - 500 = 109{,}500 \text{ Btu/hr}$$

From Table 6.2 at 95 psig (halfway between 90 and 100), the total heat of the steam is about 1188.4 Btu/lb. The steam savings, is therefore

$$\text{steam savings} = \frac{109{,}500 \text{ Btu/hr}}{1188.4 \text{ Btu/lb}}$$

Assume that condensate is returned to the boiler at around 212°F; thus the condensate has a heat content of about 180 Btu/lb. The heat required to generate 95 psig steam from this condensate is 1188.4 – 180.0 or 1008.4 Btu/lb. If the boiler is 80% efficient, then

$$\text{fuel savings} = \frac{1008.4 \text{ Btu/hr} \times 92 \text{ lb/hr}}{0.80}$$

$$= \text{approximately .116 million Btu/hr}$$

6.2.3 Properties of Superheated Steam

If additional heat is added to saturated steam with no liquid remaining, it begins to superheat and the temperature will rise. Table 6.3 shows the thermodynamic properties of superheated steam. Unlike the saturated steam of Table 6.2, in which each pressure had only a single temperature (the saturation temperature) associated with it, superheated steam may exist, for a given pressure, at any temperature above the saturation temperature. Thus the properties must be tabulated as a function of both temperature and pressure, rather than pressure alone. With this exception, the values in the superheated steam table may be used exactly like those in Table 6.2.

Table 6.3 Thermodynamic Properties of Superheated Steam

Abs. Press (psi)		Temperature (°F)														
		100	200	300	400	500	600	700	800	900	1000	1100	1200	1300	1400	1500
1	v	0.0161	392.5	452.3	511.9	571.5	631.1	690.7								
	h	68.00	1150.2	1195.7	1241.8	1288.6	1336.1	1384.5								
5	v	0.0161	78.14	90.24	102.24	114.21	126.15	138.08	150.01	161.94	173.86	185.78	197.70	209.62	221.53	233.45
	h	68.01	1148.6	1194.8	1241.3	1288.2	1335.9	1384.3	1433.6	1483.7	1534.7	1586.7	1639.6	1693.3	1748.0	1803.5
10	v	0.0161	38.84	44.98	51.03	57.04	63.03	69.00	74.98	80.94	86.91	92.87	98.84	104.80	110.76	116.72
	h	68.02	1146.6	1193.7	1240.6	1287.8	1335.5	1384.0	1433.4	1483.5	1534.6	1586.6	1639.5	1693.3	1747.9	1803.4
15	v	0.0161	0.0166	29.899	33.963	37.985	41.986	45.978	49.964	53.946	57.926	61.905	65.882	69.858	73.833	77.807
	h	68.04	168.09	1192.5	1239.9	1287.3	1335.2	1383.8	1433.2	1483.4	1534.5	1586.5	1639.4	1693.2	1747.8	1803.4
20	v	0.0161	0.0166	22.356	25.428	28.457	31.466	34.465	37.458	40.447	43.435	46.420	49.405	52.388	55.370	58.352
	h	68.05	168.11	1191.4	1239.2	1286.9	1334.9	1383.5	1432.9	1483.2	1534.3	1586.3	1639.3	1693.1	1747.8	1803.3
40	v	0.0161	0.0166	11.036	12.624	14.165	15.685	17.195	18.699	20.199	21.697	23.194	24.689	26.183	27.676	29.168
	h	68.10	168.15	1186.6	1236.4	1285.0	1333.6	1382.5	1432.1	1482.5	1533.7	1585.8	1638.8	1992.7	1747.5	1803.0
60	v	0.0161	0.0166	7.257	8.354	9.400	10.425	11.438	12.466	13.450	14.452	15.452	16.450	17.448	18.445	19.441
	h	68.15	168.20	1181.6	1233.5	1283.2	1332.3	1381.5	1431.3	1481.8	1533.2	1585.3	1638.4	1692.4	1747.1	1802.8
80	v	0.0161	0.0166	0.0175	6.218	7.018	7.794	8.560	9.319	10.075	10.829	11.581	12.331	13.081	13.829	14.577
	h	68.21	168.24	269.74	1230.5	1281.3	1330.9	1380.5	1430.5	1481.1	1532.6	1584.9	1638.0	1692.0	1746.8	1802.5
100	v	0.0161	0.0166	0.0175	4.935	5.558	6.216	6.833	7.443	8.050	8.655	9.258	9.860	10.460	11.060	11.659
	h	68.26	168.29	269.77	1227.4	1279.3	1329.6	1379.5	1429.7	1480.4	1532.0	1584.4	1637.6	1691.6	1746.5	1802.2
120	v	0.0161	0.0166	0.0175	4.0786	4.6341	5.1637	5.6831	6.1928	6.7006	7.2060	7.7096	8.2119	8.7130	9.2134	9.7130
	h	68.31	168.33	269.81	1224.1	1277.4	1328.1	1378.4	1428.8	1479.8	1531.4	1583.9	1637.1	1691.3	1746.2	1802.0
140	v	0.0161	0.0166	0.0175	3.4661	3.9526	4.4119	4.8585	5.2995	5.7364	6.1709	6.6036	7.0349	7.4652	7.8946	8.3233
	h	68.37	168.38	269.85	1220.8	1275.3	1326.8	1377.4	1428.0	1479.1	1530.8	1583.4	1636.7	1690.9	1745.9	1801.7
160	v	0.0161	0.0166	0.0175	3.0060	3.4413	3.8480	4.2420	4.6295	5.0132	5.3945	5.7741	6.1522	6.5293	6.9055	7.2811
	h	68.42	168.42	269.89	1217.4	1273.3	1325.4	1376.4	1427.2	1478.4	1530.3	1582.9	1636.3	1690.5	1745.6	1801.4
180	v	0.0161	0.0166	0.0174	2.6474	3.0433	3.4093	3.7621	4.1084	4.4505	4.7907	5.1289	5.4657	5.8014	6.1363	6.4704
	h	68.47	168.47	269.92	1213.8	1271.2	1324.0	1375.3	1426.3	1477.7	1529.7	1582.4	1635.9	1690.2	1745.3	1801.2
200	v	0.0161	0.0166	0.0174	2.3598	2.7247	3.0583	3.3783	3.6915	4.0008	4.3077	4.6128	4.9165	5.2191	5.5209	5.8219
	h	68.52	168.51	269.96	1210.1	1269.0	1322.6	1374.3	1425.5	1477.0	1529.1	1581.9	1635.4	1689.8	1745.0	1800.9
250	v	0.0161	0.0166	0.0174	0.0186	2.1504	2.4662	2.6872	2.9410	3.1909	3.4382	3.6837	3.9278	4.1709	4.4131	4.6546
	h	68.66	168.63	270.05	375.10	1263.5	1319.0	1371.6	1423.4	1475.3	1527.6	1580.6	1634.4	1688.9	1744.2	1800.2
300	v	0.0161	0.0166	0.0174	0.0186	1.7665	2.0044	2.2263	2.4407	2.6509	2.8585	3.0643	3.2688	3.4721	3.6746	3.8764
	h	68.79	168.74	270.14	375.15	1257.7	1315.2	1368.9	1421.3	1473.6	1526.2	1579.4	1633.3	1688.0	1743.4	1799.6
350	v	0.0161	0.0166	0.0174	0.0186	1.4913	1.7028	1.8970	2.0832	2.2652	2.4445	2.6219	2.7980	2.9730	3.1471	3.3205
	h	68.92	168.85	270.24	375.21	1251.5	1311.4	1366.2	1419.2	1471.8	1524.7	1578.2	1632.3	1687.1	1742.6	1798.9
400	v	0.0161	0.0166	0.0174	00162	1.2841	1.4763	1.6499	1.8151	1.9759	2.1339	2.2901	2.4450	2.5987	2.7515	2.9037
	h	69.05	168.97	270.33	375.27	1245.1	1307.4	1363.4	1417.0	1470.1	1523.3	1576.9	1631.2	1686.2	1741.9	1798.2
500	v	0.0161	0.0166	0.0174	0.0186	0.9919	1.1584	1.3037	1.4397	1.5708	1.6992	1.8256	1.9507	2.0746	2.1977	2.3200
	h	69.32	169.19	270.51	375.38	1231.2	1299.1	1357.7	1412.7	1466.6	1520.3	1574.4	1629.1	1684.4	1740.3	1796.9
600	v	0.0161	0.0166	0.0174	0.0186	0.7944	0.9456	1.0726	1.1892	1.3008	1.4093	1.5160	1.6211	1.7252	1.8284	1.9309
	h	69.58	169.42	270.70	375.49	1215.9	1290.3	1351.8	1408.3	1463.0	1517.4	1571.9	1627.0	1682.6	1738.8	1795.6
700	v	0.0161	0.0166	0.0174	0.0186	0.0204	0.7928	0.9072	1.0102	1.1078	1.2023	1.2948	1.3858	1.4757	1.5647	1.6530
	h	69.84	169.65	270.89	375.61	487.93	1281.0	1345.6	1403.7	1459.4	1514.4	1569.4	1624.8	1680.7	1737.2	1794.3
800	v	0.0161	0.0166	0.0174	0.0186	0.0204	0.6774	0.7828	0.8759	0.9631	1.0470	1.1289	1.2093	1.2885	1.3669	1.4446
	h	70.11	169.88	271.07	375.73	487.88	1271.1	1339.2	1399.1	1455.8	1511.4	1566.9	1622.7	1678.9	1736.0	1792.9
900	v	0.0161	0.0166	0.0174	0.0186	0.0204	0.5869	0.6858	0.7713	0.8504	0.9262	0.9998	1.0720	1.1430	1.2131	1.2825
	h	70.37	170.10	271.26	375.84	487.83	1260.6	1332.7	1394.4	1452.2	1508.5	1564.4	1620.6	1677.1	1734.1	1791.6
1000	v	0.0161	0.0166	0.0174	0.0186	0.0204	0.5137	0.6080	0.6875	0.7603	0.8295	0.8966	0.9622	1.0266	1.0901	1.1529
	h	70.63	170.33	271.44	375.96	487.79	1249.3	1325.9	1389.6	1448.5	1504.4	1561.9	1618.4	167~.3	1732.5	1790.3

Example: Suppose, in the preceding example, that the steam line is carrying superheated steam at 250 psia (235 psig) and 500°F (note that both temperature and pressure must be specified for superheated steam). For the same reduction in heat loss (109,500 Btu/hr), how many pounds per hour of steam is saved?

From Table 6.3 the enthalpy of steam at 250 psia and 500°F is 1263.5 Btu/lb. Thus

$$\text{steam savings} = \frac{109,500 \text{ Btu/hr}}{1263.5 \text{ Btu/lb}}$$

$$= 86.6 \text{ lb/hr}$$

Table 6.4 Orders of Magnitude of Convective Conductances

Heating Process	Order of Magnitude of h (Btu/hr ft^2 · F)
Free convection, air	1
Forced convection, air	5-10
Forced convection, water	250-1000
Condensation, steam	5000-10,000

6.2.4 Heat-Transfer Characteristics of Steam

As mentioned in Section 6.1 steam is one of the most effective heat-transfer media available. The rate of heat transfer from a fluid medium to a solid surface (such as the surface of a heat-exchanger tube or a jacketed heating vessel) can be expressed by Newton's law of cooling:

$$q = h(T_f - T_s)$$

where q is the rate of heat transfer per unit of surface area (e.g., Btu/hr/ft^2), h is a proportionality factor called the "convective conductance," T_f is the temperature of the fluid medium, and T_s is the temperature of the surface.

Table 6.4 shows the order of magnitudes of h for several heat-transfer media. Condensation of steam can be several times as effective as the flow of water over a surface for the transfer of heat, and may be 1000 times more effective than a gaseous heating medium, such as air.

In a heat exchanger, the overall effectiveness must take into account the fluid resistances on both sides of the exchanger and the conduction of heat through the tube wall. These effects are generally lumped into a single "overall conductance," U, defined by the equation

$$q = U(T_{f1} - T_{f2})$$

where q is defined as before, U is the conductance, and T_{f1} and T_{f2} are the temperatures of the two fluids. In addition, there is a tendency for fluids to deposit "fouling layers" of crystalline, particulate, or organic matter on transfer surfaces, which further impede the flow of heat. This impediment is characterized by a "fouling resistance," which, for design purposes, is usually incorporated as an additional factor in determining the overall conductance.

Table 6.5 illustrates typical values of U (not including fouling) and the fouling resistance for exchangers employing steam on the shell side versus exchangers using a light organic liquid (such as a typical heat-transfer oil). The 30 to 50% higher U values for steam translate directly into a proportionate reduction in required heat-exchanger area for the same fluid temperatures. Furthermore, fouling resistances for the steam-heated exchangers are 50 to 100% lower than for a similar service using an organic heating medium, since pure steam contains no contaminants to deposit on the exchanger surface. From the design standpoint, this means that the additional heating surface incorporated to allow for fouling need not be as great. From the operating viewpoint, it translates into energy conservation, since more heat can be transferred per hour in the exchanger for the same fluid conditions, or the same heating duty can be met with a lower fluid temperature difference if the fouling resistance is lower.

6.3 ESTIMATING STEAM USAGE AND ITS VALUE

To properly assess the worth of energy conservation improvements in steam systems, it is first necessary to determine how much steam is actually required to carry out a desired process, how much energy is being wasted through various system losses, and the dollar value of these losses. Such information will be needed to determine the potential gains achievable with insulation, repair or improvement of steam traps, and condensate recovery systems.

Table 6.5 Comparison of Steam and Light Organics as Heat-Exchange Media

Shell-Side Fluid	Tube-Side Fluid	Typical U (Btu/hr ft^2 · °F)	Typical Fouling Resistance (hr ft^2 · °F/Btu)
Steam	Light organic liquid	135-190	0.001
Steam	Heavy organic liquid	45-80	0.002
Light organic liquid	Light organic liquid	100-130	0.002
Light organic liquid	Heavy organic liquid	35-70	0.003

6.3.1 Determining Steam Requirements

Several approaches can be used to determine process steam requirements. In order of increasing reliability, they include the use of steam consumption tables for typical equipment, detailed system energy balances, and direct measurement of steam and/or condensate flows. The choice of which method is to be used depends on how critical the steam-using process is to the plant's overall energy consumption and how the data are to be used.

For applications in which a high degree of accuracy is not required, such as developing rough estimates of the distribution of energy within a plant, steam consumption tables have been developed for various kinds of process equipment. Table 6.6 shows steam consumption tables for a number of typical industrial and commercial applications. To illustrate, suppose that we wish to estimate the steam usage in a soft-drink bottling plant for washing 2000 bottles/min. From the table we see that, typically, a bottle washer uses about 310 lb of 5-psig steam per hour for each 100 bottles/min of capacity. For a washer with a 2000-bottle/min capacity we would, of course, use about 20 times this value, or 6200 lb/hr of steam. Referring to Table 6.2, we see that 5-psig steam has a total enthalpy of about 1156 Btu/lb. The hourly heat usage of this machine, therefore, would be approximately 1156 X 6200, or a little over 7 million Btu/hr. Remember that this is the heat content of the steam itself, not the fuel heat content required at the boiler, since the boiler efficiency has not yet been taken into account.

Note that most of the entries in Table 6.6 show the steam consumption "in use," not the peak steam consumption during all phases of operation. These figures are fairly reliable for equipment that operates on a more-or-less steady basis; however, they may be quite low for batch-processing operations or operations where the load on the equipment fluctuates significantly during its operation. For this reason, steam equipment manufacturers recom-

Table 6.6 Typical Steam Consumption Rates for Industrial and Commercial Equipment

Type of Installation	Description	Typical Pressure (psig)	Steam Consumption in Use (lb/hr)
Bakeries	Dough-room trough, 8 ft long	10	4
	Oven, white bread, 120-ft^2 surface	10	29
Bottle washing	Soft drinks, per 100 bottles/min	5	310
Dairies	Pasteurizer, per 100 gal heated/20 min	15-75	232 (max)
Dishwashers	Dishwashing machine	15-20	60-70
Hospitals	Sterilizers, instrument, per 100 in.^{3}, approx.	40-50	3
	Sterilizers, water, per 10 gal, approx.	40-50	6
	Disinfecting ovens, double door, 50-100 ft^3, per 10 ft^3, approx.	40-50	21
Laundries	Steam irons, each	100	4
	Starch cooker, per 10 gal capacity	100	7
	Laundry presses, per 10-in. length, approx.	100	7
	Tumblers, 40 in., per 10-in. length, approx.	100	38
Plastic molding	Each 12-15 ft^2 platen surface	125	29
Paper manufacture	Corrugators, per 1,000 ft^2	175	29
	Wood pulp paper, per 100 lb of paper	50	372
Restaurants	Standard steam tables, per ft of length	5-20	36
	Steamjacketed kettles, 25 gal of stock	5-20	29
	Steamjacketed kettles, 60 gal of stock	5-20	58
	Warming ovens, per 20 ft^3	5-20	29
Silver mirroring	Average steam tables	5	102
Tire shops	Truck molds, large	100	87
	Passenger molds	100	29

mend that estimated steam consumption values be multiplied by a "factor of safety," typically between 2 and 5, to assure that the equipment will operate properly under peak-load conditions. This can be quite important from the standpoint of energy efficiency. For example, if a steam trap is sized for average load conditions only, during startup or heavy-load operations, condensate will tend to back up into the heating vessel, reducing the effective area for condensation and hence reducing its heating capacity. For steam traps and condensate return lines, the incremental cost of slight oversizing is small, and factors of safety are used to assure that the design is adequately conservative to guarantee rapid removal of the condensate. Gross oversizing of steam traps, however, can also cause excessive steam loss. This point is discussed in more detail in the section on steam traps, where appropriate factors of safety for specific applications are given.

A second, and generally more accurate, approach to estimating steam requirements is by direct energy-balance calculations on the process. A comprehensive discussion of energy balances is beyond the scope of this section: analysis of complex equipment should be undertaken by a specialist. It is, however, possible to determine simple energy balances on equipment involving the heating of a single product.

The energy-balance concept simply states that any energy put into a system with steam must be absorbed by the product and/or the equipment itself, dissipated to the environment, carried out with the product, or carried out in the condensate. Recall that the concepts of sensible and latent heat were discussed in Section 6.2 in the context of heat absorption by water in the production of steam. We can extend this concept to consider heat absorption of any material, such as the heating of air in a dryer or the evaporation of water in the process of condensing milk.

The *sensible* heat requirement of any process is defined in terms of the "specific heat" of the material being heated. Table 6.7 gives specific heats for a number of common

Table 6.7 Specific Heats of Common Materials

Material	Btu/lb·°F	Material	Btu/lb·°F
Solids			
Aluminum	0.22	Iron, cast	0.49
Asbestos	0.20	Lead	0.03
Cement, dry	0.37	Magnesium	0.25
Clay	0.22	Porcelain	0.26
Concrete, stone	0.19	Rubber	0.48
Concrete, cinder	0.18	Silver	0.06
Copper	0.09	Steel	0.12
Glass, common	0.20	Tin	0.05
Ice, 32°F	0.49	Wood	0.32-0.48
Liquids			
Acetone	0.51	Milk	0.90
Alcohol, methyl, 60-70°F	0.60	Naphthalene	0.41
Ammonia, 104°F	1.16	Petroleum	0.51
Ethylene glycol	0.53	Soybean oil	0.47
Fuel oil, sp. gr. 86	0.45	Tomato juice	0.95
Glycerine	0.58	Water	1.00
Gases			
(Constant-Pressure Specific Heats)			
Acetone	0.35	Carbon dioxide	0.20
Air, dry, 32-392°F	0.24	Methane	0.59
Alcohol	0.45	Nitrogen	0.24
Ammonia	0.54	Oxygen	0.22

substances. The specific heat specifies the number of Btu required to raise 1 pound of a substance through a temperature rise of 1°F. Remember that for water, we stated that 1 Btu was, by definition, the amount of heat required to raise 1 lb by 1°F. The specific heat of water, therefore, is exactly 1 (at least near normal ambient conditions). To see how the specific heat can be used to calculate steam requirements for the sensible heating of products, consider the following example.

Example: A paint dryer requires about 3000 cfm of 200°F air, which is heated in a steam-coil unit. How many pounds of 50-psig steam does this unit require per hour?

The density of air at temperatures of several hundred degrees or below is about 0.075 lb/ft^3. The number of pounds of air passing through the dryer is then

3000 ft^3/min X 60 min/hr X 0.075 lb/ft^3= 13,500 lb/hr

Suppose that the air enters the steam-coil unit at 70°F. Its temperature will then be raised by 200 – 70 = 130°F. From Table 6.7 the specific heat of air is 0.24 Btu/lb °F. The energy required to provide this temperature rise is, therefore,

13,500 lb/hr X 0.24 Btu/lb °F X 130°F = 421,200 Btu/hr

Employing the energy-balance principle, whatever energy is absorbed by the product (air) must be provided by an equal quantity of steam energy, less the energy contained in the condensate.

From Table 6.2 the total enthalpy per pound of 50-psig steam is about 1179.1 Btu, and the condensate (saturated liquid) has an enthalpy of 267.5 Btu/lb. Each pound of steam therefore gives up 1179.1 – 267.5, or 911.6 Btu (i.e., its latent heat) to the air. The steam required is, therefore, just

$$\frac{\text{heat required by air per hour}}{\text{heat released per pound of steam}} = \frac{421,200}{911.6} = 462 \text{ lb/hr}$$

To illustrate how latent heat comes into play when considering the steam requirements of a process, consider another example, this time involving a steam-heat evaporator.

Example: A milk evaporator uses a steam jacketed kettle, in which milk is batch-processed at atmospheric pressure. The kettle has a 1500-lb per batch capacity. Milk is heated from a temperature of 80°F to 212°F, where 25% of its mass is then driven off as vapor. Determine the amount of 15-psig steam required per batch, not including the heating of

the kettle itself.

We must first heat the milk from 80°F to 212°F (sensible heating) and then evaporate off 0.25 X 1500 = 375 lb of water.

From Table 6.7, the specific heat of milk is 0.90 Btu/lb °F. The sensible heat requirement is, therefore,

0.90 X 1500 lb X (212 – 80)°F = 178,200 Btu/batch

In addition, we must provide the latent heat to vaporize 375 lb of water at 212°F. From Table 6.2, 970.3 Btu/lb is required. The total latent heat is, therefore,

375 X 970.3 = 363,863 Btu/batch

and the total heat input is

363,863 + 178,200 = 542,063 Btu/batch

This heat must be supplied as the latent heat of 15-psig steam, which, from Table 6.2, is about 945 Btu/lb. The total steam requirement is, then,

$$\frac{542,063 \text{ Btu/batch}}{945 \text{ Btu/lb}} = 574 \text{ lb of 15-psig steam per batch}$$

We could have also determined the startup requirement to heat the steel kettle, if we could estimate its weight, using the specific heat of 0.12 Btu/lb °F for steel, as shown in Table 6.7.

6.3.2 Estimating Surface and Leakage Losses

In addition to the steam required to actually carry out a process, heat is lost through the surfaces of pipes, storage tanks, and jacketed heater surfaces, and steam is lost through malfunctioning steam traps, and leaks in flanges, valves, and other fittings. Estimation of these losses is important, because fixing them can often be the most cost-effective energy conservation measure available.

Figure 6.2 illustrates the annual heat loss, based on 24-hr/day, 365-day/yr operation, for bare steam lines at various pressures. The figure shows, for example, that a 100-ft run of 6-in. line operating at 100 psig will lose about 1400 million Btu/yr. The economic return on an insulation retrofit can easily be determined with price data obtained from an insulation contractor.

Figure 6.3 can be used to estimate heat losses from flat surfaces at elevated temperatures, or from already insulated piping runs for which the outside jacket surface temperature is known. The figure shows the heat flow per hour per square foot of exposed surface area as a function of the

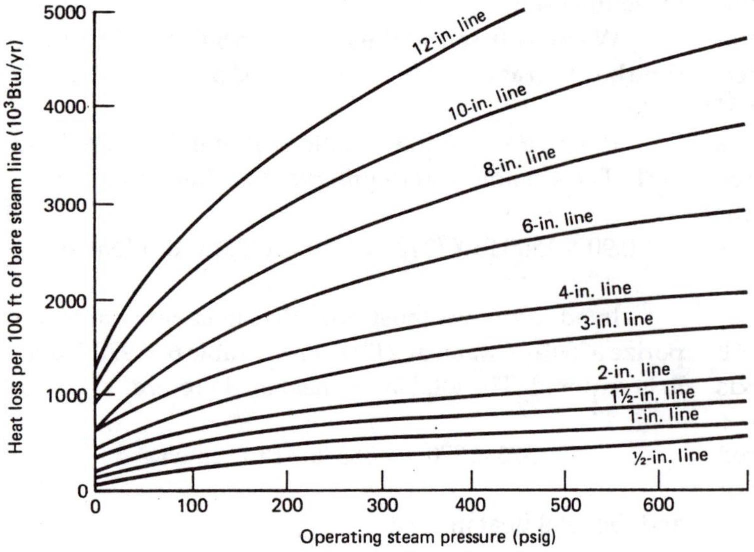

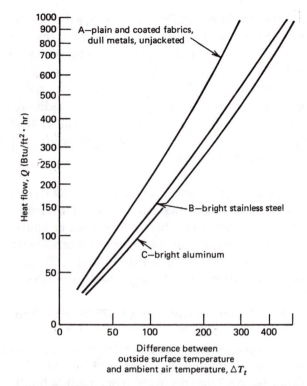

Fig. 6.2 Heat loss from bare steam lines.

Fig. 6.3 Heat losses from surfaces at elevated temperatures.

difference in temperature between the surface and the surrounding air. It will be noted that the nature of the surface significantly affects the magnitude of the heat loss. This is because thermal radiation, which is strongly dependent on the character of the radiating surface, plays an important role in heat loss at elevated temperature, as does convective heat loss to the air.

Another important source of energy loss in steam systems is leakage from components such as loose flanges or malfunctioning steam traps. Figure 6.4 permits estimation of this loss of steam at various pressures leaking through holes. The heat losses are represented in million Btu/yr, based on full-time operation. Using the figure, we can see that a stuck-open steam trap with a 1/8-in. orifice would waste about 600 million Btu/yr of steam energy when leaking from a 100-psig line. This figure can also be used to estimate magnitudes of leakage from other sources of more complicated geometry. It is necessary to first determine an approximate area of leakage (in square inches) and then calculate the equivalent hole diameter represented by that area. The following example illustrates this calculation.

Example: A flange on a 200-psig steam line has a leaking gasket. The maintenance crew, looking at the gasket, estimates that it is about 0.020 in. thick and that it is leaking from about 1/8-in. of the periphery of the flange. Estimate the annual heat loss in the steam if the line is operational 8000 hr/yr.

The area of the leak is a rectangle 0.020 in. wide and 1/8-in. in length:

$$\text{leak area} = 0.020 \times 1/8 = 0.0025 \text{ in.}^2$$

An equivalent circle will have an area of $\pi D_2/4$, so if $\pi D_2/4 = 0.0025$, then $D = 0.056$ in. From Figure 6.4, this leak, if occurring year-round (8760 hr), would waste about 200 million Btu/yr of steam energy. For actual operation

$$\text{heat lost} = \frac{8000}{8760} \times 200,000,000 = 182.6 \text{ million Btu/yr}$$

Since the boiler efficiency has not been considered, the actual fuel wastage would be about 25% greater.

6.3.3 Measuring Steam and Condensate Rates

In maintaining steam systems at peak efficiency, it is often desirable to monitor the rate of steam flow through the system continuously, particularly at points of major usage, such as steam mains. Figure 6.5 shows one of the most common types of flowmetering device, the calibrated orifice. This is a sharp-edged restriction that causes the steam flow to "neck down" and then re-expand after passing through the orifice. As the steam accelerates to pass through the restriction, its pressure drops, and this pressure drop, if measured, can be easily related to the flow rate. The calibrated orifice is one of a class of devices known as obstruction flowmeters, all of which work on the same principle of restricting the flow and producing a measurable pressure drop. Other types of obstruction meters are

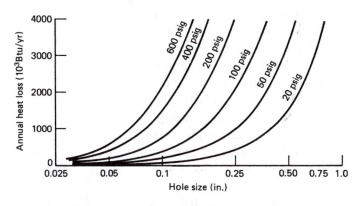

Fig. 6.4 Heat loss from steam leaks.

Fig. 6.5 Orifice flowmeter.

the ASME standard nozzle and the venturi. Orifices, although simple to manufacture and relatively easy to install, between flanges for example, are also subject to wear, which causes them, over a period of time, to give unreliable readings. Nozzles and venturis, although more expensive initially, tend to be more resistant to erosion and wear, and also produce less permanent pressure drop once the steam re-expands to fill the pipe. With all of these devices, care must be exercised in the installation, since turbulence and flow irregularities produced by valves, elbows, and fittings immediately upstream of the obstruction will produce erroneous readings.

Figure 6.6 shows another type of flowmetering device used for steam, called an annular averaging element. The annular element principle is somewhat different from the devices discussed above; it averages the pressure produced when steam impacts on the holes facing into the flow direction, and subtracts from this average impact pressure a static pressure sensed by a tube facing downstream. As with obstruction-type flowmeters, the flow must be related to this pressure difference.

A device that does not utilize pressure drop for steam metering applications is the vortex shedding flowmeter, illustrated in Figure 6.7. A solid bar extends through the flow, and as steam flows around the bar, vortices are shed alternately from one side to the other in its wake, As the vortices shift from side to side, the frequency of shedding can be detected with a thermal or magnetic detector, and this frequency varies directly with the rate of flow. The vortex shedding meter is quite rugged, since the only function of the ob-

ject extending into the flow stream is to provide an obstruction to generate vortices; hence it can be made of heavy-duty stainless steel. Also, vortex shedding meters tend to be relatively insensitive to variations in the steam properties, since they produce a pulsed output rather than an analog signal.

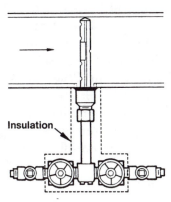

Fig. 6.6 Annular averaging element.

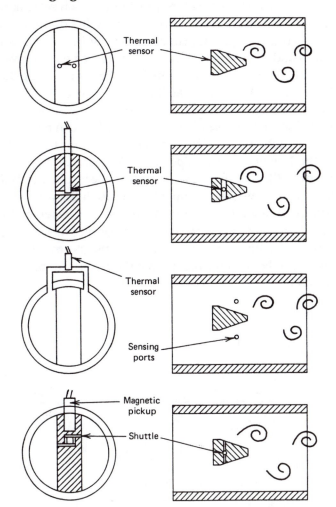

Fig. 6.7 Vortex shedding flowmeter with various methods for sensing fluctuations.

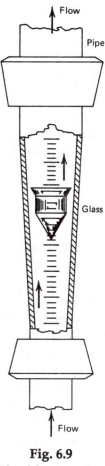

The target flowmeter, not shown, is also suitable for some steam applications. This type of meter uses a "target," such as a small cylinder, mounted on the end of a metal strut that extends into the flow line. The strut is gauged to measure the force on the target, and if the properties of the fluid are accurately known, this force can be related to flow velocity. The target meter is especially useful when only intermittent measurements are needed, as the unit can be "hot-tapped" in a pipe through a ball valve and withdrawn when not in use. The requirement of accurate property data limits the usefulness of this type of meter in situations where steam conditions vary considerably, especially where high moisture is present.

The devices discussed above are useful in permanent installations where it is desired to continuously or periodically measure steam flow; there is no simple way to directly measure steam flow on a spotcheck basis without cutting into the system. There is, however, a relatively simple indirect method, illustrated in Figure 6.8, for determining the rate of steam usage in systems with unpressurized condensate return lines, or open systems in which condensate is dumped to a drain. If a drain line is installed after the trap, condensate may be caught in a barrel and the weight of a sample measured over a given period of time. Precautions must be taken when using this technique to assure that the flash steam, generated when the condensate drops in pressure as it passes through the trap, does not bubble out of the barrel. This can represent both a safety hazard and an error in the measurements due to the loss of mass in vapor form. The barrel should be partially filled with cold water prior to the test so that flash steam will condense as it bubbles through the water. An energy balance can also be made on the water at the beginning and end of the test by measuring its temperature, and with proper application of the steam tables, a check can be made to assure that the trap is not blowing through.

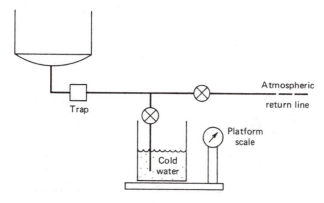

Fig. 6.8
Weigh bucket technique for condensate measurement.

Figure 6.9 illustrates another instrument which can be used to monitor condensate flow on a regular basis, the rotameter. A rotameter indicates the flow rate of the liquid by the level of a specially shaped float which rises in a calibrated glass tube, such that its weight exactly balances the drag force of the flowing condensate.

Measurements of this type can be very useful in monitoring system performance, since any unusual change in steam or condensate rate not associated with a corresponding change in production rate would tend to indicate an equipment malfunction producing poor efficiency.

6.3.4 Computing the Dollar Value of Steam

In analyzing energy conservation measures for steam systems, it is important to establish a steam value in dollars per pound: this value will depend on the steam pressure, the boiler efficiency, and the price of fuel.

Fig. 6.9
Liquid rotameter.

Steam may be valued from two points of view. The more common approach, termed the "enthalpy method," takes into account only the heating capability of the steam, and is most appropriate when steam is used primarily for process heating. The enthalpy method can be illustrated with an example.

Example: An oil refinery produces 200-psig saturated steam in a large boiler, some of which is used directly in high-temperature processes, and some of which is let down to 30 psig through regulating valves for use at lower temperatures. The feedwater is added to the boiler at about 160°F. The boiler efficiency has been determined to be 82%, and boiler fuel is priced at $2.20 million Btu. Establish the values of 200-psig and 30-psig steam ($/lb).

Using the enthalpy method, we determine the increased heat content for each steam pressure required from the boiler if feedwater enters at 160°F. From Table 6.2:

total enthalpy of steam at 200 psig
= 1199.3 Btu/lb
total enthalpy of steam at 30 psig
= 1172 Btu/lb

enthalpy of feedwater at 160°F
> = approx. 130 Btu/lb (about the same
> as saturated liquid at 160°F)

heat added per lb of 200-psig steam
> = 1199.3 − 130 = 1069.3 Btu/lb

heat added per lb of 30-psig steam
> = 1042 Btu/lb

fuel Btu required per pound at 200 psig
> = 1069.3/0.82 = 1304 Btu

fuel Btu required per lb at 30 psig
> = 1042/0.82 = 1271 Btu

With boiler fuel priced at $2.20/million Btu,
> or.22 per 1000 Btu:

value of 200-psig steam
> = 0.22 X 1.304 =.29/lb or $2.90/1000 lb

value of 30-psig steam
> = 0.22 X 1.271 =.28/lb or $2.80/1000 lb

Another approach to the valuation of steam is termed the "availability" or "entropy" method and takes into account not only the heat content of the steam, but also its power-producing potential if it were expanded through a steam turbine. This method is most applicable in plants where cogeneration (the sequential generation of power and use of steam for process heat) is practiced. The availability method involves some fairly complex thermodynamic reasoning and will not be covered here. The reader should, however, be aware of its existence in analyzing the economics of cogeneration systems, in which there is an interchangeability between purchased electric power and in-plant power and steam generation.

6.4 STEAM TRAPS AND THEIR APPLICATION

6.4.1 Functions of Steam Traps

Steam traps are important elements of steam and condensate systems, and may represent a major energy conservation opportunity (or problem, as the case may be). The basic function of a steam trap is to allow condensate formed in the heating process to be drained from the equipment. This must be done speedily to prevent backup of condensate in the system.

Inefficient removal of condensate produces two adverse effects. First, if condensate is allowed to back up in the steam chamber, it cools below the steam temperature as it gives up sensible heat to the process and reduces the effective potential for heat transfer. Since condensing steam is a much more effective heat-transfer medium than stagnant liquid, the area for condensation is reduced, and the

efficiency of the heat-transfer process is deteriorated. This results in longer cycle times for batch processes or lower throughput rates in continuous heating processes. In either case, inefficient condensate removal almost always increases the amount of energy required by the process.

A second reason for efficient removal of condensate is the avoidance of "water hammer" in steam systems. This phenomenon occurs when slugs of liquid become trapped between steam packets in a line. The steam, which has a much larger specific volume, can accelerate these slugs to high velocity, and when they impact on an obstruction, such as a valve or an elbow, they produce an impact force not unlike hitting the element with a hammer (hence the term). Water hammer can be extremely damaging to equipment, and proper design of trapping systems to avoid it is necessary.

The second crucial function of a steam trap is to facilitate the removal of air from the steam space. Air can leak into the steam system when it is shut down, and some gas is always liberated from the water in the boiling process and carried through the steam lines. Air mixed with steam occupies some of the volume that would otherwise be filled by the steam itself. Each of these components, air and steam, contributes its share to the total pressure exerted in the system; it is a fundamental thermodynamic principle that, in a mixture of gases, each component contributes to the pressure in the same proportion as its share of the volume of the space. For example, consider a steam system at 100 psia (note that in this case it is necessary to use *absolute* pressures), with 10% of the volume air instead of steam. Therefore, from thermodynamics, 10% of the pressure, or 10 psia, is contributed by the air, and only 90%, or 90 psia, by the steam. Referring to Table 6.2, the corresponding steam temperature is between 316 and 324°F, or approximately 320°F. If the air were not present, the steam pressure would be 100 psia, corresponding to a temperature of about 328°F, so the presence of air in the system reduces the temperature for heat transfer. This means that more steam must be generated to do a given heating job. Table 6.8 indicates the temperature reduction caused by the presence of air in various quantities at given pressures (shown in psig), and shows that the effective temperature may be seriously degraded.

In actual operation the situation is usually even worse than indicated in Table 6.8. We have considered the temperature reduction assuming that the air and steam are uniformly mixed. In fact, on a real heating surface, as air and steam move adjacent to the surface, the steam is condensed out into a liquid, while the air stays behind in the form of vapor. In the region very near the surface, therefore, the air occupies an even larger fraction of the volume than in the steam space as a whole, acting effectively as an insulating blanket on the surface. Suffice it it say that air is an undesirable parasite in steam systems, and its removal is important for proper operation.

Oxygen and carbon dioxide, in particular, have another adverse effect, and this is corrosion in condensate and steam lines. Oxygen in condensate produces pitting or rusting of the surface, which can contaminate the water, making it undesirable as boiler feed, and CO_2 in solution with water forms carbonic acid, which is highly corrosive to metallic surfaces. These components must be removed from the system, partially by good steam trapping and partially by proper deaeration of condensate, as is discussed in a subsequent section.

6.4.2 Types of Steam Traps and Their Selection

Various types of steam traps are available on the market, and the selection of the best trap for a given application is an important one. Many manufacturers produce several types of traps for specific applications, and manufacturers' representatives should be consulted in arriving at a choice. This section will give a brief introduction to the subject and comment on its relevance to improved energy utilization in steam systems.

Steam traps may be generally classified into three groups: mechanical traps, which work on the basis of the density difference between condensate and steam or air; thermostatic, which use the difference in temperature between steam, which stays close to its saturation temperature, and condensate, which cools rapidly; and thermodynamic, which functions on the difference in flow properties between liquids and vapors.

Figures 6.10 and 6.11 show two types of mechanical traps in common use for industrial applications. Figure 6.10 illustrates the principle of the "bucket trap." In the trap illustrated, an inverted bucket is placed over the inlet line, inside an external chamber. The bucket is attached to a lever arm which opens and closes a valve as the bucket rises and falls in the chamber. As long as condensate flows through the system, the bucket has a negative buoyancy, since liquid is present both inside and outside the bucket. The valve is open and condensate is allowed to drain continuously to the return line. As steam enters the trap it fills the bucket, displacing condensate, and the bucket rises, closing off the valve. Noncondensable gases, such as air and CO_2, bubble through a small vent hole and collect at the top of the trap, to be swept out with flash steam the next time the valve opens. Steam may also leak through the vent, but it is condensed on contact with the cool chamber walls, and collects as condensate in the chamber. The vent hole is quite small, so the rate of steam loss through this leakage action is not excessive. As condensate again begins

to enter the bucket, it loses buoyancy and begins to drop until the valve opens and again discharges condensate and trapped air.

Table 6.8 Temperature Reduction Caused by Air in Steam Systems

Pressure (psig)	Temp. of Steam, No Air Present	Temp. of Steam Mixed with Various Percentages of Air (by Volume)		
		10	20	30
10	240.1	234.3	228.0	220.9
25	267.3	261.0	254.1	246.4
50	298.0	291.0	283.5	275.1
75	320.3	312.9	204.8	295.9
100	338.1	330.3	321.8	312.4

The float-and-thermostatic (F&T) trap, illustrated in Figure 6.11, works on a similar principle. In this case, instead of a bucket, a buoyant float rises and falls in the chamber as condensate enters or is discharged. The float is

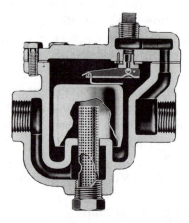

Fig. 6.10 Inverted bucket trap.

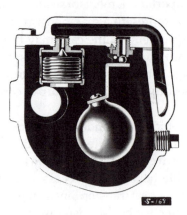

Fig. 6.11 Float and thermostatic steam trap.

attached to a valve, similar to the one in a bucket trap, which opens and closes as the ball rises and falls. Since there is no natural vent in this trap and the ball cannot distinguish between air and steam, which have similar densities, special provision must be made to remove air and other gases from the system. This is usually done by incorporating a small thermostatically actuated valve in the top of the trap. At low temperature, the valve bellows contracts, opening the vent and allowing air to be discharged to the return line. When steam enters the chamber, the bellows expands, sealing the vent. Some float traps are also available without this thermostatic air-vent feature; external provision must then be provided to permit proper air removal from the system. The F&T-type trap permits continuous discharge of condensate, unlike the bucket trap, which is intermittent. This can be an advantage in certain applications.

Figure 6.12 illustrates a thermostatic steam trap. In this trap, a temperature-sensitive bellows expands and contracts in response to the temperature of the fluid in the chamber surrounding the bellows. When condensate surrounds the bellows, it contracts, opening the drain port. As steam enters the chamber, the elevated temperature causes the bellows to expand and seal the drain. Since air also enters the chamber at a temperature lower than that of steam, the thermostatic trap is naturally self-venting, and it also is a continuous-drain-type trap. The bellows in the trap can be partially filled with a fluid and sealed, such that an internal pressure is produced which counterbalances the external pressure imposed by the steam. This feature makes the bellows-type thermostatic trap somewhat self-compensating for variations in steam pressure. Another type of thermostatic trap, which uses a bimetallic element, is also available. This type of trap is not well suited for applications in which significant variations in steam pressure might be expected, since it is responsive only to temperature changes in the system.

The thermodynamic, or controlled disk steam trap is shown in Figure 6.13. This type of trap is very simple in construction and can be made quite compact and resistant to damage from water hammer. In a thermodynamic trap, a small disk covers the inlet orifice. Condensate or air, moving at relatively low velocity, lifts the disk off its seat and is passed through to the outlet drain. When steam enters the trap, it passes through at high velocity because of its large volume. As the steam passes through the space between the disk and its seat, it impacts on the walls of the control chamber to produce a rise in pressure. This pressure imbalance between the outside of the disk and the side facing the seat causes it to snap shut, sealing off the chamber and preventing the further passage of steam to the outlet. When condensate again enters the inlet side, the disk lifts off the seat and permits its release.

An alternative to conventional steam traps, the drain orifice is illustrated in Figure 6.14. This device consists simply of an obstruction to the flow of condensate, similar to the orifice flowmeter described in an earlier section but much smaller. This small hole allows the pressure in the steam system to force condensate to drain continuously into the lower-pressure return system. Obviously, if steam enters, rather than condensate, it will also pass through the orifice and be lost. The strategy of using drain orifices is to select an orifice size that permits condensate to drain at

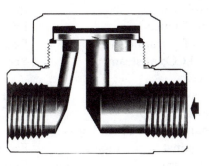

Fig. 6.13 Disk or thermodynamic steam trap.

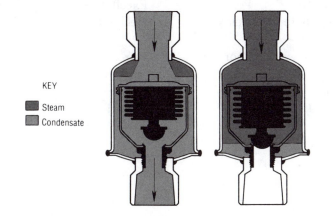

KEY

■ Steam
■ Condensate

Fig. 6.12 Thermostatic steam trap.

FLOW

Fig. 6.14 Drain orifice.

such a rate that live steam seldom enters the system. Even if steam does occasionally pass through, the small size of the orifice limits the steam leakage rate to a value much less than would be lost due to a "stuck-open" malfunction of one of the types of traps discussed above. Drain orifices can be successfully applied in systems that have a well-defined and relatively constant condensate load. They are not suited for use where condensate load may vary widely with operating conditions.

As mentioned above, a number of operating requirements must be taken into consideration in selecting the appropriate trap for a given application. Table 6.9 lists these application considerations and presents one manufacturer's ratings on the performance of the various traps discussed above. In selecting a trap for a given application, assistance from manufacturers' representatives

should be obtained, since a great body of experience in actual service has been accumulated over the years.

6.4.3 Considerations in Steam Trap Sizing

As mentioned earlier in this section, good energy conservation practice demands the efficient removal of condensate from process equipment. It is thus necessary to assure that traps are properly sized for the given condensate load. Grossly oversized traps waste steam by excessive surface heat loss and internal venting, while undersized traps permit accumulation of condensate with resultant loss in equipment heat-transfer effectiveness.

Steam traps are sized based in two specifications, the condensate load (e.g., in lbs/hr or gal/min) and the pressure differential across the trap (in psig). Section 6.3 dis-

Table 6.9 Comparison of Steam Trap Characteristics

Characteristic	Inverted Bucket	F&T	Disk	Bellows Thermostatic
Method of operation	Intermittent	Continuous	Intermittent	Continuous[a]
Energy conservation (time in service)	Excellent	Good	Poor	Fair
Resistance to wear	Excellent	Good	Poor	Fair
Corrosion resistance	Excellent	Good	Excellent	Good
Resistance to Hydraulic shock	Excellent	Poor	Excellent	Poor
Vents air and CO_2 at steam temperature	Yes	No	No	No
Ability to vent air at very low pressure (1/4 psig)	Poor	Excellent	NR[b]	Good
Ability to handle start-up air loads	Fair	Excellent	Poor	Excellent
Operation against back pressure	Excellent	Excellent	Poor	Excellent
Resistance to damage from freezing[c]	Good	Poor	Good	Good
Ability to purge system	Excellent	Fair	Excellent	Good
Performance on very light loads	Excellent	Excellent	Poor	Excellent
Responsiveness to slugs of condensate	Immediate	Immediate	Delayed	Delayed
Ability to handle dirt	Excellent	Poor	Poor	Fair
Comparative physical size	Large[d]	Large	Small	Small
Ability to handle "flash steam"	Fair	Poor	Poor	Poor
Mechanical failure (open-closed)	Open	Closed	Open[e]	Closed[f]

[a]Can be intermittent on low load.

[b] Not recommended for low-pressure operations.

[c]Cast iron traps not recommended.

[d]In welded stainless steel construction—medium..

[e]Can fail closed due to dirt.

[f]Can fail open due to wear.

cussed various methods for estimating condensate loads expected under normal operating conditions.

It is good practice to size the capacity of the trap based on this expected load times a factor of safety to account for peaks at startup and fluctuations in normal operating conditions. It is not unusual for startup condensate loads to be three to four times higher than steady operational loads, and in some applications they may range up to 10 times the steady-state load.

Table 6.10 presents typical factors of safety for condensate capacity recommended by steam trap manufacturers. This indicates typical ranges of factor of safety to consider in various applications. Although there is considerable variation in the recommended values, in both energy and economic terms the cost of oversizing is ordinarily not prohibitive, and conservative safety factors are usually used. The exception to this rule of thumb is in the sizing of disk-type traps, which may not function properly if loaded considerably below design. Drain orifices also must be sized close to normal operating loads. Again, the advice of the manufacturer should be solicited for the specific application in mind.

Table 6.10 Typical Factors of Safety for Steam Traps (Condensate Flow Basis)

Application	Factor of Safety
Autoclaves	3-4
Blast coils	3-4
Dry cans	2-3
Dryers	3-4
Dry kilns	3-4
Fan system heating service	3-4
Greenhouse coils	3-4
Hospital equipment	2-3
Hot-water heaters	4-6
Kitchen equipment	2-3
Paper machines	3-4
Pipe coils (in still air)	3-4
Platen presses	2-3
Purifiers	3-4
Separators	3-4
Steamjacketed kettles	4-5
Steam mains	3-4
Submerged surfaces	5-6
Tracer lines	2-3
Unit heaters	3-4

The other important design specification is the pressure differential over which the trap will operate. Since pressure is the driving force that moves condensate through

the trap and on to the receiver, trap capacity will increase, for a given trap size, as the pressure increases. The trap operating-pressure differential is not simply the boiler pressure. On the upstream side of the trap, steam pressure may drop through valves and fittings and through heat-transfer passages in the process equipment. Thus the appropriate upstream pressure is the pressure at the trap inlet, which to a reasonable approximation, can usually be considered to be the process steam pressure at the equipment. Back pressure on the outlet side of the trap must also be considered. This includes the receiver pressure (if the condensate return system is pressurized), the pressure drop associated with flash steam and condensate flow through the return lines, and the head of water associated with risers if the trap is located at a point below the condensate receiver. Condensate return lines are usually sized for a given capacity to maintain a velocity no greater than 5000 ft/min of the flash steam. Table 6.11 shows the expected pressure drop per 100 ft of return line which can be expected under design conditions. Referring to the table, a 60-psig system, for example, returning condensate to an unpressurized receiver (0 psig) through a 2-in. line, would have a return-line pressure drop of just under 2 psi/100-ft run, and the condensate capacity of the line would be about 2600 lb/hr. The pressure head produced by a vertical column of water is about 1 psi/2 ft of rise. These components can be summed to estimate the back pressure on the system, and the appropriate pressure for sizing the trap is then the difference between the upstream pressure and the back pressure.

Table 6.12 shows a typical pressure-capacity table extracted from a manufacturer's catalog. The use of such a table can be illustrated with the following example.

Example: A steamjacketed platen press in a plastic lamination operation uses about 500 lb/hr of 30-psig steam in normal operation. A 100-ft run of 1-in. pipe returns condensate to a receiver pressurized to 5 psig: the receiver is located 15 ft above the level of the trap. From the capacity differential pressure specifications in Table 6.12, select a suitable trap for this application.

Using a factor of safety of 3 from Table 6.10, a trap capable of handling 3 × 500 = 1500 lb/hr of condensate will be selected.

To determine the system back pressure, add the receiver pressure, the piping pressure drop, and the hydraulic head due to the elevation of the receiver.

Entering Table 6.11 at 30-psig supply pressure and 5-psig return pressure, the pipe pressure drop for a 1-in. pipe is just slightly over 2 psi at a condensate rate somewhat higher than our 1500 lb/hr; a 2-psi pressure drop is a reasonable estimate.

Table 6.11 Condensate Capacities and Pressure Drops for Return Lines[a]

Supply Press. (psig)	5	15		30			60				100					200					
Return Press. (Psig)	0	0	5	0	5	10	0	5	10	20	0	5	10	20	30	0	5	10	20	30	50
Pipe size (in.), schedule 40 pipe																					
1/2	1,425	590	1,335	360	640	1,065	235	370	535	1,010	180	270	370	615	955	115	165	215	325	450	760
	4.0	4.0	5.3	4.0	5.3	6.5	4.0	5.3	6.5	8.9	4.0	5.3	6.5	8.9	11.3	4.0	5.3	6.5	8.9	11.3	15.9
3/4	2,495	1,035	2,340	635	1,125	1,855	415	650	940	1770	310	470	645	1,085	1,675	200	285	375	570	795	1,330
	2.35	2.35	3.14	2.35	3.14	3.88	2.35	3.14	3.88	5.32	2.35	3.14	3.88	5.32	6.72	2.35	3.14	3.88	5.32	6.72	9.40
1	4,045	1,880	3,790	1,030	1,820	3,005	670	1,055	1,520	2,865	505	765	1,045	1,755	2,715	325	465	605	925	1.285	2,155
	1.53	1.53	2.04	1.53	2.04	2.51	1.53	2.04	2.51	3.44	1.63	2.04	2.51	3.44	4.36	1.53	2.04	2.51	3.44	4.36	6.15
1-1/4	7,000	2,905	6,565	1,780	3,150	5,200	1,155	1,830	2,635	4,960	875	1,320	1,810	3,035	4,695	560	800	1,050	1,600	2,225	3,735
	0.95	0.95	1.26	.95	1.26	1.55	.95	1.26	1.55	2.13	95	1.26	1.55	2.13	2.69	0.95	1.26	1.55	2.13	2.69	3.80
1-1/2	9,530	3,955	8,935	2,425	4,290	7.080	1,575	2,490	3,585	6,750	2,190	1,795	2,465	4,135	6,395	760	1,090	1,430	2,175	3,025	5,080
	0.73	0.73	0.97	0.73	0.97	1.20	0.73	0.97	1.20	1.64	0.73	0.97	1.20	1.64	2.07	0.73	0.97	1.20	1.64	2.07	2.93
2	15,710	6,525	14,725	3,995	7,070	11,670	2,595	4,105	5,910	11,125	1,985	2.960	4,060	6,810	10,540	1,255	1,800	2,355	3,585	4,990	8,375
	0.48	0.48	0.64	0.48	0.64	0.79	0.48	0.64	0.79	1.08	0.48	0.64	0.79	1.08	1.37	0.48	0.64	0.79	1.08	1.37	1.93
2-1/2	22,415	9,305	21,005	5,700	10,085	16,650	3,705	5,855	8,430	15,875	2,800	4,225	5,795	9,720	15,035	1,790	2,565	3,380	5,115	7,120	11,950
	0.36	0.36	0.48	0.36	0.48	0.59	0.36	0.48	0.69	0.81	0.36	0.48	0.59	0.81	1.03	0.36	0.48	0.59	0.91	1.03	1.45
3	34,610	14,370	32,435	8,800	15,570	25,710	5,720	9,045	13,020	24515	4,325	6,525	8,950	15,005	23,220	2,765	3,965	5.185	7,900	10,990	18,450
	0.26	0.26	0.34	0.26	0.34	0.42	0.26	0.34	0.42	0.58	0.26	0.34	0.42	0.58	0.73	0.26	0.34	0.42	0.58	0.73	1.03
3-1/2	46,285	19,220	43,380	11,765	20,825	34,385	7,650	12,095	17,410	32,785	5,785	6,725	11,970	20,070	31,050	3,695	5,300	6,940	10,565	14,700	24,675
	0.21	0.21	0.27	0.21	0.27	0.34	0.21	0.27	0.34	0.46	0.21	0.27	0.34	0.46	0.59	0.21	0.27	0.314	0.46	0.59	0.83
4	59,595	24,745	55,855	15,150	26,815	44,275	9,850	15,575	22,415	42,210	7,450	11,235	15,410	25.840	39,960	4,780	6,825	8,935	13,600	18,925	31,770
	0.17	0.17	0.23	0.17	0.23	0.28	0.17	0.23	0.28	0.38	0.17	0.23	0.28	0.38	0.49	0.17	0.23	0.28	0.36	0.49	0.25
5	93,655	38,890	87,780	23,810	42,140	69,580	15,480	24,475	35,230	66,335	11,705	17.660	24.220	40,610	62,830	7,475	10,725	14,040	21,375	29,745	49,930
	0.12	0.12	0.16	0.12	0.16	0.20	0.12	0.16	0.20	0.05	0.12	0.16	0.20	0.05	0.17	0.12	0.16	0.20	0.05	0.17	0.11
6	135,245	58,160	126,760	34,385	60,855	100,480	22,350	35,345	50,875	95,795	16,905	25,500	34,975	58,645	90,735	10,800	15,490	20,270	30,865	42,950	72,105
	0.10	0.10	0.13	0.10	0.13	0.04	0.10	0.13	0.04	0.05	0.10	0.13	0.04	0.05	0.01	0.10	0.13	0.04	0.05	0.01	0.01
8	234,195	97,245	219,505	59,540	105,380	173,995	38,705	61,205	88,095	165,880	29,270	44,160	60,565	101,650	157,115	18,700	26,820	35,105	53,450	74,175	124,855
	0.02	0.02	0.02	0.02	0.02	0.01	0.02	0.02	0.01	0.01	0.02	0.02	0.01	0.01	0.01	0.02	0.02	0.01	0.01	0.01	0.01

[a]Return-line Capacity (lb/hr) with Pressure Drop (psi) for 100 ft of Pipe at a Velocity of 5000 ft/min.

The hydraulic head due to the 15-ft riser is 2 psi/ft X 15 = 7.5 psi. The total back pressure is, therefore,

5 psi (receiver) + 7.5 psi (riser) + 2 psi (pipe) = 14.5 psi

or the differential pressure driving the condensate flow through the trap is 30 – 14.5 = 15.5 psi.

From Table 6.12 we see that a Model C trap will handle 2100 lb/hr at 15-psi differential pressure; this would then be the correct choice.

6.4.4 Maintaining Steam Traps for Efficient Operation

Steam traps can and do malfunction in two ways. They may stick in the closed position, causing condensate to back up into the steam system, or they may stick open, allowing live steam to discharge into the condensate system. The former type of malfunction is usually quickly detectable, since flooding of a process heater with condensate will usually so degrade its performance that the failure is soon evidence by a significant change in operating conditions. This type of failure can have disastrous effects on equipment by producing damaging water hammer and causing process streams to back up into other

Table 6.12 Typical Pressure-Capacity Specifications for Steam Traps

	Capacities (lb/hr)			
Model:	A	B	C	D
Differential pressure (psi)				
5	450	830	1600	2900
10	560	950	1900	3500
15	640	1060	2100	3900
20	680	880	1800	3500
25	460	950	1900	3800
30	500	1000	2050	4000
40	550	770	1700	3800
50	580	840	1 900	4100
60	635	900	2000	4400
70	660	950	2200	3800
80	690	800	1650	4000
100	640	860	1800	3600
125	680	950	2000	3900
150	570	810	1500	3500
200	—	860	1600	3200
250	—	760	1300	3500
300	—	510	1400	2700
400	—	590	1120	3100
450	—	—	1200	3200

equipment. Because of these potential problems, steam traps are often designed to fail in the open position; for this reason, they are among the biggest energy wasters in an industrial plant. Broad experience in large process plants using thousands of steam traps has shown that, typically, from 15 to 60% of the traps in a plant may be blowing through, wasting enormous amounts of energy. Table 6.13 shows the cost of wasted 100-psig steam (typical of many process plant conditions) for leak diameters characteristic of steam trap orifices. At higher steam pressures, the leakage would be even greater; the loss rate does not go down in direct proportion at lower steam pressures but declines at a rate proportional to the square root of the pressure. For example, a 1/8-in. leak in a system at 60 psig, instead of the 100 psig shown in the table, would still waste over 75% of the steam rate shown (the square root of 60/100). The cost of wasted steam far outweighs the cost of proper maintenance to repair the malfunctions, and comprehensive steam trap maintenance programs have proven to be among the most attractive energy conservation investments available in large process plants. Most types of steam traps can be repaired, and some have inexpensive replaceable elements for rapid turnaround.

A major problem facing the energy conservation manager is diagnosis of open traps. The fact that a trap is blowing through can often be detected by a rise in temperature at the condensate receiver, and it is quite easy to monitor this simple parameter. There are also several direct methods for checking trap operation. Figure 6.15 shows the simplest approach for open condensate systems where traps drain directly to atmospheric pressure. In proper normal operation, a stream of condensate drains from the line together with a lazy cloud of flash steam, produced as the condensate throttles across the trap. When the trap is blowing through, a well-defined jet of live steam will issue from the line with either no condensate, or perhaps a condensate mist associated with steam condensation at the periphery of the jet.

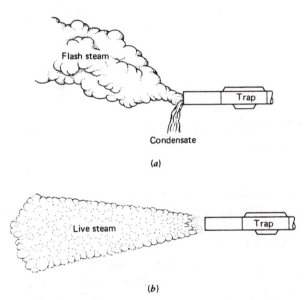

(a)

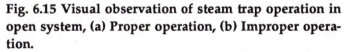

(b)

Table 6.13 Annual Cost of Steam Leaks

Leak Diameter (in.)	Steam Wasted per Month (lb)[a]	Cost per Month[b]	Cost per Year[b]
1/16	13,300	$40	$480
1/8	52,200	156	1,890
1/4	209,000	626	7,800
1/2	833,000	2,500	30,000

[a]Based on 100-psig differential pressure across the orifice.
[b]Based on steam value of $3/1000 lb. Cost will scale in direct proportion for other steam values.

Fig. 6.15 Visual observation of steam trap operation in open system, (a) Proper operation, (b) Improper operation.

Table 6.14 Operating Sounds of Various Types of Steam Traps

Trap	Proper Operation	Malfunctioning
Disk type (impulse of thermodynamic)	Opening and snap-closing of disk several times per minute	Rapid chattering of disk as steam blows through
Mechanical type (bucket)	Cycling sound of the bucket as it opens and closes	Fails open—sound of steam blowing through. Fails closed—no sound
Thermostatic type	Sound of periodic discharge if medium to high load; possibly no sound if light load; throttled discharge	Fails closed—no sound

Visual observation is less convenient in a closed condensate system, but can be utilized if a test valve is placed in the return line just downstream of the trap, as shown in Figure 6.16. This system has the added advantage that the test line may be used to actually measure condensate discharge rate as a check on equipment efficiency, as discussed earlier. An alternative in closed condensate return systems is to install a sight glass just downstream of the trap. These are relatively inexpensive and permit quick visual observation of trap operation without interfering with normal production.

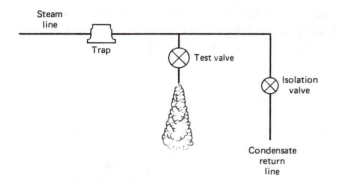

Fig. 6.16 Visual observation of steam trap operation in closed systems.

Another approach to steam trap testing is to observe the sound of the trap during operation. Table 6.14 describes the sounds made by various types of traps during normal and abnormal operation. This method is most effective with disk-type traps, although it can be used to some extent with the other types as well. An industrial stethoscope can be used to listen to the trap, although under many conditions, the characteristic sound will be masked by noises transmitted from other parts of the system. Ultrasonic detectors may be used effectively in such cases; these devices are, in effect, electronic stethoscopes with acoustic filtering to make them sensitive to sound and vibration only in the very high frequency range. Steam blowing through a trap emits a very high-pitched sound, produced by intense turbulence at the trap orifice, as contrasted with the lower-pitched and lower-intensity sound of liquid flowing through. Ultrasonic methods can, therefore, give a more reliable measure of steam trap performance than conventional "listening" devices.

A third approach to steam trap testing makes use of the drop in saturation temperature associated pressure drop across the trap. Condensate tends to cool rapidly in contact with uninsulated portions of the return line, accentuating the temperature difference. If the temperature on each side of the trap is measured, a sharp temperature

drop should be evident. Table 6.15 shows typical temperatures that can be expected on the condensate side for various condensate pressures. In practice, the temperature drop method can be rather uncertain, because of the range of temperatures the condensate may exhibit and because, in blowing through a stuck-open trap, live steam will, itself, undergo some temperature drop. For example, 85-psig saturated steam blowing through an orifice to 15 psig will drop from 328°F to about 300°F, and may then cool further by radiation and convection from uninsulated surfaces. From Table 6.15, the expected condensate-side temperature is about 215 to 238°F for this pressure. Thus although the difference is still substantial, misinterpretation is possible, particularly if accurate measurements of the steam and condensate pressures on each side of the trap are not available.

The most successful programs of steam trap diagnosis utilize a combination of these methods, coupled with a regular maintenance program, to assure that traps are kept in proper operating condition.

This section has discussed the reasons why good steam trap performance can be crucial to successful energy conservation in steam systems. Traps must be properly selected and installed for the given service and appropriately sized to assure efficient removal of condensate and gases. Once in service, expenditures for regular monitoring and maintenance easily pay for themselves in fuel savings.

Table 6.15 Typical Pipe Surface Temperatures for Various Operating Pressures

Operating Pressure (psig)	Typical Line Temperatures (°F)
0	190-210
15	215-238
45	248-278
115	295-330
135	304-340
450	395-437

6.5 CONDENSATE RECOVERY

Condensate from steam systems is wasted, or at least used inefficiently, in many industrial operations. Yet improvements in the condensate system can offer the greatest savings of any of the measures discussed in this chapter. In this section methods are presented for estimating the potential energy and mass savings achievable through good condensate recovery, and the considerations involved in condensate recovery system design are discussed. It is not possible in a brief survey to provide a comprehensive guide to the detailed design of such systems. Condensate return systems can, in fact, be quite complex, and proper design usually requires a careful engineering analysis. The energy manager can, however, define the type of system best suited to the requirements and determine whether sufficient justification exists for a comprehensive design study.

6.5.1 Estimation of Heat and Mass Losses in Condensate Systems

The saturated liquid condensate produced when steam condenses on a heating surface still retains a significant fraction of the energy contained in the steam itself. Referring to Table 6.2, for example, it is seen that at a pressure of about 80 psig, each pound of saturated liquid contains about 295 Btu, or nearly 25% of the original energy contained in the steam at the same pressure. In some plants, this condensate is simply discharged to a wastewater system, which is wasteful not only of energy, but also of water and the expense of boiler feedwater treatment. Even if condensate is returned to an atmospheric pressure receiver, a considerable fraction of it is lost in the form of flash steam. Table 6.16 shows the percent of condensate loss due to flashing from systems at the given steam pressure to a flash tank at a lower pressure. For example, if in the 80-psig

system discussed above, condensate is returned to a vented receiver instead of discharging it to a drain, nearly 12% is vented to the atmosphere as flash steam. Thus about 3% of the original steam energy (0.12 X 0.25) goes up the vent pipe. The table shows that this loss could be reduced by half by operating the flash tank at a pressure of 30 psig, providing a low-temperature steam source for potential use in other parts of the process. Flash steam recovery is discussed later in more detail.

Even when condensate is fully recovered using one of the methods to be described below, heat losses can still occur from uninsulated or poorly insulated return lines. These losses can be recovered very cost effectively by the proper application of thermal insulation as discussed in Chapter 15.

6.5.2 Methods of Condensate Heat Recovery

Several options are available for recovery of condensate, ranging in cost and complexity from simple and inexpensive to elaborate and costly. The choice of which option is best depends on the amount of condensate to be recovered, other uses for its energy, and the potential cost savings relative to other possible investments.

The simplest system, which can be utilized if condensate is presently being discharged, is the installation of a vented flash tank which collects condensate from various points of formation and cools it sufficiently to allow it to be

Table 6.16 Percent of Mass Converted to Flash Steam in a Flash Tank

Steam Pressure (psig)	Flash Tank Pressure (psig)										
	0	2	5	10	15	20	30	40	60	80	100
5	1.7	1.0	0								
10	2.9	2.2	1.4	0							
15	4.0	3.2	2.4	1.1	0						
20	4.9	4.2	3.4	2.1	1.1	0					
30	6.5	5.8	5.0	3.8	2.6	1.7	0				
40	7.8	7.1	6.4	5.1	4.0	3.1	1.3	0			
60	10.0	9.3	8.6	7.3	6.3	5.4	3.6	2.2	0		
80	11.7	11.1	10.3	9.0	8.1	7.1	5.5	4.0	1.9	0	
100	13.3	12.6	11.8	10.6	9.7	8.8	7.0	5.7	3.5	1.7	0
125	14.8	14.2	13.4	12.2	11.3	10.3	8.6	7.4	5.2	3.4	1.8
160	16.8	16.2	15.4	14.1	13.2	12.4	10.6	9.5	7.4	5.6	4.0
200	18.6	18.0	17.3	16.1	15.2	14.3	12.8	11.5	9.3	7.5	5.9
250	20.6	20.0	19.3	18.1	17.2	16.3	14.7	13.6	11.2	9.8	8.2
300	22.7	21.8	21.1	19.9	19.0	18.2	16.7	15.4	13.4	11.8	10.1
350	24.0	23.3	22.6	21.6	20.5	19.8	18.3	17.2	15.1	13.5	11.9
400	25.3	24.7	24.0	22.9	22.0	21.1	19.7	18.5	16.5	15.0	13.4

delivered back to the boiler feed tank. Figure 6.17 schematically illustrates such a system. It consists of a series of collection lines tying the points of condensate generation to the flash tank, which allows the liquid to separate from the flash steam; the flash steam is vented to the atmosphere through an open pipe. Condensate may be gravity-drained through a strainer and a trap. To avoid further generation of flash steam, a cooling leg may be incorporated to cool the liquid below its saturation temperature.

Flash tanks must be sized to produce proper separation of the flash steam from the liquid. As condensate is flashed, steam will be generated rather violently, and as vapor bubbles burst at the surface, liquid may be entrained and carried out through the vent. This represents a nuisance, and in some cases a safety hazard, if the vent is located in proximity to personnel or equipment. Table 6.17 permits the estimation of flash tank size required for a given application. Although strictly speaking, flash tanks must be sized on the basis of volume, if a typical length to

diameter of about 3 :1 is assumed, flash tank dimensions can be represented as the product of diameter times length, which has the units of square feet (area), even though this particular product has no direct physical significance. Consider, for example, the sizing of a vented flash tank for collection of 80-psig condensate at a rate of about 3000 lb/hr. For a flash tank pressure of 0 psig (atmospheric pressure), the diameter-length product is about 2.5 per 1000 lb. Therefore, a diameter times length of 7.5 ft^2 would be needed for this application. A tank 1.5 ft in diameter by 5 ft long would be satisfactory. Of course, for flash tanks as with other condensate equipment, conservative design would suggest the use of an appropriate safety factor.

As noted above, venting of flash steam to the atmosphere is a wasteful process, and if significant amounts of condensate are to be recovered, it may be desirable to attempt to utilize this flash steam. Figure 6.18 shows a modification of the simple flash tank system to accomplish this. Rather than venting to the atmosphere, the flash tank

Table 6.17 Flash Tank Sizing[a]

Steam Pressure (psig)	Flash Tank Pressure (psig)										
	0	2	5	10	15	20	30	40	60	80	100
400	5.41	4.70	3.89	3.01	2.44	2.03	1.49	1.15	0.77	0.56	0.42
350	5.14	4.45	3.66	2.84	2.28	1.91	1.38	1.07	0.70	0.51	0.37
300	4.86	4.15	3.42	2.62	2.11	1.75	1.26	0.96	0.62	0.44	0.31
250	4.41	3.82	3.12	2.39	1.91	1.56	1.11	0.85	0.52	0.37	0.25
200	3.98	3.40	2.80	2.12	1.68	1.37	0.97	0.72	0.43	0.28	0.18
175	3.75	3.20	2.61	1.95	1.57	1.26	0.87	0.64	0.38	0.23	0.15
160	3.60	3.08	2.50	1.86	1.46	1.19	0.80	0.59	0.34	0.21	0.12
150	3.48	2.98	2.41	1.80	1.40	1.14	0.77	0.56	0.31	0.19	0.10
140	3.36	2.86	2.31	1.72	1.35	1.08	0.72	0.52	0.29	0.16	0.08
130	3.24	2.76	2.23	1.65	1.29	1.02	0.67	0.49	0.26	0.14	0.07
120	3.12	2.65	2.15	1.57	1.22	0.97	0.64	0.44	0.23	0.12	0.04
110	2.99	2.52	2.05	1.50	1.15	0.91	0.58	0.40	0.20	0.09	0.02
100	2.85	2.41	1.92	1.40	1.07	0.85	0.53	0.36	0.16	0.06	
90	2.68	2.26	1.81	1.30	0.99	0.77	0.48	0.31	0.13	0.05	
80	2.52	2.12	1.67	1.18	0.90	0.68	0.42	0.25	0.09		
70	2.34	1.95	1.55	1.08	0.81	0.61	0.35	0.20	0.04		
60	2.14	1.77	1.39	0.96	0.70	0.52	0.27	0.14			
50	1.94	1.59	1.22	0.81	0.58	0.41	0.20	0.08			
40	1.68	1.36	1.02	0.67	0.44	0.30	0.11				
30	1.40	1.10	0.81	0.50	0.29	0.16					
20	1.06	0.81	0.55	0.28	0.12						
12	0.75	0.48	0.28								
10	0.62	0.42	0.23								

[a]Flash tank area (ft^2) = diameter X length of horizontal tank for discharge of 1000 lb/hr of condensate.

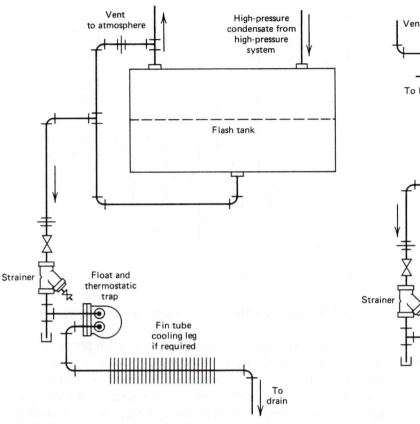

Fig. 6.17 Flash tank vented to atmosphere.

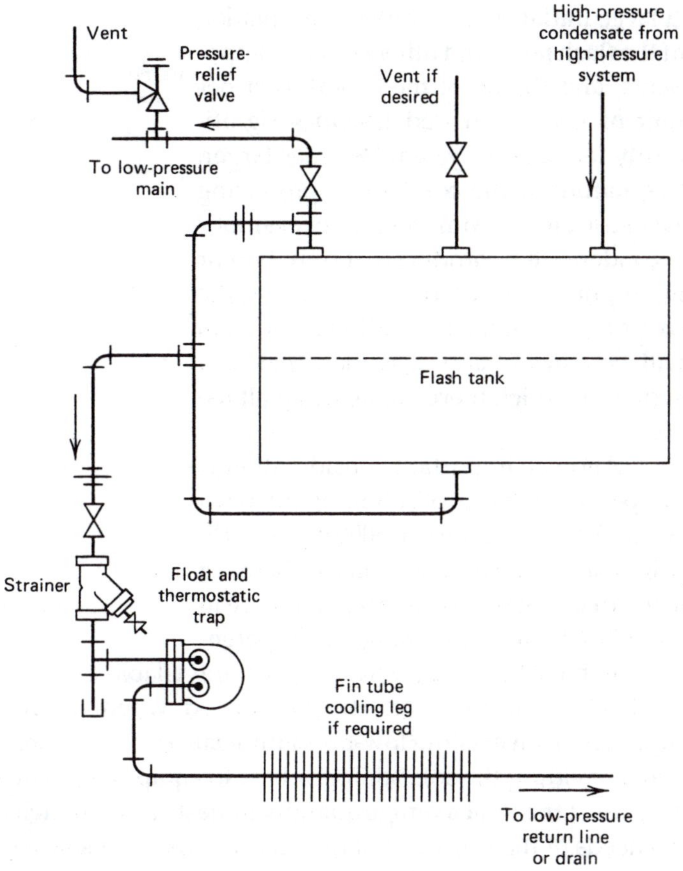

Fig. 6.18 Pressurized flash tank discharging to low-pressure steam system.

is pressurized and flash steam is piped to a low-pressure steam main, where it can be utilized for process purposes. From Table 6.17 it will be noted that the flash tank can be smaller in physical size at elevated pressure, although, of course, it must be properly designed for pressure containment. If in the example above the 80-psig condensate were flashed in a 15-psig tank, only about 2.7 ft^2 of diameter times length would be required. A tank 1 ft in diameter by 3 ft long could be utilized. Atmospheric vents are usually provided for automatic pressure relief and to allow manual venting if desired.

If pressurized flash steam is to be used, the cost of piping to set up a low-pressure steam system may be significant, particularly if the flash tank is remote from the potential low-pressure steam applications. Thus it is desirable to plan such a system to minimize these piping costs by generating the flash steam near its point of use. Figure 6.19 illustrates such an application. Here an air heater having four sections formerly utilized 100-psi steam in all sections. Because the temperature difference between the steam and the cold incoming air is larger than the difference at the exit end, the condensate load would be unevenly balanced among the four sections, with the heaviest load in the first section; lower-temperature steam could be uti-

lized here. In the revised arrangement shown, 100-psi steam is used in the last three sections; condensate is drained to a 5-psi flash tank, where low-pressure steam is generated and piped to the first section, substantially reducing the overall steam load to the heater. Note that for backup purposes, a pressure-controlled reducing valve has been incorporated to supplement the low-pressure flash steam at light-load conditions. This example shows how flash steam can be used directly without an expensive piping system to distribute it. A similar approach could apply to adjacent pieces of equipment in a multiple-batch operation.

Although the utilization of flash steam in a low-pressure system appears to offer an almost "free" energy source, its practical application involves a number of problems that must be carefully considered. These are all essentially economic in nature.

As mentioned above, the quantity of condensate and its pressure (thus yielding a given quantity of flash steam) must be sufficiently large to provide a significant amount of available energy at the desired pressure. System costs do not go up in simple proportion to capacity. Rather, there is a large initial cost for piping, installation of the flash tank, and other system components, and therefore the overall cost per unit of heat recovered becomes significantly less as the system becomes larger. The nature of the condensate-producing system itself is also important. For example, if condensate is produced at only two or three points from large steam users, the cost of the condensate collection system will be considerably less than that of a system in which there are many small users.

Another important consideration is the potential for application of the flash steam. The availability of 5000 lb/hr of 15-psig steam is meaningless unless there is a need for a heat source of this magnitude in the 250°F temperature range. Thus poten-

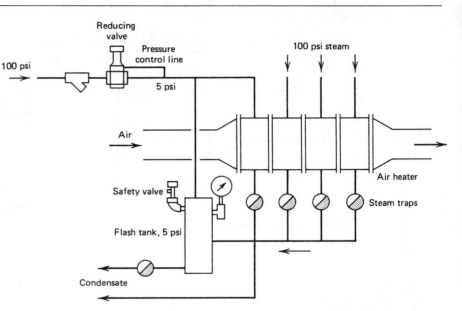

Fig. 6.19 Flash steam utilization within a process unit.

tial uses must be properly matched to the available supply. Flash steam is most effectively utilized when it can supplement an existing low-pressure steam supply rather than providing the sole source of heat to equipment. Not only must the total average quantity of flash steam match the needs of the process, but the time variations of source and user must be taken into account, since steam cannot be economically stored for use at a later time. Thus flash steam might not be a suitable heat source for sequential batch processes in which the number of operating units is small, such that significant fluctuations in steam demand exist.

When considering the possible conversion to low-pressure steam of an existing piece of equipment presently operating on high-pressure steam, it is important to recognize that steam pressure can have a significant effect on equipment operation. Since a reduction in steam pressure also means a reduction in temperature, a unit may not have adequate heating-surface area to provide the necessary heat capacity to the process at reduced pressure. Existing steam distribution piping may not be adequate, since steam is lower in density at low pressure than at high pressure. Typically, larger piping is required to transport the low-pressure vapor at acceptable velocities. Although one might expect that the heat losses from the pipe surface might be lower with low-pressure steam because of its lower temperature, in fact, this may not be the case if a large pipe (and hence larger surface area) is needed to handle the lower-pressure vapor. This requirement will also make insulation more expensive.

When flash steam is used in a piece of equipment, the resulting low-pressure condensate must still be returned to a receiver for delivery back to the boiler. Flash steam will again be produced if the receiver is vented, although

somewhat less than in the flashing of high-pressure condensate. This flash steam and that produced from the flash tank condensate draining into the receiver will be lost unless some additional provision is made for its recovery, as shown in Figure 6.20. In this system, rather than venting to the atmosphere, the steam rises through a cold-water spray, which condenses it. This spray might be boiler makeup water, for example, and hence the energy of the flash steam is used for makeup preheat. Not only is the heat content of the flash steam saved, but its mass as well, reducing makeup-water requirements and saving the incremental costs of makeup-water treatment. This system has the added advantage that it produces a deaerating effect on the condensate and feedwater. If the cold spray is metered so as to produce a temperature in the tank above about 190°F, dissolved gases in the condensate and feedwater, particularly oxygen and CO_2, will come out of

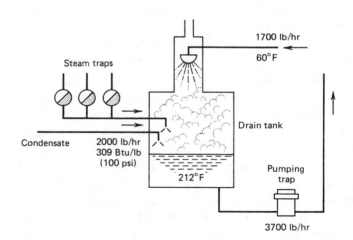

Fig. 6.20 Flash steam recovery in spray tank.

solution, and since they are not condensed by the cold-water spray, will be released through the atmospheric vent. As with flash steam systems, this system (usually termed a "barometric condenser" or "spray deaerator") requires careful consideration to assure its proper application. The system must be compatible with the boiler feedwater system, and controls must be provided to coordinate boiler makeup demands with the condensate load.

An alternative approach to the barometric condenser is shown in Figure 6.21. *In this system, condensate is cooled by passing* it through a submerged coil in the flash tank before it is flashed. This reduces the amount of flash steam generated. Cold-water makeup (possibly boiler feedwater) is regulated by a temperature-controlled valve.

The systems described above have one feature in common. In all cases, the final condensate state is atmospheric pressure, which may be required to permit return of the condensate to the existing boiler-feedwater makeup tank. If condensate can be returned at elevated pressure, a number of advantages may be realized.

Figure 6.22 shows schematics of two pressurized condensate return systems. Condensate is returned, in some cases without the need for a steam trap, to a high-pressure receiver, which routes the condensate directly back to the boiler. The boiler makeup unit and/or deaerator feeds the boiler in parallel with the condensate return unit, and appropriate controls must be incorporated to coordinate the operation of the two units. Systems such as the one shown in Figure 6.22a are available for condensate pressures up to about 15 psig. For higher pressures, the unit can be used in conjunction with a flash tank, as shown in Figure 6.22b. This system would be suitable where an application for 15-psig steam is available. These elevated-pressure systems represent an attractive option in relatively low-pressure applications, such as steam-driven absorption chillers. When considering them, care must be exercised to assure that dissolved gases in the boiler makeup are at suitable levels to avoid corrosion, since the natural deaeration effect of atmospheric venting is lost.

One of the key engineering considerations that must be accounted for in the design of all the systems described above is the problem of pumping high-temperature condensate. To understand the nature of the problem, it is necessary to introduce the concept of "net positive suction head" (NPSH) for a pump. This term means the amount of static fluid pressure that must be provided at the inlet side of the pump to assure that no vapor will be formed as the liquid passes through the pump mechanism, a phenom-

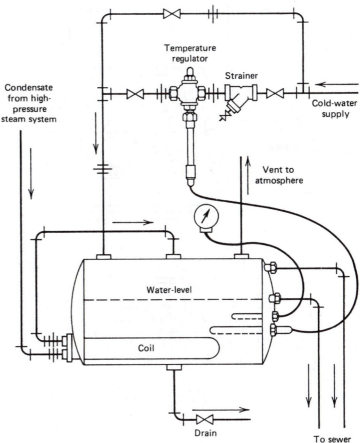

Fig. 6.21 Flash tank with condensate precooling.

enon known as cavitation. As liquid moves into the pump inlet from an initially static condition, it accelerates and its pressure drops rather suddenly. If the liquid is at or near its saturation temperature in the stationary condition, this sudden drop in pressure will produce boiling and the generation of vapor bubbles. Vapor can also be generated by air coming out of solution at reduced pressure. These bubbles travel through the pump impeller, where the fluid pressure rises, causing the bubbles to collapse. The inrush of liquid into the vapor space produces an impact on the impeller surface which can have an effect comparable to sandblasting. Clearly, this is deleterious to the impeller and can cause rapid wear. Most equipment operators are familiar with the characteristic "grinding" sound of cavitation in pumps when air is advertently allowed to enter the system, and the same effect can occur due to steam generation in high-temperature condensate pumping.

To avoid cavitation, manufacturers specify a minimum pressure above saturation which must be maintained on the inlet side of the pump, such that, even when the pressure drops through the inlet port, saturation or deaeration conditions will not occur. This minimum-pressure requirement is termed the net positive suction head.

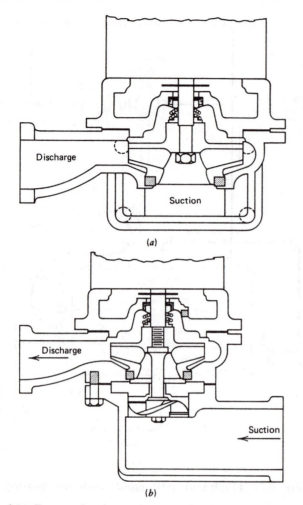

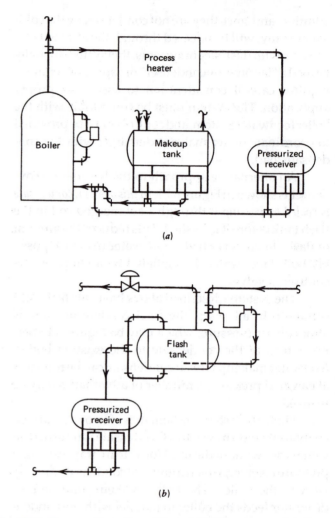

Fig. 6.22 Pressurized condensate receiver systems, (a) *Low-pressure process requirements*, (b) Flash tank for use with high-pressure systems.

Fig. 6.23 Conventional and low-NPSH pumps, (a) Conventional centrifugal pump, (b) Low-NPSH centrifugal pump.

For condensate applications, special low-NPSH pumps have been designed. Figure 6.23 illustrates the difference between a conventional pump and a low-NPSH pump. In the conventional pump (Figure 6.23a) fluid on the suction side is drawn directly into the impeller where the rapid pressure drop occurs in the entry passage. In the low-NPSH pump (Figure 6.23b) a small "preimpeller" provides an initial pressure boost to the incoming fluid, with relatively little drop in pressure at the entrance. This extra stage of pumping essentially provides a greater head to the entry passage of the main impeller, so that the system pressure at the suction side of the pump can be much closer to saturation conditions than that required for a conventional pump. Low-NPSH pumps are higher in price than conventional centrifugal pumps, but they can greatly simplify the problem of design for high-temperature condensate return, and can, in some cases, actually reduce overall system costs.

An alternative device for the pumping of condensate,

called a "pumping trap," utilizes the pressure of the steam itself as the driving medium. Figure 6.24 illustrates the mechanism of a pumping trap. Condensate enters the inlet side and rises in the body until it activates a float-operated valve, which admits steam or compressed air into the chamber. A check valve prevents condensate from being pushed back through the inlet port, and another check valve allows the steam or air pressure to drive it out through the exit side. When the condensate level drops to a predetermined position, the steam or air valve is closed, allowing the pumping cycle to start again. Pumping traps have certain inherent advantages over electrically driven pumps for condensate return applications. They have no NPSH requirement, and hence can handle condensate at virtually any temperature without regard to pressure conditions. They are essentially self-regulating, since the condensate level itself determines when the trap pumps; thus no auxiliary electrical controls are required for the system. This has another advantage in environments where

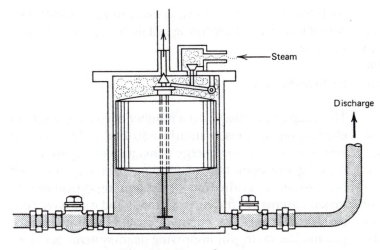

Fig. 6.24 Pumping trap.

(a)

explosion-proofing is required, such as refineries and chemical plants. Electrical lines need not be run to the system, since it utilizes steam as a driving force, and the steam line is usually close at hand. Pumping traps operate more efficiently using compressed air, if available, because when steam is introduced to the chamber, some of it condenses before its pressure can drive the condensate out. Thus for the same pressure, more steam is required to give the same pumping capacity as compressed air. The disadvantages of pumping traps are their mechanical complexity, resulting in a susceptibility to maintenance problems, and the fact that they are available only in limited capacities.

The engineering of a complete condensate recovery system from scratch can be a rather involved process, requiring the design of tanks, plumbing, controls, and pumping devices. For large systems, there is little alternative to engineering and fabricating the system to the specific plant requirements. For small- to moderate-capacity applications, however, packaged systems incorporating all of the foregoing components are commercially available. Figures 6.25a and b show examples of two such systems, the former using electrically driven low-NPSH pumps, and the latter utilizing a pumping trap (the lower unit in the figure).

6.5.3 Overall Planning Considerations in Condensate Recovery Systems

As mentioned earlier, condensate recovery systems require careful engineering to assure that they are compatible with overall plant operations, that they are safe and reliable, and that they can actually achieve their energy-efficiency potential. In this section a few of the overall planning factors that should be considered are enumerated and discussed.

(b)

Fig. 6.25
Pressurized condensate return systems.

1. **Availability of Adequate Condensate Sources.** An energy audit should be performed to collect detailed information on the quantity of condensate available from all of the various steam using sources in the plant and the relevant data associated with these sources. Such data include, for example, condensate pressure, quantity, and source location relative to other steam-using equipment and to the boiler room. Certain other information may also

be pertinent. For example, if the condensate is contaminated by contact with other process streams, it may be unsuitable for recovery. It is not valid to assume that steam used in the process automatically results in recoverable condensate. Stripping steam used in refining of petroleum and in other separation processes is a good example.

2. Survey of Possible Flash Steam Applications. The process should be surveyed in detail to assess what applications presently using first-generation steam might be adaptable to the use of flash steam or to heat recovered from condensate. Temperatures and typical heat loads are necessary but not sufficient. Heat-transfer characteristics of the equipment itself may be important. A skilled heat-transfer engineer, given the present operating characteristics of a process heater can make reasonable estimates of that heater's capability to operate at a lower pressure.

3. Analysis of Condensate and Boiler Feedwater Chemistry. To assure that conditions in the boiler are kept in a satisfactory state to avoid scaling and corrosion, a change in feedwater treatment may be required when condensate is recovered and recycled. Water samples from the present condensate drain, from the boiler blow-down, from the incoming water source, and from the outlet of the present feedwater treatment system should be obtained. With this information, a water treatment specialist can analyze the overall water chemical balance and assure that the treatment system is properly configured to maintain good boiler-water conditions.

4. Piping Systems. A layout of the present steam and condensate piping system is needed to assess the need for new piping and the adequacy of existing runs. This permits the designer to select the best locations for flash tanks to minimize the need for extensive new piping.

5. Economic Data. The bottom line on any energy recovery project, including condensate recovery, is its profitability. In order to analyze the profitability of the system, it is necessary to estimate the quantity of heat recovered, converted into its equivalent fuel usage, the costs and cost savings associated with the water treatment system, and savings in water cost. In addition to the basic capital and installation costs of the condensate recovery equipment, there may be additional costs associated with modification of existing equipment to make it suitable for flash steam utilization, and there will almost certainly be a cost of lost production during installation and checkout of the new system.

6.6 SUMMARY

This chapter has discussed a number of considerations in effecting energy conservation in industrial and commercial steam systems. Good energy management begins by improving the operation of existing systems, and then progresses to evaluation of system modifications to maximize energy efficiency. Methods and data have been presented to assist the energy manager in estimating the potential for savings by improving steam system operations and by implementing system design changes. As with all industrial and commercial projects, expenditure of capital must be justified by reductions in operating expense. With energy costs rising at a rate substantially above the rate of increase of most equipment, it is clear that energy conservation will continue to provide ever more attractive investment opportunities.

BIBLIOGRAPHY

Armstrong Machine Works, "Steam Conservation Guidelines for Condensate Drainage," *Bulletin M101*, Three Rivers, Mich., 1976.
Babcock and Wilcox Corporation, *Steam: Its Generation and Use*, 38th ed., New York, 1972.
Department of Commerce, National Bureau of Standards, *Energy Conservation Program Guide, for Industry and Commerce*, NBS Handbook 115, Washington, D.C., 1974.
Department of Energy, United Kingdom, *Utilization of Steam for Process and Heating*, Fuel Efficiency Booklet 3, London, 1977.
Department of Energy, United Kingdom, *Flash Steam and Vapour Recovery*, Fuel Efficiency Booklet 6, London, 1977.
Department of Energy, United Kingdom, *How to Make the Best Use of Condensate*, Fuel Efficiency Booklet 9, London, 1978.
IDA, MASAHIRO, "Utilization of Low Pressure Steam and Condensate Recovery," *TLV Technical Information Bulletin* 5123, TLV Company Ltd., Tokyo, 1975.
ISLES, DAVID L., "Maintaining Steam Traps for Best Efficiency," *Hydrocarbon Processing*, Vol. 56, No. 1 (1977).
JESIONOWSKI, J.M., W.E. DANEKIND, and W.R. RAGER, "Steam Leak Evaluation Technique: Quick and Easy to Use," *Oil and Gas Journal*, Vol. 74, No. 2 (1976).
KERN, DONALD Q., *Process Heat Transfer*, McGraw-Hill, New York, 1950.
MAY, DONALD, "First Steps in Cutting Steam Costs," *Chemical Engineering*, Vol. 80, No. 26 (1973).
MONROE, E.S., JR., "Select the Right Steam Trap," *Chemical Engineering*, Vol. 83, No. 1 (1976).
MOWER, JOSEPH, "Conserving Energy in Steam Systems," *Plant Engineering*, Vol. 32, No. 8 (1978).
Sarco Company, Inc., *Hook-Up Designs for Steam and Fluid Systems*, 5th ed., Allentown, Pa., 1975.

CHAPTER 7

COGENERATION

JORGE B. WONG

GE Motors & Industrial Systems
Evansville, Indiana

JOHN M. KOVACIK

GE Power Systems
Schenectady, New York

7.1 INTRODUCTION

Cogeneration is broadly defined as the coincident or simultaneous generation of combined heat and power (CHP). In a true cogeneration system a "significant" portion of the generated or recovered heat must be used in a thermal process as steam, hot air, hot water, etc. The cogenerated power is typically in the form of mechanical or electrical energy.

The power may be totally used in the industrial plant that serves as the "host" of the cogeneration system, or may be partially/totally exported to a utility grid. Figure 7.1 illustrates the potential of saving in primary energy when separate generation of heat and electrical energy is substituted by a cogeneration system. This chapter presents an overview of current design, analysis and evaluation procedures.

The combined generation of useful heat and power is not a new concept. The U.S. Department of Energy (1978) reported that in the early 1900s, 58% of the total power produced by on-site industrial plants was cogenerated. However, Polimeros (1981) stated that by 1950, on-site CHP generation accounted for only 15 percent of total U.S. electrical generation; and by 1974, this figure had dropped to about 5 percent. In Europe, the experience has been very different. The US Department of Energy, US-DOE (1978), reported that "historically, industrial cogeneration has been five to six times more common in some parts of Europe than in the U.S." In 1972, "16% of West Germany's total power was cogenerated by industries; in Italy, 18%; in France, 16%; and in the Netherlands, 10%."

Since the promulgation of the Public Utilities Regulatory Policies Act of 1978 (PURPA), however, U.S. co-generation design, operation and marketing activities have dramatically increased, and have received a much larger attention from industry, government and academy. As a result of this incentive, newer technologies such as various combined cycles have achieved industrial maturity. Also, new or improved cogeneration-related processes and equipment have been developed and are actively marketed; e.g. heat recovery steam generator and duct burners, gas-engine driven chillers, direct and indirect fired two-stage absorption chiller-heaters, etc.

The impact of cogeneration in the U.S., both as an energy conservation measure and as a means to contribute to the overall electrical power generation capacity cannot be overemphasized. SFA Pacific Inc. (1990) estimated that non-utility generators (NUGs) produce 6-7% of the power generated in the US. And that NUGs will account for about half of the 90,000 MW that will be added during the next decade. More recently, Makansi (1991) reported that around 40,000 MW of independently produced and/or cogenerated (IPP/COGEN) power was put on line since the establishment of PURPA. This constitutes about 60% of the new capacity added during the last decade. Makansi also reports that another 60,000 MW of IPP/COGEN power is in construction and in development. This trend is likely to increase since many consider CHP as a supplementary way of increasing the existing U.S. power generation capacity.

PURPA is considered the foremost regulatory instrument in promoting cogeneration, IPPs and/or NUGs. Thus, to promote industrial energy efficiency and resource conservation through cogeneration, PURPA requires electric utilities to purchase cogenerated power at fair rates to both, the utility and the generator. PURPA also orders utilities to provide supplementary and back-up power to qualifying facilities (QFs).

The Promulgation of the Energy Policy Act of 1992 EPAct (1992) have brought a "deregulated" and competitive structure to the power industry. EPAct establishes the framework that allows the possibility of "Retail Wheeling" (anyone can purchase power from any generator, utility or otherwise). This is in addition to the existing Wholesale Wheeling or electricity trade that normally occurs among U.S. utilities. EPAct's main objective is to stimulate competition in the electricity generation sector

A) Separate generation of heat and electricity in boiler and conventional condensation power plant.

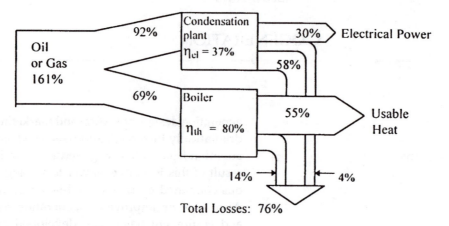

Total Losses: 76%

B) Cogeneration of heat and power.

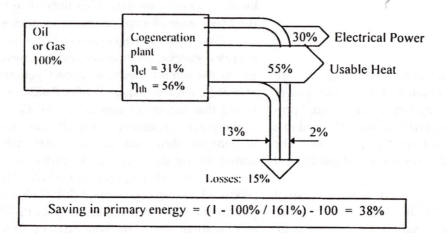

Saving in primary energy = (1 - 100% / 161%) - 100 = 38%

C) Integration of a cogeneration system and an industrial process.

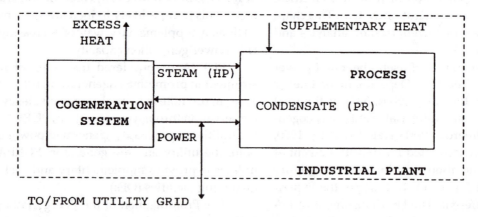

Figure 7.1 Potential of saving in primary energy when separate generation of heat and electrical energy is substituted by an industrial cogeneration system.

to and to reduce electricity costs. The impact of EPAct 1992, deregulation, and retailed wheeling on cogeneration can be tremendous, since EPAct creates a new class of generating facility called Exempt Wholesale Generators (EWGs). Economies of scale and scope—i.e. the natural higher efficiency of larger CHP plants—could trigger the grow of cogeneration based EWGs. In addition, EPAct will open the transmission grid to utilities and NUGs by ordering FERC to allow open transmission access to all approved EWGs. However, specific retail wheeling rules have not been legislated at the time of this writing. So far, only a few retail wheeling cases have been started. Thus, the actual impact of deregulation on cogeneration is considered to be uncertain. Nevertheless, some consider that retail wheeling/deregulation and cogeneration could have a synergistic effect and should be able to support each other.

7.2. COGENERATION SYSTEM DESIGN AND ANALYSIS

The process of designing and evaluating a cogeneration system has so many factors that it has been compared to the Rubik's cube—the ingenious game to arrange a multi-colored cube. The change in one of the cube faces will likely affect some other face. The most important faces are: fuel security, regulations, economics, technology, contract negotiation and financing.

7.2.1 General Considerations and Definitions

Kovacik (1982) indicates that although cogeneration should be evaluated as a part of any energy management plan, the main prerequisite is that a plant shows a significant and concurrent demand for heat and power. Once this scenario is identified, he states that cogeneration systems can be explored under the following circumstances:

1. Development of new facilities

2. Major expansions to existing facilities which increase process heat demands and/or process energy rejection.

3. When old process and/or power plant equipment is being replaced, offering the opportunity to upgrade the energy supply system.

The following terms and definitions are regularly used in the discussion of CHP systems.

Industrial Plant: the facility requiring process heat and electric and/or shaft power. It can be a process plant, a manufacturing facility, a college campus, etc. See Figure 7.1c.

Process Heat (PH): the thermal energy required in the industrial plant. This energy is supplied as steam, hot water, hot air, etc.

Process Returns (PR): the fluid returned from the industrial plant to the cogeneration system. For systems where the process heat is supplied as steam, the process returns are condensate.

Net Heat to Process (NHP): the difference between the thermal energy supplied to the industrial plant and the energy returned to the cogeneration system. Thus, NHP = PH − PR. The NHP may or may not be equal to the actual process heat demand (PH).

Plant Power Demand (PPD): the electrical power or load demanded (kW or MW) by the industrial plant. It includes the power required in for industrial processes, air-conditioning, lighting, etc.

Heat/Power Ratio (H/P): the heat-to-power ratio of the industrial plant (demand), or the rated heat-to-power ratio of the cogeneration system or cycle (capacity).

Topping Cycles: thermal cycles where power is produced prior to the delivery of heat to the industrial plant. One example is the case of heat recovered from a diesel-engine generator to produce steam and hot water. Figure 7.2 shows a diesel engine topping cycle.

Bottoming Cycle: power production from the recovery of heat that would "normally" be rejected to a heat sink. Examples include the generation of power using the heat from various exothermic chemical processes and the heat rejected from kilns used in various industries. Figure 7.3 illustrates a bottoming cycle.

Combined Cycle: this is a combination of the two cycles described above. Power is produced in a topping cycle—typically a gas-turbine generator. Then, heat exhausted from the turbine is used to produce steam; which is subsequently expanded in a steam turbine to generate more electric or shaft power. Steam can also be extracted from the cycle to be used as process heat. Figure 7.4 depicts a combined cycle.

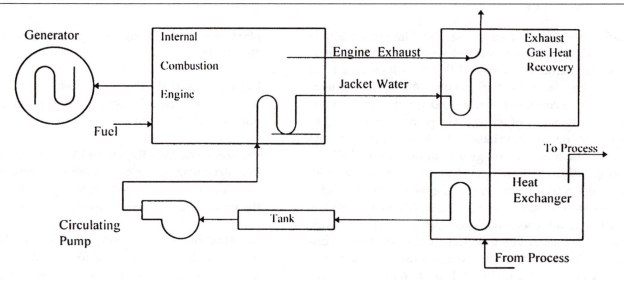

Figure 7.2 Diesel engine topping cycle.

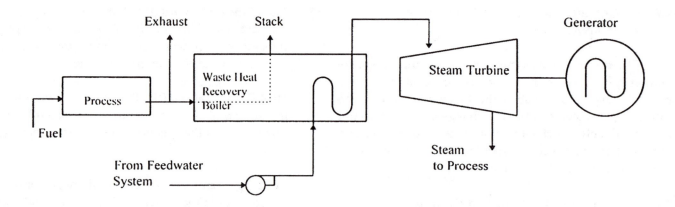

Figure 7.3 Steam turbine bottoming cycle.

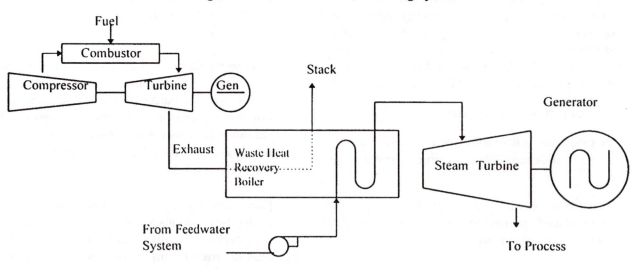

Figure 7.4 Combined cycle: gas-turbine/generator set, waste heat recovery boiler and steam-turbine/generator set.

Prime Mover: a unit of the CHP system that generates electric or shaft power. Typically, it is a gas turbine generator, a steam turbine drive or a diesel-engine generator.

7.2.2 Basic Cogeneration Systems

Most cogeneration systems are based on prime movers such as steam turbines, gas turbines, internal combustion engines and packaged cogeneration. Table 7.1 shows typical performance data for various cogeneration systems. Figures in this table (and in this chapter) are based on higher heating values, unless stated otherwise.

7.2.2.1 Steam Turbine Systems

Steam turbines are currently used as prime movers in topping, bottoming and combined cycles. There are many types of steam turbines to accommodate various heat/power ratios and loads. For limited expansion (pressure drop) and smaller loads (<4000 HP) lower cost single stage backpressure turbines are used. When several pressure levels are required (and usually for larger loads) multi-stage condensing and non-condensing turbines with induction and/or extraction of steam at intermediate pressures are generally used. Fig. 7.5 shows a variety of condensing and non-condensing turbines.

Four factors must be examined to assure that the maximum amount of power from a CHP steam plant is economically generated based on the process heat required. These factors are: (1) prime-mover size, (2) initial steam conditions, (3) process pressure levels, and (4) feedwater heating cycle.

1. Prime-Mover Type and Size. Process heat and plant electric requirements define the type and size of the steam generator. The type of CHP system and its corresponding prime mover are selected by matching the CHP system heat output to the process heat load.

If process heat demands are such that the plant power requirements can be satisfied by cogenerated power, then the size of the prime mover is selected to meet or exceed the "peak" power demand. However, cogeneration may supply only a portion of the total plant power needs. The balance has to be imported through a utility tie. In isolated plants, the balance is generated by additional conventional units. This discussion assumes that both heat and power demands remain constant all

Table 7.1 Basic cogeneration systems.

Cogeneration Systems	Unit Elec. Capacity (kW)	Heat Rate[2] (Btu/kWh)	Electrical Efficiency (%)	Thermal Efficiency (%)	Total Efficiency (%)	Exhaust Temperature (°F)	125-psig Steam Generation (.lbs/hr)
Small reciprocating Gas Engines	1-500	25,000 to 10,000	14-34	52	66-86	600-1200	0-200[1]
Large reciprocating Gas Engines	500-17,000	13,000 to 9,500	26-36	52	78-88	600-1200	100-10,000[1]
Diesel Engines	100-4,000 9,500	14,000 to	24-36	50	74-86	700-1500	100-1500[1]
Industrial Gas Turbines	800-10,000	14,000 to 11,000	24-31	50	74-81	800-1000	3,000-30,000
Utility Size Gas Turbines	10,000-150,000	13,000 to 9000	26-35	50	76-85	700-800	30,000 to 300,000
Steam-Turbine Cycles	5,000-200,000	30,000 to 10,000	10-35	28	38-63	350-1000	10,000 to 200,000

NOTES

[1]Hot water @ 250° is available at 10 times the flow of steam

[2]Heat rate is the fuel heat input (Btu/hr—higher heating value) to the cycle per kWh of electrical output at design (full load) and ISO conditions (60°F ambient temperature and sea level operation)

[3]The electrical generation efficiency in percent of a prime mover can be determined by the formula Efficiency = 3413/Heat Rate × 100

Sources: Adapted from Limaye (1985) and Manufacturers Data.

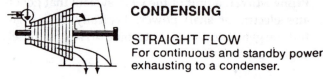

CONDENSING

STRAIGHT FLOW
For continuous and standby power exhausting to a condenser.

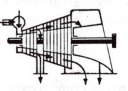

DUAL-FLOW OPPOSED EXHAUST
For large steam flow exhausting to condenser to minimize blade stresses and optimize efficiency.

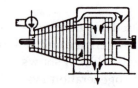

NON-AUTOMATIC EXTRACTION
To provide low flow, up to 15% of throttle flow for heating or process requirements.

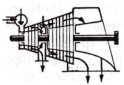

CONTROLLED EXTRACTION
Used when there is a demand for power and low pressure process steam which may be less than the steam flow required to make the power.

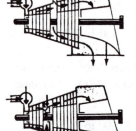

CONTROLLED INDUCTION
Enables the user to produce power from two steam pressure levels with controls which favor the less expensive lower pressure steam flow.

CONTROLLED EXTRACTION/ INDUCTION
Used where there is a variable demand for low pressure process steam which at times results in excess low pressure steam available to do work in the turbine.

BACK PRESSURE

STRAIGHT FLOW
The most economic operation where the exhaust steam is used for heating and process purposes.

NON-AUTOMATIC EXTRACTION
All the backpressure benefits plus additional higher pressure bleed for preheating.

CONTROLLED EXTRACTION/ INDUCTION
Where two low pressure steam headers are required.

Figure 7.5 Various multi-stage steam turbine systems. *Courtesy of Coppus Engineering/Murray Turbomachinery Corporation.*

times. Hence, the design problem becomes one of specifying two variables: (1) how much power should be cogenerated on-site and (2) how much power should be imported. Thus, given the technological, economical and legal constraints for a particular plant, and assuming the CHP system must be constructed at a minimum overall cost, it becomes a constrained optimization problem.

2. Initial Steam Conditions. Many industrial plants do not have adequate process steam demands to generate all the power required. Thus, it is important for the designer to examine those variables over which he has control so he can optimize the amount of power that can be economically generated. One set of these variables are the initial steam conditions, i.e. the initial pressure and temperature of the steam generated. In general, an increase in initial pressure and/or temperature will increase the amount of energy available for power generation. But the prime mover construction and cost, and the heat demand impose economical limits for the initial steam conditions. Thus, higher initial steam conditions can be economically justified in industrial plants having relatively large process steam demands.

3. Process Steam Pressure. For a given set of initial steam conditions, lowering the exhaust pressure also increases the energy available for power generation. However, this pressure is limited—totally, with non-condensing turbines, or partially, with extraction turbines—by the maximum pressure required in the industrial process. For instance, in paper mills, various pressure levels are needed to satisfy various process temperature requirements.

4. Feedwater Heating. Feedwater heating through use of steam exhausted and/or extracted from a turbine increases the power that can be generated.

Calculation of Steam Turbine Power

Given the initial steam conditions (psig, °F) and the exhaust saturated pressure (psig), Theoretical Steam Rates (TSR) specify the amount of steam heat input required to generate a kWh in an ideal turbine. The TSR is defined by

$$\text{TSR (lb/kWh)} = \frac{3412\ \text{Btu/kWh}}{h_i - h_o\ \text{Btu/lb}} \tag{7.1}$$

where $h_i - h_o$ is the difference in enthalpy from the initial steam conditions to the exhaust pressure based on an isentropic (ideal) expansion. These values can be obtained from steam tables or a Mollier chart. However, they are conveniently tabulated by the American Society of Mechanical Engineers.

The TSR can be converted to the Actual Steam Rate (ASR)

$$\text{ASR (lb/kWh)} = \frac{\text{TSR}}{\eta_{tg}} \tag{7.2}$$

where η_{tg} is the turbine-generator overall efficiency, stated or specified at "design" or full-load conditions. Some of the factors that define the overall efficiency of a turbine-generator set are: the inlet volume flow, pressure ratio, speed, geometry of turbine staging, throttling losses, friction losses, generator losses and kinetic losses associated with the turbine exhaust. Most turbine manufacturers provide charts specifying either ASR or η_{tg} values. Once the ASR has been established, the net enthalpy of the steam supplied to process (NEP) can be calculated:

NEP (Btu/lb)
$$= \text{Hi} - 3500/\text{ASR} - \text{Hc.x} - \text{Hm}(1-x) \tag{7.3}$$

where Hi   =   enthalpy at the turbine inlet conditions (Btu/lb)
       3500 =   conversion from heat to power (Btu/kWh), including the effect of 2.6% radiation, mechanical and generator losses
       Hc   =   enthalpy of condensate return (Btu/lb)
       Hm   =   enthalpy of make-up water (Btu/lb)
       x   =   condensate flow fraction in boiler feed water

(1-x)   =   make-up water flow fraction in boiler feed water

Hence, assuming a straight flow turbine (See Figure 7.5), the net heat to process (Btu/hr) defined in Section 7.2.1 can be obtained by multiplying equation 7.3 by the flow rate in lb/hr. The analysis of the overall cycle would require the replication of complete heat and mass balance calculations at part-load efficiencies. To expedite these computations, there are a number of commercially available software packages, which also produce mass/heat balance tables. See Example 8 for a cogeneration software application.

Selection of Smaller Single-Stage Steam Turbines

There exist many applications for smaller units (condensing and noncondensing), specially in mechanical drives or auxiliaries (fans, pumps, etc.); but a typical application is the replacement (or by pass) of a pressure reducing valve (PRV) by a single stage back-pressure turbine.

After obtaining the TSR from inlet and outlet steam conditions, Figure 7.6 helps in determining the approximate steam rate (ASR) for smaller (<3000HP) single-stage steam turbines. Figure 7.7 is a sample of size and speed ranges available from a turbine manufacturer. It should be noticed that there is an overlap of capacities (e.g. an ET-30 unit can operate in the ET-25 range and in portion of the ET-15/ET-20 range). Thus, the graph can help in selecting a unit subject to variable steam flows and/or loads.

Example 1. A stream of 15,500 lb of saturated steam at 250 psig (406°F) is being expanded through a PRV to obtain process steam at 50 psig. Determine the potential for electricity generation if the steam is expanded using a single-stage back-pressure 3600 RPM turbine-generator.

Data:

Steam flow (W_s) : 15,500 lb/hr

Inlet Steam   : 250 psig sat (264.7 psia), 406°F

Enthalpy, h_i : 1201.7 Btu/lb (from Mollier chart)

Outlet Steam   : 50 psig subcooled (9.67% moisture).

Enthalpy, h_o : 1090.8 Btu/lb (from Mollier chart)

Turbine Speed   : 3600 RPM

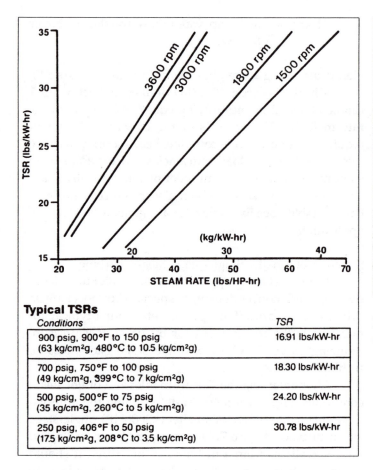

Typical TSRs

Conditions	TSR
900 psig, 900°F to 150 psig (63 kg/cm²g, 480°C to 10.5 kg/cm²g)	16.91 lbs/kW-hr
700 psig, 750°F to 100 psig (49 kg/cm²g, 399°C to 7 kg/cm²g)	18.30 lbs/kW-hr
500 psig, 500°F to 75 psig (35 kg/cm²g, 260°C to 5 kg/cm²g)	24.20 lbs/kW-hr
250 psig, 406°F to 50 psig (17.5 kg/cm²g, 208°C to 3.5 kg/cm²g)	30.78 lbs/kW-hr

Figure 7.6 Estimation of steam rates for smaller (<3,000 HP) single-stage turbines. *Courtesy Skinner Engine Co.*

a) Calculate TSR using equation 7.1:

$$TSR = \frac{3412 \, Btu/kWh}{h_i - h_o \, Btu/lb}$$

$$= \frac{3412 \, Btu/kWh}{1201.7 - 1090.8 \, Btu/lb}$$

$$= 30.77 \, lb/kWh$$

b) Obtain steam rate from Fig. 7.6, ASR = 38.5 lbs/HP-hr.

c) Calculate potential generation capacity (PGC):

$$PGC = \frac{W_s \, (0.746 \, kW/hp)}{ASR}$$

$$= \frac{(15,500 \, lb/hr) \, (0.746 \, kW/hp)}{(38.5 \, lbs/hp-hr)} = 300.3 \, kW$$

(7.4)

Figure 7.7 shows that units ET-15 or ET-20 better match the required PGC. Next, the generator would have

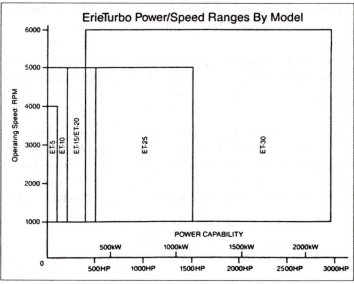

Figure 7.7 Power/speed ranges for single stage turbines. *Courtesy Skinner Engine Co.*

to be sized according to a commercially available unit size, e.g. 300 kW.

Selection of Multi-Stage Steam Turbines
Multistage steam turbines provide more flexibility to match various pressure levels and variable flow rates in larger cogeneration applications. Figure 7.5 describes a variety of back-pressure and condensing multi-stage steam turbines.

For turbine selection, a Mollier diagram should be used to explore various multi-stage turbine alternatives. In general, a preliminary analysis should include the following:

— Approximation of actual steam rates.

— Defining number of stages.

— Estimation of stage pressure and temperatures.

— Calculation of full-load and part load steam rates.

— Estimation of induction and/or extraction pressures, temperatures and flow rates for various power outputs.

For larger units (>3000 kW), Figure 7.8 shows a chart to determine the approximate turbine efficiency when the power range, speed and steam conditions are known. Figure 7.9 gives steam rate correction factors for off-design loads and speeds. However, manufacturers of multistage turbines advise that stage selection and other thermodynamic parameters must be evaluated taking into account other important factors such as speed limitations, mechanical stresses, leakage and throttling losses, windage, bearing friction and reheat. Thus, after a preliminary evaluation, it is important to compare notes with the engineers of a turbine manufacturer.

Example 2. Steam flow rate must be estimated to design a heat recovery steam generator (a bottoming cycle). The steam is to be used for power generation and is to be expanded in a 5,000 RPM multistage condensing turbine to produce a maximum of 18,500 kW. Steam inlet conditions is 600 psig/750°F and exhaust pressure is 4" HGA (absolute). Additional data are given below.

Data (for a constant entropy steam expansion @ S = 1.61 Btu/lb/°R)

PGC   : 18,500 kW

Inlet Steam : 600 psig (615 psia), 750°F

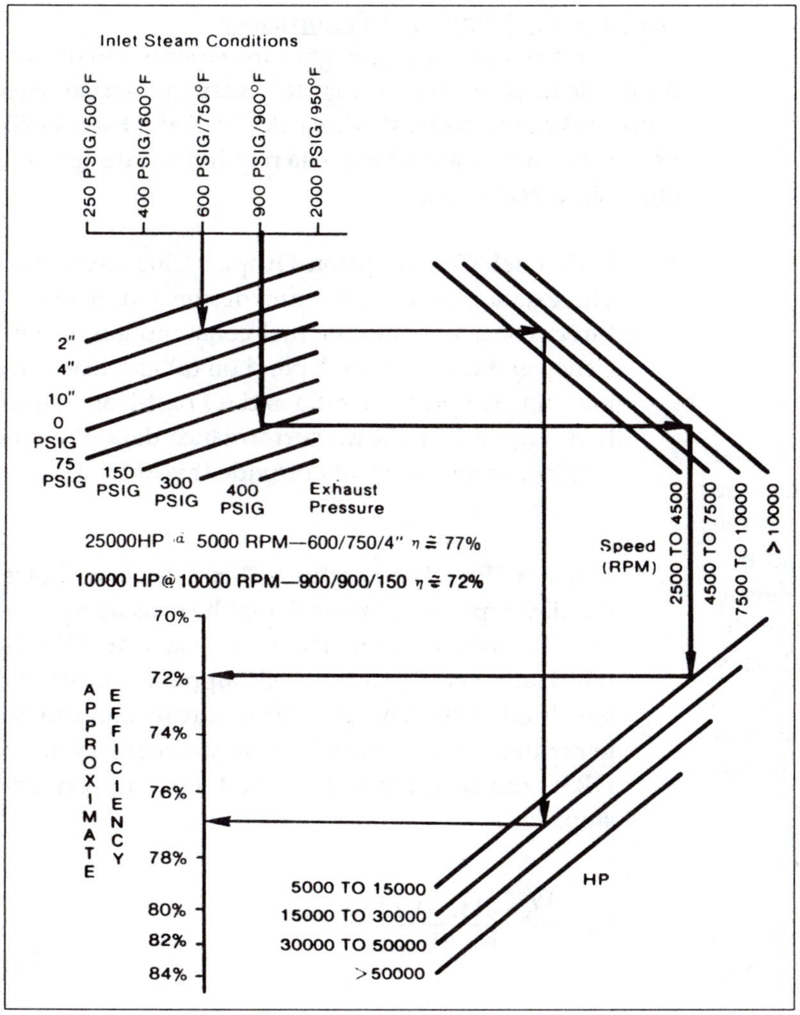

Enthalpy, h_i : 1378.9 Btu/lb (from Mollier chart or steam tables)

Outlet Steam : 4" HG absolute. (2 psia)

Enthalpy, h_o : 935 0 Btu/lb (from Mollier chart or steam tables)

Turbine Speed : 5000 RPM

a) Calculate TSR using equation 7.1 (TSR can also be obtained from ASME Tables or the Mollier chart):

$$TSR\ (lb/kWh) = \frac{3412\ Btu/kWh}{h_i - h_o\ Btu/lb}$$

$$= \frac{3412\ Btu/kWh}{1378.9 - 935.0\ Btu/lb}$$

$$= 7.68\ lb/kWh$$

b) Using the data above, Figure 7.8 gives η_{tg} = 77%.

c) Combining equations 7.2 and 7.4 and solving for W_s, the total steam flow required is

$$W_s = PGC \times TSR\ /\ \eta_{tg}$$

$$= \frac{18,500\ kW \times 7.68\ lb/kWh}{0.77}$$

$$= 184,760\ lb/hr$$

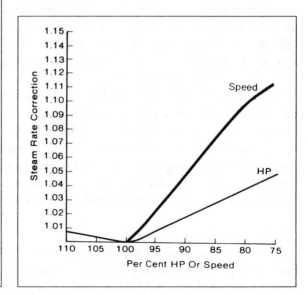

Figure 7.8 Approximate steam turbine efficiency chart for multistage steam turbines (>3,000 kW). *Courtesy Elliot Co.*

Figure 7.9 Steam turbine part-load/speed correction curves. *Courtesy Elliot Co.*

d) Hence, the waste heat recovery steam generator should be able to generate about 185,000 lb/hr. A unit with a 200,000 lb/hr nominal capacity will likely be specified.

Example 3. Find the ASR and steam flow required when the turbine of Example 2 is operated at 14,800 kW and 4,500 RPM.

% Power variation = 14,800 kW/185,500 kW = 80%

% Speed variation = 4500 RPM/5000 RPM = 90%

From Figure 7.9 the power correction factor is 1.04 and the speed correction factor is 1.05. Then, the total correction factor is:

$$1.04 \times 1.05 = 1.09.$$

Therefore, the part-load ASR is

$$= \frac{7.68 \text{ lb/kWh}}{0.77} \times 1.09 = 10.9 \text{ lb/kWh}$$

The steam flow required (@ 600 psig, 750°F) is

$$= 10.9 \text{ lb/kWh} \times 14,500 \text{ kW}$$

$$= 158,050 \text{ lb/hr.}$$

7.2.2.2 Gas Turbine Systems

Gas turbines are extensively applied in industrial plants. Two types of gas turbines are utilized: one is the lighter aircraft derivative turbine and the other is the heavier industrial gas turbine. Both industrial and aircraft engines have demonstrated excellent reliability/availability in the base load service. Due to the nature of the unit designs, aircraft derivative units usually have higher maintenance costs ($/kWh) than the industrial units.

Since gas turbines can burn a variety of liquid and gas fuels and run long times unattended, they are considered to be versatile and reliable. For a fixed capacity, they have the smallest relative foot-print (sq-ft per kW).

<u>Gas Turbine Based CHP Systems</u>

Exhaust gases from gas turbines (from 600 to 1200°F) offer a large heat recovery potential. The exhaust has been used directly, as in drying processes. Topping cycles have also been developed by using the exhaust gases to generate process steam in heat recovery steam generators

(HRSGs). Where larger power loads exist, high pressure steam is generated to be subsequently expanded in a steam turbine-generator; this constitutes the so called combined cycle (See Fig 7.4).

If the demand for steam and/or power is even higher, the exhaust gases are used (1) as preheated combustion air of a combustion process or (2) are additionally fired by a "duct burner" to increase their heat content and temperature.

Recent developments include combined cycle systems with steam injection or STIG—from the HRSG to the gas turbine (Cheng Cycle)—to augment and modulate the electrical output of the system. The Cheng Cycle allows the cogeneration system to handle a wider range of varying heat and power loads.

All these options present a greater degree of CHP generation flexibility, allowing a gas turbine system to match a wider variety of heat-to-power demand ratios and variable loads.

<u>Gas Turbine Ratings and Performance</u>

There is a wide range of gas turbine sizes and drives. Available turbines have ratings that vary in discrete sizes from 50 kW to 160,000 kW. Kovacik (1982) and Hay (1988) list the following gas turbine data required for design and off-design conditions:

1. **Unit Fuel Consumption-Output Characteristics.** These data depends in the unit design and manufacturer. The actual specific fuel consumption or efficiency and output also depend on (a) ambient temperature, (b) pressure ratio and (c) part-load operation. Figure 7.11 shows performance data of a gas turbine. Vendors usually provide this kind of information.

2. **Exhaust Flow Temperature.** This data item allows the development of the exhaust heat recovery system. The most common recovery system are HRSGs which are classified as unfired, supplementary fired and fired units. The amount of steam that can be generated in an unfired or supplementary fired HRSG can be estimated by the following relationship:

$$W_s = \frac{W_g \, C_p \left(T_1 - T_3\right) e \, L \, f}{h_{sh} - h_{sat}}$$

(7.5)

where
W_s = steam flow rate
W_g = exhaust flow rate to HRSG

C_p = specific heat of products of combustion
T_1 = gas temperature—after burner, if applicable
T_3 = saturation temperature in steam drum
L = a factor to account radiation and other losses, 0.985
h_{sh} = enthalpy of steam leaving superheater
h_{sat} = saturated liquid enthalpy in the steam drum
e = HRSG effectiveness = $(T_1\text{-}T_2)/(T_1\text{-}T_3)$, defined by Fig. 7.10.
f = fuel factor, 1.0 for fuel oil, 1.015 for gas.

3. Parametric Studies for Off-design Conditions. Varying the amount of primary or supplementary firing will change the gas flow rate or temperature and the HRSG steam output. Thus, according to the varying temperatures, several iterations of equation [7.5] are required to evaluate off-design or part load conditions. When this evaluation is carried over a range of loads, firing rates, and temperatures, it is called a parametric study. Models can be constructed or off-design conditions using gas turbine performance data provided by manufacturers (See Figure 7.11).

4. Exhaust Pressure Effects on Output and Exhaust Temperature. Heat recovery systems increase the exhaust back-pressure, reducing the turbine output in relation to simple operation (without HRSG). Turbine manufacturers provide test data about inlet and back-pressure effects, as well as elevation effects, on turbine output and efficiency (Fig 7.11).

Example 4. In a combined cycle (Fig 7.4), the steam for the turbine of Example 2 must be generated by several HRSGs (See Fig 7.10). To follow variable CHP loads and to optimize overall system reliability, each HRSG will be connected to a dedicated 10-MWe gas turbine-generator (GTG)

set. The GTG sets burn fuel-oil and each unit exhausts 140,000 kg/hr of gas at 900°F. Estimate the total gas flow to the HRSGs, if the system must produce a maximum of 160,000 lb/hr of 615 psia/750°F steam. How many HRSG-gas turbine sets are needed?

DATA (See notation of Equation 7.5 and Figure 7.10)

Gas	W_g	=	140,000 kg/hr per gas-turbine unit
Turbines	C_p	=	0.26 Btu/lb/°F, average specific heat of gases between T_1 and T_4
	T_1	=	900°F, hot gas temperature
	f	=	fuel factor, 1 for fuel oil.

The inlet steam conditions required in the steam turbine of Example 3 are: (615 psia/750°F) and h_i = 1378.9 Btu/lb (from Mollier chart). Thus,

HRSGs	W_s	=	160,000 lb/hr from all HRSGs
	T_3	=	488.8°F (temp of 615 psia sat steam)
	L	=	0.98
	e	=	HRSG effectiveness, 0.9

Therefore, h_{sh} = h_i = 1378.9 Btu/lb and,
h_{sat} = 474.7 Btu/lb (Sat. Water @ 615 psia)

From equation 7.5, the total combustion gas flow required is

$$W_g = \frac{W_s \left(h_{sh} - h_{sat}\right)}{C_p \left(T_1 - T_3\right) e\, L\, f}$$

$$= \frac{160,000\ \text{lb/hr}\ (1381.1 - 474.7\ \text{Btu/lb})}{0.26\ \text{Btu/lb/°F}\ (900 - 488.8\text{°F})\ 0.9 \times 0.98 \times 1}$$

$$= 1,537,959.3\ \text{lb/hr or } 427.2\ \text{lb/sec}.$$

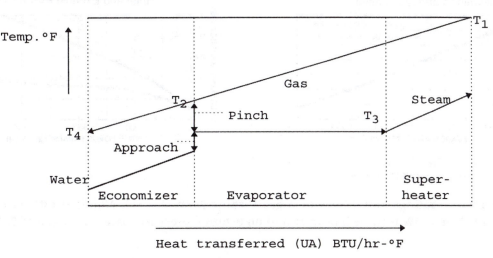

Fig 7.10 Heat recovery steam generator diagram.

FUEL	ISO RATING	POWER kW$_e$	HEAT RATE kJ/kWh (Btu/kWh)		EXHAUST FLOW kg/hr (lb/hr)		EXHAUST TEMP °C (°F)	
Natural Gas	Continuous	4370	12 870	(12 200)	75 050	(165 120)	497	(927)
	Peak	4800	12 865	(12 195)	74 970	(165 280)	534	(994)
Distillate	Continuous	4280	13 135	(12 450)	75 050	(165 120)	497	(927)
	Peak	4700	13 130	(12 445)	74 970	(165 280)	534	(994)

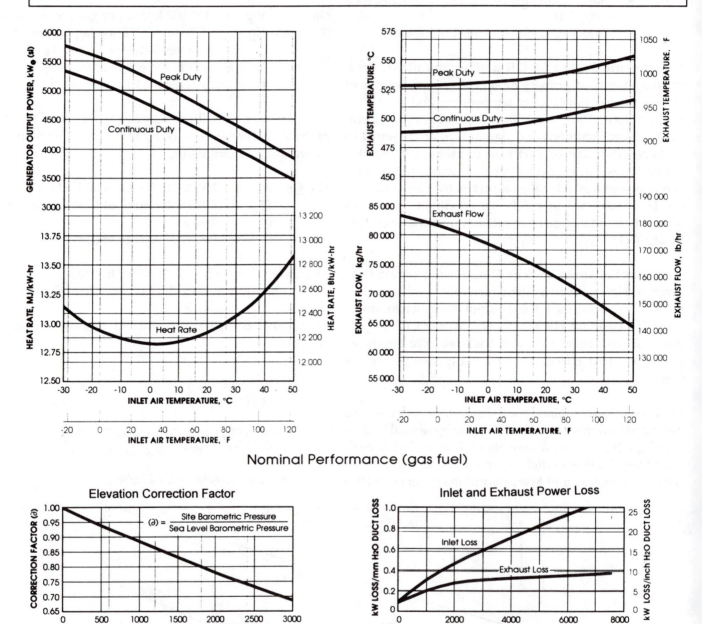

Figure 7.11 Performance ratings and curves of a gas turbine-generator set. Nominal ratings are given at ISO conditions: 59°F, sea level, 60% relative humidity and no external pressure losses. *Courtesy of Solar Turbines, Inc.*

Finally, the number of required gas turbine/HRSG sets is

= {(1,537,959 lb/hr) (1kg/2.21b)/(140,000 kg/hr/set)}

= 4.99, i.e. 5 sets.

7.2.2.3 Reciprocating Engine Systems

Reciprocating engines include a variety of internally fired, piston driven engines. Their sizes range from 10 bhp to 50,000 bhp. According to Kovacik (1982), the largest unit supplied by a U.S. manufacturer is rated at 13,500 bhp. In larger plants, several units are used to accommodate part load and to provide redundancy and better availability.

In these engines, combustion heat rejected through the jacket water, lube oil and exhaust gases, can be recovered through heat exchangers to generate hot water and/ or steam. Fig. 7.13 shows an internal combustion engine cogeneration system.

Exhaust gases have also been used directly. Reciprocating engines are classified by:

— the thermodynamic cycle: Diesel or Otto cycle.

— the rotation speed: high-speed (1200-1800 rpm) medium-speed (500-900 rpm) low speed (450 rpm or less)

— the aspiration type: naturally aspirated or turbocharged

— the operating cycle: two-cycle or four-cycle

— the fuel burned: fuel-oil fired, natural-gas fired.

Reciprocating engines are widely used to move vehicles, generators and a variety of shaft loads. Larger engines are associated to lower speeds, increased torque, and heavier duties. The total heat utilization of CHP systems based on gas-fired or fuel-oil fired engines approach 60-75%. Figure 7.12 shows the CHP balance vs. load of a diesel engine.

Example 5. Estimate the amount of 180 F water that can be produced by recovering heat; first from the jacket water and then from the exhaust of a 1200 kW diesel generator. In the average, the engine runs at a 75% load and the inlet water temperature is 70°F. The effectiveness of the jacket water heat exchanger is 90% and exhaust heat exchanger is 80%.

From Figure 7.12, at 75% load, the exhaust heat and the jacket water heat are 22% and 33%, respectively. Thus, the flow rate of water heated from 70 to 180°F is:

= (1200 kW × 75%) [(90% × 22%) + (80% × 33%)]
 (3412 Btu/kW) (1 lb.°F/Btu)(1 lb/8:33 gal)
 (1 hr/60 min)/(180 − 70°F)
= 25.8 gal per minute.

7.2.2.4 Packaged Cogeneration

Packaged cogeneration involves the development, integration and marketing of complete "off the shelf," "modular" or "total-solution" cogeneration systems. It was basically an unknown technology before the '80s. Generally speaking, the impact of packaged cogeneration systems has been in the smaller scale range (<800 kW) and mainly for commercial or institutional applications. Most of the cogeneration packages are based on reciprocating engines and nearly all produce hot water. However, skid mounted gas-turbine based combined cycle systems are currently offered in the 3 to 20 MW_e range.

Also, packaged gas-engine driven chillers and indirect fired absorption chiller-heaters are being actively marketed. Figure 7.13 is a diagram of a packaged gas-engine driven refrigeration compressor with heat recovery. In addition, there exist "custom made" packaged systems. Landers (1988) describes several "site specific

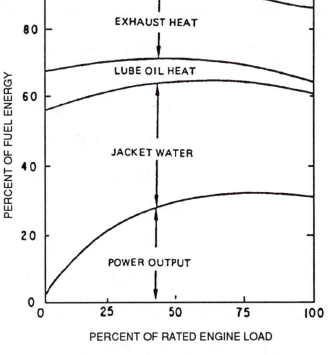

Fig. 7.12 Diesel engine heat and power balance.

Figure 7.13 This packaged cogeneration system includes a natural gas fired V-8 engine, refrigeration compressor, and heat exchangers to recover jacket water and exhaust heat (*Courtesy of Tecogen, Inc.*).

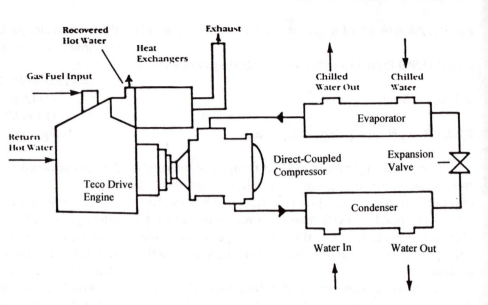

total system package" application cases. He lists the following benefits of packaged cogeneration systems:

1. The ability to test the total system (and to make any necessary corrections) at the factory.

2. Smaller foot print of the total system.

3. Reduced installation and start-up time.

4. Reduced overall costs due to reduced installation costs.

Kostrzewa and Davidson (1988) reported that industrial cogeneration applications were the first to develop due to economies of scale of larger systems, continuous operating characteristics and year-round demands for process heat.

Conversely, applications in commercial and smaller institutional facilities have been slower to develop. Among

the reasons that make packaged cogeneration systems a less attractive investment, we have: a higher unit cost ($/kWe), the comparatively small capacity, reduced operating hours and cyclic/seasonal loads. Also, energy costs for these facilities are a lesser portion of business operating costs and not always visible to the building occupant.

They remark that remaining hurdles to be overcome in the establishment of a true market for (packaged) cogeneration include consumer awareness, electric utility resistance reduction, and establishment of sales/service infrastructure, in addition to the cost issues noted previously.

Nevertheless, first costs (equipment and installation) for packaged cogeneration are considered a very significant economic factor. Mulloney et. al. (1988) reported that in commercial, residential, and speculative buildings, first cost is deemed more important than life-cycle costs. Table 7.2 lists typical packaged cogeneration installed costs.

Table 7.2 Prepackaged Cogeneration Installed Cost ($/kWe)

COST CATEGORY	AVERAGE	RANGE
F.O.B. UNIT COST	729	500 – 1,100
INSTALLATION (SITE) COST	541	100 – 1,900
INTEGRATION (SOFT) COSTS	94	0 – 500
TOTAL INSTALLED COST*	1,344	600 – 2,800

Source: Survey by Mulloney et al (1988)

*The first three cost categories are not additive to the total because some sites did not report each category but all sites reported total installed cost.

7.2.3 The Cogeneration Design Process

The following evaluation steps are suggested to carry out cogeneration system design.

1. Develop the profile of the various process steam (heat) demands at the appropriate steam pressures for the applications being studied. Also, collect data with regard to condensate returned from the process and its temperature. Data must include daily fluctuations due to normal variations in process needs, as well as seasonal weather effects; including the influence of not-working periods such as weekends, vacation periods, and holidays.

2. A profile for electric power must be developed in the same manner as the process heat demand profile. These profiles typically include hour-by-hour heat and power demands for "typical" days (or weeks) for each season or month of the year.

3. Fuel availability and present-day cost as well as projected future costs. The study should also factor process by-product fuels into the development of the energy supply system.

4. Purchased power availability and its present and expected future cost.

5. Plant discharge stream data in the same degree of detail as the process heat demand data.

6. Number and rating of major (demand and generation) equipment items. This evaluation usually establish whether spare capacity and/or supplementary firing should be installed.

7. Plant, process and CHP system economic lives.

Once this initial data bank has been established, the various alternatives that can satisfy plant heat and power demands can be identified. Subsequently, detailed technical analyses are conducted. Thus, energy balances are made, investment cost estimated, and the economic merit of each alternative evaluated. Some approaches for evaluation are discussed next.

7.2.4 Economic Feasibility Evaluation Methods

Cogeneration feasibility evaluation is an iterative process—further evaluations generally require more data. There are a number of evaluation methods using various approaches and different levels of technical detail. Most of them consider seasonal loads and equipment performance characteristics. Some of the most representative methods are discussed as follows.

7.2.4.1 General Approaches For Design and Evaluation

Hay (1988) presents a structured approach for system design and evaluation. It is a sequence of evaluation iterations, each greater than the previous and each producing information whether the costs of the next step is warranted. His suggested design process is based on the following steps.

Step 1: Site Walkthrough and Technical Screening
Step 2: Preliminary Economic Screening
Step 3: Detailed Engineering Design.

Similarly, Butler (1984) considers three steps to perform studies, engineering and construction of cogeneration projects. These are discussed as follow.

Step 1. Preliminary studies and conceptual engineering. This is achieved by performing a technical feasibility and economic cost-benefit study to rank and recommend alternatives. The determination of technical feasibility includes a realistic assessment with respect to environmental impact, regulatory compliance, and interface with a utility. Then, an economic analysis—based on the simple payback period-serves as a basis for more refined evaluations.

Step 2. Engineering and Construction Planning. Once an alternative has been selected and approved by the owner, preliminary engineering is started to develop the general design criteria. These include specific site information such as process heat and power requirements, fuel availability and pricing, system type definition, modes of operation, system interface, review of alternatives under more detailed load and equipment data, confirm selected alternative and finally size the plant equipment and systems to match the application.

Step 3. Design Documentation. This includes the preparation of project flow charts, piping and instrument diagrams, general arrangement drawings, equipment layouts, process interface layouts, building, structural and foundation drawings, electrical diagrams, and specifying an energy management system, if required.

Several methodologies and manuals have been developed to carry out Step 1, i.e. screening analysis and preliminary feasibility studies. Some of them are briefly discussed in the next sections. Steps 2 and 3 usually require ad-hoc approaches according to the characteristics of each particular site. Therefore, a general methodology is not applicable for such activities.

7.2.4.2 Preliminary Feasibility Study Approaches

AGA Manual—GKCO Consultants (1982) developed a

cogeneration feasibility (technical and economical) evaluation manual for the American Gas Association, AGA. It contains a "Cogeneration Conceptual Design Guide" that provides guidelines for the development of plant designs. It specifies the following steps to conduct the site feasibility study:

a)   Select the type of prime mover or cycle (piston engine, gas turbine or steam turbine);

b)   Determine the total installed capacity;

c)   Determine the size and number of prime movers;

d)   Determine the required standby capacity.

According to its authors "the approach taken (in the manual) is to develop the minimal amount of information required for the feasibility analysis, deferring more rigorous and comprehensive analyses to the actual concept study." The approach includes the discussion of the following "Design Options" or design criteria to determine (1) the size and (2) the operation mode of the CHP system.

Isolated Operation, Electric Load Following—The facility is independent of the electric utility grid, and is required to produce all power required on-site and to provide all required reserves for scheduled and unscheduled maintenance.

Baseloaded, Electrically Sized—The facility is sized for baseloaded operation based on the minimum historic billing demand. Supplemental power is purchased from the utility grid. This facility concept generally results in a shorter payback period than that from the isolated site.

Baseloaded, Thermally Sized—The facility is sized to provide most of the site's required thermal energy using recovered heat. The engines operated to follow the thermal demand with supplemental boiler fired as required. The authors point out that: "this option frequently results in the production of more power than is required on-site and this power is sold to the electric utility.

In addition, the AGA manual includes a description of sources of information or processes by which background data can be developed for the specific gas distribution service area. Such information can be used to adapt the feasibility screening procedures to a specific utility.

7.2.4.3 Cogeneration System Selection and Sizing.

The selection of a set of "candidate" cogeneration systems entails to tentatively specify the most appropriate prime mover technology, which will be further evaluated in the course of the study. Often, two or more alternative systems that meet the technical requirements are pre-selected for further evaluation. For instance, a plant's CHP requirements can be met by either, a reciprocating engine system or combustion turbine system. Thus, the two system technologies are pre-selected for a more detailed economic analysis.

To evaluate specific technologies, there exist a vast number of technology-specific manuals and references. A representative sample is listed as follows. Mackay (1983) has developed a manual titled "Gas Turbine Cogeneration: Design, Evaluation and Installation." Kovacik (1984) reviews application considerations for both steam turbine and gas turbine cogeneration systems. Limaye (1987) has compiled several case studies on industrial cogeneration applications. Hay (1988) discusses technical and economic considerations for cogeneration application of gas engines, gas turbines, steam engines and packaged systems. Keklhofer (1991) has written a treatise on technical and economic analysis of combined-cycle gas and steam turbine power plants. Ganapathy (1991) has produced a manual on waste heat boilers.

Usually, system selection is assumed to be separate from sizing the cogeneration equipment (kWe). However, since performance, reliability and cost are very dependent on equipment size and number, technology selection and system size are very intertwined evaluation factors. In addition to the system design criteria given by the AGA manual, several approaches for cogeneration system selection and/or sizing are discussed as follows.

Heat-to-Power Ratio

Canton et al (1987) of The Combustion and Fuels Research Group at Texas A&M University has developed a methodology to select a cogeneration system for a given industrial application using the heat to power ratio (HPR). The methodology includes a series of graphs used 1) to define the load HPR and 2) to compare and match the load HPR to the HPRs of existing equipment. Consideration is then given to either, heat or power load matching and modulation.

Sizing Procedures

Hay (1987) considers the use of the load duration curve to model variable thermal and electrical loads in system sizing, along with four different scenarios described in Figure 7.14. Each one of this scenarios defines an operating alternative associated to a system size.

Oven (1991) discusses the use of the load duration

curve to model variable thermal and electrical loads in system sizing in conjunction with required thermal and electrical load factors. Given the thermal load duration and electrical load duration curves for a particular facility, different sizing alternatives can be defined for various load factors.

Eastey et al. (1984) discusses a model (COGENOPT) for sizing cogeneration systems. The basic inputs to the model are a set of thermal and electric profiles, the cost of fuels and electricity, equipment cost and performance for a particular technology. The model calculates the operating costs and the number of units for different system sizes. Then it estimates the net present value for each one of them. Based on the maximum net present value, the "optimum" system is selected. The model includes cost and load escalation.

Wong, Ganesh and Turner (1991) have developed two statistical computer models to optimize cogeneration system size subject to varying capacities/loads and to meet an availability requirement. One model is for internal combustion engine and the other for unfired gas turbine cogeneration systems. Once the user defines a required availability, the models determine the system size or capacity that meets the required availability and maximizes the expected annual worth of its life cycle cost

7.3   COMPUTER PROGRAMS

There are several computer programs—mainly PC based—available for detailed evaluation of cogeneration systems. In opposition to the rather simple methods discussed above, CHP programs are intended for system configuration or detailed design and analysis. For these reasons, they require a vast amount of input data. Below, we examine two of the most well known programs.

7.3.1 CELCAP

Lee (1988) reports that the Naval Civil Engineering Laboratory developed a cogeneration analysis computer program known as Civil Engineering Laboratory Cogeneration Program (CELCAP), "for the purpose of evaluating the performance of cogeneration systems on a life-cycle operating cost basis. He states that "selection of a cogeneration energy system for a specific application is a complex task." He points out that the first step in the selection of cogeneration system is to make a list of potential candidates. These candidates should include single or multiple combinations of the various types of engine available. The computer program does not specify CHP systems; these must be selected by the designer. Thus, depending on the training and previous experience of the designer, different designers may select different systems

of different sizes.

After selecting a short-list of candidates, modes of operations are defined for the candidates. So, if there are N candidates and M modes of operation, then N×M alternatives must be evaluated. Lee considers three modes of operation:

1) Prime movers operating at their full-rated capacity, any excess electricity is sold to the utility and any excess heat is rejected to the environment. Any electricity shortage is made up with imports. Process steam shortages are made-up by an auxiliary boiler.

2) Prime movers are specified to always meet the entire electrical load of the user. Steam or heat demand is met by the prime mover. An auxiliary boiler is fired to meet any excess heat deficit and excess heat is rejected to the environment.

3) Prime movers are operated to just meet the steam or heat load. In this mode, power deficits are made up by purchased electricity. Similarly, any excess power is sold back to the utility.

For load analysis, Lee considers that "demand of the user is continuously changing. This requires that data on the electrical and thermal demands of the user be available for at least one year." He further states that "electrical and heat demands of a user vary during the year because of the changing working and weather conditions." However, for evaluation purposes, he assumes that the working conditions of the user—production related CHP load—remain constant and "that the energy-demand pattern does not change significantly from year to year." Thus, to consider working condition variations, Lee classifies the days of the year as working and non-working days. Then, he uses "average" monthly load profiles and "typical" 24-hour load profiles for each class.

"Average" load profiles are based on electric and steam consumption for an average weather condition at the site. A load profile is developed for each month, thus monthly weather and consumption data is required. A best fit of consumption (Btu/month or kWh/month) versus heating and cooling degree days is thus obtained. Then, actual hourly load profiles for working and non-working days for each month of the year are developed. The "best representative" profile is then chosen for the "typical working day" of the month. A similar procedure is done for the non-working days.

Next an energy balance or reconciliation is performed to make sure the consumption of the hourly load profiles agrees with the monthly energy usage. A multiplying factor K is defined to adjust load profiles that do not balance.

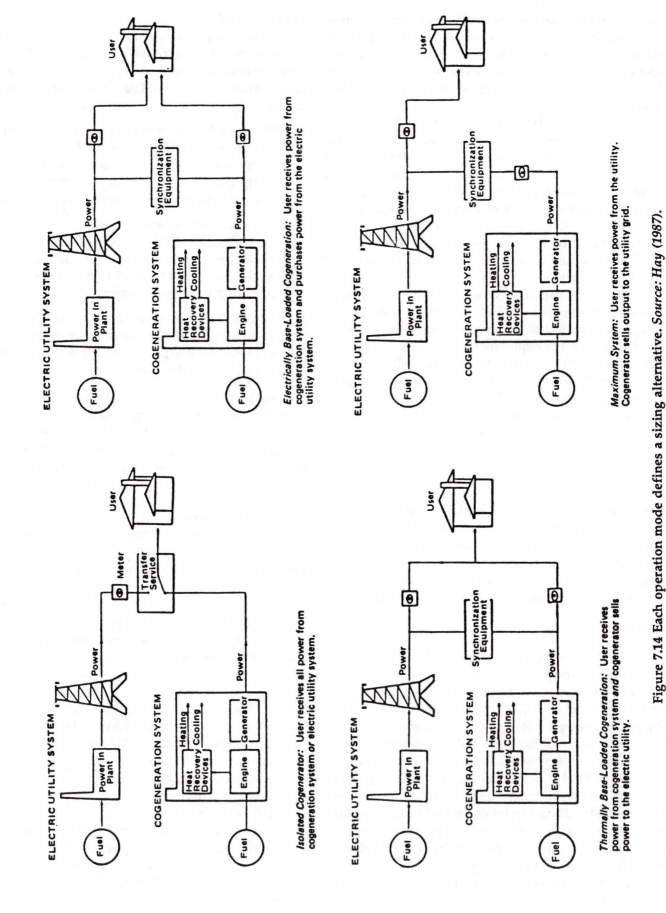

Electrically Base-Loaded Cogeneration: User receives power from cogeneration system and purchases power from the electric utility system.

Maximum System: User receives power from the utility. Cogenerator sells output to the utility grid.

Isolated Cogenerator: User receives all power from cogeneration system *or* electric utility system.

Thermally Base-Loaded Cogeneration: User receives power from cogeneration system *and* cogenerator sells power to the electric utility.

Figure 7.14 Each operation mode defines a sizing alternative. *Source: Hay (1987).*

$$K_j = E_{mj}/(AE_{wj} + AE_{nwj}) \qquad (7.9)$$

where

K_j = multiplying factor for month j

E_{mj} = average consumption (kWh) by the user for the month j selected from the monthly electricity usage versus degree day plot

AE_{wj} = typical working-day electric usage (kWh), i.e. the area under the typical working day electric demand profile for the month j

AE_{nwj} = typical non-working day usage (kWh), i.e. the area under the typical non-working day electric demand profile for the month j.

Lee suggests that each hourly load in the load profiles be multiplied by the K factor to obtain the "correct working and non-working day load profiles for the month." The procedure is repeated for all months of the year for both electric and steam demands. Lee states that "the resulting load profiles represent the load demand for average weather conditions."

Once a number of candidate CHP systems has been selected, equipment performance data and the load profiles are fed into CELCAP to produce the required output. The output can be obtained in a brief or detailed form. In brief form, the output consists of a summary of input data and a life cycle cost analysis including fuel, operation and maintenance and purchased power costs. The detailed printout includes all the information of the brief printout, plus hourly performance data for 2 days in each month of the year. It also includes the maximum hourly CHP output and fuel consumption. The hourly electric demand and supply are plotted, along with the hourly steam demand and supply for each month of the year.

Despite the simplifying assumptions introduced by Lee to generate average monthly and typical daily load profiles, it is evident that still a large amount of data handling and preparation is required before CELCAP is run. By recognizing the fact that CHP loads vary over time, he implicitly justifies the amount of effort in representing the input data through hourly profiles for typical working and non-working days of the month.

If a change occurs in the products, process or equipment that constitute the energy consumers within the industrial plant, a new set of load profiles must be generated. Thus, exploring different conditions requires sensitivity analyses or parametric studies for off-design conditions.

A problem that becomes evident at this point is that, to accurately represent varying loads, a large number of load data points must be estimated for subsequent use in the computer program. Conversely, the preliminary feasibility evaluation methods discussed previously, require very few and only "average" load data. However, criticism of preliminary methods has arisen for not being able to truly reflect seasonal variations in load analysis (and economic analysis) and for lacking the flexibility to represent varying CHP system performance at varying loads.

7.3.2 COGENMASTER

Limaye and Balakrishnan (1989) of Synergic Resources Corporation have developed COGENMASTER. It is a computer program to model the technical aspects of alternative cogeneration systems and options, evaluate economic feasibility, and prepare detailed cash flow statements.

COGENMASTER compares the CHP alternatives to a base case system where electricity is purchased from the utility and thermal energy is generated at the site. They extend the concept of an option by referring not only to different technologies and operating strategies but also to different ownership structures and financing arrangements. The program has two main sections: a Technology and a Financial Section. The technology Section includes 5 modules:

- Technology Database Module

- Rates Module

- Load Module

- Sizing Module

- Operating Module

The Financial Section includes 3 modules:

- Financing Module

- Cash Flow Module

- Pricing Module

In COGENMASTER, facility electric and thermal loads may be entered in one of three ways, depending on the available data and the detail required for project evaluation:

— A constant average load for every hour of the year.

— Hourly data for three typical days of the year

— Hourly data for three typical days of each month

Thermal loads may be in the form of hot water or steam; but system outlet conditions must be specified by the user. The sizing and operating modules permit a variety of alternatives and combinations to be considered. The system may be sized for the base or peak, summer or winter, and electric or thermal load. There is also an option for the user to define the size the system in kilowatts. Once the system size is defined, several operation modes may be selected. The system may be operated in the electric following, thermal following or constantly running modes of operation. Thus, N sizing options and M operations modes define a total of N×M cogeneration alternatives, from which the "best" alternative must be selected. The economic analysis is based on simple payback estimates for the CHP candidates versus a base case or do-nothing scenario. Next, depending on the financing options available, different cash flows may be defined and further economic analysis—based on the Net Present Value of the alternatives—may be performed.

7.4   U.S. COGENERATION LEGISLATION: PURPA

In 1978 the U.S. Congress amended the Federal Power Act by promulgation of the Public Utilities Regulatory Act (PURPA). The Act recognized the energy saving potential of industrial cogeneration and small power plants, the need for real and significant incentives for development of these facilities and the private sector requirement to remain unregulated.

PURPA of 1978 eliminated several obstacles to cogeneration so cogenerators can count on "fair" treatment by the local electric utility with regard to interconnection, back-up power supplies, and the sale of excess power. PURPA contains the major federal initiatives regarding cogeneration and small power production. These initiatives are stated as rules and regulations pertaining to PURPA Sections 210 and 201; which were issued in final form in February and March of 1980, respectively. These rules and regulations are discussed in the following sections.

Initially, several utilities—especially those with excess capacity—were reticent to buy cogenerated power and have, in the past, contested PURPA. Power (1980) magazine reported several cases in which opposition persisted in some utilities to private cogeneration. But after the Supreme Court ruling in favor of PURPA, more and more utilities are finding that PURPA can work to their advantage. Polsky and Landry (1987) report that some

utilities are changing attitudes and are even investing in cogeneration projects.

7.4.1 PURPA 201*

Section 201 of PURPA requires the Federal Energy Regulatory Commission (FERC) to define the criteria and procedures by which small power producers (SPPs) and cogeneration facilities can obtain qualifying status to receive the rate benefits and exemptions set forth in Section 210 of PURPA. Some PURPA 201 definitions are stated below.

Small Power Production Facility

A "Small Power Production Facility" is a facility that uses biomass, waste, or renewable resources, including wind, solar and water, to produce electric power and is not greater than 80 megawatts.

Facilities less than 30 MW are exempt from the Public Utility Holding Co. Act and certain state law and regulation. Plants of 30 to 80 MW which use biomass, may be exempted from the above but may not be exempted from certain sections of the Federal Power Act.

Cogeneration Facility

A "Cogeneration Facility" is a facility which produces electric energy and forms of useful thermal energy (such as heat or steam) used for industrial, commercial, heating or cooling purposes, through the sequential use of energy. A Qualifying Facility (QF) must meet certain minimum efficiency standards as described later. Cogeneration facilities are generally classified as "topping" cycle or "bottoming" cycle facilities.

7.4.2 Qualification of a "Cogeneration Facility" or a "Small Power Production Facility" under PURPA

Cogeneration Facilities

To distinguish new cogeneration facilities which will achieve meaningful energy conservation from those which would be "token" facilities producing trivial amounts of either useful heat or power, the FERC rules establish operating and efficiency standards for both topping-cycle and bottom-cycle NEW cogeneration facilities. No efficiency standards are required for EXISTING cogeneration facilities regardless of energy source or type of facility. The following fuel utilization effectiveness (FUE) values—based on the lower heating value (LHV) of the fuel—are required from QFs.

*Most of the following sections have been adapted from CFR18 (1990) and Harkins (1980), unless quoted otherwise.

For a new topping-cycle facility:

- No less than 5% of the total annual energy output of the facility must be useful thermal energy.

- For any new topping-cycle facility that uses any natural gas or oil:

  — All the useful electric power and half the useful thermal energy must equal at least 42.5% of the total annual natural gas and oil energy input; and

  — If the useful thermal output of a facility is less than 15% of the total energy output of the facility, the useful power output plus one-half the useful thermal energy output must be no less than 45% of the total energy input of natural gas and oil for the calendar.

For a new bottoming-cycle facility:

- If supplementary firing (heating of water or steam before entering the electricity generation cycle from the thermal energy cycle) is done with oil or gas, the useful power output of the bottoming cycle must, during any calendar year, be no less than 45% of the energy input of natural gas and oil for supplementary firing.

Small Power Production Facilities

To qualify as a small power production facility under PURPA, the facility must have production capacity of under 80 MW and must get more than 50% of its total energy input from biomass, waste, or renewable resources. Also, use of oil, coal, or natural gas by the facility may not exceed 25% of total annual energy input to the facility.

Ownership Rules Applying to Cogeneration and Small Power Producers

A qualifying facility may not have more than 50% of the equal interest in the facility held by an electric utility.

7.4.3 PURPA 210

Section 210 of PURPA directs the Federal Energy Regulatory Commission (FERC) to establish the rules and regulations requiring electric utilities to purchase electric power from and sell electric power to qualifying cogeneration and small power production facilities and provide for the exemption to qualifying facilities (QF) from certain federal and state regulations.

Thus, FERC issued in 1980 a series of rules to relax obstacles to cogeneration. Such rules implement sections of the 1978 PURPA and include detailed instructions to state utility commissions that all utilities must purchase electricity from cogenerators and small power producers at the utilities' "avoided" cost. In a nutshell, this means that rates paid by utilities for such electricity must reflect the cost savings they realize by being able to avoid capacity additions and fuel usage of their own.

Tuttle (1980) states that prior to PURPA 210, cogeneration facilities wishing to sell their power were faced with three major obstacles:

* Utilities had no obligation to purchase power, and contended that cogeneration facilities were too small and unreliable. As a result, even those cogenerators able to sell power had difficulty getting an equitable price.

* Utility rates for backup power were high and often discriminatory

* Cogenerators often were subject to the same strict state and federal regulations as the utility.

PURPA was designed to remove these obstacles, by requiring utilities to develop an equitable program of integrating cogenerated power into their loads.

Avoided Costs

The costs avoided by a utility when a cogeneration plant displaces generation capacity and/or fuel usage are the basis to set the rates paid by utilities for cogenerated power sold back to the utility grid. In some circumstances, the actual rates may be higher or lower than the avoided costs, depending on the need of the utility for additional power and on the outcomes of the negotiations between the parties involved in the cogeneration development process.

All utilities are now required by PURPA to provide data regarding present and future electricity costs on a cent-per-kWh basis during daily, seasonal, peak and off-peak periods for the next five years. This information must also include estimates on planned utility capacity additions and retirements, and cost of new capacity and energy costs.

Tuttle (1980) points out that utilities may agree to pay greater price for power if a cogeneration facility can:

* Furnish information on demonstrated reliability and term of commitment.

*   Allow the utility to regulate the power production for better control of its load and demand changes.

*   Schedule maintenance outages for low-demand periods.

*   Provide energy during utility-system daily and seasonal peaks and emergencies.

*   Reduce in-house on-site load usage during emergencies.

*   Avoid line losses the utility otherwise would have incurred.

In conclusion, a utility is willing to pay better "buy-back" rates for cogenerated power if it is short in capacity, if it can exercise a level of control on the CHP plant and load, and if the cogenerator can provide and/or demonstrate a "high" system availability.

PURPA further states that the utility is not obligated to purchase electricity from a QF during periods that would result in net increases in its operating costs. Thus, low demand periods must be identified by the utility and the cogenerator must be notified in advance. During emergencies (utility outages), the QF is not required to provide more power than its contract requires, but a utility has the right to discontinue power purchases if they contribute to the outage.

7.4.4 Other Regulations

Several U.S. regulations are related to cogeneration. For example, among environmental regulations, the Clean Air Act may control emissions from a waste-to-energy power plant. Another example is the regulation of underground storage tanks by the Resource Conservation and Recovery Act (RCRA). This applies to all those cogenerators that store liquid fuels in underground tanks. Thus, to maximize benefits and to avoid costly penalties, cogeneration planners and developers should become savvy in related environmental matters.

There are many other issues that affect the development and operation of a cogeneration project. For further study, the reader is referred to a variety of sources such proceedings from the various World Energy Engineering Congresses organized by the Association of Energy Engineers (Atlanta, GA). Other sources include a general compendium of cogeneration planning considerations given by Orlando (1990), and a manual—developed by Spiewak (1994)—which emphasizes the regulatory, contracting and financing issues of cogeneration.

7.5 Evaluating Cogeneration Opportunities: Case Examples

The feasibility evaluation of cogeneration opportunities for both, new construction and facility retrofit, require the comparison and ranking of various options using a figure of economic merit. The options are usually combinations of different CHP technologies, operating modes and equipment sizes .

A first step in the evaluation is the determination of the costs of a base-case (or do-nothing) scenario. For new facilities, buying thermal and electrical energy from utility companies is traditionally considered the base case. For retrofits, the present way to buy and/or generate energy is the base case. For many, the base-case scenario is the "actual plant situation" after "basic" energy conservation and management measures have been implemented. That is, cogeneration should be evaluated upon an "efficient" base case plant.

Next, suitable cogeneration alternatives are generated using the methods discussed in sections 7.2 and 7.3. Then, the comparison and ranking of the base case versus the alternative cases is performed using an economic analysis.

Henceforth, this section addresses a basic approach for the economic analysis of cogeneration. Specifically, it discusses the development of the cash flows for each option including the base case. It also discusses some figures of merit such as the gross pay out period (simple payback) and the discounted or internal rate of return. Finally, it describes two case examples of evaluations in industrial plants. The examples are included for illustrative purposes and do not necessarily reflect the latest available performance levels or capital costs.

7.5.1 General Considerations

A detailed treatise on engineering economy is presented in Chapter 4. Even so, since economic evaluations play the key role in determining whether cogeneration can be justified, a brief discussion of economic considerations and several evaluation techniques follows.

The economic evaluations are based on examining the incremental increase in the investment cost for the alternative being considered relative to the alternative to which it is being compared and determining whether the savings in annual operating cost justify the increased investment. The parameter used to evaluate the economic merit may be a relatively simple parameter such as the "gross payout period." Or one might use more sophisticated techniques which include the time value of money, such as the "discounted rate of return," on the discretionary investment for the cogeneration systems being evaluated.

Investment cost and operating cost are the expenditure categories involved in an economic evaluation. Operating costs result from the operations of equipment, such as (1) purchased fuel, (2) purchased power, (3) purchased water, (4) operating labor, (5) chemicals, and (6) maintenance. Investment-associated costs are of primary importance when factoring the impact of federal and state income taxes into the economic evaluation. These costs (or credits) include (1) investment tax credits, (2) depreciation, (3) local property taxes, and (4) insurance. The economic evaluation establishes whether the operating and investment cost factors result in sufficient after-tax income to provide the company stockholders an adequate rate of return after the debt obligations with regard to the investment have been satisfied.

When one has many alternatives to evaluate, the less sophisticated techniques, such as "gross payout," can provide an easy method for quickly ranking alternatives and eliminating alternatives that may be particularly unattractive. However, these techniques are applicable only if annual operating costs do not change significantly with time and additional investments do not have to be made during the study period.

The techniques that include the time value of money permit evaluations where annual savings can change significantly each year. Also, these evaluation procedures permit additional investments at any time during the study period. Thus these techniques truly reflect the profitability of a cogeneration investment or investments.

7.5.2 Cogeneration Evaluation Case Examples

The following examples illustrate evaluation procedures used for cogeneration studies. Both examples are based on 1980 investment costs for facilities located in the U.S. Gulf Coast area.

For simplicity, the economic merit of each alternative examined is expressed as the "gross payout period"

(GPO). The GPO is equal to the incremental investment for cogeneration divided by the resulting first-year annual operating cost savings. The GPO can be converted to a "discounted rate of return" (DRR) using Figure 7.15. However, this curve is valid only for evaluations involving a single investment with fixed annual operating cost savings with time. In most instances, the annual savings due to cogeneration will increase as fuel costs increase to both utilities and industries in the years ahead. These increased future savings enhance the economics of cogeneration. For example, if we assume that a project has a GPO of three years based on the first-year operating cost savings, Figure 7.15 shows a DRR of 18.7%. However, if the savings due to cogeneration increase 10% annually for the first three operating years of the project and are constant thereafter, the DRR increases to 21.6%; if the savings increase 10% annually for the first six years, the DRR would be 24.5%; and if the 10% increase was experienced for the first 10 years, the DRR would be 26.6%.

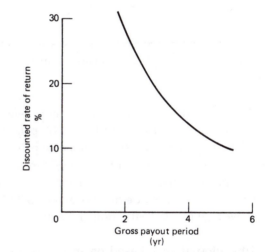

Fig. 7.15 Discounted rate of return versus gross payout period. *Basis:* (1) depreciation period, 28 years; (2) sum-of-the-years'-digits depreciation; (3) economic life, 28 years; (4) constant annual savings with time; (5) local property taxes and insurance, 4% of investment cost; (6) state and federal income taxes, 53%; (7) investment tax credit, 10% of investment cost.

Table 7.3 Plant Energy Supply System Considerations: Example 6

Process steam demands
    Net heat to process at 250 psig. 410°F—317 million Btu/hr avg.
    Net heat to process at 80 psig, 330°F—208 million Btu/hr avg. (peak requirements are 10% greater than average
        values)
Process condensate returns: 50% of steam delivered at 280°F
Makeup water at 80°F
Plant fuel is 3.5% sulfur coal
Coal and limestone for SO_2 scrubbing are available at a total cost of $2/million Btu fired
Process area power requirement is 30 MW avg.
Purchased power cost is 3.5 cents/kWh

Table 7.4 Energy and Economic Summary: Example 6

Alternative	Base Case	Case 1
Energy summary		
Boiler fuel (10^6 Btu/hr HHV)	599	714
Purchased power (MW)	33.20	4.95
Estimated total installed cost (10^6 $)	57.6	74.8
Annual operating costs (10^6 $)		
Fuel and limestone at $2/$10^6$ Btu	10.1	12.0
Purchased power at 3.5 cents/kWh	9.8	1.5
Operating labor	0.8	1.1
Maintenance	1.4	1.9
Makeup water	0.3	0.5
Total	22.4	17.0
Annual savings (10^6 $)	Base	5.4
Gross payout period (yrs)	Base	3.2

Basis: (1) boiler efficiency is 87%; (2) operation equivalent to 8400 hr/yr at Table 7-3 conditions; (3) maintenance is 2.5% of the estimated total installed cost; (4) makeup water cost for case 1 is 80 cents/1000 gal greater than Base Case water costs; (5) stack gas scrubbing based on limestone system.

Example 6: The energy requirements for a large industrial plant are given in Table 7.3. The alternatives considered include:

Base case. Three half-size coal-fired process boilers are installed to supply steam to the plant's 250-psig steam header. All 80-psig steam and steam to the 20-psig deaerating heater is pressure-reduced from the 250-psig steam header. The powerhouse auxiliary power requirements are 3.2 MW. Thus the utility tie must provide 33.2 MW to satisfy the average plant electric power needs.

Case 1. This alternative is based on installation of a noncondensing steam turbine generator. The unit initial steam conditions are 1450 psig, 950°F with automatic extraction at 250 psig and 80 psig exhaust pressure. The boiler plant has three half-size units providing the same reliability of steam supply as the Base Case. The feedwater heating system has closed feedwater heaters at 250 psig and 80 psig with a 20-psig deaerating heater. The 20-psig steam is supplied by noncondensing mechanical drive turbines used as powerhouse auxiliary drives. These units are supplied throttle steam from the 250-psig steam header. For this alternative. the utility tie normally provides 4.95 MW. The simplified schematic and energy balance is given in Figure 7.16.

    The results of this cogeneration example are tabulated in Table 7.4. Included are the annual energy requirements, the 1980 investment costs for each case,

and the annual operating cost summary. The investment cost data presented are for fully operational plants, including offices, stockrooms, machine shop facilities, locker rooms, as well as fire protection and plant security. The cost of land is not included.

    The incremental investment cost for Case 1 given in Table 7.4 is $17.2 million. Thus the incremental cost is $609/kW for the 28.25-MW cogeneration system. This illustrates the favorable per unit cost for cogeneration

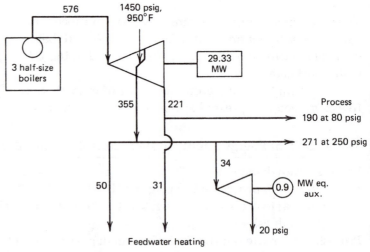

Fig. 7.16 Simplified schematic and energy-balance diagram: Example 6, Case 1. All numbers are flows in 10^3 lb/hr; Plant requirements given in Table 7.8, gross generation, 30.23 MW; powerhouse auxiliaries, 5.18 MW; net generation, 25.05 MW.

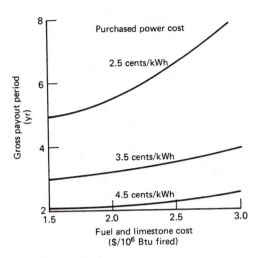

Fig. 7.17 Effect of different fuel and power costs on cogeneration profitability: Example 1. *Basis:* Conditions given in Tables 7.3 and 7.4.

systems compared to coal-fired facilities designed to provide kilowatts only, which cost in excess of $1000/kW.

The impact of fuel and purchased power costs other than Table 7.3 values on the GPO for this example is shown in Figure 7.17. Equivalent DRR values based on first-year annual operating cost savings can be estimated using Figure 7.15.

Sensitivity analyses often evaluate the impact of uncertainties in the installed cost estimates on the profitability of a project. If the incremental investment cost for cogeneration is 10% greater than the Table 7.4 estimate, the GPO would increase from 3.2 to 3.5 years. Thus the DRR would decrease from 17.5% to about 16%, as shown in Figure 7.15.

Example 7: The energy requirements for a chemical plant are presented in Table 7.5. The alternatives considered include:

Base case. Three half-size oil-fired packaged process boilers are installed to supply process steam at 150 psig. Each unit is fuel-oil-fired and includes a particulate removal system. The plant has a 60-day fuel-oil-storage capacity. A utility tie provides 30.33 MW average to supply process and boiler plant auxiliary power requirements.

Case 1. (Refer to Figure 7.18). This alternative examines the merit of adding a noncondensing steam turbine generator with 850 psig, 825°F initial steam conditions, 150-psig exhaust pressure. Steam is supplied by three half-size packaged boilers. The feedwater heating system is comprised of a 150-psig closed heater and a 20-psig deaerating heater. The steam for the deaerating heater is the

Table 7.5 Plant Energy Supply System Considerations: Example 7

Process steam demands
  Net heat to process at 150 psig sat.—158.5 million Btu/hr avg. (peak steam requirements are 10% greater than average values)
Process condensate returns: 45% of the steam delivered at 300°F
Makeup water at 80°F
Plant fuel is fuel oil
Fuel cost is $5/million Btu
Process areas require 30 MW
Purchased power cost is 5 cents/kWh

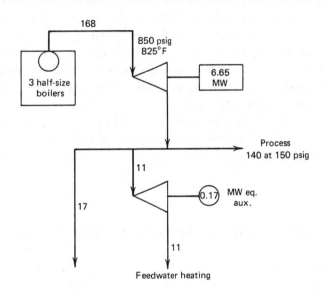

Fig. 7.18 Simplified schematic and energy-balance diagram: Example 7, Case 1. All numbers are flows in 1000 lb/hr; gross generation, 6.82 MW; powerhouse auxiliaries, 0.50 MW; net generation; 6.32 MW.

exhaust of a mechanical drive turbine (MDT). The MDT is supplied 150-psig steam and drives some of the plant boiler feed pumps. The net generation of this cogeneration system is 6.32 MW when operating at the average 150-psig process heat demand. A utility tie provides the balance of the power required.

Case 2. (Refer to Figure 7.19). This alternative is a combined cycle using the 25,000-kW gas turbine generator whose performance is given in Table 7.7. An unfired HRSG system provides steam at both 850 psig, 825°F and 150 psig sat. Plant steam requirements in excess of that available from the two-pressure level unfired HRSG system are generated in an oil fired packaged boiler. The steam supplied to the noncondensing turbine is expanded

Fig. 7.19 Simplified schematic and energy-balance diagram: Example 7, Case 2. All numbers are flows in 1000 lb/hr; gross generation, 26.77 MW; powerhouse auxiliaries, 0.23 MW: net generation; 26.54 MW.

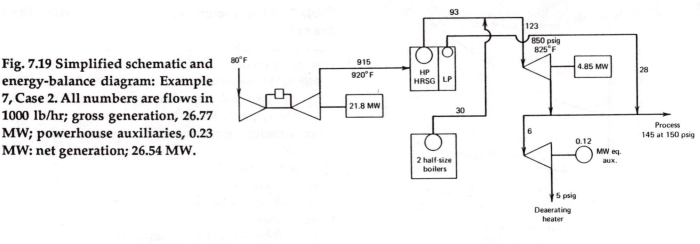

to the 150-psig steam header. The net generation from the overall system is 26.54 MW. A utility tie provides power requirements in excess of that supplied by the cogeneration system. The plant-installed cost estimates for Case 2 include two half-size package boilers. Thus full steam output can be realized with any steam generator out of service for maintenance.

The energy summary. annual operating costs. and economic results are presented in Table 7.6. The results show that the combined cycle provides a GPO of 2.5 years

based on the study fuel and purchased power costs. The incremental cost for Case 2 relative to the Base Case is $395/kW compared to $655/kW for Case 1 relative to the Base Case. This favorable incremental investment cost combined with a FCP of 5510 Btu/kWh contribute to the low GPO.

The influence of fuel and power costs other than those given in Table 7.5 on the GPO for cases 1 and 2 is shown in Figure 7.20. These GPO values can be translated to DRRs using Figure 7.15.

Table 7.6 Energy and Economic Summary: Example 7

Alternative	Base Case	Case 1	Case 2
Energy summary			
Fuel (10^6 Btu/hr HHV)			
Boiler	183	209	34
Gas turbine	—	—	297
Total	183	209	331
Purchased power (MW)	30.33	23.77	3.48
Estimated total installed cost (10^6 $)	8.3	12.6	18.9
Annual operating cost (10^6 $)			
Fuel at $5/M Btu HHV	7.7	8.8	13.9
Purchased power at 5 cents/kWh	12.7	10.0	1.5
Operating labor	0.6	0.9	0.9
Maintenance	0.2	0.3	0.5
Makeup water	0.1	0.2	0.2
Total	21.3	20.2	17.0
Annual savings (10^6 $)	Base	1.1	4.3
Gross payout period (yr)	Base	3.9	2.5

Basis: (1) gas turbine performance per Table 7-7; (2) boiler efficiency, 87%; (3) operation equivalent to 8400 hr/yr at Table 7-5 conditions; (4) maintenance, 2.5% of the estimated total installed costs; (5) incremental makeup water cost for cases 1 and 2 relative to the Base Case. $1/1000 gal.

Table 7.7 Steam Generation and Fuel Chargeable to Power: 25,000-kW ISO Gas Turbine and HRSG (Distillate Oil Fuel)[a]

Type HRSG	Unfired		Supplementary Fired		Fully Fired	
Gas Turbine						
Fuel (10^6 Btu/hr HHV)	297 ——————————————————————————→					
Output (MW)	21.8		21.6		21.4	
Airflow (10^3 lb/hr)	915 ——————————————————————————→					
Exhaust temperature (°F)	920		922		925	
HRSG fuel (10^6 Btu/hr HHV)	NA		190		769	
	Steam (10^3 lb/hr)	FCP (BTU/kWh HHV)	Steam (10^3 lb/hr)	FCP (Btu/kWh HHV)	Steam (10^3 lb/hr)	FCP (Btu/kWh HHV)
Steam conditions						
250 psig sat.	133	6560	317	5620	851	4010
400 psig, 650°F	110	7020	279	5630	751	
600 psig, 750°F	101	7340	268	5660	722	
850 psig, 825°F	93	7650	261	5700	703	
1250 psig, 900°F	—	—	254	5750	687	
1450 psig, 950°F	—	—	250	5750	675	

[a]*Basis:* (1) gas turbine performance given for 80°F ambient temperature, sea-level site; (2) HRSG performance based on 3% blowdown, 1-1/2% radiation and unaccounted losses, 228°F feedwater; (3) no HRSG bypass stack loss; (4) gas turbine exhaust pressure loss is 10 in. H_2O with unfired, 14 in. H_2O with supplementary fired, and 20 in. H_2O with fully fired HRSG; (5) fully fired HRSG based on 10% excess air following the firing system and 300°F stack. (6) fuel chargeable to gas turbine power assumes total fuel credited with equivalent 88% boiler fuel required to generate steam; (7) steam conditions are at utilization equipment; a 5% ΔP and 5°F ΔT have been assumed from the outlet of the HRSG.

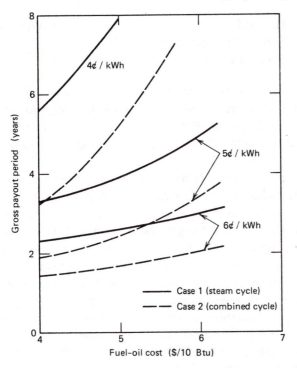

Fig. 7.20 Effect of different fuel and power cost on cogeneration profitability: Example 2. *Basis:* Conditions given in Tables 7.4 and 7.5.

Example 8. A gas-turbine and HRSG cogeneration system is being considered for a brewery to supply base-load electrical power and part of the steam needed for process. An overview of the proposed system is shown in Figure 7.21. This example shows the use of computer tools in cogeneration design and evaluation.

Base Case: Currently, the plant purchases about 3,500,000 kWh per month at $0.06 per kWh. The brewery uses an average of 24,000 lb/hr of 30 psig saturated steam. Three 300-BHP gas fired boilers produce steam at 35 psig, to allow for pressure losses. The minimum steam demand is 10,000 lb/hr. The plant operates continuously during ten months or 7,000 hr/year. The base or minimum electrical load during production is 3,200 kW. The rest of the time (winter) the brewery is down for maintenance. The gas costs $3.50/MMBtu.

Case 1: Consider the gas turbine whose ratings are given on Figure 7.11. We will evaluate this turbine in conjunction with an unfired water-tube HRSG to supply part of the brewery's heat and power loads. First, we obtain the ratings and performance data for the selected turbine, which has been sized to meet the electrical base load (3.5 MW). An air washer/evaporative cooler will be installed

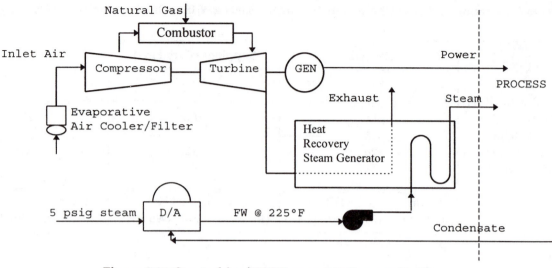

Figure 7.21 Gas turbine/HRSG cogeneration application.

at the turbine inlet to improve (reduce) the overall heat rate by precooling the inlet air to an average 70°F (80°F or less), during the summer production season Additional operating data are given below.

Operating Data
Inlet air pressure losses (filter and air
    pre-cooler): 5" H$_2$O

Exhaust Losses (ducting, by-pass valve,
    HRSG and Stack): 12" H$_2$O

Location Elevation above sea level: 850 ft

Thus, on a preliminary basis, we assume the turbine will constantly run at full capacity, minus the effect of elevation, the inlet air pressure drop and exhaust losses. Since the plant will be located at 850 ft above sea level, from Figure 7.11, the elevation correction factor is 0.90. Hence, the corrected continuous power rating (before deducting pressure losses) when firing natural gas and using 70°F inlet air is:

= (Generator Output @ 70°F)

    (Elevation correction @ 850 ft)

= 4,200 kWe × 0.9

= 3,780 kWe

Next, by using the Inlet and Exhaust Power Loss graphs in Figure 7.11, we get the exhaust and inlet losses (@ 3780 kW output): 17 and 7 kW/inch H$_2$O, respectively. So, the total power losses due to inlet and exhaust losses are:

= (17 in)(5 kW/in) + (12 in)(8 kW/in)
= 181 kW

Consequently, the net turbine output after elevation and pressure losses is

= 3,780 – 181
= 3,599 kWe

Next, from Figure 7.11 we get the following performance data for 70°F inlet air:

Heat rate	: 12,250 Btu/kWh (LHV)
Exhaust Temperature	: 935°F
Exhaust Flow	: 160,000 lb/hr

These figures have been used as input data for HGPRO—a prototype HRSG software program developed by V. Ganesh, W.C. Turner and J.B. Wong in 1992 at Oklahoma State University. The program results are shown in Fig. 7.22.

The total installed cost of the complete cogeneration plant including gas turbine, inlet air precooling, HRSG, auxiliary equipment and computer based controls is $4,500,000. Fuel for cogeneration is available on a long term contract basis (>5 years) at $2.50/MMBtu. The brewery has a 12% cost of capital. Using a 10-year after tax cash flow analysis with current depreciation and tax rates, <u>should the brewery invest in this cogeneration option</u>? For this evaluation, assume: (1) A 1% inflation for power and non-cogen natural gas; (2) an operation and maintenance (O&M) cost of $0.003/kWh for the first year after the project is installed. Then, the O&M cost should escalate at 3% per year; (3) the plant salvage value is neglected.

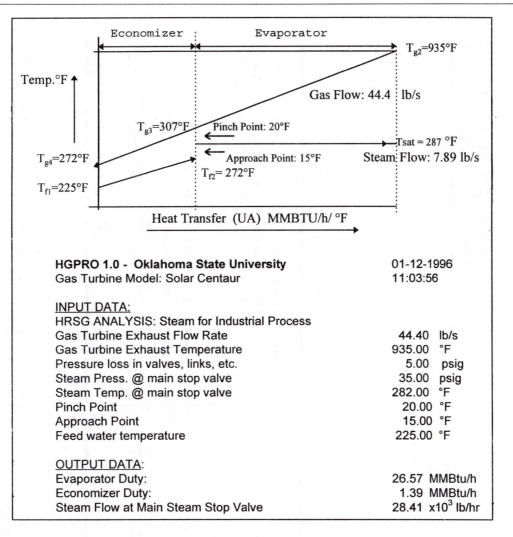

Figure 7.22 Results from HGPRO 1.0, a prototype HRSG software.

Economic Analysis

Next, we present system operation assumptions required to conduct a preliminary economic analysis.

1) The cogeneration system will operate during all the production season (7,000 hrs/year).

2) The cogeneration system will supply an average of 3.5 MW of electrical power and 24,000 lb of 35 psig steam per hour. The HRSG will be provided with an inlet gas damper control system to modulate and by-pass hot gas flow. This is to allow for variable steam production or steam load-following operation.

3) The balance of power will be obtained from the existing utility at the current cost ($0.06/kWh)

4) The existing boilers will remain as back-up units. Any steam deficit (considered to be negligible) will be produced by the existing boiler plant.

5) The cogeneration fuel (natural gas) will be metered with a dedicated station and will be available at $2.50/MMBtu during the first five years and at $2.75/MMBtu during the next five-year period. Non cogeneration fuel will be available at the current price of $3.50/MMBtu.

The discounted cash flow analysis was carried out using an electronic spreadsheet (Table 7.8). The results of the spreadsheet show a positive net present value. Therefore, when using the data and assumptions given in this case, the cogeneration project appears to be cost effective. The brewery should consider this project for funding and implementation.

7.6 CLOSURE

Cogeneration has been used for almost a century to supply both process heat and power in many large industrial plants in the United States. This technology would have

GAS TURBINE / HRSG COGENERATION TEN YEAR ECONOMIC ANALYSIS

INPUT DATA

Power & gas cost escalation rate	1%	(year 1-10)
Labor, operat. & maint. escalation rat	3%	(year 1-10)
Cogen power generation	3,500	kW
Cogen steam production	24,000	lh/hr
Existing boiler efficiency	80%	lh/hr
Present Electricity Cost	$0.06	/kWh
Present Nat. Gas Cost	$3.50	/MMBtu
Turbine plant heat rate, LHV @ 70°F	12,250 MMBtu/kW	93% LHV/HHV
Cogen Nat. Gas Cost	$2.50 /MMBtu	$2.75 /MMBtu
Cogen System O&M Cost	$0.003 /kWh	yrs: 6-10
Cogen System Installed Cost	$4,500,000	
Operation time:	7,000	hr/yr

	NPV @ 12.00%	1996	1997	1998	1999	2000	2001	2002	2003	2004	2005
Savings											
Power cost savings	8610742	1470000	1484700	1499547	1514542	1529688	1544985	1560435	1576039	1591799	1607717
Nat. Gas to steam savings	4305371	735000	742350	749774	757271	764844	772492	780217	788019	795900	803859
Total Savings	12916113	2205000	2227050	2249321	2271814	2294532	2317477	2340652	2364058	2387699	2411576
Costs											
Operating & Maint Costs	-463291	-73500	-75705	-77976	-80315	-82725	-85207	-87763	-90396	-93108	-95901
Cogen Nat. Gas Cost	-4723554	-806788	-806788	-806788	-806788	-806788	-887466	-887466	-887466	-887466	-887466
Depreciation: 5-year MACRS	-3520539	-504000	-1310400	-1290240	-773640	-579600	-435960	-146160	0	0	0
Total Costs	-8707384	-1384288	-2192893	-2175004	-1660743	-1469113	-1408633	-1121389	-977862	-980574	-983367
Net Savings	4208729	820712	34157	74317	611071	825419	908844	1219263	1386196	1407125	1428209
Cashflows											
Post-Tax Income/(Loss)	2569428	501045	20853	45370	373059	503918	554849	744360	846273	859050	871921
Investment	-4258929	-2520000	-2520000	0	0	0	0	0	0	0	0
Depreciation Add Back	3520539	504000	1310400	1290240	773640	579600	435960	146160	0	0	0
Total Cashflows	1831039	-1514955	-1188747	1335610	1146699	1083518	990809	890520	846273	859050	871921
Cumulative Cashflow	-1514955	-1514955	-2703702	-1368092	-221393	862125	1852934	2743454	3589727	4448777	5320698

NPV : $1,831,039

Table 7.8 After tax discounted cash flow economic analysis.

been applied to a greater extent if we did not experience a period of plentiful low-cost fuel and reliable low-cost electric power in the 25 years following the end of World War II. Thus economic rather than technical considerations have limited the application of this energy-saving technology.

The continued increase in the cost of energy is the primary factor contributing to the renewed interest in cogeneration and its potential benefits. This chapter discusses the various prime movers that merit consideration when evaluating this technology. Furthermore, approximate performance levels and techniques for developing effective cogeneration systems are presented.

The cost of all forms of energy is rising sharply. Cogeneration should remain an important factor in effectively using our energy supplies and economically providing goods and services in those base-load applications requiring large quantities of process heat and power.

7.6   REFERENCES

1. Butler, C.H., (1984), *Cogeneration: Engineering, Design Financing, and Regulatory Compliance*, McGraw-Hill, Inc., New York, N.Y.

2. Caton, J.A., et al., (1987), *Cogeneration Systems, Texas A&M University*, College Station, TX.

3. CFR-18 (1990): Code of Federal Regulations, Part 292 Regulations Under Sections 201 and 210 of the Public Utility Regulatory Policies Act of 1978 With Regard to Small Power Production and Cogeneration, (4-1-90 Edition).

4. Estey P.N., et al., (1984). "A Model for Sizing Cogeneration Systems," Proceedings of the 19th Intersociety Energy Conversion Engineering Conference" Vol. 2 of 4, August, 1984, San Francisco, CA.

5. Ganapathy, V. (1991). *Waste Heat Boiler Deskbook*, The Fairmont Press, Inc., Lilburn, GA.

6. Harkins H.L., (1981), "PURPA New Horizons for Electric Utilities and Industry," IEEE Transactions, Vol. PAS-100, pp 2784-2789.

7. Hay, N., (1988), *Guide to Natural Gas Cogeneration*, The Fairmont Press, Lilburn, GA.

8. Kehlhofer, R., (1991). *Combined-Cycle Gas & Steam Turbine Power Plants*, The Fairmont Press, Inc. Lilburn, Ga.

9. Kostrzewa, L.J. & Davidson, K.G., (1988). "Packaged Cogeneration," *ASHRAE Journal*, February 1988.

10. Kovacik, J.M., (1982), "Cogeneration," in *Energy Management Handbook*, ed. by W.C. Turner, Wiley, New York, N.Y.

11. Kovacik, J.M., (1985), "Industrial Cogeneration: System Application Consideration," *Planning Cogeneration Systems*, The Fairmont Press, Lilburn, Ga.

12. Lee, R.T.Y., (1988), "Cogeneration System Selection Using the Navy's CELCAP Code," *Energy Engineering*, Vol. 85, No. 5, 1988.

13. Limaye, D.R. and Balakrishnan, S., (1989), "Technical and Economic Assessment of Packaged Cogeneration Systems Using Cogenmaster," *The Cogeneration Journal*, Vol. 5, No. 1, Winter 1989-90.

14. Limaye, D.R., (1985), *Planning Cogeneration Systems*, The Fairmont Press, Atlanta, GA.

15. Limaye, D.R., (1987), *Industrial Cogeneration Applications*, The Fairmont Press, Atlanta, GA.

16. Mackay, R. (1983). "Gas Turbine Cogeneration: Design, Evaluation and Installation." The Garret Corporation, Los Angeles, CA— The Association Of Energy Engineers, Los Angeles CA, February, 1983.

17. Makansi, J., (1991). "Independent Power/Cogeneration, Success Breeds New Obligation—Delivering on Performance," *Power*, October 1991.

18. Mulloney, et. al., (1988). "Packaged Cogeneration Installation Cost Experience," Proceedings of The 11th World Energy Engineering Congress, October 18-21, 1988.

19. Orlando, J.A., (1991). *Cogeneration Planners Handbook*, The Fairmont Press, Atlanta, GA.

20. Oven, M., (1991), "Factors Affecting the Financial Viability Applications of Cogeneration," XII Seminario Nacional Sobre El Uso Racional de La Energia," Mexico City, November, 1991.

21. Polimeros, G., (1981), *Energy Cogeneration Handbook*, Industrial Press Inc, New York.

22. Power (1980), "FERC Relaxes Obstacles to Cogeneration," *Power*, September 1980, pp 9-10.

23. SFA Pacific Inc (1990). "Independent Power/Cogeneration, Trends and Technology Update," *Power*, October 1990.

24. Somasundaram, S., et al., (1988), *A Simplified Self-Help Approach To Sizing of Small-Scale Cogeneration Systems*, Texas A&M University, College Station, TX.

25. Spiewak, S.A. and Weiss L., (1994) *Cogeneration & Small Power Production Manual*, 4th Edition, The Fairmont Press, Inc. Lilburn, GA.

26. Turner, W.C. (1982). *Energy Management Handbook*, John Wiley & Sons, New York, N.Y.

27. Tuttle, D.J., (1980), PURPA 210: New Life for Cogenerators," *Power*, July, 1980.

28. Williams, D. and Good, L., (1994) *Guide to the Energy Policy Act of 1992*, The Fairmont Press, Inc. Lilburn, GA.

29. Wong, J.B., Ganesh, V. and Turner, W.C. (1991), "Sizing Cogeneration Systems Under Variable Loads," 14th World Energy Engineering Congress, Atlanta, GA.

30. Wong, J.B. and Turner W.C. (1993), "Linear Optimization of Combined Heat and Power Systems," Industrial Energy Technology Conference, Houston, March, 1993.

CHAPTER 8
WASTE-HEAT RECOVERY

WESLEY M. ROHRER, JR.

Emeritous Associate Professor of
    Mechanical Engineering
University of Pittsburgh
Pittsburgh, Pennsylvania

8. 1 INTRODUCTION

8.1.1 Definitions

Waste heat, in the most general sense, is the energy associated with the waste streams of air, exhaust gases, and/or liquids that leave the boundaries of a plant or building and enter the environment. It is implicit that these streams eventually mix with the atmospheric air or the groundwater and that the energy, in these streams, becomes unavailable as useful energy. The absorption of waste energy by the environment is often termed thermal pollution.

In a more restricted definition, and one that will be used in this chapter, waste heat is that energy which is rejected from a process at a temperature high enough above the ambient temperature to permit the economic recovery of some fraction of that energy for useful purposes.

8.1.2 Benefits

The principal reason for attempting to recover waste heat is economic. All waste heat that is successfully recovered directly substitutes for purchased energy and therefore reduces the consumption of and the cost of that energy. A second potential benefit is realized when waste-heat substitution results in smaller capacity requirements for energy conversion equipment. Thus the use of waste-heat recovery can reduce capital costs in new installations. A good example is when waste heat is recovered from ventilation exhaust air to preheat the outside air entering a building. The waste-heat recovery reduces the requirement for space-heating energy. This permits a reduction in the capacity of the furnaces or boilers used for heating the plant. The initial cost of the heating equipment will be less and the overhead costs will be reduced. Savings in capital expenditures for the primary conversion devices can be great enough to com-

pletely offset the cost of the heat-recovery system. Reduction in capital costs cannot be realized in retrofit installations unless the associated primary energy conversion device has reached the end of their useful lives and are due for replacement.

A third benefit may accrue in a very special case. As an example, when an incinerator is installed to decompose solid, liquid, gaseous or vaporous pollutants, the cost of operation may be significantly reduced through waste-heat recovery from the incinerator exhaust gases.

Finally, in every case of waste-heat recovery, a gratuitous benefit is derived: that of reducing thermal pollution of the environment by an amount exactly equal to the energy recovered, at no direct cost to the recoverer.

8.1.3 Potential for Waste-Heat Recovery in Industry

It had been estimated[1] that of the total energy consumed by all sectors of the U.S. economy in 1973, that fully 50% was discharged as waste heat to the environment. Some of this waste is unavoidable. The second law of thermodynamics prohibits 100% efficiency in energy conversion except for limiting cases which are practically and economically unachievable. Ross and Williams,[2] in reporting the results of their second-law analysis of U.S. energy consumption, estimated that in 1975, economical waste-heat recovery could have saved our country 7% of the energy consumed by industry, or 1.82 $\times$ 10^{16} Btus (1.82 quads.)

Roger Sant[3] estimated that in 1978 industrial heat recovery could have resulted in a national fuel savings of 0.3%, or 2.65 $\times$ 10^{16} Btus quads. However, his study included only industrial furnace recuperators.* In terms of individual plants in energy-intensive industries, this percentage can be greater by more than an order of magnitude.

The Annual Energy Review 1991[4] presents data to show that although U.S. manufacturing energy intensity increased by an average of 26.7% during the period 1980 to 1988, the manufacturing sector's energy use efficiency, for all manufacturing, increased by an average of 25.1%.

---

*Recuperators are heat exchangers that recover waste heat from the stacks of furnaces to preheat the combustion air. Section 8.4.2 subjects this device to more detailed scrutiny.

In reviewing the Annual Energy Reviews over the years, it becomes quite clear that during periods of rising fuel prices energy efficiency increases, while in periods of declining fuel prices energy efficiency gains are eroded. Although the average gain in energy use efficiency, in the 7-year period mentioned above, is indeed impressive, several industrial groups accomplished much less than the average or made no improvements at all during that time. As economic conditions change to favor investments in waste-heat recovery there will be further large gains made in energy use efficiency throughout industry.

8.1.4 Quantifying Waste Heat

The technical description of waste heat must necessarily include quantification of the following characteristics: (1) quantity, (2) quality, and (3) temporal availability.

The quantity of waste heat available is ordinarily expressed in terms of the enthalpy flow of the waste stream, or

$$\dot{H} = \dot{m}h \qquad (8.1)$$

where $\dot{H}$ = total enthalpy flow rate of waste stream, Btu/hr

$\dot{m}$ = mass flow rate of waste stream, lb/hr

h = specific enthalpy of waste stream, Btu/lb

The mass flow rate, m, can be calculated from the expression

$$\dot{m} = \rho Q \qquad (8.2)$$

where ρ = density of material, lb/ft^3

Q = volumetric flow rate, ft^3/hr

The potential for economic waste-heat recovery, however, does not depend as much on the quantity available as it does on whether its quality fits the requirements of the potential heating load which must be supplied and whether the waste heat is available at the times when it is required.

The quality of waste heat can be roughly characterized in terms of the temperature of the waste stream. The higher the temperature, the more available the waste heat for substitution for purchased energy. The primary source of energy used in industrial plants are the combustion of fossil fuels and nuclear reaction, both

occurring at temperatures approaching 3000°F. Waste heat, of any quantity, is ordinarily of little use at temperatures approaching ambient, although the use of a heat pump can improve the quality of waste heat economically over a limited range of temperatures near and even below ambient. As an example, a waste-heat stream at 70°F cannot be used directly to heat a fluid stream whose temperature is 100°F. However, a heat pump might conceivably be used to raise the temperature of the waste heat stream to a temperature above 100°F so that a portion of the waste-heat could then be transferred to the fluid stream at 100°F. Whether this is economically feasible depends upon the final temperature required of the fluid to be heated and the cost of owning and operating the heat pump.

8.1.5 Matching Loads to Source

It is necessary that the heating load which will absorb the waste heat be available at the same time as the waste heat. Otherwise, the waste heat may be useless, regardless of its quantity and quality. Some examples of synchrony and non-synchrony of waste-heat sources and loads are illustrated in Figure 8.1. Each of the graphs in that figure shows the size and time availability of a waste-heat source and a potential load. In Figure 8.1a the size of the source, indicated by the solid line, is an exhaust stream from an oven operating at 425°F during the second production shift only. One possible load is a water heater for supplying a washing and rinsing line at 135°F. As can be seen by the dashed line, this load is available only during the first shift. The respective quantities and qualities seem to fit satisfactorily, but the time availability of the source could not be worse. If the valuable source is to be used, it will be necessary to (1) reschedule either of the operations to bring them into time correspondence, (2) generate the hot water during the second shift and store it until needed at the beginning of the first shift the next day, or (3) find another heat load which has an overall better fit than the one shown.

In Figure 8.1b we see a waste-heat source (solid line) consisting of the condenser cooling water of an air-conditioning plant which is poorly matched with its load (dashed line)—the ventilating air preheater for the building. The discrepancy in availability is not diurnal as before, but seasonal.

In Figure 8.1c we see an almost perfect fit for source and load, but the total availability over a 24 hour period is small. The good fit occurs because the source, the hot exhaust gases from a heat-treat furnace, is used to preheat combustion air for the furnace burner. However, the total time of availability over a 24-hour period

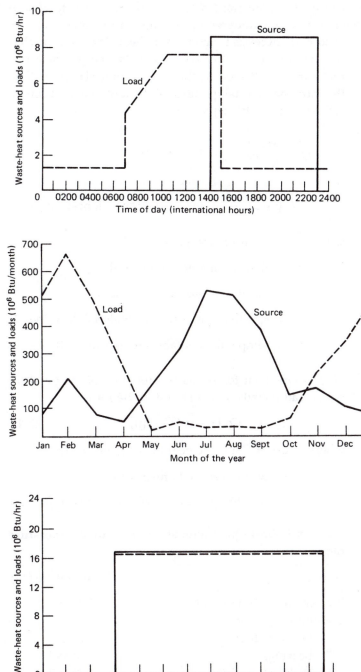

Figure 8.1 Matching waste-heat sources and loads.

is so small as to cast doubt on the ability to pay off the capital costs of this project.

8.1.6 Classifying Waste-Heat Quality

For convenience, the total range of waste-heat temperatures, 80 to 3000°F, is broken down into three subranges: high, medium, and low. These classes are de-signed to match a similar scale which classifies commercial waste-heat-recovery devices. The two systems of classes allow matches to be made between industrial process waste heat and commercially available recovery equipment. Subranges are defined in terms of temperature range as:

High range	$1100 \leq T \leq 3000$
Medium range	$400 \leq T < 1100$
Low range	$80 \leq T < 400$

Waste heat in the high-temperature range is not only the highest quality but is the most useful, and costs less per unit to transfer than lower-quality heat. However, the equipment needed in the highest part of the range requires special engineering and special materials and thus requires a higher level of investment. All of the applications listed in Table 8.1 result from direct-fired processes. The waste heat in the high range is available to do work through the utilization of steam turbines or gas turbines and thus is a good source of energy for cogeneration plants.*

Table 8.2 gives the temperatures of waste gases primarily from direct-fired process equipment in the medium-temperature range. This is still in the temperature range in which work may be economically

Table 8.1 Waste-heat sources in the high-temperature range.

Type of Device	Temperature (°F)
Nickel refining furnace	2500-3000
Aluminum refining furnace	1200-1400
Zinc refining furnace	1400-2000
Copper refining furnace	1400-1500
Steel heating furnaces	1700-1900
Copper reverberatory furnace	1650-2000
Open hearth furnace	1200-1300
Cement kiln (dry process)	1150-1350
Glass melting furnace	1800-2800
Hydrogen plants	1200-1800
Solid waste incinerators	1200-1800
Fume incinerators	1200-2600

*The waste heat generates high-pressure steam in a waste-heat boiler which is used in a steam turbine generator to generate electricity. The turbine exhaust steam at a lower pressure provides process heat. Alternatively, the high-temperature gases may directly drive a gas turbine generator with the exhaust generating low-pressure steam in a waste-heat boiler for process heating.

Table 8.2 Waste-heat sources in the medium-temperature range.

Type of Device	Temperature (°F)
Steam boiler exhausts	450-900
Gas turbine exhausts	700-1000
Reciprocating engine exhausts	600-1100
Reciprocating engine exhausts (turbocharged)	450-700
Heat treating furnaces	800-1200
Drying and baking ovens	450-1100
Catalytic crackers	800-1200
Annealing furnace cooling systems	800-1200
Selective catalytic reduction systems for NO_x control	525-750

extracted using gas turbines in the range 15 to 30 psig or steam turbines at almost any desired pressure. It is an economic range for direct substitution of process heat since requirements for equipment are reduced from those in the high-temperature range.

The use of waste heat in the low-temperature range is more problematic. It is ordinarily not practical to extract work directly from the waste-heat source in this temperature range. Practical applications are generally for preheating liquids or gases. At the higher temperatures in this range air preheaters or economizers can be utilized to preheat combustion air or boiler make-up water, respectively. At the lower end of the range heat pumps may be required to raise the source temperature to one that is above the load temperature. An example of an application which need not involve heat pump assistance would be the use of 95°F cooling water from an air compressor to preheat domestic hot water from its ground temperature of 50°F to some intermediate temperature less than 95°F. Electric, gas-fired, or steam heaters could then be utilized to heat the water to the temperature desired. Another application could be the use of 90°F cooling water from a battery of spot welders to preheat the ventilating air for winter space heating. Since machinery cooling can't be interrupted or diminished, the waste-heat recovery system, in this latter case, must be designed to be bypassed or supplemented when seasonal load requirements disappear. Table 8.3 lists some waste-heat sources in the low-temperature range.

8.1.7 Storage of Waste Heat

Waste heat can be utilized to adapt otherwise mismatched loads to waste-heat sources. This is possible be-

cause of the inherent ability of all materials to absorb energy while undergoing a temperature increase. The absorbed energy is termed stored heat. The quantity that can be stored is dependent upon the temperature rise that can be achieved in the storage material as well as the intrinsic thermal qualities of the material, and can be estimated from the equation

$$Q = \int_{T_1}^{T_2} mC \, dT = \int_{T_1}^{T_2} \rho \, VC \, dT$$

$$= \rho \, VC \, (T - T_0) \quad \text{for constant specific heat} \quad (8.3)$$

where m = mass of storage material, lb_m

ρ = density of storage material, lb/ft^3

V = volume of storage material, ft^3

C = specific heat of storage material, Btu/lb_m °R

T = temperature in absolute degrees, °R

The specific heat for solids is a function of temperature which can usually be expressed in the form

$$C_0 = C_0 [1 + \alpha (T - T_0)] \quad (8.4)$$

where C_0 = specific heat at temperature T_0

T_0 = reference temperature

α = temperature coefficient of specific heat

Table 8.3 Waste-heat sources in the low-temperature range.

Source	Temperature (°F)
Process steam condensate	130-190
Cooling water from:	
Furnace doors	90-130
Bearings	90-190
Welding machines	90-190
Injection molding machines	90-190
Annealing furnaces	150-450
Forming dies	80-190
Air compressors	80-120
Pumps	80-190
Internal combustion engines	150-250
Air conditioning and refrigeration condensers	90-110
Liquid still condensers	90-190
Drying, baking, and curing ovens	200-450
Hot-processed liquids	90-450
Hot-processed solids	200-450

It is seen from equation 8.3 that storage materials should have the properties of high density and high specific heat in order to gain maximum heat storage for a given temperature rise in a given space. The rate at which heat can be absorbed or given up by the storage material depends upon its thermal conductivity, k, which is defined by the equation

$$\frac{\delta Q}{\delta t} = -kA \frac{dT}{dx}\bigg|_{x=0} = Q \tag{8.5}$$

where t = time, hr
       k = thermal conductivity, Btu-ft/hr ft^2 °F
       A = surface area

$$\frac{dT}{dx}\bigg|_{x=0} = \text{temperature gradient at the surface}$$

Thus additional desirable properties are high thermal conductivity and large surface area per unit mass (specific area). This latter property is inversely proportional to density but can also be manipulated by designing the shape of the solid particles. Other important properties

for storage materials are low cost, high melting temperature, and a resistance to spalling and cracking under conditions of thermal cycling. To summarize: the most desirable properties of thermal storage materials are (1) high density, (2) high specific heat, (3) high specific area, (4) high thermal conductivity, (5) high melting temperature, (6) low coefficient of thermal expansion, and (7) low cost.

Table 8.4 lists the thermophysical properties of a number of solids suitable for heat-storage materials.

The response of a storage system to a waste-heat stream is given approximately by the following expression due to Rummel[4]:

$$\frac{Q}{A} = \frac{\Delta T_{l,m} / (\theta' + \theta'')}{1/h'' \, \theta'' + 1/h' \, \theta' + 1/2.5C_s\rho_s R_B + R_B/k \, (\theta' + \theta'')} \tag{8.6}$$

where $T_{l,m}$ = logarithmic mean temperature difference based upon the uniform inlet temperature of each stream and the average outlet temperatures

       C_s = specific heat of storage material, Btu/lb °F

Table 8.4 Common refractory materials[a,b].

Name	Formula	Density (lbm/ft^3)	Specific Heat (Btu/lb$_m$)	Mean Thermal Conductivity (Btu/ft hr °F) (to 1000°)	Coefficient of Cubical Expansion (per °F)	Maximum Use Temperature (°F)	Melting Point (°F)
Alumina	Al_2O_3	230	0.24	2.0	8×10^{-6}	3300	3700
Beryllium oxide	BeO	190	0.24	—	9×10^{-6}	4000	4600
Calcium oxide	CaO	200	0.18	4.5	13×10^{-6}	4200	4700
Carbon, graphite	C	120	0.36	7	3×10^{-6}	4000	6500[c]
Chrome	40% Cr_2O_3	200	0.20	1.0	8×10^{-6}	3200	3800
Corundum	90% Al_2O_3	200	0.22	1.5	7×10^{-6}	3200	—
Forsterite	2 MgO SiO_2	160	0.23	1.2	10×10^{-6}	3000	3300
Magnesia	MgO	210	0.25	2.3	11×10^{-6}	4000	5000
Magnesium oxide	MgO	175	0.25	2.0	10×10^{-6}	3500	5000
Mullite	3 Al_2O_3 SiO_2	160	0.23	1.2	5×10^{-6}	3000	3350
Silica	SiO_2	110	0.24	1.0	7×10^{-6}	2800	3100
Silicon carbide	SiC	170	0.23	8	3×10^{-6}	3000	4000[d]
Spinel	MgO Al_2O_3	220	0.23	5	7×10^{-6}	3300	—
Titanium oxide	TiO_2	260	0.17	2.2	8×10^{-6}	3000	3300
Zircon	ZrO_2 SiO_2	220	0.15	1.3	5×10^{-6}	3500	4500
Zirconium oxide	ZrO_2	360	0.13	1.3	4×10^{-6}	4400	4800

[a]Most of these materials are available commercially as refractory tile, brick, and mortar. Properties will depend on form, purity, and mixture. Temperatures given should be considered as high limits.
[b]For density in kg/m^3, multiply value in lb/ft^3 by 16.02. For specific heat in J/kg K, multiply value in Btu/lb$_m$ °F by 4184. For thermal conductivity in W/m K, multiply value in Btu/ft hr °F by 1.73.
[c]Sublimes.
[d]Dissociates.

$$\rho_S = \text{density of storage material, lb/ft3}$$

k = conductivity of storage material, Btu/hr ft °F

R_B = volume per unit surface area for storage material, ft

h = coefficient of convective heat transfer of gas streams, Btu/hr ft^2 °F

θ = time cycle for gas stream flows, hr

The primed and double-primed values refer, respectively, to the hot and cold entering streams. In cases where the fourth term in the denominator is large compared to the other three terms, this equation should not be used. This will occur when the cycle times are short and the thermal resistance to heat transfer is large. In those cases there exists insufficient time for the particles to get heated and cooled. Additional equations for determining the rise and fall in temperatures, and graphs giving temperature histories for the flow streams and the storage material, may be found in Rohsenow and Hartnett.[5]

8.1.8 Enhancing Waste Heat with Heat Pumps

Heat pumps offer only limited opportunities for waste-heat recovery simply because the cost of owning and operating the heat pump may exceed the value of the waste heat recovered.

A heat pump is a device that operates cyclically so that energy absorbed at low temperature is transformed through the application of external work to energy at a higher-temperature which can be absorbed by an existing load. The commercial mechanical refrigeration plant can be utilized as a heat pump with small modifications, as indicated in Figure 8.2. The coefficient of performance (COP) of the heat pump cycle is the simple ratio of heat delivered to work required:

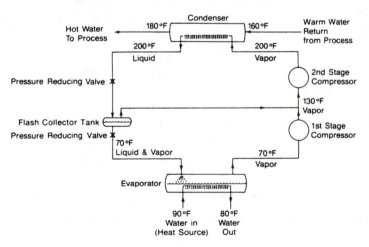

Figure 8.2 Heat pump.

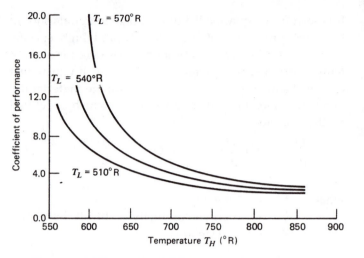

Figure 8.3 Theoretical COP vs. load temperature.

$$\text{COP}_{HP} = \frac{Q_H}{Q_{net}} = \frac{Q_H}{W_{net}} \qquad (8.7)$$

Since the work requirement must be met by a prime mover that is either an electric motor or a liquid-fueled engine, the COP must be considerably greater than 3.0 in order to be an economically attractive energy source. That is true because the efficiency of the prime movers used to drive the heat pump, or to generate the electrical energy for the motor drive, have efficiencies less than 33%. The maximum theoretical COP for an ideal heat pump is given by

$$\text{COP}_{HP} = \frac{1}{(1 - T_L/T_H)}$$

where T_L = temperature of energy source

T_H = temperature of energy load

The ideal cycle, however, uses an ideal turbine as a vapor expander instead of the usual throttle valve in the expansion line of the mechanical refrigeration plant.

Figure 8.3 is a graph of the theoretical COP versus load temperature for a number of source temperatures. Several factors prevent the actual heat pump from approaching the ideal:

1. The compressor efficiency is not 100%, but is rather in the range 65 to 85%.
2. A turbine expander is too expensive to use in any but the largest units. Thus the irreversible throttling process is used instead of an ideal expansion through a turbine. All of the potential turbine work is lost to the cycle.

3. Losses occur from fluid friction in lines, compressors, and valving.

4. Higher condenser temperatures and lower evaporator temperatures than the theoretical are required to achieve practical heat flow rates from the source and into the load.

An actual two-stage industrial heat pump installation showed[7] an annual average COP of 3.3 for an average source temperature of 78°F and a load temperature of 190°F. The theoretical COP is 5.8. Except for very carefully designed industrial units, one can expect to achieve actual COP values ranging from 50% to 65% of the theoretical.

An additional constraint on the use of heat pumps is that high-temperature waste heat above 230°F cannot be supplied directly to the heat pump because of the limits imposed by present compressor and refrigerant technology. The development of new refrigerants might raise the limit of heat pump use to 400°F.

8.1.9 Dumping Waste Heat

It cannot be emphasized too strongly that the interruption of a waste heat load, either accidentally or intentionally, may impose severe operating conditions on the source system, and might conceivably cause catastrophic failures of that system.

In open system cooling the problem is easier to deal with. Consider the waste-heat recovery from the cooling water from an air compressor. In this case the cooling water is city tap water which flows serially through the water jackets and the intercooler and is then used as makeup water for several heated treatment baths. Should it become necessary to shut off the flow of makeup water to the baths, it would be necessary to valve the cooling water flow to a drain so that the compressor cooling continues with no interruption. Otherwise, the compressor would become overheated and suffer damage.

In a closed cooling system supplying waste heat to a load requires more extensive safeguards and provisions for dumping heat rather than fluid flow. Figure 8.4 is the schematic of a refrigeration plant condenser supplying waste heat for space heating during the winter. Since the heating load varies hourly and daily, and disappears in the summer months, it is necessary to provide an auxiliary heat sink which will accommodate the entire condenser discharge when the waste-heat load disappears. In the installation shown, the auxiliary heat sink is a wet cooling tower which is placed in series with the waste-heat exchanger. The series arrangement is preferable to the alternative parallel arrangement for several reasons. One is that fewer additional controls are needed. Using the parallel arrangement would require that the flows through the two paths be carefully controlled to maintain required condenser temperature and at the same time optimize the waste-heat recovery.

In the above examples the failure to absorb all of the available waste heat had serious consequences on the system supplying the waste heat. A somewhat different waste-heat dumping problem occurs when the effect of excessive waste-heat availability has an adverse affect on the heat sink. An example would be the use of the cooling air stream from an air-cooled screw-type compressor for space heating in the winter months. During the summer months all of the compressor cooling air would have to be dumped to the outdoors in order to prevent overheating of the work space.

8.1.10 Open Waste-Heat Exchangers

An open heat exchanger is one where two fluid streams are mixed to form a third exit stream whose energy level (and temperature) is intermediate between the two entering streams. This arrangement has the advantage of extreme simplicity and low fabrication costs with no complex internal parts. The disadvantages are that (1) all flow streams

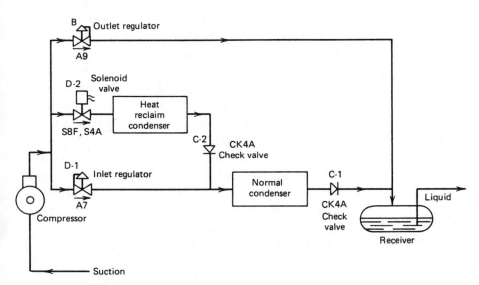

Figure 8.4 System with cold weather condenser pressure control.

must be at the same pressure, and (2) the contamination of the exit fluids by either of the entrance flows is possible. Several effective applications of open waste-heat exchangers are listed below:

1.   The exhaust steam from a turbine-driven feedwater pump in a boiler plant is used to preheat the feedwater in a deaerating feedwater heater.

2.   The makeup air for an occupied space is tempered by mixing it with the hot exhaust products from the stack of a gas-fired furnace in a plenum before discharge into the space. This recovery method may be prohibited by codes because of the danger of toxic carbon monoxide; a monitor should be used to test the plenum gases.

3.   The continuous blowdown stream from a boiler plant is used to heat the hot wash and rinse water in a commercial laundry. A steam-heated storage heater serves as the open heater.

8.1.11 Serial Use of Process Air and Water

In some applications, waste streams of process air and water can be directly used for heating without prior mixing with other streams. Some practical applications include:

1. Condenser cooling water from batch coolers used directly as wash water in a food-processing plant;

2. steam condensate from wash water heaters added directly to wash water in the bottling section of a brewery;

3. air from the cooling section of a tunnel kiln used as the heating medium in the drying rooms of a refractory; and

4. condensate from steam-heated chemical baths returned directly to the baths;

5. the exhaust gases from a waste-heat boiler used as the heating medium in a lumber kiln.

In all cases, the possibility of contamination from a mixed or a twice-used heat-transport medium must be considered.

8.1.12 Closed Heat Exchangers

As opposed to the open heat exchanger, the closed heat exchanger separates the stream containing the heating fluid from the stream containing the heated fluid, but allows the flow of heat across the separating boundaries. The reasons for separating the streams may be:

1.   A pressure difference may exist between the two streams of fluid. The rigid boundaries of the heat exchanger are designed to withstand the pressure differences.

2.   One stream could contaminate the other if allowed to mix. The impermeable, separating boundaries of the heat exchanger prevents mixing.

3.   To permit the use of an intermediate fluid better suited than either of the principal exchange media for transporting waste heat through long distances. While the intermediate fluid is often steam, glycol and water mixtures and other substances can be used to take of their special properties.

Closed heat exchangers fall into the general classification of industrial heat exchangers, however they have many pseudonyms related to their specific form or to their specific application. They can be called recuperators, regenerators, waste-heat boilers, condensers, tube-and-shell heat exchangers, plate-type heat exchangers, feedwater heaters, economizers, and so on. Whatever name is given all perform one basic function: the transfer of heat across rigid and impermeable boundaries. Sections 8.3 and 8.4 provide much more technical detail concerning the theory, application, and commercial availability of heat exchangers.

8.1.13 Runaround Systems

Whenever it is necessary to ensure isolation of heating and heated systems, or when it becomes advantageous to use an intermediate transfer medium because of the long distances between the two systems, a runaround heat recovery system is used. Figure 8.5 shows the schematic of a runaround system which recovers heat from the exhaust stream from the heating and ventilating system of a building. The circulating medium is a water-glycol mixture selected for its low freezing point. In winter the exhaust air gives up some energy to the glycol in a heat exchanger located in the exhaust air duct. The glycol is circulated by way of a small pump to a second heat exchanger located in the inlet air duct. The outside air is preheated with recovered waste-heat that substitutes for heat that would otherwise be added in the main heating coils of the building's air handler. During the cooling season the heat exchanger in the exhaust duct heats the exhaust air, and the one in the inlet duct precools the outdoor air prior to its passing through the cooling coils of the air handler. The principal reason for using a runaround system in this application is the long

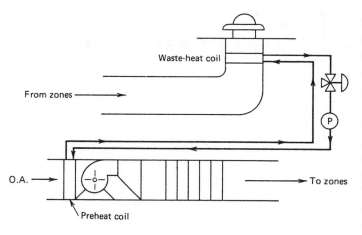

Figure 8.5 Runaround heat-recovery system.

separation distance between the inlet air and the exhaust air ducts. Had these been close together, one air-to-air heat exchanger (with appropriate ducting) could have been more economical.

Figure 8.6 is the schematic diagram of a runaround system used to recover the heat of condensation from a chemical bath steam heater. In this case the bath is a highly corrosive liquid. A leak in the heater coils would cause the condensate to become contaminated and thus do damage to the boiler. The intermediate transport fluid isolates the boiler from a potential source of contamination and corrosion. It should be noted that the presence of corrosive chemicals in the bath, which dictated the choice of the runaround system, are also in contact with one side of the condensate heat exchanger. The materials of construction for that heat exchanger should be carefully selected to withstand the corrosion from that chemical.

8.2 THE WASTE-HEAT SURVEY

8.2.1 How to Conduct the Survey

The survey should be carried out as an integral part of the energy audit of the plant. The survey consists of a systematic study of the sources of waste heat in the plant and of the opportunities for its use. The survey is carried out on three levels. The first step is the identification of every energy-containing nonproduct that flows from the plant. Included are waste streams containing sensible heat (substances at elevated temperatures); examples are hot inert exhaust products from a furnace or cooling water from a compressor. Also to be included are waste streams containing chemical energy (waste

fuels), such as carbon monoxide from a heat-treatment furnace or cupola, solvent vapors from a drying oven, or sawdust from a planing mill: Figure 8.7 is an example of a survey form used for listing waste streams leaving the plant.

The second step is to learn more about the original source of the waste-heat stream. Information should be gathered that can lead to a complete heat balance on the equipment or the system that produces it. Since both potential savings and the capital costs tend to be large in waste-heat recovery situations, it is important that the data be correct. Errors in characterization of the energy-containing streams can either make a poor investment look good, or conversely cause a good investment to look bad and be rejected. Because of the high stakes involved, the engineering costs of making the survey may be substantial. Adequate instrumentation must be installed for accurately metering flow streams. The acquired data are used for designing waste-heat-recovery systems. The instruments are then used for monitoring system operation after the installation has been completed. This is to ensure that the equipment is being operated correctly and maintained in optimum condition so that full benefits will be realized from the capital investment. Figure 8.8 is a survey form used to gather information on each individual system or process unit.

8.2.2 Measurements

Because waste-heat streams have such variability, it is difficult to list every possible measurement that might be required for its characterization. Generally speaking, the characterization of the quantity, quality, and temporal availability of the waste energy requires that volumetric flow rate, temperature, and flow intervals be measured. Chapter 6 of the NBS Handbook 121[8] is devoted to this topic. A few further generalizations are sufficient for planning the survey operation:

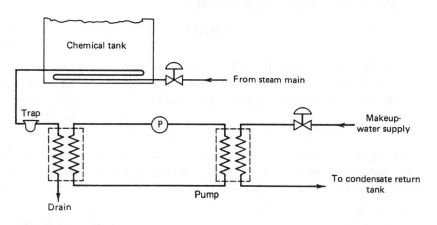

Figure 8.6 Runaround heat-recovery-system process steam source.

Designation	Location	Composition	Flow Rate	Exit Temperature	Heat Rate	Comments

Figure 8.7 Waste-Heat Source Inventory

WASTE HEAT SURVEY

SURVEY FORM FOR INDUSTRIAL PROCESS UNITS

NAME OF PROCESS UNIT <u>Tube Reheat Furnace</u> INVENTORY NUMBER <u>DE-37</u>

LOCATION OF PROCESS UNIT, PLANT NAME <u>Plattsburg Works</u> BUILDING <u>B</u>

MANUFACTURER <u>Symonds Eng'ng Ltd</u> MODEL <u>50-DE-2</u> SERIAL NUMBER <u>18031</u>

		NAME	FIRING RATE	HHV	TEMPERATURE OF			FLUE GAS COMPOSITION % VOLUME				
					COMB. AIR	FUEL	STACK	CO₂	O₂	CO	CH₄	N₂
FOR DIRECT FIRED UNITS	PRIMARY FUEL	Nat.gas	408.8	1030	100 F	100 F	2200 F	7.80	6.30	0.50		85.40
	FIRST ALTERNATIVE	#2 oil	52.729/hr	131,500	100 F	100 F	2200 F	9.30	7.72	0.79		82.19
	SECOND ALTERNAT.											

		FLOW PATH 1	FLOW PATH 2	FLOW PATH 3	FLOW PATH 4
FOR ALL PROCESS UNITS	FLUID COMPOSITION	Steel Tubes	Water		
	FLOW RATE	50 Tons/hr	600 gpm		
	INLET TEMPERATURE	600°F	80°F		
	OUTLET TEMPERATURE	2000°F	191°F		
	DESCRIPTION	Product	Cooling Water		

ANNUAL HOURS OPERATION <u>5700</u> ANNUAL CAPACITY FACTOR, % _____

ANNUAL FUEL COMSUMPTION: PRIMARY FUEL <u>1.39x10⁶ MCF</u>, FIRST ALTERN. <u>4.43x10⁴ gal</u> SEC. ALTERN. _____

PRESENT FUEL COST: PRIMARY FUEL <u>$ 3.21/MCF</u>, FIRST ALTERN. <u>$ 0.535/gal</u> SEC. ALTERN. _____

ANNUAL ELECTRICAL ENERGY COMSUMPTION, KWHR. <u>2,047,000</u>

PRESENT ELECTRICAL ENERGY RATE <u>$ 0.0278</u>

Figure 8.8

1. Flow continuity requires that the mass flow rate of any flow stream under steady-state conditions be constant everywhere in the stream; that is,

$$\dot{m}_{inlet} = \dot{m}_{outlet} \quad \text{or} \quad \rho_{in}A_{in}V_{in} = \rho_{out}A_{out}V_{out}$$

where Q is the material density, A the cross-sectional flow area, and V the velocity of flow normal to that area. The equation can also be written

$$\rho_{in}Q_{in} = \rho_{out}Q_{out} \tag{8.9}$$

where Q is the volumetric flow rate. It is safer, more convenient, and usually more accurate to measure low-temperature flows than those at higher temperatures. Thus in many cases the volumetric flow rate of the cold flow and the temperature of the inlet and outlet flows are sufficient to infer the characteristics of the waste-heat stream.

2. Fuel flows in direct-fired equipment are easily measured with volumetric rate meters. Combustion air and exhaust gas flows are at least an order of magnitude greater than the associated fuel flows. This effectively precludes the use of the volumetric meter because of the expense. However ASME orifice meters, using differential pressure cells are often used for that purpose if the associated pressure drop can be tolerated. It is even cheaper, although less convenient, to determine the volumetric proportions of the flue-gas constituents. Using this data, the air flow quantity and the flue gas flow rate can be calculated from the combustion equation and the law of conservation of mass*.

---

*See Appendix 1, Section 1.2.6.

3. The total energy flux of the fluid streams can be determined from the volumetric flow rate and the temperature using the equation

$$\dot{H} = \rho Q h \qquad (8.10)$$

where $\dot{H}$ = enthalpy flux, Btu/hr
        ρ = density, lb_m/ft^3
        h = specific enthalpy, Btu/lb_m
        Q = volumetric flow rate, ft^3/hr

The density, the specific enthalpy, and many other thermal and physical properties as a function of temperature and chemical species is given in tables and graphs found in a number of engineering handbooks[9,10] and other specialized volumes.[11-13] Appendix II also contains some of that data.

4. In order to complete the waste-heat survey for the plant it is not necessary to completely and permanently instrument all systems of interest. One or more gas meters can be temporarily installed and then moved to other locations. In fact, portable instruments can be used for all measurements. While equipment monitoring, should be carried out with permanently installed instruments, compromises may be necessary to keep survey costs reasonable. However, permanently installed instruments should become a part of every related capital-improvement program.

5. For steady-state operations a single temperature can be assigned to each outlet flow stream. But for a process with preprogrammed temperature profiles in time, an average over each cycle must be carefully determined. For a temperature-zoned device, averages of firing rate must be determined carefully over the several burners.

8.2.3 Estimation without Measurement

It is risky to base economic predictions used for decisions concerning expensive waste-heat-recovery systems on guesswork. However, when measurements are not possible, it becomes necessary to rely on the best approximations available. The approximations must be made taking full advantage of all relevant data at hand. This should include equipment nameplate data; installation, operating, and maintenance literature; production records; fuel and utility invoices; and equipment logs. The energy auditor must attempt to form a consensus

among those most knowledgeable about the system or piece of equipment. This can be done by personally interviewing engineers, managers, equipment operators, and maintenance crews. By providing iterative feedback to these experts, a consensus can be developed. However, estimations are very risky. This author, after taking every possible precaution, using all available data and finding a plausible consensus among the experts, has found his economic projections to be as much as 100% more favorable than actual measurements proved. Those errors would have been avoided by accurate measurements.

8.2.4 Constructing the Heat-Balance Diagram

The first law of thermodynamics (see Appendix I) as applied to a steady flow-steady state system is conveniently written

$$\dot{Q} = \sum_{i=1}^{n} m_i h_i + \dot{W} = \sum_{1}^{n} \rho_i Q_i h_i + \sum_{1}^{n} Q_i h_i + \dot{W} \qquad (8.11)$$

where

$\dot{Q}$ = net rate of heat loss or gain, Btu/hr

p_i = density of ith inflow or outflow, lb/ft^3

Q_i = volumetric flow rate of ith inflow or outflow, scfh

h_i = specific enthalpy of ith inflow or outflow, Btu/lb_m

h'_i = specific enthalpy of ith inflow or outflow, Btu/scf

W = net rate of mechanical or electric work being transferred to or from the system, Btu/hr

n = total number of inlet or outlet paths penetrating system boundaries

Equation 8.11 constitutes the theoretical basis and the mathematical model of the heat-balance diagram shown in Figure 8.9. It corresponds to the data requirements of the survey form shown in Figure 8.8. Using the data taken from that form, we compute the separate terms of the heat balance for a hypothetical furnace as follows:

$$\dot{H}_f \text{ (fuel energy rate)} = \text{firing rate} \times \text{HHV} \qquad (8.12)$$

where HHV is the higher heating value of the fuel in Btu/ft^3 or Btu/gal.

$$\dot{H}_{f,1} = 403.8 \times 10^3 ft^3/hr \times 1030 Btu/ft^3$$

$$= 415.9 \times 10^6 Btu/hr$$

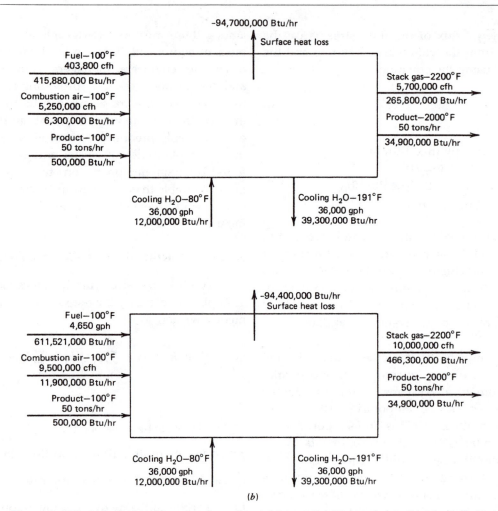

Figure8.9 Heat-balance diagram for reheat furnace(a) Natural gas.(b) No.2 fuel oil.

Because this is a dual-fuel installation, we can also construct a second heat-balance diagram for the alternative fuel:

$$\dot{H}_{f2} = 3162.6 \text{ gph} \times 131,500 \text{ Btu/gal}$$

$$= 415.9 \times 10^6 \text{Btu/hr}$$

Writing the combustion equation on the basis of 100 ft³ of dry flue gas from the flue-gas analysis* for natural gas (fuel no. 1):

$$\alpha CH_4 + \beta O_2 + \gamma N_2 = 7.8CO_2 + 6.3O_2$$
$$+ 0.5CO + 85.4N_2 + 2\alpha H_2O$$

Because the chemical atomic species are conserved, we can solve for the relative quantities of air and fuel:

---

*For simplicity the assumption is made that this natural gas is pure methane. This assumption is often valid for high-Btu fuel gas. For more precision and with other fuel gases, the exact composition should be determined and used in the combustion equation.

For carbon: $\alpha = 7.8 + 0.5 = 8.3$

For nitrogen: $\lambda = 85.4$

For oxygen: $\beta = 7.8 + 6.3 + \dfrac{0.5}{2} + \dfrac{2 \times 8.3}{2} = 22.65$

$$\frac{A}{F} = \frac{\text{volume of air}}{\text{volume of fuel}} = \frac{22.65 + 85.4}{8.3} = 13.0 \frac{\text{ft}^3 \text{ air}}{\text{ft}^3 \text{ gas}}$$

$$\dot{H}_{\text{comb. air}} = \frac{A}{F} \times \text{fuel firing rate} \times h'_{\text{air}}$$

h'_{air} at 100°F is found to be 1.2 Btu/ft³.

$$\dot{H}_{\text{comb. air}} = 13.0 \times 403.8 \times 10^3 \times 1.2$$

$$= 6.3 \times 10^6 \text{ Btu/hr}$$

The specific enthalpy of each of the flue-gas components at 2200°F is found from Figure 8.10 to be

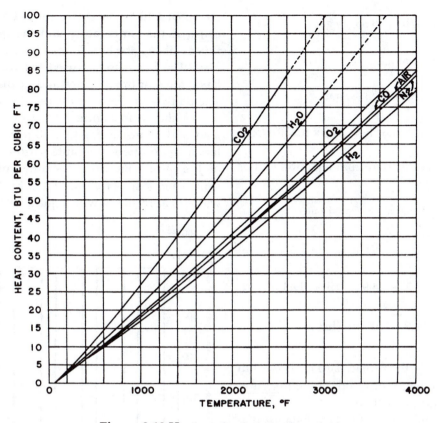

Figure 8.10 Heat content vs. temperature.

$h'_{CO_2} = 69.3 \text{ Btu/scf}$

$h'_{O_2} = 45.4 \text{ Btu/scf}$

$h'_{CO} = 44.1 \text{ Btu/scf}$

$h'_{N_2} = 43.4 \text{ Btu/scf}$

$h'_{H_2O} = 54.7 \text{ Btu/scf}$

$\dot{H}_{\text{stack gas}} = \text{fuel firing rate } (Q_{CO_2} \times h'_{CO_2}$

$+ Q_{O_2} \times h'_{O_2} + Q_{CO} \times h'_{CO} Q_N \times h'_{N_2}$

$+ Q_{H_2O} \times h'_{H_2O})$

$= 403.8 \times 10^3 \left(\dfrac{7.8}{8.3} \times 69.3 + \dfrac{6.3}{8.3} \times 45.4 \right.$

$\left. + \dfrac{0.5}{8.3} \times 44.1 + \dfrac{85.4}{8.3} \times 43.4 + \dfrac{2 \times 8.3}{8.3} \times 54.7\right) \text{btu/hr}$

$= 265.8 \times 10^6 \text{ Btu/hr}$

Because each fuel has its own chemical composition, the stack-gas composition will be different for each fuel, as will the enthalpy flux: thus the calculation should be repeated for each fuel.

Flow path 1 represents the flow of product through the furnace.

$\dot{H}_{\text{prod. in}} = m_{\text{prod}} C_{\text{prod}} T_{\text{prod. in}}$

$= 50 \text{ tons} \times 2000 \text{ lb/ton} \times 0.115 \text{ Btu/lb} \cdot °F \, (100 - 60) \, °F$

$= 0.5 \times 10^6 \text{ Btu/hr}$

$\dot{H}_{\text{prod. out}} = m_{\text{prod}} C_{\text{prod}} T_{\text{prod. out}}$

$= 50 \text{ tons} \times 2000 \text{ lb/ton}$

$\times 0.180 \text{ Btu/lb} \cdot °F \, (2000 - 60) \, °F$

$= 34.9 \times 10^6 \text{ Btu/hr}$

Flow path 2 is the cooling-water flow for the conveyor.

$\dot{H}_{\text{CW. in}} = \text{gpm} \times 60 \text{ min/hr} \times 8.33 \text{ lb/gal}$

$\times 1 \text{ Btu/lb} \cdot F° \times T_{\text{CW. in}}$

$= 600 \text{ gpm} \times 60 \text{ min/hr} \times 8.33 \text{ lb/gal}$

$\times 1 \text{ Btu/lb} \cdot °F \, (100 - 60) \, °F = 12.0 \times 10^6 \text{ Btu/hr}$

$$\dot{H}_{CW.out} = gpm \times 60 \text{ min/hr} \times 8.33 \text{ lb/gal}$$

$$\times 1 \text{ Btu/lb} \cdot {}^\circ F \times T_{CW.\,out}$$

$$= 600 \text{ gpm} \times 60 \text{ min/hr} \times 8.33 \text{ lb/gal}$$

$$\times 1 \text{ Btu/lb} \cdot {}^\circ F \, (191 - 60) \, {}^\circ F = 39.3 \times 10^6 \text{ Btu/hr}$$

The heat losses are then estimated from equation 8.11 as

$$\dot{Q} = \dot{H}_{stack \; gas} - \dot{H}_{f1} - \dot{H}_{comb.\,air} + \dot{H}_{prod.\,out}$$

$$- \dot{H}_{prod.\,in} + \dot{H}_{CW,\,out} - \dot{H}_{CW,\,in}$$

$$= (265.8 - 415.9 - 6.3 + 34.9 - 0.5 + 39.3$$

$$- 12.0)10^6 = -94.7 \times 10^6 \text{ Btu/hr}$$

where $\dot{Q}$ includes not only the furnace surface losses but any unaccounted for enthalpy flux and all the inaccuracies of measurement and calculation.

Figure 8.9 shows the completed heat-balance diagrams for the reheat furnace. From that diagram we identify three waste-heat streams, as listed in Table 8.5.

8.2.5 Constructing Daily Waste-Heat Source and Load Diagrams

The normal daily operating schedule for the furnace analyzed in Section 8.2.4 is plotted in Figure 8.11. It is shown that the present schedule shows furnace operation over two shifts daily, 6 days/week. It is necessary to identify one or more potential loads for each of the two sources.

If high-temperature exhaust stack streams from direct-fired furnaces are recuperated to heat the combustion air stream, then the load and source diagrams are identical. This kind of perfect fit enhances the economics of waste-heat recovery. In order to use the cooling-water

stream it will be necessary to find a potential load. Because of the temperature and the quantity of enthalpy flux available, the best fit may be the domestic hot-water load. The hot water is used for wash facilities for the labor force at break times and at the end of each shift. The daily load diagram for the domestic hot-water system is shown in Figure 8.12. Time coincidence for the load and source is for two 10-minute periods prior to lunch and for two 30-minute wash-up periods at the end of the shifts. Because the time coincidence between source and load is for only 1-1/2 hours in each 16 hours, waste-heat recovery will require heat storage.

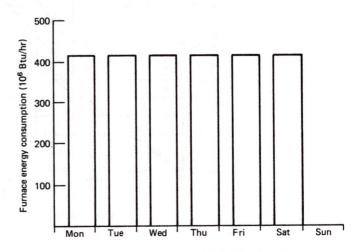

Figure 8.11 Furnace operating schedule.

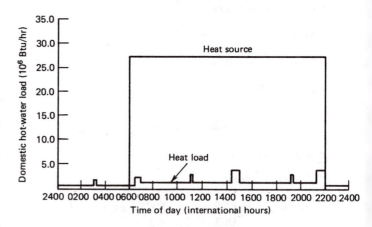

Figure 8.12 Daily domestic water load.

Table 8.5 Waste-Heat Streams from Reheat Furnace Fired with Natural Gas

Composition	Exhaust Stack: Combustion Products	Cooling Water	Product: Steel Castings
Temperature (°F)	2200	191	2000
Flow rate	57,000,000 cfh	36,000 gph	50 tons/hr
Enthalpy rate (Btu/hr)	265,800,000	39,300,000	34,900,000
Percent fuel energy rate	64	9	8

8.2.6 Conceptual Design of the Waste-Heat-Recovery System

Prior to equipment design and before a detailed economic analyses is performed, it is necessary to develop one or more conceptual designs which can serve as a model for the future engineering work, This approach is illustrated by the analyses done in Sections 8.2.4 and 8.2.5 for the two waste-heat streams. An excellent reference text which is useful for the design of waste-heat recovery systems is Hodge's *Analysis and Design of Energy Systems*.[14]

Stack-Gas Stream

Clearly, recuperation is the most promising candidate for heat recovery from high-temperature exhaust gas streams. In the application pictured in Figure 8.13 the hot exhaust gases will be cooled by the incoming combustion air. Because of the temperature of the gases leaving the furnace, the heat exchanger to be selected is a radiation recuperator. This is a concentric tube heat exchanger which replaces the present stack. The incoming combustion air is needed to cool the base of the recuperator and thus parallel flow occurs. Figure 8.13 includes a sketch of the temperature profiles for the two streams. It is seen that in the parallel flow exchanger, heat recovery ceases when the two streams approach a common exit temperature. For a well-insulated recuperator the conservation of energy is expressed by the equation

$$Q_{\text{stack gas}}\left(h'_{\text{stack gas, in}} - h'_{\text{stack gas, out}}\right)$$
$$= Q_{\text{comb. air}}\left(h'_{\text{comb. air, out}} - h'_{\text{comb. air, in}}\right) \quad (8.14)$$

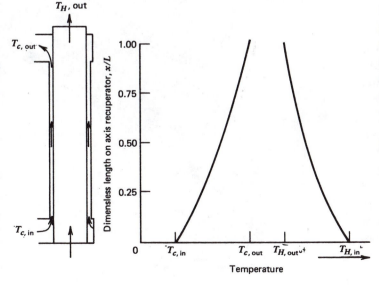

Figure 8.13 Temperature distribution in recuperator.

Both the right- and left-hand terms represent the heat-recovery rate as well as the decrease in fuel energy required. If the burners or associated equipment have maximum temperature limitations, those temperatures also become the high limits for the combustion air, which in turn fixes the maximum allowable enthalpy for the combustion air. Thus the maximum rate of heat recovery is also fixed. Otherwise, the final temperatures of the two stream are based on an optimization of the economic opportunity. This is so because increased heat recovery implies increased recuperator area and thus increased cost. It may also imply higher combustion air temperatures with the resultant increase in fan operating costs and additional investment costs in high-temperature burners, combustion air ducts, and larger fans. In this case we assume that a 100°F temperature difference will occur between the preheated combustion air and the stack gases leaving the recuperator. Equation 8.14 cannot be used directly because the volume rates of fuel and air required are reduced with recuperation. However, if the air/fuel ratio is maintained constant, then $Q_{\text{stack/gas}}/Q_{\text{comb.air}}$ remains almost constant; then equation 8.14 can be written

$$\frac{Q_{\text{stack gas}}}{Q_{\text{comb. air}}} = \text{const.} = \frac{h'_{\text{comb. air, out}} - h'_{\text{comb. air, in}}}{h'_{\text{stack gas, in}} - h'_{\text{stack gas, out}}}$$
$$(8.15)$$

This equation can be solved with the help of data from Figure 8.9 and the temperature relationship

$$T_{\text{stack gas.out}} = T_{\text{comb air out}} + 100°F \quad (8.16)$$

Separate solutions are required for the primary and alternative fuels. The solutions are found by iterating equations 8.15 and 8.16 through a range of temperatures. The value of preheat temperature found for natural gas is 1260°F. For fuel oil the temperature is only slightly different. The annual heat recovery for each fuel was assumed to be proportional to the total consumption of each fuel, so that the heat recovered was found as

$$\text{heat recovered} = 1.39 \times 10^9 \frac{\text{ft}^3\text{gas}}{\text{yr}} \times 12.95 \frac{\text{ft}^3\text{ air}}{\text{ft}^3\text{ gas}}$$

$$\times (24.6 - 1.2) \frac{\text{Btu}}{\text{ft}^3} + 4.43 \times 10^6 \frac{\text{gal}}{\text{yr}}$$

$$\times 2040.8 \frac{\text{ft}^3\text{ air}}{\text{gal oil}} \times (24.6 - 1.2) \frac{\text{Btu}}{\text{ft}^3}$$

$$= 4.2 \times 10^{11} + 2.1 \times 10^{11} = 6.3 \times 10^{11} \text{ Btu/yr}$$

This is an energy savings of

$$\frac{6.3 \times 10^{11}}{1.39 \times 10^9 \times 1030 + 4.43 \times 10^6 \times 131,500} = 0.31 \text{ or } 31\%$$

The predicted cost savings is

$1,308,900	for natural gas	
950,200	for No. 2 fuel oil	
$2,259,100	total	

The complete retrofit installation is estimated to cost less than $2,000,000, and the payback period is less than one year. Two points must be emphasized. The entire retrofit installation must be well engineered as a system. This includes the recuperator itself as well as modifications and/or replacement of burners and fans, and the system controls. Only then can the projected system life span be attained and the capital payback actually realized. The cost of lost product must also be factored into the economic analysis if the installation is planned at a time that will cause a plant shutdown. Economics may dictate a delay for the retrofit until the next scheduled or forced maintenance shutdown.

8.3 WASTE-HEAT EXCHANGERS

8.3.1 Transient Storage Devices

The earliest waste-heat-recovery devices were "regenerators." These consisted of extensive brick work, called "checkerwork," located in the exhaust flues and inlet air flues of high-temperature furnaces in the steel industry. Regenerators are still used to a limited extent in open hearth furnaces and other high-temperature furnaces burning low-grade fuels. It is impossible to achieve steel melt temperature unless regenerators are used to boost the inlet air temperature. In the process vast amounts of waste heat are recovered which would otherwise be supplied by expensive high-Btu fuels. Pairs of regenerators are used alternately to store waste heat from the furnace exhaust gases and then give back that heat to the inlet combustion air. The transfer of exhaust-gas and combustion-air streams from one regenerator to the other is accomplished by using a four-way flapper valve. The design of and estimates of the performance of recuperators follows the principles presented in Section 8.1.7. One disadvantage of this mode of operation is that heat-exchanger effectiveness is maximum only at the beginning of each heating and cooling cycle and falls to almost zero at the end of the cycle. A second disadvantage is that the tremendous mass of the checkerwork and

the volume required for its installation raises capital costs above that for the continuous-type air preheaters.

An alternative to the checkerwork regenerator is the heat wheel. This device consists of a permeable flat disk which is placed with its axis parallel to a pair of split ducts and is slowly rotated on an axis parallel to the ducts. The wheel is slowly rotated as it intercepts the gas streams flowing concurrently through the split ducts. Figure 8.14 illustrates those operational features.

As the exhaust-gas stream in the exhaust duct passes through one-half of the disk it gives up some of its heat which is temporarily stored in the disc material. As the disc is turned, the cold incoming air passes through the heated surfaces of the disk and absorbs the energy. The materials used for the disks include metal alloys, ceramics and fiber, depending upon the temperature of the exhaust gases. Heat-exchanger efficiency for the heat wheel has been measured as high as 90% based upon the exhaust stream energy. Further details concerning the heat wheel and its applications are given in Section 8.4.3.

8.3.2 Steady-State Heat Exchangers

Section 8.4 treats heat exchangers in some detail. However, several important criteria for selection are listed below.

1. **Flow Arrangements.** These are characterized as:

Parallel flow	Crossflow
Counterflow	Mixed flow

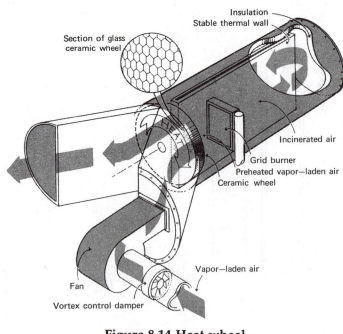

Figure 8.14 Heat wheel.

The flow arrangement helps to determine the overall effectiveness, the cost, and the highest achievable temperature in the heated stream. The latter effect most often dictates the choice of flow arrangement. Figure 8.15 indicates the temperature profiles for the heating and heated streams, respectively. If the waste-heat stream is to be cooled below the load stream exit, a counterflow heat exchanger must be used.

2. **Character of the Exchange Fluids.** It is necessary to specify the heated and cooled fluids as to:

    Chemical composition
    Physical phase (i.e., gaseous, liquid, solid, or
        multiphase)
    Change of phase, if any, such as evaporating or
        condensing

    These specifications may affect the optimum flow arrangement and/or the materials of construction.

8.3.3 Heat-Exchanger Effectiveness

The effectiveness of a heat exchanger is defined as a ratio of the actual heat transferred to the maximum possible heat transfer considering the temperatures of two streams entering the heat exchanger. For a given flow arrangement, the effectiveness of a heat exchanger is directly proportional to the surface area that separates the heated and cooled fluids. The effectiveness of typical heat exchangers is given in Figure 8.16 in terms of the parameter AU/C_{min} where A is the effective heat-transfer area, U the effective overall heat conductance, and C_{min} the mass flow rate times the specific heat of the fluid with minimum mc. The conductance is the heat rate per unit area per unit temperature difference. Note that as AU/C_{min} increases, a linear relation exists with the effectiveness until the value of AU/C_m approaches 1.0. At this point the curve begins to knee over and the increase in effectiveness with AU is drastically reduced. Thus one sees a relatively early onset of the law of diminishing returns for heat-exchanger design. It is implied that one pays heavily for exchangers with high effectiveness.

8.3.4 Filtering or Fouling

One of the important heat-exchanger parameters related to surface conditions is termed the fouling factor. The fouling of the surfaces can occur because of film deposits, such as oil films; because of surface scaling due

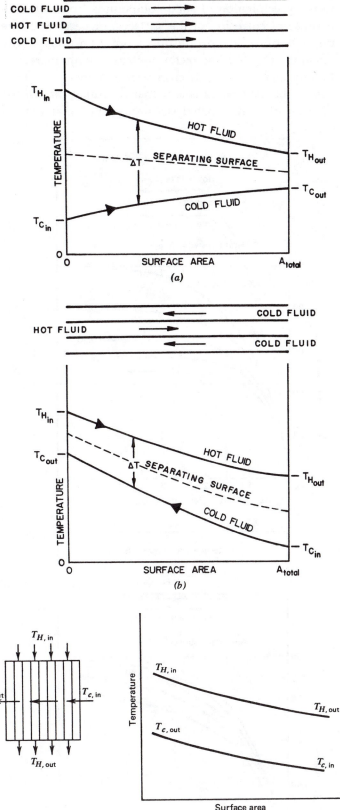

Figure 8.15 Cross-flow heat exchanger.

to the precipitation of solid compounds from solution; because of corrosion of the surfaces; or because of the deposit of solids or liquids from two-phase flow streams. The fouling factor increases with increased fouling and causes a drop in heat exchanger effectiveness. If heavy fouling is anticipated, it may call for the filtering of contaminated streams, special materials of construction, or a mechanical design that permits easy access to surfaces for frequent cleaning.

8.3.5 Materials and Construction

These topics have been reviewed in previous sections. In summary:
1. High temperatures may require the use of special materials.
2. The chemical and physical properties of exchange fluids may require the use of special materials.
3. Contaminated fluids may require special materials and/or special construction.
4. The additions of tube fins on the outside, grooved surfaces or swaged fins on the inside, and treated or coated surfaces inside or outside may be required to achieve compactness or unusually high effectiveness.

8.3.6 Corrosion Control

The standard material of construction for heat exchangers is mild steel. Heat exchangers made of steel are the cheapest to buy because the material is the least expensive of all construction materials and because it is so easy to fabricate. However, when the heat transfer media are corrosive liquids and/or gases, more exotic ma-

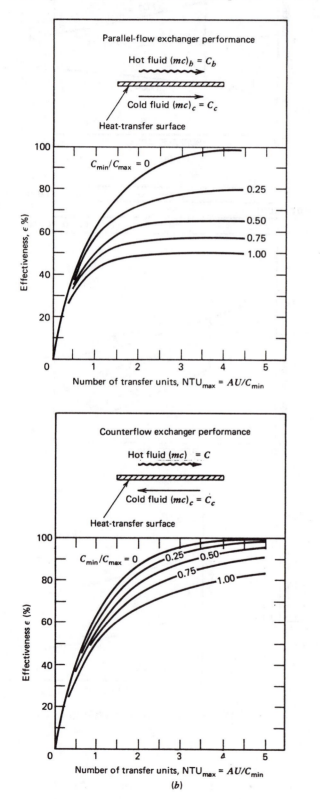

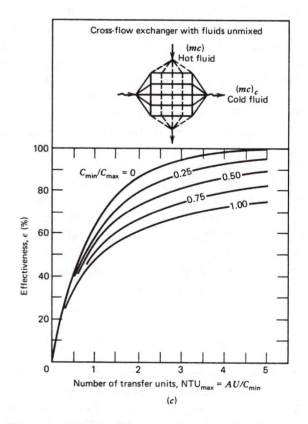

Figure 8.16 Typical heat-exchanger effectiveness.

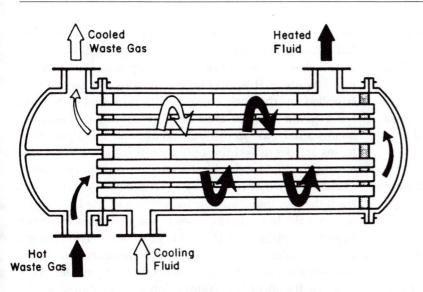

Figure 8.17. Shell heat exchanger.

terials may have to be used. Corrosion tables[15] give the information necessary to estimate the life of the heat exchanger and life-cycle-costing studies allow valid comparisons of the costs of owning the steel heat exchanger versus one constructed of exotic materials. The problem is whether it will be cheaper to replace the steel heat exchanger at more frequent intervals or to buy a unit made of more expensive materials, but requiring less frequent replacement. Mechanical designs which permit easy tube replacement lower the cost of rebuilding and favor the use of mild steel heat exchangers.

Corrosion-resisting coatings, such as the TFE plastics, are used to withstand extremely aggressive liquids and gases. However, the high cost of coating and the danger of damaging the coatings during assembly and during subsequent operation limit their use. One disadvantage of using coatings is that they almost invariably decrease the overall conductance of the tube walls and thus necessitate an increase in size of the heat exchanger. The decision to use coatings depends first upon the availability of alternate materials to withstand the corrosion as well as the comparative life-cycle costs, assuming that alternative materials can be found.

Among the most corrosive and widely used materials flowing in heat exchangers are the chlorides such as hydrochloric acid and saltwater. Steel and most steel alloys have extremely short lives in such service. One class of steel alloys that have shown remarkable resistance to chlorides and other corrosive chemicals is called duplex steels[16] and consists of half-and-half ferrite and austenitic microstructures. Because of their high tensile strength, thinner tube walls can be used and this offsets some of the higher cost of the material.

8.3.7 Maintainability

Provisions for gaining access to the internals may be worth the additional cost so that surfaces may be easily cleaned, or tubes replaced when corroded. A tube-and-shell heat exchanger with flanged and bolted end caps which are easily removed for maintenance is shown in Figure 8.17. Economizers are available with removable panels and multiple one-piece finned, sepentine tube elements, which are connected to the headers with standard compression fittings. The tubes can be removed and replaced on site, in a matter of minutes, using only a crescent wrench.

8.4 COMMERCIAL OPTIONS IN WASTE-HEAT-RECOVERY EQUIPMENT

8.4.1 Introduction

It is necessary to completely specify all of the operating parameters as well as the heat exchange capacity for the proper design of a heat exchanger, or for the selection of an off-the-shelf item. These specifications will determine the construction parameters and thus the cost of the heat exchanger. The final design will be a compromise among pressure drop (which fixes pump or fan capital and operating costs), maintainabilty (which strongly affects maintenance costs), heat exchanger effectiveness, and life-cycle cost. Additional features, such as the on-site use of exotic materials or special designs for enhanced maintainability, may add to the initial cost. That design will balance the costs of operation and maintenance with the fixed costs in order to minimize the life-cycle costs. Advice on selection and design of heat exchangers is available from manufacturers and from T.E.M.A.* *Industrial Heat Exchangers* [17] is an excellent guide to heat exchanger selection and includes a directory of heat exchanger manufacturers.

The essential parameters that should be known and specified in order to make an optimum choice of wasteheat-recovery devices are:

Temperature of waste-heat fluid
Flow rate of waste-heat fluid
Chemical composition of waste-heat fluid
Minimum allowable temperature of waste-heat fluid

---

*Tubular Equipment Manufacturers Association, New York, NY

Amount and type of contaminants in the waste-heat fluid
Allowable pressure drop for the waste-heat fluid
Temperature of heated fluid
Chemical composition of heated fluid
Maximum allowable temperature of heated fluid
Allowable pressure drop in the heated fluid
Control temperature, if control required

In the remainder of this section, some common types of commercially available waste-heat-recovery devices are discussed in detail.

8.4.2 Gas-to-Gas Heat Exchangers: Recuperators

Recuperators are used in recovering waste heat to be used for heating gases in the medium- to high-temperature range. Some typical applications are soaking ovens, annealing ovens, melting furnaces, reheat furnaces, afterburners, incinerators, and radiant-heat burners. The simplest configuration for a heat exchanger is the metallic radiation recuperator, which consists of two concentric lengths of metal tubing, as shown in Figure 8.18. This is most often used to extract waste heat from the exhaust gases of a high-temperature furnace for heating the combustion air for the same furnace. The assembly is often designed to replace the exhaust stack.

The inner tube carries the hot exhaust gases while the external annulus carries the combustion air from the

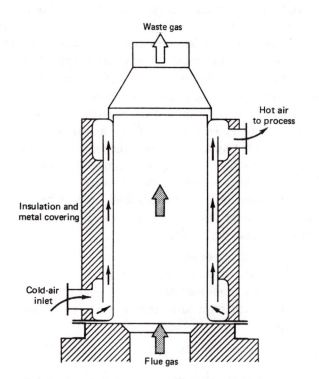

Figure 8.18 Metallic radiation recuperator.

atmosphere to the air inlets of the furnace burners. The hot gases are cooled by the incoming combustion air, which then carries additional energy into the combustion chamber. This is energy that does not have to be supplied by the fuel; consequently, less fuel is burned for a given furnace loading. The saving in fuel also means a decrease in combustion air, and therefore stack losses are decreased not only by lowering the stack exit temperatures, but also by discharging smaller quantities of exhaust gas. This particular recuperator gets its name from the fact that a substantial portion of the heat transfer from the hot exhaust gases to the surface of the inner tube takes place by radiative heat transfer. The cold air in the annulus, however, is almost transparent to infrared radiation so that only convection heat transfer takes place to the incoming combustion air. As shown in the diagram, the two gas flows are usually parallel, although the configuration would be simpler and the heat transfer more efficient if counterflow were used. The reason for the use of parallel flow is that the cold air often serves the function of cooling the hottest part of the exhaust duct and consequently extends its service life.

The inner tube is often fabricated from high-temperature materials such as high-nickel stainless steels. The large temperature differential at the inlet causes differential expansion, since the outer shell is usually of a different and less expensive material. The mechanical design must take this effect into account. More elaborate designs of radiation recuperators incorporate two sections; the bottom operating in parallel flow, and the upper section using the more efficient counterflow arrangement. Because of the large axial expansions experienced and the difficult stress conditions that can occur at the bottom of the recuperator, the unit is often supported at the top by a freestanding support frame and the bottom is joined to the furnace by way of an expansion joint.

A second common form for recuperators is called the tube-type or convective recuperator. As seen in the schematic diagram of a combined radiation and convective type recuperator in Figure 8.19, the hot gases are carried through a number of small-diameter parallel tubes, while the combustion air enters a shell surrounding the tubes and is heated as it passes over the outside of the tubes one or more times in directions normal to the tubes. If the tubes are baffled as shown so as to allow the air to pass over them twice, the heat exchanger is termed a two-pass convective recuperator; if two baffles are used, a three-pass recuperator; and so on. Although baffling increases the cost of manufacture and also the pressure drop in the air path, it also increases the effectiveness of heat exchange. Tube-type recuperators are generally more compact and have a higher effectiveness than do radiation recuperators, because of the larger effective heat-transfer area made possible through the use of multiple tubes and multiple passes of the air. For maximum effectiveness of heat transfer, combinations of the two types of recuperators are used, with the convection type always following the high-temperature radiation recuperator.

The principal limitation on the heat recovery possible with metal recuperators is the reduced life of the liner at inlet temperatures exceeding 2000°F. This limitation forces the use of parallel flow to protect the bottom of the liner. the temperature problem is compounded when furnace combustion air flow is reduced as the furnace loading is reduced. Thus the cooling of the inner shell is reduced and the resulting temperature rise causes rapid surface deterioration. To counteract this effect, it is necessary to provide an ambient air bypass to reduce the temperature of the exhaust gases. The destruction of a radiation recuperator by overheating is a costly accident. Costs for rebuilding one are about 90% of the cost of a new unit.

To overcome the temperature limitations of metal recuperators, ceramic-tube recuperators have been developed whose materials permit operation to temperatures of 2800°F and on the preheated air side to 2200°F, although practical designs yield air temperatures of 1800°F. Early ceramic recuperators were built of tile and joined with furnace cement. Thermal cycling caused cracking of the joints and early deterioration of the units. Leakage rates as high as 60% were common after short service periods. Later developments featured silicon carbide tubes joined by flexible seals in the air headers. This kind of design, illustrated in Figure 8.20, maintains the seals at a relatively low temperature and the life of seals has been much improved, as evidenced by leakage rates of only a few percent after two years of service.

An alternative design for the convective recuperator is one in which the cold combustion air is heated in a bank of parallel tubes extending into the flue-gas stream normal to the axis of flow. This arrangement is shown in Figure 8.21. The advantages of this configuration are compactness and

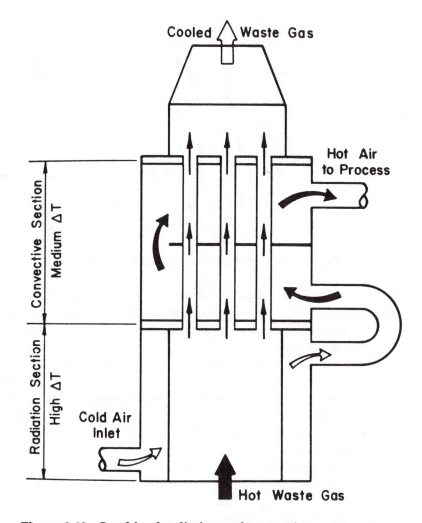

Figure 8.19. Combined radiation and convective recuperator.

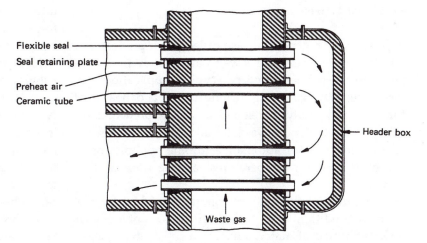

Figure 8.20. Silicon-carbide-tube ceramic recuperator.

the ease of replacing individual units. This can be done during full-load operation and minimizes the cost, inconvenience, and possible furnace damage due to a forced shutdown from recuperator failure.

Recuperators are relatively inexpensive and they

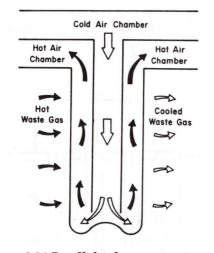

Figure 8.21 Parallel-tube recuperator.

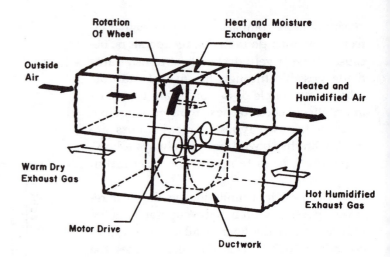

Figure 8.22 Rotary regenerator (heat wheel).

do reduce fuel consumption. However, their use may require extensive capital improvements. Higher combustion air temperatures may require:

- burner replacement
- larger-diameter air lines with flexible expansion fittings
- cold-air piping for cooling high-temperature burners
- modified combustion controls
- stack dampers
- cold air bleeds
- recuperator protection systems
- larger combustion air fans to overcome the additional pressure drops in the system.

8.4.3 Heat Wheels

A rotary regenerator, also called an air preheater or a heat wheel, is used for low- to moderately high-temperature waste-heat recovery. Typical applications are for space heating, curing, drying ovens and heat-treat furnaces. Originally developed as an air preheater for utility steam boilers, it was later adapted, in small sizes, as a regenerator for automotive turbine applications. It has been used for temperatures ranging from 68°F to 2500°F.

Figure 8.22 illustrates the operation of a heat wheel in an air conditioning application. It consists of a porous disk, fabricated of material having a substantial specific heat. The disk is driven to rotates between two side-by-side ducts. One is a cold-gas duct and the other is a hot-gas duct. Although the diagram shows a counterflow configuration, parallel flow can also be used. The axis of the disk is located parallel to and on the plane of the partition between the ducts. As the disk slowly rotates, sensible heat (and in some cases, moisture-containing la-

tent heat) is transferred to the disk by the hot exhaust gas. As the disk moves into the area of the cold duct, the heat is transferred from the disk to the cold air. The overall efficiency of heat transfer (including latent heat) can be as high as 90%.

Heat wheels have been built as large as 70 ft in diameter with air capacities to 40,000 cfm. Multiple units can be used in parallel. This modular approach may be used to overcome a mismatch between capacity requirements and the limited number of sizes available in commercial units.

The limitations on the high-temperature range for the heat wheel are primarily due to mechanical difficulties introduced by uneven thermal expansion of the rotating wheel. Uneven expansion can cause excessive deformations of the wheel that result in the loss of adequate gas seals between the ducts and the wheel. The deformation can also result in damage due to the wheel rubbing against its retaining enclosure.

Heat wheels are available in at least four types: 1) A metal frame packed with a core of knitted mesh stainless steel, brass, or aluminum wire, 2) A so-called laminar wheel fabricated from corrugated materials which form many small diameter parallel-flow passages, 3) A laminar wheel constructed from a high-temperature ceramic honeycomb, and 4) A laminar wheel constructed of a fibrous material coated with a hygroscopic so that latent heat can be recovered.

Most gases contain some water vapor since it is a natural component of air and it is also a product of hydrocarbon combustion. Water vapor, as a component of a gas mixture, carries with it its latent heat of evaporation. This latent heat may be a substantial part of the energy contained within the exit-gas streams from air-conditioned spaces or from industrial processes. To recover some of the latent heat in the gas stream, using a heat wheel, the sheet must be coated with a hygroscopic material such as lithium chloride (LiCl) which readily absorbs water vapor to form a hydrate, which in the case of lithium chloride is the hydrate LiCl•H2O. The hydrate consists of one mole of lithium chloride chemically combined with one mole of water vapor. Thus the weight ratio of water to lithium-chloride is 3:7. In a hygroscopic heat wheel, the hot gas stream gives up some part of its water vapor to the lithium-chloride coating; the gases to be heated are dry and absorb some of the water held in the hydrate. The latent heat in that water vapor adds directly to the total quantity of recovered heat. The efficiency of recovery of the water vapor in the exit stream may be as high as 50%.

Because the pores or passages of heat wheels carry small amounts of gas from the exhaust duct to the intake duct, cross-contamination of the intake gas can occur. If the contamination is undesirable, the carryover of exhaust gas can be partially eliminated by the addition of a purge section located between the intake and exhaust ducts, as shown in Figure 8.23. The purge section allows the passages in the wheel to be cleared of the exhaust gases by introducing clean air which discharges the contaminant to the atmosphere. Note that additional gas seals are required to separate the purge ducts from the intake and exhaust ducts and consequently add to the cost of the heat wheel.

Common practice is to use six air changes of clean air for purging. This results in a reduction of cross-contamination to a value as little as 0.04% for the gas and 0.2% for particulates in laminar wheels, and less than 1.0% total contaminants in packed wheels.

If the heated gas temperatures are to be held constant, regardless of heating loads and exhaust gas temperatures, the heat wheel must be driven at variable speed. This requires a variable-speed drive with a speed-controller with an air temperature sensor as the control element. When operating with outside air in periods of sub-zero temperatures and high humidity, heat wheels may frost up requiring the protection of an air-preheat system. When handling gases containing water-soluble, greasy, or large concentrations of particulates, air filters may be required in the exhaust system upstream from the heat wheel. These features, however, add to the complexity and the cost of owning and operating the system.

Contaminant buildup on ceramic heat wheels can often be removed by raising the temperature of the exhaust stream to exceed the ignition temperature of the contaminant. However, heat wheels are inherently self-cleaning, because materials entering the wheel from the hot-gas stream tend to be swept out by the reverse flow of the cold-gas stream.

8.4.4 Passive Air Preheaters

Passive gas-to-gas regenerators are available for applications where cross-contamination cannot be tolerated. One such type of regenerator, the plate-type, is shown in Figure 8.24. A second type, the heat pipe array is shown in Figure 8.25. Passive air preheaters are used in the low- and medium-temperature applications. Those include drying, curing, and baking ovens; air preheaters in steam boilers; air dryers; waste heat recovery from exhaust steam; secondary recovery from refractory kilns and reverbatory furnaces; and waste heat recovery from conditioned air.

The plate-type regenerator is constructed of alternate channels which separate adjacent flows of heated and heating gases by a thin wall of conducting metal. Although their use eliminates cross-contamination, they are bulkier, heavier, and more expensive than a heat wheel of similar heat-recovery and flow capacities. Furthermore, it is difficult to achieve temperature control of the heated gas, while fouling may be a more serious problem.

The heat pipe is a heat-transfer element that is assembled into arrays which are used as compact and effi-

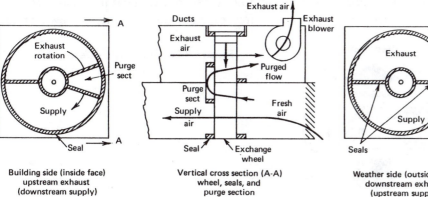

Building side (inside face)
upstream exhaust
(downstream supply)

Vertical cross section (A-A)
wheel, seals, and
purge section

Weather side (outside face)
downstream exhaust
(upstream supply)

Figure 8.23
Heat wheel with purge section.

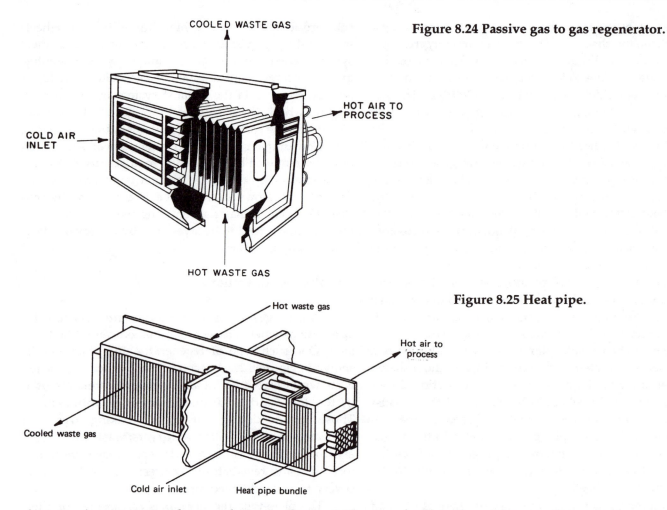

Figure 8.24 Passive gas to gas regenerator.

Figure 8.25 Heat pipe.

cient passive gas-to-gas heat exchangers. Figure 8.25 shows how the bundle of finned heat pipes extend through the wall separating the inlet and exhaust ducts in a pattern that resembles the conventional finned tube heat exchangers. Each of the separate pipes, however, is a separate sealed element. Each consists of an annular wick on the inside of the full length of the tube, in which an appropriate heat-transfer fluid is absorbed. Figure 8.26 shows how the heat transferred from the hot exhaust gases evaporates the fluid in the wick. This causes the vapor to expand into the center core of the heat pipe. The latent heat of evaporation is carried with the vapor to the cold end of the tube. There it is removed by

transferral to the cold gas as the vapor is recondensed. The condensate is then carried back in the wick to the hot end of the tube. This takes place by capillary action and by gravitational forces if the axis of the tube is tilted from the horizontal. At the hot end of the tube the fluid is then recycled.

The heat pipe is compact and efficient for two reasons. The finned-tube bundle is inherently a good configuration for convective heat transfer between the gases and the outside of the tubes in both ducts. The evaporative-condensing cycle within the heat tubes is a highly efficient method of transferring heat internally. This design is also free of cross-contamination. However, the

Figure 8.26 Heat pipe operation.

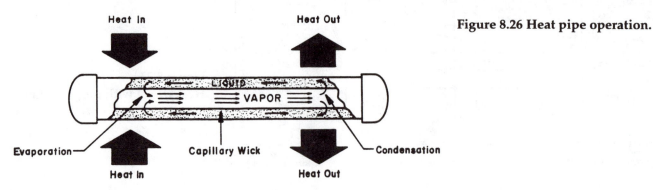

Table 8.6 Temperature Ranges for Heat-Transfer Fluids Used in Heat Pipes

Fluid	Temperature Range (°F)			Compatible Metals
Nitrogen	− 300	to	− 110	Stainless steel
Ammonia	− 95	to	+ 140	Nickel, aluminum, stainless steel
Methanol	− 50	to	+ 240	Nickel, copper, stainless steel
Water	40	to	425	Nickel, copper
Mercury	375	to	1000	Stainless steel
Sodium	950	to	1600	Nickel, stainless steel
Lithium	1600	to	2700	Alloy of niobium and zirconium
Silver	2700	to	3600	Alloy of tantalum and tungsten

temperature range over which waste heat can be recovered is severely limited by the thermal and physical properties of the fluids used within the heat pipes. Table 8.6 lists some of the transfer fluids and the temperature ranges in which they are applicable.

8.4.5 Gas or Liquid-to-Liquid Regenerators: The Boiler Economizer

The economizer is ordinarily constructed as a bundle of finned tubes, installed in the boiler's breeching. Boiler feedwater flows through the tubes to be heated by the hot exhaust gases. Such an arrangement is shown in Figure 8.27. The tubes are usually connected in a series arrangement, but can also be arranged in series-parallel to control the liquid-side pressure drop. The air-side pressure drop is controlled by the spacing of the tubes and the number of rows of tubes. Economizers are available both prepackaged in modular sizes and designed and fabricated to custom specifications from standard components. Materials for the tubes and fins can be selected to withstand corrosive liquids and/or exhaust gases.

Temperature control of the boiler feedwater is necessary to prevent boiling in the economizer during low-steam demand or in case of a feedwater pump failure. This is usually obtained by controlling the amount of exhaust gases flowing through the economizer using a damper, which diverts a portion of the gas flow through a bypass duct.

The extent of heat recovery in the economizer may be limited by the lowest allowable exhaust gas temperature in the exhaust stack. The exhaust gases contain water vapor both from the combustion air and from the

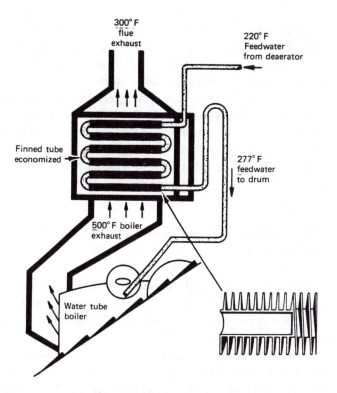

Figure 8.27 Boiler economizer.

combustion of the hydrogen that is contained in the fuel. If the exhaust gases are cooled below the dew point of the water vapor, condensation will occur and cause damage to the structural materials. If the fuel also contains sulfur, the sulfur-dioxide will be absorbed by the condensed water to form sulfuric acid. This is very corrosive and will attack the breeching downstream of the economizer and the stack lines. The dew point of the exhaust gases from a natural-gas-fired boiler varies from approximately 138°F for a stoichiometric fuel/air mixture, to 113°F for 100% excess air. Because heat-transmission losses through the stack cause axial temperature gradients from 0.2 to 2°F/ft, and because the stack liner may exist at a temperature 50 to 75°F lower than the gas

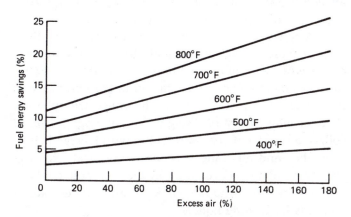

Figure 8.28 Fuel savings from a gas-fired boiler using economizer.

bulk temperature, it is considered prudent to limit minimum stack temperatures to 300°F, or no lower than 250°F when burning natural gas. When using the fuels containing sulfur, even greater caution is taken. This means that the effectiveness of an economizer is limited unless the exhaust gases from the boiler are relatively hot. Figure 8.28 is a graph of the percent fuel saved plotted against percent excess air for a number of stack gas temperatures using natural gas as a boiler fuel. The plots are based on a 300°F hot-gas temperature leaving the economizer.

8.4.6 Shell-and-Tube or Concentric-Tube Heat Exchangers

Shell-and-tube and concentric-tube heat exchangers are used to recover heat in the low and medium range from process liquids, coolants, and condensates of all kinds for heating liquids.

When the medium containing waste heat is either a liquid or a vapor that heats a liquid at a different pressure, a totally exclosed heat exchanger must be used. The two fluid streams must be separated so as to contain their respective pressures. In the shell-and-tube heat exchanger, the shell is a cylinder that contains the tube bundle. Internal baffles may be used to direct the fluid in the shell over the tubes in multiple passes. Because the shell is inherently weaker than the tubes, the higher-pressure fluid is usually circulated in the tubes while the lower-pressure fluid circulates in the shell. However, when the heating fluid is a condensing vapor, it is almost invariably contained within the shell. If the reverse were attempted, the condensation of the vapor within the small-diameter parallel tubes would cause flow instabilities. Tube-and-shell heat exchangers are produced in a wide range of standard sizes with many combinations of materials for the tubes and the shells. The overall conductance of these heat exchangers range to a maximum of several hundred Btu/hr ft^2 °F.

A concentric-tube exchanger is used when the fluid pressures are so high that a shell design is uneconomical, or when ease of dissembly is paramount. The hotter fluid is almost invariably contained in the inner tube to minimize surface heat losses. The concentric-tube exchanger may consist of a single straight length, a spiral coil, or a bundle of concentric tubes with hairpin bends.

Shell-and-tube and concentric-tube heat exchangers are used to recover heat in the low and medium range from process liquids, coolants, and condensates of all kinds for heating liquids.

8.4.7 Waste-Heat Boilers

Waste-heat boilers are water tube boilers in which hot exhaust gases are used to generate steam. The exhaust gases may be from a gas turbine, an incinerator, a diesel engine, or any other source of medium- to high-temperature waste heat. Figure 8.29 shows a conventional, two-pass waste-heat boiler. When the heat source is in the medium-temperature range, the boiler tends to become bulky. The use of finned tubes extends the heat transfer areas and allows a more compact size. Table 8.7 gives specifications for a typical waste-heat boiler. If the quantity of waste heat is insufficient for generating a needed quantity of steam, it is possible to add auxiliary burners to the boiler or an afterburner to the ducting upstream of the boiler. The conventional waste-heat boiler cannot generate super-heated steam so that an external superheater is required if superheat is needed.

A more recently designed waste-heat boiler utilizes

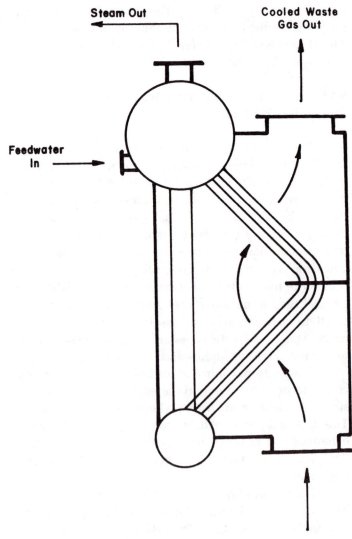

Figure 8.29 Two-pass waste-heat boiler.

a finned-tube bundle for the evaporator, an external drum, and forced recirculation of the feedwater. The design, which is modular, makes for a compact unit with high boiler efficiency. Additional tube bundles can be added for superheating the steam and for preheating the feedwater. The degree of superheat which can be achieved is limited by the waste-heat temperature. The salient features of the boiler are shown on the schematic diagram in Figure 8.30.

Waste-heat boilers are commercially available in capacities from less than 1000 up to 1 million cfm of exhaust gas intake.

8.4.8 Input-Output Matrix for Waste-Heat-Recovery Devices

Table 8.8 presents the significant attributes of the most common types of industrial heat exchangers. This matrix is useful in making selections from competing types of heat exchangers for waste-heat recovery.

8.5 ECONOMICS OF WASTE-HEAT RECOVERY

8.5.1 General

Economic analysis techniques used for analyzing investment potential for waste-heat-recovery systems are no different from those used for the analysis of any other industrial capital project. These techniques are thoroughly discussed in Chapter 4 of this volume and in Chapter 3 of the NBS Handbook 121.[18] The economic potential for this class of systems is often limited by factors that are crucial yet overlooked. Although the capital cost of these systems is proportional to the peak rate of heat recovery, the capital recovery depends principally on the annual fuel savings. These savings depend on a number of factors, such as the time distribution of waste-heat source availability, the time distribution of heat-load availability, the availability of waste-heat-recovery equipment that can perform at the specified thermal conditions, and the current and future utility rates and prices of fuel. The inability to accurately predict these factors can make the normal investment decision-making process ineffectual.

There is another important distinction to be made about waste-heat recovery investment. When capital projects involve production-related equipment, the rate of capital recovery can be adjusted by manipulating product selling prices. When dealing with waste-heat-recovery systems, this is generally not an option. The rate of capital recovery is heavily influenced by utility rates and fuel prices which are beyond the influence of

Table 8.8 Operation and Application Characteristics of Industrial Heat Exchangers

Commercial Heat-Transfer Equipment	Low temperature: subzero –250°F	Intermediate temp: 250-1200°F	High temperature: 1200-2000°F	Recovers moisture	Large temperature differentials permitted	Packaged units available	Can be retrofit	No cross-contamination	Compact size	Gas-to-gas heat exchange	Gas-to-liquid heat exchanger	Liquid-to-liquid heat exchanger	Corrosive gases permitted with special construction
Radiation recuperator			×		×	a	×	×		×			×
Convection recuperator		×	×		×	×	×	×		×			×
Metallic heat wheel	×	×	b			×	×	c	×	×			×
Hygroscopic heat wheel	×			×		×	×	c	×	×			
Ceramic heat wheel		×	×		×	×	×		×	×			×
Passive regenerator	×	×			×	×	×	×		×			×
Finned-tube heat exchanger	×	×			×	×	×	×	×		×		d
Tube shell-and-tube exchanger	×	×			×	×	×	×	×		×	×	
Waste-heat boilers	×	×	×			×	×	×			×		d
Heat pipes	×	×	×		e	×	×	×	×	×			×

[a] Off-the-shelf items available in small capacities only.
[b] Controversial subject. Some authorities claim moisture recovery. Do not advise depending on it.
[c] With a purge section added, cross-contamination can be limited to less than 1% by mass.
[d] Can be constructed of corrosion-resistant materials, but consider possible extensive damage to equipment caused by leaks or tube ruptures.
[e] Allowable temperatures and temperature differential limited by the phase-equilibrium properties of the internal fluid.

Figure 8.30 A Recirculation waste-heat boiler.

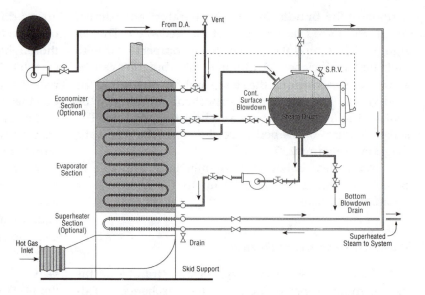

the investor. This points out the importance of gathering solid data, and using the best available predictions in preparing analyses for the investment decision process.

8.5.2 Effect of Utility Rates and Fuel Prices

Good investment potential exists in an economic climate where energy costs are high relative to equipment costs. However, the costs of goods are sharply influenced by energy costs. This would seem to indicate that the investment potential for waste-heat-recovery projects is not going to improve any faster than energy-cost escalations occur. In one sense this is true of future projects. But because capital projects involve one-time expenditures, which are usually financed by fixed-rate loans, the worth of a present investment will benefit from the rising costs of energy.

8.5.3 Effect of Load and Use Factors

The load factor is defined as the ratio of average annual load to rated capacity and the use factor as the fractional part of a year that the equipment is in use. It is clear that the capital recovery rate is directly proportional to these factors.

8.5.4 Effects of Reduced System Life

Waste-heat recovery equipment is susceptible to damage from natural and human-made environmental conditions. Damage can result from overheating, freezing, corrosion, collision, erosion, and explosion. Furthermore, capital recovery can never be completed if the equipment fails to achieve its expected life. One must either factor the risks of equipment damage into the eco-

nomic analysis, or insist that sufficient provision for equipment safety be engineered into the systems.

References

1. E. Cook, "Energy Flow in Industrial Societies," *Scientific America*, Vol. 225 (Sept. 1971), p. 135.
2. M. H. Ross and R. H. Williams, "The Potential for Fuel Conservation." *Technology Review*, Vol. 79. No. 4 (1977), p. 43.
3. R. W. Sant, "The Least Cost Energy Strategy: Minimizing Consumer Costs through Competition," Report of Energy Productivity Center, CMU, 1979, pp. 35-36.
4. *Annual Energy Review, 1991*, U.S. Energy Information Administration, Washington, D.C.
5. W.M. Rohsenow and J.P. Hartnett, Eds. *Handbook of Heat Transfer*, McGraw-Hill, New York, 1978, pp. 18-73.
6. B.C. Langley, *Heat Pump Technology*, 2nd Ed., Prentice-Hall, Englewood Cliffs, N.J., 1989.
7. K. Kreider and M. McNeil, Eds., *Waste Heat Management Guidebook*, NBS Handbook 121, U.S. Government Printing Office, Washington, D.C., 1976, pp. 121-125.
8. K. Kreider and M. McNeil, Eds., *Waste Heat Management Guidebook*, NBS Handbook 121, U.S. Government Printing Office, Washington, D.C., 1976, pp. 155-184.
9. Baumeister and Marks, Eds., *Standard Handbook for Mechanical Engineers*, 7th ed. McGraw-Hill, New York, 1967.
10. J. K. Salisbury, *Kent's Mechanical Engineers' Handbook, Power*, 12th ed., Wiley, New York, 1967.
11. *Thermodynamic and Transport Properties of Steam*, ASME, New York, 1967.
12. *Selected Values of Chemical Thermodynamic Properties*, NBS Technical Note 270-3, NBS, Washington, D.C., 1968.
13. *Gas Engineers Handbook*, American Gas Association, New York, 19.
14. B.K. Hodge, *Analysis and Design of Energy Systems*, Prentice-Hall, Englewood Cliffs, NJ, 1985.
15. Philip A. Schweitzer, Ed., *Corrosion Resistance Tables, Vols, I and II*, 3rd Edition, Marcel Dekker, New York City, 1991.
16. Weymueller, C.R., Duplex Steels Fight Corrosion, Welding Design and Fabrication, August 1990, pp 30-33, Penton, Cleveland, OH.
17. G. Walker, *Industrial Heat Exchangers*, 2nd Ed., Hemisphere, New York, NY, 1990.
18. K. Kreider and M. McNeil, Eds., *Waste Heat Management Guidebook*, NBS Handbook 121, U.S. Government Printing Office, Washington, D.C., 1976, pp. 25-51.

CHAPTER 9

BUILDING ENVELOPE

KEITH E. ELDER, P.E.

Iverson Elder Inc.
Bothell, Washington

9.1 INTRODUCTION

Building "Envelope" generally refers to those building components that enclose conditioned spaces and through which thermal energy is transferred to or from the outdoor environment. The thermal energy transfer rate is generally referred to as "heat loss" when we are trying to maintain an indoor temperature that is greater than the outdoor temperature. The thermal energy transfer rate is referred to as "heat gain" when we are trying to maintain an indoor temperature that is lower than the outdoor temperature. While many principles to be discussed will apply to both phenomena, the emphasis of this chapter will be upon heat loss.

Ultimately the success of any facility-wide energy management program requires an accurate assessment of the performance of the building envelope. This is true even when no envelope-related improvements are anticipated. Without a good understanding of how the envelope performs, a complete understanding of the interactive relationships of lighting and mechanical systems cannot be obtained.

In addition to a good understanding of basic principles, seasoned engineers and analysts have become aware of additional issues that have a significant impact upon their ability to accurately assess the performance of the building envelope.

1. The actual conditions under which products and components are installed, compared to how they are depicted on architectural drawings.

2. The impact on performance of highly conductive elements within the building envelope; and

3. The extent to which the energy consumption of a building is influenced by the outdoor weather conditions, a characteristic referred to as *thermal mass*.

It is the goal of this chapter to help the reader develop a good qualitative and analytical understanding of the thermal performance of major building envelope components. This understanding will be invaluable in better understanding the overall performance of the facility as well as developing *appropriate* energy management projects to improve performance.

9.1.1 Characteristics of Building Energy Consumption

Figure 9.1 below shows superimposed plots of average monthly temperature and fuel consumption for a natural gas-heated facility in the Northwest region of the United States.

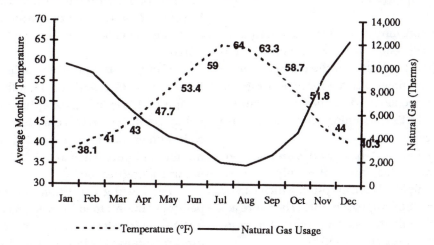

Figure 9.1

Experienced energy analysts recognize the distinctive shape of the monthly fuel consumption profile and can often learn quite a bit about the facility just from inspection of this data. For example, the monthly energy consumption is inversely proportional to the average monthly temperature. The lower the average monthly temperature, the more natural gas appears to be consumed.

Figure 9.1 also indicates that there is a period during the summer months when it appears no heating should be required, yet the facility continues to consume some energy. For natural gas, this is most likely that which is consumed for the heating of domestic hot water, but it could also be due to other sources, such as a gas range in a kitchen. This lower threshold of monthly energy consumption is often referred to as the "base," and is characterized by the fact that its magnitude is independent of outdoor weather trends. The monthly fuel consumption which exceeds the "base" is often referred to as the "variable" consumption, and is characterized by the fact that its magnitude is dependent upon the severity of outdoor environmental conditions. The distinction between the base and the variable fuel consumption is depicted below in Figure 9.2.

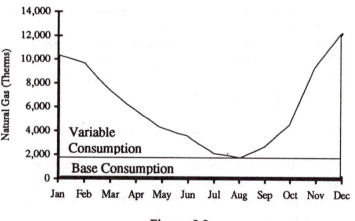

Figure 9.2

Often the base consumption can be distinguished from the variable by inspection. The lowest monthly consumption often is a good indicator of the base consumption. However there are times when a more accurate assessment is necessary. Section 9.9 describes one technique for improving the analyst's discrimination between base and variable consumption.

The distinction between base and variable consumption is an important one, in that it is only the variable component of annual fuel consumption which can be saved by building envelope improvements. The more accurate the assessment of the fuel consumption, the more accurate energy savings projections will be.

9.1.2 Quantifying Building Envelope Performance

The rate of heat transfer through the building envelope will be found to be related to the following important variables:

1.   Indoor and outdoor temperature;

2.   Conductivity of the individual envelope components; and

3.   The square footage of each of the envelope components.

For a particular building component exposed to a set of indoor and outdoor temperature conditions, these variables are often expressed in equation form by the following:

$$q = UA(T_i - T_o) \tag{9.1}$$

Where:

q =   the component heat loss, Btu/hr

U =   the overall heat transfer coefficient, Btu/ (hr-ft^2-°F)

A =   the area of the component, ft^2

T_i =   the indoor temperature, °F

T_o =   the outdoor temperature, °F

9.1.3 Temperatures for Instantaneous Calculations

The indoor and outdoor temperatures for equation (9.1) are those conditions for which the heat loss needs to be known. Traditionally interest has been in the design heat loss for a building or component, and is determined by using so-called "design-day" temperature assumptions.

Outdoor temperature selection should be made from the 1993 ASHRAE Handbook of Fundamentals, Chapter 24, for the geographic location of interest. The values published in this chapter are those that are statistically known not to be exceeded more than a prescribed number of hours (such as 2-1/2 %) of the respective heating or cooling season.

For heating conditions in the winter, indoor temperatures maintained between 68 and 72°F result in comfort to the greatest number of people. Indoor temperatures maintained between 74 and 76°F result in the greatest comfort to the most people during the summer (cooling) period.

We will see in later sections that other forms of temperature data collected on an annual basis are useful for evaluating envelope performance on an annual basis.

The next section will take up the topic of how the U-factors for various envelope components are determined.

9.2   PRINCIPLES OF ENVELOPE ANALYSIS

The successful evaluation of building envelope performance first requires that the analyst be well-versed in the use of a host of analytical tools which adequately address the unique way heat is transferred through each component. While the heat loss principles are similar, the calculation will vary somewhat from component to component.

9.2.1 Heat Loss Through Opaque Envelope Components

We have seen from Equation (9.1) that the heat loss through a component, such as a wall, is proportional to the area of the component, the indoor-outdoor temperature difference, and U, the proportionality constant which describes the temperature-dependent heat transfer through that component. U, currently described as the "U-factor" by ASHRAE[2], is the reciprocal of the total thermal resistance of the component of interest. If the thermal resistance of the component is known, U can be calculated by dividing the total thermal resistance into "1" as shown:

$$U = 1/R_t \tag{9.2}$$

The thermal resistance, R_t, is the sum of the individual resistances of the various layers of material which comprise the envelope component. R_t is calculated by adding them up as follows:

$$R_t = R_1 + R_2 + R_n +... \tag{9.3}$$

R_1, R_2 and R_n represent the thermal resistance of each of the elements in the path of the "heat flow." The thermal resistance of common construction materials can be obtained from the 1993 ASHRAE Handbook of Fundamentals, Chapter 22. Other physical phenomenon, such as convection and radiation, are typically included as well. For instance, free and forced convection are treated as another form of resistance to heat transfer, and the "resistance" values are tabulated in the 1993 ASHRAE Fundamentals Manual for various surface orientations and wind velocities. For example, the outdoor resistance due to forced convection (winter) is usually taken as 0.17 hr/(Btu–ft^2–°F) and the indoor resistance due to free con-

vection of a vertical surface is usually taken to be 0.68 hr/(Btu–ft^2–°F).

To calculate the overall U-factor, one typically draws a cross-sectional sketch of the building component of interest, assigns resistance values to the various material layers, sums the resistances and uses the reciprocal of that sum to represent U. The total calculation of a wall U-factor is demonstrated in the example below.

Example

Calculate the heat loss for 10,000 square feet of wall with 4-inch face brick, R-11 insulation and 5/8-inch sheet rock when the outdoor temperature is 20°F and the indoor temperature is 70°F.

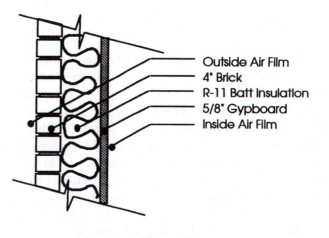

Outside Air Film
4" Brick
R-11 Batt Insulation
5/8" Gypboard
Inside Air Film

Figure 9.3

In the ASHRAE Handbook we find that a conservative resistance for brick is 0.10 per inch. Four inches of brick would therefore have a resistance of 0.40. Sheet rock (called gypsum board by ASHRAE) has a resistance of 0.90 per inch, which would be 0.56 for 5/8-inch sheet rock. Batt insulation with a rating of R-11 will have a resistance of 11.0, if expanded to its full rated depth.

The first step is to add all the resistances.

	R_i
Outdoor Air Film (15 m.p.h.)	0.17
4-inch Face Brick	0.40
R-11 Batt Insulation	11.00
5/8-inch Gypsum Board	0.56
Indoor Air Film (still air)	0.68

$$R_t = 12.81 \; \frac{\text{hr–ft}^2 \, °F}{\text{Btu}}$$

The U-factor is the reciprocal of the total resistance

$$U_o = \frac{1}{R_t} = \frac{1}{12.81} = 0.078 \; \frac{\text{Btu}}{\text{hr–ft}^2 \, °F}$$

The heat loss is calculated by multiplying the U-factor by the area and the indoor-outdoor temperature difference.

$$q = 0.078 \times 10,000 \times (70 - 20)$$

$$= 39,000 \text{ Btu/hr}$$

The accuracy of the previous calculation is dependent on at least two important assumptions. The calculation assumes:

1. The insulation is not compressed; and

2. The layer(s) of insulation has not been compromised by penetrations of more highly conductive building materials.

9.2.2 Compression of Insulation

The example above assumed that the insulation is installed according to the manufacturer's instructions. Insulation is always assigned its R-value rating according to a specific standard thickness. If the insulation is compressed into a smaller space than it was rated under, the performance will be less than that published by the manufacturer. For example, R-19 batt insulation installed in a 3-1/2-inch wall might have an effective rating as low as R-13. Table 9.1[7] is a summary of the performance that can be expected from various levels of fiberglass batt insulation types installed in different envelope cavities.

9.2.3 Insulation Penetrations

One of the assumptions necessary to justify the use of the one-dimensional heat transfer technique used in equation 9-1 is that the component must be thermally homogeneous. Heat is transferred from the warm side of the component to the colder side and through each individual layer in a series path, much like current flow through simple electrical circuit with the resistances in series. No lateral or sideways heat transfer is assumed to take place within the layers. For this to be true, the materials in each layer must be continuous and not penetrated by more highly conductive elements.

Unfortunately, there are very few walls in the real world where heat transfer can truly be said to be one-dimensional. Most common construction has wood or metal studs penetrating the insulation, and the presence of these other materials must be taken into consideration.

Traditionally studs are accounted for by performing separate U-factor calculations through both wall sections, the stud and the cavity. These two separate U-factors are then combined in parallel by "weighting" them by their respective wall areas. The following example (Figure 9.4) shows how this would typically be done for a wall whose studs, plates and headers constituted 23% of the total gross wall area.

	R-Cavity	R-Frame
Outdoor Air Film (15 m.p.h.)	0.17	0.17
4-Inch Face Brick	0.40	0.40
R-11 Batt Insulation	11.00	—
3-1/2-Inch Wood Framing	—	3.59
5/8-Inch Gypsum Board	0.56	0.56
Indoor Air Film (still air)	0.68	0.68
R_t =	12.81	5.40 $\frac{\text{hr–ft}^2\text{–°F}}{\text{Btu}}$
U_i =	0.078	0.185 $\frac{\text{Btu}}{\text{hr–ft}^2\text{–°F}}$

Combining the two U-factors by weighted fractions:

$$U_o = 0.77 \times 0.078 + 0.23 \times 0.185 = 0.103 \frac{\text{Btu}}{\text{hr–ft}^2\text{–°F}}$$

Insulation R-Value at Standard Thickness								
R-Value	38	30	22	21	19	15	13	11
Standard Thickness	12"	9-½"	6-¾"	5-½"	6-¼"	3-½"	3-5/8"	3-½"
Nominal Lumber Sizes, Inches / Actual Depth of Cavity, Inches	Insulation R-Values when Installed in a Confined Cavity							
2 x 12 / 11-1/4	37	--	--	--	--	--	--	--
2 x 10 / 9-1/4	32	30	--	--	--	--	--	--
2 x 8 / 7-1/4	27	26	--	--	--	--	--	--
2 x 6 / 5-1/2	--	21	20	21	18	--	--	--
2 x 4 / 3-1/2	--	--	14	--	13	15	13	11

Table 9.1 R-Value of fiberglass batts compressed within various depth cavities.

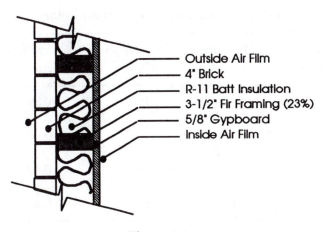

Outside Air Film
4" Brick
R-11 Batt Insulation
3-1/2" Fir Framing (23%)
5/8" Gypboard
Inside Air Film

Figure 9.4

In the absence of advanced framing construction techniques, wood studs installed 16 inches on-center comprise 20-25% of a typical wall (including window framing, sill, etc.). Wood studs installed 24 inches on center comprise approximately 15-20% of the gross wall.

The use of this method is appropriate for situations where the materials in the wall section are sufficiently similar that little or no lateral, or "sideways" heat transfer takes place. Of course a certain amount of lateral heat transfer does take place in every wall, resulting in some error in the above calculation procedure. The amount of error depends on how thermally dissimilar the various elements of the wall are. The application of this procedure to walls whose penetrating members' conductivities deviate from the insulation conductivities by less than an order of magnitude (factor of 10) should provide results that are sufficiently accurate for most analysis of construction materials.

Because the unit R-value for wood is approximately 1.0 per inch and fiberglass batt insulation is R 3.1 per inch, this approach is justified for wood-framed building components.

9.3   METAL ELEMENTS IN ENVELOPE COMPONENTS

Most commercial building construction is not wood-framed. Economics as well as the need for fire-rated assemblies has increased the popularity of metal-framing systems over the years. The conductivity of metal framing is significantly more than an order of magnitude greater than the insulation it penetrates. In some instances it is several thousand times greater. However, until recent years, the impact of this type of construction on envelope thermal performance has been ignored by much of the design industry. Yet infrared photography in the field and hot-box tests in the laboratory have demon-

strated the severe performance penalty paid for this type of construction.

The introduction of the metal stud framing system into a wall has the potential to nearly double its heat loss! Just how is this possible, given that a typical metal stud is only about 1/20 of an inch thick? The magnitude of this effect is counter-intuitive to many practicing designers because they have been trained to think of heat transfer through building elements as a one-dimensional phenomenon, (as in the previous examples). But when a highly conductive element such as a metal stud is present in an insulated cavity, two and three-dimensional considerations become extremely important, as illustrated below.

Figure 9.5 shows the temperature distribution centered about a metal stud in a section of insulated wall. The lines, called "isotherms," represent regions with the same temperature. Each line denotes a region that is one degree Fahrenheit different from the adjacent line. These lines are of interest because they help us to visualize the

Warm Side

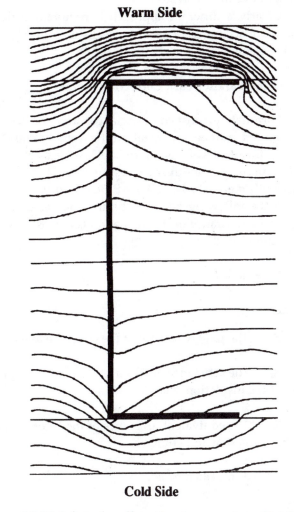

Cold Side

Figure 9.5 Metal stud wall section temperature distribution.

characteristics of the heat flow through the wall section. The direction of heat flow is perpendicular to the lines of constant temperature, and the heat flow is more intense through regions where the isotherms are closer together. It is possible to visualize both the direction and intensity of heat flow through the wall section by observing the isotherms alone.

If the heat flow was indeed one-dimensional and occurring through a thermally homogeneous material, the lines would be horizontal and would show an even and linear temperature-drop progression from the warm side of the wall to the cold side. Notice that the isotherms are far from horizontal in certain regions around the metal stud. Figure 9.5 shows that the heat flow is not parallel, nor does it move directly through the wall section. Also note that the area with the greatest amount of heat flow is not necessarily restricted to the metal part of the assembly. The metal stud has had a negative influence on the insulation in the adjacent region as well. This is the reason for the significant increase in heat loss reported above for metal studs.

Clearly a different approach is required to determine more accurate U-factors for walls with highly conductive elements.

9.3.1 Using Parallel-Path Correction Factors

Fortunately, most commercial construction consists of a limited number of metal stud assembly combinations, so the results of laboratory "hot box" tests of typical wall sections can be utilized to estimate the U-factors of walls with metal studs. ASHRAE Standard 90.1 recommends the following equation be used to calculate the equivalent resistance of the insulation layer installed between metal studs:

$$R_t = R_i + R_e \qquad (9.4)$$

Where:

R_t = the total resistance of the envelope assembly

R_i = the resistance of all series elements except the layer of the insulation & studs

R_e = the equivalent resistance of the layer containing the insulation and studs

$$= R_{insulation} \times F_c$$

Where:

F_c = the correction factor from the table following:

Table 9.2 Parallel path correction factors.

Size of Members	Framing	Insulation R-Value	Correction Factor, F_c
2 × 4	16 in. O.C.	R-11	0.50
2 × 4	24 in. O.C.	R-11	0.60
2 × 6	16 in. O.C.	R-19	0.40
2 × 6	24 in. O.C.	R-19	0.45

The use of the above multipliers in wall U-factor calculation is demonstrated in the example given below:

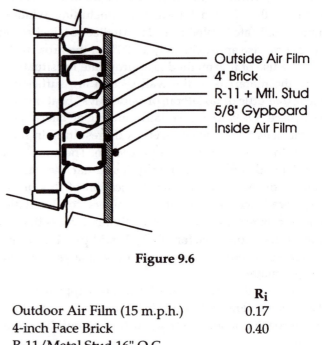

Outside Air Film
4" Brick
R-11 + Mtl. Stud
5/8" Gypboard
Inside Air Film

Figure 9.6

	R_i
Outdoor Air Film (15 m.p.h.)	0.17
4-inch Face Brick	0.40
R-11/Metal Stud 16" O.C.	
= 11.0 × 0.50 =	5.50
5/8-inch Gypsum Board	0.56
Indoor Air Film (still air)	0.68

$$R_t = 7.31 \ \frac{hr\text{–}ft^2\text{–}°F}{Btu}$$

$$U_o = \frac{1}{R_t} = \frac{1}{7.31} = 0.137 \ \frac{Btu}{hr\text{–}ft^2\text{–}°F}$$

Notice that only one path is calculated through the assembly, rather than two. The insulation layer is simply corrected by the ASHRAE multiplier and the calculation is complete. Also notice that it is only the metal stud/insulation layer that is corrected, not the entire assembly.

This does not mean that only this layer is affected by the metal studs, but rather that this is the approach ASHRAE Standard 90.1 intends for the factors to give results consistent with tested performance.

The above example shows that the presence of the metal stud has increased the wall heat loss by 75%! The impact is even more severe for R-19 walls. The importance of accounting for the impact of metal studs in envelope U-factor calculations cannot be overstated.

9.3.3 The ASHRAE "Zone Method"

The ASHRAE Zone Method should be used for U-factor calculation when highly conductive elements are present in the wall that do not fit the geometry or spacing criteria in the above parallel path correction factor table. The Zone Method, described on page 22.10 of the 1993 ASHRAE Handbook of Fundamentals, is an empirically derived procedure which has been shown to give reasonably accurate answers for simple wall geometries. It is a structured way to calculate the heat transmission through a wall using both series and parallel paths.

The Zone Method takes its name from the fact that the conductive element within the wall influences the heat transmission of a particular region or "zone." The zone of influence is typically denoted as "Zone A," which experiences a significant amount of lateral conduction. The remaining wall section that remains unaffected is called "Zone B." The width of Zone A, which is denoted "W," can be calculated using the following equation:

$$W = m + 2d \qquad (9.5)$$

Where "m" is the width of the metal element at its widest point and "d" is the shortest distance from the metal surface to the outside surface of the entire component. If the metal surface is the outside surface of the component, then "d" is given a minimum value of 0.5 (in English units) to account for the air film.

For example a 3-1/2-inch R-11 insulation is installed between 1.25-inch wide metal studs 16 inches on-center. The assembly is sheathed on both sides with 1/2-inch gypsum board. The variable "m" is the widest part of the stud, which is 1.25 inches. The variable "d" is the distance from the metal element to the outer surface of the wall, which in this case, is the thickness of the gypsum wall board, 0.5 inches. The width of Zone A is:

$$W_A = 1.25 + 2 \times 0.5 \quad = \quad \underline{2.25 \text{ inches}}$$

The width of Zone B is:

$$W_B = 16 - 2.25 \quad = \quad \underline{13.75 \text{ inches}}$$

Figure 9.7 shows the boundaries of the zones described above.

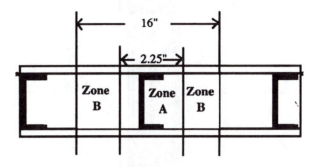

Figure 9.7 Zones A and B.

Actual step-by-step procedures for performing the calculations for Zone A and Zone B can be found on page 22.10 of the 1993 ASHRAE Handbook of Fundamentals as well as other publications.[5]

9.3.4 Improving the Performance of Envelope Components with Metal Elements

The impact of metal elements can be mitigated by:

1. Installing the insulation outside of the layer containing metal studs.

2. Installing interior and exterior finished materials on horizontal "hat" sections, rather than directly on the studs themselves.

3. Using expanded channel (thermally improved) metal studs.

4. Using non-conductive thermal breaks at least 0.40 to 0.50 inches thick between the metal element and the inside and outside sheathing. The R-value of the thermal break conductivity should be at least a factor of 10 less than the metal element in the envelope component.

9.3.5 Metal Elements in Metal Building Walls

Many metal building walls are constructed from a corrugated sheet steel exterior skin, a layer of insulation, and sometimes a sheet steel V-rib inner liner. The inner liner can be fastened directly to the steel frame of the building. The exterior cladding is attached to the inner liner and the structural steel of the building through cold formed sheet steel elements called Z-girts. Fiber glass batt insulation is sandwiched between the inner liner and the exterior cladding. Figures 9.8 and 9.9 show details of a typical sheet steel wall.

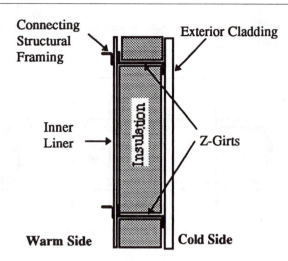

Figure 9.8 Vertical section view.

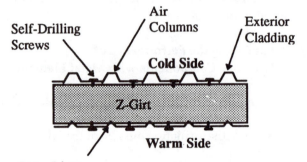

Figure 9.9 Horizontal section view.

The steel framing and metal siding materials provide additional opportunities for thermal short circuits to occur. The highly conductive path created by the metal in the girts, purlins and frames connected directly to the metal siding can result in even greater thermal short-circuiting than that discussed in Section 9.3.4 for metal studs in insulated walls. Because of this unobstructed high conductivity path, the temperature of the metal element is nearly the same through the entire assembly, rather than varying as heat flows through the assembly.

In laboratory tests, the introduction of 16 gauge Z-Girts spaced 8 feet apart reduced the R-value of a wall with no girts from 23.4 to 16.5 hr–ft^2–°F/Btu.[4] Substitution of Z-Girts made of 12-gauge steel reduced the R-value even further, to 15.2 hr–ft^2–°F/Btu. The extent of this heat loss will be dictated by the spacing of the Z-girts, the gauge of metal used and the contact between the metal girts and the metal wall panels.

9.3.6 Metal Wall Performance

While the performance of metal building walls will vary significantly, depending on construction features, Tables 9.3-9.5 will provide a starting point in predicting the performance of this type of construction.[6]

9.3.7   Strategies for Reducing Heat Loss in Metal Building Walls

9.3.7.1 Maintain Maximum Spacing Between Z-Girts

Tests on a 10-inch wall[8] with a theoretical R-value of 36 hr-ft^2-°F/Btu demonstrated a measured R-value of 10.2 hr-ft^2-°F/Btu with the girts spaced 2 feet apart. Increasing the spacing to 8 feet between the girts had a corresponding effect of increasing the R-value to 16.5 hr-ft^2-°F/Btu. While typical girt spacing in the wall of a metal building is 6 feet on-center, this study does emphasize the desirability of maximizing the girt spacing wherever possible.

9.3.7.2 Use "Thermal" Girts

The thermal performance of metal building walls can be improved by the substitution of thermally improved Z-girts. The use of expanded "thermal" Z-Girts has the potential to increase the effective R-value by 15-20% for a wall with Girts spaced 8 feet apart.[4] A significant amount of the metal has been removed from these

Table 9.3 Wall insulation installed between 16 gauge Z-girts.

Unbridged R-Value	Nominal Thickness (Inches)	R-Value 2' O.C.	R-Value 4' O.C.	R-Value 5' O.C.	R-Value 6' O.C.	R-Value 8' O.C.
6	2	4.9	5.7	5.8	5.9	6.0
10	3	6.4	7.9	8.3	8.5	8.7
13	4	7.4	9.4	9.9	10.2	10.6
19	5-1/2	9.0	12.0	12.9	13.4	14.2
23	7	10.2	13.8	14.9	15.6	16.5
36	10	14.1	19.8	21.5	22.8	24.4

Table 9.4 Wall insulation installed between 12 gauge Z-girts.

Unbridged R-Value	Nominal Thickness (Inches)	R-Value 2' O.C.	R-Value 4' O.C.	R-Value 5' O.C.	R-Value 6' O.C.	R-Value 8' O.C.
6	2	4.8	5.6	5.8	5.9	6.0
10	3	6.1	7.7	8.1	8.3	8.6
13	4	6.9	9.0	9.5	9.9	10.3
19	5-1/2	8.1	11.1	12.0	12.6	13.4
23	7	8.7	12.3	13.4	14.1	15.2

thermal girts, which increases the length of the heat flow path and reduces the girt's heat flow area. Figure 9.10 indicates the basic structure of such a girt.

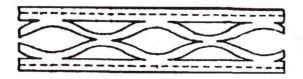

Figure 9.10 Expanded "thermal" girt.

Table 9.5 combines test results with calculations to estimate the benefits that may be realized utilizing "thermal" girts.

9.3.7.3 Increase Z-Girt-to-Metal Skin Contact Resistance

While the Z-Girt itself does not offer much resistance to heat flow, testing has found that up to half of the resistance of the girt-to-metal skin assembly is attributable to "poor" contact between the girt and the metal skin. Quantitative estimates of the benefit of using less conductive materials to "break" the interface between the two metals are not readily available, but laboratory tests have shown some improvement in thermal performance, just by using half the number of self-tapping screws to fasten the metal skin to the girts.

9.3.7.4 Install Thermal Break Between Metal Elements

The impact of this strategy will vary with the flexibility offered by the geometry of the individual wall component, the on-center spacing of the wall girts, thermal break material and contact resistance between the girt and metal wall before consideration of the thermal break. Figure 9.11 illustrates the impact of thermal breaks of varying types and thicknesses installed between a corrugated metal wall and Z-Girts mounted 6 feet on-center.

To be effective, reduction by a factor of 10 or more in thermal conductivity between the thermal break and the metal may be required for a significant improvement in performance. The nominal thickness of the break also plays an important role. Tests of metal panels indicate that, regardless of the insulator, inserts less than 0.40 inches will not normally be sufficient to prevent substantial loss of their insulation value.

9.3.8 Calculating U-factors for Metal Building Walls

ASHRAE Standard 90.1 recommends The Johannesson-Vinberg Method[7] for determining the U-factor for sheet metal construction, internally insulated with a metal structure bonded on one or both sides with a metal skin. This method is useful for the prediction of U-factors for metal girt/purlin constructions which can be defined by the geometry describedin Figure 9.12.

Table 9.5 Wall insulation installed between 16 gauge "thermal" Z-girts.

Unbridged R-Value	Nominal Thickness (Inches)	R-Value 2' O.C.	R-Value 4' O.C.	R-Value 5' O.C.	R-Value 6' O.C.	R-Value 8' O.C.
6	2	5.6	6.2	6.3	6.4	6.4
10	3	8.3	9.4	9.6	9.8	10.0
13	4	10.2	11.7	12.1	12.3	12.6
19	5-1/2	14.0	16.4	16.9	17.3	17.8
23	7	16.4	19.5	20.2	20.7	21.3

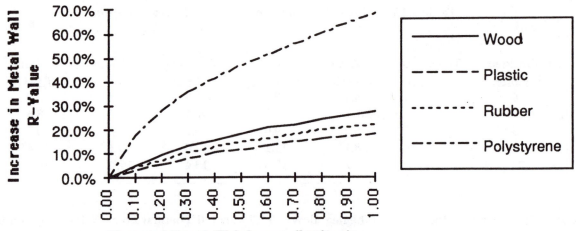

Figure 9.11[5] Impact of thermal break on metal wall R-value corrugated metal with Z-girts 6'0" on-center.

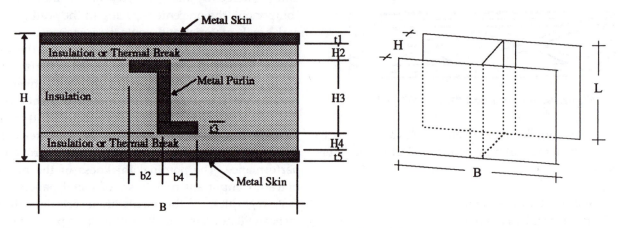

Figure 9.12 Metal skinned structure w/internal metal element.

The dimensions of the elements shown, as well as their conductivities are evaluated in a series of formulations, which are in turn combined in two parallel paths to determine an overall component resistance. The advantage of this method is that metal gauge as well as thermal break thickness and conductivity can be taken into account in the calculation. Caution is advised when using this method to evaluate components with metal-to-metal contact. The method assumes complete thermal contact between all components, which is not always the case in metal building construction. To account for this effect, values for contact resistance can be introduced in the calculation in place of or in addition to the thermal break layer.

9.4 ROOFS

In many cases the thermal performance of roof structures is similar to that of walls, and calculations can be performed in a similar way to that described in the previous examples. As was the case with walls, metal penetrations through the insulation will exact a penalty. With roofs these penetrations will usually take one of several forms. The first form is the installation of batt insulation between z-purlins, similar to that discussed in Section 9.3. The performance of these assemblies will be similar to that of similarly constructed walls.

9.4.1 Insulation Between Structural Trusses

Another common penetration is that which occurs as a result of installing the insulation between structural trusses with metal components, as shown in Figure 9.13. Table 9.6 is a summary of the derated R-values that might be expected as a result of this type of installation compared to the unbridged R-values published by insulation manufacturers.

9.4.2 Insulation Installed "Over-the-Purlin"

One of the most economical methods of insulating the roof of a metal building, shown in Figure 9.14, is to

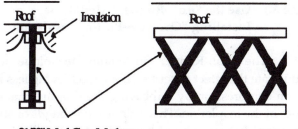

21/32" Metal Cross Members
4, 6 or 8 feet on-center

Figure 9.13

**Table 9.6 Roof insulation installed
between metal trusses.[5]**

Unbridged R-Value	R-Value 4 Ft O.C.[10]	R-Value 6 Ft O.C.[11]	R-Value 8 Ft O.C.[11]
5	4.8	4.9	4.9
10	9.2	9.6	9.7
15	13.2	14.0	14.3
20	17.0	18.4	18.9
25	20.3	22.4	23.2
30	23.7	26.5	27.6

stretch glass fiber blanket insulation over the purlins or trusses, prior to mounting the metal roof panels on top.

The roof purlin is a steel Z-shaped member with a flange width of approximately 2-1/2 inches. The standard purlin spacing for girts supporting roofs in the metal building industry is 5 feet. Using such a method, the insulation is compressed at the purlin/panel interface, resulting in a thermal short circuit. The insulation thickness averages 2-4 inches, but can range up to 6 inches. While thicker blanket insulation can be specified to mitigate this, the thickness is limited by structural considerations. With the reduced insulation thickness at the

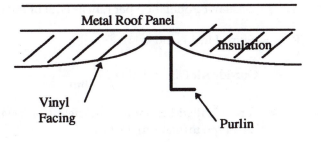

Figure 9.14

purlin, there is a diminishing return on the investment in additional insulation. Table 9.7 is a summary of the installed R-values that might be expected using this installation technique.

9.4.2.1 Calculating Roof Performance for "Over-the-Purlin" Installations

Traditional methods of calculating assembly R-values will not result in accurate estimates of thermal performance for "over-the-purlin" installations. The insulation is not installed uniformly, nor at the thickness required to perform at the rated R-value, and it is penetrated with multiple metal fasteners. The Thermal Insulation Manufacturers Association (TIMA) has developed an empirical formula based on testing of insulation meeting the TIMA 202 specification. Insulation not designed to be laminated (such as filler insulation) will show performance up to 15% less than indicated by the formula below. Assuming the insulation meets the TIMA specification, the U-factor of the completed assembly can be estimated as:

$$U = 0.012 + \frac{0.255}{(0.31 \times R_f + t)} \times \left(1 - \frac{N}{L}\right) + N \times \frac{0.198 + 0.065 \times n}{L}$$

(9.6)

Where:
L = Length of Building Section, feet
N = Number of Purlins or Girts in the L Dimension

Table 9.7 Roof insulation installed compressed over-the-purlin[5].

Unbridged R-Value	Nominal Thickness (Inches)	R-Value 2' O.C.	R-Value 3' O.C.	R-Value 4' O.C.	R-Value 5' O.C.	R-Value 6' O.C.
10	3	5.4	6.5	7.2	7.7	8.1
13	4	5.7	7.1	8.0	8.7	9.3
16	5-1/2	5.9	7.4	8.6	9.5	10.2
19	6	6.0	7.7	9.0	10.0	10.9

n   =   Fastener Population per Linear Foot of Purlin

R_f = Sum of Inside and

Outside Air Film R–Values, $\dfrac{hr - ft^{2\circ}F}{Btu}$

t   =   Pre-installed insulation thickness (for TIMA 202 type insulation), inches

Example Calculation Using the TIMA Formula

Assume a metal building roof structure with the following characteristics:

Length of Building Section   = 100 feet
Number of Purlins   = 21 purlins
Fastener Population per Foot of Purlin = 1 per foot

Sum of Air Film R–Values = 0.61 + 0.17 = 0.78 $\dfrac{hr–ft^{2\circ}F}{Btu}$

Insulation thickness   = 5 inches

$$U = 0.012 + \frac{0.255}{(0.31 \times 0.78 + 6)} \times \left(1 - \frac{21}{105}\right) + 21$$

$$\times \frac{0.198 + 0.065 \times 1}{105} = \mathbf{0.0973} \ \frac{Btu}{hr–ft^{2\circ}F}$$

9.4.3 Strategies for Reducing Heat Loss in Metal Building Roofs

Many of the strategies suggested for metal walls are applicable to metal roofs. In addition to strategies that minimize thermal bridging due to metal elements, other features of metal roof construction can have a significant impact on the thermal performance.

Tests have shown significant variation in the thermal performance of insulation depending upon the facing used, even though the permeability of the facing itself has no significant thermal effect. The cause of this difference is in the flexibility of the facing itself. The improvement in permeability rating produces a higher U-factor rating due to the draping characteristics. During installation, the insulation is pulled tightly from eave to eave of the building. The flexibility of vinyl facing allows it to stretch and drape more fully at the purlins than reinforced facing. Figures 9.15 and 9.16 show a generalized illustration of the difference in drape of the vapor retarders and the effect on effective insulation thickness.

The thermal bridging and insulation compression issues discussed above make the following recommendations worth considering.

9.4.3.1 Use the "Roll-Runner" Method of Installing "Over-the-Purlin" Insulation

The "Roll-Runner" installation technique helps to mitigate the effect described above, achieving less insulation compression by improving the draping characteristics of the insulation. This is done using bands or straps to support the suspended batt insulation, as illustrated in Figure 9.17.

9.4.3.2 Use Thermal Spacers Between Purlin and Standing Seam Roof Deck

Figure 9.18 depicts a thermal spacer installed between the top of the purlin and the metal roof deck. The

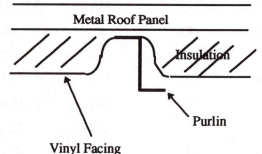

Figure 9.15 Vinyl facing.

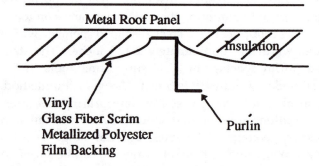

Figure 9.16 Vinyl facing with glass fiber scrim.

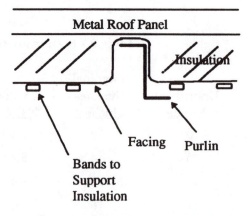

Figure 9.17 Roll Runner Method

impact of this strategy will vary with the geometry of the individual component as well as the location of the spacer.

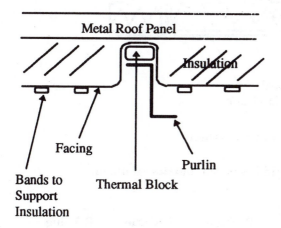

Figure 9.18 Insulated purlin with "roll-runner" installation.

Table 9.8 summarizes the comparative performance of the two installation techniques discussed above. In all cases, purlins were installed 5 feet on-center. While improvements of 8 to 19 percent in effective R-value are achieved using the Roll-Runner method, performance of 60 to 80 percent can be realized when the purlins were also insulated from direct contact with the metal structure.

9.4.3.3 Install Additional Uncompressed Insulation Between Purlins or Bar Joists

This insulation system illustrated in Figure 9.19, is sometimes referred to as "Full Depth/Sealed Cavity," referring to the additional layer of insulation that is installed between purlins or bar joists, and which is not compressed as is the over-the-purlin insulation above it. In this configuration, the main function of the over-the-purlin insulation above is to act as a thermal break.

9.4.3.4 Add Rigid Insulation Outside of Purlins

The greatest benefit will be derived where insula-

tion can be added which is neither compressed nor penetrated by conductive elements, as shown in Figure 9.20.

Table 9.9 comparatively summarizes the performance of the above installation for varying levels of insulation installed over purlins with varying on-center spacing.

9.5   FLOORS

Floors above grade and exposed to outdoor air can be calculated much the same way as illustrated previously for walls, except that the percentages assumed for floor joists will vary somewhat from that assumed for typical wall constructions.

9.5.1 Floors Over Crawl Spaces

The situation is a little different for a floor directly over a crawl space. The problem is that knowledge of the temperature of the crawl space is necessary to perform the calculation. But the temperature of the crawl space is dependent on the number of exposed crawl space surfaces and their U-factors, as well as the impact of crawl space venting, if any. For design-day heat loss calculations, it is usually most expedient to assume a crawl space temperature equal to the outdoor design temperature. This will very nearly be the case for poorly insulated or vented crawl spaces.

When actual, rather than worst-case heat loss is needed, it is necessary to perform a heat balance on the crawl space. The process is described on page 25.9-25.10 in the 1993 ASHRAE Handbook of Fundamentals. Below is a brief summary of the approach.

Heat loss, q_{floor} from a Floor to a Crawl Space:

$$q_{floor} = q_{perimeter} + q_{ground} + q_{air\ exchange}$$

$$U_fA_f(t_i-t_c) = U_pA_p(t_c-t_o) + U_gA_g(t_c-t_g)$$
$$+ 0.67H_cV_c(t_c-t_o) \qquad (9.7)$$

Where:

t_i = indoor temperature, °F
t_o = outdoor temperature, °F

Table 9.8 "Roll runner" method

Mfgr's Rated R-Value	Nominal Thickness (Inches)	Over the Purlin Method	Roll Runner Method	Roll-Runner with Insulated Purlin
10	3	7.1	7.7	12.5
13	4	8.3	9.1	14.3
19	6	11.1	12.5	20.0

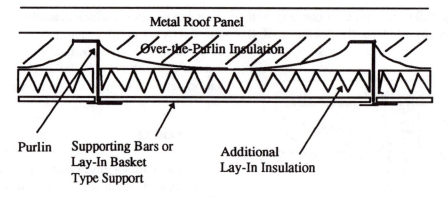

Figure 9.19 Insulation suspension system.

Table 9.9 Insulation "over-the-purlin" w/rigid insulation outside of purlin.

Uninstalled Unbridged R-Value	Thickness (Inches)	R-Value 2' O.C.	R-Value 3' O.C.	R-Value 4' O.C.	R-Value 5' O.C.	R-Value 6' O.C.
13	2	12.2	12.9	13.3	12.5	13.8
17	3	12.6	13.7	14.4	14.3	15.3
20	4	12.9	14.3	15.2	15.5	16.5
23	5-1/2	13.1	14.6	15.8	16.7	17.4
26	6	13.2	14.9	16.2	18.3	18.1

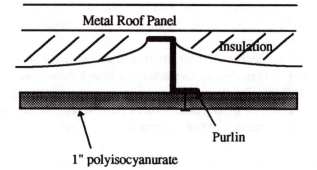

Figure 9.20 Insulation suspension system.

t_g = ground temperature, °F
t_c = crawl space temperature, °F
A_f = floor area, ft^2
A_p = exposed perimeter area, ft^2
A_g = ground area ($A_g = A_f$), ft^2
U_f = floor heat transfer coefficient, Btu/(hr–ft^2–°F)
U_g = ground coefficient, Btu/hr–ft^2°F
U_p = perimeter heat transfer coefficient, Btu/(hr-ft^2-°F)
V_c = volume of the crawl space, ft^3
H_c = volumetric air heat capacity (0.018 Btu/(ft^3–°F)
0.67 = assumed air exchange rate (volume/hour)

The above equation must be solved for t_c, the crawl space temperature. Then the heat loss from the space above to the crawl space, using the floor U-factor can be calculated.

9.5.2 Floors On Grade

A common construction technique for commercial buildings is to situate the building on a concrete slab right on grade. The actual physics of the situation can be quite complex, but methods have been developed to simplify the problem. In the case of slab-on-grade construction, it has been found that the heat loss is proportional to the perimeter length of the slab, rather than the floor area. Rather than using a U-factor, which is normally associated with a wall or roof <u>area,</u> we use an "F-factor," which is associated with the number of linear <u>feet</u> of slab perimeter. The heat loss is given by the equation below:

$$q_{slab} = F \times Perimeter \times (T_{inside} - T_{outside}) \qquad (9.8)$$

F-factors are published by ASHRAE, and are also available in many state energy codes. As an example, the F-factor for an uninsulated slab is 0.73 Btu/(hr-ft-°F). A slab with 24 inches of R-10 insulation installed inside the foundation wall would be 0.54 Btu/(hr-ft-°F).

9.5.3 Floors Below Grade

Very little performance information exists on the performance of basement floors. What does exist is more relevant to residential construction than commercial. Fortunately, basement floor loss is usually an extremely small component of the overall envelope performance. For every foot the floor is located below grade, the magnitude of the heat loss is diminished dramatically. Floor heat loss is also affected somewhat by the shortest dimension of the basement. Heat loss is not directly proportional to the outside ambient temperature, as other above grade envelope components. Rather, basement floor loss has been correlated to the temperature of the soil four inches below grade. This temperature is found to vary sinusoidally over the heating season, rather on a daily cycle, as the air temperature does.

A method for determining basement floor and wall heat loss is described, with accompanying calculations, in Chapter 25 of the 1993 ASHRAE Handbook of Fundamentals.

9.6 FENESTRATION

The terms "Fenestration," "window," and "glazing" are often used interchangeably. To describe the important aspects of performance in this area requires that terms be defined carefully. "Fenestration" refers to the design and position of windows, doors and other structural openings in a building. When we speak of *windows*, we are actually describing a system of several components. *Glazing* is the transparent component of glass or plastic windows, doors, clerestories, or skylights. The *sash* is a frame in which the glass panes of a window are set. The *Frame* is the complete structural enclosure of the glazing and sash system. *Window* is the term we give to an entire assembly comprised of the sash, glazing, and frame.

Because a window is a thermally nonhomogeneous system of components with varying conductive properties, the thermal performance cannot be accurately approximated by the one-dimensional techniques used to evaluate common opaque building envelope components. The thermal performance of a window system will vary significantly, depending on the following characteristics:

- The number of panes
- The dimension of the space between panes
- The type of gas between the panes
- The emissivity of the glass
- The frame in which the glass is installed
- The type of spacers that separate the panes of glass

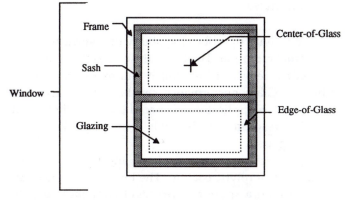

Figure 9.21 Window components.

9.6.1 Multiple Glass Panes

Because of the low resistance provided by the glazing itself, the major contribution to thermal resistance in single pane glazing is from the indoor and outdoor air films. Assuming 0.17 outdoor and 0.68 indoor air film resistances, a single paned glazing unit might be expected to have an overall resistance of better than 0.85 (hr-ft^2-°F)/Btu, or a U-factor of 1.18 Btu/(hr-ft^2-°F). The addition of a second pane of glass creates an additional space in the assembly, increasing the glazing R-value to 1.85, which results in a U-factor of approximately 0.54 Btu/(hr-ft^2-°F). Similarly, the addition of a third pane of glass might increase the overall R-value to 2.85 (hr-ft^2°F)/Btu, yielding a U-factor of 0.35 Btu/(hr-ft^2-°F). This one-dimensional estimate does not hold true for the entire window unit, but only in the region described by ASHRAE as the "center-of-glass" (see Figure 9.21). Highly conductive framing or spacers will create thermal bridging in much the same fashion as metal studs in the insulated wall evaluated in Section 9.31.

9.6.2 Gas Space Between Panes

Most multiple-paned windows are filled with dry air. The thermal performance can be improved by the substitution of gases with lower thermal conductivities. Other gases and gas mixtures used besides dry air are Argon, Krypton, Carbon Dioxide, and Sulfur Hexafluoride. The use of Argon instead of dry air can improve the "center-of-glass" U-factor by 6-9%, depending on the distance between panes, and CO_2 filled units achieve similar performance to Argon gas. For spaces up to 0.5 inches, the mixture of Argon and SF_6 gas can produce the same performance as Argon, and Krypton can provide superior performance to that of Argon.

9.6.3 Emissivity

Emissivity describes the ability of a surface to give off thermal radiation. The lower the emissivity of a warm surface, the less heat loss that it will experience due to radiation. Glass performance can be substantially improved by the application of special low emissivity coatings. The resulting product has come to be known as "Low-E" glass.

Two techniques for applying the Low-E film are sputter and pyrolytic coating. The lowest emittances are achieved with a sputtering process by magnetically depositing silver to the glass inside a vacuum chamber. Sputter coated surfaces must be protected within an insulated glass unit and are often called "soft coat." Pyrolytic coating is a newer method which applies tin oxide to the glass while it is still somewhat molten. The pyrolytic process results in higher emittances than sputter coating, but surfaces are more durable and can be used for single glazed windows. While normal glass has an emissivity of approximately 0.84, pyrolytic coatings can achieve emissivities of approximately 0.40 and sputter coating can achieve emissivities of 0.10 and lower. The emittance of various Low-E glasses will vary considerably between manufacturers.

9.6.4 Window Frames

The type of frame used for the window unit will also have a significant impact on the performance. In general, wood or vinyl frames are thermally superior to metal. Metal frame performance can be improved significantly by the incorporation of a "thermal-break." This usually consists of the thermal isolation of the cold side of the frame from the warm side by means of some low-conducting material. Estimation of the performance of a window due to the framing elements is complicated by the variety of configurations and combinations of materials used for sash and frames. Many manufacturers combine materials for structural or aesthetic purposes. Figure 9.22 below illustrates the impact that various framing schemes can have on glass performance schemes.

The center-of-glass curve illustrates the performance of the glazing system without any framing. Notice that single pane glass is almost unaffected by the framing scheme utilized. This is due to the similar order-of-magnitude thermal conductivity between glass and metal. As additional panes are added, emissivities are lowered and low conductivity gases are introduced, the impact of the framing becomes more pronounced, as shown by the increasing performance "spread" toward the right hand side of the chart. Notice that a plain double pane window with a wood or vinyl frame actually has similar performance to that of Low-E glass (hard coating) with Argon gas fill and a metal frame. Also note how flat the curve is for metal-framed windows with no thermal break. It should be clear that first priority should be given to framing systems before consideration is given to Low-E coatings or low conductivity gases.

9.6.5 Spacers

Double and triple pane window units usually have continuous members around the glass perimeter to sepa-

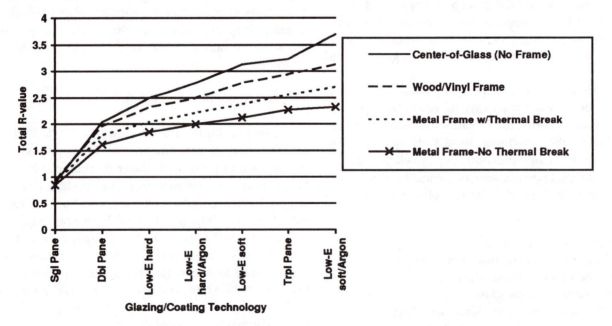

Figure 9.22 Window system performance comparison.

rate the glazing lites and provide an edge seal. These spacers are often made of metal, which increases the heat transfer between the panes and degrades the performance of the glazing near the perimeter. ASHRAE reports this conductive region to be limited to a 2.5-inch band around the perimeter of the glazing unit, and appropriately describes it as the "edge of glass." Obviously, low conductivity spacers, such as plastic, fiberglass or even glass, are to be preferred over high conductivity spacers, such as metal. The impact of highly conductive spacers is only felt to the extent that other high performance strategies are incorporated into the window. For example, ASHRAE reports no significant performance difference between spacers types when incorporated into window systems with thermally unimproved frames. When thermally improved or thermally broken frames are incorporated, the spacer material becomes a factor in overall window performance.

The best performing windows will incorporate thermally broken frames, non-metallic spacers between the panes, low-emissivity coatings, or low conductivity gases, such as argon or krypton, between the panes.

9.6.6 Advanced Window Technologies

9.6.6.1 Interstitial Mylar Films

One product going by the trade name Heat Mirror™ incorporates a Low-E coated mylar or polyester film between the inner and outer glazing of the window unit. High performance is achieved by the addition of a second air space, another Low-E surface, but without the weight of an additional pane of glass.

9.6.6.2 Advanced Spacers

Significant improvements in spacer design have been made in the past 10-15 years. The "Swiggle Strip" spacer sandwiches a thin piece of corrugated metal between two thicker layers of butyl rubber combined with a desiccant. The conductivity is reported to be one-quarter to one-half of the conductivity of a normal aluminum spacer. Other spacers with conductivities only 6 to 11 percent as great as aluminum have been designed using silicon foam spacers with foil backing or by separating two conventional aluminum spacers with a separate strip of polyurethane foam.

9.6.6.3 Future Window Enhancement Technologies

Electrochromic glass systems, referred to as "smart windows" are currently under development in a number of laboratories, and have been incorporated in limited numbers in some automobiles. These systems sandwich indium tin oxide, amorphous tungsten trioxide, magne-

sium fluoride and gold between layers of heat-resistant Pyrex. Imposing a small voltage to the window changes its light and heat transmission characteristics, allowing radiation to be reflected in the summer and admitted in the winter. These systems are expected to be commercially available in the late 1990's and following the turn of the century.

Windows with R-values of R-16 are theoretically possible if the air or other insulating gas between window panes is replaced with a vacuum. Until recently a permanent vacuum has not been achievable due to the difficulty of forming an airtight seal around the edges of the unit. Work is presently under way at the Solar Research Institute on a technique to laser-weld the edges of the glass panes at more than 1000°F to create a leakproof seal. Tiny glass beads are used as spacers between the panes to offset the effect of atmospheric pressure and keep the glass apart.

9.6.7 Rating the Performance of Window Products

Window energy performance information has not been made available in any consistent form, until recently. Some manufacturers publish R-values that rival many insulation materials. Usually the center-of-glass U-factor is published for glazing, independent of the impact of the framing system that will eventually be used. While the center-of-glass rating alone may be impressive, it does not describe the performance of the entire window. Other manufacturers provide performance data for their framing systems, but they do not reflect how the entire product performs. Even when manufacturers provide ratings for the whole product different methods are used to determine these ratings, both analytical and in laboratory hot box tests. This has been a source of confusion in the industry, requiring some comprehensive standard for the reporting of window thermal performance.

The National Fenestration Rating Council (NFRC), sanctioned by the federal government under the Energy Policy Act of 1992 was established to develop a national performance rating system for fenestration products. The NFRC has established a program where the factors that affect window performance are included in published performance ratings. NFRC maintains a directory of certified products. Currently many local energy codes require that NFRC ratings be used to demonstrate compliance with their window performance standards.

9.6.8 Doors

In general, door U-factors can be determined in a similar manner to the walls, roofs and exposed floor demonstrated above. Softwoods have R-values around 1.0 to 1.3 per inch, while hardwoods have R-values that range from 0.80 to 0.95 per inch. A 1-3/8-inch panel door has a U-factor of approximately 0.57 Btu/(Hr-ft^2-°F), while a 1-3/4 solid core flush door has a U-factor of approximately 0.40.

As has been shown to be the case with some walls and windows, metal doors are a different story. The same issues affecting windows, such as framing and thermal break apply to metal doors as well. The U-factor of a metal door can vary from 0.20 to 0.60 Btu/(Hr-ft^2-°F) depending on the extent to which the metal-to-metal contact can be "broken." In the absence of tested door U-factor data, Table 22.6 in Chapter 22 of the 1993 ASHRAE Handbook of Fundamentals can be used to estimate the U-factor of typical doors used in residential and commercial construction.

9.7 INFILTRATION

Infiltration is the uncontrolled inward air leakage through cracks and interstices in a building element and around windows and doors of a building, caused by the effects of wind pressure and the differences in the outdoor/indoor air density. The heat loss due to infiltration is described by the following equation:

$$q_{infiltration} = 0.019 \times Q \times (T_{inside} - T_{outside}) \quad (9.9)$$

Where Q is the infiltration air flow in cubic feet per hour.

The determination of Q is an extremely imprecise undertaking, in that the actual infiltration for similar buildings can vary significantly, even though observable parameters appear to be the same.

9.7.1 Estimating Infiltration for Residential Buildings

ASHRAE suggests that the infiltration rate for a residence can be estimated as:

$$Q = L [(A(T_i - T_o) + (Bv^2)]^{1/2} \quad (9.10)$$

Where:

Q = The infiltration rate, ft^3/hr
A = Stack Coefficient, CFM^2/[in^4-°F]
T$_i$ = Average indoor Temperature, °F
T$_o$ = Average outdoor Temperature, °F
B = Wind Coefficient, CFM^2/[in^4-(mph)^2]
v = Average wind speed, mph

The method is difficult to apply because:

1. L, the total crack area in the building is difficult to determine accurately;

2. The determination of the stack and wind coefficient is subjective;

3. The average wind speed is extremely variable from one micro-climate to another; and

4. Real buildings, built to the same standards, do not experience similar infiltration rates.

ASHRAE reports on the analysis of several hundred public housing units where infiltration varied from 0.5 air changes per hour to 3.5 air changes per hour. If the real buildings experience this much variation, we cannot expect a high degree of accuracy from calculations, unless a significant amount of data is available. However, the studies reported provided some useful guidelines.

In general, older residential buildings without weather-stripping experienced a median infiltration rate of 0.9 air changes per hour. Newer buildings, presumably built to more modern, tighter construction standards, demonstrated median infiltration rates of 0.5 air changes per hour. The structures were unoccupied during tests. It has been estimated that occupants add an estimated 0.10 to 0.15 air changes per hour to the above results.

9.7.2 Estimating Infiltration for Complex Commercial Buildings

Infiltration in large commercial buildings is considerably more complex than small commercial buildings or residential buildings. It is affected by both wind speed and "stack effect." Local wind speed is influenced by distance from the reporting meteorological station, elevation, and the shape of the surrounding terrain. The pressure resulting from the wind is influenced by the local wind velocity, the angle of the wind, the aspect ratio of the building, which face of the building is impinged and the particular location on the building, which in turn is effected by temperature, distance from the building "neutral plane," the geometry of the building exterior envelope elements, and the interior partitions, and all of their relationships to each other. If infiltration due to both wind velocity (Q$_w$) and stack effect (Q$_s$) can be determined, they are combined as follows to determine the overall infiltration rate.

$$Q_{ws} = \sqrt{Q_w^2 + Q_s^2} \quad (9.11)$$

Where:

Q_{ws} = The combined infiltration rate due to wind and stack effect, ft^3/hr

Q_w = The infiltration rate due to wind, ft^3/hr

Q_s = The infiltration rate due to stack effect, ft^3/hr

While recent research has increased our understanding of the basic physical mechanisms of infiltration, it is all but impossible to accurately calculate anything but a "worst-case" design value for a particular building. Techniques such as the air-change method, and general anecdotal findings from the literature are often the most practical approaches to the evaluation of infiltration for a particular building.

9.7.2.1 Anecdotal Infiltration Findings

General studies have shown that office buildings have air exchange rates ranging from 0.10 to 0.6 air changes per hour with no outdoor intake. To the extent outdoor air is introduced to the building and it is pressurized relative to the local outdoor pressure, the above rates will be reduced.

Infiltration through modern curtain wall construction is a different situation than that through windows "punched" into walls. Studies of office buildings in the United States and Canada have suggested the following approximate leakage rates per unit wall area for conditions of 0.30 inches of pressure (water gauge):

Tight Construction — 0.10 CFM/ft^2 of curtainwall area

Average Construction — 0.30 CFM/ft^2 of curtainwall area

Leaky Construction — 0.60 CFM/ft^2 of curtainwall area

9.7.2.2 The Air Change Method

While it is difficult to quantitatively predict actual infiltration, it is possible to conservatively predict infiltration that will give us a sense of "worst-case." This requires experience and judgment on the part of the engineer or analyst. Chapter 5 of the 1972 ASHRAE Fundamentals Manual published guidelines for estimating infiltration on the basis of the number of "air changes." That is, based on the volume of the space, how many complete changes of air are likely to occur within the space of an hour?

Table 9.10 is a summary of the recommended air changes originally published by ASHRAE.

These guidelines have been found to be sound over the years and are still widely used by many practicing professionals. The values represent a good starting point for estimating infiltration. They can be modified as required for local conditions, such as wind velocity or excessive building stack effect.

The above values are based on doors and operable windows that are not weather-stripped. ASHRAE originally recommended that the above factors be reduced by 1/3 for weather-stripped windows and doors. This would be a good guideline applicable to modern buildings, which normally have well-sealed windows and doors. Fully conditioned commercial spaces are slightly pressurized and often do not have operable sash. This will also tend to reduce infiltration.

9.7.3 An Infiltration Estimate Example

Assume a 20' × 30' room with a 10' ceiling and one exposed perimeter wall with windows. What would the worst-case infiltration rate be?

The total air volume is 20 × 30 × 10 = 6,000 ft^3. The table above recommends using 1.0 air changes per hour. However, assuming modern construction techniques, we follow ASHRAE's guideline of using 2/3 of the table values.

$$2/3 \times 6,000 \ ft^3/hr = 4,000 \ ft^3/hr$$

The heat loss due to infiltration for this space, if the indoor temperature was 70°F and the outdoor temperature was 25°F would be:

$$q_{infiltration} = 0.019 \times 4,000 \times (70 - 25)$$
$$= 3,420 \ Btu/hr$$

Table 9.10 Recommended air changes due to infiltration.

Type of Room	Number of Air Changes Taking Place per Hour
Rooms with no windows or exterior doors	1/2
Rooms with windows or exterior doors on one side	1
Rooms with windows or exterior doors on two sides	1-1/2
Rooms with windows or exterior doors on three sides	2
Entrance Halls	2

9.8 SUMMARIZING ENVELOPE PERFORMANCE WITH THE BUILDING LOAD COEFFICIENT

In Section 9.1.2 it was shown that building heat loss is proportional to the indoor-outdoor temperature difference. Our review of different building components, including infiltration has demonstrated that each individual component can be assigned a proportionality constant that describes that particular component's behavior with respect to the temperature difference imposed across it. For a wall or window, the proportionality constant is the U-factor times the component area, or "UA." For a slab-on-grade floor, the proportionality constant is the F-factor times the slab perimeter, or "FP." For infiltration, the proportionality constant is 0.019 time the air flow rate in cubic feet per hour. While not attributed to the building envelope, the effect of ventilation must be accounted for in the building's overall temperature-dependent behavior. The proportionality constant for ventilation is the same for infiltration, except that it is commonly expressed as 1.10 times the ventilation air flow in cubic feet per minute.

The instantaneous temperature-dependent performance of the total building envelope is simply the sum of all the individual component terms. This is sometimes referred to as the *Building Load Coefficient* (BLC). The BLC can be expressed as:

$$BLC = \Sigma UA + \Sigma FP + 0.018\, Q_{INF} + 1.1\, Q_{VENT} \quad (9.12)$$

Where:

ΣUA = the sum of all individual component "UA" products

ΣFP = the sum of all individual component "FP" products

Q_{INF} = the building infiltration volume flow rate, in cubic feet per hour

Q_{VENT} = the building ventilation volume flow rate, in cubic feet per minute

From the above and Equation (9.1), it follows that the total instantaneous building heat loss can be expressed as:

$$q = BLC(T_{indoor} - T_{outdoor}) \quad (9.13)$$

9.9 THERMAL "WEIGHT"

Thermal weight is a qualitative description of the extent to which the building energy consumption occurs in "lock-step" with local weather conditions. Thermal "weight is characterized by the mass and specific heat (heat capacity) of the various components which make up the structure, as well as the unique combination of internal loads and solar exposures which have the potential to offset the temperature-dependent heat loss or heat gain of the facility.

Thermally "Light" Buildings are those whose heating and cooling requirements <u>are</u> proportional to the weather. Thermally "Heavy" Buildings are those Buildings whose heating and cooling requirements are <u>not</u> proportional to the weather. The "heavier" the building, the less temperature-dependent the building's energy consumption appears to be, and the less accuracy that can be expected from simple temperature-dependent energy consumption calculation schemes.

The concept of thermal weight is an important one when it comes to determining the energy-saving potential of envelope improvements. The "heavier" the building, the less savings per square foot of improved envelope can be expected for the same "UA" improvement. The "lighter" the building, the greater the savings that can be expected. A means for characterizing the thermal "weight" of buildings, as well as analyzing the energy savings potential of "light" and "heavy" buildings will be taken up in Section 9.10.

9.10 ENVELOPE ANALYSIS FOR EXISTING BUILDINGS

9.10.1 Degree Days

In theory, if one wanted to predict the heat lost by a building over an extended period of time, Equation 9.12 could be solved for each individual hour, taking into account the relevant changes of the variables. This is possible because the change in the value BLC with respect to temperature is not significant and the indoor temperature (T_i) is normally controlled to a constant value (such as 70°F in winter). That being the case, the total energy transfer could be predicted by knowing the summation of the individual deviations of outdoor temperature (T_o) from the indoor condition (T_i) over an extended period of time.

The summation described has come to be known as "Degree-Days" and annual tabulations of Degree-Days for various climates are published by NOAA, ASHRAE, and various other public and military organizations. Historically an indoor reference point of 65°F has been used to account for the fact that even the most poorly constructed building is capable of maintaining comfort conditions without heating when the temperature is at least 65°F.

Because of the impracticality of obtaining hour-by-hour temperature data for a wide variety of locations, <u>daily</u> temperature averages are often used to represent 24-hour blocks of time. The daily averages are calculated by taking the average of daily maximum and minimum temperature recordings, which in turn are converted to Degree Days. Quantitatively this is calculated as:

$$Degree - Days = \frac{\left(T_{reference} - T_{average}\right)n}{24}$$ (9.14)

Where "n" represents the number of hours in the period for which the Degree Days are being reported, and $T_{reference}$ is a reference temperature at which no heating is assumed to occur. Typically, a reference temperature of 65°F is used. Units of degree hours can be obtained by multiplying the results of any Degree-Day tabulation by 24.

Figure 9.23 shows the monthly natural gas consumption of a metropolitan newspaper building overlaid on a plot of local heating Degree Days.

Because of the clear relationship between heating Degree Days and heating fuel consumption for many buildings, such as the above, the following formula has been used since the 1930's to predict the future heating fuel consumption.

$$E = \frac{q \times DD \times 24}{\Delta t \times H \times Eff}$$ (9.15)

Where:

E   = Fuel consumption, in appropriate units, such as Therms natural gas

q   = The design-day heat loss, Btu/hr

DD = The annual heating Degree-Days (usually referenced to 65°F)

24 = Converts Degree-Days to Degree-Hours

Δt = The design day indoor-outdoor temperature difference, °F

H   = Conversion factor for the type of fuel used

Eff = The annualized efficiency of the heating combustion process

Note that if the BLC is substituted for q/Δt, Equation 9.15 becomes:

$$E = \frac{BLC \times DD \times 24}{H \times Eff}$$ (9.16)

Ignoring the conversion terms, we can see that this equation has the potential to describe the relationship between monthly weather trends and the building heating fuel consumption shown in Figure 9.23.

As building construction techniques have improved over the years, and more and more heat-producing equipment has found its way into commercial and even residential buildings, a variety of correction factors have been introduced to accommodate these influences. The Degree-Day method, in its simpler form, is not presently considered a precise method of estimating future building energy consumption. However it is useful for indicating the severity of the heating season for a region and it can prove to be a very powerful tool in the analysis of existing buildings (i.e. ones whose energy consumption is already known).

It is possible to determine an effective Building

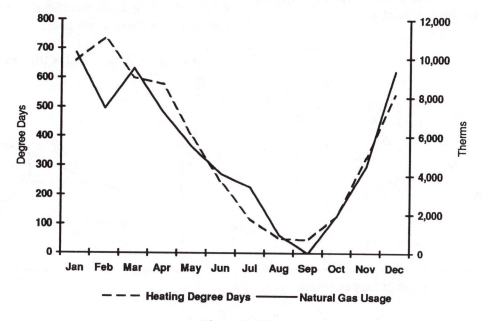

Figure 9.23.

Load Coefficient for a building by analyzing the monthly fuel consumption and corresponding monthly Degree Days using the technique of linear regression

9.10.2 Analyzing Utility Billings

If we were to plot the monthly natural gas consumption and corresponding Degree-Days for a building, the result might look something like Figure 9.24. It can be seen that the monthly energy consumption appears to follow the weather in an indirect fashion.

If we could draw a line that represented the closest "fit" to the data, it might look like the Figure 9.25.

As a general rule, the simpler the building and its heating (or cooling) system, the better the correlation. Buildings that are not mechanically cooled will show an even better correlation. To put it in terms of our earlier discussion regarding thermal "weight," the "lighter" the building, the better the correlation between building energy consumption and Degree Days. To the extent this line fits the data, it gives us two important pieces of information discussed in Section 9.1.1.

• The Monthly Base Consumption

• The Relationship Between the Weather and Energy Consumption Beyond the Monthly Base, which we have been calling the Building Load Coefficient.

9.10.3 Using Linear Regression for Envelope Analysis

The basic idea behind linear regression is that any physical relationship where the value of a result is linearly dependent on the value of an independent quantity can be described by the relationship:

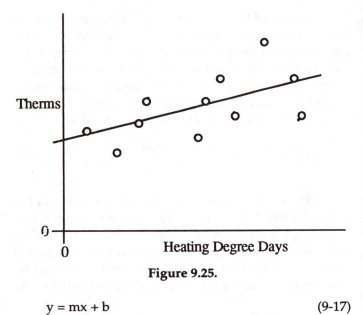

Figure 9.25.

$$y = mx + b \qquad (9\text{-}17)$$

where "y" is the dependent variable, x is the independent variable, m is the slope of the line that describes the relationship, and b is the y-intercept, which is the value of y when x = 0.

In the case of monthly building energy consumption, the monthly Degree Days, can be taken as the independent variable, and the monthly fuel consumption as the dependent variable. The slope of the line that relates the monthly consumption to monthly Degree Days is the Building Load Coefficient (BLC). The monthly base fuel consumption is the intercept on the fuel axis (or the monthly fuel consumption when there are no heating Degree Days). The fuel consumption for any month can be determined by multiplying the monthly Degree Days by the Building Load Coefficient and adding it to the monthly base fuel consumption.

$$\text{Fuel Consumption} = \text{BLC} \times \text{DD} + \text{base energy} \qquad (9.18)$$

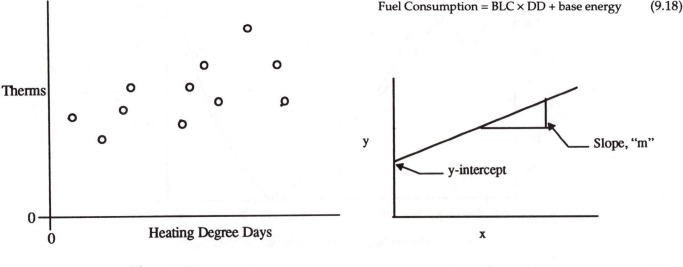

Figure 9.24.

Figure 9.26.

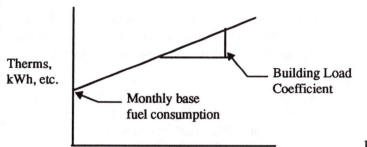

Figure 9.27.

The value of the above is that once the equation is determined, projected fuel savings can be made by recalculating the monthly fuel consumption using a "conservation-modified" BLC.

The determination of the Building Load Coefficient is useful in the analysis of building envelope for a number of reasons.

- Often the information necessary for the analysis of building envelope components is not available.

- If envelope information is readily available and the Building Load Coefficient has been determined by analysis of all the individual components, the regression derived BLC gives us a "real-world" check on the calculated BLC.

- Linear regression also gives feedback in terms of how "good" the relationship is between fuel consumption and local climate, giving a good indication as to the thermal "weight" of the building, discussed in Section 9.9.

The following is a brief overview of the regression procedure as it can be applied to corresponding pairs of monthly Degree-Day and fuel consumption data. The Building Load Coefficient which describes the actual performance of the building can be determined with the following relationship:

$$BLC = \frac{n\Sigma D_i E_i - \Sigma D_i \Sigma E_i}{n\Sigma D_i^2 - (\Sigma D_i)^2}$$

(9.19)

Where:

n   =   the number of degree-day/fuel consumption pairs

D_i =   the degree days accumulated for an individual month

E_i =   the energy or fuel consumed for an indi-

vidual month The monthly base fuel consumption can be calculated as:

$$Base = \frac{\Sigma E_i}{n} - BLC\frac{\Sigma D_i}{n}$$

(9.20)

The units of the BLC will be terms of the fuel units per degree day. The BLC will be most valuable for additional analysis if it is converted to units of Btu/(hr-°F) or Watts/°C.

The following example demonstrates how the above might be used to evaluate the potential for envelope improvement in a building.

Example

A proposal has been made to replace 5,000 ft² of windows in an electrically heated building. The building envelope has been analyzed on a component-by-component basis and found to have an overall analytical BLC = 15,400 Btu/(Hr-°F). Of this total, 6,000 Btu/(Hr-°F) is attributable to the single pane windows, which are assumed to have a U-factor of 1.2 Btu/(hr-ft²-°F). A linear regression is performed on the electric utility data and available monthly Degree-Day data (65°F reference). The BLC is found to be 83.96 kWh/degree-day, which can be converted to more convenient units by the following:

$$BLC = 83.96\frac{kWh}{°F - Days} \times 3413\frac{Btu}{kWh} \times \frac{1\ Day}{24\ hours}$$

$$= 11,940\ Btu/(hr-°F)$$

What is the reason for the discrepancy between the BLC calculated component-by-component and the BLC derived from linear regression of the building's performance?

The explanation comes back to the concept of thermal weight. Remember, the more thermally "heavy" the building, the more independent of outdoor conditions it is. The difference between the value above and the calculated BLC is due in part to internal heat gains in the building that offset some of the heating that would normally be required.

This has significant consequences for envelope retrofit projects. If the potential savings of a window conservation retrofit is calculated on the basis of the direct "UA" improvement, the savings will be overstated in a building such as the above. A more conservative (and realistic) estimate of savings can be made by de-rating the theoretical UA improvement by the ratio of the regression UA to the calculated UA.

9.10.4 Evaluating the Usefulness of the Regression Results

The Correlation Coefficient, R, is a useful term that describes how well the derived linear equation accounts for the variation in the monthly fuel consumption of the building. The correlation coefficient is calculated as follows:

$$R = BLC \sqrt{\frac{\sum D_i^2 - \dfrac{\sum D_i}{n}}{\sum F_i^2 - \dfrac{\sum F_i}{n}}} \qquad (9.21)$$

The square of the Correlation Coefficient, R^2 provides an estimate of the number of independent values (fuel consumption) whose variation is explained by the regression relationship. For example, an R^2 value of 0.75 tells us that 75% of the monthly fuel consumption data points evaluated can be accounted for by the linear equation given by the regression analysis. As a general rule of thumb, an R^2 value of 0.80 or above describes a thermally "light" building, and a building whose fuel consumption indicates an R^2 value significantly less than 0.80 can be considered a thermally "heavy" building.

9.10.5 Improving the Accuracy of the Building Load Coefficient Estimate

As discussed elsewhere in Section 9.9, one of the reasons a building might be classified as thermally "heavy" would be the presence of significant internal heat gains, such as lighting, equipment and people. To the extent these gains occur in the perimeter of the building, they will tend to offset the heat loss predicted using the tools described previously in this chapter. The greater the internal heat, relative to the overall building temperature dependence (BLC), the lower the outside temperature has to be before heating is required.

The so-called "building balance temperature" is the theoretical outdoor temperature where the total building heat loss is equal to the internal gain. We discussed earlier that the basis of most published Degree-Day data is a reference (or balance) temperature of 65°F. If data of this sort is used for analysis, the lower the actual building balance temperature than 65°F, the more error will be introduced into the analysis. This accounts for the low R^2 values encountered in thermally heavy buildings.

While most degree data is published with 65°F as the reference point, it is possible to find compiled sources with 55°F and even 45°F reference temperatures. The use of this monthly data can improve the accuracy of the analysis significantly, if the appropriate data is utilized.

Example

A regression is performed for a library building with electric resistance heating. The results of the analysis indicate an R^2 of 0.457. In other words, only 46% of the variation in monthly electricity usage is explained by the regression. The figure below is a plot of the monthly electric consumption versus Degree Days calculated for a 65°F reference.

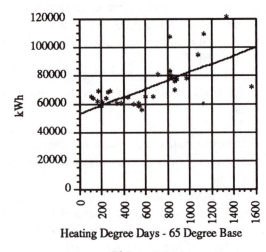

Figure 9.28.

The plot following shows the result of the regression run using monthly Degree Day referenced to 55°F. Notice in the figure that there are less data points showing. This is explained by the observation that all the days with average temperatures greater than 55°F are not included in the data set. The resulting R^2 has increased to 0.656.

The next figure shows the result of the regression run using monthly Degree Day referenced to 45°F. The resulting R^2 has increased to 0.884. Notice how few data points are left. Normally in statistical analysis great emphasis is placed on the importance of having an adequate data sample for the results to be meaningful. While we would like to have as many points as possible in our analysis, the concern is not so great with this type of analysis. Statistical analysis usually concerns itself with <u>whether</u> there is a relationship to be found. The type of analysis we are advocating here assumes that the relationship <u>does</u> exist, and we are merely attempting to discover the most accurate form of the relationship. Another way of saying this is that we are using a trial-and-error technique to discover the building balance point.

The final figure shows the result of the regression run with monthly degree days referenced to 40°F. The resulting R^2 has <u>decreased</u> to 0.535.

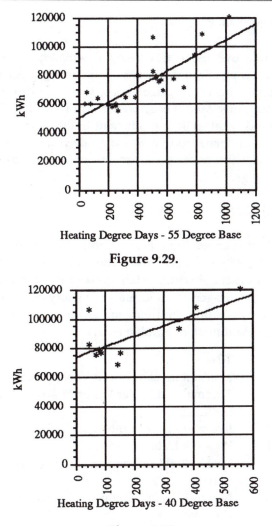

Figure 9.29.

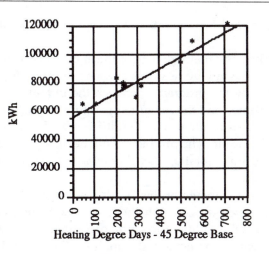

Figure 9.31.

Figure 9.30.

We have learned from the above that the balance point of the building analyzed is somewhere between 40° and 50°F, with 45°F probably being pretty close. Additional iterations can be made if more precision is desired, and if the referenced Degree Day data is available.

If the published Degree Day data desired is not available, it is a relatively simple matter to construct it from the average daily temperatures published for the local climate. If this is done with an electronic spreadsheet, a table can be constructed in such a way that the degree days, BLC, monthly base and R^2 values are all linked to an assumed balance temperature. This balance temperature is modified iteratively until R^2 is maximized. Of course any energy saving calculations should consistently use both the derived BLC and degree days that accompany the correlation with the highest R^2.

9.11   ENVELOPE ANALYSIS FOR NEW BUILDINGS

Envelope analysis of new buildings offers a different kind of challenge to the analyst. While new construc-

tion offers much greater opportunity for economical improvements to the envelope, the analysis is much more open-ended than that for existing buildings. In other words, we do not know the monthly energy consumption (the answer), and are left totally at the mercy of tools designed to assist us in <u>predicting</u> the future energy consumption of an as-yet unconstructed building, the necessary details of which constitute thousands of unknowns.

We say the above to emphasize the extreme difficulty of the task, and the inadvisability of harboring any illusions that this can be done reliably, short of using costly hour-by-hour computer simulation techniques. Even with the powerful programs currently available, accuracy's no better than 10 to 20 percent should be expected.

Nevertheless, we are often called upon to quantify the benefits of using one envelope strategy in place of another. We are able to quantify the *difference* in annual energy consumption between two options much more reliably than the absolute consumption. A number of techniques have been developed to assist in this process. One of the more popular and useful tools is the *Temperature Bin Method*.

9.11.1 The Temperature Bin Method

The Temperature Bin Method requires that instantaneous energy calculations be performed at many different outdoor temperature conditions, with the results multiplied by the number of hours expected at each temperature condition. The "bins" referred to represent the number of hours associated with groups of temperatures, and are compiled in 5°F increments. The hour tabulations are available in annual, monthly and sometimes 8-hour shift totals. All hourly occurrences in a bin are assumed to take place at the bin "center-temperature." For example, it is assumed that 42°F quantitatively represents the 40-44°F bin.

The basic methodology of this method requires calculating the unique heat loss at each bin by multiplying the BLC by the difference between the indoor temperature assumed and the center-temperature for each respective bin. This result is in turn multiplied by the number of hours in the bin. The products of all the bins are summed to arrive at the total predicted annual heating energy for the building.

The advantage of this "multiple-measure" method over a "single-measure" method, such as the degree day method, is the ability to accommodate other temperature-dependent phenomena in the analysis. For example, the power requirement and capacity of an air-to-air heat pump are extremely temperature dependent. This dependency is easily accommodated by the bin method.

However, just as was the case with the Degree-Day method, energy savings predicted by the bin method may vary significantly from actual, depending on the building balance temperature (building thermal weight). While the balance temperature cannot be predicted for a new building with the techniques previously discussed, it can be estimated with the following equation.

$$T_{balance} = T_{indoor} - \frac{q_{internal}}{BLC} \tag{9.22}$$

Where:

$T_{balance}$ = The predicted balance temperature, °F
T_{indoor} = The assumed indoor conditioned space temperature, °F
$q_{internal}$ = The assumed internal heat gain in building temperature control zones adjacent to the envelope, Btu/hr
BLC = The Building Load Coefficient, BLC, Btu/(hr-°F)

More accurate energy predictions will result by omitting calculations for bins whose center temperature exceeds the assumed balance temperature. For example, no calculation should be performed for the 50-54°F bin, if the predicted balance temperature is 50°F. The center temperature of 47°F for the 45-49°F bin indicates that a 3°F temperature difference is appropriate for that bin (50-47°F).

A complete description of the bin method can be found in Chapter 28 of the 1993 ASHRAE Handbook of Fundamentals.

9.12 SUMMARY

While the above discussion of envelope components has emphasized the information needed to perform rudimentary heat loss calculations, you'll find that the more you understand these basics, the more you begin to understand what makes an efficient building envelope. This same understanding will also guide you in deciding how to prioritize envelope improvement projects in existing buildings.

9.13 ADDITIONAL READING

As you can see from this brief introduction, the best source of comprehensive information on building envelope issues is the ASHRAE Handbook of Fundamentals. You are encouraged to continue your study of building envelope by reading the following chapters in the 1993 ASHRAE Handbook of Fundamentals.

Chapter	Title
20,21	Thermal Insulation and Vapor Retarders
22	Thermal and Water Vapor Transmission Data
23	Infiltration and Ventilation
24	Weather Data
25	Residential Cooling and Heating Calculations
26	Nonresidential Cooling and Heating Calculations
27	Fenestration

9.13.1 References
1. American Society of Heating, Refrigerating, and Air Conditioning Engineers, Inc., *Heat Transmission Coefficients for Walls, Roofs, Ceilings, and Floors*, 1993.
2. ASHRAE Handbook of Fundamentals, American Society of Heating, Refrigerating and Air Conditioning Engineers, Inc., Atlanta, GA, 1993.
3. ASHRAE Standard 901.-1989, American Society of Heating, Refrigerating, and Air Conditioning Engineers, Inc., Atlanta, GA, 1989.
4. Brown, W.C., *Heat-Transmission Tests on Sheet Steel Walls*, ASHRAE Transactions, 1986, Vol. 92, part 2B.
5. Elder, Keith E., *Metal Buildings: A Thermal Performance Compendium*, Bonneville Power Administration, Electric Ideas Clearinghouse, August 1994.
6. Elder, Keith E., *Metal Elements in the Building Envelope. A Practitioner's Guide*, Bonneville Power Administration, Electric Ideas Clearinghouse, October 1993.
7. Johannesson, Gudni, *Thermal Bridges in Sheet Metal Construction*, Division of Building Technology, Lund Institute of Technology, Lund, Sweden, Report TVHB-3007, 1981 .
8. Loss, W., *Metal Buildings Study: Performance of Materials and Field Validation*, Brookhaven National Laboratory, December 1987.
9. Washington State Energy Code, Washington Association of Building Officials April 1, 1994

CHAPTER 10

HVAC SYSTEMS

ERIC NEIL ANGEVINE, P.E.

Associate Professor
School of Architecture
Oklahoma State University
Stillwater, Oklahoma

JENNIFER S. FAIR, P.E.

Vice President
PSA Consulting Engineers
Oklahoma City, Oklahoma

10.1 INTRODUCTION

The mechanical heating or cooling load in a building is dependent upon the various heat gains and losses experienced by the building including solar and internal heat gains and heat gains or losses due to transmission through the building envelope and infiltration (or ventilation) of outside air. The primary purpose of the heating, ventilating, and air-conditioning (HVAC) system in a building is to regulate the dry-bulb air temperature, humidity and air quality by adding or removing heat energy. Due to the nature of the energy forces which play upon the building and the various types of mechanical systems which can be used in non-residential buildings, there is very little relationship between the heating or cooling load and the energy consumed by the HVAC system.

This chapter outlines the reasons why energy is consumed and wasted in HVAC systems for non-residential buildings. These reasons fall into a variety of categories, including energy conversion technologies, system type selection, the use or misuse of outside air, and control strategies. Following a review of the appropriate concerns to be addressed in analyzing an existing HVAC system, the chapter discusses the aspects of human thermal comfort. Succeeding sections deal with HVAC system types, energy conservation opportunities and domestic hot water systems.

10.2 SURVEYING EXISTING CONDITIONS

As presented in Chapter 3, the first stage of any effective energy management program is an energy audit of the facility in question. In surveying the HVAC system(s) in a facility, the first step is to find out what you have to work with: what equipment and control systems exist. It is usually beneficial to divide the HVAC systems into two categories: equipment and systems which provide heating and cooling, and equipment and systems which provide ventilation. It is essential to fully document the type and status of all equipment from major components including boilers, chillers, cooling towers and air-handling units to the various control systems: thermostats, valves and gauges, whether automated or manual; in order to later determine what elements can be replaced or improved to realize a saving in energy consumed by the system.

The second step is to determine how the system is operating. This requires that someone measure the operating parameters to determine whether the system actually operates as it was specified to operate. Determine the system efficiency under realistic conditions. This may be significantly different from the theoretical, or full-load efficiency. Determine how the system is operated. What are the hours of operation? Are changes in system controls manual or automatic? Find out how the system is actually operated, which may differ from how the system was designed to be operated. It is best to talk to operators and/or users of the system who know a lot more about how the system operates than the engineers or managers.

If the system is no longer operating at design conditions, it is extremely useful to determine what factors are responsible for the change. Potential causes of operational changes are modifications in the building or system and lapses in maintenance. Have there been structural or architectural changes to the building without corresponding changes to the HVAC system? Have there been changes in building operations? Is the system still properly balanced? Has routine maintenance been performed? Has scheduled preventative maintenance been performed?

Finally, it is useful to determine whether the system can or should be restored to its initial design conditions. If practical, it may be beneficial to carry out the needed maintenance *before* proceeding to analyze the system for further improvements. However, some older systems are so obviously inefficient that bringing them back to original design parameters is not worth the time or expense.

Before continuing with an analysis of the system, it is also useful to determine future plans for the building and the HVAC system which can seriously effect the energy efficiency of system operation. Are there plans to remodel the building or parts of the building? How extensive are proposed changes? Are changes in building operations planned?

Document everything. Only when you have a full record of what the system consists of, how it is operating and how it is operated, and what changes have been made and will be made in the future, can you properly evaluate the benefit of energy conservation techniques which may be applicable to a particular building system.

10.3 HUMAN THERMAL COMFORT

The ultimate objective of any heating, cooling and ventilating system is typically to maximize human thermal comfort. Due to the prevalence of simple thermostat control systems for residential and small-scale commercial HVAC systems, it is often believed that human thermal comfort is a function solely, or at least primarily, of air temperature. But this is not the case.

Human thermal comfort is actually maximized by establishing a heat balance between the occupant and his or her environment. Since the body can exchange heat energy with its environment by conduction, convection and radiation, it is necessary to look at the factors which affect these heat transfer processes along with the body's ability to cool itself by the evaporation of perspiration.

All living creatures generate heat by burning food, a process known as metabolism. Only 20 percent of food energy is converted into useful work; the remainder must be dissipated as heat. This helps explain why we remain comfortable in an environment substantially cooler than our internal temperature of nearly 100°F (37°C).

In addition to air temperature, humidity, air motion and the surface temperature of surroundings all have a significant influence on the rate at which the human body can dissipate heat. At temperatures below about 80°F (27°C) most of the body's heat loss is by con-

vection and radiation. Convection is affected mostly by air temperature, but it is also strongly influenced by air velocity. Radiation is primarily a function of the relative surface temperature of the body and its surroundings. Heat transfer by conduction is negligible, since we make minimal physical contact with our surroundings which is not insulated by clothing.

At temperatures above 80°F (27°C) the primary heat loss mechanism is evaporation. The rate of evaporation is dependent on the temperature and humidity of the air, as well as the velocity of air which passes over the body carrying away evaporated moisture.

In addition to these environmental factors, the rate of heat loss by all means is affected by the amount of clothing, which acts as thermal insulation. Similarly, the amount of heat which must be dissipated is strongly influenced by activity level. Thus, the degree of thermal comfort achieved is a function of air temperature, humidity, air velocity, the temperature of surrounding surfaces, the level of activity, and the amount of clothing worn.

In general, when environmental conditions are cool the most important determinant of human thermal comfort is the radiant temperature of the surroundings. In fact, a five degree increase in the mean-radiant temperature of the surroundings can offset a seven degree reduction in air temperature.

When conditions are warm, air velocity and humidity are most important. It is not by accident that the natural response to being too warm is to increase air motion. Similarly, a reduction in humidity will offset an increase in air temperature., although it is usually necessary to limit relative humidity to no more than 70% in summer and no less than 20% in winter.

There is, of course, a human response to air temperature, but it is severely influenced by these other factors. The most noticeable comfort response to air temperature is the reaction to drift, the change of temperature over time. A temperature drift of more than one degree Fahrenheit per hour (0.5°C/hr) will result in discomfort under otherwise comfortable conditions. Temperature stratification can also cause discomfort, and temperature variation within the occupied space of a building should not be allowed to vary by more than 5 degrees F (3°C).

Modern control systems for HVAC systems can respond to more than just the air temperature. One option which has been around for a long time is the humidistat, which senses indoor humidity levels and controls humidification. However, state-of-the-art control systems can measure *operative temperature*, which is the air temperature equivalent to that affected by radiation and convection conditions of an actual environment.[*] Another useful construct is that of *effective temperature,* which is a computed temperature that includes the effects of humidity and radiation.[†]

The location and type of air distribution devices play a role equal in importance to that of effective controls in achieving thermal comfort. The discomfort caused by stratification can be reduced or eliminated by proper distribution of air within the space.

In general terms, thermal comfort can be achieved at air temperatures between about 68°F and 80°F, and relative humidities between 20% and 70%, under varying air velocities and radiant surface temperatures. Figure 10.1 shows the generalized "comfort zone" of dry bulb temperatures and humidities plotted on the psychrometric chart. However it should not be forgotten that human thermal comfort is a complex function of temperature, humidity, air motion, thermal radiation from local surroundings, activity level and amount of clothing.

10.4 HVAC SYSTEM TYPES

The energy efficiency of systems used to heat and cool buildings varies widely but is generally a function of the details of the system organization. On the most simplistic level the amount of energy consumed is a function of the source of heating or cooling energy, the amount of energy consumed in distribution, and whether the working fluid is simultaneously heated and cooled. System efficiency is also highly dependent upon the directness of control, which can sometimes overcome system inefficiency.

HVAC system types can be typically classified according to their energy efficiency as highly efficient, moderately efficient or generally inefficient. This terminology indicates only the comparative energy consumption of typical systems when compared to each other. Using these terms, those system types classified as generally inefficient will result in high energy bills for the building in which they are installed, while an equivalent building with a system classified as highly efficient will usually have lower energy bills. However, it is important to recognize that there is a wide range of efficiencies within each category, and that a specific energy-efficient example of a typically inefficient system might have lower energy bills than the least efficient example of a moderately, or even highly efficient type of system.

Figure 10.2 shows the relative efficiency of the more commonly used types of HVAC systems discussed below. The range of actual energy consumption for each system type is a function of other design variables including how the system is configured and installed in a particular building as well as how it is controlled and operated.

To maximize the efficiency of any type of HVAC system, it is important to select efficient equipment, minimize the energy consumed in distribution and avoid simultaneous heating and cooling of the working fluid. It is equally important that the control system directly control the variable parameters of the system.

Most HVAC systems include zones, which are areas within the building which may have different climatic and/or internal thermal loads and for which heat can be supplied or extracted independent of other zones.

The four-volume *ASHRAE Handbook,* published sequentially in a four-year cycle by the American Society of Heating, Refrigerating and Air-Conditioning Engineers, Inc., provides the most comprehensive and authoritative reference on HVAC systems for buildings. The reader is specifically referred to the *1992 ASHRAE Handbook: HVAC Systems and Equipment* for additional information on any of the systems described below.[3] Additional information regarding the most appropriate types of HVAC system for specific applications can be found in the *1995 ASHRAE Handbook: HVAC Applications.*[2]

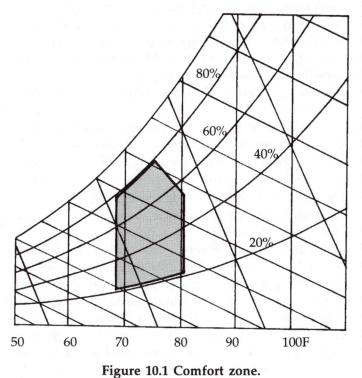

Figure 10.1 Comfort zone.

[*]Operative temperature is technically defined as the uniform temperature of an imaginary enclosure with which an individual exchanges the same heat by radiation and convection as in the actual environment.
[†]Effective temperature is an empirical index which attempts to combine the effect of dry bulb temperature, humidity and air motion into a single figure related to the sensation of thermal comfort at 50% relative humidity in still air.

Figure 10.2 Relative energy efficiency of air-conditioning systems.

Least Efficient   Most Efficient

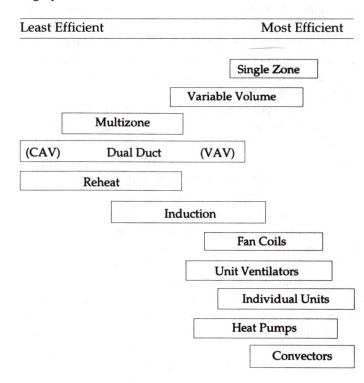

10.4.1 All-Air Systems

The most common types of systems for heating and cooling buildings are those which moderate the air temperature of the occupied space by providing a supply of heated or cooled air from a central source via a network of ducts. These systems, referred to as all-air systems, increase or decrease the space temperature by altering either the volume or temperature of the air supplied.

Recalling that the most important determinant of thermal comfort in a warm environment is air velocity, most buildings which require cooling employ all-air systems. Consequently, all-air systems are the system of choice when cooling is required. All-air systems also provide the best control of outside fresh air, air quality, and humidity control. An added benefit of forced air systems is that they can often use outside air for cooling interior spaces while providing heating for perimeter spaces. (see §10.5.5; Economizers) The advantages of all-air systems are offset somewhat by the energy consumed in distribution.

All-air systems tend to be selected when comfort cooling is important and for thermally massive buildings which have significant internal cooling loads which coincide with heating loads imposed by heat loss through the building envelope.

The components of an all-air HVAC system in-

clude an air-handling unit (AHU) which includes a fan, coils which heat and/or cool the air passing through it, filters to clean the air, and often elements to humidify the air. Dehumidification, when required, is accomplished by cooling the air below the dew-point temperature. The conditioned air from the AHU is supplied to the occupied spaces by a network of supply-air ducts and air is returned from conditioned spaces by a parallel network of return-air ducts. (Sometimes the open plenum above a suspended ceiling is used as part of the return-air path.) The AHU and its duct system also includes a duct which supplies fresh outside air to the AHU and one which can exhaust some or all of the return air to the outside. Figure 10.3 depicts the general arrangement of components in an all-air HVAC system.

Single Duct Systems

The majority of all-air HVAC systems employ a single network of supply air ducts which provide a continuous supply of either warmed or cooled air to the occupied areas of the building.

Single Zone - The single duct, single-zone system is the simplest of the all-air HVAC systems. It is one of the most energy-efficient systems as well as one of the least expensive to install. It uses a minimum of distribution energy,* since equipment is typically located within or immediately adjacent to the area which it conditions. The system is directly controlled by a thermostat which turns the AHU on and off as required by the space temperature. The system shown in Figure 10.3 is a single zone system.

Single zone systems can provide either heating or cooling, but provide supply air at the same volume and temperature to the entire zone which they serve. This limits their applicability to large open areas with few windows and uniform heating and cooling loads. Typical applications are department stores, factory spaces, arenas and exhibit halls, and auditoriums.

Variable Air Volume - The variable air volume (VAV) HVAC system shown in Figure 10.4 functions much like the single zone system, with the exception that the temperature of individual zones is controlled by a thermostat which regulates the volume of air that is discharged into the space. This arrangement allows a high degree of local temperature control at a moderate cost. Both installation cost and operating costs are only slightly greater than the single-zone system.

The distribution energy consumed is increased slightly over that of a single-zone system due to the fric-

*Distribution energy includes all of the energy used to move heat within the system by fans and pumps. Distribution energy is typically electrical energy.

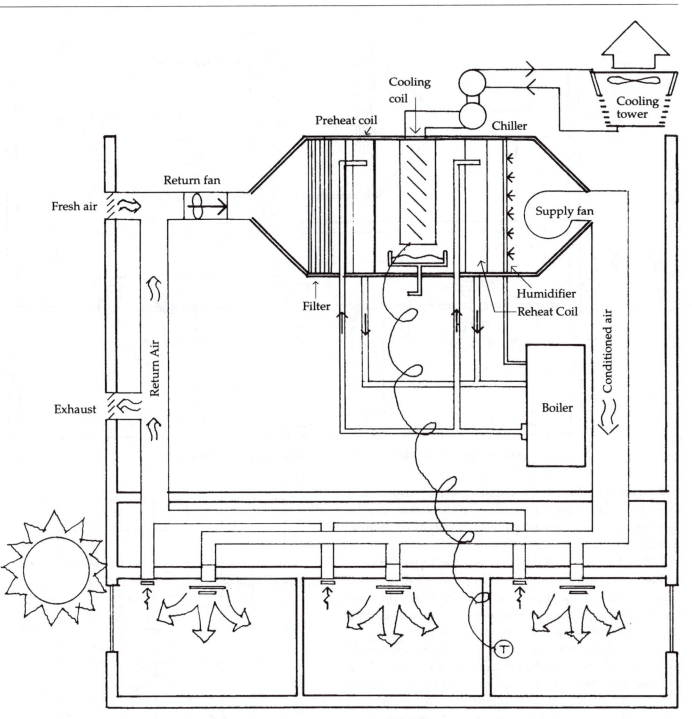

Figure 10.3 Elements of an all-air air-conditioning system.

tion losses in VAV control devices, as well as the fact that the fan in the AHU must be regulated to balance the overall air volume requirements of the system. Fan regulation by inlet vanes or outlet dampers forces the fan to operate at less than its optimum efficiency much of the time (see Figure 10.5) Consequently a variable speed fan drive is necessary to regulate output volume of the fan. For the system to function properly, it is necessary that air be supplied at a constant temperature, usually about 55°F (13°C). This requires indirect control of the supply air temperature with an accompanying decrease in control efficiency.

Single-duct VAV systems can often provide limited heating by varying the amount of constant temperature air to the space. By reducing the cooling airflow, the space utilizes the lights, people and miscellaneous

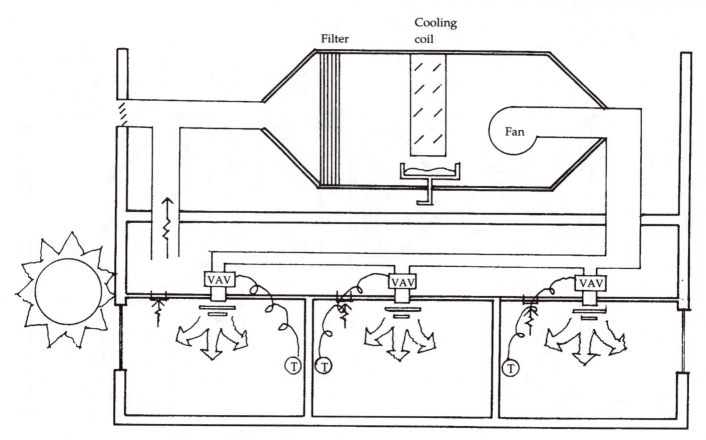

Figure 10.4 Variable air volume system schematic diagram.

equipment to maintain the required space temperature. However, if the space requires more heat than can be supplied by internal heat gains, a separate or supplemental heating system must be employed.

Single-duct VAV systems are the most versatile and have become the most widely used of all systems for heating and cooling large buildings. They are appropriate for almost any application except those requiring a high degree of control over humidity or air exchange.

Reheat systems - Both the single-zone and single duct VAV systems can be modified into systems which provide simultaneous heating and cooling of multiple zones with the addition of reheat coils for each zone (Figure 10.6). These systems are identical in design to the foregoing systems up to the point where air enters the local ductwork for each zone. In a reheat system supply air passes through a reheat coil which usually contains hot water from a boiler. In a less efficient option, an electrical resistance coil can also be used for reheat. (See comments regarding the efficiency of electric resistance heating in §10.5.6.)

A local thermostat in each zone controls the temperature of the reheat coil, providing excellent control of the zone space temperature. Constant air volume (CAV) reheat systems are typically used in situations which re-

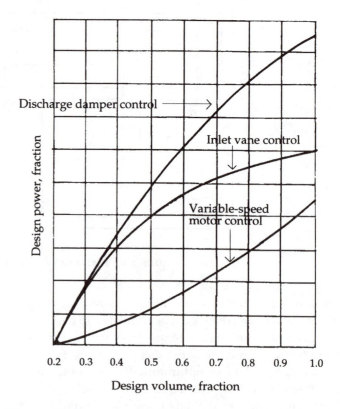

Figure 10.5 Fan power vs discharge volume characteristics.

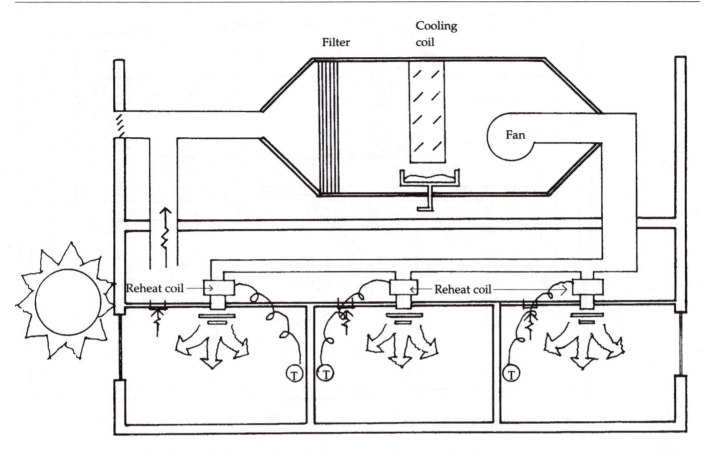

Figure 10.6 Reheat system schematic diagram.

quire precise control of room temperature and/or humidity, often with constant airflow requirements, such as laboratories and medical facilities.

Both the CAV and VAV reheat systems are inherently inefficient, representing the highest level of energy consumption of the all-air systems. This is due to the fact that energy is consumed to cool the supply air and then additional energy is consumed to reheat it. In VAV reheat systems, the reheat coil is not activated unless the VAV controls are unable to meet local requirements for temperature control, and they are therefore somewhat more energy efficient than CAV reheat systems.

Both CAV and VAV reheat systems can also be used with specialized controls to condition spaces with extremely rigid requirements for humidity control, such as museums, printing plants, textile mills and industrial process settings.

Multizone - Although commonly misused to indicate any system with thermostatically controlled air-conditioning zones, the multizone system is actually a specific type of HVAC system which is a variation of the single-duct CAV reheat system. In a multizone system, each zone is served by a dedicated duct which connects it directly to a central air handling unit (Figure 10.7).

In the most common type of multizone system, the

AHU produces warm air at a temperature of about 100°F (38°C) as well as cool air at about 55°F (13°C) which are blended with dampers to adjust the supply air temperature to that called for by zone thermostats. In a variation of this system, a third neutral deck uses outside air as an economizer to replace warm air in the summer or cool air in the winter. In another variation, the AHU produces only cool air which is tempered by reheat coils located in the fan room. In this case, the hot deck may be used as a preheat coil.

Multizone systems are among the least energy efficient, sharing the inherent inefficiency of reheat systems since energy is consumed to simultaneously heat and cool air which is mixed to optimize the supply air temperature. Since a constant volume of air is supplied to each zone, blended conditioned air must be supplied even when no heating or cooling is required.

In addition, multizone systems require a great deal of space for ducts in the proximity of the AHU which restricts the number of zones. They also consume a great deal of energy in distribution, due to the large quantity of constant volume air required to meet space loads. These drawbacks have made multizone systems nearly obsolete except in relatively small buildings with only a few zones and short duct runs.

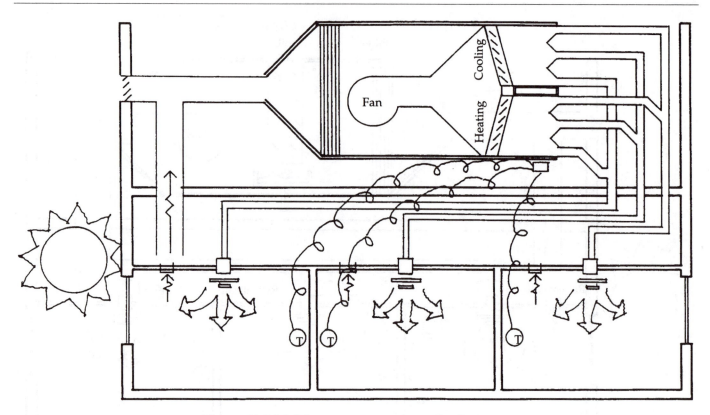

Figure 10.7 Multizone system schematic diagram.

Dual Duct Systems

Dual duct systems are similar to the multizone concept in that both cool supply air and warm supply air are produced by a central AHU. But instead of blending the air in the fan room, separate hot-air ducts and cold-air ducts run parallel throughout the distribution network and air is mixed at terminal mixing boxes in each zone (Figure 10.8). The mixing boxes may include an outlet for delivering air directly to the space, or a duct may connect a branch network with air mixed to a common requirement.

Dual duct systems require the greatest amount of space for distribution ductwork. In order to offset the spatial limitations imposed by this problem, dual duct systems often employ high velocity/high pressure supply ducts, which reduce the size (and cost) of ductwork, as well as the required floor-to-floor height. However this option increases the fan energy required for distribution. Their use is usually limited to buildings with very strict requirements for temperature and or humidity control.

Constant Volume Dual Duct - For a long time, the only variation of the dual duct system was a CAV system, which functioned very much like the multizone system. This system exhibits the greatest energy consumption of any all-air system. In addition to the energy required to mix conditioned air even when no heating or cooling is required, it requires a great amount of distribution energy even when normal pressure and low air velocities are used. For these reasons it has become nearly obsolete, being replaced with dual duct VAV or other systems.

Dual Duct VAV - Although the dual duct VAV system looks very much like its CAV counterpart, it is far more efficient. Instead of providing a constant volume of supply air at all times, the primary method of responding to thermostatic requirements is through adjusting the volume of either cool or warm supply air.

The properly designed dual duct VAV system functions essentially as two single duct VAV systems operating side by side; one for heating and one for cooling. Except when humidity control is required it is usually possible to provide comfort at all temperatures without actually mixing the two air streams. Even when humidity adjustment is required a good control system can minimize the amount of air mixing required.

The dual-duct VAV system still requires more distribution energy and space than most other systems. The level of indirect control which is necessary to produce heated and cooled air also increases energy consumption. Consequently its use should be restricted to applications which benefit from its ability to provide exceptional temperature and humidity control and which do not require a constant supply of ventilation air.

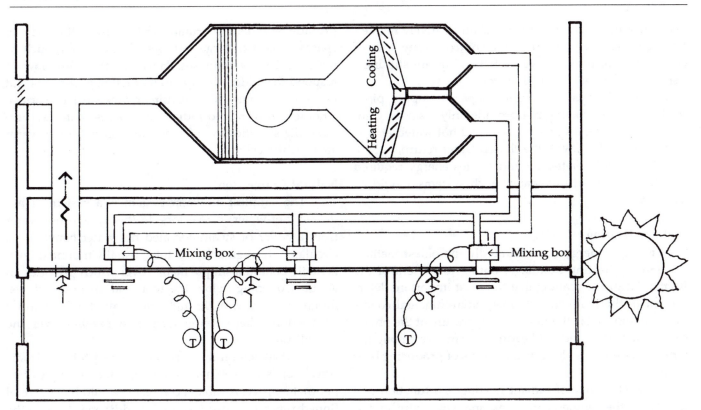

Figure 10.8 Dual duct system schematic diagram.

10.4.2 All-Water Systems

Air is not a convenient medium for transporting heat. A cubic foot of air weighs only about 0.074 pounds (0.34 kg) at standard conditions (70°F; 1 atm.). With a specific heat of about 0.24 Btu/lb°F (0.14 joule/C), one cubic foot can carry less than 0.02 Btu per degree Fahrenheit temperature difference. By comparison, a cubic foot of water weighs 62.4 pounds and can carry 62.4 Btu/ft^3.

Water can be used for transporting heat energy in both heating and cooling systems. It can be heated in a boiler to a temperature of 160 to 250°F (70-120°C) or cooled by a chiller to 40 to 50°F (4-10°C), and piped throughout a building to terminal devices which take in or extract heat energy typically through finned coils.

Steam can also be used to transport heat energy. Steam provides most of its energy by releasing the latent heat of vaporization (about 1000 Btu/lb or 2.3 joules/kg). Thus one mass unit of steam provides as much heating as fifty units of water which undergo a 20°F (11°C) temperature change. However, when water vaporizes, it expands in volume more than 1600 times. Consequently liquid water actually carries more energy per cubic foot than steam and therefore requires the least space for piping.

All-water distribution systems provide flexible zoning for comfort heating and cooling and have a relatively low installed cost when compared to all-air systems. The minimal space required for distribution piping makes them an excellent choice for retrofit installation in existing buildings or in buildings with significant spatial constraints. The disadvantage to these systems is that since no ventilation air is supplied, all-water distribution systems provide little or no control over air quality or humidity and cannot avail themselves of some of the energy conservation approaches of all-air systems.

Water distribution piping systems are described in terms of the number of pipes which are attached to each terminal device:

One-pipe systems use the least piping by connecting all of the terminal units in a series loop. Since the water passes through each terminal in the system, its ability to heat or cool becomes progressively less at great distances from the boiler or chiller. Thermal control is poor and system efficiency is low.

Two-pipe systems provide a supply pipe and a return pipe to each terminal unit, connected in parallel so that each unit (zone) can draw from the supply as needed. Efficiency and thermal control are both high, but the system cannot provide heating in one zone while cooling another.

Four-pipe systems provide a supply and return pipe for both hot water and chilled water, allowing simultaneous heating and cooling along with relatively high efficiency and excellent thermal control. They are, of course, the most expensive to install, but are still inexpensive compared to all-air systems.

Three-pipe systems employ separate supply pipes for heating and cooling but provide only a single, common return pipe. Mixing the returned hot water, at perhaps 140°F (60°C), with the chilled water return, at 55°F (13°C), is highly inefficient and wastes energy required to reheat or recool this water. Such systems should be avoided.

Radiant Heating

Radiant energy is undoubtedly the oldest method of centrally heating buildings, dating to the era of the Roman Empire. Recalling that the most important determinant of thermal comfort when environmental conditions are too cool is the radiant temperature of the physical surroundings, radiant heating systems are among the most economical, so long as the means of producing heat is efficient.

The efficiency of radiant heating is a function primarily of the temperature, area and emissivity of the heat source and the distance between the radiant source and the observer. It is therefore essential that radiant heat sources be located so that they are not obstructed by other objects. Emissivity is an object's ability to absorb and emit thermal radiation, and is primarily related to color. Dark objects absorb and emit radiation better than light colored objects.

There are three categories of radiant heating devices, classified according to the temperature of the source of heat. All may employ electric resistance heating elements, but are energy-efficient only if they employ combustion as a heat source.

Low temperature radiant floors employ the entire floor area as a radiating surface by embedding hot water coils in the floor. The water temperature is typically less than 120°F (50°C). By distributing the heat energy uniformly though the floor, surface temperature is normally below 100°F (40°C).

By increasing the temperature of the radiant surface its area can be reduced. In medium temperature radiant panels, hot water circulates through metal panels, heating them to a temperature of about 140°F (60°C). Consequently the panels must be located out-of-reach, usually on the ceiling or on upper walls.

High temperature *infrared* heaters are typically gas-fired or oil-fired and are discussed below under packaged systems.

Because they are not dependent upon maintaining a static room air temperature, radiant heating systems provide excellent thermal comfort and efficiency in spaces subject to large influxes of outside air, such as factories and warehouses. However they are slow to respond to sudden changes in thermal requirements and malfunctions may be difficult or awkward to correct. Another drawback to radiant systems is that they promote the stratification of room air, concentrating warm air near the ceiling.

Natural Convection

The simplest all-water system is a system of hydronic (hot-water) convectors. In this system hot water from a boiler or steam-operated hot water converter is circulated through a finned tube, usually mounted horizontally behind a simple metal cover which provides an air inlet opening below the tube and an outlet above. Room air is drawn through the convector by natural convection where it is warmed in passing over the finned tube.

A variation on the horizontal finned-tube hydronic convector is the cabinet convector, which occupies less perimeter space. A cabinet convector would have several finned tubes in order to transfer additional heat to the air passing through it. When this is still insufficient a small electric fan can be added, converting the convector to a *unit heater*. Although an electric resistance element can be used in place of the finned tube, the inefficiency of electric resistance heating should eliminate this option.

Hydronic convectors are among the least expensive heating systems to operate as well as to install. Their use is limited, however, to heating only and they do not provide ventilation, air filtration, nor humidity control.

Hydronic convectors and unit heaters may be used alone in buildings where cooling and mechanical ventilation is not required or to provide heating of perimeter spaces in combination with an all-air cooling system. They are the most suitable type of system for providing heat to control condensation on large expanses of glass on exterior wall systems.

Fan-coils

A fan-coil terminal is essentially a small air-handling unit which serves a single space without a ducted distribution system. One or more independent terminals are typically located in each room connected to a supply of hot and/or chilled water. At each terminal, a fan in the unit draws room air (sometimes mixed with outside air) through a filter and blows it across a coil of hot water or chilled water and back into the room. Condensate which forms on the cooling coil must be collected in a drip pan and removed by a drain (Figure 10.9).

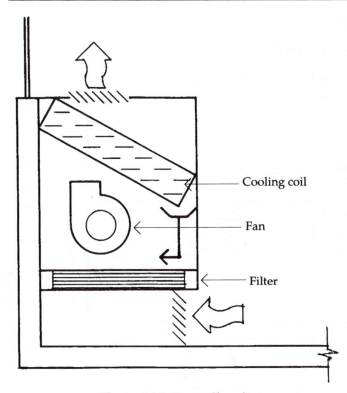

Figure 10.9 Fan-coil unit.

Although most fan-coil units are located beneath windows on exterior walls, they may also be mounted horizontally at the ceiling, particularly for installations where cooling is the primary concern.

Technically, a fan-coil unit with an outside air inlet is called a *unit ventilator*. Unit ventilators provide the capability of using cool outside air during cold weather to provide free cooling when internal loads exceed the heat lost through the building envelope. See the discussion of economizers, §10.5.5.

Fan-coil units and unit ventilators are directly controlled by local thermostats, often located within the unit, making this system one of the most energy efficient. Drawbacks to their use is a lack of humidity control and the fact that all maintenance must occur within the occupied space.

Fan-coil units are typically used in buildings which have many zones located primarily along exterior walls, such as schools, hotels, apartments and office buildings. They are also an excellent choice for retrofitting air-conditioning into buildings with low floor-to-floor heights. Although a four-pipe fan-coil system can be used for a thermally massive building with high internal loads, it suffers the drawback that the cooling of interior zones in warm weather must be carried out through active air-conditioning, since there is no supply of fresh (cool) outside air to provide free cooling. They are also utilized to control the space temperature in laboratories where con-

stant temperature make-up air is supplied to all spaces.

Closed-Loop Heat Pumps

Individual heat pumps (§10.4.4) have a number of drawbacks in nonresidential buildings. However, closed-loop heat pumps, more accurately called water-to-air heat pumps, offer an efficient option for heating and cooling large buildings. Each room or zone contains a water-source heat pump which can provide heating or cooling, along with air filtration and the dehumidification associated with forced-air air-conditioning.

The water source for all of the heat pumps in the building circulates in a closed piping loop, connected to a cooling tower for summer cooling and a boiler for winter heating. Control valves allow the water to bypass either or both of these elements when they are not needed (Figure 10.10). The primary energy benefit of closed-loop heat pumps is that heat removed from overheated interior spaces is used to provide heat for underheated perimeter spaces during cold weather.

Since the closed-loop heat pump system is an all-water, piped system, distribution energy is low, and since direct, local control is used in each zone, control energy is also minimized, making this system one of the most efficient. Although the typical lack of a fresh-air supply eliminates the potential for an economizer cycle, the heat recovery potential discussed above more than makes up for this drawback.

Heat pump systems are expensive to install and maintenance costs are also high. Careful economic analysis is necessary to be sure that the energy savings will be great enough to offset the added installation and maintenance costs. Closed-loop heat pumps are most applicable to buildings such as hotels which exhibit a wide variety of thermal requirements along with simultaneous heating requirements in perimeter zones and large internal loads or chronically overheated areas such as kitchens and assembly spaces.

10.4.3 Air & Water systems - Induction

Once commonly used in large buildings, induction systems employ terminal units installed at the exterior perimeter of the building, usually under windows. A small amount of fresh outside ventilation air is filtered, heated or cooled, and humidified or dehumidified by a central AHU and distributed throughout the building at high-velocity by small ducts.

In each terminal unit, this *primary air* is discharged in such a way that it draws in a much larger volume of *secondary air* from the room, which is filtered and passed through a coil for additional heating or cooling (Figure

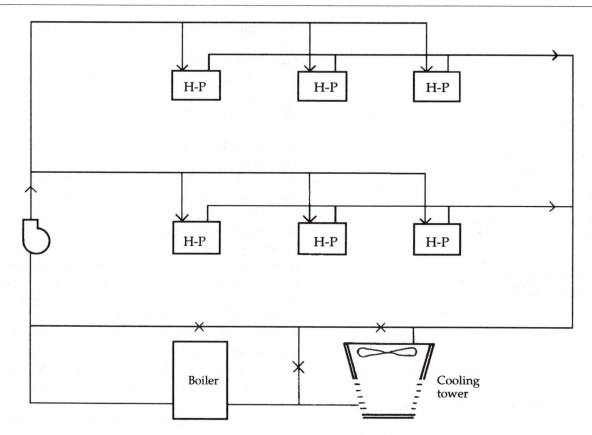

Figure 10.10 Closed-loop heat pump system schematic diagram.

10.11). The use of primary air as the motive force eliminates the need for a fan in the induction unit. The cooling coil is often deliberately kept at a temperature greater than the dew point temperature of the room air which passes through it, eliminating the need for a condensate drain. Although the standard air-water induction system is a cooling-only system, room terminals can employ reheat coils to heat perimeter zones.

Despite the high pressures and velocities required for the primary air distribution, distribution energy is minimized by the relatively small volume of primary air. But the energy saved in primary air distribution is more than offset by the energy consumed in the indirect control and distribution of cooling water, making air-water induction systems among the least energy efficient.

Air-water induction units tend to be noisy and the system provides negligible control of humidity. The applicability of these systems is limited to buildings with widely varying cooling or heating loads where humidity control is not necessary, such as office buildings. Concerns about indoor air quality limits their use as well.

10.4.4 Packaged systems

All of the systems described above may be classified as *central* air-conditioning systems in that they contain certain central elements, typically including a boiler, chiller and cooling tower. Many large buildings provide heating and cooling with distributed systems of unitary or packaged systems, where each package is a stand-alone system which provides all of the heating and cooling requirements for the area of the building which it serves. Individual units derive their energy from raw energy sources typically limited to electricity and natural gas.

Since large pieces of equipment usually have higher efficiencies than smaller equipment, it might be thought that packaged systems are inherently inefficient when compared to central air-conditioning systems. Yet almost all packaged systems actually use much less energy. There are several reasons for this.

First, there is much less energy used in distribution. Fans are much smaller and pumps are essentially non-existent. In addition, control of the smaller packaged units is local and direct. Typically, the unit is either on or off, which can be a disadvantage when the space use requires that ventilation air not be turned off. However there are some advantages associated with this control flexibility. It allows individual thermal control and accurate metering of use. In addition, if equipment failure occurs it does not affect the entire building.

A third reason for the energy efficiency of pack-

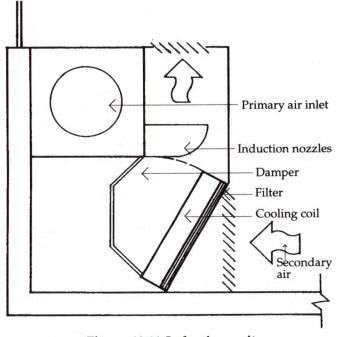

Figure 10.11 Induction unit.

Primary air inlet
Induction nozzles
Damper
Filter
Cooling coil
Secondary air

aged systems involves the schedule of operation. While large equipment is more efficient overall, it only operates at this peak efficiency when it is running at full-load. Small packaged units, due to their on/off operation, run at full load or not at all. In a central air-conditioning system, the central equipment must run whenever any zone requires heating or cooling, often far from its peak load, optimum efficiency conditions.

A secondary advantage to the use of packaged systems is the advantage of diversity. The design of a large central air-conditioning system sometimes requires that a compromise be made between the ideal type of system for one part of a building and a different type of system for another. When packaged systems are employed, parts of a building with significantly different heating and cooling requirements can be served by different types of equipment. This will always provide improved thermal comfort, and often results in improved efficiency as well.

Packaged Terminal Air-Conditioners

The most common type of packaged equipment is the packaged terminal air-conditioner, often called a PTAC or incremental unit, due to the fact that increases in equipment can be made incrementally. Examples of PTAC's are through-the-wall air-conditioners and single-zone rooftop equipment. Their use is limited to about 500 square feet per unit.

Individual air-to-air (air-source) heat pumps can also be installed as a packaged system. A heat pump is essentially a vapor-compression air-conditioner which can be reversed to extract heat from the outdoor environment and discharge it into the occupied space. A significant drawback to air-source heat pumps is that vapor compression refrigeration becomes inefficient when the evaporator is forced to extract heat from a source whose temperature is 30°F (0°C) or below.

In large systems, heat pumps can utilize a source of circulating water from which to extract heat during cold weather, so that the evaporator temperature never approaches 30°F (0°C). The circulating water would be heated in the coldest weather, and could be cooled by a cooling tower to receive rejected heat during warm weather. These closed-loop heat pumps are discussed under all-water systems above.

Unit Heaters

Packaged heating-only units typically utilize electricity or natural gas as their primary source of energy. As discussed in a later section, electricity is the most expensive source of heat energy and should be avoided. However, natural gas (or liquefied propane) provides an economical source of heat when used in packaged unit heaters.

Fan-forced unit heaters can disperse heat over a much larger area than packaged air-conditioners. They can distribute heat either vertically or horizontally and respond rapidly to changes in heating requirements.

High temperature infrared radiant heaters utilize a gas flame to produce a high-temperature (over 500°F, 260°C) source of radiant energy. Although they do not respond rapidly to changes in heating requirements, they are essentially immune to massive intrusion of cold outside air. Because they warm room surfaces and physical objects in the space, thermal comfort returns within minutes of an influx of cold air.

HVAC systems may be central or distributed; all-air, all-water, or air-water (induction). Each system type has advantages and disadvantages, not the least important of which is its energy efficiency. An economic analysis should be conducted in selecting an HVAC system type and in evaluating changes in HVAC systems in response to energy concerns.

10.5 ENERGY CONSERVATION OPPORTUNITIES

The ultimate objective of any energy management program is the identification of energy conservation opportunities (ECO's) which can be implemented to produce a cost saving. However, it is important to recognize that the fundamental purpose of an HVAC system is to provide human thermal comfort, or the equivalent environmental conditions for some specific process. It is therefore necessary to examine each ECO in the context of its effect on indoor air quality, humidity and thermal comfort standards, air velocity and ventilation requirements, and requirements for air pressurization.

10.5.1 Thermal Comfort, Air Quality and Airflow

It is not wise to undertake modifications of an HVAC system to improve energy efficiency without considering the effects on thermal comfort, air quality, and airflow requirements.

Thermal Comfort

One of the most serious errors which can be made in modifying HVAC systems is to equate a change in dry bulb air temperature with energy conservation. It is worthwhile to recall that dry bulb air temperature is not the most significant determinant of thermal comfort during either the heating season or cooling season.

Since *thermal comfort in the cooling season is most directly influenced by air motion* cooling energy requirements can be reduced by increasing airflow and/or air motion in occupied spaces without decreasing dry bulb air temperature. During the *heating season thermal comfort is most strongly influenced by radiant heating.* Consequently, changes from forced-air heating to radiant heating can improve thermal comfort while decreasing heat energy requirements.

The total energy, or enthalpy, associated with a change in environmental conditions includes both sensible heat and latent heat. *Sensible* heat is the heat energy required to increase dry bulb temperature. The heat energy associated with a change in moisture content of air is known as *latent* heat. (See Figure 10.12)

Changes in HVAC system design or operation which reduce sensible heating or cooling requirements may increase latent heating or cooling energy which offsets any energy conservation advantage. This is particularly true of economizer cycles (§10.5.5). The use of cool, but humid outside air in an economizer can actually increase energy consumption if dehumidification requirements are increased. For this reason, the only reliable type of economizer control is enthalpy control which prevents the economizer from operating when latent cooling requirements exceed the savings in sensible cooling.

Air velocity and airflow

The evaluation of energy conservation opportunities often neglects previously established requirements for air velocity and airflow. As discussed above, the volume of air supplied and its velocity have a profound influence on human thermal comfort. ECO's which reduce airflow can inadvertently *decrease* thermal comfort. Even more important, airflow cannot be reduced below the volume of outdoor air required by codes for ventilation.

The design of an all-air or air-water HVAC system is much more complex that just providing a supply air duct and thermostat for each space. The completed system must be *balanced* to assure adequate airflow to each space, not only to offset the thermal loads, but also to provide the appropriate *pressurization* of the space.

It is a common practice for supply air to exceed return air in selected spaces to create positive pressurization, which minimizes infiltration and prevents the intrusion of odors and other contaminants from adjacent spaces. Similarly, negative pressurization can be achieved by designing exhaust or return airflow to exceed supply air requirements in order to maintain a sterile field or to force contaminants to be exhausted. Any alterations in air supply or return requirements upset the relationship between supply and return airflows requiring that the system be rebalanced.

Indoor Air Quality

Another factor which is often neglected in the application of ECO's is the effect of system changes on indoor air quality. Indoor air quality requirements are most commonly met with ventilation and filtration provided by an all-air or air-water HVAC system. When outdoor air quantities are altered significantly, the effect

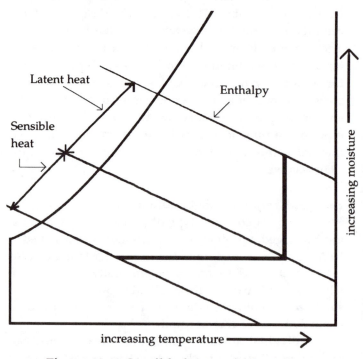

Figure 10.12 Sensible heat and latent heat.

on indoor air quality is unknown and must be determined.

In a polluted environment, increasing outdoor air volume, for example with an economizer cycle, either increases filtration requirements or results in a deterioration of indoor air quality. On the other hand, reducing airflow to conditioned spaces due, for example, to a conversion to a variable air volume system, reduces the filtration of recycled indoor air which can likewise produce a reduction in air quality. The concerns regarding indoor air quality are discussed in more detail in Chapter 17.

10.5.2 Maintenance

The first place to begin in the investigation of any existing HVAC system is to review the condition and maintenance of the system. Inadequate or improper maintenance is a major cause of system inefficiency. The system may have been designed to be energy-efficient and some ECO's may have been previously implemented, but due to a lack of maintenance the system no longer functions in an energy-efficient mode.

A more comprehensive review of maintenance procedures is included in Chapter 14, particularly with regard to boiler maintenance. What follows here a brief summary of some of the things to watch for in inspecting the system for inadequate maintenance.

The control system (see §10.5.6) should be performing its function. Thermostats should be properly calibrated. Time clocks, setback devices, and automatic controls should be intact and operating. Bearings and belts should be in good shape and belts should be tight. Bearings and other moving parts should be properly lubricated with the appropriate lubricant. Lubricant selection can actually have a significant effect on system performance and efficiency.

Dirt build-up should be periodically removed especially from heat transfer surfaces, particularly the condenser coils of air-cooled condensers. Filters should be cleaned or replaced regularly rather than waiting for them to appear so dirty as to need cleaning. Dirty filters increase fan horsepower while reducing airflow. If routine and preventative maintenance has not been performed on the system, complete the required maintenance before proceeding to determine other applicable ECO's.

10.5.3 Demand Management

Demand management (see also Chapter 11) involves shutting off or deferring the operation of equipment to minimize the peak electrical load which occurs at a given time. This strategy can be extended to shutting off or reducing HVAC equipment requirements when they are not needed. The most common demand management strategies for HVAC systems are scheduled operation and night set-back.

Operating Schedule

One of the greatest causes of energy waste is unnecessary operation. The most energy-efficient equipment will consume excess energy if it runs when it is not needed. Unnecessary operation tends to be caused by large systems which condition entire buildings or sections of buildings and a lack of control to turn off equipment when it is not needed.

When buildings are conditioned by large central systems, off-hours use of one area may require the operation of the entire system to condition a single space. Sometimes the installation of a local packaged system serving a specialized area can prevent the unnecessary operation of a larger system. For example, a radio station located in a large office building installed a packaged air-conditioner to provide cooling for its studios which operated 24 hours a day while the offices remained on the building system which operated only from 7 a.m. until 5 p.m. Despite the fact that the packaged system was less efficient than the central building system, the energy saved was substantial.

Exhaust fans, dust collectors, and other small equipment which serve specific rooms are often left running continuously due solely to the lack of local control to turn them off. Local switching alone is seldom adequate to save energy since it is difficult to get occupants to turn equipment off and sometimes to turn it on. Passive switching or other automatic devices can eliminate the need to run equipment continuously.

For example an exhaust fan for a seldom-occupied store room can be controlled to turn on with an occupancy sensor or when the lights are turned on and to remain on for a set period of time after the lights are turned off or occupants leave the room. Spring-wound interval timers are useful for controlling small devices which are turned on manually and then left running. Exhaust fans can be interlocked with the central air-conditioning system to shut down when the central system is turned off.

Turning off equipment, particularly central air-conditioning systems, in buildings which are unoccupied at night is another obvious energy saver. Simple time clocks can be used where more sophisticated computer controlled systems are not used. Even when it is necessary to keep the system operating to provide minimal cooling during the night or on weekends, exhaust fans and fresh-air ventilation can be shut down when the building is essentially unoccupied.

Some equipment loads are *deferrable* in that they can be turned off for a short interval and then turned back on with no appreciable impact on operations. Water heating and cooling are two examples, but even space heating can be deferred for periods of up to thirty minutes with no perceptible effect on thermal comfort. During extreme weather outside air can be reduced for short periods in order to improve heating or cooling efficiency. See the discussion of warm-up and cool-down cycles which follows.

Computerized controls can be used to prevent large pieces of equipment from operating simultaneously which increases peak demand and the expense which accompanies increased demand. This is particularly attractive in large systems of distributed packaged equipment.

With adequate storage both heat and "cool" can be stored in order to take advantage of off-peak utility rates and to reduce instantaneous heating or cooling requirements during peak periods. See the discussion of thermal storage in section 10.6.4 and in chapter 19.

Night Set-back

The principle of night set-back is to reduce the amount of conditioning provided at night by allowing the interior temperature to drift naturally to a marginal temperature during the night and then to recondition it to normal conditions in the morning. With the possible exception of electric heat pumps, night set-back will always save energy. The energy saved can be computed from the equation:

$$E = [(A_{per} \times U) + q_{vent}] \Delta T \times hr \qquad (10.1)$$

where

E	=	energy saved, Btu/yr
A_{per}	=	surface area of perimeter envelope, ft^2
U	=	effective U-value of thermal envelope, Btu/hr ft^2°F
q_{vent}	=	ventilation load, Btu/hr ft^2
ΔT	=	night setback temperature difference, degrees F
hr	=	heating season unoccupied hours

In a temperate climate this can amount to a saving of 40 percent of the heating energy for a building with little thermal mass which is occupied only ten hours per day. However the amount of energy saved will not always be large and local utility rate structures may be such that the actual cost of the energy saved is less than the cost of the energy to recondition the building in the morning.

For example, many electric utility companies offer time-of-day rates in which energy used during the night is charged at a rate much lower than that charged for electricity consumed during the day. If an air-conditioned building allowed to heat up during the night must be recooled at the day rate in the morning, night set-back could end up costing more than cooling the building all night.

Similarly, the more energy-efficient the building is the smaller the energy saving will be from night set-back and the less attractive this option will be from a financial standpoint. Night set-back is marginal at best for cooling, and will be only slightly better for heating unless the building has little thermal mass and significant infiltration heat loss.

Where night set-back is a viable option, a secondary question is whether to turn the system off or simply turn the thermostat set-point down (or up). This is not a question which can be answered simply. The answer is dependent on the length of time the system will be set-back and how much temperature variation will occur if the system is turned off. The question of whether it is even advisable to turn the system off is also influenced by humidification and pressurization requirements, air quality concerns, etc. See §10.5.1.

Warm-up and Cool-down

When a building is operated with night set-back, it is not necessary to condition the building until the last person leaves in the evening. But the system must be restarted in the morning so that optimum conditions are reached before the beginning of the business day. It is not difficult to determine an appropriate time to let the system begin to drift in the afternoon, but morning warm-up (or cool-down) can require substantial time and energy.

One of the best approaches to re-conditioning a building after night set-back is with the use of an *optimum-time start* device. This microprocessor-based thermostat compares the outside and inside temperatures along with the desired set-point during the operating cycle. It determines how long it will take to re-condition the building to the set-point based on previous data and turns the system on at the appropriate time to reach the set-point temperature just in time. Since time clocks must be set to turn the system on with adequate time to re-condition the building on even the coldest morning, the optimum-time start device can save more than half of the warm-up time on days when the outdoor ambient temperature is moderate.

A second method of saving energy during warm-up or cool down is with a *warm-up/cool-down cycle*. During the warm-up/cool-down cycle the system recirculates building return air until a temperature within one or two degrees of the set-point is reached, saving the energy which would be required to heat or cool outside ventilation air. Since outside ventilation air is provided primarily to meet human fresh-air needs there is no harm in reducing or eliminating it during warm-up or cool-down. However modern building codes prohibit eliminating all ventilation air unless the building is truly unoccupied.

Warm-up/cool-down cycles provide a secondary advantage in times of extreme weather. Should the heating and cooling equipment be unable to maintain the set-point temperature due to a lack of capacity to condition the ventilation air because of its extreme temperature, the warm-up/cool-down cycle could close the outside-air damper and reduce the volume of outdoor air until interior conditions were restored to near the set-point. Since the system would run continuously in either case, this procedure technically doesn't save energy but it does maintain thermal comfort without expending additional energy.

10.5.4 Heat Recovery

Heat recovery systems are used during the heating season to extract waste heat and humidity from exhaust air which is used to preheat cold fresh air from outside. In warm weather, they may be used to extract heat and humidity from warm fresh air and dispel it into the exhaust air stream. Their use is particularly appropriate to buildings with high outside air requirements. Heat recovery systems are discussed in more detail in Chapter 8.

One special type of heat recovery, dubbed bootstrap heating, recovers waste heat from warm condenser water to provide nearly free heat to meet perimeter heating needs. This is most applicable to buildings in which interior zone cooling exceeds the requirements for perimeter heating. Care must be taken in utilizing bootstrap heating to avoid reducing the condenser water temperature below the chiller equipment's entering water effective temperature.

10.5.5 Economizers

One advantage of all-air HVAC systems is that they can utilize outside air to condition interior spaces when it is at an appropriate temperature. *The use of outside air to actively cool interior spaces is referred to as an economizer, or economizer cycle.*

Whenever the outside air is cooler than the cooling set-point temperature only distribution energy is required to provide cooling with outside air. When the outside air temperature is above the set-point temperature but less than the return air temperature, it requires less cooling energy to utilize 100 percent outdoor air for supply air than to condition recycled indoor air. An economizer cycle is simply a control sequence that adjusts outside air and exhaust dampers to utilize 100 percent outside air when its temperature makes it advantageous to do so.

The energy consumed in heating outside air may be computed from the equation

$$q = 1.08 \ Q \cdot \Delta T \qquad (10.2a)$$

where

q = hourly ventilation heating load, Btu/hr
Q = volumetric flow rate of air, cfm
ΔT = temperature increase, °F

The economizer cycle is most appropriate for thermally massive buildings which have high internal loads and require cooling in interior zones year round. It is ineffective in thermally light buildings and buildings whose heating and cooling loads are dominated by thermal transmission through the envelope. Economizer cycles will provide the greatest benefit in climates having more than 2000 heating degree days per year, since warmer climates will have few days cold enough to permit the use of outside air for cooling.

In theory 100 percent outside air would be utilized for cooling whenever the outside air temperature is below the return air temperature. But this assumption disregards the fact that outside air typically has a higher relative humidity than conditioned indoor air and contains significant *latent heat*. Consequently, the economizer is seldom activated at this temperature.

The simplest type of economizer utilizes a *dry-bulb temperature control* which activates the economizer at a predetermined outside dry-bulb temperature, usually the supply air temperature, or about 55°F (13°C). Above 55°F (13°C), minimum outdoor air is supplied for ventilation. Below 55°F (13°C), the quantity of outdoor air is gradually reduced from 100 percent and blended with return air to make 55°F (13°C) supply air.

Eq. 10.2a predicts that the energy required to cool room (return) air 20°F (13°C) from 75°F (24°C) to a supply air temperature of 55°F (13°C) is 1.08 x Q x 20 = 21.6 Q. The energy saved [E, Btu/hr] from the use of a simple economizer which blends outside air with return to provide a supply air temperature is therefore:

$$E = 21.6 \ Q \qquad (10.2b)$$

Because it is possible to utilize outdoor air at temperatures above 55°F (13°C) to save cooling energy, a *modified dry-bulb temperature control* can be used. This is identical to the simpler dry-bulb temperature control except that when the outside temperature is between 55°F (13°C) and a preselected higher temperature based on the typical humidity, 100 percent outdoor air is used, but cooled to 55°F for supply air.

The third and most efficient type of control for an economizer cycle is *enthalpy control*, which can instantaneously determine and compare the amount of energy required to cool 100 percent outdoor air with that required to cool the normal blend of return air. It then selects the source which requires the least energy for cooling.

Air-handling units which lack adequate provision for 100 percent outside air can utilize an alternative "wet-side" economizer. This energy-saving technique is a potential retrofit for existing older buildings. Several variations on the wet-side economizer exist, all of which allow the chiller to be shut down. The simplest water-side economizer is a coil in the air-handling unit through which cooling tower water can be circulated to provide free cooling. An alternative arrangement for large chilled water systems interconnects the chilled water circuit directly with the cooling tower, to allow the chiller to be bypassed. To reduce contamination of the chilled water, the cooling tower water should only be coupled to the chilled water circuit through a heat exchanger (Figure 10.13).

The energy saved [E, kWh] due to the use of a wet-side economizer can be determined from a knowledge of the hours [hr] of chiller operation during when the outside temperature is below 40°F (4°C), the capacity of chiller in tons and the kW/ton rating of the chiller:

$$E = hr \times tons \times kW/ton \qquad (10.3)$$

A word of caution is advised. A wet-side economizer uses condenser water at 40-45°F (4-7°C). But the chiller typically operates with condenser water much higher than this. In some cases, the lowest condenser water temperature at which the chiller will operate is 65°F (18°C). This means that when the chiller is turned off there would be no useful coolant until the cooling tower cooled to 45°F (7°C). This "cool down" period is typically no more than 30 minutes. For additional discussion of the wet-side economizer, see §10.6.6.

10.5.6 Control Strategies

Control systems play a large part in the energy conservation potential of a HVAC system. To be effective at controlling energy use along with thermal comfort they must be used appropriately, work properly and be set correctly. Overheated or overcooled spaces not only waste energy, they are uncomfortable. An easy way to spot such areas is to walk around the outside of the building and look for open windows which usually indicate overheated (or overcooled) spaces.

The efficiency of control systems is mostly related to the *directness of control*. Simple controls which turn the system off and on when needed provide the most direct control. Systems in which air and/or water temperatures are controlled by parameters other than the actual need for heating and cooling may be said to be indirectly controlled, and are inherently less efficient.

Direct control systems most commonly employ primarily thermostats which turn equipment on or off or adjust the volume of air supplied to a space on the basis of room temperature. Direct thermostat control is most effective in thermally light buildings whose energy requirements are directly proportional to the exterior weather conditions. Where thermostats are an appropriate choice, individual zones must be used for spaces with different heating or cooling requirements.

Many older HVAC systems, particularly heating-only systems, were created as single-zone systems for which the HVAC system is either on throughout the building or entirely off. Self-contained thermostatic valves can be used to modify single-zone convection heating systems and single-zone forced air systems can be readily modified into zoned VAV or reheat air-conditioning systems without replacing the costly central equipment.

The selection of set points, even for directly-controlled systems, is important. Thermostats should be set to maximize thermal comfort rather than relying on previously established design values. The need for humidification varies with outdoor temperature and humidistats should not be set at constant levels and forgotten. The need for humidification is also a function of occupancy and should be adjusted in assembly spaces when they are unused or underutilized. The use of outside air for ventilation when this air must be conditioned likewise can be reduced with proper controls to adjust outside-air dampers. Modern microprocessor-based energy management control systems (EMCS) allow for vastly improved direct control of HVAC systems.

A useful concept made possible by modern electronics is the use of *enthalpy control*. Enthalpy control relates the energy required to cool air at a given dry-bulb temperature to the energy required to remove humidity from moist outdoor air. Enthalpy controls are most commonly used to adjust economizer cycles, determining when it is

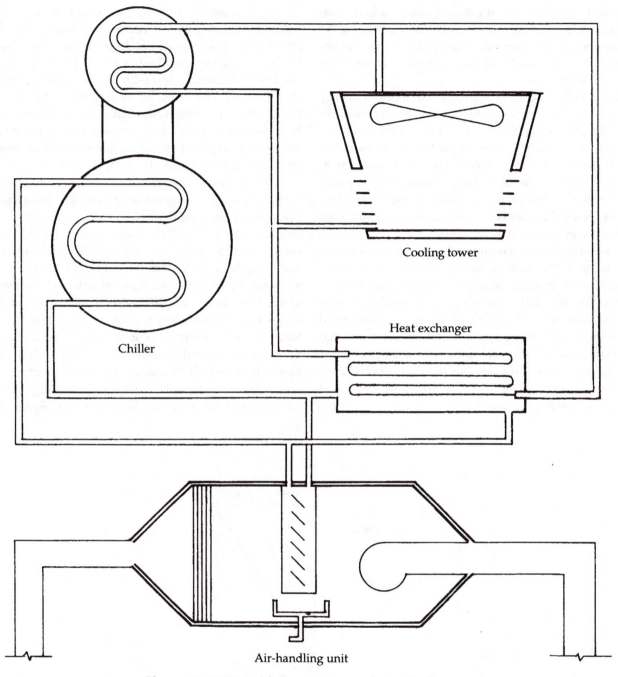

Figure 10.13 Wet-side economizer schematic diagram.

more efficient to recycle warm indoor air than to condition humid outside air for "free" cooling.

Control systems which are not well-understood or simple to use are subject to misuse by occupants or building service personnel. For example the controls on a unit ventilator include a room temperature thermostat which controls the valve on the heating or cooling coil, a damper control which adjusts the proportion of fresh air mixed with recirculated room air, and a low-limit thermostat which prevents the temperature of outside air from dropping below a preset temperature (usually 55 to 60°F; 13 to 16°C). A common error of occupants or building custodians in response to a sense that the air supplied by the unit ventilator is too cold is to increase the set-point on the low-limit thermostat, which prevents free cooling from outside air or, on systems without a cooling coil, prevents cooling altogether. Controls which are subject to misadjustment by building occupants should be placed so that they cannot be tampered with.

The energy consumption of thermally-massive buildings is less related to either the inside or outside air temperature. Both the heating and cooling loads in thermally-massive buildings are heavily dependent on the heat generated from internal loads and the thermal energy stored in the building mass which may be dissipated at a later time.

In an indirect control system the amount of energy consumed is not a function of human thermal comfort needs, but of other factors such as outdoor temperature, humidity, or enthalpy. Indirect control systems determine the set points for cool air temperature, water temperatures, etc. As a result indirect control systems tend to adjust themselves for peak conditions rather than actual conditions. This leads to overheating or overcooling of spaces with less than peak loads.

One of the most serious threats to the efficiency of any system is the need to heat and cool air or water simultaneously in order to achieve the thermal balance required for adequate conditioning of spaces. Figure 10.14 indicates that 20 percent of the energy consumed in a commercial building might be used to reheat cooled air, offsetting another 6 percent that was used to cool the air which was later reheated. For the example building the energy used to cool reheated air approaches that actually used for space cooling.

Following the 1973 oil embargo federal guidelines encouraged everyone to reduce thermostat settings to 68°F (20°C) in winter and to increase thermostat settings in air-conditioned buildings to 78°F (26°C) in summer. [In 1979, the winter guideline was reduced farther to 65°F (18°C).] The effect of raising the air-conditioning thermostat on a reheat, dual-duct, or multizone system is actually to *increase* energy consumption by increasing the energy required to reheat air which has been mechanically cooled (typically to 55°F; 13°C).

To minimize energy consumption on these types of systems it makes more sense to raise the discharge temperature for the cold-deck to that required to cool perimeter areas to 78°F (26°C) under peak conditions. If the system was designed to cool to 75°F (24°C) on a peak day using 55°F (13°C) air, the cold deck discharge could be increased to 58°F (14.5°C) to maintain space temperatures at no more than 78°F (26°C), saving about $5 per cfm per year. Under less-than-peak conditions these systems would operate *more* efficiently if room temperatures were allowed to fall below 78°F (26°C) than to utilize reheated air to maintain this temperature.

A more extensive discussion of energy management control systems may be found in Chapter 12.

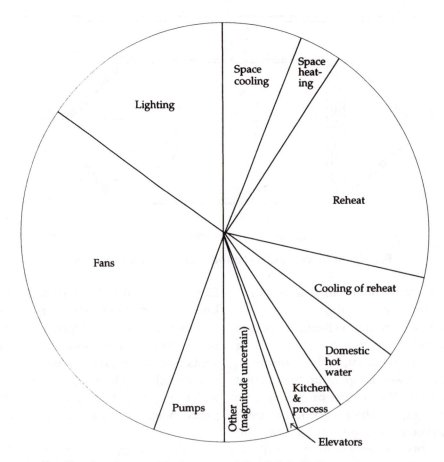

Figure 10.14 Energy cost distribution for a typical non-residential building using an all-air reheat HVAC system.

10.5.7 HVAC Equipment

The elements which provide heating and cooling to a building can be categorized by their intended function. HVAC equipment is typically classified as heating equipment, including boilers, furnaces and unit heaters; cooling equipment, including chillers, cooling towers and air-conditioning equipment; and air distribution elements, primarily air-handling units (AHU's) and fans. A more lengthy discussion of boilers may be found in Chapter 6, followed by a discussion of steam and condensate systems in Chapter 7. Cooling equipment is discussed in section 10.6, below. What follows here relates mostly to air-handling equipment and distribution systems.

Figure 10.14 depicts the typical energy cost distribution for a large commercial building which employs an all-air reheat-type HVAC system. Excluding the energy costs associated with lighting, kitchen and miscellaneous loads which are typically 25-30 percent of the total, the remaining energy can be divided into two major categories: the energy associated with heating and cooling and the energy consumed in distribution. The total energy consumed for HVAC systems is therefore dependent on the efficiency of individual components, the efficiency of distribution and the ability of the control system to accurately regulate the energy consuming components of the system so that energy is not wasted.

The size (and heating, cooling, or air-moving capacity) of HVAC equipment is determined by the mechanical designer based upon a calculation of the peak internal and envelope loads. Since the peak conditions are arbitrary (albeit well-considered and statistically valid) and it is likely that peak loads will not occur simultaneously throughout a large building or complex requiring all equipment to operate at its rated capacity, it is common to specify equipment which has a total capacity slightly less than the peak requirement. This diversity factor varies with the function of the space. For example, a hospital or classroom building will use a higher diversity multiplier than an office building.

In sizing heating equipment however, it is not uncommon to provide a total heating capacity from several units which exceeds the design heating load by as much as fifty percent. In this way it is assured that the heating load can be met at any time, even in the event that one unit fails to operate or is under repair.

The selection of several boilers, chillers, or air-handling units whose capacities combine to provide the required heating and cooling capability instead of single large units allows one or more components of the system to be cycled off when loads are less than the maximum.

This technique also allows off-hours use of specific spaces without conditioning an entire building.

Equipment Efficiency

Efficiency, by definition, is the ratio of the energy output of a piece of equipment to its energy input, in like units to produce a dimensionless ratio. Since no equipment known can produce energy, efficiency will always be a value less than 1.0 (100%).

Heating equipment which utilizes electric resistance appears at first glance to come closest to the ideal of 100 percent efficiency. In fact, every kilowatt of electrical power consumed in a building is ultimately converted to 3414 Btu per hour of heat energy. Since this is a valid unit conversion it can be said that electric resistance heating is 100 percent efficient. What is missing from the analysis however, is the inefficiency of producing electricity, which is most commonly generated using heat energy as a primary energy source.

Electricity generation from heat is typically about 30 percent efficient, meaning that only 30 percent of the heat energy is converted into electricity, the rest being dissipated as heat into the environment. Energy consumed as part of the generation process and energy lost in distribution use up about ten percent of this, leaving only 27 percent of the original energy available for use by the consumer. By comparison, state-of-the-art heating equipment which utilizes natural gas as a fuel is more than eighty percent efficient. Distribution losses in natural gas pipelines account for another 5 percent, making natural gas approximately three times as efficient as a heat energy source than electricity.

The relative efficiency of cooling equipment is usually expressed as a *coefficient of performance* (COP), which is defined as the ratio of the heat energy extracted to the mechanical energy input in like units. Since the heat energy extracted by modern air conditioning far exceeds the mechanical energy input a COP of up to 6 is possible.

Air-conditioning equipment is also commonly rated by its *energy efficiency ratio* (EER) or seasonal energy efficiency ratio (SEER). EER is defined as the ratio of heat energy extracted (in Btu/hr) to the mechanical energy input in watts. Although it should have dimensions of Btu/hr/watt, it is expressed as a dimensionless ratio and is therefore related to COP by the equation

$$EER = 3.41 \cdot COP \qquad (10.4)$$

Although neither COP nor EER is the efficiency of a chiller or air-conditioner, both are measures which allow the comparison of similar units. The term *air-condi-*

tioning efficiency is commonly understood to indicate the extent to which a given air-conditioner performs to its maximum capacity. As discussed below, most equipment does not operate at its peak efficiency all of the time. For this reason, the *seasonal energy efficiency ratio* (SEER), which takes varying efficiency at partial load into account, is a more accurate measure of air-conditioning efficiency than COP or EER.

In general, equipment efficiency is a function of size. Large equipment has a higher efficiency than small equipment of similar design. But the rated efficiency of this equipment does not tell the whole story. Equipment efficiency varies with the load imposed. All equipment operates at its optimum efficiency when operated at or near its design full-load condition. Both overloading and under-loading of equipment reduces equipment efficiency.

This fact has its greatest impact on system efficiency when large systems are designed to air-condition an entire building or a large segment of a major complex. Since air-conditioning loads vary and since the design heating and cooling loads occur only rarely under the most severe weather or occupancy conditions, most of the time the system must operate under-loaded. When selected parts of a building are utilized for off-hours operation this requires that the entire building be conditioned or that the system operate far from its optimum conditions and thus at far less than its optimum efficiency.

Since most heating and cooling equipment operates at less than its full rated load during most of the year, its part-load efficiency is of great concern. Because of this, most state-of-the-art equipment operates much closer to its full-load efficiency than does older equipment. A knowledge of the actual operating efficiency of existing equipment is important in recognizing economic opportunities to reduce energy consumption through equipment replacement.

Distribution Energy

Distribution energy is most commonly electrical energy consumed to operate fans and pumps, with fan energy typically being far greater than pump energy except in all-water distribution systems. The performance of similar fans is related by three fan laws which relate fan power, airflow, pressure and efficiency to fan size, speed and air density. The reader is referred to the 1992 *ASHRAE Handbook: HVAC Systems and Equipment* for additional information on fans and the application of the fan laws.[3]

Fan energy is a function of the quantity of airflow moved by the fan, the distance over which it is moved, and the velocity of the moving air (which influences the pressure required of the fan). Most HVAC systems, whether central or distributed packaged systems, all-air, all-water, or a combination are typically oversized for the thermal loads that actually occur. Thus the fan is constantly required to move more air than necessary, creating inherent system inefficiency.

One application of the third fan law describes the relationship between fan horsepower (energy consumed) and the airflow produced by the fan:

$$W_1 = W_2 \times (Q_1/Q_2)^3 \qquad (10.5)$$

where
W = fan power required, hp
Q = volumetric flow rate, cfm

Because fan horsepower is proportional to the cube of airflow, reducing airflow to 75 percent of existing will result in a reduction in the fan horsepower by the cube of 75 percent, or about 42 percent: $[(0.75)^3 = 0.422]$ Even small increases in airflow result in disproportional increases in fan energy. A ten percent increase in airflow requires 33 percent more horsepower $[1.10^3 = 1.33]$, which suggests that airflow supplied solely for ventilation purposes should be kept to a minimum.

All-air systems which must move air over great distances likewise require disproportionate increases in energy as the second fan law defines the relationship between fan horsepower [W] and pressure [p], which may be considered roughly proportional to the length of ducts connected to the fan:

$$W_1 = W_2 \times (p_1/p_2)^{3/2} \qquad (10.6)$$

The use of supply air at temperatures of less than 55°F (13°C) for primary cooling air permits the use of smaller ducts and fans, reducing space requirements at the same time. This technique requires a complex analysis to determine the economic benefit and is seldom advantageous unless there is an economic benefit associated with space savings.

System Modifications

In examining HVAC systems for energy conservation opportunities, the less efficient a system is, the greater is the potential for significant conservation to be achieved. There are therefore several "off-the-shelf" opportunities for improving the energy efficiency of selected systems.

All-air Systems - Virtually every type of all-air system can benefit from the addition of an economizer cycle, particularly one with enthalpy controls. Systems with substantial outside air requirements can also benefit from heat recovery systems which exchange heat between exhaust air and incoming fresh air. This is a practical retrofit only when the inlet and exhaust ducts are in close proximity to one another.

Single zone systems, which cannot provide sufficient control for varying environmental conditions within the area served can be converted to variable air volume (VAV) systems by adding a VAV terminal and thermostat for each new zone. In addition to improving thermal comfort this will normally produce a substantial saving in energy costs.

VAV systems which utilize fans with inlet vanes to regulate the amount of air supplied can benefit from a change to variable speed or variable frequency fan drives. Fan efficiency drops off rapidly when inlet vanes are used to reduce airflow.

In terminal reheat systems, all air is cooled to the lowest temperature required to overcome the peak cooling load. Modern "discriminating" control systems which compare the temperature requirements in each zone and cool the main airstream only to the temperature required by the zone with the greatest requirements will reduce the energy consumed by these systems. Reheat systems can also be converted to VAV systems which moderate supply air volume instead of supply air temperature, although this is a more expensive alteration than changing controls.

Similarly, dual-duct and multizone systems can benefit from "smart" controls which reduce cooling requirements by increasing supply air temperatures. Hot-deck temperature settings can be controlled so that the temperature of warm supply air is just high enough to meet design heating requirements with 100 percent hot-deck supply air and adjusted down for all other conditions until the hot-deck temperature is at room temperature when outside temperatures exceed 75°F (24°C). Dual duct terminal units can be modified for VAV operation.

An economizer option for multizone systems is the addition of a third "bypass" deck to the multizone air-handling unit. This is not appropriate as a retrofit although an economizer can be utilized to provide cold-deck air as a retrofit.

All-water systems - Wet-side economizers are the most attractive common energy conservation measure appropriate to chilled water systems. Hot-water systems benefit most from the installation of self-contained thermostat valves, to create heating zones in spaces formerly operated as single-zone heating systems.

Air-water Induction - Induction systems are seldom installed anymore but many still exist in older buildings. The energy-efficiency of induction systems can be improved by the substitution of fan-powered VAV terminals to replace the induction terminals.

10.6 COOLING EQUIPMENT

The most common process for producing cooling is vapor-compression refrigeration, which essentially moves heat from a controlled environment to a warmer, uncontrolled environment through the evaporation of a refrigerant which is driven through the refrigeration cycle by a compressor.

Vapor compression refrigeration machines are typically classified according to the method of operation

Table 10.1 Summary of HVAC System Modifications for Energy Conservation

System type	Energy Conservation Opportunities
All-air systems (general):	economizer
	heat recovery
Single zone systems	conversion to VAV
Variable air volume (VAV) systems	replace fan inlet vane control with variable frequency drive fan
Reheat systems	use of discriminating control systems
	conversion to VAV
Constant volume dual-duct systems	use of discriminating control systems
	conversion to dual duct VAV
Multizone systems	use of discriminating control systems
	addition of by-pass deck*
All-water systems:	
hydronic heating systems	addition of thermostatic valves
chilled water systems	wet-side economizer
Air-water induction systems	replacement with fan-powered VAV terminals

*Requires replacement of air-handling unit

of the compressor. Small air-to-air units most commonly employ a reciprocating compressor which is combined with an air-cooled condenser to form a condensing unit. This is used in conjunction with a direct-expansion (DX) evaporator coil placed within the air-handling unit.

Cooling systems for large non-residential buildings typically employ chilled water as the medium which transfers heat from occupied spaces to the outdoors through the use of chillers and cooling towers.

10.6.1 Chillers

The most common type of water chiller for large buildings is the centrifugal chiller which employs a centrifugal compressor to compress the refrigerant, which extracts heat from a closed loop of water which is pumped through coils in air-handling or terminal units within the building. Heat is rejected from the condenser into a second water loop and ultimately rejected to the environment by a cooling tower.

The operating fluid used in these chillers may be either a CFC or HCFC type refrigerant. Currently eighty to ninety percent of centrifugal chillers use CFC-11 or CFC-12 refrigerants, the manufacture and use of which is being eliminated under the terms of the Montreal Protocol. New refrigerants HCFC-123 and HCFC-134a are being used to replace the CFC refrigerants but refrigerant modifications to existing equipment will reduce the overall capacity of this equipment by 15 to 25 percent.

Centrifugal chillers can be driven by open or hermetic electric motors or by internal combustion engines or even by steam or gas turbines. Natural gas engine-driven equipment sized from 50 to 800 tons of refrigeration are available and in some cases are used to replace older CFC-refrigerant centrifugal chillers. These engine-driven chillers use the same cooling towers and pumps usually, but take advantage of cost savings in fuel costs. Part-load performance modulates both engine speed and compressor speed to match the load profile, maintaining close to the peak efficiency down to 50 percent of rated load. They can also use heat recovery options to take advantage of the engine jacket and exhaust heat.

Turbine-driven compressors are typically used on large equipment with capacities of 1200 tons or more. The turbine may be used as part of a cogeneration process but this is not required. (For a detailed discussion of cogeneration, see Chapter 7.) If excess steam is available, in industry or a large hospital, a steam turbine can be used to drive the chiller. However the higher load on the cooling tower due to the turbine condenser must be considered in the economic analysis.

Small water chillers, up to about 200 tons of capac-

ity, may utilize reciprocating compressors and are typically air-cooled instead of using cooling towers. An air-cooled reciprocating chiller uses a single or multiple reciprocating compressors to operate a DX liquid cooler. Air-cooled reciprocating chillers are widely used in commercial and large-scale residential buildings.

Other types of reciprocating refrigeration systems include liquid overfeed systems, flooded coil systems and multi-stage systems. These systems are generally used in large industrial or low-temperature applications.

10.6.2 Absorption Chillers

An alternative to vapor-compression refrigeration is absorption refrigeration which uses heat energy to drive a refrigerant cycle, extracting heat from a controlled environment and rejecting it to the environment (Figure 10.15). Thirty years ago absorption refrigeration was known for its low coefficient of performance and high maintenance requirements. Absorption chillers used more energy than centrifugal chillers and were economical only if driven by a source of waste heat.

Today, due primarily to the restriction on the use of CFC and HCFC refrigerants, the absorption chiller is making a comeback. Although new and improved, it still uses heat energy to drive the refrigerant cycle and typically uses aqueous lithium bromide to absorb the refrigerant and water vapor in order to provide a higher coefficient of performance.

The new absorption chillers can use steam as a heat source or be direct-fired. They can provide simultaneous heating and cooling which eliminates the need for a boiler. They do not use CFC or HCFC refrigerants, which may make them even more attractive in years to come. Improved safety and controls and better COP (even at part load) have propelled absorption refrigeration back into the market.

The most effective use of refrigeration equipment in a central-plant scenario is to have some of each type, comprising a hybrid plant. From a mixture of centrifugal and absorption equipment the operator can determine what equipment will provide the lowest operating cost under different conditions. For example a hospital that utilizes steam year round, but at reduced rates during summer, might use the excess steam to run an absorption chiller or steam-driven turbine centrifugal chiller to reduce its summertime electrical demand charges.

10.6.3 Chiller Performance

Most chillers are designed for peak load and then operate at loads less than the peak most of the time. Many chiller manufacturers provide data that identifies

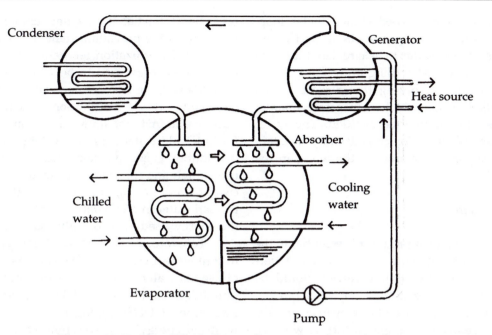

Figure 10.15 Simplified absorption cycle schematic diagram.

a chiller's part-load performance as an aid to evaluating energy costs. Ideally a chiller operates at a desired temperature difference (typically 45-55 degrees F; 25-30 degrees C) at a given flow rate to meet a given load. As the load requirement increases or decreases, the chiller will load or unload to meet the need. A reset schedule that allows the chilled water temperature to be adjusted to meet thermal building loads based on enthalpy provides an ideal method of reducing energy consumption.

Chillers should not be operated at less than 50 percent of rated load if at all possible. This eliminates both surging and the need for hot-gas bypass as well as the potential that the chiller would operate at low efficiency. If there is a regular need to operate a large chiller at less than one-half of the rated load it is economical to install a small chiller to accommodate this load.

10.6.4 Thermal Storage

Thermal storage can be another effective way of controlling electrical demand by using stored chilled water or ice to offset peak loads during the peak demand time. A good knowledge of the utility consumption and/or load profile is essential in determining the applicability of thermal storage. See Chapter 19 for a discussion of thermal storage systems.

10.6.5 Cooling Towers

Cooling towers use atmospheric air to cool the water from a condenser or coil through evaporation. In gen-

eral there are three types of cooling tower, named for the relationship between the fan-powered airflow and the flow of water in the tower: counterflow induced draft, crossflow induced draft and counterflow forced draft.

The use of variable-speed, two-speed or three-speed fans is one way to optimize the control of the cooling tower in order to reduce power consumption and provide adequate water cooling capacity. As the required cooling capacity increases or decreases the fans can be sequenced to maintain the approach temperature difference. For most air-conditioning systems this usually varies between 5 and 12 degrees F (3 to 7 degrees C).

When operated in the winter, the quantity of air must be carefully controlled to the point where the water spray is not allowed to freeze. In cold climates it may be necessary to provide a heating element within the tower to prevent freeze-ups. Although electric resistance heaters can be used for this purpose it is far more efficient to utilize hot water or steam as a heat source if available.

10.6.6 Wet-side Economizer

The use of "free-cooling" using the cooling tower water to cool supply air or chilled water is referred to as a wet-side economizer. The most common and effective way of interconnecting the cooling tower water to the chilled water loop is through the use of a plate-and-frame heat exchanger which offers a high heat transfer rate and low pressure drop. This method isolates the cooling tower water from the chilled water circuit maintaining the integrity of the closed chilled water loop. Another method is to use a separate circuit and pump that allows cooling tower water to be circulated through a coil located within an air-handling unit.

The introduction of cooling tower water, even through a so-called strainer cycle, can create maintenance nightmares and should be avoided. The water treatment program required for chilled water is intensive due to the required cleanness of the water in the chilled water loop.

10.6.7 Water treatment

A good water treatment program is essential to the maintenance of an efficient chilled water system. Filtering make-up water for the cooling tower should be evaluated. In some cases, depending on water quality, this can save the user a great deal of money in chemicals. Pretreating new systems prior to initial start-up will also provide longer equipment life and insure proper system performance.

Chiller performance is based on given design parameters and listed in literature provided by the chiller manufacturer. The performance will vary with building load, chilled water temperature, condenser water temperature and fouling factor. The fouling factor is the resistance caused by dirt, scale, silt, rust and other deposits on the surface of the tubes in the chiller and significantly affects the overall heat transfer of the chiller.

10.7 DOMESTIC HOT WATER

The creation of domestic hot water (DHW) represents about 4 percent of the annual energy consumption in typical non-residential buildings. In buildings where sleeping or food preparation occur, including hotels, restaurants, and hospitals, DHW may account for as much as thirty percent of total energy consumption.

A typical lavatory faucet provides a flow of 4 to 6 gal/min (0.25 to 0.38 l/s). Since hand washing is a function more of time than water use, substantial savings can be achieved by reducing water flow. Reduced-flow faucets which produce an adequate spray pattern can reduce water consumption to less than 1 gal/min (0.06 l/s). Flow reducing aerator replacements are also available.

Reducing DHW temperature has also been shown to save energy in non-residential buildings. Since most building users accept water at the available temperature regardless of what it is water temperature can be reduced from the prevailing standard of 140°F (60°C) to a 105°F (40°C) utilization temperature saving up to one-half of the energy used to heat the water.

Many large non-commercial buildings employ recirculating DHW distribution systems in order to reduce or eliminate the time required and water wasted in flushing cold water from hot water piping. Recirculating distribution is economically attractive only where DHW use is high and/or the cost of water greatly exceeds the cost of water heating. In most cases the energy required to keep water in recirculating DHW systems hot exceeds the energy used to heat the water actually used.

To overcome this waste of energy there is a trend to convert recirculating DHW systems to localized point-of-use hot water heating, particularly in buildings where plumbing facilities are widely separated. In either case insulation of DHW piping is essential in reducing the waste of energy in distribution. One-inch of insulation on DHW pipes will result in a 50% reduction in the distribution heat loss.

One often-overlooked energy conservation opportunity associated with DHW is the use of solar-heated hot water. Unlike space-heating, the need for DHW is relatively constant throughout the year and peaks during hours of sunshine in non-residential buildings. Year-round use amortizes the cost of initial equipment faster than other active-solar options.

Many of the techniques appropriate for reducing energy waste in DHW systems are also appropriate for energy consumption in heated service water systems for industrial buildings or laboratories.

10.8 ESTIMATING HVAC ENERGY CONSUMPTION

The methods for estimating building heating and cooling loads and the consumption of energy by HVAC systems are described in Chapter 9.

References
1. *ASHRAE Handbook: Fundamentals*, American Society of Heating, Refrigerating and Air-Conditioning Engineers, Inc., Atlanta, 1993.
2. *ASHRAE Handbook: HVAC Applications*, American Society of Heating, Refrigerating and Air-Conditioning Engineers, Inc., Atlanta, 1995.
3. *ASHRAE Handbook: HVAC Systems and Equipment*, American Society of Heating, Refrigerating and Air-Conditioning Engineers, Inc., Atlanta, 1992.
4. *ASHRAE Handbook: Refrigeration*, American Society of Heating, Refrigerating and Air-Conditioning Engineers, Inc., Atlanta, 1994.

CHAPTER 11

ELECTRIC ENERGY MANAGEMENT

K.K. LOBODOVSKY
BSEE & BSME
Certified Energy Auditor
State of California

11.1 INTRODUCTION

Efficient use of electric energy enables commercial, industrial and institutional facilities to minimize operating costs, and increase profits to stay competitive.

The majority of electrical energy in the United States is used to run electric motor driven systems. Generally, systems consist of several components, the electrical power supply, the electric motor, the motor control, and a mechanical transmission system.

There are several ways to improve the systems efficiency. The cost effective way is to check each component of the system for an opportunity to reduce electrical losses. A qualified individual should oversee the electrical system since poor power distribution within a facility is a common cause of energy losses.

Technology Update Ch. 18[1] lists 20 items to help facility management staff identify opportunities to improve drive system efficiency.

1. Maintain Voltage Levels.
2. Minimize Phase Imbalance.
3. Maintain Power Factor.
4. Maintain Good Power Quality.
5. Select Efficient Transformers.
6. Identify and Fix Distribution System Losses.
7. Minimize Distribution System Resistance.
8. Use Adjustable Speed Drives (ASDs) or 2-Speed Motors Where Appropriate.
9. Consider Load Shedding.
10. Choose Replacement Before a Motor Fails.
11. Choose Energy-Efficient Motors.
12. Match Motor Operating Speeds.
13. Size Motors for Efficiency.
14. Choose 200 Volt Motors for 208 Volt Electrical Systems.
15. Minimize Rewind Losses.
16. Optimize Transmission Efficiency.
17. Perform Periodic Checks.
18. Control Temperatures.
19. Lubricate Correctly.
20. Maintain Motor Records.

Some of these steps require the one-time involvement of an electrical engineer or technician. Some steps can be implemented when motors fail or major capital changes are made in the facility. Others involve development of a motor monitoring and maintenance program.

11.2 POWER SUPPLY

Much of this information consists of standards defined by the National Electrical Manufacturers Association (NEMA).

The power supply is one of the major factors affecting selection, installation, operation, and maintenance of an electrical motor driven system. Usual service conditions, defined in NEMA Standard Publication MG1-1987, *Motors and Generators*,[2] include:

- Motors designed for rated voltage, frequency, and number of phases.

- The supply voltage must be known to select the proper motor.

- Motor nameplate voltage will normally be less then nominal power system voltage.

Nominal Power System Voltage (Volts)	Motor Utilization (Nameplate) Voltage Volts
208	200
240	230
480	460
600	575
2400	2300
4160	4000
6900	6600
13800	13200

- Operation within tolerance of ±10 percent of the rated voltage.

- Operation from a sine wave of voltage source (not to exceed 10 percent deviation factor).

- Operation within a tolerance of ±5 percent of rated frequency.

- Operation within a voltage unbalance of 1 percent or less.

Operation at other than usual service conditions may result in the consumption of additional energy.

11.3 EFFECTS OF UNBALANCED VOLTAGES ON THE PERFORMANCE OF POLYPHASE SQUIRREL-CAGE INDUCTION MOTORS (MG 1-20.56)

When the line voltages applied to a polyphase induction motor are not equal, unbalanced currents in the stator windings result. A small percentage of voltage unbalance results in a much larger percentage current unbalance. Consequently, the temperature rise of the motor operating at a particular load and percentage voltage unbalance will be greater than for the motor operating under the same conditions with balanced voltages.

Voltages should be evenly balanced as closely as they can be read on a voltmeter. If the voltages are unbalanced, the rated horsepower of polyphase squirrel-cage induction motors should be multiplied by the factor shown in Figure 11.1 to reduce the possibility of damage to the motor. Operation of the motor with more than a 5-percent voltage unbalance is not recommended.

When the derating curve of Figure 11.1 is applied for operation on balanced voltages, the selection and setting of the overload device should take into account the combination of the derating factor applied to the motor and the increase in current resulting from the unbalanced voltages. This is a complex problem involving the variation in motor current as a function of load and voltage unbalance in the addition to the characteristics of the overload device relative to $I_{MAXIMUM}$ or $I_{AVERAGE}$. In the absence of specific information it is recommended that overload devices be selected and/or adjusted at the minimum value that does not result in tripping for the derating factor and voltage unbalance that applies. When the unbalanced voltages are unanticipated, it is recommended that the overload devices be selected so as to be responsive to $I_{MAXIMUM}$ in preference to overload devices responsive to $I_{AVERAGE}$.

11.4 EFFECT ON PERFORMANCE— GENERAL (MG 1 20.56.1)

The effect of unbalanced voltages on polyphase induction motors is equivalent to the introduction of a "negative-sequence voltage" having a rotation opposite to that

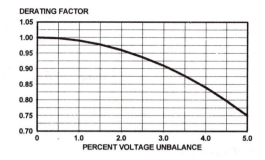

Figure 11.1 Polyphase squirrel-cage induction motors derating factor due to unbalanced voltage.

occurring the balanced voltages. This negative-sequence voltage produces an air gap flux rotating against the rotation of the rotor, tending to produce high currents. A small negative-sequence voltage may produce current in the windings considerably in excess of those present under balanced voltage conditions.

11.4.1 Unbalanced Defined (MG 1 20.56.2)

The voltage unbalance in percent may be defined as follows:

$$\text{Percent Voltage Unbalance} = 100 \times \frac{\text{Maximum voltage deviation from average voltage}}{\text{average voltage}}$$

Example—With voltages of 220, 215 and 210, the average is 215, the maximum deviation from the average is 5

PERCENT VOLTAGE UNBALANCE
= 100 * 5/215 = 2.3 PERCENT

11.4.2 Torque (MG 1 20.56.3)

The locked-rotor torque and breakdown torque are decreased when the voltage is unbalanced. If the voltage unbalance is extremely severe, the torque might not be adequate for the application.

11.4.3 Full-Load Speed (MG 1 20.56.4)

The full-load speed is reduced slightly when the motor operates at unbalanced voltages.

11.4.4 Currents (MG 1 20.56.5)

The locked-rotor current will be unbalanced but the locked rotor kVA will increase only slightly.

The currents at normal operating speed with unbalanced voltages will be greatly unbalanced in the order of 6 to 10 times the voltage unbalance.

11.5 MOTOR

The origin of the electric motor can be traced back to 1831 when Michael Faraday demonstrated the fundamental principles of electromagnetism. The purpose of an electric motor is to convert electrical energy into mechanical energy.

Electric motors are efficient at converting electric energy into mechanical energy. If the efficiency of an electric motor is 80%, it means that 80% of electrical energy delivered to the motor is directly converted to mechanical energy. The portion used by the motor is the difference between the electrical energy input and mechanical energy output.

A major manufacturer estimate that US annual sales exceed 2 million motors. Table 11.1 lists sales volume by motor horsepower. Only 15% of these sales involve high-efficiency motors.[3]

Table 11.1 Polyphase induction motors annual sales volume.

hp	Units
1-5	1,300,000
7.5-20	500,000
25-50	140,000
60-100	40,000
125-200	32,000
250-500	11,000
Total	2,023,000

Motor terms are used quite frequently, usually on the assumption that every one knows what they mean or imply. Such is far too often not the case. The following section is a list of motor terms.

11.6 GLOSSARY OF FREQUENTLY OCCURRING MOTOR TERMS[4]

Amps
Full Load Amps
The amount of current the motor can be expected to draw under full load (torque) conditions is called Full Load Amps. It is also known as nameplate amps.
Locked Rotor Amps
Also known as starting inrush, this is the amount of current the motor can be expected to draw under starting conditions when full voltage is applied.
Service Factor Amps
This is the amount of current the motor will draw when it is subjected to a percentage of overload equal to the service factor on the nameplate of the motor. For example, many motors will have a service factor of 1.15, meaning that the motor can handle a 15% overload. The service factor amperage is the amount of current that the motor will draw under the service factor load condition.

Code Letter
The code letter is an indication of the amount of inrush current or locked rotor current that is required by a motor when it is started. Motor code letters usually applied to ratings of motors normally started on full voltage (chart below).

Code letter	Locked rotor* kVA per horsepower	Horsepower Single-phase	Horsepower Three-phase
F	5.0 to 5.6		15 up
G	5.6 to 6.3	5	7.5 to 10
H	6.3 to 7.1	3	5
J	7.1 to 8.0	1.5 to 2	3
K	8.0 to 9.0	0.75 to 1.00	1.5 to 2
L	9.0 to 10.0	0.5	1

*Locked rotor kVA is equal to the product of the line voltage times motor current divided by 1,000 when the motor is not allowed to rotate; this corresponds to the first power surge required to start the motor. Locked-rotor kVA per horsepower range includes the lower figure up to but not including the higher figure.

Design

The design letter is an indication of the shape of the torque speed curve. Figure-11.2 shows the typical shape of the most commonly used design letters. They are A, B, C, and D. Design B is the standard industrial duty motor which has reasonable starting torque with moderate starting current and good overall performance for most industrial applications. Design C is used for hard to start loads and is specifically designed to have high starting torque. Design D is the so-called high slip motor which tends to have very high starting torque but has high slip RPM at full load torque. In some respects, this motor can be said to have a 'spongy' characteristic when loads are changing. Design D motors particularly suited for low speed punch press, hoist and elevator applications. Generally, the efficiency of Design D motors at full load is rather poor and thus they are normally used on those applications where the torque characteristics are of primary importance. Design A motors are not commonly specified but specialized motors used on injection molding applications have characteristics similar to Design B. The most important characteristic of Design A is the high pull out torque.

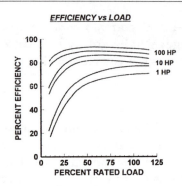
EFFICIENCY vs LOAD

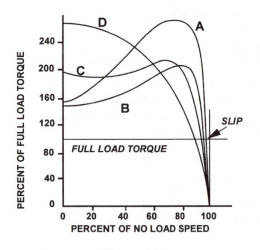

Figure-11.2

Efficiency

Efficiency is the percentage of the input power that is actually converted to work output from the motor shaft. Efficiency is now being stamped on the nameplate of most domestically produced electric motors. See the section 11.14.

$$\text{Efficiency} = \text{EFF} = \frac{746 \times \text{HP Output}}{\text{Watts Input}}$$

Frame Size

Motors, like suits of clothes, shoes and hats, come in various sizes to match the requirements of the applications. In general, the frame size gets larger with increasing horsepower or with decreasing speeds. In order to promote standardization in the motor industry, NEMA (national electrical manufacturers association) prescribes standard frame sizes for certain horsepower, speed, and enclosure combinations. Frame size pins down the mounting and shaft dimension of standard motors. For example, a motor with a frame size of 56, will always have a shaft height above the base of 3- 1/2 inches.

Frequency

This is the frequency for which the motor is designed. The most commonly occurring frequency in this country is 60 cycles but, on an international basis, other frequencies such as 25, 40, and 50 cycles can be found.

Full Load Speed

An indication of the approximate speed that the motor will run when it is putting out full rated output torque or horsepower is called full load speed.

High Inertia Load

These are loads that have a relatively high fly wheel effect. Large fans, blowers, punch presses, centrifuges, industrial washing machines, and other similar loads can be classified as high inertia loads.

Insulation Class

The insulation class is a measure of the resistance of the insulating components of a motor to degradation from heat. Four major classifications of insulation are used in motors. They are, in order of increasing thermal capabilities, A, B, F, and H.

Class of Insulation System	Temperature, Degrees C
A	75
B	905
F	115
H	130

Load Types

Constant Horsepower

The term constant horsepower is used in certain types of loads where the torque requirement is reduced as the speed is increased and vice-versa. The constant horsepower load is usually associated with metal removal applications such as drill presses, lathes, milling machines, and similar types of applications.

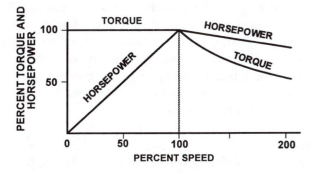

Constant Torque

Constant torque is a term used to define a load characteristic where the amount of torque required to drive the machine is constant regardless of the speed at which it is driven. For example, the torque requirement of most conveyors is constant.

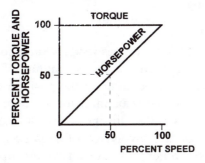

Variable Torque

Variable torque is found in loads having characteristics requiring low torque at low speeds and increasing values of torque required as the speed is increased. Typically examples of variable torque loads are centrifugal fans and centrifugal pumps.

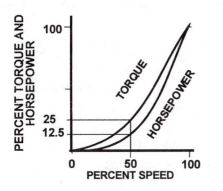

Phase

Phase is the indication of the type of power supply for which the motor is designed. Two major categories exist: single phase and three phase. There are some very spotty areas where two phase power is available but this is very insignificant.

Poles

This is the number of magnetic poles within the motor when power is applied. Poles are always an even number such as 2, 4, 6. In an AC motor, the number of poles work in conjunction with the frequency to determine the synchronous speed of the motor. At 50 and 60 cycles, common arrangements are:

Synchronous speed

Poles	60 Cycles	50 Cycles
2	3600	3000
4	1800	1500
6	1200	1000
8	900	750
10	720	600

Power Factor

Percent power factor is a measure of a particular motor's requirements for magnetizing amperage. For more information see section 11.7.

$$\text{Power Factor} = pf = \frac{\text{Watts Input}}{\text{Volts} \times \text{Amps} \times 1.73}$$

(3 phase)

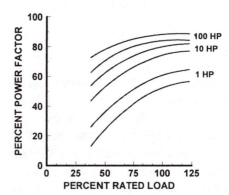

Service Factor

The service factor is a multiplier that indicates the amount of overload a motor can be expected to handle. For example, a motor with a 1.0 service factor cannot be expected to handle more than its nameplate horsepower on a continuous basis. Similarly, a motor with a 1.15 service factor can be expected to safely handle intermittent loads amounting to 15% beyond its nameplate horsepower.

Slip

Slip is used in two forms. One is the slip RPM which is the difference between the synchronous speed and the full load speed. When this slip RPM is expressed as a percentage of the synchronous speed, then it is called percent slip or just 'slip.' Most standard motors run with a full load slip of 2% to 5%.

$$\% \, \text{Slip} = \frac{\text{Synchronous speed—Running speed} \times 100}{\text{Synchronous speed}}$$

$$\% \, \text{Slip} = \frac{1800 - 1750 \times 100}{1800} = 2.8\% \, \text{Slip}$$

Synchronous Speed

This is the speed at which the magnetic field within the motor is rotating. It is also approximately the speed that the motor will run under no load condition. For example, a 4 pole motor running in 60 cycles would have a magnetic field speed of 1800 RPM. The no load speed of that motor shaft would be very close to 1800, probably 1798 or 1799 RPM. The full load speed of the same motor might be 1745 RPM. The difference between the synchronous speed of the full load speed is called the slip RPM of the motor.

$$\text{RPM} = \frac{(120) \times (\text{Frequency, HZ})}{\text{Number of Poles in Winding}}$$

$$\text{RPM} = \frac{(120) \times (60)}{4} = 1800$$

Temperature
Ambient Temperature.

Ambient temperature is the maximum safe room temperature surrounding the motor if it is going to be operated continuously at full load. In most cases, the standardized ambient temperature rating is 40°C (104°F). This is a very warm room. Certain types of applications such as on board ships and in boiler rooms, may require motors with a higher ambient temperature capability such as 50°C or 60°C.

Temperature Rise.

Temperature rise is the amount of temperature change that can be expected within the winding of the mo-

tor from non-operating (cool condition) to its temperature at full load continuous operating condition. Temperature rise is normally expressed in degrees centigrade.

Time Rating

Most motors are rated in continuous duty which means that they can operate at full load torque continuously without overheating. Motors used on certain types of applications such as waste disposal, valve actuators, hoists, and other types of intermittent loads, will frequently be rated in short term duty such as 5 minutes, 15 minutes, 30 minutes or 1 hour. Just like a human being, a motor can be asked to handle very strenuous work as long as it is not required on a continuous basis.

Torque

Torque is the twisting force exerted by the shaft or a motor. Torque is measured in inch, pounds, foot pounds, and on small motors, in terms of inch ounces.

Full Load Torque

Full load torque is the rated continuous torque that the motor can support without overheating within its time rating.

Peak Torque

Many types of loads such as reciprocating compressors have cycling torque where the amount of torque required varies depending on the position of the machine. The actual maximum torque requirement at any point is called the peak torque requirement. Peak torque are involved in things such as punch presses and other types of loads where an oscillating torque requirement occurs.

Pull Out Torque

Also known as breakdown torque, this is the maximum amount of torque that is available from the motor

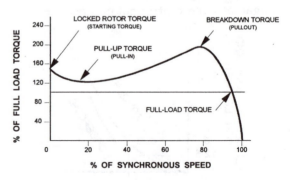

Figure 11.3 Typical speed—torque curve.

shaft when the motor is operating at full voltage and is running at full speed. The load is then increased until the maximum point is reached. Refer to Figure-11.3.

Pull Up Torque

The lowest point on the torque speed curve for a motor accelerating a load up to full speed is called pull up torque. Some motors are designed to not have a value of pull up torque because the lowest point may occur at the locked rotor point. In this case, pull up torque is the same as locked rotor torque.

Starting Torque

The amount of torque the motor produces when it is energized at full voltage and with the shaft locked in place is called starting torque. This value is also frequently expressed as 'Locked rotor torque.' It is the amount of torque available when power is applied to break the load away and start accelerating it up to speed.

Voltage

This would be the voltage rating for which the motor is designed. Section 11.2.

11.7 POWER FACTOR

<u>WHAT IS POWER FACTOR (pf)?</u>

It is the mathematical ratio of *ACTIVE POWER* (W) to *APPARENT POWER* (VA)

$$pf = \frac{Active\,power}{Apparent\,power} = W = Cos\,\theta$$

pf angle in degrees = $cos^{-1}\,\theta$
ACTIVE POWER = **W** = "real power" = supplied by the power system to actually turn the motor.

REACTIVE POWER = **VAR** = **(W)tan** θ = is used strictly to develop a magnetic field within the motor.

or **(VA)^2** = **(W)^2** + **(VAR)^2**

NOTE: Power factor may be "leading" or "lagging" depending on the direction of VAR flow.

CAPACITORS can be used to improve the power factor of a circuit with a large inductive load. Current through capacitor LEADS the applied voltage by 90 electrical degrees (VAC), and has the effect of "opposing"

the inductive "LAGGING" current on a "one-for-one" (VAR) basis.

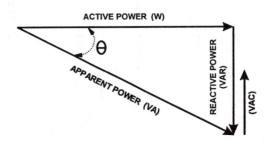

<u>WHY RAISE POWER FACTOR (pf)?</u>

Low (or "unsatisfactory") power factor is caused by the use of inductive (magnetic) devices and can indicate possible low system electrical operating efficiency. These devices are:

* non-power factor corrected fluorescent and high intensity discharge lighting fixture ballasts (40%-80% pf)

* arc welders (50%-70% pf)

* solenoids (20%-50% pf)

* induction heating equipment (60%-90% pf)

* lifting magnets (20%-50% pf)

* small "dry-pack" transformers (30%-95% pf)

* and most significantly, induction motors (55%-90% pf)

Induction motors are generally the principal cause of low power factor because there are so many in use, and they are usually <u>not fully loaded</u>. The correction of the condition of LOW power factor is a problem of vital economic importance in the generation, distribution and utilization of a-c power.

<u>MAJOR BENEFITS OF</u>
<u>POWER FACTOR MPROVEMENT ARE</u>:

* increased plant capacity,

* reduced power factor "penalty" charges for the electric utility,

* improvement of voltage supply,

* less power losses in feeders, transformers and distribution equipment.

WHERE TO CORRECT POWER FACTOR?

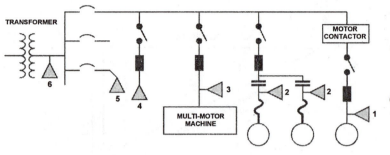

Capacitor correction is relatively inexpensive both in material and installation costs. Capacitors can be installed at any point in the electrical system, and will improve the power factor between the point of application and the power source. However, the power factor between the utilization equipment and the capacitor will remain unchanged. Capacitors are usually added at each piece of offending equipment, ahead of groups of small motors (ahead of motor control centers or distribution panels) or at main services. Refer to the National Electrical Code for installation requirements.

The advantages and disadvantages of each type of capacitor installation are listed below:

Capacitor on each piece of equipment (1,2)

ADVANTAGES
- increases load capabilities of distribution system.

- can be switched with equipment; no additional switching is required.

- better voltage regulation because capacitor use follows load.

- capacitor sizing is simplified

- capacitors are coupled with equipment and move with equipment if rearrangements are instituted.

DISADVANTAGES
- small capacitors cost more per KVAC than larger units (economic break point for individual correction is generally at 10 HP).

Capacitor with equipment group (3)

ADVANTAGES
- increased load capabilities of the service,

- reduced material costs relative to individual correction

- reduced installation costs relative to individual correction

DISADVANTAGES
- switching means may be required to control amount of capacitance used.

Capacitor at main service (4,5, & 6)

ADVANTAGES

- low material installation costs.

DISADVANTAGES

- switching will usually be required to control the amount of capacitance used.

- does not improve the load capabilities of the distribution system.

OTHER CONSIDERATIONS

Where the loads contributing to power factor are relatively constant, and system load capabilities are not a factor, correcting at the main service could provide a cost advantage. When the low power factor is derived from a few selected pieces of equipment, individual equipment correction would be cost effective. Most capacitors used for power factor correction have built-in fusing; if not, fusing must be provided.

The growing use of ASD's (nonlinear loads) has increased the complexity of system power factor and its corrections. The application of pf correction capacitors without a thorough analysis of the system can aggravate rather than correct the problem, particularly if the fifth and seventh harmonics are present.

POWER QUALITY REQUIREMENTS[6]

The electronic circuits used in ASDs may be susceptible to power quality related problems if care is not taken during application, specification and installation. The most common problems include transient overvoltages, voltage sags and harmonic distortion. These power quality problems are usually manifested in the form of nuisance tripping.

TRANSIENT OVERVOLTAGES—Capacitors are devices used in the utility power system to provide power factor correction and voltage stability during periods of heavy loading. Customers may also use capacitors for power factor correction within their facility. When capacitors are energized, a large transient overvoltage may develop causing the ASD to trip.

VOLTAGE SAGS—ASDs are very sensitive to temporary reductions in nominal voltage. Typically, voltage sags are caused by faults on either the customer's or the utilities electrical system.

HARMONIC DISTORTION—ASDs introduce harmonics into the power system due to nonlinear characteristics of power electronics operation. Harmonics are components of current and voltage that are multiples of the normal 60Hz ac sine wave. ASDs produce harmonics which, if severe, can cause motor, transformer and conductor overheating, capacitor failures, misoperation of relays and controls and reduce system efficiencies.

Compliance with IEEE-519 "Recommended Practices and Requirements for Harmonic Control in Electrical Power Systems" is strongly recommended.

11.8 HANDY ELECTRICAL FORMULAS & RULES OF THUMB

Conversion formulas

REQUIRED	DIRECT CURRENT	ALTERNATING CURRENT	
		SINGLE-PHASE	THREE-PHASE
AMPERES WHEN H.P. IS KNOWN	$\dfrac{H.P. \times 746}{E \times EFF.}$	$\dfrac{H.P. \times 746}{E \times EFF. \times P.F.}$	$\dfrac{H.P. \times 746}{1.73 \times E \times EFF. \times P.F.}$
AMPERES WHEN KILOWATTS ARE KNOWN	$\dfrac{KW \times 1000}{E}$	$\dfrac{KW \times 1000}{E \times P.F.}$	$\dfrac{KW \times 1000}{1.73 \times E \times P.F.}$
AMPERES WHEN KVA IS KNOWN		$\dfrac{KVA \times 1000}{E}$	$\dfrac{KVA \times 1000}{1.73 \times E}$
KILOWATTS	$\dfrac{I \times E}{1000}$	$\dfrac{I \times E \times P.F.}{1000}$	$\dfrac{I \times E \times P.F. \times 1.73}{1000}$
KVA		$\dfrac{I \times E}{1000}$	$\dfrac{I \times E \times 1.73}{1000}$
HORSEPOWER OUTPUT	$\dfrac{I \times E \times EFF.}{746}$	$\dfrac{I \times E \times EFF. \times P.F.}{746}$	$\dfrac{I \times E \times 1.73 \times EFF. \times P.F.}{746}$

I = AMPERES   H.P. = HORSEPOWER   EFF. = EFFICIENCY (EXPRESSED IN A DECIMAL)
E = VOLTS   P.F. = POWER FACTOR   KW = KILOWATTS   KVA = KILOVOLT-AMPERES

Rules of thumb.
At 3600 RPM, a motor develops 1.5 lb.-ft. per HP.
At 1800 RPM, a motor develops 3 lb.-ft. per HP.
At 1200 RPM, a motor develops 4.5 lb.-ft. per HP.
At 550 & 575 Volts, a 3 phase motor draws 1 amp per HP.
At 440 & 460 Volts, a 3 phase motor draws 1.25 amp per HP.
At 220 & 230 Volts, a 3 phase motor draws 2.5 amp per HP.

11.9 ELECTRIC MOTOR OPERATING LOADS

Most electric motors are designed to operate at 50 to 100 percent of their rated load. One reason is the motors optimum efficiency is generally 75 percent of the rated load, and the other reason is motors are generally sized for the starting requirements.

Several surveys of installed motors reveal that large portion of motors in use are improperly loaded. Underloaded motors, those loaded below 50 percent of rated load, operate inefficiently and exhibit low power factor. Low power factor increases losses in electrical distribution and utilization equipment, such as wiring, motors, and transformers, and reduces the load-handling capability and voltage regulation of the building's electrical system. Typical part-load efficiency and power factor characteristics are shown in Figure 11.4.

POWER SURVEY

Power surveys are conducted to compile meaningful records of energy usage at the service entrance, feeders and individuals loads. These records can be analyzed

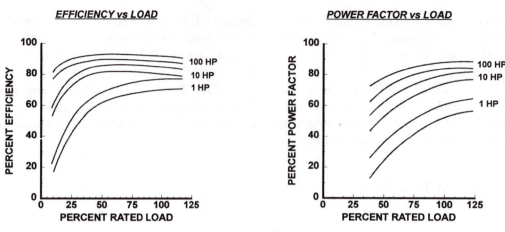

Figure 11.4 Typical part-load efficiency and power factor characteristics

to prioritize those areas yielding the greatest energy savings. Power surveys also provide information for load scheduling to reduce peak demand and show operational characteristics of loads that may suggest component or system replacement to reduce energy consumption. Only through the measurement of AC power parameters can true cost benefit analysis be performed.[7]

11.10 DETERMINING ELECTRIC MOTOR OPERATING LOADS

Determining if electric motors are properly loaded enables a manager to make informed decisions about when to replace them and which replacement to choose. There are several ways to determine motor loads. The best and the simplest way is by direct electrical measurement using a Power Meter. Slip Measurement or Amperage Readings methods can be used to **estimate** the actual load.

11.11 POWER METER

To understand the electrical power usage of a facility, load or device, measurements must be taken over a time span to have a profile of the unit's operation. Digital power multimeters, measure Amps, Volts, kWatts, kVars, kVA, Power Factor, Phase Angle and Firing Angle. The GENERAL TEST FORM Figure 11.5 provides a format for documentation with corresponding connection diagrams for various power circuit configurations.

*Such measurements should only be
performed by trained personnel*

Selection of Equipment for Power Measurement or Surveys

When choosing equipment to conduct a power survey, many presentation formats are available including indicating instruments, strip chart recorders and digital devices with numeric printout. For most survey applications, changing loads makes it mandatory for data to be compiled over a period of time. This period may be an hour, day, week or month. Since it is not practical to write down varying readings from an indicating device for a long period of time, a chart recorder or digital device with numeric printout is preferred. If loads vary frequently, an analog trend recording will be easier to analyze than trying to interpret several numeric reports. Digital power survey monitors are typically less expensive than analog recordings systems. Complete microprocessor based power survey systems capable of measuring watts, VARs, kVA, power factor, watt hours, VAR hours and demand including current transformers are available for under $3000. With prices for memory and computers going down, digital devices interfaced to disk or cassette storage will provide a cost effective method for system analysis.[7]

Loads

When analyzing polyphase motors, it is important to make measurement with equipment suited for the application. Watt measurements or var measurements should be taken with a two element device. Power factor should be determined from the readings of both measurements. When variable speed drives are encountered, it is always preferable to take measurements on the line side of the controller. When measurements are required on the load side of the controller, the instrument specifi-

TO MEASURE	1 Phase, 2 Wire	1 Phase, 3 Wire	3 Phase, 3 Wire	3 Phase, 4 Wire	3 Phase, 4 Wire TAP
VOLTAGE (V)	L - 1 to N	L - 1 to N	A to B Phase	A Phase to N	B Phase to N
		L - 2 to N	C to B Phase	B Phase to N	A Phase to N
				C Phase to N	C Phase to N
CURRENT (A)	L - 1	L - 1	A Phase	A Phase	B Phase
		L - 2	C Phase	B Phase	A Phase
				C Phase	C Phase
POWER (KW)	L - 1 to N	L - 1 to N		A Phase to N	
		L - 2 to N		B Phase to N	
				C Phase to N	
		Total KW	Total KW	Total KW	Total KW
VOLT-AMPERS REACTIVE (KVAR)	L - 1 to N	L - 1 to N		A Phase to N	
		L - 2 to N		B Phase to N	
				C Phase to N	
		Total KVAR	Total KVAR	Total KVAR	Total KVAR
VOLT-AMPERS (KVA)	L - 1 to N	L - 1 to N		A Phase to N	
		L - 2 to N		B Phase to N	
				C Phase to N	
		Total KVA	Total KVA	Total KVA	Total KVA
POWER FACTOR PF %	L - 1 to N	L - 1 to N		A Phase to N	
		L - 2 to N		B Phase to N	
				C Phase to N	
		Combined PF %	Combined PF %	Combined PF %	Total PF %

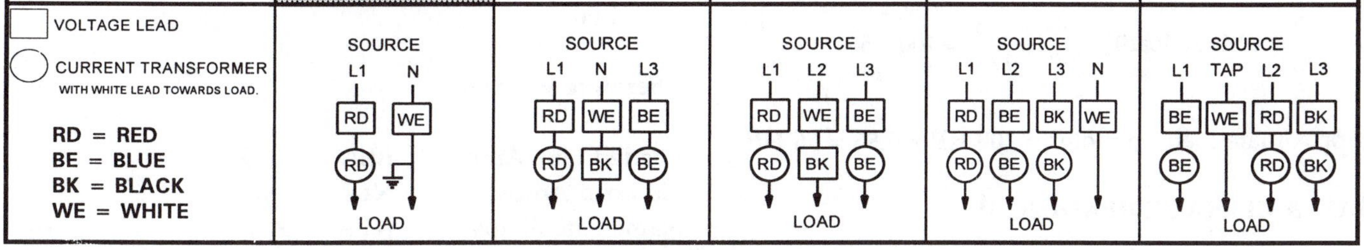

Legend:
☐ VOLTAGE LEAD
◯ CURRENT TRANSFORMER WITH WHITE LEAD TOWARDS LOAD.
RD = RED
BE = BLUE
BK = BLACK
WE = WHITE

SOURCE (1 Phase, 2 Wire): L1 N — RD, WE — RD — LOAD

SOURCE (1 Phase, 3 Wire): L1 N L3 — RD, WE, BE — RD, BK, BE — LOAD

SOURCE (3 Phase, 3 Wire): L1 L2 L3 — RD, WE, BE — RD, BK, BE — LOAD

SOURCE (3 Phase, 4 Wire): L1 L2 L3 N — RD, BE, BK, WE — RD, BE, BK — LOAD

SOURCE (3 Phase, 4 Wire TAP): L1 TAP L2 L3 — BE, WE, RD, BK — BE, RD, BK — LOAD

Figure 11.5 General test form (for use with power meter).

cations should be reviewed and if there is a question on the application the manufacturer should be contacted.[7]

11.12 SLIP MEASUREMENT

Conditions
1. Applied voltage must be within 5% of nameplate rating.
2. Should not be used on rewound motors.
3. Motors should be operating under steady load conditions.
4. Should be performed by trained personnel.

Note: Values used in this analysis are subject to rounding errors. For example, full load speed often rounded to the nearest 5 RPM.

Procedure
1. Read and record the motors nameplate Full Load Speed. (RPM)
2. Determine Synchronous speed No Load Speed (RPM) (900, 1200, 1800, 3600)
3. Measure and record Operating Load Speed with tachometer. (RPM)
4. Insert the recorded values in the following formula and solve.

$$(\% \text{ Motor load}) = \frac{\text{NLS} - \text{OLS}}{\text{NLS} - \text{FLS}} \times 100$$

Where:
    NLS = No load or synchronous speed
    OLS = Operating load speed
    FLS = Full load speed

Example:
    Consider a 100 HP, 1800 RPM Motor
    FLS = 1775 RPM, OLS = 1786 RPM

$$(\% \text{ Motor load}) = \frac{1800 - 1786}{1800 - 1775} \times 100 = 56$$

Approximate load on motor = **100 HP × 0.56 = 56 HP**

11.13 AMPERAGE READINGS[4]

Conditions
1. Applied voltage must be within 5% of nameplate rating.
2. You must be able to disconnect the motor from the load. (By removing V-belts or disconnecting a coupling).

3. Motor must be 7 1/2 HP or larger, 3450, 1725 or 1140 RPM.
4. The indicated line amperage must be below the full load nameplate rating.

Procedure
1. Measure and record line amperage with load connected and running.
2. Disconnect motor from load. Measure and record the line amperage when the motor is running without load.
3. Read and record the motors nameplate amperage for the voltage being used.
4. Insert the recorded values in the following formula and solve.

$$(\% \text{ Rated HP}) = \frac{2 \times \text{LLa} - \text{NLA}}{2 \times \text{NPA} - \text{NLA}} \times 100$$

Where:
    LLA = Loaded Line Amps
    NLA = No Load Line Amps (Motor disconnected from load)
    NPA = Nameplate Amperage (For operating voltage)

Please note: This procedure will generally yield reasonably accurate results when motor load is in the 40 to 100% range and deteriorating results at loads below 40%.

Example:

- A 20 HP motor driving a pump is operating on 460 volts and has a loaded line amperage of 16.5.

- When the coupling is disconnected and the motor operated at no load the amperage is 9.3.

- The motor nameplate amperage for 460 volts is 24.0.

Therefore we have:

Loaded Line Amps	LLA	= 16.5
No Load Amps	NLA	= 9.3
Nameplate Amps	NPA	= 24.0

$$(\% \text{ Rated HP}) = \frac{2 \times 16.5 - 9.3}{2 \times 24.0 - 9.3} \times 100 = \frac{23.7}{38.7} \times 100 = 61.2\%$$

Approximate load on motor = 20 HP × 0.612 = 12.24
                                or slightly over 12 HP

11.14 ELECTRIC MOTOR EFFICIENCY

The efficiency of a motor is the ratio of the mechanical power output to the electrical power input. It may be expressed as:

$$\text{Efficiency} = \frac{\text{Output}}{\text{Input}} = \frac{\text{Input} - \text{Losses}}{\text{Input}} = \frac{\text{Output}}{\text{Output} + \text{Losses}}$$

Design changes, better materials, and manufacturing improvements reduce motor losses, making premium or energy-efficient motors more efficient than standard motors. Reduced losses mean that an energy-efficient motor produces a given amount of work with less energy input than a standard motor.[3]

In 1989, the National Electrical Manufacturers Association (NEMA) developed a standard definition for energy-efficient motors.[2]

How should we interpret efficiency labels?

Efficiencies and Different Standards

Standard	7.5 HP motor	20 HP motor
International (IEC 34-2)	82.3%	89 4%
British (BS-269)	82.3%	89.4%
Japanese (IEC-37)	85.0%	90.4%
U.S. (IEEE -112 Method B)*	80.3%	86.9%

The critical part of the efficiency comparison calculations is that the efficiencies used **must be comparable**.

The Arthur D Little report contained the following interesting statement: "Reliable and consistent data on motor efficiency is not available to motor appliers. Data published by manufacturers appears to range from very conservative to cavalier."

Recognizing that less than a 10 percent spread in losses is statistically insignificant NEMA has set up efficiency bands. Any motor tested by IEEE - 112, Method B, will carry the nominal efficiency of the highest band for which the average full load efficiency for the model is equal to or above that nominal.

The NEMA nominal efficiency is defined as the average efficiency of a large population of motors of the same design. The spread between nominal efficiency in the table based on increments of 10 percent losses. The spread between the nominal efficiency and the associated minimum is based on an increment of 20 percent losses.

Nominal Efficiency		Minimum Efficiency
93.6	—— 20% Greater Losses →	92.4
|		
10% Greater Losses		
↓		
93.0		

11.14 THE FOLLOWING IS REPRINTED FROM NEMA MG 1-1987 EFFICIENCY (MG 1-12.54)

Determination of Motor Efficiency and Losses (MG 1-12.54.1)

Efficiency and losses shall be determined in accordance with IEEE Std 112 Standard Test Procedures for Polyphase Induction Motors and Generators*. The efficiency shall be determined at rated output, voltage, and frequency.

Unless otherwise specified, horizontal polyphase squirrel-cage medium motors rated 1 to 125 horsepower shall be tested by dynamometer (Method B) as described in par. 5.2.2.4 of IEEE Std 112. Motor efficiency shall be calculated using MG 1-12.57 in lieu of Form E of IEEE Std 112. Vertical motors in this horsepower range shall also be tested by Method B if bearing construction permits; otherwise they shall be tested by segregated losses (Method E) as described in par. 5.2.3.1 of IEEE Std 112, including direct measurement of stray-load loss.

The following losses shall be included in determining the efficiency:

1. Stator I^2R.
2. Rotor I^2R.
3. Core Loss.
4. Stray load loss.
5. Friction & windage loss.[†]
6. Brush contact loss of wound-rotor machines

---

*See Referenced Standards, MG 1-1.01

†In the case of motors which are furnished with thrust bearings, only that portion of the thrust bearing loss produced by the motor itself shall be included in the efficiency calculation. Alternatively, a calculated value of efficiency, including bearing loss due to external thrust load, shall be permitted to be specified.

In the case of motors which are furnished with less than a full set of bearing, friction and windage losses which are representative of the actual installation shall be determined by (1) calculations or (2) experience with shop tested bearings and shall be included in the efficiency calculations.

Power required for auxiliary items, such as external pumps or fans, that are necessary for the operation of the motor shall be stated separately.

In determining I^2R losses, the resistance of each winding shall be corrected in a temperature equal to an ambient temperature of 25°C plus the observed rated load temperature rise measured by resistance. When the rated load temperature rise has not been measured, the resistance of the winding shall be corrected to the following temperature:

Class of Insulation System	Temperature, Degrees C
A	75
B	95
F	115
H	130

This reference temperature shall be used for determining I^2R losses at all loads. If the rated temperature rise is specified as that of a lower class of insulation system, the temperature for resistance correction shall be that of the lower insulation class.

NEMA Standard 5-12-1975, revised 6-21-1979; 11-12-1981; 11-20-1986; 1-11-1989.

Efficiency Of Polyphase Squirrel-cage Medium Motors with Continuous Ratings (MG 1-12.54.2)

The full-load efficiency of Design A and B single-speed polyphase squirrel-cage medium motors in the range of 1 through 125 horsepower for frames assigned in accordance with NEMA Standards Publication No. MG 13, Frame Assignments for Alternating Current Integral-horsepower Induction Motors, (see MG1-1.101) and equivalent Design C ratings shall be identified on the nameplate by a nominal efficiency selected from the Nominal Efficiency column in Table 11.2 (NEMA Table 12.6A) which shall be not greater than the average efficiency of a large population of motors of the same design.

The efficiency shall be identified on the nameplate by the caption "NEMA Nominal Efficiency" or "NEMA Nom. Eff."

The full load efficiency, when operating at rated voltage and frequency, shall be not less than the minimum value indicated in Column B of Table 11.2 (NEMA Table 12.6A) associated with the nominal value in Column A.

Suggested Standard for Future Design 3-16-1977, NEMA Standard 1-17-1980, revised 3-8-1983; 3-14-1991.

The full-load efficiency, when operating at rated voltage and frequency, shall be not less than the minimum value indicated in Column C of Table 11.2 (NEMA Table 12.6A) associated with the nominal value in Column A.

Suggested Standard for Future Design 3-14-1991.

Variations in materials, manufacturing processes, and tests result in motor-to-motor efficiency for a large population of motors of a single design is not a unique efficiency but rather a band of efficiency. Therefore, Table 11.2 (NEMA Table 12.6A) has been established to indicate a logical series of nominal motor efficiencies and the minimum associated with each nominal. The nominal efficiency represents a value which should be used to compute the energy consumption of a motor or group of motors.

Authorized Engineering Information 3-6-1977, revised 1-17-1980;1-11-1989.

Efficiency Levels of Energy Efficient Polyphase Squirrel-Cage Induction Motors (MG 1-12.55)

The nominal full-load efficiency determined in accordance with MG 1-12.54.1, identified on the nameplate in accordance with MG 1.12.54.2, and having corresponding minimum efficiency in accordance with Column B of Table 11.2 (NEMA Table 12.6A) shall equal or exceed the values listed in Table 11.3 (NEMA Table 12.6B) for the motor to be classified as "energy efficient."

NEMA Std 1-11-1989;3-14-1991.

Efficiency Levels of Energy Efficient Polyphase Squirrel-Cage Induction Motors (MG 1-12.55A)
(SUGGESTED STANDARD FOR FUTURE DESIGN)

The nominal full-load efficiency determined in accordance with MG 1-12.54.1, identified on the nameplate in accordance with MG 1-12.54.2 and having minimum efficiency in accordance with Column C of Table 11.2 (NEMA Table 12.6A) shall equal or exceed the values listed in Table 11.4 (NEMA Table 12.6C) for the motor to be classified as "energy efficient."

Suggested Standard for future design 9-5-1991.

Table 11.2 (NEMA Table 12.6A)

Column A Nominal Efficiency	Column B* Minimum Efficiency based on 20% Loss Difference	Column C† Minimum Efficiency Based on 10% Loss Difference
99.0	98.8	98.9
98.9	98.7	98.8
98.8	98.6	98.7
98.7	98.5	98.6
98.6	98.4	98.5
98.5	88.2	98.4
98.4	98.0	98.2
98.2	97.8	98.0
98.0	97.6	97.8
97.8	97.4	97.6
97.6	97.1	97.4
97.4	96.8	97.1
97.1	96.5	96.8
96.8	96.2	96.5
96.5	95.8	96.2
96.2	95.4	95.8
95.8	95.0	95.4
95.4	94.5	95.0
95.0	94.1	94.5
94.5	93.6	94.1
94.1	93.0	93.6
93.6	92.4	93.0
93.0	91.7	92.4
92.4	91.0	91.7
91.7	90.2	91.0
91.0	89.5	90.2
90.2	88.5	89.5
89.5	87.5	88.5
88.5	86.5	87.5
87.5	85.5	86.5
86.5	84.0	85.5
85.5	82.5	84.0
84.0	81.5	82.5
82.5	80.0	81.5
81.5	78.5	80.0
80.0	77.0	78.5
78.5	75.5	77.0
77.0	74.0	75.5
75.5	72.0	74.0
74.0	70.0	72.0
72.0	68.0	70.0
70.0	66.0	68.0
68.0	64.0	66.0
66.0	62.0	64.0
64.0	59.5	62.0
62.0	57.5	59.5
59.5	55.0	57.5
57.5	52.5	55.0
55.0	50.5	52.5
52.5	48.0	50.5
50.5	46.0	48.0

*Column B approved as NEMA Standard 3/14/1991
†Column C approved as Suggested Standard for Future Designs 3/14/ 1991

Table 11.3 (NEMA Table 12.6B) Full-load efficiencies of energy efficient motors.

OPEN MOTORS

HP	2 POLE (3600)		4 POLE (1800)		6 POLE (1200)		8 POLE (900)	
	NOMINAL EFFICIENCY	MINIMUM EFFICIENCY	NOMINAL EFFICIENCY	MINIMUM EFFICIENCY	NOMINAL EFFICIENCY	MINIMUM EFFICIENCY	NOMINAL EFFICIENCY	MINIMUM EFFICIENCY
1.0	-------	-------	82.5	80.0	77.0	74.0	72.0	68.0
1.5	80.0	77.0	82.5	80.0	82.5	80.0	75.5	72.0
2.0	82.5	80.0	82.5	80.0	84.0	81.5	85.5	82.5
3.0	82.5	80.0	86.5	84.0	85.5	82.5	86.5	84.0
5.0	85.5	82.5	86.5	84.0	86.5	84.0	87.5	85.5
7.5	85.5	82.5	88.5	86.5	88.5	86.5	88.5	86.5
10.0	87.5	85.5	88.5	86.5	90.2	88.5	89.5	87.5
15.0	89.5	87.5	90.2	88.5	89.5	87.5	89.5	87.5
20.0	90.2	88.5	91.0	89.5	90.2	88.5	90.2	88.5
25.0	91.0	89.5	91.7	90.2	91.0	89.5	90.2	88.5
30.0	91.0	89.5	91.7	90.2	91.7	90.2	91.0	89.5
40.0	91.7	90.2	92.4	91.0	91.7	90.2	90.2	88.5
50.0	91.7	90.2	92.4	91.0	91.7	90.2	91.7	90.2
60.0	93.0	91.7	93.0	91.7	92.4	91.0	92.4	91.0
75.0	93.0	91.7	93.6	92.4	93.0	91.7	93.6	92.4
100.0	93.0	91.7	93.6	92.4	93.6	92.4	93.6	92.4
125.0	93.0	91.7	93.6	92.4	93.6	92.4	93.6	92.4
150.0	93.6	92.4	94.1	93.0	93.6	92.4	93.6	92.4
200.0	93.6	92.4	94.1	93.0	94.1	93.0	93.6	92.4

ENCLOSED MOTORS

HP	2 POLE (3600)		4 POLE (1800)		6 POLE (1200)		8 POLE (900)	
	NOMINAL EFFICIENCY	MINIMUM EFFICIENCY	NOMINAL EFFICIENCY	MINIMUM EFFICIENCY	NOMINAL EFFICIENCY	MINIMUM EFFICIENCY	NOMINAL EFFICIENCY	MINIMUM EFFICIENCY
1.0	-------	-------	80.5	77.0	75.5	72.0	72.0	68.0
1.5	78.5	75.5	81.5	78.5	82.5	80.0	75.5	72.0
2.0	81.5	78.5	82.5	80.0	82.5	80.0	82.5	80.0
3.0	82.5	80.0	84.0	81.5	84.0	81.5	81.5	78.5
5.0	85.5	82.5	85.5	82.5	85.5	82.5	84.0	81.5
7.5	85.5	82.5	87.5	85.5	87.5	85.5	85.5	82.5
10.0	87.5	85.5	87.5	85.5	87.5	85.5	87.5	85.5
15.0	87.5	85.5	88.5	86.5	89.5	87.5	88.5	86.5
20.0	88.5	86.5	90.2	88.5	89.5	87.5	89.5	87.5
25.0	89.5	87.5	91.0	89.5	90.2	88.5	89.5	87.5
30.0	89.5	87.5	91.0	89.5	91.0	89.5	90.2	88.5
40.0	90.2	88.5	91.7	90.2	91.7	90.2	90.2	88.5
50.0	90.2	88.5	92.4	91.0	91.7	90.2	91.0	89.5
60.0	91.7	90.2	93.0	91.7	91.7	90.2	91.7	90.2
70.0	92.4	91.0	93.0	91.7	93.0	91.7	93.0	91.7
100.0	93.0	91.7	93.6	92.4	93.0	91.7	93.0	91.7
125.0	93.0	91.7	93.6	92.4	93.0	91.7	93.6	92.4
150.0	93.0	91.7	94.1	93.0	94.1	93.0	93.6	92.4
200.0	94.1	93.0	94.5	93.6	94.1	93.0	94.1	93.0

Table 11.4 (NEMA Table 12.6C) (Suggested standard for future design)
Full-load efficiencies of energy efficient motors.

OPEN MOTORS

HP	2 POLE (3600)		4 POLE (1800)		6 POLE (1200)		8 POLE (900)	
	NOMINAL EFFICIENCY	MINIMUM EFFICIENCY	NOMINAL EFFICIENCY	MINIMUM EFFICIENCY	NOMINAL EFFICIENCY	MINIMUM EFFICIENCY	NOMINAL EFFICIENCY	MINIMUM EFFICIENCY
1.0	-------	-------	82.5	81.5	80.0	78.5	74.0	72.0
1.5	82.5	81.5	84.0	82.5	84.0	82.5	75.5	74.0
2.0	84.0	82.5	84.0	82.5	85.5	84.0	85.5	84.0
3.0	84.0	82.5	86.5	85.5	86.5	85.5	86.5	85.5
5.0	85.5	84.0	87.5	86.5	87.5	86.5	87.5	86.5
7.5	87.5	86.5	88.5	87.5	88.5	87.5	88.5	87.5
10.0	88.5	87.5	89.5	88.5	90.2	89.5	89.5	88.5
15.0	89.5	88.5	91.0	90.2	90.2	89.5	89.5	88.5
20.0	90.2	89.5	91.0	90.2	91.0	90.2	90.2	89.5
25.0	91.0	90.2	91.7	91.0	91.7	91.0	90.2	89.5
30.0	91.0	90.2	92.4	91.7	92.4	91.7	91.0	90.2
40.0	91.7	91.0	93.0	92.4	93.0	92.4	91.0	90.2
50.0	92.4	91.7	93.0	92.4	93.0	92.4	91.7	91.0
60.0	93.0	92.4	93.6	93.0	93.6	93.0	92.4	91.7
75.0	93.0	92.4	94.1	93.6	93.6	93.0	93.6	93.0
100.0	93.0	92.4	94.1	93.6	94.1	93.6	93.6	93.0
125.0	93.6	93.0	94.5	94.1	94.1	93.6	93.6	93.0
150.0	93.6	93.0	95.0	94.5	94.5	94.1	93.6	93.0
200.0	94.5	94.1	95.0	94.5	94.5	94.1	93.6	93.0

ENCLOSED MOTORS

HP	2 POLE (3600)		4 POLE (1800)		6 POLE (1200)		8 POLE (900)	
	NOMINAL EFFICIENCY	MINIMUM EFFICIENCY	NOMINAL EFFICIENCY	MINIMUM EFFICIENCY	NOMINAL EFFICIENCY	MINIMUM EFFICIENCY	NOMINAL EFFICIENCY	MINIMUM EFFICIENCY
1.0	75.5	74.0	82.5	81.5	80.0	78.5	74.0	72.0
1.5	82.5	81.5	84.0	82.5	85.5	84.0	77.0	75.5
2.0	84.0	82.5	84.0	82.5	86.5	85.5	82.5	81.5
3.0	85.5	84.0	87.5	86.5	87.5	86.5	84.0	82.5
5.0	87.5	86.5	87.5	86.5	87.5	86.5	85.5	84.0
7.5	88.5	87.5	89.5	88.5	89.5	88.5	85.5	84.0
10.0	89.5	88.5	89.5	88.5	89.5	88.5	88.5	87.5
15.0	90.2	89.5	91.0	90.2	90.2	89.5	88.5	87.5
20.0	90.2	89.5	91.0	90.2	90.2	89.5	89.5	88.5
25.0	91.0	90.2	92.4	91.7	91.7	91.0	89.5	88.5
30.0	91.0	90.2	92.4	91.7	91.7	91.0	91.0	90.2
40.0	91.7	91.0	93.0	92.4	93.0	92.4	91.0	90.2
50.0	92.4	91.7	93.0	92.4	93.0	92.4	91.7	91.0
60.0	93.0	92.4	93.6	93.0	93.6	93.0	91.7	91.0
75.0	93.0	92.4	94.1	93.6	93.6	93.0	93.0	92.4
100.0	93.6	93.0	94.5	94.1	94.1	93.6	93.0	92.4
125.0	94.5	94.1	94.5	94.1	94.1	93.6	93.6	93.0
150.0	94.5	94.1	95.0	94.5	95.0	94.5	93.6	93.0
200.0	95.0	94.5	95.0	94.5	95.0	94.5	94.1	93.6

11.15   COMPARING MOTORS

It is essential that motor comparison be done on the same basis as to type, size, load, cost of energy, operating hours and most importantly the efficiency values such as nominal vs. nominal or guaranteed vs. guaranteed.

The following equations are used to compare the two motors.

For loads not sensitive to motor speed—

Note: Replacing a standard motor with an energy-efficient motor in a centrifugal pump or fan application can result in increased energy consumption if energy-efficient motor operates at a higher RPM.

Same horsepower—different efficiency.

$$kW_{saved} = hp \times 0.746 \times \left(\frac{100}{E_{STD}} - \frac{100}{E_{EE}} \right)$$

Same horsepower and % load—different efficiency

$$kW_{saved} = hp \times 0.746 \times L \times \left(\frac{100}{E_{STD}} - \frac{100}{E_{EE}} \right)$$

Annual $ savings due to difference in efficiency

$$S = hp \times 0.746 \times L \times C \times N \times \left(\frac{100}{E_{STD}} - \frac{100}{E_{EE}} \right)$$

			Example
S	=	$ Savings (annual)	100
hp	=	Horsepower	100
L	=	% Load	100
C	=	Energy cost ($/kWh)	0.08
N	=	Operating house (annual)	4000
E_{STD}	=	% Efficiency of standard motor	91.7
E_{EE}	=	% Efficiency of energy eff. motor	95.0
RPM_{STD}	=	Speed of standard motor	1775
RPM_{EE}	=	Speed of energy eff. motor	1790

For loads sensitive to motor speed

Above equations should be multiplied by speed ratio correction factor.

SRCF = Speed Ratio Correction Factor

$$\left(\frac{RPM_{EE}}{RPM_{STD}} \right)^3$$

Example:

$$S = 100 \times 0.746 \times 1 \times 0.080 \times 4000$$
$$\times (100/91.7 - 100/95.0) = \$904$$
$$S = 100 \times 0.746 \times 1 \times 0.080 \times 4000$$
$$\times (100/91.7 - 100/95.0) \times (1790/1775)3 = \$262$$
$$\$642 \text{ reduction in expected savings.}$$

Relatively minor, 15 RPM, increase in a motor's rotational speed results in a 2.6 percent increase in the load placed upon the motor by the rotating equipment.

11.16   SENSITIVITY OF LOAD TO MOTOR RPM

When employing electric motors, for air moving equipment, it is important to remember that the performance of fans and blowers is governed by certain rules of physics. These rules are known as "The Affinity Law" or "The Fan Law." There are several parts to it, and are all related to each other in a known manner and when one changes, all others change. For centrifugal loads, even a minor change in the motor's speed translates into significant change in energy consumption and is especially troublesome when the additional air flow is not needed or useful. Awareness of the sensitivity of load and energy requirements to motor speed can help effectively identify motors with specific performance requirements. In most cases we can capture the full energy conservation benefits associated with an energy efficient motor retrofits.

Terminology of Load to Motor RPM

CFM   Fan capacity (Cubic Feet per Minute)
Volume of air moved by the fan per unit of time.

P   Pressure. Pressure produced by the fan that can exist whether the air is in motion or confined in a closed duct.

HP   Horsepower. The power required to drive an air moving device.

RPM   Revolutions Per Minute. The speed at which the shaft of air moving equipment is rotating.

Affinity Laws or Fan Laws

$$\text{Law \# 1} \quad \frac{CFM_2}{CFM_1} = \frac{RPM_2}{RPM_1}$$

Quantity (CFM) varies as fan speed (RPM)

$$\text{Law \# 2} \quad \frac{P_2}{P_1} = \frac{(RPM_2)^2}{(RPM_1)^2}$$

Pressure (P) varies as the square of fan speed (RPM)

$$\text{Law \# 3} \quad \frac{HP_2}{HP_1} = \frac{(RPM_2)^3}{(RPM_1)^3}$$

Horsepower (HP) varies as the <u>cube</u> of fan speed (RPM)

Example

Fan system <u>32,000</u> CFM
Motor <u>20</u> HP <u>1750</u> RPM (existing)
Motor <u>20</u> HP <u>1790</u> RPM (new EE)

kW = 20 × 0.746 = 14.92 kW
New CFM with new motor = 1790/1750 × 32,000 = 32,731 or 2.3% increase
New HP = (1790/1750)3 × 20 × = 21.4 HP or 7% increase.
New kW = 21.4 × 0.746 = 15.96 kW 7% increase in kW and work performed by motor.

Replacing a standard motor with an energy efficient motor in centrifugal pump or a fan application can result in increased energy consumption if the energy efficient motor operates at a higher RPM. Table 11.5 shows how a 10 RPM increase can negate any savings associated with a high efficiency motor retrofit.

11.17 THEORETICAL POWER CONSUMPTION

Figure 11.5 illustrates the energy saving potential of the application of *Adjustable Speed Drive* to an application that traditionally uses throttling control, such as *Discharge Damper, Variable Inlet Vane* or *Eddy Current Drive.*

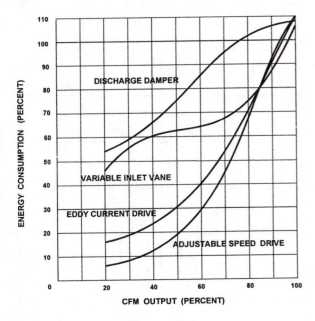

Figure 11.5

From the standpoint of maximum energy conservation, the most optimal method to reduce fan CFM is to reduce the fan's speed (RPM). This can be accomplished by changing either, the sheaves of the motor, the sheaves of the fan or by varying fan motor speed.

Applications Involving Extended Periods of Light Load Operation[2]

A number of methods have been proposed to reduce the voltage applied to the motor in response to the applied load, the purpose of this being to reduce the magnetizing losses during the periods when the full torque capability of the motor is not required. Typical of these devices is the power factor controller. The power factor controller is a device that adjusts the voltage applied to the motor to approximate a preset power factor.

These power factor controllers may, for example, be beneficial for use with small motors operating for extended periods of light loads where the magnetization losses are a relatively high percentage of the total loss. Care must be exercised in the application of these controllers. Savings are achieved only when the controlled motor is operated for extended periods at no load or light load.

Particular care must be taken when considering their use with other than small motors. A typical 10 horsepower motor will have idle losses in the order of 4 or 5 percent of the rated output. In this size range the magnetization losses that can be saved may not be equal to the losses added by the controller plus the additional motor losses caused by the distorted voltage wave form induced by the controller.

Applications Involving Throttling or By-pass Control.[2]

Many pump and fan applications involve the control of flow or pressure by means of throttling or bypass devices. Throttling and bypass valves are in effect series and parallel power regulators that perform their function by dissipating the difference between source energy supplied and the desired sink energy.

These losses can be dramatically reduced by controlling the flow rate or pressure by controlling the speed of the pump or fan with a variable speed drive.

Figure 11.5 illustrates the energy saving potential of the application of variable speed drive to an application that traditionally uses throttling control.

A simplistic example will serve to illustrate the savings to be achieved by the use of this powerful energy conservation tool.

Table 11.5 Hourly operating costs.

HOURLY OPERATING COSTS
100% Load

A MINOR RPM INCREASE IN A MOTOR'S ROTATIONAL SPEED RESULTS IN AN INCREASE IN THE LOAD PLACED UPON THE MOTOR BY ROTATING EQUIPMENT.

HP	AVERAGE STANDARD MOTOR FL RPM EFF. %	ENERGY EFFICIENT MOTOR SAME RPM EFF. %	ENERGY EFFICIENT MOTOR + 10 RPM EFF. %	$/kWh $0.01	AVERAGE ELECTRICITY PRICE ($/kWh)						
					$0.05	$0.06	$0.07	$0.08	$0.09	$0.10	$0.11
5	83.8			0.0445	$0.22	$0.27	$0.31	$0.36	$0.40	$0.45	$0.49
5		86.5		0.0431	$0.22	$0.26	$0.30	$0.34	$0.39	$0.43	$0.47
5			86.5	0.0439	$0.22	$0.26	$0.31	$0.35	$0.39	$0.44	$0.48
7.5	85.3			0.0656	$0.33	$0.39	$0.46	$0.52	$0.59	$0.66	$0.72
7.5		88.5		0.0632	$0.32	$0.38	$0.44	$0.51	$0.57	$0.63	$0.70
7.5			88.5	0.0643	$0.32	$0.39	$0.45	$0.51	$0.58	$0.64	$0.71
10	87.2			0.0856	$0.43	$0.51	$0.60	$0.68	$0.77	$0.86	$0.94
10		88.5		0.0843	$0.42	$0.51	$0.59	$0.67	$0.76	$0.84	$0.93
10			88.5	0.0857	$0.43	$0.51	$0.60	$0.69	$0.77	$0.86	$0.94
15	87.6			0.1277	$0.64	$0.77	$0.89	$1.02	$1.15	$1.28	$1.41
15		90.2		0.1241	$0.62	$0.74	$0.87	$0.99	$1.12	$1.24	$1.36
15			90.2	0.1262	$0.63	$0.76	$0.88	$1.01	$1.14	$1.26	$1.39
20	88.4			0.1688	$0.84	$1.01	$1.18	$1.35	$1.52	$1.69	$1.86
20		91.0		0.1640	$0.82	$0.98	$1.15	$1.31	$1.48	$1.64	$1.80
20			91.0	0.1667	$0.83	$1.00	$1.17	$1.33	$1.50	$1.67	$1.83
25	89.2			0.2091	$1.05	$1.25	$1.46	$1.67	$1.88	$2.09	$2.30
25		91.7		0.2034	$1.02	$1.22	$1.42	$1.63	$1.83	$2.03	$2.24
25			91.7	0.2068	$1.03	$1.24	$1.45	$1.65	$1.86	$2.07	$2.28
30	89.2			0.2509	$1.25	$1.51	$1.76	$2.01	$2.26	$2.51	$2.76
30		91.7		0.2441	$1.22	$1.46	$1.71	$1.95	$2.20	$2.44	$2.68
30			91.7	0.2482	$1.24	$1.49	$1.74	$1.99	$2.23	$2.48	$2.73
40	90.2			0.3308	$1.65	$1.98	$2.32	$2.65	$2.98	$3.31	$3.64
40		92.4		0.3229	$1.61	$1.94	$2.26	$2.58	$2.91	$3.23	$3.55
40			92.4	0.3284	$1.64	$1.97	$2.30	$2.63	$2.96	$3.28	$3.61
50	90.1			0.4140	$2.07	$2.48	$2.90	$3.31	$3.73	$4.14	$4.55
50		92.4		0.4037	$2.02	$2.42	$2.83	$3.23	$3.63	$4.04	$4.44
50			92.4	0.4105	$2.05	$2.46	$2.87	$3.28	$3.69	$4.11	$4.52
60	91.0			0.4919	$2.46	$2.95	$3.44	$3.93	$4.43	$4.92	$5.41
60		93.0		0.4813	$2.41	$2.89	$3.37	$3.85	$4.33	$4.81	$5.29
60			93.0	0.4895	$2.45	$2.94	$3.43	$3.92	$4.41	$4.89	$5.38
75	91.9			0.6088	$3.04	$3.65	$4.26	$4.87	$5.48	$6.09	$6.70
75		93.6		0.5978	$2.99	$3.59	$4.18	$4.78	$5.38	$5.98	$6.58
75			93.6	0.6079	$3.04	$3.65	$4.26	$4.86	$5.47	$6.08	$6.69
100	91.7			0.8135	$4.07	$4.88	$5.69	$6.51	$7.32	$8.14	$8.95
100		93.6		0.7970	$3.99	$4.78	$5.58	$6.38	$7.17	$7.97	$8.77
100			93.6	0.8106	$4.05	$4.86	$5.67	$6.48	$7.30	$8.11	$8.92
125	91.7			1.0169	$5.08	$6.10	$7.12	$8.14	$9.15	$10.17	$11.19
125		93.6		0.9963	$4.98	$5.98	$6.97	$7.97	$8.97	$9.96	$10.96
125			93.6	1.0132	$5.07	$6.08	$7.09	$8.11	$9.12	$10.13	$11.15
150	92.9			1.2045	$6.02	$7.23	$8.43	$9.64	$10.84	$12.05	$13.25
150		94.1		1.1892	$5.95	$7.13	$8.32	$9.51	$10.70	$11.89	$13.08
150			94.1	1.2094	$6.05	$7.26	$8.47	$9.68	$10.88	$12.09	$13.30
200	93.1			1.6026	$8.01	$9.62	$11.22	$12.82	$14.42	$16.03	$17.63
200		94.1		1.5855	$7.93	$9.51	$11.10	$12.68	$14.27	$15.86	$17.44
200			94.1	1.6125	$8.06	$9.68	$11.29	$12.90	$14.51	$16.13	$17.74

Assumptions:

Line Power Required by Fan at full CFM
    without flow control device   100 HP
Full CFM required   1000 hours per year
75% CFM required   3000 hours per year
50% CFM required   2000 hours per year
Cost of energy $.06 per kWhr

% Power consumption with various flow control methods per Figure 11.5.

Annual cost of Energy

$$\$ = \frac{hrs \times hp \times 746 \times \% \, energy \, consumption \times \$/kW \, hw}{1000 \times 100}$$

Annual Cost of Energy Summary:

Discharge Damper	$24,600
Variable Inlet Vane	$19,600
Eddy Current Drive	$15,900
Adjustable Speed Drive	$13,900

11.18 MOTOR EFFICIENCY MANAGEMENT

Many think that when one is saying *Motor Efficiency* the logical word to follow is *improvement*. Where are we going? How far can we push manufacturers in our quest for the perfect motor?

During the 102nd Congress, the Markey Bill, H.R. 2451, was introduced. The bill mandated component efficiency standards for such products as lighting, distribution transformers and electric A.C. motors.

This plan was met with opposition by NEMA and other interested groups. They called for a system approach that would recognize the complex nature of the product involved under the plan. The bill passed by the Energy & Power Subcommittee on the theory that the elimination of the least efficient component from the market would ensure that consumers would purchase and use the most efficient products possible.

Although motors tend to be quite efficient in themselves, several factors can contribute to cost-effective replacement or retrofit alternatives to obtain efficiency gain in motors. We are well aware that the electric motor's primary function is to convert electrical energy into mechanical work. It is also important to remember that *good energy management requires a consideration of the total system of which the motor is a part.*

Experience indicates that despite heightened awareness and concern with energy efficiency, the electric motor is either completely neglected or decisions are made on the basis of incomplete information. At this point I would like to quote me.

"Motors Don't Waste Energy, People Do."

What this really means is that we must start managing efficiency and not just improving the motor. This is what will improve your corporate bottom line.

11.19 MOTORS ARE LIKE PEOPLE

Motors can be managed the same way and with the same skills as people. There are amazing similarities. I have spent years managing both and find there is very little difference between the two.

The expectations are the same for one as for the other. The employee's performance is evaluated to identify improvement opportunities linking them to organizational goals and business objectives. The manager measures the performance of an employee as an individual and as a member of a team. Why then would it not work the same with your motors? Motors employed just as people are and they work as an individual or as a team. They will perform their best if cared for, maintained, evaluated and rewarded.

An on going analysis of motor performance prevents major breakdown. Performance evaluation of a motor should be done as routinely as it is done on an employee. Both the motor and an employee are equally important. Applied motor maintenance will keep the building or plant running smoothly with minimal stress on the system or downtime due to failure.

MAXIMIZE YOUR EFFECTIVENESS WITH MOTOR EVALUATION SYSTEM

11.20 MOTOR PERFORMANCE MANAGEMENT PROCESS (MPMP)

The Motor Performance Management Process (MPMP) is designed to be the Motor Manager's primary tool to evaluate, measure and most importantly *manage electric motors.* MPMP focuses on building a stronger relationship between Motor Manager and the electric motor *employed* to perform a task. Specifically, it is a logical, systematic and structured approach to reduce energy waste. Energy waste reduction is fundamental in becoming more efficient in an increasingly competitive market. The implementation of MPMP is more than a good business practice it is an intelligent management resource.

→NEGLECTING YOUR MOTORS CAN BE A COSTLY MISTAKE←

Motor Managers must understand motor efficiency, how it is achieved and how to conduct an economic evaluation. Timely implementation of MPMP would be an effective way to evaluate existing motor performance in a system and identify improvement opportunities linking them to organizational goals. The following *Motor Manager* model defines the task of managing and enabling motors to operate in the ways required to achieve business objectives.

$$\frac{\text{Motor}}{\text{Manager}} \Rightarrow \text{Motors} \Rightarrow \frac{\text{Desired}}{\text{Operation}} \Rightarrow \frac{\text{Business}}{\text{Objectives}}$$

It is vital to evaluate the differences between motors offered by various manufacturers and only choose those that clearly meet your operating criteria. An investment of 20 to 25% for an Energy Efficient motor, over the cost of a standard motor, can be recovered in a relatively short period of time. Furthermore, with some motors the cost of wasted energy exceeds the original motor price even in the first year of operation.

11.21 HOW TO START MPMP.

Begin by conscientiously gathering information on each motor used in excess of 2000 hours per year.

Complete a <u>MOTOR RECORD FORM</u> (Figure 11.6) for each motor. (*A detailed profile of your existing Motor.*)

* This form must be established for each motor to serve as a performance record. It will help you to understand WHEN, WHERE, and HOW your motors are used, and identify opportunities to improve drive system efficiency.

* Each item in this form must be addressed, paying particular attention to the following items: *Motor Location, Application, Energy Cost $/kWh, Operating Speed, Operating Load* and *Nameplate information.*

* Motors with special electrical designs or mechanical features should be studied carefully. Some applications require special attention, such as fans, compressors and pumps.

* Motors with a history of repeated repair should be of special interest.

* Examine your completed MOTOR RECORD FORM and select the best candidates for possible retrofit or future replacement.

* Check with your local utility regarding the availability of financial incentives or motor rebates.

Finding the right motor for the application, and calculating its energy and cost savings can be done with the <u>MotorMaster</u>[5] software and WHAT IF motor comparison form Figure 11.7. This form has the capability to compare several motors and analyze potential savings.

K.K.LOBODOVSKY 11/1993
(MM_MRF.DOC & XLS)

MOTORMANAGER

MOTOR RECORD FORM

CUSTOMER NAME:		PREPARED BY:
MOTOR LOCATION:		DATE PREPARED:
APPLICATION:		ENERGY COST $/KWH

MOTOR MANUFACTURER	HP	OPERATING HOURS	% EFF 100% LOAD	% EFF 75% LOAD	% EFF 50% LOAD	% EFF 25% LOAD
MODEL NO.	DESIGN	RPM (FULL LOAD)	VOLTS	AMPS (FULL LOAD)	HZ	SERVICE FACTOR
SERIAL NO.	YEAR PURCHASED	FRAME	INSULATION CLASS	TYPE	CODE	PHASE
NAME:	YEAR REWOUND	DUTY	POWER FACTOR			

NO LOAD DATA FROM MOTOR MANUFACTURER	KW	AMPS	VOLTS	CORE LOSS	F. & W. LOSS	STATOR RESIST.

STATOR RESISTANCE DATA	A - B	B - C	C - A	CONDUCTOR	MCM:	AWG :
MEASURED RESISTANCE ->				MATERIAL	CU:	AL:
MOTOR SURFACE TEMPERATURE->		<-DEG F DEG C ->		DISTANCE POINT. TO MOTOR IN FEET ->		
AMBIENT TEMPERATURE->		<-DEG F DEG C ->		WIRE RESISTANCE IN OHMS / 1000FT->		

	LOAD TEST			NO LOAD TEST		
TEST DATE						
CURRENT TRANSFORMER						
SPEED (RPM)						
VOLTS A - B						
VOLTS B - C						
AMPS A						
AMPS C						
KW TOTAL						
KVAR TOTAL						
KVA TOTAL						
PF TOTAL						

BE = BLUE
RD = RED
BK = BLACK
WE = WHITE

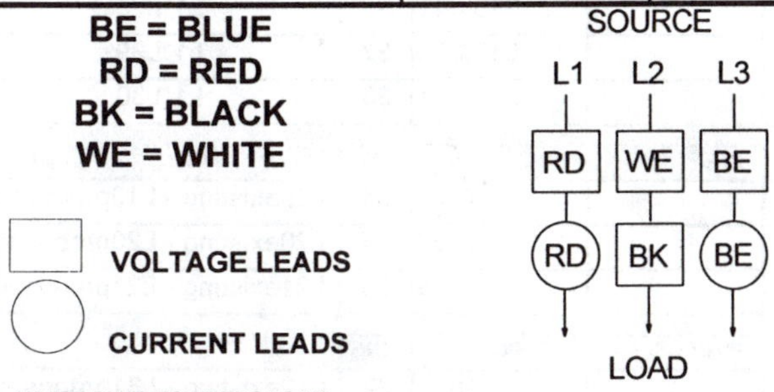

☐ VOLTAGE LEADS

◯ CURRENT LEADS

WHITE PLUG TOWARDS LOAD

SOURCE

L1 L2 L3

RD WE BE

RD BK BE

LOAD

REMARKS:

Figure 11.6 Motor record form.

WHAT IF

$$S = hp \times 0.746 \times L \times C \times N \times \left(\frac{100}{E_{STD}} - \frac{100}{E_{EE}} \right)$$

S=$ SAVINGS hp= HORSEPOWER L=%LOAD C=ENERGY COST N=OPERATING HRS E=%EFFICIENCY STD. OR EE.

	EXISTING	PROPOSAL 1	PROPOSAL 2	PROPOSAL 3	PROPOSAL 4	PROPOSAL 5	LINE	FORMULA (L1 = Line 1…)
ENERGY COST $/KWH	$0.080					EXAMPLE	1	C
OPERATING HRS (PER YEAR)	4000					Baldor	2	N
MOTOR HP	100					EM4400T-8	3	hp
NO-LOAD SPEED RPM	1800						4	NL RPM
FULL-LOAD SPEED (SEE Note)	1775					1780	5	FL RPM
Note: FULL-LOAD SPEED information is important for loads sensitive to motor speed.								
EFFICIENCY @ 100% LOAD	91.7					95.0	6	EFFICIENCY @ 100% LOAD
EFFICIENCY @ 75% LOAD	91.7					95.4	7	EFFICIENCY @ 75% LOAD
EFFICIENCY @ 50% LOAD	91.7					95.5	8	EFFICIENCY @ 50% LOAD
INVESTMENT / SALVAGE	$0					$7,180	9	INVESTMENT / SALVAGE
SLIP	25					20	10	L4-L5
CALCULATED SPEED @ 75% LOAD	1,781					1,785	11	L5+(L10*0.25)
CALCULATED SPEED @ 50% LOAD	1,788					1,790	12	L5+(L10*0.5)
FOR LOADS NOT SENSITIVE TO MOTOR SPEED		PROPOSAL 1	PROPOSAL 2	PROPOSAL 3	PROPOSAL 4	PROPOSAL 5	LINE	
KW @ 100% EFF & LOAD	81.4					78.5	13	L3*0.746*100/L6
KW @ 75% EFF & LOAD	61.0					58.6	14	L3*0.746*0.75*100/L7
KW @ 50% EFF & LOAD	40.7					39.1	15	L3*0.746*0.5*100/L8
KWh @ 100% EFF & LOAD	325,409					314,105	16	L2*L13
KWh @ 75% EFF & LOAD	244,057					234,591	17	L2*L14
KWh @ 50% EFF & LOAD	162,704					156,230	18	L2*L15
OPERATING COST 100%LOAD	$26,033					$25,128	19	L1*L16
OPERATING COST 75%LOAD	$19,525					$18,767	20	L1*L17
OPERATING COST 50%LOAD	$13,016					$12,498	21	L1*L18
FOR LOADS SENSITIVE TO MOTOR SPEED		PROPOSAL 1	PROPOSAL 2	PROPOSAL 3	PROPOSAL 4	PROPOSAL 5	LINE	
SPEED RATIO OF 100%LOAD						1.0085	22	(L5prop./L5exist)cubed
(MOTOR B / MOTOR A) CUBE 75%LOAD						1.0063	23	(L11prop./L11exist)cubed
(FAN OR AFFINITY LAW) 50%LOAD						1.0042	24	(L12prop./L12exist)cubed
KW @ 100% EFF & LOAD						79.2	25	L3*0.746*L22*100/L6
KW @ 75% EFF & LOAD						59.0	26	L3*0.746*0.75*L23*100/L7
KW @ 50% EFF & LOAD						39.2	27	L3*0.746*0.5*L24*100/L8
KWh @ 100% EFF & LOAD						316,767	28	L2*L25
KWh @ 75% EFF & LOAD						236,076	29	L2*L26
KWh @ 50% EFF & LOAD						156,887	30	L2*L27
OPERATING COST 100%LOAD						$25,341	31	L1*L28
OPERATING COST 75%LOAD						$18,886	32	L1*L29
OPERATING COST 50%LOAD						$12,551	33	L1*L30
FOR LOADS NOT SENSITIVE TO MOTOR SPEED		PROPOSAL 1	PROPOSAL 2	PROPOSAL 3	PROPOSAL 4	PROPOSAL 5	LINE	
$ SAVINGS (LOSS) @ 100%LOAD						$904	34	L19existing - L19proposed
$ SAVINGS (LOSS) @ 75%LOAD						$757	35	L20existing - L20proposed
$ SAVINGS (LOSS) @ 50%LOAD						$518	36	L21existing - L21proposed
FOR LOADS SENSITIVE TO MOTOR SPEED		PROPOSAL 1	PROPOSAL 2	PROPOSAL 3	PROPOSAL 4	PROPOSAL 5	LINE	
$ SAVINGS (LOSS) @ 100%LOAD						$691	37	L19existing - L31proposed
$ SAVINGS (LOSS) @ 75%LOAD						$638	38	L20existing - L32proposed
$ SAVINGS (LOSS) @ 50%LOAD						$465	39	L21existing - L33proposed

Figure 11.7 WHAT IF motor comparison form.

HOW TO GET AROUND IN THE 'WHAT IF' FORM.

COLUMN	LINE	EXPLANATION
EXISTING	1-9	Information can be taken from the Motor Record Form if previously generated. If not available, generate data.
EXISTING	1	ENERGY COST $/kWh Self Explanatory
EXISTING	2	OPERATING HOURS PER YEAR is very important to be as accurate as possible
EXISTING	3	MOTOR HORSEPOWER Self Explanatory
EXISTING	4	NO LOAD SPEED RPM (synchronous speed) is usually within 5% of Full Load Speed i.e. 900, 1200, 1800, or 3600 rpm.
EXISTING	5	FULL LOAD SPEED is found on the nameplate. This information is important for loads sensitive to motor speed.
EXISTING	6	EFFICIENCY @ 100% LOAD NEMA % efficiency at full load. If motor is relatively new this will be found on the nameplate, if older, it will be necessary to contact the manufacturer or the MotorMaster data base. (WSEO)
EXISTING	7-8	EFFICIENCY @ 75% AND 50% LOAD This information can be obtained from the manufacturer or the MotorMaster data base. (WSEO)
EXISTING	9	INVESTMENT/SALVAGE This should include total cost associated with purchase of a motor, such as cost of motor installation, balancing alignment and disposition of existing motor.
PROPOSAL 1-5	5-9	Information is acquired from the motor manufacturers catalogs or the MotorMaster data base which contains information on more than 10,000 motors from various manufacturers.
EXISTING	10-21	The data is calculated using the formulas in the column entitled FORMULA or is automatically calculated if using the What If spreadsheet.
PROPOSAL 1-5	10-39	The data is calculated using the formulas in the column entitled FORMULA or is automatically calculated if using the What If spreadsheet.
FORMULA	10-39	The formulas used for calculating. These formulas may also be used to create a spreadsheet similar to 'What If.'

11.22 NAMEPLATE GLOSSARY

HP—The number of, or fractional part of a horsepower, the motor will produce at rated speed.

RPM—An indication of the approximate speed that the motor will run when it is putting out full rated output torque or horsepower is called full load speed.

VOLTS—Voltage at which the motor may be operated. Generally, this will be 115 Volts, 230 Volts, 115/230 V, or 220/440 V.

AMPS—The amount of current the motor can be expected to draw under full load (torque) conditions is called full load amps. It is also known as nameplate amps.

HZ—Frequency at which the motor is to be operated. This will almost always be 60 Hertz.

SERVICE FACTOR—The service factor is a multiplier that indicates the amount of overload a motor can be expected to handle. For example, a motor with a 1.15 service factor can be expected to safely handle intermittent loads amounting to 15% beyond its nameplate horsepower.

DESIGN—The design letter is an indication of the shape of the torque speed curve. They are A. B. C and D. Design B is the standard industrial duty motor which has reasonable starting torque with moderate starting current and good overall performance for most industrial applications. Design C is used for hard to start loads and is specifically designed to have high starting torque. Design D is the so called high slip motor which tends to have very high starting torque but has high slip rpm at full load torque. Design A motors are not commonly specified but specialized motors used on injection molding applications.

FRAME—Motors, like suits of clothes, shoes and hats, come in various sizes to match the requirements of the application.

INSULATION CLASS—The insulation class is a measure of the resistance of the insulating components of a motor to degradation from heat. Four major classifications of insulation are used in motors. They are, in order of increasing thermal capabilities, A, B, C, and F.

TYPE—Letter code that each manufacturer uses to indicate something about the construction and the power the motor runs on. Codes will indicate split phase, capacitor start, shaded pole, etc.

CODE—This is a NEMA code letter designating the locked rotor kVA per horsepower.

PHASE—The indication of the type of power supply for which the motor is designed. Two major categories exist; single phase and three phase.

AMB. DEG. C.—Ambient temperature is the maximum safe room temperature surrounding the motor if it is going to be operated continuously at full load. In most cases, the standardized ambient temperature rating is 40 degrees C (104 degrees F).

TEMPERATURE RISE—Temperature rise is the amount of temperature change that can be expected within the winding of the motor from non-operating (cool condition) to its temperature at full load continuous operating condition. Temperature rise is normally expressed in degrees centigrade.

DUTY—Most motors are rated for continuous duty which means that they can operate at full load torque continuously without overheating. Motors used on certain types of applications such as waste disposal, valve actuators, hoists, and other types of intermittent loads, will frequently be rated for short term duty such as 5 minutes, 15 minutes, 30 minutes, or 1 hour.

EFF. INDEX OR NEMA %—Efficiency is the percentage of the input power that is actually converted to work output from the motor shaft. Efficiency is now being stamped on the nameplate of most domestically produced electric motors.

Summary

Over 60 percent of electricity in the United States is consumed by electric motor drive systems. Generally, a motor drive system consists of several components; a power supply, controls, the electric motor and the mechanical transmission system. The function of an electric motor is to convert electric energy into mechanical work.

During the conversion the only power consumed by the electric motor is the energy losses within the motor. Since the motor losses are in the range of 5-30% of the input power, it is important to consider the total system of which the motor is a part.

This chapter deals mostly with electric motor drive systems and provides practical methods for managing motors.

One of the "MotorManager" methods is the Motor Performance Management Process (MPMP) which effectively evaluates the performance of existing motors and identifies opportunities to link them to organizational goals. It is a primary tool to evaluate, measure and manage motors and a logical, systematic, structured approach in reducing energy waste, fundamental to efficiency in a competitive market. With minor changes, this process can be used to evaluate other electrical and mechanical equipment.

Considerable attention must be paid to the efficiencies of all electric equipment being purchased today. This is true not only for motors but for transformers of all types and other electrical devices.

To guard against the waste of electrical energy, manufacturers of dry type transformers are designing them with lower than normal conductor and total losses. The reduction in these losses also lowers the temperature rise of the transformer resulting in improved life expectancies as well as a reduction in the air conditioning requirements.

References

1. Technology Update Produced by the Washington State Energy Office for the Bonneville Power Administration.
2. NEMA Standards publication No. MG 10. *Energy Management Guide for Selection and Use of Polyphase Motors.* National Electrical Manufacturers Association, Washington, D.C. 1989
3. McCoy, G., A. Litman, and J. Douglass. Energy Efficient Electric Motor Selection Handbook. Bonneville Power Admin. 1991
4. Ed Cowern Baldor Electric Motors.
5. **MotorMaster** software developed by the Washington State Energy Office (WSEO) puts information on more than 10,000 three-phase motors at your fingertips. The **MotorMaster** lets you review features and compare efficiency, first cost, and operating cost. The U.S. Department of Energy (DOE) and the Bonneville Power Administration (BPA) funded the development of the software. You may obtain this valuable software or more detailed information by contacting WSEO (360) 956-2000
6. Financing for Energy Efficiency Measures, Virginia Power, POSTAND, REV(0)08-17-93.
7. William E. Lanning and Steven E. Strauss Power Survey Techniques for electrical loads Esterline Angus Instrument Corp.

APPENDIX

ELECTRONIC ADJUSTABLE SPEED DRIVES: ISSUES AND APPLICATIONS

CLINT D. CHRISTENSON

Oklahoma Industrial Assessment Center
School of Industrial Engineering and Management
322 Engineering North
Oklahoma State University
Stillwater, OK 74078

INTRODUCTION

Electric motors are used extensively to drive fans, pumps, conveyors, printing presses, and many other processes. A majority of these motors are standard, 3-phase, AC induction motors that operate at a single speed. If the process (fan, pump, etc.) is required to operate at a speed different than the design of the motor, pulleys are applied to adjust the speed of the equipment. If the process requires more than one speed during its operation, various methods have been applied to allow speed variation of a single speed motor. These methods include variable pitch pulley drives, motor-generator sets, inlet or outlet dampers, inlet guide vanes, and Variable Frequency Drives. The following section will briefly discuss each of these speed control technologies.

Changing the pulleys throughout the day to follow demand is not feasible, but the use of a "Reeves" type variable pitch pulleys drive was a common application. These drives utilized a wide belt between two pairs of opposing conical pulleys. As the conical pulleys of the driven shaft were brought together (moved apart) the pulley diameter would increase (decrease) and decrease (increase) the belt speed and the process speed. This type of system is still extensively in use in the food and chemical industries where mixing speeds can dramatically effect product quality. These systems are a mechanical speed adjustment system which has inherent function losses and require routine maintenance.

Motor-Generator sets were used in the past to convert incoming electricity to a form required in the process including changes from AC to DC. The DC output could then be used to synchronize numerous dc motors at the required speed. This was the common type of speed control in printing presses and other "web" type systems. The use of an Eddy-current clutch would vary the output of the generator to the specific needs of the system. The windage and other losses associated with motors are at least doubled with the generator and the efficiency of the system drops drastically at low load situations.

A majority of the commercial and industrial fan and pump speed control techniques employed do not involve speed control at all. These systems utilize inlet dampers, outlet dampers (valves) with or without by-pass, or inlet guidevanes to vary the flow to the process. Inlet dampers, used in fan applications, reduce the amount of air supplied to the process by reducing the inlet pressure. Outlet dampers (valves) maintain system pressure (head) seen by the fan (pump) while reducing the actual volume of air (liquid) flowing. Inlet guide vanes are used in fans similar to inlet dampers but the guidevanes are situated such that as air flow is reduced, the circular motion of the fan is imparted upon the incoming air. Each of these control methods operates to reduce the amount of flow with some reduction in energy required.

Variable Frequency Drives (VFD) change the speed of the motor by changing the voltage and frequency of the electricity supplied to the motor based upon system requirements. This is accomplished by converting the AC to DC and then by various switching mechanisms invert the DC to a synthetic AC output with controlled voltage and frequency [Phipps, 1994]. If this process is accomplished properly, the speed of the motor can be controlled over a wide variation in shaft speed (0 rpm through twice name plate) with the proper torque characteristics for the application. The remainder of this paper will discuss the various issues and applications of VFD's. Figure 1 shows the percent power curves versus percent load for simplified centrifugal air handling fan application.

VARIABLE FREQUENCY DRIVE TYPES

In order to maintain proper power factor and reduce excessive heating of the motor, the name plate volts per hertz ratio must be maintained. This is the main function of the variable frequency drive (VFD). The four main components that make up AC variable frequency drives (VFD's) are the Converter, Inverter, the DC circuit which links the two, and a control unit, shown in Figure 2. The converter contains a rectifier and other circuitry which converts the fixed frequency AC to DC. The inverter converts the DC to an adjustable frequency, adjustable voltage AC (both must be adjustable to maintain a constant volts to hertz ratio). The DC circuit filters the DC and conducts the DC to the inverter. The control unit controls the output voltage and frequency based upon feedback from the process (e.g. pressure sensor). The three main types of inverter designs are voltage source

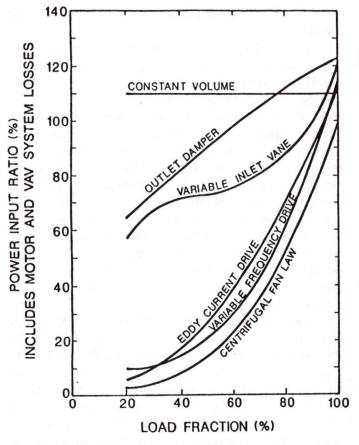

Figure 1. Typical power consumption of various fan control systems. (Source: Moses et. al., 1989)

inverters, current source inverters, and pulse width modulation inverters. Each will be briefly discussed in the next section.

Inverter Types

The voltage source inverters (VSI) use a silicon controlled rectifier (SCR) to rebuild a pseudosine wave form for delivery to the motor. This is accomplished with a six-step voltage inverter with a voltage source converter and a variable voltage DC bus. As with any SCR system, troublesome harmonics are reflected to the power source. Also the six-step wave form sends current in pulses which can cause the motor to cog at low frequencies, which can damage keyways, couplings, pump impellers, etc. [Phipps, 1994].

The current source inverters (CSI) also use an SCR input form the power source but control the current to the motor rather than the voltage. This is accomplished with a six-step current inverter with a voltage source converter and a variable voltage DC bus. The CSI systems have the same problems with cogging and harmonics as the VSI systems. Many manufacturers only offer VSI or CSI VFD's for larger horsepower sizes (over 300 HP).

The pulse width modulation (PWM) VFD's have become the state of the art concept in the past several years, starting with the smaller hp sizes, and available up to 1500 hp from at least one manufacturer [Phipps, 1994]. PWM uses a simple diode bridge rectifier for power input to a constant voltage DC bus. The PWM inverter develops the output voltage by producing pulses of varying widths which combine to synthesize the desired wave form. The diode bridge significantly reduces the harmonics from the power source. The PWM system produces a current wave form that more closely matches the power line wave form, which reduces adverse heating. The PWM drives also have the advantage of virtually constant power factor at all speeds. Depending on size, it is possible to have power factors over 95% [Phipps, 1994]. Another advantage of the PWM VFD's is that sufficient current frequency (~200+ Hz) is available to operate multiple motors on a single drive, which would be advantageous for the printing press example discussed earlier. The next section will discuss the types of loads that require adjustable speeds that may be controlled by variable frequency drives.

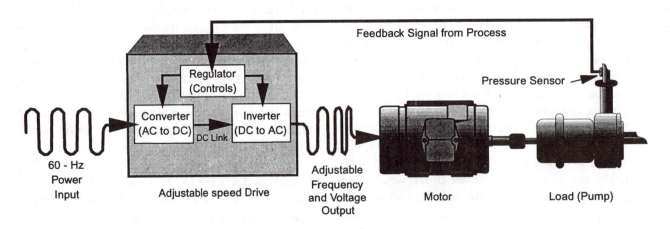

Figure 2. Typical VFD system. (Source: Lobodovsky, 1996)

VARIABLE SPEED LOADS

The three common types of adjustable speed loads are variable torque, constant torque, and constant horsepower loads. A variable torque load requires much lower torque at low speeds than at high speeds. With this type of load, horsepower varies approximately as the cube of speed and the torque varies approximately as the square of the speed. This type of load is used in applications such as centrifugal fans, pumps, and blowers. A constant torque load requires the same amount of torque at low speed as at high speed. The torque remains constant throughout the speed range, and the horsepower increases or decreases in direct proportion to the speed. A constant torque load is used in applications such as conveyors, positive displacement pumps, some extruders, and for shock loads, overloads, or high inertia loads. A constant horsepower load requires high torque at low speeds, and low torque at high speeds, and therefore constant horsepower at any speed. Constant horsepower loads are encountered in most metal cutting operations, and some extruders [Lobodovsky, 1996]. The savings available from non-centrifugal (constant torque or constant horsepower) loads are based primarily on the VFD's high efficiency (when compared to standard mechanical systems), increased power factor, and reduced maintenance costs. The next section will discuss several applications of VFD's and the issues involved with the application.

VARIABLE FREQUENCY DRIVE APPLICATIONS

Variable frequency drive systems offer many benefits that result in energy savings through efficient and effective use of electric power. The energy savings are achieved by eliminating throttling, performance, and friction losses associated with other mechanical or electromechanical adjustable speed technologies. Efficiency, quality, and reliability can also be drastically improved with the use of VFD technology. The application of a VFD system is very load dependent and a thorough understanding of the load characteristics is necessary for a successful application. The type of load (i.e. Constant torque, variable torque, constant horsepower) should be known as well as the amount of time that the system operates (or could operate) at less than full speed. Figures 3 and 4 show the energy savings potential for a variable speed fan and pump, respectively. The VFD pump is compared with a valve control system which would be adjusted to maintain a constant pressure in the system. The VFD fan is compared with both the damper control and the inlet guidevane controls. These curves

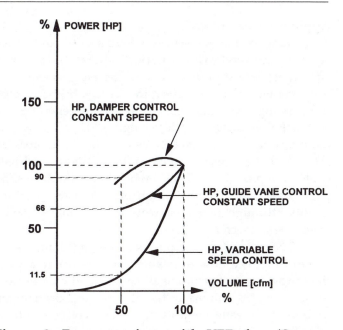

Figure 3. Energy savings with VFD fan. (Source: Lobodovsky, 1996)

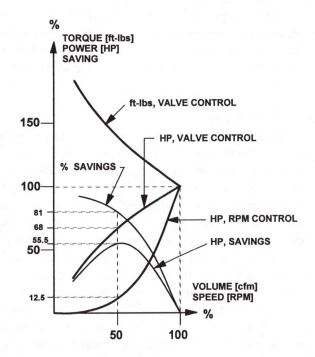

Figure 4. Savings with VFD pump. (Source: Lobodovsky, 1996)

do not account for system characteristics (i.e., head or static pressure), which would need to be included in an actual design. These figures show that the amount of savings achievable from the VFD is based upon the percent volume flow for both cases. The consumption savings would be determined by the percent of time at a particular load multiplied by the amount of time at that particular load.

Application Identification

There are many instances where a VFD can be successfully applied. The situation where the existing equipment already utilizes one of the older speed control technologies is the easiest to identify. In the case of the printing press that utilizes an Motor-generator set with a Eddy-current clutch, a single VFD and new AC motors could replace the MG-set and the DC motors. A mixer utilizing variable pitch pulleys ("Reeves" drive) could be replaced with a VFD which would reduce slippage losses and could dramatically improve product quality (through better control) and reliability (solid state versus mechanical).

Equipment (fans or pumps) utilizing dampers or valves to reduce flow is another instance where a VFD may provide a better means of control and energy savings. A variable volume air handler that utilizes damper or inlet guidevane controls could be replaced with a VFD drive controller. This would reduce the amount of energy required to supply the required amount of air to the system. Chiller manufacturers have utilized VFD controllers to replace the standard butterfly inlet valves on centrifugal compressors. This can significantly reduce the part load power requirements of a chiller which occur for a majority of the operating cycle in most applications.

Application Analysis

Project evaluation methods depend in large part on the size of the project. Smaller and less complex projects may only require reviewing specifications, installation sketches and vendors' quotes. Larger, more complex projects require more detailed engineering drawings and a drive system specialist will need to review the plan [Lobodovsky, 1996]. Once a possible application for a VFD is identified, the load profile (percent load versus time) should be determined. This curve(s) could be developed with the use of demand metering equipment or process knowledge (less desirable). This curve will be used to determine the available savings to justify the project as well as assure that the motor is properly sized. The proper load profile (constant torque, variable torque, etc.) can be compared to that loads corresponding VFD load profile in order to develop savings potential. The motor must be evaluated to assure that the VFD is matched to the motor and load, determine the motor temperature requirements, that the minimum motor speeds are met, among others. Many VFD manufacturers will require that the existing motor be replaced with a new model to assure that the unit is properly sized and not affected by earlier motor misuse or rewinding. At this stage the expertise of a VFD analyst or sales representative should be brought into the project for further design issues and costs. Phipps includes comprehensive chapters on applying drives to various applications where several check lists are included. The next section includes two case studies of applications of VFD's in industry.

CASE STUDIES

The following case studies are included as an example of possible VFD applications and the analysis procedures undertaken in the preliminary systems analysis.

Boiler Combustion Fan

This application involves the use of a VFD to vary the speed of a centrifugal combustion air intake fan (50 nameplate horsepower) on a scotch marine type high pressure steam boiler. The existing system utilizes an actuator to simultaneously vary the amount of gas and air that enters the burner. The air is controlled with the use of inlet dampers (not guidevanes). As the amount of "fire" is reduced, the damper opening is reduced and visa versa. The centrifugal fan and continuous variation in fire rate make this a feasible VFD application. The load profile of the boiler and corresponding motor demand (measured with demand metering equipment) is listed in Table 1. This table also includes the corresponding VFD demand requirements (approximated from Figure 4), kW savings, and kWh savings.

The annual savings for this example totaled 88,000 kWh, which would equate to an annual savings of $4,400 (based upon a cost of energy of $0.05/kWh). Lobodovsky provides an average estimated installed cost of VFD's in this size range at around $350 per horsepower, or an installed cost of $17,500 (50 hp * $350/hp). This would yield a simple payback of around 4 years. This example does not take into account demand savings which may result if the demand reduction corresponds with the plant peak demand. The control of the fan VFD would be able to utilize the same output signal that the existing actuator does.

Industrial Chiller Plant

A different calculation procedure will be used in the following example. A malting plant in Wisconsin uses seven 550 ton chillers to provide cold water for process cooling. Three of the chillers work all of the time and the other four are operated according to the plant's varying demand for cooling. The chillers are currently controlled by variable inlet vanes (VIV).

The typical load diversity and the power input re-

Table 1. Boiler combustion fan load profile and VFD Savings

Operating Time at load (hours)	Percent Full Load Power[2]	Existing Load w/ Damper (kW)[1]	Load with VFD Control (kW)[2]	Kilowatt Savings	Kilowatt-Hour Savings
1000	100	37	37	0	0
2000	105	40	30	10	20,000
3000	95	35	20	15	45,000
1000	90	33	10	23	23,000

[1]Measured with a Demand Meter

[2]Approximated using Figure 4

88,000 kWh

quired for one of the chillers under varying load were obtained from the manufacturer. In addition, the data in Table 2 gives the power input of a proposed variable frequency drive (also referred to as adjustable speed drive (ASD)) for one chiller that operates about 6,000 hours per year.

A weighted average of the load and duty-cycle fractions gives a load diversity factor (Idf) of 64.2 percent. This means that, on the average, the chiller operates at 64 percent of its full load capacity. A duty-cycle fraction weighted average of the savings attainable by a VFD can be estimated, which is 26.6 percent savings per year. The energy savings due to avoided cost of electricity usage are computed as follows: Savings = (0.266)(550 ton)(0.7 kW/ton)(0.642)(6,000 hr/yr) = Idf= 394,483

Table 2. Centrifugal chiller load and power input (Source: updated from Moses et Al., 1989)

Load Fraction (1)	VIV % kW (2)	ASD % kW (3)	Savings % kW (4)	Duty Cycle Fraction (5)
0.3	41	17	24	0.08
0.4	53	22	31	0.10
0.5	66	33	33	0.13
0.6	75	40	35	0.18
0.7	82	54	28	0.22
0.8	87	63	24	0.15
0.9	92	78	14	0.09
1.0	100	100	0	0.05

Column 2. From typical compressor performance with inlet and vane guide control (York Division of Borg-Warner Corp).

Column 3. From Carrier Corporation's Handbook of Air Conditioning Design: Comparative Performance of Centrifugal Compressor Capacity Control Methods.

Column 5. Actual performance data.

kWh/year The dollar savings at $0.04 per kWh are (394,483 kWh/year)($0.04/kWh) = $15,779/year. For a 500 horsepower variable frequency drive, the installed cost is estimated at $75,000, based upon Lobodovsky average installed cost of $150 per ton for units of this size. Therefore, the payback period is:

($75,000)/($15,779/year) = 4.75 years.

VFD ATTRIBUTES

These are just a few of the examples to show savings calculations. The advantages of VFD's include other aspects beyond energy savings, including improved productivity, reduced maintenance costs, and improved product quality, among others. The application of VFDs is relatively straight forward but a thorough analysis of the existing system and design of the future system is necessary to assure successful application. The load profile of the existing system is necessary to both determine savings as well as assure the system is properly sized. The specification of the actual VFD should only be done by VFD suppliers and/or experts. The list of references is a good source of more detailed discussions of each of the points discussed in this paper.

Attached are some additional forms and information on variable speed drives. These are reproduced from a forthcoming book by Mr. Konstantin Lobodovsky.

References

Lobodovsky, Konstantin K., *Motor Efficiency Management & Applying Adjustable Speed Drives*, April 1996.

Moses, Scott A., Wayne C. Turner, Jorge Wong, and Mark Duffer, "Profit Improvement With Variable Frequency Drives, *Energy Engineering. Vol. 86, No. 3, 1989.

Phipps, Clarence A. *Variable Speed Drive Fundamentals* Lilburn, GA, The Fairmont Press, 1 994.

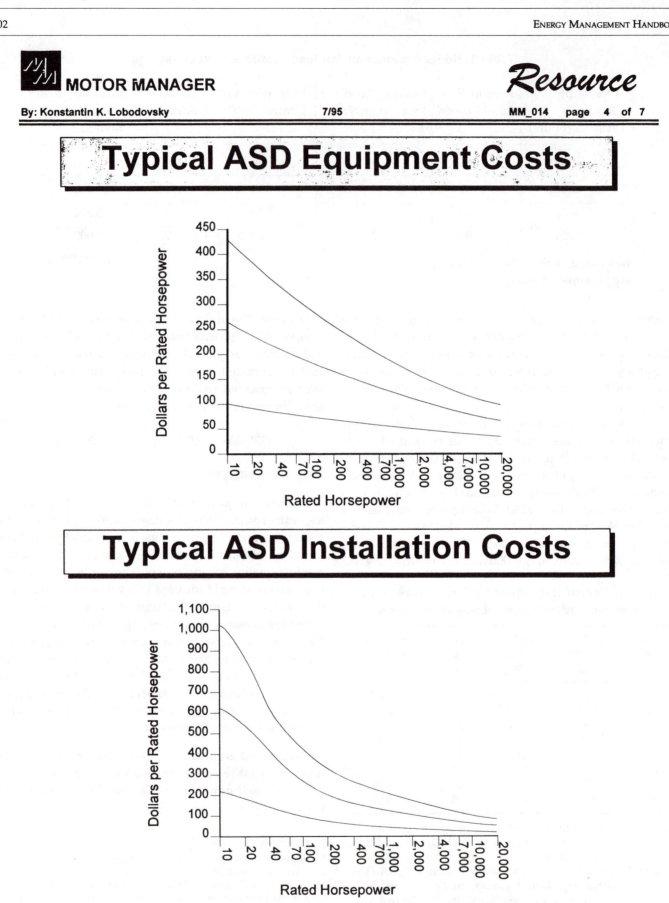

Figure 4.7: Typical ASD Equipment and Installation Costs

 MOTOR MANAGER

TECHNICAL FEASIBILITY CHECKLIST

Drive Identification and Location:_____

_____

1. How would changing the speed of the driving motor cause a change in the process or its rate?

2. Will product quality be improved or impaired?

3. What effects will the improvement or impairment have?

4. In what way can the machinery operate at other than its current speed?

5. In what way can any speed-changing mechanism (such as step pulleys, gears, or fluid drives) be installed to provide suitable electrical signal(s)?

6. If the existing process is mechanically modulated (dampers, valves, gates), how would new sensor(s) be installed to provide suitable electrical signal(s)?

7. If the existing process is electrically modulated (dc motors or wound-rotor induction motors), how would squirrel cage induction motors be adapted to the equipment?

8. Describe how a process modulating control has been or would be, applied to a drive system of this type.

9. Describe the physical space for installing a new or additional electrical motor controller, (Conventional induction motor ASD electronics need at least twice the space of existing starters; synchronous motor ASDs may need more.)

10. If the existing constant speed motor is a totally enclosed fan cooled induction motor, how would additional ventilation be provided if needed, when operating at lower speeds using an ASD?

11. How much derating of the existing motor would be necessary for heating caused by harmonics? Is this derating acceptable?

12. If the machinery is a pump, fan, or compressor, what data sheets or test means are available to estimate the operating characteristics of the unit?

13. Are data, drawings or other means available to estimate the torque requirements at various speeds for machinery other than pumps, fans, or compressors?

14. What drawings or other means available to validate the construction or installation details of the motor and machinery involved?

MOTOR MANAGER   *Resource*

ECONOMIC FEASIBILITY CHECKLIST

Drive Identification and Location_____

_____

1. How will a change in the speed of the driving motor result in a lower energy requirements?

2. What are your electrical energy costs in terms of your utility bill (consumed kWh, demand charges, etc.) or in terms of product costs?

3. If the existing process is mechanically modulated, what portion of the operating time is at other than maximum flow (or load)?

4. How many hours per week does the equipment operate?

5. How will the use of an ASD improve quality (through better speed control, elimination of waste, product reversion, etc.) and/or ultimately result in lower product costs?

6. What costs associated with drive inefficiencies (friction heat, cooling water, etc.) can be reduced by using an energy-efficient ASD system?

7. What are the costs for maintaining existing mechanical speed-changing equipment (transmissions, etc.)? are they obsolete and in need of replacement?

8. What are the costs of maintaining existing electrical speed-changing equipment (dc and wound-rotor motor, or reduced-voltage starting)?   Are they obsolete and in need of replacement?

9. Are there other problems of equipment reliability that cause production delays and higher product costs? Who can they be eliminated by ASDs with self-diagnostic features?

10. What opportunity is there to create additional space by removing large mechanical equipment (transmissions, etc.) with the installation of an ASD controller?

11. Can plant noise be reduced through lessening of the mechanical noise by installing an ASD control, or will the noise of the ASD be excessive?

12. Describe the shutdown arrangements required to provide time for an ASD to be installed.

CHAPTER **12**

ENERGY MANAGEMENT CONTROL SYSTEMS

WILLIAM E. CRATTY
ALFRED R. WILLIAMS, P.E.

Energy Conservation Management Company
Bethel, Connecticut
(203)797-1274

DALE A. GUSTAVSON, C.E.M.

D.A. Gustavson Workshops
Orange, California
(714)639-6100

Competitive economic pressures on owners to reduce building operating expenses are challenging the traditional design and control of heating, ventilating, air conditioning (HVAC) and lighting functions. Facility owners and operators have strong financial incentives to match more closely control, zoning and HVAC equipment sizing to the use of building spaces and outside environmental conditions. This must be done without sacrificing comfort and safety. Energy management systems play a key roll in meeting this challenge.

12.1 ENERGY MANAGEMENT SYSTEMS

Energy management is the control of energy consuming devices for the purpose of minimizing energy demand and consumption. Manually toggling on and off devices based upon need is a rudimentary form of energy management. The advent of mechanical devices such as time clocks for automatic toggling and bimetallic strip thermostats to control the output of heating and cooling devices along with pneumatic and electrical transmission systems provided means for developing early energy management systems in the form of automatic temperature controls. The advent of solid state electronic control devices and the increasing power of the microprocessor based personal computer have led to dramatic advances in energy management and what today is termed the energy management control system (EMCS). The primary difference between early automatic temperature control systems and EMCS is the application of a broad base of variables through programmable logic controllers to optimize the use of

energy. While EMCS is used to control building environmental conditions and industrial processes, this chapter will deal only with EMCS applications for control of building HVAC and lighting functions.

12.1.1 Direct Digital Control

HVAC building control system manufacturers have greatly enhanced EMCS by incorporating direct digital control (DDC). DDC is defined as a digital computer that measures particular variables, processes this data via control algorithms and controls a terminal device to maintain a given setpoint or the on/off status of an output device. The term "digital" refers only to the fact that input/output information is processed digitally and not that input or output devices are digital. Inputs and outputs relative to a DDC EMCS can be either digital or analog. Typically, most inputs are analog signals converted to digital signals by the computer while the greater portion of outputs are likely to be digital (zero or full voltage).

DDC systems use software to program microprocessors, therefore providing tremendous flexibility for controlling and modifying sophisticated control applications. Changing control sequences by modifying software allows the user continually to improve performance of control systems throughout a building.

DDC EMCSs can be programmed for customized control of HVAC and lighting systems and perform facility wide energy management routines such as electrical peak demand limiting, ambient condition lighting control, start/stop time optimization, sitewide chilled water and hot water reset, time-of-day scheduling and outdoor air free cooling control. An EMCS using DDC can integrate automatic temperature control functions with energy management functions to ensure that HVAC systems operate in accord with one another for greater energy savings.

The most significant benefits DDC systems offer are: a) the ability to customize the scheduling of equipment and their component devices to react to ever changing conditions of use and weather, and b) additional control modes (integral and derivative) that result in quicker, more accurate control when compared to pneumatic systems. Pneumatic systems inherently offer only proportional (how far is the input from setpoint) control in which the terminal control device linearly varies the output as an input variable changes relative to setpoint. Linear control results in offset or "hunting" by a terminal control device (valve) as it throttles to control to setpoint. By adding the integral (how long has the input been away from setpoint) mode to a proportional controller, offset is minimized as the control point is automatically reset while the controller continually shifts the throttling range. By adding the derivative (how fast is the input approaching/moving away from setpoint) mode the controller can achieve setpoint much quicker and more accurately whenever load varies. Generally, proportional-integral (PI) control is sufficient for building HVAC applications.

The primary objective of an EMCS using DDC is to optimize the control and sequencing of mechanical systems. A DDC EMCS also allows centralized and remote monitoring, supervision and programming maintenance of the HVAC and lighting functions. Additionally, such systems can lead to improved indoor environmental comfort and air quality.

Importantly, EMCS can also perform valuable non-energy related tasks (NERTs). These often overlooked DDC applications, or "extra-standard" functions, can be business specific or facility specific such as follows:

A design/build EMCS specialist for shopping malls adds people counting, vehicle counting, and precipitation logs to normal EMCS monitoring to help mall management lease space and track the success of mall promotions.

During a facility appraisal of a year-old drug store distribution center, a West Coast EMCS consultant/contractor discovered an opportunity to control the facility's conveyor system. The application not only resulted in energy savings, but also streamlined shipping.

An imaginative plant engineer commissioned the writing of custom front-end software for a brand name EMCS to perform noise level control in a factory. This 'stretch' of the EMCS now helps protect the hearing of employees.

DDC EMCSs·are being used to control environments in mushroom farms and banana distribution centers, to control pivot-irrigation systems, and even to control snow-making equipment at ski resorts. The limiting of DDC EMCS application strictly to HVAC and lighting control diminishes the financial benefit of this technology to system owner.

Sometimes a business specific or facility specific

application will have a greater impact on EMCS justification than does the energy saving component. For example, a 1990 law change in California requires banks to maintain specified levels of lighting near their automatic teller machines (ATMs). The law was passed in response to muggings and killings at poorly lit ATMs. Some banks have responded by randomly installing additional lighting fixtures. This shortsighted approach still begs the question, what happens if the lights go out?

One possible DDC solution would be to measure and log an ATM's light level. When lumens drop below a specified level, that facility's EMCS could simultaneously activate an emergency backup lighting system, notify an alarm monitoring company, and automatically dial out to a central EMCS monitoring station to record and verify that the notification function worked. An alarm light or tone also could be activated at the site to alert the facility personnel to the problem the next morning. All of this by the same microprocessor that controls the HVAC and lighting? Easily! Would it be cost effective? That question might best be asked of the bank's legal department.

For more discussion on NERTs and other benefits of EMCSs, see Section 12.2.

12.1.2 Hardware

An EMCS can range from a very simple stand-alone unitary microprocessor based controller with firmware routines (control software logic that the user cannot modify except for setpoints) that provides control of a terminal unit such as a heat pump to a very sophisticated large building DDC EMCS that interfaces with fire and security systems. Although this chapter will generally cite examples from applications in medium to large facilities, there have been numerous documented successes with stand-alone controllers in small buildings. Fast food and dinner houses, auto dealerships, retail stores, bowling centers, super markets, branch banks, small commercial offices, etc., have all benefited from this technology. Typically, EMCS installations at small buildings are 'design/build'. That is, an EMCS manufacturer, distributor, installing contractor, end user, or some combination thereof select the EMCS hardware and decides how it is to be applied.

Much study yet remains to be done to determine the extent by which firmware routines themselves contribute to the economics in small building EMCS projects. Empirical data suggests that while EMCSs in small facilities do improve the accuracy and response of mechanical system controls, the energy management routines are responsible only partially for the savings which are achieved. In applications where unitary controllers have been treated simply as 'devices', by-in-large savings have proven to be less dramatic than where they have been treated as 'systems' and used as 'tools for saving energy'. Better applied, installed, documented, supported, and maintained EMCS projects have yielded better results than those in which 'black boxes' were hung on walls in broom closets and left alone to work their 'magic'.

What is it that would lead one restaurant chain to bypass or disconnect hundreds of small EMCSs and return to time clock and manual control while at another chain, using nearly identical systems and strategies, a 40% return on investment, better comfort, and more rapid HVAC service response is realized? The opposite results suggest that the "human dimension" is a factor which must be more clearly understood and weighed. This is true for small design/build projects as well as for very large, sophisticated, 'engineered' projects. The "people" factor must not be ignored. Different building managers and occupants have different needs and occupancy habits. Also, no matter how well a system is designed, if the building operators and occupants are not properly taught how to use and maintain it, the EMCS will never live up to its full potential. For more discussion on the importance of training and the "human dimension," see Section 12.3.8.

The selection of EMCS type and sophistication for any given application should balance management and control desires with economics. An all encompassing DDC EMCS will provide the best overall control and management capability, but it is also the most expensive. On the other hand, stand-alone controllers are the least expensive for individual control applications; they also restrict control strategy and management capability. In terms of energy savings, an all encompassing system may not provide significantly greater savings than using stand-alone controllers for all HVAC functions at a given building, but it surely provides for better and easier management of the mechanical systems. However, with appropriate telecommunications software a system of unitary stand-alone controllers connected to modems could prove to be a very effective EMCS for a chain of widely dispersed small facilities.

There are two basic economic opportunities for application of an EMCS. The first is a retrofit of existing buildings with functioning automatic temperature control systems. The second is in new construction. It could be very difficult to justify removing a functioning automatic temperature control system in an existing building to install an all encompassing DDC EMCS solely for energy conservation purposes. For example, an air han-

dling unit is typically managed by a controller that monitors signals for outside air temperature, mixed air temperature, discharge air temperature, and return air temperature, compares the signals to setpoint conditions and in turn transmits a signal to a terminal control device to position the outside air dampers, exhaust dampers, return air dampers, and heating valve or chilled water valve to maintain a predetermined setpoint. If the unit has a variable speed drive, humidifier or dehumidifier there are even more conditions to monitor and control. To remove all of the existing controls and install an all encompassing DDC EMCS could be very costly. An economical alternative is to eliminate the traditional time clock function from the automatic temperature control system and interface a DDC EMCS in a "supervisory" mode over the existing controller by connecting key electric/pneumatic relays (EPs) to DDC outputs and installing strategically placed input sensors. A representative sampling of conditions such as a space temperature or discharge air temperature in conjunction with global inputs and well designed software in most cases would provide adequate management to maintain comfort and save a significant amount of energy.

Even though the first cost of a DDC EMCS installation has decreased dramatically in the past several years, the traditional pneumatic control system currently has a first cost advantage over a DDC EMCS in new construction; however, the pneumatic system still just performs temperature control. When comparing maintenance costs, energy efficiency and the convenience of the building management features EMCS offers it is not difficult to justify an all encompassing DDC EMCS in new construction. Some of the advantages of the DDC system over the pneumatic system are: 1) more precise control, 2) unlimited customization of control schemes for energy management and comfort, 3) centralizes and integrates control and monitoring of HVAC and lighting, 4) easier to maintain, and 5) easier to expand and 'grow' with building size and use. It is conceivable though, that due to the proliferating nature of the industry, DDC EMCS will have a first cost advantage by 1995.

Current day DDC EMCS hardware configuration varies from manufacturer to manufacturer but has a common hierarchical configuration of microprocessor based digital controllers as well as a front-end personal computer. This configuration can be categorized by three levels: 1) terminal equipment level controllers, 2) system level controllers, and 3) the operator interface level. The three levels are networked together via a communications trunk which allows information to be shared between other terminal equipment controllers and all level controllers.

ENERGY MANAGEMENT HANDBOOK

The most important part of a DDC EMCS are the system controllers because they monitor and control most mechanical equipment. A major feature of system level controllers is their ability to handle multiple control loops and functions such as proportional-integral-derivative (PID) control, energy management routines, and alarms while the terminal equipment controllers are single control loop controllers with specific firmware. The operator interface level is generally a personal computer that serves as the primary means to monitor the network for specific data and alarms, customize the control software for downloading to specific system controllers, maintain time-of-day scheduling, and generate management reports. However, it is not uncommon for a single personal computer to serve both as the system controller and operator interface.

The specific method of communications within an EMCS is significant because of the amount of data being processed simultaneously. While methods of data transfer between DDC controllers and the operator interface level vary from manufacturer to manufacturer, communications protocols can be simplified into two distinct categories: 1) poll/response, and 2) peer-to-peer or token-ring-passing network.

A peer-to-peer network does not have a communication master or center point as does the poll/response system because every trunk device, be it a terminal level controller attached to a systems level controller or a system level controller networked to other system level controllers, at some point, has a time slot allowing it to operate as the master in its peer grouping. However, the terminal level controller in a peer-to-peer network usually is a poll/response device. A system using peer-to-peer communication can offer distinct advantages over poll/response communication when redundancy of critical global data is accommodated. These advantages are: 1) communication is not dependent on one device, 2) direct communication between controllers does not require communication through the operator interface level, and 3) global information can be communicated to all controllers quickly and easily.

The speed of system communication in building control usually is not an issue with today's DDC technology. However, the response time for reaction to control parameters can be an issue with the application of a centralized poll/response system where control and monitoring point densities are very high or alarm response time is critical. Generally, peer-to-peer communication distributive systems that network system level controllers and terminal equipment level controllers provide the quickest response time. However, the ever increasing speed of the personal computer

is allowing centralized poll/response systems effectively to expand their control and monitoring horizon.

Manufacturers of panelized DDC EMCSs with peer-to-peer communication have not been able to keep pace with the rapid advances in processing power of the personal computer technology. As a result a DDC EMCS with poll/response communication using the latest personal computer as a front-end will have greater memory and ability to implement more complex, customized control logic, monitoring and analysis routines than a system with peer-to-peer communication.

Be it new construction or a retrofit when selecting an EMCS the actual needs and requirements of the particular building or campus of buildings must be considered. While facility layout and construction type will influence EMCS configuration, software is the most critical element in any EMCS application.

12.1.3 Software

The effectiveness of the software control logic is what provides the building operator with the benefits of an energy management system. While color graphics and other monitoring enhancements are nice features to have, the control logic in the system and terminal controllers is what improves building systems efficiency and produces the energy savings.

Networking DDC devices has provided building operators with the ability to customize the traditional energy management strategies such as time-of-day/holiday scheduling, demand shedding, duty cycling, optimum start/stop and temperature control as well as the ability to implement energy management techniques such as occupied/unoccupied scheduling of discrete building areas, resulting in reduced airflow volumes and unoccupied period setback strategies that greatly reduce operating cost. PID control provides for more accurate, precise and efficient control of building HVAC systems. However, software that provides for adaptive or self tuning control not only enhances savings but also addresses environmental quality. Adaptive control software monitors the performance of a particular control loop and automatically adjusts PID parameters to improve performance. This feature improves control loop response to more complicated and dynamic processes.

While a DDC EMCS with peer-to-peer communication may be preferred over a system with poll/response communication from a hardware standpoint, in a particular application the DDC EMCS with a personal computer and poll/response communication serving as the system controller could prove to be the system of choice because of the complexity of the software requirements.

If the energy management industry has an Achilles heel it is the specifier who specifies an EMCS by reference to a particular manufacturer's general hardware specifications rather than detailing the specifics of software sequencing logic and system configuration. Specification by manufacturer reference is not a complement to a hardware manufacturer rather it is an indication of EMCS ignorance on the part of the specifier. HVAC mechanical equipment operation sequences are becoming increasingly complex. This is a result of stricter criteria established by national codes (American Society of Heating, Refrigerating and Air Conditioning Engineers' Minimum Outdoor Air Requirements for example) and by building operators who require highly accurate control systems, energy use monitoring and accounting by zone of use. To meet the challenges posed by these stricter codes and building operator demands the EMCS specifier must be thoroughly knowledgeable about all aspects of EMCSs.

As suggested in the earlier discussion of the human dimension phenomenon, it is essential that the EMCS specifier look beyond technology or, as the old saying goes, look beyond the trees to the forest. That is, a specifier must be knowledgeable about not only the building construction and mechanical and lighting systems to be controlled but also the businesses and people which occupy the building. And, of equal importance is the need to assess the level of training and ongoing support that the EMCS owner/operator will need to realize the anticipated benefit.

Also, the EMCS specifier must be knowledgeable about the willingness of DDC EMCS manufacturers to serve the customer. For instance, in order to secure customers for their unitary stand-alone controllers which manufacturers would be willing to provide user friendly telecommunications software to a bank that has multiple branch offices to facilitate the management of a widely dispersed EMCS?

12.1.4 Control Strategies

A DDC EMCS can serve six basic functions: 1) manage the demand or need for energy at any given time, 2) manage the length of time that devices consume energy, 3) set alarms when devices fail or malfunction, 4) facilitate monitoring of HVAC system performance and the functioning of other building systems, 5) assist the building operator to administer equipment maintenance, and 6) provide the building/business owner/operator with non-energy related tasks (NERTs), i.e., extra-standard functions to make the EMCS more effective or advantageous. There are financial benefits for each of these six functions, yet traditionally, only the first two are quantified in an economic analysis. The industry still has work to do to develop new, accurate, reliable models for predicting energy savings by an EMCS. But, to limit the inquiry to just energy savings falls far short of what is necessary for EMCS technology to realize anywhere close to its truly staggering potential.

It is the industry's responsibility to develop models for predicting the value of all EMCS functions. It is too often left to sales and marketing people to explore and, even more importantly, assert the full range of benefits of EMCS technology. The energy and engineering communities tend to stop where science ends and speculation begins. Speculation is a healthy exercise. Here are some questions which might be worth asking:

What is the annual dollar value of an efficient office building 95% leased as compared to a less efficient building 75% leased?

What is the annual dollar value of being able to provide basic comfort troubleshooting by telephone using a personal computer and remote communication software?
What is the annual dollar value of increased productivity in a building, which suffers little or no discomfort?

What is the value of lowering absenteeism by using EMCS to maintain indoor air quality monitoring and preventive action?

What if one of the above benefits is the difference between justifying an EMCS project and not going ahead? The key to making the most of NERTs, or extra-standard EMCS functions is recognizing that: 1) quantified benefits have more impact than unquantified 'hopes' when buying decisions are made, 2) it is the industry's duty to continually observe, track and create new EMCS 'byproducts', and 3) only imagination limits the range of extra-standard functions which can be incorporated into the EMCS design. The extent and proficiency of performing these functions depends on the software programming.

The energy management industry has evolved standard time-proven software routines that can be used as a starting point to develop effective programming. These routines are commonly referred to as control strategies. While these routines can be applied in virtually all EMCS installations they must be customized for each building. A description of some of these routines follows.

Daily Scheduling: This routine provides for individual, multiple start and stop schedules for each piece of equipment, for each day of the week.

Holiday Programming: This routine provides for multiple holiday schedules which can be configured up to a year in advance. Typically, each holiday can be programmed for complete shutdown where all zones of control are maintained at setback levels, or for special days requiring the partial shutdown of the facility. In addition, each holiday generally can be designated as a single date or a range of dates.

Yearly Scheduling: Typically, any number of control points can be assigned to special yearly scheduling routines. The system operator usually can enter schedules for yearly scheduled control points for any date during the year. Depending on the particular EMCS software operating system, yearly scheduled dates may be erased once the dates have passed and the schedules have been implemented; or, the scheduled dates may repeat in the following year as do scheduled holiday dates.

Demand Limiting or Load Shedding: Demand limiting can be based on a single electric meter or multiple meters. Generally, loads are assigned to the appropriate meters in a specified order in which equipment is to be shed. Usually, a control point that is allowed to be shed is given a status as a "round-robin" demand point (first off-first on), a "priority" demand point (first off-last on), or a "temperature" demand shed point (load closest to setpoint is shed first). Other parameters per control point include Rated kW, Minimum Shed Time, Maximum Shed Time and Minimum Time Between Shed.

Minimum On-Minimum Off Times: Normally a system turns equipment on and off based on temperatures, schedules, duty cycles, demand limits and other environmental parameters. However, mechanical equipment often is specified to run for a minimum amount of time once started and/or remain off for a minimum amount of time once shut down. Therefore, this routine gives the operator the ability to enter these minimum times for each piece of equipment controlled.

Duty Cycling: This routine provides the ability for a control point to be designated for either temperature compensated (cycle a piece of equipment on and off to maintain a setpoint within a dead band) or straight time dependent (cycle a piece of equipment on and off for distinct time intervals) duty cycling. Control parameters for temperature compensated duty cycling include Total Cycle Lengths, Long and Short Off Cycles and Hi and Low Temperatures. There are separate sets of parameters for both heating and cooling. Time dependent duty cycling uses Total Cycle Length and Off Cycle Lengths. Cycles for each load can be programmed in specific minute increments and based on a selectable offset from the top of the hour.

Optimum Start/Stop: For both heating and cooling seasons DDC systems can provide customized optimum start/stop routines which take into account outside temperature and inside zone temperatures when preparing the building climate for the occupant or shutting the facility down at the end of the day. During unoccupied hours (typically at night) the software tracks the rate of heat loss or gain and then utilizes this data to decide when equipment will be enabled in order to regain desired climatic conditions by the scheduled time of occupation. The same logic is used in reverse for optimum stop.

Night Setback: This routine allows building low temperature limits for nighttime, weekend and holiday hours, as well as parameters and limits for normal occupied operation to be user selectable. Night setback is usually programmed to evaluate outside air temperature in the algorithm.

Hot Water Reset: This routine varies the temperature of the hot water in a loop such that the water temperature is reduced as the heating requirement for the building decreases. The reset temperature for night setback usually is less than the day reset temperature. Typically, reset is accomplished by controlling a three-way valve in the hot water loop; however, depending on boiler type, reset can be accomplished by "floating" the aquastat setpoint. The primary variable in the reset algorithm is outside air temperature.

Boiler Optimization: In a facility that has multiple boilers this routine schedules the boilers to maximize plant efficiency by staging the units to give preference to the most efficient boiler, by controlling the burner firing mode when desirable, and by minimizing partial loading.

Chilled Water Reset: This routine varies the temperature of the chilled water in a loop such that the water temperature is increased as the cooling requirement for the building decreases. This is typically done by controlling a three-way valve in the chilled water loop. Chilled water reset also can be accomplished by interfacing the EMCS with the chiller controls to reduce the maximum available cooling capacity such as during demand limiting. The primary variable in the reset algorithm is outside air temperature.

Chiller Optimization: In a facility that has multiple chillers this routine schedules the chillers to maximize plant efficiency by staging the units to give preference to the most efficient chiller. Unlike the boiler optimization routine, however, it can be more efficient to operate a chiller plant with units sharing the load rather than fully loading a unit.

Chiller Demand Limiting: This routine limits chiller demand by interfacing the EMCS with the chiller controls to reduce the maximum available cooling capacity in several fixed steps. The primary variable in the demand limiting algorithm is kW demand.

Free Cooling: Free cooling is the use of outside air to augment air conditioning or to ventilate a building when the enthalpy (total heat content) of the outside air is less than the enthalpy of the internal air and there is a desire to cool the building environment.

Recirculation: This routine provides for rapid warm-up during heating and rapid cool-down during cooling by keeping outside air dampers fully closed and return air dampers fully opened during system start-up.

Hot Deck/Cold Deck Temperature Reset: This routine selects the zones or areas with the greater heating and cooling requirements and establishes the minimum hot and cold deck temperature differential which will meet the requirements in order to maximize system efficiency.

Motor Speed Control: This routine will vary the speed of fan and pump motors to reduce air and water velocity as loads decrease. Speed control can be accomplished either by controlling a two speed motor with a digital output or a variable speed drive with an analog output.

Manual Override: This routine provides separate manual override schedules to allow for direct control of equipment for specific periods of time.

Duty Logs: DDC systems can track and display various types of information such as last time on, last time off, daily equipment runtimes, temperatures (or other analog inputs) kWh and even remote panel hardware performance. DDC systems can also accumulate and display monthly scheduled and unscheduled runtimes for each piece of equipment controlled or monitored.

Temperature and hardware performance information can be monitored and displayed. Change of state can be logged for all loads and can be reviewed on a daily basis for all loads. In addition, a system can produce line graphics or historical trends of input data and hardware systems information.

Energy consumption logs can be kept separately for every pulse meter point. For electricity these energy logs can record data (by the day) such as daily kilowatt hour consumption, kilowatt demand, time the peak occurred, selected demand limit, time that any load was

shed, minimum, maximum and average outside temperatures and degree day information.

Alarm Monitoring and Reporting: DDC systems register and display alarms for conditions such as 1) manual override of machinery at remote locations, 2) equipment failures, 3) high temperatures, 4) low temperatures, invalid temperatures (sensor is being tampered with), and 5) communication problems.

12.2 JUSTIFICATION OF EMCSs

Traditionally, a commercial building is constructed by one party (a developer) owned by a second party (investors) and operated by a third party (a facilities management firm). Typically, the building will have one electric meter and a central heating plant. Therefore, total energy cost must be included in the rental structure. Before the 'first energy crisis' and energy was cheap, lessees typically paid a flat rental charge based on square footage rented and nobody paid much attention to operating hours of the HVAC and lighting systems.

When energy prices started their rapid upward spiral building owners and operators implemented measures to pass the cost increase directly on to the lessee. However, now because of the volume of space available and other competitive pressures building owners and operators are being forced to find means of reducing the energy cost component of the base rent in order to reduce operating costs and entice customers with lower base rentals in the lease market.

Depending on the type of HVAC and lighting control systems installed in a building, energy consumption and operating costs can be dramatically different. Therefore, it should go without saying that since operating costs are a key ingredient when establishing competitive rentals, the type of control system installed can have a significant impact on the owner's income. And, to follow this line of reasoning to its conclusion, since rental income is a key ingredient in the appraisal of a commercial facility, the type of control systems installed can have a significant impact on the value placed on the building should the owner want to sell.

A well-designed and properly operated EMCS is an investment which offers the building owner/manager a multitude of benefits, which can oftentimes be bundled and quantified. These benefits include, yet go beyond, impressive payback and return-on-investment figures. In addition to the immediate financial benefits, there are long-range benefits to managing energy and demand as well. Such benefits can be wide-ranging, positively affecting society in general. Consider the following EMCS functions, along with the related benefits:

Manage energy consumption and demand—Lower operating expenses, higher profits and increased competitiveness, keep energy prices affordable in the long term. Less smog, reduced acid rain and less global warming lead to stronger national security.

Optimize operating efficiencies of energy consuming equipment—Above plus extended equipment life which further increases profits and competitiveness.

Improve comfort—Increases occupant productivity and concentration. Helps occupants feel more alert, rested, make fewer mistakes and perform better. Leads to a more vital U.S. economy with better educated and less stressed populace.

Improve indoor air quality—Saves lives, increases productivity, prevents lawsuits, lowers insurance costs and reduces absenteeism. Contributes to better economical and educational systems.

Activate alarms when equipment malfunctions—Provides for less costly disruptions of productivity due to faster response, increased profitability, extended equipment life and increased competitiveness.

Assist on- or off-site operator administer service/maintenance—Results in better records, histories for more efficient, accurate work, less downtime, more productivity and lower cost for service. Keeps polluting vehicles off the street. Leads to higher profits and increased competitiveness.

Monitor/log building equipment performance and energy use—Resulting, accumulated data leads to improved performance of all EMCS functions, thereby increasing magnitude of all benefits itemized above. Also confirms cost avoidance benefit of project.

Perform NERTs (non-energy related tasks)—Sometimes the value of NERTs exceeds the value of all the other benefits. (Refer back to Section 12.1.1 for specific examples of NERT applications.)

While some of these benefits may seem far-reaching, different building owners and operators have different needs and requirements. Energy management professionals should always be on the alert for specific ways an EMCS can benefit a facility, as each case is unique.

12.2.1 An EMCS Opportunity

For purposes of discussion, consider a ten story, multiple tenant, commercial office building configured and occupied as follows as an example of an opportunity for an EMCS application (Figure 12.1).

a.   One boiler room in the basement with 2 boilers and 2 redundant sets of circulating pumps (4 in all) to supply hot water to perimeter radiation and heating coils.

b.   One domestic hot water system for the entire building.

c.   Four mechanical rooms—a north and a south mechanical room on the 5th floor each with 2 air handling units (AHUs) supplying the first 5 floors and a north and a south mechanical room on the 10th floor each with 2 AHUs supplying the top 5 floors. Each AHU has a preheat coil on the inlet side of the supply fan and a chilled water coil and a reheat coil on the discharge side to maintain desired hot deck and cold deck temperatures for a double duct distribution system. The north mechanical room on the 10th floor has a chiller and a redundant set of circulators to service the north side AHUs while the south mechanical room on the 10th floor has a chiller and a redundant set of circulators to service the south side AHUs.

d.   One roof top mounted cooling tower serving both chillers.

e.   The building is zoned by AHU (8 zones) and each room has a mixing box that is controlled by its own thermostat to allow heating and cooling as required.

f.   Toilet exhaust is provided by 2 roof mounted exhaust fans that are switched in one of the 10th floor mechanical rooms.

g.   Corridor and lobby lighting is switched from a central location while space lighting is operated by a wall switch in each space.

h.   The building is normally occupied between 8:00 am and 5:00 p.m. Monday thru Friday and sporadically occupied a various times after normal hours until midnight on work days and occasionally on weekends.

An opportunity for use of DDC EMCS in either retrofits or new construction can be found when considering building use. One of the keys to effective utilization of an EMCS is zoning. Through flexible program scheduling to control the energy consuming systems by zone the building operator can realize substantial energy savings.

Most likely an existing building configured and occupied as the above exampled building will have manual lighting control and a mechanical time clock based pneumatic system controlling HVAC functions as follows:

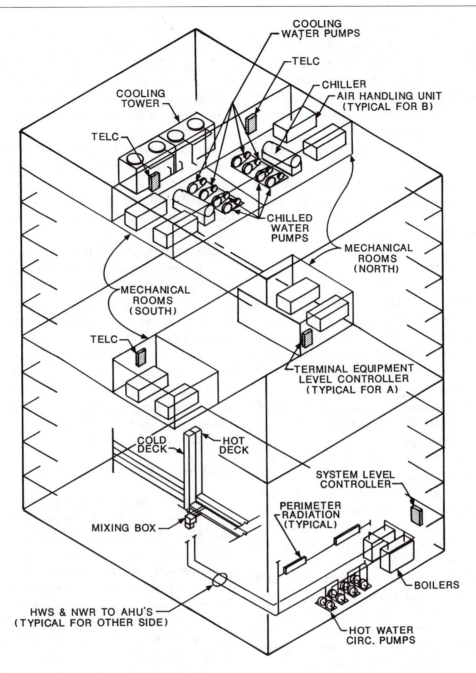

Figure 12.1 Ten-story office building.

a. Each mechanical room has a time clock activated control panel for occupied/unoccupied switch over of day/night air pressures for space thermostats and start/stop of the mechanical equipment.

b. Each AHU has a mixed air temperature controller, a cold deck temperature controller and a hot deck temperature controller to maintain supply air temperatures at preset levels.

c. Outside air dampers are calibrated to maintain desired minimum air intake.

d. Each zone has a centrally located night thermostat for its respective AHU.

e. In the boiler room there is a summer/winter change over of the heating and cooling functions.

f. Corridor and lobby lighting is controlled by a mechanical time clock in the electrical control room.

To accommodate after normal hours occupancy the temperature control system provides for occupied conditions from 7:00 am to midnight on weekdays. On weekends the system provides for occupied conditions from 8:00 am to 5:00 pm. Toilet exhaust fans run continuously. The corridor and lobby lighting controller is set to maintain lights on from 6:00 am to 1:00 am on weekdays and from 7:00 am to 6:00 pm on weekends. Cleaning crews and tenants are responsible for control of lighting in the tenant spaces and toilets.

In multiple tenant, high rise commercial office buildings with HVAC and lighting systems configured and operated in a manner similar to the exampled facility the DDC EMCS provides an opportunity to save a significant amount of energy and recover costs associated with after normal hours tenant use.

12.2.2 The EMCS Retrofit

Without the benefit of a DDC EMCS the operator of the exampled building must incur after normal hours operating costs and allocate them over the entire leasing cost. This unfairly spreads the costs generated by a few tenants over the lease costs of all tenants. The overall effect are higher leasing charges for all.

The most cost effective way to retrofit a DDC EMCS in a building controlled by a reasonably maintained pneumatic control system is to interface it with the existing pneumatics in a supervisory mode by eliminating night thermostats, replacing time clock contacts with DDC outputs, direct hard wiring for start/stop control of the equipment not under the influence of the pneumatic system (i.e. toilet exhaust fans, domestic service hot water circulators, and corridor, lobby and exterior lighting); direct hard wiring boilers, chillers and hot water and chilled water circulators for start/stop control, and direct hard wiring to enable/disable the equipment under control of the pneumatic system (i.e. the AHUs and dampers). Enabled equipment is allowed to operate under the control of the pneumatic system. Disabled equipment is shut down (Figure 12.2).

With a supervisory DDC system the building operator can effect a dramatic reduction in energy usage by incorporating a simple timed on/off control strategy for all HVAC and lighting zones that segregates the operation of the equipment into discrete tenant zones. The control strategy would shut down the tenant zones when outside of normal lease times. If a tenant wished to utilize a space after hours, he/she could initial an override (using a simple automated telephone interface from his/her office) to signal the system to preserve his/her working environment. Additionally, on weekends and holidays the entire building can be kept in the unoccupied mode unless a tenant requires extended occupancy of a particular zone. That tenant could initiate

an override from any outside telephone to ensure occupied conditions for his/her space upon arrival at the building. Whenever an override is initiated, an operator interface level personal computer could track the energy consumption and or/runtimes of that zone's HVAC equipment for the purpose of billing the tenant an after hours usage charge.

Experience at a multiple tenant, high rise commercial office building on Long Island, New York, has shown that this "tenant strategy" not only results in a dramatic reduction in energy usage but also allows the building operator to reduce the base rental and recover after normal hours operating cost directly from those tenants responsible for the use.

By installing a DDC system as a supervisor of and existing pneumatic control system the building operator can implement traditional control strategies (see section 12.1.4) to improve building operating efficiency and achieve additional energy cost reductions. In addition to optimizing heating and cooling functions, the toilet exhaust fans, domestic service hot water and corridor and lobby lighting can be interlocked with zone occupied/unoccupied status to shut down these systems when not needed. Pay backs for this type of EMCS retrofit are typically very attractive.

12.2.3 New Construction EMCS

If the exampled building were under construction as a new building, the developer would be remiss not to consider a DDC EMCS as part of the original design. In addition to all of the benefits described above, the all encompassing system would provide added benefit in the form of PI control of heating and cooling valves and dampers. The design engineer also would be remiss not to consider improvements in the mechanical design that along with DDC control could significantly improve the energy efficiency of the building.

By combining more aggressive zoning with DDC EMCS in new construction the building designer has the opportunity to better match control of the HVAC with building usage and optimize equipment sizing to better fit loads thus improving equipment operating efficiency. However, the bulk of the EMCS opportunity lies in the thousands of existing buildings which are controlled with pneumatic and electrical/mechanical automatic temperature control systems. Surprisingly enough there are also buildings still being controlled by hand!

12.3 SYSTEMS INTEGRATION

The performance of an EMCS is directly proportional to the quality of the designer's systems integration

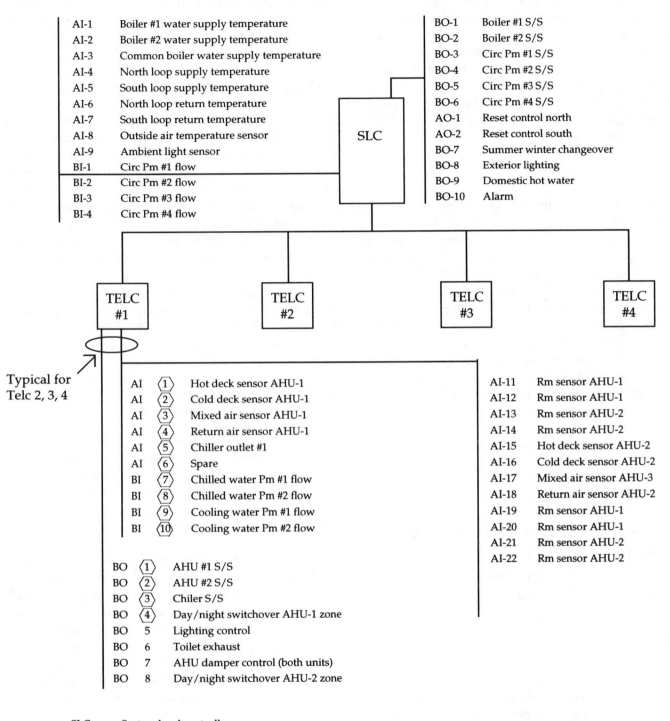

AI-1	Boiler #1 water supply temperature	
AI-2	Boiler #2 water supply temperature	
AI-3	Common boiler water supply temperature	
AI-4	North loop supply temperature	
AI-5	South loop supply temperature	
AI-6	North loop return temperature	
AI-7	South loop return temperature	
AI-8	Outside air temperature sensor	
AI-9	Ambient light sensor	
BI-1	Circ Pm #1 flow	
BI-2	Circ Pm #2 flow	
BI-3	Circ Pm #3 flow	
BI-4	Circ Pm #4 flow	

BO-1	Boiler #1 S/S
BO-2	Boiler #2 S/S
BO-3	Circ Pm #1 S/S
BO-4	Circ Pm #2 S/S
BO-5	Circ Pm #3 S/S
BO-6	Circ Pm #4 S/S
AO-1	Reset control north
AO-2	Reset control south
BO-7	Summer winter changeover
BO-8	Exterior lighting
BO-9	Domestic hot water
BO-10	Alarm

SLC

TELC #1   TELC #2   TELC #3   TELC #4

Typical for Telc 2, 3, 4

AI	(1)	Hot deck sensor AHU-1
AI	(2)	Cold deck sensor AHU-1
AI	(3)	Mixed air sensor AHU-1
AI	(4)	Return air sensor AHU-1
AI	(5)	Chiller outlet #1
AI	(6)	Spare
BI	(7)	Chilled water Pm #1 flow
BI	(8)	Chilled water Pm #2 flow
BI	(9)	Cooling water Pm #1 flow
BI	(10)	Cooling water Pm #2 flow

AI-11	Rm sensor AHU-1
AI-12	Rm sensor AHU-1
AI-13	Rm sensor AHU-2
AI-14	Rm sensor AHU-2
AI-15	Hot deck sensor AHU-2
AI-16	Cold deck sensor AHU-2
AI-17	Mixed air sensor AHU-3
AI-18	Return air sensor AHU-2
AI-19	Rm sensor AHU-1
AI-20	Rm sensor AHU-1
AI-21	Rm sensor AHU-2
AI-22	Rm sensor AHU-2

BO	(1)	AHU #1 S/S
BO	(2)	AHU #2 S/S
BO	(3)	Chiler S/S
BO	(4)	Day/night switchover AHU-1 zone
BO	5	Lighting control
BO	6	Toilet exhaust
BO	7	AHU damper control (both units)
BO	8	Day/night switchover AHU-2 zone

SLC -	System level controller
Telc -	Terminal equipment level controller
AI -	Analog input
BI -	Binary input
AO -	Analog output
BO -	Binary output
S/S -	Stop/start

Figure 12.2 Office building supervisory EMCS configuration,

effort. Systems integration can be defined as those activities required to accomplish the precise monitoring and control desired when incorporating an EMCS into a building's HVAC and lighting systems. This definition implies that systems integration includes not only selecting the appropriate EMCS and detailing monitoring points and control interfaces, but also, customizing the software algorithms to meet the user's needs and commissioning the system.

12.3.1 Facility Appraisal

An appraisal of the job site is the first step of EMCS systems integration. The appraisal is generally done by interviewing the building users to determine building usage patterns (be careful to differentiate between normally scheduled usage and after hours usage), to identify any comfort complaints (i.e. are there 'hot' or 'cold' spots in the building, is ventilation adequate, etc.), to assess the attitude that the building maintenance staff has towards an EMCS, and to determine exactly what the building manager expects the system to accomplish.

During the facility appraisal, the following people should be interviewed:

Tenants (by company, department, job function)

Prospective tenants

Delivery services

Security personnel

Service vendors (including janitorial, HVAC, lighting, electrical, life safety)

Building engineers/facility managers

Corporate energy manger, if any

Building owner/CEO/CFO

Property manager and staff

Leasing agents.

Perhaps the most overlooked above are leasing agents. These professionals must understand the benefits of energy efficiency, or they will not be able to sell it to perspective tenants.

12.3.2 Equipment List

Next, the designer should develop a complete and accurate equipment list. In a retrofit situation the list should include all of the equipment and devices to be controlled, the present method of control for these items, the operating condition of the items and their controls, and the area each item services. Do not forget to include exterior and interior lighting on the list.

12.3.3 Input/Output Point Definition

The third step is to define the input and output points required to achieve the monitoring and control desired in the most cost-effective or efficient manner. This step requires an in-depth appraisal of such items as the equipment list, layout of building spaces, method of building construction, HVAC physical plant makeup and layout, methods of heating and cooling, types of secondary HVAC distribution systems and controls, and the condition of the mechanical equipment and existing controls.

The definition of a control point should include a listing of all inputs that will influence the functioning of that point. The installation specifications should detail this information and indicate the nature of the influence as follows:

Output Point	Analog Input Point	Notes
AHU #1	Outside Air Temperature	
	Room #1027 Temperature-------------------	
	Room #935 Temperature	
	Room #837 Temperature	Room of greatest
	Room #715 Temperature	Demand
	Room #625 Temperature--------------------	

12.3.4 Systems Configuration

Once the inputs and outputs have been defined, the designer needs to define any constraints that might be placed on the EMCS hardware. Building size, building design, building location, building use, design of the physical plant, type and layout of the secondary distribution systems, design of the electrical system, and type and condition of existing controls in retrofits as well as the nature of the personnel operating the building are some of the items to be considered when configuring an EMCS.

Cost of installation can vary significantly depending on the type of system architecture (centralized versus distributive), the method of transmitting information between system controllers and the input/output devices (direct hardwire versus multiplexing versus powerline carrier), the types of system controllers and input/output devices selected (two way versus one way communicators) and the amount of programming required (customized algorithms versus firmware).

Knowing who is going to maintain the software and perform the necessary administrative functions (i.e. inputting schedule changes, monitoring alarms, and producing reports) not only affects how many but also where the operator interface level personal computer is to be located, and the type of monitoring and alarming that will be needed. This will also determine training requirements and the extent of support services needed.

12.3.5 Specifics of Software Logic

The most critical phase of systems integration is the software specification. The designer must take care to be as explicit as he/she can when detailing the nature of the algorithms to be installed. This can be done by structuring the software specifications in a manner that clearly defines the algorithms and their interrelationships. The designer who simply includes a manufacturer's general hardware specifications rather than detailing the specifics of software sequencing logic most assuredly dooms the system to failure.

APPENDIX A "EMCS Software Specifications" is a sample of what can be described as a minimally acceptable software specification. Ideally, in addition to a descriptive specification, the designer would leave little doubt as to exactly what he/she has in mind by providing a listing, in easy to understand English format, of all system application programs associated with each piece of equipment. Following is an example of what a listing might look like for control of boilers.

HOT WATER BOILER CONTROL

Where there are multiple boilers in a set the software shall change the lead-lag status of boilers 1 and 2 in the set on a biweekly basis (Note the call to the lead-lag algorithm). This specific algorithm calls the general CENTHEAT algorithm which assumes that the heating circulation pump is stage one of the central heating plant.

For the following control points CONTRACTOR shall apply this HOT WATER BOILER CONTROL algorithm:

1. Boiler 1
2. Boiler 2

VARIABLE DEFINITION:

User defined variables (operating parameters):

HEATSET:	The heating target temperature (72°F).
SETBACKTEMP:	The minimum set back temperature allowed (55°F).
MAXOATHEAT:	The maximum outside air temperature for heating to be utilized (65°F).
OCCLOWOAT:	The low outside air temperature parameter to be used for hot water reset during occupied time of day (0°F).
OCCHIGHOAT:	The high outside air temperature parameter to be used for hot water reset during occupied time of day (60°F).
OCCLOWSWT:	The low hot water supply temperature parameter to be used for hot water reset during occupied time of day (120°F).
OCCHIGHSWT:	The high hot water supply temperature parameter to be used for hot water reset during occupied time of day (190°F).
UNOCCLOWOAT:	The low outside air temperature parameter to be used for hot water reset during unoccupied time of day (0°F).
UNOCCHIGHOAT:	The high outside air temperature parameter to be used for hot water reset during unoccupied time of day (40°F).
UNOCCLOWSWT:	The low hot water supply temperature parameter to be used for hot water reset during unoccupied time of day (120°F).
UNOCCHIGHSWT:	The high hot water supply temperature parameter to be used for hot water reset during unoccupied time of day (160°F).

Measured variables:

ACTUALTEMP:	The temperature being controlled by the I/0 point in question. In this case of a central boiler, the ACTUALTEMP shall represent the coolest of all of the zone temperatures. Refer to the Analog Temperature Assignment List to relate the temperature inputs to the control outputs.
OATEMP:	Outside air temperature.

Calculated variables:

MANOVRFLAG: Holds the cue for a manual override of normal program logic, 0 = manually override off, 1 manually override on, 2 = no manual override in effect. This is calculated by the MANOVR sub program.

MINFLAG: Holds the cue for whether a point is currently being controlled by its minimum on or minimum off time, 0 = it is not currently affected by minimum on/off, 1 = it is currently affected by minimum on/off times.

TOD: Holds a value of 1 if the point is within its time of day schedule, 0 if the point is not within its time of day schedule.

MODETEST: Holds the cue for the System Mode, 0 = heating season, 1 = non-heating season. Its contents are determined by the MODETEST algorithm that is called from this algorithm.

TURNON: Holds a value of 1 if the point should be turned on, 0 if the point should be turned off.

PROGRAM CODE BLOCK

```
' identify point characteristics

500   COOLFLAG = 0          ' not a cooling point.
      HEATFLAG = 1          ' heating point.
      TODFLAG = 1           ' time of day point.
      OPTSTFLAG = 1         ' optimum start point.
      TCFLAG = 1            ' temperature control point.
      OALFLAG = 1           ' low outside air temperature point.
      SETBACKFLAG = 1       ' night setback point.
      FREEZEFLAG = 1        ' freeze protection point.

1000  TURNON = 0            ' assume that it can be turned off.
      CALL LEAD-LAG         ' see if it is the lead or lag boiler.
      CALL MANOVR           ' manual override.
        IF MANOVRFLAG = 1 THEN TURNON = 1 : GOTO 3000
      CALL MINTEST          ' minimum on/off test.
        IF MINFLAG = 1 THEN GOTO 3000 ' TURNON is set in MINTEST ' when MINFLAG is = 1
      CALL FREZPROT         ' freeze protection override.
        IF TURNON = 1 THEN GOTO 3000
      CALL MODETEST         ' check to see if in heating or non-heating.
        IF SYSMODE = 1 THEN TURNON = 0 : GOTO 3000

' outside air heat lockout.
        IF OATEMP > MAXOAHEAT AND OALFLAG = 1 THEN TURNON = 0 : GOTO 3000
      CALL SNOWDAY — IF SNOWDAYFLAG = 1 THEN TOD = 0 : GOTO 2010
      CALL OPENHOUSE — IF OPHOUSEFLAG = 1 THEN TOD = 1 : GOTO 2000
      CALL ACTIVITY — IF ACTIVITYFLAG = 1 THEN TOD = 1 : GOTO 2000
      CALL TOD               ' see if its scheduled now.

' load appropriate hot water parameters, call outside air
' reset program to calculate target water temps then call
' the appropriate algorithm to set TURNON variable.
2000    IF TOD = 1 THEN
        LOWOAT = OCCLOWOAT : HIGHSWT = OCCHIGHSWT
        HIGHOAT = OCCHIGHOAT : LOWSWT = OCCLOWSWT
      CALL OARESET

' if there is no call for heat, leave boiler off.
        IF ACTUALTEMP > HEATSET THEN TURNON = 0 : GOTO 3000
      CALL CENTHEAT
        END IF

2010    IF TOD = O THEN
        LOWOAT = UNOCCLOWOAT : HIGHSWT = UNOCCHIGHSWT
        HIGHOAT = UNOCCHIGHOAT : LOWSWT = UNOCCLOWSWT
      CALL OARESET

' if there is no call for heat, leave boiler off
        IF ACTUALTEMP > SETBACKTEMP THEN TURNON = 0 : GOTO 3000

      CALL CENTHEAT
        END IF

3000  END BLOCK
```

12.3.6 Control Point Tie-Ins

To ensure that the EMCS is interfaced to a device to be controlled exactly as desired, the integrator should include sketches detailing the specifics of the interface at the control circuit of that device as exampled in Figure 12.3.

12.3.7 Commissioning

The final step of systems integration is commissioning. Commissioning should include both the hardware and software. All of the hardware should be physically inspected for correct installation and the wiring tested for continuity. Inputs should be checked for accuracy against a standard. The operation of all outputs should be demonstrated from the operator interface personal computer or system controller if no personal computer is installed. This inspection should not be taken lightly. The system will not function as designed if any hardware component is incorrectly installed or does not function as desired.

The software should be checked before the system is put on the line. This requires a careful review of all algorithms, setpoints and schedules for compliance with the detailed specifications.

APPENDIX B "EMCS Installation Requirements" provides the basic parameters to formalize specifications to bid the installation of a DDC EMCS. Commissioning procedures are detailed in sections 15, 16 and 17.

12.3.8 Tackling the Human Dimension in EMCS Specifications

Training should also be considered a part of the commissioning. In fact, training and long-term support are often the most overlooked aspects of an EMCS project. In what form and for whom training should be provided differs for every project. This is because the "people" involved are different every time.

In the end, the performance of an EMCS depends on the operating personnel. Those who are to work with the system must understand it. The most common cause for the failure of an EMCS to achieve intended results is ignorance. If a system is overridden frequently, or if controls are disconnected, or if parameters and schedules are not properly maintained the EMCS becomes useless. APPENDIX B includes training and documentation instructions in sections 10, 11, 12, 13, and 14.

Training is not something that can be adequately described in a few short paragraphs in a specification. In reality, there needs to be Comprehensive, Explicit, and Relevant Training and Support (CERTs). CERTs can be used to prevent misunderstandings and dissatisfaction on the part of the building owner, operators and tenants. CERTs also prevent system disconnections and "vapor-watts" (energy savings that never materialize).

In 12.B11.2, the boiler plate reads, "Contractor shall orient the training specifically to the system installed rather than a general training course." A case can be made for expanding this to include business specific and facility specific instructions. In new construction, these would be determined by peering as far into the future as possible to determine what might be required. The best approach here is to ask copious questions about how the facility will be used, staffed, serviced, and marketed (in the case of an office building).

In retrofit applications, the depth and quality of the questions asked during the facility appraisal (refer back to Section 12.3.8) heavily influence the detail spelled out in the specification. For example, for the EMCS RETROFIT discussed in 12.2.2, a list of facility specific instructions might also include the following:

Program user-friendly, menu-prompted, after-hours access setup routines.

Program user-friendly, menu-prompted, monthly after-hours access bill printing routines. Set up ten (10) test access codes which demonstrate options tenants will have for naming after hours zones. Print out bill examples using hourly rates which will be billed in this building.

Provide one 2-hour training session in after-hours access code programming and bill printing for building engineer.

Provide two 2-hour training sessions including system overview, control strategies, and temperature and status data interpretation, for property manager.

Provide two 2-hour training sessions in after-hours access code logic, benefits, options, and programming for property manager.

Provide one 2-hour training session in after-hours bill printing for property manager and property management secretary.

Provide color copies of access code maps for tenants to easily determine what areas will be illuminated and conditioned when accessed after hours.

Conduct a meeting with all tenants' designated representatives to explain how to request HVAC

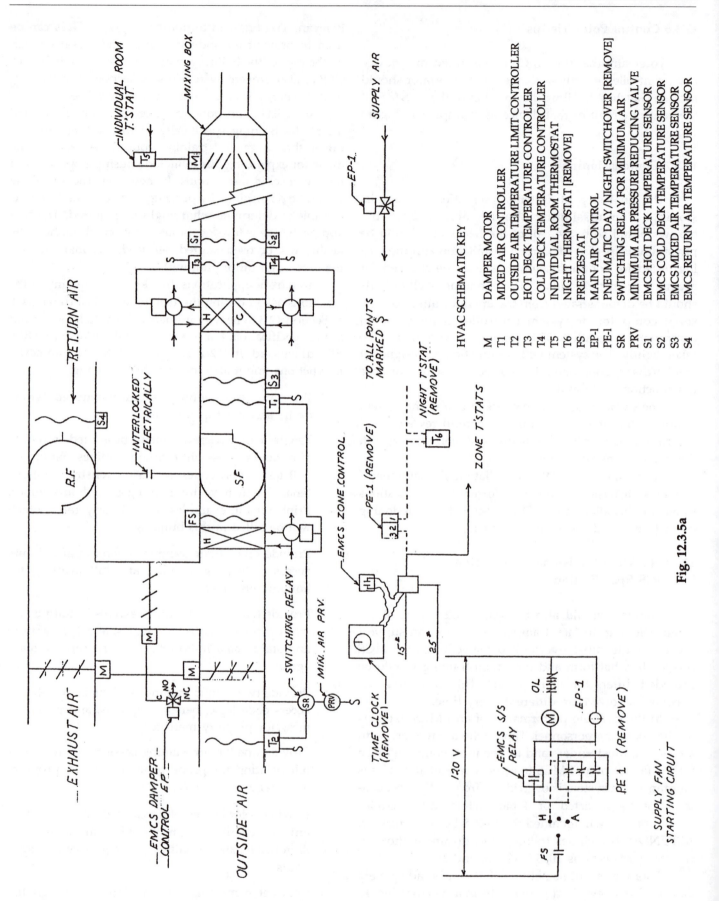

Fig. 12.3.5a

and/or lighting by telephone and to determine the need for additional access code sub-zones.

Conduct follow-up meeting with all tenants' designated representatives to review first month's billing and deliver access code maps documenting sub-zones requested in first meeting.

Conduct a meeting with janitorial service management personnel to explain the use of access codes (lighting only).

Provide two 2-hour training sessions for outside leasing agents in EMCS benefits (i.e. temperature control, efficiency, improved maintenance and after-hours fairness), why a building with an EMCS is different and better than a building without an EMCS.

Attend up to two lease negotiation meetings during first 12 months after start-up to explain after-hours and other EMCS benefits to prospective tenants.

Act as guest speaker at a tenant meeting to explain EMCS benefits beyond the after-hours fairness already explained in private meetings.

Conduct one 2-hour meeting to explain EMCS operation, provide basic documentation, and establish communication and service-call protocol with all service vendors (i.e. security, HVAC, lighting, and electrical).

The above are only a few of what might be a long list of services which could be included in an EMCS specification for a particular building owner to ensure that projected benefits are realized.

Whether one is a consulting engineer, a contractor doing a design/build project, or a facility director writing a guide specification, every effort should be made to anticipate future needs. EMCS projects priced, negotiated, or bid without the benefit of 'complete' specifications are doomed to less than stellar performance at best and total failure at worse. The human dimension may be the most challenging aspect of all of the factors to weigh and incorporate in an EMCS design.

12.3.9 Equipment and Contractor Qualification and Selection

Many EMCS projects have gone astray because decisions as to system manufacturer and installer (or bid list) were overly influenced by the features of EMCS equipment. There is a big difference between 'capabilities' and 'effective application of appropriate control and facility specific strategies'.

Too much emphasis on a 'brand' of EMCS overlooks the vital role of the contractor in the success or failure of a project. Even when protected by the most detailed of specifications, it pays to look beyond the low bid and conduct a thorough investigation of the installers being considered. Several of the major EMCS manufacturers also have contracting divisions. Talent and experience often vary immensely from branch office to branch office. Just because a company has been in business for 50 to 100 years does not automatically mean it is the best choice for a particular project. After all, none of the 'people' who will be involved in the project have worked for that company for 50 to 100 years.

Thorough interviews, inspections of previous job sites, conversations with references, and visits to potential contractors' offices to meet 'the delivery team' are all good ways to increase the chances of quality installation. The delivery team will have a greater impact on the outcome of a project than the EMCS equipment. A thorough investigation of the entire team should be made and only the qualified allowed to compete. The same argument could easily be made for the selection of a consulting engineer.

EMCS technology does not save energy, not on its own at any rate. People save energy by using EMCS technology as a tool. By seeking out the crafts-people and the people most focused on saving energy, the chances of a project paying off are significantly greater.

APPENDIX A

EMCS SOFTWARE SPECIFICATIONS

This Appendix provides a sample of the descriptive portion of an EMCS software specification for a new construction junior high school in Connecticut. The school is constructed of red brick, has a pitched, well insulated roof and comprises 140,000 sq. ft. Entry ways are double doored. Windows comprise a moderate percentage of outside wall area and are of the double pane construction. The building is fully air conditioned. The HVAC systems consists of a central boiler/chiller plant and single duct distribution without static air pressure control. Equipment is as follows: 1) a bank of 10 modular oil fired hot water boilers controlled by the manufacturer's microprocessor controller that stages the units on as needed to maintain a temperature reset schedule as determined by the outside air temperature, 2) 2 two stage reciprocating chillers connected in parallel, 3) a single cell cooling tower with a two speed fan, 4) one con-denser water circulator, 5) a pair of space heating circulators to circulate hot water, 6) a pair of space cooling circu-lators to circulate chilled water, 7) 17 air handling units—10 of which contain a preheat coil, a supply air fan, an outside air damper and a return air damper supply air to terminal units—7 of the units which contain a cooling coil, a heating coil, a supply air fan, an outside air damper and a return air damper supply air directly to the space, 8) one makeup air unit with an outside air damper, 9) 100 space terminal units each containing a reheat coil, a chilled water coil and a damper, 10) an oil fired domestic service hot water heater and associated circulator to supply heat energy to a heat exchanger in the domestic service hot water storage tank, 11) one circulator to circulate domestic service hot water through the building, 12) 21 attic exhaust fans, 13) 14 cabinet unit heaters, and 14) 27 toilet exhaust fans all to be controlled by a distributive DDC EMCS with peer-to-peer communication. There are several exhaust fans for ventilating mechanical rooms, fume hoods, science rooms, the auditorium stage and storage rooms. These units are locally controlled on an as needed basis. Occupancy sensors perform space lighting control.

The reader should note that sequencing logic requirements are organized into three sections: 1) administrative, 2) system, and 3) operating.

12.A1.0 GENERAL

12.A1.1 The listings in this section are meant to provide a logical guide for the sequence of operations to be used for each point of control. There are many considerations other than those listed in this section which must be incorporated into a proper energy management system control program.

12.A1.2 CONTRACTOR must provide programs in the language that is appropriate for the hardware that is used. The programs written in the primary language of the controller must conform in operation and functionality to the sequences listed in these specifications. CONTRACTOR IS LIABLE FOR PROOF OF COMPLIANCE WITH THESE SPECIFICATIONS. BASIS FOR PROOF OF COMPLIANCE WILL BE COMPARING CONTRACTOR'S WRITTEN PROGRAM CODE WITH THE SEQUENCES PROVIDED IN THESE SPECIFICATIONS AND FIELD TESTING OF ACTUAL SYSTEM OPERATION. EACH CONTROL POINT IN THE SYSTEM MUST BE EXERCISED IN THE PRESENCE OF OWNER'S REPRESENTATIVE TO PROVE PROPER OPERATION.

12.A2.0 ADMINISTRATIVE SEQUENCE LOGIC

12.A2.1 BASIC SCHEDULE: Install school occupancy schedules for the building as follows:

a. School Year: Provide capability for the user to easily and conveniently schedule the start and stop dates for the normal school term (School Year = 1). For the current School Year install August 28 as the start date and June 18 as the end date.

b. School Day: Provide capability for the user to easily and conveniently schedule a day of the week during the School Year for normal school activity (School Day = 1). For the current School Year install the 1992-1993 calendar.

e. School Hours: Provide capability for the user to easily and conveniently schedule occupancy hours during the School Day for the normal daily school session (School Hours = 1). For the current year install 7:00 am as the start time and 3:00 pm as the end time.

c. Regular Office Hours: Provide capability for the user to easily and conveniently schedule regular occupancy hours during the School Day for the central administrative office (Regular Office Hours = 1). For the current School Year install 7:00 am as the start time and 5:00 pm as the end time.

d. Vacation Office Hours: Provide capability for the user to easily and conveniently schedule administrative office occupancy hours for days other than a School Day (Vacation Office Hours = 1). For the current year install 8:00 am as the start time and 3:00 pm as the end time.

12.A2.2 ACTIVITY SCHEDULING: Install capability for the user to easily and conveniently schedule for up to twelve months in advance two (2) occupancy periods per day in addition to the Basic Schedule for an individual area activity (XXX Activity = 1). For the current year install the activities currently scheduled on the 1992-1993 school calendar.

a. Whenever an activity is scheduled, all equipment necessary to maintain normal occupancy conditions for that area at the date and time of that activity shall be enabled for occupancy.

b. The user must be able to easily and conveniently initiate, set time parameters and cancel each of the below listed scheduled activities as a separate function.

1. Clrm East Activity
2. Clrm S.E. Activity
3. Clrm South Activity
4. Clrm S.W. Activity
5. Clrm West Activity
6. Clrm North Activity
7. Media Activity
8. Auditorium Activity
9. Admin Activity
10. Main Gym Activity
11. Aux Gym Activity
12. Music Area Activity
13. Cafeteria Activity
14. Industrial Arts Activity
15. Open House

c. The following are equipment assignments for the above scheduled activity areas.

1. Points affected by CLRM EAST ACTIVITY:
   Boilers—Chillers—AHU-1

2. Points affected by CLRM S.E. ACTIVITY:
   Boilers—Chillers—AHU-2

3. Points affected by CLRM SOUTH ACTIVITY:
   Boilers—Chillers—AHU-3

4. Points affected by CLRM S.W. ACTIVITY:
   Boilers—Chillers—AHU-4

5. Points affected by CLRM WEST ACTIVITY:
   Boilers—Chillers—AHU-5

6. Points affected by CLRM NORTH ACTIVITY:
   Boilers—Chillers—AHU-6

7. Points affected by MEDIA ACTIVITY:
   Boilers—Chillers—AHU-7

8. Points affected by AUDITORIUM ACTIVITY:
   Boilers—Chillers—AHU-8—AHU-15 Domestic Hot Water

9. Points affected by ADMIN ACTIVITY:
   Boilers—Chillers—AHU-9

10. Points affected by MAIN GYM ACTIVITY:
    Boilers—Chillers—AHU-10
    Domestic Hot Water

11. Points affected by AUX GYM ACTIVITY:
    Boilers—Chillers—AHU-11
    Domestic Hot Water

12. Points affected by MUSIC ACTIVITY:
    Boilers—Chillers—AHU-12

13. Points affected by CAFETERIA ACTIVITY:
    Boilers—Chillers—AHU-13—FCU-2
    Domestic Hot Water

14. Points affected by INDUSTRIAL ARTS ACTIVITY:
    Boilers—Chillers—AHU-14—MAU-1

15. Points affected by OPEN HOUSE OVERRIDE:
    All control points

12.A2.3 INDIVIDUAL OVERRIDES: Install timed overrides for each output point as follows:

a. At any time the user must be able to easily and conveniently Enable/on or Disable/off any output point for a user selected time period.

b. At the end of the override period that point overridden shall return to its normal programmed control and the override period shall be reset to zero.

c. An Individual Override shall have priority over the Snow Day Override.

12.A2.4 SNOW DAY OVERRIDE: Install an easy and convenient method to override off any scheduled occupancy and keep the entire building in the unoccupied mode (Snow Day =1).

a. A Snow Day override shall automatically terminate at 2359 hours each day.

12.A2.5 GLOBAL SOFTWARE POINTS: The user variable software points identified by bold print in this specification shall be referenced in the algorithms such that the user need only to enter a new value for a named software point to effect a value change in all algorithms which include that software variable.

12.A3.0 SYSTEM SEQUENCE LOGIC

12.A3.1 SYSTEM MODE: Install capability for the user to easily and conveniently set and vary annual beginning and ending dates for the heating season to prevent simultaneous heating and cooling. Heating equipment operation will be determined by the time of year (i.e. space heating boilers and circulators will only operate during the heating season And be locked out during the rest of the year regardless of outside air temperature). Install October 15th to May 15th as initial parameters for the heating season. Heating season System Mode = 0, non-heating season System Mode = 1.

12.A3.2 SUMMER/WINTER: Install user adjustable variables for heating and cooling operations during building occupancy as follows:

a.   Whenever System Mode = 0 and the outside air temperature is less than or equal to the user variable Winter Occupancy Setpoint (i.e. 65°F), summer/winter shall go to winter. If the outside air temperature is greater than the Winter Occupancy Setpoint and the highest occupied space temperature + greater than some user variable setpoint (i.e. 74°F) and Chiller Plant Operation is Enabled/on then summer/winter shall go to summer to allow the chillers to function. If the outside air temperature is greater than the Winter Occupancy Setpoint and Chiller Plant Operation is Disable/off, summer/winter shall remain in winter to allow maximum ventilation with outside air.

b.   Whenever System Mode = 1 and the outside air temperature is greater than or equal to the user variable Summer Occupancy Setpoint (i.e. 60°F), summer/winter shall go to summer to allow the chillers to function. If the outside air temperature is less than the Summer Occupancy Setpoint and the lowest occupied space temperature is greater than some user variable setpoint (i.e. 72°F) and is Disable/off, summer/winter shall go to winter to allow maximum ventilation with outside air. If the outside air temperature is less than the Summer Occupancy Setpoint and Chiller Plant Operation is Enabled/on, summer/winter shall remain in summer.

12.A3.3 OPTIMUM START: Install an adaptable optimum start of all control points for both heating and cooling functions as follows:

a.   During optimum start for both heating and cooling, outside air dampers shall go fully closed; and, classroom and return air dampers shall go fully open; and, all exhaust fans shall be Disabled/off.

b.   Optimum start shall be limited to a maximum number of hours and be user adjustable.

12.A3.4 CHILLER PLANT OPERATION: Install capability for the user to easily and conveniently set and vary annual beginning and ending dates for operation of the chiller plant. Install May 1 to November 10 as initial parameters for Chiller Plant Operation.

12.A3.5 DESIGNATED SPACE TEMPERATURE SETPOINTS: Install user variable space temperature setpoints as follows:

a.   The occupied setpoint for all Classrooms, all Offices, the Media Center, the Music Area, the Cafeteria and Kitchen, The Industrial Arts Area, and the Audito-

rium shall be 72°F for both heating and cooling functions.

b.   The occupied setpoint for the Locker Rooms shall be 74°F for both heating and cooling functions.

c.   The occupied setpoint for the Main Gymnasium and the Auxiliary Gymnasium shall be 68°F for heating and 74°F for cooling.

d.   The unoccupied heating Setback Temperature shall be 55°F.

e.   The unoccupied cooling Setup Temperature shall be 80°F.

12.A3.6 FREE COOLING: Install free cooling into all algorithms that control air handling unit fans, attic exhaust fans, outside air dampers, return air dampers and space terminal unit dampers as follows:

a.   Free cooling shall be Enabled/on to ventilate the building during unoccupied periods whenever the outside air temperature is greater than the user variable Free Cool Minimum (i.e. 45°F) and the enthalpy of the outside air is less than the enthalpy of the inside air and the average building temperature is greater than the user variable Free Cool Setpoint (i.e. 72°F).

b.   Whenever free cooling is Enabled/on all outside air dampers, return air dampers and space terminal unit dampers shall go to the fully open position; and, all air handling units and attic exhaust fans shall be Enabled/on; and, the chiller plant and boiler plant shall be Disabled/off locked out.

12.A3.7 MINIMUM ON/OFF: Install a user adjustable Minimum On/Off time period (i.e. 5 minutes) in the control program for each piece of mechanical equipment.

12.A4.0 OPERATING SEQUENCE LOGIC

12.A4.1 AHU START/STOP: Install Enable/Disable of Air Handling Units as follows:

a.   Whenever School Hours = 1 and Snow Day = 0 all Air Handling Units shall be Enabled/on to start the supply air fans.

b.   Whenever Regular Office Hours = 1 or Vacation Office Hours = 1 AHU-9 shall be Enabled/on.

c. Whenever an individual area activity is scheduled on (i.e. a certain XXX Activity = 1) and Snow Day = 0 the air handling unit serving that area shall be Enabled/on.

d. Whenever School Hours = 0 and the activity flag for an individual area (i.e. XXX Activity = 0) the air handling unit serving that area shall be Disabled/off.

e. Whenever Regular Office Hours = 0 or Vacation Office Hours = 0 and Admin Activity = 0 AHU-9 shall be Disabled/off.

12.A4.2 AUDITORIUM AHU: Install control of the Air Handling Unit for the Auditorium as follows:

a. Whenever School Day = 1 and Snow Day = 0 and the Auditorium is unoccupied as determined by the Auditorium occupancy sensors AHU-8 shall be duty cycled as a standby load to maintain the Designated Space Temperature for the Auditorium within the user variable Space Dead Band (i.e. 6°F).

b. If, during the standby mode, the Auditorium should become occupied as determined by the Auditorium occupancy sensors for a user variable time period (i.e. 5 minuets) AHU-8 shall be Enabled/on to maintain the Designated Space Temperature for the Auditorium.

c. Whenever Auditorium Activity = 1 and Snow Day = 0 AHU-8 shall be Enabled/on to maintain the Designated Space Temperature for the Auditorium.

12.A4.3 MAIN GYMNASIUM AHU: Install control of the Air Handling Unit for the Main Gymnasium as follows:

a. Whenever School Day = 1 and Snow Day = 0 AHU-10 shall be duty cycled to maintain the Designated Space Temperature for the Main Gymnasium within the Space Dead Band.

b. Whenever Main Gym Activity = 1 and Snow Day = 0 AHU-10 shall be Enabled/on to maintain the Designated Space Temperature for the Main Gymnasium.

12.A4.4 MAKEUP AIR UNIT (MAU -1): Install Enable/Disable control of MAU-1 such that MAU-1 will be allowed to operate during School Hours and an Industrial Arts Activity. Actual control of on/off while MAU-1 is Enable/on will be by local toggle switch.

12.A4.5 AHU OUTSIDE AIR & RETURN AIR DAMPERS: Install control of the Air Handling Unit dampers as follows:

a. Whenever a unit is in normal operation (the supply air fan is functioning and free cool or optimum start is not in effect) and the outside air damper for that unit is not under a lockout the dampers shall be calibrated to maintain the user variable Minimum Fresh Air Intake Volume (i.e. 20%).

b. Whenever a unit is not operating the outside air damper and return air damper for that unit shall go fully closed.

c. Whenever in System Mode = 0 and the outside air temperature is equal to or greater than the user variable Winter Lockout Setpoint (i.e. 20°F) outside air dampers and return air dampers on operating Air Handling Units shall be Enabled/on and modulate under the control of a user adjustable PID loop to maintain the user variable Mixed Air Temperature (i.e. 55°F) for supply air to the preheat or heating coil.

d. Whenever in System Mode = 1 and the outside air temperature is equal to or less than the user variable Summer Lockout Setpoint (i.e. 85°F) outside air dampers shall be Enabled/on and go to minimum position; and, return air dampers shall go to the fully open position.

e. Whenever the outside air temperature is less than the Winter lockout Setpoint or greater than the Summer Lockout Setpoint outside air dampers shall be Disabled/off and go fully closed; and, return air dampers shall go to the fully open position.

12.A4.6 AHU PREHEAT COIL CONTROL VALVES: Install control of preheat coil control valves for the Air Handling Units as follows:

a. The AHU preheat coil control valve shall go fully open whenever air flow to the coil stops.

b. Whenever System Mode = 0 and there is air flow to an AHU preheat coil that coil control valve shall modulate under the control of a user adjustable PID loop to maintain the user variable Discharge Air Temperature (i.e. 60°F).

12.A4.7 AHU SPACE HEATING COIL CONTROL VALVES: Install control of heating coil control valves for the Air Handling Units as follows:

a. The AHU heating coil control valve shall go fully open whenever air flow to the coil stops.

b. Whenever System Mode = 0 and there is air flow to the heating coil that coil control valve shall modulate under the control of a user adjustable PID loop to maintain the designated Space Temperature Setpoint for that space.

12.A4.8 AHU COOLING COIL CONTROL VALVES: Install control of cooling coil control valves for the Air Handling Units as follows:

a. Whenever System Mode = 0 the cooling coil control valve shall go fully open.

b. Whenever System Mode = 1 and there is air flow to the cooling coil that coil control valve shall modulate under the control of a user adjustable PID loop to maintain the designated Space Temperature Setpoint for that space.

c. Whenever System Mode = 1 and air flow to a cooling coil stops that AHU cooling control valve shall go fully closed.

12.A4.9 SPACE TERMINAL UNIT DAMPERS & CONTROL VALVES:

a. During an occupancy mode whenever a space is occupied as determined by that space's motion sensor, the damper for the terminal unit serving that space shall go fully open and remain open for the duration of the occupancy; and, the hot water heating valve or chilled water cooling valve, as the season may be, shall modulate under the control of a user adjustable PID loop to maintain the designated Space Temperature Setpoint for that space.

b. During an occupancy mode whenever a space is unoccupied as determined by that space's motion sensor that space's terminal unit damper and heating or chilled water cooling valve, as the season may be, shall go fully closed.

c. If, during an occupancy mode, a space is unoccupied as determined by that space's motion sensor and that space's temperature should exit the user variable Space Dead Band the damper for that space's terminal unit shall go fully open; and, the hot water heating valve or chilled water cooling valve, as the season may be, shall modulate under the control of the PID loop to re-

turn that space's temperature to within the Space Dead Band.

d. When a space is in the unoccupied mode and System Mode = 1 and that space's temperature is less than the Setup Temperature that space's terminal unit damper and chilled water cooling valve shall go fully closed.

e. If, during the unoccupied mode and System Mode = 1, the temperature in a space should exceed the Setup Temperature, then the terminal unit damper for that space shall go fully open; and, the chilled water cooling valve for that space shall modulate under the control of the PID loop to cool that space to a temperature that is less than the Setup Temperature by an amount equal to the user variable Setback Differential (i.e. 2°F).

f. If, during the unoccupied mode, the temperature in a space should fall below the Setback Temperature, then the terminal unit damper for that space shall go fully open; and, the hot water heating valve for that space shall modulate under the control of the PID loop to heat that space until the temperature in that space exceeds the Setback Temperature by an amount equal to the Setback Differential.

12.A4.10 SPACE HEATING CIRCULATORS: Install control of space heating circulators as follows:

a. Whenever the building is occupied during System Mode = 0 and the outside air temperature is less than or equal to the Winter Occupancy Setpoint the lead space heating circulator shall be Enabled/on.

b. Whenever the building is occupied during System Mode = 0 and the outside air temperature is greater than the Winter Occupancy Setpoint the lead space heating circulator shall be Disabled/off.

c. Whenever the building is unoccupied during System Mode = 0 and the outside air temperature is greater than or equal to the user variable Freeze Point (i.e. 35°F) the space heating circulators shall be Disabled/off.

d. Whenever the building is unoccupied during System Mode = 0 and the outside air temperature is greater than or equal to the Freeze Point and the lowest space temperature falls below the Setback Temperature the lead space heating circulator shall be Enabled/on until all space temperatures exceed the Setback Temperature by an amount equal to the Setback Differential.

e. Whenever the outside air temperature is less than the Freeze Point the lead space heating circulator shall be Enabled/on.

f. The circulators will alternate lead/lag every 14 days and the then current lagging circulator will Enable/on in the event that the lead circulator should fail. Circulator failure is to be determined by a flow switch at each circulator discharge.

12.A4.11 BOILERS: Install control of boiler operation as follows:

a. Unless boiler override is on, boiler operation shall be allowed only if System Mode = 0.

b. Whenever a space heating circulator is Enabled/on the boilers shall be Enabled/on to maintain the appropriate supply water temperature as determined by the boiler reset controller.

c. Whenever space heating circulators are Disabled/off the boilers shall be Disabled/off.

12.A4.12 CHILLED WATER CIRCULATORS: Install control of chilled water circulators as follows:

a. Whenever the building is occupied and summer/winter is in summer and Chiller Plant Operation is Enabled/on the lead circulator shall be Enabled/on.

b. Whenever the building is unoccupied and the highest space temperature in a cooling area is less than or equal to the Setup Temperature the chilled water circulators shall be Disabled/off.

c. Whenever the building is unoccupied and the highest space temperature in a cooling area is greater than the Setup Temperature the lead chilled water circulator shall be Enabled/on until all space temperatures are less than the Setup Temperature by an amount equal to the Setback Differential.

d. The circulators will alternate lead/lag every 14 days and the then current lagging circulator will Enable/on in the event that the lead circulator should fail. Circulator failure is to be determined by a flow switch at each circulator discharge.

12.A4.13 CHILLERS: Install control of chillers as follows:

a. Whenever a chilled water circulator is Enabled/on the lead chiller shall be Enabled/on.

b. A chiller shall operate to maintain chilled water discharge (supply to loop) temperature as determined by that chiller's controller.

c. During School Hours only, whenever the lead chiller is Enabled/on and the outside air temperature is greater than the user variable Chiller Maximum Output Point (i.e. 85°F) the lag chiller shall be Enabled/on.

d. Whenever chilled water circulators are Disabled/off the chillers shall be Disabled/off.

e. The chillers will alternate lead/lag every 14 days and the then current lagging chiller will Enable/on in the event that the lead chiller should fail.

12.A4.14 ATTIC EXHAUST FANS: Install control for operation of attic exhaust fans as follows:

a. Whenever the attic air humidity is greater than the user variable Attic Maximum Humidity Point (i.e. 60%) attic exhaust fans shall be Enabled/on.

12.A4.15 TOILET EXHAUST FANS: Install control for operation of toilet exhaust fans as follows:

a. Whenever the building is occupied all toilet exhaust fans shall be Enabled/on.

b. Whenever and activity is scheduled on the toilet exhaust fans serving the building area associated with that activity shall be Enabled/on.

c. Whenever the building is unoccupied toilet exhaust fans shall be Disabled/off.

12.A4.16 CABINET UNIT HEATERS: Install control of the cabinet unit heaters as follows:

a. Whenever the building is occupied during System Mode = 0 and the outside air temperature is less than or equal to the Winter Occupancy Setpoint the unit cabinet heaters shall be Enabled/on; and, the hot water heating valve shall modulate under the control of a user adjustable PID loop to maintain space temperature at a user variable setpoint (i.e. 65°F).

b.   Whenever the building is <u>occupied</u> during System Mode = 0 <u>and</u> the outside air temperature is greater than the Winter Occupancy Setpoint the unit cabinet heaters shall be Disabled/off.

c.   Whenever the building is <u>unoccupied and</u> the outside air temperature is equal to or greater than the Freeze Point the unit cabinet heaters shall be Disabled/off.

d.   Whenever the building is <u>unoccupied and</u> the outside air temperature is less than the Freeze Point the unit cabinet heaters shall be Enabled/on; and, the hot water heating valve shall modulate under the control of the PID loop to maintain space temperature at a user variable setpoint (i.e. 55°F).

12.A4.17 <u>DOMESTIC SERVICE HOT WATER</u>: Install control of the domestic service hot water system as follows:

a.   During School Hours <u>and</u> scheduled activities for the Cafeteria, Main Gymnasium and Auxiliary Gymnasium the domestic service hot water heater and loop circulators and burner shall be Enabled/on. The local aquastat will cycle the heater circulator and burner on and off as required to maintain the desired loop temperature.

b.   During <u>unoccupied</u> periods the domestic service hot water system shall be Disabled/off.

12.A4.18 <u>COOLING TOWER</u>: The cooling tower shall be interlocked with the chillers such that whenever a chiller is Enabled/on the condenser water circulator and cooling tower fan are Enable/on.

a.   Cooling tower fan speed shall be determined by the cooling tower controller.

12.A4.19 <u>AHU FREEZE PROTECTION</u>: Install freeze protection for the Air Handling Units and the Make-up Air Unit as follows:

a.   Whenever in System Mode = 0 <u>and</u> the air temperature as measured at the face of the discharge side of a preheat or heating coil is less than the user variable Freeze Stat Point (i.e. 45°F) the supply fan for that unit shall be Disabled/off and the outside air damper shall be Disabled/off and go fully closed.

b.   If a unit has been Disabled/off because of freeze

protection, should the air temperature as measured at the face of the discharge side of a preheat or heating coil increase to a value greater than the Freeze Stat Point plus some user variable differential (i.e. 50°F) that unit shall be returned to its normal programmed control.

APPENDIX B

EMCS INSTALLATION REQUIREMENTS

In addition the Software Specifications and appropriate drawings and sketches necessary to integrate a DDC EMCS into a building, the following can be used as a guide in developing specifications to select a supplier and installer of the desired system.

12.B1.0 GENERAL

12.B1.1 The system within each specific building shall be configured as a distributive (or central as the case may be) control and monitoring system. All computing devices, as designed in accordance with FCC Rules and Regulations Part 15, shall be verified to comply with the requirements for Class A computing devices. The system shall perform all data consolidation, control, timing, alarm, feedback status verification and calculation for the system.

12.B1.2 The system controller (or controllers as the case may be) shall function as the overall system coordinator, perform automatic energy management functions, control peripheral devices and perform calculations associated with operator interactions, alarm reporting, and event logging.

12.B1.3 The system shall have at least 48 hours of battery backup to maintain all volatile memory and the real time clock.

12.B1.4 The system shall have a battery backed uninterruptible real time clock with an accuracy within 10 seconds per day.

12.B1.5 The system shall operate on 120 VAC, 60 Hz power. Operation shall continue in accordance with the Detailed Specifications from line voltages as low as 95 VAC. Line voltages below the operating range of the system shall be considered outages.

12.B1.6 The system shall be capable of interfacing with no additional hardware, other than the appropriate cabling, to industry standard terminals/PCs. The interface shall support as a minimum: RS232 EIA port, ASCII character code with baud rate selectable between 300-9600 baud.

12.B1.7 The system shall have the capability of connecting to a local terminal/PC or printer to perform all programming, modification, monitoring and reporting requirements of the Detailed Specifications.

12.B1.8 The system shall support a Hayes Smartmodem with full RS232 control signals to communicate with remote terminals and other systems via dial-up voice grade phone lines at a minimum of 1200 baud.

12.B1.9 The system shall accept control programs in accordance with the Software Specifications, perform automated energy management functions, control peripheral devices and perform all necessary mathematical calculations.

12.B1.10 The system central processing unit shall be a microcomputer of modular design. The word size shall be 8 bits or larger, with a memory cycle time less than 1 micro second. All chips shall be second sourced.

12.B1.11 Any control program of a system may affect control of another program if desired. Each program shall have full access to all input/output facilities of the system.

12.B1.12 The system shall bi-directionally transfer data to input/output units and local control units at a minimum of 720 baud.

12.B1.13 The system shall provide signal conditioning to insure integrity.

12.B1.14 The system shall be user friendly and shall allow OWNER to designate appropriate nomenclature for all inputs, outputs programs, routines and menus.

12.B2.0 INSTALLATION PARTICULARS

12.B2.1 CONTRACTOR shall install all equipment, devices, cabling, wiring, etc., in compliance with manufacturer's recommendations.

12.B2.2 The system shall have key lock protected access to internal circuitry and any output override switches.

12.B2.3 The system controller(s) shall be properly located to provide easy access, and to ensure the ambient temperature range is +32 degrees F to +95 degrees F,

10% to 95% relative humidity non-condensing. Location and accessibility for removal of covers for repair shall be sufficient to facilitate required access by service personnel. Location shall be approved by OWNER'S REPRESENTATIVE before installation.

12.B2.4 The load on any output shall not exceed 50% of that output's contact rating.

12.B2.5 Only one (1) wire pair per input/output point shall terminate within any input/output unit enclosure. All relays, wire junctions, terminal strips, transformers, modems and other devices not an integral part of the input/output unit shall be located external to the input/output unit in a separate suitable enclosure(s) properly mounted.

12.B2.6 Each processing unit and controller shall have a separate dedicated electric service of the appropriate amperage.

12.B2.7 All electrical relays, electric/pneumatic relays (EP's) and other control devices shall be mounted in an appropriate enclosure, securely fastened and clearly labeled as to purpose and item controlled. All enclosures shall be of adequate size to allow easy access. A minimum of 70 cubic inches shall be allowed per device and the minimum size enclosure shall be 6" × 6" × 4". All relays shall be base mounted and face out towards their enclosure opening. Relays not applicable to base mounting shall have screw type wire connections.

12.B2.8 Where stranded wire is used all connections shall be made with solderless connections.

12.B2.9 All wiring and cabling shall be labeled with a logical numbering system. The labels shall remain consistent from the input/output unit to termination with labels being clearly visible at all terminations and penetrations.

12.B2.10 For all new work, temperature sensors on piping shall be installed in wells.

12.B2.11 Where strap-on sensors are used in existing piping, sensors shall be bedded in heat conducting compound with a shield and securely fastened to pipe under insulation with strapping as recommended by manufacturer. If not present, insulation as approved by OWNER'S REPRESENTATIVE shall be installed to extend 12" on both sides of sensor. Sensors shall not be installed on PVC pipe.

12.B2.12 Sensors in ductwork shall protrude into air stream.

12.B2.13 Where a system output directly or indirectly controls equipment previously controlled by a time clock, that time clock shall be disconnected or otherwise removed from the control circuit of the equipment.

12.B2.14 Where a system output directly or indirectly controls equipment which has in its control circuit a Hand On-Off-Auto controller, the Hand On function shall be taken out of the control circuit.

12.B2.15 All control and sensor wiring in boiler rooms, equipment rooms, electrical rooms and unfinished areas shall be enclosed in metal conduit.

12.B2.16 All control and sensor wiring in finished areas shall be enclosed in metal wire mold.

12.B2.17 All control and sensor wiring in non-accessible spaces such as ceilings and tunnels shall be tied in a good workmanship manner acceptable to OWNER'S REPRESENTATIVE (maximum 3' spacing) to a suitable support with approved ties. Said wiring shall not lie on ceiling tiles. All wiring in plenum spaces shall be listed for use in same.

12.B2.18 Control of boilers, fans, pumps, chillers, compressors or any other equipment which utilizes electric or fossil fuel motors as a power source shall be totally by electrical means.

   a. Control points identified as "Electric" control points shall be wired such that the digital signal sent by a control unit is received directly by the device being controlled. Unless otherwise specified, any use of pneumatics to interface an "Electric" control point with the device to be controlled is not acceptable.

12.B2.19 Use of pneumatics for control of heating/cooling valves, hot/chilled water reset valves, dampers or similar modulating devices is acceptable.

   a. In the event that CONTRACTOR should utilize any component of existing pneumatic controls to interface "Pneumatic" control points with the device(s) to be controlled, CONTRACTOR shall be responsible for continuity of the pneumatic signal such that the signal from the control unit is transmitted undiminished and as intended to each device to be controlled.

12.B2.20 All conduit, boxes, and enclosures shall be metal.

12.B2.21 All wiring carrying 120 volts or more shall be #12 AWG minimum.

12.B2.22 All enclosures shall be labeled on the outside stating purpose and relays or devices enclosed. All relays and other devices shall be labeled inside the enclosures to clearly designate their use. Relays are to be labeled at the base or adjacent to the base so that the relay can be replaced without losing the label.

12.B2.23 Labels shall be of engraved phenolic type riveted to the enclosure. Minimum letter size shall be 1/4". Labels shall be approved by OWNER'S REPRESENTATIVE.

12.B2.24 All control and sensor wiring shall be a minimum of #18 gauge wire. All wire in plenums shall be plenum rated.

12.B3.0 GLOBAL DATA EXCHANGE

12.B3.1 Each control unit and input/output unit in the system shall be able to exchange information between other control units, input/output units and the system controller(s) each network scan.

12.B3.2 Any network structure shall be transparent such that each control unit in the system may store and reference any global variable available in the network for use in the controller's calculations or programs.

12.B4.0 INPUT/OUTPUT UNITS

12.B4.1 Input/output units shall be used to connect the data environment to the system and contain all necessary input/output functions to read field sensors and operate controlled equipment based on instructions from the central processing unit.

12.B4.2 Input/output units shall be fully supervised to detect failures. The input/output units shall report the status of all points in its data environment at the rate of at least once every 60 seconds for temperature control. Slightly longer times may be acceptable for non-critical loads on an exception basis.

12.B4.3 Upon failure of the input/output unit (including transmission failure), the input/output unit shall automatically force outputs to a predetermined state.

12.B4.4 For system voltages below 95 VAC the input/output unit shall totally shut down and de-energize its outputs.

12.B5.0 INPUTS

12.B5.1 The system shall be capable of receiving the following input signals:

a. <u>Analog Inputs (AI)</u>: The AI function shall monitor each analog input, perform analog to digital (A/D) conversion, and hold the digital value in a buffer for interrogation.

b. <u>Digital Inputs (DI)</u>: The DI function shall accept dry contact closures.

c. <u>Temperature Inputs</u>: Temperature inputs originating from a thermister, shall be monitored and buffered as an AI, except that automatic conversion to degrees F shall occur without any additional signal conditioning. Cable for temperature input wiring from sensors that incorporate RFI circuitry in the sensor do not require shielded cable. Cable for temperature input wiring that does not incorporate RFI circuitry in the sensor shall be shielded.

d. <u>Pulse Accumulators</u>: The pulse accumulator function shall have the same characteristics as the DI, except that, in addition, a buffer shall be required to totalize pulses between interrogations. The pulse accumulator shall accept rates up to 10 pulse/second.

12.B6.0 OUTPUTS

12.B6.1 The system shall be capable of outputting the following system signals:

a. <u>Digital Outputs (DO)</u>: The DO output function shall provide dry contact closures for momentary and maintained programmable operation of field devices. Any manual overrides shall be reported to the system at each update.

b. <u>Analog Output (AO)</u>: The AO output function may be a true analog output providing a 0-l0VDC and/or 0-20ma DC signal; or, accept digital data, perform digital to analog (D/A) conversion and output a pulse within the range of 0.1 to 3276.6 seconds in order to perform pulse width modulation of modulating equipment.

12.B7.0 DEVICES

12.B7.1 <u>Temperature Sensors</u>: Temperature sensors shall have a range suitable to their purpose. For example, typical degree ranges should be as follows: outside air -

$40°F$ to $+120°F$, indoor spaces $+40°F$ to $+100°F$, hot water $+80°F$ to $+260°F$, chilled water $+20°F$ to $+90°F$, cold duct air $+35°F$ to $+80°F$, and hot duct air $+60°F$ to $+200°F$. RTDs, thermocouples and sensors based on the National Semiconductor LM 34CZ chip are permitted for sensing temperature. Thermisters of nominal resistance of at least 10K ohms may be used for temperature sensing also. Thermisters shall be encapsulated in epoxy or series 300 stainless steel.

12.B7.2 <u>Thermowells</u>: Thermowells shall be series 300 stainless steel for use in steam or fuel oil lines, and monel for use in water lines. Thermowells shall be wrought iron for measuring flue gases.

12.B7.3 <u>Pressure Sensors</u>: Pressure sensors shall withstand up to 150% of rated pressure. Accuracy shall be plus or minus 3% of full scale. Pressure sensors shall be either capsule, diaphragm, bellows or bourdon.

12.B7.4 <u>Pressure Switches</u>: Pressure switches shall have a repetitive accuracy plus or minus 3% of range and withstand up to 150% of rated pressure. Sensors shall be diaphragm or bourdon tube. Switch operation shall be adjustable over the operating pressure range. The switch shall have an application-rated for C, snapacting wiping contact of platinum alloy, silver alloy or gold plating.

12.B7.5 <u>Watt-hour transducers</u>: Watt-hour transducers shall have an accuracy of plus or minus 1 percent for kW and kWh and shall be internally selectable without requiring the changing of current or potential transformers.

12.B7.6 <u>Potential Transformers</u>: Potential transformers shall be in accordance with ANSI C57.13.

12.B7.7 <u>Current Transformers</u>: Current transformers shall be in accordance with ANSI C57.13.

12.B7.8 <u>Watt-hour Meters with Diamond Register</u>: Meters shall be in accordance with ANSI C12 and have pulse initiators for remote monitoring of watt-hour consumption and instantaneous demand. Pulse initiators shall utilize light emitting diodes and photo-detectors. Pulse initiators shall consist of form C contacts with a current rating of 2 amps, 10 VA maximum and a life rating of one billion operations.

12.B7.9 <u>Flow Meters</u>: Flow meter outputs shall be compatible with the system. Accuracy shall be plus or minus 3% of full scale.

12.B7.10 <u>Control Relays</u>: Control relay contacts shall be rated for the application, with form C contacts, enclosed in a dustproof enclosure. Relays shall have silver cadmium contacts with a minimum life span rating of one million operations. When required, provide relays on H-0-A board with LEDs and relay status feedback.

12.B7.11 <u>Reed Relays</u>: Reed relays shall have been encapsulated in a glass type container housed in a plastic or epoxy case. Contacts shall be rated for the application. Operating and release times shall be five milliseconds or less. Reed relays shall have a minimum life span rating of 10 million operations.

12.B7.12 <u>Solid State Relays (SSRs)</u>: Input/output isolation shall be greater than 10E9 ohms with a breakdown voltage of 1500V root mean square or greater at 60 Hz. The contact life shall be 10 x 10E6 operations or greater. The ambient temperature range of SSRs shall be minus 20 to plus 140 degrees F. Input impedance shall not be less than 500 ohms. Relays shall be rated for the application. Operating and release time shall be 100 milliseconds or less. Transient suppression shall be provided as an integral part of the relay.

12.B7.13 <u>Surge Protection</u>: Surge protection shall meet the following minimum criteria.

a. Solid state silicon avalanche junction diode device

b. Response time less than or equal to 5ns (nanoseconds)

c. Voltage protection level 300 volts maximum

d. Service frequency 60 Hertz

e. Service voltage 110/120 volts AC

f. Life expectancy > 10 years

g. Ambient operating temperatures 0 to 40 degrees Celsius

h. Suppression power 12,000 watts (1 × 1000 microsecond pulse)

i. Instant automatic reset

j. UL Listed

12.B8.0 OPERATING SYSTEM

12.B8.1 The operating system shall be real-time operating system requiring no operator interaction to initiate and commence operations. The programming shall include:

a. Operations and management of all devices;

b. Error detection and recovery from arithmetic and logical functions;

c. Editing software to allow the user to develop and alter application programs; and,

d. System self-testing.

12.B9.0 SOFTWARE

12.B9.1 The software shall be designed to run twenty four hours a day, three hundred sixty five days a year. It shall continually compile and act upon acquired input and display pertinent information on system operation.

12.B9.2 The software shall be user friendly, user interactive and provide complete environmental and machine status information at the touch of a key. It shall allow the user to change operation parameters through easily understood display prompts.

12.B9.3 The software shall be flexible and adaptable. All displays must be clear and concise, using the actual names of the equipment and areas in question rather than employing a cryptic computer code to identify locations and machinery.

12.B9.4 All operator input shall require a minimum of keystrokes in response to prompted messages clearly identifying the required options for entries.

12.B9.5 The software modules and data base shall be stored in nonvolatile or battery hacked memory to prevent loss of the operating parameters in the event of a power failure. When the system starts, it shall first reload all the parameters and decide what should be on or off. The system shall then scan the facility, see what equipment is actually running and update all temperatures. If, in fact, the status is no longer what it should be, the system shall stagger start all appropriate equipment.

12.B9.6 The software shall provide telecommunication capability for remote support at a minimum of 1200 baud. After logging onto the system with a remote terminal, the remote user must have access to all user defined variables (operational parameters) as described hereinafter.

12.B9.7 The software shall provide the capability to upload and download all data, programs, I/O point databases and user configurable parameters from the remote terminal.

12.B9.8 The software shall be capable of autodialing a minimum of three different phone numbers, in pulse or dial mode.

12.B9.9 The software shall update the display of data with the most current data available within 5 seconds of operator request.

12.B9.10 The software shall allow the user to assign unique identifiers of his choice to each connected point. Identifiers shall have at least eight alpha/numeric characters. All reference to these points in programs, reports and command messages shall be by these identifiers.

12.B9.11 The software shall provide a simple method for the qualified systems integrator to designate control and monitoring characteristics of each control/monitor point. There shall be a menu selection to print the I/O point listings with current configuration data.

12.B9.12 The software shall be custom configured for the project and shall conform to the sequences that are explicitly specified in the Software Specifications. All configuration must be completed by CONTRACTOR prior to acceptance. GENERIC SOFTWARE TO BE TOTALLY CONFIGURED BY THE END USER IS UNACCEPTABLE. The operator must be able to easily change operating setpoints and parameters from legible, comprehensive menus, but input/output point definition, control algorithms and machinery interlocks must be custom tailored by CONTRACTOR to conform to the control sequences outlined. These algorithms and interlocks are designed to coordinate the operation of all controlled machinery for optimum energy conservation. Access to input/output point definition and algorithm selection/alternation shall be available to a qualified systems integrator through password protected routines and configuration file manipulation, however, these functions shall be invisible to the system operator.

12.B9.13 Software shall be modular in nature and easily expandable for both more control points and/or more control features.

12.B9.14 All input gathered by remote sensors shall be globally accessible to the system controller(s). Software designation rather than hardware configuration shall define the interrelationships of all input/output points.

12.B9.15 The software shall provide at least 3 password access codes that can be user configurable for leveled access to menu functions. All password protection shall be enabled or disabled through simple menu selections.

12.B9.16 The execution of control functions shall not be interrupted due to normal user communications including: interrogation, program entry, printout of the program for storage, etc.

12.B9.17 In the dynamic display mode the software shall have the ability to activate real time logic tracers. These tracers shall display the logical sequences being executed by the program as they occur.

12.B10.0 ENERGY MANAGEMENT FUNCTIONS

12.B10.1 Scheduling: The software shall provide for up to 5 (user configurable) multiple start and stop schedule(s) for each piece of machinery, for each day of the week. In addition, it shall provide for twenty-six (26) special holiday schedules which can be configured a year in advance. The operator shall have the ability to designate each holiday as either a single date or a range of dates. In addition, each of these 26 holidays shall be configurable as either a complete shutdown day where all zones of control are maintained at setback levels, or a special holiday allowing for the partial shut down of the facility. Each piece of equipment shall have a unique schedule for each of these holidays and the schedules must be identical in format and execution to the normal day-of-week schedules.

12.B10.2 Demand Shedding (if applicable): The operator shall be allowed to select the maximum electrical demand limit for each monitored electrical service in the facility that the system will strive to protect. The operator must have the ability to set the order of priority in which equipment is to be shed. The operator must also have the ability to identify any given control point as a "round-robin" shed point (first off-first on), a "priority" demand point (first off-last on) or a "temperature" demand point (load closest to setpoint shed first). Special user selectable demand parameters for each load shall be shed type, shed priority, minimum shed time, maximum shed time, minimum time between shed. There shall be a programmable demand shedding schedule that will determine the time of day during which demand shedding will be active.

12.B10.3 Minimum on-minimum off times: The operator shall have the ability to enter minimum on and minimum off times for each piece of machinery controlled. This feature shall override all other control parameters to prevent equipment short cycling.

12.B10.4 Straight Time Duty Cycling: The operator shall be able to establish any given control point as a straight time duty cycled point and to set the cycling parameters. For each cycled load the operator shall be able to designate any two minute segment of the hour as either a cycled off or cycled on period. The program shall always begin the programmed cycle periods at the designated offset from the top of the hour. Cycling parameters shall be able to be copied from point to point, or from one point to a group of points. The parameter copy feature shall automatically stagger the cycle period offset from the top of the hour. There shall be a programmable duty cycling schedule that will determine the time of day during which duty cycling will be active.

12.B10.5 Temperature Compensated Duty Cycling: The operator shall be able to establish any given control point as a temperature compensated duty cycled point and to set the cycling parameters for both heating and cooling ramps (if applicable) with a temperature Dead Band. For each cycled load the operator shall be able to designate high and low temperatures, long off and short off times for both heating and cooling ramps, and a total cycle length. All time parameters shall be programmable in minutes and the software shall verify data entry to prohibit conflicting time parameters. There shall be a programmable duty cycling schedule that will determine the time of day during which duty cycling will be active.

12.B10.6 Optimum Start/Stop: For both heating and cooling seasons, the system shall provide customized optimum start/stop routines which take into account outside temperature and inside zone temperatures when preparing the building climate for occupation or shutting the facility down at the end of the day. The software shall track the rate of heat loss or gain in each zone and utilize this data when activating optimum start/stop routines.

12.B10.7 Temperature Control: Building target temperatures for night time, weekend and holiday hours, as well as parameters and limits on normal occupied operation shall be user selectable. The system will strive to maintain these setpoint temperatures in consideration of other energy management functions such as demand shedding, duty cycling, optimum start/stop, etc. Temperature setpoint parameters for on/off points shall be separate temperatures for heating and cooling (if applicable) with Dead Band. Temperature setpoints for analog control points shall include minimum output position (in %), heating full, heating start, cooling start and cooling full.

12.B10.8 Air Conditioning Optimization: Air conditioning shall be balanced and controlled considering the following factors:

a. Date (user selectable start and stop dates);

b. Time of day;

c. Temperature (outside and inside zone temperatures);

d. Electrical load (staggered start ups & cycling);

e. Short cycle protection (user selectable minimum on and minimum off times); and,

f. Economizer control of equipment configured to support this type of operation.

12.B10.9 Freeze Protection: Any control point can be assigned to a special freeze protection menu for use in cold climate areas. Points so assigned shall be programmed to run based on an outside temperature setpoint. This freeze protection algorithm will supersede all other temperature control parameters.

12.B10.10 Manual Override: Separate software manual overrides shall be available for each controlled load that allow the operator to turn on any load for a specific period of time.

12.B10.11 Daylight Savings Time: Start and stop dates for Daylight Savings Time shall be programmable and automatically implemented by the system.

12.B10.12 Alarms: The software shall register and record alarms for the following conditions:

a. Failure of controlled equipment to respond to computer commands (where positive feedback is used);

b. High analog value (user selectable limits); and,

c. Low analog value (user selectable limits).

All alarms must be visible and history shall be maintained for a user selectable number of events with a minimum storage of 10 events. In addition, the system must have the capability to automatically dial out of the building to alert user determined authorities in the event of specific alarms. The operator shall have the ability to enable and disable automatic alarm printing. Alarm logs shall contain the date, time, location and type of alarm registered. Also, the software must record the same data for the time when the alarm cleared.

12.B10.13 Monitoring: The software shall track and display various types of information on monitored or controlled points including but not limited to:

a. Last time on

b. Last time off

c. Temperature

d. Pressure

e. Alarms

The software shall have the capability to track analog values on the basis of a user defined time period and store the information for a user configurable number of time periods with a minimum storage of 24 periods.

12.B11.0 TRAINING

12.B11.1 CONTRACTOR shall provide the services of competent instructors who will give full instruction to designated personnel in the operation, maintenance and programming of the system.

12.B11.2 CONTRACTOR shall orient the training specifically to the system installed rather than a general training course. Instructors shall be thoroughly familiar with the subject matter they are to teach.

12.B11.3 CONTRACTOR shall provide a training manual for each student which describes in detail the data included on each training program. CONTRACTOR shall provide equipment and material required for classroom training.

12.B11.4 Training on the functional operation of the system shall include:

    a. Operation of equipment;
    b. Programming;
    c. Diagnostics;
    d. Failure recovery procedure;
    e. Alarm formats (where applicable);
    f. Maintenance and calibration; and,
    g. Trouble shooting, diagnostics and repair instructions.

12.B12.0 SYSTEMS MANUAL

12.B12.1 CONTRACTOR shall provide three (3) system manuals describing programming and testing, starting with a system overview and proceeding to a detailed description of each software feature. The manual shall instruct the user on programming or reprogramming any portion of the system. This shall include all control programs, algorithms, mathematical equations, variables, setpoints, time periods, messages, and other information necessary to load, alter, test and execute the system. The manual shall include:

12.B12.2 Complete description of programming language, including commands, editing and writing control programs, algorithms, printouts and logs, mathematical calculations and passwords.

12.B12.3 Instructions on modifying any control algorithm or parameter, verifying errors, status, changing passwords and initiating or disabling control programs.

12.B12.4 Complete point identification, including terminal number, symbol, engineering units and control program reference number.

12.B12.5 Field information including location, device, device type and function, electrical parameters and installation drawing number.

12.B12.6 Location identification of system control hardware.

12.B12.7 For each system function, a listing of digital and/or analog hardware required to interface the system to the equipment.

12.B12.8 Listing of all system application programs associated with each piece of equipment. This listing shall include all control algorithms and mathematical equations. The listing shall be in easy to understand English format. All application programs must be submitted. No proprietary control algorithms will be accepted.

12.B13.0 MAINTENANCE MANUAL

12.B13.1 CONTRACTOR shall provide three (3) maintenance manuals containing descriptions of maintenance on all system components, including sensors and controlled devices. The manual shall cover inspection, periodic preventive maintenance, fault diagnosis and repair or replacement of defective components.

12.B14.0 CONTROL DRAWINGS

12.B14.1 CONTRACTOR shall submit the following Shop Drawings for the new energy management system for approval prior to start of the work.

12.B14.2 Wiring diagram indicating general system layout for each system.

12.B14.3 Control drawings including outline of systems being controlled and location of devices. Include floor plans showing locations of EMS panels, wiring runs, junction boxes, EP's, PE's, etc.

12.B15.0 CONTRACTOR'S VERIFICATION OF SYSTEM INSTALLATION

12.B15.1 CONTRACTOR shall provide to OWNER written certification in the form prescribed by OWNER'S REPRESENTATIVE that installation of the system is substantially complete as to all respects of these specifications prior to final checkout pursuant to section 12.B17.0 hereof. This certification shall be submitted to OWNER'S REPRESENTATIVE at least ten (10) calendar days prior to the then current Contract completion date. As part of the certification, CONTRACTOR shall perform as listed below.

12.B15.2 CONTRACTOR shall verify correct temperature readings and scaling of all temperature inputs by comparing actual space temperature to remote temperature indication as displayed by the system video monitor. Both actual and displayed temperatures shall be recorded in table form and certified as to authenticity by CONTRACTOR. Actual space temperature shall be measured by a device traceable to the National Bureau of Standards. Said device will be noted in the certification.

12.B15.3 CONTRACTOR shall verify operation of all outputs by operating each output over its full range of operation and observing for correct response at the controlled equipment.

12.B15.4 CONTRACTOR shall provide to OWNER a list of all existing nonoperational equipment that is interfaced with the system.

12.B15.5 CONTRACTOR shall perform an inspection of the completed work for compliance with all provisions of these specifications and the Software Specifications.

12.B16.0 CONTRACTOR'S VERIFICATION OF SYSTEM PROGRAMMING

12.B16.1 CONTRACTOR shall provide to OWNER written certification that all system programming is complete as to all respects of these specifications and the Software Specifications prior to final checkout pursuant to section 12.B17.0 hereof. This certification shall be submitted to OWNER'S REPRESENTATIVE at least twenty (20) calendar days prior to the then current Contract completion date. As part of the certification, CONTRACTOR shall perform as listed below.

12.B16.2 CONTRACTOR shall provide to OWNER a printed copy of all software programming, point designations and schedules installed into the system.

12.B16.3 CONTRACTOR shall demonstrate to

OWNER'S REPRESENTATIVE all software programming as detailed in the Software Specifications. The demonstration shall include but not be limited to verification of the following:

a. Individual equipment control programs;
b. Activity schedule/programs;
c. Holiday schedule;
d. Specific area activity overrides;
e. Freeze protection;
f. Global override;
g. Data logging (trending, last time on, last time off); and,
h. Software interlocks.

12.B17.0 FINAL SYSTEMS CHECKOUT

12.B17.1 CONTRACTOR shall arrange with OWNER'S REPRESENTATIVE a final systems checkout to be performed at least five (5) calendar days prior to the then current Contract completion date.

12.B17.2 CONTRACTOR shall provide all materials and equipment necessary for final systems checkout.

12.B17.3 CONTRACTOR shall provide all personnel necessary for final systems checkout. OWNER'S REPRESENTATIVE will act only as an observer for verification and not actively participate in this checkout procedure.

12.B17.4 CONTRACTOR shall demonstrate operation of the following system capabilities and components in the presence of OWNER'S REPRESENTATIVE:

a. Remote terminal access to the central system controller and all of its peers via telephone modem;
b. Remote uploading and downloading of all programs and data points on the energy management system (to include the central system controller and any peers);
c. Local terminal access to the central system controller and any peers;
d. Operation of each output point as per subparagraph 12.B17.8 below; and,
e. Listing of the status of all input points with verification of dynamic response to changing conditions as per subparagraph 12.B17.7 below.

12.B17.5 CONTRACTOR shall demonstrate proper operation of each input point to OWNER'S REPRESENTATIVE. Input points will be verified for accuracy, location, scaling and operation. Operation will be verified by including a dynamic change in the ambient condition sensed by input sensor and tracking the sensors response and subsequent return to the static condition.

12.B17.6 CONTRACTOR shall demonstrate proper operation of each output point to OWNER'S REPRESENTATIVE. Each output point will be verified in four (4) parts as follows:

a. Building automation system to the interface device (i.e. relay, EP, etc.); b. Orientation of interface device (normally open, normally closed, etc.); c. Interface device to the specified equipment; and, d. Actual control/operation of specified equipment.

12.B17.7 OWNER'S REPRESENTATIVE will conduct an inspection of the physical installation to include, but not be limited to, the following:

a. Mounting of the system controller and peers;
b. Labeling of wires and components;
c. Neat and orderly wiring and terminations;
d. Mounting/location of modem;
e. Mounting/location of auxiliary enclosures;
f. Access to relays and equipment mounted in auxiliary enclosures;
g. Mounting/location of temperature and pressure sensors; and,
h. Completion of associated electrical work including disconnecting all time clocks associated with HVAC equipment and disconnecting "Hand On" position on all associated Hand On-Off-Auto switches.

12.B18.0 AS BUILT DRAWINGS

12.B18.1 CONTRACTOR shall provide to OWNER "as built" drawings in conjunction with CONTRACTOR'S systems installation certification.

12.B18.2 As built drawings shall include floor layout drawings including but not limited to:

a. Location of all control devices;
b. Location of all controlled equipment;
c. Location of all system controllers and terminal controllers;
d. Location of all wire runs and number of conductors;
e. All spare wires;
f. Location of all circuit and feeds including panel

designations and circuit breaker numbers;
g. Room numbers and or space designations;
h. Location of all sensors; and,
i. Wire labeling at terminal points.

12.B18.3 As built drawings shall include all wiring and pneumatic schematics showing the systems interfacing for each piece of controlled equipment.

APPENDIX C

THE DDC DICTIONARY

ALGORITHM—An assortment of rules or steps presented for solving a problem.

ANALOG INPUT—A variable input that is sensed, like temperature, humidity or pressure and is sent to a computer for processing.

ANALOG OUTPUT—A variable signal such as voltage, current pneumatic air pressure which would in turn operate a modulating motor which will drive valves, dampers, etc.

ANALOG TO DIGITAL A/D CONVERTER—An integrated circuit that takes data such as a temperature sensor and converts it into digital logic that the microprocessor can recognize. All analog inputs of a DDC system are fed into the microprocessor through an A/D converter.

BAUD RATE—The speed, in bits per second, at which information travels over a communications channel.

BIT—The smallest representation of digital data there is. It has two states; 1 (on) or 0 (off).

BITS OF RESOLUTION—A representation of how finely information can be represented. Eight bits of resolution means information may be divided into 256 different states, ten bits allows division into 1024 states and twelve bits allows division into 4096 states.

BYTE—Eight bits of digital information. Eight bits can represent up to 256 different states of information.

CONTROL LOOP—The strategy that is used to make a piece of equipment operate properly. The loop receives the appropriate inputs and sets the desired condition.

DAISY CHAIN—A wiring scheme where units must follow one another in a specific order.

DEAD BAND—An area around a setpoint where there is no change of state.

DEAD TIME—A delay deliberately placed between two related actions in order to avoid an overlap in operation that could cause equipment problems.

DERIVATIVE CONTROL—A system which changes the output of a controller based on how fast a variable is moving from or to a setpoint.

DIGITAL INPUT—An input where there are only two possibilities, such as on/off, open/closed.

DIGITAL COMMUNICATION BUS—A set of wired, usually a twisted pair for DDC controllers. Information is sent over this bus using a digital value to represent the value of the information.

DIGITAL OUTPUT—An output that has two states, such as on/off, open/closed.

DIGITAL TO ANALOG D/A CONVERTER—An integrated circuit that takes data from the microprocessor and converts it to analog data that is represented by a voltage or current. All analog outputs of a DDC system go through a D/A converter.

DISTRIBUTED CONTROL—A system where all intelligence is not in a single controller. If something fails, other controllers will take over to control a unit.

EPROM—Erasable Programmable Read Only Memory. An integrated circuit that is known as firmware, where software instructions have been "burned" into it.

EEPROM—Electrically Erasable Programmable Read Only Memory. An integrated circuit like the EPROM above except that the information may be altered electronically with the chip installed in a circuit.

FCU—Fan Coil Unit. A type of terminal unit.

FEEDBACK—The signal or signals which are sent back from a controlled process telling the current status.

FIRMWARE—Software that has been programmed into an EPROM.

FLOATING POINT—Numerical data that is displayed or manipulated with automatic positioning of the decimal point.

HP—Heat Pump. A type of terminal unit.

INPUT—Data which is supplied to a computer or control system for processing.

INTEGRAL CONTROL—A system which changes the output of a controller based on how long a variable has been offset from a setpoint. This type of control reduces the setpoint/variable offset.

INTELLIGENT BUILDING—A building that is controlled by a DDC system.

LAG—The delay in response of a system to a controlled change.

LAN—Local Area Network. A communication line through which a computer can transmit and receive information from other computers.

MICROPROCESSOR—The brains of all DDC systems and, for that matter, personal computers. An integrated circuit which has logic and math functions built into it. If fed the proper instructions, it will perform the defined functions.

MODEM—A device which allows the computer to communicate over phone lines. It allows the DDC system to be viewed, operated and programmed through a telephone system.

OFFSET—The difference between a variable and its setpoint.

OPERATOR'S TERMINAL—The computer at which the operator can control a motor or to a relay to start/stop a unit.

OUTPUT—Processed information that is sent by a controller to an actuator to control a motor or to a relay to start/stop a unit.

PI CONTROL—A combination of proportional and integral control. Adequate for almost all HVAC control applications. Control loops using PI control to look at both how far an input is from setpoint and how long it has been away from setpoint.

PID CONTROL—Proportional plus Integral plus Derivative Control. A system which directs control loops to look at how far away an input is from setpoint, how long it has been at setpoint and how fast it is approaching/moving away from setpoint.

PROPORTIONAL CONTROL—A system which linearly varies the output as an input variable changes relative to setpoint. The farther the input is from setpoint, the larger the change in the output.

PULSE WIDTH MODULATION—A means of proportionally modulating an actuator using digital outputs. One output opens the motor, another closes it. Used extensively in VAV boxes for driving the damper motor on the box.

RAM—Random Access Memory. A computer chip that information can be written to and read from. Used to store information needed in control loop calculations.

SETPOINT—A value that has been assigned to a controlled variable. An example would be a cooling setpoint of 76 degrees F.

SOFTWARE—The list of instructions written by an engineer or programmer that makes a controller or computer operate as it does.

TERMINAL UNIT—Part of the mechanical system that serves an individual zone, such as VAV box, hydronic heat pump, fan coil, etc.

TRANSDUCER—A unit that converts one type of signal to another. An electronic signal to a pneumatic signal.

TWISTED PAIR—Two wires in one cable that are twisted the entire length of the cable. DDC systems often use a twisted pair for communication link between controllers.

TWO POSITION—Something that has only two states. A two position valve has two states, either open or closed.

USER INTERFACE—The operator's terminal is usually referred to as the user interface. If allows the operator to interrogate the system and perform desired functions. Most interfaces today are personal computers or minicomputers.

VAV—Variable Air Volume. A terminal unit that varies the amount of air delivered to a space, depending upon the demand for cooling.

WORD—Sixteen bits of digital information. Sixteen bits can represent up to 65,536 states of information.

APPENDIX D

EMCS MANUFACTURERS DIRECTORY

A.E.T. Systems, Inc., 77 Accord Executive Park Drive, Norwell, MA 02061
617-871-4801

Alerton Technologies, 2475 140th Ave. N.E., Bldg. A, Bellevue, WA 98005
206-644-9500

American Auto-Matrix, One Technology Drive, Export, PA 15632
412-733-2000

Andover Controls Corp., 300 Brickstone Square, Andover, MA 01810
508-470-0555

Automated Logic Corp., 1283 Kennestone Circle, Marietta, GA 30066
404-423-7474

Barber-Coleman Co., 1354 Clifford Ave., Rockford, IL 61132
815-877-0241

Broadmoor Electric Co., 1947 Republic Ave., San Leandro, CA 94577
415-483-2000

Carrier Corp., One Carrier Place, Farmington, CT 06034
203-674-3000

Chrontrol Corp., 9707 Candida Street, San Diego, CA 92826
619-566-5656

CSI Control Systems International, Inc., 1625 W. Crosby Road, Carrollton, TX 75006
214-323-1111

D.K. Enterprises, Inc., 7361 Ethel Avenue, Suite 1, North Hollywood, CA 91605
818-764-0819

Danfoss-EMC, Inc., 5650 Enterprise Pkwy., Ft. Myers, FL 33905
813-693-5522

Delta Controls, Inc., 13520 78th Avenue, Surrey, B.C. Canada V5W836
604-590-8184

Dencor, Inc., 1450 W. Evans Avenue, Denver, CO 80223
303-922-1888

EDA Controls Corp., 6645 Singletree Drive, Columbus, OH 43229
614-431-0694

Electronic Systems USA, Inc., 1014 East Broadway, Louisville, KY 40204
502-589-1000

Elemco Building Controls, 1324 Motor Parkway, Hauppauge, NY 11788
516-582-8266

Energy Control Systems, Inc., 2940 Cole Street, Norcross, GA 30071
404-448-0651

Functional Devices, Inc., 310 S. Union Street, Russiaville, IN 46979
317-883-5538

Grasslin Controls Corp., 45 Spear Road, Ramsey, NJ 07446
201-825-9696

Honeywell, Inc., Comm. Bldg. Group, Honeywell Plaza, Minneapolis, MN 55408
612-870-5200

Honeywell, Inc., Res. & Bldg. Controls, 1985 Douglas Drive North, Minneapolis, MN 55422
612-870-5200

Integrated Energy Controls (Enercon Data), 7464 78th St. West, Minneapolis, MN 55435
612-829-1900

ISI Wireless, 3000-D South Highland Drive, Las Vegas, NV
702-733-6500

ISTA Energy Systems Corp., 407 Hope Ave., P.O. Box 618, Roselle, NJ 07203
908-241-8880

Johnson Controls, Inc., 507 E. Michigan Street, P.O. Box 423, Milwaukee, WI 53201
414-274-4128

Kreuter, Mfg., 5000 Peachtree Ind. Blvd., #125, Norcross, GA 30071
404-662-5720

Landis & Gyr Powers, 1000 Deerfield Parkway, Buffalo Grove, IL 60089
708-215-1000

Logic Power, 13500 Wright Circle, Tampa, FL 33626
813-854-1588

Microcontrol Systems, Inc., 6579 N. Sidney Pl., Milwaukee, WI 53209
414-262-3143

North American Technologies, Inc., 1300 N. Florida Mango Road, Ste 32, W. Palm Beach, FL 33409
407-687-3051

Novar Controls Corp., 24 Brown Street, Barberton, OH 44203
216-745-0074

Paragon Electric Company, 606 Parkway Blvd., Two Rivers, WI 54241
414-793-1161

Phonetics, Inc., 901 Tryens Road, Aston, PA 19014
215-558-2700

Process Systems, Inc., 100-A Forsythe Hall Dr, Box 240451, Charlotte, NC 28224
704-588-4660

Robertshaw Controls, 1800 Glenside Drive, Richmond, VA 23261
804-289-4200

Scientific Atlanta, 4300 Northeast Expressway, Atlanta, GA 30340
404-449-2902

Snyder General, Inc., 13600 Industrial Park Blvd., Minneapolis, MN 55440
612-553-5330

Solidyne Corp., 1202 Carnegie Street, Rolling Meadows, IL 60008
619-427-9640

Staeta Control Systems Inc., 8515 Miralani Drive, San Diego, CA 92126
619-530-1000

Teletrol Systems, Inc., 324 Commercial Street, Manchester, NH 03101
603-645-6061

The Watt Stopper, Inc., 296 Brokaw Road, Santa Clara, CA 95050
408-988-5331

Trane Company, The, 20 Yorkton Court, St. Paul, MN 55117
612-490-3900

Triangle Microsystems, Inc., 456 Bacon Street, Dayton, OH 45402
513-223-0373

Unity Systems, Inc., 2606 Spring Street, Redwood City, CA 94063
415-369-3233

Xencom Systems, 12015 Shiloh Road, Suite 155, Dallas, TX 75228
214-991-1643

Zeta Engineering Corp., 797 Industrial Court, Bloomfield Hills, MI 48013
313-332-2828

The authors wish to thank the following organizations for making this directory available:

Electric Power Research Institute, 3412 Hillview Ave., Palo Alto, CA, 94303
Plexus Research, Inc., 289 Great Road, Acton, MA 01720

REFERENCES

1. Cratty, William E., "Energy Conservation Design and Services With Ventana," *APEC Journal*, Spring 1990.
2. Crosby, Philip B., *Quality Is Free*, Penguin Books, 1979.
3. Elemco Building Controls, Inc., *Elemco Energy Executive II Control Algorithms*, 1989.
4. Gustavson and Gustavson, *The Business of Energy Management/ Doing It Right And Making Money At It*, Gannam/Kubat Publishing, Inc. 1990.
5. Honeywell Inc., *Engineering Manual of Automatic Control For Commercial Buildings*, 1988.
6. Kirkpatrick, Bill, "Computers In EMS/Controls," *Contracting Business*, September 1990.
7. Kirkpatrick, Bill, "Overlay Controls," *Contracting Business*, November 1990.
8. Mehta and Thumann, *Handbook of Energy Engineering*, Fairmont Press, Inc., 1989.
9. Patel, Jayant R., "Case Study: Energy Conservation/A Continual Process At Defense Contractor's Manufacturing Facility," *Energy Engineering*, Vol. 88, No. 4, 1991.
10. Rosenfeld, Shlomo I., "Worker Productivity: Hidden HVAC Cost," *Heating/Piping/Air Conditioning*, September 1991.
11. Rosenfeld, Shlomo I., "Worker Productivity: Hidden HVAC Cost," *Heating/Piping/Air Conditioning*, September 1990.
12. Rosenfeld, Shlomo I., "Worker Productivity: Hidden HVAC Cost," *Heating/Piping/Air Conditioning*, September 1989.
13. Silveria, Kent B., "The New EMCS Generation Has A Lot To Offer," *Consulting-Specifying Engineering*, January 1992.
14. Stern and Aronson., *Energy Use/The Human Dimension*, W.H. Freeman and Company, 1984.
15. Thumann, Albert. *Plant Engineer and Managers Guide To Energy Conservation*, Fairmont Press, Inc. 1989.

CHAPTER 13

LIGHTING

ERIC A. WOODROOF

School of Industrial Engineering and Management
Oklahoma State University

LESLIE "LES" L. PACE, CEM, CLEP

Lektron
Tulsa, Oklahoma

13.1 INTRODUCTION

In today's cost-competitive, market-driven economy, everyone is seeking technologies or methods to reduce energy expenses and environmental impact. Because nearly all buildings have lights, opportunities for lighting retrofits are very common and generally offer an attractive return on investment. Electricity used to operate lighting systems represents a significant portion of total electricity consumed in the United States. In 1991, lighting systems consumed approximately 20% of the electricity generated in the United States. U.S. consumers spent approximately $9 billion/year on lighting equipment and $38 billion/year for the associated electricity.[1]

One driving force behind energy management is to reduce energy costs. Energy costs are usually divided between demand charges (kW) and consumption charges (kWh). An attractive feature of lighting energy management is that many lighting retrofits can provide savings in both demand and consumption charges. Thus, the potential for dollar savings is maximized.

The distinguishing element of lighting retrofits is that in addition to reducing energy expenses, they can also improve the visual environment and worker productivity. Conversely, if a lighting retrofit reduces lighting quality, worker productivity may drop and the energy savings could be overshadowed by reduced profits. This was the case with the lighting retrofits of the 1970s, when employees where left "in the dark" due to massive de-lamping initiatives. However, due to substantial advances in technologies, today's lighting retrofits can reduce energy expenses while *improving* lighting quality and worker productivity.

This chapter will provide the energy manager with a good understanding of lighting fundamentals, so that he/she can ensure lighting upgrades will not compromise visual comfort. The Example Section (near the end of this chapter) contains analyses of a few common lighting retrofits. The Schematics Section contains illustrations of many lamps, ballasts and lighting systems.

13.2 LIGHTING FUNDAMENTALS

Energy managers who desire to identify good opportunities for lighting retrofits must understand a few fundamentals about lighting design and light quality. This section will introduce the important concepts about lighting, and the two objectives of the lighting designer: (1) to provide the right quantity of light, and (2) provide the right quality of light.

13.2.1 Lighting Quantity

Lighting quantity is the amount of light provided to a room. Unlike light quality, light quantity is easy to measure and describe.

13.2.1.1 Units

Lighting quantity is related to three types of units: watts, lumens and foot-candles (fc). Figure 13.1 shows the relationship between each unit. The watt is the unit for measuring electrical power. It defines the rate of energy consumption by an electrical device when it is in operation. The amount of watts consumed represents the electrical input to the lighting system.

The output of a lighting system is measured in lumens. The amount of lumens represents the brightness of the lamp. For example, one standard four-foot fluorescent lamp would provide 2,900 lumens in a standard office system. The amount of lumens can also be used to describe the output of an entire fixture (comprising several lamps). Thus, the number of lumens describes how much light is being produced by the lighting system.

The number of foot-candles shows how much light

is actually reaching the workplane (or task). Foot-candles are the end result of watts being converted to lumens, the lumens escaping the fixture and traveling through the air to reach the workplane. In an office, the workplane is the desk level. You can measure the amount of foot-candles with a light meter when it is placed on the work surface where tasks are performed. Foot-candle measurements are important because they express the "result" and not the "effort" of a lighting system. The Illuminating Engineering Society (IES) recommends light levels for specific tasks using foot-candles, not lumens or watts.

Efficacy

Similar to efficiency, efficacy describes an output/input ratio, the higher the output (while input is kept constant), the greater the efficacy. Efficacy is the amount of lumens per watt from a particular energy source. A common misconception is that lamps with greater wattage provide more light. However, light sources with

high efficacy can provide more light with the same amount of power (watts), when compared to light sources with low efficacy.

Figure 13.2 presents the spectrum of efficacies available from different types of lighting systems. These systems will be discussed in greater detail in Section 13.2.3; Lighting System Components.

13.2.1.2 IES Recommended Light Levels

The Illuminating Engineering Society (IES) is the largest organized group of lighting professionals in the United States. Since 1915, IES has prescribed the appropriate light levels for many kinds of visual tasks. Although IES is highly respected, the appropriate amount of light for a given space can be subjective. For many years, lighting professionals applied the philosophy that "more light is better," and light levels recommended by IES generally increased until the 1970s. However, recently it has been shown that occupant comfort decreases when a space has too much light. Numerous experiments have confirmed that some of IES's light levels were excessive and worker productivity was decreasing due to poor visual comfort. Due to these findings, IES revised their Handbook and reduced the recommended light levels for many tasks.

The lighting designer must avoid over-illuminating a space. Unfortunately, over-illuminated spaces have become the "norm" in many buildings. The tradition of excessive illumination is easily continued due to habit. To correct this trend, the first step for a lighting retrofit should be to examine the existing system to determine if it is over-illuminated. Appropriate light levels for all types of visual tasks can be found in the most recent IES

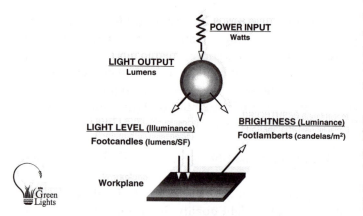

Figure 13.1 Units of measurement.

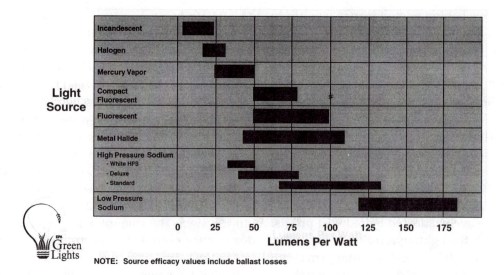

Figure 13.2 Spectrum of efficacies for different lighting systems.

Handbook. Table 13.1 is a summary of some of the IES recommendations.

It is important to remember that IES light levels correspond to particular visual tasks. In an office, there are many tasks: walking around the office, viewing computer screens, reading and writing on paper. Each task requires a different light level. In the past, lighting designers would identify the task that required the most light and design the lighting system to provide that level of illumination for the entire space. However, as previously stated, these design methods often lead to envi-

Table 13.1 Recommended light levels for visual tasks.

Building/Space Type	Guideline Illuminance Range (footcandles)
Commercial interiors	
Art galleries	30-100
Banks	50-150
Hotels (rooms and lobbies)	10-50
Offices	30-100
-Average reading and writing	50-75
-Hallways	10-20
-Rooms with computers	20-50
Restaurants (dining areas)	20-50
Stores (general)	20-50
Merchandise	100-200
Institutional interiors	
Auditoriums/assembly places	15-30
Hospitals (general areas)	10-15
Labs/treatment areas	50-100
Libraries	30-100
Schools	30-150
Industrial interiors	
Ordinary tasks	50
Stockroom storage	30
Loading and unloading	20
Difficult tasks	100
Highly difficult tasks	200
Very difficult tasks	300-500
Most difficult tasks	500-1000
Exterior	
Building security	1-5
Floodlighting	
(low/high brightness or surroundings)	5-30
Parking	1-5

ronments with excessive brightness and poor worker productivity.

If the IES tables are applied for each task ("Task Lighting"), a superior lighting system is constructed. For example, in an office with computers, there should be up to 30 fc for ambient lighting. Small task lights on desks could provide the additional foot-candles needed to achieve a total illuminance of 50 to 75 fc for reading and writing. Task lighting techniques are discussed in greater detail in Section 13.3.

13.2.2 Lighting Quality

Lighting quality is a more subjective issue for the lighting designer. However, it needs to be addressed because good lighting quality can be extremely important. Lighting quality can have a dramatic influence on the attitude and performance of occupants. In fact, different "moods" can be created by a lighting system. Consider the behavior of people when they eat in different restaurants. If the restaurant is a fast-food restaurant, the space is usually is illuminated by bright white lights, with a significant amount of glare from shiny tables. Occupants rarely spend much time there partly because the space creates an uncomfortable mood and the atmosphere is "fast" (eat and leave). In contrast, consider an elegant restaurant with a candle-lit tables and a "warm" atmosphere. Occupants tend to relax and take more time to eat. Although occupant behavior is also linked to interior design and other factors, lighting quality represents a significant influence. Occupants perceive and react to a space's light color. It is important that the lighting designer be able to recognize and create the subtle aspects of an environment that define the image of the space. For example, drug and grocery stores use white lights to create a "cool" and "clean" environment. Imagine if these spaces were illuminated by the same color lights as in an elegant restaurant. How would the image of the store change?

Occupants can be influenced to work more effectively if they are in an environment that promotes a "work-like" atmosphere. The goal of the lighting designer is to provide the appropriate quality of light for a particular task to create the right "mood" for the space.

Employee comfort and performance are worth more than energy savings. Although the cost of energy for lighting ($.50-$1.00/year/ft^2) is substantial, it is relatively small compared to the cost of labor ($100-$300/year/ft^2). Improvements in lighting quality can yield high dividends for businesses because gains in worker productivity are common when lighting quality is improved. Conversely, if a lighting retrofit reduces lighting quality, occupant performance may decrease, quickly offsetting any savings in energy costs. Good energy managers should remember that buildings were not designed to save energy; they exist to create an environment where people can work efficiently. Occupants should be able to see clearly without glare, excessive shadows or other uncomfortable features of an illuminated space.

Lighting quality can be divided into four main considerations: Uniformity, Glare, Color Rendering Index and Coordinated Color Temperature.

13.2.2.1 Uniformity

The uniformity of illuminance describes how evenly light spreads over an area. Creating uniform illumination requires proper luminaire spacing. Non-uniform illuminance creates bright and dark spots, which can distract and discomfort some occupants.

Traditionally, lighting designers have specified uniform illumination. This option eliminates the problems with non-uniform illumination and provides excellent flexibility for changes in the work environment. Unfortunately, uniform lighting applied over large areas can waste large amounts of energy. For example, in a manufacturing building, 20% of the floor space may require high levels of illumination (100 fc) for a specific visual task. The remaining 80% of the building may only require 40 foot candles. Uniform illumination over the entire space would require 100 fc at any point in the building. Clearly, this is a tremendous waste of energy and money.

13.2.2.2 Glare

Glare is a sensation caused by relatively bright objects in an occupant's field of view. The key word is *relative*, because glare is most probable when bright objects are located in front of dark environments. For example, a car's high beam headlights cause glare to oncoming drivers at night, yet create little discomfort during the day. *Contrast* is the relationship between the brightness of an object and its background. Although most visual tasks generally become easier with increased contrast, too much brightness causes glare and makes the visual task more difficult. Glare in certain work environments is a serious concern because it usually will cause discomfort and reduce worker productivity.

Visual Comfort Probability (VCP)

The Visual Comfort Probability is a rating given to a luminaire which indicates the percent of people who are comfortable with the glare. Thus, a luminaire with a VCP = 80 means that 80% of occupants are comfortable with the amount of glare from that luminaire. A minimum VCP of 70 is recommended for general interior

spaces, while luminaires with VCPs exceeding 80 are recommended in computer areas and high-profile executive office environments.

To completely correct a lighting system with excessive glare, a lighting designer should be consulted. However there are some basic "rules of thumb" which can assist the energy manager. A high-glare environment is characterized by either excessive illumination and reflection, or the existence of very bright areas typically around luminaires. To minimize glare, the energy manager can try to obscure the bare lamp from the occupant's field of view, relocate luminaires or replace the luminaires with ones that have a high VCP.

Reducing glare is commonly achieved by using indirect lighting, using deep cell parabolic troffers, or special lenses. Although these measures will reduce glare, luminaire efficiency will be decreased because more light will be "trapped" in the luminaire. Alternatively, glare can be minimized by reducing ambient light levels and using task lighting techniques.

Visual Display Terminals (VDTs)

Today's office environment contains a variety of special visual tasks, including the use of computer monitors or visual display terminals (VDTs). Occupants using VDTs are extremely vulnerable to glare and discomfort. When reflections of ceiling lights are visible on the VDT screen, the occupant has difficulty reading the screen. This phenomena is also called "discomfort glare," and is very common in rooms that are uniformly illuminated by luminaires with low a VCP. Therefore, lighting for VDT environments must be carefully designed, so that occupants remain comfortable. Because the location of VDTs can be frequently changed, lighting upgrades should also be designed to be adjustable. Moveable task lights and luminaires with high VCP are very popular for these types of applications. Because each VDT environment is unique, each upgrade must be evaluated on a case-by-case basis.

13.2.2.3 Color

Color considerations have an incredible influence on lighting quality. Light sources are specified based on two color-related parameters: the Color Rendering Index (CRI) and the Coordinated Color Temperature (CCT).

Color Rendering Index (CRI)

In simple terms, the CRI provides an evaluation of how colors appear under a given light source. The index range is from 0 to 100. The higher the number, the easier to distinguish colors. Generally, sources with a CRI > 75 provide excellent color rendition. Sources with a CRI <

55 provide poor color rendition. To provide a "base-case," offices illuminated by most T12 Cool White lamps have a CRI = 62.

It is extremely important that a light source with a high CRI be used with visual tasks that require the occupant to distinguish colors. For example, a room with a color printing press requires illumination with excellent color rendition. In comparison, outdoor security lighting for a building may not need to have a high CRI, but a large quantity of light is desired.

Coordinated Color Temperature (CCT)

The Coordinated Color Temperature (CCT) describes the color of the light source. For example, on a clear day, the sun appears yellow. On an over-cast day, the partially-obscured sun appears to be gray. These color differences are indicated by a temperature scale. The CCT (measured in degrees Kelvin) is a close representation of the color that an object (black-body) would radiate at a certain temperature. For example, imagine a wire being heated. First it turns red (CCT = 2000K). As it gets hotter, it turns white (CCT = 5000K) and then blue (CCT = 8000K). Although a wire is different from a light source, the principle is similar.

CCT is not related to CRI, but it can influence the atmosphere of a room. Laboratories, hospitals and grocery stores generally use "cool" (blue-white) sources, while expensive restaurants may seek a "warm" (yellow-red) source to produce a candle-lit appearance. Traditionally, office environments have been illuminated by Cool White lamps, which have a CCT = 4100K. However, a more recent trend has been to specify 3500K triphosphor lamps, which are considered neutral. Table 13.2 illustrates some common specifications for different visual environments.

13.2.3 Lighting System Components

After determining the quantity and quality of illumination required for a particular task, most lighting designers specify the lamp, then the ballast, and finally the luminaire to meet the lighting needs. The Schematics Section (near the end of this chapter) contains illustrations of many of the lamps and systems described in this section.

13.2.3.1 Lamps

The lamp is the first component to consider in the lighting design process. The lamp choice determines the

Table 13.2 Sample design considerations for a commercial building.

Office Areas		Light Levels	CRI	Color Temperature	Glare
Executive	General	100FC	≥80	3000K	VCP≥70
	Task	≥50FC	≥80	3000K	VCP≥70
Private	General	30-50FC	≥70	3000-3500K	VCP≥70
	Task	≥50FC	≥70	3000-3500K	VCP≥70
Open Plan Computers	General	30-50FC	≥70	3000-3500K	VCP≥90
	Task	≥50FC	≥70	3000-3500K	VCP≥90
Hallways	General	10-20FC	≥70	3000-3500K	VCP≥70
	Task	10-20FC	≥70	3000-3500K	VCP≥70
Reception/ Lobby	General	20-50FC	≥80	3000-5000K	VCP≥90
	Task	≥50FC	≥80	3000-3500K	VCP≥90
Conference	General	10-70FC	≥80	3000-4100K	VCP≥90
	Task	10-70FC	≥80	3000-4100K	VCP≥90
Open Plan General	General	30-70FC	≥70	3500-4100K	VCP≥80
	Task	30-70FC	≥70	3500-4100K	VCP≥80
Drafting	General	70-100FC	≥70	4100-5000K	VCP≥90
	Task	100-150FC	≥780	4100-5000K	VCP≥90

light quantity, CRI, CCT, relamping time interval and operational costs of the lighting system. This section will only cover the most popular types of lamps. Table 13.3 summarizes the differences between the primary lamps and lighting systems.

Incandescent

The oldest electric lighting technology is the incandescent lamp. Incandescent lamps are also the least efficacious (have the lowest lumens per watt) and have the shortest life. They produce light by passing a current through a tungsten filament, causing it to become hot and glow. As the tungsten emits light, it gradually evaporates, eventually causing the filament to break. When this happens, the lamps is said to be "burned-out."

Although incandescent sources are the least efficacious, they are still sold in great quantities because of market barriers. Many fixtures originally designed for incandescent bulbs have not been re-designed to enclose more efficient sources, which may be larger in size. In addition, lamp manufacturers continue to aggressively market incandescent bulbs because they are easily produced. Consumers still purchase incandescent bulbs because they have low initial costs. However, if life-cycle cost analyses are used, incandescent lamps are usually more expensive than other lighting systems with higher efficacies.

Compact Fluorescent Lamps (CFLs)

Compact Fluorescent Lamps (CFLs) are energy efficient, long lasting replacements for some incandescent lamps. For simplicity, this chapter refers to a CFL as a lamp and ballast system. CFLs are available in many styles and sizes. They can be purchased as self-ballasted units or as discrete lamps and ballasts. A self-ballasted unit "screws in" to an existing incandescent socket.

CFLs are composed of two parts, the lamp and the ballast. The short tubular lamps (commonly called PL or SLS lamps) can last longer than 8,000 hours. The ballasts (plastic component at the base of tube) usually last longer than 60,000 hours. When a CFL is not working, often only the lamp needs to be replaced.

Self-ballasted CFLs are common replacements for incandescent lamps in table lamps, downlights, surface lights, pendant luminaires, task lights, wall sconces, and flood lights. Although CFLs have been known to be too large and bulky to fit into fixtures designed for incandescent lamps, the physical size of CFLs has decreased considerably in recent years, making them more applicable for retrofits. Due to the benefits of CFLs, numerous manufacturers now produce fixtures that are specially designed for CFLs, which are commonly installed in new facilities.

Improvements to CFL technologies have been occurring every year since they became commercially available. In comparison to first-generation CFLs, products available today provide 20% higher efficacies as well as instant starting, reduced lamp flicker, quiet operation, smaller size and lighter weight. Dimmable CFLs are now available, and it can be expected that their per-

Table 13.3 Lamp characteristics.

Wattages (lamp only)	Incandescent Including Tungsten Halogen 15-1500	Fluorescent 15-219	Compact Fluorescent 4-40	Mercury Vapor (Self-ballasted) 40-1000	Metal Halide 175-1000	High-Pressure Sodium (Improved Color) 70-1000	Low-Pressure Sodium 35-180
Life (hr)	750-12,000	7,500-24,000	10,000-20,000	16,000-15,000	1,500-15,000	24,000 (10,000)	18,000
Efficacy (lumens/W) lamp only	15-25	55-100	50-80	50-60 (20-25)	80-100	75-140 (67-112)	Up to 180
Lumen maintenance	Fair to excellent	Fair to excellent	Fair	Very good (good)	Good	Excellent	Excellent
Color rendition	Excellent	Good to excellent	Good to excellent	Poor to excellent	Very good	Fair	Poor
Light direction control	Very good to excellent	Fair	Fair	Very good	Very good	Very good	Fair
Relight time	Immediate	Immediate	Imm- 3 seconds	3-10 min.	10-20 min.	Less than 1 min.	Immediate
Comparative fixture cost	Low: simple	Moderate	Moderate	Higher than fluorescent	Generally higher than mercury	High	High
Comparative operating cost	High	Lower than incandescent	Lower than incandescent	Lower than incandescent	Lower than mercury	Lowest of HID types	Low

formance will increase with time.

In most applications, CFLs are excellent replacements for incandescent lamps. CFLs provide similar light quantity and quality while only requiring about 30% of the energy of comparable incandescent lamps. In addition, CFLs last 7-10 times longer than their incandescent counterparts. In many cases, it is cost-effective to replace an entire incandescent fixture with a fixture specially designed for CFLs.

Problems with CFLs

Because CFLs are not point sources, (like incandescent lamps), they are not as effective in projecting light over distance. The light is more diffuse and difficult to focus on intended targets in directional lighting applications. As a subset of fluorescent lamps, CFLs don't last as long when rapidly switched on and off, and lumen output decreases over time. However, if a larger CFL is installed, light levels can be greater than the incandescent system while retaining most of the energy savings.

The light output of CFLs can also be reduced when used in luminaires that trap heat near the lamp. The orientation of the lamp can also affect lumen output. Depending on the lamp design and ambient temperature, the light output in the base-down orientation may be 15% less than in the base-up position.

CFL ballasts are not all the same, some are magnetic with a one year warranty, while others are electronic or hybrids with up to a 10 year warranty. Some have difficulty starting when ambient temperature drops below 40°F, while others start at temperatures below freezing. Some CFL units can produce as much as 110% Total Harmonic Distortion, although systems exist with less than 20% THD. THD is explained in the section titled Electrical Considerations.

Despite these difficulties, most CFLs provide comparable light quality to incandescents while significantly reducing energy consumption. Even if life or light output is reduced slightly, the benefits of CFLs are large enough that they usually remain cost-effective investments.

Fluorescent

Fluorescent lamps are the most common light source for commercial interiors in the U.S. They are repeatedly specified because they are relatively efficient, have long lamp lives, and are available in a wide variety of styles. The most common fluorescent lamp used in offices is the four-foot F40T12 lamp, which is usually used with a magnetic ballast. However, these lamps are being rapidly replaced by F32T8 and F40T10 tri-phosphor lamps with electronic ballasts.

The labeling system used by manufacturers may appear complex, however it is actually quite simple. For example, with an F40T12 lamp, the "F" stands for fluorescent, the "40" means 40 watts, and the "T12" refers to the tube thickness. Since tube thickness (diameter) is measured in 1/8 inch increments, a T12 is 12/8 or 1.5 inches in diameter. A T8 lamp is 1 inch in diameter. Some lamp labels include additional information, indicating the CRI and CCT. Usually, CRI is indicated with one digit, like "8" meaning CRI = 80. CCT is indicated by the two digits following, "35" meaning 3,500K. For example, a F32T8/841 label indicates a lamp with a CRI = 80 and a CCT = 4,100K. Alternatively, the lamp manufacturer might label a lamp with a letter code referring to a specific lamp color. For example, "CW" to mean Cool White lamps with a CCT = 4,100K.

Some lamps have "ES," "EE" or "EW" printed on the label. These acronyms attached at the end of a lamp label indicate that the lamp is an energy-saving type. These lamps consume less energy than standard lamps, however they also produce less light.

Tri-phosphor lamps have a coating on the inside of the lamp which improves performance. Tri-phosphor lamps usually provide greater color rendition. A bi-phosphor lamp (T12 Cool White) has a CRI= 62. By upgrading to a tri-phosphor lamp with a CRI = 75, occupants will be able to distinguish colors better. Tri-phosphor lamps are commonly specified with T8 or T10 systems using electronic ballasts. Lamp flicker and ballast humming are also significantly reduced with electronically-ballasted systems. For these reasons, the visual environment and worker productivity are likely to be improved.

There are many options to consider when choosing fluorescent lamps. Carefully check the manufacturers specifications and be sure to match the lamp and ballast to the application. Table 13.4 shows some of the specifications that vary between different lamp types.

High Intensity Discharge (HID)

High-Intensity Discharge (HID) lamps are similar to fluorescent lamps because they produce light by discharging an electric arc through a tube filled with gases. HID lamps generate much more light, heat and pressure within the arc tube than fluorescent lamps, hence the title "high intensity" discharge. Like incandescent lamps, HIDs are physically small light sources, (point sources) which means that reflectors, refractors and light pipes can be effectively used to direct the light. Although originally developed for outdoor and industrial applications, HIDs are also used in office, retail and other indoor applications. With a few exceptions, HIDs

Table 13.4 Sample fluorescent lamp specifications.

| | MANUFACTURERS' INFORMATION | | |
| | F40T12CW | F40T10 | F32T8 |
	Bi-phosphor	Tri-phosphor	Tri-phosphor
CRI	62	83	83
CCT (K)	4,150	4,100 or 5,000	4,100 or 5,000
Initial lumens	3,150	3,700	3,050
Maintained lumens	2,205	2,960	2,287
Lumens per watt	55	74	71
Rated life (hrs)	24,000	48,000[†]	20,000
Service life (hrs)	16,800	33,600[†]	14,000

[†]This extended life is available from a specific lamp-ballast combination. Normal T10 lamp lives are approximately 24,000 hours. Contact Mr. Les Pace for information on this specialty system. Service life refers to the typical lamp replacement life.

require time to warm up and should not be turned on and off for short intervals. They are not ideal for certain applications because, as point sources of light, they tend to produce more defined shadows than non-point sources such as fluorescent tubes, which emit diffuse light.

Most HIDs have relatively high efficacies and long lamp lives, (5,000 to 24,000+ hours) reducing maintenance re-lamping costs. In addition to reducing maintenance requirements, HIDs have many unique benefits. There are three types of HID sources (listed in order of increasing efficacy): Mercury Vapor, Metal Halide and High Pressure Sodium. A fourth source, Low Pressure Sodium, is not technically a HID, but provides similar quantities of illumination and will be referred to as an HID in this chapter. Table 13.3 shows that there are dramatic differences in efficacy, CRI and CCT between each HID source type.

Mercury Vapor

Mercury Vapor systems were the "first generation" HIDs. Today they are relatively inefficient, provide poor CRI and have the most rapid lumen depreciation rate of all HIDs. Because of these characteristics, other more cost-effective HID sources have replaced mercury vapor lamps in nearly all applications. Mercury Vapor lamps provide a white-colored light which turns slightly green over time. A popular lighting upgrade is to replace Mercury Vapor systems with Metal Halide or High Pressure Sodium systems.

Metal Halide

Metal Halide lamps are similar to mercury vapor lamps, but contain slightly different metals in the arc tube, providing more lumens per watt with improved color rendition and improved lumen maintenance. With nearly twice the efficacy of Mercury Vapor lamps, Metal Halide lamps provide a white light and are commonly used in industrial facilities, sports arenas and other spaces where good color rendition is required. They are the current best choice for lighting large areas that need good color rendition.

High Pressure Sodium (HPS)

With a higher efficacy than Metal Halide lamps, HPS systems are extremely popular for most outdoor or industrial applications where excellent color rendition is not required. HPS is common in parking lots and produces a light golden color that allows some color rendition. Although HPS lamps do not provide the best color rendition, they are adequate for indoor applications at some industrial facilities. The key is to apply HPS in an area where there are no other light source types available for comparison. Because occupants usually prefer "white light," HPS installations can result with some occupant complaints. However, when HPS is installed at a great distance from metal halide lamps or fluorescent systems, the occupant will have no reference "white light" and he/she will accept the HPS as "normal." This technique has allowed HPS to be installed in countless indoor gymnasiums and industrial spaces with minimal complaints.

Low Pressure Sodium

Although LPS systems have the highest efficacy of any commercially available HID, the monochromic light source produces the poorest color rendition of all lamp types. With a low CCT, the lamp appears to be "pumpkin orange," and all objects illuminated by its light appear white, shades of gray or black. Applications are limited to security or street lighting. The lamps are physically long (up to 3 feet) and not considered to be point sources. Thus optical control is poor, making LPS less effective for extremely high mounting heights.

LPS has become popular because of its extremely high efficacy. With up to 60% greater efficacy than HPS, LPS is economically attractive. Several cities, such as San Diego, California, have installed LPS systems on streets. Despite the poor CRI, many parking lots in San Diego have also switched to LPS, with few complaints. Although there are many successful applications, LPS installations must be carefully considered. Often lighting quality can be improved by supplementing the LPS system with other light sources (with a greater CRI).

13.2.3.2 Ballasts

With the exception of incandescent systems, nearly all lighting systems (fluorescent and HID) require a ballast. A ballast controls the voltage and current supplied to lamps. Because ballasts are an integral component of the lighting system, they have a direct impact on light output. The ballast factor is the ratio of a lamp's light output to a reference ballast. General purpose fluorescent ballasts have a ballast factor that is less than one (typically .88 for most electronic ballasts). Special ballasts may have higher ballast factors to increase light output, or lower ballast factors to reduce light output. Naturally, a ballast with a high ballast factor also consumes more energy than a general purpose ballast.

Fluorescent Ballasts

Specifying the proper ballast for fluorescent lighting systems has become more complicated than it was 20 years ago, when magnetic ballasts were practically the only option. Electronic ballasts for fluorescent lamps have been available since the early 1980s, and their introduction has resulted in a variety of options.

During the infancy of electronic ballast technology, reliability and harmonic distortion problems hampered their success. However, most electronic ballasts available today have a failure rate of less than one percent, and many distort harmonic current less than their magnetic counterparts. Electronic ballasts are superior to magnetic ballasts because they are typically 30% more energy efficient. They also produce less lamp flicker, ballast noise, and waste heat.

There are two types of fluorescent ballasts: magnetic and electronic.

Magnetic

There are three primary types of magnetic ballasts.

- Standard core and coil
- High-efficiency core and coil (Energy-Efficient Ballasts)
- Cathode cut-out or Hybrid

Standard core and coil magnetic ballasts are essentially core and coil transformers that are relatively inefficient at operating fluorescent lamps. Although these types of ballasts are no longer sold in the US, they still exist in many facilities. The "high-efficiency" magnetic ballast can replace the "standard ballast," improving the system efficiency by approximately 10%.

"Cathode cut-out" or "hybrid" ballasts are high-efficiency core and coil ballasts that incorporate electronic components that cut off power to the lamp cathodes after the lamps are operating, resulting in an additional 2-watt savings per lamp.

Electronic

In nearly every fluorescent lighting application, electronic ballasts can be used in place of conventional magnetic core and coil ballasts. Electronic ballasts improve fluorescent system efficacy by converting the standard 60 Hz input frequency to a higher frequency, usually 25,000 to 40,000 Hz. Lamps operating on these frequencies produce about the same amount of light, while consuming up to 30% less power than a standard magnetic ballast. Other advantages of electronic ballasts include less audible noise, less weight, virtually no lamp flicker and dimming capabilities.

There are three primary types of electronic ballasts: T12, T10 and T8 ballasts. T12 electronic ballasts are designed for use with conventional (T12) fluorescent lighting systems. Although most T10 ballasts consume the same (or more) energy as T12 ballasts, T10 systems provide more light output. In fact, some T10 systems are more efficient than comparable T8 systems. Because T10s provide more light per lamp, fewer lamps can be installed in a fixture.

T8 ballasts offer some distinct advantages over other types of electronic ballasts. They are generally more efficient, have less lumen depreciation, are available with more options. T8 ballasts can operate one, two, three or four lamps. Most T12 and T10 ballasts can only operate one, two or three lamps. Therefore, one T8 ballast can replace two T12 ballasts in a 4 lamp fixture.

Some electronic ballasts are parallel-wired, so that when one lamp burns out, the remaining lamps in the fixture will continue to operate. In a typical magnetic, (series-wired system) when one component fails, all lamps in the fixture shut off. Before maintenance personnel can relamp, they must first diagnose which lamp failed. Thus the electronically-ballasted system will reduce time to diagnose problems, because maintenance personnel can immediately see which lamp failed.

Parallel-wired ballasts also offer the option of reducing lamps per fixture (after the retrofit) if an area is over-illuminated. This option allows the energy manager to experiment with different configurations of lamps in different areas. However, each ballast operates best when controlling the specified number of lamps.

Due to the advantages of electronically-ballasted systems, they are produced by many manufacturers and prices are very competitive. Due to their market penetration, T8 systems (and replacement parts) are more likely to be available, and at lower costs.

HID Ballasts

As with fluorescent systems, High Intensity Discharge lamps also require ballasts to operate. Although there are not nearly as many specification options as with fluorescent ballasts, HID ballasts are available in dimmable and bi-level light outputs. Instant restrike systems are also available.

Capacitive Switching HID Luminaires

Capacitive switching or "bi-level" HID luminaires are designed to provide either full or partial light output based on inputs from occupancy sensors, manual switches or scheduling systems. Capacitive-switched dimming can be installed as a retrofit to existing luminaires or as a direct luminaire replacement. Capacitive switching HID upgrades can be less expensive than installing a panel-level variable voltage control to dim the lights, especially in circuits with relatively few luminaires.

The most common applications of capacitive switching are athletic facilities, occupancy-sensed dimming in parking lots, and warehouse aisles. General purpose transmitters can be used with other control devices such as timers and photo sensors to control the bi-level luminaires. Upon detecting motion, the occupancy sensor signals the bi-level HID ballasts. The system rapidly brings light levels from a standby reduced level to about 80 percent of full output, followed by the normal warm-up time between 80 and 100 percent of full light output.

Depending of the lamp type and wattage, the standby lumens are roughly 15-40 percent of full output and the standby wattage is 30-60 percent of full wattage. When the space is unoccupied and the system is dimmed, you can achieve energy savings of 40-70 percent. However, these types of applications must be carefully designed. Lamp manufacturers do not recommend dimming below 50% of the rated input power.[2]

13.2.3.3 Luminaires

A luminaire (or light fixture) is a unit consisting of the lamps, ballasts, reflectors, lenses or louvers and housing. The main function is to focus or spread light emanating from the lamp(s). Without luminaires, lighting systems would appear very bright and cause glare.

Luminaire Efficiency

Luminaires block or reflect some of the light exiting the lamp. The efficiency of a luminaire is the percentage of lamp lumens produced that actually exit the fixture in the intended direction. Efficiency varies greatly among different luminaire and lamp configurations. For example, using four T8 lamps in a luminaire will be more efficient than using four T12 lamps because the T8 lamps are thinner, allowing more light to "escape" between the lamps and out of the luminaire. Understanding luminaires is important because a lighting retrofit may involve changing some components of the luminaire to improve the efficiency and deliver more light to the task.

The Coefficient of Utilization (CU) is the percent of lumens produced that actually reach the work plane. The CU incorporates the luminaire efficiency, mounting height, and reflectances of walls and ceilings. Therefore, improving the luminaire efficiency will improve the CU.

Reflectors

Installing reflectors in most luminaires can improve its efficiency because light leaving the lamp is more likely to "reflect" off interior walls and exit the luminaire. Because lamps block some of the light reflecting off the luminaire interior, reflectors perform better when there are less lamps (or smaller lamps) in the luminaire. Due to this fact, a common luminaire upgrade is to install reflectors and remove some of the lamps in a luminaire. Although the luminaire efficiency is improved, the overall light output from each luminaire is likely to be reduced, which will result in reduced light levels. In addition, reflectors will redistribute light (usually more light is reflected down), which may create bright and dark spots in the room. Altered light levels and different distributions may be acceptable, however these changes need to be considered.

To ensure acceptable performance from reflectors, conduct a trial installation and measure "before" and "after" light levels at various locations in the room. Don't compare an existing system, (which is dirty, old and contains old lamps) against a new luminaire with half the lamps and a clean reflector. The light levels may appear to be adequate, or even improved. However, as the new system ages and dirt accumulates on the surfaces, the light levels will drop.

A variety of reflector materials are available: highly reflective white paint, silver film laminate, and anodized aluminum. Silver film laminate usually has the highest reflectance, but is considered less durable.

In addition to installing reflectors within luminaires, light levels can be increased by improving the reflectiveness of the room's walls, floors and ceilings. For example, by covering a brown wall with white paint, more light will reflect back into the workspace, and the Coefficient of Utilization is increased.

Lenses and Louvers

Most indoor luminaires use either a lens or louver to prevent occupants from directly seeing the lamps. Light that is emitted in the shielding angle or "glare zone" (angles above 45° from the fixture's vertical axis) can cause glare and visual discomfort, which hinders the occupant's ability to view work surfaces and computer screens. Lenses and louvers are designed to shield the viewer from these uncomfortable, direct beams of light. Lenses and louvers are usually included as part of a luminaire when purchased, and they can have a tremendous impact on the VCP of a luminaire.

Lenses are sheets of hard plastic (either clear or milky white) that are located on the bottom of a luminaire. Clear, prismatic lenses are very efficient because they allow most of the light to pass through, and less light to be trapped in the fixture. Milky-white lenses are called "diffusers" and are the least efficient, trapping a lot of the light within the fixture. Although diffusers have been routinely specified for many office environments, they have one of the lowest VCP ratings.

Louvers provide superior glare control and high VCP when compared to most lenses. As Figure 13.3 shows, a louver is a grid of plastic "shields" which blocks some of the horizontal light exiting the luminaire. The most common application of louvers is to reduce the

luminaire glare in sensitive work environments, such as in rooms with computers. Parabolic louvers usually improve the VCP of a luminaire, however efficiency is reduced because more light is blocked by the louver. Generally, the smaller the cell, the greater the VCP and less the efficiency. Deep-cell parabolic louvers offer a better combination of VCP and efficiency, however deep-cell louvers require deep luminaires, which may not fit into the ceiling plenum space.

Table 13.5 shows the efficiency and VCP for various lenses and louvers. VCP is usually inversely related to luminaire efficiency. An exception is with the milky-white diffusers, which have low VCP and low efficiency.

Table 13.5 Luminaire efficiency and VCP.

Shielding Material	Luminaire Efficiency (%)	Visual Comfort Probability (VCP)
Clear Prismatic Lens	60-70	50-70
Low Glare Clear Lens	60-75	75-85
Deep-Cell Parabolic Louver	50-70	75-95
Translucent Diffuser	40-60	40-50
Small-Cell Parabolic Louver	35-45	99

Light Distribution/Mounting Height

Luminaires are designed to direct light where it is needed. Various light distributions are possible to best suit any visual environment. With "direct lighting," 90-100% of the light is directed downward for maximum use. With "indirect lighting," 90-100% of the light is directed to the ceilings and upper walls. A "semi-indirect" system distributes 60-90% down, with the remainder upward. Designing the lighting system should incorporate the different light distributions of different luminaires to maximize comfort and visual quality.

Fixture mounting height and light distribution are presented together since they are interactive. HID systems are preferred for high mounting heights since the lamps are physically small, and reflectors can direct light downward with a high degree of control. Fluorescent lamps are physically long and diffuse sources, with less ability to control light at high mounting heights. Thus fluorescent systems are better for low mounting heights and/or areas that require diffuse light with minimal shadows.

Generally, "high-bay" HID luminaires are designed for mounting heights greater than 20 feet high. "High-bay" luminaires usually have reflectors and focus most of their light downward. "Low-bay" luminaires are

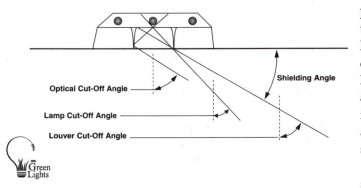

Figure 13.3 Higher shielding angles for improved glare control.

designed for mounting heights less than 20 feet and use lenses to direct more light horizontally.

HIDs are potential sources of direct glare since they produce large quantities of light from physically small lamps. The probability of excessive direct glare may be minimized by mounting fixtures at sufficient heights. Table 13.6 shows the minimum mounting height recommended for different types of HID systems.

Table 13.6 Minimum mounting heights for HIDs

Lamp Type	feet above ground
400 W Metal Halide	16
1000 W Metal Halide	20
200 W High Pressure Sodium	15
250 W High Pressure Sodium	16
400 W High Pressure Sodium	18
1000 W High Pressure Sodium	26

13.2.3.4 Exit Signs

Recent advances in exit sign systems have created attractive opportunities to reduce energy and maintenance costs. Because emergency exit signs should operate 24 hours per day, energy savings quickly recover retrofit costs. There are generally two options, buying a new exit sign, or retrofitting the existing exit sign with new light sources.

Most retrofit kits available today contain adapters that screw into the existing incandescent sockets. Installation is easy, usually requiring only 15 minutes per sign. However, if a sign is severely discolored or damaged, buying a new sign might be required in order to maintain illuminance as required by fire codes.

Basically, there are five upgrade technologies: Compact Fluorescent Lamps (CFLs), incandescent assemblies, Light Emitting Diodes (LED), Electroluminescent panels, and Self Luminous Tubes.

Replacing incandescent sources with compact fluorescent lamps was the "first generation" exit sign upgrade. Most CFL kits must be hard-wired and can not simply screw into an existing incandescent socket. Although CFL kits are a great improvement over incandescent exit signs, more technologically advanced upgrades are available that offer reduced maintenance costs, greater efficacy and flexibility for installation in low (sub-zero) temperature environments.

As Table 13.7 shows, LED upgrades are the most cost-effective because they consume very little energy, and have an extremely long life, practically eliminating maintenance. The LED retrofit consists of a pair of LED strips that adhere to the inside of the exit sign enclosure, providing a soft, glowing exit sign.

Another low-maintenance upgrade is to install a "rope" of incandescent assemblies. These low-voltage "luminous ropes" are an easy retrofit because they can screw into existing sockets like LED retrofit kits. However, the incandescent assemblies create bright spots which are visible through the transparent exit sign and the non-uniform glow is a noticeable change. In addition, the incandescent assemblies don't last nearly as long as LEDs.

Although electroluminescent panels consume less than one watt, light output rapidly depreciates over time. These self-luminous sources are obviously the most energy-efficient, consuming no electricity. However the spent tritium tubes, which illuminate the unit, must be disposed of as a radioactive waste, which will increase over-all costs.

13.2.3.5 Lighting Controls

Lighting controls offer the ability for systems to be turned on and off either manually or automatically. There are several control technology upgrades for lighting systems, ranging from simple (installing manual switches in proper locations) to sophisticated (installing occupancy sensors).

Table 13.7 Exit sign upgrades.

Light Source	Watts	Life	Replacement
Incandescent (Long Life)	40	8 months	lamps
Compact Fluorescent	10	1.7 years	lamps
Incandescent Assembly	8	3 + years	light source
Light Emitting Diode (LED)	<4	>25	light source
Electroluminescent	1	8+ years	panel
Self luminous (Tritium)	0	10-20 years	luminous tubes

Switches

The standard manual, single-pole switch was the first energy conservation device. It is also the simplest device and provides the least options. One negative aspect about manual switches is that people often forget to turn them off. If switches are far from room exits or are difficult to find, occupants are more likely to leave lights on when exiting a room.[3] Occupants do not want to walk through darkness to find exits. However, if switches are located in the right locations, with multiple points of control for a single circuit, occupants find it easier to turn systems off. Once occupants get in the habit of turning lights off upon exit, more complex systems may not be necessary. The point is: switches can be great energy conservation devices as long as they are installed in proper locations so that occupants can conveniently use them.

Another opportunity for upgrading controls exists when lighting systems are designed such that all circuits in an area are controlled from one switch, when only some circuits need to be activated. For example, a college football stadium's lighting system is designed to provide enough light for TV applications. However, this intense amount of light is not needed for regular practice nights or other non-TV events. Because the lights are all controlled from one switch, every time the facility is used all the lights are turned on. By dividing the circuits and installing one more switch to allow the football stadium to use only 70% of its lights during practice nights, significant energy savings are possible.

Generally, if it is not too difficult to re-circuit a poorly designed lighting system, additional switches can be added to optimize the lighting controls.

Time Clocks

Time clocks can be used to control lights when their operation is based on a fixed operating schedule. Time clocks are available in electronic or mechanical styles. However, regular check-ups are needed to ensure that the time clock is controlling the system properly. After a power loss, electronic timers without battery backups can get off schedule and cycle on and off at the wrong times. It requires a great deal of maintenance time to reset isolated time clocks if many are installed.

Photocells

For most outdoor lighting applications, photocells (which turn lights on when it gets dark, and off when sufficient daylight is available) offer a low-maintenance alternative to time clocks. Unlike time clocks, photocells are seasonally self-adjusting and automatically switch on when light levels are low, such as during rainy days.

A photocell is inexpensive and can be installed on each luminaire, or can be installed to control numerous luminaires on one circuit. Photocells can also be effectively used indoors, if daylight is available through skylights.

Photocells have worked well in almost any climate, however they should be cleaned when luminaires are re-lamped. Otherwise, dust will accumulate on the photodiode aperture, causing the controls to always perceive it is a cloudy day, and the lights will stay on.

Occupancy Sensors

Occupancy sensors save energy by turning off lights in spaces that are unoccupied. When the sensor detects motion, it activates a control device that turns on a lighting system. If no motion is detected within a specified period, the lights are turned off until motion is sensed again. With most sensors, sensitivity (the ability to detect motion) and the time delay (difference in time between when sensor detects no motion and lights go off) are adjustable. Occupancy sensors are produced in two primary types: Ultrasonic (US) and Passive Infrared (PIR). Dual-Technology (DT) sensors, that have both ultrasonic and passive infrared detectors, are also available. Table 13.8 shows the estimated percent energy savings from occupancy sensor installation for various locations.

Table 13.8 Estimated % savings from occupancy sensors.

Application	Energy Savings
Offices (Private)	25-50%
Offices (Open Spaces)	20-25%
Rest Rooms	30-75%
Corridors	30-40%
Storage Areas	45-65%
Meeting Rooms	45-65%
Conference Rooms	45-65%
Warehouses	50-75%

US and PIR sensors are available as wall-switch sensors, or remote sensors such as ceiling mounted or outdoor commercial grade units. With remote sensors, a low-voltage wire connects each sensor to an electrical relay and control module, which operates on common voltages. With wall-switch sensors, the sensor and control module are packaged as one unit. Multiple sensors and/or lighting circuits can be linked to one control module allowing flexibility for optimum design.

Wall-switch sensors can replace existing manual

swi..ches in small areas such as offices, conference rooms, and some classrooms. However, in these applications, a manual override switch should be available so that the lights can be turned off for slide presentations and other visual displays. Wall-switch sensors should have an unobstructed coverage pattern (absolutely necessary for PIR sensors) of the room it controls.

Ceiling-mounted units are appropriate in corridors, rest rooms, open office areas with partitions and any space where objects obstruct the line of sight from a wall-mounted sensor location. Commercial grade outdoor units can also be used in indoor warehouses and large aisles. Sensors designed for outdoor use are typically heavy duty, and usually have the adjustable sensitivities and coverage patterns for maximum flexibility. Table 13.9 indicates the appropriate sensors for various applications.

Ultrasonic Sensors (US)

Ultrasonic sensors transmit and receive high-frequency sound waves above the range of human hearing. The sound waves bounce around the room and return to the sensor. Any motion within the room distorts the sound waves. The sensor detects this distortion and signals the lights to turn on. When no motion has been detected over a user-specified time, the sensor sends a signal to turn the lights off. Because ultrasonic sensors need enclosed spaces (for good sound wave echo reflection), they can only be used indoors and perform better if room surfaces are hard, where sound wave absorption is minimized. Ultrasonic sensors are most sensitive to motion toward or away from the sensor. Applications

include rooms with objects that obstruct the sensor's line of sight coverage of the room, such as restroom stalls, locker rooms and storage areas.

Passive Infrared Sensors (PIR)

Passive Infrared sensors detect differences in infrared energy emanating in the room. When a person moves, the sensor "sees" a heat source move from one zone to the next. PIR sensors require an unobstructed view, and as distance from the sensor increases, larger motions are necessary to trigger the sensor. Applications include open plan offices (without partitions), classrooms and other areas that allow a clear line of sight from the sensor.

Dual-Technology Sensors (DT)

Dual-Technology (DT) sensors combine both US and PIR sensing technologies. DT sensors can improve sensor reliability and minimize false switching. However, these types of sensors are still only limited to applications where ultrasonic sensors will work.

Occupancy Sensors' Effect on Lamp Life

Occupancy Sensors can cause rapid on/off switching which reduces the life of certain fluorescent lamps. Offices without occupancy sensors usually have lights constantly on for approximately ten hours per day. After occupancy sensors are installed, the lamps may be turned on and off several times per day. Several laboratory tests have shown that some fluorescent lamps lose 25% of their life if turned off and on every three hours. Although occupancy sensors may cause lamp life to be

Table 13.9 Occupancy sensor applications.

Sensor Technology	Private Office	Large Open Office Plan	Partitioned Office Plan	Conference Room	Rest Room	Closets/ Copy Room	Hallways Corridors	Warehouse Aisles Areas
US Wall Switch	•			•	•	•		
US Ceiling Mount	•			•	•	•		
IR Wall Switch	•			•		•		
IR Ceiling Mount	•	•	•	•		•		
US Narrow View							•	
IR High Mount Narrow View							•	•
Corner Mount Wide		•						
View Technology Type								

reduced, the annual burning hours also decreases. Therefore, in most applications, the time period until re-lamp will not increase. However, due to the laboratory results, occupancy sensors should be carefully evaluated if the lights will be turned on and off rapidly. The longer the lights are left off, the longer lamps will last.

The frequency at which occupants enter a room makes a difference in the actual percent time savings possible. Occupancy sensors save the most energy when applied in rooms that are not used for long periods of time. If a room is frequently used and occupants re-enter a room before the lights have had a chance to turn off, no energy will be saved. Therefore, a room that is occupied every other hour will be more appropriate for occupancy sensors than a room occupied every other minute, even though the percent vacancy time is the same.

Although occupancy sensors were not primarily developed for HIDs, some special HID ballasts (bi-level) offer the ability to dim and re-light lamps quickly. Another term for bi-level HID technology is Capacitive Switching HID Luminaires, which are discussed in the HID ballast section.

13.3 LIGHTING ENERGY MANAGEMENT STEPS

Three basic steps to lighting energy management:

1.   Identify necessary light quantity and quality to perform visual task.

2.   Increase light source efficiency if occupancy is frequent.

3.   Optimize lighting controls if occupancy is infrequent.

Step 1, identifying the proper lighting quantity and quality is essential to any illuminated space. However, steps 2 & 3 are options that can be explored individually or together. Steps 2 & 3 can both be implemented, but often the two options are economically mutually exclusive. If you can turn off a lighting system for the majority of time, the extra expense to upgrade lighting sources is rarely justified. Remember, light source upgrades will only save energy (relative to the existing system) when the lights are on.

13.3.1 Identify necessary light quantities and qualities to perform tasks.

Identifying the necessary light quantities for a task is the first step of a lighting retrofit. Often this step is overlooked because most energy managers try to mimic the illumination of an existing system, even if it is over-illuminated and contains many sources of glare. For many years, lighting systems were designed with the belief that no space can be over-illuminated. Recently, the IES retracted it's "more light is better" philosophy. Between 1972 and 1987, light levels recommended by the IES declined by 15% in hospitals, 17% in schools, 21% in office buildings and 34% in retail buildings.[4] Even with IES's adjustments, there are still many excessively illuminated spaces in use today. Energy managers can reap remarkable savings by simply redesigning a lighting system so that the proper illumination levels are produced.

Although the number of workplane footcandles are important, the occupant needs to have a contrast so that he can perform a task. For example, during the daytime your car headlights don't create enough contrast to be noticeable. However, at night, your headlights provide enough contrast for the task. The same amount of light is provided by the headlights during both periods, but daylight "washes out" the contrast of the headlights.

The same principle applies to offices, and other illuminated spaces. For a task to appear relatively bright, objects surrounding that task must be relatively dark. For example, if ambient light is excessive (150 fc) the occupant's eyes will adjust to it and perceive it as the "norm." However when the occupant wants to focus on something he/she may require an additional light to accent the task (at 200 fc). This excessively illuminated space results in unnecessary energy consumption. The occupant would see better if ambient light was reduced to 30-40 fc and the task light was used to accent the task at 50 fc. As discussed earlier, excessive illumination is not only wasteful, but it can reduce the comfort of the visual environment and decrease worker productivity.

After identifying the proper quantity of light, the proper quality must be chosen. The CRI, CCT and VCP must be specified to suit the space.

13.3.2 Increase Source Efficacy

Increasing the source efficacy of a lighting system means replacing or modifying the lamps, ballasts and/or luminaires to become more efficient. In the past, the term "source" has been used to imply only the lamp of a system. However, due to the inter-relationships between components of modern lighting systems, we also consider ballast and luminaire retrofits as "source upgrades." Thus increasing the efficacy simply means getting more lumens per watt out of lighting system. For example, to increase the source efficacy of a T12 system with a magnetic ballast, the ballast and lamps could be replaced with T8 lamps and an electronic ballast, which is a more efficacious (efficient) system.

Another retrofit that would increase source efficacy would be to improve the luminaire efficiency by installing reflectors and more efficient lenses. This retrofit would increase the lumens per watt, because with reflectors and efficient lenses, more lumens can escape the luminaire, while the power supplied remains constant.

Increasing the efficiency of a light source is one of the most popular types of lighting retrofits because energy savings can almost be guaranteed if the new system consumes less watts than the old system. With reduced lighting load, electrical demand savings are also usually obtained. In addition, lighting quality can be improved by specifying sources with higher CRI and improved performance. These benefits allow capital improvements for lighting systems that pay for themselves through increased profits.

Task lighting

As a subset of Increasing Source Efficacy, "Task lighting" or "Task/Ambient" lighting techniques involve improving the efficiency of lighting in an entire workplace, by replacing and relocating lighting systems. Task lighting means retrofitting lighting systems to provide appropriate illumination for each task. Usually, this results in a reduction of ambient light levels, while maintaining or increasing the light levels on a particular task. For example, in an office the light level needed on a desk could be 75 fc. The light needed in aisles is only 20 fc. Traditional uniform lighting design would create a workplace where ambient lighting provides 75 fc throughout the entire workspace. Task lighting would create an environment where each desk is illuminated to 75 fc, and the aisles only to 20 fc. Figure 13.4 shows a typical application of Task/Ambient lighting.

Task lighting upgrades are a model of energy efficiency, because they only illuminate what is necessary. Task lighting designs are best suited for office environ-

Figure 13.4 Task/ambient lighting.

ments with VDTs and/or where modular furniture can incorporate task lighting under shelves. Alternatively, moveable desk lamps may be used for task illumination. Savings result when the energy saved from reducing ambient light levels exceeds the energy used for task lights.

In most work spaces, a variety of visual tasks are performed, and each employee has optimal lighting preferences. Most workers prefer lighting systems designed with task lighting because it is flexible and allows individual control. For example, older workers may require greater light levels than young workers. Modular task lighting would allow older workers to increase their light levels on specific tasks.

Task lighting techniques are also applicable in industrial facilities, or any area where ambient lighting can be reduced/replaced by task lights. For example, high intensity task lights can be installed on fork trucks (to supplement headlights) for use in rarely occupied warehouses. With this system, the entire warehouse's lighting can be turned off (or reduced), saving a large amount of energy.

Identifying task lighting opportunities may require some creativity, but the potential dollar savings can be enormous.

13.3.3 Optimize Lighting Controls

The third step of lighting energy management is to investigate optimizing lighting controls. As shown earlier, improving the efficiency of a lighting system can save a percentage of the energy consumed *while the system is operating.* However, sophisticated controls can turn entire systems off when they are not needed, allowing energy savings to accumulate quickly. The Electric Power Research Institute (EPRI) reports that spaces in an average office building may only be occupied 60-75% of the time, although the lights may be on for the entire 10-hour day[5]. Lighting controls include switches, time clocks, occupancy sensors and other devices that regulate a lighting system. These systems are discussed in Section 13.2.3, Lighting System Components.

13.4 MAINTENANCE

13.4.1 Isolated Systems

Most lighting manuals prescribe specialized technologies to efficiently provide light for particular tasks. An example is dimmable ballasts. For areas that have sufficient daylight, dimmable ballasts can be used with integrated circuitry to reduce energy consumption during peak periods. However, although there may be some shedding of lighting load along the perimeter, these energy cost savings may not represent a great percentage of the building's total lighting load. Furthermore, applications of specialized technologies (such as dimmable ballasts) may be dispersed and isolated in several buildings.

The applications of specialized technologies in isolated locations presents an inventory challenge for maintenance personnel. In many cases, maintenance costs may escalate as personnel spend more time attempting to identify the location of a system needing repair. If maintenance needs to make additional site visits to get the right equipment to re-lamp or "fine-tune" special systems, the labor costs may exceed the energy cost savings.

In facilities with low potential for energy cost savings, facility managers may not want to spend a great deal of time monitoring and "fine-tuning" a lighting system if other maintenance concerns need attention. If a specialized lighting system malfunctions, repair may require special components, that may be expensive and more difficult to install. If maintenance cannot effectively repair the complex technologies, the systems will fail and occupant complaints will increase. Thus the isolated, complex technology that appeared to be a unique solution to a particular lighting issue is often replaced with a system that is easy to maintain.

In addition to the often eventual replacement of technologies that are difficult to maintain, well intended repairs to the system may accidentally result in "snap-

back." "Snap-back" is when a specialized or isolated technology is accidentally replaced with a common technology within the facility. For example, if dimmable ballasts only represent 10% of the building's total ballasts, maintenance probably won't keep them in stock. When replacement is needed, the maintenance personnel may accidentally install a regular ballast. Thus, the lighting retrofit has "snapped back" to its original condition.

The above arguments are not meant to "shoot down" the application of all new technologies. However new technologies usually bring new problems. The authors ask that the energy manager carefully consider the maintenance impact when evaluating an isolated technology. Once again, all lighting systems depend on regular maintenance.

13.4.2 Maintaining System Performance

As with most manufactured products, lighting systems lose performance over time. This degradation can be the result of Lamp Lumen Depreciation (LLD), Luminaire Dirt Depreciation (LDD), Room Surface Dirt Depreciation (RSDD), and many other factors. Several of these factors can be recovered to maintain performance of the lighting system. Figure 13.5 shows the LLD for various types of lighting systems

Lamp Lumen Depreciation occurs because as the lamp ages, its performance degrades. LLD can be accelerated if the lamp is operated in harsh environments, or the system is subjected to conditions for which it was not designed. For example, if a fluorescent system is turned

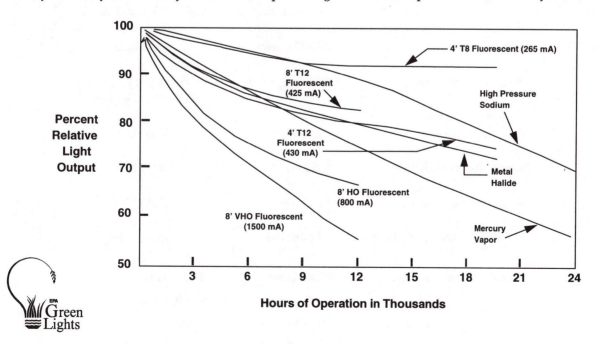

Figure 13.5 Lamp lumen depreciation (LLD).

on and off every minute, the lamps and ballasts will not last as long. Light loss due to lamp lumen depreciation can be recovered by re-lamping the fixture.

Luminaire Dirt Depreciation and Room Surface Dirt Depreciation block light and can reduce light levels. However, these factors can be minimized by cleaning surfaces and minimizing dust. The magnitude of these factors is dependent on each room, thus recommended cleaning intervals can vary. Generally it is most economical to clean fixtures when re-lamping.

13.4.3 Group Re-lamping

Most companies replace lamps when someone notices a lamp is burned out. In a high rise building, this could become a full-time job, running from floor to floor, office to office, disrupting work to open a fixture and replace a lamp. However, those astute facility managers know it is less costly to group re-lamp on a pre-determined date. The rule of thumb is: group relamp at 50% to 70% of the lamp's rated life. However, depending on site-specific factors and the lumen depreciation of the lighting system, relamping interval may vary.

The facility manager must evaluate their own building, and determine the appropriate relamp interval by observing when lamps start to fail. Due to variations in power voltages (spikes, surges and low power), lamps may have different operating characteristics and lives from one facility to another. It is important to maintain records on lamp and ballast replacements and determine the most appropriate relamping interval. This also helps keep track of maintenance costs, labor needs and budgets.

Group relamping is the least costly method to relamp due to reduced time and labor costs. For example, Table 13.10 shows the benefits of group relamping. As more states adopt legislation requiring special disposal of lighting systems, group relamping in bulk may offer reduced disposal costs due to large volumes of material.

13.4.4 Disposal Costs

Disposal costs and regulations for lighting systems vary from state to state. These expenses should be included in an economic analysis of any retrofit. If proper disposal regulations are not followed, the EPA could impose fines and hold the violating company liable for environmental damage in the future.

13.5 NEW TECHNOLOGIES & PRODUCTS

Currently, the energy efficient lighting market is extremely competitive, forcing manufacturers to develop

Table 13.10 Group relamping example: 1,000 3-Lamp T8 Lensed troffers

	Spot Relamping (on burn-out)	Group Relamping (@ 70% rated life)
Relamp cycle	20,000 hours	14,000 hours
Avg. relamps/year	525 relamps/yr	750 relamps/yr (group) 52 relamps/yr (spot)
Avg. material cost/year	$1,050/yr	$1,604/yr
Lamp disposal @ 0.50 ea.	$236/yr	$375/yr
Avg. labor cost/year	$3,150/yr	$1,437/yr
MAINTENANCE BUDGET:	$4,463/yr	$3,416/yr
Assumptions: Labor:	$6.00/lamp	$1.50/lamp
Material:	$2.00/lamp	$2.00/lamp
Operation:	3,500 hr/yr	3,500 hr/yr

new products to survive. The development is so rapid, this chapter cannot possibly announce all of the most recent lighting products. However we acknowledge that new savings opportunities are constantly arriving in the marketplace, and it is worthwhile for the energy manager to be aware of the advancements.

On the technology side, we expect benefits from further development of HIDs, the sulfur lamp, light pipes and light shelves. Because consumers have realized the benefits of natural illumination, daylighting interior spaces may also increase. Although the exact future is unknown, as lighting manufacturers diversify and improve their products, opportunities arise for creative applications. Because there are so many variations of systems possible, the ability to custom design your lighting system is becoming easier and less expensive.

13.6 SPECIAL CONSIDERATIONS

13.6.1 EPACT

The National Energy Policy Act of 1992 (EPACT) was designed to dramatically reduce energy consumption via more competitive electricity generation and more efficient buildings, lights and motors. Because lighting is common in nearly all buildings, it is a primary focus of EPACT. The 1992 legislation bans the production of lamps that have low efficacy or CRI. Table 13.11 indicates which lamps are banned and a few options for replacing the banned systems. From left to right, the table shows several options, for each banned system, ranging from the most efficient substitute to the minimum compliance substitute. Generally, the minimum compliance substitute has the lowest initial cost, but after energy costs have been included, the most efficient upgrades have the lowest life-cycle costs.

Often the main expense with a lighting upgrade is the labor cost to install new products. However the incremental labor cost of installing high-efficiency equipment is minimal. Therefore, it is usually beneficial to install the most efficient technologies because they will have the lowest operational and life-cycle costs. EPACT only eliminates the "bottom of the barrel" in terms of available lighting technology. To keep "one step ahead" of future lamp bans, it is a good idea to consider upgrades with greater efficiencies than the minimum acceptable substitute.

Electrical Considerations

Due to the increasingly complex lighting products available today, concern about effects on power distribution systems have risen. In certain situations, lighting retrofits can reduce the power quality of an electrical system. Poor power quality can waste energy and the capacity of an electrical system. In addition, it can harm the electrical distribution system and devices operating on that system.

Electrical concerns peaked when the first generation electronic ballasts for fluorescent lamps caused power quality problems. Due to advances in technology, electronic ballasts available today can *improve* power quality when replacing magnetically-ballasted systems in almost every facility. However, some isolated problems may still occur in electronically sensitive environments such as intensive-care units in hospitals. In these types of areas, special electromagnetic shielding devices are available, and are usually required.

The energy manager should ensure that a new system will improve the power quality of the electrical system. The following sub-sections will provide some basic electrical considerations relevant to lighting systems.

Power Quality

Certain lighting systems and other electric devices can cause distortion in the electrical current, which can affect power quality. Most incandescent lighting systems do not reduce the power quality of a distribution system because they have sinusoidal current waveforms that are in phase with the voltage waveform (the current and voltage both increase and decrease at the same time). Fluorescent, HID and low-voltage systems, which use ballasts or transformers, may have distorted current waveforms. Devices with heavily distorted current waveforms draw current in short bursts (instead of drawing it smoothly), which creates distortion in the voltage. These devices' current waveforms may also be out of phase with the voltage waveform. Such a "phase displacement" can reduce the efficiency of the alternating current circuit. In Figure 13.6 the current wave lags behind the voltage wave. The device produces work only during the time represented by the non-shaded parts of the cycle, which represent the circuit's "active power." However, during the shaded periods, the current is positive while the voltage is negative (or vice versa). During this period, the current and voltage work against each other, creating "reactive power." Reactive power does not distort the voltage. However, it is an important power quality concern because utilities' distribution systems must have the capacity to carry reactive power even though it accomplishes no useful work.[6]

Harmonics

A harmonic is a higher multiple of the primary frequency (usually 60 Hertz) superimposed on the alternat-

Table 13.11 EPACT's effect: lamp bans and options.

F96T12 SLIMLINE (EFFECTIVE MAY 1, 1994)

NONCOMPLYING LAMPS	MOST EFFICIENT (Ballast change required)	GOOD RETROFIT	MINIMUM COMPLIANCE (See note above)
F96T12/CW (75 W, Cool White)	F96T8/41K-85CRI (4100°K, 85 CRI) F96T8/41K-75CRI (4100°K, 75 CRI)	F96T12/41K-80CRI/ES (60 W, 4100°K, 80 CRI) F96T12/41K-70CRI/ES (60 W, 4100°K, 70 CRI	F96T12/CW/ES (60 W, Cool white, 62 CRI) F96T12/41K-80CRI (75 W, 4100°K, 80 CRI) F96T12/41K-70CRI (75 W, 4100°K, 70 CRI)
F96T12/W (75 W, White)	F96T835K-85CRI (3500°K, 85 CRI) F96T8/35K-75CRI (3500°K, 75 CRI)	F96T12/35K-80CRI/ES (60 W, 3500°K, 80 CRI) F96T12/35K-70CRI/ES (60 W, 3500°K, 70 CRI)	F96T12/W/ES (60 W, White, 57 CRI) F96T12/35K-80CRI (75 W, 3500°K, 80 CRI) F96T12/35K-70CRI (75 W, 3500°K, 70 CRI)
F96T12/WW (75 W, Warm white)	F96T8/30K-85CRI (3000°K, 85 CRI) F96T8/30K-75CRI (3000°K, 75 CRI)	F96T12/30K-80CRI/ES (60 W, 3000°K, 80 CRI) F96T12/30K-70CRI/ES (60 W, 3000°K, 70 CRI)	F96T12/WW/ES (60 W, Warm white, 52 CRI) F96T12/30K-80CRI (75, W, 3000°K, 80 CRI) F96T12/30K-70CRI (75, W, 3000°K, 70 CRI)
F96T12/D (75 W, Daylight)		F96T12/64-80CRI/ES (60 W, 6400°K, 80 CRI)	F96T12/64K-80CRI (75 W, 6400°K, 80 CRI)

F96T12/HO HIGH OUTPUT (EFFECTIVE MAY 1, 1994)

F96T12/CW/HO (110 W, Cool white)		F96T12/41K-70CRI/HO/ES (95 W, 4100°K, 70 CRI)	F96T12/CW/HO/ES (95, W, Cool white, 62 CRI) F96T12/LW/HO/ES (95 W, Lite white, 48 CRI) F96T12/41K-70CRI/HO (110 W, 4100°K, 70 CRI
F96T12/W/HO (110 W, White)		F96T12/35K-70CRI/HO/ES (95 W, 3500°K, 70 CRI)	F96T12/35K-70CRI/HO (110 W, 3500°K, 70 CRI)
F96T12/WW/HO (110 W, Warm white)		F96T12/30K-70CRI/HO/ES (95 W, 3000°K, 70 CRI)	F96T12/WW/HO/ES (95 W, Warm white, 52 CRI) F96T12/30K-70 CRI/HO (110 W, 3000°K, 70 CRI)
F96T12/D/HO (110 W, Daylight)		F96T12/64K-80CRI/HO/ES (95 W, 6400°K, 80 CRI)	F96T12/64K-80CRI/HO (110 W, 6400°K, 80 CRI)

F40T12 (EFFECTIVE NOVEMBER 1, 1995)

F40/CW (40 W, Cool white)	F32T8/41K-85CRI/ES (4100°K, 85 CRI) F32T8/41K-75CRI (4100°K, 75 CRI)	F40/41K-80CRI/ES (32 W, 4100°K, 80 CRI) F40/41K-70CRI (34 W, 4100°K, 70 CRI)	F40/CW/ES (34 W, Cool white, 62 CRI) F40/41K-80CRI (40 W, 4100°, 80 CRI) F40/41K-70CRI (40 W, 4100°K, 70 CRI)
F40/W (40 W, White)	F32T8/35K-85CRI (3500°K, 85 CRI) F32T8/35K-75CRI (3500°K, 75 CRI)	F40/35K-80CRI/ES (34 W, 3500°K, 80 CRI F40/35K-70CRI/ES (34 W, 3500°K, 70 CRI)	F40/W/ES (34 W, White, 57 CRI) F40/35K-80CRI (40 W, 3500°K, 80 CRI) F40/35K-70CRI (40 W, 3500°K, 70 CRI)
F40/WW (40 W, Warm white) F40/WWX (40 W, Warm white deluxe) F40/WWX/ES (34 W, Warm white deluxe)	F32T8/30K-85CRI (3000°K, 85 CRI) F32T8/30K-75CRI (3000°K, 75 CRI)	F40/30K-80CRI/ES (34 W, 3000°K, 80 CRI) F40/30K-70CRI/ES (34 W, 3000°K, 70 CRI)	F40/WW/ES (34 W, Warm white, 52 CRI) F40/30K-80CRI (40 W, 3000°K, 80 CRI) F40/30K-70CRI (40 W, 3000°K, 70 CRI)
F40/D (40 W, Daylight) F40/D/ES (34 W, Daylight)	F32T8/50K-75CRI (5000°K, 75 CRI)	F40/64K-80CRI (34 W, 6400°K, 80 CRI)	F40/64K-80CRI (40 W, 6400°K, 80 CRI)

FB40/6 CURVALUME "U-LAMP" (EFFECTIVE NOVEMBER 1, 1995)

F40/U/6/CW	F32T8/U/6/41K-85CRI	F40/U/6/41K-70CRI/ES	F40/U/6/CW/ES

(40 W, Cool white)	(4100°K, 85 CRI) **F32T8/U/6/41K-75CRI** (4100°K, 75 CRI)	(34 W, 4100°K, 70 CRI)	(34 W, Cool white, 62 CRI) **F40/U/6/41K-70CRI** (40 W, 4100°K, 70 CRI)
F40/U/6W (40 W, White)	**F32T8/U/6/35K-85CRI** (3500°K, 85 CRI) **F32T8/U/6/35K-75CRI** (3500°K, 75 CRI)	**F40/U/6/35K-70CRI/ES** (34 W, 3500°K, 70 CRI)	**F40/U/6/W/ES** (34 W, White, 57 CRI) **F40/U/6/35K-70CRI** (40 W, 3500°K, 70 CRI)
F40/U/6/WW (40 W, Warm white) **F40/U/6/WWX** (40 W, Warm white deluxe)	**F32T8/U/6/30K-85CRI** (3000°K, 85 CRI) **F32T8/U/6/30K-75CRI** (3000°K, 75 CRI)	**F40/U/6/30K-70CRI/ES** (34 W, 3000°K, 70 CRI)	**F40/U/6/WW/ES** (34 W, Warm white, 52 CRI) **F40/U/30K-80CRI** (40 W, 3000°K, 80 CRI) **F40/U/30K-70CRI** (40 W, 3000°K, 70 CRI)

INCANDESCENT REFLECTOR LAMPS (EFFECTIVE NOVEMBER 1, 1995)

NONCOMPLYING LAMPS	BEST COMPLYING RETROFIT	SUITABLE SUBSTITUTE	EXEMPT
75PAR38	45PAR/Halogen	50PAR30/Halogen	Colored ty
100PAR38	75PAR/Halogen	75ER30	Rough ser
150PAR38	90PAR/Halogen	75PAR/Halogen	ER Shape
75/65PAR38	45PAR/Halogen	50PAR30/Halogen	
100/80PAR38	75PAR/Halogen	75ER30	
150/120PAR38	90PAR/Halogen	75PAR/Halogen	
75R30	50PAR30/Longneck/Halogen	50ER30	
75R40	45PAR/Halogen/Very Wide Flood	50ER30	
100R40	75PAR/Halogen	75ER30	
150R40	90PAR/Halogen	120ER40	**Source:**

When a device's current waveform is out of phase with the voltage waveform, the difference between the two is the phase displacement. The shaded areas represent the reactive power that results.

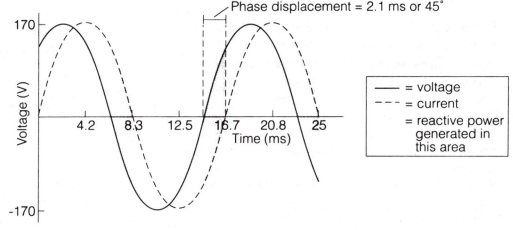

Figure 13.6 Phase displacement and reactive power.

ing current waveform. A distorted 60 Hz current wave may contain harmonics at 120 Hz, 180 Hz and so on. The harmonic whose frequency is twice that of the fundamental is called the "second-order" harmonic. The harmonic whose frequency is three times the fundamental is the "third-order" harmonic.

Highly distorted current waveforms contain numerous harmonics. The even harmonics (second-order, fourth order, etc.) tend to cancel each other's effects, but the odd harmonics tend to add in a way that rapidly increases distortion because the peaks and troughs of their waveforms coincide. Lighting products usually indicate a common measurement of distortion percentage: Total Harmonic Distortion (THD). Table 13.12 shows the

% THD for various types of lighting and office equipment.

Although lighting systems can have high % THD, the Total Harmonic Current (THC) is what is really important.

(THC) = (%THD)*(current draw of the lighting system)

Therefore, although an electronic ballast might have a slightly higher % THD than a magnetic ballast, the THC is often less because the electronic ballast consumes less power.

Both lighting manufacturers and building owners can take steps to improve power quality and reduce harmonic current. Most electronic ballasts for full-size

Table 13.12 Power quality characteristics for different electric devices.

	Active Power (W)	Power Factor	Current THD (%)
Compact fluorescent lighting systems			
13-W quad-tube compact fluorescent lamp w/ NPF magnetic ballast	16	0.54	13
13-W quad-tube compact fluorescent lamp w/ NPF electronic ballast	13	0.50	153
16-W quad-tube compact fluorescent lamp w/ HPF electronic ballast	16	0.91	20
Full-size fluorescent lighting systems (two lamps per ballast)			
T12 40-W lamps w/ energy-efficient magnetic ballast for T12 lamps	87	0.98	17
T12 34-W lamps w/ energy-efficient magnetic ballasts for T12 lamps	72	0.94	22
T10 40-W lamps w/ energy-efficient magnetic ballast for T12 lamps	93	0.98	22
T12 40-W lamps w/ electronic ballast for T12 lamps	72	0.99	5
T12 34-W lamps w/ electronic ballast for T12 lamps	62	0.99	5
T10 40-W lamps w/ electronic ballast for T12 lamps	75	0.99	5
T9 34-W lamps w/ electronic ballasts for T12 lamps	79	0.99	5
T9 32-W lamps w/ electronic ballast for T8 lamps	61	0.98	6
T8 32-W lamps w/ electronic ballast for T8 lamps	63	0.98	6
High-intensity discharge lighting systems			
400-W high-pressure sodium lamp w/ magnetic transformer	425	0.99	14
400-W metal halide lamp w/ magnetic transformer	450	0.94	19
Incandescent lighting systems			
100-W incandescent A lamp	101	1.00	1
50-W MR16 low-voltage halogen lamp w/ magnetic transformer	62	0.97	6
50-W MR16 low-voltage halogen lamp w/ electronic transformer	51	0.99	10
Office equipment			
Desktop computer without monitor	33	0.56	139
13" high-resolution color monitor for desktop computer	49	0.56	138
Laser printer while in standby	29	0.40	224
Laser printer while printing	799	0.98	15
External fax/modem	5	0.73	47
Electric pencil sharpener	85	0.41	33

*NLPIP measured specific products and reported their characteristics. These characteristics may vary substantially for similar products; specifiers should check with product manufacturers for specific information.

NPF = Normal Power Factor

HPF = High Power Factor

fluorescent lamps have filters to reduce current distortion. Generally, if the THC for a new system is less than the old system, harmonics should not be a problem. For example, as Table 13.13 shows, an electronic ballast with 20% current THD may produce less THC than a magnetic ballast with 16% THD because the electronic ballast uses only .21 amps of current, as compared to the .32 amp magnetic ballast. As Table 13.13 shows, THC and input power is reduced if the magnetic ballast is replaced with the electronic ballast.

If a facility manager believes that a problem with harmonics exists, he/she should ask an electric utility representative to meter the harmonics on a given circuit, and then on the entire service to diagnose the problem.

Crest Factor

The crest factor is basically a ratio of peak current to the Root Mean Square (RMS) current values. Crest factor is one of the criteria used by lamp manufacturers to estimate fluorescent rapid-start lamp life. Manufacturers design ballasts based on a 1.44 crest factor. The closer a ballast operates to the 1.44 crest factor, the longer the lamp life and the better the lamp/ballast efficiency. Current from a ballast with a higher crest factor may cause materials to be eroded from the lamp electrodes, shortening lamp life.

Power Factor

Power factor is a measure of how effectively a device converts input current and voltage into useful electric power. It describes the combined effects of current THD and reactive power from phase displacement. A device with a power factor of unity has 0% current THD and a current draw that is synchronized with the voltage. Resistive loads such as incandescent lamps have power factors of unity.

A device is said to have "normal power factor" (NPF) if the power factor is between 0.5 and 0.9. "High

power factor" (HPF) refers to a device with power factor greater than 0.9. Magnetic and electronic ballasts for fluorescent lamps may be either NPF or HPF. HPF ballasts usually have filters to reduce harmonics and capacitors to reduce phase displacement. Generally, HPF ballasts should be chosen to preserve power quality. Building owners also can install capacitors in their building distribution systems to compensate for electrical systems with low power factor.

13.6.2 HVAC Effects

Nearly all energy consumed by lighting systems is converted to light, heat and noise, which dissipate into the building. Therefore, if the amount of energy consumed by a lighting system is reduced, the amount of heat energy going into the building will also be reduced, and less air-conditioning will be needed. Consequently, the amount of winter-time heating may be increased to compensate for a lighting system that dissipates less heat.

Because most offices use air-conditioning for more months per year than heating, a more efficient lighting system can significantly reduce air-conditioning costs. In addition, air conditioning (usually electrically-powered) is much more expensive that heating (usually gas-powered). Therefore, the air-conditioning electricity savings are usually worth more dollars than the additional gas cost.

13.6.3 The Human Aspect

Regardless of the method selected for achieving energy savings, it is important to consider the human aspect of energy conservation. Buildings and lighting systems should be designed to help occupants work in comfort, safety and enjoyment. Retrofits that improve the lighting quality (and the performance of workers)

Table 13.13 Power quality characteristics: comparing magnetic and electric ballasts.

Ballast Type	# Lamps per Ballast	Watts per Ballast	Line Amps	THD	THC
Magnetic Ballast	2	86	.32	16%	.05
Electronic Ballast	2	58	.21	20%	.042

THD = Total Harmonic Distortion
THC = Total Harmonic Current
Reference: Advance Transformer Specification Manual

should be installed, especially when they save money. The recent advances in electronic ballast technology offer an opportunity for energy conservation to actually improve worker productivity. High frequency electronic ballasts and tri-phosphor lamps offer improved CRI, less audible noise and lamp flicker. These benefits have been shown to improve worker productivity and reduce headaches, fatigue and absenteeism.

Implementation Tactics

In addition to utilizing the appropriate lighting products, the implementation method of a lighting upgrade can have a serious impact on its success. To ensure favorable reaction and support from employees, they must be involved in the lighting upgrade. Educating employees and allowing them to participate in the decision process of an upgrade will reduce the resistance of change to a new system. Of critical importance is the maintenance department, because they will have an important role in the future upkeep of the system.

Once it has been decided to upgrade the lighting in an area, and a trial installation has been approved, a complete retrofit should be completed as soon as possible. Due to economies of scale and minimal employee distraction, an all-at-once retrofit is usually optimal.

If the lighting quality is not significantly increased in a particular retrofit, an over-night or over-the-weekend installation might be preferred. This method would avoid possible criticisms from side-by-side comparisons of the old and new systems. For example, a task lighting retrofit may appear darker than a uniformly-illuminated space adjacent to it. The average worker who believes "more light is better" might protest the retrofit. However, if the upgrade is done over the weekend, the worker may not easily notice the changes.

13.6.4 Lighting Waste and the Environment

Upgrading any lighting system will require disposal of lamps and ballasts. Some of this waste may be hazardous and/or require special management. Contact your state to identify the regulations regarding the proper disposal of lighting equipment in your area.

Mercury

With the exception of incandescent bulbs, all gaseous discharge lamps (fluorescent and HIDs) contain small quantities of mercury that end up in the environment, unless recycled. Mercury is also emitted as a byproduct of electricity generation from some fossil-fueled power plants.

Although compact fluorescent lamps contain the most mercury per lamp, they save a great deal of energy when compared to incandescent sources. Because they reduce energy consumption, (and avoid power plant emissions) CFLs introduce to the environment less than half the mercury of incandescents.[7] Mercury sealed in glass lamps is also much less available to ecosystems than mercury dispersed throughout the atmosphere. Mercury in lamps can be recycled, and regulations may soon require it.

PCB Ballasts

Ballasts produced prior to 1979 may contain polychlorinated biphenyls (PCBs). Human exposure to these possible carcinogens can cause skin, liver, and reproductive disorders. Fluorescent and HID ballasts contain high concentrations of PCBs. These chemical compounds were widely used as insulators in electrical equipment such as capacitors, switches and voltage regulators until 1979. The proper method for disposing used PCB ballasts depends on the regulations in the state where the ballasts are removed or discarded. Generators of PCB containing ballast wastes may be subject to notification and liability provisions under the Comprehensive Environmental Response, Compensation and Liability Act of 1980 (CERCLA)—also known as "Superfund."[8]

Generally, the PCB ballast is considered to be a hazardous waste only when the ballast is leaking PCBs. An indication of possible PCB leaking is an oily, tar-like substance emanating from the ballast. If the substance contains PCBs, the ballast and all materials it contacts are considered PCB waste, and are subject to state regulations. Leaking PCB ballasts must be disposed of at an EPA approved high-temperature incinerator.

Energy Savings and Reduced Power Plant Emissions

When appliances use less electricity, power plants don't need to produce as much electricity. Because most power plants use fossil fuels, a reduction in electricity generation results in reduced fossil fuel combustion and airborne emissions. Considering the different types of power plants (and the different fuels used) in different geographic regions, the Environmental Protection Agency has calculated the reduced power plant emissions by saving one kWh. Table 13.14 shows the reduction of CO_2, SO_x and NO_x per kWh saved in different regions of the US[9].

13.7 DAYLIGHTING

Human beings developed with daylight as their primary light source. For thousands of years humans evolved to the frequency of natural diurnal illumination.

Table 13.14

POLLUTION PREVENTION					
To estimate pollution prevention of an energy conservation project, use the following formulas and factors.					
CO_2:	kWh/yr saved	x	emission factor	=	lbs/yr
SO_2:	kWh/yr saved	x	emission factor	=	g/yr
NOx	kWh/yr	x	emission	=	g/yr

EPA Regional Emission Factors

REGION 1: CT, MA, ME, NH, RI, VT		
Emission per	CO_2	SO_2	NOx
kWh saved:	1.1	4.0	1.4

REGION 2: NJ, NY, PR, VI		
Emission per	CO_2	SO_2	NOx
kWh saved:	1.1	3.4	1.3

REGION 3: DC, DE, MD, PA, VA, WV		
Emission per	CO_2	SO_2	NOx
kWh saved:	1.6	8.2	2.6

REGION 4: AL, FL, GA, KY, MS, NC, SC, TN		
Emission per	CO_2	SO_2	NOx
kWh saved:	1.5	6.9	2.5

REGION 5: IL, IN, MI, MN, OH, WI		
Emission per	CO_2	SO_2	NOx
kWh saved:	1.8	10.4	3.5

REGION 6: AR, LA, NM, OK, TX		
Emission per	CO_2	SO_2	NOx
kWh saved:	1.7	2.2	2.5

REGION 7: IA, KS, MO, NE		
Emission per	CO_2	SO_2	NOx
kWh saved:	2.0	8.5	3.9

REGION 8: CO, MT, ND, SD, UT, WY		
Emission per	CO_2	SO_2	NOx
kWh saved:	2.2	3.3	3.2

REGION 9: AZ, CA, HI, NV, Guam, Am Samoa		
Emission per	CO_2	SO_2	NOx
kWh saved:	1.0	1.1	1.5

REGION 10: AK, ID, OR, WA		
Emission per	CO_2	SO_2	NOx
kWh saved:	0.1	0.5	0.3

Note: State pollution emission factors are aggregated by EPA region.

Daylight is a flicker-free source, generally with the widest spectral power distribution and highest comfort levels. With the twentieth century's trend towards larger buildings and dense urban environments, the development and wide spread acceptance of fluorescent lighting allowed electric light to become the primary source in offices.

Daylighting interior spaces is making a comeback because it can provide good visual comfort, and it can save energy if electric light loads can be reduced. New control technologies and improved daylighting methods allow lighting designers to conserve energy and optimize employee productivity.

There are three primary daylighting techniques available for interior spaces: Utilizing Skylights, Building Perimeter Daylighting and Building Core Daylighting. Skylights are the most primitive and are the most common in industrial buildings. Perimeter daylighting is defined as using natural daylight (when sufficient) such that electric lights can be dimmed or shut off near windows at the building perimeter. Traditionally, the amount of dimming depends on the interior distance from fenestration. However, ongoing research and application of "Core Daylighting Techniques" can stretch daylight penetration distance further into a room. Core daylighting techniques include the use of light shelves, light pipes, active daylighting systems and fiber optics. These technologies will likely become popular in the near future

Perimeter and core daylighting technologies are technologies which are being further developed to function in the modern office environment. The modern office has many visual tasks which require special considerations to avoid excess illumination or glare. It is important to properly control daylighting in offices, so that excessive glare does not reduce employee comfort or the ability to work on VDTs. A poorly designed or poorly managed daylit space reduces occupant satisfaction and can increase energy use if occupants require additional electric light to balance excessive daylight-induced contrast.

Windows and daylighting typically cause an increased solar heat gain and additional cooling load for HVAC systems. However, development of new glazings and high performance windows has allowed designers to use daylighting without severe heat gain penalties. With dynamic controls, most daylit spaces can now have lower cooling loads than non-daylit spaces with identical fenestration. "The reduction in heat-from-lights due to daylighting can represent a 10% down-sizing in perimeter zone cooling and fans.[10]" However, because there are several parameters, daylighting does not al-

ways reduce cooling loads any time it displaces electric light. As window size increases, the maximum necessary daylight may be exceeded, creating additional cooling loads.

Whether interior daylighting techniques can be economically utilized depends on several factors. However the ability to significantly reduce electric lighting loads during utility "peak periods" is extremely attractive.

13.8 Common Retrofits

Although there are numerous potential combinations of lamps, ballasts and lighting systems, a few retrofits are very common.

Offices

In office applications, popular and profitable retrofits involve installing electronic ballasts and energy efficient lamps, and in some cases, reflectors. Table 13.15 shows how a typical system changes with the addition of reflectors and the removal or substitution of lamps and ballasts. Notice that thin lamps allow more light to exit the luminaire, thereby increasing luminaire efficiency. Reflectors improve efficiency by greater amounts when there are less lamps (or thinner lamps) to block exiting light beams.

The expression (Lumens)/(Luminaire watt) is an indicator of the overall efficiency of the lighting system. It is similar to the efficacy of a lamp.

Indoor/Outdoor Industrial

In nearly all applications with significant annual operating hours, mercury vapor systems can be replaced by metal halide systems. This retrofit will improve CRI, reduce operating and relamping costs. In applications where CRI is not critical, HPS systems (which have a higher efficacy than metal halide systems) can be used.

Almost Anywhere

In nearly all applications where incandescent lamps are on for more than 5 hours per day, switching to CFLs will be cost-effective.

13.8.1 Sample Retrofits

This section provides the equations to calculate several different types of retrofits. For each type of retrofit, the calculations shown are based on average conditions and costs, which vary from location to location. For example, annual air conditioning hours will vary from building to building and from state to state. The energy

Table 13.15 Fluorescent lighting upgrade options.

ORIGINAL SYSTEM
Energy Efficient Magnetic Ballast

	40W lamps (T-12)	34W lamps (T-12)
Number of lamps	4	4
Total Watts	176	144
Ballast Factor	0.94	0.87
Available Lumens	12000	9700
Luminaire Efficiency	0.65	0.65
Lumens/Luminaire	7800	6300
Lumens/Luminaire watt	44.3	43.8

POTENTIAL RETROFITS

	EE Magnetic Ballast 42W lamps (T-10)	Electronic Rapid Start Ballast 42W lamps (T-10)
Number of lamps	2	2
Total Watts	92	63
Ballast Factor	0.95	0.73
Available Lumens	7000	5400
Luminaire Efficiency	0.77	0.77
Lumens per Luminaire	5400	4200
Lumens/Luminaire watt	58.7	66.7
ADD SILVER REFLECTOR		
Luminaire Efficiency	0.83	0.83
Lumens/Luminaire	5800	4500
Lumens/Luminaire watt	63.0	71.4

Electronic Instant Start Ballast

	32W lamps (T-8)	32W lamps (T-8)	32W lamps (T-8)
Number of lamps	4	3	2
Total Watts	112	90	60
Ballast Factor	0.88	0.88	0.88
Available Lumens	10200	7700	5100
Luminaire Efficiency	0.74	0.76	0.78
Lumens/Luminaire	7500	5900	4000
Lumens/Luminaire watt	67.0	65.6	66.7
ADD SILVER REFLECTOR			
Luminaire Efficiency	0.78	0.81	0.84
Lumens per Luminaire	8000	6200	4300
Lumens/Luminaire watt	71.4	68.9	71.7

References:
U.S. EPA Green Lights Program, Lighting Upgrade Manual, Apr. 94.
Advance Transformer Specification Guide
Magnetek Specification Guide

NOTES:
40W lamps are rated at 3200 lumens per lamp.
34W lamps are rated at 2800 lumens per lamp.
42W lamps are rated at 3700 lumens per lamp.
32W lamps are rated at 2900 lumens per lamp.
New luminaires may have greater efficiencies, due to highly reflective paints.

costs used in the following examples were based on $4/MCF, $10/kW month and $.05/kWh. In the following examples demand savings would likely occur in all except for examples # 4 and # 6. To accurately estimate the cost and savings from these types of retrofits, simply insert local values into the equations.

EXAMPLE 1:
  UPGRADE T12 LIGHTING SYSTEM TO T8

A hospital had 415 T12 fluorescent fixtures, which operate 24 hours/day, year round. The lamps and ballasts were replaced with T8 lamps and electronic ballasts, which saved about 30% of the energy, and provided higher quality light. Although the T8 lamps cost a little more (resulting in additional lamp replacement costs), the energy savings quickly recovered the expense. In addition, because the T8 system produces less heat, air conditioning requirements during summer months will be reduced. Conversely, heating requirements during winter months will be increased.

Calculations
= (# fixtures) [(Present input watts/fixture) – (Proposed input watts/fixture)]
= (415)[(86 watts/T12 fixture)–(60 watts/T8 fixture)]
= 10.8 kW

kWh Savings
= (kW savings)(Annual Operating Hours)
= (10.8 kW)(8,760 hours/year)
= 94,608 kWh/year

Air Conditioning Savings
= (kW savings)(Air Conditioning Hours/year) (1/Air Conditioner's COP)
= (10.8 kW)(2000 hours)(1/2.6)
= 8,308 kWh/year

Additional Gas Cost
= (kW savings)(Heating Hours/year)(.003413 MCF/kWh)(1/Heating Efficiency)(Gas Cost)
= (10.8 kW)(1,500 hours/year)(.003413 MCF/kWh) (1/0.8)($4.00/MCF)
= $ 276/year

Lamp Replacement Cost
= [(# fixtures)(# lamps/fixture)][((annual operational hours/proposed lamp life)(proposed lamp cost)) – ((annual hours operation/present lamp life) (present lamp cost))]

= [(415 fixtures)(2 lamps/fixture)][((8,760 hours/ 20,000 hours)($ 3.00/T8 lamp)) – ((8,760 hours/20,000 hours)($ 1.50/T12 lamp))]
= $ 545/year

Total Annual Dollar Savings
= (kW Savings)(kW charge)+[(kWh savings)+ (Air Conditioning savings)](kWh cost) –(Additional gas cost) – (lamp replacement cost)
= (10.8 kW)($ 120/kW year)+[(94,608 kWh)+ (8,308 kWh)]($ 0.05/kWh) –($ 276/year) – ($ 545/year)
= $ 5,621/year

Implementation Cost
= (# fixtures) (Retrofit cost per fixture)
= (415 fixtures) ($ 45/fixture)
= $ 18,675

Simple Payback
= (Implementation Cost)/ (Total Annual Dollar Savings)
= ($ 18,675)/($ 5,621/year)
= 3.3 years

EXAMPLE 2:
  UPGRADE T12 SYSTEM TO T10 WITH A REFLECTOR

Often it is beneficial to retrofit a T12 lighting systems to a T10 system. In this case, the existing fixture was a four lamp T12 system, that consumed 192 watts per fixture. The existing room was over-illuminated. The fixtures were retrofitted by installing T10 lamps, electronic ballasts and a custom-made specular aluminum reflector. Light levels were lowered by 12% and the retrofitted fixtures only consumed 68 watts per fixture.

EXAMPLE 3:
  REPLACE INCANDESCENT LIGHTING WITH
  COMPACT FLUORESCENT LAMPS

A power plant has 111 incandescent fixtures which operate 24 hours/day, year round. The incandescent lamps were replaced with compact fluorescent lamps, which saved over 70% of the energy, and last over ten times as long. Because the lamp life is so much longer, there is a maintenance relamping labor savings. Air conditioning savings or heating costs were not included because these fixtures are located in a high-bay building which is not heated or air-conditioned.

Calculations

Watts Saved Per Fixture

= (Present input watts/fixture) –
  (Proposed input watts/fixture)

= (150 watts/fixture) – (30 watts/fixture)

= 120 watts saved/fixture

kW Savings

= (# fixtures)(watts saved/fixture)(1 kW/1000 watts)

= (111 fixtures)(120 watts/fixtures)(1/1000)

= 13.3 kW

kWh Savings

= (Demand savings)(annual operating hours)

= (13.3 kW)(8,760 hours/year)

= 116,508 kWh/year

Lamp Replacement Cost

= [(Number of Fixtures)(cost per CFL Lamp)
  (operating hours/lamp life)] – [(Number of
  existing incandescent bulbs)(cost per bulb)
  (operating hours/lamp life)]

= [(111 Fixtures)($10/CFL lamp)
  (8,760 hours/10,000 hours)]
  – [(111 bulbs)($1.93/type "A" lamp)
  (8,760 hours/750 hours)]

= $ – 1,530/year§   §Negative cost indicates savings.

Maintenance Relamping Labor Savings

= [(# fixtures)(maintenance relamping cost per fixture)]
  [((annual hours operation/present lamp life))–
  ((annual hours operation/proposed lamp life))]

= [(111 fixtures)($1.7/fixture)][((8,760/750))–
  ((8,760/10,000))]

= $ 2,039/year

Total Annual Dollar Savings

= (kWh savings)(kWh cost)+ (kW savings)(kW cost)
  – (lamp replacement cost) +
  (maintenance relamping labor savings)

= (116,508 kWh)($.05/kWh)+(13.3)($120/kW year) –
  (–1,530/year) + (2,039/year)

= $ 10,999/year

Total Implementation Cost

= [(# fixtures)(cost/CFL ballast and lamp)]
  + (retrofit labor cost)]

= (111 fixtures)($45/fixture)

= $ 4,995

Simple Payback

= (Total Implementation Cost)/

(Total Annual Dollar Savings)

= ($ 4,995)/(10,999/year)

= 0.5 years

EXAMPLE 4:
INSTALL OCCUPANCY SENSORS

In this example, an office building has many individual offices that are only used during portions of the day. After mounting wall-switch occupancy sensors, the sensitivity and time delay settings were adjusted to optimize the system. The following analysis is based on an average time savings of 35% per room. Air conditioning costs and demand charges would likely be reduced, however these savings are not included.

Calculations

kWh Savings

= (# rooms)(# fixtures/room)(input watts/fixture)
  (1 kW/1000 watts)
  (Total annual operating hours)
  (estimated % time saved/100)

= (50 rooms)(4 fixtures/room)(144 watts/
  fixture)(1/1000)
  (4,000 hours/year)(.35)

= 40,320 kWh/year

Total Annual Dollar Savings ($/Year)

= (kWh savings/year)(kWh cost)

= (40,320 kWh/year)($.05/kWh)

= $ 2,016/year

Implementation Cost

= (# occupancy sensors needed)[(cost of occupancy
  sensor)+ (installation
  time/room)(labor cost)]

= (50)[($ 75)+(1 hour/sensor)($20/hour)]

= $ 4,750

Simple Payback

= (Implementation Cost)/(Total Annual Dollar Savings)

= ($ 4,750)/($ 2,016/year)

= 2.4 years

EXAMPLE 5:
RETROFIT EXIT SIGNS WITH L.E.D. EXIT WANDS

A hospital had 117 exit signs, which used incandescent bulbs. The exit signs were retrofitted with LED wands, which saved 90% of the energy. Even though the existing incandescent bulbs were "long-life" models, (which are expensive) material and maintenance savings were significant. Basically, the hospital should not have to relamp exit signs for 25 years!

Calculations

Input Wattage – Incandescent Signs
= (Watt/fixture) (number of fixtures)
= (40 Watts/fix) (117 fix)
= 4.68 kW

Input Wattage – LED Signs
= (Watt/fixture) (number of fixtures)
= (3.6 Watts/fix) (117 fix)
= .421 kW

kW Savings
= (Incandescent Wattage) – (LED Wattage)
= (4.68 kW) – (.421 kW)
= 4.26 kW

kWh Savings
= (kW Savings)(operating hours)
= (4.26 kW)(8,760 hours)
= 37,318 kWh/yr

Lamp Replacement Cost
= [(Number of LED Exit Fixtures)(cost per
  LED Fixture)(operating hours/Fixture life)]
 – [(Number of existing Exit lamps)(cost per Exit
  lamp)(operating hours/lamp life)]
= [(117 Fixtures)($ 60/lamp kit)(8,760 hours/219,000
  hours)] – [(234 Exit lamps)($5.00/lamp)
  (8,760 hours/8,760 hours)]
= –$ 889/year§ §Negative cost indicates savings.

Maintenance Relamping Labor Savings
= (# signs)(Number of times each fixture is
  relamped/yr)(time to
  relamp one fixture)(Labor Cost)
= (117 signs)(1 relamp/yr)(.25 hours/sign)($20/hour)
= $585/year

Annual Dollar Savings
= [(kWh savings)(electrical consumption cost)] +
  [(kW savings)(kW cost)] + [Maintenance Cost
  Savings]–[lamp replacement cost]
= [(37,318 kWh)($.05/kWh)] + [(4.26 kW)

($120/kW yr)]+ [$585/yr]– [– $889/yr]
= $ 3,851/year

Implementation Cost
= [# Proposed Fixtures][(Cost/fixture +
  Installation Cost/fixture)]
= [117][$60/fixture + $5/fixture]
= $ 7,605

Simple Payback
= (Implementation Cost)/(Annual Dollar Savings)
= ($7,605)/($3,851/yr)
= 2 years.

EXAMPLE 6:
  REPLACE OUTSIDE MERCURY VAPOR LIGHTING
  SYSTEM WITH HIGH AND LOW PRESSURE
  SODIUM LIGHTING SYSTEM

A parking lot is illuminated by mercury vapor lamps, which are relatively inefficient. The existing fixtures were replaced with a combination of High Pressure Sodium (HPS) and Low Pressure Sodium (LPS) lamps. The LPS provides the lowest-cost illumination, while the HPS provides enough color rendering ability to distinguish the colors of cars. By replacing the fifty 400 watt Mercury Vapor lamps with ten 250 watt HPS and forty 135 watt LPS fixtures, the company saved approximately $ 2,750/year with an installed cost of $12,500 and a payback of 4.6 years.

EXAMPLE 7:
  REPLACE "U" LAMPS WITH STRAIGHT T8 TUBES

The existing fixtures were 2' by 2' Lay-In Troffers with two F40T12CW "U" lamps, with a standard ballast consuming 96 watts per fixture. The retrofit was to remove the "U" lamps and install three F017T8 lamps with an electronic ballast, which had only 47 watts per fixture.

13.9 SCHEMATICS

References for all schematics are in Reference Section.

Lamps

Fluorescent

Bulb Shapes (Not Actual Sizes)

The size and shape of a bulb is designated by a letter or letters followed by a number. The letter indicates the shape of the bulb while the number indicates the diameter of the bulb in eighths of an inch. For example, "T-12" indicates a tubular shaped bulb having a diameter of ¹²/₈ or 1½ inches. The following illustrations show some of the more popular bulb shapes and sizes.

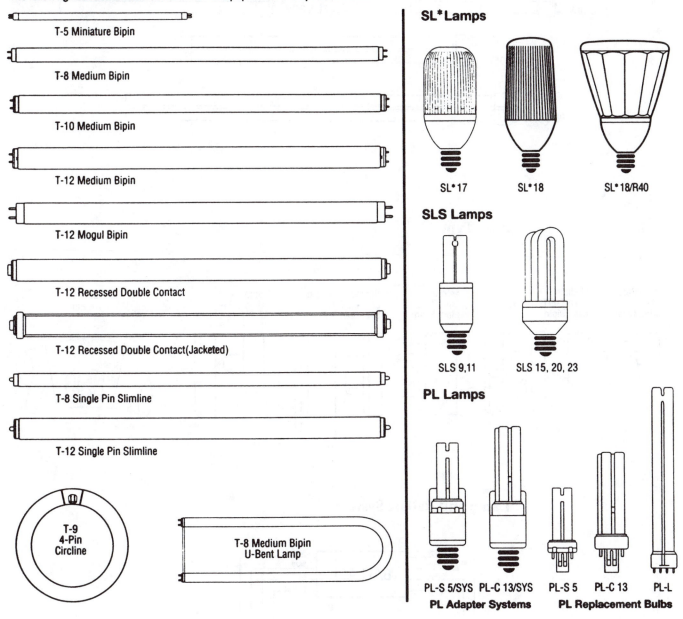

Fluorescent (continued)

Biax Lamps

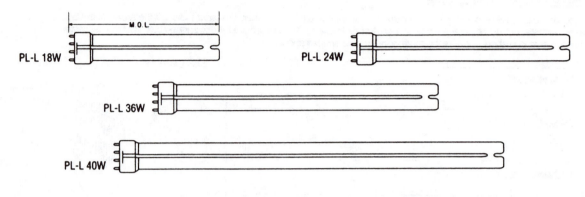

PL-L 18W

PL-L 24W

PL-L 36W

PL-L 40W

Base Types

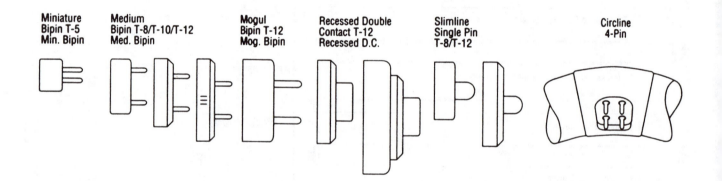

Miniature
Bipin T-5
Min. Bipin

Medium
Bipin T-8/T-10/T-12
Med. Bipin

Mogul
Bipin T-12
Mog. Bipin

Recessed Double
Contact T-12
Recessed D.C.

Slimline
Single Pin
T-8/T-12

Circline
4-Pin

Lamp and Ballast System

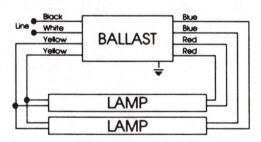

HID

Mercury Vapor

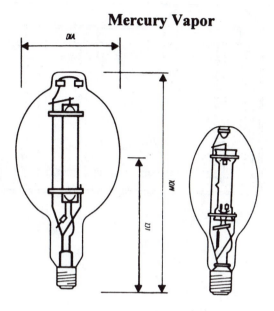

Metal Halide

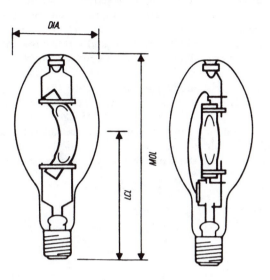

High Pressure Sodium

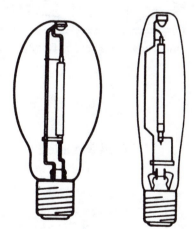

Low Pressure Sodium

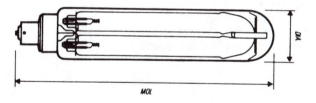

Exit Signs

LED

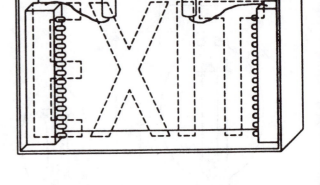

- **1.8-3.6 Input Watts/Fixture. (Replaces standard 20-25 watt lamps.)**
- Convert existing incandescent EXIT signs to use energy efficient LED liehg strips.
- Each kit contains two LED light strips and a reflective backing to provide even light distribution and a new red lens for the fixture.
- Estimated life is 25 years.
- Complies with OSHA and NFPA requirements.
- Available in four base styles to fit existing sockets or as a hard wire kit.
- LED light strips emit a bright red light and are not recommended for use with green signs.
- In addition to DGSC standard warranty, manufacturer's 25-year warranty applies.
- UL approved.

CFL

Quick connecting adapter screws into existing incandescent socket. For use with medium screw base societs.

Two lamp EXIT sign retrofit system, backup lamp will take over if the primary lamp fails.

UL approved.
- Lamp: 9 watt, twin tube compact fluorescent
- Lumens: 600
- Lamp Avg Life: 10,000 hours
- Ballast Losses: 2 watts/ballast
- System Input Watts: 11 watts
- Minimum Starting Temperature: 0°F

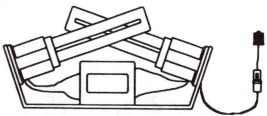

Unit Dimensions: 8"L X 4-5/8"H X 1"H

Occupancy Sensors
Ceiling Mounted

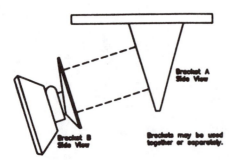

Wall Mounted

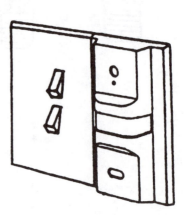

13.10 SUMMARY

In summary, this chapter will help the energy manager make informed decisions about lighting. The following "recipe" reviews some of the main points that influence the effectiveness of lighting retrofits.

A Recipe for Successful Lighting Retrofits

1. Identify visual task—Distinguish between tasks that involve walking and tasks that involve reading small print.

2. Identify lighting needs for each task-Use IES tables to determine target light levels.

3. Research available products and lighting techniques—Talk to lighting manufacturers about your objectives, let them help you select the products. Perhaps they will offer a demonstration or trial installation. Be aware of the relative costs, especially the costs associated with specialized technologies.

4. Identify lamps to fulfill lighting needs—Pick the lamp that has the proper CRI, CCT, lamp life and lumen output.

5. Identify ballasts and luminaires to fulfill lighting needs. Select the proper ballast factor, % THD, voltage, luminaire light distribution, lenses or baffles, luminaire efficiency.

6. Identify the optimal control technology: Decide whether to use IR, US or DT Occupancy Sensors. Know when to use time clocks or install switches.

7. Consider system variations to optimize:
    -Employee performance—Incorporate the importance of lighting quality into the retrofit process.
    -Energy savings—Pick the most efficient technologies that are cost-effective.
    -Maintenance—Installing common systems for simple maintenance, group re-lamping, and maintenance training.
    -Ancillary effects—Consider effects on the HVAC system, security, safety, etc.

8. Publicize results—As with any energy management program, your job depends on demonstrating progress. By making energy cost savings known to the employees and upper-level management, all people contributing to the program will know that there is a benefit to their efforts.

9. Continually look for more opportunities—The lighting industry is constantly developing new products that could improve profitability for your company. Keep in-touch with new technologies and methods to avoid "missing the boat" on a good opportunity. Table 13.16 offers a more complete listing of energy saving ideas for lighting retrofits.

Table 13.16 Energy saving checklist.

Lighting Needs

*	Visual tasks: specification	Identify specific visual tasks and locations to determine recommended illuminances for tasks and for surrounding areas.
*	Safety and aesthetics	Review lighting requirements for given applications to satisfy safety and aesthetic criteria.
*	Over-illuminated application	In existing spaces, identify applications where maintained illumination is greater than recommended. Reduce energy by adjusting illuminance to meet recommended levels.
*	Groupings: similar visual tasks	Group visual tasks having the same illuminance requirements, and avoid widely separated workstations.
*	Task lighting	Illuminate work surfaces with luminaires properly located in or on furniture; provide lower ambient levels.
*	Luminance ratios	Use wall-washing and lighting of decorative objects to balance brightness.

(Continued)

Space Design and Utilization

* Space plan   When possible, arrange for occupants working after hours to work in close proximity to one another.

* Room surfaces   Use light colors for walls, floors, ceilings and furniture to increase utilization of light, and reduce connected lighting power to achieve required illuminances. Avoid glossy finishes on room and work surfaces to limit reflected glare.

* Space utilization   Use modular branch circuit wiring to allow for flexibility in moving, relocating or adding
  branch circuit wiring   luminaires to suit changing space configurations.

* Space utilization: Light building for occupied periods only, and when required for security or cleaning purposes
  occupancy   (see chapter 31, Lighting Controls).

Daylighting

* Daylight compensation   If daylighting can be used to replace some electric lighting near fenestration during substantial periods of the day, lighting in those areas should be circuited so that it may be controlled manually or automatically by switching or dimming.

* Daylight sensing   Daylight sensors and dimming systems can reduce electric lighting energy.

* Daylight control   Maximize the effectiveness of existing fenestration-shading controls (interior and exterior) or automatically by switching or dimming.

* Space utilization   Use daylighting in transition zones, in lounge and recreational areas, and for functions where the variation in color, intensity and direction may be desirable. Consider applications where daylight can be utilized as ambient lighting, supplemented by local task lights.

Lighting Sources: Lamps and Ballasts

* Source efficacy   Install lamps with the highest efficacies to provide the desired light source color and distribution requirements.

* Fluorescent lamps   Use T8 fluorescent and high-wattage compact fluorescent systems for improved source efficacy and color quality.

* Ballasts   Use electronic or energy efficient ballasts with fluorescent lamps.

* HID   Use high-efficacy metal halide and high-pressure sodium light sources for exterior floodlighting.

* Incandescent   Where incandescent sources are necessary, use reflector halogen lamps for increased efficacy.

* Compact fluorescent   Use compact fluorescent lamps, where possible, to replace incandescent sources.

* Lamp wattage
  reduced-wattage lamps   Use reduced-wattage lamps where illuminance is too high.

* Control compatibility   If a control system is used, check compatibility of lamps and ballasts with the control device.

* System change   Substitute metal halide and high-pressure sodium systems for existing mercury vapor lighting systems.

(Continued)

Luminaires

* Maintained efficiency — Select luminaires which do not collect dirt rapidly and which can be easily cleaned.

* Improved maintenance — Improved maintenance procedures may enable a lighting system with reduced wattage to provide adequate illumination throughout systems or component life.

* Luminaire efficiency replacement or relocation — Check luminaire effectiveness for task lighting and for overall efficiency; if ineffective or inefficient, consider replacement or relocation.

* Heat removal — When luminaire temperatures exceed optimal system operating temperatures, consider using special luminaires to improve lamp performance and reduce heat gain to the space.

* Maintained efficiency — Select a lamp replacement schedule for all light sources, to more accurately predict light loss factors and possibly decrease the number of luminaires required.

Lighting controls

* Switching; local control — Install switches for local and convenient control of lighting by occupants. This should be in combination with a building-wide system to turn lights off when the building is unoccupied.

* Selective switching — Install selective switching of luminaires according to groupings of working tasks and different working hours.

* Low-voltage switching systems — Use low-voltage switching systems to obtain maximum switching capability.

* Master control system — Use a programmable low-voltage master switching system for the entire building to turn lights on and off automatically as needed, with overrides at individual areas.

* Multipurpose spaces — Install multi-circuit switching or preset dimming controls to provide flexibility when spaces are used for multiple purposes and require different ranges of illuminance for various activities. Clearly label the control cover plates.

* "Tuning" illuminance — Use switching and dimming systems as a means of adjusting illuminance for variable lighting requirements.

* Scheduling — Operate lighting according to a predetermined schedule, based on occupancy.

* Occupant/motion sensors — Use occupant/motion sensors for unpredictable patterns of occupancy.

* Lumen maintenance — Fluorescent dimming systems may be utilized to maintain illuminance throughout lamp life, thereby saving energy by compensating for lamp-lumen depreciation and other light loss factors.

* Ballast switching — Use multilevel ballasts and local inboard-outboard lamp switching where a reduction in illuminances is sometimes desired.

Operation and Maintenance

* Education — Analyze lighting used during working and building cleaning periods, and institute an education program to have personnel turn off incandescent lamps promptly when the space is not in use, fluorescent lamps if the space will not be used for 10 min. or longer, and HID lamps (mercury, metal halide, high-pressure sodium) if the space will not be used for 30 min. or longer.

(Continued)

* Parking	Restrict parking after hours to specific lots so lighting can be reduced to minimum security requirements in unused parking areas.
* Custodial service	Schedule routine building cleaning during occupied hours.
* Reduced illuminance	Reduce illuminance during building cleaning periods.
* Cleaning schedules	Adjust cleaning schedules to minimize time of operation, by concentrating cleaning activities in fewer spaces at the same time and by turning off lights in unoccupied areas.
* Program evaluation	Evaluate the present lighting maintenance program, and revise it as necessary to provide the most efficient use of the lighting system.
* Cleaning and maintenance	Clean luminaires and replace lamps on a regular maintenance schedule to ensure proper illuminance levels are maintained.
* Regular system checks	Check to see if all components are in good working condition. Transmitting or diffusing media should be examined, and badly discolored or deteriorated media replaced to improve efficiency.
* Renovation of luminaries	Replace outdated or damaged luminaires with modern ones which have good cleaning capabilities and which use lamps with higher efficacy and good lumen maintenance characteristics.
* Area maintenance	Trim trees and bushes that may be obstructing outdoor luminaire distribution and creating unwanted shadow.

GLOSSARY[11]

Ampere: The standard unit of measurement for electric current that is equal to one coulomb per second. It defines the quantity of electrons moving past a given point in a circuit during a specific period. Amp is an abbreviation.

ANSI: Abbreviation for American National Standards Institute.

ARC TUBE: A tube enclosed by the outer glass envelope of a HID lamp and made of clear quartz or ceramic that contains the arc stream.

ASHRAE: American Society of Heating, Refrigerating and Air-Conditioning Engineers.

AVERAGE RATED LIFE: The number of hours at which half of a large group of product samples have failed.

BAFFLE: A single opaque or translucent element used to control light distribution at certain angles.

BALLAST: A device used to operate fluorescent and HID lamps. The ballast provides the necessary starting voltage, while limiting and regulating the lamp current during operation.

BALLAST CYCLING: Undesirable condition under which the ballast turns lamps on and off (cycles) due to the overheating of the thermal switch inside the ballast. This may be due to incorrect lamps, improper voltage being supplied, high ambient temperature around the fixture, or the early stage of ballast failure.

BALLAST EFFICIENCY FACTOR: The ballast efficiency factor (BEF) is the ballast factor (see below) divided by the input power of the ballast. The higher the BEF—within the same lamp ballast type—the more efficient the ballast.

BALLAST FACTOR: The ballast factor (BF) for a specific lamp-ballast combination represents the percentage of the rated lamp lumens that will be produced by the combination.

CANDELA: Unit of luminous intensity, describing the intensity of a light source in a specific direction.

CANDELA DISTRIBUTION: A curve, often on polar coordinates, illustrating the variation of luminous intensity of a lamp or luminaire in a plane through the light center.

CANDLEPOWER: A measure of luminous intensity of a light source in a specific direction, measured in candelas (see above).

COEFFICIENT OF UTILIZATION: The ratio of lumens from a luminaire received on the work plane to the lumens produced by the lamps alone. (Also called "CU")

COLOR RENDERING INDEX (CRI): A scale of the effect of a light source on the color appearance of an object compared to its color appearance under a reference light source. Expressed on a scale of 1 to 100, where 100 indicates no color shift. A low CRI rating suggests that the colors of objects will appear unnatural under that particular light source.

COLOR TEMPERATURE: The color temperature is a specification of the color appearance of a light source, relating the color to a reference source heated to a particular temperature, measured by the thermal unit Kelvin. The measurement can also be described as the "warmth" or "coolness" of a light source. Generally, sources below 3200K are considered "warm;" while those above 4000K are considered "cool" sources.

COMPACT FLUORESCENT: A small fluorescent lamp that is often used as an alternative to incandescent lighting. The lamp life is about 10 times longer than incandescent lamps and is 3-4 times more efficacious. Also called PL, Twin-Tube, CFL, or BIAX lamps.

CONTRAST: The relationship between the luminance of an object and its background.

DIFFUSE: Term describing dispersed light distribution. Refers to the scattering or softening of light.

DIFFUSER: A translucent piece of glass or plastic sheet that shields the light source in a fixture. The light transmitted throughout the diffuser will be directed and scattered.

DIRECT GLARE: Glare produced by a direct view of light sources. Often the result of insufficiently shielded light sources. (SEE GLARE)

DOWNLIGHT: A type of ceiling luminaire, usually fully recessed, where most of the light is directed downward. May feature an open reflector and/or shielding device.

EFFICACY: A metric used to compare light output to energy consumption. Efficacy is measured in lumens per watt. Efficacy is similar to efficiency, but is expressed in dissimilar units. For example, if a 100-watt source produces 9000 lumens, then the efficacy is 90 lumens per watt.

ELECTRONIC BALLAST: A ballast that uses semi-conductor components to increase the frequency of fluorescent lamp operation-typically, in the 20-40 kHz range. Smaller inductive components provide the lamp current control. Fluorescent system efficiency is increased due to high frequency lamp operation.

ENERGY-SAVING BALLAST: A type of magnetic ballast designed so that the components operate more efficiently, cooler and longer than a "standard magnetic" ballast. By US law, standard magnetic ballasts can no longer be manufactured.

ENERGY-SAVING LAMP: A lower wattage lamp, generally producing fewer lumens.

FLUORESCENT LAMP: A light source consisting of a tube filled with argon, along with krypton or other inert gas. When electrical current is applied, the resulting arc emits ultraviolet radiation that excites the phosphors inside the lamp wall, causing them to radiate visible light.

FOOTCANDLE (FC): The English unit of measurement of the illuminance (or light level) on a surface. One footcandle is equal to one lumen per square foot.

FOOTLAMBERT: English unit of luminance. One footlambert is equal to 1/p candelas per square foot.

GLARE: The effect of brightness or differences in brightness within the visual field sufficiently high to cause annoyance, discomfort or loss of visual performance.

HARMONIC: For a distorted waveform, a component of the wave with a frequency that is an integer multiple of the fundamental.

HID: Abbreviation for high intensity discharge. Generic term describing mercury vapor, metal halide, high pressure sodium, and (informally) low pressure sodium light sources and luminaires.

HIGH-BAY: Pertains to the type of lighting in an industrial application where the ceiling is 20 feet or higher. Also describes the application itself.

HIGH OUTPUT (HO): A lamp or ballast designed to operate at higher currents (800 mA) and produce more light.

HIGH PRESSURE SODIUM LAMP: A high intensity discharge (HID) lamp whose light is produced by radiation from sodium vapor (and mercury).

HVAC: Heating, ventilating and air conditioning systems.

ILLUMINANCE: A photometric term that quantifies light incident on a surface or plane. Illuminance is commonly called light level. It is expressed as lumens per square foot (footcandles), or lumens per square meter (lux).

INDIRECT GLARE: Glare produced from a reflective surface.

INSTANT START: A fluorescent circuit that ignites the lamp instantly with a very high starting voltage from the ballast. Instant start lamps have single-pin bases.

LAMP LUMEN DEPRECIATION FACTOR (LLD): A factor that represents the reduction of lumen output over time. The factor is commonly used as a multiplier to the initial lumen rating in illuminance calculations, which compensates for the lumen depreciation. The LLD factor is a dimensionless value between 0 and 1.

LAY-IN-TROFFER: A fluorescent fixture; usually a 2' × 4' fixture that sets or "lays" into a specific ceiling grid.

LED: Abbreviation for light emitting diode. An illumination technology used for exit signs. Consumes low wattage and has a rated life of greater than 80 years.

LENS: Transparent or translucent medium that alters the directional characteristics of light passing through it. Usually made of glass or acrylic.

LIGHT LOSS FACTOR (LLF): Factors that allow for a lighting system's operation at less than initial conditions. These factors are used to calculate maintained light levels. LLFs are divided into two categories, recoverable and non-recoverable. Examples are lamp lumen depreciation and luminaire surface depreciation.

LOUVER: Grid type of optical assembly used to control light distribution from a fixture. Can range from small-cell plastic to the large-cell anodized aluminum louvers used in parabolic fluorescent fixtures.

LOW-PRESSURE SODIUM: A low-pressure discharge lamp in which light is produced by radiation from so-

dium vapor. Considered a monochromatic light source (most colors are rendered as gray).

LUMEN: A unit of light flow, or luminous flux. The lumen rating of a lamp is a measure of the total light output of the lamp.

LUMINAIRE: A complete lighting unit consisting of a lamp or lamps, along with the parts designed to distribute the light, hold the lamps, and connect the lamps to a power source. Also called a fixture.

LUMINAIRE EFFICIENCY: The ratio of total lumen output of a luminaire and the lumen output of the lamps, expressed as a percentage. For example, if two luminaires use the same lamps, more light will be emitted from the fixture with the higher efficiency.

MEAN LIGHT OUTPUT: Light output in lumens at 40% of the rated life.

MERCURY VAPOR LAMP: A type of high intensity discharge (HID) lamp in which most of the light is produced by radiation from mercury vapor. Emits a blue-green cast of light. Available in clear and phosphor-coated lamps.

METAL HALIDE: A type of high intensity discharge (HID) lamp in which most of the light is produced by radiation of metal halide and mercury vapors in the arc tube. Available in clear and phosphor-coated lamps.

OCCUPANCY SENSOR: Control device that turns lights off after the space becomes unoccupied. May be ultrasonic, infrared or other type.

PHOTOCELL: A light sensing device used to control luminaires and dimmers in response to detected light levels.

POWER FACTOR: Power factor is a measure of how effectively a device converts input current and voltage into useful electric power. Power factor is the ratio of kW/kVA.

RAPID START (RS): The most popular fluorescent lamp/ballast combination used today. This ballast quickly and efficiently preheats lamp cathodes to start the lamp. Uses a "bi-pin" base.

REACTIVE POWER: Power that creates no useful work; it results when current is not in phase with voltage. Calculated using the equation:

reactive power = V × A × sinφ

where φ is the phase displacement angle.

RECESSED: The term used to describe the luminaire that is flush mounted into a ceiling.

RETROFIT: Refers to upgrading a fixture, room, or building by installing new parts or equipment.

ROOT-MEAN-SQUARE (rms): The effective average value of a periodic quantity such as an alternating current or voltage wave, calculated by averaging the squared values of the amplitude over one period, and taking the square root of that average.

SPACE CRITERION: A maximum distance that interior fixtures may be spaced that ensures uniform illumination on the work plane. The luminaire height above the work plane multiplied by the spacing criterion equals the center-to-center luminaire spacing.

SPECULAR: Mirrored or polished surface. The angle of reflection is equal to the angle of incidence. This word describes the finish of the material used in some louvers and reflectors.

T12 LAMP: Industry standard for a fluorescent lamp that is 12 one-eighths (1.5 inches) in diameter.

TANDEM WIRING: A wiring option in which a ballast is shared by two or more luminaires. This reduces labor, materials, and energy costs. Also called "master-slave" wiring.

TOTAL HARMONIC DISTORTION (THD): For current or voltage, the ratio of a wave's harmonic content to its fundamental component, expressed as a percentage. Also called "harmonic factor," it is a measure of the extent to which a waveform is distorted by harmonic content.

VCP: Abbreviation for visual comfort probability. A rating system for evaluating direct discomfort glare. This method is a subjective evaluation of visual comfort expressed as the percent of occupants of a space who will be bothered by direct glare. VCP allows for several factors: luminaire luminances at different angles of view, luminaire size, room size, luminaire mounting height, illuminance, and room surface reflectiveness. VCP tables are often provided as part of photometric reports for specific luminaires.

VERY HIGH OUTPUT (VHO): A fluorescent lamp that operates at a "very high" current (1500 mA), producing more light output than a "high output" lamp (800 mA) or standard output lamp (430 mA).

WATT (W): The unit for measuring electrical power. It defines the rate of energy consumption by an electrical device when it is in operation. The energy cost of operating an electrical device is calculated as its wattage times the hours of use. In single phase circuits, it is related to volts and amps by the formula: Volts × Amps × PF = Watts. (Note: For AC circuits, PF must be included.)

WORK PLANE: The level at which work is done and at which illuminance is specified and measured. For office applications, this is typically a horizontal plane 30 inches above the floor (desk height).

References

1. Mills, E., Piette, M., 1993, "Advanced Energy-Efficient Lighting Systems: Progress and Potential," *Energy-The International Journal*, 18 (2) pp. 75.
2. United States EPA Green Lights Program, 1995, *Lighting Upgrade Manual-Lighting Upgrade Technologies*, p. 16.
3. Woodroof, E., 1995, *A LECO Screening Procedure for Surveyors*, Oklahoma State University Press-Thesis, p. 54.
4. Bartlett, S., 1993, *Energy—The International Journal*, 18 p. 99.
5. Flagello, V, 1995, "Evaluating Cost-Efficient P.I.R. Occupancy Sensors: A Guide to Proper Application," *Strategic Planning for Energy and the Environment*, 14 (3) p. 52.
6. National Lighting Product Information Program, 1995, *Lighting Answers*, 2(2) p. 2.
7. Mills, E., Piette, M., 1993, "Advanced Energy-Efficient Lighting Systems: Progress and Potential," *Energy-The International Journal*, 18 (2), p. 90.
8. United States EPA Green Lights Program, 1995, *Lighting Upgrade Manual—Lighting Waste Disposal*, p. 2.
9. United States EPA Green Lights Program, 1995, "Green Lights Update" p. 2, March 1995.
10. R. Rundquist, 1991, "Daylighting controls: Orphan of HVAC design," *ASHRAE Journal*, p. 30, November 1991.
11. United States EPA Green Lights Program, 1995, *Lighting Upgrade Manual—Lighting Fundamentals*.

Tables

13.1— United States EPA Green Lights Program, 1995, *Green Lights Workshop Slide Hard Copies* p. 16. AND Illuminating Engineering Society of North America, 1993, Lighting Handbook, 8th Edition, pp. 460-475, IES, New York.
13.2—Phillips Lighting, 1994, Phillips Office Lighting Application Guide p. 24.
13.3—Turner, W., 1993, *The Energy Management Handbook*, p. 311, The Fairmont Press, Lilburn, GA.
13.5—United States EPA—Green Lights Program, 1995, *Green Lights Workshop Slide Hard Copies* p. 18.
13.6 —Lindsay, J., 1991, *Applied Illuminating Engineering*, p. 291, The Fairmont Press, Lilburn, GA.
13.7 —United States EPA Green Lights Program, 1995, *Green Lights Workshop Slide Hard Copies* p. 28.

13.8 —United States EPA Green Lights Program, 1995, *Green Lights Workshop Slide Hard Copies* p. 62.

13.9—United States EPA Green Lights Program, 1995, *Green Lights Workshop Slide Hard Copies* p. 87.

13.10—United States EPA Green Lights Program, 1995, *Green Lights Workshop Slide Hard Copies* p. 67.

13.11—Yarnell, K., 1995, "The Impact of EPACT on Lighting Specification," *Consulting Specifying Engineer*, p. 31, April 1995.

13.12—National Lighting Product Information Program, 1995, *Lighting Answers*, 2(2)p. 5.

13.13—Advance Transformer, 1995, *Advance Transformer Specification Guide.*]

13.14—United States EPA Green Lights Program, 1995, "Green Lights Update," June 1995.

13.16—Illuminating Engineering Society of North America, 1993, *Lighting Handbook, 8th Edition*, p. 855-857, IES, New York.

Figures

13.1— United States EPA Green Lights Program, 1995, *Green Lights Workshop Slide Hard Copies* p. 14.

13.2— United States EPA Green Lights Program, 1995, *Green Lights Workshop Slide Hard Copies* p. 23.

13.3— United States EPA Green Lights Program, 1995, *Green Lights Workshop Slide Hard Copies* p. 19.

13.4— United States EPA Green Lights Program, 1995, *Green Lights Workshop Slide Hard Copies* p. 49.

13.5— United States EPA Green Lights Program, 1995, *Green Lights Workshop Slide Hard Copies* p. 65.

13.6— National Lighting Product Information Program, 1995, *Lighting Answers*, 2(2)p. 2.

Schematics

Fluorescent lamps—Phillips Lighting Company, 1994, *Lamp Specification and Application Guide*, p. 62, 78.

HID lamps—General Electric Company, 1993, *Spectrum 9200 Lamp Catalog*, p. 3-1 to 3-18.

Exit Signs—Defense General Supply Center, 1994, *Energy Efficient Lighting*, p. 13-16.

Occupancy Sensors—Defense General Supply Center, 1994, *Energy Efficient Lighting*, p. 61-64.

CHAPTER 14
ENERGY SYSTEMS MAINTENANCE

W.J. KENNEDY, P.E.

Department of Industrial Engineering
Clemson University
Clemson, SC 29634-0920

Energy systems maintenance is the maintenance of all systems that use or affect the use of energy. These systems are found in every kind of organization that uses energy, whether a hospital, a church, a store, a university, a warehouse, or a factory. Energy systems maintenance includes such routine maintenance tasks as lubrication, examination, and cleaning of electrical contacts and calibrating thermostats, and such non-routine tasks as repainting walls to increase the effective lighting, cleaning fins on compressors, and cleaning damper blades and linkages.

A good energy systems maintenance program can save a company substantial amounts of money in wasted steam and wasted electricity and in the lost production and additional expense caused by preventable equipment breakdowns. Other benefits include general cleanliness, improved employee morale, and increased safety. In a good maintenance program, planning, scheduling, and monitoring are all carried out in a predictable and well-organized manner. This chapter is designed to assist the reader to develop such a program. In addition, a special section is included on some recent developments and on some of the special problems and opportunities associated with material handling systems.

The maintenance part of this chapter is divided into three sections. The first section gives a four-step procedure for developing a maintenance program for energy management. This plan relies on a detailed knowledge of the systems within any plant and the components of these systems. The systems are described and the main components of each system are listed. The second section provides a more detailed description of the maintenance of these components and constitutes the main body of this chapter. The final section describes some of the instruments useful in the maintenance associated with energy management.

14.1 DEVELOPING THE MAINTENANCE PROGRAM

There are four steps in the development of a maintenance program. Step 1 is to determine the present condition of the existing facility. This step includes a detailed examination of each of the major energy-consuming systems. The output from this examination is a list of the motors, lights, transformers, and other components that make up each system, together with a report of the condition of each. Step 2 is the preparation of a list of routine maintenance tasks with an estimate of the number of times that each task must be performed. This list should also include, for each task, the craft, the needed material, and the appropriate equipment. These data are then incorporated in step 3 into a regular schedule for the accomplishment of the desired maintenance. Step 4 is the monitoring necessary to keep the program in force once it has been initiated. These four steps are discussed in more detail in the following sections.

14.1.1 Determine the Present Condition of the Facility

The purpose of this step is to create a starting point. The questions to be answered are: (1) What equipment and systems are in the building? (2) What material is available to describe each system and/or its components? (3) What needs to be done to get the energy-related systems into working condition and to keep them that way?

This step should incorporate (1) vendor data and operating specifications for as much of the installed equipment as possible, kept in a notebook, a file cabinet, or a computer data base (depending upon the size and complexity of the facility); (2) a diagram of each major system, showing the location of all important equipment and the direction of all fluid flows; (3) a complete list of all the equipment in the building, showing the name, location, and condition of each item; and (4) a comprehensive list of maintenance tasks required for each piece of equipment. This information constitutes an equipment reference source unique to the equipment in your facility. It should be kept current, with each addition initialed and dated, and its location should be known to all maintenance personnel and their supervisors.

The preparation of the notebook and the compilation of the other data is made much easier if separate energy-related systems are defined within the plant so that the systems can be examined in turn. A suggested classification of these systems would constitute (1) the building envelope, including all surfaces of the facility exposed to the outside; (2) the boiler and steam distribution system; (3) the heating, ventilating, and air-conditioning system, together with its controls; (4) the electrical system; (5) the lights, windows, and adjacent reflective walls, ceilings, and floors; (6) the hot-water distribution system; (7) the compressed-air distribution system; and (8) the manufacturing system, consisting of motors and specialized energy-consuming equipment used in the creation of products. Each of these systems should be inspected, the condition of each part noted, and a diagram drawn if appropriate. The following descriptions and tables have been prepared to assist you in this inspection.

Building Envelope

This consists of all parts of the facility that can leak air into or out of any building. Its components and appropriate initial maintenance measures are given in Table 14.1. The envelope should be described by a blueprint of the building, showing locations and compositions of all outside walls, windows, ceilings, and floors, and the locations of all outside doors. The primary malfunction of this system is leakage of air, and this leakage can often be detected by sight—looking for cracks— or by noting the presence of a draft. Infrared scanning from the outside can also be helpful. The benefits from maintaining this system are (1) a reduction in the amount of air that must be heated or cooled, and (2) increased comfort due to decreased drafts.

Boilers and Steam Distribution Systems

(These systems are described in greater detail in Chapters 5 and 6). A boiler is often the largest consumer of fuel in a factory or building. Any improvements that maintenance can make in its operation are therefore immediately reflected in decreased energy consumption and decreased energy cost. If the steam distribution system has leaks or is not properly insulated, these faults cause the boiler to generate more steam than is needed; eliminating these problems saves money. But boilers can malfunction, and steam leaks can cause severe burns if maintenance is performed by untrained personnel. The first step in proper boiler maintenance is usually to get the boiler and the steam distribution system inspected by a licensed professional. It is possible, however, to examine your own boiler and determine whether your system has some of the more conspicuous problems. To estimate the value of repairs to this system, assume that any boiler that has not been adjusted for two years can have its efficiency increased by 25% by a suitable adjustment, with a corresponding decrease in fuel consumption and in projected fuel costs. In steam distribution, a defective steam trap can typically waste 50,000,000 Btu/yr, at a cost of between $100 and $1000, depending upon the source of fuel. The savings from boiler and steam distribution system maintenance measures are thus worth pursuing. The system components and an appropriate set of initial maintenance measures are given in Table 14.2.

Heating, Ventilating, and Air-Conditioning Systems

(These systems are described in more detail in Chapter 10.) The purposes of these systems are to supply enough air of the right temperature to keep people comfortable and to exhaust harmful or unpleasant air contaminants. A complete description of this system should include a blueprint with the location of all dampers,

Table 14.1 Problems and solutions: the building envelope.

System Component	Problem	Initial Maintenance Action
Door	Loose fitting	Weatherstripping, new threshold, or frame repair
	Does not close	If problem is caused by air pressure, balance intake and exhaust air; correct door fit; check door-closing hardware
Window	Air leakage	Weatherstrip, caulk, or add storm window
	Broken	Replace broken panes
Wall	Drafts from wall openings	Caulk or seal openings on outside of wall
	Cracks	Patch or seal if air is entering or leaving
Ceiling	Drafts around exhaust piping	Caulk or repair flashing
Roof	Holes	Patch or cover

Table 14.2 Problems and solutions: the boiler and steam distribution system.

System Component	Problem	Initial Maintenance Action
Boiler	Inoperable gauges	Overhaul boiler controls as soon as possible
	Most recent boiler adjustment at least two years ago	Have boiler adjusted for most efficient firing
	Scale deposits on water side of shutdown boiler	Remove scale; check water-softening system
	Boiler stack temperature more than 150°F above steam or water temperature	Clean tubes and adjust fuel burner (Ref. 1)
	Fuel valves leak	Repair
	Stack shows black smoke or no haze when boiler is operating	Check combustion controls
	Rust in water gauge	Check return line for evidence of corrosion (Ref. 2)
	Safety valves not checked or tagged	Have inspection performed immediately
Steam trap	Leaks	Have inspection performed; repair or replace
Steam valve	Leaks	Repair
Steam line	Lines uninsulated	Have insulation installed
	Water hammer noted	Fix steam trap
Condensate return	Uninsulated	Insulate if hot to touch
Condensate tank	Steam plumes at tank vents	Check and repair leaking steam traps
	No insulation	Install insulation
Condensate pumps	Excessive noise	Check and repair
	Leaks	Replace packing: overhaul or replace pump if necessary

fans, and ducts, a complete diagram of the control system showing the location of all gauges, thermostats, valves, and other components, and a list of the correct operating ranges for each dial and gauge. The descriptive material should also include any vendor-supplied manuals and the engineering diagrams and reports prepared when each system was installed or modified. Since many systems have been modified since their installation, it is also desirable to have someone prepare a diagram of your system as it is now.

Determining the present condition of the system includes preparing a diagram of the existing system based on an actual survey of the system; placing labels on each valve, gauge, and piece of equipment; and examining each system component to see if it is working. Table 14.3 gives a list of some of the more expensive troubles that may be encountered.

Significant amounts of money may be saved by proper maintenance of this system. These savings come from three sources: reducing the energy used by the system and its associated cost, decreasing the amount of unanticipated repair that is necessary in the absence of

good maintenance, and reducing downtime caused when the system does not work and conditions become either uncomfortable or unsafe. Since energy maintenance on this system usually includes maintaining the temperature established by company policy, and since company policies have increasingly favored a high temperature threshold for cooling and a low threshold for heating, energy maintenance is directly responsible for realizing the considerable savings made possible by these policies. It is not uncommon for a good maintenance policy to cause a decrease of 50% in the energy consumption of a building. In addition to the energy cost savings, a good energy maintenance program can spot deterioration of equipment and can enable repair to be scheduled at a time that will not cause extensive disruption of work. Finally, there have been cases when a buildup of poisonous gas was caused when exhaust fans ceased to function. Such problems can often be avoided with a good maintenance program aimed at the ventilation system.

The controls for a heating, ventilating, and air-conditioning system can range from a simple thermostat to

Table 14.3 Problems and solutions: the heating, ventilating, and air-conditioning system.

Component	Problem	Initial Maintenance Action
Filter	Excessively dirty	Replace or clean
Damper	Blocked open or linkage disconnected	Check damper controls
	Leaks badly	Clean and overhaul
Ductwork	Open joints	Repair with duct tape
	Loose insulation in duct work	Replace and attach firmly
	Water leakage or rust spots	Repair
	Crushed	Replace
Grillwork	Air flow impossible due to dirt	Remove and clean
	Blocked by equipment	Remove equipment
Fan	Motor not hooked up to fan	Disconnect motor or install fan belts
	Excess noise	Check bearings, belt tension mountings, dirty blades
	Insufficient ventilation	Check fan and surrounding duct work and grill work
	Belt too tight or too loose	Adjust motor mount
	Pulleys misaligned	Correct alignment
Pump	Hot-water pump is cold (or vice versa)	Inspect valving; check direction of flow
Blower	Not moving air in acceptable quantities	Check direction of rotation and change wiring if needed; clean if dirty
	Excessive noise	Check bearings, coupler
	Rotation wrong direction	Check wiring
	Shaft does not turn freely by hand	Check lubricant; repair pump
Chiller	Leaks	Repair
Cooling tower	Scaling on spray nozzles	Remove by chipping or by chemical means
	Leaks	Repair
	Cold water too warm	Check pumps, fans, and wood fill (if used); clean louvers and fill
	Excessive water drift	Check drift elimination, metering orifices, and basins (for leaks); check for overpumping
Compressor		See Table 14.7
Thermostat	Temperature reading not accurate	Calibrate
	Leaks water or oil from mounting	Check pneumatic control lines

a modern computer-controlled network. The amount of troubleshooting that you can do will be directly proportional to your knowledge of the system. If you know a lot about it, you may be able to accomplish a great deal, even initially. If your knowledge is limited, get advice from someone who has installed such systems, obtain all the operating manuals that pertain to your system, and get estimates of the cost needed to bring your control system up to its designed operation. You can save larger amounts of money by simply adjusting controls—for example, by using a lower temperature on weekends and when no one is in the facility—than you can by any other comparable expense of effort. It pays to get your control system in operating condition and to keep it that way.

Electrical System

Many industrial electrical distribution systems are being used in ways not foreseen by the designers. These changes in use can cause some problems in energy consumption and in safety. If a motor is operating at a lower voltage than it was designed for, it is probably using more amperage than was intended and is causing unnecessary losses in transmission lines. If the wires are too small for the load, line losses can be large, and fire hazards increase significantly. Other problems that can create unnecessary energy loss are voltage imbalance in three-phase motors and leaks from voltage sources to ground.[1] Most of these problems are safety hazards as well as expensive in energy costs, and it is imperative that they be checked.

It is desirable to have a qualified electrician or electrical contractor examine your facility to find safety problems. At the same time, a load survey should be performed to determine whether your wiring, transformers, and switch gear are appropriately sized for the load they are currently carrying. To supplement this formal examination, Table 14.4 gives a list of trouble indicators you should look for as you are ascertaining the current condition of your facility. This list should be used in conjunction with the formal survey.

Another problem that may be costing money is a low power factor. The power factor is the ratio between the resistive component of ac power and the total (resistive and inductive) supplied. Because it costs more to provide electricity with a low power factor, it is common for electrical utilities to make an additional charge if the power factor is less than some value, typically 75%. If your electrical bill includes a charge for a low power factor, it may be worthwhile to have a power-factor meter installed and to focus management attention on the problem of increasing the power factor. Equipment that contributes to a low power factor includes welding machines, induction motors, power transformers, electric arc furnaces, and fluorescent light ballasts. A low power factor can be corrected with the installation of capacitors or with a separate power supply for machinery causing the problem. The maintenance of this equipment then becomes another item on the list of scheduled energy management maintenance. Power-factor management is covered in Chapter 11.

Lights, Windows, and Reflective Surfaces

Maintenance of the lighting adjacent to reflective walls, ceilings, and floors serves several purposes in energy management. The energy consumed by lights is significant, as is the energy used by the heating, ventilating, and air-conditioning system to remove the heat put into a building by the lights. But the psychological value may have a greater impact. If lights are conspicuously absent from corridors and, where not needed, from management offices, people tend to look upon energy conservation as a program that is being taken seriously, and they begin to take it more seriously themselves. Similarly, a plant where energy management has been encouraged but where no attention has been paid to lighting is often seen as a plant where energy management is not taken seriously. (For a more complete discussion of lighting, see Chapter 13.)

Many factors modify the effectiveness of the lighting system. Of particular importance are the condition of the lights, the cleanliness of the luminaires, and the cleanliness of the walls, ceiling, and floors. When determining the condition of this system, a light meter should be carried as standard equipment. This will show the rooms where the IES (Illuminating Engineering Society) or other lighting standards have been exceeded or where not enough light is present. There may also be some additional problems, which are described in Table 14.5. Windows can also be a source of light. If they are used this way in your facility, they should be cleaned. If they are not used as a light source, consider boarding them up to avoid having to remove the heat that they allow in from the sun. The lighting systems in a facility are important, both in energy cost and as a morale factor. They deserve your attention.

Hot-Water Distribution System

Hot water can have several uses within a facility. It can be used for sterilization, for industrial cleaning, as a source of process heat, or for washing hands. The maintenance principles for all of these uses are the same, but the temperatures that must be maintained may differ. The purpose of energy systems maintenance in this area is to keep the temperatures as low as possible, to prevent leaks, to keep insulation in repair, and to keep heat-transfer surfaces clean. To put this into perspective, a

Table 14.4 Problems and solutions: the electrical system.

Component	Problem	Initial Maintenance Action
Transformer	Leaking oil	Have electric company check at once
	Not ventilated	Install ventilation or provide for natural ventilation
	Dirt or grease in transformer and control room	Install air filtering system to insure clean contacts
	Water on control room floor	Install drainage or stop leaks into control room
Contact	Burned spots	Indicates shorting; repair immediately (Ref. 4)
	Frayed wire	May cause shorting; use tape to secure frayed ends
Switch	Sound of arcing, lights flicker	Replace

Table 14.5 Problems and solutions: lights, windows, and reflective surfaces.

Component	Problem	Initial Maintenance Action
Light	Illuminate unused space	Remove light and store for later use elsewhere
	Flickers (fluorescents)	Replace quickly
	Too little light	Increase lighting to acceptable levels
	Ballasts buzz	Adjust voltage or change ballast types
	Smoking	Replace ballast; check contacts and electrical wiring; do not use until condition is remedied
Wall	Dirty or greasy	Clean
	Painted with dark paint	On next painting use brighter paint
Floor	Hard to keep clean	Examine possibility of changing floor surface
Window	Dirty	Clean, if used for light; otherwise, consider boarding over to prevent solar gain and heat loss

leak of 1 cup/min of water that has been heated from 55°F to 180°F uses about 30 million Btu/yr. This is approximately equivalent to 30,000 ft^3 of natural gas.

Part of the description of this system should include hot-water temperatures throughout the facility. Hot-water temperatures should be kept at a low level unless there is some good reason for keeping them high. It is also unlikely that water needs to be kept hot during weekends. A further area for improved maintenance is in the insulation of hot-water tanks and lines. If a line or tank is hot to the touch, it should probably be insulated, both for the energy saved and for safety. Adding 1 in. of insulation to a water main carrying 150°F water can save as much as $1.60 per foot per year, depending upon the fuel used as a heat source. Heat-transfer surfaces should also be examined—any fins or radiators that are plugged up with debris or dirt are causing the hotwater heater to consume more energy. Table 14.6 gives some additional troubleshooting suggestions for the initial maintenance.

Air Compressors and the Air Distribution System

Compressed air usually serves one of three functions: as a control medium, for cleaning, or as a source of energy for tools or machines. As a control medium, it serves to regulate various parts of heating, ventilating, and air-conditioning systems. In cleaning, compressed air can be used to dry materials or blow away various kinds of dirt. And it can be a convenient source of energy for tools or for various kinds of hydraulic equipment. All three uses are affected badly by line leaks and by poor compressor performance. When a pneumatic control system for building temperature develops leaks, the usual result is that only hot air comes into a room. Thus this kind of leak creates two kinds of energy waste: excess running of the compressor and excess heating. If the air is used for cleaning, the affect of a moderate leak

Table 14.6 Problems and solutions: hot-water distribution system.

Component	Problem	Initial Maintenance Action
Faucet	Leaks	Fix; replace with spring-actuated units
	Water too hot for washing	Turn down thermostat
Piping	Hot to touch	Lower water temperature or add insulation
	Leaking	Replace
	Scale buildup	Consider installation of water-softening unit
Water storage tank	Hot to touch	Insulate or lower water temperature
	Leaks	Repair or replace
Electric boiler	Scale buildup	Install water-softening unit or start regular flushing operation
	On during periods when no people are in facility	Install time clocks to regulate use
Radiator	Finned surface badly fouled	Clean with soap, water, and a brush
	Obstructed	Remove obstructions

Table 14.7 Problems and solutions: air compressors and the air distribution system.

Component	Problem	Initial Maintenance Action
Compressor (Ref. 5)	Low suction pressures	Look for leaks on low-pressure side; overhaul if necessary
	Gauges do not work	Repair or replace
	Excess vibration	Check mountings and initial installation instructions
	Cold crankcase heater during compressor operation	Check for lubrication problems
	Loose wiring or frayed wires	Repair
	Leaks on high-pressure side	Examine compressor closely; check gaskets, connections, etc., and replace if necessary
Air line	Leaks	Tape or replace
	Water in line	Look for leak at compressor
	Oil in line	Look for lubrication leak in compressor

may only be that more air is used; this depends upon the individual application. If air is used as an energy source for motors or tools, the speed of the motor or tool usually depends upon the air pressure, and any decrease in this pressure will affect the performance of the motor. Another thing—controls are particularly affected by oil, water, and dust particles in control air, so it is of utmost importance that the compressor be checked regularly to see that the air cleaner is working as intended. More detail on compressors can be found in Chapter 12 of Ref. 5. Table 14.7 gives a guide to the initial maintenance procedures for air compressors and air distribution systems.

The Manufacturing Equipment System

This section would not be complete without a discussion of some of the more common equipment used in manufacturing. Some of this equipment has unique maintenance requirements, and these must be obtained from the manufacturer. Much equipment, however, is common to many companies or facilities, and this includes motors, ovens, and time clocks.

Motors. Motors can consume excess amounts of energy if they are improperly mounted, if they are not hooked up to their load, or, in the case of three-phase motors, if the voltages in the opposing legs are different. In this last case, the *ASHRAE Equipment Handbook* states:

With three-phase motors, it is essential that the phase voltages be balanced. If not, a small voltage imbalance can produce a greater current unbalance and a much greater temperature rise which can result in nuisance overload trips or motor failures. Motors should not be operated where the voltage unbalance is greater than 1% without first consulting the manufacturer.[5]

Bearings. Bearing wear is one of the more significant failure types in a motor, and this can be avoided by a procedure devised by Harold Tornberg, a plant manager and former industrial engineer for Safeway Stores, Inc. The Tornberg procedure is to connect a stethoscope to a decibel meter and to take readings at both ends of a motor. The reading on the driving end of the motor will usually be 2 to 3 dB above the reading on the inactive end. If there is no difference, the bearings on the inactive end probably need to be lubricated or replaced. If the difference is 5 to 6 dB, the bearings on the active end need to be replaced. If the difference is 7 to 9 dB, the bearing is turning inside the housing, and the motor should be overhauled as soon as possible. If the difference in readings is more than 9 dB, the motor is on the verge of failure and should be replaced immediately. These procedures were developed on motors in use at Safeway Stores and are in use throughout the Safeway system. To adapt these procedures to your facility, buy or make an instrument that allows you to determine the noise level at a particular point. Then survey the motors in your facility using the criteria discussed above. Take one of the motors that the test has indicated as needing repair into your shop, and tear it down. If you find that the failure was worse than these standards indicated, lower the decibel limit needed to indicate a problem. If you find that the failure was not as bad as you thought, raise the standards. In any case, this procedure is one that can be adapted to your needs to indicate bearing failures before they occur.

Ovens. Ovens use much energy, and many standard operating practices are available to help decrease associated waste.[6] In describing the initial maintenance actions that must be performed, check the seals, controls, refractory, and insulation. The possible problems are shown in Table 14.8. Correcting these will save energy and improve operating efficiency at the same time. Remember also that heat lost from an oven must be removed by the air-conditioning system.

Time Clocks. Time clocks can be used to significant advantage in the control of equipment that can be turned off at regular intervals. But two problems may occur to eliminate any savings that might otherwise be generated. First, people can wire around time clocks or otherwise obstruct their operation. Or maintenance personnel can forget to reset them after power outages. If either of these happens, all the potential energy savings possible with their use will have been lost. These items are included in Table 14.8.

14.1.2   Prepare a List of Routine Maintenance Actions with Time Estimates, Materials, and Frequency for Each

This is the second step in the four-step procedure for developing a maintenance program. The first step was to determine the present condition of each system in the facility, using Tables 14.1 to 14.8 to help locate trouble spots. The products of this step are: (1) a list of all the equipment in the facility by system; (2) a list of the major one-time maintenance problems associated with this equipment; and (3) a notebook with these lists and with the diagrams for each of the major systems. The next major step is to augment the notebook by a list of preventive energy maintenance actions for each system together with an estimate of the materials needed, the time required, and the maintenance frequency for each piece of equipment. Table 14.9 gives representative inspection intervals to help you develop your own procedures. For convenience, the items of equipment are arranged in alphabetical order, and the maintenance actions are described only briefly. A more detailed descrip-

tion of the required maintenance for some items is given in Section 14.2 and necessary instruments are discussed in Section 14.5.

Incidentally, the lists described above should be maintained in a form that maintenance personnel can use them. If your personnel are familiar with database searching, or if you have a computer system that you have used, understand, and like, then by all means use a computer system. Such systems have information retrieval capabilities that can be most helpful, they offer fast access to information, and they are easy to update. They also don't take up much room. On the other hand, they can be a barrier to getting work done. If the software is unfamiliar, if the maintenance people don't like to use it, or if keeping it up to date is difficult, then a manual system is better. The objective is to get the maintenance done, and done right, not to demonstrate another use for a computer.

Time Standards. Accurate time standards for maintenance are difficult to obtain except where the maintenance actions are the same whenever they are repeated. In this case, predetermined time standard systems such as MTM and MOST are claimed to work, as are some detailed standards such as those developed by the Navy.[7] But most maintenance actions include troubleshooting. Troubleshooting depends upon the condition of the equipment, the maintenance history, and the skill of the maintenance personnel. In general, hiring journeymen rather than apprentices is an investment that is well worthwhile. To estimate the amount of maintenance time that will be needed, consult equipment manufacturers first, then other users of the same kind of equipment. Modify these estimates to include the experience of your personnel, the present condition of the equipment, and the availability of necessary repair parts. Then record your time estimates, compare these estimates with actual experience, and revise the estimates to conform with your actual experience.

Table 14.8 Motors, ovens, and time clocks.

Component	Problem	Initial Maintenance Action
Motor	Noisy	Check bearings
	Too hot	Check voltage on both legs of three-phase input
	Vibrates	Check mounting
Oven	Door gaskets worn	Replace
	Insulation or refractory brick missing	Replace
	Thermostats out of calibration	Replace or recalibrate
Time clock	Does not work	Repair or replace
	Time incorrect	Adjust clock to correct time

Table 14.9 Table of preventive maintenance actions.

	Frequency
Air lines	
1. Check for leaks	Annually
Blowers	
1. Inspect belts for tension and alignment	
2. Inspect pulley wheels	
3. Inspect for dirt and grease	
Boilers (see Section 14.2.1)	
1. Check temperature and pressure	Daily
2. Clean tubes and other heating surfaces	As needed
3. Check water gauge glass	Daily
4. Remove scale	Annually
5. Perform flue-gas analysis	Monthly
6. Calibrate controls	Annually
Chillers (Ref. 5)	Annually
1. Clean condenser and oil cooler	
2. Calibrate controls	
3. Check electrical connections	
4. Inspect valves and bearings	
Condensate return system	
1. Check valves, pumps, and lines	Annually
Cooling coils	
1. Brush and wash with soap	Quarterly or when needed
2. Clean drip pan drain	When needed
Compressors	
1. Check oil levels	Monthly
2. Check wiring	Annually
3. Visual check for leaks	Monthly
4. Log oil temperature and pressure	Monthly
5. Remove rust with wire brush	Annually
6. Replace all drive belts	Annually
Condenser	
1. Clean fan	Annually
2. Brush off coil	Monthly
Controls	
1. Calibrate thermostats	Semiannually (Ref. 4)
2. Get professional inspection of control system	According to equipment specifications
3. Check gauges to see that readings are in correct range	Monthly
4. Examine control tubing for leaks	Monthly
Cooling towers (Ref. 3)	
1. Inspect for clogging and unusual noise	Daily
2. Check gear reducer oil	Level—weekly. For sludge and water— monthly
3. Inspect for leakage	Semiannually
4. Tighten loose bolts	Semiannually
5. Clean suction screen	Weekly
Dampers	
1. Check closure	Every 6 weeks
2. Clean with brush	Semiannually

Table 14.9 (*Continued*)

	Frequency
Ductwork	Semiannually
1. Inspect and refasten loose insulation	
2. Check and repair leaks	
3. Inspect for crushed or punctured ducts; repair	
Electrical system	
1. Inspect equipment for frayed or burned wiring	Semiannually
2. Perform electrical load analysis	Whenever major equipment changes occur, or every 2 years
Fans	
1. Check fan blades and clean if necessary	Semiannually
2. Check fan belts for proper tension and wear	Monthly
3. Check pulleys for wear and alignment	Semiannually
4. Check for drive noise, loose belts, and excessive vibration	Semiannually
Faucets	
1. Check for leaks; replace washers if needed	Annually
Filters, air	
1. Replace	When dirty, or monthly
2. Check for gaps around filters	When replacing
3. Inspect electrical power equipment (for roll filters; from Ref. 6)	Monthly
Filter, oil	
1. Clean and oil	Whenever compressor oil is changed
Gauges	
1. Check calibration	Annually
2. Check readings	As needed
Grillwork	
1. Remove dirt, grease, bugs	Monthly
2. Remove obstructions from in front of grill	Monthly
3. Check air direction	Monthly
4. Recaulk seams	As required
Leaks	
1. Check for refrigerant leaks	Annually
Lights, inside	
1. Perform group relamping	See Chapter 13
2. Perform survey of lighting in actual use	Semiannually
3. Clean luminaires	Office area: every 6 months; laboratories: every 2 months; maintenance shops: every 6 months; heavy manufacturing areas: every 3 months; warehouses: every 12 months
4. Replace flickering lights	Immediately
Lights, outside	See Chapter 13
Motors	
1. Lubricate bearings	When needed
2. Check alignment	Semiannually
3. Check mountings	Semiannually
Ovens	Semiannually
1. Check insulation	

Table 14.9 (*Continued*)

	Frequency
2. Check controls	Annually
3. Check firebrick and insulation	Annually
Piping	
1. Check mountings	Annually
2. Check for leaks	Semiannually
Pumps	
1. Check lubrication	Monthly
2. Examine for leaks	Semiannually
Steam traps, (Ref. 8), high pressure (250 psig or more)	
1. Test	Daily-weekly
Steam traps, medium pressure (30-250 psig)	
1. Test	Weekly-monthly
Steam traps, low pressure	
1. Test	Monthly-annually
Steam traps, all (see Section 14.2.3)	
1. Take apart and check for dirt and corrosion	If test shows problems
Thermostats	
1. Check calibration	Annually
Time clocks	
1. Reset	After every power outage
2. Check and clean	Annually
Transformers (Ref. 6)	
1. Inspect gauges and record readings	Monthly
2. Remove debris	Monthly
3. Sample and test dielectric	Annually
4. Remove cover and check for water	Every 10 years
Walls and windows	
1. Check for air infiltration	Semiannually
2. Clean	Regularly, when dirty

14.1.3 Prepare a Maintenance Schedule

Section 14.1.2 gave a list of maintenance actions needed for many equipment items, together with approximate maintenance frequencies. To develop your own maintenance schedule, first list the equipment you have in your facility. Then, using the table, estimate the maintenance or inspection frequencies for each unit. A section of such a table is shown in Table 14.10. This table should go into the workbook with manufacturers' manuals described earlier. This master table and the workbook are used to develop two files: an equipment file and a tickler file.

The equipment file is a list of cards describing the maintenance for each unit. A unit may have more than one card, depending upon the complexity of its maintenance. These cards contain troubleshooting directions, a description of how to perform specific maintenance on each piece of equipment, and a record of individual equipment maintenance. The cards should be updated every time maintenance is performed or whenever a new troubleshooting procedure is devised. Figure 14.1 shows such a card. The tickler file has a separate card for each day and gives the maintenance to be performed on that day. This file has two parts. The first part is for actions to be done monthly, quarterly, semiannually, or annually and is arranged by months. The second part of the file is for actions to be done weekly and is arranged by day of the week. In preparing this tickler file, items that require the same skills should be grouped together where possible to save time and to take advantage of any economies of scale. (An alternative is to do all tasks in the same area at the same time.) Figure 14.2 shows an example of one card from the tickler file. Ideally, all of

Table 14.10 Example of equipment maintenance and frequencies.

Equipment	Location	Action	Standard, Frequency
Air-handling units	Bldg 211	Clean filters	0.7 hr, monthly
		Clean fan blades	1.2 hr, annually
		Check ducts	1.4 hr, annually
		Replace fan belts	0.6 hr, annually
		Check belt tension	0.1 hr, monthly

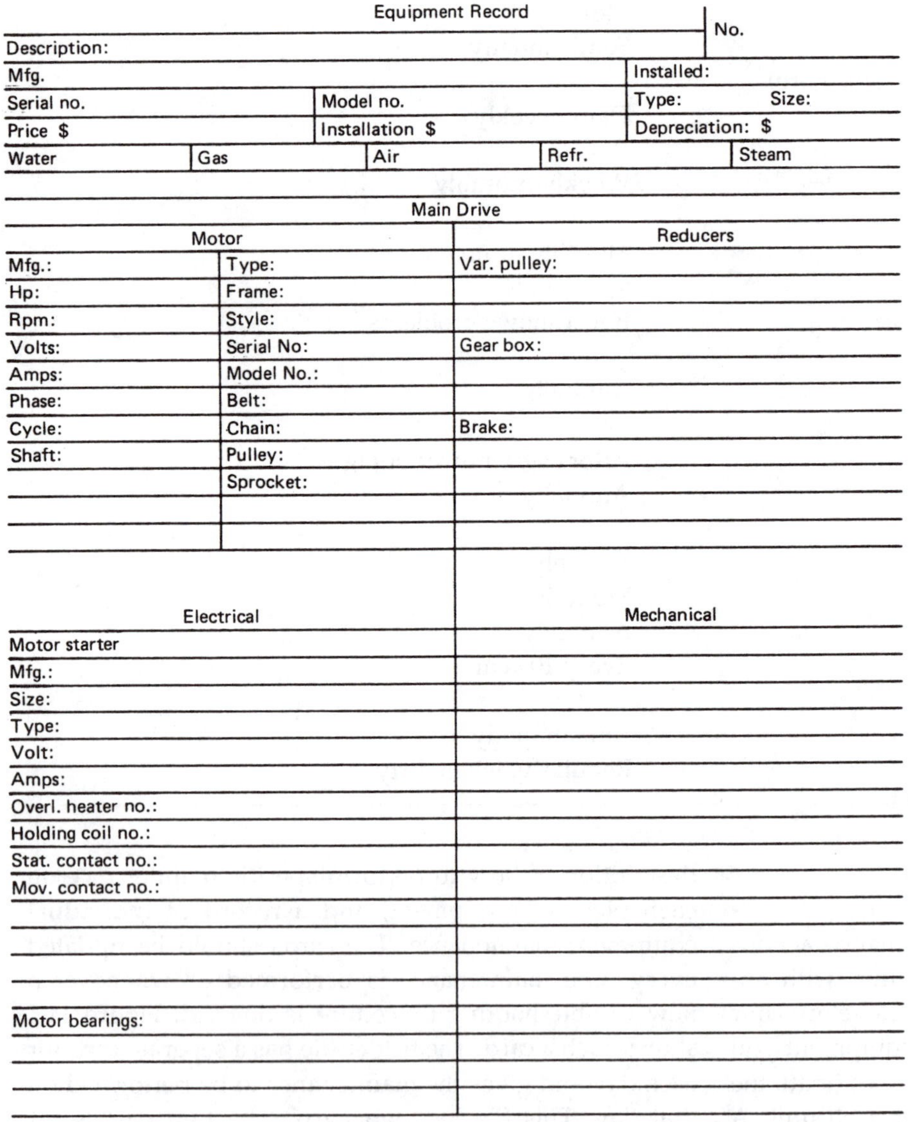

Equipment Record

			No.
Description:			
Mfg.		Installed:	
Serial no.	Model no.	Type:	Size:
Price $	Installation $	Depreciation: $	
Water	Gas	Air	Refr.	Steam

Main Drive

Motor		Reducers	
Mfg.:	Type:	Var. pulley:	
Hp:	Frame:		
Rpm:	Style:		
Volts:	Serial No:	Gear box:	
Amps:	Model No.:		
Phase:	Belt:		
Cycle:	Chain:	Brake:	
Shaft:	Pulley:		
	Sprocket:		

Electrical	Mechanical
Motor starter	
Mfg.:	
Size:	
Type:	
Volt:	
Amps:	
Overl. heater no.:	
Holding coil no.:	
Stat. contact no.:	
Mov. contact no.:	
Motor bearings:	

Fig. 14.1 Equipment record. (Courtesy of Safeway Stores, Inc.)

these files should be on a computer, but see the note in Section 14.1.2.

The final step is to use the tickler and equipment files to set up an operating schedule by assigning each task to a particular person or crew.

14.1.4 Follow-up and Monitoring

Management action is necessary to see that the prescribed maintenance policies are being followed. This step, the fourth in the development of an energy maintenance program, has three objectives. First, to ensure that maintenance is being performed as scheduled. This objective can be accomplished by periodic inspections of the records and of the action recorded. The second objective is for the management to get an update on the condition of the facility so as to anticipate and plan for capital expenditures relating to maintenance. This planning requires knowledge of the current condition of the plant. The third objective is to update all the files.

The energy equipment notebook should be changed every time a new kind of equipment is installed or whenever a manufacturer announces improved maintenance procedures for equipment already on hand. The equipment card file should be changed at the same time, with abbreviated versions of any new maintenance instructions. The equipment file should also be changed whenever someone discovers an improved way of maintaining an item on a card. The tickler file should be changed to reflect the actual needs of the facility. If, for example, the original standard on filter changing was 0.5 hr per change, and a new method is developed that enables these to be changed in 0.3 hr, the new method should be noted on the equipment card file and the new standard on the tickler card files. Similarly, if the frequency of filter changes was originally listed as every month, and experience shows that 3 months is a better interval, the three months should be noted in the tickler file. By following this procedure, the time standards and the maintenance frequencies can be kept current, and the maintenance plan can be adapted to each individual facility.

14.2 DETAILED MAINTENANCE PROCEDURES

Section 14.1 gave an outline for the development of a maintenance program for use in energy management. To maintain some of the more complex equipment properly, it is necessary to have more detail than could be presented in Table 14.9. This section is therefore devoted to a more complete discussion of boilers, pumps, and steam traps.

14.2.1 Boilers

Money spent in proper boiler maintenance is one of the best investments a company can make. The benefits are substantial. Suppose, for example, that your gas bill is $1,500,000 per year and that 90% of this, or $1,350,000, is for your central heating plant. Improving your boiler efficiency 5% saves $67,500 per year, in addition to any benefits that you may realize through decreased downtime, decreased repair costs, and other nontrivial expenses. To gain this benefit requires proper boiler maintenance, either by your own personnel or by a boiler service company hired to provide this service on a regular basis. The following section is designed to help you decide what must be done, how often, and why. Two warnings must be observed. First, safety precautions must be known and observed at all times. *Make sure that all safety interlocks are functioning before doing any work on a boiler.* Second, nothing in the following section should be construed as superseding manufacturers' instructions or local safety and environmental regulations. Keeping these warnings in mind, consider the following problems, their effects, and the best means of correcting them.

The basic reference for the following material is *Guidelines for Industrial Boiler Performance Improvement*, published by the Environmental Protection Agency.[9]

Excess O_2 in Stack Gas

The amount of oxygen mixed with fuel is directly related to the amount of excess air introduced to the boiler, as shown in Figure 14.3. The problem of excess O_2 is simply this: If there is too little air available to combine with fuel, incomplete combustion takes place, the boiler smokes, and hazards of boiler malfunction increase dramatically (in an oil- or coal-fired boiler) or the carbon monoxide (CO) concentration builds up (in a gas-fired boiler). If too much air is introduced, a great deal of the energy in the fuel is used to heat up the excess air, and the efficiency of the boiler decreases. These inefficiencies are shown in Figures 14.4 and 14.5. The two curves in each figure reflect the range within which the curve of a typical boiler would be expected to fall. The exact curve for your boiler should be established by a

Maintenance Department

Weekly Lubrication Schedule

Day of week   Tuesday, 11/13/79   _____Plant

Date_____ 19_____ By_____

Alto Roll Slicer		Tail Off Conveyor		
Clean and oil all moving parts; clean and		Clean and oil drive chain, #1 oil; check		
oil all drive chains, #1 oil; check all gear-		gearmotor oil level, #8 oil.		
motors oil level, #8 oil; check variable-				
speed internal drive belt for wear;		All Cooling Conveyors		
check out machine.		Oil all transfer roller bearings, #30 oil.		
United Roll Bagger		Model-K-Roll Machine		
Clean and oil all moving parts; clean and		Check head oilers; oil sifter linkage,		
oil all drive chains, #1 oil; check gear		sifter bushings, chain take up idler		
reducer oil level, #8 oil; check drive belts		sprocket, #1 oil; check vacuum com-		
for wear; check out, wipe off machine.		pressor oil level, compressor oil.		
Kwik-Lok		Pan-O-Mat		
Clean and oil all moving parts, #1 oil;		Oil all sifter linkage, sifter bushings, vari-		
check out, wipe off machine.		able speed adjusting pulley, chain take up		
		idler sprocket, control arm linkage, cam		
Brush Unit		rollers, gate bushings, tail off conveyor		
Remove side cover; clean and oil all		pulleys, #1 oil.		
chains, #1 oil; check out, wipe off				
		Total time		
The following items need attention:				
machine.				

Fig. 14.2 Tickler file card. (Courtesy of Safeway Stores, Inc.)

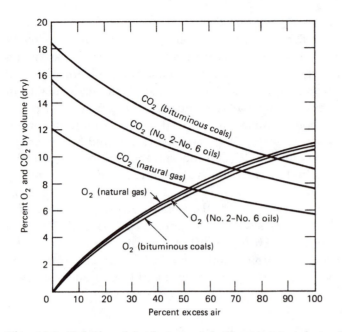

Fig. 14.3 Relationship between boiler excess air and stack-gas concentration of excess oxygen. (Adapted from Ref. 9.)

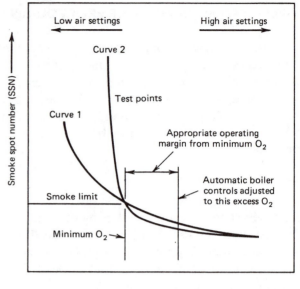

Fig. 14.4 Typical smoke-O_2 characteristic curves for coal- or oil-fired industrial boilers. Curve 1, gradual smoke/O_2 characteristic; curve 2, steep smoke/O_2 characteristic. (Adapted from Ref. 9.)

careful test at each of the firing rates you anticipate.

The combustion controls on the boiler should be used to decrease the amount of excess air to a point where the amount of excess O_2 represents a compromise between the optimal minimum and the point at which actual waste of energy occurs. The effect of reducing excess oxygen toward the optimum can be dramatic; if a reduction in excess air from 80% to 20% can be accompanied by a reduction in stack-gas temperature from 500°F to 400°F, the resulting improvement in boiler efficiency can be 6 to 8% or more. On an annual gas bill of $1,000,000, this is a reduction of $60,000 to $80,000 in gas alone.

But it is not usually desirable to operate at the minimum indicated on Figures 14.4 and 14.5, for several reasons. The consequences of exceeding the minimum O_2 percentage by a small amount are not as damaging as those of being less than the minimum by a corresponding amount. If the minimum is exceeded slightly, fuel is wasted. If, however, there is insufficient excess air and thus insufficient excess oxygen, smoking or CO buildup can occur, the tubes can become fouled, and the boiler can malfunction. Since there is usually some play in the combustion control system, a setting greater than the minimum helps to guard against these problems. Typical values are shown in Table 14.11.

The amount of O_2 in stack gas can be measured continuously, by a recording gas analyzer, or periodically, by an Orsat analyzer. The advantages of a continuous, mounted unit are that readings are always available

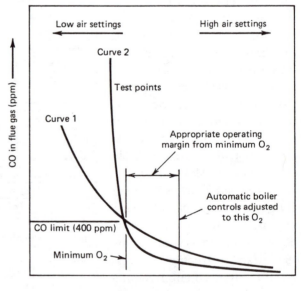

Fig. 14.5 Typical CO-O_2, characteristic curves for gas-fired industrial boilers. Curve 1, gradual CO/O_2 characteristic: curve 2, steep CO/O_2 characteristic. (Adapted from Ref. 9.)

to boiler operating personnel for daily boiler maintenance and for troubleshooting. The cost of an installed gas analyzing unit will vary from application to application, but typical costs are $200 to $1000. The readings should be logged every hour.

Table 14.11 Typical values for minimum excess O_2.

Fuel Type	Typical Minimum Values of Excess O_2 at High Firing Rates (%)
Natural gas	0.5-3.0
Oil fuels	2.0-4.0
Pulverized coal	3.0-6.0
Coal stoker	4.0-8.0

Source: Ref. 9.

Sudden changes in the percentage of excess oxygen from hour to hour are unlikely but should be analyzed immediately if they occur. Some of the more common causes for changes in excess O_2 are tube fouling, changes in fuel or in atmospheric conditions, or damage to the control system. The base for comparison should be established by the boiler service personnel or the boiler contractor at the time of initial installation of the boiler, and he or she should be consulted immediately if any unusual conditions are observed.

Excessive Stack-Gas Temperature

This problem can be caused by water-side or fire-side fouling. If fire-side fouling (i.e., a buildup of soot, ash, or other particles) is taking place, the rate of heat transfer from the boiler to the water or steam it is heating is impeded, and the stack gas is correspondingly hotter. The effect of soot on fuel consumption is noticeable, according to the American Boiler Manufacturers Association (ABMA).[10] Efficiency losses due to soot are approximately as given in Table 14.12. For this reason, the ABMA recommends that tubes be cleaned once every shift where practical. Where such cleaning is not practical, it is recommended that tubes be cleaned whenever the stack temperature rises 75°F.

Table 14.12 Typical fuel losses due to soot.

Soot Layer on Heating Surfaces (in.)	Increase in Fuel Consumption (%)
1/32	2.5
1/16	4.4
1/8	8.5

Source: Ref. 10.

If the fouling is on the water side because of scale buildup, accumulation of mud or slime, or for some other reason, heat transfer will be impeded as described

Table 14.13 Effect of water-side sealing.

Thickness of Scale (in.)	Loss of Heat (%) Soft Carbonate	Hard Carbonate	Hard Sulfate
1/50	3.5	5.2	3.0
1/32	7.0	8.3	6.0
1/25	8.0	9.9	9 0
1/20	10.0	11.2	11.0
1/16	12.5	12.6	12.6
1/11	15.0	14.3	14.3
1/9	n.a.	16.0	16.0

Source: Ref. 10.

above, and increased stack temperature will result. The effect of this scaling is seen in Table 14.13. The detrimental effect can be severe.

As with excess oxygen, it is advisable to monitor stack temperature on an hourly basis and to record the readings. By graphing the readings, it is possible to determine how much variation is caused by shifting load, when given temperature indicates that something unusual has happened, and when a gradual temperature rise indicates that it is time to clean the water side or the fire side of the boiler. If soot blowers are being used, the stack temperature should drop immediately after the tubes are cleaned. If the temperature does not drop, the soot blowers may not be working or the thermometer may have become fouled—in any case, something is wrong.

Smoking or Excess CO

Excess CO, in the case of natural gas, or smoking for coal or oil fuels, gives an indication that something has changed. Changes in the fuel composition or wear in some component of the burner can cause these problems, or there may be a change in the air supply. In any case, the problem should be corrected immediately. A stack-gas analyzer can be used to monitor the percentage of CO on a continuous basis.

Flame Appearance

The appearance of the flame can give some valuable information. If the pattern is unusual, there may have been changes in the burner tips or in other parts of the burner, or there may be a malfunction in a related part of the boiler. Also, examining the flame pattern can show if part of the boiler is getting overheated. At the same time the flame is being examined, the inside of the boiler can be examined as far as possible to see that everything is in order—the stoker (if it is a coal-fired boiler), the refractory, the burners, and so on. The flame check is quick and should not replace the other inspections noted above, but it can provide additional information. Such a check should be performed every hour.

Record Keeping

The Environmental Protection Agency guidelines recommend that a log be kept on each boiler with the information shown in Table 14.14.[9] In addition to recording this information, plotting it on a graph can give early indications of unusual trends or cycles in the data; these patterns can then be incorporated into the general system used to indicate when something is about to go wrong.

Maintenance Actions and Frequencies

Table 14.15 gives a list of the most common boiler maintenance actions that need to be performed annually. Table 14.16 gives a checklist of other routine maintenance items. If your staff is trained in boiler operation and maintenance this table can be used to help define a pattern of boiler maintenance. If you decide to hire boiler maintenance done by an outside firm, the table can help you to determine what they should do and to determine whether you would gain more from doing this work with your own personnel.

Ultimately, your own maintenance personnel will do much of the routine maintenance. To be sure they know what to expect and how to perform these tasks, have a boiler service representative or contractor train each person in the more routine kinds of boiler operation, such as reading a sight glass, inspecting the flame, and blowing down the boiler. In particular, they should learn enough about the boiler that they know how to operate it safely. Keep in mind that many maintenance people are undertrained as well as underpaid and that they are not familiar with procedures for handling live steam. Without proper training, they may also let sight glasses go dry, and they may miss such facts as that the low-water cutoff valve is not working. These maintenance people can be the most expensive ones you hire—if their lack of training causes you to lose a boiler.

Another note—the given maintenance intervals are averages. If your boiler uses a great deal of makeup water, more sludge can develop, and this will have to be removed by blowing down more often than indicated. If your boiler is a closed system with very few leaks, it is possible that water quality will not be a problem. In that case, the blowdowns can be much less frequent. You must adapt the procedures to your own boiler, preferably with the help of a local professional and the vendor.

Table 14.14 Boiler information to be logged.

General data to establish unit output
    Steam flow, pressure
    Superheated steam temperature (if applicable)
    Feedwater temperature
Firing system data
    Fuel type (in multifuel boilers)
    Fuel flow rate
    Oil or gas supply pressure
    Pressure at burners
    Fuel temperature
    Burner damper settings
    Windbox-to-furnace air pressure differential
    Other special system data unique to particular installation
Air flow indication
    Air preheater inlet gas O_2
    Stack gas O_2
    Optional: air flow pen, forced-draft fan damper position, forced-draft fan amperes
Flue-gas and air temperature
    Boiler outlet gas
    Economizer or air heater outlet gas
    Air temperature to air heater
Unburned combustion indication
    CO measurement
    Stack appearance
    Flame appearance
Air and flue-gas pressures
    Forced-draft fan discharge
    Furnace boiler outlet
    Economizer differential
    Air heater and gas-side differential
Unusual conditions
    Steam leaks
    Abnormal vibration or noise
    Equipment malfunctions
    Excessive makeup water
Blowdown operation
Soot-blower operation

Source: Ref. 9.

14.2.2 Package Boilers

The foregoing discussion applies to the operator of a large boiler with complete controls and one or more persons directly responsible for the boiler operation. In many cases, however, needs are met by a package boiler—a self-contained unit that generally requires little maintenance. Depending upon the quantity of fresh water used annually, water treatment may or may not be

Table 14.15 Boiler checklist for annual maintenance.

Check	Examine for:
Safety interlocks	Operability—must work
Boiler trip circuits	Operability
Burners	
1. Oil tip openings	Erosion or deposits
2. Oil temperatures	Must meet manufacturer's specifications
3. Atomizing steam pressure	Must meet manufacturer's specifications
4. Burner diffusers	Burned or broken, properly located in burner throat
5. Oil strainers	In place, clean
6. Throat refractory	In good condition
Gas injection system	
1. Orifices	Unobstructed
2. Filters and moisture traps	In place, clean, and operating
3. Burner parts	Missing or damaged
Coal burners	
1. Burner components	Working properly
2. Coal	Fires within operating specifications
3. Grates	Excessive wear
4. Stokers	Location and operation
5. Air dampers	Unobstructed, working
6. Cinder reinjection system	Working, unobstructed
Combustion Controls	
1. Fuel valves	Move readily, clean
2. Control linkages and dampers	Excessive "play"
3. Fuel supply inlet pressures on atomizing	Meet manufacturer's specifications steam or air systems
4. Controls	Smooth response to varying loads
5. Gauges	Functioning and calibrated
Furnace	
1. Fire-side tube surfaces	Soot and fouling
2. Soot blowers	Operating properly
3. Baffling	Damaged; gas leaks
4. Refractory and insulation	Cracks, missing insulation
5. Inspection ports	Clean
Water treatment	
1. Gauges	Working properly
2. Blowdown valves	Working properly
3. Water tanks	Sludge
4. Water acidity	Within specifications

Source: Ref. 9.

Table 14.16 Boiler checklist for routine maintenance.

Action	Frequency
Check safety controls	Daily
Check stack-gas analysis	Daily or more often
Blow down water in gauges	Weekly
Blow down sludge from condensate tanks	Whenever needed; frequency depends upon amount of makeup water used
Have water chemistry checked	Quarterly
Perform combustion efficiency check; log results	Daily or weekly
Check and record pressures and readings from boiler gauges	Daily or weekly

critically important. Most of the important maintenance procedures are covered in the operating manual that comes with the boiler, and this manual should be kept available and up to date. Some of the more important procedures for keeping a package boiler running are described below.

1. *Safety and relief valves* should be checked occasionally to see that they work and that they will reseat properly. They should be checked carefully to avoid excessive steam loss and scaling. If any such valve fails to work properly, the fact should be noted and the valve fixed by a boiler service representative at the first opportunity.
2. *Air supply* should be kept open so that the boiler can have enough combustion air at all times. Restricting combustion air by blocking the boiler air openings or by blocking all air openings into the boiler room can create a buildup of carbon monoxide and/or cause the boiler to operate inefficiently.
3. *Low-water and high-water gauges and controls* should be flushed periodically to remove sludge. If sludge builds up in a float-operated valve, the valve can fail to operate, with expensive consequences to the boiler. The gauges should be flushed periodically and should not be allowed to get rusty or clogged.
4. *Combustion controls* should be inspected regularly and adjusted if sooting or burner wear is taking place. Sooting causes a great drop in boiler efficiency (see Table 14.12), and burner wear can cause irregular firewall wear, inefficient combustion, and the need for increased boiler maintenance. It is generally worthwhile to have your combustion controls inspected at least annually.
5. *Tubes* can be cleaned if scaling occurs by removing head plates at either end and running a special brush through each tube. For detailed procedures and necessary safety precautions, see your local boiler service representative. Tubes can be replaced if necessary, but the job calls for special skills and should be done by someone trained for this job.

14.2.3 Steam Traps[8]

A steam trap is a mechanical device used to remove air, carbon dioxide, and condensed steam and to prevent steam from flowing freely into the outside air from steam distribution systems. Steam traps are necessary for several reasons. If air is not removed from steam, the oxygen can dissolve in low-temperature steam condensate and help cause corrosion of the valves, pipes, and coils in the steam distribution system. If carbon dioxide is not removed, it can combine with steam condensate to form carbonic acid, another major source of corrosion in the steam distribution system. Air and carbon dioxide also act as insulators to impede heat transfer from the steam; their presence creates a partial pressure that lowers the steam temperature and the heat-transfer rate.

Perhaps the main function of steam traps, however, is to permit the removal of steam condensate from a system while simultaneously preventing the free escape of steam. In this last function, the energy of the steam is kept within the system, and the amount of the live steam within the facility is controlled.

Steam traps occur as parts of nearly every steam distribution system. They are often not maintained, and this lack of maintenance can create a hidden cost that is significant. The cost of a steam trap failure is dependent upon the failure mode, with a strong dependence upon the original design. The two main failure modes are failing open and failing shut, and the design consideration of most maintenance interest is proper drainage. Consider these problems in turn.

Problems if Trap Fails Open

If a trap fails open, live steam flows directly from the steam system through the trap, a pressure buildup is caused in the condensate return system and the condensate return lines are heated unnecessarily, or the steam is released directly to the air. In the first case, the back pressure in the condensate lines may cause other steam traps to fail, and the condensate may not be removed from the steam distribution system. In the second case, steam discharging to air costs as much as any other kind of steam leak and creates pressure losses. These costs can be estimated using Table 14.17 and the formula

$$C = \text{waste} \times \frac{1100\,\text{Btu/lb}}{\text{boiler eff.}} \times \frac{\text{energy cost/million Btu}}{1,000,000}$$

where
C = energy cost per month due to steam loss
waste = number of pounds of steam wasted per month, from Table 14.17
1100 Btu/lb = approximate value for total heat in saturated steam at 100 psi
boiler eff. = boiler efficiency, usually 60 to 80%
energy cost/million Btu = cost that can be obtained from your fuel supplier

These figures are close estimates for steam at 100-psia pressure; for other pressures between 50 and 200 psi, the cost is nearly proportional to the pressure, using 100 psi and Table 14.17.

Table 14.17 Cost of Steam Leaks at 100 psi Assuming a Boiler Efficiency of 80% and Input Energy Cost of $2 per Million Btu

Size of Orifice (in.)	Steam Wasted per Month (lb)	Total Cost per Month	Total Cost per Year
1/2	835,000	$2,480	$29,760
7/16	637,000	1,892	22,704
3/8	470,000	1,396	16,752
5/16	325,000	965	11,580
1/4	210,000	624	7,488
3/16	117,000	347	4,164
1/8	52,500	156	1,872

Problems if Trap Fails Shut

If a steam trap fails shut, it acts like a plug in that part of the steam system. Condensate builds up and cools, heat transfer stops, and noncondensable gases dissolve in the water and create corrosion in the pipes and in the equipment served by the trap. If a blocked trap causes condensation to build up in a line containing active steam, the steam may push water ahead of it to form a water hammer. To visualize the effects of water hammer, think of water at 90 miles/hr colliding with the inside of a pipe. Severe damage can result.

Problems of Improper Drainage

If steam condensate is not drained properly, weight accumulates, heat transfer stops, and freezing may result. Water weighs 62.4 lb/ft^3, steam at 100 psi about 0.26 lb/ft^3. If a 6-in. pipe is filled with condensate rather than steam, it is carrying 12.25 lb per linear foot rather than 0.05 lb. The heat transfer problem has been presented earlier. Another problem may be the freezing of pipes, particularly if the system depends upon proper operation of the steam trap to prevent freezing.

There are four basic types of steam traps, and each requires a different troubleshooting procedure. The following sections show how each trap type operates, what can go wrong with it, and what troubleshooting procedures are recommended.

Open Bucket Traps

In the trap shown in Figure 14.6, the condensate enters through the top. Since the entering hole is at the side of the bucket, condensate flows into the body of the trap, raising the bucket and causing the valve to plug the condensate relief hole. When the bucket has raised as far as the trap frame will allow, condensate fills up the trap and finally overflows into the bucket, decreasing its buoyancy. When enough water has flowed into the bucket, it drops, opening the condensate return valve. Steam pressure then forces the condensate into the condensate return system. Because of the design of the bucket, some of the condensate within the bucket is also emptied at this time, but the bucket is never completely empty, thus maintaining a water seal over the outlet and preventing steam from blowing through the trap. The cycle then repeats.

The advantages of this kind of trap are its reliability and its general ease of servicing. It can develop leaks in the bucket, and pivot wear or the accumulation of dirt can make it unworkable.

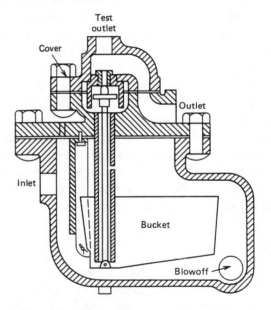

Fig. 14.6 Open bucket steam trap. (Courtesy of Armstrong Machine Works.)

Open Bucket Traps, Troubleshooting and Maintenance. The operation of an open bucket trap is characterized by its regular opening and closing. This creates a noise which can be detected with an industrial stethoscope. When a trap is first installed, the frequency of clicking should be noted on the body of the trap. If this frequency changes drastically without a corresponding change in steam pressure, the trap should be taken apart (after isolating it from live steam !) and overhauled. If the trap is not making noise and is cold, determine whether condensate is getting to the trap. If no condensate or steam is reaching the trap, check strainers, valves, and lines upstream of the trap to discover where the blockage is occurring. If condensate is reaching the trap, check the pressure of the steam coming into the trap. If this pressure is too great, the trap will not work. If the pressure is correct and the trap is not working, valve off the trap so that no live steam is coming from either the steam distribution system or the condensate return and take off the cover or remove the trap from the line. Look for dirt, worn parts, or worn or plugged orifices. It is also possible that some fluctuations in steam pressure caused the trap to lose the small amount of water in the bucket necessary for its operation (the *priming*). This condition can be detected if the trap is blowing live steam. Close the inlet valve for a few minutes to let condensate build up, then let this condensate into the trap and it should work.

Another problem that this trap can have is caused by the need to have the priming water in the trap at all times. The trap is automatically vulnerable to freezing if the external temperature gets cold enough and if the supply of steam is shut off. If such freezing occurs, many traps have provisions for the admission of live steam to thaw the ice.

Note: There is a difference between *flash steam* and *live steam*. Flash steam is formed when condensate at a high temperature is exposed to air at a lower temperature, and part of it "flashes" to steam. Flash steam has very little pressure and gives an irregular, undirected flow pattern. Live steam however, is steam under pressure and has a well-defined, consistent pattern. The difference is significant, since the presence of flash steam is

expected, whereas live steam gives an indication that something is wrong.

Inverted Bucket Traps

The inverted bucket trap (Figure 14.7) works like the open bucket trap described above, with two exceptions. First, the trap is opened and closed by steam lifting a bucket rather than by condensate lowering it. Second, the design of the trap is self-cleaning—dirt is scoured from the bottom of the trap automatically as a result of the trap design. Thus there is less likelihood that the trap will become plugged up with dirt, although a large dirt particle could still keep the bucket lifted up or the valve jammed open. Both of these problems can be alleviated by installing and maintaining a strainer upstream of the trap. This trap is maintained using the procedures outlined for the maintenance of the bucket trap described above. This trap can also freeze or lose its prime.

Float and Thermostatic Traps

The float and thermostatic (F&T) steam trap is often used where steam pressures vary over a wide range and where gravity may be the only force available to remove condensate. This trap, shown in Figure 14.8, has a float-controlled valve to let condensate out and a thermostatically controlled valve, shut at condensate temperatures but open at lower temperatures to let out air and other nondissolved gases. It can be used where continuous condensate flow is desirable, unlike the bucket

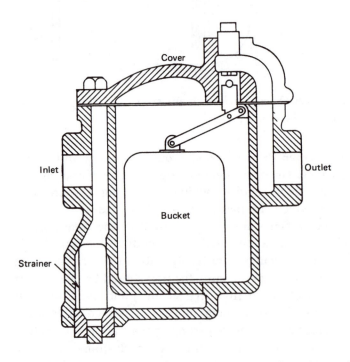

Fig. 14.7 Inverted bucket steam trap. (Courtesy of Armstrong Machine Works)

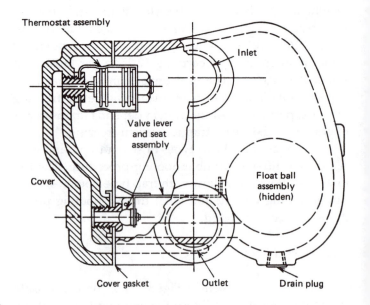

Fig. 14.8 Float and thermostatic steam trap. (Courtesy of Armstrong Machine Works.)

traps, which yield only intermittent flow.

Float and Thermostatic Traps, Troubleshooting and Maintenance Procedures. In case of no steam flow, the float may have developed a leak or have collapsed, or the air valve may be plugged. The mechanism may be worn and in need of replacement. If the trap is blowing live steam, the air hole may be blocked open because of wear or by a piece of scale, or the trap may be filled with dirt. To check any of these, valve off the trap and examine or remove the trap. Freezing can also be a problem.

Disk Traps

Another common type of steam trap is the disk trap, one example of which is shown in Figure 14.9. When condensate or air enters from the left, it pushes up the disk and flows out through the passage on the right. When steam enters the trap, it passes around the edge of the disk and creates a downward pressure on the top of the trap. When this pressure is great enough, the disk drops and the trap closes. When the trap closes, the disk is no longer heated by the steam, and the steam above the trap cools, giving lower pressure against the top of the disk. When this pressure is lower than the pressure from underneath, the trap opens and the process repeats itself.

Disk Traps, Troubleshooting and Maintenance Procedures. The contact area between the disk and its seat can become corroded or blocked, and this can cause a problem. Other problems may include installation backward to the proper direction of steam flow. To maintain this trap, first check the installation to see if the manufacturer's directions have been followed. Then, if installation is correct, valve off the trap from live steam, remove the top, and check on the condition of the disk and its contact surface. Unlike the bucket traps, this trap has no prime, so losing its prime is not a problem.

Thermostatic Traps

This trap operates on the general principles illustrated in Figure 14.10. In this trap, the bellows or bimetallic strips are extended when hot and contracted when cold. Since condensate is cold (relative to steam), its pressure causes the tap to contract, opening the outlet and allowing condensate and noncondensable gases such as air and carbon dioxide to escape. When these have left the trap, steam comes in and heats up the bellows or bimetallic strip. This expands, closing the outlet. As steam condenses, the quantity of condensate builds up, the valve opens again, and the cycle repeats. This trap can be used to give a continuous flow of condensate if desired.

Thermostatic Traps, Troubleshooting and Maintenance. Depending upon the mounting of the trap, dirt and scale may not be a problem in the outlet. If the bellows leaks and fills with condensate, the trap may fail instead. If a bimetallic strip is used instead of bellows, corrosion can be a problem. Like most other traps, failure can be caused in this trap if scale or dirt gets between the valve and its seat. Also, the dirt screen can get plugged up, causing the trap to fail closed.

Steam Distribution System, Maintenance Frequency

The steam distribution system, including piping, steam-heated equipment, steam traps, and the condensate return system, should be checked regularly for leaks and to ensure correct operation. The frequency of inspection should vary with the age of the installation—in a new system, dirt and other foreign matter in the pipes will be deposited in strainers, steam traps, and the bottom of steam-heated equipment. After the installation of a new system, it thus makes sense to check steam traps, strainers, and drains frequently, perhaps every 3 months. As the system gets older, scale can become a problem if the water is not properly treated. If scaling is a problem, the steam distribution system will have to be maintained more often than without scaling—semiannually rather than annually, for example, on steam trap

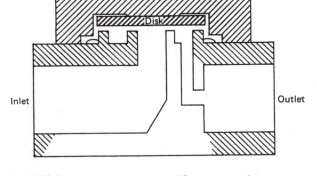

Fig. 14.9 Disk-type steam trap. (Courtesy of Armstrong Machine Works.)

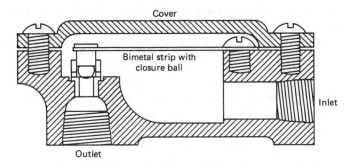

Fig. 14.10 Thermostatic steam trap (bimetal).

inspection. The entire steam system should be inspected carefully at least once each year for leaks and to keep steam traps and steam-heated equipment operating correctly. When this inspection and maintenance procedure is followed regularly, the amount of steam needed will be reduced, and the cost savings will be substantial.

14.3 MATERIALS HANDLING MAINTENANCE

Much energy is used in materials handling. Proper maintenance can reduce this amount of energy in the operation of lift trucks, conveyors, and the ancillary batteries, motors, fans, and tires. The proper maintenance actions are described in great detail in the operating and maintenance manuals of equipment manufacturers, and much of this material is condensed from those publications. For detailed instructions on the maintenance of a particular equipment item, refer to the manual for that item. The material in this chapter can, however, be used to develop check-off lists and to give a good idea of the types and amounts of maintenance required.

14.3.1 Electric Lift Trucks[11,12]

The general maintenance principles that apply to an electric lift truck are (1) keep the battery charged, (2) make sure that all fluid levels are where they should be, (3) check the brakes and keep them working correctly, (4) inspect the electrical equipment and keep it running, and (5) lubricate the truck according to manufacturer's specifications as follows.

Battery
The battery should be inspected at least weekly with specific attention paid to the specific gravity, the contacts, and the electrolyte level. The specific gravity should be between 1.250 and 1.275 and should not vary more than 0.020 between cells. Special care must be taken to see that the charger is connected up with correct polarity and the correct voltage, and smoking should be prohibited near the charging so as not to ignite the hydrogen gas liberated during charging. If the battery fails to hold a charge, a test with a hydrometer can often detect whether a particular cell is at fault; if no cell is clearly bad, have a load test performed by someone who knows the necessary safety precautions.

The battery contacts should be inspected daily for corrosion. At the same time, the wires leading to the contacts should also be inspected for cracked insulation and for other signs of wear. The level of electrolyte should be inspected daily and any deficiencies made up using distilled water.

Hydraulic Components
The level of hydraulic oil should be inspected daily and filled if it is below the lower mark on the dipstick. It should be changed at least every 2000 hours. As the daily hydraulic level inspection is being performed, the operator should also check all hydraulic lines for leaks and fix them (or have them fixed) immediately.

The brakes should also be checked daily to see that the shoes are not dragging and that the brake pedal has the right amount of play and resistance to foot pressure. Brake shoes should be inspected and adjusted at least once a year.

Tires
If the tires on a particular lift truck are pneumatic, it is important that they be kept inflated to the pressures recommended in the operating manual and that the inflation be checked at least weekly Underinflation can lead to excessive wear. Underinflation also increases electrical consumption. If the tires are made of hard rubber, they should be inspected daily so that the operator knows their condition. If a tire has large chunks out of it, it may constitute a safety hazard, and the irregular ride that this problem generates may be damaging to the material being transported. If tires show irregular wear, the front end may be out of line—this should be checked and fixed.

General Lubrication
General lubrication should follow the schedule and procedures laid out in the operation manual for the particular lift truck being maintained. This lubrication should also include inspection, adjustment, and oiling of the hoist chain.

14.3.2 Lift Trucks, Non-electric

The main points to maintain on a lift truck powered by diesel, gasoline, or LPG are the engine, the hydraulic system, the transmission, and the braking system. Batteries are important, but their care is not as central to the maintenance of these lift trucks as it is to the maintenance of electric-powered trucks.

Engine
Special attention should be paid to the oil level, the radiator, and fan belts. The oil level should be inspected daily, and the oil and oil filter should be changed every 150 hours (unless the environment is very dirty or muddy, in which case the oil should be changed more often). The oil level on the dipstick should be between the high and low marks—if it is too low, more should be

added. At the same time this check is being made, the operator should look for any oil leaks. The operator should also try to keep track of the frequency of adding oil as warning of unseen leaks. The coolant level in the radiator should be checked before each day's work, either by looking at the coolant level in the recovery bottle or by carefully removing the radiator cap to check the level there. Fan belts should be checked periodically. If they are too tight, bearing wear will be excessive; if they are too loose, insufficient cooling and/or generator power will take place. More details on the maintenance of diesel engines are given in Section 14.4.

The Hydraulic System

If the hydraulic system fails on a lift truck, people can get hurt. Any hydraulic pumps that are run without fluid can burn out, causing a significant and usually avoidable expense. The way that such an expense can be avoided is by careful checking of the hydraulic levels daily and by daily inspection of all hydraulic connections for leaks.

Transmission

The fluid level in the transmission should also be checked daily. This minor inconvenience can save the cost of a new transmission and the loss of the vehicle during the repair time.

Brakes

Part of the daily check should be a quick check of the foot brake and the parking brake. The footbrake pedal should move a short distance—1/4 to 1/2 in., depending upon the truck model—before the brake engages, and the brake should respond firmly to pressure. The parking brake should be checked to see if it provides the braking necessary to act as a backup for the foot brakes. Air brakes should have the drain cock opened until no more water escapes with the air.

14.3.3 Conveyor Systems

Conveyor belts, in-floor towlines, and similar chain- or belt-driven equipment have three important areas where proper maintenance can save energy and money. These are in controls, drive motors and gear, and the actual moving belt or line.

Conveyor Controls

One way to save energy on controls is to install and maintain controls that cause the conveyor to move only when there is material to be moved. This measure has saved significant amounts of energy in coal mines and in

pneumatic conveying systems, for example, where in one case the total energy consumed was reduced by 90%. The additional wear and tear on starting motors was not found to be significant when these motors were maintained according to the original procedures specified at the time of purchase of the equipment. Adding a load-related starting control saves on conveyor wear and on each part of the equipment that is no longer operating continuously.

Drive Motors and Gears

Drive motors should be checked regularly using the Tornberg procedure for motors described in Section 14.1.1. When the motor is being inspected, the rest of the driving gear should also be inspected, with the details being dependent upon the power transmission system being used by the conveyor.

Conveyor Belts or Lines

The belt, towline, or other equipment used to move the material should be checked regularly. The kinds of problem to include in the inspection are worn belts; noisy rollers, indicating possible bearing problems; belts that are too tight or too loose; and any place that shows conspicuous wear. Such an inspection should be performed regularly and coupled with regular lubrication of appropriate points. Manufacturing specifications should be consulted as a source for specific lubrication directions for all conveying equipment.

14.4 TRUCK OPERATION AND MAINTENANCE[13]

Most organizations use trucks in some form, whether for coal haulage, parcel delivery, or earth moving. Wherever trucks are used, there are opportunities for saving energy and money. The three general rules for achieving these savings are: (1) match the equipment to the load, (2) keep the equipment tuned up and in good repair, and (3) turn it off if you do not need to have it running. The following section gives more details on these rules as they relate to general operation and to the operation of diesel and gasoline engines. The information presented here has been distilled from manuals produced by Cummins, Caterpillar, Detroit Diesel, and Hyster.

14.4.1 Management Decisions

The person supervising a company with a vehicle fleet can have a large impact upon the costs of running that fleet in at least five different areas: (1) matching the vehicle to the job, (2) maintenance of the vehicle yard, (3) instituting a system of fuel accountability, (4) taking advantage of unavoidable delays to create more efficient operation, and (5) keeping all vehicles tuned up.

Matching the Vehicle to the Job

This takes two forms. First is the policy that assigns small passenger or pickups to personnel on one-person trips; delivery vans for local deliveries, where the demand calls for more capacity than a pickup provides; and semi-trucks only where justified by the amount of material to be moved. Within each category the same rationale is applied—use the larger vehicles only when their use is necessitated by the load. Once the vehicles have been chosen or assigned based on their intended use, the second rule can come into play—if you do not need all the power of the vehicle now, see if it can be modified so that the maximum power used is lowered. This can happen where large earth-moving or transporting equipment has been designed for a 10% grade and where actual conditions do not have more than a 7% grade. If the actual requirement is less than the equipment was designed for, the governor or the fuel pump can be modified to limit the horsepower available to the operator.

Maintaining the Vehicle Yard

Mud resists vehicle travel and makes maintenance into a chore. The effect of mud, whether it is on a haul road or in a vehicle yard, can be seen in Figure 14.11. If a significant part of your fleet travel is along roads under your maintenance or in a yard you control, this figure can help you decide whether to surface the road or yard or how often to grade the road. Particularly where your vehicles have 30% or more of their travel off the road, the savings can be substantial.

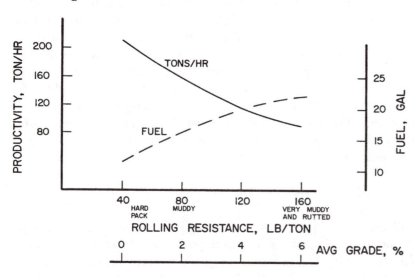

Fig. 14.11 Gasoline consumption as a function of rolling resistance.

Instituting Fuel Accountability

If your company does not have a system accounting for the use of every gallon of fuel, you probably could benefit from instituting such a program. The program has two benefits: It enables you to determine which of your vehicles is using the most fuel, and it decreases losses to employee vehicles. If you can locate the vehicles that are least efficient, you can start the process to find out whether something is wrong with the vehicle, whether some operating procedures need to be changed, or whether the particular design is inefficient. In any case, you have a better idea of where your vehicle fuel is going.

Taking Advantage of Bottlenecks

If your facility has certain inherent bottlenecks, such as a power shovel that has less capacity than the trucks serving it, it may be possible to have these work to your advantage. In the example of the shovel, fuel can be saved by slowing down on the return trip from the dumping area, using the time that would be wasted waiting for the shovel in the additional time along the haul road. Make adjustments to the vehicle engine if operators are unwilling to conform to these practices.

Keep Vehicles Tuned Up

When a 1% improvement in vehicle efficiency can create a savings of $1 per day for each vehicle, it makes sense to keep vehicles tuned up. Keeping a vehicle fleet in tune saves three ways: in energy consumption, in frequency of downtime and the repairs involved, and in the cost of repairs that do have to be made. The tuneup procedure for each vehicle is prescribed by the manufacturer.

14.4.2 Daily Maintenance Practices

In addition to the general maintenance practices described above, daily procedures performed by the operators can create substantial savings in fuel and operating costs. In addition to the mandatory inspection of such safety items as brakes, the horn, and lights (if appropriate), a daily inspection should be made of tires, batteries, air hoses, fuel and lubricant levels, and air filters.

Tires

Proper inflation of tires reduces rolling resistance and thus reduces energy consumption. Figure 14.11 shows a typical curve relating inflation and fuel efficiency. In addition to the fuel savings, proper inflation increases tire life and decreases the total cost per mile of running each vehicle.

Batteries

If one cell of a battery is weaker than the other cells, the generator uses more power than if all cells were in good condition. This additional power cuts into fuel efficiency. To avoid this problem, the water level in each cell should be checked daily and filled with distilled water whenever the level is below the prescribed fill mark, and the battery voltages should be checked frequently.

Air Hoses

These should be checked for leaks as soon as they are under pressure. Air leaks cause the air compressor to use more power than necessary and create another waste of fuel.

Fuel and Lubricant Levels

It is generally recommended that fuel tanks not be filled to more than 95% of the tank capacity. The extra 5% is to allow for expansion of the fuel as it heats up. If the tanks are filled to the top, this expansion can cause an overflow of fuel, a waste and possible safety hazard. Lubricant levels should be checked routinely to detect leaks and to keep proper lubrication in the engine, transmission, and other systems.

Air Filters

An engine uses many gallons of air for every gallon of fuel burned, so keeping the air clean is critical. If the air filter gets clogged, the turbocharger (if the vehicle has one) must use additional amounts of energy and thus extra fuel is consumed; if the air filter gets too dirty, the engine will not run. Also, particles of the filter or dirt particles may be drawn into various parts of the engine, causing unnecessary wear.

14.4.3 Operating Practices

In addition to the daily maintenance check described above, each operator can do several other things to decrease the fuel consumption of his or her vehicle. These measures include eliminating unnecessary idling, warming the engine correctly, choosing the most economical speeds for a given load, keeping the hydraulic system above stall speed, and keeping the cooling temperature within specified limits. The following paragraphs discuss these actions in more detail.

Eliminating Unnecessary Idling

The three largest makers of diesels are unanimous in condemning unnecessary idling as a waste of energy and as hard on the engine, although some idling is necessary.

If equipment will be idled for more than 10 to 15 minutes, the cost for shutdown and restarting the engine will generally be less than the cost of fuel for idling. The engine should not be shut down immediately after a hard run, however, but should be allowed to idle for a few minutes to allow engine temperatures to drop. Otherwise, a surge of hot coolant to the heads can cause damage. Also, turbocharger damage can occur if coolant oil and air flow are shut off too abruptly.

Warming Up the Engine

When starting, the engine should be warmed so that the coolant is in the recommended operating range before putting a load on the vehicle. This is recommended for diesel engines. For gasoline engines, check with a reputable dealer to find whether this warm-up period is necessary.

Choosing the Most Economical Speed

Every vehicle operator should know the most economical rpm range to achieve good fuel economy while delivering the desired horsepower. The best rpm range can be determined from performance curves, such as Figure 14.12, and from the fuel consumption map, such as Figure 14.13. If these curves describe the engine operated by a particular operator, that person should know

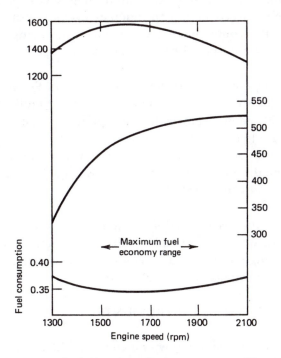

Fig. 14.12 Typical diesel performance curve. (Courtesy of Cummins Engine Company.)

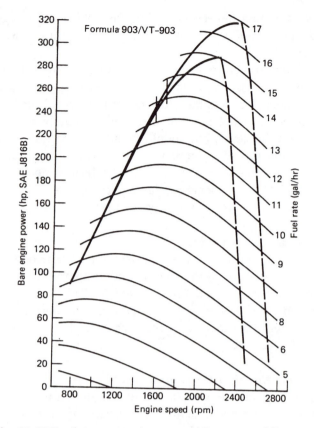

Fig. 14.13 Fuel consumption map. (Courtesy of Cummins Engine Company.)

that operating at 1700 rpm rather than 2300 rpm saves approximately 8% in fuel. Similarly, mechanical drive equipment should be operated in the highest gear practical. Since the economical operation of a vehicle fleet may enable the company to stay in business, saving this amount of fuel should be important to any operator.

Operating the Hydraulic System at the Right Speed

Care should be taken to operate the hydraulic system with enough power that the pumps do not stall. Stalling a hydraulic pump creates heat but does not get work done. Similarly, if operating a converter, the operator should downshift to get more power rather than allow the converter to stall.

Maintaining Recommended Coolant Temperatures

The operator should keep the coolant temperature in the range recommended by the manufacturers in order to achieve maximum fuel economy.

14.5 MEASURING INSTRUMENTS

Instruments serve two functions in maintenance—checking equipment condition and on-line monitoring. It is often impossible to know whether some equipment is

functioning properly without measuring a temperature, pressure, or other parameter associated with the operation of the equipment. For example, the only indication that a steam trap is not working properly may be the absence of a regular clicking as the trap door opens and closes, and this noise may be detectable only with a stethoscope. A similar use of instruments is checking on the effectiveness of repairs. Another function of instrumentation is to observe major system operating parameters so that equipment can be operated with the minimum effective use of energy. Instrumentation for this purpose is generally mounted permanently, with gauges readable either at the equipment or at some central location.

Measuring instruments used in maintenance can be classified by type or by function—by type according to what they measure, by function according to what system they serve. A classification of instruments into these categories is shown in Table 14.18, followed by descriptions of each instrument in the table.

14.5.1 Electrical Measurements

Electrical measurements can be used to assess the condition of individual items of equipment and to analyze the energy consumption patterns of the entire facility. Troubleshooting and monitoring of individual units can be accomplished with portable wattmeters, multimeters, and power-factor meters. Many of these instruments are available in recording models also for permanent installations for monitoring changes in overall consumption. Safety precautions must be known and carefully observed in the installation and use of all equipment, particularly that involving electricity.

Multimeters

A multimeter is capable of measuring voltage, current, and resistance. These meters come in many ranges. One of the more common types is the clip-on meter capable of measuring 0 to 300 A, 0 to 600 V, and 0 to 1000 Ω. This meter typically sells for approximately $50 and is easily carried in a belt holster. It can be used to check many parts of the electrical system and to make electrical checks on motors.

Power-Factor Meter

This meter is used to determine the ratio between the resistive component and the total (resistive and inductive) electric power supplied. It can be used to check the power factor on equipment and to determine whether a power-factor-improvement program is working.

Table 14.18 Instruments for use in energy management.

System	Instrument	Portable	Mounted	Parameter Measured
Building envelope	Infrared photography	×		Heat loss
Boilers	See Chapter 5 and Section 14.5			
Steam traps	Thermometer (sensitive)	×		Temperature differences between inlet and outlet
	Stethoscope	×		Opening and closing noise
Heating, ventilating, and air conditioning	Flow hood	×		Air flow rates
	Pitot tube	×		Air flow rates
	Inclined-tube manometer		×	Pressure differential between two points
	Bourdon tube	×	×	Pressure
	Pocket thermometer	×		Temperature of rooms, pipes, etc.
	Orifice flowmeters		×	Flow rates of air or steam
	Psychrometer	×	×	Humidity
Electrical	Ammeter, recording		×	Electricity usage, peaks
	Ammeter, clip-on, with probes	×		Voltage, current, resistance
	Wattmeter, recording		×	Power consumption
	Power-factor meter	×	×	Electrical power factor
Lighting	Industrial lightmeter	×		Light levels
Hot water	Thermometer	×	×	Temperature
Air compressors	Pressure gauges		×	Oil pressure, air pressure
	Stethoscope with gauge	×		Bearing wear on motors
	Multimeters	×		Voltage balance on three-phase motors
	Strobe light	×		Motor vibration
	Infrared film and camera	×		Bearing wear

Recording Ammeter

This records the electrical consumption by hour for a prescribed period. Its most important function is to enable someone to determine the magnitude and timing of peak load demand and the magnitude of the underlying base load (the equipment that is on all the time). A recording ammeter can also help determine the effectiveness of an ongoing energy management program. It can be used to assist operating personnel in shifting peak loads and in determining the magnitude and timing of any secondary peaks. Such a meter, installed, costs $200 to $1000 and can provide a payback period of one to two months. Similar meters are available to record watts, volts, and power factor.

Wattmeter

A wattmeter is used to determine directly the amount of power used by a piece of equipment or by a facility. It can be used to analyze electrical consumption by enabling maintenance personnel to first determine the total power being used in an area, then find what equipment is using the power, then turn off machines and shift loads from peak periods to off-peak periods.

14.5.2 Light Measurements

Reducing unnecessary lighting has psychological impacts that can far exceed any direct impact on energy consumption. To convince people that their areas are overlit or underlit, however, it is necessary to have well-established lighting standards and a way to compare existing conditions against those standards. The standards have been provided by the Illuminating Engineering Society in the *IES Handbook* and are discussed in Chapter 13 of this book.

Industrial Light Meter

This instrument is typically designed to measure incident light directly and can be used to measure reflected light and light transmittance as well. An industrial light meter differs from a photographic light meter in two ways: (1) it reads directly in footcandles in scales that are approximately linear rather than the logarithmic scales used in photographic light meters; (2) it is designed to receive light from wide angles rather than focusing on a particular part of the viewing field. Such a meter is small, light, and inexpensive, and sufficiently useful that it should be part of the equipment of every maintenance supervisor.

14.5.3 Pressure Measurements

If equipment is not operating in its proper pressure range, it can be damaged or the equipment it serves can be damaged. This is true whether the pressure being measured comes from oil (compressors or lift trucks), fuel (boilers), steam (boilers), or other sources. Measuring pressure using portable equipment generally requires that the equipment to be measured be equipped with a fitting specifically designed for pressure-sensing equipment. With such fittings in place, it is possible to use a Bourdon tube or a diaphragm gauge for routine inspections. Where information is to be available at any time, or when the readings are to be taken so often that portable pressure-sensing equipment becomes a nuisance, it may be desirable to permanently install a Bourdon tube or a diaphragm gauge, or to rely on a manometer.

Bourdon Gauge

This common pressure gauge consists of a curved tube closed at one end with the other end connected to the pressure to be measured. When the pressure inside the tube is greater than the pressure outside the tube, the tube tends to straighten, and the amount of change in length or curvature can be translated directly into a gauge reading. Such gauges are available in many pressure ranges and accuracies.

Diaphragm Gauge

If the pressure inside a bellows or on one side of a diaphragm is greater than the pressure outside the bellows or on the other side of the diaphragm, the bellows or the diaphragm will move. The amount of movement is related to the pressure between the inside and outside the bellows (or on the different sides of the diaphragm). These pressure gauges are also very common.

Manometer

If a glass tube has liquid in it, and if one end is open to the air and the other end is exposed to a pressure other than air pressure, the end with the higher pressure will have a lower liquid level. This kind of

gauge can be mounted across a filter bank to indicate when a clogged filter is causing pressure to build up, or it can be mounted in a place where a much higher pressure is to be measured. If the tube is inclined, a smaller pressure difference is detected, as in the inclined-tube manometer. These gauges are easy to read, easy to maintain (the glass must be kept clean, and the inlet and outlet holes must be kept clear of debris), and inexpensive.

14.5.4 Stack-Gas Analysis

As explained in Section 14.2.1, proper maintenance of boilers depends heavily on knowledge of stack-gas composition. If there is too much molecular oxygen, the boiler is operating inefficiently; if there is too much carbon monoxide or too much smoke, the boiler is operating inefficiently and creating an operating hazard. Thus it is important to keep track of O_2, CO, and smoke. Three types of monitoring equipment meet these needs: Orsat kits, permanently mounted meters, and smoke detectors.

Orsat Apparatus

The Orsat analysis kit consists of three tubes filled with potassium hydroxide, potassium pyrogallate, and cuprous chloride, respectively. Flue gas is introduced into all three tubes, and the amount of carbon dioxide, oxygen, and carbon monoxide removed in the first, second, and third tubes, respectively, indicates the proportion of those gases in the flue gas. Any gas remaining is assumed to be nitrogen. This apparatus is capable of accurate readings when representative samples of flue gas have been taken from different points in the flue stream, when the apparatus does not leak, and when the operator is well trained in its use.[14]

Smoke Detectors

Smoke detectors work by comparing the amount of light going through a sample of smoke with some standard shades. If a Ringlemann scale is used, the smoke number is between 1 and 4, if the Bacharach scale, between 1 and 9. Smoke detectors can be either portable or permanently mounted.

14.5.5 Temperature Measurement

Temperature measurements are essential to energy management and proper maintenance in at least four situations: for comfort, to determine where heat is leaking from a building, to define abnormally hot areas in a machine, and to use in the analysis of boiler operations and of industrial operations using process heat. Temperatures in these situations call for instruments such as a pocket thermometer, infrared photography, and permanently mounted devices possibly using thermocouples.

Pocket Thermometers

Every person who must set thermostats in a building and listen to the complaints of uncomfortable people needs one of these. It is possible to get a rugged thermometer, small enough to fit into a shirt pocket and accurate to within $\pm 1/2°F$ between 0 and 220°F, for $12 to $16. Such a thermometer can be used to calibrate thermostats, to check complaints, and to add to the professional image of the wearer. By placing the heat-sensing end on a hot pipe, it is possible to estimate the temperature of the material in the pipe or to determine that it is beyond the temperature range of the thermometer. This thermometer can also be used to check the temperature of water coming from various points of a culinary hot-water system. The thermometer is small and inexpensive, and its value as a tool in energy management makes it indispensable.

Infrared Equipment

This equipment works by sensing infrared radiation, a kind of radiation given off by warm materials. Infrared equipment, whether photographic or using a television-like device, senses differences in temperature. It can therefore detect heat leaks that are not revealed by a visual inspection. Infrared examinations have been useful in detecting areas of major heat loss in buildings or between buildings where steam leaks were occurring. Such inspections have also proven valuable in detecting areas of unseen friction in motors and thus finding problem areas before the problems have become major.

Thermocouples[15]

To create a thermocouple, two wires or strips of different material are joined at both ends. Then one end is kept at a constant temperature and the other end is used in the temperature probe. If the ends are at different temperatures, a voltage difference will occur between the ends. Measuring this voltage provides a way to estimate the temperature difference between the two ends. Instruments using this principle are found in many places, particularly where a permanent meter is desired for temperature in a remote or inaccessible place.

14.5.6 Velocity and Flow-Rate Measurement

In many situations, it is necessary to know the flow rate of some substance, such as air or steam, to determine where energy is being used. For example, a factory with several buildings and a central steam system may not be metered so that the steam consumption of each building can be determined. As another example, it is generally difficult to estimate the amount of air being moved by a fan in an air-conditioning system without making some measurement of air velocity. Three types of measuring instruments used are flow hoods, pitot tubes, and orifice plates.

Flow Hoods

A flow hood resembles an inverted pyramid with the top replaced by a small cube. The inverted pyramid is made of cloth treated to minimize air leaks. The small cube is a turbine that generates current which is measured by an attached meter. In practice, the opening of the hood is placed over the grill emitting the air. Air is forced to the base of the pyramid, turns the turbine, and generates electricity. The meter is calibrated in ft/min (or m/sec). Since the cross-sectional area is known at the point where the velocity is being measured, the number of ft^3/min (or m^3/sec) can be immediately calculated.

Pitot Tubes

The pitot tube operates on the principle that air flow across the end of an open tube creates a pressure drop, and a measurement of this pressure drop can be converted into a measurement of the air velocity at the end of the tube. Pitot tubes have been used in applications ranging from air flow rates from a duct to determining the air speed of an airplane. They can be used to estimate the air flow velocity across a duct and may be more convenient than flow hoods for some applications. Care must be taken, however, that enough readings are made to give a representative velocity profile—one reading is not enough.

Orifice Plates[15]

An orifice plate consists of a disk with a hole of known diameter mounted in a pipe or duct with a manometer attached to the pipe upstream and downstream from the orifice plate. Since the hole in the orifice plate is always smaller than the inside diameter of the pipe, the pressure downstream of the orifice plate is smaller than the pressure upstream of the plate. This pressure difference can be used together with the diameter of the orifice and the inside diameter of the pipe and the density of the material to give a value for the velocity of the material flow.

14.5.7 Vibration Measurement

Vibration is found in most mechanical devices that move. Sometimes this noise is helpful, as in the case of the noise caused by a bucket steam trap opening and closing. Often, however, an increase in vibration of a machine is an indication that something is going wrong. Among the many instruments that can be used to check vibrations, two are of particular value in the maintenance associated with energy management: the stethoscope and the stroboscope.

Stethoscope

A stethoscope brings noise from a particular place to the ears of the observer and by isolating the source seems to amplify the noise. Two valuable uses for this are in the Tornberg procedure for monitoring motor bearings, described in Section 14.1, 1, and in the procedure for monitoring steam trap operation of steam traps described in Section 14.2.3.

Stroboscope

A stroboscope is a flashing light whose flashes occur at regular intervals that can be changed at the will of the operator. When the flashes occur at the same time that a particular part of rotating machinery passes one point, the flashes make the machinery appear to stand still. Thus any lateral vibrations of the equipment appear in slow motion and can be examined in detail. This procedure also shows cracks that open up only when the machinery is in motion. By detecting incipient trouble before it causes the equipment to be shut down, the stroboscope enables maintenance personnel to order replacement parts and to schedule corrective maintenance at convenient times rather than at the times dictated by equipment breakdowns.

14.6 SAVING ENERGY DOLLARS IN MATERIALS HANDLING AND STORAGE

The earlier parts of this chapter have dealt with various topics in energy-efficient maintenance of equipment and facilities. Two other topics of interest not covered elsewhere in this handbook are energy management in materials handling and new devices for energy cost savings. This section covers the first of those topics.

Section 14.3 discussed the maintenance of certain kinds of materials handling equipment, specifically lift trucks and conveyor systems. In addition to proper maintenance, however, there are many operating changes that together can save a large fraction of the materials handling energy cost. These cost savings can be realized if a systematic approach is followed. One approach is to answer these four questions:

1. What is energy being used for now, and is all of this use productive?

2. How much electricity, gas, and oil is used now, and can this use or its cost be reduced by modifying equipment usage?

3. Can the working hours or the locations of people be changed to reduce energy requirements without adverse impacts?

4. How can the energy management program be monitored? These questions are discussed in Sections 14.6.1 to 14.6.4.

14.6.1 Analyzing Present Energy Usage

Energy is used in three ways in materials handling and storage: to move material, to condition spaces for material, and to condition spaces for people who are moving or storing material. Energy used to move material includes fuel or electricity for lift trucks, diesel or electricity for conveyor power, or cranes and electricity for automatic storage and retrieval systems. Examples of energy to condition spaces for material includes refrigeration for vegetables, controlled temperature storage for electric components, and materials and ventilation of areas used to store volatile liquids. The energy used to condition spaces for people includes heating, cooling, and ventilating a warehouse to make it comfortable for persons working there, air conditioning a cab of a large crane, and providing a comfortable working area for the secretaries supporting the materials handling and storage functions. These three functions also interact—if gasoline is used in a warehouse lift truck, the amount of ventilation air required for personnel is 8000 ft^3/min per lift truck, a requirement that is not present for electric lift trucks. (Nonelectric trucks may, however, have offsetting advantages for individual applications.) Energy used for conditioning material storage spaces can also interact in other ways with energy used to condition space for people—for example, waste heat from a refrigerator has been used to heat office space.

14.6.2 Walk-Through Audit

The first step in analyzing energy usage in your materials handling and storage system is to find out what equipment is being used to move material and, when it is turned on, to check the conditions that are being maintained for material. Then determine where the people are located in the system and what conditions are being maintained for their benefit. To do this, three walk-through audits are recommended. The first audit is performed during working hours in an attempt to find practices that can be improved. For this audit, you should be equipped with an industrial light meter (about $20 to $30) and a pocket thermometer ($10 to $15). As you walk around the facility, you should write or record potential improvements for later analysis. These improvements can be found in answer to questions such as these:

1.   Does equipment need to be idling when no one is using it, and, if so, what is a reasonable maximum for an idling period?
2.   Is the temperature being maintained unnecessarily high or low—is it necessary to air-condition this space in the summer or to heat it in the winter?
3.   Is all the lighting necessary? Standard guidelines for warehouse lighting in the *IES Handbook*[16] range from 5 to 50 footcandles, depending upon whether the storage is in active use and upon the amount of visual effort needed to distinguish one item from another.
4.   Is a great deal of conditioned air lost whenever trucks unload at the facility?

The second walk-through audit is performed when office personnel have left for the day. The intent of this audit is to discover equipment and lights that are on but are not needed for the security of the building or its contents. This audit should focus on equipment, lights, and unnecessary heating and cooling of office spaces. The use of a night setback—setting the thermostat to 55°F at night—can save as much as 20% of the bill for office heating and cooling.

The third audit is the 2 A.M. audit, performed sometime after most people have left the building. In this audit, look for motors that are on unnecessarily, for lights that are on but are not needed for security purposes, and for temperatures that are higher or lower than they need to be.

Completion of the three walk-through audits gives a qualitative survey of the equipment being used, lighting levels and temperatures being maintained, and some operating improvements that can be instituted. This is a good start. More information is necessary, however, before an energy management program can achieve its potential, and this is provided by the next step.

14.6.3 Finding and Analyzing Improvements that Cost Money

Walk-through audits can uncover operating practices that can be improved, but a different kind of analysis is needed to discover possible capital-intensive improvements. This analysis has three parts: examination of past energy bills, use of a checklist, and economic justification of the most promising improvements. (The economic analysis is discussed in Chapter 4.)

Examining Past Bills

One purpose in examining energy bills is to determine the total amount that can be saved. If your bill for fuel for vehicles is $25,000 per year, then $25,000 is the upper annual limit on savings. Another purpose for examining bills is to find what factors are significant in the billing for your facility. If, for example, you are being charged for a low power factor on your electrical bill, power-factor improvement may be worthwhile. If your electrical bill shows a factor labeled "power" or "demand," you are being charged for the peak power you use. To find whether peak electrical demand presents an opportunity for cost savings, compute the load factor by the formula

$$\text{load factor} = \frac{\text{kWh used in billing period}}{\text{peak demand} \times \text{hours in billing period}}$$

(The billing period in days is given on the bill.) If the load factor is less than 70 to 80%, there are significant periods of high electrical usage. To determine when these occur, have your electrical utility install a recording ammeter or install one yourself. When you find the peak usage times, examine your equipment to find what causes the peaks and see if some of this can be rescheduled at off-peak times.

A third purpose for examining past bills is to establish a base for comparison for your energy management program. For your continued economic health it is essential that you manage your energy consumption. In order to find the measures that have worked for your facility, however, you need to be able to show that these measures have caused you to fall below previous energy usage. If measures you institute do not cause the consumption to change significantly, you know that you need to do more.

Possible Areas for Energy Consumption Improvement

As soon as your bills have been examined, you are in a position to examine possible areas where money invested can have large returns in energy cost containment. (Much of this material was taken, with permission, from Ref. 17.) These areas, in abbreviated form, are presented in Tables 14.19 to 14.22.

14.6.4 Monitoring

When the energy consumption base has been established and the energy management plan has begun to

Table 14.19 Materials handling energy savings: doors.

Problem	Solution
People use truck bay doors to enter and leave building	Install personnel doors Benefits include less loss of heated or cooled air
Open passages between heated and cooled areas	Use strip curtain doors if lift trucks use the passageway, unless there is a pressure difference between the areas
Truck bays open to outside air	Install adjustable dock cushions

Source: Ref. 17.

Table 14.20 Materials handling energy savings: industrial trucks.

Problem	Solution
Ventilation is used primarily to get rid of truck fumes	1. Modify trucks to reduce ventilation requirements if possible 2. Replace diesel or propane trucks by electric trucks 3. Replace with more efficient trucks
Excessive demand charge for electric power	1. Have electric trucks charged during off-peak hours 2. Install drop-in dc generators
Excessive fuel consumption	1. Reduce truck idle time to manufacturer's specifications 2. Use smaller trucks where possible 3. Use a communication system between dispatcher and trucks to reduce unloaded travel time 4. Use improved ignition and carburetion devices 5. Use more energy-efficient truck tires

Source: Ref. 17.

Table 14.21 Materials handling energy savings: hoists and cranes.

Problem	Solution
Cranes powered by compressed air	Use electrical hoists and cranes (5 hp on compressor = 1 hp on hoist)
Idling	Use switches to turn crane off when not in use for a preset time (5-10 min)
Too many moves	Package in larger unit loads
Most loads too small for full use of hoist or crane capacity	Size the crane or hoist to load moved most often
Cranes running during off hours	Turn off as part of regular closing check list

Source: Ref. 17.

Table 14.22 Materials handling energy savings: conveyors.

Problem	Solution
Excessive idling	1. Install controls to keep conveyors running only when loaded 2. Wire conveyors to light switch
Motors running at less than top load	1. Downsize motors and install slow-start controls
Motors burn out	1. Institute preventive maintenance program for motors (see Section 14.1.1) 2. Replace two-phase by three-phase motors

take effect, energy consumption must be monitored. Monitoring serves three functions: to evaluate progress, to show which measures work, and to show which measures do not work.

The first step in a monitoring program is to choose some measure of energy consumption to monitor. Each kind of energy used in the facility should be monitored, with separate graphs for gasoline, propane, gas, and electric usage (kWh) and electric demand (kW). These should be plotted for the 12 months previous to the installation of the energy management plan, where possible, to establish a base against which to judge improvement. If your facility does not have a separate electric meter or separate accounting procedures for fuel, it is probably worthwhile to install them so you can know exactly what effect your program is having on your energy consumption.

Once the energy management plan has gone into effect, the next step is to graph the consumption, month by month, to see whether the program has helped and, if so, how much. If the program has not decreased energy consumption, you can use the monitoring equipment to analyze your energy use in more detail—to find what areas or pieces of equipment are using the energy. Then concentrate on these, and start again.

When this monitoring has shown that a particular measure is particularly effective, find out why, and copy its best features elsewhere. If a measure did not work, find out why and make sure that the bad features of it are not being duplicated elsewhere.

14.7 RECENT DEVELOPMENTS

Many new devices and concepts are being marketed with the promise of saving money and energy. Some of these have the status of proven technology and have been presented earlier—infrared portable thermometers, stethoscope checks on steam traps, and transparent plastic door strips. These devices and concepts represent technology that can be of substantial value to the readers of this handbook. Other useful ideas that are in this category but have not been presented earlier in this chapter are presented alphabetically in Section 14.7.1. In addition to this list, a reader should have some criteria available to assist him or her in evaluating new techniques other than those in the list. These criteria are in the form of six questions in Section 14.7.2. Section 14.7.3 gives some credible sources of information.

14.7.1 Glossary of Recent Devices

Casablanca Fans (also known as *Panama* fans): These are the large slow-moving fans that formerly adorned general stores, before the advent of refrigerated air conditioning. They blow warm air down from the ceilings to where it is needed for comfort. They thus save thermal energy at a moderate or low cost in electrical energy.

Cost-Center Metering: How can a manager of a department know how effective his or her energy management program has been without knowing the energy reduction in his or her department? Metering at cost centers is designed to give this feedback. It has proven very effective and can be well worth its cost.

Decal Temperature Indicators: If a maintenance person can tell how hot the outlet of a steam trap is by simple visual inspection, the time required for monitoring can be reduced significantly. This is the promise of decal temperature indicators. They may have other problems—consult with your local vendors and with similar users of these decals before making a large investment in them.

Dock Cushions and Seals: The amount of air that leaks around the back of an open trailer can represent a large cost in heating or cooling. To decrease this leakage, a number of dock shelters, seals, and cushions have appeared on the market recently. If properly installed, they can reduce the leakage of conditioned air to a minimum with a corresponding savings in money.

Floating Ball Covers: The cover referred to is a collection of light balls designed to float on the top of hot liquid and thereby prevent the escape of hot vapor, These balls have been used successfully in several applications, but they can create maintenance problems.

Flue Dampers: Even when a furnace is turned off, heated air may be escaping out the flue. A flue damper is designed to overcome this problem. Only those tested and approved by the American Gas Association or such groups as Factory Mutual should be used, however, and the installation should be by a licensed contractor—*if a damper is incorrectly designed or incorrectly installed, carbon monoxide can build up, and people can be killed.*

Gas-Fired Infrared Heaters: These have been installed where mechanics needed to be kept comfortable on otherwise cold floors with much movement of cold air, and they have worked. They heat only the substances that absorb radiant heat and do not heat the surrounding air. One disadvantage may be either the initial cost or the gas cost.

Greenhouse Tube: A greenhouse tube is a tube mounted along the top of a gabled ceiling to collect warm air and to reintroduce it at ground level. Like Casablanca fans, these reduce the amount of energy required to heat air at the expense of increasing the amount of energy needed to move the air.

Layout Concepts: This is a new field in which you can expect to see worthwhile changes taking place quickly in the future. Some ideas that will be developed include the use of landscaping to provide shading from the sun in the summer and from the wind in the winter, the use of different layout techniques to enable a company to take more advantage of waste heat generated within a facility, and the layout of storage according to the heat requirements of the stored material.

Microprocessors to Control Loads and Equipment Operation: Microprocessors have been developed with the capability to perform many of the routine logical operations currently being performed by people, and to do them better. Microprocessors do not get bored, and they do not forget to turn machines on or off. Many controls are being made utilizing these devices, and they are an integral part of energy management.

Polarized Lighting Panels: These are claimed to make far more efficient use of light than do regular luminaires. They do, however, have their detractors, and potential buyers should speak to past users before buying these en masse.

Turbulators: Turbulators are designed to increase the heat-transfer efficiency within boilers by creating turbulence after the heated gas leaves the flame. It is claimed that they increase efficiency several percentage points and decrease the fouling on tubes. If so, they are a worthwhile investment.

Underground Facilities: If the temperature of the ground (below 36 in. deep) is 55°F, it seems more reasonable to have the ground as a surrounding medium than to have the cold or warm air provided by nature. Also, losing heat to a medium with a constant temperature of 55°F creates less heat flow than if the outside temperature is 5 to 10°F, temperatures not uncommon in winter in most of the United States. The advantages in the summer are even more obvious. This is also a recent idea and one that is receiving much research effort.

14.7.2 New Technology: Five Questions

The last section provided brief descriptions of some new technology. The reader will probably be del-

uged with many other new devices, and the function of this section is to provide criteria for use in evaluating these gimmicks. These criteria are in the form of five questions.

1. **Does the Concept or Device Make Sense?** An attempt has been made to sell a water-softening device consisting of two magnets on either side of a hose containing the water coming into a business. The sellers claim that the magnets work "although the principle behind its operation is beyond the realms of present science." Such claims do not enhance the technical attractiveness of the magnetic water softener.

2. **Do You Need it?** Another black box shown to the author was a timer based on microcircuits and batteries rather than on a clock run by 60-Hz ac current. Since one of the weaknesses of the 60-Hz timer is its vulnerability to power outages, and since a timer seems to be a good way to reduce peak electrical demand, this black box appears to be a good way to meet the need for a timer.

3. **Does it Come from a Reliable Source?** If a reputable local dealer or a national company known for its customer service sponsors a gimmick intended to save energy, the company's reputation is at stake and the device is more likely to be worthwhile. Similarly, if a device has the backing of a strong utility group, such as the American Gas Association, or the Electric Power Research Institute, the concept has been tested and some individual models will work.

4. **Is it Worthwhile if it Saves Half as Much Money as it Claimed?** The energy management market has its share of glib-tongued sales personnel—people who base their energy-saving claims on escalating utility costs without taking into account the increasing price of everything else. It is also generally true that installing two devices into one system will not save as much as installing each in a separate system. The interaction between devices can be significant.

5. **Does it have Non-economic Advantages that Would Make it Worthwhile?** Some recent concepts have significant advantages in reducing pollution. An example of this is fluidized-bed combustion, an idea that has been studied for at least 20 years. This combustion system may save money directly, but its strongest advantages seem to be its low SO_x and NO_x emission characteristics and that it can burn a wide range of fuels.

Sources of Information

Credible information on new products can be obtained from one or more of the following sources: (1)

Department of Energy office (in each state); (2) the U.S. Department of Energy; (3) your local electrical or gas utility; and (4) Factory Mutual.

The mechanical or industrial engineering departments of your state university usually have several people who also keep current on new technology and can serve as a source of referrals to other knowledgeable persons.

References

1. National Electrical Contractors Association and National Electrical Manufacturers Association, *Total Energy Management: a Practical Handbook on Energy Conservation and Management,* National Electrical Contractors Association, 1976.
2. *Operation and Maintenance Manual for Steel Boilers* (Including Water Treatment), The Hydronics Institute, 34 Russo Place, Berkeley Heights, N.J. 07922, 1964.
3. "Operation and Maintenance Instructions, Class 600 Industrial Cross-Flow Cooling Towers," The Marley Co., 5800 Foxridge Drive, Mission, Kans. 66202.
4. New *HVAC Energy Saving Developments,* Air Distribution Associates, Inc., 955 Lively Boulevard, Wood Dale, Ill. 60191.
5. *ASHRAE Handbook and Product Directory, 1979, Equipment,* American Society of Heating, Refrigerating and Air Conditioning Engineers, Inc., New York, 1979.
6. Thomas F. Sack, A *Complete Guide to Building and Plant Maintenance,* Prentice-Hall, Englewood Cliffs, N.J., 1971.
7. *Unit Price Standards Handbook,* Depts. of the Navy, the Army, and the Air Force, NAVFAC P-716.
8. *Steam Conservation Guidelines for Condensate Drainage,* Armstrong Machine Works, Three Rivers, Mich. 49093.
9. Dale E. Shore and Michael W. McElroy, *Guidelines for Industrial Boiler Performance Improvement,* EPA Report EPA-600/8-77-003a, Jan. 1977, U.S. Environmental Protection Agency, Industrial Environmental Research Laboratory, Research Triangle Park, N.C.
10. W.H. Axtman, *Boiler Fuel Management and Energy Conservation,* American Boiler Manufacturers Association, Suite 700, 1500 Wilson Boulevard, Arlington, Va. 22209.
11. *Owner's and Operator's Guide, E-Series Sit Drive,* Hyster Company, Danville, Il. 61832, 1976.
12. *Owner's and Operator's Guide, Challenger,* Hyster Company, Danville, Il. 61832, 1975.
13. *Off-Highway Fuel Savings Guide,* Cummins Engine Co., Inc., Columbus, Ind., 1977.
14. *Automotive Marketing Bulletin 78-25,* Cummins Engine Co., Inc., Columbus, Ind.
15. *ASHRAE Handbook and Product Directory, 1977 Fundamentals,* American Society of Heating, Refrigerating and Air Conditioning Engineers, Inc., New York, 1978.
16. "IES Recommended Levels of Illumination," *IES Lighting Handbook,* 5th ed., Illuminating Engineering Society of North America, New York, 1977.
17. Wayne C. Turner, "The Impact of Energy on Material Handling," presented at the 1980 Materials Handling Institute Seminar, St. Louis, Mo.

Chapter 15
Industrial Insulation

JAVIER A. MONT

Oklahoma State University
Stillwater, Oklahoma

MICHAEL R. HARRISON

Manager, Engineering and Technical Services
Johns-Manville Sales Corp.
Denver, Colorado

Thermal insulation plays a key role in the overall energy management picture. In fact, the use of insulation is mandatory for the efficient operation of any hot or cold system. It is interesting to consider that by using insulation, the entire energy requirements of a system are reduced. Most insulation systems reduce the unwanted heat transfer, either loss or gain, by at least 90% as compared to bare surfaces.

Since the insulation system is so vital to energy-efficient operations, the proper selection and application of that system is very important. This chapter describes the various insulation materials commonly used in industrial applications and explores the criteria used in selecting the proper products. In addition, methods for determining the proper insulation thickness are developed, taking into account the economic trade-offs between insulation costs and energy savings.

15.1 FUNDAMENTALS OF THERMAL INSULATION DESIGN THEORY

The basic function of thermal insulation is to retard the flow of unwanted heat energy either to or from a specific location. To accomplish this, insulation products are specifically designed to minimize the three modes of energy transfer. The efficiency of an insulation is measured by an overall property called thermal conductivity.

15.1.1 Thermal Conductivity

The thermal conductivity, or k value, is a measure of the amount of heat that passes through 1 square foot of 1-inch-thick material in 1 hour when there is a temperature difference of 1°F across the insulation thickness. Therefore, the units are Btu-in./hr ft^2 °F. This property relates only to homogeneous materials and has nothing to do with the surfaces of the material. Obviously, the lower the k value, the more efficient the insulation. Since products are often compared by this property, the measurement of thermal conductivity is very critical. The American Society of Testing Materials (ASTM) has developed sophisticated test methods that are the standards in the industry. These methods allow for consistent evaluation and comparison of materials and are frequently used at manufacturing locations in quality control procedures.

Conduction

Energy transfer in this mode results from atomic or molecular motion. Heated molecules are excited and this energy is physically transferred to cooler molecules by vibration. It occurs in both fluids (gas and liquid) and solids, with gas conduction and solid conduction being the primary factors in insulation technology.

Solid conduction can be controlled in two ways: by utilizing a solid material that is less conductive and by utilizing less of the material. For example, glass conducts heat less readily than steel and a fibrous structure has much less through-conduction than does a solid mass. Gas conduction does not lend itself to simple modification. Reduction can be achieved by either reducing the gas pressure by evacuation or by replacing the air with a heavy-density gas such as Freon®. In both cases, the insulation must be adequately sealed to prevent reentry of air into the modified system. However, since gas conduction is a major component of the total thermal conductivity, applications requiring very low heat transfer often employ such gas-modified products.

Convection

Energy transfer by convection is a result of hot fluid rising in a system and being replaced by a colder, heavier fluid. This fluid is then heated, rises, and carries more heat away from the heat source. Convective heat transfer is minimized by the creation of small cells within which the temperature gradients are small. Most thermal insulations are porous structures with enough density to block radiation and provide structural integrity. As such, convection is virtually eliminated within the insulation except for applications where forced convection is being driven through the insulation structure.

Radiation

Electromagnetic radiation is responsible for much of the energy transferred through an insulation and increases in its significance as temperatures increase. The radiant energy will flow even in a vacuum and is governed by the emittance and temperature of the surfaces involved. Radiation can be controlled by utilizing surfaces with low emittance and by inserting absorbers or reflectors within the body of the insulation. The core density of the material is a major factor, with radiation being reduced by increased density. The interplay between the various heat-transfer mechanisms is very important in insulation design. High density reduces radiation but increases solid conduction and material costs. Gas conduction is very significant, but to alter it requires permanent sealing at additional cost. In addition, the temperature in which the insulation is operating changes the relative importance of each mechanism. Figure 15.1 shows the contribution of air conduction, fiber conduction, and radiation in a glass fiber insulation at various densities and mean temperatures.

15.1.2 Heat Transfer

There are many texts dedicated to the physics of heat transfer, some of which are listed in the references.

In its simplest form, however, the basic law of energy flow can be stated as follows:

A steady flow of energy through any medium of transmission is directly proportional to the force causing the flow and inversely proportional to the resistance to that force.

In dealing with heat energy, the forcing function is the temperature difference and the resistance comes from whatever material is located between the two temperatures.

$$\text{heat flow} = \frac{\text{temperature difference}}{\text{resistance to heat flow}}$$

This is the fundamental equation upon which all heat-transfer calculations are based.

Temperature Difference

By definition, heat transfer will continue to occur until all portions of the system are in thermal equilibrium (i.e., no temperature difference exists). In other words, no amount of insulation is able to provide enough resistance to totally stop the flow of heat as long as a temperature difference exists. For most insulation applications, the two temperatures involved are the operating temperature of the piping or equipment and the surrounding ambient air temperature.

Thermal Resistance

Heat flow is reduced by increasing the thermal resistance of the system. The two types of resistances commonly encountered are mass and surface resistances. Most insulations are homogeneous and as such have a thermal conductivity or k value. Here the insulation resistance, $R_I = tk/k$, where tk represents the thickness of the insulation. In cases of nonhomogeneous products such as multifoil metallic insulations, the thermal properties of the products at their actual finished thicknesses are expressed as conductances rather than conductivities based on a 1-in. thickness. In this case the resistance $R_I = 1/C$, where C represents the measured conductance.

The other component of insulation resistance is the surface resistance, $R_S = 1/f$, where f represents the surface film coefficient. These values

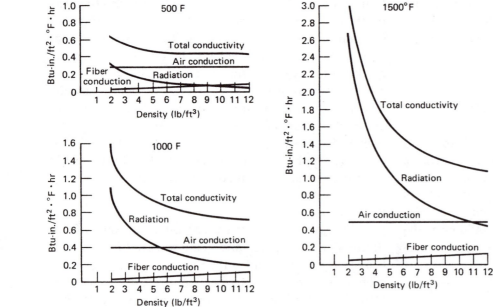

Fig. 15.1 Contribution of each mode of heat transfer. (From Ref. 15.)

are dependent on the emittance of the surface and the temperature difference between the surface and the surrounding environment.

Thermal resistances are additive and as such are the most convenient terms to deal with. Following are several expressions for the heat-transfer equation, showing the relationships between the commonly used R, C, and U values. For a single insulation with an outer film:

$$Q = \frac{\Delta t}{R_I + R_s}$$

$$= \frac{\Delta t}{tk/k + 1/f}$$

$$= \frac{\Delta t}{1/C_I + 1/f} \quad \text{Where the insulation is not homogeneous}$$

$$= U \, \Delta t \quad \text{where } U = \frac{1}{R_{\text{total}}} = \frac{1}{R_I + R_s}$$

U is termed the overall coefficient of heat transmission of the insulation system.

15.2 INSULATION MATERIALS

Marketplace needs, in conjunction with active research and development programs by manufacturers, are responsible for a continuing change in insulation materials available to industry. Some products have been used for decades, whereas others are relatively new and are still being evaluated. The following sections describe the primary insulation materials available today, but first, the important physical properties will be discussed.

15.2.1 Important Properties

Each insulation application has a unique set of requirements as it relates to the important insulation properties. However, certain properties emerge as being the most useful for comparing different products and evaluating their fitness for a particular application. Table 15.1 lists the insulation types and product properties that are discussed in detail below. One area that will not be discussed is industrial noise control. Thermal insulations

Table 15.1 Industrial insulation types and properties.

Insulation Type and Forms(a)	Temp. Range (°F)	Density (lb/cu.ft)	Thermal Conductivity [Btu-in/hr-ft2-°F at Tmean (°F)]											Compressive Strength (psi) at % Deformation	Fire Hazard Classification Flame-Spread- Smoke Developed	Cell Structure (Permeability and Moisture Absorption)
			-300	-100	0	75	100	200	300	500	700	800	900	1400			
Calcium silicate blocks, shapes and P/C	to 1200	11-15					0.38	0.41	0.44	0.52	0.62		0.72		100-250 at 5%	Non-combustible	Open cell
Glass fiber blankets	to 1000	1				0.24	0.25	0.34	0.46	0.78							
		2				0.22	0.22	0.30	0.36	0.57					0.02-3.5 at 10%	Non-combustible to 25/50	Open cell
Glass fiber boards	to 850	3				0.22	0.23	0.27	0.32	0.49							
Glass fiber P/C	-20 to 850	3				0.22	0.23	0.31	0.39	0.62							
Mineral fiber blocks, boards and P/C	to 1800	15-24					0.32	0.37	0.42	0.52	0.62		0.74		1-18 at 10%	Non-combustible to 25/50	Open cell
Cellular glass blocks, boards and P/C	to 900	8	0.18	0.24	0.29	0.33	0.34	0.41	0.49	0.70					100 at 5%	Non-combustible	Closed cell
Expanded perlite blocks, shapes and P/C	to 1500	13-15					0.4	0.45	0.5	0.6	0.71		0.83		90 at 5%	Non-combustible	Open cell
Urethane foam blocks and P/C	(-100 to -450) to 225	to 1.5				0.16-0.18									16-75 at 10%	25-75 to 140-400	95% closed cell
Polyisocyanurate foam blocks and P/C	to 250	2				0.14	0.15								17-25 at 10%	25-55 to 100	85-90% closed cell
Phenolic foam P/C	-40 to 250	2-3				0.22	0.26								13-22 at 10%	25/50	Open cell
Elastomeric closed cell sheets and P/C	to 400	8.5-9.5				0.29	0.32								40 at 10%	25-75 to 115-490	Closed cell
Ceramic fiber blankets	to 2600	6-8							0.47-0.50		0.70-0.60			1.20-1.80	0.5-1 at 10%	Non-combustible	Open cell

(a) P/C means pipe covering.

Sources: Refs. 17, 18 and manufacturers' literature.

are often used as acoustical insulations for their absorption or attenuation properties. Many texts are available for reference in this area.

Temperature-Use Range

Since all products have a point at which they become thermally unstable, the upper temperature limit of an insulation is usually quite important. In some cases the physical degradation is gradual and measured by properties such as high-temperature shrinkage or cracking. In such cases, a level is set for the particular property and the product is rated to a temperature at which that performance level is not exceeded. Occasionally, the performance levels are established by industry standards, but frequently, the manufacturers establish their own acceptance levels based on their own research and application knowledge.

In other cases, thermal instability is very rapid rather than gradual. For example, a product containing an organic binder may have a certain temperature at which an exothermic reaction takes place due to a too-rapid binder burnout. Since this type of reaction can be catastrophic, the temperature limit for such a product may be set well below the level at which the problem would occur.

Low-end temperature limits are usually not specified unless the product becomes too brittle or stiff and, as such, unusable at low temperatures. The most serious problem with low-temperature applications is usually vapor transmission, and this is most often related to the vapor-barrier jacket or coating rather than to the insulation. In general, products are eliminated from low-temperature service by a combination of thermal efficiency and cost.

Thermal Conductivity

This property is very important in evaluating insulations since it is the basic measure of thermal efficiency, as discussed in Section 15.1.1. However, a few points must be emphasized. Since the k value changes with temperature, it is important that the insulation *mean* temperature be used rather than the operating temperature. The mean temperature is the average temperature within the insulation and is calculated by summing the hot and cold surface temperatures and dividing by 2: $(t_h + t_s)/2$. Thermal conductivity data are always published per mean temperature, but many users incorrectly make comparisons at operating temperatures.

A second concern relates to products which have k values that change with time. In particular, foam products often utilize an agent that fills the cells with a gas heavier than air. Shortly after manufacture, some of this gas migrates out, causing an increase in thermal conductivity. This new value is referred to as an "aged k" and is more realistic for design purposes.

Compressive Strength

This property is important for applications where the insulation will see a physical load. It may be a full-time load, such as in buried lines or insulation support saddles, or it may be incidental loading from foot traffic. In either case, this property gives an indication of how much deformation will occur under load. When comparing products it is important to identify the percent compression at which the compressive strength is reported. Five and 10% are the most common, and products should be compared at the same level.

Fire Hazard Classification

Insulation materials are involved with fire in two ways: fire hazard and fire protection. Fire protection refers to the ability of a product to withstand fire exposure long enough to protect the column, pipe, or vessel it is covering. This topic is discussed in Section 15.3.1.

Fire hazard relates to the product's contribution to a fire by either flame spread or smoke development. The ASTM E-84 tunnel test is the standard method for rating fire hazard and compares the FS/SD (flame spread/smoke developed) to that of red oak, which has a 100/100 rating. Typically, a 25/50 FHC is specified where fire safety is an important concern. Certain concealed applications allow higher ratings, while the most stringent requirements require a noncombustible classification.

Cell Structure

The internal cell structure of an insulation is a primary factor in determining the amount of moisture the product will absorb as well as the ease in which vapor will pass through the material. Closed-cell structures tend to resist both actions, but the thickness of the cell walls as well as the base material will also influence the long-term performance of a closed-cell product. In mild design conditions such as chilled-water lines in a reasonable ambient, closed-cell products can be used without an additional vapor barrier. However, in severe conditions or colder operating temperatures, an additional vapor barrier is suggested for proper performance.

Available Forms

An insulation material may be just right for a specific application, but if it is not manufactured in a form compatible with the application, it cannot be used. Insulation is available in different types (Ref. 19).

Loose-fill insulation and insulating cements. Loose-fill insulation consists of fibers, powders, granules or nodules which are poured or blown into walls or other irregular spaces. Insulating cements are mixtures of a loose material with water or other binder which are blown on a surface and dried in place.

Flexible, semirigid and rigid insulation. Flexible and semirigid insulation, which are available in sheets or rolls, are used to insulate pipes and ducts. Rigid insulation is available in rectangular blocks, boards or sheets and are also used to insulate pipes and other surfaces. The most common forms of insulation are flexible blankets, rigid boards and blocks, pipe insulation half-sections and full-round pipe sections.

Formed-in-place insulation. This type of insulation can be poured, frothed or sprayed in place to form rigid or semirigid foam insulation. They are available as liquid components, expandable pellets or fibrous materials mixed with binders.

Removable-reusable insulation covers. Used to insulate components that require routine maintenance (like valves, flanges, expansion joints, etc.). These covers use belts, Velcro or stainless steel hooks to reduce the installation time.

Other Properties

For certain applications and thermal calculations, other properties are important. The pH of a material is occasionally important if a potential for chemical reaction exists. Density is important for calculating loads on support structures and occasionally has significance with respect to the ease of installation of the product. The specific heat is used together with density in calculating the amount of heat stored in the insulation system, primarily of concern in transient heat-up or cool-down cycles.

15.2.2 Material Description

Calcium Silicate

These products are formed from a mixture of lime and silica and various reinforcing fibers. In general, they contain no organic binders, so they maintain their physical integrity at high temperatures. The calcium silicate products are known for exceptional strength and durability in both intermediate- and high-temperature applications where physical abuse is a problem. In addition, their thermal performance is superior to other products at the higher operating temperatures.

Glass Fiber

Fiberglass insulations are supplied in more forms, sizes, and temperature limits than are other industrial insulations. All of the products are silica-based and range in density from 0.6 to 12 lb/ft^3. The binder systems employed include low-temperature organic binders, high-temperature organic/antipunk binders, and needled mats with no binders at all. The resulting products include flexible blankets, semirigid boards, and preformed one-piece pipe covering for a very wide range of applications from cryogenic to high temperature. In general, the fiberglass products are not considered load bearing.

Most of the organic binders used begin to oxidize (burn out) in the range 400 to 500°F. The loss of binder somewhat reduces the strength of the product in that area, but the fiber matrix composed of long glass fibers still gives the product good integrity. As a result, many fiberglass products are rated for service above the binder temperature, and successful experience indicates that they are completely suitable for numerous applications.

Mineral Fiber/Rock Wool

These products are distinguished from glass fiber in that the fibers are formed from molten rock or slag rather than silica. Most of the products employ organic binders similar to fiberglass but the very high temperature, high-density blocks use inorganic clay-type binders. The mineral wool fibers are more refractory (heat resistant) than glass fibers, so the products can be used to higher temperatures. However, the mineral wool fiber lengths are much shorter than glass and the products do contain a high percentage of unfiberized material. As a result, after binder burnout, the products do not retain their physical integrity very well and long-term vibration or physical abuse will take its toll.

Cellular Glass

This product is composed of millions of completely sealed glass cells, resulting in a rigid insulation that is totally inorganic. Since the product is closed cell, it will not absorb liquids or vapors and thus adds security to cryogenic or buried applications, where moisture is always a problem. Cellular glass is load bearing, but also somewhat brittle, making installation more difficult and causing problems in vibrating or flexing applications. At high temperatures, thermal-shock cracking can be a problem, so a cemented multilayer construction is used. The thermal conductivity of cellular glass is higher than for most other products, but it has unique features that make it the best product for certain applications.

Expanded Perlite

These products are made from a naturally occurring mineral, perlite, that has been expanded at a high temperature to form a structure of tiny air cells surrounded by vitrified product. Organic and inorganic binders together with reinforcing fibers are used to hold the structure together. As produced, the perlite materials have low moisture absorption, but after heating and oxidizing the organic material, the absorption increases dramatically. The products are rigid and load bearing but have lower compressive strengths and higher thermal conductivities than the calcium silicate products and are also much more brittle.

Plastic Foams

There are three foam types finding some use in industrial applications, primarily for cold service. They are all produced by foaming various plastic resins.

Polyurethane/Isocyanurate Foams. These two types are rigid and offer the lowest thermal conductivity since they are expanded with fluorocarbon blowing agents. However, sealing is still required to resist the migration of air and water vapor back into the foam cells, particularly under severe conditions with large differentials in vapor pressure. The history of urethanes is plagued with problems of dimensional stability and fire safety. The isocyanurates were developed to improve both conditions, but they still have not achieved the 25/50 FHC (fire hazard classification) for a full range of thickness. As a result, many industrial users will not allow their use except in protected or isolated areas or when covered with another fire-resistant insulation. The advantage of these foam products is their low thermal conductivity, which allows less insulation thickness to be used, of particular importance in very cold service.

Phenolic Foam. These products have achieved the required level of fire safety, but do not offer k values much different from fiberglass. They are rigid enough to eliminate the need for special pipe saddle supports on small lines. However, the present temperature limits are so restrictive that the products are primarily limited to plumbing and refrigeration applications.

Polyimide Foams

Polyimide foams are used as thermal and acoustical insulation. This material is fire resistant (FS/SD) of 10/10) and lightweight, so they requires fewer mechanical fastening devices. Thermal insulation is available in open-cell structure. Temperature stability limits its application to chilled water lines and systems up to 100°F.

Elastomeric Cellular Plastic

These products combine foamed resins with elastomers to produce a flexible, closed-cell material. Plumbing and refrigeration piping and vessels are the most common applications, and additional vapor-barrier pro-

tection is not required for most cold service conditions. Smoke generation has been the biggest problem with the elastomeric products and has restricted their use in 25/50 FHC areas. To reduce installation costs, elastomeric pipe insulation is available in 6-ft long, pre-split tubular sections with a factory-applied adhesive along the longitudinal joint.

Refractories

Insulating refractories consist primarily of two types, fiber and brick.

Ceramic Fiber. These alumina-silica products are available in two basic forms, needled and organically bonded. The needled blankets contain no binders and retain their strength and flexibility to very high temperatures. The organically bonded felts utilize various resins which provide good cold strength and allow the felts to be press cured up to 24 lbm3 density. However, after the binder burns out, the strength of the felt is substantially reduced. The bulk ceramic fibers are also used in vacuum forming operations where specialty parts are molded to specific shapes.

Insulating Firebrick. These products are manufactured from high-purity refractory clays with alumina also being added to the higher temperature grades. A finely graded organic filler which burns out during manufacture provides the end product with a well-designed pore structure, adding to the product's insulating efficiency. Insulating firebrick are lighter and therefore store less heat than the dense refractories and are superior in terms of thermal efficiency.

Protective Coatings and Jackets

Any insulation system must employ the proper covering to protect the insulation and ensure long-term performance. Weather barriers, vapor barriers, rigid and soft jackets, and a multitude of coatings exist for all types of applications. It is best to consult literature and representatives of the various coating manufacturers to establish the proper material for a specific application. Jackets with reflective surfaces (like aluminum and stainless steel jackets) have low emissivity (ε). For this reason, reflective jackets have lower heat loss than plain or fabric jackets (high emissivity). In hot applications, this will result in higher surface temperatures and increase the risk of burning personnel. In cold applications, surface temperatures will be lower, which could cause moisture condensation. Regarding jacketing material, existing environment and abuse conditions and desired esthetics usually dictate the proper material. Section 15.3.3 will discuss jacketing systems typically used in industrial work.

15.3 INSULATION SELECTION

The design of a proper insulation system is a two-fold process. First, the most appropriate material must be selected from the many products available. Second, the proper thickness of material to use must be determined. There is a link between these two decisions in that one product with superior thermal performance may require less thickness than another material, and the thickness reduction may reduce the cost. In many cases, however, the thermal values are so close that the same thickness is specified for all the candidate materials. This section deals with the process of material selection. Section 15.4 addresses thickness determination.

15.3.1 Application Requirements

Section 15.2.1 discussed the insulation properties that are of most significance. However, each application will have specific requirements that are used to weigh the importance of the various properties. There are three items that must always be considered to determine which insulations are suitable for service. They are operating temperature, location or ambient environment, and form required.

Operating Temperature

This parameter refers to the hot or cold service condition that the insulation will be exposed to. In the event of operating design temperatures that may be exceeded during overrun conditions, the potential temperature extremes should be used to assure the insulation's performance after the excess exposure.

Cryogenic (–455 to –150°F). Cryogenic service conditions are very critical and require a well-designed insulation system. This is due to the fact that if the system allows water vapor to enter it will not only condense to a liquid but will subsequently expand and destroy the insulation. Proper vapor barrier design is critical in this temperature range. Closed-cell products are often used since they provide additional vapor resistance in the event that the exterior barrier is damaged or inadequately sealed.

For the lowest temperatures where the maximum thermal resistance is required, vacuum insulations are often employed. These insulations are specially designed to reduce all the modes of heat transfer. Multiple foil sheets (reduced radiation) are separated by a thin mat filler of fiberglass (reduced solid conduction) and are then evacuated (reduced convection and gas conduction). These "super insulations" are very efficient as long as the vacuum is maintained, but if a vacuum failure

occurs, the added gas conduction drastically reduces the efficiency.

Finely divided powders are also used for bulk, cavity-fill insulation around cryogenic equipment. With these materials, only a moderate vacuum is required, and in the event that the vacuum fails, the powder still acts as an insulation. It is, however, very important to keep moisture away from the powders, as they are highly absorbent and the ingress of moisture will destroy the system.

Some plastic foams are suitable for cryogenic service, whereas others become too brittle to use. They must all have additional vapor sealing since high vapor pressures can cause moisture penetration of the cell walls. Closed-cell foamed glass (cellular glass) is quite suitable for this service in all areas except those requiring great thermal efficiency. Since it is not evacuated and has solid structure, the thermal conductivity is relatively high.

Because of the critical nature of much cryogenic work, it is very common to have the insulation system specifically designed for the job. The increased use of liquified gases (natural and propane) together with cryogenic fluids in manufacturing processes will require continued use and improvement of these systems.

Low Temperature (–150 to 212°F). This temperature range includes the plumbing, HVAC, and refrigeration systems used in all industries from residential to aerospace. There are many products available in this range, and the cost of the installed thermal efficiency is a large factor. Products typically used are glass fiber, plastic foams, phenolic foam, elastomeric materials, and cellular glass. In below-ambient conditions, a vapor barrier is still required, even though as the service approaches ambient temperature, the necessary vapor resistance becomes less. Above-ambient conditions require little special attention, with the exception of plastic foams, which approach their temperature limits around 200°F.

Because of the widespread requirement for the plumbing and HVAC services within residential and commercial buildings, the insulations are subject to a variety of fire codes. Many codes require a flame spread rating less than 25 for exposed material and smoke ratings from 50 to 400, depending on location. A composite rating of 25/50 FHC is suitable for virtually all applications, with a few applications requiring non-combustibility.

Intermediate Temperature (212 to 1000°F). The great majority of steam and hot process applications fall within this operating range. Refineries, power plants, chemical plants, and manufacturing operations all require insulation for piping and equipment at these temperatures. The products generally used are calcium silicate, glass fiber, mineral wool, and expanded perlite. Most of the fiberglass products reach their temperature limit somewhere in this range, with common breakpoints at 450, 650, 850, and 1000°F for various products.

There are two significant elements to insulation selection at these temperatures. First, the thermal conductivity values change dramatically over the range of mean temperatures, especially for light-density products under 8 lbm3. This means, for example, that fiberglass pipe covering will be more efficient than calcium silicate for the lower temperatures, with the calcium silicate having an advantage at the higher temperatures. A thermal conductivity comparison is of value in making sure that the insulation mean temperature is used rather than the operating temperature.

The second item relates to products that use organic binders in their manufacture. All the organics will burn out somewhere within this temperature range, usually between 400 and 500°F. Many products are designed to be used above that temperature, whereas others are not. This is mentioned here only to call attention to the fact that some structural strength is usually lost with organic binder.

High Temperature (1000 to 1600°F). Superheated steam, boiler exhaust ducting, and some process operations deal with temperatures at this level. Calcium silicate, mineral wool, and expanded perlite products are commonly used together with the lower-limit ceramic fibers. Except for a few clay-bonded mineral wool materials, these products reach their temperature limits in this range. Thermal instability, as shown by excessive shrinkage and cracking, is usually the limiting factor.

Refractory (1600 to 3600°F). Furnaces and kilns in steel mills, heat treating and forging shops, as well as in brick and tile ceramic operations, operate in this range. Many types of ceramic fiber are used, with alumina-silica fibers being the most common. Insulating firebrick, castables, and bulk-fill materials are all necessary for meeting the wide variety of conditions that exist in refractory applications. Again, thermal instability is the controlling factor in determining the upper temperature limits of the many products employed.

Location

The second item to consider in insulation selection is the location of the system. Location includes many factors that are critical to choosing the most cost-effective product for the life of the application. Material selection based on initial price only without regard to location can be not only inefficient, but dangerous under certain conditions.

Surrounding Environment. For an insulation to remain effective, it must maintain its thickness and thermal conductivity over time. Therefore, the system must either be protected from or able to withstand the rigors of the environment. An outdoor system needs to keep water from entering the insulation, and in most areas, the jacketing must hold up under radiant solar load. Indoor applications are generally less demanding with regard to weather resistance, but there are washdown areas that see a great deal of moisture. Also, chemical fumes, atmospheres, or spillage may seriously affect certain jacketing materials and should be evaluated prior to specifications. Direct burial applications are normally severe, owing to soil loading, corrosiveness, and moisture. It is imperative that the barrier material be sealed from groundwater and resistant to corrosion. Also, the insulation must have a compressive strength sufficient to support the combined weight of the pipe, fluid, soil backfill, and potential wheel loads from ground traffic.

Another concern is insulation application on austenetic stainless steel, a material subject to chloride stress-corrosion cracking. There are two specifications most frequently used to qualify insulations for use on these stainless steels: MIL-1-24244 and Nuclear Regulatory Commission NRC Reg. Guide 1.36. The specifications require, first, a stress-corrosion qualification test on actual steel samples; then, on each manufacturing lot to be certified, a chemical analysis must be performed to determine the amount of chlorides, fluorides, sodium, and silicates present in the product. The specific amounts of sodium and silicates required to neutralize the chlorides and fluorides are stated in the specifications.

There are many applications where vibration conditions are severe, such as in gas turbine exhaust stacks. In general, rigid insulations such as calcium silicate withstand this service better than do fibrous materials, especially at elevated temperatures. If the temperature is high enough to oxidize the organic binder, the fibrous products lose much of their compressive strength and resiliency. On horizontal piping, the result can be an oval-shaped pipe insulation which is reduced in thickness on the top of the pipe and sags below the underside of the pipe, thus reducing the thermal efficiency of the system. On vertical piping and equipment with pinned-on insulation, the problem of sag is reduced, but the vibration can still tend to degrade the integrity of the insulation.

Location in a fire-prone area can affect the insulation selection in two ways. First, the insulation system cannot be allowed to carry the fire to another area; this is fire hazard. Second, the insulation can be selected and designed to help protect the piping or equipment from the fire. There are many products available for just fireproofing such areas as structural steel columns, but in general they are not very efficient thermal insulations. When an application requires both insulation during operation and protection during a fire, calcium silicate is probably the best selection. This is due to the water of hydration in the product, which must be driven off before the system will rise above the steam temperature. Other high-density, high-temperature products are used as well. With all the products it is important that the jacketing system be designed with stainless steel bands and/or jacketing since the insulation must be maintained on the piping in order to protect it. Figure 15.2 shows fire test results for three materials per the ASTM E-119 fire curve and indicates the relative level of fire protection provided by each material.

A final concern deals with the transport of volatile fluids through piping systems. When leaks occur around flanges or valves, these fluids can seep into the insulation. Depending on the internal insulation structure, the surface area may be increased significantly, thus reducing the fluid's flash point. If this critical temperature drops below the operating temperature of the system, autoignition can occur, thus creating a fire hazard. In areas where leaks are a problem, either a leakage drain must be provided to remove the fluid or else a closed-cell material such as cellular glass should be used, since it will not absorb the fluid.

The previous discussion is intended to draw attention to specific application requirements, not necessarily to determine the correct insulation to be used. Each situation should be evaluated for its own requirements, and in areas of special concern (auto-ignition, fire, etc.) the manufacturer's representative should be called upon to answer questions specific to the product.

Resistance to Physical Abuse. Although this issue is related to location, it is so important that it needs its own discussion. In commercial construction and many light industrial facilities, the pipe and equipment insulation is either hidden or isolated from any significant abuse. In such cases, little attention need be given to this issue. However, in most heavy industrial applications, physical abuse and the problems caused by it are matters of great concern.

Perhaps by definition, physical abuse differs from physical loading in that loading is planned and designed for, whereas abuse is not. For example, with cold piping, pipe support saddles are often located external to the

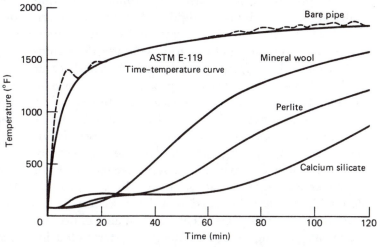

Fig. 15.2 Fire resistance test data for pipe insulation. (Used by permission from Ref. 6.)

insulation and vapor barrier. This puts the combined weight of the pipe and fluid onto the lower portion of the insulation. This is a designed situation, and a rigid material is inserted between the pipe and the saddle to carry the load. However, if a worker decides to use the insulated pipe as a scaffold support, a walkway, or a hoist support, the insulation may not be designed to support such a load and damage will occur. A quick walk-through of any industrial facility will show much evidence of "unusual" or "unanticipated" abuse. In point of fact, many users have seen so much of this that they now design for the abuse, having determined that it is "usual" for their facility.

The effects of abuse are threefold. First, dented and creased aluminum jacketing is unsightly and lowers the overall appearance of the plant. Nonmetal jackets may become punctured and torn. The second point is that wherever the jacketing is deformed, the material under it is compressed and as a result is a much poorer insulation since the thickness has been reduced. Finally, on outside lines some deformation will undoubtedly occur at the jacketing overlap. This allows for water to enter the system, further degrading the insulation and reducing the thermal efficiency.

In an effort to deal with the physical abuse problem, some specifications call for all horizontal piping to be insulated with rigid material while allowing a fibrous option on the vertical lines. Others modify this specification by requiring rigid insulation to a height of 6 to 10 ft on vertical lines to protect against lateral abuse. Still, in facilities that have a history of a rough environment, it is most common to specify the rigid material for all piping and equipment except that which is totally enclosed or isolated.

As previously mentioned, the primary insulation material choice is between rigid and nonrigid materials. Calcium silicate, cellular glass, and expanded perlite products fit the rigid category, whereas most mineral wool and fiberglass products are nonrigid. Over the years, the calcium silicate products have become the standard for rigorous services, combining good thermal efficiency with exceptional compressive strength and abuse resistance. The maintenance activities associated with rigid insulations are significantly less than the maintenance and replacement needs of the softer insulations in abuse areas. The costs associated with this are discussed in a later section. However, it is also recognized that often, maintenance activities are lacking, which results in a deteriorated insulation system operating at reduced efficiency for a long period of time.

Form Required

The third general category to consider is the insulation form required for the application. Obviously, pipe insulation and flat sheets are manufactured for specific purposes, and the lack of a specific form eliminates that product from consideration. However, there are subtle differences between form that can make a significant difference in installation costs and system efficiency.

On flat panels, the two significant factors are panel size and the single-layer thickness available. A fibrous 4 × 8 ft sheet is applied much more rapidly than four 2 × 4 ft sheets, and it is possible that the number of pins required might be reduced. In regard to thickness, if one material can be supplied 4 in. thick as a single layer as opposed to two 2-in.-thick layers, the first option will result in significant labor savings. The same holds true for 18-in.- vs. 12-in.-wide rigid block installation.

Fibrous pipe insulations have three typical forms: one-piece hinged snap-on, two piece halfsections, and flexible blanket wraparound. For most pipe sizes, the one-piece material is the fastest to install and may not require banding if the jacket is attached to the insulation and secured to itself. Two-piece products must be wired in place and then subsequently jacketed in a separate operation. Wrap-around blankets are becoming more popular, especially for large-diameter pipe and small vessels. They come in standard roll lengths and are cut to length on the job site.

Rigid insulations also have different forms, which vary with the manufacturer. The two-piece half-section pipe insulation is standard. However, these sections can be supplied prejacketed with aluminum, which in effect gives a one-piece hinged section that does not need a separate application of insulation and jacket. Also, thicknesses up to 6 in. are available, eliminating the need for double-layer applications where they are not required for expansion reasons. The greatest diversity comes in the large-diameter pipe sizes. Quads (quarter sections) are available and are both quicker to install and thermally more efficient than is scored block bent around the pipe. Similarly, curved radius blocks are available for sizes above quads and provide a better fit than does flat beveled block or scored block.

The important point is that the available forms of insulation may well affect the decision as to which material to select and which manufacturer to purchase that material from. It is unwise to assume that all manufacturers offer the same sizes and forms or that the cost to install the product is not affected by its form.

15.3.2 Cost Factors

Section 15.3.1 dealt with the process of selecting the materials best suited for a specific application. In some cases the requirements are so stringent that only one specific material is acceptable. For most situations, however, more than one insulation material is suitable, even though they may be rank-ordered by anticipated performance. In these cases, several cost factors should be considered to determine which specific material (and/or manufacturer) should be selected to provide the best system for the lowest cost.

Initial Cost

In new construction, the owner is usually interested primarily in the installed cost of the insulation system. As long as the various material options provide similar thermal performance, they can be compared on an equal basis. The contractor, on the other hand, is much more concerned about the insulation form and its effect on installation time. It may be of substantial benefit to the contractor to utilize a more costly material that can be installed more efficiently for reduced labor costs. In a highly competitive market, these savings need to be passed through to the owner for the contractor to secure the job. The point is that the lowest-cost material does not necessarily become the lowest-cost installed material, so all acceptable alternatives should be evaluated.

Maintenance Cost

To keep their performance and appearance at acceptable levels, all insulation systems must be maintained. This means, for example, that outdoor weather protection must be replaced when damaged to prevent deterioration of the insulation. If left unattended, the entire system may need to be replaced. In a high-abuse area, a nonrigid insulation may need to be replaced quite frequently in order to maintain performance. Aesthetics often play an important part in maintenance activities, depending on the type of operation and its location. In such cases there is benefit in utilizing an insulation that maintains its form and if possible aids the jacketing or coating in resisting abuse.

The trade-off comes between initial cost and maintenance cost in that a less costly system may well require greater maintenance. Unfortunately, the authorities for initial construction and ongoing maintenance are often split, so the owner may not be aware of the future consequences of the initial system selection. It is imperative that both aspects be viewed together.

Lost Heat Cost

If the various suitable insulations are properly evaluated, a more thermally efficient product should require less thickness to meet the design parameters However, if a common thickness is specified for all products, there can be a substantial difference in heat loss or gain between the systems In such a case, the more efficient product should receive financial credit for transferring less heat, and this should be considered in the overall cost calculations

Referring to the previous discussion on maintenance costs, there was an underlying assumption that maintenance would be performed to the extent that the original thermal efficiency would be maintained. In reality, maintenance is usually not performed until the situation is significantly deteriorated and sometimes not even then. The result of this is reduced thermal efficiency for much of the life of a maintenance-intensive system. It is very difficult to assign a figure to the amount of additional heat transfer due to deterioration. In a wet climate, for example, a torn jacket will allow moisture into the system and drastically affect the performance. Conversely, in a dry area, the insulation might maintain its performance for quite some time. Still, when dealing with maintenance costs, it is a valid concern that systems in need of maintenance generally are transferring more heat at greater cost than are systems requiring less maintenance.

Design Life

The anticipated project life is the foundation upon which all costs are compared. Since there are trade-offs between initial cost and ongoing maintenance and heat-loss costs, the design life is important in determining the total level of the ongoing costs. To illustrate, consider the difference between designing a 40-year power plant and a two-year experimental process. Assuming that the in-

sulation in the experimental process will be scrapped at the close of the project, it makes no sense to use a more costly insulation that has lower maintenance requirements, since those future benefits will never be realized. Similarly, utilizing a less costly but maintenance-intensive system when the design life is 40 years makes little sense, since the additional front-end costs could be remained in only a few years of reduced maintenance costs.

15.3.3 Typical Applications

This section is designed to give a brief overview of commonly used materials and application techniques. For a detailed study of application, techniques, and recommendations, see Ref. 7, as well as the guide specifications supplied by most insulation manufacturers.

The Heat Plant

Boilers are typically insulated with fiberglass or mineral wool boards, with some usage of calcium silicate block when extra durability is desired. Powerhouse boilers are normally insulated on-site with the fibrous insulation being impaled on pins welded to the boiler. Box-rib aluminum is then fastened to the stiffeners or buckstays as a covering for the insulation. In most commercial and light industrial complexes, package boilers are normally used. These are insulated at the factory, usually with fiberglass or mineral wool.

Breechings and other high-temperature duct work are insulated with calcium silicate (especially where traffic patterns exist), mineral wool, and high-temperature fiberglass. On very large breechings, prefabricated panels are used, as discussed in the following paragraph. H-bar systems supporting the fibrous materials are common, with the aluminum lagging fastened to the outside of the H-bar members. Also, many installations utilize roadmesh over the duct stiffeners, creating an air space, and then wire the insulation to the mesh substrate. Indoors, a finish coat of cement may be used rather than metal lagging.

Precipitators are typically insulated with prefabricated panels filled with mineral wool or fiberglass blankets. For large, flat areas, such panels provide very efficient installations, as the panels are simply secured to the existing structure with self-tapping screws. H-bar and Z-bar systems are also used to contain the fibrous boards.

Steam piping insulation varies with temperature and location, as discussed earlier. Calcium silicate wired in place and then jacketed with corrugated or plain aluminum is very widely used. The jacketing is either screwed at the overlap or banded in place. Fiberglass is used extensively in low-pressure steam work in areas of limited abuse. Mineral wool and expanded perlite can also be used for higher-temperature steam, but calcium silicate is the standard.

Process Work

Hot process piping and vessels are typically insulated with calcium silicate, mineral wool, or high-temperature fiberglass. Horizontal applications are generally subject to more abuse than vertical and as such have a higher usage of calcium silicate. Many vessels have the insulation banded in place and then the metal lagging banded in place separately. Most vessel heads have a cement finish and may or may not be subsequently covered with metal. Recent product developments have provided a fiberglass wraparound product for large-diameter piping and vessels. This flexible material conforms to the curvature and need only be pinned at the bottom of a horizontal vessel. Banding is then used to secure both the insulation and the jacketing. In areas of chemical contamination, stainless steel jacketing is frequently used.

Cold process vessels and piping also use a variety of insulations, depending on the minimum temperature and the thermal efficiency required. Cellular glass is widely used in areas where the closed-cell structure is an added safeguard (the material is still applied with a vapor-barrier jacket or coating). It is also used wherever there is a combined need for closed-cell structure and high compressive strength. However, the polyurethane materials are much more efficient thermally, and in cryogenic work, maximum thermal resistance is often required. Multiple vapor barriers are used with the urethanes to prevent the migration of moisture throughout the entire system. In all cold work, the workmanship, particularly on the outer vapor barrier, is extremely critical. There are many other specially engineered systems for cryogenic work, as discussed in Section 15.3.1.

Fluid storage tanks located outdoors are typically insulated with fiberglass insulation. Prefabricated panels are either installed on studs or banded in place. Also, the jacketing can be banded on separately over the insulation. A row of cellular glass is placed along the base of the tank to prevent moisture from wicking up into the fiberglass. Sprayed urethane is also used on tanks that will not exceed 200°F but a trade-off exists between cost efficiency and the long-term durability of the system.

Tank roofs are a problem because of the need for a rigid walking surface as well as a lagging system that will shed water. Many tops are left bare for this reason, whereas others utilize a spray coating of cork-filled mastic, which provides only minimal insulation. Rigid fiberglass systems can be made to work with a well-designed covering system that drains properly. The most secure system is to use a built-up roofing system similar to those used on flat-top buildings. The installation is generally more costly, but acceptable long-term performance is much more probable.

HVAC System

Duct work constructed of sheet metal is usually wrapped with light-density fiberglass with a preapplied foil and kraft facing. The blanket is overlapped and then stapled, with tape or mastic being applied if the duct flow is cold and a vapor barrier is required. Support pins are required to prevent sag on the bottom of horizontal ducts. Fiberglass duct liner is used inside sheet-metal ducts to provide better sound attenuation along the duct; this provides a thermal benefit as well. For exposed duct work, a heavier-density fiberglass board may be used as a wrap to provide a more acceptable appearance. In all cases, the joints in the sheet metal ducts should be sealed with tape or caulking to minimize air leakage and allow the transport of air to the desired location, rather than losing much of it along the run.

Rigid fiberglass duct board and round duct are also used in many low-pressure applications. These products form the duct itself as well as providing the thermal, acoustical, and vapor-barrier requirements most often needed. The closure system used to join the duct sections also acts to seal the system for minimum air leakage.

Chillers and chilled water expansion tanks are usually insulated with closed-cell elastomeric sheet to prevent condensation on the equipment. The joints are sealed and a finish may or may not be applied to the outside, depending on location.

Piping for both hot and cold service is normally insulated with fiberglass pipe insulation. On cold work, the vapor-barrier jacket is sealed at the overlap with either an adhesive or a factory-applied self-seal lap. If staples are used, they should be dabbed with mastic to secure the vapor resistance. Aluminum jacketing is often used on outside work with a vapor barrier applied beneath if it is cold service. Domestic hot- and cold-water plumbing and rain leaders are also commonly insulated with fiberglass. Insulation around piping supports takes many forms, depending on the nature of the hanger or support. On cold work, the use of a clevis hanger on the outside of the insulation requires a high-density insert to support the weight of the piping. This system eliminates the problem of adequately sealing around penetrations of the vapor barrier.

15.4 INSULATION THICKNESS DETERMINATION

This section presents formulas and graphical procedures for calculating heat loss, surface temperature, temperature drop, and proper insulation thickness. Although the overall objective is to determine the right amount of insulation that should be used, some of the equations use thickness as an input variable rather than solving for it. However, all the calculations are simply manipulations or further refinements of the equation in Section 15.1.2:

$$Q = \frac{\Delta t}{R_I + R_s}$$

Following is a list of symbols, definitions, and units to be used in the heat-transfer calculations.

t_a = ambient temperature, °F

t_s = surface temperature of insulation next to ambient, °F

t_h = hot surface temperature, normally operating temperature (cold surface temperature in cold applications), °F

k = thermal conductivity of insulation always determined at mean temperature, Btu-in./hr ft^2 °F

t_m = $(t_h + t_s)/2$ = mean temperature of insulation, °F

t_h = $(t_{in} + t_{out})/2$ = average hot temperature when fluid enters at one temperature and leaves at another, °F

tk = thickness of insulation, in.

r_1 = actual outer radius of steel pipe or tubing, in.

r_2 = $(r_1 + tk)$ = radius to outside of insulation on piping, in.

Eq tk = $r_2 \ln (r_2/r_1)$ = equivalent thickness of insulation on a pipe, in.

f = surface air film coefficient, Btu/hr ft^2 °F

R_s = $1/f$ = surface resistance, hr ft^2 °F/Btu

R_I = tk/k = thermal resistance of insulation, hr ft^2 °F/Btu

Q_F = heat flux through a flat surface, Btu/hr ft^2

Q_p = $Q_F(2\pi r_2/12)$ = heat flux through a pipe, Btu/hr lin. ft

A = area of insulation surface, ft^2

L = length of piping, lin. ft

Q_T = $Q_F \times A$ or $Q_p \times L$ = total heat loss, Btu/hr

H = time, hr

C_p = specific heat of material. Btu/lb · °F

ρ = density, lb/ft^3

M = mass *flow* rate of a material, lb/hr

Δ = difference by subtraction, unit less

RH = relative humidity, %

DP = dew-point temperature, °F

15.4.1 Thermal Design Objective

The first step in determining how much insulation to use is to define what the objective is. There are many reasons for using insulation, and the amount to be used will definitely vary based on the objective chosen. The four broad categories, which include most applications, are (1) personnel protection, (2) condensation control, (3) process control, and (4) economics. Each of these is discussed in detail, with sample problems leading through the calculation sequence.

15.4.2 Fundamental Concepts

Thermal Equilibrium

A very important law in heat transfer is that under steady-state conditions, the heat *flow* through any portion of the insulation system is the same as the heat *flow* through any other part of the system. Specifically, the heat *flow* through the insulation equals the heat *flow* from the surface to the ambient, so the temperature difference for each section is proportional to the resistance for each section:

$$\text{Heat flow} = \frac{\text{temperature difference}}{\text{resistance to heat flow}}$$

$$Q = \frac{t_h - t_a}{R_I + R_s} = \frac{t_h - t_s}{R_I} = \frac{t_s - t_a}{R_s}$$

Because all of the heat flows Q are equal, this relationship is used to check surface temperature or other interface temperatures. For an analysis concerned with the inner surface film coefficient, the same reasoning applies.

$$Q = \frac{t_h - t_a}{R_{s1} + R_I + R_{s2}} = \frac{t_h - t_{s1}}{R_{s1}} = \frac{t_{s1} - t_{s2}}{R_I} = \frac{t_{s2} - t_a}{R_{s2}}$$

Or for a system with two insulation materials involved, the interface temperature t_{if} between the materials is involved.

$$Q = \frac{t_h - t_a}{R_{I1} + R_{I2} + R_s} = \frac{t_h - t_{if}}{R_{I1}} = \frac{t_{if} - t_s}{R_{I2}} = \frac{t_s - t_a}{R_s}$$

$$= \frac{t_{if} - t_a}{R_{I2} + R_s}$$

It should be apparent that the heat flow Q is also equal for any combination of Δt and R values, as shown by the last equivalency above, which utilized two parts of the system instead of just one.

Finally, it is of critical importance to calculate the R_I values using the insulation mean temperature, not the operating temperature. The mean temperature is the sum of the temperatures on either side of the insulation divided by 2. Again for the last set of equivalencies:

$$t_m \text{ for } R_{I1} = \frac{t_h + t_{if}}{2}$$

$$t_m \text{ for } R_{I2} = \frac{t_{if} + t_s}{2}$$

Pipe vs. Flat Calculations—Equivalent Thickness

Because the radial heat flows in a path from a smaller-diameter pipe, through the insulation, and then off a larger-diameter surface, a phenomenon termed "equivalent thickness" (Eq tk) occurs. Because of the geometry and the dispersion of the heat to a greater area, the pipe really "sees" more insulation than is actually there. When the adjustment is made to enter a greater insulation thickness into the calculation, the standard flat geometry formulas can be used by substituting Eq tk for tk into the equations.

The formula for equivalent thickness is

$$\text{Eq tk} = r_2 \ln \frac{r_2}{r_1}$$

where r_1 and r_2 are the inner and outer radii of the insulation system. For example, an 8-in. IPS with 3-in.

insulation would lead to an equivalent thickness as follows (8-in. IPS has 8.625 in. actual outside diameter):

$$r_1 = \frac{8.625}{2} = 4.31 \quad \text{(Table 15.2)}$$

$$r_2 = r_1 + \text{tk} = 4.31 + 3 = 7.31$$

$$\text{Eq tk} = 7.31 \ln \frac{7.31}{4.31} = 7.31 \ln 1.70$$

$$= 3.86 \quad \textit{actual outside diameter}$$

This Eq tk can then be used in the flat geometry equation by substituting Eq tk for tk.

$$Q = \frac{t_h - t_a}{\text{Eq tk}/k + R_s}$$

The example above used an even insulation thickness of 3 in. Some products are manufactured to such even thicknesses, and Table 15.2 lists the Eq tk for such products. However, many products are manufactured to "simplified" thicknesses, which allow a proper fit when nesting double-layer materials. ASTM-C-5859 lists these standard dimensions, and Table 15.3 shows Eq tk values for the simplified thicknesses. Figure 15.3 also shows the conversion for any thickness desired and will be used later in the reverse fashion.

Surface Resistance

There is always diversity of opinion when it comes to selecting the proper values for the surface resistance R_s. The surface resistance is affected by surface emittance, surface air velocity, and the surrounding environment. Heat-transfer texts have developed procedures for calculating R_s values, but they are all based on speculated values of emittance and air velocity. In actuality, the emittance of a surface often changes with time, temperature, and surface contamination, such as dust. As a result, it is unnecessary to labor over calculating specific R_s values, when the conditions are estimates at best.

Table 15.4 lists a series of R_s values based on three different surface conditions and the temperature difference between the surface and ambient air. Also included are single-point R_s values for three different surface air velocities. See the note at the bottom of Table 15.4 relating to the effect of R_s on heat-transfer calculations.

15.4.3 Personnel Protection

Workers need to be protected from high-temperature piping and equipment in order to prevent skin burns. Before energy conservation analyses became commonplace, many insulation systems were designed simply to maintain a "safe-touch" temperature on the outer jacket. Now, with energy costs so high, personnel protection calculations are generally limited to temporary installations or waste-heat systems, where the energy being transferred will not be further utilized.

Normally, safe-touch temperatures are specified in the range 130 to 150°F, with 140°F being used most often. It is important to remember that the surface temperature is directly related to the surface resistance R_S, which in turn depends on the emittance of the surface. As a result, an aluminum jacket will be hotter than a dull mastic coating over the same amount of insulation. This is demonstrated below.

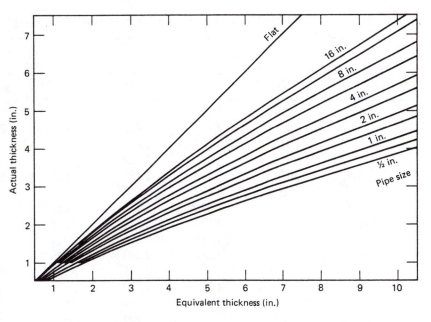

Fig. 15.3 Equivalent thickness chart. (From Ref. 16.)

Calculation

The objective is to calculate the amount of insulation required to attain a specific surface temperature. As

Table 15.2 Equivalent thickness values for even insulation thicknesses.

Nominal Pipe Size (in.)	r_1	Actual Thickness (in.)						
		1	1-1/2	2	2-1/2	3	3-1/2	4
1/2	0.420	1.730	2.918	4.238	5.662	7.172	8.755	10.402
3/4	0.525	1.626	2.734	3.966	5.297	6.712	8.199	9.747
1	0.658	1.532	2.563	3.711	4.953	6.275	7.665	9.117
1-1/4	0.830	1.447	2.405	3.472	4.626	5.856	7.153	8.507
1-1/2	0.950	1.403	2.321	3.342	4.449	5.629	6.872	8.171
2	1.188	1.337	2.195	3.148	4.177	5.276	6.436	7.648
2-1/2	1.438	1.287	2.099	2.997	3.968	5.001	6.093	7.234
3	1.750	1.242	2.012	2.858	3.771	4.742	5.768	6.840
3-1/2	2.000	1.217	1.959	2.772	3.649	4.582	5.564	6.592
4	2.250	1.194	1.916	2.704	3.549	4.448	5.396	6.386
4-1/2	2.500	1.178	1.880	2.645	3.464	4.337	5.253	6.211
5	2.781	1.163	1.846	2.590	3.388	4.231	5.118	6.043
6	3.313	1.138	1.799	2.510	3.270	4.071	4.911	5.790
7	3.813	1.120	1.761	2.453	3.184	3.956	4.759	5.604
8	4.313	1.108	1.737	2.407	3.116	3.863	4.644	5.452
9	4.813	1.097	1.714	2.369	3.056	3.783	4.541	5.330
10	5.375	1.088	1.693	2.333	3.007	3.714	4.450	5.214
11	5.875	1.079	1.675	2.305	2.972	3.663	4.383	5.123
12	6.375	1.076	1.662	2.286	2.936	3.619	4.321	5.048
14	7.000	1.069	1.647	2.265	2.900	3.569	4.258	4.969
16	8.000	1.059	1.639	2.231	2.858	3.504	4.178	4.866
18	9.000	1.053	1.622	2.206	2.822	3.449	4.110	4.776
20	10.000	1.048	1.608	2.188	2.789	3.411	4.051	4.711
24	12.000	1.040	1.589	2.163	2.736	3.347	3.971	4.598
30	15.000	1.032	1.572	2.122	2.704	3.281	3.874	4.497

Source: Ref. 16.

Table 15.3 Equivalent thickness values for simplified insulation thicknesses.

Nominal Pipe Size (in.)	Actual Thickness (in.)							
	r^1	1	1-1/2	2	2-1/2	3	3-1/2	4
1/2	0.420	1.730	3.053	4.406	6.787	8.253	9.972	12.712
3/4	0.523	1.435	2.660	3.885	5.996	7.447	8.965	10.642
1	0.638	1.715	2.770	4.013	5.358	6.702	8.112	9.581
1-1/4	0.830	1.281	2.727	3.333	4.552	5.777	7.070	8.420
1-1/2	0.950	1.457	2.382	4.025	5.253	6.476	7.759	9.179
2	1.188	1.438	2.367	3.398	4.446	5.561	6.733	8.027
2-1/2	1.438	1.383	2.765	3.657	4.737	5.815	7.015	8.195
3	1.750	1.286	2.114	2.968	3.889	4.868	5.965	7.046
3-1/2	2.000	1.625	2.459	3.258	4.166	5.251	6.266	7.256
4	2.230	1.281	2.010	2.806	3.659	4.059	5.577	6 543
4-1/2	2.300	1.564	2.351	3.152	4.905	4.962	5.907	7 080
5	2.781	1.202	1.893	2.639	3.489	4.339	5.230	6.461
6	3.313	1.138	1.799	2.555	3.317	4.122	5.237	6.015
7	3.813		1.804	2.495	3.230	4.153	4.969	5.821
8	4.313		1.776	2.445	3.391	4.010	4.842	5.768
9	4.813		1.752	2.579	3.232	3.971	4.786	5.583
10	5.375		1.810	2.457	3.108	3.850	4.591	5.361
11	5.875		1.793	2.428	3.140	3.793	4.519	5.271
12	6.375		1.777	2.405	3.103	3.745	4.456	5.241
14	7.000		1.647	2.265	2.900	3.569	4.258	4.969
16	8.000		1.639	2.231	2.858	3.504	4.178	4.866
18	9.000		1.622	2.206	2.822	3.449	4.110	4.776
20	10.000		1.608	2.188	2.789	3.411	4.051	4.711
24	12.000		1.589	2.163	2.736	3.347	3.971	4.598
30	15.000		1.572	2.122	2.704	3.281	3.874	4.497

Source: Ref. 16.

Table 15.4 R_s Values[a] (hr $\cdot$ ft^2 °F/Btu).

$t_s - t_a$ (°F)	Still Air Plain, Fabric, Dull Metal: $\varepsilon = 0.95$	Aluminum: $\varepsilon = 0.2$	Stainless Steel: $\varepsilon = 0.4$
10	0.53	0.90	0.81
25	0.52	0.88	0.79
50	0.50	0.86	0.76
75	0.48	0.84	0 75
100	0.46	0.80	0 72
With Wind Velocities			
Wind Velocity (mph)			
5	0.35	0.41	0.40
10	0.30	0.35	0.34
20	0.24	0.28	0.27

Source: Courtesy of Johns-Manville, Ref. 16.
[a]For heat-loss calculations, the effect of R_s is small compared to R_I, so the accuracy of R_s is not critical. For surface temperature calculations, Rs is the controlling factor and is therefore quite critical. The values presented in Table 15.4 are commonly used values for piping and flat surfaces. More precise values based on surface emittance and wind velocity can be found in the references.

noted earlier,

$$\frac{t_h - t_s}{R_I} = \frac{t_s - t_a}{R_s}$$

Therefore,

$$R_I = R_s \left(\frac{t_h - t_s}{t_s - t_a}\right) = \frac{tk}{k} \text{ (flat)} = \frac{Eq\ tk}{k} \text{ (pipe)}$$

Therefore,

$$tk\ or\ Eq\ tk = kR_s \left(\frac{t_h - t_s}{t_s - t_a}\right)$$

Example. For a 4-in. pipe operating at 700°F in an 85°F ambient temperature with aluminum jacketing over the insulation, determine the thickness of calcium silicate that will keep the surface temperature below 140°F.

Since this is a pipe, the equivalent thickness must first be calculated and then converted to actual thickness.

STEP 1. Determine k at $t_m = (700 + 140)/2 = 420°F$. $k = 0.49$ from appendix Figure 15.A1 for calcium silicate.

STEP 2. Determine R_s from Table 15.4 for aluminum. $t_s - t_a = 140 - 85 = 55$. So $R_s = 0.85$.

STEP 3. Calculate Eq tk:

$$Eq\ tk = (0.49)(0.85)\frac{700 - 140}{140 - 85}$$

$$= 4.24\ in.$$

STEP 4. Determine the actual thickness from Figure 15.3. The effect of 4.24 in. on a 4-in. pipe can be accomplished by using 2.9 in. of insulation.

STEP 5. The thickness recommended for this application is 3 in. of calcium silicate pipe insulation.

Note: Thickness recommendations are always increased to the next 1-in. increment. If a surface temperature calculation happens to fall precisely on an even increment (such as 3 in.), it is advisable to be conservative and increase to the next increment (such as 3-1/2 in.). This reduces the criticality of the R_s number used. In the preceding example, it would not be unreasonable to recommend 3-1/2 in. of insulation, since it was found to be so close to 3 in.

To illustrate the effect of surface type, consider the same example with a mastic coating.

Example. From Table 15.4, $R_s = 0.50$, so

$$Eq\ tk = (0.49)(0.50)\frac{700 - 140}{140 - 85}$$

$$= 2.49\ in.$$

This corresponds to an actual thickness requirement on a 4-in. pipe of 1.8 in., which rounds up to 2 in. This compares with 3 in. required for an aluminum-jacketed system. It is of interest to note that even though the aluminum system has a higher surface temperature, the actual heat loss is less because of the higher surface resistance value.

Graphical Method

The calculations illustrated above can also be carried out using graphs which set the heat loss through the insulation equal to the heat loss off the surface, following the discussion in Section 15.4.2.

Figure 15.4 will be used for several different calculations. The following example gives the four-step procedure for achieving the desired surface temperature for personnel protection. The accompanying diagram outlines this procedure.

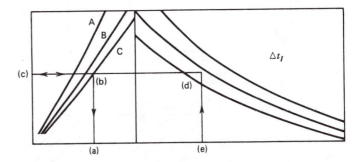

Example. We follow the procedure of the first example, again using aluminum jacketing.

STEP 1. Determine $t_s - t_a$, $140 - 85 = 55°F$.

STEP 2. In the diagram, proceed vertically from (a) of $\Delta t = 55$ to the curve for aluminum jacketing (b).

STEP 2a. Although not required, read the heat loss $Q = 65$ Btu/hr ft^2) (c).

STEP 3. Proceed to the right to (d), the appropriate curve for $t_h - t_s = 700 - 140 = 560°F$. Interpolate between lines as necessary.

STEP 4. Proceed down to read the required insulation resistance $R_t = 8.6$ at (e). Since $R = tk/k$ or $Eq\ tk/k$, tk or Eq tk $= R_lk$

$$t_m = \frac{700 + 140}{2} = 420°F$$

$k = 0.49$ from appendix Figure 15.A1 and

tk or Eq tk $= (8.6)(0.49) = 4.21$ in.

which compares well with the 4.24 in. from the earlier calculation.

The conversion of Eq tk to actual thickness required for pipe insulation is done in the same manner, using Figure 15.3.

A better understanding of the procedure involved in utilizing this quick graphical method will be obtained after working through the remainder of the calculations in this section.

15.4.4 Condensation Control

On cold systems, either piping or equipment, insulation must be employed to prevent moisture in the warmer surrounding air from condensing on the colder surfaces. The insulation must be of sufficient thickness to keep the insulation surface temperature above the dew point of the surrounding air. Essentially, the calculation procedures are identical to those for personnel protection except that the dew-point temperature is substituted for the desired surface temperature. (*Note:* The surface temperature should be kept 1 or 2° above the dew point to prevent condensation at that temperature.)

Dew-Point Determination

The condensation (saturation) temperature, or dew point, is dependent on the ambient dry-bulb and wet-bulb temperatures. With these two values and the use of a psychrometric chart, the dew point can be determined. However, for most applications, the relative humidity is more readily attainable, so the dew point is determined using dry-bulb temperature and relative humidity rather

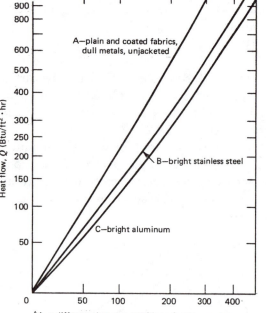

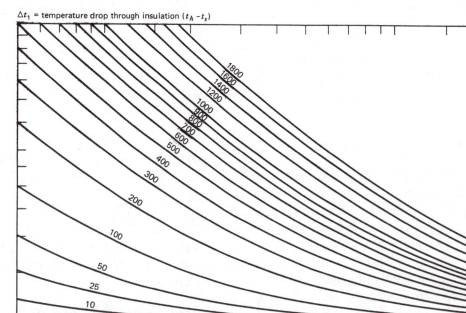

Fig. 15.4 Heat loss and surface temperature graphical method. (From Ref. 16.)

than wet-bulb. Table 15.5 is used to find the proper dew-point temperature.

Calculation

This equation is identical to the previous surface-temperature problem except that the surface temperature t_s now takes on the value of the dew point of the ambient air. Also, t_h now represents the cold operating temperature.

$$tk \text{ or Eq } tk = kR$$

Example. For a 6-in.-diameter chilled-water line operating at 35°F in an ambient of 90°F and 85% RH, determine the thickness of fiberglass pipe insulation with a composite kraft paper jacket required to prevent condensation.

STEP 1. Determine the dew point (DP) using either a psychrometric chart or Table 15.5. DP at 90°F and 85% RH = 85°F.

STEP 2. Determine k at $t_m = (35 + 85)/2 = 60°F$. k at 60°F = 0.23, from appendix Figure 15.A2.

STEP 3. Determine R_s from Table 15.4. Δt here is $(t_a, - t_s)$ rather than $(t_s - t_a)$, $t_a - t_s = 90 - 85 = 5°F$, $R_s = 0.54$.

STEP 4. Calculate Eq tk.

$$\text{Eq } tk = (0.23)(0.54)\frac{35-85}{85-90}$$

$$= 1.24 \text{ in.}$$

STEP 5. Determine the actual thickness from Figure 15.3 for 6-in. pipe, 1.24 in. Eq tk. The actual thickness is 1.1 in.

STEP 6. The thickness recommended for this application is 1-1/2 in. of fiberglass pipe covering.

Table 15.5 Dew-point temperature.

Dry-Bulb Temp. (°F)	Percent Relative Humidity																		
	10	15	20	25	30	35	40	45	50	55	60	65	70	75	80	85	90	95	100
5	−35	−30	−25	−21	−17	−14	−12	−10	−8	−6	−5	−4	−2	−1	1	2	3	4	5
10	−31	−25	−20	−16	−13	−10	−7	−5	−3	−2	0	2	3	4	5	7	8	9	10
15	−28	−21	−16	−12	−8	−5	−3	−1	1	3	5	6	8	9	10	12	13	14	15
20	−24	−16	−8	−4	−2	2	4	6	8	10	11	13	14	15	16	18	19	20	
25	−20	−15	−8	−4	0	3	6	8	10	12	15	16	18	19	20	21	23	24	25
30	−15	−9	−3	2	5	8	11	13	15	17	20	22	23	24	25	27	28	29	30
35	−12	−5	1	5	9	12	15	18	20	22	24	26	27	28	30	32	33	34	35
40	−7	0	5	9	14	16	19	22	24	26	28	29	31	33	35	36	38	39	40
45	−4	3	9	13	17	20	23	25	28	30	32	34	36	38	39	41	43	44	45
50	−1	7	13	17	21	24	27	30	32	34	37	39	41	42	44	45	47	49	50
55	3	11	16	21	25	28	32	34	37	39	41	43	45	47	49	50	52	53	55
60	6	14	20	25	29	32	35	39	42	44	46	48	50	52	54	55	57	59	60
65	10	18	24	28	33	38	40	43	46	49	51	53	55	57	59	60	62	63	65
70	13	21	28	33	37	41	45	48	50	53	55	57	60	62	64	65	67	68	70
75	17	25	32	37	42	46	49	52	55	57	60	62	64	66	69	70	72	74	75
80	20	29	35	41	46	50	54	57	60	62	65	67	69	72	74	75	77	78	80
85	23	32	40	45	50	54	58	61	64	67	69	72	74	76	78	80	82	83	85
90	27	36	44	49	54	58	62	66	69	72	74	77	79	81	83	85	87	89	90
95	30	40	48	54	59	63	67	70	73	76	79	82	84	86	88	90	91	93	95
100	34	44	52	58	63	68	71	75	78	81	84	86	88	91	92	94	96	98	100
105	38	48	56	62	67	72	76	79	82	85	88	90	93	95	97	99	101	103	105
110	41	52	60	66	71	77	80	84	87	90	92	95	98	100	102	104	106	108	110
115	45	56	64	70	75	80	84	88	91	94	97	100	102	105	107	109	111	113	115
120	48	60	68	74	79	85	88	92	96	99	102	105	107	109	112	114	116	118	120
125	52	63	72	78	84	89	93	97	100	104	107	109	111	114	117	119	121	123	125

Graphical Method

The graphical procedures are as described in Section 15.4.3. As the applications become colder, it is apparent that the required insulation thicknesses will become larger, with R_I values toward the right side of Figure 15.4. It is suggested that the graphical procedure not be used when the resulting R_I values must be determined from a very flat portion of the $(t_h - t_s)$ curve. It is difficult to read the graph with sufficient accuracy, particularly in light of the simplicity of the mathematical calculation.

Thickness Chart for Fiberglass Pipe Insulation

Table 15.6 gives the thickness requirements for fiberglass pipe insulation with a white, all-purpose jacket in still air. The calculations are based on the lowest temperature in each temperature range. Three temperature/humidity conditions are depicted.

15.4.5 Process Control

Included under this heading will be all the calculations other than those for surface temperature and economics. It is often necessary to calculate the heat flow through a given insulation thickness, or conversely, to calculate the thickness required to achieve a certain heat flow rate. The final situation to be addressed deals with the temperature drop in both stagnant and flowing systems.

Heat Flow for a Specified Thickness

Calculation Equations. Again, the basic equation for a single insulation material is

$$Q_F = \frac{t_h - t_a}{R_I + R_s}$$

Example. For an 850°F boiler operating indoors in an 80°F ambient temperature insulated with 4 in. of calcium silicate covered with 0.016 in. aluminum jacketing, determine the heat loss per square foot of boiler surface and the surface temperature.

STEP 1. Find k for calcium silicate at tm. Assume that $ts = 140°F$. Then $tm = (850 + 140)/2 = 495°F$, k at 495°F = 0.53, from appendix Figure 15.A1.

STEP 2. Determine R_s for aluminum from Table 15.4. $ts - ta = 140 - 80 = 60°F$, so $R_s = 0.85$.

STEP 3. Calculate $R_I = 4/0.53 = 7.5$.

STEP 4. Calculate

$$Q_F = \frac{850 - 80}{7.5 + 0.85} = 92 \text{ Btu/hr} \cdot \text{ft}^2$$

STEP 5. Calculate the surface temperature ts, as follows:

$$R_s \times Q_F = t_s - t_a$$
$$(R_s \times Q_F) + ta = ts$$
$$t_s = (0.85 \times 92) + 80$$
$$= 158°F$$

Table 15.6 Fiberglass pipe insulation: minimum thickness to prevent condensation[a].

Operating Pipe Temperature (°F)	80°F and 90% RH		80°F and 70% RH		80°F and 50% RH	
	Pipe Size (in.)	Thickness (in.)	Pipe Size (in.)	Thickness (in.)	Pipe Size (in.)	Thickness (in.)
0-34	Up to 1	2	Up to 8	1	Up to 8	1/2
	1-1/4 to 2	2-1/2	9-30	1-1/2	9-30	1
	2-1/2 to 8	3				
	9-30	3-1/2				
35-49	Up to 1-1/2	1-1/2	Up to 4	1/2	Up to 30	1/2
	2-8	2	412-30			
	9-30	3				
50-70	Up to 3	1-1/2	Up to 30	1/2	Up to 30	1/2
	3-1/2 to 20	2				
	21-30	2-1/2				

Source: Courtesy of Johns-Manville, Ref. 16.
[a]Based on still air and AP Jacket.

STEP 6. Calculate t_m to check assumption and to check the k value used.

$$t_m = \frac{850 + 158}{2}$$

$$= 504°F$$

As k at $504°F = k$ at $495°F$ (assumed) $= 0.53$, the assumption is okay. A check on R_s can also be made based on the calculated surface temperature.

STEP 7. If the assumption is not okay, recalculate using a new k value based on the new t_m.

The QF used above is for flat surfaces. In determining heat flow from a pipe, the same equations are used with Eq tk substituted for tk in the R_I calculation as discussed in Section 15.4.2. Often, it is desired to express pipe heat losses in terms of Btu/hr-lin.-ft rather than Btu/hr ft². This is termed Q_P, with

$$Q_P = Q_F \left(\frac{2\pi r_2}{12}\right)$$

Graphical Method. Figure 15.4 may again be used in lieu of calculations. The main difference from the previous chart usage is that surface temperature is now an unknown, and must be determined such that thermal equilibrium exists.

Example. Determine the heat loss from the side walls of a vessel operating at 300°F in an 80°F ambient temperature. Two inches of 3-lb/ft³ fiberglass is used with aluminum lagging.

STEP 1. Assume a surface temperature $t_s = 120°F$.

STEP 2. Calculate

$$t_m = \frac{t_h + t_s}{2} = \frac{300 + 120}{2} = 210°F$$

Determine k from appendix Figure 15.A3 at 210°F. $k = 0.27$.

STEP 3. Calculate $R_I = tk/k = 2/0.27 = 7.41$.

STEP 4. Go to position (a) on the chart shown for $RI = 7.41$ and read vertically to (b), where $t_h - t_s = 180°F$.

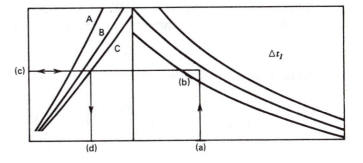

STEP 5. Read to the left to (c) for heat loss $Q = 24$ Btu/hr ft².

STEP 6. Read down from the proper surface curve to (d), which represents $ts - ta$, to check the surface-temperature assumption. For aluminum, $ts - ta$ (chart) is 21°F, compared with the $120 - 80 = 40°F$ assumption.

STEP 7. Calculate a new surface temperature $80 + 21 = 101°F$; then calculate a new $tm, = (300 + 101)/2 = 200.5°F$. Then find a new $k = 0.26$, which gives a new $RI = 2/126 = 7.69$.

STEP 8. Return to step 4 with the new RI and proceed. This example shows the insensitivity of heat loss to changes in surface temperature since the new $Q = 22$ Btu/hr ft².

For pipe insulation, the same procedure is followed except that RI is calculated using the equivalent thickness. Also, conversion to heat loss per linear foot must be done separately after the square-foot loss is determined.

Thickness for a Specified Heat Loss

Again, a surface temperature t_s must first be assumed and then checked for accuracy at the end of the calculation.

From Section 15.4.2,

$$Q = \frac{t_h - t_s}{R_I} = \frac{t_h - t_s}{tk/k}$$

$$tk \text{ (or Eq tk)} = k\left(\frac{t_h - t_s}{Q}\right)$$

where k is determined at $t_m = (t_h + t_s)/2$.

Example. How much calcium silicate insulation is required on a 650°F duct in an 80°F ambient temperature if the maximum heat loss is 50 Btu/hr ft²? The insulation will be finished with a mastic coating.

STEP 1. Assume that $ts = 105°F$. So $t_m = (650 + 105)/2 = 377°F$. k from appendix Figure 15.A1 at 377°F $= 0.46$.

STEP 2. Find tk as follows:

$$\text{tk} = k\left(\frac{t_h - t_s}{Q}\right) = 0.46\left(\frac{650 - 105}{50}\right)$$

$$= 5.01 \text{ in.}$$

STEP 3. Check surface temperature assumption by $t_s = (Q \times R_s) + t_a$ using $R_s = 0.52$. From Table 15.4 for a mastic finish,

$$t_s = 50(0.52) + 80$$
$$= 106°F$$

(Note that this in turn changes the $ts - ta$ from 40 to 25, which changes R_s from 0.49 to 0.51, which is insignificant.)

For a graphical solution to this problem, Figure 15.4 is again used. It is simply a matter of reading across the desired Q level and adjusting the t_s and R_I values to reach equilibrium. Thickness is then determined by $tk = kR_I$.

Temperature Drop in a System

The following discussion is quite simplified and is not intended to replace the service of the process design engineer. The material is presented to illustrate how insulation ties into the process design decision.

Temperature Drop in Stationary Media over Time. The procedure calls for standard heat-flow calculations now tied into the heat content of the fluid. To illustrate, consider the following example.

Example. A water storage tank is calculated to have a surface area of 400 ft² and a volume of 790 ft³. How much will the temperature drop in a 72-hr period with an ambient temperature of 0°F, assuming that the initial water temperature is 50°F? The tank is insulated with 2-in. fiberglass with a mastic coating.

Before proceeding, realize that the maximum heat transfer will occur when the water is at 50°F. As it drops in temperature, the heat-transfer rate is reduced due to a smaller temperature difference. As a first approximation, it is reasonable to use the maximum heat transfer based on 50°F. Then if the temperature drop is significant, an average water temperature can be used in the second iteration.

STEP 1. Assume a surface temperature, calculate the mean temperature, find the k factor from appendix Figure 15A.3, and determine R_S from Table 15.4. With $t_S = 10°F$, $t_m = 30°F$, $k = 0.22$, and $R_S = 0.53$.

STEP 2. Calculate heat loss with fluid at 50°F.

$$Q_F = \frac{50 - 0}{(2/0.22) + 0.53} = 5.2 \text{ Btu/hr} \cdot \text{ft}^2$$

$$Q_T = Q_F \times A = 5.2 (400) = 2080 \text{ Btu/hr}$$

STEP 3. Calculate the amount of heat that must be lost for the entire volume of water to drop 1°F.

Available heat per °F

$$= \text{volume} \times \text{density} \times \text{specific heat}$$
$$= 790 \text{ ft}^3 \times 62.4 \text{ lb/ft}^3 \times 1 \text{ Btu/lb °F}$$
$$= 49,296 \text{ Btu/°F}$$

STEP 4. Calculate the temperature drop in 72 hr by determining the total heat flow over the period: $Q = 2080 \times 72 = 149,760$ Btu. Divide this by the available heat per 1°F drop:

$$\frac{149,760 \text{ Btu}}{49,296 \text{ Btu/°F}} = 3.04°F \text{ drop}$$

This procedure may also be used for fluid lying stationary in a pipeline. In this case it is easiest to do all the calculations for 1 linear foot rather than for the entire length of pipe.

One conservative aspect of this calculation is that the heat capacity of the metal tank or pipe is not included in the calculation. Since the container will have to decrease in temperature with the fluid, there is actually more heat available than was used above.

Temperature Drop in Flowing Media. There are two common situations in this category, the first involving flue gases and the second involving water or other fluids with a thickening or freezing point. This section discusses the flue-gas problem and the following section, freeze protection.

A problem is encountered with flue gases that have fairly high condensation temperatures. Along the length of a duct run, the temperature will drop, so insulation is added to control the temperature drop. This calculation is actually a heat balance between the mass flow rate of energy input and the heat loss energy outlet.

For a round duct of radius r_1 and length L, gas enters at t_h, and must not drop below t_{min} (the dew point). The flow rate is M lb/hr and the gas has a specific heat of Cp Btu/lb °F. Therefore, the maximum allowable heat loss in Btu/hr is

$$Q_t = MC_{p,}\Delta t = MCp(t_h - t_{min})$$

Also,

$$Q_T = Q_P \times L \frac{t_h - t_a}{R_I + R_s} \times \frac{2\pi r_2}{12} \times L$$

where

$$t_h = \frac{t_{in} + t_{out}}{2}$$

= average gas temperature along the length

(A conservative simplification would be to set $t_h = t_{in}$ since the higher temperature, t_{in}, will cause a greater heat loss.)

To simplify on large ducts, assume that $r_1 = r_2$ (ignore the insulation-thickness addition to the surface area). Therefore,

$$\frac{t_h - t_a}{R_I + R_s} \times \frac{2\pi r_1}{12} \times L = MC_p\left(t_h - t_{min}\right)$$

and

$$R_I + R_s = \frac{t_h - t_a}{t_h - t_{min}} \times \frac{2\pi r_1}{12} \times L$$

Therefore,

$$R_t = \left[\frac{t_h - t_a}{MC_p\left(t_h - t_{min}\right)} \times \frac{2\pi r_1}{12} \times L\right] - R_s$$

$$= \frac{tk}{k} \quad \therefore \quad tk = R_1 \times k$$

Example. A 48-in.-diameter duct 90 ft long in a 60°F ambient temperature has gas entering at 575°F and 15,000 cfm. The standard conditions gas density is 0.178 lb/ft^3 and the gas outlet must not be below 555°F. $Cp = 0.18$ Btu/lb °F. Determine the thickness of calcium silicate required to keep the outlet temperature above 565°F, giving a 10°F buffer to account for the interior film coefficient. A more sophisticated approach calculates an interior film resistance R_s (interior) instead of using a 10°F or larger buffer. The resulting equation for Q_p would be

$$Q_p = \frac{t_h - t_a}{R_s \text{ (interior)} + R_I + R_s} \times \frac{2\pi r_2}{12}$$

This equation, however, will not be used.

STEP 1. Determine t_h the average gas temperature, = (575 + 565)/2 = 570°F. (A logarithmic mean could be calculated for more accuracy, but it is usually not necessary.)

STEP 2. Determine M lb/hr. The flow rate is 15,000 cfm of hot gas (570°F). At standard conditions (1 atm, 70°F), the flow rate must be determined by the absolute temperature ratio:

$$t_h = \frac{t_{start} + 32}{2}$$

$$\frac{t_h + 460}{70 + 460} = \frac{15,000}{\text{std. flow}}$$

$$\text{Std. flow} = 15,000\left(\frac{70 + 460}{570 + 460}\right)$$

$$= 6262 \text{ cfm std. gas (or scfm)}$$

$$M = 6262 \text{ cfm} \times 0.178 \text{ lb/ft}^3 \times 60 \text{ min/hr}$$

$$= 66,878 \text{ lb/hr}$$

STEP 3. Determine Rs from Table 15.4 assuming t_s = 80°F and a dull surface $R_S = 0.52$.

STEP 4. Calculate R_I.

$$R_I = \frac{570 - 60}{(66,878)(0.18)(575 - 565)} \times \frac{2\pi 24}{12} \times 90 - 0.52$$

$$= 4.79 - 0.52$$

$$= 4.27$$

STEP 5. Calculate the thickness. Assume that t_s = 80°F.

$$t_m = \frac{565 + 80}{2} = 323°F$$

k at 323°F = 0.45 for calcium silicate from Appendix Figure 15.A1.

tk = $R_I \times k = 4.27 \times 0.45$
= 1.93 in.

STEP 6. The thickness required for this application is 2 in. of calcium silicate. Again, a more conservative recommendation would be 2-1/2 in.

Note: The foregoing calculation is quite complex. It is, however, the basis for many process control and freeze-prevention calculations. The two equations for Q, can be manipulated to solve for the following:

Temperature drop, based on a given thickness and flow rate.

Minimum flow rate, based on given thickness and temperature drop.

Minimum length, based on thickness, flow rate, and temperature drop.

Freeze Protection. Four different calculations can be performed with regard to water-line freezing (or the unacceptable thickening of any fluid).

1. Determine the time required for a stagnant, insulated water line to reach 32°F.
2. Determine the amount of heat tracing required to prevent freezing.
3. Determine the flow rate required to prevent freezing of an insulated line.
4. Determine the insulation required to prevent freezing of a line with a given flow rate.

Calculations 1 and 2 relate to Section 15.4.5, where we dealt with stationary media. To apply the same principles to the freeze problems, the following modifications should be made.

a. In calculation 1, the heat transfer should be based on the average water temperature between the starting temperature and freezing:

$$\dot{t}_h = \frac{t_{start} + 32}{2}$$

b. Rather than solving for temperature drop, given the number of hours, the hours are determined based on

$$\text{hours to freeze} = \frac{\text{available heat}}{\text{heat loss/hr}} \frac{\text{Btu}}{\text{Btu/hr}}$$

where available heat is $WCp \ \Delta t$, with

$\quad W \quad = \text{lb of water}$
$\quad Cp \quad = \text{specific heat of water (1 Btu/lb °F)}$
$\quad \Delta t \quad = t_{start} - 32$

c. In calculation 2, the heat-loss value should be calculated based upon the minimum temperature at which the system should stay, for example, 35°F. The heat tracing should provide enough heat to the system to offset the naturally occurring losses of the pipe. Heat-trace calculations are quite complex

and many variables are involved. References 8 and 10 should be consulted for this type of work.

Calculations 3 and 4 relate to Section 4.5.3, dealing with flows. In the case of water, the minimum temperature can be set at 32°F and the heat-transfer rate is again on an operating average temperature

$$\dot{t}_h = \frac{t_{start} + 32}{2}$$

The equations given can be manipulated to solve for flow rate or insulation thickness.

As an aid in estimating the amount of insulation for freeze protection, Table 15.7 shows both the hours to freezing and the minimum flow rate to prevent freezing based on different insulation thicknesses. These figures are based on an initial water temperature of 420°F, an ambient temperature of − 10°F, a surface resistance of 0.54, and a thermal conductivity for fiberglass pipe insulation of $k = 0.23$.

15.4.6 Operating Conditions

Like all other calculations, heat-transfer equations yield results that are only as accurate as the input variables used. The operating conditions chosen for the heat-transfer calculations are critical to the result, and very misleading conclusions can be drawn if improper conditions are selected.

The term "operating conditions" refers to the environment surrounding the insulation system. Some of the variable conditions are operating temperature, ambient temperature, relative humidity, wind velocity, fluid type, mass flow rate, line length, material volume, and others. Since many of these variables are constantly changing, the selection of a proper value must be made on some logical basis. Following are three suggested methods for determining the appropriate variable values.

1. **Worst Case.** If a severe failure might occur with insufficient insulation, a worst-case approach is probably warranted. For example, freeze protection should obviously be based on the historical temperature extremes rather than on yearly averages. Similarly, exterior condensation control should be based on both ambient temperature and humidity extremes in addition to the lowest operating temperature. The *ASHRAE Handbook of Fundamentals* as well as U.S. Weather Bureau data give proper design conditions for most locales. In process areas, an appropriate example involves flue-gas condensation. Here the minimum flow rate is the most critical and should be used in the calculation.

Table 15.7 Hours to freeze and flow rate required to prevent freezing[a].

Nominal Pipe Size (in.)	1 in		2 in.		3 in	
	Hours to Freeze	gpm/100 ft	Hours to Freeze	gpm/100 ft	Hours to Freeze	gpm/100 ft
1/2	0.30	0.087	0.42	0.282	0.50	0.053
3/4	0.47	0.098	0.66	0.070	0.79	0.058
1	0.66	0.113	0.96	0.078	1.16	0.065
1-1/2	0.90	0.144	1.35	0.096	1.67	0.078
2	1.72	0.169	2.64	0.110	3.31	0.088
2-1/2	2.13	0.195	3.33	0.124	4.24	0.098
3	2.81	0.228	4.50	0.142	5.80	0.110
4	3.95	0.279	6.49	0.170	8.49	0.130
5	5.21	0.332	8.69	0.199	11.54	0.150
6	6.48	0.386	10.98	0.228	14.71	0.170
7	7.66	0.437	13.14	0.255	17.75	0.189
8	8.89	0.487	15.37	0.282	20.89	0.207

Source: Ref. 16.

[a]Calculations based on fiberglass pipe insulation with k = 0.23, initial water temperature of 42°F, and ambient air temperature of −10°F. Flow rate represents the gallons per minute required in a 100-ft pipe and may be prorated for longer or shorter lengths.

As a general rule, worst-case conditions will result in greater insulation thickness than will average conditions. In some cases the difference is very substantial, so it is important to determine initially if a worst-case calculation is required.

2. **Worst Season Average.** When a heating or cooling process is only operating part of a year, it is sensible to consider the average conditions only during that period of time. However, in year-round operations, a seasonal average is also justified in many cases. For example, personnel protection requires a maximum surface temperature that is dependent on the ambient air temperature. Taking the average summer daily maximum temperature is more practical than taking the absolute maximum ambient that could occur. The following example illustrates this.

Example. Consider an 8-in.-diameter, 600°F waste-heat line operating indoors with an average daily high of 80°F (but occasionally it will be 105°F). To maintain the surface below 135°F, 2 in. of calcium silicate is required with the 80°F ambient, whereas 3-1/2 in. is required with the 105°F ambient. The difference is significant and must be weighed against the benefit of the additional insulation in terms of worker safety.

3. **Yearly Average.** Economic calculations for continuously operating equipment should be based on yearly average operating conditions rather than on worst-case design conditions. Since the intent is to maximize the owner's financial return, an average condition will not overstate the savings as the worst case or worst season might. A good approach to process work is to calculate the economic thickness based on yearly averages and then check the sufficiency of that thickness under the worst-case design conditions. That way, both criteria are met.

15.4.7 Bare-Surface Heat Loss

It is often desirable to determine if any insulation is required and also to compare bare surface losses with those using insulation. Table 15.8 gives bare-surface losses based on the temperature difference between the surface and ambient air. Actual temperature conditions between those listed can be arrived at by interpolation. To illustrate, consider a bare, 8-in.-diameter pipe operating at 250°F in an 80°F ambient temperature. $\Delta t = 250 - 80 = 170°F$. Q for Δt of 150°F = 812.5 Btu/hr-lin.-ft; Q for Δt of 200°F = 1203 Btu/hr lin. ft. Interpolating between 150 and 200°F gives

$$Q_{170} = Q_{150} + (2/5)(Q_{200} - Q_{150})$$

$$= 812.5 + 0.4(1203 - 812.5)$$

$$= 968.7 \text{ Btu/hr lin. ft}$$

Table 15.8 Heat loss from bare surfaces[a].

Normal Pipe Size (in.)	Temperature Difference (°F)															
	50	100	150	200	250	300	350	400	450	500	550	600	700	800	900	1000
1/2	22	47	79	117	162	215	279	355	442	541	650	772	1,047	1,364	1,723	2,123
3/4	27	59	99	147	203	269	349	444	552	677	812	965	1,309	1,705	2,153	2,654
1	34	75	124	183	254	336	437	555	691	846	1,016	1,207	1,637	2,133	2,694	3,320
1-1/4	42	94	157	232	321	425	552	702	873	1,070	1,285	1,527	2,071	2,697	3,406	4,198
1-1/2	49	107	179	265	367	487	632	804	1,000	1,225	1,471	1,748	2,371	3,088	3,899	4,806
2	61	134	224	332	459	608	790	1,004	1,249	1,530	1,837	2,183	2,961	3,856	4,870	6,002
2-1/2	74	162	271	401	556	736	956	1,215	1,512	1,852	2,224	2,643	3,584	4,669	5,896	7,267
3	89	197	330	489	677	897	1,164	1,480	1,841	2,256	2,708	3,219	4,365	5,685	7,180	8,849
3-1/2	102	225	377	558	773	1,024	1,329	1,690	2,102	2,576	3,092	3,675	4,984	6,491	8,198	10,100
4	115	254	424	628	869	1,152	1,496	1,901	2,365	2,898	3,479	4,135	5,607	7,304	9,224	11,370
4-1/2	128	282	471	698	965	1,280	1,662	2,113	2,628	3,220	3,866	4,595	6,231	8,116	10,250	12,630
5	142	313	524	776	1,074	1,424	1,848	2,350	2,923	3,582	4,300	5,111	6,931	9,027	11,400	14,050
6	169	373	624	924	1,279	1,696	2,201	2,799	3,481	4,266	5,121	6,086	8,254	10,750	13,580	16,730
7	195	430	719	1,064	1,473	1,952	2,534	3,222	4,007	4,910	5,894	7,006	9,501	12,380	15,630	19,260
8	220	486	813	1,203	1,665	2,207	2,865	3,643	4,531	5,552	6,666	7,922	10,740	13,990	17,670	21,780
9	246	542	907	1,343	1,859	2,464	3,198	4,066	5,057	6,197	7,440	8,842	11,990	15,620	19,720	24,310
10	275	606	1,014	1,502	2,078	2,755	3,576	4,547	5,655	6,930	8,320	9,888	13,410	17,470	22,060	27,180
11	300	661	1,106	1,638	2,267	3,005	3,901	4,960	6,169	7,560	9,076	10,790	14,630	19,050	24,060	29,660
12	326	718	1,202	1,779	2,463	3,265	4,238	5,338	6,701	8,212	9,859	11,720	15,890	20,700	26,140	32,210
14	357	783	1,319	1,952	2,703	3,582	4,650	5,912	7,354	9,011	10,820	12,860	17,440	22,710	28,680	35,350
16	408	901	1,508	2,232	3,090	4,096	5,317	6,759	8,407	10,300	12,370	14,700	19,940	25,970	32,790	40,410
18	460	1,015	1,698	2,514	3,480	4,612	5,987	7,612	9,467	11,600	13,930	16,550	22,450	29,240	36,930	45,510
20	510	1,127	1,885	2,790	3,862	5,120	6,646	8,449	10,510	12,880	15,460	18,380	24,920	32,460	40,990	50,520
24	613	1,353	2,263	3,350	4,638	6,148	7,980	10,150	12,620	15,460	18,570	22,060	29,920	38,970	49,220	60,660
30	766	1,690	2,827	4,186	5,795	7,681	9,971	12,680	15,770	19,320	23,200	27,570	37,390	48,700	61,500	75,790
Flat	98	215	360	533	738	978	1,270	1,614	2,008	2,460	2,954	3,510	4,760	6,200	7,830	9,650

Source: Ref. 16.

[a]Losses given in Btu/hr lin. ft of bare pipe at various temperature differences and Btu/hr-ft^2 for flat surfaces. Heat losses were calculated for still air and $\varepsilon = 0.95$ (plain, fabric or dull metals).

15.5 INSULATION ECONOMICS

Thermal insulation is a valuable tool in achieving energy conservation. However, to strive for maximum energy conservation without regard for economics is not acceptable. There are many ways to manipulate the cost and savings numbers, and this section explains the various approaches and the pros and cons of each.

15.5.1 Cost Considerations

Simply stated, if the cost of insulation can be recouped by a reduction in total energy costs, the insulation investment is justified. Similarly, if the cost of additional insulation can be recouped by the additional energy-cost reduction, the expenditure is justified. There is a significant difference between the "full thickness" justification and the "incremental" justification. This is discussed in detail in Section 15.5.3. The following discus-

sions will generally use the incremental approach to economic evaluation.

Insulation Costs

The insulation costs should include everything that it takes to apply the material to the pipe or vessel and to properly cover it to finished form. Certainly, it is more costly to install insulation 100 ft in the air than it is from ground level, and metal jackets are more costly than all-purpose indoor jackets. Anticipated maintenance costs should also be included based on the material and application involved. The variations in labor costs due to both time and base rate should be evaluated for each particular insulation system design and locale. In other words, insulation costs tend to be job specific as well as being differentiated by product.

Lost Heat Costs

Reducing the amount of unwanted heat loss is the

function of insulation, and the measurement of this is in Btu. The key to economic analyses rests in the dollar value assigned to each Btu that is wasted. At the very least, the energy cost must include the raw-fuel cost, modified by the conversion efficiency of the equipment. For example, if natural gas costs $2.50/million Btu and it is being converted to heat at 70% efficiency, the effective cost of the Btu is 2.50/0.70 = $3.57/million Btu.

The cost of the heat plant is always a point of discussion. Many calculations ignore this capital cost on the basis that a heat plant will be required whether insulation is used or not. On the other hand, the only purpose of the heat plant is to generate usable Btus. So the cost of each Btu should reflect the capital plant cost ammortized over the life of the plant. The recent trend that seems most reasonable is to assign an incremental cost to increases in capital expenditures. This cost is stated as dollars per 1000 Btu per hour. This gives credit to a well-insulated system that requires less Btu/hr capacity.

Other Costs

As the economic calculations become more sophisticated, other costs must be included in the analysis. The major additions are the cost of money and the tax effect of the project. Involving the cost of money recognizes the real fact that many projects are competing for each investment dollar spent.

Therefore, the money used to finance an insulation project must generate a sufficient after-tax return or the money will be invested elsewhere to achieve such a return. This topic, together with an explanation of the use of discount factors, is discussed in detail in Chapter 4.

The effect of taxes can also be included in the analysis as it relates to fuel expense and depreciation. Since both of these items are expensed annually, the after-tax cost is significantly reduced. The final example in Section 15.5.3 illustrates this.

15.5.2. Energy Savings Calculations

The following procedure shows how to estimate the energy cost savings resulting from installing thermal insulation.

Procedure

STEP 1. Calculate present heat losses (Q_{Tpres}). You can use one of the following methods to calculate the heat losses of the present system:

- Heat flow equations. These equations are in Section 15.4.2.

- Graphical method. To calculate heat losses, follow

Steps 1, 2 and 2a of the graphical method presented in Section 15.4.3.

- Table values. Table 15.8 presents heat losses values for bare surfaces (dull metals).

STEP 2. Determine insulation thickness (tk). Using Section 15.4, you can determine the insulation thickness according to your specific needs. Depending on the pipe diameter and temperature, the first inch of insulation can reduce bare surface heat losses by approximately 85-95% (Ref. 20). Then, for a preliminary economic evaluation, you can use tk = 1-in. If the evaluation is not favorable, you will not be able to justify a thicker insulation. On the other hand, if the evaluation is favorable, you will need to determine the appropriate insulation thickness and reevaluate the investment.

STEP 3. Calculate heat losses with insulation (Q_{Tins}). Use the equations from Section 15.4.5.

STEP 4. Determine heat loss savings ($Q_{Tsavings}$). Subtract the heat losses with insulation from the present heat losses ($Q_{Tsavings} = Q_{Tpres} - Q_{Tins}$)

STEP 5. Estimate fuel cost savings. Estimate the amount of fuel used to generate each Btu wasted and use this value to calculate the energy cost savings. With this savings, you can evaluate the insulation investment using any appropriate financial analysis method (see Section 15.5.3).

Example. For the example presented in section 15.4.3, determine the fuel cost savings resulting from insulating the pipe.
Data
- Pipe data: 4-in pipe operating at 700°F in an 85°F ambient temperature.

- Jacket type: Aluminum.

- Pipe length: 100-ft.

- Operating hours: 4,160 hr/yr

- Fuel data: Natural gas, burned to heat the fluid in the pipe at $3/MCF. Efficiency of combustion is approximately 80%

STEP 1. Determine present heat loss. From Table 15.8 (4-in. pipe, temperature difference = $t_s - t_a$ = 700 – 85 = 615°F), heat loss = 4,356 Btu/hr-lin.ft. Then,

QT_{pres} = (heat loss/lin.ft)(length)
= (4,356 Btu/hr-lin.ft.)(100 ft)
= 435,600 Btu/hr

STEP 2. Determine insulation thickness. In this example, the surface temperature has to be below 140°F, which is accomplished with an insulation thickness = tk = 3.5-in.

STEP 3. Determine heat losses with insulation. For this example, we need to calculate the heat losses for tk = 3.5-in following the procedure outlined in Section 15.4.5.

1) From the example in Section 15.4.3, ts = 140 °F, k = 0.49 and R_s = 0.85.

2) From Table 15.2, Eq tk for 3-1/2-in insulation on a 4-in. pipe = 5.396 in. Then,

R_I =Eq tk/k =5.396/0.49= 11

3) Calculate heat loss Q_F:

$$Q_F = \frac{700-85}{11+0.85} = 52 \text{ Btu/hr ft}^2$$

4) Calculate surface temperature t_s:
$t_s = t_a + R_s \times Q_F$= 85 + (0.85 × 52) = 129°F

5) Calculate t_m = (700+129)/2 = 415°F. The insulation thermal conductivity at 415°F is 0.49, which is close enough to the assumed value (see Appendix 15.1). Then, Q_F = 52 Btu/hr ft².

6) Determine the outside area of insulated pipe. From Table 15.2, pipe radius = rl = 2.25-in., then, outside insulated area (ft²):

= 2π (rl+tk)(length)/(12 in./ft)

= 2π (2.25 in+3.5 in.)(100 ft)/(12 in./ft)

=301 ft²

7) Calculate heat losses with insulation:

QT_{ins} = (Q_F)(outside area)
= (52 Btu/hr ft²)(301 ft²)
= 15,652 Btu/hr

STEP 4. Determine heat losses savings $QT_{savings}$:

$QT_{savings}$ = (QT_{pres} − QT_{ins})(hr/yr)

= (435,600–15,652 Btu/hr)(4,160 hr/yr)
(1 MMBtu/10^6 Btu)

= 1,747 MMBtu/yr

STEP 5. Determine fuel cost savings. Assuming 1 MCF = 1 MMBtu:

Fuel savings

= ($QT_{savings}$)(conversion factor)/
(combustion efficiency)

= (1,747 MMBtu/yr)(1 MCF/MMBtu)/(0.8)

= 2,184 MCF/yr

Then,

Fuel cost savings = (fuel savings) (fuel cost)
= (2,184 MCF/yr) ($3/MCF)
= $6,552/yr

15.5.3 Financial Analysis Methods—Sample Calculations

Chapter 4 offers a complete discussion of the various types of financial analyses commonly used in industry. A review of that material is suggested here, as the methods discussed below rely on this basic understanding.

To select the proper financial analysis requires an understanding of the degree of sophistication required by the decision maker. In some cases, a quick estimate of profitability is all that is required. At other times, a very detailed cash flow analysis is in order. The important point is to determine what level of analysis is desired and then seek to communicate at that level. Following is an abbreviated discussion of four primary methods of evaluating an insulation investment: (1) simple payback; (2) discounted payback; (3) minimum annual cost using a level annual equivalent; and (4) present-value cost analysis using discounted cash flows.

Economic Calculations

Basically, a simple payback period is the time required to repay the initial capital investment with the operating savings attributed to that investment. For example, consider the possibility of upgrading a present insulation thickness standard.

Thickness

	Current Standard	Upgraded Thickness	Difference
Insulation			
investment ($)	225,000	275,000	50,000
Annual fuel cost ($)	40,000	30,000	10,000

$$\text{Simple payback} = \frac{\text{investment difference}}{\text{annual fuel saving}} = \frac{50,000}{10,000} = 5.0 \text{ years}$$

This calculation represents the incremental approach, which determines the amount of time to recover the additional $50,000 of investment.

In the following table, the full thickness analysis is similar except that the upgraded thickness numbers are now compared to an uninsulated system with zero insulation investment.

	Uninsulated System	Upgraded Thickness	Difference
Insulation			
investment ($)	0	275,000	275,000
Annual fuel cost ($)	340,000	30,000	310,000

$$\text{Simple payback} = \frac{275,000}{310,000} = 0.89 \text{ year}$$

The magnitude of the difference points out the danger in talking about payback without a proper definition of terms. If in the second example, management had a payback requirement of 3 years, the full insulation investment easily complies, whereas the incremental investment does not. Therefore, it is very important to understand the intent and meaning behind the payback requirement.

Although simple payback is the easiest financial calculation to make, its use is normally limited to rough estimating and the determination of a level of financial risk for a certain investment. The main drawback with this simple analysis is that it does not take into account the time value of money, a very important financial consideration.

Time Value of Money

Again, see Chapter 4. The significance of the cost of money is often ignored or underestimated by those who are not involved in their company's financial mainstream. The following methods of financial analysis are all predicated on the use of discount factors that reflect the cost of money to the firm. Table 15.9 is an abbreviated table of present-value factors for a steady income stream over a number of years. Complete tables are found in Chapter 4.

Table 15.9 Present-Value Discount Factors for an Income of $1 Per Year for the Next n Years

Years	Cost of Money at:				
	5%	10%	15%	20%	25%
1	.952	.909	.870	.833	.800
2	1.859	1.736	1.626	1.528	1.440
3	2.723	2.487	2.283	2.106	1.952
4	3.546	3.170	2.855	2.589	2.362
5	4.329	3.791	3.352	2.991	2.689
10	7.722	6.145	5.019	4.192	3.571
15	10.380	7.606	5.847	4.675	3.859
20	12.460	8.514	6.259	4.870	3.954

Discounted Payback

Although similar to simple payback, the utilization of the discount factor makes the savings in future years worth less in present-value terms. For discounted payback, then, the annual savings times the discount factor must now equal the investment to achieve payback in present-value dollars. Using the same example:

Thickness

	Current Standard	Upgraded Thickness	Difference
Insulation			
investment ($)	225,000	275,000	50,000
Annual fuel cost ($)	40,000	30,000	10,000

Now, payback occurs when:

$$\text{investment} = \text{discount factor} \times \text{annual savings}$$

$$50,000 = (\text{discount factor}) \times 10,000$$

so solving for the discount factor,

$$\text{discount factor} = \frac{\text{investment}}{\text{annual savings}} = \frac{50,000}{10,000} = 5.0$$

For a 15% cost of money, read down the 15% column of Table 15.9 to find a discount factor close to 5. The corresponding number of years is then read to the left, approximately 10 years in this case. For a cost of money of only 5%, the payback is achieved in about 6 years. Obviously, a 0% cost of money would be the same as the simple payback calculation of 5 years.

Minimum Annual Cost Analysis

As previously discussed, an insulation investment must involve a lump-sum cost for insulation as well as a stream of fuel costs over the many years. One method of putting these two sets of costs into the same terms is to spread out the insulation investment over the life of the project. This is done by dividing the initial investment by the appropriate discount factor in Table 15.9. This produces a "level annual equivalent" of the investment for each year which can then be added to the annual fuel cost to arrive at a total annual cost.

Utilizing the same example with a 20-year project life and 10% cost of money:

	Thickness	
	Current Standard	Upgraded Thickness
Insulation investment ($)	225,000	275,000

For 20 years at 10%, the discount factor is 8.514 (Table 15.9), so

	Thickness	
	Current Standard	Upgraded Thickness
Equivalent annual insulation costs	$\frac{225,000}{8.514}$	$\frac{275,000}{8.514}$
=	26,427	32,300
Annual fuel cost ($)	40,000	30,000
Total annual cost ($)	66,427	62,300

Therefore, on an annual cost basis, the upgraded thickness is a worthwhile investment because it reduces the annual costs by $4127.

Now, to illustrate again the importance of using a proper cost of money, change the 10% to 20% and recompute the annual cost. The 20% discount factor is 4.870.

	Thickness	
	Current Standard	Upgraded Thickness
Equivalent annual insulation cost ($)	$\frac{225,000}{4.870}$	$\frac{275,000}{4.870}$
=	46,201	56,468
Add the annual fuel cost ($)	40,000	30,000
Total annual cost	86,201	86,468

In this case, the higher cost of money causes the upgraded annual cost to be greater than the current cost, so the project is not justified.

Present-Value Cost Analysis

The other method of comparing project costs is to bring all the future costs (i.e., fuel expenditures) back to today's dollars by discounting and then adding this to the initial investment. This provides the total present-value cost of the project over its entire life cycle, and projects can be chosen based on the minimum present-value cost. This discounted cash flow (DCF) technique is used regularly by many companies because it allows the analyst to view a project's total cost rather than just the annual cost and assists in prioritizing among many projects.

	Thickness	
	Current Standard	Upgraded Thickness
Annual fuel cost ($)	40,000	30,000

For 20 years at 10% the discount factor is 8.514 (Table 15.9), so

	Thickness	
	Current Standard	Upgraded Thickness
Present value of fuel cost over 20 years	40,000 × 8.514 =340,560	30,000 × 8.514 255,420
Insulation investment ($)	225,000	275,000
Total present-value cost of insulation project	$565,460	$530,420

Again, the lower total project cost with the upgraded thickness option justifies that project.

So far, the effect of taxes and depreciation has been ignored so as to concentrate on the fundamentals. However, the tax effects are very significant on the cash flow to the company and should not be ignored. In the case where the insulation investment is capitalized utilizing a 20-year straight-line depreciation schedule and a 48% tax rate, the following effects are seen (see table at top of next page).

This illustrates the significant impact of both taxes and depreciation. In the preceding analysis, the PV benefit of upgrading was (565,460 − 530,420) = $35,040. In this case, the cash flow benefit is reduced to (356,116 − 351,626) = $4490.

The final area of concern relates to future increases in fuel costs. So far, all the analyses have assumed a constant stream of fuel costs, implying no increase in the base cost of fuel. This assumption allows the use of the PV factor in Table 15.9. To accommodate annual fuel-price increases, either an average fuel cost over the project life is used or each year's fuel cost is discounted separately to PV terms. Computerized calculations permit this, whereas a manual approach would be extremely laborious.

	Thickness Current Standard	Upgraded Thickness
1. Annual energy cost ($)	40,000	30,000
2. After-tax energy cost ($) ((1) × (1.0 − 0.48))	20,800	15,600
3. Insulation depreciation ($ tax benefit)		
(225,000/20 yr)(0.48)	5,400	
(275,000/20 yr)(0.48)		6,600
4. Net annual cash costs [$; (2) − (3)]	15,400	9,000
5. Present-value factor for 20 years at 10% = 8.514		
6. Present value of annual cash flows [$; (4) × (5)]	131,116	76,626
7. Present value of cash flow for insulation purchase ($)	225,000	275,000
8. Present-value cost of project [$; (6) + (7)]	356,116	351,626

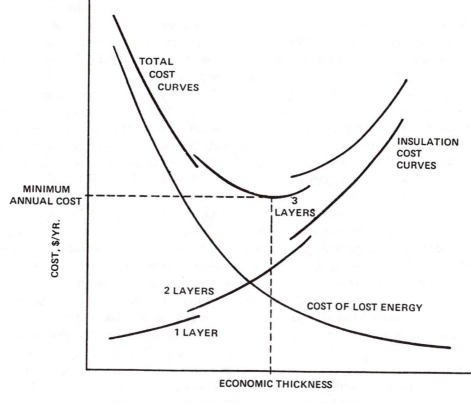

Fig. 15.5 Economic thickness of insulation (ETI) concept.

15.5.4 Economic Thickness (ETI) Calculations

Section 15.5.3 developed the financial analyses often used in evaluating a specific insulation investment. As presented, however, the methods evaluate only two options rather than a series of thickness options. Economic thickness calculations are designed to evaluate each 1/2-in. increment and sum the insulation and operating costs for each increment. Then the option with the lowest total annual cost is selected as the economic thickness. Figure 15.5 graphically illustrates the optimi-

zation method. In addition, it shows the effect of additional labor required for double- and triple-layer insulation applications.

Mathematically, the lowest point on the total-cost curve is reached when the incremental insulation cost equals the incremental reduction in energy cost. By definition, the economic thickness is:

that thickness of insulation at which the cost of the next increment is just offset by the energy savings due to that increment over the life of the project.

Historical development

A problem with the McMillan approach was the large number of charts that were needed to deal with all the operating and financial variables. In 1949, Union Carbide Corp. in a cooperation with West Virginia University established a committee headed by W.C. Turner to establish practical limits for the many variables and to develop a manual for performing the calculations. This was done, and in 1961, the manual was published by the National Insulation Manufacturers Association (previously called TIMA and now NAIMA, North American Insulation Manufacturers Association). The manual was entitled *How to Determine Economic Thickness of Insulation* and employed a number of nomographs and charts for manually performing the calculations.

Since that time, the use of computers has greatly changed the method of ETI calculations. In 1973, TIMA released several programs to aid the design engineer in selecting the proper amount of insulation. Then in 1976, the Federal Energy Administration (FEA) published a nomograph manual entitled *Economic Thickness of Industrial Insulation* (Conservation Paper #46). In 1980, these manual methods were computerized into the "Economic Thickness of Industrial Insulation for Hot and Cold Surfaces." Through the years, NAIMA developed a version for personal computers; the newest program (1992) was renamed 3E and calculates the ETI thickness of insulation.

Perhaps the most significant change occurring is that most large owners and consulting engineers are developing and using their own economic analysis programs, specifically tailored to their needs. As both heat-transfer and financial calculations become more sophisticated, these programs will continue to be upgraded and their usefulness in the design phase will increase.

Nomograph Methods

A nomograph methods is not presented here, but the interested reader can review the following references:

- FEA manual (see Ref. 12). This manual provides a fairly complete but time-consuming nomograph method.

- 1972 ASHRAE Handbook of Fundamentals, Chapter 17 (see Ref. 13) which provides a simplified, one-page nomograph. This approach is satisfactory for a quick determination, but it lacks the versatility of the more complex approach. The nomograph has been eliminated in the latest edition and reference is made to the computer analyses and the FEA manual.

Computer Programs

As mentioned earlier, many companies are developing ETI programs of their own. This is an excellent trend because it indicates the greater degree of confidence that management has in its internally developed methods. For the convenience of owners that do not have this in-house capability, NAIMA offers 3E, a program that uses an after-tax annual analysis (see Ref 14). The program is very flexible and allows the simulation of a large number of conditions.

All the insulation owning costs are expressed on an equivalent uniform annual cost basis. This program uses the ASTM C680 method for calculating the heat loss and surface temperatures. Each commercially available thickness is analyzed, and the thickness with the lowest annual cost is the economic thickness (ETI).

Figure 15.6 shows the output generated by the NAIMA 3E program. The first several lines are a readout of the input data. The different variables used in the program allow to simulate virtually any job condition. The same program can be used for retrofit analyses and bare-surface calculations. There are two areas of input data that are not fully explained in the output. The first is the installed insulation cost. The user has the option of entering the installed cost for each particular thickness or using an estimating procedure developed by the FEA (now DOE).

The second area that needs explanation is the insulation choice, which relates to the thermal conductivity of the material. the example in Figure 15.6 shows the insulation as Glass Fiber Blanket. The program includes the thermal conductivity equations of several generic types of thermal insulation, which were derived from ASTM materials specifications. The user has the option of supplying thermal conductivity data for other materials.

The lower portion of the output supplies seven columns of information. The first and second columns are input data, while the others are calculated output. The program also calculates the reduction in CO_2 emissions by insulating to economic thickness. The meaning of columns two to seven of the output are explained below.

Annual Cost ($/yr). This is the annual operating cost including both energy cost and the amortized insulation cost. Tax effects are included. This value is the one that determines the economic thickness. As stated under the columns, the lowest annual cost occurs with 2.50 in. of insulation which is the economic thickness.

Payback period (yr). This value represents the discounted payback period of the specific thickness as compared to the reference thickness. In this example, the reference thickness variable was input as zero, so the payback is compared to the uninsulated condition.

Present Value of Heat Saved ($/ft). This gives the energy cost savings in discounted terms as compared to the uninsulated condition. As discussed earlier, the first incre-

Figure 15.6 NAIMA 3E computer program output.

Project Name =	Date = 11-13-1995
Project Number =	Engineer =
System =	Contact =
Location =	Phone =

Fuel Type =	Gas
First Year Price =	3.36 $ per mcf
Heating Value =	1000 Btu per cf
Efficiency =	80.0%
Annual Fuel Inflation Rate =	6.0%
Annual hours of operation =	8320 hours

ECONOMIC DATA

Interest rate or Return on investment =	10.0%
Effective Income Tax Rate =	30%
Physical Plant Depreciation Period =	7 years
New Insulation Depreciation Period =	7 years
Incremental Equipment Investment Rate =	3.47 $/MMBtu/hr
Percent of New Insulation Cost for	
Annual Insulation Maintenance =	2%
Percent of Annual Fuel Bill for	
Physical Plant Maintenance =	1%
Ambient temperature =	75 F
Emittance of outer jacketing =	0.10
Wind speed =	0 mph
Emittance of existing surface =	0.80
Reference thickness for payback calculations =	0.0 inches

Insulation material = GLASS FIBER BLANKET
Horizontal Pipe

Pipe Size =	5 inch
Average Installation Complexity factor =	1.20
Performance Service factor =	1.00

Insulation costs estimated by FEA method

Labor rate =	38.35 $/hr
Productivity factor =	100
Price of 2x2 pipe insulation =	4.97 $/ln ft
Price of 2 inch block =	1.71 $/sqft

Operating Temperature 450 F

Insulation Thick Inches	Cost $/ft	Annual Cost $/ft	Payback Period Years	Pres Value Heat Saved $/ft	Heat Loss Btu/ft	Surf Temp F
Bare	57.37				1834	450
1.0	10.18	9.25	0.2	1133.90	226	193
1.5	12.10	8.07	0.2	1170.00	174	163
2.0	14.96	7.76	0.3	1190.76	145	146
2.5	17.27	7.60	0.4	1205.83	124	133
3.0	19.68	7.71	0.4	1214.97	111	125
4.0	25.45	8.36	0.6	1228.33	92	113
Double layer						
3.0	22.30	8.27	0.5	1214.97	111	125
4.0	29.22	9.18	0.7	1228.33	92	113
5.0	36.14	10.34	0.9	1235.90	81	106
6.0	43.07	11.63	1.1	1240.32	75	102
Triple layer						
6.0	67.48	16.91	1.7	1240.32	75	102
7.0	77.39	18.86	2.0	1244.51	69	99
8.0	87.00	20.79	2.3	1247.76	64	96

The Economic Thickness is single layer 2.5 inches.

The savings for the economic thickness is 49.77 $/ln ft/yr and the reduction in Carbon Dioxide emissions is 1608 lbs/lnft/yr.

ment has the most impact on energy savings, but the further incremental savings are still justified, as evidenced by the reduction of annual cost to the 2.50-in. thickness.

Heat Loss (Btu/ft). This calculation allows the user to check the expected heat loss with that required for a specific process. It is possible that under certain conditions a thickness greater than the economic thickness may be required to achieve a necessary process requirement.

Surface Temperature (°F). This final output allows the user to check the resulting surface temperature to assure that the level is within the safe-touch range. The ETI program is very sophisticated. It employs sound methods of both thermal and financial analysis and provides output that is relevant and useful to the design engineer and owner. NAIMA makes this program available to those desiring to have it on their own computer systems. In addition, several of the insulation manufacturers offer to run the analysis for their customers and send them a program output.

APPENDIX 15.1
Typical Thermal Conductivity Curves Used in Sample Calculations*

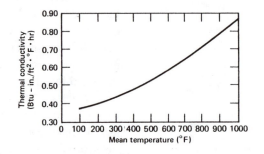

Fig. 15.A1 Calcium cilicate.

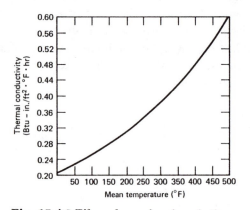

Fig. 15.A2 Fiberglass pipe insulation.

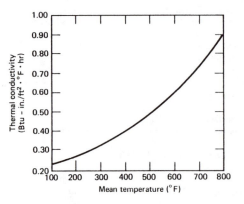

Fig. 15.A3 Fiberglass board, 3 lb/ft³.

References

1. American Society for Testing and Materials, Annual Book of ASTM Standards: Part 18—Thermal and Cryogenic Insulating Materials; Building Seals and Sealants; Fire Test.;.Building Constructions;Environmental Acoustics;Part 17—Refractories, Glass and Other Ceramic Materials; Manufactured Carbon and Graphite Products.
2. W.H. McADAMS, Heat Transmission, McGraw-Hill, New York, 1954.
3. E.M. SPARROW and R.D. CESS, Radiation Heat Transfer, McGraw-Hill, New York, 1978.
4. L.L. BERANEK, Ed., Noise and Vibration Control, McGraw-Hill, New York, 1971.
5. F.A. WHITE, Our Acoustic Environment, Wiley, New York, 1975.
6. M. KANAKIA, W. HERRERA, and F. HUTTO, JR., "Fire Resistance Tests for Thermal Insulation," Journal of Thermal Insulation, Apr. 1978, Technomic, Westport, Conn.
7. Commercial and Industrial Insulation Standards, Midwest Insulation Contractors Association, Inc., Omaha, Neb., 1979.
8. J.F. MALLOY, Thermal Insulation, Reinhold, New York, 1969.
9. American Society for Testing and Materials, Annual Book of ASTM Standards, Part 18, STD C-585.
10. J.F. MALLOY, Thermal Insulation, Reinhold, New York, 1969, pp. 72-77, from Thermon Manufacturing Co. technical data.
11. L.B. McMILLAN, "Heat Transfer through Insulation in the Moderate and High Temperature Fields: A Statement of Existing Data," No. 2034, The American Society of Mechanical Engineers, New York, 1934.

12. Economic Thickness of Industrial Insulation, Conservation Paper No. 46, Federal Energy Administration, Washington, D.C., 1976. Available from Superintendent of Documents, U.S. Government Printing Office, Washington, D.C. 20402 (Stock No. 041-018-00115-8).
13. ASHRAE Handbook of Fundamentals, American Society of Heating, Refrigerating and Air Conditioning Engineers, Inc., New York, 1972, p. 298.
14. NAIMA 3 E's Insulation Thickness Computer Program, North American Insulation Manufacturers Association, 44 Canal Center Plaza, Suite 310, Alexandria, VA 22314.
15. P. Greebler, "Thermal Properties and Applications of High Temperature Aircraft Insulation," American Rocket Society, 1954. Reprinted in Jet Propulsion, Nov.-Dec. 1954.
16. Johns-Manville Sales Corporation, Industrial Products Division, Denver, Colo., Technical Data Sheets.
17. ASHRAE Handbook of Fundamentals, American Society of Heating, Refrigerating and Air Conditioning Engineers, Inc., Atlanta, GA, 1992, p.22.16.
18. W.C.Turner and J.F. Malloy, Thermal Insulation Handbook, Robert E. Krieger Publishing Co. And McGraw Hill, 1981.
19. Ahuja, A., "Thermal Insulation: A Key to Conservation," Consulting-Specifiying Engineer, January 1995, p. 100-108.
20. U.S. Department of Energy, "Industrial Insulation for Systems Operating Above Ambient Temperature," Office of Industrial Technologies, Bulletin ORNL/M-4678, Washington, D.C., September 1995.

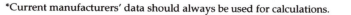

*Current manufacturers' data should always be used for calculations.

CHAPTER 16
USE OF ALTERNATIVE ENERGY

JERALD D. PARKER

Professor Emeritus, Oklahoma State University
Stillwater, Oklahoma
Professor Emeritus, Oklahoma Christian University
Oklahoma City, Oklahoma

16.1 INTRODUCTION

Any energy source that is classified as an "alternative energy source" is that because, at one time it was not selected as the best choice. If the original choice of an energy source was a proper one the use of an alternative energy source would make sense only if some condition has changed. This might be:

1. Present or impending nonavailability of the present energy source
2. Change in the relative cost of the present and the alternative energy
3. Improved reliability of the alternative energy source
4. Environmental or legal considerations

To some, an alternative energy source is a nondepleting or renewable energy source, and, for many it is this characteristic that creates much of the appeal. Although the terms "alternative energy source" and "renewable energy sources" are not intended by this writer to be synonymous, it will be noted that some of the alternative energy sources discussed in this section are renewable.

It is also interesting that what we now think of as alternative energy sources, for example solar and wind, were at one time important conventional sources of energy. Conversely, natural gas, coal, and oil were, at some time in history, alternative energy sources. Changes in the four conditions listed above, primarily conditions 2 and 3, have led us full circle from the use of solar and wind, to the use of natural gas, coal, and oil, and back again in some situations to a serious consideration of solar and wind.

In a strict sense, technical feasibility is not a limitation in the use of the alternative energy sources that will be discussed. Solar energy can be collected at any rea-

sonable temperature level, stored, and utilized in a variety of ways. Wind energy conversion systems are now functioning and have been for many years. Refuse-derived fuel has also been used for many years. What is important to one who must manage energy systems are the factors of economics, reliability, and in some cases, the nonmonetary benefits, such as public relations.

Government funding for R&D as well as tax incentives in the alternative energy area has dropped sharply during the decade of the eighties and early nineties. This has caused many companies with alternative energy products to go out of business, and for others to cut back on production or to change into another product or technology line. Solar thermal energy has been hit particularly hard in this respect, but solar powered photovoltaic cells have had continued growth both in space and in terrestrial applications. Wind energy systems have continued to be installed but not at the rapid rate some have predicted. The burning of refuse has met with some environmental concerns and strict regulations. Recycling of some refuse materials such as paper and plastics have given an alternative to burning.

Surviving participants in the alternative energy business have in some cases continued to grow and to improve their products and their competitiveness. As some or all of the four conditions listed above change, we will see rising or falling interest on the part of the government, industry and private individuals in particular alternative energy systems.

Fuel cells were not discussed in the previous edition because they are not a new *source* of energy. Continuing widespread use of these alternative energy conversion systems in space applications and interest and trial use of these devices by electric utilities have justified inclusion of material on them in this chapter.

16.2 SOLAR ENERGY

16.2. 1 Availability

"Solar energy is free!" states a brochure intended to sell persons on the idea of buying their solar products. "There's no such thing as a free lunch" should come to mind at this point. With a few exceptions, one must invest capital in a solar energy system in order to reap the benefits of this alternative energy source. In addition to the cost of the initial capital investment, one is usually faced with additional periodic or random costs due to operation and maintenance. Provided that the solar system does its expected task in a reasonably reliable manner, and presuming that the conventional energy source is available and satisfactory, the important question usually is: Did it save money compared to the conventional system? Obviously, the cost of money, the cost of conventional fuel, and the cost and performance of the solar system are all important factors. As a first step in looking at the feasibility of solar energy, we will consider its availability.

Solar energy arrives at the outer edge of the earth's atmosphere at a rate of about 428 Btu/hr ft^2 (1353 W/m^2). This value is referred to as the solar constant. Part of this radiation is reflected back to space, part is absorbed by the atmosphere and re-emitted, and part is scattered by atmospheric particles. As a result, only about two-thirds of the sun's energy reaches the surface of the earth. At 40° north latitude, for example, the noontime radiation rate on a flat surface normal to the sun's rays is about 300 Btu/hr · ft^2 on a clear day. This would be the approximate maximum rate at which solar energy could be collected at that latitude. A solar collector tracking the sun so as to always be normal to the sun's rays could gather approximately 3.6×10^3 Btu/ft^2 · day as an absolute upper limit. To gather 1 million Btu/day, for example, would require about 278 ft^2 (26 m^2) of movable collectors, collecting all the sunlight that would strike them on a clear day.

Since no collector is perfect and might collect only 70% of the energy striking it, and since the percent sunshine might also be about 70%, a more realistic area would be about 567 ft^2 (53 m^2) to provide 1 million Btu of energy per day. In the simplest terms, would the cost of constructing, operating, and maintaining a solar system consisting of 567 ft^2 of tracking solar collectors justify a reduction in conventional energy usage of 1 million Btu/day? Fixed collectors might be expected to deliver approximately 250,000 Btu/yr for each square foot of surface.

A most important consideration which was ignored in the discussion above was that of the system's ability to use the solar energy when it is available. A space-heating system, for example, cannot use solar energy in the summer. In industrial systems, energy demand will rarely correlate with solar energy availability. In some cases, the energy can be stored until needed, but in most systems, there will be some available solar energy that will not be collected. Because of this factor, particular types of solar energy systems are most likely to be economically viable. Laundries, car washes, motels, and restaurants, for example, need large quantities of hot water almost every day of the year. A solar water-heating system seems like a natural match for such cases. On the other hand, a solar system that furnishes heat only during the winter, as for space heating, may often be a poor economic investment.

The amount of solar energy available to collect in a system depends upon whether the collectors move to follow or partially follow the sun or whether they are fixed. In the case of fixed collectors, the tilt from horizontal and the orientation of the collectors may be significant. The remainder of this section considers the energy available to fixed solar collecting systems.

Massive amounts of solar insolation data have been collected over the years by various government and private agencies. The majority of these data are hourly or daily solar insolation values on a horizontal surface, and the data vary considerably in reliability. Fixed solar collectors are usually tilted at some angle from the horizontal so as to provide a maximum amount of total solar energy collected over the year, or to provide a maximum amount during a particular season of the year. What one needs in preliminary economic studies is the rate of solar insolations on tilted surfaces.

Figure 16.1 shows the procedure for the conversion of horizontal insolation to insolation on a tilted surface. The measured insolation data on a horizontal surface consist of direct radiation from the sun and diffuse radiation from the sky. The total radiation must be split into these two components (step A) and each component analyzed separately (steps B and C). In addition, the

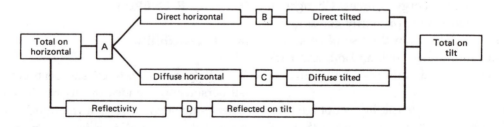

Fig. 16.1 Conversion of horizontal insolation to insolation on tilted surface.

solar energy reflected from the ground and other sur-roundings must be added into the total (step D). Proce-dures for doing this are given in Refs. 1 to 4.

A very useful table of insolation values for 122 cit-ies in the United States and Canada is given in Ref. 5. These data were developed from measured weather data using the methods of Refs. 2 and 3 and are only as reli-able as the original weather data, perhaps ± 10%. A sum-mary of the data for several cities is given in Table 16.1.

One of the more exhaustive compilations of U.S. solar radiation.data is that compiled by the National Climatic Center in Asheville, North Carolina, for the Department of Energy. Data from 26 sites were rehabili-tated and then used to estimate data for 222 stations, shown in Figure 16.2. A summary of these data is tabu-lated in a textbook by Lunde.[6] It should be remembered that measured data from the past do not predict what will happen in the future. Insolation in any month can be quite variable from year to year at a given location.

Another approach is commonly used to predict insolation on a specified surface at a given location. This method is to first calculate the clear-day insolation, us-ing knowledge of the sun's location in the sky at the given time. The clear-day insolation is then corrected by use of factors describing the clearness of the sky at a given location and the average percent of possible sun-shine.

The clear-sky insolation on a given surface is readily found in references such as the *ASHRAE Hand-book of Fundamentals*. A table of percent possible sun-shine for several cities is given in Table 16.2.

16.2.2 Solar Collectors

A wide variety of devices may be used to collect solar energy. A general classification of types is given in Figure 16.3. Tracking-type collectors are usually used where relatively high temperatures (above 250°F) are required. These types of collectors are discussed at the end of this section. The more common fixed, flat-plate collector will be discussed first, followed by a discussion of tube-type or mildly concentrating collectors.

The flat-plate collector is a device, usually faced to the south (in the northern parts of the globe) and usually at some fixed angle of tilt from the horizontal. Its pur-pose is to use the solar radiation that falls upon it to raise the temperature of some fluid to a level above the ambient conditions. That heated fluid, in turn, may be used to provide hot water or space heat, to drive an engine or a refrigerating device, or perhaps to remove moisture from a substance. A typical glazed flat-plate solar collector of the liquid type is shown in Figure 16.4.

The sun's radiation has a short wavelength and easily passes through the glazing (or glazings), with only about 10 to 15% of the energy typically reflected and absorbed in each glazing. The sunlight that passes through is almost completely absorbed by the absorber surface and raises the absorber temperature. Heat loss

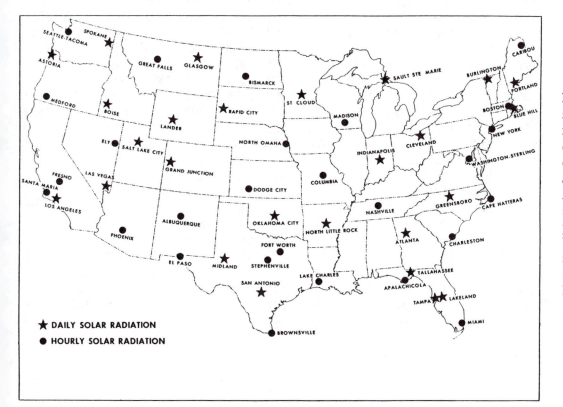

Fig. 16.2 Weather stations for which rehabilitated measured (asterisks) and derived data have been collected. [From *SOLMET, Volume 1*, and *Input Data for Solar Sys-tems*, Nov. 1978, pre-pared by NOAA for DOE, Interagency agree-ment E (49-26)-1041. Some data are given in Ref. 6.]

Table 16.1 Average Daily Radiation on Tilted Surfaces for Selected Cities

Average Daily Radiation (Btu/day ft^2).

City	Slope	Jan.	Feb.	Mar.	Apr.	May	June	July	Aug.	Sept.	Oct.	Nov.	Dec.
Albuquerque, NM	hor.	1134	1436	1885	2319	2533	2721	2540	2342	2084	1646	1244	1034
	30	1872	2041	2295	2411	2346	2390	2289	2318	2387	2251	1994	1780
	40	2027	2144	2319	2325	2181	2182	2109	2194	2369	2341	2146	1942
	50	2127	2190	2283	2183	1972	1932	1889	2028	2291	2369	2240	2052
	vert.	1950	1815	1599	1182	868	754	795	1011	1455	1878	2011	1927
Atlanta, GA	hor.	839	1045	1388	1782	1970	2040	1981	1848	1517	1288	975	740
	30	1232	1359	1594	1805	1814	1801	1782	1795	1656	1638	1415	1113
	40	1308	1403	1591	1732	1689	1653	1647	1701	1627	1679	1496	1188
	50	1351	1413	1551	1622	1532	1478	1482	1571	1562	1679	1540	1233
	vert.	1189	1130	1068	899	725	659	680	811	990	1292	1332	1107
Boston, MA	hor.	511	729	1078	1340	1738	1837	1826	1565	1255	876	533	438
	30	830	1021	1313	1414	1677	1701	1722	1593	1449	1184	818	736
	40	900	1074	1333	1379	1592	1595	1623	1536	1450	1234	878	803
	50	947	1101	1322	1316	1477	1461	1494	1448	1417	1254	916	850
	vert.	895	950	996	831	810	759	791	857	993	1044	842	820
Chicago, IL	hor.	353	541	836	1220	1563	1688	1743	1485	1153	763	442	280
	30	492	693	970	1273	1502	1561	1639	1503	1311	990	626	384
	40	519	716	975	1239	1425	1563	1544	1447	1307	1024	662	403
	50	535	723	959	1180	1322	1341	1421	1363	1274	1034	682	415
	vert.	479	602	712	746	734	707	754	806	887	846	610	373
Ft. Worth, TX	hor.	927	1182	1565	1078	2065	2364	2253	2165	1841	1450	1097	898
	30	1368	1550	1807	1065	1891	2060	2007	2097	2029	1859	1604	1388
	40	1452	1601	1803	1020	1755	1878	1845	1979	1995	1907	1698	1488
	50	1500	1614	1758	957	1586	1663	1648	1820	1914	1908	1749	1549
	vert.	1315	1286	1196	569	728	679	705	890	1185	1459	1509	1396
Lincoln, NB	hor.	629	950	1340	1752	2121	2286	2268	2054	1808	1329	865	629
	30	958	1304	1605	1829	2004	2063	2088	2060	2092	1818	1351	1027
	40	1026	1363	1620	1774	1882	1909	1944	1971	2087	1894	1450	1113
	50	1068	1389	1597	1679	1724	1720	1763	1838	2030	1922	1512	1170
	vert.	972	1162	1156	989	856	788	828	992	1350	1561	1371	1100
Los Angeles, CA	hor.	946	1266	1690	1907	2121	2272	2389	2168	1855	1355	1078	905
	30	1434	1709	1990	1940	1952	1997	2138	2115	2066	1741	1605	1439
	40	1530	1776	1996	1862	1816	1828	1966	2002	2037	1788	1706	1550
	50	1587	1799	1953	1744	1644	1628	1758	1845	1959	1791	1762	1620
	vert.	1411	1455	1344	958	760	692	744	918	1230	1383	1537	1479
New Orleans, LA	hor.	788	954	1235	1518	1655	1633	1537	1533	1411	1316	1024	729
	30	1061	1162	1356	1495	1499	1428	1369	1456	1490	1604	1402	1009
	40	1106	1182	1339	1424	1389	1309	1263	1371	1451	1626	1464	1058
	50	1125	1174	1292	1324	1256	1170	1137	1259	1381	1610	1490	1082
	vert.	944	899	847	719	599	546	548	647	843	1189	1240	929
Portland, OR	hor.	578	872	1321	1495	1889	1992	2065	1774	1410	1005	578	508
	30	1015	1308	1684	1602	1836	1853	1959	1830	1670	1427	941	941
	40	1114	1393	1727	1569	1746	1739	1848	1771	1680	1502	1020	1042
	50	1184	1442	1727	1502	1622	1594	1702	1673	1651	1539	1073	1116
	vert.	1149	1279	1326	953	889	824	890	989	1172	1309	1010	1109

Source: Ref. 5.

Table 16.2 Mean percentage of possible sunshine for selected U.S. cities.

Station	Jan.	Feb.	Mar.	Apr.	May	June	July	Aug.	Sept.	Oct.	Nov.	Dec.	Annual
Albuquerque, NM	70	72	72	76	79	84	76	75	81	80	79	70	76
Atlanta, GA	48	53	57	65	68	68	62	63	64	67	60	47	60
Boston, MA	47	56	57	56	59	62	64	63	61	58	48	48	57
Chicago, IL	44	49	53	56	63	69	73	70	65	61	47	41	59
Ft. Worth, TX	56	57	65	66	67	75	78	78	74	70	63	58	68
Lincoln, NB	57	59	60	60	63	69	76	71	67	66	59	55	64
Los Angeles, CA	70	69	70	67	68	69	80	81	80	76	79	72	73
New Orleans, LA	49	50	57	63	66	64	58	60	64	70	60	46	59
Portland, OR	27	34	41	49	52	55	70	65	55	42	28	23	48

Source: Ref. 7.

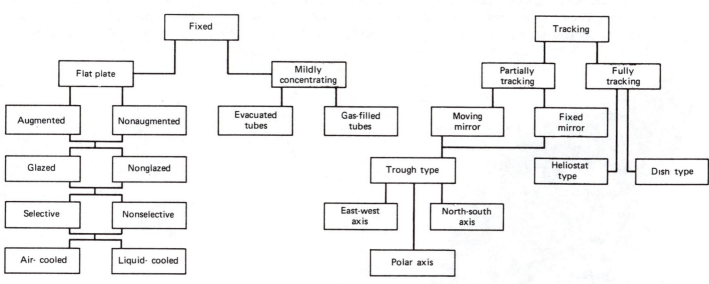

Fig. 16.3 Types of solar collectors.

out the back from the absorber plate is minimized by the use of insulation. Heat loss out the front is decreased somewhat by the glazing, since air motion is restricted. The heated absorber plate also radiates energy back toward the sky, but this radiation is longer-wavelength radiation and most of this radiation not reflected back to the absorber by the glazing is absorbed by the glazing. The heated glazing, in turn, converts some of the absorbed energy back to the air space between it and the absorber plate. The trapping of sunlight by the glazing and the consequent heating is known as the "greenhouse effect."

Energy is removed from the collector by the coolant fluid. A steady condition would be reached when the absorber temperature is such that losses to the coolant and to the surroundings equal the energy gain from the solar input. When no energy is being removed from the collectors by the coolant, the collectors are said to be at *stagnation.* For a well-designed solar collector, that stagnation temperature may be well above 300°F. This must be considered in the design of solar collectors and solar systems, since loss of coolant pumping power might be expected to occur sometime during the system lifetime. A typical coolant flow rate for flat-plate collectors is about 0.02 gpm/ft^2 of collector surface.

The fraction of the incident sunlight that is collected by the solar collector for useful purposes is called the collector efficiency. This efficiency depends upon several variables, which might change for a fixed absorber plate design and fixed amount of back and side insulation. These are:

1. Rate of insolation
2. Number and type of glazing
3. Ambient air temperature
4. Average (or entering) coolant fluid temperature

A typical single-glazed flat-plate solar collector efficiency curve is given in Figure 16.5. The measured performance can be approximated by a straight line. The left intercept is related to the product $\tau\alpha$, where τ is the transmittance of the glazing and α is the absorptance of the absorber plate. The slope of the line is related to the magnitude of the heat losses from the collector, a flatter line representing a collector with reduced heat-loss characteristics.

A comparison of collector efficiencies for unglazed, single-glazed, and double-glazed flat-plate collectors is shown in Figure 16.6. Because of the lack of glazing reflections, the unglazed collector has the highest efficiencies at the lower collector temperatures. This factor, combined with its lower cost, makes it useful for swimming pool heating. The single-glazed collector also performs well at lower collector temperatures, but like the unglazed collector, its efficiency drops off at higher collec-

tion temperatures because of high front losses. The double-glazed collector, although not performing too well at lower temperatures, is superior at the higher temperatures and might be used for space heating and/or cooling applications. The efficiency of an evacuated tube collector is also shown in Figure 16.6. It can be seen that it performs very poorly at low temperatures, but because of small heat losses, does very well at higher temperatures.

A very important characteristic of a solar collector surface is its selectivity, the ratio of its absorptance α_s for sunlight to its emittance ε for long-wavelength radiation. A collector surface with a high value of α_s/ε is called a selective surface. Since these surfaces are usually formed by a coating process, they are sometimes called *selective coatings*. The most common commercial selective coating is *black chrome*. The characteristics of a typical black

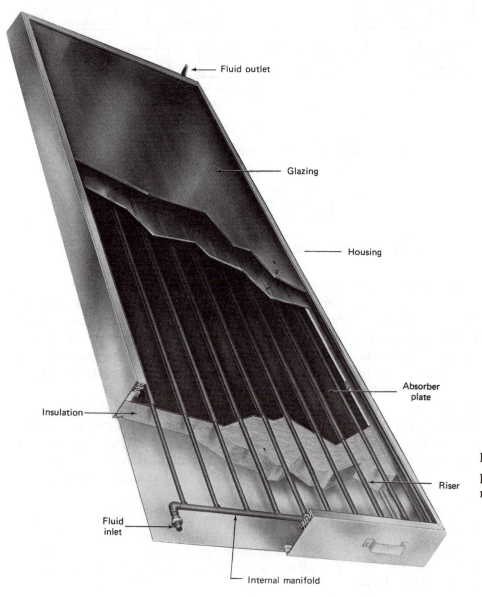

Fig. 16.4 Typical double-glazed flat-plate collector, liquid type, internally manifolded. (Courtesy LOF.)

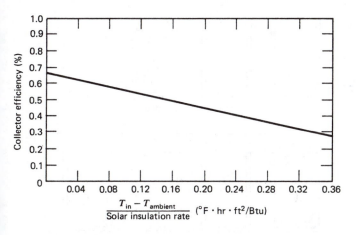

Fig. 16.5 Efficiency of a typical liquid-type solar collector panel.

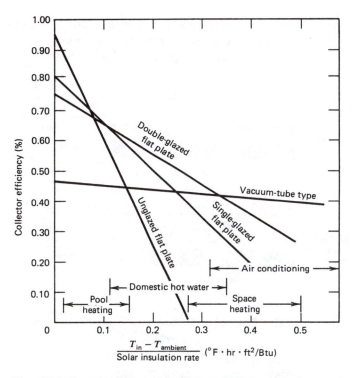

Fig. 16.6 Comparison of collector efficiencies for various liquid type collectors.

chrome surface are shown in Figure 16.7, where $\alpha\lambda = \varepsilon\lambda$, the monochromatic absorptance and monochromatic emittance of the surface. Note that at short wavelengths (~ 0.5 μ), typical of sunlight, the absorptance is high. At the longer wavelengths (~2 μ and above), where the absorber plate will emit most of its energy, the emittance is high. Selective surfaces will generally perform better than ordinary blackened surfaces. The performance of a flat black collector and a selective coating collector are compared in Figure 16.8. The single-glazed selective collector performance is very similar to the double-glazed nonselective collector. Economic considerations usually lead one to pick a single-glazed, selective or a double-glazed, nonselective collector over a double-glazed, selective collector, although this decision depends heavily upon quoted or bid prices.

Air-type collectors are particularly useful where hot air is the desired end product. Air collectors have distinct advantages over liquid-type collectors:

1. Freezing is not a concern.
2. Leaks, although undesirable, are not as detrimental as in liquid systems.
3. Corrosion is less likely to occur.

Air systems may require large expenditures of fan power if the distances involved are large or if the delivery ducts are too small. Heat-transfer rates to air are typically lower than those to liquids, so care must be taken in air collectors and in air heat exchangers to provide sufficient heat-transfer surface. This very often in-

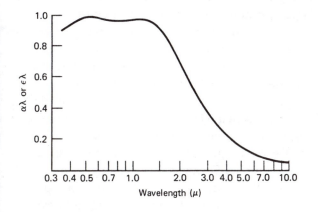

Fig. 16.7 Characteristics of a typical selective (black chrome) collector surface.

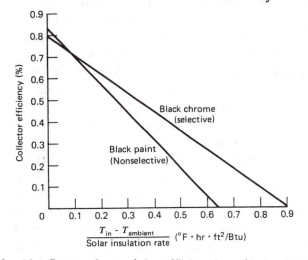

Fig. 16.8 Comparison of the efficiencies of selective and nonselective collectors.

volves the use of extended surfaces or fins on the sides of the surface, where air is to be heated or cooled. Typical air collector designs are shown in Figure 16.9.

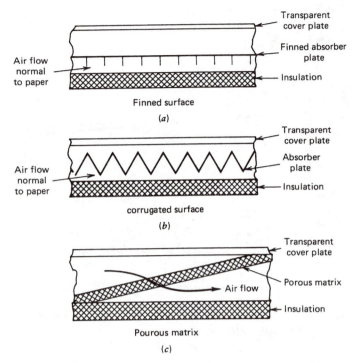

Fig. 16.9 Typical air collector designs. (a) Finned surface. (b) Corrugated surface. (c) Porous matrix.

Flat-plate collectors usually come in modules about 3 ft wide by 7 ft tall, although there is no standard size. Collectors may have internal manifolds or they may be manifolded externally to form collector arrays (Figure 16.10). Internally manifolded collectors are easily connected together, but only a small number can be hooked together in a single array and still have good flow distribution. Small arrays (5 to 15) are often piped together with similar arrays in various series and parallel arrangements to give the best compromise between nearly uniform flow rates in each collector, and as small a pressure drop and total temperature rise as can be attained. Externally manifolded collectors are easily connected in balanced arrays if the external manifold is properly designed. These types of arrays require more field connections, however, have more exposed piping to insulate, and are not as neat looking.

The overall performance of a collector array, measured in terms of the collector array efficiency, may be quite a bit less than the collector efficiency of the individual collectors. This is due primarily to unequal flow distribution between collectors, larger temperature rises in series connections than in single collectors, and heat losses from the connecting piping. A good array design will minimize these factors together with the pumping

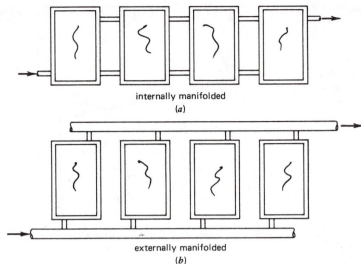

Fig. 16.10 Examples of collectors hooked in parallel. (a) Internally manifolded. (b) Externally manifolded.

requirements for the array.

Concentrating collectors provide relatively high temperatures for applications such as air conditioning, power generation, and the furnishing of industrial or process heat above 250°F (121°C). They generally cannot use the diffuse or scattered radiation from the sky and must track so that the sun's direct rays will be concentrated on the receiver. The theory is simple. By concentrating the sun's rays on a very small surface, heat losses are reduced at the high temperature desired. An important point to make is that concentrating collectors do not increase the amount of energy above that which falls on the mirrored surfaces; the energy is merely concentrated to a smaller receiver surface.

A typical parabolic trough-type solar collector array is shown in Figure 16.11. Here the concentrating surface or mirror is moved, to keep the sun's rays concentrated as much as possible on the receiver, in this case a tube through which the coolant flows. In some systems the tube moves and the mirrored surfaces remain fixed.

This type collector can be mounted on an east-west axis and track the sun by tilting the mirror or receiver in a north-south direction (Figure 16.12a). An alternative is to mount the collectors on a north-south axis and track the sun by rotating in an east-west direction (Figure 16.12b). A third scheme is to use a polar mount, aligning the trough and receiver parallel to the earth's pole, or inclined at some angle to the pole, and tracking east to west (Figure 16.12c). Each has its advantages and disadvantages and the selection depends upon the application. A good discussion of concentrating collectors is given in Ref. 8.

Fully tracking collectors may be a parabolic disk

Fig. 16.11 Typical parabolic trough-type solar collector array (Suntec, Inc.).

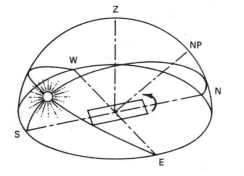

(a)

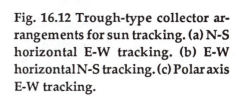

Fig. 16.12 Trough-type collector arrangements for sun tracking. (a) N-S horizontal E-W tracking. (b) E-W horizontal N-S tracking. (c) Polar axis E-W tracking.

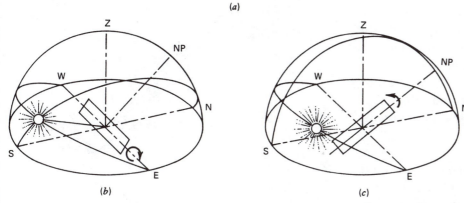

(b)   (c)

with a "point source" or may use a field of individual nearly flat moving mirrors or heliostats, concentrating their energy on a single source, such as might be installed on a tower (a power tower). Computers usually control the heliostat motion. Some trough-type collectors are also fully tracking, but this is not too common. All partial and fully tracking collectors must have some device to locate the sun in the sky, either by sensing or by prediction. Tracking motors, and in some cases flexible or movable line connections, are additional features of tracking systems. Wind loads can be a serious problem for any solar collector array that is designed to track. Ability to withstand heavy windloads is perhaps the biggest single advantage of the flat-plate, fixed collector array.

16.2.3 Thermal Storage Systems

Because energy demand is almost never tied to solar energy availability, a storage system is usually a part of the solar heating or cooling system. The type of storage may or may not depend upon the type of collectors used. With air-type collectors, however, a rock-bed type of storage is sometimes used (Figure 16.13). The rocks are usually in the size range 3/4 to 2 in. in diameter to give the best combination of surface area and pressure drop. Air flow must be down for storing and up for removal if this type system is to perform properly. Horizontal air flow through a storage bed should normally be avoided. An air flow rate of about 2 cfm/ft^2 of collector is recommended. The amount of storage required in any solar heating system is tied closely to the amount of collector surface area installed, with the optimum amount being determined by a computer calculation. As a rule of thumb, for rough estimates one should use about 75 lb of rock per square foot of air-type collectors. If the storage is too large, the system will not be able to attain sufficiently high temperatures, and in addition, heat losses will be high. If the storage is too small,

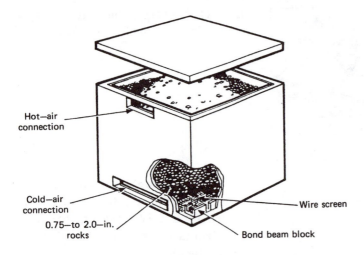

Fig. 16.13 Rock-bed-type storage system.

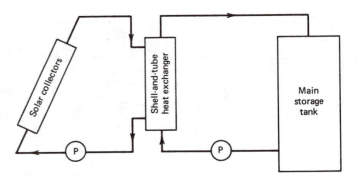

Fig. 16.14 External heat exchanger between collectors and main storage.

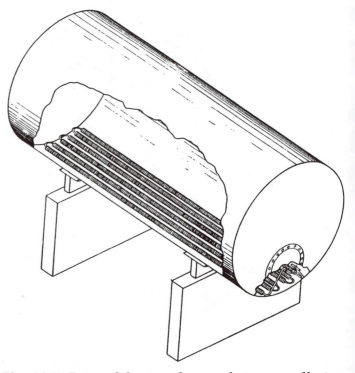

Fig. 16.15 Internal heat exchanger between collector and storage medium.

the system will overheat at times and may not collect and store a large enough fraction of the energy available.

The most common solar thermal storage system is one that uses water, usually in tanks. As a rule the water storage tank should contain about 1.8 gal/ft^2 of collector surface. Water has the highest thermal storage capability of any common single-phase material per unit mass or per unit volume. It is inexpensive, stable, nontoxic, and easily replaced. Its main disadvantage is its high vapor pressure at high temperatures. This means that high pressures must be used to prevent boiling at high temperatures.

Water also freezes, and therefore in most climates, the system must either (1) drain all of the collector fluid back into the storage tank, or (2) use antifreeze in the collectors and separate the collector fluid from the storage fluid by use of a heat exchanger.

Drain-down systems must be used cautiously because one failure to function properly can cause severe damage to the collectors and piping. It is the more usual practice in large systems to use a common type of heat exchanger, such as a shell-and-tube exchanger, placed external to the storage tank, as shown in Figure 16.14. Another method, more common to small solar systems, is to use coils of tubing around the tank or inside the tank, as shown in Figure 16.15.

In any installation using heat exchangers between the collectors and storage, the exchanger must have sufficient surface for heat transfer to prevent impairment of system performance. Too small a surface area in the exchanger causes the collector operating temperature to be higher relative to the storage tank temperature, and the collector array efficiency decreases. As a rough rule of thumb, the exchanger should be sized so as to give an effectiveness of at least 0.60, where the effectiveness is

the actual temperature decrease of the collector fluid passing through the exchanger to the maximum possible temperature change. The maximum possible would be the difference between the design temperature of the collector fluid entering the exchanger and the temperature entering from the storage tank.

Stratification normally occurs in water storage systems, with the warmest water at the top of the tank. Usually, this is an advantage, and flow inlets to the tank should be designed so as not to destroy this stratification. The colder water at the bottom of the tank is usually pumped to the external heat exchanger and the warmer, returning water is placed at the top or near the center of the tank. Hot water for use is usually removed from the top of the tank.

Phase-change materials (PCMs) have been studied extensively as storage materials for solar systems. They depend on the ability of a material to store thermal energy during a phase change at constant temperature. This is called latent storage, in contrast to the sensible storage of rock and water systems. In PCM systems large quantities of energy can be stored with little or no change in temperature. The most common PCMs are the eutectic salts. Commercial PCMs are relatively expensive and, to a certain extent, not completely proven as to lifetime and reliability. They offer distinct advantages, however, particularly in regard to insulation and space requirements, and will no doubt continue to be given attention.

16.2.4 Control Systems

Solar systems should operate automatically with little attention from operating personnel. A good control system will optimize the performance of the system with reliability and at a reasonable cost. The heart of any solar thermal collecting system is a device to turn on the collector fluid circulating pump (and other necessary devices) when the sun is providing sufficient insolation so that energy can be collected and stored, or used. With flat-plate collectors it is common to use a differential temperature controller (Figure 16.16), a device with two temperature sensors. One sensor is normally located on the collector fluid outlet and the other in the storage tank near the outlet to the heat exchanger (or at the level of the internal heat exchanger). When the sun is out, the fluid in the collector is heated. When a prescribed temperature difference (about 20°F) exists between the two sensors, the controller turns on the collector pump and other necessary devices. If the temperature difference drops below some other prescribed difference (about 3 to 5°F), the controller turns off the necessary devices. Thus clouds or sundown will cause the system to shut down and prevent not only the unnecessary loss of heat to the collectors but also the unnecessary use of electricity. The distinct temperature difference to start and to stop is to prevent excessive cycling.

Differential temperature controllers are available with adjustable temperature difference settings and can also be obtained to modulate the flow of the collector fluid, depending upon the solar energy available.

Controllers for high-temperature collectors, such as evacuated tubes and tracking concentrators, sometimes use a light meter to sense the level of sunlight and turn on the pumps. Some concentrating collectors are inverted for protection when light levels go below a predetermined value.

In some systems the storage fluid must be kept above some minimum value (e.g., to prevent freezing). In such cases a *low-temperature controller* is needed to turn on auxiliary heaters if necessary. A *high-temperature controller* may also be needed to bypass the collector fluid or to turn off the system so that the storage fluid is not overheated.

Figure 16.17 shows a control diagram for a solar-heated asphalt storage system (Figure 16.18) in which the fluid must be kept between two specified temperatures. Solar heat is used whenever it is available (collector pump on). If the storage temperature drops below the specified minimum, the pump *and* an electric heater are turned on to circulate electrically heated fluid to the tank. If the tank fluid gets too warm, the system shuts off. Almost any required control pattern can be developed for solar systems using the proper arrangement of a differential temperature controller, high- and low-temperature controllers, relays, and electrically operated valves.

16.2.5 Sizing and Economics

An article on how to identify cost-effective solar-thermal applications is given in the *ASHRAE Journal*[9]. In almost any solar energy system the largest single expense are the solar collector panels and support structure. For this reason the system is usually "sized" in terms of collector panel area. Pumps, piping, heat exchangers, and storage tanks are then selected to match.

Very rarely can a solar thermal system pro-

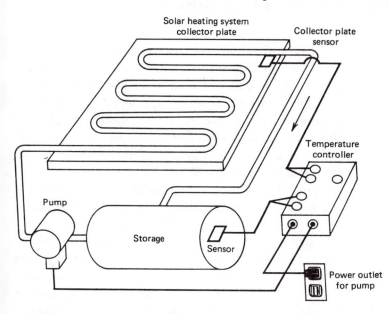

Fig. 16.16 Installation of a differential temperature controller in a liquid heating system.

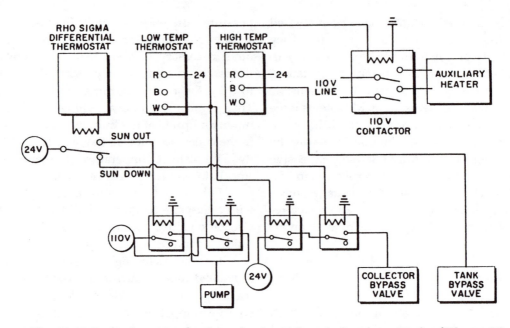

Fig. 16.17 Control system for the solar-heated asphalt storage tank of Figure 16.

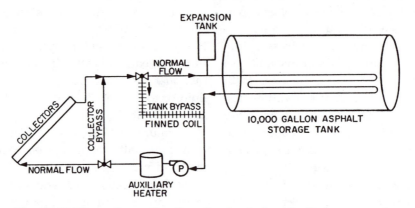

Fig. 16.18 Flow schematic of a solar-heated asphalt storage system.

vide 100% of the energy requirements for a given application. The optimum-size solar system is the one that is the most economical on some chosen basis. The computations may be based on (1) lowest life-cycle cost, (2) quickest payout, (3) best rate of return on investment, and, (4) largest annual savings. All of these computations involve the initial installed cost, the operating and maintenance costs, the life of the equipment, the cost of money, the cost of fuel, and the fuel escalation rate, in addition to computations involving the amount of energy furnished by the solar system.

When done by hand calculation, the demand and solar values are usually computed on a monthly basis using averaged weather data for each month. Annual values are determined by summation and the economic factors are then calculated. If a computer is used, hourly or daily simulations and computations may be made, supposedly with greater accuracy.

A typical set of calculations might lead to the results shown in Figure 16.19, the net annual savings per year versus the collector area, with the present cost of fuel as a parameter.[10] Curve a represents a low fuel cost, the net savings is negative, and the system would cost rather than save money. Curve b represents a slightly higher fuel cost where a system of about 800 ft^2 of collectors would break even.

Curves c and d, representing even higher fuel costs, show a net savings, with optimum savings occurring at about 1200 and 2000 ft^2, respectively.

High interest rates tend to reduce the economic viability of solar systems. High fuel costs obviously have the opposite effect, as does a longer life of the equipment. Federal and state tax credits would also have an important effect on the economics of solar energy as an alternative energy source. Technical improvements and lower first costs can obviously have an important effect

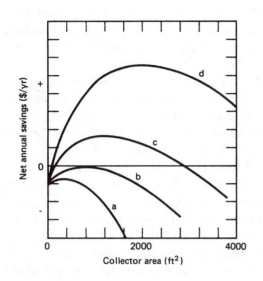

Fig. 16.19 Collector area optimization curves for a typical solar heating system. Ref. 10.

on the economics, but contributions of these two factors have been relatively slow in coming.

16.2.6 Solar Cells

Solar cells use the electronic properties of semiconductor material to convert sunlight directly into electricity. They are widely used today in space vehicles and satellites, and in terrestrial applications requiring electricity at remote locations. Since the conversion is direct, solar cells are not limited in efficiency by the Carnot principle. An excellent text by Martin[11] is devoted entirely to the operating principles, technology and system applications of solar cells.

Most solar cells are very large area p-n junction diodes. Figure 16.20a. A p-n junction has electronic asymmetry. The n-type regions have large electron densities but small hole densities. Electrons flow readily through the material but holes find it very difficult. P-type material has the opposite characteristic. Excess electron-hole pairs are generated throughout the p-type material when it is illuminated. Electrons flow from the p-type region to the n-type and a flow of holes occurs in the opposite direction. If the illuminated p-n junction is electrically short circuited a current will flow in the short-circuiting lead. The normal rectifying current-voltage characteristic of the diode is shown in Figure 16.20b. When illuminated (insulated) the current generated by the illumination is superimposed to give a characteristic where power can be extracted.

The characteristic voltage and current parameters of importance to utilizing solar cells are shown in Figure 16.20b. The short-circuit current I_{sc} is, ideally, equal to the light generated current I_L. The open-circuit voltage V_{oc} is determined by the properties of the semiconductor. The particular point on the operating curve where the power is maximum, the rectangle defined by V_{mp} and I_{mp} will have the greatest area. The fill factor FF is a measure of how "square" the output characteristics are. It is given by:

$$FF = \frac{V_{mp}I_{mp}}{V_{oc}I_{sc}}$$

Ideally FF is a function only of the open-circuit voltage and in cells of reasonable efficiency has a value in the range of 0.7 to 0.85.

Most solar cells are made by doping silicon, the second most abundant element in the earth's crust. Sand

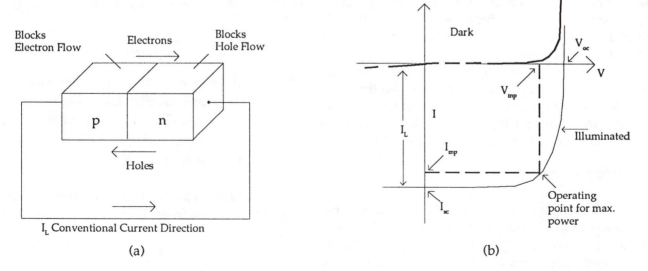

Figure 16.20 Nomenclature of solar cells.

is reduced to metallurgical-grade silicon, which is then further purified and converted to single-crystal silicon wafers. The wafers are processed into solar cells which are then encapsulated into weatherproof modules. Boron is used to produce p-type wafers and phosphorus is the most common material used for the n-type impurity. Other types of solar cells include CdS, CdTe, and GaAs. Theoretical maximum conversion efficiencies of these simple cells are around 21 to 23 percent.

Major factors which, when present in real solar cells, prohibit the attainment of theoretical efficiencies include reflection losses, incomplete absorption, only partial utilization of the energy, incomplete collection of electron-hole pairs, a voltage factor, a curve factor and internal series resistance.

Thin-film solar cells have shown promise in reducing the cost of manufacturing and vertical junction cells have been shown to have high end-of-life efficiencies. Solar cells are subject to weathering and radiation damage. Care must be taken in solar arrays to avoid poor interconnection between cells, and increased series resistance due to deterioration of contacts.

Solar cells are arranged in a variety of series and parallel arrangements to give the voltage-current characteristics desired and to assure reliability in case of individual cell failure. Fixed arrays are placed at some optimal slope and usually faced due south in the northern hemisphere. Large arrays are usually placed on a structure allowing tracking of the sun similar to those used for concentrating solar thermal collectors. In some arrays the sunlight is concentrated before it is allowed to impinge on the solar cells. Provision must be made for thermal energy removal since the solar cell typically converts only a small fraction of the incident sunlight into electrical power. Increasing temperature of the cell has a dominant effect on the open circuit voltage, causing the power output and efficiency to decrease. For silicon cells the power output decreases by 0.4 to 0.5 % per degree Kelvin increase.

Provisions must usually be made for converting the direct current generated by the array into the more useful alternating current at suitable frequency and voltage. In many systems where 24 hour/day electricity is needed some type of storage must be provided.

16.3 WIND ENERGY

Wind energy to generate electricity is most feasible at sites where wind velocities are consistently high and reasonably steady. Ideally these sites should be remote from densely populated areas, since noise generation, safety, and disruption of TV images may be problems.

On the other hand the generators must be close enough to a consumer that the energy produced can be utilized without lengthy transmission. With these considerations it is no surprise to find most of the existing wind energy systems are installed in mountain passes in California and along the seacoast of Europe. In 1990 California produced 85 to 90 percent of the world's wind power, its 13,500 windmills capable of producing enough electricity to meet all of San Francisco's residential needs.[12] Interest in wind power has continued even after the dropping of tax credits in the mid 1980's. Enough interest exists in the private sector to cause several manufacturing firms to continue production and development of new models. The wind industry has had good luck attracting private capital to develop wind farms for the production of electricity to sell to electric utilities.

16.3. 1 Availability

A panel of experts from NSF and NASA estimated that the power potentially available across the continental United States, including offshore sites and the Aleutian arc, is equivalent to approximately 10^5 GW of electricity.[13] This was about 100 times the electrical generating capacity of the United States. Figure 16.21 shows the areas in the United States where the average wind velocities exceed 18 miles/hr (6 meters/sec) at 150 ft (45.7 m) above ground level. As an approximation, the wind velocity varies approximately as the 1/7 power of distance from the ground.

The power that is contained in a moving air stream per unit area normal to the flow is proportional to the cube of the wind velocity. Thus small changes in wind velocity lead to much larger changes in power available. The equation for calculating the power density of the wind is

$$\frac{P}{A} = \frac{1}{2}\rho V^3$$

where P = power contained in the wind

  A = area normal to the wind velocity

  ρ = density of air (about 0.07654 lbm/ft3 or 1.23 kg/m^3)

  V = velocity of the air stream

Consistent units should be selected for use in equation 16.1. It is convenient to rewrite equation 16.1 as

$$\frac{P}{A} = KV^3$$

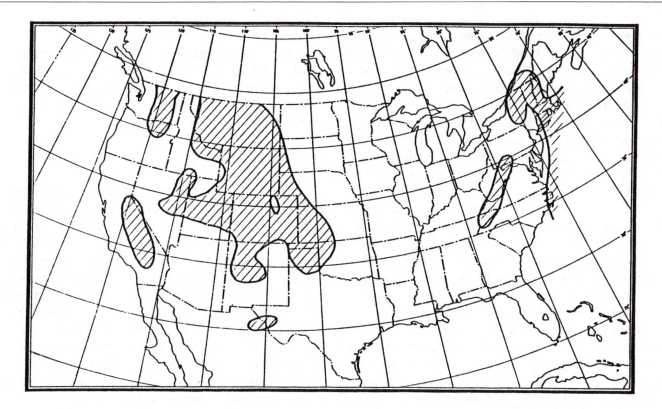

Fig. 16.21 Areas in the United States where average wind speeds exceed 18 miles/hr (8 miles/sec) at 150 ft (45.7 m) elevation above ground level. (From Ref. 13.)

If the power density P/A is desired in the units W/ft^2, then the value of K depends upon the units selected for the velocity V. Values of K for various units of velocity are given in Table 16.3.

The fraction of the power in a wind stream that is converted to mechanical shaft power by a wind device is given by the power coefficient Cp.

It can be shown that only 16/27 or 0.5926 of the power in a wind stream can be extracted by a wind machine, since there must be some flow velocity downstream from the device for the air to move out of the way. This upper limit is called the Betz coefficient (or Glauert's limit). No wind device can extract this theoretical maximum. More typically, a device might extract some fraction, such as 70%, of the theoretical limit. Thus a real device might extract approximately $(0.5926)(0.70) = 41\%$ of the power available. Such a device would have an aerodynamic efficiency of 0.70 and a power coefficient of 0.41. The power conversion capability of such a device could be determined by using equation 16.2 and Table 16.3. Assume a 20-mile/hr wind. Then

$$\left(\frac{P}{A}\right)_{\text{actual}} = \left(5.08 \times 10^{-3}\right)(20)^3 (0.41) = 16.7\,\text{W/ft}^2$$

Notice that for a 30-mile/hr wind the power conversion capability would be 56.2 W/ft^2, or more than three times as much.

Table 16.3 Values of K to Give P/A (W/ ft^2) in Equation 16.2^a.

Units of V	K
ft/sec	1.61×10^{-3}
miles/hr	5.08×10^{-3}
km/hr	1.22×10^{-3}
m/sec	5.69×10^{-2}
knots	7.74×10^{-3}

^{a}To convert w/ft^2 to W/m^2, multiply by 10.76.

Because the power conversion capability of a wind device varies as the cube of the wind velocity, one cannot predict the annual energy production from a wind device using mean wind velocity. Such a prediction would tend to underestimate the actual energy available.

16.3.2 Wind Devices

Wind conversion devices have been proposed and built in a very wide variety of types. The most general types are shown in Figure 16.22. The most common type is the horizontal-axis, head-on type, typical of conventional farm windmills. The axis of rotation is parallel to the direction of the wind stream. Where the wind direction is variable, the device must be turned into the wind, either by a tail vane or, in the case of larger systems, by a servo device. The rotational speed of the single-, double-, or three-bladed devices can be controlled by feathering of the blades or by flap devices or by varying the load.

In most horizontal-axis wind turbines, the generator is directly coupled to the turbine shaft, sometimes through a gear drive. In the case of the bicycle multi-bladed type, the generator may be belt driven off the rim, or the generator hub may be driven directly off the rim by friction. In the later case there is no rotational speed control except that imposed by the load.

In the case of a vertical-axis wind turbine (VAWT) such as the Savonius or the Darrieus types, the direction of the wind is not important, which is a tremendous advantage. The system is more simple and there are no stresses created by yawing or turning into the wind as occurs on horizontal-axis devices. The VAWT are also lighter in weight, require only a short tower base, and can have the generator near the ground. VAWT enthusiasts claim much lower costs than for comparable horizontal-axis systems.

The side wind loads on a VAWT are accommodated by guy wires or cables stretched from the ground to the upper bearing fixture.

The Darrieus-type VAWT can have one, two, three, or more blades, but two or three are most common. The curved blades have an airfoil cross section with very low starting torque and a high tip-to-wind speed.

The Savonius-type turbine has a very high starting torque but a relatively low tip-to-wind speed. It is primarily a drag-type device, whereas the Darrieus type is primarily a lift-type device. The Savonius and the Darrieus types are sometimes combined in a single turbine to give good starting torque and yet maintain good performance at high rotational speeds.

Figure 16.23 shows the variation of the power coefficient Cp as the ratio of blade tip speed to wind speed varies for different types of wind devices. It can be seen that two-blade types operating at relatively high speed ratios have the highest value of Cp, in the range of 0.45, which is fairly close to the limiting value of the Betz coefficient (0.593). The Darrieus rotor is seen to have a slightly lower maximum value, but like the two-blade type, performs best at high rotational speeds. The American (bicycle) multi-blade type is seen to perform best at lower ratios of tip to wind speed, as does the Savonius.

For comparison, in a 17-mile/hr (7.6-meters/sec) wind, a 2000-kW horizontal-axis wind turbine would have a diameter of 220 ft (67 m) and a 2000-kW Darrieus type would have a diameter of 256 ft (78 m) and would stand about 312 ft (95 m) tall.[12]

16.3.3 Wind Systems

Because the typical wind device cannot furnish energy to exactly match the demand, a storage system and a backup conventional energy source may be made a part of the total wind energy system (Figure 16.24). The storage system might be a set of batteries and the backup system might be electricity from a utility. In some cases the system may be designed to put electrical power into the utility grid whenever there is a surplus and to draw power from the utility grid whenever there is a deficiency of energy. Such a system must be synchronized with the utility system and this requires either rotational speed control or electronic frequency control such as might be furnished by a field-modulated generator.

Economics favors the system that feeds surplus power into the utility grid over the system with storage, but the former does require reversible metering devices and a consenting utility. Some states have and others probably will pass laws that require public utilities to accept such power transfers.

16.3.4 Wind Characteristics—Siting

The wind characteristics given in Figure 16.21 are simple average values. The wind is almost always quite variable in both speed and direction. Gusting is a rapid up-and-down change in wind speed and/or direction. An important characteristic of the wind is the number of hours that the wind exceeds a particular speed. This information can be expressed as speed-duration curves, such as those shown in Figure 16.25 for three sites in the United States. These curves are similar to the load-duration curves used by electric utilities.

Because the power density of the wind depends on the cube of the wind speed, the distribution of annual average energy density of winds of various speeds will be quite different for two sites with different average wind speeds. A comparison between sites having average velocities of 13 and 24 miles/hr (5.8 and 10.7 meters/sec) is given in Figure 16.25. The area under the curve is the total energy available per unit area per year for each case.

Sites should be selected where the wind speed is as high and steady as possible. Rough terrain and the pres-

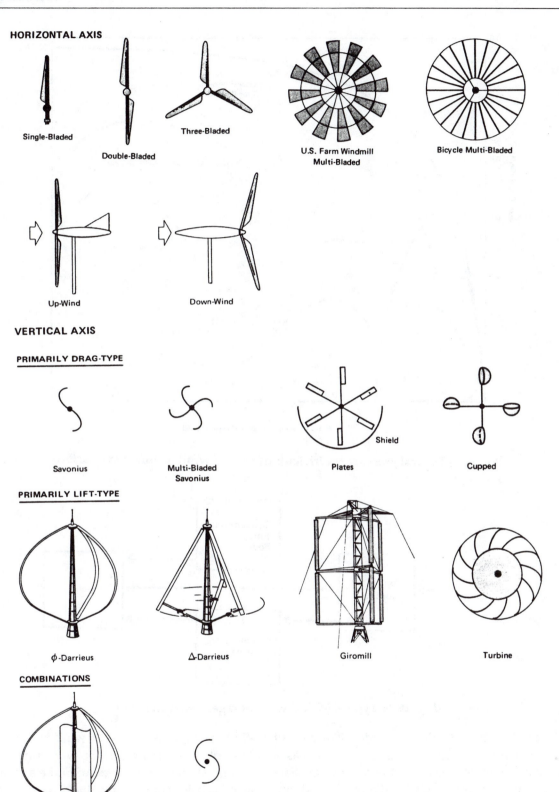

HORIZONTAL AXIS

Single-Bladed

Double-Bladed

Three-Bladed

U.S. Farm Windmill
Multi-Bladed

Bicycle Multi-Bladed

Up-Wind

Down-Wind

VERTICAL AXIS

PRIMARILY DRAG-TYPE

Savonius

Multi-Bladed
Savonius

Plates

Shield

Cupped

PRIMARILY LIFT-TYPE

ϕ-Darrieus

Δ-Darrieus

Giromill

Turbine

COMBINATIONS

Savonius/ϕ-Darrieus

Split Savonius

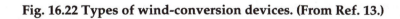

Fig. 16.22 Types of wind-conversion devices. (From Ref. 13.)

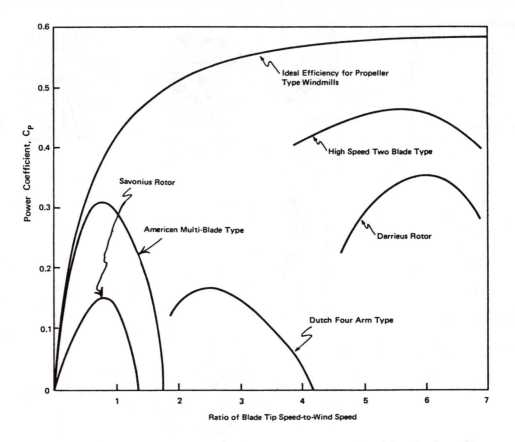

Fig. 16.23 Typical pressure coefficients of several wind turbine devices. (From Ref. 13.)

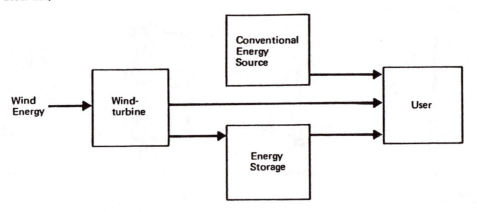

Fig. 16.24 Typical WECS with storage. (From Ref. 13.)

ence of trees or building should be avoided. The crest of a well-rounded hill is ideal in most cases, whereas a peak with sharp, abrupt sides might be very unsatisfactory, because of flow reversals near the ground. Mountain gaps that might produce a funnelling effect could be most suitable.

16.3.5 Performance of Turbines and Systems

There are three important wind speeds that might be selected in designing a wind energy conversion system (WECS). They are (1) cut-in wind speed, (2) rated

wind speed, and (3) cut-off wind speed. The names are descriptive in each case. The wind turbine is kept from turning at all by some type of brake as long as the wind speed is below the cut-in value. The wind turbine is shut off-completely at the cut-off wind speed to prevent damage to the turbine. The rated wind speed is the lowest speed at which the system can generate its rated power. If frequency control were not important, a wind turbine would be permitted to rotate at a variable speed as the wind speed changed. In practice, however, since frequency control must be maintained, the wind turbine rotational speed might be controlled by varying the load

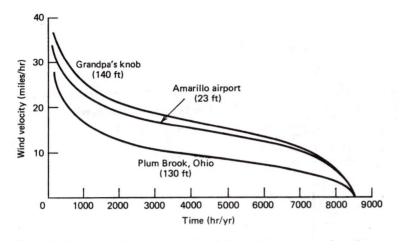

Fig. 16.25 Annual average speed-duration curves for three sites. (From Ref. 13.)

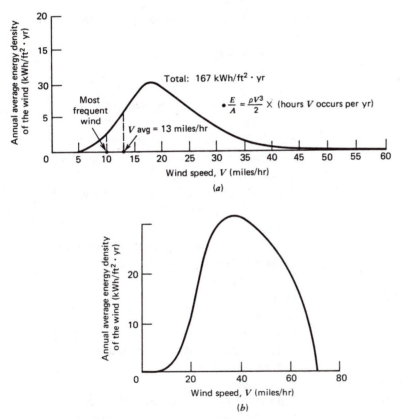

Fig. 16.26 Comparison of distribution of annual average energy density at two sites. (From Ref. 13.) (a) V_{avg} = 13 miles/hr. (b) V_{avg} = 24 miles/hr.

on the generator when the wind speed is between the cut-in and rated speed. When the wind speed is greater than the rated speed but less than cut-out speed, the spin can be controlled by changing the blade pitch on the turbine. This is shown in Figure 16.27 for the 100-kW DOE/NASA system at Sandusky, Ohio. A system such

as that shown in Figure 16.27 does not result in large losses of available wind power if the average energy content of the wind at that site is low for speeds below the cut-in speed and somewhat above the rated speed.

Another useful curve is the actual annual power density output of a WECS (Figure 16.28). The curve shows the hours that the device would actually operate and the hours of operation at full rated power. The curve is for a system with a rated wind speed of 30 miles/hr (13.4 meters/ sec), a cut-in velocity of 15 miles/hr (6.7 meters/ sec) and a cut-off velocity of 60 miles/hr (26.8 meters/ sec) with constant output above 30 miles/hr.

16.3.6 Loadings and Acoustics

Blades on wind turbine devices have a variety of extraneous loads imposed upon them. Rotor blades may be subject to lead-lag motions, flapping, and pitching. These motions and some of their causes are shown in Figure 16.29. These loads can have a serious effect on the system performance, reliability, and lifetime.

Acoustics can be a serious problem with wind devices, especially in populated areas. The DOE/ NASA device at Boone. North Carolina caused some very serious low-frequency (~1 Hz) noises and was taken out of service.

The most promising wind systems from an economic standpoint appear to be mid-size, propeller-type systems, located in large numbers at one site and controlled from a central terminal. Most systems will likely be owned by an electric utility or sell their power to a utility.

16.4 REFUSE-DERIVED FUEL

16.4.1 Process Wastes

Typical composition of solid waste is shown in Table 16.4. It can be seen that more than 70% by weight is combustible. More important, more than 90% of the volume of typical solid waste can be eliminated by combustion. Burning waste as fuel has the advantage of not only replacing scarce fossil fuels but also greatly reducing the problem of waste disposal.

Solid wastes affect public health, the environment, and also present an opportunity for reuse or recycling of the material. Managing this in an optimum way is some-

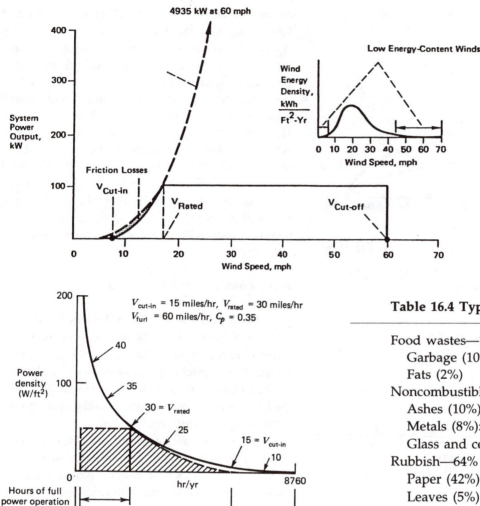

Fig. 16.27 Power output of a 100-kW WECS at various wind speeds. (From Ref. 13.)

Fig. 16.28 Actual annual power density output of a WECS. (From Ref. 13.)

Table 16.4 Typical composition of solid waste.

Food wastes—12% by weight
   Garbage (10%)
   Fats (2%)
Noncombustibles—24% by weight
   Ashes (10%)
   Metals (8%): cans, wire, and foil
   Glass and ceramics (6%): bottles primarily
Rubbish—64% by weight
   Paper (42%): various types, some with fillers
   Leaves (5%) Grass (4%)
   Street sweepings (3%)
   Wood (2.4%): packaging, furniture, logs, twigs
   Brush (1.5%)
   Greens (1.5%)
   Dirt (1%)
   Oil, paints (0.8%)
   Plastics (0.7%): polyvinyl chloride, polyethyl-
     ene, styrene, etc., as found in packaging,
     housewares, furniture, toys, and nonwoven
     synthetics
   Rubber (0.6%): shoes, tires, toys, etc.
   Rags (0.6%): cellulose, protein, and woven syn-
     thetics
   Leather (0.3%): shoes, tires, toys, etc.
   Unclassified (0.6%)

Source: Ref. 14.

times called *integrated solid waste management.* A textbook on that subject (15) provides much more detail than can be furnished in this brief handbook discussion. That reference estimates that between 2500 and 7750 pounds of waste is generated per person each year in the United States. Included in this is typically 2225 pounds of municipal waste, 750 pounds of industrial waste, and between 250 and 3000 pounds of agricultural waste per person each year.

The total mass of solid wastes in the United States reached more than 41 billion tons in 1971[14] and is probably more than double that amount today. It was estimated that each person in the United States consumed 660 lb of packaging material in 1976, and uses over 5 lb/day of waste products.

The heating value of the refuse would be an important consideration in any refuse-derived fuel (RDF) application. Typical heating values of solid waste refuse components are given in Table 16.5. Other values are given in Ref. 16.

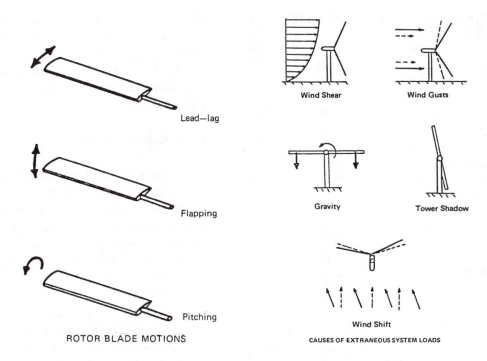

Fig. 16.29 Rotor blade motions and their causes. (From Ref. 13.)

16.4.2 Refuse Preparation

There are several routes by which waste can be used to generate steam or electricity. The possible paths for municipal wastes are shown in Figure 16.30. In the past the most common method was for the refuse to be burned unprepared in a waterwall steam generator. This technology is simple and well developed and costs can be accurately predicted. Another approach is for the refuse to be placed in a landfill and gas formed from decomposition of the organic material recovered and burned. This is not efficient and requires land for use in the landfill. Refuse may be given some treatment, such as shredding and separation, and then burned in a waterwall steam generator.

The more sophisticated methods involve some type of treatment after shredding to change the refuse into a more desirable fuel form. This may involve converting the shredded refuse into a gas or liquid or into solid pellets.

The shredding, which can be done wet or dry (as received), converts the refuse into a relatively homogeneous mixture. The shredding is usually done by hammermills or crushers. This shredding operation is costly both in terms of energy and maintenance. One reference gives 1977 maintenance costs of 60 cents/ton of waste.[16]

Problems with fire, explosions, vibrations, and noise are common in the shredding operation.

Density separation increases the fuel's heating value, minimizes wear on transporting and boiler heat-transfer surfaces, and makes the ash more usable. Resource recovery can be an important by-product, with separation of metals and glass for resale.

16.4.3 Pyrolysis and Other Processes

Pyrolysis is the thermal decomposition of material in the absence of oxygen. The product can be a liquid or a gas suitable for use as a fuel. There are many pyrolysis projects in the research and development stage. An apparently successful pyrolytic heat-recovery system is described in Ref. 17. The new plant saved $53,000 the first year while disposing of 90% of the firm's waste. It was expected that the system would pay out in approximately three years. Emissions were said to be below standards set by EPA. A second pyrolytic heat recovery is being installed to take care of expanding needs.

Anaerobic digestion processes, similar to those used in wastewater treatment facilities, can also be used to convert the shredded, separated waste into a fuel. About 3 scf of methane can be produced from about 1 lb of refuse. In this process the shredded organic material is mixed with nutrients in an aqueous slurry, heated to about 140°F, and circulated through a digester for several days. The off-gas has a heating value of about 600 Btu/scf but can be upgraded to nearly pure methane.

Solid fuel pellets can also be prepared from refuse

Table 16.5 Typical Heating Values of Solid Waste Refuse Components[a]

	MJ/kg	Btu/lb
Domestic refuse		
Garbage	4.23	1,820
Grass	8.88	3,820
Leaves	It 39	4,900
Rags (cotton, linen)	14.97	6,440
Brush, branches	16.60	7,140
Paper, cardboard, cartons, bags	17.81	7,660
Wood, crates, boxes, scrap	18.19	7,825
Industrial scrap and plastic refuse		
Boot, shoe trim, and scrap	19.76	8,500
Leather scrap	23.24	10,000
Cellophane	27.89	12,000
Waxed paper	27.89	12,000
Rubber	28.45	12,240
Polyvinyl chloride	40.68	17,500
Tires	41.84	18,000
Oil, waste, fuel-oil residue	41.84	18,000
Polyethylene	46.12	19,840
Agricultural		
Bagasse	8.37-15.11	3,600-6,500
Bark	10.46-12.09	4,500-5,200
Rice hulls	12.15-15.11	5,225-6,500
Corncobs	18.60-19.29	8,000-8,300
Composite		
Municipal	10.46-15.11	4,500-6,500
Industrial	15.34-16.97	6,600-7,300
Agricultural	6.97-13.95	3,000-6,000

Source: Ref. 14.

[a] Calorific value in MJ/kg (Btu/lb) as fired.

which are low in inorganics and moisture and with heating values around 7500 to 8000 Btu/lbm (17,000 to 19,000 kJ/kg). Some pellets have been found to be too fibrous to be ground in the low- and medium-speed pulverizers that might normally be found in coal-fired plants.

16.4.4 Refuse Combustion

The major problems in firing refuse in steam generators seems to be fouling of heat-transfer surfaces and corrosion. Fouling is caused by slag and fly-ash deposition. It is reduced by proper sizing of the furnace, by proper arrangement of heat-transfer surfaces, and by proper use of boiler cleaning equipment.[17]

Corrosion in RDF systems is usually due to:

1. Reducing environment caused by stratification or improper distribution of fuel and air.
2. Halogen corrosion caused by presence of polyvinyl chloride (PVC) in the refuse.
3. Low-temperature corrosion caused when some surface in contact with the combustion gases is below the dew-point temperature of the gas.

It appears that many existing coal-fired boilers can be modified to use suitably prepared refuse as a fuel.

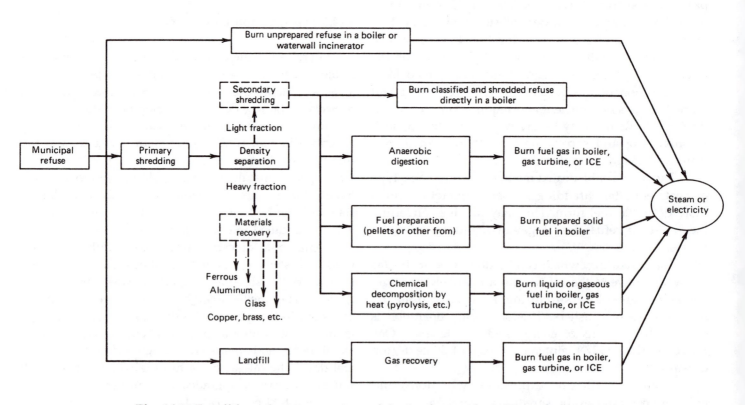

Fig. 16.30 Possible paths to convert municipal refuse to fuel. (From Ref. 15.)

One type of refuse that is both abundant and readily usable in certain types of furnaces and steam generators is tires. Paper mills, cement plants and electric utilities are among those who have discovered this abundant and readily available fuel, with a higher heating value than coal and with less production of pollutants such as nitrous oxides and ash.[18]

Old tires have been a problem in landfills because their shape permits the trapping of water and methane. Tire piles which catch fire are difficult to extinguish and may burn for days, spewing black smoke into the air and oozing oil into the ground. Burning tires as fuel solves the difficult problems created by the U.S. inventory of old tires, estimated to be about 850 million, and the 250 million tires added to the piles each year. Added incentives using old tires are offered by most state governments and tire collectors typically charge a tipping fee for taking a used tire. The company or utility operating the furnace can assume responsibility for direct collection or hire that done by a vendor.

Most of the experience in using tires as a fuel in the U.S. electric utility industry has been with cyclone-fired boilers, which make up less than 10 percent of utility capacity. Little modification is required in these type boilers except for the conveyer system bringing the fuel to the combustion chambers. The metal bead wire around the rim of the tires must be removed prior to combustion and any other metal left after burning must be removed magnetically. The tires are typically burned in cyclone-fired boilers as chips about 1 inch square and as a small percentage of the total fuel. Slightly larger sizes have been successfully burned in stoker-fired units where the fuel sits on a moving grate near the bottom of the boiler.

Tires are more difficult to burn successfully in pulverized-coal boilers except as whole tires in wet-bottom furnaces operating at high temperatures, around 3200°F. In one project whole tires were fed into the boiler at 10 second intervals by a conveyor and lock hopper system. There appears to be good promise of burning tires in fluidized-bed combustion systems, particularly if they are designed in advance to handle such fuel. In the near future it is likely that all tires being discarded will find their way directly into use in some material manufacturing, such as road surfacing, or will be used as a fuel. The split between the various uses will be determined primarily by the economics.

16.5 FUEL CELLS

All energy conversion processes that utilize the concept of a heat engine are limited in their thermal ef-

ficiency by the Carnot principle. The fuel cell, since it is a direct conversion device, has the advantage of not being limited by that principle. The fuel cell is an electrochemical device in which the chemical energy of a conventional fuel is converted directly and efficiently into low voltage direct-current electrical energy. It can be thought of as a primary battery in which the fuel and oxidizer are stored external to the battery and are fed to it as needed. Fundamentals of fuel cell operation are described in texts on direct energy conversion, such as the one by Angrist.[19]

A schematic of a fuel cell is given in Figure 16.31. The fuel, which is in gaseous form, diffuses through the anode and is oxidized. This releases electrons to the external circuit. The oxidizer gas diffuses through the cathode where it is reduced by the electrons coming from the anode through the external circuit. The resulting oxidation products are carried away. In contrast to the combustion in a heat engine, where electrons pass directly from the fuel molecules to the oxidizer molecules, the fuel cell keeps the fuel molecules from mixing with the oxidizer molecules. The transfer of electrons takes place through the path containing the load.

The theoretical efficiency of a simple fuel cell is the maximum useful work we can obtain (the change in Gibbs free energy) divided by the heat of reaction. Since the change in Gibbs free energy is given by $\Delta G = \Delta H - T\Delta S$, the efficiency of the simple cell is:

$$\eta = \frac{\Delta G}{\Delta H} = 1 - \frac{T\Delta S}{\Delta H}$$

If any heat is being rejected by the cell $T\Delta S$ is not zero and the cell efficiency will be less than zero. Other factors that must be considered in real fuel cells are losses associated with attendant accessories, undesirable reactions taking place in the cell, some hindrance of the reaction at the anode or cathode, a concentration gradient in the electrolyte, and Joule heating in the electrolyte. The difference between the maximum useful work and the actual work must appear as rejected heat, in keeping with the first law of thermodynamics. A voltage efficiency is defined by:

$$\eta_v = \frac{Vac}{V}$$

The Faradaic or current efficiency η_F is also defined for the fuel cell, and is the fraction of the reaction which is occurring electrochemically to give current. The part of the chemical free energy that actually results in electrical

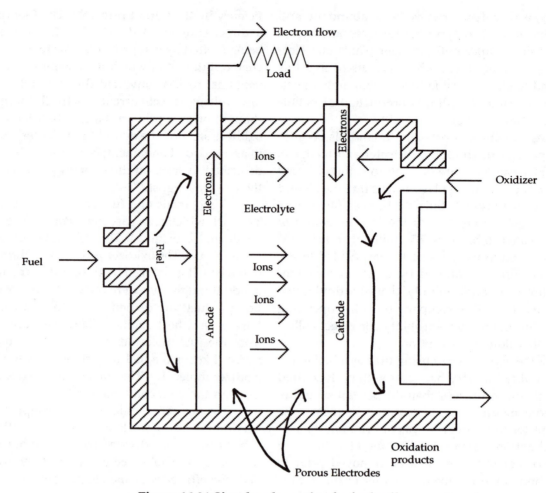

Figure 16.31 Simple schematic of a fuel cell.

energy is the product of $\eta_\upsilon \, \eta_F$.

The ideal fuel would be hydrogen, which if used with pure oxygen as the oxidizer, would give pure water as the product of combustion.

Rarely publicized or discussed outside the space and electric industries, the fuel cell has much to offer those sectors, and has developed rapidly into an economically competitive energy conversion system. In October of 1995 an article by Hirschenhofer and McClelland stated that there were approximately 200 fuel cell units operating in 15 countries.[20] The authors concluded that the future of fuel cells looks bright since they promise reliability, high conversion efficiency and low emission of pollutants to the environment.

Fuel cell modules of identical design can be stacked to accommodate a variety of capacity needs. The efficiency is relatively independent of size, and fuel cell systems are easy to put in place because of their extremely low environmental intrusion. They can use a variety of fuels and change out between fuels can be accomplished rapidly. They can provide VAR control, they have a quick ramp rate and they can be remotely controlled in unattended operation.

In non-space (utility) operation the four most promising types of fuel cells may be classified by their electrolytes and the corresponding temperatures at which they operate. These are:

Electrolyte	Temperature
phosphoric acid fuel cell (PAFC)	200°C
proton exchange membrane (PEM)	80°C
molten carbonate fuel cells (MCFC)	650°C
solid oxide fuel cell (SOFC)	1000°C

Operating temperature has a very significant effect on materials of construction, fabrication methods, and the way in which a unit may be applied.

The phosphoric acid (PAFC) is the only type of fuel cell that is in commercial production at present. The other three are in various stages of research and development but are of interest since they promise higher efficiencies and lower first costs that the PAFC. As of 1995 commercial production is a few years into the future.

The most common PAFC applications today are cogeneration units placed on site and fired with natural gas. Promising fuels for the near future applications include methanol and coal.

The proton exchange membrane (PEM) fuel cells are a recent development, and have a solid polymer membrane electrolyte that can be operated at 80°C. A major developer has been Ballard Power Systems of Canada. Ballard is utilizing plastic for the gas manifolds and other components in an attempt to reduce costs. The Chicago Transit Authority has agreed on a program to use Ballard's PEM cells in three of their buses starting in 1996.

The molten carbonate fuel cells (MCFC) operate at 650°C, high enough to allow the cell's rejected heat to be used in a thermal steam cycle. Another advantage of the temperature is that reforming can take place within the cell, using a reforming catalyst, thus improving the cell efficiency. The cell operation does not require the use of precious metal catalysts and major cell-stack components can be stamped from less expensive metals. Corrosion prevention and selection of materials for construction are the major technical challenges.

Developments now taking place (1995) with major government funding show promising results both in performance and cost. Early demonstration tests have pinpointed problems that seem to be solvable by improved control strategies, operating procedure and equipment design. Several demonstration plants are underway or planned for the near future.

The solid oxide fuel cell (SOFC) can be fabricated in a variety of shapes due to the use of a solid ceramic electrolyte. Corrosion problems are alleviated and natural gas can be used directly without external reforming. The high temperatures do lead to problems in thermal expansion mismatches and sealing between cells.

Development on SOFCs has taken place continuously for a longer period of time than any of the other cell types mentioned above, Westinghouse has been a primary developer and has shown their ability to operate these cells for long periods of time. Encouraged by the results of their demonstrations and operations of relatively small (20-25 kW) units, Westinghouse is looking ahead to larger installations.

Although many improvements can be foreseen for fuel cells to operate commercially there does not seem to be any serious technical barriers. Most effort will be directed to making the cells more economically competitive and reliable with suitable lifetimes of operation.

References

1. S.E. Fuller, "The Relationship between Diffuse, Total and Extraterrestrial Solar Radiation," *Journal of Solar Energy and Technology*, Vol. 18, No. 3 (1976), pp. 259-263.
2. B.V.H. Liu and R.C. Jordan, "Availability of Solar Energy for Flat-Plate Solar Heat Collectors," *Application of Solar Energy for Heating and Cooling of Buildings*, ASHRAE GRP-170, American Society of Heating, Refrigerating and Air Conditioning Engineers, Inc., New York, 1977.
3. S.A. Klein, "Calculations of Monthly Average Insolation on Tilted Surfaces," Vol. 1, *Proceedings of the ISES and SESC Joint Conference*, Winnipeg, 1976.
4. D.W. Ruth and R.E. Chant, "The Relationship of Diffuse Radiation to Total Radiation in Canada," *Journal of Solar Energy and Technology*, Vol. 18, No. 2 (1976), pp. 153-154.
5. "Introduction to Solar Heating and Cooling Designing and Sizing," DOE/CS-0011 UC-59a, b, c, Aug. 1978.
6. Peter J. Lunde, *Solar Thermal Engineering*, Wiley, New York, 1980.
7. *Climatic Atlas of the United States*, U.S. Government Printing Office, Washington, D.C., 1968.
8. A.B. Meinel and M.P. Meinel, *Applied Solar Energy, an Introduction*, Addison Wesley, Reading, Mass., 1976.
9. Maria Telkes, "Solar Energy Storage," *ASHRAE Journal*, Sept. 1974, pp. 38-44.
10. *ERDA Facilities Solar Design Handbook*, ERDA 77-65, Aug. 1977.
11. Martin A. Green, *Solar Cells*, Prentice-Hall, Englewood Cliffs, N.J.
12. John Flinn, "Wind farms start to pay off," *San Francisco Examiner*, January 14, 1990.
13. F.R. Eldridge, *Wind Machines*, NSF-RA-N-75-051, prepared for NSF by the Mitre Corporation, Oct. 1975, U.S. Government Printing Office, Washington, D.C. (Stock No. 038000-00272-4).
14. R.F. Rolsten et al., "Solid Waste as Refuse Derived Fuel," *Nuclear Technology*, Vol. 36 (mid-Dec. 1977), pp. 314-327.
15. G. Tchobanoglous, H. Theisen, and S.A. Vigil, *Integrated Solid Waste Management*, McGraw-Hill, New York, 1993.
16. "Power from Wastes—A Special Report," *Power*, Feb. 1975.
17. A.W. Morgenroth, "Solid Waste as a Replacement Energy Source," *ASHRAE Journal*, June 1978.
18. Leslie Lamarre, "Tapping the tire pile," *EPRI Journal*, Sept/Oct 1995, pages 28-34, EPRI, Palo Alto, CA.
19. Stanley W. Angrist, *Direct Energy Conversion*, Fourth Edition, Allyn and Bacon, Boston, 1982.
20. J.H. Hirschenhofer and R.H. McClelland, "The coming age of fuel cells," *Mechanical Engineering*, October 1995, pp 84-88.

CHAPTER **17**

INDOOR AIR QUALITY

H.E. BARNEY BURROUGHS

Atlanta, Ga.

17.1 INTRODUCTION AND BACKGROUND

IAQ and Energy Management are in an undeniably linked relationship. The purpose of this chapter is to provide the energy manager with a general and overall understanding of the IAQ issue. From this understanding will come the knowledge of the delicate balance between IAQ and energy management and, especially, the confidence that they can work together synergistically and positively to create buildings that operate both healthfully and efficiently.

Many authors and experts focus on the mid-70s with its energy crisis as the spawning ground of today's IAQ problems. Skyrocketing energy costs led to tight building construction which resulted in drastic reductions in ventilation air and infiltration. This yielded the early expression "Tight Building Syndrome." Unfortunately, this label was not only misleading, it was inadequate in providing a full explanation for the issue. It resulted in the focus of blame on *ventilation* and its inherent *energy* cost as being the "cause all"—"fix all"— of indoor air quality. This is simply not the case, then or now.

There were a number of other trends and changes that were happening at the same time. These combined and intermingled to cause today's situation of many existing or potential problem buildings. Let's examine these issues so that we have a better understanding of how the IAQ situation developed.

17.1.1 Impact of "Least Cost" Design and Construction

"Value Engineering" has lost its once positive connotation. Now we know it really means devalued and *"cheapened."* This "least cost" mentality led to many design/construction decisions and tactics that save on first cost. However, both the owner and the occupants too often pay a deferred price in increased life/cycle costs, discomfort, illness and health costs, and low productivity. In the extreme, first cost savings return their dubious "value" multifold in the form of lost lease income and even litigation.

17.1.2 Technical Advances in Office Equipment and Processes

What modern office could survive without the high speed/volume copier, the fax, the laser printer, and the desktop computer terminal? In addition to their heat load, these bring ozone, Volatile Organic Compounds (VOCs), and submicron respirable particles from the print toner into the indoor environment.

17.1.3 Changes in Office Layout and Furnishings

High density office layout is now possible through "office systems." In addition to the VOC contaminant outgassing of their fabric and fiberboard components, work stations can add unpredicted barriers to air flow and ventilation effectiveness. Further, they facilitate increased occupancy density which can exceed original design parameters.

17.1.4 Shift of Materials used in Construction

Polymers and plastics are now widely used in construction components as glues, binders, and soiling retardants. Typical are formaldehyde (HCHO) and other aldehydes which are ubiquitous components of furniture, fabric, particle board, and plastic surfaces. There is a widening usage of fabrics and fleecy materials in the space. Polymeric glues and high solvent-loaded adhesives were also widely used to attach them. The usage of carpeting is now widespread in schools, for example, where hard surfaces previously prevailed. Porous wall dividers and wall coverings add to this fabric burden and contribute to both outgassing and "sinks." An example is 4PC (4-phenyl-cyclo-hexane) which is a component in the carpet backing. Exposure to high concentration peaks or prolonged exposure over time to these chemicals can trigger a health response risk known as MCS (Multiple-Chemical-Sensitivity).

17.1.5 Focus on Energy Conservation in HVAC Systems

Many well-meaning energy managers and engineers designated energy conservation above comfort (remember the "Building Emergency Thermostat Settings" [BETS] of the Carter era). This "feeding frenzy" of energy saving brought on VAV (Variable Air Volume) distribution that was designed to *completely* shutdown supply air to zones with low demand. This resulted in lowered or nonexistent ventilation effectiveness. This meant that conditioned dilution air was not being delivered to the occupied zone. This allows air within the space to become "aged" or stale and stuffy and the system to go out of proper air balance. Unventilated space also becomes a "sink" for contaminants. Contaminants remaining in the space deposit onto and into fleecy and porous surfaces. Then, they re-emerge over time by outgassing due to changes in temperature, humidity, and airflow.

17.1.6 Focus on Ventilation

Because of the energy concern and its resulting predominance, ventilation, and the initial label of "Tight Building Syndrome" received undue attention. This is not to say that ventilation is not integrally involved in many IAQ problems. However, this early focus overshadowed the roles of source control and filtration as control mechanisms. It also minimized the many other aspects of good IAQ such as maintenance, moisture control, housekeeping, commissioning, and other facility management practices.

17.1.7 Prevailing National Priorities

The early focus on energy and the overall environment lessened under the Reagan Administration with the de-emphasis on regulation. However, the Federal regulators focused on Radon and Asbestos as being serious indoor environmental threats to the general public health and welfare. Heavy regulations were created regarding asbestos in public buildings such as schools and commercial buildings. The early focus of the regulations was mitigation through removal. This created tremendous financial burden on both the private and public building management segment. Of course, the regulations and their interpretation have since softened. Their impact, however, was to financially burden facility and plant management teams and to sensitize decision makers against environmental concerns.

17.1.8 "No Maintenance" Mentality

The traditional "if it ain't broke, don't fix it" atti-

tude matched with the "least cost" perspective has contributed heavily to the current status of building environmental quality. Inadequate or non-existent maintenance can destroy the finest of designs. Most IAQ experts agree that buildings that do not work properly due to poor upkeep are the most likely candidates for Sick Building Syndrome. EPA researchers claim over half of the complaint buildings they visit are plagued with poor maintenance.

17.1.9 Time Marches On

Few energy managers today remember the OPEC crisis that began the energy panic. It was in 1974—over 20 years ago. Buildings constructed in the early stages of this period have considerable age. Systems, controls, decorations, roofs, elevators, and HVAC systems, have been subjected to two decades of use and abuse. Poor maintenance, poor filtration, low ventilation rates, water leaks and least cost purchasing decisions have taken their toll. It was revealed, for example, in the trial concerning the EPA headquarters building in Washington, D.C. that the ductwork had only 1/32" of uniform dirt buildup. Yet, this equated to *tons* of particulate contamination that had to be cleaned out of the building distribution system. This had accumulated over years of operating with low efficiency blanket media particulate filters. The aging process alone is a major factor to deal with when considering IAQ problems.

17.1.10 Low Filtration is No Filtration

Low efficiency disposable type furnace filters or blanket media were and are widely used in commercial buildings. They represent the least cost, however, they also represent least performance in providing contaminant control and protection of the systems, the building, and the occupants. The resulting poor filtration allowed contamination accumulation on cooling coils, ductwork, and the facility. This robs efficiency and energy because of operating with fouled inefficient systems like dirty coils. Poor filtration also provides nutrients for microbial growth as well as a source of odor and VOC sinks. Furthermore, the conditioned and occupied space is not protected from the pollutants generated in the outdoor environment.

17.1.11 "It's No Problem Boss!"

The natural response by Facility Managers to IAQ complaints has been "frustration mixed with skepticism" and a "tendency to downplay complaints"[1].

Handling of IAQ problems has therefore been slow or non-existent. Thus, the real problems are complicated by the perception or lack of concern, lack of response, lack of communications and just poor human relations. This inappropriate response adds human anger to the health effects equation and it usually equals litigation.

17.1.12 Public Awareness prevails

IAQ issues have been in all the pertinent trade publications, most network TV news shows and many hometown newspapers. The general public was conditioned to environmental issues with the focus on radon, asbestos, PCB, insecticides, fertilizers, and now lead. Problem buildings like EPA buildings and numerous high profile public courthouses make great headlines.

Thus, your tenants, your employees, even your family and children know about air quality in their space. The general public is therefore suspicious of "places that make them sick" and "chemical smells" and "black stuff that comes out of the ductwork." This awareness has raised the expectation of occupants regarding the quality of the air in their personal space.

17.1.13 Summary

It should be obvious that there are a number of issues and related trends that worked together to bring us the current situation. The interesting point is that many of these influences are totally under the control of the facility and plant management team. Like the philosopher of the Okefenokee (Pogo) once stated. *"We has found the enemy—and he is us."*

17.2 WHAT IS THE CURRENT SITUATION?

Have we learned anything over the last two decades of experience? This discussion is intended to brief the reader on our current knowledge of the IAQ issue.

17.2.1 Symptoms of IAQ

We still lack a full understanding of the health effect consequences of indoor air contaminant exposure. This is especially true of the more recently identified ailments, such as MCS (Multiple Chemical Sensitivity). However, we do have a good working understanding of the health related symptoms that are reported in a sick building. Listed in Table 17.1[2], are the health effects that are typical of occupant complaints. What makes them symptoms of "Sick Building Syndrome" is that they occur only when the individual is exposed to the building environment. When no longer in the exposure environ-

ment, the symptoms tend to go away or lessen dramatically. If the symptoms persist even after leaving the space and they fit an identifiable medical diagnostic pattern, then the ailment is likely to be a "Building Related Illness" such as Legionnaire's Disease or Hypersensitivity Pneumonitis. Unfortunately, some of the symptoms are vague and are easily confused with other causes, such as the common cold, preexisting allergies or the prevailing "flu bug." This fuels the response of facility managers to treat the occupant complaints as "walk-in" problems rather than caused by the building space.

Table 17.1 Typical IAQ Symptoms

Burning eyes
Headache & sinusitis
Runny nose
Tightness in chest
Cough & hoarseness
Lethargy & fatigue
Sore throat
Itching skin
Various combinations of the above

17.2.2 Causes of IAQ Problems

The more common and obvious causes of IAQ problems are known and are listed in Table 17.2. However, they seldom represent a *single* causative element because most problem buildings exhibit multiple and intermingled causes. Further, all of the involved causes have to be brought under control to assure that the building returns to a nonproblem status. We still lack clear linkages between specific cause and the resulting health effect. Yet, we know that the ultimate and overriding cause of IAQ problems is the unusual or elevated concentration of components in the air supplied to the space. The typical air components noted in Table 17.3 are prevalent in most all indoor air and exposure to them in routine concentrations does not bring about health effects or discomfort. However, it is the elevated concentrations of these same components that contaminate the space and bring on health complaints. Thus, it is *contaminant amplification* that is the primary IAQ cause. Therefore, the search for the specific cause and effect in a specific building site should focus on those elements that impact the contaminant amplification and retention, such as water and dirt as amplifiers of fungal growth. When we fully understand these factors, we can then diagnose building related complaints. They also provide insight regarding what makes one building "sick" and another "healthy."

Table 17.2 Typical IAQ Causes

Low ventilation rates
Inadequate ventilation efficiency
Poor quality makeup air
Poor filtration
Inadequate humidity control
Water incursion
Poor maintenance
Microbiological accumulation and growth
Internal pollutants
Ground gas
Space usage and activity
Concurrent stressors

Table 17.3 Typical Air Contaminants

Respirable particles
Viable particles
Allergens
Spores
Odors
Corrosive gases
VOCs & Formaldehyde
Hydrocarbons
Reactive inorganic compounds
Oxidants

17.2.3 No Clear Definition of "Acceptable" Air

We lack a clearly stated and measurable definition of "acceptable indoor air quality." The primary current parameter is tied to an outdoor air ventilation rate prescribed by the ASHRAE Ventilation Standard 62-1989[4]. This rate is, in turn, tied to the human occupancy level of the building and is stated in CFM *per occupant*. This has proven to be an inadequate yardstick because it ignores the contaminant load of the building, its furnishings, and the activities within the space. Further, it promotes the philosophy that air quality and comfort is solely a function of the ventilation rate. This is definitely not the case as has been discussed previously. Thus, designers, building owners, and facility managers need a better target guideline than furnished by Standard 62 for defining and determining acceptable indoor air quality. The Standard specifically defines it as "air in which there are no known contaminants at harmful concentrations as determined by cognizant authorities and with which a substantial majority (80% or more) of the people exposed do not express "dissatisfaction." Thus, the issue becomes a *complaint* based and a *comfort* based issue. The 20% dissatisfaction factor also becomes irrelevant when

we realize that just one highly vocal or highly affected individual can trigger an IAQ incident. The current ASHRAE Standing Standard Project Committee (SSPC 62-R) is revising Standard 62 and is dealing with the definition issue. They are also wrestling with the rationale for establishing a more meaningful and appropriate ventilation rate. In the meantime, it is very important for the IAQ manager to be aware of actual ventilation rates and related occupancy populations within the facility.

17.2.4 Litigation as the Driver

With the failure of repeated legislative initiatives and the sidetrack of proposed OSHA regulations, the remaining driver of IAQ is the risk of litigation. Lawsuits in both the worker's compensation and civil courts abound. Often, they are highly publicized and involve millions of dollars in liability and mitigation. The specter of litigation is, therefore, a powerful and destruction force at work in the IAQ arena. A legacy of the proposed OSHA regulation[5] persists, however, in the almost universal response of the building management community to the smoking issue. After only the announcement of a proposed ruling that would ban smoking, the voluntary control of tobacco smoking in public space was widespread. This provides some indication of the power of regulation and the influence that it could impose on the indoor air quality issue. This powerful influence provides some guidance to the potential outcome of the proposed ASHRAE Standard 62 Revision Draft as it emerges for public review. This may be particularly true for this standard is being drafted in mandatory code language. Managers of energy and IAQ should track standards and regulations pertaining to IAQ very closely and carefully.

17.3 SOLUTIONS AND PREVENTION OF IAQ PROBLEMS

The following discussions provide both practical and philosophical aspects to the prevention and mitigation of the causes of IAQ complaints.

17.3.1 Understand the Issue!

Some experts in the field have re-expressed the issue in the broader term "IEQ" (Indoor Environmental Quality). This recognizes that the perceived quality of the conditioned space also includes the broader issues of lighting, sound, ergonomics, and psycho/social issues.

Even as a complex and multidisciplined issue, IAQ control and prevention reduces to simple principles.

These principles are inherent in the definition of air *conditioning* as "the control of temperature, humidity, distribution, and cleanliness of air supplied to occupied space." When a building HVAC system fails to address any aspect of this definition, then complaints, discomfort, and health effects can result. Cleanliness of the air is probably the most overlooked but perhaps the most crucial aspect in commercial buildings. Thus, an overriding cause of complaints and adverse health effects from IAQ is an abnormal or excessive exposure to accumulated airborne contaminants.

The tactics to control excessive contaminant buildup in conditioned space are: dilution, extraction, and source control.[4]

Dilution is the reduction of contaminant load by *addition* of clean or at least lesser contaminated air, i.e., *ventilation.*

Extraction is the reduction of contaminant load by *subtraction* of contaminants by means of *filtration.*

Source Control is the reduction of contaminant load by *elimination* by means of exhaust or selection of low emitting components and practices.

Most of the management, control, or prevention suggestions reflect attention to all three of these control tactics. Of these, however, the most cost effective is source control and the least cost effective is ventilation. Thus, filtration and source control should receive the priority when analyzing mitigation alternatives from a cost effectiveness standpoint.

17.3.2 Treat the Issue as Reality!

An IAQ complaint may be real or it may be imagined—but it is *real* to the individual reporting the complaint. Thus, potential IAQ complaints require prompt and positive response from the facility management team. This sends a proactive message to employees and tenants that their comfort, health, and feelings are important to you. This avoids frustration and anger experienced by occupants and avoids a legitimate charge of negligence on your part. Complaints should be logged and follow-up documented which aids in providing information that is helpful in the diagnostics process. Frequency, nature, location, timing and affected individuals provide guidance in determining the causal factors of the incidents. This also helps in the determination of the severity and urgency of the situation.

17.3.3 Put Someone in Charge and Make Them Smart!

Appoint an experienced member of the facility management team as your IAQ Manager. Many times this is the Energy Manager because of the close relationship between IAQ and energy management issues. This is consistent with the recommendation in the EPA Building Owners Manual. This individual should be the clearing house for IAQ complaints, investigation, problem resolutions, and record keeping. This also becomes the individual who receives training to become the in-house "guru." The IAQ manager should become very familiar with causes, prevention, and mitigation tactics. ASHRAE, IFMA, BOMA, EPA, AIHA, and AWMA are associations or organizations that are sources of information and training on IAQ matters. In addition, the AEE (Association of Energy Engineers) offers a course on IAQ Fundamentals that leads to a certification (Certified Indoor Air Quality Professional—CIAQP).

17.3.4 Establish Proactive Monitoring!

I now hear building managers say "Sure, we respond to IAQ complaints"! That is not proactive but "*reactive response.*" Proactive response takes effort, time, and money just like preventative maintenance. Periodic facility inspections and assessments will provide baseline air quality performance data. It can also reveal imminent problems that can be proactively mitigated before occupant complaints occur. This entails a physical inspection walk-through of the entire facility to identify potential IAQ causes and practices before they manifest into complaints. It also may be advisable to perform baseline air diagnostics on key environmental parameters such as temperature and relative humidity, Carbon Dioxide, Carbon Monoxide, Total Volatile Organics, Airborne Microbials, and Formaldehyde. This will enable the Facility Manager to establish "norms" for the facility that can be used for comparative purposes if complaints occur and performance levels alter.

17.3.5 Use ASHRAE Standard 62-89!

Really use it, which means to go beyond Table II (see Table 17.4).[4] This table numerically designates outdoor air levels for ventilation using the Ventilation Rate Procedure of Standard 62-1989. Most designers stop at this table and never go beyond. Thus, they don't realize that the Standard also covers such issues as outdoor air quality, ventilation effectiveness, monitoring, maintenance, filtration, humidity control, and source control. The Indoor Air Quality alternative procedure allows for consideration of energy management concerns through outdoor air reduction and air cleaning. Because of this broad coverage, Standard 62 becomes an excellent overall protocol for addressing many of the issues discussed above, including energy management.

17.3.6 Test, Adjust and Balance Again!

Unfortunately, "TAB" work is usually considered a one-time event—done at system start-up. Then the report is filed away. However, a building is a dynamic model, changing as it matures through its life cycle of changing layouts, tenants, and usage. TAB becomes an effective tool to assure ongoing maximized system performance and ventilation effectiveness.

17.3.7 Exploit Filtration!

Perhaps the most under-utilized of the prevention tactics is filtration, both gas phase and particulate. High efficiency filtration addresses contaminant control by extraction, can cost-effectively substitute for ventilation to save energy, and can act as source control. More often than not, urban outdoor air does not even meet the am-

Table 17.4 Excerpts from Table II - ASHRAE Standard 62-1989

Application	Estimated Maximum Occupancy P/1000 ft^2 or 100 m^2	Outdoor Air Requirements cfmi person	cfm/ft^2	Comments
Offices				
Office space	7	20		Some office equipment may require local exhaust.
Reception areas	60	15		
Telecommunication centers and data entry areas	60	20		
Conference rooms	50	20		Supplementary smoke-removal equipment may be required.
Public Spaces				
Corridors and utilities			0.05	
Public restrooms, cfm/wc or urinal		50		Mechanical exhaust with no recirculation is recommended. Normally supplied by transfer air, local mechanical exhaust: with no recirculation recommended
Locker and dressing rooms			0.5	
Smoking lounge	70	60		
Elevators			1.00	Normally supplied by transfer air.
Retail Stores, Sales Floors, and Show Room Floors				
Basement and street	30		0.30	
Upper floors	20		0.20	

bient air quality levels prescribed by Standard 62. Though Standard 62-1989 is largely silent on filtration, the revision draft will incorporate a minimum particulate filter efficiency. This will be set at the particle extraction performance of not less than the ASHRAE 30% (ADSM) medium efficiency filter. Filtration can cleanse this outdoor dilution air to eliminate it as a contaminant source and enhance its effectiveness as a space dilutant. Further, self-contained air cleaners can provide localized air quality improvement using both gaseous and particulate filter components. Tables 17.5 and 17.6[3] provide additional usages and benefits from more extensive usage of filtration.

Table 17.5 Usages of Filtration

Protect mechanical equipment such as heat exchange coils

Protect systems such as air distribution ductwork, outlets, and porous components

Protect occupied space such as wall and ceiling surfaces

Protect occupants from contaminant exposure

Protect processes such as pharmaceuticals and electronic chip fabrication

Provide clean makeup air to use as high quality dilution air

Protect environment from contaminated exhaust

Provide source control to avoid reintrainment into the return public air

Augment ventilation air through treatment of the return air

Table 17.6 Benefits of Filtration

Increased system efficiency from clean components

Increase system life

Lower maintenance costs

Lower housekeeping costs

Avoidance of product failure

Enhanced energy management

Lower health risk and costs

Increased productivity and reduced absenteeism

17.3.8 Use Lifelong Commissioning!

Thought by many to be just system start-up, commissioning is emerging as an essential component of the successful quality construction or renovation process. The commissioning process really commences with the first vision of a construction process, and then proceeds throughout the life of the building. (See ASHRAE Guideline #1, Commissioning).[7] There is a very strong role for the commissioning agent to assure that air quality is built into and retained within the building construction process.

17.3.9 Leverage the Use of Source Control!

Controlling the contaminant, whether it be an odor or a toxic substance, is most effectively and efficiently done at the source. This can mean localized exhaust of contaminants to atmosphere. It can mean localized spot filtration using air cleaners. It can mean a rigorous review of all MSDS records and restricting introduction of purposeful, but harmful, products and chemicals to the space. This includes pesticides and routine housekeeping chemicals. It can also mean a review of construction and buildout components for outgassing properties and/or potential. Specific control tactics are scheduled in Table 17.7.[8]

17.3.10 Look at Controls!

As the nerve center of the building, controls play a major role in effective and efficient building management… getting the right air to the right place at the right time. They both monitor and control response to the varying conditions within and without the building envelope. They can impact air quality in the way that outdoor air levels are managed, distribution systems respond, and mechanical equipment reacts to both the changing dynamics of the building and the needs of the occupants.

17.3.11 Diligent Maintenance!

One of the most frequently cited deficiencies in air quality, proper operation and maintenance can be an effective deterrent to IAQ problems. Cleanliness and avoidance of water in the systems are critical to maintaining high air quality. Poor maintenance can unfortunately negate the finest of designs and construction efforts. Likewise, housekeeping and janitorial protocols should be examined. Cleaning chemicals, vacuuming techniques, and other purposeful practices such as pest control, should all be scrutinized to make sure that they are lowering and not aggravating contaminant levels.

17.3.12 Perform Life Cycle Costing!

As opposed to "least cost" thinking, (see Chapter 4) life cycle costing must re-emerge as the real cost decision model. For example, high efficiency filters are more expensive first cost than cheap furnace filters. Yet, they are

Table 17.7 Typical Control Tactics

Contaminant	Sources	Control Techniques
Asbestos	Furnace, pipe, wall ceiling insulation fireproofing, acoustical and floor tiles	1. Enclose: shield 2. Encapsulate; seal 3. Remove 4. Label ACM 5. Use precautions against breathing when disturbed
Bioaerosols	Wet insulation, carpet, ceiling tile, wall coverings, furniture, air conditioners, dehumidifiers, cooling towers, drip pans, and cooling coils. People, pets, plants, insects, and soil	Use effective filters. Check and clean any areas with standing water. Be sure condensate pans drain and are clean. Treat with algicides. Maintain humidifiers and dehumidifiers. Check & clean duct linings. Keep surfaces clean.
Carbon Monoxide (CO)	Vehicle exhaust, esp. attached garages; unvented kerosene heaters and gas appliances; tobacco smoke; malfunctioning furnaces	Check and repair furnaces, flues, heat exchangers, etc. for leaks. Use only vented combustion appliances. Be sure exhaust from garage does not enter air intake.
Combustion Products (NO$_x$ and CO)	Incomplete combustion process	Use vented appliances and heaters. Avoid air from loading docks and garages entering air intake. Check HVAC for leaks regularly, repair promptly. Use plants as air cleaners. Filter particulates.
Environmental Tobacco Smoke (ETS)	Passive smoking; sidestream and mainstream smoke	Eliminate smoking; confine smokers to designated areas; or isolate smokers with direct outside exhaust. Increase ventilation. Filter contaminants.
Formaldehyde (HCHO)	Building products; i.e., paneling, materials with lower particleboard, plywood urea-formaldehyde insulation as well as fabrics and furnishings	Selective purchasing of materials with lower formaldehyde emissions. Barrier coatings and sealants. New construction commissioning.
Radon	Soil around basements and slab on grade	Ventilate crawl spaces. Ventilate sub-slab. Seal cracks, holes, around drain pipes. Positive pressure in tight basements. NOTE: Increased ventilation does not necessarily reduce radon levels.
Volatile Organic Compounds (VOCs)	Solvents in adhesives cleaning agents, paints, fabrics, tobacco smoke, linoleum, pesticides, gasoline, photocopying materials, refrigerants, building material	Avoid use of solvents and pesticides indoors. If done, employ time of use and excessive ventilations. Localized exhaust near source when feasible. Selective purchasing. Increase ventilation.

more cost effective when the energy impact, health impact, and long life cycle issues are considered. Recognize the real "life" cycle and the ecological and environmental cost of a product from its cradle to its grave. And the "Green Building" folks would like us to think in terms of cradle to cradle—including the cost and ecological impact of recycling. For example, in some urban communities, furnace filters have to be demanufactured at great cost to be disposed in solid waste handling sites. Likewise, air quality concerns and their resulting impact on the human asset base must be considered when design/construct decisions are made. Shortcuts made for reasons of budget, time, or convenience, will demand payment in due course.

17.3.13 Prepare for Regulation and Litigation Now!

The proactive Facility Team will prepare for this eventuality by bringing your facilities and your practices into compliance with current "state of the art" and to document your efforts.

It is never too soon to prepare for the eventuality of litigation. The best defense against either litigation or regulation is to proactively follow the recommendations in this chapter. Then document, document, document! Document your design path! Document your decision path!! Document your baseline performance path using proactive response to the IAQ issue!!!

References

1.   Building Operating Management Magazine, August, 1994
2.   Burroughs, H.E. *The Role of Risk Assessment as a Mitigation Protocol for IAQ Causes.* World Energy Engineering Congress, Association of Energy Engineers Conference, 1991.
3.   Burroughs, H.E. *Air Filtration. A New Look At An Existing IAQ Asset.* Carrier Global Engineering Conference, 1994.
4.   ASHRAE Standard 62-1989, Ventilation for Acceptable Indoor Air Quality. American Society of Heating, Refrigerating, and Air-Conditioning Engineers, Inc., 1791 Tullie Circle, NE, Atlanta, GA. 30329
5.   *Department of Labor. Federal Register (OSHA)*, Occupational Safety and Health Administration, Pages 1 5969 -1 6039.
6.   *Building Air Quality: A Guide For Building Owners And Facility Managers.* US EPA, 401 M St., Washington, DC 20460.
7.   ASHRAE Guideline 1-1989, Guideline for Commissioning of HVAC Systems. American Society of Heating, Refrigerating, and Air-Conditioning Engineers, Inc. 1791 Tullie Circle, NE, Atlanta, GA. 30329.
8.   Hansen, Shirley J., *Managing Indoor Air Quality.* 1991. The Fairmont Press, Lilburn, GA.

CHAPTER 18

ELECTRIC AND GAS UTILITY RATES FOR COMMERCIAL AND INDUSTRIAL CUSTOMERS

LYNDA J. WHITE
RICHARD A. WAKEFIELD
JAIRO A. GUTIÉRREZ

CSA Energy Consultants
Arlington, VA

18.1 INTRODUCTION

Purpose and Limitations

The main focus of this chapter on rates is to provide information on how an average commercial or industrial customer can identify potential rate-related ways of reducing its energy costs. The basic costs incurred by electric and gas utilities are described and discussed. How these costs are reflected in the final rates to commercial and industrial customers is illustrated. Some examples of gas and electric rates and how they are applied are included. In addition, this chapter identifies some innovative rates that were developed by electric and gas utilities as a response to the increasing pressure for the development of a more competitive industry.

Because of the breadth and complexity of the subject matter, the descriptions, discussions and explanations presented in this chapter can not cover every specific situation. Energy consumption patterns are often unique to a particular commercial and/or industrial activity and therefore case by case evaluations are strongly suggested. The purpose is to present some general cost background and guidelines to better understand how to identify potential energy cost savings measures.

18.1.2 General Information

Historically, electric and gas utility rate structures were developed by the utilities themselves within a much less complex regulatory environment, by simply considering market factors (demand) as well as cost factors (supply). Today the increasing pressures to develop more competitive markets have forced utilities to recon-sider their traditional pricing procedures. Other factors affecting today's electric and gas markets include rising fuel prices, environmental concerns, and energy conservation mandates. These factors and pressures have affected gas and electric utility costs and hence their rates to their final customers.

In general, electric and gas rates differ in structure according to the type and class of consumption. Differences in rates may be due to actual differences in the costs incurred by a utility to serve one specific customer vs. another. Utility costs also vary according to the time when the service is used. Customers using service at off-peak hours are less expensive to serve than on-peak users. Since electricity cannot be stored, and since a utility must provide instantaneous and continuous service, the size of a generation plant is determined by the aggregate amount of service taken by all its customers at any particular time.

The main cost elements generally included in rate-making activities are: energy costs, customer costs, and demand costs. Each of these is discussed in the next section.

18.2 UTILITY COSTS

Utilities perform their activities in a manner similar to that of any other privately-owned company. The utility obtains a large portion of its capital in the competitive money market to build its system. It sells a service to the public. It must generate enough revenues to cover its operating expenses and some profit to stay in business and to attract capital for future expansions of its system.

In general there are two broad types of costs incurred by a utility in providing its service. First, there are the fixed capital costs associated with the investment in the facilities needed to produce (or purchase) and deliver the service. Some of the expenses associated with fixed capital costs include interest on debts, depreciation, insurance, and taxes. Second, there are the expenses associated with the operation and maintenance of those same facilities. These expenses include such things as salaries and benefits, spare parts, and the purchasing, handling, preparing, and transporting of energy resources. The rates paid by utility customers are designed to generate the necessary revenues to recover both types of costs. Both capital and operation and maintenance costs are allocated between the major cost elements incurred by a utility.

18.2.1 Cost Components

The major costs to a utility can be separated into three components. These include customer costs, energy/commodity costs, and demand costs. These cost components are briefly described below.

Customer Costs

Customer costs are those costs incurred in the connection between customer and utility. They vary with the number of customers, not with the amount of use by the customer. These costs include the operating and capital costs associated with metering (original cost and on-going meter-reading costs), billing, and maintenance of service connections.

Energy/Commodity Costs

Energy and commodity costs consist of costs that vary with changes in consumption of kilowatt-hours (kWh) of electricity or of cubic feet of gas. These are the capital and operating costs that change only with the consumption of energy, such as fuel costs and production supplies. They are not affected by the number of customers or overall system demand.

Demand Costs

Electric utilities must be able to meet the peak demand—the period when the greatest number of customers are simultaneously using service. Gas utilities must be responsive to daily or hourly peak use of gas. In either case, the utility will need to generate or purchase enough power to cover its firm customers' needs at all times. Demand-related costs are dependent upon overall system requirements. Demand costs can be allocated in many different ways, but utilities tend to allocate on-peak load. Included in these costs are the capital and operating costs for production, transmission, and storage (in the case of gas utilities) that vary with demand requirements.

18.2.2 Allocation of Costs

Once all costs are identified, the utility must decide how to allocate these costs to its various customer classes. How much of each cost component is directly attributable to serving a residential, a lighting, or a manufacturing customer? In answer to this question, each utility performs a cost-of-service study to devise a set of allocation factors that will allow them to equitably divide these costs to the various users. After the costs are allocated, the utility devises a rate structure designed to collect sufficient revenue to cover all its costs, plus a fair rate of return (currently, this is running between 10 and 14% of the owners' equity.)

18.3 RATE STRUCTURES

18.3.1 Basic Rate Structure

The rate tariff structure generally follows the major cost component structure. The rates themselves usually consist of a customer charge, an energy charge, and a demand charge. Each type of charge may consist of several individual charges and may be varied by the time or season of use.

Customer Charge

This is generally a flat fee per customer ranging from zero to $25 for a residential customer to several thousand dollars for a large industrial customer. Some utilities base the customer charge to large industrial customers on the level of maximum annual use.

Energy Charge

This is a charge for the use of energy, and is measured in dollars per kilowatt-hour for electricity, or in dollars per therm or cubic foot of gas. The energy charge often includes a fuel adjustment factor that allows the utility to change the price allocated for fuel cost recovery on a monthly, quarterly, or annual basis without resorting to a formal rate hearing. This passes the burden of variable fuel costs (either increases or decreases) directly to the consumer. Energy charges are direct charges for the actual use of energy.

Demand Charge

The demand charge is usually not applied to residential or small commercial customers, though it is not always limited to large users. The customer's demand is generally measured with a demand meter that registers the maximum demand or maximum average demand in any 15-, 30-, or 60-minute period in the billing month. For customers who do not have a demand meter, an approximation may be made based on the number of kilowatt hours consumed. Gas demand is determined over an hour or a day and is usually the greatest total use in the stated time period.

Another type of demand charge that may be included is a reactive power factor charge; a charge for kilovolt reactive demand (kVAR). This is a method used to charge for the power lost due to a mismatch between the line and load impedance. Where the power-factor charge is significant, corrective action can be taken, for example by adding capacitance to electric motors.

Demand may be "ratcheted" back to a period of greater use in order to provide the utility with revenues to maintain the production capabilities to fulfill the greater-use requirement. In other words, if a customer uses a maximum demand of 100 kW or 100 MMBtu one month, then uses 60 kW or MMBtu for the next six months, he/she may have to pay for 100 kW or MMBtu each month until the ratchet period (generally 12 months) is over.

18.3.2 Variations

Utilities use a number of methods to tailor their rates to the needs of their customers. Some of the different structures used to accomplish this include: seasonal pricing; block pricing; riders; discounts; and innovative rates.

Seasonal Pricing

Costs usually vary by season for most utilities. These variations may be reflected in their rates through different demand and energy charges in the winter and summer. When electric utilities have a seasonal variation in their charges, usually the summer rates are higher than the winter rates, due to high air conditioning use. Gas utilities will generally have winter rates that are higher than summer rates, reflecting increased space-heating use.

Block Pricing

Energy and demand charges may be structured in one of three ways: 1) a declining block structure; 2) an inverted block structure; or 3) a flat rate structure. An inverted block pricing structure increases the rate as the consumption increases. A declining block pricing method decreases the rate as the user's consumption increases. When a rate does not vary with consumption levels it is a "flat" structure. With the declining or inverted block structures, the number of kWh, MMBtu, or therms used is broken into blocks. The unit cost (cents per kWh or cents per MMBtu or therm) is lower or higher for each succeeding block.

A declining block reflects the fact that most utilities can generate additional electricity or provide additional gas for lower and lower costs—up to a point. The capital

costs of operation are spread over more usage. The inverted block structure reflects the fact that the incremental cost of production exceeds the average cost of energy. Hence, use of more energy will cause a greater cost to the utility.

Most utilities offer rates with more than one block pricing structure. A utility may offer some combination of inverted, declining and flat block rates, often reflecting seasonal energy cost differentials as well as use differentials. For example, a gas utility may use an inverted block pricing structure in the winter that reflects the higher energy costs in that period, but use either a flat or declining block pricing structure in the summer when energy costs are lower.

Riders

A "Rider" modifies the structure of a rate based on specific qualifications of the customer. For example, a customer may be on a general service rate and subscribe to a rider that reduces summer energy charges where the utility is granted physical control of the customers air conditioning load.

Discounts

The discount most often available is the voltage discount offered by electric utilities. A voltage discount provides for a reduction in the charge for energy and/or demand if the customer receives service at voltages above the standard voltage. This may require the customer to install, operate and maintain the equipment necessary to reduce the line voltage to the appropriate service voltage. Each customer must evaluate the economics of the discounts against the cost of the equipment they will have to provide.

Innovative Rates

Increased emphasis on integrated resource planning, demand-side management and the move to a more competitive energy marketplace has focused utility attention on innovative rates. Those rates designed to change customer load use, help customers maintain or increase market share, or provide the utility with a more efficient operating arena are innovative. Most rates offered today fit into the innovative category.

18.4 INNOVATIVE RATE TYPES

Utilities have designed a variety of rate types to accomplish different goals. Some influence the customer to use more or less energy or use energy at times that are helpful to the utility. Others are designed to retain or attract customers. Still others are designed to encourage efficient use of energy. The following are some of the innovative rate types that customers should know about.

Time-of-Use Rates are used for the pricing of electricity only. The primary purpose of the time-of-use (TOU) rate is to send the proper pricing signals to the consumer regarding the cost of energy during specific times of the day. Generally, a utility's daytime load is higher than its nighttime load, resulting in higher daytime production costs. Proper TOU price signals will encourage customers to defer energy use until costs are lower. TOU rates are usually offered as options to customers, though some utilities have mandatory TOU rates.

End-Use Rates, these rates include air-conditioning, all-electric, compressed natural gas, multi-family, space-heating, thermal energy storage, vehicle fuel and water-heating rates. These rates are all intended to encourage customers to use energy for a specific end-use.

Financial Incentive Rates include rates such as residential assistance, displacement, economic development, and surplus power rates. Assistance rates provide discounts to residential customers who meet specific low-income levels, are senior citizens or suffer from some physical disability. Displacement rates are offered by electric utilities to customers who are capable of generating their own electricity. The price offered to these customers for utility-provided power is intended to induce the customer to "displace" its own generated electricity with utility-provided electricity. Economic development rates are generally offered by utilities to provide economic incentives for businesses to remain, locate, or expand into areas which are economically distressed. This type of rate is an attempt to attract new customers into the area and to get existing customers to expand until the area is revitalized. Surplus power rates are offered to large commercial and industrial customers. They are offered gas or electric capacity at greatly reduced prices when the utility has an excess available for sale.

Interruptible rates generally apply to commercial and industrial customers. The utilities often offer several options with respect to the customer's ability to interrupt. Prices vary based on the amount of capacity that is interruptible, the length of the interruption, and the notification time before interruption. Such interruptions are generally, but not always, customer controlled. In addition, the total number of interruptions and the maximum annual hours of interruption may be limited.

18.5 CALCULATION OF A MONTHLY BILL

Following is the basic formula used for calculating the monthly bill under a utility rate. The sum of these components will result in the monthly bill.

1) Customer Charge
- Customer charge = fixed monthly charge
2) Energy Charge
- Energy charge = dollars × energy use
- Energy/Fuel Cost Adjustment = dollars × energy use
3) Demand Charge
- Demand charge = dollars × demand
- Reactive Demand Charges (electric only) = dollars × measured kilovolt-ampere reactive demand
4) Tax/Surcharge
- Tax/surcharge = either sum of one or more of items 1-3 above multiplied by tax percentage, dollars × energy use, or dollars × demand

Some examples illustrating the calculation of a monthly bill follow. These examples are actual rate schedules used by the utilities shown. The information regarding the consumption of electricity used in the sample calculations are based on the typical figures shown in Table 18-1.

18.5.1 Commercial General Service with Demand Component

The commercial general service rate often involves the use of demand charges. Table 18.2 provides rate data from Public Service Electric & Gas Company. A sample bill is calculated using the data from Table 18.1 for a convenience store.

Energy Usage – 17,588 kWh; Billing Demand – 30 kW; Season – Summer (June)

Customer Charge:	$3.74
Energy Charge:	$1,661.01
17,588 kWh ×.09444	
Energy Cost Adjustment:	–$302.37
17,588 kWh × <.017192>	
Demand Charge:	$262.34
(1 kW × 4.53) + (29 kW × 8.89)	
Total Monthly Charge:	**$1,624.72**

Energy Usage Measured in kWh per kW Demand
The standard measure of electric energy blocks is in kilowatt hours. An alternative measure used by electric utilities for commercial and industrial rates is energy per unit of demand (e.g., kWh per kW). This block measurement is illustrated by Virginia Power's Schedule GS-

Table 18.1 Typical usage patterns.

	Convenience Store		Office Building		Shopping Center	
	kW	kWh	kW	kWh	kW	kWh
JAN	28	12,049	660	194,500	1,935	892,000
FEB	30	13,097	668	215,500	1,905	795,000
MAR	31	15,001	668	223,500	1,740	719,000
APR	32	13,102	600	177,500	1,515	633,000
MAY	30	14,698	345	101,000	1,425	571,000
JUN	30	17,588	293	95,000	1,455	680,000
JUL	32	17,739	375	118,000	1,440	661,000
AUG	35	17,437	420	156,500	1,395	667,000
SEP	31	18,963	555	130,000	1,455	733,000
OCT	32	16,003	540	207,000	1,455	646,000
NOV	30	17,490	570	169,500	1,545	675,000
DEC	29	12,684	600	172,500	1,740	768,000

Table 18.2 Commercial general service with demand component.

Company: Public Service Electric & Gas Company
Rate Class: Commercial
Rate Type: General Service
Rate Name: Schedule GLP
Service: Electric
Effective Date: 01/01/94
Qualifications: General purposes where demand is less then 150 kW.

	Winter Oct-May	Summer Jun-Sep
Customer Charge:	$3.74	$3.74
Minimum Charge:	$3.74	$3.74
Energy Cost Adjustment:	-$0.019037	-$0.017192
Tax Rate:	$0.00	$0.00
Surcharge:	$0.00	$0.00
# of Energy Blocks:	1	1
Block 1 Size:	> 0	> 0
Block 1 Energy Charge:	$0.09444	$0.09444
No. of Demand Blocks:	2	2
Block 1 Size:	1	1
Block 2 Size:	> 1	> 1
Block 1 Demand Charge:	$4.53	$4.53
Block 2 Demand Charge:	$7.84	$8.89

BILLING DEMAND: Estimated by dividing kWh usage by 100 or, if metered, greatest average 30-minute demand in the month.

2 Intermediate rate in Table 18.3. We are calculating the bill for October use. According to Table 18.1, the convenience store used 16,003 kWh in October and had a billing demand of 32 kW. The bill for this use is as follows.

Customer Charge:	$21.17
Energy Charge:	
Block 1 - (150 kWh × 32 kW),	
or 4,800 kWh @ $0.04840 =	$232.32
Block 2 - (150 kWh × 32 kW),	
or 4,800 kWh @ $0.02712 =	$130.18
Block 3 - (all remaining kWh),	
or 6,403 kWh @ $0.01173 =	$75.11
Energy Cost Adjustment: (16,003 × $0.01418)	$226.92
Demand Charge: (32 kW × $5.023)	$160.74
Subtotal:	$846.44
Tax Rate:	
25% of first $50 of subtotal	$12.50
12% on remainder of subtotal	$95.57
Total Monthly Charge:	**$954.51**

Time-of-Use Rates

Time-of-use rates are calculated very differently from general service rates. The customer's use must be recorded on a time-of-use meter, so that billing can be calculated on the use in each time period. In the following sample calculation from Long Island Lighting Company (Table 18.4), we assume the customer has responded to the price signal and has relatively little on-peak usage.

Energy Usage - 1,200 kWh; Season - Summer; On-Peak Period - 10:00 AM-8:00 PM, Monday-Friday; On-Peak Usage - 12.5%

Customer Charge:	$9.79
Energy Charge:	$176.53
150 on-peak kWh @ $0.3739 = $56.09	
1050 off-peak kWh @ $0.1147 = $120.44	
Energy Cost Adjustment:	$2.88
Tax:	$10.02
Total Monthly Charge:	**$199.22**

Table 18.3 Energy usage measured in kWh per kW of demand.

Company:	Virginia Electric & Power Company
Rate Class:	Commercial
Rate Type:	General Service
Rate Name:	Schedule GS-2 - Intermediate
Effective Date:	1/01/94
Qualifications:	Non-residential use with at least three billing demands => 30 kW in the current and previous 11 billing months; but not more than two billing months of 500 kW or more.

	OCT-MAY	JUN-SEP
Customer Charge:	$21.17	$21.17
Minimum Charge:	See Note	See Note
Energy Cost Adjustment:	$0.01418	$0.01418
Tax Rate:	See Note	See Note
No. of Energy Blocks:	3	3
Block 1 Size:	150 kWh/kW demand	150 kWh/kW demand
Block 2 Size:	150 kWh/kW demand	150 kWh/kW demand
Block 3 Size:	> 300 kWh/kW demand	> 300 kWh/kW demand
Block 1 Energy Charge:	$0.0484	$0.0484
Block 2 Energy Charge:	$0.02712	$0.02712
Block 3 Energy Charge:	$0.01173	$0.01173
No. of Demand Blocks:	1	1
Block 1 Size:	> 0	> 0
Block 1 Demand Charge:	$5.023	$6.531

MINIMUM CHARGE: Greater of: 1) contract amount; or 2) sum of customer charge, energy charge and adjustments, plus $1.604 times the maximum average 30-minute demand measured in the month. TAX: Tax is 25% of the first $50, and 12% of the excess.

BILLING DEMAND: Maximum average 30-minute demand measured in the month, but not less than the maximum demand determined in the current and previous 11 months when measured demand has reached 500 kW or more.

Table 18.4 Electric time-of-use rate.

Company:	Long Island Lighting Company	
Rate Class:	Residential	
Rate Type:	Time-of-Use	
Rate Name:	Schedule SC 1-VMRP, Rate 2	
Effective Date:	04/11/95	
Qualifications:		

Use for all residential purposes where consumption is 39,000 kWh or less for year ending September 30, or under 12,600 kWh for June through September.

	OCT-MAY	JUN-SEP
Customer Charge:	$9.79	$9.79
Minimum Charge:	$16.53	$16.53
Energy Cost Adjustment:	$0.0024	$0.0024
Tax Rate:	5.29389%	5.29389%
No. of Energy Blocks:	1	1
Block 1 Size:	> 0	> 0
Block 1 Energy Charge:		
On-Peak:	$0.1519	$0.3739
Off-Peak:	$0.0978	$0.1147

TAX: applied to total bill
PEAK PERIOD:
  On-Peak Hours: 10 a.m.-8 p.m., MON-FRI.
  Off-Peak Hours: All remaining hours.

18.5.2 Gas General Service Rates

Gas general service rates are calculated in a similar manner to the electric general service rates. Gas rates are priced in dollars per MMBtu, or in dollars per MCF, depending upon the individual utility's unit of measurement.

Commercial General Service
Rate with Demand Component

Gas rates for the commercial or industrial customer may involve a demand charge. The method for calculating this type of bill is done in the same manner as an electric rate with a demand charge. The example in Table 18.5 comes from Northern Illinois Gas Company.

Energy Usage - 11,500 MMBtu; Billing Demand - 400 MMBtu; Season - Winter

Customer Charge:	$325.00
Energy Charge:	$4,899.00
Purchased Gas Adjustment	
(Energy Cost Adjustment):	$28,221.00
Demand Charge:	$2,828.00
Tax:	$1,849.92
Total Charge:	**$38,122.92**

18.6 CONDUCTING A LOAD STUDY

Once a customer understands how utility rates are implemented, he can perform a simple load study to make use of this information. A load study will help the energy user to identify his load patterns, amount and time of occurrence of maximum load, and the load factor. This information can be used to modify use in ways that can lower electric or gas bills. It can also help the customer to determine the most appropriate rate to use.

The first step is to collect historical load data. Past bills are one source for this information. One year of data is necessary to identify seasonal patterns; two or more years of data is preferable. Select a study period that is fairly representative of normal consumption conditions.

Table 18.5 Commercial gas rate with demand component.

Company:	Northern Illinois Gas Company
Rate Class:	Commercial/Industrial Large
Rate Type:	General Service
Rate Name:	Schedule 6
Effective Date:	09/01/89
Qualifications:	General commercial use. All charges are shown per MMBtu.

	Winter		Summer
Customer Charge:	$325.00		$325.00
Minimum Charge:	$3,000.00		$3,000.00
Purchased Gas Adjustment:	$2.454		$2.61
Tax Rate:	5.1%		5.1%
No. of Energy Blocks:	1		1
Block 1 Size:	> 0	>	0
Block 1 Energy Charge:	$0.426		$0.426
No. of Demand Blocks:	1		1
Block 1 Size:	> 0	>	0
Block 1 Demand Charge:	$7.07		$5.388

TAX: applied to the total bill and is the lower of 5% of revenues or $0.24 per MMBtu, plus 0.10%.

BILLING DEMAND: per MMBtu of customer's Maximum Daily Contract Quantity.

The PGA and Demand charges shown here change monthly. Winter charges applied in January 1994; and Summer charges applied in July 1993.

Table 18.6 Basic steps for conducting a load analysis.
1) Collect historical load data
   - compile data for at least one year
2) Organize data by month for
   - kWh consumption
   - maximum kW demand
   - load factor
3) Review data for
   - seasonal patterns of use
   - peak demands
4) Determine what demand or use can be eliminated or reduced
5) Review load data with utility

The next step is to organize the data so that use patterns are evident. One way to analyze the data is to plot the kWh usage, the maximum demand, and the load factor. The load factor is the ratio of the average demand to the maximum demand. The average demand is determined by the usage in kWh divided by the total number of hours (24 × number of days) in the billing period. The number of days in the billing period may vary depending on how often the meter is read. (See Table 18.7.)

Table 18.7 Load factor calculation.

$$\text{Load Factor} = \frac{\text{average demand (kW)}}{\text{maximum demand (kW)}}, \text{ where}$$

$$\text{Average Demand} = \frac{\text{kWh usage}}{(24) \times (\text{number of days})}$$

Example: December office building load from Table 18.1

$$\text{Average Demand} = \frac{172,500 \text{ kWh}}{(24) \times (31)} = 231.9 \text{ kW}$$

$$\text{Load Factor} = \frac{231.9 \text{ kW}}{600 \text{ kW}} = 0.39$$

Next, review the data. Seasonal variations will be easily pinpointed. For example, most buildings will show a seasonal trend with two peaks. One will occur in winter and another during summer, reflecting seasonal heating and cooling periods. There may be other peaks due to some aspect of some industrial process, such as a cannery where crops are processed when they are harvested.

In Figure 18.1, kWh, maximum kW, and average kW are plotted from the data in Table 18.1 for the shopping center and the office building.

Note that the shopping center has a dominant winter peak—it is in the winter that the maximum kWh and kW are used for the year. The load factor ranges between 0.54 and 0.70. Overall, the average demand for this customer is about 60% of the peak demand.

The office building shows a different pattern. The load factor ranges between 0.35 and 0.52. This reflects the fact that the office building is really used less than half the time. Generally, working hours span from 8 a.m. to 6 p.m. Although some electrical load continues during the night hours, it is not as intense as during the normal office hours.

If the load factor were 1, this would imply uniform levels of use—in effect, a system that was turned on and left running continuously. This may be the case with some manufacturing processes such as steel mills and refineries.

The fourth step is to determine what demand or use can be eliminated, reduced, or redirected. How can the shopping center reduce its energy costs? By reducing or shifting the peak demand it can shave demand costs. Although overall consumption is not necessarily reduced, the demand charge is reduced. Where demand ratchets are in place, shaving peak demand may result in savings over a period of several months, not just the month of use. One way to shift peak demand is to install thermal storage units for space cooling purposes; this will shift day time load to night time, giving the customer an overall higher load factor. This may qualify the customer for special rates from the utility as well.

Where there may not be much that can be done about the peak demand, (in a high load factor situation) more emphasis should be placed on methods to reduce usage. Some examples: turn up the thermostat at night during the summer, down during the winter; install motion detectors to turn off unnecessary lights; turn off other equipment that is not in use.

Where the customer is charged for electric service on a time-of-use basis, a more sophisticated load study

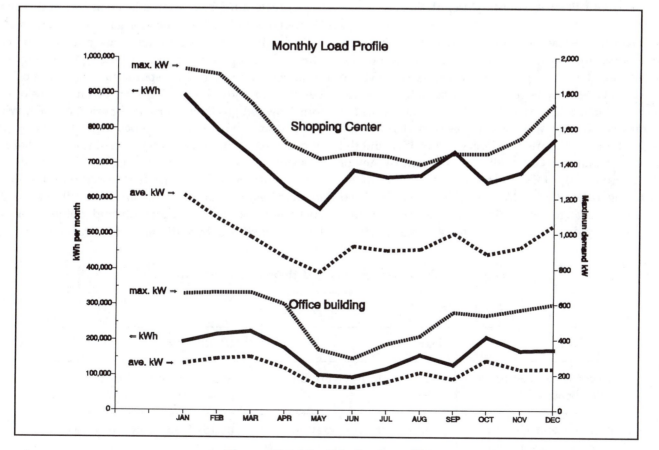

Figure 18.1 Monthly load profile.

should be performed. The data collected should consist of hourly load data over at least one year. This data can be obtained through the use of recording meters. Once acquired, the data should be organized to show use patterns on a monthly basis with Monday through Friday (or Saturday, depending on the customer's uses) use plotted separately from weekend use. Review of this data should show where shaving or shifting energy or demand can lower overall electricity bills.

Once the customer has obtained a better understanding of his energy usage patterns, he can discuss with his utility how to best benefit from them. The utility most likely will be interested because it will also receive some benefits. The customer can consider implementing certain specific measures to better fit in the utility's load pattern, and at the same time improve his energy use. The customer's benefit will generally be associated with less energy-related costs. Table 18.8 contains some examples of options that can be taken by commercial and industrial customers and the effect of those options on the utility.

18.7 EFFECTS OF DEREGULATION ON CUSTOMER RATES

18.7.1 Gas and Electric Supply Deregulation

Over the last two decades, many changes have either occurred or begun to occur in the structure of the nation's electric and gas supply industries. These changes have already begun to affect the rate types and structures for U.S. gas and electricity consumers. In the natural gas industry, well-head prices were deregulated as a result of the Natural Gas Policy Act of 1978 and the subsequent Natural Gas Well-Head Decontrol Act of 1989. Subsequently, FERC introduced a number of restructuring rules (Order Nos. 436, 500, and 636) that dramatically change the regulation of the nation's pipelines and provide access for end-users to transport gas

purchased at the well-head. In the electric industry, supply deregulation commenced with passage of the Public Utility Regulatory Policies Act of 1978, which encouraged electric power generation by certain non-utility producers. The Energy Policy Act of 1992 further deregulated production and mandated open transmission access for wholesale transfers of electricity between qualified suppliers and wholesale customers.

These legal and regulatory changes will have a significant and lasting effect on the rate types and rate structures experienced by end-users. In the past, most gas and electric customers paid a single bundled rate that reflected all costs for capacity and energy, storage, delivery, and administration. Once customers are given the opportunity to purchase their gas and electric resources directly from producers, it then becomes necessary to unbundle the costs associated with production from the costs associated with transportation and delivery to end users. This unbundling process has already resulted in separate rates for many services whose costs were previously combined in the single unit price for either gas or electricity.

18.7.2 Effect on Gas Rates

Much of the discussion in Section 18.3 of this chapter pertains to bundled rates for gas. However, as a result of unbundling, many utilities are now offering customers four separate services, including balancing, procurement, storage, and transportation of gas. Gas balancing rates provide charges for over- or under-use of customer-owned gas over a specified period of time. When the customer has the utility procure gas for transportation to the customer, gas procurement rates are charged. Gas storage rates are offered to customers for the storage of customer-owned gas. Gas transportation rates are offered to commercial, industrial and non-utility generator customers for the transportation and delivery of customer-owned gas. In addition to these rates, there is the

Table 18.8 Customer options and their effects on utility.

OPTIONS		
Commercial	*Industrial*	*Utility Effect*
Accept direct control of water heaters	Subscribe to interruptible rates	Reduction of load during peak periods
Store hot water to increase space heating	Add nighttime operations	Builds load during off-peak periods

actual cost of purchasing the gas to be transported. Gas procurement, balancing, storage and transportation rates have increased in usage as the structure of the gas industry has evolved.

Two other types of gas rates also are evolving as a result of industry deregulation. These include negotiated gas rates and variable gas rates. The former refers to rates that are negotiated between individual customers and the utility. Such rates are often subject to market conditions. The latter, variable gas service rates, refer to rates that vary from month to month. A review of all of the gas service rates collected by the Gas Research Institute (GRI) in 1994 indicates that 52% of the gas utilities surveyed offered at least one type of variable pricing. Such rates are often indexed to an outside factor, such as the price of gasoline or the price of an alternative fuel, and they usually vary between established floor and ceiling prices. The most common types of variable rates are those offered for transportation services.

18.7.3 Effect on Electric Rates

In the past, most U.S. electric customers have paid a single bundled rate for electricity. Many of these customers purchased from a utility that produced, transmitted, and delivered the electricity to their premises. In other cases, customers purchased from a distribution utility that had itself purchased the electricity at wholesale from a generating and transmitting utility. In both of these cases, the customer paid for electricity at a single rate that did not distinguish between the various services required to produce and deliver the power. In the future, as a result of the deregulation process already underway, there is a far greater likelihood that initially large customers, and later many smaller customers, will have the ability to select among a number of different suppliers. In most of these cases, however, the transmission and delivery of the purchased electricity will continue to be a regulated monopoly service. Consequently, future electricity consumers are likely to receive separate bills for:

* electric capacity and energy;
* transmission; and
* distribution.

In some cases, a separate charge may also be made for system control and administrative services, depending on exact industry structure in the given locality. For each such charge, a separate rate structure will apply. At present, it appears likely that there will be significant regional and local differences in the way these rates evolve and are implemented.

GLOSSARY

There are a few terms that the user of this document needs to be familiar with. Below is a listing of common terms and their definitions.

Billing Demand: The billing demand is the demand that is billed to the customer. The electric billing demand is generally the maximum demand or maximum average measured demand in any 15-, 30-, or 60-minute period in the billing month. The gas billing demand is determined over an hour or a day and is usually the greatest total use in the stated time period.

British Thermal Unit (Btu): Quantity of heat needed to bring one pound of water from 58.5 to 59.5 degrees Fahrenheit under standard pressure of 30 inches of mercury.

Btu Value: The heat content of natural gas is in Btu per cubic foot. Conversion factors for natural gas are:
* Therm = 100,000 Btu;
* 1 MMBtu = 1,000,000 Btu = 1 Decatherm.

Contract Demand: The demand level specified in a contractual agreement between the customer and the utility. This level of demand is often the minimum demand on which bills will be determined.

Controllable Demand: A portion or all of the customer's demand that is subject to curtailment or interruption directly by the utility.

Cubic Foot: Common unit of measurement of gas volume; the amount of gas required to fill one cubic foot.

Curtailable Demand: A portion of the customer's demand that may be reduced at the utility's direction. The customer, not the utility, normally implements the reduction.

Customer Charge: The monthly charge to a customer for the provision of the connection to the utility and the metering of energy and/or demand usage.

Demand Charge: The charge levied by a utility for metered demand of the customer. The measurement of demand may be either in kW or kVA.

Dual-Fuel Capability: Some interruptible gas rates require the customer to have the ability to use a fuel other than gas to operate their equipment.

Energy Blocks: Energy block sizes for gas utilities are either in MCFs or in MMBtus. The standard measures of energy block sizes for electric utilities are kWhs. However, several electric utilities also use an energy block size based on the customers' demand level (i.e. kWh per kW). Additionally, some electric utilities combine the standard kWh value with the kWh per kW value.

Energy Cost Adjustment (ECA): A fuel cost factor

charged for energy usage. This charge usually varies on a periodic basis, such as monthly or quarterly. It reflects the utilities' need to recover energy related costs in a volatile market. It is often referred to as the fuel cost adjustment, purchased power adjustment or purchased gas adjustment.

Excess or Non-Coincidental Demand: Some utilities charge for demands in addition to the on- or off-peak demands in time-of-use rates. An excess demand is demand used in off-peak time periods that exceeds usage during on-peak hours. Non-coincidental demand is the maximum demand measured any time in a billing period. This charge is usually in addition to the on- or off-peak demand charges.

Firm Demand: The demand level that the customer can rely on for uninterrupted use.

Interruptible Demand: All of the customer's demand may be completely interrupted at the utility's direction. Either the customer or the utility may implement the interruption.

MCF: Thousand (1000) cubic feet.

MMCF: Million (1,000,000) cubic feet.

Minimum Charge: The minimum monthly bill that will be charged to a customer. This generally is equal to the customer charge, but may include a minimum demand charge as well.

Off-Peak Demand: Greatest demand measured in the off-peak time period.

On-Peak Demand: Greatest demand measured in the on-peak time period.

Ratchet: A ratchet clause sets a minimum billing demand that applies during peak and/or non-peak months. It is usually applied as a percentage of the peak demand for the preceding season or year.

Reactive Demand: In electric service, some utilities have a special charge for the demand level in kilovolt-amperes reactive (kVAR) that is added to the standard demand charge. This value is a measure of the customer's power factor.

Surcharge: A charge levied by utilities to recover fees or imposts other than taxes.

Therm: A unit of heating value equal to 100,000 Btu.

Transportation Rates: Rates for the transportation of customer-owned gas. These rates do not include purchase or procurement of gas.

Voltage Discounts: Most electric utilities offer discounted rates to customers who will take service at voltages other than the general distribution voltages. The voltages for which discounts are generally offered are Secondary, Primary, Sub-transmission and Transmission. The actual voltage of each of these levels vary from utility to utility.

CHAPTER 19

THERMAL ENERGY STORAGE

CLINT CHRISTENSON
Research Associate

Oklahoma State University

19.1 INTRODUCTION

A majority of the technology developed for energy management has dealt with the more efficient consumption of electricity, rather than timing the demand for it. Variable frequency drives, energy efficient lights, electronic ballasts and energy efficient motors are a few of these consumption management devices. This side of electricity management effects only a portion of the actual electric cost, and often a small portion, compared to that dealing with the demand charge for electricity. Strangely enough the conservation of demand charges deals very little with the consumption of energy, but mainly with the ability of the utility to supply energy when needed. It is this timing of consumption that is the basis of demand management.

Experts agree that demand management is actually not a form of energy **conservation** but a form of cost **management.** Energy management has entered a new stage of maturation with the introduction of demand management. Utilities often charge more for energy used during periods when electricity is used most, i.e. on-peak periods, and levy penalties for high demand during certain periods in the form of peak and ratchet clauses. This has prompted many organizations to implement some form of demand management. Companies often control the demand of electricity by utilizing some of the techniques listed above and other consumption management actions which also reduce demand. More recently the ability to shift the **time** when electricity is needed has provided a means of balancing or shifting the demand for electricity to "off-peak" hours. This technique is often called demand balancing or demand shedding. This demand balancing may best be seen with the use of an example 24-hour chiller consumption plot during the peak day, Figure 19.1 and Table 19.1. This facility exhibits a typical office building load profile. The utility rate schedule has a summer on-peak demand period from 10 a.m. to 5:59 p.m., an 8-hour period.

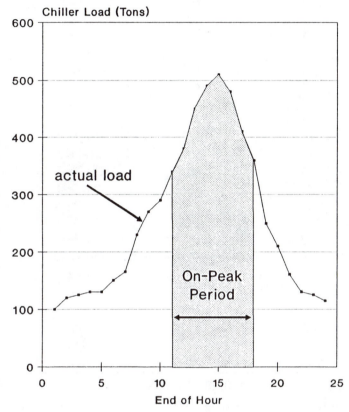

Figure 19.1 Typical office building chiller consumption profile.

Moving load from the on-peak rate period to the off-peak period can both balance the demand and reduce residual ratcheted peak charges. Thermal energy storage is one method available to accomplish just that.

Thermal energy storage (TES) is the concept of generating and storing energy in the form of heat or cold for use during peak periods. For the profile in Figure 19.1, a cooling storage system could be implemented to reduce or eliminate the need to run the chillers during the on-peak rate period. By running the chillers during off-peak hours and storing this capacity for use during the on-peak hours, a reduction in energy costs can be realized. If this type of system is implemented during new construction or when equipment is being replaced, smaller capacity chillers can be installed, since the chiller can spread the production of the total load over the entire day, rather than being sized for peak loads.

Table 19.1 Example chiller consumption profile

Chiller Consumption Profile

End of Hour	Chiller Load (Tons)	Rate
1	100	Reg
2	120	Reg
3	125	Reg
4	130	Reg
5	130	Reg
6	153	Reg
7	165	Reg
8	230	Reg
9	270	Reg
10	290	Reg
11	340	On-Peak
12	380	On-Peak
13	450	On-Peak
14	490	On-Peak
15	510	On-Peak
16	480	On-Peak
17	410	On-Peak
18	360	On-Peak
19	250	Reg
20	210	Reg
21	160	Reg
22	130	Reg
23	125	Reg
24	115	Reg

Daily Total	0123	Ton-Hrs
Daily Avg.	255.13	Tons
Peak Total	3420	Ton-Hrs
Peak Demand	510	Tons

Often the chiller load and efficiency follow the chiller consumption profile, in that the chiller is running at high load, i.e. high efficiency, only a small portion of the day. This is due to the HVAC system having to produce cooling when it is needed as well as to be able to handle instantaneous peak loads. With smaller chiller systems designed to handle the base and peak loads during off-peak hours, the chillers can run at higher average loads and thus higher efficiencies. Appendix A following this chapter lists several manufacturers of thermal energy storage systems.

19.2 STORAGE SYSTEMS

There are two general types of storage systems, ones that shut the chiller down during on-peak times and run completely off the storage system during that time are known as "full storage systems." Those designed to have the chiller run during the on-peak period supplementing the storage system are known as "partial storage systems." The full storage systems have a higher first cost since the chiller is off during peaking times and the cooling load must be satisfied by a larger chiller running fewer hours and a larger storage system storing the excess. The full storage systems do realize greater savings than the partial system since the chillers are completely turned off during on-peak periods. Full storage systems are often implemented in retrofit projects since a large chiller system may already be in place.

A partial storage system provides attractive savings with less initial cost and size requirements. New construction projects will often implement a partial storage system so that the size of both the chiller and the storage system can be reduced. Figures 19.2 and 19.3 and Tables 19.2 and 19.3 demonstrate the chiller load required to satisfy the cooling needs of the office building presented in Figure 19.1 for the full and partial systems, respectively. Column 2 in these tables represents the building cooling load each hour, and column 3 represents the chiller output for each hour. Discussion of the actual calculations that are required for sizing these different systems is included in a subsequent section. For simplicity sake, these numbers do not provide for any system losses, which will also be discussed in a later section.

The full storage system has been designed so that the total daily chiller load is produced during the off-peak hours. This eliminates the need to run the chillers during the on-peak hours, saving the increased rates for demand charges during this period and as well as any future penalties due to ratchet clauses. The partial storage system produces 255.13 tons per hour during the entire day, storing excess capacity for use when the building demand exceeds the chiller production. This provides the ability to control the chiller load, limit the peak chiller demand to 255.13 kW,[*] and still take advantage of the off-peak rates for a portion of the on-peak chiller load.

An advantage of partial load systems is that they can provide a means of improving the performance of a system that can handle the cumulative cooling load, but not the instantaneous peak demands of the building. In such a system, the chiller could be run nearer optimal load continuously throughout the day, with the excess cooling tonnage being stored for use during the peak

---

[*]assuming $COP = 3.5 => \dfrac{12,000\,\text{Btu/Hr}}{(X\,\text{kW}/3,412\,\text{Btu})} => X = 1 = \text{kW/Ton}$

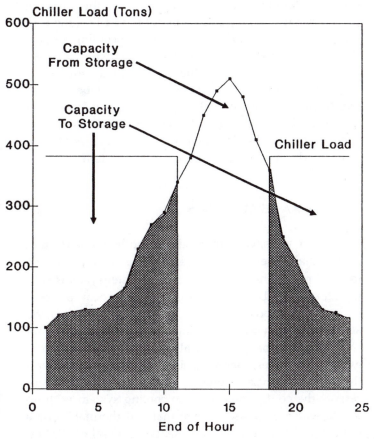

Figure 19.2 Full storage chiller consumption profile.

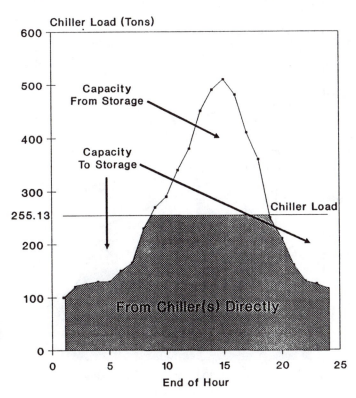

Figure 19.3 Partial storage chiller consumption
profile.

Table 19.2 Full storage chiller consumption profile.

Chiller Consumption Profile
Full Storage System

	1	2	3	4
End of Hour (Tons)	Cooling Load (Tons)	Chiller Load[2]	Rate	
1	100	382.69	Reg
2	120	382.69	Reg
3	125	382.69	Reg
4	130	382.69	Reg
5	130	382.69	Reg
6	153	382.69	Reg
7	165	382.69	Reg
8	230	382.69	Reg
9	270	382.69	Reg
10	290	382.69	Reg
11	340	0	On-Peak
12	380	0	On-Peak
13	450	0	On-Peak
14	490	0	On-Peak
15	510	0	On-Peak
16	480	0	On-Peak
17	410	0	On-Peak
18	360	0	On-Peak
19	250	382.69	Reg
20	210	382.69	Reg
21	160	382.69	Reg
22	130	382.69	Reg
23	125	382.69	Reg
24	115	382.69	Reg

Daily Total (Ton-Hrs) 6123   6123
Daily Avg (Tons):   255.13[1] 255.13

Peak Total (Ton-Hrs)   3420[3] 0[4]
Peak Demand (Tons)   510[3]   0[4]

[1] $\dfrac{6123\,\text{Ton--Hr}}{24\,\text{Hours}} = 255.13\,\text{Avg Tons}$

[2] $\dfrac{6123\,\text{Ton--Hr}}{16\,\text{Hours}} = 382.69\,\text{Avg Tons}$

[3]This load is supplied by the TES, not the chiller
[4]This is the chiller load and peak during on-peak periods

Table 19.3 Partial storage chiller consumption profile.

Chiller Consumption Profile
Partial Storage System

1	2	3	4
Hour of Day	Cooling Load (Tons)	Chiller Load (Tons)[1]	Rate
1	100	255.13	Reg
2	120	255.13	Reg
3	125	255.13	Reg
4	130	255.13	Reg
5	130	255.13	Reg
6	153	255.13	Reg
7	165	255.13	Reg
8	230	255.13	Reg
8	270	255.13	Reg
10	290	255.13	Reg
11	340	255.13	On-Peak
12	380	255.13	On-Peak
13	450	255.13	On-Peak
14	490	255.13	On-Peak
15	510	255.13	On-Peak
16	480	255.13	On-Peak
17	410	255.13	On-Peak
18	360	255.13	On-Peak
19	250	255.13	Reg
20	210	255.13	Reg
21	160	255.13	Reg
22	130	255.13	Reg
23	125	255.13	Reg
24	115	255.13	Reg

Daily Total (Ton-Hrs)	6123	6123	
Daily Avg (Tons):	255.13	255.13	
Peak Total (Ton-Hrs):	3420[2]	2041[3]	
Peak Demand (Tons):	510[2]	255.13[3]	

[1]
$$\frac{6123 \text{ Ton–Hr}}{24 \text{ Hours}} = 255.13 \text{ Avg Tons}$$

[2]This load is supplied by the TES supplemented by the chiller
[3]This is the chiller load and peak during on-peak period.

periods. An optional method for utilizing partial storage is a system that already utilizes two chillers. The daily cooling load could be satisfied by running both chillers during the off-peak hours, storing any excess cooling capacity, and running only one chiller during the on-peak period, to supplement the discharge of the storage system. This also has the important advantage of offering a reserve chiller during peak load times. Figure 19.4

shows the chiller consumption profile for this optional partial storage arrangement and Table 19.4 lists the consumption values. Early and late in the cooling season, the partial load system could approach the full load system characteristics. As the cooling loads and peaks begin to decline, the storage system will be able to handle more of the on-peak requirement, and eventually the on-peak chiller could also be turned off. A system such as this can be designed to run the chillers at optimum load, increasing efficiency of the system.

19.3 STORAGE MEDIUMS

There are several methods currently in use to store cold in thermal energy storage systems. These are water, ice, and phase change materials. The water systems simply store chilled water for use during on-peak periods. Ice systems produce ice that can be used to cool the actual chilling water, utilizing the high latent heat of fusion. Phase change materials are those materials that exhibit properties, melting points for example, that lend themselves to thermal energy storage. Figure 19.5a represents the configuration of the cooling system with either a water or phase change material thermal storage system and Figure 19.5b represents a general configura-

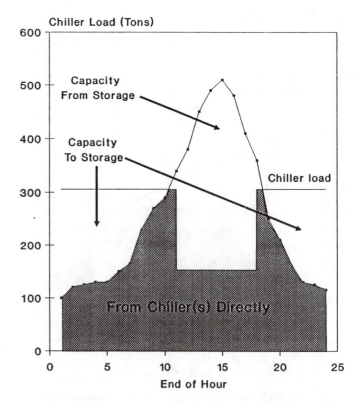

Figure 19.4 Optional partial storage chiller profile.

tion of a TES utilizing ice as the storage medium The next few sections will discuss these different mediums.

19.3.1 Chilled Water Storage

Chilled water storage is simply a method of storing chilled water generated during off-peak periods in a large tank or series of tanks. These tanks are the most commonly used method of thermal storage. One factor

Table 19.4 Partial storage chiller consumption profile.

1 Hour of Day	2 Cooling Load (Tons)	3 Chiller Load[1,2] (Tons)	4 Rate
1	100	306	Reg
2	120	306	Reg
3	125	306	Reg
4	130	306	Reg
5	130	306	Reg
6	153	306	Reg
7	165	306	Reg
8	230	306	Reg
9	270	306	Reg
10	290	306	Reg
11	340	153	On-Peak
12	380	153	On-Peak
13	450	153	On-Peak
14	490	153	On-Peak
15	510	153	On-Peak
16	480	153	On-Peak
17	410	153	On-Peak
18	360	153	On-Peak
19	250	306	Reg
20	210	306	Reg
21	160	306	Reg
22	130	306	Reg
23	125	306	Reg
24	115	306	Reg

Daily Total (Ton-Hrs):	6123	6123
Daily Avg (Tons):	255.13	255.13
On-Peak (Ton-Hrs):	3420[3]	1225[4]
Peak Demand (Tons):	510[3]	153[4]

[1]
$$\frac{(6123\,\text{Ton–Hr})(2\,\text{Chillers Operating})}{(16\,\text{Hours})(2\,\text{Chillers}) + (8\,\text{Hours})(1\,\text{Chiller})} = 306\,\text{Tons}$$

[2]
$$\frac{(6123\,\text{Ton–Hr})(1\,\text{Chiller Operating})}{(16\,\text{Hours})(2\,\text{Chillers}) + (8\,\text{Hours})(1\,\text{Chiller})} = 153\,\text{Tons}$$

[3]This load is supplied by the TES supplemented by the chiller.
[4]This is the chiller load and peak during the on-peak period.

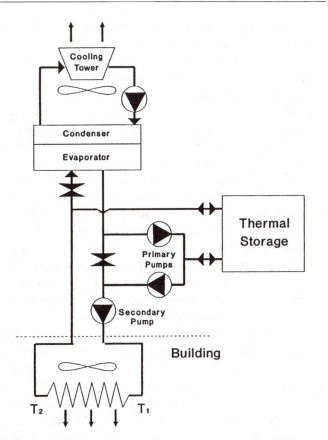

Figure 19.5a Water & eutectic storage system configuration.

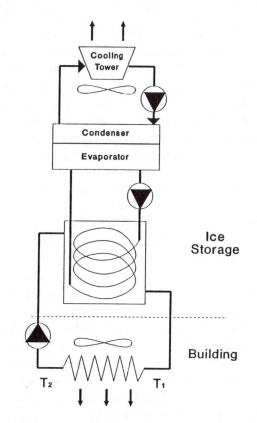

Figure 19.5b Ice storage system configuration.

to this popularity is the ease to which these water tanks can be interfaced with the existing HVAC system. The chillers are not required to produce chilled water any colder than presently used in the system so the system efficiency is not sacrificed. The chiller system draws warmer water from one end of the system and this is replaced with chilled water in the other. During the off-peak charge cycle, the temperature of the water in the storage will decline until the output temperature of the chiller system is approached or reached This chilled water is then withdrawn during the on-peak discharge cycle, supplementing or replacing the chiller(s) output.

Facilities that have a system size constraint such as lack of space often install a series of small insulated tanks that are plumbed in series. Other facilities have installed a single, large volume tank either above or below ground. The material and shape of these tanks vary greatly from installation to installation. These large tanks are often designed very similar to municipal water storage tanks. The main performance factors in the design of these tank systems, either large or multiple, is location and insulation. An Electric Power Research Institute's (EPRI) Commercial Cool Storage Field Performance Monitoring Project (RP-2732-05) Report states that the storage efficiencies of tanks significantly decrease if tank walls were exposed to sunlight and outdoor ambient conditions and/or had long hold times prior to discharging[7]. To minimize heat gain, tanks should be out of the direct sun whenever possible. The storage efficiency of these tanks is also decreased significantly if the water is stored for extended periods.

One advantage to using a single large tank rather than the series of smaller ones is that the temperature differential between the warm water intake and the chilled water outlet can be maintained. This is achieved utilizing the property of thermal stratification where the warmer water will migrate to the top of the tank and the colder to the bottom. Proper thermal stratification can only be maintained if the intake and outlet diffusers are located at the top and bottom of the tank and the flow rates of the water during charge and discharge cycles is kept low This will reduce a majority of the mixing of the two temperature waters. Another method used to assure that the two temperature flows remain separated is the use of a movable bladder, creating a physical partition. One top/bottom diffuser tank studied in the EPRI study used a thermocouple array, installed to measure the chilled water temperature at one foot intervals from top to bottom of the tank. This tank had a capacity of 550,000 gallons and was 20 feet deep but had only a 2.5 foot blend zone over which the temperature differential was almost 20 degrees[7].

The advantages of using water as the thermal storage medium are:

1.   Retrofitting the storage system with the existing HVAC system is very easy,

2.   Water systems utilize normal evaporator temperatures,

3.   With proper design, the water tanks have good thermal storage efficiencies,

4.   Full thermal stratification maintains chilled water temperature differential, increasing chiller loading and efficiencies, and

5.   Water systems have lower auxiliary energy consumption than both ice and phase change materials since the water has unrestricted flow through the storage system.

19.3.2 Ice Storage

Ice storage utilizes water's high latent heat of fusion to store cooling energy. One pound of ice stores 144 Btu's of cooling energy while chilled water only contains 1 Btu per pound $-°F$[7,8]. This reduces the required storage volume approximately 75%[7] if ice systems are used rather than water. Ice storage systems form ice with the chiller system during off-peak periods and this ice is used to generate chilled water during on-peak periods.

There are two main methods in use to utilize ice for on-peak cooling. The first is considered a static system in which serpentine expansion coils are fitted within a insulated tank of cooling water. During the charging cycle, the cooling water forms ice around the direct expansion coil as the cold gases pass through it (see Figure 19.5b). The thickness of the ice varies with the ice building time (charge time) and heat transfer area. During the discharge cycle, the cooling water contained in the tank is used to cool the building and the warmer water returned from the building is circulated through the tank, melting the ice, and using its latent heat of fusion for cooling.

The second major category of thermal energy storage systems utilizing ice can be considered a dynamic system. This system has also been labeled a plate ice maker or ice harvester. During the charging cycle the cooling water is pumped over evaporator "plates" where ice is actually produced. These thin sheets of ice are fed into the cooling water tank, dropping the temperature. During on-peak periods, this chilled water is circulated through the building for cooling. This technology is considered dynamic due to the fact that the ice is removed from the evaporator rather than simply remaining on it.

Static ice storage systems are currently available in factory-assembled packaged units which provide ease of installation and can provide a lower initial capital cost. When compared to water storage systems, the size and weight reduction associated with ice systems makes them very attractive to facilities with space constraints One main disadvantage to ice systems is the fact that the evaporator must be cold enough to produce ice. These evaporator temperatures usually range from 10° to 25° while most chiller evaporator temperatures range from 42° to 47°[9]. This required decrease in evaporator temperature results in a higher energy demand per ton causing some penalty in cooling efficiency. The EPRI Project reported that chillers operating in chilled water or eutectic salt (phase change material) used approximately 20% less energy than chillers operating in ice systems (0.9 vs. 1.1 kW/ton)[7]. The advantages of using ice as the thermal storage medium are:

1. Retrofitting the storage system with the existing HVAC chilled water system is feasible,

2. Ice systems require less space than that required by the water systems,

3. Ice systems have higher storage but lower refrigeration efficiencies than those of water, and

4. Ice systems are available in packaged units, due to smaller size requirements

19.3.3 Phase Change Materials

The benefit of capturing latent heat of fusion while maintaining evaporating temperatures of existing chiller systems can be realized with the use of phase change materials. There are materials that have melting points higher than that of water that have been successfully used in thermal energy storage systems. Several of these materials fall into the general category called "eutectic salts" and are salt hydrates which are mixtures of inorganic salts and water. Some eutectic salts have melting (solidifying) points of 47°[7], providing the opportunity for a direct retrofit using the current chiller system since this is at or above the existing evaporator temperatures. In a thermal storage system, these salts are placed in plastic containers, which are immersed within an insulated chilled water tank. During the charging cycle, the chilled water flows through the gaps between the con-

tainers, freezing the salts within them. During the on-peak discharge, the warmer building return water circulates through the tank, melting the salts and utilizing the latent heat of fusion to cool the building. These salt solutions have latent heat of fusion around 40 Btu/lb[9].

This additional latent heat reduces the storage volume by 66% of that required for an equivalent capacity water storage system[9]. Another obvious benefit of using eutectic salts is that the efficiency of the chillers is not sacrificed, as stated earlier, since the phase change occurs around normal evaporator suction temperatures. One problem with the eutectic salt systems is that the auxiliary energy consumption is higher since the chilled water must be pumped through the array of eutectic blocks. The auxiliary energy consumption of the ice systems is higher than both the water and eutectic salt systems since the chilled water must be pumped through the ice system coils, nozzles, and heat exchangers. The EPRI study found that the chilled water systems had an average auxiliary energy use of 0.43 kWh/Ton-Hr compared to the phase change systems (eutectic and ice) average auxiliary energy use of 0.56 kWh/Ton-Hr[5]. The advantages of using eutectic salts as the thermal storage medium are that they:

1. can utilize the existing chiller system for generating storage due to evaporator temperature similarity,

2. require less space than that required by the water systems, and

3. have higher storage and equivalent refrigeration efficiencies to those of water.

19.4 SYSTEM CAPACITY

The performance of thermal storage systems depends upon proper design. If it is sized too small or too large, the entire system performance will suffer. The following section will explain this sizing procedure for the example office building presented earlier. The facility has a maximum load of 510 tons, a total cooling requirement of 6,123 Ton-Hours, and a on-peak cooling requirement of 3,420 Ton-Hours. This information will be analyzed to size a conventional chiller system, a partial storage system, a full storage system, and the optional partial storage system. These results will then be used to determine the actual capacity needed to satisfy the cooling requirements utilizing either a chilled water, a eutectic salt, or an ice thermal storage system. Obviously some greatly simplifying assumptions are made.

19.4.1 Chiller System Capacity

The conventional system would need to be able to handle the peak load independently, as seen in Figure 19.1. A chiller or series of chillers would be needed to produce the peak cooling load of 510 tons. Unfortunately, packaged chiller units usually are available in increments that mandate excess capacity but for simplicity one 600-ton chiller will be used for this comparison. The conventional chiller system will provide cooling as it is needed and will follow the load presented in Figure 19.1 and Table 19.1.

To determine the chiller system requirement of a cooling system utilizing partial load storage, further analysis is needed. Table 19.1 showed that the average cooling load of the office building was 255.13 tons per hour. The ideal partial load storage system will run at this load (see Figure 19.3 and Table 19.3). The chiller system would need to be sized to supply the 255.13 tons per hour, so one 300-ton chiller will be used for comparison purposes. Table 19.5 shows how the chiller system would operate at 255.13 tons per hour, providing cooling required for the building directly and charging the storage system with the excess. Although the storage system supplements the cooling system for 2 hours before the peak period, the cooling load is always satisfied.

Comparing the peak demand from the bottoms of columns 2 and 3 of Table 19.5 shows that the partial storage system reduced this peak load almost 50% (510 − 255.13 = 254.87 Tons). Column 4 shows the tonnage that is supplied to the storage system and column 5 shows the amount of cooling contained in the storage system at the end of each hour of operation. This system was design so that there would be zero capacity remaining in the thermal storage tanks after the on-peak period. The values contained at the bottom of Table 19.5 are the total storage required to assure that there is no capacity remaining and the maximum output required from storage. These values will be utilized in the next section to determine the storage capacity required for each of the different storage mediums.

The full storage system also requires some calculations to determine the chiller system size. Since the chillers will not be used during the on-peak period, the entire daily cooling requirement must be generated during the off-peak periods. Table 19.1 listed the total cooling load as 6,123 Ton-Hours for the peak day. Dividing this load over the 16 off-peak hours yields that the chillers must generate 383 tons of cooling per hour (6,123 Ton-Hours/16 hours). A 450-ton chiller will be utilized in this situation for comparison purposes. Table 19.6 shows how the chiller system would operate at 383 tons per hour, providing cooling required for the building directly and charging the storage system with the excess.

Comparing the peak demand from the bottoms of columns 2 and 3 of Table 19.6 shows that the full storage system eliminated all load from the on-peak period. Column 4 shows the tonnage that is supplied to the storage system and column 5 shows the amount of cooling contained in the storage system at the end of each hour of operation. This system was designed so that there would be 0 capacity remaining in the thermal storage tanks after the on-peak period, as shown at the bottom of Table 19.6. The values in Table 19.6 will be utilized in the next section to determine the storage capacity required for each of the different storage mediums.

The optional partial storage system is a blend of the two systems presented earlier. Values given in Table 19.7 and Figure 19.4 are one combination of several possibilities that would drop the consumption and peak demand during the on-peak period. Once again this system has been designed to run both chillers during off-peak hours and run only one during on-peak hours. Benefits of this arrangement are that the current chiller system could be used in combination with the storage system and that the storage system does not require as much capacity as the full storage system. Also, a reserve chiller is available during peak-load times.

Comparing the peak demand from the bottoms of columns 2 and 3 of Table 19.7 shows that the optional partial storage system reduces the peak load from 510 tons to 153 tons, or approximately 70% during the on-peak period Column 4 shows the tonnage that is supplied to the storage system and column 5 shows the amount of cooling contained in the storage system at the end of each hour of operation. This system was designed so that there would be zero capacity remaining in the thermal storage tanks after the on-peak period. The values contained at the bottom of Table 19.7 are the total storage capacity required and the maximum output required from storage. These values will be utilized in the next section to determine the storage capacity required for each of the different storage mediums. Table 19.8 summarizes the performance parameters for the three configurations discussed above. The next section summarizes the procedure used to determine the size of the storage systems required to handle the office building.

19.4.2 Storage System Capacity

Each of the storage mediums has different size requirements to satisfy the needs of the cooling load. This section will describe the procedure to find the actual volume or size of the storage system for the partial load system for each of the different storage mediums. The

Table 19.5 Partial storage operation profile.

Thermal Storage Operation Profile
Partial Storage System

1 End of Hour	2 Cooling Load (Tons)	3 Chiller Load (Ton-Hrs)	4 Capacity to Storage (Ton-Hrs)	5 Capacity In Storage	6 Storage Cycle
1	100	255.13	155	696	Charge
2	120	255.13	135	831	Charge
3	125	255.13	130	961	Charge
4	130	255.13	125	1086	Charge
5	130	255.13	125	1211	Charge
6	153	255.13	102	1314	Charge
7	165	255.13	90	1404	Charge
8	230	255.13	25	1429	Charge
9	270	255.13	-15	1414	Discharge
10	290	255.13	-35	1379	Discharge
11	340	255.13	-85	1294	Discharge
12	380	255.13	-125	1169	Discharge
13	450	255.13	-195	974	Discharge
14	490	255.13	-235	740	Discharge
15	510	255.13	-255	485	Discharge
16	480	255.13	-225	260	Discharge
17	410	255.13	-155	105	Discharge
18	360	255.13	-105	0	Discharge
19	250	255.13	5	5	Charge
20	210	255.13	45	50	Charge
21	160	255.13	95	145	Charge
22	130	255.13	125	271	Charge
23	125	255.13	130	401	Charge
24	115	255.13	140	541	Charge

Daily Total (Ton-Hrs):	6123	6123			
Daily Avg (Tons):	255.13	255.13			

Peak Total (Ton-Hrs):	3420	2041	Storage Total =	1429	
Peak Demand (Tons):	510	255.13	Peak Storage Output =	255	

Column 4 = Column 3 − Column 2

Column 5(n) = Column 5(n−1) + Column 4(n)

design of the chiller and thermal storage system must provide enough chilled water to the system to satisfy the peak load, so particular attention should be paid to the pumping and piping. Table 19.9 summarizes the size requirement of each of the three different storage options.

To calculate the capacity of the partial load storage system, the relationship between capacity (C), mass (M), specific heat of water (Cp), and the coil temperature differential (T_2-T_1) shown in Figure 19.5a will be used:

$$C = M \, Cp \, (T_2-T_1)$$

where: M = lbm

Cp = 2 Btu/lbm °R

(T_2-T_1) = °R

The partial load system required that 1,429 Ton-Hrs be stored to supplement the output of the chiller during on-peak periods. This value does not allow for any thermal loss which normally occurs. For this discussion, a conservative value of 20% is used, which is an average suggested in the EPRI report[7]. This will increase the storage requirements to 1,715 Ton-Hrs and chilled water storage

Table 19.6 Full storage operation profile.

Thermal Storage Operation Profile
Full Storage System

1 Hour of Day	2 Cooling Load (Tons)	3 Chiller Load (Tons)	4 Capacity to Storage (Ton-Hrs)	5 Capacity In Storage (Ton-Hrs)	6 Storage Cycle
1	100	383	283	1589	Charge
2	120	383	263	1852	Charge
3	125	383	258	2109	Charge
4	130	383	253	2362	Charge
5	130	383	253	2615	Charge
6	153	383	230	2844	Charge
7	165	383	218	3062	Charge
8	230	383	153	3215	Charge
9	270	383	113	3327	Charge
10	290	383	93	3420	Charge
11	340	0	-340	3080	Discharge
12	380	0	-380	2700	Discharge
13	450	0	-450	2250	Discharge
14	490	0	-490	1760	Discharge
15	510	0	-510	1250	Discharge
16	480	0	-480	770	Discharge
17	410	0	-410	360	Discharge
18	360	0	-360	0	Discharge
19	250	383	133	133	Charge
20	210	383	173	305	Charge
21	160	383	223	528	Charge
22	130	383	253	781	Charge
23	125	383	258	1038	Charge
24	115	383	268	1306	Charge

Daily Total (Ton-Hrs):	6123	6123			
Daily Avg (Tons):	255.13	255.13			
Peak Total (Ton-Hrs):	3420	0	Storage Total =	3420	
Peak Demand (Tons):	510	0	Peak Storage Output =	510	

Column 4 = Column 3 – Column 2
Column 5(n) = Column 5(n–1) + Column 4(n)

systems in this size range cost approximately $200/Ton-Hr including piping and installation[5]. Assuming that there are 12,000 Btu's per Ton-Hr, this yields:

C = (1,715 Ton-Hrs)*(12,000 Btu/Ton-Hr)
    = 20.58 × 10^6 Btu's.

Assuming $(T_2 - T_1) = 12°$ and $C_p = 1$ Btu/lbm °R, the relation becomes:

$$M = \frac{C}{C_p(T_2 - T_1)} = 1\frac{20.58 \times 10^6 \text{ Btu's}}{(\text{Btu/lbm °R})(12 \text{ °R})} = 1.72 \times 10^6 \text{ lbm H}_2\text{O}$$

$$\text{Volume of Water} = \text{Mass/Density} = \frac{1.72 \times 10^6 \text{ lbm}}{62.5 \text{ lbm/Ft}^3}$$

$$= 27,520 \text{ Ft}^3 \text{ or}$$

$$\frac{1.72 \times 10^6 \text{ lbm}}{8.34 \text{ lbm/gal}} = 206,235 \text{ gal.}$$

Sizing the storage system utilizing ice is completed in a very similar fashion. The EPRI study states that the

Table 19.7 Optional partial storage operation profile.

Thermal Storage Operation Profile
Optional Partial Storage System

1 End of Hour	2 Cooling Load (Tons)	3 Chiller Load (Tons)	4 Capacity to Storage (Ton-Hrs)	5 Capacity In Storage (Ton-Hrs)	6 Storage Cycle
1	100	306	206	1053	Charge
2	120	306	186	1239	Charge
3	125	306	181	1420	Charge
4	130	306	176	1597	Charge
5	130	306	176	1773	Charge
6	153	306	153	1926	Charge
7	165	306	141	2067	Charge
8	230	306	76	2143	Charge
9	270	306	36	2179	Charge
10	290	306	16	2195	Charge
11	340	153	-187	2008	Discharge
12	380	153	-227	1782	Discharge
13	450	153	-297	1485	Discharge
14	490	153	-337	1148	Discharge
15	510	153	-357	791	Discharge
16	480	153	-327	464	Discharge
17	410	153	-257	207	Discharge
18	360	153	-207	0	Discharge
19	250	306	56	56	Charge
20	210	306	96	152	Charge
21	160	306	146	298	Charge
22	130	306	176	475	Charge
23	125	306	181	656	Charge
24	115	306	191	847	Charge

Daily Total (Ton-Hrs):	6123	6123			
Daily Avg (Tons):	255.13	255.12			

Peak Total (Ton-Hrs):	3420	1225	Storage Total =	2195	
Peak Demand (Tons):	510	153	Peak Storage Output =	357	

Column 4 = Column 3 - Column 2
Column 5(n) = Column 5(n-1) + Column 4(n)

ice storage tanks had average daily heat gains 3.5 times greater than the chilled water and eutectic systems due to the higher coil temperature differential (T_2–T_1) To allow for these heat gains a conservative value of 50% will be added to the actual storage capacity, which is an average suggested in the EPRI report[7] This will increase the storage requirements to 2,144 Ton-Hrs. Assuming that there are 12,000 Btu's per Ton-Hr, this yields: (2,144 Ton-Hrs)*(12,000 Btu's/Ton-Hr) = 25.73 × 10^6 Btu's. The ice systems utilize the latent heat of fusion so the C_1 now

becomes

$$C_1 = \text{Latent Heat} = 144 \text{ Btu/lbm.}$$

Because the latent heat of fusion, which occurs at 32°F, is so large compared to the sensible heat, the sensible heat (C_p) is not included in the calculation. The mass of water required to be frozen becomes:

$$M = C/C_1 = \frac{25.73 \times 10^6 \text{ Btu's}}{(144 \text{ Btu/lbm})} = 1.79 \times 10^5 \text{ lbm } H_2O$$

Table 19.8 System performance comparison.

	SYSTEM			
PERFORMANCE PARAMETERS	Conventional No Storage	Partial Storage	Full Storage	Optional Partial
Overall Peak Demand (Tons)	510	255.13	383	306
On-Peak, Peak Demand (Tons)	510	255.13	0	153
On-Peak Chiller Consumption (Ton-Hrs)	3,420	2,041	0	1,225
Required Storage Capacity[1] (Ton-Hrs)	—	1,379	3,420	2,195
MAXIMUM STORAGE OUTPUT[1] (Tons)	—	255	510	357

[1]Values from Table 19.5, 19.6, and 19.7. Represent the capacity required to be supplied by the TES.

$$\text{Volume of Ice} = \frac{\text{Mass}}{\text{Density}} = \frac{1.79 \times 10^5 \, \text{lbm}}{62.5 \, \text{lbm/Ft}^3}$$

$$= 2.864 \, \text{Ft}^3$$

This figure is conservative since the sensible heat has been ignored but calculates the volume of ice needed to be generated. The actual volume of ice needed will vary and the total amount of water contained in the tank around the ice coils will vary greatly. The ability to purchase pre-packaged ice storage systems makes their sizing quite easy For this situation, two 1,080 Ton-Hr ice storage units will be purchased for approximately \$150/Ton-Hr including piping and installation[4] (note that this provides 2,160 Ton-Hrs compared to the needed 2,144 Ton-Hrs).

Sizing the storage system utilizing the phase change materials or eutectic salts is completed just as the ice storage system. The EPRI study states that the eutectic salt storage tanks had average daily heat gains approximately the same as that of the chilled water systems. To allow for these heat gains a conservative value of 20% is added to the actual storage capacity[5]. This increases storage requirements to 1,715 Ton-Hrs. Assuming there are 12,000 Btu's per Ton-Hr, this yields: (1,715 Ton-Hrs)*(12,000 Btu's/Ton-Hr) = 20.58 × 10^6 Btu's.

The eutectic system also utilizes the latent heat of fusion like the ice system and the temperature differen-

tial shown in Figure 19.5a is not used in the calculation. The C, now becomes:

$$C_1 = \text{Latent Heat} = 40 \, \text{Btu/lbm}$$

$$M = C/C_1 = \frac{20.58 \times 10^6 \, \text{Btu's}}{(40 \, \text{Btu/lbm})} = 5.15 \times 10^5 \, \text{lbm}$$

$$\text{Volume of Eutectic Salts} = \frac{\text{Mass}}{\text{Density}} = \frac{5.15 \times 10^5 \, \text{lbm}}{62.5 \, \text{lbm/Ft}^3}$$
(assuming density = water)

$$= 8,232 \, \text{Ft}^3$$

The actual volume of eutectic salts needed would need to be adjusted for density differences in the various combinations of the salts. Eutectic systems have not been studied in great detail and factory sized units are not yet readily available. The EPRI report[7] studied a system that required 1,600 Ton-Hrs of storage which utilized approximately 45,000 eutectic "bricks" contained in an 80,600 gallon tank of water. For this situation, a similar eutectic storage unit will be purchased for approximately \$250/Ton-Hr including piping and installation. The ratio of Ton-Hrs required for partial storage and the required tank size will be utilized for sizing the full and optional partial storage systems.

Table 19.9 summarizes the sizes and costs of the different storage systems and the actual chiller systems

Table 19.9 Complete system comparison.

Performance Parameters	Conventional No Storage	Partial Storage	Full Storage	Optional Partial
CHILLER				
SIZE (# and Tons)	1 @ 600	1 @ 300	1 @ 450	2 @ 175
COST($)	180,000	90,000	135,000	105,000
WATER STORAGE				
Capacity (Ton-Hrs)	—	1,715	4,104	2,634
Volume (cubic feet)	—	27,484	65,769	42,212
Volume (gallons)	—	205,635	492,086	315,827
Cost per Ton-Hr ($)	—	200	135	165
Storage cost ($)	—	343,000	554,040	434,610
ICE STORAGE				
Capacity (Ton-Hrs)		2,144	5,130	3,293
# and size (Ton-Hrs)	—	2 @ 1,080	4 @ 1,440	3 @ 1,220
Ice volume (cubic feet)	—	2,859	6,840	4,391
Cost per Ton-Hr (S)	—	150	150	150
Storage cost ($)[1]	—	324,000	864,000	549,000
EUTECTIC STORAGE				
Capacity (Ton-Hrs)		1,715	4,104	2,634
Eutectic vol (cubic feet)	—	8,232	19,699	12,643
Cost per Ton-Hr ($)	—	250	200	230
Storage cost ($)	—	428,750	820,000	605,820

[1](2 units)(1,080 Ton-Hrs/units)($150/Ton-Hr) = $324,000
Note: The values in this table vary slightly from those in the text from additional significant digits.

for each of the three storage arrangements. The values presented in this example are for a specific case and each application should be analyzed thoroughly. The cost per ton hour of a water system dropped significantly as the size of the tanks rises as will the eutectic systems since the engineering and installation costs are spread over more capacity.

19.5 ECONOMIC SUMMARY

Table 19.9 covered the approximate costs of each of the three system configurations utilizing each of the three different storage mediums. Table 19.8 listed the various peak day performance parameters of each of the systems presented. To this point, the peak day chiller consumption has been used to size the system. To analyze the savings potential of the thermal storage systems, much more information is needed to determine daily cooling and chiller loads and the respective storage

system performance. To calculate the savings accurately, a daily chiller consumption plot is needed for at least the summer peak period. These values can then be used to determine the chiller load required to satisfy the cooling demands. Only the summer months may be used since most of the cooling takes place and a majority of the utilities "time of use" charges (on-peak rates) are in effect during that time. There are several methods available to estimate or simulate building cooling load. Some of these methods are available in a computer simulation format or can also be calculated by hand.

For the office building presented earlier, an alternative method will be used to estimate cooling savings. An estimate of a monthly, average day cooling load will be used to compare the operating costs of the respective cooling configurations. For simplicity, it is assumed that the peak month is July and that the average cooling day is 90% of the cooling load of the peak day. The average cooling day for each of the months that make up the summer cooling period are estimated based upon July's average cooling load. These factors are presented in Table 19.10 for June through October[11]. These factors are applied to the hourly chiller load of the average July day to determine the season chiller/TES operation loads. The monthly average day, hourly chiller loads for each of the three systems are presented in Table 19.11. The first column for each month in Table 19.11 lists the hourly cooling demand. The chiller consumption required to satisfy this load utilizing each of the storage systems is also listed. This table does not account for the thermal efficiencies used to size the systems but for simplicity, these values will be used to determine the rate and demand savings that will be achieved after implementing the system. The formulas presented for the peak day thermal storage systems operations have been used for simplicity. These chiller loads do not represent the optimum chiller load since some of partial systems approach full storage systems during the early and late cooling months. The bottom of the table contains the totals for the chiller systems. These totaled average day values will now be used to calculate the savings. The difference between the actual cooling load and the chiller load is the approximate daily savings for each day of that month.

A hypothetical southwest utility rate schedule will be used to apply economic terms to these savings. The electricity consumption rate is $0.04/kWh and the demand rate during the summer is $3.50/kW for the peak demand during the off-peak hours and $5.00/kW for the peak demand during the on-peak hours. These summer demand rates are in effect from June through October. This rate schedule only provides savings from balancing the demand, although utilities often have cheaper off-peak consumption rates. It can be seen that the off-peak demand charge assures that the demand is leveled and not merely shifted. This rate schedule will be applied to

the total values in Table 19.11 and multiplied by the number of days in each month to determine the summer savings. These savings are contained in Table 19.12. The monthly average day loads in Table 19.11 are assumed to be 90% of the actual monthly peak billing demand, and are adjusted accordingly in Table 19.12. The total monthly savings for each of the chiller/TES systems is determined at the bottom of each monthly column.

These cost savings are not the only monetary justification for implementing TES systems. Utilities often extend rebates and incentives to companies installing thermal energy storage systems to shorten their respective payback period. This helps the utility reduce the need to build new generation plants. The southwest utility serving the office building studied here offers $200 per design day peak kW shifted to off-peak hours up to $200,000. Table 19.13 summarizes the incentives available for the various systems using this subsidy plan. Appendix B following this chapter contains a short list of the incentives provided by several utilities across the nation.

19.6 CONCLUSIONS

Thermal energy storage will play a large role in the future of demand side management programs of both private organizations and utilities. The success of the storage system and the HVAC system as a whole depend on many factors:

- The chiller load profile,
- The utility rate schedules and incentive programs,
- The condition of the current chiller system,
- The space available for the various systems,
- The selection of the proper storage medium, and
- The proper design of the system and integration of this system into the current system.

Table 19.10 Average summer day cooling load factors.

MONTH	kW FACTOR[1]	PEAK TONS[2]	kWh FACTOR[1]	Ton-Hrs/day[3]
JUNE	0.8	360	0.8	4,322
JULY	1	450	1	5,403
AUGUST	0.9	405	0.9	4,863
SEPT	0.7	315	0.7	3,782
OCT	0.5	225	0.5	2,702

[1]kW and kWh factors were estimated to determine utility cost savings.
[2]The average day peak load is estimated to be 90% of the peak day. The kW factor for each month is multiplied by the peak months average tonnage. For JUNE: PEAK TONS = (0.8)*(450) = 360
[3]The average day consumption is estimated to be 90% of the peak day. The kWh factor for each month is multiplied by the peak months average consumption. For JUNE: CONSUMPTION = (0.8)*(5,403) = 4,322

Table 19.11 Monthly average day chiller load profiles.

END OF HOUR	JUNE (in tons)				JULY (in tons)				AUGUST (in tons)				SEPTEMBER (in tons)				OCTOBER (in tons)			
	Actual	partial	full	optional	Actual	partial	full	optional	Actual	partial	full	optional	Actual	partial	full	optional	Actual	partial	full	optional
1	71	180	270	216	88	225	338	270	79	203	304	243	62	158	236	189	44	113	169	135
2	85	180	270	216	106	225	338	270	95	203	304	243	74	158	236	189	53	113	169	135
3	88	180	270	216	110	225	338	270	99	203	304	243	77	158	236	189	55	113	169	135
4	92	180	270	216	115	225	338	270	103	203	304	243	80	158	236	189	57	113	169	135
5	92	180	270	216	115	225	338	270	103	203	304	243	80	158	236	189	57	113	169	135
6	108	180	270	216	135	225	338	270	122	203	304	243	95	158	236	189	68	113	169	135
7	116	180	270	216	146	225	338	270	131	203	304	243	102	158	236	189	73	113	169	135
8	162	180	270	216	203	225	338	270	183	203	304	243	142	158	236	189	101	113	169	135
9	191	180	270	216	238	225	338	270	214	203	304	243	167	158	236	189	119	113	169	135
10	205	180	270	216	256	225	338	270	230	203	304	243	179	158	236	189	128	113	169	135
11	240	180	0	108	300	225	0	135	270	203	0	122	210	158	0	95	150	113	0	68
12	268	180	0	108	335	225	0	135	302	203	0	122	235	158	0	95	168	113	0	68
13	318	180	0	108	397	225	0	135	357	203	0	122	278	158	0	95	199	113	0	68
14	346	180	0	108	432	225	0	135	389	203	0	122	303	158	0	95	216	113	0	68
15	360	180	0	108	450	225	0	135	405	203	0	122	315	158	0	95	225	113	0	68
16	339	180	0	108	424	225	0	135	381	203	0	122	296	158	0	95	212	113	0	68
17	289	180	0	108	362	225	0	135	326	203	0	122	253	158	0	95	181	113	0	68
18	254	180	0	108	318	225	0	135	286	203	0	122	222	158	0	95	159	113	0	68
19	176	180	270	216	221	225	338	270	199	203	304	243	154	158	236	189	110	113	169	135
20	148	180	270	216	185	225	338	270	167	203	304	243	130	158	236	189	93	113	169	135
21	113	180	270	216	141	225	338	270	127	203	304	243	99	158	236	189	71	113	169	135
22	92	180	270	216	115	225	338	270	103	203	304	243	80	158	236	189	57	113	169	135
23	88	180	270	216	110	225	338	270	99	203	304	243	77	158	236	189	55	113	169	135
24	81	180	270	216	101	225	338	270	91	203	304	243	71	158	236	189	51	113	169	135
TOTALS:	JUNE	partial	full	optional	JULY	partial	full	optional	AUG	partial	full	optional	SEPT	partial	full	optional	OCT	partial	full	optional
TOTAL: (ton-hrs)	4,322	4,322	4,322	4,322	5,403	5,403	5,403	5,403	4,862	4,862	4,862	4,862	3,782	3,782	3,782	3,782	2,701	2,701	2,701	2,701
AVG: (tons)	180	180	180	180	225	225	225	225	203	203	203	203	158	158	158	158	113	113	113	113
OFF-PEAK MAX: (tons)	205	180	270	216	256	225	338	270	230	203	304	243	179	158	236	189	128	113	169	135
ON-PEAK MAX: (tons)	360	180	0	108	450	225	0	135	405	203	0	122	315	158	0	95	225	113	0	68
ON-PEAK CONSUMP: (ton-hrs)	2,414	1,441	0	864	3,018	1,801	0	1,081	2,716	1,621	0	972	2,112	1,261	0	756	1,509	900	0	540

For June Partial Storage: (4,322 Ton–Hrs)/(24 Hrs) = 180 Tons
For June Full Storage: (4,322 Ton–Hrs)/(16 Hrs) = 270 Tons

Thermal storage is a very attractive method for an organization to reduce electric costs and improve system management. New installation projects can utilize storage to reduce the initial costs of the chiller system as well as savings in operation. Storage systems will become easier to justify in the future with increased mass production, technical advances, and as more companies switch to storage.

References

1. Cottone, Anthony M., "Featured Performer: Thermal Storage," in *Heating Piping and Air Conditioning*, August 1990, pp. 51-55.
2. Hopkins, Kenneth J., and James W. Schettler, "Thermal Storage Enhances Heat Recovery," in *Heating Piping and Air Conditioning*, March 1990, pp. 45-50.
3. Keeler, Russell M., "Scrap DX for CW with Ice Storage," in *Heating Piping and Air Conditioning*, August 1990, pp. 59-62.
4. Lindemann, Russell, Baltimore Aircoil Company, Personnel Phone Interview, January 7, 1992.
5. Mankivsky, Daniel K., Chicago Bridge and Iron Company, Personnel Phone Interview, January 7, 1992.
6. Pandya, Dilip A., "Retrofit Unitary Cool Storage System," in *Heating Piping and Air Conditioning*, July 1990, pp. 35-37.
7. Science Applications International Corporation, *Operation Performance of Commercial Cool Storage Systems* Vols. 1 & 2, Electric Power Research Institute (EPRI) Palo Alto, September 1989.
8. Tamblyn, Robert T., "Optimizing Storage Savings," in *Heating Piping and Air Conditioning*, August 1990 pp. 43-46.
9. Thumann, Albert, *Optimizing HVAC Systems* The Fairmont Press, Inc., 1988.
10. Thumann, Albert and D. Paul Mehta, *Handbook of Energy Engineering*, The Fairmont Press, Inc., 1991.
11. Wong, Jorge-Kcomt, Dr. Wayne C. Turner, Hemanta Agarwala, and Alpesh Dharia, *A Feasibility Study to Evaluate Different Options for Installation of a New Chiller With/Without Thermal Energy Storage System*, study conducted for the Oklahoma State Office Buildings Energy Cost Reduction Project, Revised 1990.

Table 19.12 Summer monthly system costs and TES savings.

	JUNE (30 days)				JULY (31 days)				AUGUST (31 days)				SEPTEMBER (30 days)				OCTOBER (31 days)			
	Actual	Partial	Full	Optional	Actual	Partial	Full	Optional	Actual	Partial	Full	Optional	Actual	Partial	Full	Optional	Actual	Partial	Full	Optional
ON–PEAK, PEAK[1] (kW)	400	200	0	120	500	250	0	150	450	226	0	136	350	176	0	106	250	126	0	76
OFF–PEAK, PEAK[1] (kW)	228	200	300	240	284	250	376	300	256	226	338	270	199	176	262	210	142	126	188	150
CONSUMPTION[1] (kW–Hr)	4,322	4,322	4,322	4,322	5,403	5,403	5,403	5,403	4,862	4,862	4,862	4,862	3,782	3,782	3,782	3,782	2,701	2,701	2,701	2,701
DEMAND COST[2] ($)	2,797	1,700	1,050	1,440	3,496	2,125	1,314	1,800	3,144	1,917	1,182	1,623	2,446	1,492	918	1,263	1,748	1,067	657	903
CONSUMPTION COST[3] ($)	5,186	5,186	5,186	5,186	6,700	6,700	6,700	6,700	6,029	6,029	6,029	6,029	4,538	4,538	4,538	4,538	3,349	3,349	3,349	3,349
TOTAL COST[4] ($)	7,984	6,886	6,236	6,626	10,195	8,825	8,014	8,500	9,173	7,946	7,211	7,652	6,985	6,031	5,456	5,801	5,097	4,416	4,006	4,252
SAVINGS[5] ($)		1,097	1,747	1,357		1,371	2,181	1,696		1,227	1,962	1,522		954	1,528	1,183		681	1,091	845

Table 19.13 Available demand management incentives.

Performance Parameters	System			
	Conventional No Storage	Partial Storage	Full Storage	Optional Partial
Actual On-Peak Demand[1] (kW)	510	255	0	153
On-Peak Demand Shifted[2] (kW)		255	510	357
Utility Subsidy[3] ($)		51,000	102,000	71,400

[1]Yearly design peak demand from Table 19.8.
[2]Demand shifted from design day on-peak period. For partial: 510 kW - 255 kW = 255 kW.
[3]Based upon $200/kW shifted from design day on-peak period. For partial: 255 kW * $200/kW = $51,000.

APPENDIX 19-A
Partial list of manufacturers of thermal storage systems.
Source: *Energy User News*, December 1991.

Company:	Address:
Baltimore Aircoil Co.	P.O. Box 7322 Baltimore, MD 21227 (301) 799-6146
Belyea Company, Inc.	45 Howell St. Jersey City, NJ 07306 (201) 653-3334
Carrier Corporation	One Carrier Place Farmington, CT 06034-4015 (203) 674-3139

Control Pak Corporation	23840 Industrial Park Drive Farmington Hills, MI 48335 (313) 471-0337
Control Systems Intl.	1625 West Crosby Rd. #400 Carrollton, TX 75006 (214) 323-1111
The Trane Company	3600 Pammel Creek Rd. La Crosse, WI 54601-7599 (608) 787-2000

APPENDIX 19-B

Partial list of Utility Cash Incentive Programs.
Source: Dan Mankivsky, Chicago Bridge & Iron, August 1991.

STATE - Electric Utility	CASH INCENTIVE $/kW Shifted	Maximum
ARIZONA		
- Arizona Public Service	75-125	no limit
- Salt River Project	60-250	no limit
CALIFORNIA		
- American Public Utilities Dept.	60	50,000
- L.A. Dept of Water & Power	250	40% cost
- Pacific Gas & Electric	300	50%-70%
- Pasadena Public Utility	300	no limit
- Riverside Public Utility	200	no limit
- Sacramento Municipal Util Dist.	200	no limit
- San Diego Gas & Electric	50-200	no limit
- Southern California Edison	100	300,000
DISTRICT OF COLUMBIA		
- Patomac Electric Power Co.	200-250	no limit
FLORIDA		
- Florida Power & Light Co.	250/ton	no limit
- Florida Power Corp.	160-180	25%
- Tampa Electric Co.	200	no limit
INDIANA		
- Indianapolis Power & Light	200	no limit
- Northern Indiana Public Service	200/ton	
MARYLAND		
- Baltimore Gas & Electric	200	no limit
- Patomac Electric Power Co.	200-250	no limit
MINNESOTA		
- Northern States Power	400/ton	no limit
NEVADA		
- Nevada Power	100-150	no limit
NEW JERSEY		
- Atlantic Electric	150	200,000
- Jersey Central Power & Light	300	250,000
- Orange & Rockland Utilities	250	no limit
- Public Service Electric & Gas	125-250	no limit

(Continued)

STATE - Electric Utility	CASH INCENTIVE $/kW Shifted	Maximum
NEW YORK		
- Central Edison Gas & Electric	25/Ton-Hr	equip cost
- Consolidated Edison Co.	600	no limit
- Long Island Lighting Co.	300-500	no limit
- New York State Electric & Gas	113	no limit
- Orange & Rockland Utilities	250	no limit
- Rochester Gas & Electric	200-300	70,000
NORTH DAKOTA		
- Northern States Power	400/ton	no limit
OHIO		
- Cincinnati Gas L Electric	150	no limit
- Toledo Edison	200-250	—
OKLAHOMA		
- Oklahoma Gas & Electric	125-200	225,000
PENNSYLVANIA		
- Metropolitan Edison	100-250	40,000
- Orange & Rockland Utilities	250	no limit
- Pennsylvania Electric	250	no limit
- Pennsylvania Power & Light	100	no limit
- Philadelphia Electric	100-200	25,000
SOUTH DAKOTA		
- Northern States Power	400/ton	no limit
TEXAS		
- Austin Electric Department	300	150,000
- El Paso Electric Company	200	no limit
- Gulf States Utilities	250	—
- Houston Lighting & Power	350	—
- Texas Utilities (Dallas Power, Texas Electric Service, and Texas Power & Light)	125-250	no limit
WISCONSIN		
- Madison Gas & Electric	60 - 80	no limit
- Northern States Power	175	no limit
- Wisconsin Electric Power	350	no limit

*Note: Some states have additional programs not listed here and some of the listed programs have additional limitations.

CHAPTER 20

CODES, STANDARDS & LEGISLATION

ALBERT THUMANN, P.E., CEM

Association of Energy Engineers
Atlanta, GA

This chapter presents an historical perspective on key codes, standards, and regulations which have impacted energy policy and are still playing a major role in shaping energy usage. The Energy Policy Act of 1992 is far reaching and its implementation is impacting electric power deregulation, building codes and new energy efficient products. Sometimes policy makers do not see the far reaching impact of their legislation. The Energy Policy Act for example has created an environment for retail competition. Electric utilities will drastically change the way they operate in order to provide power and lowest cost. This in turn will drastically reduce utility sponsored incentive and rebate programs which have influenced energy conservation adoption.

20.1 THE ENERGY POLICY ACT OF 1992

This comprehensive legislation is far reaching and impacts energy conservation, power generation, and alternative fuel vehicles as well as energy production. The federal as well as private sectors are impacted by this comprehensive energy act. Highlights are described below:

Energy Efficiency Provisions
Buildings

- Requires states to establish minimum commercial building energy codes and to consider minimum residential codes based on current voluntary codes.

Utilities

- Requires states to consider new regulatory standards that would: require utilities to undertake integrated resource planning; allow efficiency programs to be at least as profitable as new supply options; and encourage improvements in supply system efficiency.

Equipment Standards

- Establishes efficiency standards for: commercial heat-

ing and air-conditioning equipment; electric motors; and lamps.

- Gives the private sector an opportunity to establish voluntary efficiency information/labeling programs for windows, office equipment and luminaires, or the Dept. of Energy will establish such programs.

Renewable Energy

- Establishes a program for providing federal support on a competitive basis for renewable energy technologies. Expands program to promote export of these renewable energy technologies to emerging markets in developing countries.

Alternative Fuels

- Gives Dept. of Energy authority to require a private and municipal alternative fuel fleet program starting in 1998. Provides a federal alternative fuel fleet program with phased-in acquisition schedule; also provides state fleet program for large fleets in large cities.

Electric Vehicles

- Establishes comprehensive program for the research and development, infrastructure promotion, and vehicle demonstration for electric motor vehicles.

Electricity

- Removes obstacles to wholesale power competition in the Public Utilities Holding Company Act by allowing both utilities and non-utilities to form exempt wholesale generators without triggering the PUHCA restrictions.

Global Climate Change

- Directs the Energy Information Administration to establish a baseline inventory of greenhouse gas emissions and establishes a program for the voluntary reporting of those emissions. Directs the Dept. of Energy to prepare a report analyzing the strategies for mitigating global climate change and to develop a least-cost energy strategy for reducing the generation of greenhouse gases.

Research and Development

- Directs the Dept. of Energy to undertake research and development on a wide range of energy technologies, including: energy efficiency technologies, natural gas end-use products, renewable energy resources, heating and cooling products, and electric vehicles.

20.2 STATE CODES

More than three quarters of the states have adopted ASHRAE Standard 90-80 as a basis for their energy efficiency standard for new building design. The ASHRAE Standard 90-80 is essentially "prescriptive" in nature. For example, the energy engineer using this standard would compute the average conductive value for the building walls and compare it against the value in the standard. If the computed value is above the recommendation, the amount of glass or building construction materials would need to be changed to meet the standard.

Most states have initiated "Model Energy Codes" for efficiency standards in lighting and HVAC. Probably one of the most comprehensive building efficiency standards is California Title 24. Title 24 established lighting and HVAC efficiency standards for new construction, alterations and additions of commercial and noncommercial buildings.

ASHRAE Standard 90-80 has been updated into two new standards:

ASHRAE 90.1-1989 Energy Efficient Design of New Buildings Except New Low-Rise Residential Buildings

ASHRAE 90.2 Energy Efficient Design of New Low Rise Residential Building

The purposes of ASHRAE Standard 90.1-1989 are:

(a) set minimum requirement for the energy efficient design of new buildings so that they may be constructed, operated, and maintained in a manner that minimizes the use of energy without constraining the building function nor the comfort or productivity of the occupants.
(b) provide criteria for energy efficient design and methods for determining compliance with these criteria.
(c) provide sound guidance for energy efficient design.

In addition to recognizing advances in the performance of various components and equipment, the Standard encourages innovative energy conserving designs. This has been accomplished by allowing the building designer to take into consideration the dynamics that exist between the many components of a building through use of the System Performance Method or the Building Energy Cost Budget Method compliance paths. The standard, which is cosponsored by the Illuminating Engineering Society of North America, includes an extensive section on lighting efficiency, utilizing the Unit Power Allowance Method.

20.3 MODEL ENERGY CODE

The American Building Officials (ABO) has upgraded their Model Energy Code (MEC). The MEC is a widely accepted model energy standard that is developed in an open process involving code officials, builders, manufacturers, government agencies and researchers. In a July 1991 report by the Alliance to Save Energy it found 34 states currently using less stringent codes and standards. The report recommended:

• States should review their existing building code standards. It is recommend that states use the MEC as a minimum baseline for upgrading their energy standards for new housing.

• The federal government should help states upgrade their energy standards. The Department of Energy should increase its efforts to provide technical assistance to states, and it should do further studies of the benefit of various model building energy standards and codes.

• Model Energy Code standards should be strengthened in the southern region. The studies found that in many southern states the MEC's thermal requirements were too lax, especially considering the heavy use of electricity for air conditioning in new homes.

• The MEC's standards for multifamily homes should be reassessed. The study found that MEC standards for multifamily housing, while beneficial overall, showed less cost-effectiveness and affordability benefits than single-family housing. This suggests that the MEC's multifamily standards should be refined and strengthened where necessary.

20.4 NATIONAL ENERGY CONSERVATION POLICY ACT (NECPA)—1978.

This act is probably the most important legislation passed. Programs initiated by NECPA are still in progress today. One of the first successful aspects of NECPA is the: "Matching Grants Program for Schools & Hospitals."

The Matching Grants Program for Schools and Hospitals is a voluntary program to assist these non-profit institutions in saving energy.

Authorized by Part 1, Title III of the National Energy Conservation Policy Act of 1978 (NECPA) and administered by the Department of Energy's Institutional Conservation Programs (ICP) through state energy offices, the ICP provides grants for detailed energy analyses and for installation of energy-saving capital improvements to eligible institutions. Participation in the grant program requires a 50% match of funds from recipient institutions, except in hardship cases.

Program Activities

NECPA divided the program into two main phases: Phase I, Preliminary Energy Audits and Energy Audits, and Phase II, Technical Assistance and Energy Conservation Measures. Funding for Phase I activities has expired; however, the program still administers the Technical Assistance/Energy Conservation Measure phase.

The Preliminary Energy Audit (PEA) was a data-gathering activity to determine basic information about the number, size, type, and rate of energy use in eligible buildings and the actions taken to conserve energy. The Energy Audit (EA) consisted of an on-site visit which confirmed and expanded upon information gathered in the PEA and provided further information to establish a "building profile."

The Technical Assistance (TA) Audit consists of detailed engineering analyses and reports specific costs, payback periods, and projected energy savings possible from the purchase and installation of devices or systems, or from modification to building operations and the building shell. Examples of this type of activity are the installation of storm windows, insulation, solar hot water or space heating systems, or automatic setback devices.

Energy Conservation Measure (ECM) grants provide for the actual purchase and installation of energy measures recommended as a result of the TA Audit.

20.5 INDOOR AIR QUALITY (IAQ) STANDARDS[1]

Indoor Air Quality (IAQ) is an emerging issue of concern to building managers, operators, and designers.

Recent research has shown that indoor air is often less clean than outdoor air and federal legislation has been proposed to establish programs to deal with this issue on a national level. This, like the asbestos issue, will have an impact on building design and operations. Americans today spend long hours inside buildings, and building operators, managers and designers must be aware of potential IAQ problems and how they can be avoided.

IAQ problems, sometimes termed "Sick Building Syndrome," have become an acknowledged health and comfort problem. Buildings are characterized as sick when occupants complain of acute symptoms such as headache, eye, nose and throat irritation, dizziness, nausea, sensitivity to odors and difficulty in concentrating. The complaints may become more clinically defined so that an occupant may develop an actual building-related illness that is believed to be related to IAQ problems.

The most effective means to deal with an IAQ problem is to remove or minimize the pollutant source, when feasible. If not, dilution and filtration may be effective.

Dilution (increased ventilation) is to admit more outside air to the building ASHRAE's 1981 standard recommended 5 CFM/person outside air in an office environment. The new ASHRAE ventilation standard, 62-1989, now requires 20 CFM/person for offices if the prescriptive approach is used.

Increased ventilation will have an impact on building energy consumption However, this cost need not be severe. If an airside economizer cycle is employed and the HVAC system is controlled to respond to IAQ loads as well as thermal loads, 20 CFM/person need not be adhered to and the economizer hours will help attain air quality goals with energy savings at the same time. Incidentally, it was the energy cost of treating outside air that led to the 1981 standard. The superseded 1973 standard recommended 15-25 CFM/person.

Energy savings can be realized by the use of improved filtration in lieu of the prescriptive 20 CFM/person approach. Improved filtration can occur at the air handler, in the supply and return ductwork, or in the spaces via self-contained units. Improved filtration can include enhancements such as ionization devices to neutralize airborne biological matter and to electrically charge fine particles, causing them to agglomerate and be more easily filtered.

The Occupational Safety and Health Administration (OSHA) announced a proposed rule on March 25, 1994 that would regulate indoor air quality (IAQ) in workplaces across the nation. The proposed rule addresses all indoor contaminants but a significant step

[1]Source: Indoor Air Quality: Problems & Cures, M. Black & W. Robertson, Presented at 13th World Energy Engineering Congress.

would ban all smoking in the workplace or restrict it to specially designed lounges exhausted directly to the outside. The smoking rule would apply to all workplaces while the IAQ provisions would impact "non-industrial" indoor facilities.

There is growing consensus that the most promising way to achieve good indoor air quality is through contaminant source control. Source control is more cost effective than trying to remove a contaminant once it has disseminated into the environment. Source control options include chemical substitution or product reformulating, product substitution, product encapsulation, banning some substances or implementing material emission standards. Source control methods except emission standards are incorporated in the proposed rule.

20.6 REGULATIONS & STANDARDS IMPACTING CFCs

For years, chlorofluorocarbons (CFCs) have been used in air-conditioning and refrigeration systems designed for long-term use. However, because CFCs are implicated in the depletion of the earth's ozone layer, regulations will require the complete phaseout of the production of new CFCs by the turn of the century. Many companies, like DuPont, are developing alternative refrigerants to replace CFCs. The need for alternatives will become even greater as regulatory cutbacks cause continuing CFC shortages.

Air-conditioning and refrigeration systems designed to operate with CFCs will need to be retrofitted (where possible) to operate with alternative refrigerants so that these systems can remain in use for their intended service life.

DuPont and other companies are commercializing their series of alternatives—hydrochlorofluorocarbon (HCFC) and hydrofluorocarbon (HFC) compounds. See Table 20.1.

The Montreal Protocol which is being implemented by the United Nations Environment Program (UNEP) is a worldwide approach to the phaseout of CFCs. A major revision to the Montreal Protocol was implemented at the 1992 meeting in Copenhagen which accelerated the phaseout schedule.

The reader is advised to carefully consider both the "alternate" refrigerants entering the market place *and* the alternate technologies available. Alternate refrigerants come in the form of HCFCs and HFCs. HFCs have the attractive attribute of having no impact on the ozone layer (and correspondingly are not named in the Clean Air Act). Alternative technologies include absorption and ammonia refrigeration (established technologies

since the early 1900's), as well as desiccant cooling.

Taxes on CFCs originally took effect January 1, 1990. The Energy Policy Act of 1992 revised and further increased the excise tax effective January 1, 1993.

Another factor to consider in ASHRAE Guidelines 3-1990—Reducing Emission of Fully Halogenated Chlorofluorocarbon (CFC) Refrigerants in Refrigeration and Air-Conditioning Equipment and Applications:

> The purpose of this guideline is to recommend practices and procedures that will reduce inadvertent release of fully halogenated chlorofluorocarbon (CFC) refrigerants during manufacture, installation, testing, operation, maintenance, and disposal of refrigeration and air-conditioning equipment and systems.

The guideline is divided into 13 sections. Highlights are as follows:

The Design Section deals with air-conditioning and refrigeration systems and components and identifies possible sources of loss of refrigerants to atmosphere. Another section outlines refrigerant recovery reuse and disposal. The Alternative Refrigerant section discusses replacing R11, R12, R113, R114, R115 and azeotropic mixtures R500 and R502 with HCFCs such as R22.

20.7 CLEAN AIR ACT AMENDMENT

On November 15, 1990, the new Clean Air Act (CAA) was signed by President Bush. The legislation includes a section entitled Stratospheric Ozone Protection (Title VI). This section contains extraordinarily comprehensive regulations for the production and use of CFCs, halons, carbon tetrachloride, methyl chloroform, and HCFC and HFC substitutes. These regulations will be phased in over the next 40 years, and they will impact every industry that currently uses CFCs.

The seriousness of the ozone depletion is such that as new findings are obtained, there is tremendous political and scientific pressure placed on CFC end-users to phase out use of CFCs. This has resulted in the U.S., under the signature of President Bush in February 1992, to have accelerated the phaseout of CFCs.

20.8 REGULATORY AND LEGISLATIVE ISSUES IMPACTING COGENERATION & INDEPENDENT POWER PRODUCTION[2]

Federal, state and local regulations must be addressed when considering any cogeneration project. This

---

[2]Source: *Georgia Cogeneration Handbook*, published by the Governor's Office of Energy Resources.

Table 20.1 Candidate Alternatives for CFCs in Existing Cooling Systems

CFC	Alternative	Potential Retrofit Applications
CFC-11	HCFC-123	Water and brine chillers; process cooling
CFC-12	HFC-134a or Ternary Blends	Auto air conditioning; medium temperature commercial food display and transportation equipment; refrigerators/freezers; dehumidifiers; ice makers; water fountains
CFC-114	HCFC-124	Water and brine chillers
R-502	HFC-125	Low-temperature commercial food equipment

section will provide an overview of the federal regulations that have the most significant impact on cogeneration facilities.

20.8.1 Federal Power Act

The Federal Power Act asserts the federal government's policy toward competition and anti-competitive activities in the electric power industry. It identifies the Federal Energy Regulatory Commission (FERC) as the agency with primary jurisdiction to prevent undesirable anti-competitive behavior with respect to electric power generation. Also, it provides cogenerators and small power producers with a judicial means to overcome obstacles put in place by electric utilities.

20.8.2 Public Utility Regulatory Policies Act (PURPA)

This legislation was part of the 1978 National Energy Act and has had perhaps the most significant effect on the development of cogeneration and other forms of alternative energy production in the past decade. Certain provisions of PURPA also apply to the exchange of electric power between utilities and cogenerators.

PURPA provides a number of benefits to those cogenerators who can become Qualifying Facilities (QFs) under the act. Specifically, PURPA

- Requires utilities to purchase the power made available by cogenerators at reasonable buy-back rates. These rates are typically based on the utilities' cost.

- Guarantees the cogenerator or small power producer interconnection with the electric grid and the availability of backup service from the utility.

- Dictates that supplemental power requirements of cogenerator must be provided at a reasonable cost.

- Exempts cogenerators and small power producers

from federal and state utility regulations and associated reporting requirements of these bodies.

In order to assure a facility the benefits of PURPA, a cogenerator must become a Qualifying Facility. To achieve Qualifying Status, a cogenerator must generate electricity and useful thermal energy from a single fuel source. In addition, a cogeneration facility must be less than 50% owned by an electric utility or an electric utility holding company. Finally, the plant must meet the minimum annual operating efficiency standard established by FERC when using oil or natural gas as the principal fuel source. The standard is that the useful electric power output plus one half of the useful thermal output of the facility must be no less than 42.5% of the total oil or natural gas energy input. The minimum efficiency standard increases to 45% if the useful thermal energy is less than 15% of the total energy output of the plant.

20.8.3 Natural Gas Policy Act (NGPA)

The major objective of this legislation was to create a deregulated national market for natural gas. It provides for incremental pricing of higher cost natural gas supplies to industrial customers who use gas, and it allows the cost of natural gas to fluctuate with the cost of fuel oil. Cogenerators classified as Qualifying Facilities under PURPA are exempt from the incremental pricing schedule established for industrial customers.

20.8.4 Resource Conservation and Recovery Act of 1976 (RCRA)

This act requires that disposal of non-hazardous solid waste be handled in a sanitary landfill instead of an open dump. It affects only cogenerators with biomass and coal-fired plants. This legislation has had little, if any, impact on oil and natural gas cogeneration projects.

20.8.5 Public Utility Holding Company Act of 1935

The Public Utility Holding Company Act of 1935 (the 35 Act) authorizes the Securities and Exchange Commission (SEC) to regulate certain utility "holding companies" and their subsidiaries in a wide range of corporate transactions.

The Energy Policy Act of 1992 creates a new class of wholesale-only electric generators—"exempt wholesale generators" (EWGs)—which are exempt from the Public Utility Holding Company Act (PUHCA). The Act dramatically enhances competition in U.S. wholesale electric generation markets, including broader participation by subsidiaries of electric utilities and holding companies. It also opens up foreign markets by exempting companies from PUHCA with respect to retail sales as well as wholesale sales.

20.8.6 Moving towards a deregulated electric power marketplace

The Energy Policy Act set into motion a widespread movement for utilities to become more competitive. Retail wheeling proposals were set into motion in states such as California, Wisconsin, Michigan, New Mexico, Illinois and New Jersey. There are many issues involved in a deregulated power marketplace and public service commission rulings and litigation will certainly play a major role in the power marketplace of the future. Deregulation has already brought about several important developments:

• Utilities will need to become more competitive. Downsizing and minimization of costs including elimination of rebates are the current trend. This translates into lower costs for consumers. For example Southern California Edison announced that the system average price will be reduced from 10.7 cents/kWh to lower than 10 cents by the year 2000. This would be a 25% reduction after adjusting for inflation.

• Utilities will merge to gain a bigger market share. Wisconsin Electric Power Company recently announced a merger with Northern States Power; this is the largest merger of two utilities of its kind in the nation resulting in a savings of $2 billion over 10 years.

• Utilities are forming new companies to broaden their services. Energy service companies, financial loan programs and purchasing of related companies are all part of the new utility strategy.

• In 1995 one hundred power marketing companies have submitted applicants to FERC. Power marketing companies will play a key role in brokering power between end users and utilities in different states and in purchasing of new power generation facilities.

• Utilities will need to restructure to take advantage of deregulation. Generation Companies may be split away from other operating divisions such as transmission and distribution. Vertical disintegration will be part of the new utility structure.

• Utilities will weigh the cost of repowering and upgrading existing plants against purchasing power from a third party.

20.9 OPPORTUNITIES IN THE SPOT MARKET[3]

Basics of the Spot Market

A whole new method of contracting has emerged in the natural gas industry through the spot market. The market has developed because the Natural Gas Policy Act of 1978 (NGPA) guaranteed some rights for end-users and marketers in the purchasing and transporting of natural gas. It also put natural gas supplies into a more competitive position with deregulation of several categories.

The Federal Energy Regulatory Commission (FERC) provided additional rulings that facilitated the growth of the spot market. These rulings included provisions for the Special Marketing Programs in 1983 (Order 2346) and Order 436 in 1985, which encouraged the natural gas pipelines to transport gas for end-users through blanket certificates.

The change in the structure of markets in the natural gas industry has been immense in terms of both volumes and the participants in the market. By year-end 1986, almost 40% of the interstate gas supply was being transported on a carriage basis. Not only were end-users participating in contract carriage, but local distribution companies (LDCs) were accounting for about one half of the spot volumes on interstate pipelines.

The "spot market" or "direct purchase" market refers to the purchase of gas supplies directly from the producer by a marketer, end-user or LDC. (The term "spot gas" is often used synonymously with "best efforts gas," "interruptible gas," "direct purchase gas" and

[3]Source: *New Opportunities for Purchasing Natural Gas*, Fairmont Press, Atlanta, GA

"self-help gas.") This type of arrangement cannot be called new because the pipelines have always sold some supplies directly to end-users.

The new market differs from the past arrangements in terms of the frequency in contracting and the volumes involved in such contracts. Another characteristic of the spot market is that contracts are short-term, usually only 30 days, and on an interruptible basis. The interruptible nature of spot market supplies is an important key to understanding the operation of the spot market and the costs of dealing in it. On both the production and transportation sides, all activities in transportation or purchasing supplies are on a "best efforts" basis. This means that when a cold snap comes the direct purchaser may not get delivery on his contracts because of producer shutdowns, pipeline capacity and operational problems or a combination of these problems. The "best efforts" approach to dealing can also lead to problems in transporting supplies when demand is high and capacity limited.

FERC's Order No. 436

The impetus for interstate pipeline carriage came with FERC's Order No. 436, which provided more flexibility in pricing and transporting natural gas. In passing the 1986 ruling, FERC was attempting to get out of the day-to-day operations of the market and into more generic rule making. More significantly, FERC was trying to get interstate pipelines out of the merchant business into the transportation business—a step requiring a major restructuring of contracting in the gas industry.

FERC has expressed an intent to create a more competitive market so that prices would signal adjustments in the markets. The belief is that direct sales ties between producers and end-users will facilitate market adjustments without regulatory requirements clouding the market. As more gas is deregulated, FERC reasoned that natural gas prices will respond to the demand: Lower prices would assist in clearing excess supplies; then as markets tightened, prices would rise drawing further investment into supply development.

FERC Order No. 636

Order 636 required significant "Restructuring" in interstate pipeline services, starting in the fall of 1993. The original Order 636:
- Separates (unbundles) pipeline gas sales from transportation
- Provides open access to pipeline storage
- Allows for "no notice" transportation service
- Requires access to upstream pipeline capacity

- Uses bulletin boards to disseminate information
- Provides for a "capacity release" program to temporarily sell firm transportation capacity
- Pregrants a pipeline the right to abandon gas sales
- Bases rates on straight fixed variable (SFV) design
- Passes through 100% of transition costs in fixed monthly charges to firm transport customers

FERC Order No. 636A

Order 636A makes several relatively minor changes in the original order and provides a great deal of written defense of the original order's terms. The key changes are:

- Concessions on transport and sales rates for a pipeline's traditional "small sales" customers (like municipalities).
- The option to "release" (sell) firm capacity for less than one month—*without posting it on a bulletin board system or bidding.*
- Greater flexibility in designing special transportation rates (i.e., offpeak service) while still requiring overall adherence to the straight fixed variable rate design.
- Recovery of 10% of the transition costs from the interruptible transportation customers (Part 284).

Court action is still likely on the Order. Further, each pipeline will submit its own unique tariff to comply with the Order. As a result, additional changes and variations are likely to occur.

20.10 THE CLIMATIC CHANGE ACTION PLAN

The Climatic Change Action Plan was established April 21, 1993 and includes the following:

- Returns U.S. greenhouse gas emissions to 1990 levels by the year 2000 with cost effective domestic actions.

- Includes measures to reduce all significant greenhouse gases, carbon dioxide, methane, nitrous oxide, hydrofluorocarbons and other gases.

20.11 SUMMARY

The dynamic process of revisions to existing codes plus the introduction of new legislation will impact the energy industry and bring a dramatic change. Energy conservation and creating new power generation supply options will be required to meet the energy demands of the twenty-first century.

CHAPTER 21

NATURAL GAS PURCHASING

CAROL FREEDENTHAL

Jofree Corporation
Houston, TX

21.1 PREFACE

When the editors of this book asked for chapter authors to review and revise their individual works, it became apparent in revising this chapter of Natural Gas Purchasing, more than a mere revision was necessary. A major remodeling from front to back was required because of the changes occurring in the U.S. natural gas industry from the time the first edition was written and published. Something greater than a simple revision of the previous text was necessary to match the changes that have occurred in the industry in the last five years!

The U.S. natural gas industry from 1985 through 1995—10 years—a decade—has gone through a transformation of tremendous magnitude. The roots of this change started earlier than 1985. Significant events prior to this time played a major role in the transformation. But, the actual changes and transformation occurred in the decade from 1985 through 1995. While more will be covered in the chapter on these changes and the impact on both sellers and buyers of natural gas, it is important before starting the chapter to understand the magnitude of this transition.

The transition was to convert the country's natural gas industry—big industry with total revenues in 1994 close to $80 Billion—from a government economically regulated, sales dominated industry to a free market, service business. Prior to the transformation, interstate natural gas prices were regulated by the Federal government. The transportation sector served as both the means of getting the gas from the field to the city gate and the merchant for selling the commodity to local distributors and large gas users. Gas was sold at the wellhead to transporters who sold the gas to major consumers and distribution companies for resales to the bulk of the consumers. "Wellhead" natural gas pricing was essentially insensitive to market demands and did not respond when gas supplies were short such as the conditions in the mid-1970s.

The big change was to remove the merchant function from the transporters and open up the marketing of gas to everyone. Further, the additional services the transporters supplied like gathering, processing and storage and were a part of the transportation tariff whether a buyer or seller needed the additional services. These were "unbundled" so the user only paid for transportation. Transportation services became transportation only. Further, transportation is available to any buyer or seller, so a large gas consumer in New York can make arrangements with a marketing company or a producer to bring gas to his facility. This and the wellhead force decontrol were the transitions and changes occurring in the 1985-1995 period.

Some new elements coming out of this broad spectrum of change include a wide range of participants and new instruments in the buying and selling of natural gas. The most outstanding new participant is the group called "marketers." Where before the merchant function was a part of the transportation system whether it be the pipelines bringing gas from the production field to the city gate or the distribution company transporting the gas to the consumer, now a new group handles the marketing of natural gas. The marketer buys and sells gas, arranges transportation, and handles risk management for either buyer or seller. In addition to marketers, the industry now has specific companies handling processing, storage, and hub management. In addition to these, the industry now has certain instruments to help with risk management. The change in price stability coming from a regulated to a competitive market place gives need for some type of instrument to handle the price risk function. Futures and options in the gas industry make it possible to hedge against price volatility.

There is still fine tuning to be done in this big a reorganization. The handling of gas at the city gate and sales by distribution companies is still going through the same "open access" transition the transporting pipelines went through in the last decade. Some of the distributors will allow the use of their pipes for transportation by anyone, while some areas may restrict the use to the distribution company only. The agencies to rule on this are the countless state and local Public Utility Commissions having jurisdiction over the distribution companies. With hundreds of distribution companies in the U.S., the transition will not come overnight in this end of the business.

With the change of the gas industry from a controlled, non-responsive business to the market oriented, service supplying industry additional changes occurred. Now there is an interchange with financial service companies to make gas available as a commodity. Futures markets are operating already with Henry Hub in Louisiana and Permian Basin in Texas as the contract locations. Additional financial services are available from other companies serving this market. Computer and systems companies are taking an active role in modernizing the gas industry and providing the system so gas now can be sold on as short a basis as a few hours.

The purpose of this section is to give the gas or fuel buyer for any operation, whether a first time experience or on-going job, the background and information to seek out the necessary supplies of natural gas at the lowest possible price consistent with the reliability needed. Buying natural gas whether for heating a large apartment complex or fuel for a manufacturing facility does require a good background and solid information on the natural gas industry.

21.2 INTRODUCTION

Natural gas is the nation's second largest source of fuel and hydrocarbon feedstock. With plentiful supplies and many economic and environmental advantages, natural gas is the fuel of choice for many applications. As a fuel for industry, for heating and as a chemical feedstock, there is nothing better than clean burning natural gas. For residential and commercial uses, the security of supply and efficiency in supplying, natural gas makes it the ideal fuel. Natural gas has tremendous potential to gain even greater use in the generation of electricity as nuclear plants age and are retired. Petroleum products have already lost major market share to other fuels for electric generation. Even the economically and supply-stable coal will lose some market share in electric generation to natural gas because of its clean burning efficiency and easier transportation.

In recent years, use of natural gas has gained momentum because of changes in the way gas is marketed. With economic regulations removed from the wellhead pricing of natural gas, the industry has become market sensitive. Supplies increase or decrease in tune with economic considerations. With no wellhead pricing rules,

when gas supplies become tight, the increase in gas value quickly generates additional supplies. While at times there is over supply, the decrease in supply also follows through to the natural gas producer. The natural gas industry has come into its prime as a major supplier of energy for residential, business, industry and electric generating demands.

Today, with the plentiful supply, ease of transportation, and marketing and financial systems in place to make gas competitive with other fuels, natural gas markets are taking advantage of the changed conditions. New markets are developing to expand the use of the commodity. The previous inordinate amount of government regulation on natural gas at the wellhead severely impacted the availability of gas at times making severe shortages as seen in the mid-1970s. This is no longer the case. As the gas supply today is free to respond to economic incentives, the fears of supply shortages because of the lack of sufficient field development have diminished.

The expanding new markets include the use of natural gas for transportation fuels, expanded use in the electric generation industry, and use of natural gas for industrial and commercial air conditioning. As environmental concerns increase and the reliability and availability of supply continue to improve, the future looks bright for natural gas.

To take advantage of the uses of natural gas, what are the important elements for the fuels' buyer in obtaining natural gas supplies? Of the primary fuels available to consumers, natural gas does have some distinct attributes making natural gas purchasing different from other fuels. A big example of this difference can be seen in just one illustration. From the time the natural gas is produced, when it is brought out from the ground from depths anywhere from a couple of hundred feet to two to three miles, to the time it is consumed at either the burner tip, engine or chemical reactor, it is never seen, nor smelled, nor touched. From the time natural gas is produced until the time it is used, it travels in enclosed pipes through the full journey!

Not only will it help the gas buyer to know where and how natural gas is produced and moved from wellhead to market, but a knowledge of the economic parameters of the gas business will help in working with marketing and trading people to get needed gas supplies. Further, today with active futures markets for natural gas, depending on the nature of the buyer's needs and requirements for fuel, knowledge of the financial markets and commodity trading can be helpful as well as in securing natural gas at the lowest possible price consistent with supply security and service requirements.

To help current and future gas buyers, this chapter will cover how natural gas is produced, the origin of the gas industry, moving natural gas from field to burner tip, the impact of remaining economic and operational regulatory control on the natural gas industry, natural gas prices, sources of information on the natural gas industry, buying natural gas, and an outlook for how the natural gas industry will evolve with the changing picture of energy supply in the U.S. Especially important to the gas buyer will be the information on gas prices and the discussion on gas production, transporting gas from wellhead to city gate, gas distribution and marketing. The section on gas buying will include information on natural gas contracts. Gas marketing companies will be discussed and major ones listed to help the buyer in seeking gas supplies. Also, within the pricing section, discussion on natural gas futures and forms of risk management will aid the buyer in developing programs to cover the risk management of gas purchasing.

The natural gas business has made the change from a simple, sales oriented commodity business to a highly sophisticated, market oriented and financially directed industry. Figure 21.1 is a simplified schematic showing the natural gas industry from field to burner tip. This is the "map" of the industry and shows the many parts making up the $80 Billion (B) business. The major sectors of the industry—production, marketing, transportation, and distribution revenues generated by each sector are shown graphically in Figure 21.2. The amounts shown for the marketing, transportation, and distribution are the value added by the sector and is not the full revenue produced by this sector since the gas itself has developed its own value. Many opportunities arise in the business for the savvy gas trader to increase his margin either as a gas buyer, trader, or consumer. Knowledge of the industry—knowing the commodity, the way it is transported and "packaged," how financial areas impact pricing—all go hand-in-hand in making natural gas purchases.

21.3 ORIGIN OF THE U.S. NATURAL GAS INDUSTRY

The term " natural gas industry" includes the people, equipment, and systems starting at the wells where the natural gas is produced, gathered and processed to storage or to the pipelines for transporting the gas to the market area storage or to the distribution systems to deliver the gas to the consuming burner tip,

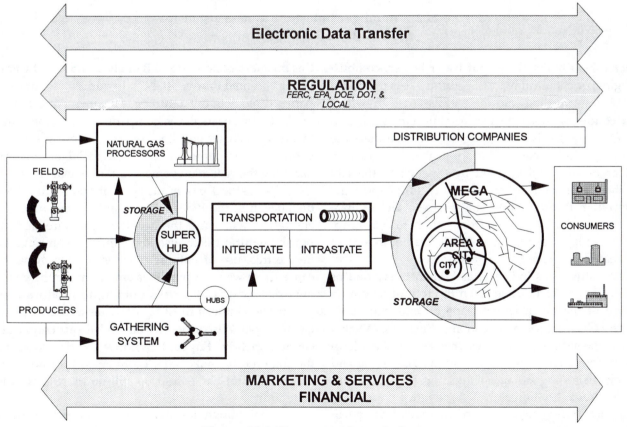

Figure 21.1 New natural gas industry.

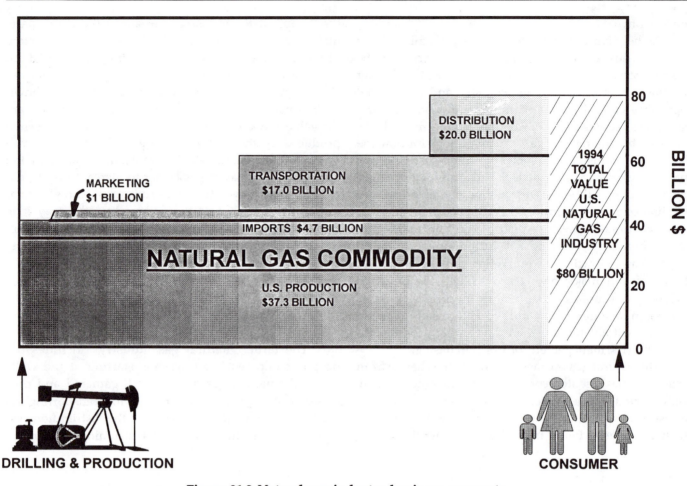

Figure 21.2 Natural gas industry business segments.
(1992 U.S. Gas Market)

whether it be a boiler flame, hot water heater, combustion engine, or chemical reactor. Natural gas is produced in the field by drilling into the earth's crust anywhere from a couple of thousand feet to almost five miles in depth. Once the gas is found and the wells completed to bring the gas to the earth's surface, it is treated or processed as needed and shipped by pipeline to the city gate where the distribution company takes over the transportation through additional pipes to the ultimate consumer. As stated earlier, from origin to consumption~ the gas is always in a pipeline if it is not in a plant being treated or processed or in storage.

The natural gas industry, as known today and as shown in Figure 21.1, while often thought of as only a step-child of the much larger crude oil and petroleum industry, is in itself a major economic business. What is known as the natural gas business, consisting of exploration and production companies, treating, processing and gathering businesses, storage facilities, and pipeline companies for transportation as well as separate companies for local distribution has grown to be a major business in the U.S. industrial spectrum. In recent years, the value of

this business is close to $80 B or about 1.5% of the annual Gross National Product (GNP). In addition to those companies directly in the natural gas business, credit must be given to those ancillary companies taking part in the business such as gas liquids companies, systems and computer companies and the financial organizations handling the risk management and financial areas.

The natural gas industry as it has matured to today, began in the 1930s when seamless steel pipe strong enough to withstand the necessary pressures for transporting natural gas thousands of miles became available. Before this, use of natural gas was on a local basis because there were no pipelines able to move the gas over long distances. Today, it is nothing to produce natural gas in the southern most tip of Texas and transport the gas to upper Wisconsin or to the northeast corner of the country in New England. Without the development of the material to make safe, economical transportation a reality, natural gas would be still no more than a local commodity.

Even though the modern gas industry started in the 1930s, actually, the U.S. gas industry is one of the oldest businesses on record. Its roots begin early in American history when synthetic gas for fuel and city lighting called "manufactured" gas was produced in many of the young country's large municipalities. Using coal and steam, a low Btu (British thermal units, a term used in the measurement of energy and representing the amount of energy required to raise the temperature of a pound of water at 60 degrees Fahrenheit one degree; in practicality one Btu is about the amount of energy released when a single wooden match is lit!) gas was produced for centralized city heating and lighting of street lamps.

This industry was one of the earliest of what is commonly known as "public utilities." Only water companies predate the gas industry as public utilities. In 1816, in Baltimore, MD, the first public utility, Gas Light Company of Baltimore, was formed to manufacture and supply gas service. This was five years before the first natural gas well was drilled in 1821 and 14 years before the first railroad company was formed in 1830.

When natural gas became available in the late 19th century in the U.S., it was easy to replace the temperamental mixture sold as manufactured gas. The synthetic gas was both explosive and poisonous while the new "natural gas," a by-product from the oil production fields was safe, non-poisonous and non-explosive. The main problem was moving the natural gas from producing locations to burner tip. The early stages of the industry saw the gas used only close to production sites.

One of the oldest known uses of natural gas was to bring it from the New York oil fields to Fedonia, NY, in 1821. This original use of gas included using a pipeline made from a three quarter inch wooden tube to transport the gas. The first iron pipeline was reportedly built in 1872 in Pennsylvania, to bring associated gas from the oil fields to Titusville, PA.

The long distance, thousand-mile or longer, large pipelines did not begin to come into place until the 1930s after the technology was developed for making large diameter, seamless welded steel pipe in the 1920s. The need for fuels forced the development of long distance natural gas facilities. The growth of the industry was stopped in the forties because of the second world war but picked up with greater intensity afterwards. Today, the country is crossed by a large number of pipelines not only to bring the large resources of natural gas from the Southwest to the consuming regions of the North Atlantic and North Central U.S. but also, to bring the large resources of natural gas in the North from Canada across the country and into the markets of the U.S.

The U.S. gas business has grown from the local supplier near the oil production into a major energy supplier across the nation. With the growth of the Canadian gas industry and the potential of growth for Mexican resources, the U.S. gas industry is really a North American business. And it is a total business in the modern sense. Not only is it a business for supplying energy through the commodity natural gas, but it is complete with ancillary groups supplying electronic data systems, financial instruments, and associated businesses to compliment the basic industry.

21.4 NATURAL GAS PRODUCTION— HOW IT'S PRODUCED

Natural gas, as it is commercially known, is composed mostly of the hydrocarbon gas "methane," the simplest hydrocarbon available. Methane is a naturally occurring gas consisting of a central carbon atom (C) surrounded by four hydrogen atoms (H) with the chemical formula, $CH4$. Its chemical simplicity makes it easily burned with almost complete combustion. The resulting major products are carbon dioxide and water, both environmentally acceptable chemicals. Under normal operating temperatures and pressures, the physical state of the methane is a gas. Since gases are volatile and will diffuse into the air if not contained in an enclosed system, natural gas is confined to a pipeline or storage vessels of varying kinds from the time of coming out of the ground at the well site until being consumed at the burner tip or reactor vessel.

The methane making up natural gas is the product of decay of organic matter when there is not sufficient oxygen available to oxidize the carbon. When sufficient oxygen is available for the chemical combination of methane for full oxidation to occur, the methane is converted to water, H_2O, and carbon dioxide, CO_2. Theory is natural gas deep in the ground is a product of decaying organic matter from past thousands of years of the earth's history. Without sufficient oxygen in the ground, the decaying organic or carbon containing material combined with hydrogen available from the material itself or from water in the ground to form methane. With the other elements available as the organic material decomposed, other compounds like hydrogen sulfide, nitrogen and all kinds of other compounds were formed. In those cases where some oxygen was available, the conversion was to carbon dioxide and many gas streams are found containing from trace amounts of carbon dioxide to quantities as high as 90 to 95 percent of carbon dioxide. Natural gas is produced from well depths as shallow as a few thousand feet to wells in the 20 to 25 thousand foot depths.

Natural gas produced from the ground for commercial use in the U.S. comes from both associated wells, where it is found in combination with crude oil, and non-associated wells where only natural gas is produced from the wells. Roughly 40% of the natural gas produced in the United States comes from associated wells. The remainder of the domestically supplied gas comes from non-associated or gas-only producing wells. The only major difference between gas produced from associated wells and that coming from non-associated wells is the priority granted by state regulatory bodies. Operators are allowed to produce gas from associated wells without any prorationing to insure maximum crude oil production, while states might put restrictions on the production from non-associated wells because of market, conservation, or other conditions.

Natural gas quantities are measured in two sets of units. The volume of the gas at standard conditions is one measure. Basically, at standard conditions of temperature and pressure, how much volume does the gas occupy? In the U.S., this is usually expressed in terms of cubic feet, abbreviated as "cf." Since a single cubic foot is a relatively small measure, when talking of volumes of gas, the usual term is "thousand cubic feet" where a capital "M" is used to signify the thousand. The next larger unit of volume measurement would be designated, million cubic feet and be abbreviated as "MMcf," which is one thousand, thousand cubic feet (Mcf). Larger volumes, a thousand MMcf would be a billion cubic feet (bcf). A trillion cubic feet (tcf) would be a thousand bcf.

Since natural gas is not a pure compound but a mixture of many decaying products, the energy content of each cubic foot of gas at standard conditions would still have varying quantities of energy when burned. The typical unit of measure of energy in the English system is the British thermal units (Btu) which physically is the amount of energy required to raise the temperature of one pound of water one degree Fahrenheit when the water is at 60 degrees Fahrenheit under standard conditions of pressure. A cubic foot of pure natural gas, if there were such a commodity, would contain about 1,000 Btu and likewise, a Mcf of theoretically pure natural gas would contain a MMBtu or a million British thermal units. Natural gas as it comes out of the well might contain anywhere from a couple of hundred Btu to a couple of thousand Btu per cubic foot because the compound known as natural gas is not a pure material. Statistics for an "average" cubic foot of gas produced in the U.S. for commercial markets as dry natural gas would have an average 1,028 Btu/cf. The gas is treated and processed to remove the contaminants which both raise and lower the energy content per cubic foot depending on their chemical identity and oxidation state. Pipeline quality natural gas will contain from 950 to 1150 Btu per cubic foot.

When natural gas is sold at the local distribution level, such as residential or commercial users, another term is frequently used to describe the energy content of the gas. The term is "dekatherm" abbreviated "Dt" and is equivalent to 100 Btu. Ten Dt would make a thousand Btu or one MBtu.

Major natural gas producing areas in the U.S. are Texas, Louisiana, Oklahoma, and New Mexico. These states, including the offshore areas along the Gulf Coast stretching from Alabama to the southern tip of Texas account for over 70% of the gas produced in the country. Figure 21.3 shows the major gas producing areas of the U.S. and Canada for 1994. On a state basis, other states with significant gas production are California, Wyoming, Colorado, New York, Pennsylvania, Alabama, Mississippi, and Michigan. In total, approximately 18 states in the U.S. produce commercial quantities of natural gas based on information supplied by the Federal government's Energy Information Administration (EIA).

The other major supplier of natural gas for U.S. markets is Canada. Major production areas for Canada are British Columbia and Alberta Provinces which in 1994 supplied 12.4 percent of the gas consumed in the U.S. and amounted to 2.69 tcf. This was about 60% of the gas produced in total in Canada. Imports into the U.S. from Canada have grown since 1985, almost tripling by 1994. Canada went through its own "open access" decontrol of natural gas in the mid-1980s and one of the benefits of this was additional gas supplies for the U.S. Canadian gas comes into the United States at seven major locations. These are shown in Figure 21.4. More will be discussed about Canadian imports in the transportation section as the competitiveness of Canadian gas on a geographic basis is dependent on certain aspects of natural gas transportation.

Natural gas produced in the field, either from associated or gas only wells, may be of sufficient quality to be put directly into commercial pipelines for sale and distribution. Some large wells, especially those in the offshore producing areas, are piped directly from the producing platforms to onshore terminals where the gas goes into major transportation pipelines. Rudimentary mechanical equipment is used to knock out entrained gas liquids such as condensate and contaminants such as water at the wellhead. This cleanup is part of the wellhead production and typically, gas from these operations is collected in a "gathering" system for transport to either further treating and processing facilities or to an interstate pipeline for transportation.

After gathering, the gas may be "treated" to re-

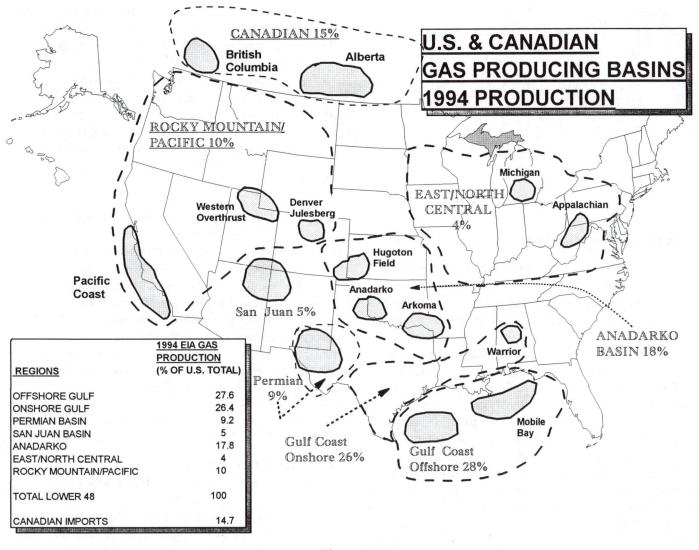

REGIONS	1994 EIA GAS PRODUCTION (% OF U.S. TOTAL)
OFFSHORE GULF	27.6
ONSHORE GULF	26.4
PERMIAN BASIN	9.2
SAN JUAN BASIN	5
ANADARKO	17.8
EAST/NORTH CENTRAL	4
ROCKY MOUNTAIN/PACIFIC	10
TOTAL LOWER 48	100
CANADIAN IMPORTS	14.7

Figure 21.3

move acid portions of the gas such as carbon dioxide or hydrogen sulfide. Natural gas with these contaminants is considered "acid gas" and if the acid components were left in the gas, they could seriously corrode the gas-carrying pipeline. Also, hydrogen sulfide is highly poisonous and must be removed from the gas in case of leakage during transportation. Water must be removed to protect the pipelines from corrosion also. A further threat from having water in the gas is the possibility of it freezing in the winter and stopping the flow of gas to the consumer. Further, moisture, carbon dioxide, and other inert contaminants such as nitrogen or in some rare cases, helium or argon, act to lower the heating value of the gas reducing its effectiveness and requiring greater volumes of gas for the same quantity of heat. The inert gases helium, argon, neon and xenon, which are always in relatively small quantities are removed when economics and content in the stream justify marketing these products.

Even if there is no need to treat the gas to remove the contaminants like carbon dioxide, hydrogen sulfide, and moisture, the gas may need to be "processed" to remove higher carbon chain hydrocarbons known as "natural gas liquids" (NGL). Entrained NGL must be removed from the gas stream because they might condense in the pipeline, interfering with transportation and because of their own economic value. The NGL consists of the higher hydrocarbons ethane (C_2H_6), propane (C_3H_8), butanes (C_4H_{10}), pentanes (C_5H_{12}) and higher hydrocarbons. Each product has specific markets where the ensuing physical and chemical attributes make the individual hydrocarbon more valuable for a specific end use than leaving it in the natural gas stream. Sometimes ethane markets are very low in prices and then the gas producer might reinject the ethane back into the gas stream for sale with the natural gas.

Figure 21.4 Canadian natural gas entry locations.

Once the natural gas is treated and/or processed, it is ready for gathering, if still at the field level, or transportation to the consuming areas. Sometimes the gas is gathered in the field and a combined stream is treated and/or processed with the residue gas then being pipeline quality and ready to go into a pipeline for inter or intrastate shipment to a major consumer or distribution company for further transportation to individual users. Many times, if the gas has no hydrogen sulfide in it, it is moved directly to the transporting pipeline from the wellhead or gathering system and the full stream of that pipeline is processed to remove liquid hydrocarbons and contaminants like moisture and carbon dioxide. Plants treating a full transportation pipeline stream are known as "straddle" plants. Natural gas going into pipelines for transportation must meet certain quality standards to be pipeline quality. Treating and/or processing plants are used to bring the gas to these standards.

When the natural gas stream goes through a treating and/or processing plant, there is a loss in both gas volume and heating value. This is called the "fuel and shrinkage" loss. The gas plant owner charges a fee to the gas owner for the treating and/or processing. Each of the gas owners receives his share of the proceeds from the processing and sale of the extracted NGL less the cost of processing. Usually, the fee for processing is covered by the share of gas liquids the processor is allowed to keep. There is no single standard for charges. Each plant processor negotiates with the gas owners wanting to use its facilities for processing charges to meet pipeline specifications for the gas coming into the plant. Under some conditions, the lower hydrocarbon liquids like ethane and propane cannot be economically justified to be removed and sold as separate products. In these cases, if the quantity contained in the gas stream is not too great to raise the heat content of the gas too high, these products might be left in the gas stream.

Natural gas to be considered pipeline quality must meet certain standards. One of the criteria of natural gas is the heat content of the gas per unit of volume. Typically, pipeline quality gas will run around 1,000 Btu per cubic foot of gas or more typically expressed as 1,000,000 Btu per thousand cubic feet (1 MMBtu/Mcf). Pipeline quality gas can range from 950 to 1150 Btu/cf. The exact amount is measured for the stream as the gas is sold on a Btu basis. Typical other specifications for an interstate pipeline for the transmission of natural gas are given in Table 21.1.

Table 21.1 Specification for natural gas quality.

Contaminants may not exceed the following levels:

20 grains of elemental sulfur per 100 cubic feet
1 grain of hydrogen sulfide per 100 cubic feet
7 pounds of water per million cubic feet
3 percent of carbon dioxide by volume
Other impurities (i.e. oxygen, nitrogen, dirt, gum) if their levels *exceed amounts that the buyer must incur costs to make the gas meet pipeline specifications.*

Source: *Handbook on Gas Contracts*, Thomas G. Johnson, IED Press, Inc., Oklahoma City, OK. 1982, page 63.

21.5 REGULATORY IMPACT ON THE NATURAL GAS INDUSTRY

Until recently, the natural gas industry was tightly controlled with economic regulations on the wellhead price of gas going into interstate and intrastate commerce. Now that the wellhead price is unregulated, there are still considerable federal controls on the interstate shipment of natural gas and on the local distribution to consumers by state and local public utility commissions or similar agencies. These are strictly transportation and storage regulatory controls on natural gas and are basically economic and operational in nature. In addition to these economic controls, there are countless other regulatory agencies impacting the production, transportation, and consumption of the commodity. Some knowledge of the history and the influence regulations have had on the industry is helpful in understanding the business today. Further, some knowledge of the continuing regulatory control aimed directly at natural gas and those regulations associated with it will be helpful also.

The industry expanded from its start-up at the turn of the century supplying local consumers to a major source of energy for the country by shipping gas thousands of miles in the early 1930s. In a 1935 Federal Trade

Commission (FTC) report, recommendations were made for federal regulatory law to be enacted regarding interstate pipelines transporting natural gas. The FTC said the interstate pipelines monopoly of power was damaging to consumers. The Federal agency also claimed individual states could do little to regulate pipelines because of protection by the commerce clause of the United States Constitution.

Congress responded by enacting the Natural Gas Act (NGA) in 1938. This Act, 58 years later, is still the basic law regarding federal natural gas regulations. The act covers the interstate transportation of natural gas, the sale for resale of natural gas in interstate commerce, and the companies engaging in either the transportation and/or resale of gas in interstate commerce. The production and/or gathering of natural gas, transportation and sale of intrastate natural gas were specifically excluded from the Act. Distribution is also excluded because local public service commissions or other state agencies regulate these industries.

The Federal Power Commission (FPC) was given the authority to administer the Act giving it very broad power over the emerging industry. In addition to its authority to establish "just and reasonable" rates for interstate transportation and sales of natural gas, the FPC also had the authority to grant the certificates of "public convenience and necessity" needed for companies to transport and sell gas in interstate commerce. Along with this went the authority for construction and operation of facilities to transport and sell natural gas. Lastly, the FPC had the authority to terminate or change jurisdictional services by requiring abandonment authority from the Commission before a company subject to the authority of the FPC could abandon service. These laws to a great degree are still in effect and play a role in the industry even though tremendous strides were taken in the last 25 years to deregulate the industry.

In the early stages of the Act, it only applied to transportation and downstream sales of the gas. In 1954, a Supreme Court ruling (Phillips Petroleum vs. Wisconsin), changed this jurisdiction interpretation to include wellhead regulation of gas pricing in interstate commerce. The Commission now had the authority to regulate prices of natural gas going into interstate commerce and set up a two-tiered market of interstate and intrastate sales. Intrastate pipelines were under the jurisdiction of state agencies, and wellhead prices for gas in intrastate commerce were not regulated by the federal government.

Controlling the price of interstate natural gas at the wellhead became a very difficult and time consuming operation for the regulatory body, the FPC. It severely restricted the industry and kept natural gas supplies from growing. Each year, a bill was introduced in the U.S. Congress to decontrol natural gas, but none ever got far. Around 1970, the Permian rate case was resolved by the FPC which set the price producers could charge for gas produced in the Permian Basin. Gas prices were set very low by the Commission. So low, many recognized it would be uneconomic to continue to develop gas reserves. This was the start of the serious effort on all fronts to decontrol the price of natural gas at the wellhead.

The problems from controlling the price of gas at the wellhead became serious when the first oil embargo of the early 1970s was put into effect by the Organization of Exporting Countries (OPEC). Crude oil itself escalated in prices from a few dollars per barrel to the mid-teens. Natural gas, because of the wellhead control, could not follow the price increases for gas in interstate commerce. Demand for more gas existed in these years, but the economics forced on the industry by wellhead price control prevented further expansion of the industry. The move to decontrol gas prices went into full force.

The need for the market place to impact gas production at the wellhead was obvious. Operating the industry under the control of the federal agency was like operating in a vacuum. Without the market impacting the price of gas at the wellhead, as it did in the intrastate market where prices went into the $4/MMBtu range, while interstate prices stayed below a $1/MMBtu, industries, schools, and residences faced gas shortages. Natural gas prices were given a boost because supply was so low and drilling had been cut back by producers. In 1976, the FPC passed Orders 770 and 770A which practically overnight tripled the price producers could obtain for new natural gas supplies.

Congress passed the Natural Gas Policy Act of 1978 (NGPA) to start the orderly process of decontrolling natural gas prices. Congress also recognized that to make it a free market, the pipelines had to give up the merchant function and be transporters only. Further, transportation had to be available to everyone. The passage of the NGPA put into effect a series of events to decontrol the price and provide open access transportation. The NGPA also set up the Federal Energy Regulatory Commission (FERC) as the government body to regulate the natural gas industry as well as other energy industries.

The FERC in turn mandated a series of orders to implement Congress' wishes as spelled out in the NGPA. The last of these orders, FERC Order 636, was passed in late 1993 and was the major order changing the industry into the form it is today.

While the government through Congress and the regulatory agencies was active in restructuring the natural gas industry, the market place was really the trend setter. The regulatory agencies and the government are always reacting to changes demanded by the market. Many times the regulatory changes come after the market has made the innovations to meet the business demands. The government has never taken the initiative to change industry; it is always reacting to conditions.

While industry has seen the removal of government controls of wellhead prices for natural gas and unified interstate and intrastate markets as well as open transportation, there are still many economic rules and regulations impacting the gas industry at both the Federal and state levels. To build new interstate pipelines or new terminals, to set transportation rates, and a host of other activities still require, what one could call, economic regulatory approval. The tremendous strides made in the last 25 years are still clouded with some economic regulations. In addition, the industry is still subject to the multitude of regulations coming from the Environmental Protection Agency, Department of Transportation, Commodity Futures Exchange Commission and a host of other regulatory bodies.

21.6 MOVING NATURAL GAS FROM FIELD TO BURNER TIP

Perhaps, except for the sales or marketing source itself, the most important segment of the natural gas industry to someone interested in buying gas for use in a commercial or industrial application is transportation. Coincidentally, this is the part of the business seeing the most changes in the last decade. Many times, the transportation will be covered by the selling or marketing group the customer has gone to for buying the gas but, it is still important for the buyer to know something about the transportation industry. Again, as a reminder, natural gas is only transported in pipelines in the United Sates (with a little exception in some Northeast and Northcentral areas of the country where liquefied natural gas (LNG) is used for storage purposes and is transported as LNG).

Before going into the mechanics of moving natural gas from the field to the plant gate, a review of where the markets are for natural gas is important. U.S. natural gas consumption was the greatest in 1972 just after the first embargo on crude oil by the Organization of Petroleum Exporting Countries (OPEC) began to make its impact on the world oil market. Gas consumption peaked then, not because of a decline in demand, but because of a decline in gas availability for the interstate

markets. Natural gas consumption would have increased in future years from the 22.8 trillion cubic feet (tcf) consumed in 1972 if more gas had been available for the interstate market. At that time, based on current regulatory law, gas sources could serve only certain markets. Gas produced in a given state and consumed in that state was known as intrastate gas and came only under the sole jurisdiction of the state where it was produced. Gas produced in one state and transported to another state for consumption was termed interstate gas and was under Federal pricing and use jurisdiction.

Gas demand increased as the oil embargo decreased crude oil supplies and as crude oil prices increased multi-fold. The lack of available gas for non-producing states forced curtailment by interstate pipelines in the mid-1970s. Natural gas as reserves in the ground was available for development and production, but the economic incentive to produce gas for the interstate market was not there. Natural gas prices for the commodity in interstate commerce were regulated by the Federal Power Commission. To get a change in price took so long that gas prices for interstate gas lagged prices seen by crude oil and intrastate gas. Interstate gas prices for many years after the embargo began and ended were still under a dollar per MMBtu while intrastate gas prices took off and were in the three to four dollar per MMBtu range by the mid-1970s.

Supply curtailment in the interstate market continued from 1973 through close to the end of the decade. The passage of the various regulations including the Natural Gas Policy Act of 1978 (NGPA) by Congress changed the natural gas supply situation. Gas became more available towards the end of the 1970s with supply exceeding demand in the early 1980s. Later in the 1980s, the surplus was so great, serious declines in gas prices occurred because of gas-on-gas competition and an excess of all energy supplies. From the late 1970s into the early 1990s, the industry has experienced what is called the "gas bubble" when supply exceeds demand. Further, even though the bubble might "evaporate" at times, as suppliers became more optimistic and produced new rounds of drilling and exploration, the bubble returns with the resulting drop in gas prices. Figure 21.5 gives the U.S. consumption of natural gas for the period 1970 through 1995.

Once natural gas pipelines linked the country, the most extensive use of natural gas was for residential and space heating in commercial facilities. When natural gas is used in an industrial location for space heating, it is considered a commercial application rather than industrial one. Natural gas quickly replaced coal in home and other space heating applications and has slowly replaced even fuel oil in a great many locations for this use. Natu-

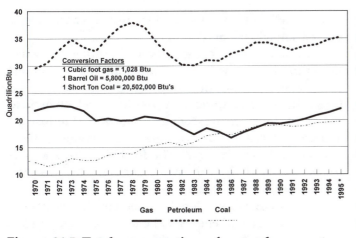

Figure 21.5 Total consumption of natural gas, petroleum, and coal.

ral gas is clean, economical, and easily suitable for all types of space heating.

In areas where gas distribution companies have laid pipes to individual homes and places of business, getting gas supplies is easy. Residential uses of natural gas for space heating, hot water, and cooking and commercial space heating made up 27.6 percent of the natural gas used in the U.S. in 1950. By 1975, 38 percent of the natural gas used in the country went to residential and commercial use and represented 16.4 percent of all the fuel used in the U.S. In 1994, residential and commercial applications consumed 37 percent of the U.S. gas supply. The major factor in residential and commercial use of gas is basically weather and economy. How cold or hot the various seasons are plays an important role in how much is used along with the country's economic condition. Amazing, when the economy is bad and winter hits, how many users will find ways to economize to save heating dollars.

The industrial sector is the next largest user of natural gas. This is the sector showing the greatest growth over the last few years since gas demand sank to its low levels of the mid-1980s. Gas is used in heating industrial processes as well as a feedstock for certain chemical manufacturing operations. Heating for industrial processes is either where gas is directly burned in the process or where gas is used in industrial boilers to produce steam for heating and further processing. Either way, natural gas makes an ideal fuel for industrial applications, especially where the fuel is burned in contact with the process goods such as fibers, food, paper and pigments. The clean burning qualities of natural gas make it the only fuel used in these applications. Weather and economy play a major role in gas used in industrial applications. Also, price plays a significant factor. Recent increases in gas consumption in industrial applications

stem from the increased availability of gas and its lower prices. Some industrial applications are very sensitive to gas prices such as gas used in ammonia and related products manufacturing.

Electric generation is the last of the major uses for natural gas. Gas accounted for 23.8% of the fossil fuels used in the United States in 1990 for electric generation, making it the number two fuel choice after coal at 28.6%. Over the last few years, natural gas consumption in electric generating equipment has increased. Still, the use is considerably less than was used in the early 1970s when the Organization of Petroleum Exporting Countries (OPEC) made its first embargo of crude oil to the free world. This was also the time when the greatest quantities of gas were consumed in the U.S.

The major typical use for fuels where gas has played only a minor role is in the transportation sector. The major use of natural gas for transportation is in fuel for the compressors used to transport natural gas in the interstate and intrastate pipe systems. In recent years, attention was generated for using natural gas for automobile and other vehicular transportation. The need for special fueling stations and the additional requirements for tankage on the vehicle has kept the use of natural gas limited in vehicular transportation. Some fleet operations have converted to gas but overall, liquid fuel is still the most popular form for vehicle use.

The only current economic means of moving natural gas in the United States from production site to consumer is by using pipelines. This is both a boon and a bane. On the positive side is the ease of handling the gas in pipelines from production to combustion. The commodity in the pipeline is a non-corrosive, easy to move, gaseous form that can be economically transported to the market place. On the other side of the ledger, the fact natural gas can only be transported by pipeline and the difficulties encountered in shipping because of this have lowered the value of natural gas in the U.S. industrial sector. Even though natural gas production and pricing have been freed of many of the regulatory restrictions imposed on it at the wellhead, the natural gas transportation and distribution industries remain heavily regulated. Natural gas moving from production to consumer goes through many stages before coming to its end. Figure 21.1 is a schematic of the activities to bring natural gas from the field to burner tip.

21.7 MARKETING NATURAL GAS

Marketing natural gas is what has changed in the U.S. gas industry because of wellhead decontrol and transportation services unbundling. This is the crux of the industry's change through the last 25 or so years. A very important aspect to understanding is what the term "marketing" means, especially as opposed to the word "selling."

Selling is moving a commodity from the producer or secondary holder of the commodity to the buyer. It implies little service to go with the movement of the commodity. Marketing is a broader term and covers all the activities the producer or marketer does, including many services, to move the commodity in the market place. Selling is the bare action of putting the commodity at the dock and letting people buy as needed. Marketing is the all inclusive action including services, advertising, technical assistance, and special conditions needed to move the commodity. These additional services are as much a part of the price the buyer pays for the commodity as the commodity itself

The old natural gas industry was a sellers' market. Producers developed and produced natural gas which the pipelines took at the wellhead or close to it and sold to the distribution companies who made the final sale to the consumer. The only service function was at the last move, the sell to the consumer and that was fairly limited also. By opening up the market and allowing marketing companies to be the merchants, the transformation has occurred to make the gas industry a service oriented business. The services offered by the seller might include many things from variable amounts of gas available on the contract, risk management, and any other service needed to secure the selling of gas to the buyer.

In looking at the markets, one of the first ways of classifying the market is by knowing the buyers' purpose in seeking gas supplies. Is the buyer looking for gas to resell such as in the case of the distribution companies or is he looking for the gas for his own consumption. By definition, buyers looking for gas for resale are considered wholesale buyers regardless of the level they are at in the system or the quantity of gas purchased. Those buyers looking for gas because they are end users—because essentially they have a "burner tip" for the consumption of the gas, are considered retail buyers. The wholesale buyer is both a buyer and seller of the gas while the retail buyer is strictly a buyer except for the few times he might have excess supplies and needs to resell some of his purchased gas. Even if this should occur, the original intent of the retail buyer was purchasing gas for his own use.

This gives a group of marketers who are predominantly either wholesale or retail sellers. There is an additional group called agency marketers who help a buyer find the best sources of gas for the buyer's needs. They assist in the buying function for the buyer whether

he is a reseller or consumer and are used because of their buying knowledge and ability.

Another major classification of markets is where the natural gas used. There are four major markets which cover the uses of natural gas outside of gas used in the field for compressor operations and other field uses. The four are residential, commercial, industrial, and electric utility. The volume of gas going to these markets is shown in Figure 21.6 for 1975-1994. The electric utility market is as the name implies, where electricity is generated for resale.

Electricity is considered a secondary fuel because it takes other forms of energy to be the feedstock for electric generation. The major fuel for electric generation including nuclear, is coal which has a 56 percent share of the market electricity generation. Natural gas has increased its market share of the electric generation business. Expectations are it will continue to gain market share as older coal plants become obsolete and nuclear facilities either are phased out or outlive their license. Natural gas is an ideal fuel for electric plants for a number of reasons. The most important are the clean burning aspects of gas and the lower capital needed to build new gas fired generating facilities.

Electricity that is generated by independent power producers, which includes cogeneration plants in the country, and which use natural gas as fuel are not included in the consumption data on electric utilities. Data on gas consumption for these facilities goes under the category of industrial. Natural gas in the industrial sector goes for process fuel and as feedstock in many chemical applications. Process fuel would include applications like product drying, paint ovens, high temperature reactors and other processing equipment used in the chemical, paper, cement, iron and other metals, and other industries. Natural gas is the feedstock for major chemicals as it is the raw material (as well as fuel) for the

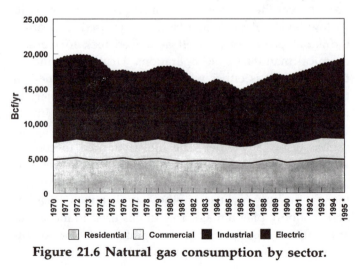

Figure 21.6 Natural gas consumption by sector.

production of ammonia and its by products, ammonium nitrate and urea.

The commercial user of natural gas is any business using gas for space heating or for heating water. This definition includes the use of gas in industrial facilities where it is categorized under the commercial group if it is used for space heating. Only when the gas is used for process fuel or feedstock would it be included in the industrial category. Natural gas is the major fuel for home or residential heating and has over 50 percent of the market. Fuel oil and electricity are the other residential fuels. Natural gas has the competitive edge over electricity except in certain areas where between the heating load and the cost of electricity, an all electric home can be competitive to using natural gas for fuel. Oil is used in certain areas where gas supplies do not reach and in certain areas where the price of natural gas is higher than other fuels by the time the gas reaches that market. This is still true today in certain areas of the Northeast like Connecticut and Massachusetts.

There is one additional market where a small amount of natural gas is consumed and that is transportation. The major use for natural gas at the time in the transportation area is for pipeline compressors used to move the natural gas. A small amount of gas in recent years has moved into the motor fuels market but, to date, this is rather limited. While promising, the amount of gas going to vehicular transportation is small and will take time and economic incentives to grow significantly.

In the mid-1980s as the natural gas industry started into the decontrol of wellhead prices and open access transportation, natural gas production capability greatly exceeded demand requirements. The pipelines were still the merchants for the industry and as this excess capacity grew, or what was termed the gas "bubble" expanded, the pipelines began to curtail taking gas under their long term contracts. Not only did demand decrease during these years, but the producers found better ways to stimulate the gas wells making significant production increases. This only led to additional curtailments and contract problems.

While many producers sought contract enforcement, both pipelines and producers began to form separate marketing companies to start selling the natural gas on a short term or spot market basis. This was the beginning of the marketing companies as the merchants in the gas business. In the early days, the marketers had little formality and essentially, anyone with a telephone and knowledge of gas marketing could be a marketer.

By the time FERC Order 636 came about with specific changes directed at the pipeline to open transportation access and to unbundle pipeline services, marketing companies were more formalized, significant business centers. With one or two exceptions, all the marketing companies had significant capital investments in either pipelines, production, or processing areas to back financially their marketing efforts. In addition to supplying natural gas to customers either for resale or for consumption, the marketing companies supplied additional services like balancing, transportation supervision, risk management, and in some cases, fuel management. The marketers have even branched out today into wholesale electric power marketing.

Most marketers today come from either of two major sources. They are either subsidiaries of production companies where the marketing is a necessary adjunct to moving the gas from production into sales or nonregulated subsidiaries of pipelines, either those engaged in transportation or in the distribution of the gas. Some marketing companies are subsidiaries of companies in other areas of the natural gas business like gas gathering, processing, or storage. Only a few of the marketers today are strictly marketers and have no other major connection to the natural gas industry. The need for financial support and backup capital almost makes it a necessity that the marketing company has assets whether it is a subsidiary of a company basically in production, transportation, processing or some other area of the gas business.

The organization of the marketing companies has been in constant change since the first ones were organized. As the industry has changed and as the sophistication increased in the marketing area, many changes have occurred. Some of the latest changes have been the formation of "strategic alliances" between marketing companies to strengthen their ability to gain market share. Many of the strategic alliances are actually acquisitions of one marketing company by another to gain additional market coverage. Many of these acquisitions are between marketing companies where one of the marketing companies is a subsidiary of a producer. Classic examples of this in 1996 were the major acquisitions where the Natural Gas Clearinghouse, renamed NGC, purchased the marketing subsidiary of Chevron Corporation and PanEnergy's marketing subsidiary acquired the marketing subsidiary of Mobil Oil. Both of these were megamergers for the natural gas marketing business as each saw companies with over a hundred employees and moving significant quantities of natural gas absorbed into equally large marketing subsidiaries of either a processing company, NGC, or a pipeline, PanEnergy, previously Panhandle Eastern Pipeline Company.

If this trend continues, where marketing subsidiaries of producers combine with other marketing companies so that the producers are back to "selling" natural gas rather than "marketing" the commodity, many of the producers will no longer have marketing subsidiaries. For the small natural gas producer, this is not unique since they have such small volumes of gas it might not be economically feasible for them to continue to market gas. But, for the large producer, those having over a billion cubic feet per day of production, this outsourcing of the marketing function represents a critical strategic move. In the end, it all comes down to who can market most efficiently and successfully.

Major natural gas marketers today are those moving over a billion cubic feet of gas per day on an annual average basis. The total gas marketed by the company divided by the number of days in the year gives this annual average. In reality, natural gas marketing swings considerably between the lows of late summer and the peaks of the winter heating season. For 1995, the average movement of natural gas was 57.5 Bcf/d which comes from taking all the gas consumed and dividing by the number of days in the year. On peak demand days close to 90 Bcf/d of gas will be consumed while during the summer, the consumption might fall to around 40 Bcf/d.

Gas produced at the wellhead is sold by the producer through its own marketing affiliate or by a company given the responsibility for selling the producer's gas. Usually, when the producer's marketing affiliate has responsibility for marketing the gas, in addition to the gas from its own production, commonly called "equity gas," the marketer might also market natural gas coming from interest owners in the same field or well as the producer, or it might seek other gas from other producers it can market to increase its size of operations and efficiency of doing business. The gas at the wellhead might be sold to another marketing company who will resell it again. It might be sold at the wellhead to a distribution company for resale to the final consumer. The number of times an individual cubic foot of gas is handled by a party is at best, a "guesstimate." For planning purposes, on an annual basis, an average of four times is assumed. This would mean on an annual basis, each day about 240 Bcf of gas is moved by the marketing companies as physical sales. Appendix 21.2 Table A is a list of the major marketing companies in the spring of 1996. The estimate of the annualized amount of gas each company is moving is given in the table. Since the largest company is moving approximately 10 Bcf/d, it has about a four percent share of the market. Table A lists 32 companies as the major marketers at a time when there are about a total of 400 to 500 marketers. The top 50 might include all companies moving at least a Bcf/d with the remaining companies moving from 900 MMcf/d down

to less than a million cubic feet per day (MMcf/d). The smaller ones might have niche markets where the margins are sufficient to sustain profitably the marketing effort made by the company in moving the small amount of gas.

In moving the natural gas from the wellhead to the first marketer and subsequent ones, either party can be responsible for the transportation or other services that might be needed. Some of the marketing companies have contracted for firm transportation with the pipelines while almost all of the distribution companies still have contracts with pipelines for firm transportation. This is a changing part of the business and major changes will occur in the next five years. When wellhead decontrol was being put into effect, in some of the regulatory orders the distribution companies were released from their contractual obligations to take gas from the pipelines. Some of their transportation obligations were also modified, but the bulk of the firm transportation contracts stayed in place. The pipelines under the new regulations still offered firm and interruptible transportation.

With firm transportation, the transportation buyer is obligated to pay a portion of the transportation charge regardless of its using the transportation. This is the demand charge of firm transportation tariffs. The methodology of the rate making procedure used to recover pipeline transportation costs which includes a rate of return for the pipeline or actually its profit, is such that when a pipeline sells all of its firm transportation, it has maximized its rate of return. Interruptible transportation carries no guarantee to the buyer of space in the pipeline. During periods of high gas demand, pipelines will sometimes cut off the interruptible transportation.

Typically, the pipeline uses about 80% of its capacity for firm transportation and the rest for interruptible. All of the interstate pipelines went through a rate case when FERC Order 636 was imposed in 1992-93 which included the firm contracts they held with individual distribution companies, marketing companies, and others. Many of these contracts will terminate around the year 2000 and pipelines will have to renegotiate them. A question arises whether the transportation buyers will want firm transportation again and many pipelines may be hard pressed to renew sufficient quantities of firm transportation to maintain their desired rate of return. The first of the contract renegotiations started with Transwestern Pipeline, El Paso Pipeline, and Natural Gas Pipeline of America (NGPL) in 1996. The patterns and precedent could be set by these early negotiations.

The importance to the gas buyer of this discussion is the knowledge of how the transportation industry is still unstable and open to change. This will have an impact on gas marketing as the amount of available transportation will play a role in delivered gas prices. Further opening up of transportation from a contract standpoint will also impact the availability of transportation to the marketers so they can sell a bundled product of delivered gas to a given destination.

At the current time, most of the gas marketing companies are basically wholesale sellers of gas. This means most marketing companies sell gas to buyers who are going to resell the gas. Most of the retail trade is done by the individual distribution companies who buy gas at the city gate. The distribution sector has yet to make the unilateral decision to offer open access transportation as in the interstate transportation sector. The change to open access on a state basis will be regulated by the individual Public Service or Utility Commission in that state. Different from the interstate transportation where only one regulatory agency acted to enforce open access, each state and in some cases districts within the state, will be dependent on the individual state or district regulatory agency to direct the open access activities. Some states like Connecticut, New York, and Wisconsin have already taken the initiative and have made the distribution companies in their state offer open access.

When and if open access is available at the state level, the industrial and utility users will be the first to contact marketing companies for their gas purchases. The industrial and utility users are large enough to deal directly with marketers for direct purchase of gas supplies and use the distributor only for transportation. Large commercial users will be the next to deal directly with marketing companies. In all likelihood, the residential user will continue to buy gas through the distribution company. Open access, allowing marketing companies to sell directly beyond the city gate, will make available risk management instruments to buyers which are not available when buying gas through a regulated company like the distribution companies are now. The other side, whether the distribution companies will be allowed to use risk management tools, is something to be considered also. Most likely, as the distribution companies move to building non-regulated marketing subsidiaries to handle the down stream merchant function, this will be the company to handle risk management for the distribution company.

21.8 NATURAL GAS PRICES & TRANSPORTATION COSTS

Natural gas has become a commodity. Like any other commodity, its price reflects the parameters of supply and demand and to a lesser extent, the actual costs in finding, developing and producing the gas. The pricing of natural gas supplies in the United States is unique because the U.S. is one of the few major gas consuming countries where natural gas prices at the wellhead are not close to crude oil prices for the equivalent energy content. Today, natural gas is the second largest source of fuel in the country. Possibly, it is the best fuel, yet its pricing is low compared to its major competitor, oil products.

In the early days of the oil and gas industry, gas was produced only in conjunction with crude oil. Many times, natural gas was flared at the production site to get rid of it so crude oil from the well could be produced without restraint. Further, the only practical way to move natural gas from the field to consumer is by hooking the well to a pipeline. This took additional time and money and slowed the pace of producing the crude oil. Crude oil can be produced and stored at the well site. Tank trucks can pick up the oil and deliver it to the refinery if necessary. In the early days, gas was more a nuisance than an asset.

The U.S. did not begin to understand the importance of energy conservation and the impact of prices until the early 1970s. The first oil embargo by the OPEC countries quickly brought home the impact of cost and prices for fuels. The "energy is almost free" mentality of the U.S. came to a rapid halt with the first embargo in the early 1970s. Crude oil prices went from a few dollars per barrel to the teens quickly. Figure 21.7 shows oil and gas prices from 1970 through 1995 in both current and constant dollars. The higher crude oil prices of the first embargo moderated within a few years as supply exceeded demand. The higher prices brought more supply than demand could consume. Crude prices were coming down by the mid-1970s. The country began to think of energy products as almost free again.

Natural gas prices responded only partially to the increased crude oil prices in the early 1970s because gas in the interstate market was under federal price controls. Intrastate gas supplies had prices three to four times those of interstate supplies by mid-decade. In 1978, before passage of the NGPA, interstate new contract gas prices were in the $1.50/MMBtu range, while some intrastate contracts were as high as $4.00/MMBtu. In 1978, crude oil wellhead prices in the U.S. were $9.00/B. Expressed in energy values, crude oil at the time was $1.55/MMBtu since a barrel of crude oil is equivalent to 5.8 MMBtu.

Decontrolling natural gas at the wellhead started a new and different two-tiered market system in the gas industry. Further, it opened the door for a futures market so buyer and seller could hedge the risk of price

Figure 21.7 Wellhead price of crude oil and natural gas. (Expressed in current and constant dollars [1994])

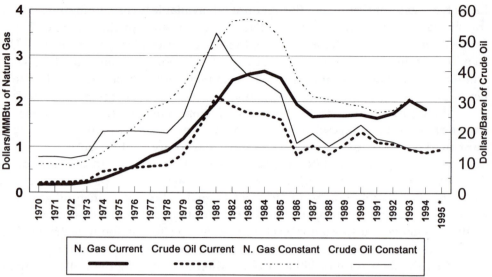

fluctuations. Like other commodities, natural gas needs all three types of markets. The physical markets are the secure supply, longer term market. Another is the market which reflects the short term and spot demand. Both of these are for the physical trading of the commodity. The third is a futures market where the commodity itself is usually not physically moved, but allows buyer and seller risk management opportunities.

The short term market started in the mid-1980s when pipelines stopped taking contract gas and producers sought other avenues for moving their gas supplies and maintaining cash flows. Interestingly, there was already a short term gas market in effect for many years before the mid-1980s. In the mid-1970s, when natural gas supplies for the interstate market were short, the FPC endorsed moves allowing a spot market for natural gas. The highest bidder should have been able to get supply. Under the FPC plan, consumers in Ohio could go to Texas and contract directly with producers for gas supplies under the new rules. In addition to the new ruling, pipelines could always make "off-systems" sales which were a form of spot sales.

But again, economics were the barrier to success. There was no incentive for a producer to sell natural gas directly because prices were still federally controlled for interstate gas. Price bidding was impossible because of price controls. The mechanics were in place in the mid-1970s for a short term market but, the regulation of gas at the wellhead prevented the spot market from becoming a reality.

In the passage of the NGPA in 1978, the new categories of gas pricing caused an important change in the way gas deliverability was handled. Under the old system, producers sold a field's reserves when they did a contract with the pipeline company to take the gas. Prior

to the NGPA, the deliverability was based on the reserves in the field under contract. Each year, the buyer was obligated to take a certain percentage of the field's reserves. The NGPA called for a new way where the well was to be tested and the annual takes would be based on the flowing gas available. This prompted the producers to find better ways to fracture a well to increase the well's deliverability so that it would recover its costs sooner. Time is money. Taking the gas faster and reducing the payout time of the well were nonmonetary ways of giving the producer added economic benefit and an incentive to sell its gas reserves to one pipeline or another. If the pipeline did not take the gas at the contracted rate based on the deliverability, then it was obligated to pay the producer for the untaken gas at the end of the year. This provision of the contract was known as the "take or pay" clause. Such clauses impacted the amount of gas available in the early 1980s and contributed to the excess supply problems of that time.

The outgrowth of this was the start of the short term or spot market in natural gas. By early 1985, a spot market was in progress and through the next decade the market evolved into three physical markets—spot, short term and long term. Spot sales are essentially anything from about an hour of gas availability to a 30-day or month's supply. Short term would be up to six months' contracted supply and long term would typically be up to five years. In some cases such as an independent power producer building a cogeneration plant, the contract might be for 15 to 20 years. The evolution of the gas business has come around to today's system where prices reflect the current supply/demand parameters and to a lesser degree, the cost of finding and producing the gas. Competition with other fuels, especially distillate and residual fuel oil and coal, also interact with gas prices.

The pricing discussed to this point was in reference to the physical trading and sales of the commodity, natural gas. When gas prices were deregulated, it made it possible to start a "futures" market for the commodity, natural gas. There are futures markets for other commodities like grain, metals, and other fuels. When a buyer buys a futures contract on a commodity, he is purchasing the commodity today at the estimated price for the later delivery date and delivery location as set by the futures contract sales point. In the case of natural gas, the first futures market was established on the New York Mercantile Exchange (NYMEX) for the sales of gas at the Henry Hub in Louisiana. In 1995, a second market was established by the Kansas City Board of Trade (KCBT) for sales out of the Waha terminal in West Texas. The NYMEX was granted in early 1996, a second contract for sales out of the Permian Basin in West Texas. Futures trading under this contract will most likely, start before the end of the year. The NYMEX is planning contracts for other cities, but has not yet sought agency approval of the planned contracts.

A futures market allows for buyers and sellers to "hedge" the risk in timing that can occur when buying or selling a commodity. As an example, a copper wire converter who buys copper in bulk form and makes it into wire could sustain significant losses if when he bought copper it was at a regular price—say $1.20/pound and by the time he had converted the wire bar into wire, the value of copper had gone up to $1.40/pound. To replace the copper he had used in making the initial order, maybe two to three months later when the finished wire is delivered, he would have had to pay significantly more to replace the copper used in the first order. To protect himself from such a situation, the wire maker would buy a futures contract for copper for delivery when he completes the first order and now needs new raw material to process new orders. The price of the wire is based on the raw material copper being priced at the futures price plus a margin for the conversion and profit of the wire maker. When the wire maker delivers the first order, he is able to buy copper at the price he used in making the first order and is protected from price escalation. He has hedged his risk or managed the risk in buying raw material because of changing times and changing prices.

Natural gas buyers and sellers face the same risk. A buyer will need natural gas over the next five years for use in making a certain product. To hedge his risk, he will buy futures contracts to insure getting his natural gas at an appropriate price. The NYMEX and KCBT fu-

tures markets in natural gas are both 36 months forward. Other financial institutions making a market in natural gas go out into time further.

The futures market opens the door for other financial instruments in trading of natural gas. The most common are options and derivatives, but there are many more. It also opens the door for "speculators," traders who do not need the commodity, but only play the market as a way of making money. The futures trading area is a very sophisticated financial business and large sums of money can be made or lost easily. At the same time, when it is used only for hedging purposes, it can protect buyers and sellers from the volatility of the market place. The closest month to the time of trading is called the nearest month and even though the trading is on a time period a month away, many traders use this as a guide for current physical buying and selling of natural gas. The further out months are an estimate on a given day of what the future price will be including the cost of money and other variables and is only an indicator on the day of the trading what the future price is expected to be then. Futures prices for natural gas on the NYMEX are listed in various newspapers and *The Wall Street Journal* daily when the market is open. Trade publications and the wire services shown in Appendix 21.1 contain summaries of this information also with the wire services giving instantaneous information from the trading marts.

When referring to natural gas prices, it is important to know where in the gas delivery system the quoted price applies. The first point of gas sales could be at the wellhead, a field gathering point, or the tailgate of a treating and/or processing plant. The price would be known as the wellhead or field price for the natural gas. The next major point of sales would be into the intrastate or interstate pipeline. This is the "into pipeline" price. The alternative to this could be a "hub" price where gas is brought to a central facility that has the ability to dispatch the gas among a number of pipelines. After the hub, the "into pipeline price" would be applicable as the gas goes into one of the pipelines working out of the hub.

The next pricing stage could be the city-gate where the gas is transferred from the transporting pipeline to a local distribution company for dissemination to consumers. In some instances, in today's market, the gas could go directly to the consumer from the transporting pipeline so the buyer's plant gate might be the pricing point. The major consuming sectors are the residential, commercial, industrial, and electric utility markets. Since the progression from each of the stages from production to market carries a cost factor, it is important to know where in the delivery chain the price quoted applies. Figure 21.8 is a comparison of prices at each of these major market points for December 1994 and 1995.

In prior years, the differential in price between gas produced in the various basins was relatively small. Louisiana has almost always been the leader because of its access to markets within the state and the big markets of the Northeast and North Atlantic states. Gas produced in East Texas might carry a five cents less differential from Louisiana and gas from Oklahoma might be 10 cents/MMBtu less. Gas from West Texas was a little less than gas from East Texas. The gas in West Texas is influenced by the California market with some consideration for the Eastern markets. The limited transportation available to bring Texas gas from the West to the East hampers the free movement of the gas to the highest paying markets.

Gas produced in the Appalachian and Michigan areas might carry a premium over Louisiana gas because of the proximity of these locations to the market place. As a general rule, the closer to the market area the gas is produced or where the market area is easily accessible, the higher the value of the gas. The difference between locations for gas prices is referred to as the basis differential. The winter of 1995-96 saw this differential increase significantly from the five to 15 cents in prior years to as much as a dollar to $1.50/MMBtu. The reason for this was the inability to move the gas from the producing area to the markets with the greatest demand. Gas produced in basins with close access or easy access to the Northeast and North central and then to the midcontinent received the highest prices because these were the areas with the greatest demand for natural gas.

Gas pricing statistics are available from a number of sources. Data on the average price by state for gas going to residential, commercial, industrial, and utility markets are reported by the federal government in its publication, DOE/EIA-0130, *Natural Gas Monthly*. The same publication also publishes the monthly average wellhead price for all gas produced in the U.S. regardless of whether it is sold under long term or short term contracts. The prices for the various markets for natural gas are based on the prices consumers pay for the commodity at the plant gate or residential site. In making a pricing analysis for industrial and utility users of different fuels, it might be more appropriate to measure the cost of fuel at the smoke stack of the consuming system. As clean-up costs increase, especially for a fuel like coal, the final costs after combustion and including the clean-up costs would give the most representative value of the fuel being used.

Another element to be considered in looking at gas prices is the seasonal aspect. In recent years, markets have not followed the seasonal aspects and sometimes the gas was more expensive in the summer than in midwinter, the coldest part of the year when gas demand is usually the greatest. To compare prices from year to year, daily prices are usually averaged into monthly prices and an average of the monthly prices reported as the averaged yearly price. Comparing annual averages is the only way to make valid comparison of the prices for natural gas and other fuels because of the influence of seasonal aspects.

For those buying and trading natural gas, the government price data is not specific enough nor timely enough to be practical. Many of the trade publications now have a pricing section with prices for gas in the field, into pipelines or at hubs, and at the city gate quoted on a daily and weekly basis. Also, pricing information from the wire services is available on a continu-

Figure 21.8 Gas price range for market served. December 94 and December 95.

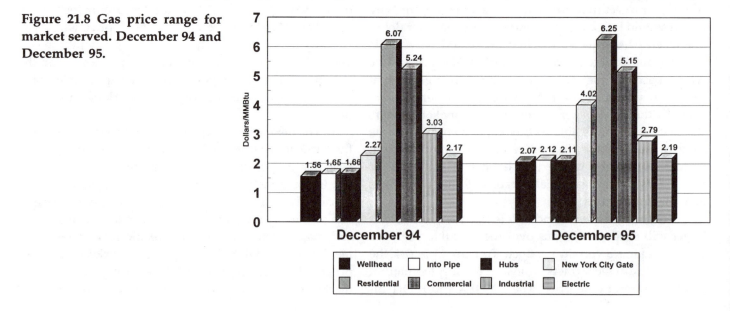

ous basis. Those actively trading natural gas might want to take advantage of these services. The pricing services from both the wire services and the trade publications are listed in Appendix 21.1.

Wellhead and other field prices are developed by the publications by polling gas buyers and sellers on a daily basis. These prices are usually short term or spot prices and might be exclusive of any longer term contracts between sellers and buyers. The data collected by the government for wellhead prices makes no distinction for the type of contract the gas is sold under and is more representative of the total value for all the gas moving from a given state or, collectively, within the U.S. in total. The into pipeline prices are a combination of the prices quoted by the individual pipelines much like a posted price in the crude oil refining business and a polling of traders as well. Pipelines, being out of the merchant function now, purchase only a small amount of gas for resale. The FERC did make a provision for some pipelines to continue to sell small amounts of gas to certain buyers where the retail function was not readily available from other sources.

Natural gas prices, like other commodities, vary for many reasons, some of which are not always understandable. On a macro scale, the greatest influence of gas prices, especially at any of the stages in the chain from production to use, is the supply/demand balance at the time. As gas supplies become greater, prices reflect the additional supplies and decline. This is true at the wellhead as well as at the city gate or plant gate. Many times, the fears of shortages will raise the prices quickly. The 1995-96 winter, which was one of the coldest winters in recent years and that brought cold weather early, saw gas prices escalate in mid-December when supply tightness occurred in the Northeast and North central states. The price escalation was not uniform across the basins where gas is produced. The greatest increases came in those states providing supply to the country's regions where the weather was most severe.

There was no supply shortness at the wellhead. Many of the basins where transportation to the high demand section of the country was limited saw prices only rise minimally. Basins supplying the high demand sectors of the country had prices shoot up dramatically. The problem was a shortage of firm transportation. Again, when prices escalated widely for gas coming into the Chicago area in late January, early February, 1996 it was a combination of transportation and wells in the Oklahoma, North Texas areas freezing and being shut-in. Again, prices were impacted by the current supply/demand situation, but more local than in the whole market.

In addition to supply/demand impacting prices,

other elements to a lesser degree will have an impact also. The lower costs for finding and developing natural gas reserves did much to bring down the prices for natural gas in mid-1994, early 1995. The technology improvements allowing the lower costs of finding, developing, and producing gas played two roles in impacting natural gas prices. First, it made gas cheaper at the wellhead as producers would still seek new supplies even though prices were lower. Secondly, because it made lower cost gas available, many producers drilled more wells to get the low cost gas and only added to the excess supply and lowered prices for the commodity.

Natural gas prices at the wellhead or in the field are a function of the plant gate or burner tip price. The price paid for natural gas is solely a function of the market place—the demand and competition for the fuel. Wellhead prices reflect the buyers' costs for the fuel less the expense to bring the gas from the production location to the plant gate or burner tip. Typically, the plant gate is the point of pricing for industrial users. The price for natural gas will be set by the many factors affecting the fuel at the plant gate. The cost of transporting the gas from the production area to the gate will be deducted to reach a "net-back" price to the wellhead. The competitive value in relation to other fuels such as coal, fuel oil, electricity, or other sources of natural gas will set the value of natural gas at the plant gate or burner tip. In looking at natural gas pricing, the plant gate and burner tip prices are essentially the same since there are no storage or fuel preparation costs associated with natural gas. Both oil and coal have storage costs and many times coal has a preparation cost making a difference between the plant gate and burner tip values.

In addition to the actual heating values of the competing fuels, certain intangibles play a role in setting the fuel's prices. Natural gas is clean burning, easy to use, burns efficiently, and is not needed to be stored at the consuming site. These elements add to the value of natural gas compared to other fuels. At the same time, natural gas has certain disadvantages, some real and some only in perception. Setting up transportation, the degree of regulation of the gas industry, and until recently, the difficulty in buying supplies all tended to lower the value of natural gas in relationship to crude oil products competing with it. For the three fossil fuels in the United States—oil products, natural gas, and coal—oil products are usually the most expensive while coal is the cheapest with natural gas being in the middle on an energy equivalent basis. Crude oil is rarely used as fuel and the products coming from the refining of crude oil are used for fuel. Figure 21.9 gives the costs of fuels to electric utilities for fossil fuels for 1980 through 1995.

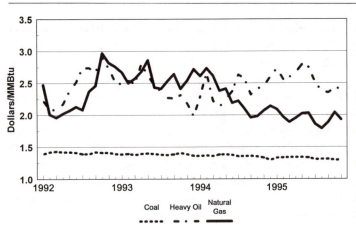

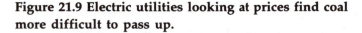

Figure 21.9 Electric utilities looking at prices find coal more difficult to pass up.

The measure of energy for fuels is British thermal units, Btu. Natural gas is sold directly in energy units and the price is usually quoted in $/MMBtu. Fuel oil is usually sold in gallons if it is the cleaner, better grade number two fuel oil (No. 2 FO) or in barrels if it is number six, residual fuel oil (No. 6 FO). To convert NO. 2 FO given in gallons to barrels, divide the number of gallons by 42, the number of gallons in a barrel of oil. The heating equivalent of No. 2 FO is 5.8 MMBtu/B while No. 6 FO has a heating equivalent of 6.3 MMBtu/B. NO. 2 FO is used in heating homes and in industrial applications where the contaminants of No. 6 FO would be detrimental. In recent years, some states have put a tax on NO. 6 FO to limit its use for boiler fuel applications because of the environmental problems coming from the dirty fuel. Coal energy content is found by multiplying the short tons of coal by 20.502 MMBtu when the coal is used by the electric utility industry or by 20.852 MMBtu when it is for consumption with no designated industry area. Figure 21.5 shows the U.S. energy consumption in energy units for coal, oil, and natural gas.

If natural gas prices were on parity with oil, the price of gas in MMBtu would be about one sixth the value of oil when the oil is expressed in $/B. If expressed in energy units and gas and oil were at parity, the values would be the same. In 1994, the average short term wellhead gas price was $1.66/MMBtu and the overall average wellhead price for all gas was $1.84/MMBtu. The average gas price paid by electric generating utilities in 1994 was $2.230/MMBtu which makes comparisons with oil and coal prices easy. Average price of heavy oil (NOS. 4, 5 & 6 FO) was $2.409/MMBtu. Average price for NO. 2 FO would have been considerably higher in the range of $3.60/MMBtu. Average price for coal was $1.355/MMBtu. Even though natural gas has some decided advantages over both these fuels for

specific applications, it is easily seen how gas compares price wise among the three basic fossil fuels in the U.S.

A very important aspect of natural gas prices is the cost of transportation to get the fuel from producing field to plant gate. With natural gas, there is only one form of transportation—pipeline. Other fuels have the advantage that while pipeline is the preferred method, the buyer can resort to other means to insure his supply during difficult times. Gas buyers do not have this luxury. The best hedge for insuring transportation is not a hindrance to supply is to have more than one pipeline source to the plant or using facility. Buyers with a multiplicity of supplying pipelines have a distinct advantage since many times one distributor cannot get gas supplies while another may have plenty during difficult times.

"Net-back pricing" is a term used to compare gas prices at the plant gate or city gate on a wellhead basis. The price for the gas at either location less the transportation costs to get to that location makes up the net-back price. Transportation costs play a significant role in finding the net-back gas price. In looking at a net-back price for gas at the wellhead and starting, say at the plant gate, the typical transportation and gathering factors would include the following:

- Transportation and distribution costs from the city gate or mainline sales point to the plant gate by the distribution company.

- Transportation charges from the pipeline pick-up or terminal point to the city gate by either an interstate or intrastate pipeline.

- Transportation from the wellhead or field processing plant tailgate to the pipeline pick-up or terminal which is usually a gathering company charge.

- There could be an initial transportation charge from the wellhead to a field gathering point or to a processing plant by a gathering company.

Pipeline transportation might include more than one pipeline. Also, any additional charges, such as for treating, processing, and or storage, would enter into the net-back pricing. Exactly who pays for the transportation at each step is open to negotiation. Usually, the producer is responsible for the gathering and field costs of getting the gas to the transportation pipeline's inlet, either on its pipeline or at a terminal point, sometimes designated as a "hub." Many times when the transportation pipeline's pipe goes through a producing field, the producer will only be responsible for the gathering charges to get the gas from the wellhead to the field's central point for discharge into the pipeline's pipe. The gathering and field charges along with the transportation to the transportation pipeline inlet is what makes the difference between wellhead gas prices and into pipe gas prices.

Who pays for the transportation charges from the transportation pipeline's inlet to the city gate or distribution company's inlet, even if it includes more than one transporting pipeline, is negotiable between the marketing company selling the gas and the buyer. The marketing company selling the gas might quote a delivered price to the buyer, especially if the seller is holding transportation rights with the pipeline handling the transportation. If the buyer has the transportation, he might take the gas FOB (Free on Board, the point where title transfers and where transportation charges to that point are included in the sales price) at the transportation pipeline's inlet. These are all part of the marketing and negotiating in moving the gas from field to city gate or consumer.

The transporting pipeline in a few cases may actually deliver the gas to the consumer, especially for very large users like electric generating or chemical plants. Most of the time, the gas is delivered at the city gate, which means to the inlet of the distribution company serving a certain area. The final transportation from the city gate to the consumer is usually a part of the total gas price charged the consumer by the local distribution company.

This final transportation is the area open to change. In some states, distribution companies have already made the change and gone to open access transportation. In those areas where open access is part of the distribution company's tariffs, the transportation of the gas from city gate to consumer could be a separate charge and negotiable between the natural gas marketer and the buyer. Here, the distribution company will be selling transportation just like the interstate or intrastate pipelines do for moving the gas from the field to the city gate.

Transportation by pipelines is either on a firm or interruptible basis. Firm transportation means the pipeline guarantees the transportation buyer or seller space for a certain gas volume or energy content. When firm transportation is held, the transportation buyer will pay a two part rate; a demand charge guaranteeing space and a variable or transportation charge to cover the cost of moving the natural gas. The demand charge is a fixed cost for the volume of gas space guaranteed by the pipeline and is payable to the pipeline regardless of whether or not the transportation buyer actually uses the space to

transport his gas. The variable or transportation charge usually includes a deduction in the volume of the gas received for transportation as payment for the fuel used to transport the gas in addition to the monetary charge for the transportation.

In interruptible transportation, space is on a first come, first served basis and there is a single charge for the actual transportation. This transportation charge usually includes a deduction from the gas transported to pay for the fuel used in the transportation, the same as in the variable charge of the firm transportation. The fuel deduction for recompressing the gas along the transportation route and normal losses associated with transportation.

FERC Order 636 initiated a new type of transportation charge, commonly called a "brokered capacity" rate. Holders of firm transportation on a given pipeline have the opportunity if they are not going to use their guaranteed space to turn the space back to the pipeline for it to "broker" this returned space to natural gas buyers and sellers needing transportation. The transportation seller can take back its brokered capacity under certain conditions but, this makes a way for a transportation buyer or seller to get discounted transportation rates. The risk is there may not be any brokered capacity on a given line, or if demand becomes very tight, the original transportation "owner" might take back its capacity. For brokered capacity transportation, where the tariff for transportation might be 75 cents/MMBtu to transport the gas, the brokered charge might only be 15 to 25 percent of this transportation charge. Many times when gas demand is low, the change in transportation charges associated with brokered capacity charges makes a significant price differential in the gas. This can make the delivered price of natural gas vary considerable depending on the marketer and his ability to get brokered capacity transportation.

Another type of charge where considerably lower transportation charges are available is when the pipeline can "back haul" the gas. An example of this would be for the pipeline to be transporting gas from the Gulf coast to the West coast. On the way to the West coast, the pipeline also picks up gas in New Mexico bound for the West coast market also. The pipeline can "back haul" gas from New Mexico to the Gulf coast by shipping less Gulf coast gas to the West coast and instead picking up a larger volume of New Mexico gas to be moved to the West coast. This additional volume then takes the place of gas which would have been shipped from the Gulf coast to the West coast and the gas from the Gulf coast which did not get shipped west is handled as gas coming from New Mexico instead. The New Mexico gas gets to the Gulf coast without ever being shipped and is a paper transfer in reality. This is back hauling and the pipeline is able to offer very attractive transportation rates.

Much of the gas coming to the United States from Canada through the Northern Border pipeline to the Midwest is back hauled with gas from the Gulf coast destined for the Chicago market. The natural gas from Canada goes to the Chicago market and takes the place of actual shipments from the Gulf coast to Chicago. When this is done, gas from Canada is back hauled from Chicago to the Southwestern markets and can compete effectively against Southwestern produced gas. Some marketers in recent years have been able to put Canadian gas in Florida as cheap as putting Southwestern in Florida by back hauling the Canadian gas.

Charges for transportation by natural gas pipelines in interstate commerce are regulated by the FERC. Pipeline charges for transportation can be obtained from the pipeline or from the FERC by getting a copy of its tariffs. There are also industry reporting services which make tariffs available as part of their consulting services. Intrastate pipeline transportation is regulated by the state regulatory agencies. Currently, gathering transpor-

tation charges are regulated by the state agencies but this is a rather new area for these agencies. Before open access, much of the gathering in the field from individual wells and leases to a central pick up point for the pipeline was done by the pipeline as part of its transportation services and the charges were based on the pipeline's allowed rate of return. Since the pipeline unbundled these services and now either independent companies or non-regulated subsidiaries of the pipelines handle the gathering transportation, the regulatory burden has fallen on the state regulatory agencies.

An appreciation for the costs of transportation at each level can be seen in Figure 21.10 where the cost of gas at the wellhead, into pipe, city gate, and consumer's receipt point are given on a generalized basis. One should remember that these costs are heavily influenced by the amount of service needed to complete the transportation involved at each level and some levels include the addition of local taxes and other charges.

21.9 BUYING NATURAL GAS

Buying natural gas now is not too different from buying any other commodity. Like all commodities, a

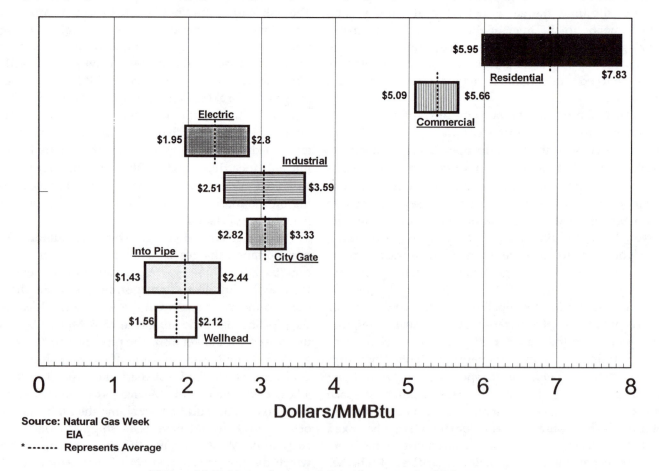

Figure 21.10 Natural gas pricing elements for U.S. 1994.

knowledge of the commodity and how it is transported and delivered are all a part of the buying scene. As discussed earlier, some of the current changes going on in the industry will continue to impact the business. Any knowledgeable buyer will keep on top of the changes and be aware of the significance of them to change his own purchasing activities. If one is a large buyer, it might be beneficial for the buyer to subscribe to one of the trade publications listed previously to be more knowledgeable of the gas industry.

Service, security of supply, and price play a major role in selecting the source of gas for the buyer. Certain suppliers, marketing companies and distribution companies retaining the merchant function have a greater ability to aggregate supplies and insure the buyer of supply and/or transportation. Prices will play a role as secure supplies could cost more than buying gas on the short term or spot market. Service such as storage and additional swing gas play an important role also. The buyer will have to take all of these aspects into consideration when looking to buying gas supplies and developing suppliers. Each of these aspects has an important role in determining who will be the marketer in bringing gas to the consumer.

Next to knowing the natural gas industry, an equally important aspect of buying gas is to know the parameters where the fuel is to be used. Knowing what are the fuel requirements on an annual, monthly, and hourly basis are crucial. Knowing what are the swings— how little and how much more than the norm will be required—is essential in buying natural gas. Knowing what the impact would be if fuel is curtailed or how many days can the company afford to be without natural gas and what other fuels can serve as back up if supplies should be curtailed are all important in developing a gas buying program. Security of supply is an economic consideration in gas buying and knowing what can be important and what is not will be instrumental in developing a cost/service quality buying program.

The buyer should be knowledgeable of the current fuel uses in his facilities. If natural gas is already in use and the buyer is looking to insure he is getting the best service for the price of the fuel, he should have knowledge of the current supplier, costs, delivery points, and supply considerations like percent swing, guaranteed takes, and any other supply considerations. Knowledge of this information is a necessity in developing an analysis of the lowest cost suppliers for the natural gas.

In buying and selling natural gas, a contract between the two parties will spell out the details of the transactions. Even though much of today's gas deals are done over the telephone or by computer, whenever there is an on going relationship of supplying gas over a period of time, there will be a contract to cover the buying and selling conditions. There might be a contract to supply gas for as little as an hour or for as long as 20 years. Prior to the deregulation of the industry, most contracts for the sale of gas at the wellhead were for periods of 10 to 20 years. When the spot market evolved in the 1980s and some sales were for as little as an hour, long term contracts became obsolete as they were known previously. Typically today, anything longer than two years is considered a long term contract with the exception of gas being supplied to independent power producers where a long term contract of 15 to 20 years is required in the financing of the project.

The contract, among many other things, will usually have the following major areas of consideration. Since today, sellers and buyers will vary considerably by their position in the gas industry, the contract needs to be tailored specifically between the two or more parties involved. A contract buying gas from a distribution company will be different in some aspects than the contract between a marketing company and the buyer.

1. **Purpose & Scope of the Agreement**—What is to be accomplished by the contract.

2. **Definitions**—Lists the standard and special terms used in the contract, which is especially important in natural gas contracts because of the uniqueness of the industry and commodity.

3. **Term of the Agreement**—A statement giving how long the agreement will be in force and what conditions will terminate the agreement. Some contracts will include information on extending the contract past the initial term of the contract.

4. **Quantity**—This section will detail the total quantity of gas to be sold and delivered by the seller. Information on the daily contract quantity (DCQ) as well as any other specific periodic quantities will be included in this section. Included are the penalties the seller is willing to accept for the buyer's failure to take the gas as well as penalties the buyer is willing to accept for the seller's non-performance. This is the section where take-or-pay language, if any, will be in the new contracts. Also, a "best efforts" clause on the part of either party may be included in this section. The newer contracts today often have additional language protecting either party from the vagaries of today's market

place. While these in effect reduce the coverage of the contract, often both parties are willing to have a contract with a legal means of changing the conditions. This section of the contract might also include ways for the buyer to nominate volumes of gas for delivery as well as any means for testing well deliveries or operational procedures related to quantity depending on whether the seller is a producer or reseller. This section might also include penalties for either over or under takes and balancing conditions.

5. **Price**—The price to be paid for the gas to be delivered as well as statements on any price escalators or means to renegotiate the price will be stated in this section. Any language needed for agreements on price indexing or other means of adjusting the price to current market conditions will be included also. Good contracts today will include price changes up or down depending on changing market conditions. The Price Section will cover any additional expenses or costs the buyer is willing to undertake in addition to the direct cost of the gas. If the contract is with the producer or an interest owner in a gas well, it might state who will be responsible for any gathering or processing charges. Again, for a contract with the seller being a producer, provisions will be in this section for who has responsibility for severance and other taxes, royalties or other charges the producer might be liable for payment.

6. **Transportation**—Transportation must be covered in the contract. The buyer must insure he is covered in case transportation is unobtainable or ceases after deliveries have started. This section will include any special conditions on either the buyer's or seller's part to take into account any special situations either party could have to interfere with the gas transportation. The contract should also cover who has responsibility with the transporter for overages, measurement, disagreements on quantities, and payment of transportation and associated charges.

7. **Delivery Point(s)**—Since the delivery point is different in each situation, the contract needs to state the delivery or alternate delivery points in very specific language.

8. **Measurement & Quality**—Conditions for the measurement and the quality of the gas need to be stated as well as how these conditions will be tested. Alternatives should be given in those cases where the gas fails to meet the quality requirements.

9. **Billing**—Billing terms and who is responsible for payment are included in this section. How payment is to be paid and terms for payment are included also.

10. **Force Majeure**—This clause allows either party affected by a totally unforeseeable occurrence which is beyond the control of the party (an act of God) to get out of the contract.

11. **Warranty of Title**—This clause guarantees the seller has title to the gas and can sell it. Included are allowances for the buyer to recover damages if there is a failure of title. The clause includes language stating the gas is free of all liens.

12. **Regulatory**—All necessary permits, licenses, etc. must be obtained according to FERC regulations and any state or local authorities having jurisdiction over the gas. Who has responsibility for obtaining these should be covered.

13. **Assignments**—Specifications for the transfer of rights under the contract would be covered in this section.

While in prior times, a typical contract would be of the order of magnitude of 10 to 30 pages, today with short supply times some contracts are no longer than a page or two. Since a contract is a binding legal instrument, the buyer should be knowledgeable enough to understand the general terms of the contract. Especially in early days of his or her buying gas, he should consult a lawyer for guidance and review of the contract. During the days of changing times in the gas industry, sometimes a contract looked worthless. In the end though, many millions of dollars were paid as damages when parties breached or did not live up to the conditions of the contract. One should be sure the contract covers what is wanted in buying the natural gas. Because of the terminology and precedent in the industry, legal aid is a must. Most short term agreements are on a "best efforts" basis. In these cases, contracts can be short and simple to reflect the actual sales agreements. The buyer must recognize what he wants in the purchase and the degree of supply/price security he is comfortable in obtaining. Greater degrees of security are actually non-monetary pricing conditions. The cost of the gas to the buyer will reflect these conditions.

Depending on who the consumer is and from whom he is buying the gas, the responsibility for transportation will vary. In most situations where the seller is the producer, the buyer will be responsible for transportation. In recent years, marketing companies representing producers have made firm contracts with pipelines to insure having transportation capacity. At times, the marketer will supply the transportation. In buying natural gas at the local level, the distribution company could be both, the merchant and the transporter. As more states adopt open access rules for local distribution companies, they might only supply the transportation and a marketer will be the gas merchant. In all transactions, the seller should at least assist in developing or be knowledgeable of the transportation to insure the gas gets to the desired delivery point.

A big issue to the buyer is the question of whether to buy the natural gas on a short term or a longer term contract. Spot sales, sometimes called "swing sales" can be for as short a duration as hours up to 30 days. Short term purchases are usually for 30 days one month or up to six months. Each period, the buyer will seek supplies. Longer term contracting could be for six months or an annual basis calling for gas for two to three years. Even though spot or short term contracts are usually relatively simple, they are manpower intensive because of the time needed for the continuing negotiations on supply and sometimes, transportation. Short term buying has little supply and price security and puts the buyer at risk for future supplies.

The supply security is less a risk. In an open market with no price regulations, like now, supply is almost always available at some price. If the buyer cannot find gas at, say $2.50/MMBtu, by raising his bidding price, he will eventually find supplies. The price risk is the greater problem on short term buying. There are risk management tools for hedging including the use of alternate fuels if gas prices temporarily rise too high for the buyer's budget. The prudent buyer will have looked at the necessary needs for gas and made his decision on the contract term and pricing protection needed.

For longer term purchases, the parties will spend considerably more time in developing the original contract. But, on a unit time basis, there is considerable economy in the longer term purchases. Sometimes, when the marketer negotiates a longer term agreement, there will be a premium on the gas price for the longer agreement because of the security of supply it affords. This might be in the form of an added charge to the price of the gas or it might be included in having to pay an

overall higher price for the gas. Even in longer term contracts, in today's markets the contract will have a pricing provision where the price is "indexed" to changing market conditions and can vary either up or down on a periodic basis, usually month to month. The index is the reference for changing the price as periodically set.

Selecting an index for natural gas contracts is a negotiation in itself. There are two major ways the buyer can seek to negotiate. The price changes can be triggered by changes in natural gas prices or changes in the buyer's business conditions. The more typical way is for the changes in natural gas prices to trigger the price changes paid by the buyer. Many contracts call for the price each month to be set by prices published in the trade publication, *Inside FERC Gas Market Report*, the pricing edition from the publisher of *Inside FERC*, which is published twice a month. In the first issue of the month, a listing of prices for natural gas into the major transportation pipelines at different locations is listed. This is called the "index" price for the coming month. The data for making this index price is obtained by the publication by polling selected gas sellers and buyers to estimate the coming month's index prices based on prices received during the last portion of the closing month. When contract prices are set by this index, the price is constant for the month as listed by the index even though during the month gas prices might vary significantly. Price might go up or down but the price paid by the buyer is the same as listed in the index. The indexed price will usually be given for a certain pipeline and specific pipeline pick-up point. This is just one of the models for setting or indexing gas prices to meet the gas' price volatility.

Other references of gas prices can be used for indexing. Sometimes, the price paid on the New York Mercantile Exchange (NYMEX) for the near month is used as an index of current prices. Other publications publish price summaries and one of these might be used as the price index also. This is a negotiable point between buyer and seller. Sometimes a "basket" of indexes will be used. This is more common in international gas contracts but is sometimes seen in special occasions in the U.S.

Many times, when the buyer is a large consumer of gas and the price of the gas plays a major role in the consumer's business such as in the ammonia industry, the buyer will seek to index the natural gas price he pays against the changing value of his own product. Ammonia and agriculture products made from ammonia have volatile prices on an annual basis. Natural gas, used for fuel and feedstock in making ammonia and the other products using the ammonia, constitutes a major portion of the manufacturing cost of the products. The buyer would like to "hedge" his risk in the market place by indexing the raw material price to his own product prices. Again, this is an area of negotiation between buyer and seller.

Contract prices can be set for a fixed price for a given period. In this case, the seller, buyer, or marketer making the offer will hedge the fixed price risk by going to the futures market and buying or selling futures contracts to hedge the price paid or to be received for the gas. No one will make a fixed price contract without the proper risk management protection.

Once the buyer knows the quantities of natural gas needed and conditions that need to be covered in buying the gas, the big question is from whom to buy the gas'? In large purchases by industrials and utilities, this is easier to answer, even though there are more choices, than in the smaller applications whether it be for commercial, industrial or utility users. Under today's market and regulatory conditions, the residential user has little choice and usually buys from the local distribution company. In some cities, more than one distributor is available. Then even the residential buyer has an option.

Under a plan for open access at the distribution level, over which individual states will have regulatory control, everyone would have an opportunity to buy gas from the marketer of his choice and the distributor will be only the transporter. In all likelihood, even with open access at the distributor level, the distributor in the local area will still be a merchant as well as the transporter and might be the best choice for gas supply under certain conditions.

As this is the situation, the following discussion on buying natural gas is applicable to the larger gas buyer regardless of his specific use for the gas. The large buyer might be an industrial, utility, or commercial user. Usually in today's gas industry, in a typical urban location, gas requirements over a few million cubic feet per day are required to justify seeking suppliers other than the local distribution company.

In looking to who might be the supplier, it is not unusual to have more than one especially if the facility receiving the gas has multiple connections from different pipelines or distribution companies. Large gas consumers might have more than one connection allowing for picking multiple suppliers or to allow for competition between suppliers. Many large consumers have put in additional facilities just for these reasons.

A list of the major gas marketers is given in Appendix 21.2, Table A. This list was current at the time of publication, but with the rapid pace of mergers and strategic alliances between marketers, the list changes constantly. All of the major interstate pipelines have non-regulated marketing companies, as do many of the natural gas producing companies. All of the companies supplying natural gas liquids have marketing subsidiaries selling natural gas along with the natural gas liquids. Most of the major marketing companies are mainly sellers in the wholesale business selling to resellers. For the retail business, sales to a consumer, many of the marketers are non-regulated affiliates of the distribution companies or some of the smaller marketing companies. Table B in Appendix 21.2 lists some of the current marketing companies having strong retail marketing operations, including some of those affiliated with local distribution companies.

A crucial factor in selecting a marketing company is to insure the company can provide the service needed to insure the security of supply at the price developed during the negotiations. As the gas marketing group began to materialize, and the merchant function transferred from the interstate pipelines to the marketing companies, it became apparent financial strength of the marketing company was an essential criteria for doing business. Most producers in selling their gas will require a letter of credit or some other instrument to insure payment for the gas. The buyer should be equally concerned about the credit worthiness of the marketing company he is dealing with to supply his gas. Working with the subsidiary of a large company offers some protection, but still credit consciousness is important.

In addition to the reputation and security of the marketing company is important relating to financial matters, the company's reputation in its ability to provide gas as promised is also important. Even though there may be contractual provisions insuring the buyer against non-performance, the important function of the buying company's purchasing agent is to insure getting the gas when needed. Shutting down a furnace or having to stop an assembly line because of low fuel supplies can be the success or failure of the buyer. Supplies are critical and during certain times of the year when gas supplies might be tight because of field supply problems or transportation failures, the value of the marketer in getting gas to the buyer can well be worth the penny or two more spent in selecting a reputable and known supplier.

Since natural gas prices are usually seasonally impacted, buyers wanting longer than short term purchases should pick a time when gas prices are low to make the longer contract. Even though most contracts contain a clause to adjust the price on a regular basis, buying longer term supplies will be influenced by the current gas buying atmosphere. Negotiating a three to four year contract for natural gas on a sub-zero day during a winter month is not a prudent time. The buyer would be better to pay the higher prices for short term supplies and seek a better time for the longer term contracting.

Risk management of the commodity's pricing is an essential element for the buyer as it is for the producer and marketer. Like all major commodities such as grains and metals, natural gas has an active futures market. Currently, there are two contracts available, one covering natural gas at the Henry Hub in Louisiana supplied by the New York Mercantile Exchange (NYMEX) and the other contract for natural gas at the Waha, Texas hub from the Kansas City Board of Trade. A third contract is planned to start in mid to late 1996 offered by the NYMEX for natural gas at a location in Western Texas. Additional contracts at other natural gas hubs or terminals are planned for the future because of the large difference in gas prices at different producing basins in recent times. In all likelihood there will be one out of Canada and maybe one closer to the north central U.S. markets. In addition to the exchanges offering futures contracts, there are many banks and other financial institutions making a market in gas futures. While the commodity exchange contracts currently go out for 36 months, some of the financial institution markets go out for much longer times.

Hedging in commodities is an art in itself. Today there are options and many other types of instruments for working the futures market. Derivatives trading, though very speculative in nature, is another out growth of the risk management arena. The buyer's management should set very strong parameters if the buying company wants to hedge the natural gas price. Hedging allows both the buyer and seller market insurance from price fluctuations. When the futures buyer is not an active user of the commodity, any buying or selling of futures is done on a speculative basis. Many companies have been financially hurt because the hedging function became a speculative one and considerable losses were incurred before management could stop the activities.

Hedging can be a good business practice when done properly and within restraints imposed by management. The volatility of gas prices, where in a 12-month period they might double or go even higher, makes risk management an important part of the buying and purchasing activities. Especially when the fuel or feedstock requirements are a major portion of the cost of business operations, either as a part of the product costs or facilities operations, risk management of fuel prices can be critical to profitability. With a good hedging program, natural gas supplies can be purchased over a relatively long term at fairly constant prices. Almost all of the major marketing companies have groups well acquainted with hedging operations. Either through their own sales program or working with the buyer, the marketer can work out a relatively constant purchase price using the futures markets for up to three years and using other financial institutions making a market in natural gas for longer periods.

Appendix 21.1

Natural Gas Industry Publications & Wire Services Having Current Natural Gas Prices & Statistics

Publications

Bloomberg Natural Gas Report	**Gas Daily**	**Inside FERC**
P.O. Box 888 Princeton, NJ 08542-0888	600 Travis Street Suite 2340, Houston, TX 77002	McGraw Hill Publications
A/C 609/279-4261 Weekly Publication	A/C 713/225-6035 Daily Publication	1221 Ave. of the Americas
		New York, NY 10020
		A/C 800/223-6180
Natural Gas Intelligence	**Natural Gas Week**	
8425 B Carlise Dr. Herndon, VA 22070	1401 New York Ave. NW, Washington, DC 20005	Bi-monthly Pub'n
A/C 703/318-8848 Weekly Publication	A/C 202/662-7000 Weekly Publication	

Wire Services

Bloomberg Energy Services	**McGraw Hill**	**Reuters**
Princeton, NJ 08542-0888	New York, NY 10020	New York, NY 10021
A/C 609/279-4264	A/C 800/438-6992	A/C 212/603-3300

Appendix 21.2 Table A

U.S. Major Natural Gas Marketing Companies—Spring 1996

Rank	Company	Parent Cos.	Address	Estimated Current Avg. Annual Volume Bcfd
1	NGC Corporation	Chevron Corp. and NGC	1301 McKinney Houston, TX 77010	10.0
2	Enron Capital and Trade	Enron Corp.	1400 Smith Street Houston, TX 77002	9.2
3	Pan Energy	Mobil Oil Corp Panenergy Corp.	5718 Westeimer Suite 1250 Houston, TX 77057	8.0
4	Amoco Gas Marketing	Amoco Corp.	550 Westlake Park Blvd Houston, TX 77079-2696	5.5
5	Williams Energy Services Co.	The Williams Companies Tulsa OK 74172	One Williams Center	4.5
6	Coral Energy Resources	Shell Oil & Tejas Pipeline L.P.	200 North Dairy Ashford Houston, TX 77079	4.4
7	Transcanada Gas Services	Transcanada Pipeline	111-Fifth Avenue S.W. Calgary, Alberta T2P 3Y6 Canada	3.6
8	Coastal Gas Services	Coastal Corp.	7373 East Doubletree Ranch Road, Suite 200 Scottsdale, AZ 85258	3.5
9	Texaco Natural Gas	Texaco Oil Co.	1111 Bagby Street Houston, TX 77002-0200	3.5
10	Columbia Gas Marketing	Columbia Gas System	20 Montchanin Road Wilmington, DE 19807	3.2
11	El Paso Gas Marketing	El Paso Natural Gas Co.	100 North Stanton El Paso, TX 79901	3.2
12	Hadson Gas Service Inc.	Louisville Gas & Electric Co.	2777 Stemons Freeway Suite 700 Dallas, TX 75207	3.0
13	Noram Energy	Noram Energy	1600 Smith Suite 1220 Houston, TX 77002	2.5
14	Valero Natural Gas	Valero Energy	1200 Smith Street Suite 900 Houston 77002	2.5

15	Vastar/Arco	Arco Oil & Gas	1601 Bryan Street Dallas, TX 75201-3499	2.5
16	Western Gas Resources, Inc.		12200 North Pecos Street Denver, CO 80234-3439	2.5
17	Conoco	Dupont Inc.	600 North Dairy Ashford Houston, TX 77079	2.5
18	Aquila Energy	Utilicorp Corp	2533 North 117th Avenue Suite 200 Omaha, NE 68164-8618	2.2
19	Sonat Marketing	Sonat & Atlanta Gas Light	5 Greenway Plaza 18th Floor Houston, TX 77046	2.1
20	Exxon Gas Marketing	Exxon	800 Bell Street Houston, TX 77002-7426	2.0
21	Proenergy (Producers Energy Marketing, LLC)	Apache, Arker Parsley& Oryx	2000 Post Oak Suite 100 Houston, TX 77056-4400	2.0
22	Koch Gas Services	Koch Industries	600 Travis 14th Floor Houston, Tx 77002	1.8
23	Union Pacific Fuels, Inc.	Union Pacific Resources	801 Cherry Street MS 3200 Fort Worth, TX 76102	1.8
24	KN Energy Gas Marketing	KN Energy	P.O. Box 281304 Lakewood, CO 80228-8304	1.6
25	Catex-Vitol		470 Atlantic Avenue Boston, MA 02210-2208	1.5
26	CNG Energy Services	Consolidated Natural Gas Co.	Four Greenspoint Plaza 16945 Northchase Dr., #1750 Houston, TX 77060	1.5
27	Midcon Gas Services	Occidental Petroleum Co.	3200 Southwest Freeway Suite 3000 Houston, TX 77027	1.5
28	Tenneco	Tenneco Energy	1010 Milam Houston, Tx 77002	1.3

(Continued)

Appendix 21.2 Table A

U.S. Major Natural Gas Marketing Companies—Spring 1996

Rank	Company	Parent Cos.	Address	Estimated Current Avg. Annual Volume Bcfd
29	Amerada Hess	Amerada Hess	One Allen Center 500 Dallas Street, Level Two Houston, TX 77002	1.2
30	Meridian Oil, Inc	Burlington Resources	5051 Westeimer Suite 1400 Houston, TX 77056-2124	1.2
31	Penn Union Energy Services	Pennzoil Corporation Brooklyn Union Gas Co	1330 Post Oak Boulevard 20th Floor Houston, TX 77056	1.2
32	Unocal Gas Marketing	Unocal	1201 West Fifth Street Los Angeles, CA 90017	1.2

Appendix 21.2 Table B

Natural Gas Industry Marketing Companies
Significant Retail Sales Operations

COMPANY	ADDRESS	PARENT COMPANY
AQUILA ENERGY	2533 North 117th Avenue #200, Omaha NE 68164-8618	UTILICORP CORPORATION
CATEX-VITOL	470 Atlantic Avenue, Boston MA 02210-2208	CALTEX-VITOL
COKINOS	5718 Westeimer #900, Houston TX. 77057	
ENRON CAPITAL & TRADE	1400 Smith Street, Houston TX. 77002	ENRON CORPORATION
HADSON GAS SERVICES	2777 Stemons Freeway #700, Dallas Tx. 75207	LOUISVILLE GAS & ELECTRIC
NORAM ENERGY	1600 Smith #1220, Houston TX 77002	NORAM ENERGY CO.
PEOPLES ENERGY SERVICES	122 South Michigan Avenue, Chicago IL 60603	PEOPLES GAS (Chicago)
PINNICAL-OLYMPIC	5100 Westeimer#320, Houston TX 77056	PINNACLE GAS
RESOURCE ENERGY SERVICES	1915 Rexford Road,Charlotte NC 28211	PIEDMONT NATURAL GAS
TECO	13355 Noel Road, Dallas, TX 75240	TECO CORPORATION
TEXLA	1100 Louisiana, Houston, TX 77002	
VESTA	320 ONEOK Plaza 100 West Fifth St., Tulsa OK 74103	
WICKFORD	2323 South Shepherd #810, Houston TX 77019	
WILLIAMS ENERGY SERVICES	One Williams Center, Tulsa OK 74172	WILLIAMS COMPANIES
YANKEE GAS SERVICES	599 Research Parkway, Meridian CT 06450	YANKEE GAS COMPANY

CHAPTER 22

CONTROL SYSTEMS

ALAN J. ZAJAC
JAMES R. SMITH

Johnson Controls Inc.
Milwaukee, Wisconsin

22.1 INTRODUCTION

Economic pressures on building managers continue to force reductions in operating expenses, while at the same time social pressures from building occupants and government regulation require improved indoor air quality, better temperature control comfort and improved lighting quality. Controls on HVAC systems as well as the quality, type and design of the lighting systems impact these issues. Chapter 12 discusses the use of EMCS. Later within this chapter we will discuss how controls are used in the EMCS.

Controls manage the use and demand of the HVAC and Lighting systems. Without controls, life as we know it would change dramatically. You couldn't turn on a light or adjust the temperature in your home. From a simple on-off switch to complex computerized proportional integrated control sequence we are continually impacted by the control of energy. Fundamentally a control system does four things: 1) Establishes a final condition, 2) Provides safe operation of equipment, 3) Eliminates the need of ongoing human attention, and 4) Assures economical operation. Hardware, software, installation materials (wire, tubing, etc.), HVAC processes and the final condition under control, must all play together to insure occupant comfort.

22.2 THE FUNDAMENTAL CONTROL LOOP

Control systems can be simplified by breaking the system into functional blocks. When examined in smaller manageable pieces any system becomes simpler to understand. For that reason this chapter will begin with an overview of a control system from a functional block perspective. See Figure 22.1.

Final conditions at the end of the block diagram represent the fruit of the controls system's efforts. This is the reason control systems exist. Most HVAC systems are designed for occupant comfort and will control the final conditions of temperature, humidity and pressure.

Controls system types include:
- Self Contained Controls Systems
- Pneumatic Controls Systems
- Electric Control System
- Electronic Control System
- Digital Control System

22.2.1 Self Contained Controls Systems

Self-contained contained control systems combine the controller and controlled device into one unit. In this system, a change in the controlled medium is used to actuate the control device. A self-contained valve with a vapor, gas, or liquid temperature sensing element that uses the displacement of the sensing fluid to position the valve, is one example of a self-contained control system. A steam or water pressure control valve that uses a slight pressure change in the sensing medium to actuate the controlled device is an example.

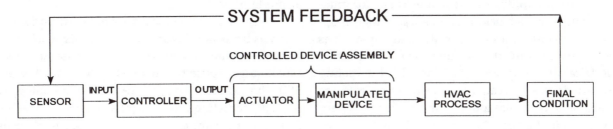

Figure 22.1

22.2.2 Pneumatic Controls Systems

Pneumatic control systems use compressed air to modulate the controlled device. In this system, air is applied to the controller at a constant pressure, and the controller regulates the output pressure of the controlled device according to the rate of load change. Typically, compressed air at 20 PSI is used. However, the controlled device can be operated by air pressures as high as 60 PSI. Advantages of pneumatics: actuators for valve and damper controls are inexpensive, easily maintained, and cost effective. The technology is mature, controls are reliable, and different manufacturer components can be used interchangeably. Disadvantages are: they require clean dry air, calibration of the controls on a regular basis, and customized complex control panels for advanced temperature control systems.

22.2.3 Electric Control Systems

Electric control systems use electricity as the power source of a control device. This system can have two position action in which the controller switches an electric motor, resistance heating element, or solenoid coil directly or through microprocessor based electro-mechanical means. Alternatively, the system can be proportional so that the controlled device is modulated by an electric motor. Advantages are the two-position controls are simple and reliable and use simple low voltage electrical technology. Disadvantages are the controls cannot modulate and actuators can be expensive.

22.2.4 Electronic Control Systems

Electronic control systems use solid state components in electronic circuits to create control signals in response to sensor information. Advantages are that modulated controls are reliable and require less calibration and use electricity. Disadvantages are actuators and controllers are expensive.

22.2.5 Digital Control System

Digital systems controllers utilize electronic technology to detect, amplify, and evaluate sensor information. The evaluation can include sophisticated logical operations and results in a output command signal. It is often necessary to convert this output command signal to an electrical or pneumatic signal capable of operating a controlled device. Advantages are that controls are highly reliable and require minimal maintenance. Disadvantages are initial costs which may be high.

22.3 SENSORS

Certain basic field hardware is necessary for a control system to function properly. Sensors provide appropriate information concerning the HVAC control system. Communications paths must be available to transmit sensor and control information. Often referred to as inputs, sensed signals convey either analog or binary information. Analog Inputs convey variable signals such as outdoor air temperature.

Binary Inputs convey status signals such as fan or pump status, ON or OFF.

This network of field hardware must function properly if the building control system is to be effective. It is a distinction of professional building management for the entire network of sensors, controllers and communications to remain functioning and accurate. This necessitates an investment in effective preventive maintenance and continuous fault monitoring and correction, but pays rich dividends in the ability to provide a well controlled, cost effective environment. See Figure 22.2.

Sensors types include:
- Temperature
- Humidity
- Pressure
- Air Quality

22.3.1 Temperature Sensors

Temperature sensing elements are a critical part of any building control system. Various temperature sensors are available, to meet the requirements of the full range of applications.

A bulb and capillary element contains thermally sensitive fluid (in the bulb) which expands through the capillary as the temperature increases. The familiar mercury thermometer is one example. Most control system applications connect a diaphragm to the capillary, so that fluid expansion changes the internal pressure and therefore the relative position of the diaphragm. This type of sensor is used in ducts (insertion or averaging) or piping (immersion) applications. Figure 22.3 shows an example of each.

A bimetal element consists of two dissimilar metals (such as brass and invar) fused together, Each metal has a different rate of thermal expansion, so that temperature changes cause the metal strip to bend. The bending of the bimetal strip may cause an electric contact to open or close, actuating the manipulated device. The bimetallic element is simple and common, and is often used in

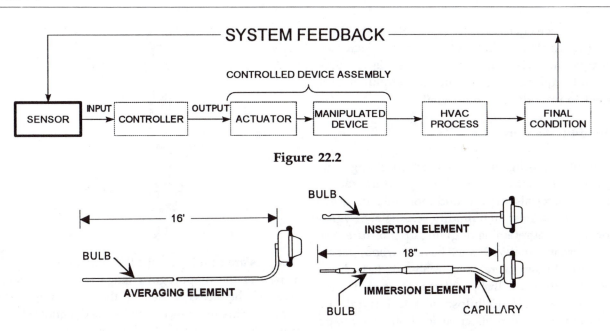

Figure 22.2

Figure 22.3

pneumatic room mounted thermostats. An example of a bimetal element installed on a pneumatic room thermostat is shown in Figure 22.4.

A rod and tube element consists of a high expansion tube inside of which is a low expansion rod, attached to one end of the tube. The high expansion tube changes length as the temperature changes, causing displacement of the rod. This type of sensor is sometimes used in immersion thermostats.

A sealed bellows element is either a vapor, gas, or liquid filled element. The fluid changes in pressure and volume as temperature changes, forcing a movement that may make an electrical contact, adjust an orifice, or react against a constant spring pressure to activate a control. This type of sensing element is used in room thermostats and remote bulb sensing thermostats. An example of a vapor charged element is shown in Figure 22.5. Here a corresponding temperature pressure relationship will establish a given motion.

A thermocouple is a union of two dissimilar metals (e.g., copper and constantan) that generate an electrical voltage at the point of union. The voltage is a nonlinear function of temperature. This sensor can be sensitive to noise, as voltage levels are typically in the millivolt range, and changes per degree are in the microvolt range. The sensing element of a gas oven is typically of this type.

A resistance element or resistance temperature detector is a wire of nickel platinum or silicon chip in which electrical resistance increases with rising temperature. Therefore, if a constant voltage is placed across the wire, the current would decrease. This change is then

transformed by the controller to a suitable output signal. These type of sensing elements are popularly used in electronic control systems.

A thermistor is a semiconductor device in which

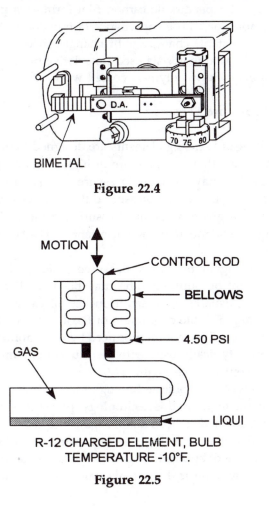

Figure 22.4

R-12 CHARGED ELEMENT, BULB TEMPERATURE -10°F.

Figure 22.5

electrical resistance changes with temperature. It differs from a resistance element in that its resistance decreases as the temperature rises. It is usually used in an electronic circuit, and its output must be amplified and transmitted to provide a usable signal.

22.3.2 Humidity Sensors

Humidity sensing elements react to changes in relative humidity within a given temperature range. Two types of materials, organic and inorganic are used. Whatever the type most all humidity sensors are hygroscopic, or capable of retaining or giving up moisture. For many years organic materials such as hair, wood, paper, or animal membranes were used. Materials have been developed to eliminate the problems associated with these fragile materials. Cellulose acetate butyrate or C.A.B. sensors have been effective for both pneumatic and electric controls.

More sophisticated, electronic sensors use capacitive sensing circuits. A typical material used for sensing relative humidity is a polymer whose dielectric constant changes with the number of water molecules present in the material. The polymer film is coated on both sides with a water penetrable carbon film forming a parallel plate capacitor whose capacitance (a function of the dielectric constant) changes with changes in relative humidity. Typical materials used for relative humidity applications include polymers coated with carbon.

22.3.3 Pressure Sensors

Pressure sensing elements are designed to measure pressure in either low pressure or high pressure ranges. The device may measure pressure relative to atmospheric pressure, or the pressure difference between two points in a given medium. Pressure sensing in HVAC piping, ducts and tanks is important to HVAC system control.

For higher pressures measured in PSI, a bellows, diaphragm, or Bourdon tube may be used. For lower pressures, usually expressed in inches of water column, WG, a large flexible diaphragm or flexible metal bellows is often used. The motion produced by the diaphragm or bellows is typically used with mechanical pneumatic or electric controls. See Figure 22.6.

Newer electronic technologies are using piezo-resistive sensing. In this technology pressure sensitive micro-machined, silicon diaphragms are used. Silicon can be described as being a perfect spring which is ideal for this type of application. Unlike the large diaphragms used in mechanical controls micro-machined dia-

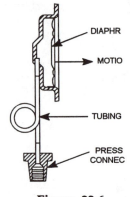

Figure 22.6

phragms are small, perhaps 0.1" square. Resistors on the surface of the diaphragm change resistance when subjected to the stress caused by the deflection of the diaphragm. This resistance is the sensed signal which communicates the pressure to the controller.

22.3.4 Air Quality

Air quality sensing elements are designed to measure components of air, be it Carbon Dioxide (CO_2), Carbon Monoxide (CO), Oxygen (O_2) levels or Volatile Organic Compounds (VOCs). Various electronic means are used to sense these, from infrared sensing to heated electronic elements. Today the costs are high and the units many times are high maintenance items. A definite return on investment analysis or air quality risk assessment should be made before implementing the use of these. In the future the quality will go up and the cost will go down and more ventilation uses will be created for these sensors.

22.4 CONTROLLERS

Controller types include:

*   Two Position
*   Proportional Action
*   Proportional plus Integral (PI)
*   Proportional Integral Derivative (PID)

Controllers are devices which create changes, known as system response according to sensor information. Controllers play the critical role of maintaining the desired building conditions.

Controllers produce five distinct types of control action to control a buildings environment at desired settings. These types of control action will be presented, beginning with the simplest and progressing through the most sophisticated. Other types of control action are available.

22.4.1 Two Position Controls

22.4.1.1 Two position controls

Simple Two position controllers turn the heating or cooling fully on and off, as the temperature varies. The function is the same as simple home thermostats. When heat is required the thermostat turns on the furnace. The furnace continues to run until the temperature has risen a few degrees above where it was when the furnace turned on. It then shuts off. This difference between the lower temperature (furnace on) and the higher temperature (furnace off) is referred to as the thermostat's temperature "differential." Two position controls have limited use in commercial buildings because of the relatively crude form of control. Applications are usually fan coils in vestibules or unit heaters in shipping areas.

An example is shown in Figure 22.7.

It is characteristic of two position control systems to overshoot the set point, the setting of the controller or thermostat. The overshoot is usually greater during low load conditions. For example, consider the home heating thermostat set at 75°F with a four degree operating differential. As the temperature falls to 73°F, the system heat input is less than the building heat losses because the furnace is off. Before the rate of heat input from the furnace can equal and surpass the heat losses, the system must operate for a period of time during which the temperature will continue to fall below the 73°F turn on temperature. Conversely, when the temperature rises to 77°F, the thermostat will turn off the heating system. Since the system heat input continues for a short time, more heat is provided than required to maintain 77°F. As a result space temperatures will continue to rise and over shoot.

22.4.1.2 Two Position with Anticipation

To partially offset the overshoot phenomenon just mentioned, two position controllers with anticipation were developed. This involves placing a heating element inside a heating thermostat which is activated when the thermostat activates. The heat from this element falsely loads the thermostat, causing it to deactivate before the controlled space overshoots the thermostat setting.

22.4.1.3 Floating

Floating control responses are similar to two position control responses, except that the controller produces a gradual continuous action in the controlled device. The controlled device is normally a reversible electric motor. This type of control action produces an output signal which causes a movement of the controlled device toward its open stem down or closed stem up position until the controller is satisfied, or until the controlled device reaches the end of its travel. Generally there is a dead band or a neutral zone in which no motion stops as is of the controlled device is required by the controller. In a heating or cooling system using floating control, the controller modulates a source of heating or cooling (e.g., air, water, or steam) that is maintained at a constant temperature. The controller for floating action systems normally has a dead band differential of 1 to 2 above and below the set point.

Assume the controller controlling a steam valve is set at 70°F and has a dead band differential of 2 (69 to 71). As the discharge temperature falls below 69, the controller will energize the steam valve motor to open the valve gradually until the temperature at the controller's sensing element has risen into the dead band range. When the controlled temperature continues to rise above 71, the controller will energize the steam valve motor to close the valve gradually. The motor's speed for this type of control is slow, and it may take a minute or two for the valve to move from fully closed to a fully open position. The motor speed must be compatible with the desired rate of temperature change in the controlled area. This type of control system will function satisfactorily in a heating or cooling system with slow changes in load. Figure 22.8 shows a graphic response.

22.4.2 Proportional Action

In this form of control the controller's output signal

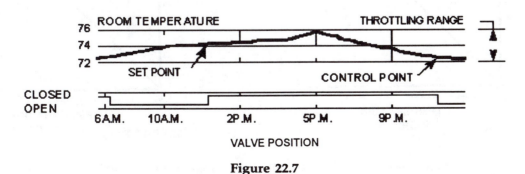

VALVE POSITION

Figure 22.7

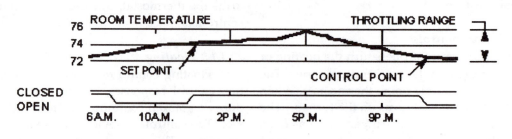

Figure 22.8

varies in proportion to the change in control variable measured, from zero 0% (no value, closed) to maximum 100% (full value, open) or design value. The output signal varies proportionately throughout its output range. A proportional controller's output variation is usually adjusted to produce a given number of units of change, such as one or more pounds per square inch (PSI) change, or one or more volts change, or one or more milliamps change per degree change in temperature. Common terms for the controller settings which accomplish a unit change per degree are *sensitivity* or *throttling range adjustment* for pneumatic controls and bandwidth for analog electronic controls. Another way to express the controllers response is by *proportional band*, which is the amount of change in the control variable required to cause the controlled device to travel from 0-100%.

The variation in output signal is either a direct or reverse relationship to the variation in temperature change measured. The set point of the controller is established by producing an output signal in the mid-range (50%) of the device or devices it is controlling. An example is shown in Figure 22.9a. When modulating devices from open to closed, the controller has done so because of a change in load. Many controllers operate with a set point aligned to the middle of the actuators

operating range. Some controllers work from a 0% or closed position, opening only on a call for cooling or heating. See Figure 22.9b.

Proportional action describes the relationship between a controller and the controlled device. In this system, the controlled device assumes a position that is proportional to the magnitude of the load sensed by the controller. Proportional systems have an operating range called the throttling range or proportional band, which is the change in the controlled variable required to move the controlled device from open to closed. Such a system tends to reach a balance within its throttling range. The balance point is related to the magnitude of the load at a given time. If the throttling range of the control is 4, and the set point is 55°F, the load and control system should balance at 57°F under maximum cooling requirements, and at 53°F under minimum cooling requirements. See Figure 22.10.

Much like an automobile without cruise control, proportional control will not maintain a desired setting. Proportional control systems, like your automobile, are subject to variations in the load much like the road. As the terrain changes, so does the control point. The difference between setpoint and control point is known as offset. See Figure 22.11.

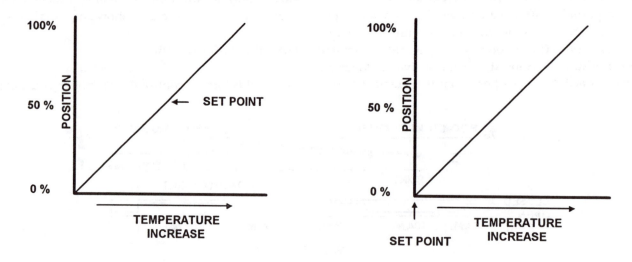

Figure 22.9a & b Temperature output signal relationships.

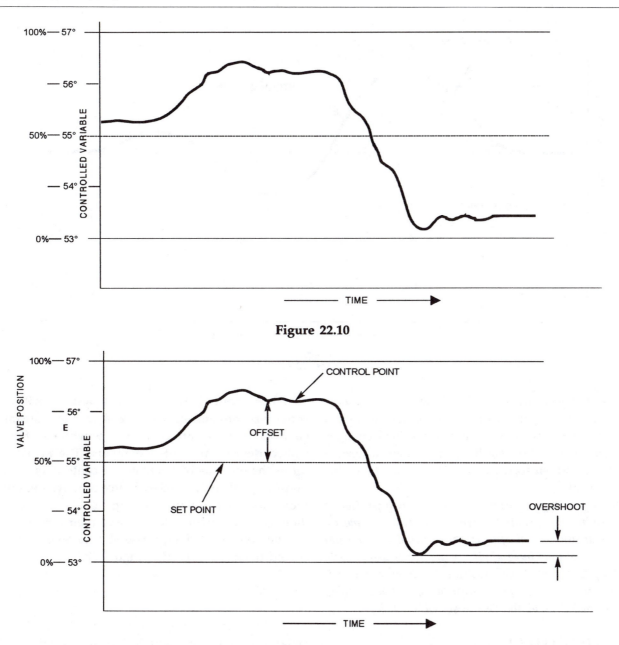

Figure 22.10

Figure 22.11

22.4.3 Proportional Plus Integral Action

Many control strategies require a control response that will maintain set point. Proportional plus integral action (PI) provides this feature. Upon a load change, the control response will cause the controlled device to be positioned so that set point is maintained. Sometimes referred to as automatic reset this control scheme has gained much popularity in recent years, largely due to the simplicity of adding this feature due to digital controls.

Have you ever noticed how the cruise control takes over and over time eventually restores the desired

speed? Proportional plus integral action does exactly the same thing. The controller continues to drive the output signal until setpoint is established. Cruise control does this by taking over control of the throttle until the speed is back to the set point! See Figure 22.12.

22.4.4 Proportional Plus Integral Action plus derivative

PID Control works similar to PI control except that derivative control has been added. This is applied to systems that experience rapid and sometimes erratic changes in their load. Derivative action reacts to the rate of change in the load and works quickly to prevent large

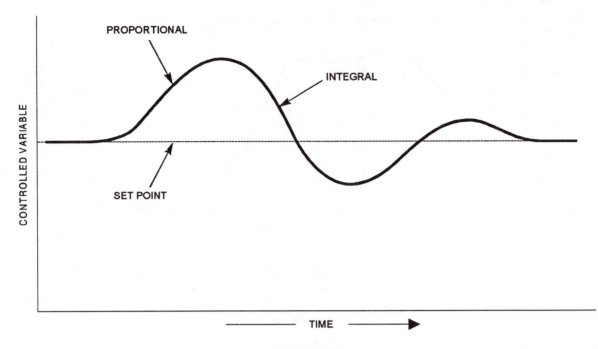

Figure 22.12

differences between the setpoint and the controlled variable. It is a difficult parameter to setup properly, so it is rarely used in HVAC systems where the load changes are relatively slow. Large auditoriums and areas susceptible to large sudden solar gains or losses might use this type of control.

Of the four types of control mentioned, two position is largely used in electric control systems. Proportional is mostly used in pneumatic control systems. Proportional plus integral and proportional plus integral plus derivative is used mostly in direct digital control (DDC) systems. The intelligence, ability to perform complex control algorithms, and networking capabilities makes DDC the control system of choice.

22.5 CONTROLLED DEVICES

Controlled Devices include:

- Valves
- Dampers
- Actuators for Valves and Dampers

Just about all HVAC control systems will require some type of controlled device. Water and steam flow controlled devices are called valves while air flow controlled devices are called dampers. The actuator performs the function of receiving the controllers command output signal and produces a force or movement used to move the manipulated device usually the valve or damper. Manipulated devices and their actuators make up the controlled device assembly known as the controlled device as shown in Figure 22.13.

2.5.1 Valves

Valves for automatic temperature control are classified in a number of ways. Valves are classified by body style, either two-way or three-way. Two-way valves control the flow rate to the heating or cooling equipment. Three-way valves control also the flow rate to the heating or cooling equipment with the added advantage of maintaining a constant flow rate in the primary piping system. Figure 22.14 gives an example of each.

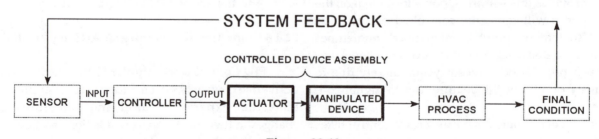

Figure 22.13

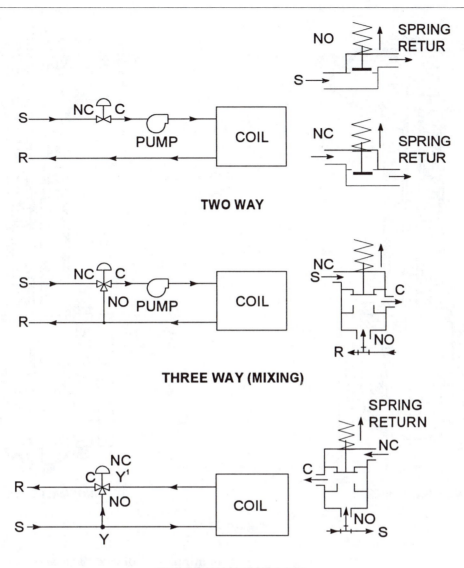

Figure 22.14

Other valve classifications exist, such as by flow characteristic, body pressure rating as well as subtle internal differences. Aside from the two body styles discussed earlier, another classification is by normal position, normally open or normally closed. The normal position is the fail-safe which will occur upon loss of the control signal. This fail-safe is achieved by a spring on those valves so equipped. Given a loss in the control signal the spring will return the plug (the part within the valve that controls the flow) to it's normal position. Figure 22.15 gives an example of both normally open and normally closed two-way valves.

22.5.2 Dampers

Dampers for automatic temperature control systems are much simpler than valves. Different classifications are based upon damper blade arrangement, leakage ratings, application, and flow characteristics. The two different blade arrangements are shown in Figure 22.16.

Like valves, dampers have a fail-safe. The fail-safe is established by the actuator and the way it is mounted to the damper. A typical fail-safe on an outdoor air damper is typically normally closed. Fail-safe on the adjoining return air damper would be normally open. The fail-safe positions of both dampers are arranged so as to prevent freezing outdoor air from entering the air handling system when the unit is off.

22.5.3 Actuators for Valves and Dampers

Actuators, sometimes called motors or operators, provide the force and movement required to stroke the

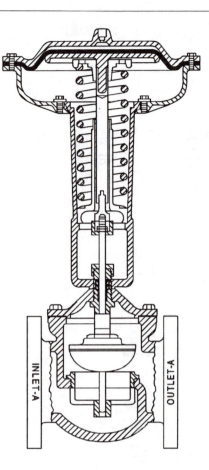

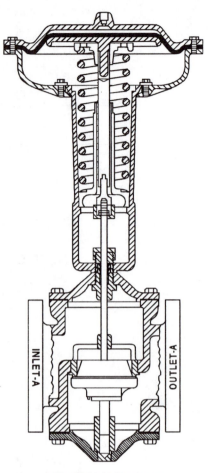

NORMALLY OPEN

NORMALLY CLOSED

Figure 22.15

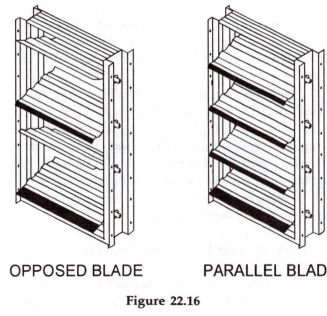

OPPOSED BLADE   PARALLEL BLAD

Figure 22.16

manipulated device. The actuator must be powerful enough to overcome the fluid pressure differences against which the valve or damper must close. Also the actuator must be able to overcome any frictional forces such as valve packings, damper blade bearings and linkages.

Actuators may be pneumatic or electric. Pneumatic actuators consist of a pressure head, diaphragm, diaphragm plate or piston, and the associated connecting linkage. Pneumatic actuators produce a linear or straight line motion. Little has to be done to covert this motion for into a form acceptable to valves or dampers. Since valve stems require linear movement the installation of pneumatic actuators is usually a direct mount. See Figure 22.17. Damper blades require a rotary motion which is easily accomplished by a crank arm.

Electric actuators are typically electric motors. Some actuators use slip-clutch mechanisms which engage or disengage to create or stop movement of the manipulated device. Other types of electric motors have a limited rotation of 270°. Figure 22.18 is one example.

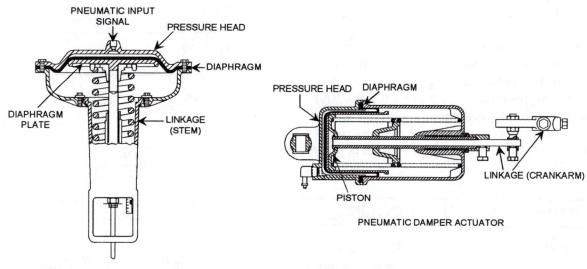

Figure 22.17

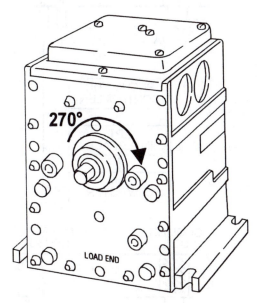

Figure 22.18

22.6 HVAC PROCESSES

HVAC Processes include:
* Control Agents
* Operations

Most HVAC systems have the capability for year-round air conditioning. The geography and climatic conditions have much effect on a systems design and the control strategies used. Regardless of the area it is safe to say that most systems will have to provide control over both sensible and latent heat. Temperature and humidity control and pressurization regulation or automatic controls insure occupant comfort. Automatic controls regulate by controlling HVAC processes within predetermined limits as defined by ASHRAE. Figure 22.19 shows HVAC processes as part of the overall functional block diagram.

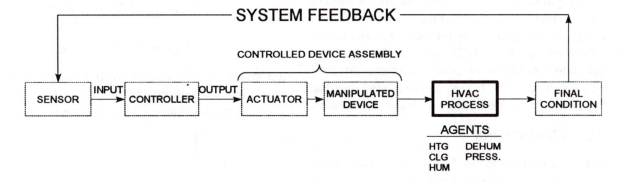

Figure 22.19

22.6.1 Control Agents

Control Agents are the representative of the source under control, cooling, heating, dehumidification, humidification and pressurization. Further breakdown of these processes reveals more detail on the exact type of process. For example, heating may be accomplished by hot water, steam, electric heat, heat recovery or even return air. A detailed look at possible sources is shown in Figure 22.20.

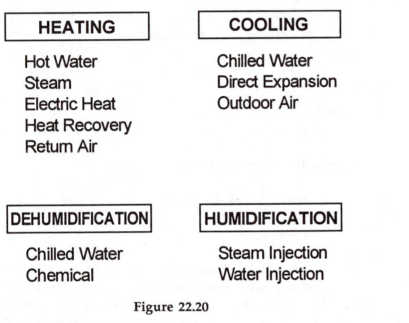

HEATING	COOLING
Hot Water	Chilled Water
Steam	Direct Expansion
Electric Heat	Outdoor Air
Heat Recovery	
Return Air	

DEHUMIDIFICATION	HUMIDIFICATION
Chilled Water	Steam Injection
Chemical	Water Injection

Figure 22.20

22.6.2 Operations

Operating the HVAC equipment is key to achieving optimal comfort levels. Equipment designed for worse case conditions, will operate at maximum load less than 2% of the time. In order to turn down the HVAC equipment, from full load design conditions, automatic controls are used to regulate flows, positions and temperatures.

Valves, dampers and variable speed devices are often used to regulate equipment. The equipment used is controlled at a level in response to the loads, on the system. Several seasonal scenarios for temperature control are presented in Figure 22.21. The system scenarios represent possible conditions of boilers, chillers, air handlers, interior zone VAV boxes, exterior zone heating valves and are based upon outdoor and zone conditions.

22.7 FINAL CONDITIONS

Final conditions at the end of the functional block diagram, Figure 22.22, represent the fruit of the control systems efforts. The control strategies which achieve the

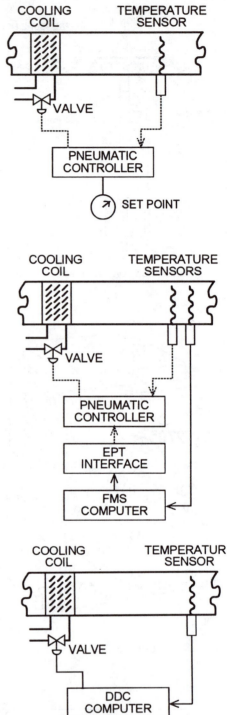

Figure 22.21a, b, & c

final conditions produced by the control systems are varied, usually falling within certain tolerances established for the equipment under control.

22.8 FEEDBACK

Feedback Systems are
* Closed Loop Systems
* Open Loop System

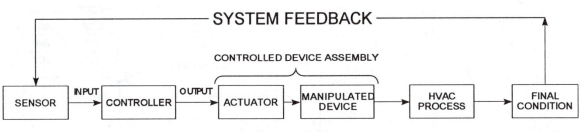

Figure 22.22

Feedback, sometimes called system feedback, is transmitting the results of an action or operation back to its origin.

22.8.1 Closed Loop Systems

Figure 22.23 shows a typical closed loop system controlling discharge air temperature. Closed loop systems use feedback for accurate control of HVAC processes. The controller in this system and its sensor measures the actual changes in the final conditions and actuates the controlled device to bring about a corrective change, which is again measured by the controller. Without feedback this system would not control.

22.8.2 Open Loop Systems

An open loop system is used to correct for load changes on final conditions. A typical example is shown in Figure 22.24.

Here an outdoor air sensor and its controller are arranged so as to cause an inverse relationship between outdoor temperature and hot water temperature. As the outdoor temperature decreases the hot water gets progressively hotter. Notice that the outdoor sensor is in an open loop while the hot water sensor is in a closed loop. Open loop systems simply sense, they do not control, as in the case of the outdoor air sensor. One can only sense outdoor air, it cannot be controlled. Yet sensing of outdoor air is critical in the proper function of this system.

22.9 CONTROL STRATEGIES

22.9.1 Zone Control

An advantage of *zone control* is that the actual load in the space is sensed and balanced by the controllers response. The only time the controller reacts is when it detects a load change. A load consists of heat transfer to

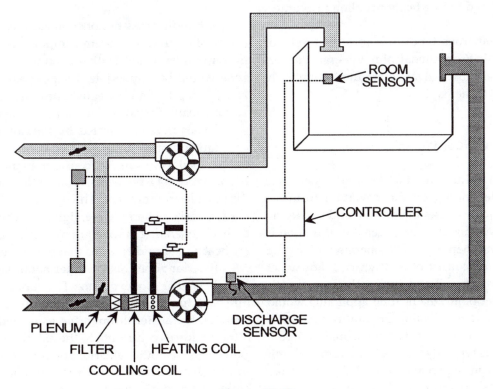

Figure 22.23

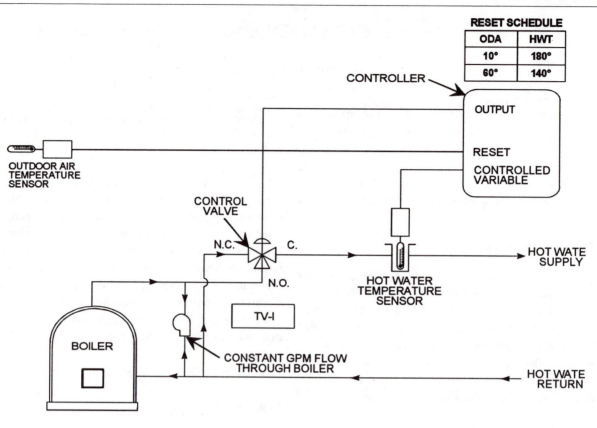

RESET SCHEDULE	
ODA	HWT
10°	180°
60°	140°

Figure 22.24

or from the space. When this occurs the space sensor and controller react by requesting heat or cooling to counterbalance the load.

There are numerous types of control methods, of these; Proportional, Proportional plus integral, Proportional plus integral plus derivative are most commonly applied to HVAC Systems

22.9.2 How Zone Control Works

A thermostat or humidistat in the zone senses zone conditions and depending on the deviation from set point, the control logic causes the heating and or cooling apparatus to balance the zone requirement. If a zone is below set point, the thermostat will operate the heating apparatus. This may consist of a steam coil, hot water coil or electric heating coil and their valves or electric switches. Above set point, the heating apparatus would be closed or off and the cooling apparatus such as a chilled water valve or cold air damper would be open.

Zone control is typically handled by a terminal unit such as the reheat coil shown in Figure 22.25, which may serve one or several rooms, or partitioned areas.

22.9.3 Zoning

A building may be zoned in various ways. Zoning is a way of dedicating system components, including controls, to similar loads. Without zoning, comfort in a given zone would be impossible. Imagine one system, in one zone, trying to simultaneously heat and cool because of varying loads. Obviously, it would be impossible!

Zones may be selected by interior, exterior, or by orientation; north, south, east or west. Each considers a unique load. For example, consider an interior area. Interior zones don't have walls, roofs or floor area exposed to the outside environment. The only load changes that occur are variations in people, lights, machinery or electrical heat generating office equipment. These loads actually are heat and humidity gains that only require cooling.

Exterior zones on the other hand have at least one outside wall or roof exposure. The variations in temperature and humidity of the outside environment causes conductive transfer through the walls and windows. Infiltration losses contribute to the exterior convective heat gain or loss. Another exterior zoning consideration is the sun's radiation on the north, east, west or south sides of a building. Of course, the sun affects just the opposite sides

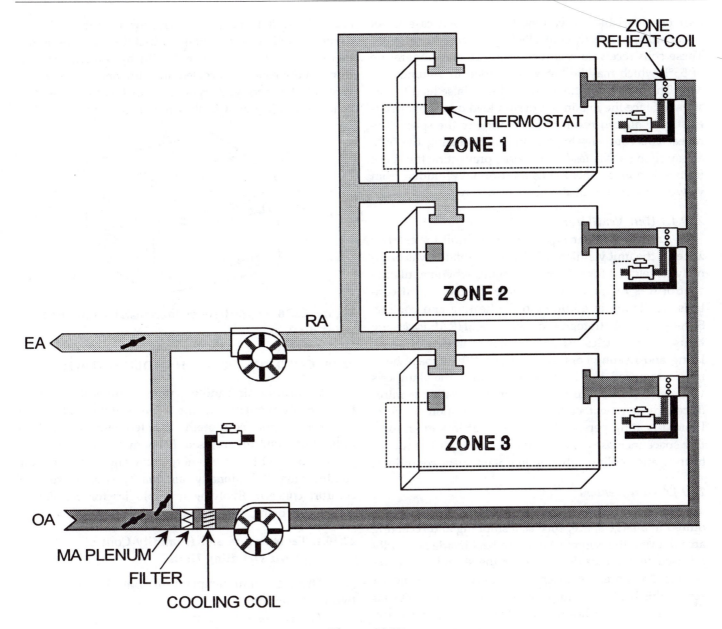

Figure 22.25

of a building in the southern hemisphere.

Unlike interior spaces which are purely heat gain, exterior zones are subject to both heat gains and losses.

22.9.4 Terminal Equipment Controlled From the Zone

Each of the following terminal unit control strategies applies to a particular piece of equipment.

22.9.4.1 Baseboard Radiation

Baseboard radiation or finned tube radiation, provides the blanket of heat for exterior wall surfaces. The radiation system along the exterior walls, radiates outward and convects upward, usually along the window areas, to replace the heat which flows to the outside.

This prevents extreme variations of the existing heat in the space. Heat is required when the thermostat senses a drop in space temperature. The heat lost through the wall and windows must be balanced by an equal amount of heat input in order for the space temperature to be maintained. Steam or hot water is modulated through the finned tube radiation by a valve. This valve is controlled directly by a room thermostat. If electric heaters are used, an electric thermostat senses this heat loss and energizes one or more stages to balance the load.

22.9.4.2 Reheat Coils

Reheat coils are installed close to the zone in the ducts of either constant volume systems or as an integral part of a variable air volume box. In each case a hot water valve is directly controlled by the zone thermostat. These coils receive a relatively constant air temperature of 55°F which may be heated or varied in volume.

The reason for the reheat coil is to false load the air stream. Since the maximum design load does not always exist 55°F air can result in subcooling of the space. The reheat coil compensates for this by reheating the 55°F air, applying a *false load* to the zone, preventing the space from subcooling. These coils are normally hidden from view in the ceiling of the controlled zone.

22.9.4.3 Unit Ventilator

Another popular type of terminal unit for exterior zones is the unit ventilator. The unit ventilator was originally developed for school classrooms, when ventilation control was required by law. Here the thermostat controls ventilation in addition to the heating and cooling. Several control strategies known as *ASHRAE cycles* are in use. These cycles use various combinations of ventilating and heating control strategies. When space heating is required the thermostats control the damper's volume, decrease it to a minimum, and heating is introduced by a reheat coil, baseboard radiation or both. From a heating perspective, as the heat loss in an exterior space increases, the terminal units modulate the air heating and cooling valves to maintain space conditions.

22.9.4.4 Unit Heaters

Unit heaters are used where high output is required in a large space, such as a shipping and receiving area. During the winter season the heat inside is rapidly released to the outside whenever the shipping and receiving doors are open. Generally, an electric thermostat senses this heat loss and turns on a fan which blows air through a steam or hot water coil to warm up the space. The unit heater continues to run even after the doors have been shut until the space air temperature returns to the thermostat set point!

22.9.4.5 Variable Air Volume Boxes

Variable air volume boxes can be applied to either interior or exterior zones. From a cooling perspective, upon a change in load the thermostat's output varies to modulate the variable air volume box damper. The volume of air varies from it's minimum to it's maximum position. This increasing quantity of air, generally at 55°F balances the heat gain in the space.

22.9.4.6 Thermostat Placement and Tampering

Accurate placement of the thermostat and sensors is critical for proper sensing. Occasionally, the thermostat may provide an inaccurate measure of load. For example, if it is next to a window or behind heavy drapes. Another disadvantage is that the set point on the thermostat may be tampered with by anyone passing through the room. Concealed adjustments will help prevent this tampering. A typical room thermostat with a concealed adjustment is shown in Figure 22.26.

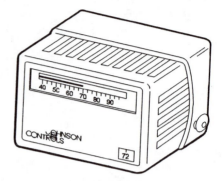

Figure 22.26 Typical room thermostat (concealed set point dial).

22.10 CONTROL OF AIR HANDLING UNITS

Control of air handling units ensures that the air being made available to the zone is delivered at the proper condition. Four methods of temperature and two of humidity control are used. Economizer cycles are employed to select the most energy efficient air streams for conditioning. Additionally, air quality is an increasing comfort concern. Evolving technologies for ventilation control are being introduced into the marketplace.

22.10.1 Temperature and Humidity Control of Air Handling Units

There are four temperature control methods and two for humidity.

Temperature F
1.   Zone/room temperature control
2.   Return air temperature control
3.   Discharge air temperature control
4.   Room reset of discharge temperature

Humidity
1.   Room/return air humidity control
2.   Dew point control of discharge air

22.10.1.1 Zone/room temperature control of a single zone unit.

The thermostat located in the zone sends it's control signal to the unit to position the heating, outside air dampers and cooling apparatus, so as to provide the desired air temperature to the zone. See Figure 22.27.

22.10.1.2. Return Air Control

Return air control can be described as a controller

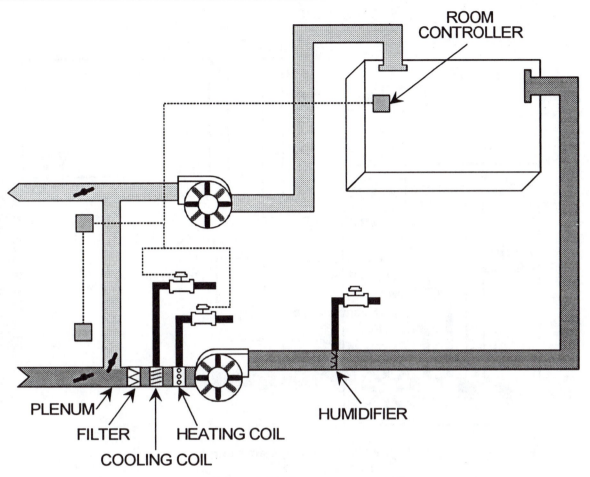

ROOM
CONTROLLER

PLENUM

FILTER

COOLING COIL

HEATING COIL

HUMIDIFIER

Figure 22.27 Zone room control.

receiving its signal from the temperature sensor located in the return air stream. It is actually a control strategy which uses an average temperature of a large area or number of rooms. This strategy can be used to control heating, cooling or humidification apparatus. This type of control is utilized when a single space sensor location is not representative of the entire area to be controlled. See Figure 22.28.

22.10.1.3. Discharge Air Control

Discharge air control of an entire unit is common to systems with multiple zones. In this configuration the sensor is located in the unit discharge and the controller signal positions the heating, outside air dampers and cooling apparatus to precondition the air so as to send a constant temperature (typically 55°F) to the zones. This air temperature is often the minimum temperature delivered to the zones. This strategy has been used with constant volume systems for many years. The zones have thermostats which control reheat coils which can add heat to prevent subcooling. Also, this strategy works well with the energy efficient variable air volume system, which today is the all air system of choice. The

zones have thermostats which control variable air volume boxes to modulate the quantity of 55 degree air delivered to the zone.. See Figure 22.29.

22.10.1.4 Room Reset of Discharge

Room reset of discharge of entire unit. An example

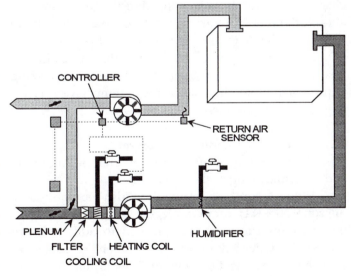

CONTROLLER

RETURN AIR
SENSOR

PLENUM

FILTER

COOLING COIL

HEATING COIL

HUMIDIFIER

Figure 22.28 Return air control.

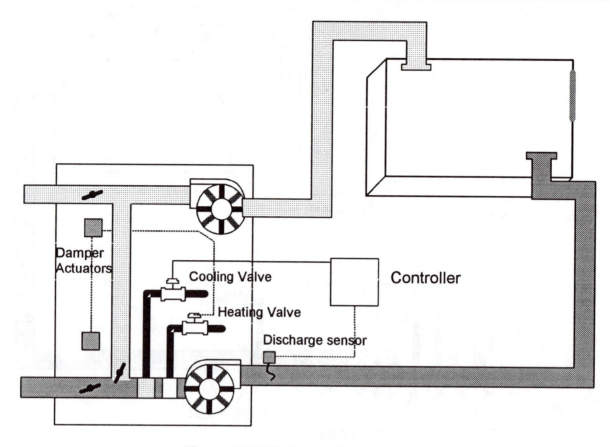

Figure 22.29 Discharge air control.

of this type of control for a single zone unit is shown in Figure 22.30. The use of two sensors, one in the zone and one in the unit discharge allows for closer control of the temperature within the space. The control of the unit is actually the combination of room and discharge control methods. Two feedback loops are utilized. The discharge control sensor's set point is adjusted higher or lower to compensate for changes in room temperature. The discharge sensor controls the unit directly so no undesirable variations in temperature reach the zone. Room reset of discharge is also used where multiple room temperature zones provide reset of the discharge. In this case the warmest room temperature resets the cooling requirements and the coldest room temperature resets the heating requirements.

22.10.1.5 Room or Return Air Control

Room or return air control of units humidifying and dehumidifying apparatus.

When a constant relative humidity is required in the zone, 50% R.H. for example, the humidifier and cooling (dehumidification) is controlled to add moisture during the winter (low moisture content season), and cool the air during the summer. This arrangement requires a continuous use of energy to provide a single set point space value. See Figure 22.31.

A more acceptable and energy saving concept of humidity control is the two set point method. Again whether sensed in the room or return air, the low moisture content season (usually mid-winter) will be controlled at a low relative humidity, such at 35% R.H. If the zone humidity should go below 35%, the zone humidity

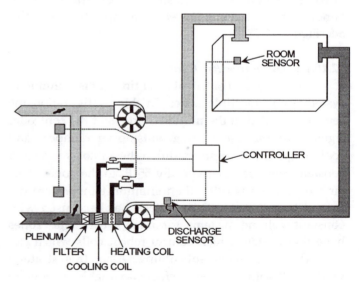

Figure 22.30 Reset control.

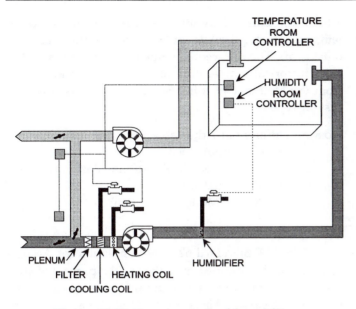

Figure 22.31 Zone control of humidity.

controller would add moisture by modulating the humidifier. If the humidity increases above 65% R.H., the zone humidity controller dehumidification signal would override the temperature control signal to the cooling/dehumidifying apparatus to prevent the zone R.H. from rising too high. As a result of the dehumidification process the space temperature may drop or be "subcooled." This would require the temperature control to reheat the sub-cooled air. The unit will need a heating coil located on the downstream side of the cooling/dehumidification apparatus. The zone will fluctuate between 35% and 65% R.H. throughout the intermediate seasons of spring and fall. During this time no energy to add or remove moisture will be used.

22.10.1.6 *Dew point Control of the Discharge Air*

This strategy is specialized for certain process operations such as textile and tobacco production. In these types of manufacturing facilities precise humidity control is required for quality control. Dehumidification or humidification can also be accomplished by controlling the cooling apparatus via a dry bulb temperature sensor located down stream of a cooling coil which has a high efficiency rating. The high efficiency coil will cause the leaving air of the cooling coil to be very close to saturation (100% R.H.). By sensing and controlling for dew point, achieved by dew forming on a dry bulb sensor, this arrangement will control temperature at saturation. This provides an extremely predictable moisture content, since dew point can be equated to an exact humidity ratio. When this air is reheated to a required value the relative humidity will be precise. Air washers because of their air washing capabilities can also be used

instead of high efficiency coils in dew point control applications. This is common in manufacturing facilities such as textile mills.

22.10.2 Economizer Cycles

The term *economizer cycle* has been used to define the control strategy which allows *free cooling* from outside air, thus reducing cooling load. This control strategy is applied to mixed air systems where the outside air or return air may be used to economize the cooling requirements at the cooling coil. The reduced load on the refrigeration equipment results in tremendous savings in electrical energy.

There are three types of economizer switch-over cycles:
*   Dry Bulb
*   Enthalpy
*   Floating differential adjustable switch-over

All three types of economizer switch-over cycles choose between outdoor or return air streams. The way that economizer cycles choose which air stream to use is what distinguishes them from one another.

22.10.2.1 *Dry Bulb Switch-over*
The Dry Bulb Switch-over Cycle chooses whether the mixed air system should be using outdoor air for free cooling, or return air based on the outdoor air dry bulb temperature.

The Dry Bulb Economizer Switch-over Cycle is a control strategy that saves energy all year round. In the winter mode of operation, it saves cooling energy by taking in free cooling from the outdoor air. In the summer mode of operation, by removing heat from the return air that has already passed through the cooling coil, therefore lessening the latent load on the cooling coil.

Winter Mode of Operation: When the outdoor air dry bulb temperature is below the switch-over temperature (dependent upon locality), the temperature control system will have the ability to modulate open the outdoor damper upon a call for cooling. **Summer Mode of Operation:** When the outdoor air dry bulb temperature is above the switch-over temperature, the outdoor damper is placed to its minimum position providing minimum ventilation as required by code. In this mode of operation, the control system is not able to modulate the outdoor damper and remain at minimum position. The primary source of air is the return which will have a lower total cooling (sensible and latent) load. **Economizer Switchpoint**—Since this strategy senses only the sensible load (temperature only) care must be taken when selecting the dry bulb economizer switchpoint. The switchpoint for a given geographic area must consider the latent loads that exist at the dry bulb switchpoint. For example Denver or similar "dry climates" will have somewhat higher dry bulb switch-over temperatures while coastal areas such as San Francisco will be subject to lower switch-over temperatures due to the higher moisture content of the air.

> Dry Bulb Switch-over Logic:
> OA Temperature > Switch-over Temperature
>   = Summer Mode (Minimum Position)
> OA Temperature < Switch-over Temperature
>   = Winter Mode (Free Cooling)

22.10.2.2 Enthalpy Switch-over

The Enthalpy Economizer Switch-over Cycle chooses whether the mixed air system should be using outdoor air for free cooling or return air by measuring the total heat content or enthalpy of each air stream. Enthalpy economizer is sometimes referred to as "true economizer" because it can sense both the sensible and latent components of the air. Dry bulb temperature and relative humidity is measured in both outdoor air and return air streams. This economizer will choose the air stream that will impose the least load on the cooling coil.

> Enthalpy Logic:
> OA Enthalpy > RA Enthalpy OR OA Temp. >
>   RA Temp. = Summer Mode (Min. Position)
> OA Enthalpy < RA Enthalpy AND OA Temp. <
>   RA Temp. = Winter Mode (Free Cooling)

Winter Mode (Free Cooling)

Enthalpy is a much more accurate measure of the load on the cooling coil. To maximize the efficient use of energy in a system, enthalpy should be used. Devices available today for sensing moisture content in air streams, especially those with wide temperature and humidity variations, require periodic maintenance. This must be considered in making the decision to use enthalpy switch-over. If the potential for proper maintenance is not good, then the best choice may be a dry bulb economizer or the floating switch-over cycle, discussed next.

22.10.2.3. Floating Switch-over - Adjustable Differential

The Floating Switch-over - Adjustable Differential economizer switch-over cycle chooses whether the mixed air system should be using outdoor air for free cooling or return air by measuring both outdoor and return air dry bulb temperatures. This Economizer provides a differential temperature between outdoor and return air. This differential indicates the difference in temperature required to obtain free cooling from the outside air. The differential value is computer generated from historical temperature and relative humidity data from the National Weather Service and is based on the system type.

> Floating Differential Adjustable Logic:
> OA + Differential > RA
>   = Summer Mode (Minimum Position)
> OA + Differential < RA
>   = Winter Mode (Free Cooling)

Ventilation

ASHRAE Standard 62-1989 Ventilation for Acceptable Air Quality defines *ventilation* as the process of supplying and removing air by natural or mechanical means to or from a space. Air is provided at a specified quantity known as the *ventilation rate*. The quality of the outdoor air used is subject to air quality standards for outdoor air as set by regulatory agencies such as the Environmental Protection Agency in the United States.

Ventilation rates may vary, a lobby area may require 15 CFM per person while a public rest room 50 CFM per toilet fixture. Local, state codes, ASHRAE Standard 62-1989 Ventilation for Acceptable Air Quality, or job specifications provide guidance to the designer and commissioning personnel.

22.11   CONTROL OF PRIMARY EQUIPMENT

Unique control strategies for primary equipment; boilers, heat exchangers and converters, chillers and cooling towers exist for each particular piece of equipment. Common approaches to control these primary equipment systems are discussed. There are many more which are beyond the scope of this text.

22.11.1 Hot Water Systems

Boilers are controlled by packaged controls installed at the point of manufacture. Hot water boilers are controlled at a fixed temperature, generally around 180°F. They operate around this fixed temperature in a two position (on-off) manner, or some combination of low fire, modulating or high fire rates. This helps the boiler maintain a high efficiency and a long life.

To avoid overheating problems associated with this fixed boiler hot water temperature, cooler return water is mixed with water leaving the boiler to obtain the desired hot water system temperature. This hot water system temperature is inversely reset from ODA temperature, commonly known as *hot water outdoor air reset*.

See Figure 22.32. As outdoor air temperature goes down to 10°F, the hot water will be readjusted to a maximum heating value such as 180°F. Conversely, when the outdoor air is 60°F, the hot water is readjusted to a light load condition, such as 140°F.

The reason for hot water reset via outdoor air temperature is to provide a more controllable water temperature at the perimeter zone valve. This prevents overheating in the zone.

Colder outdoor air temperature increases the heat transfer through the walls and windows. The perimeter zone thermostat will sense this heat loss and proportionally open its valve.

If the zone thermostat opens the valve too far to balance the heat loss, overheating might occur. By having the proper hot water temperature available to the terminal unit the possibility of overheating is eliminated.

Another benefit is the energy conservation which results from optimizing hot water temperatures to match the load.

22.11.2 Chilled Water Systems

Chillers, much like boilers, generally come with control packages installed at the point of manufacture. Constant chilled water temperatures ranging from 42°F

to 45°F are usually regulated by various capacity controls, such as inlet vanes or unloaders, which command the refrigerating effect of the machine. Chilled water is required for two basic reasons:

1. To provide a minimum of 55°F air temperatures to the zones.

2. To lower the dew point of the primary air so as to dehumidify the air in the zone.

When temperature rises in the zones a conditioned source of air, lower than the zones temperature and moisture, must be fed to the zone. This cooler, drier air absorbs the excess heat and moisture. The warm return air, carries the sensible heat from the space, mixes with outdoor air being brought in for ventilation. This mixed air then rejects its heat to the cooling coil. This causes the chilled water to increase as much as 10°Fs. This chilled water returns to the chiller at 50°F to 55°F (worst case design load) to be cooled down to 43°F to 45°F, before once again repeating the process. See Figure 22.33.

22.11.3 Cooling Towers

Cooling towers have the important job of rejecting heat from the building. Whether there is one cooling tower or several, control is usually done in any of four

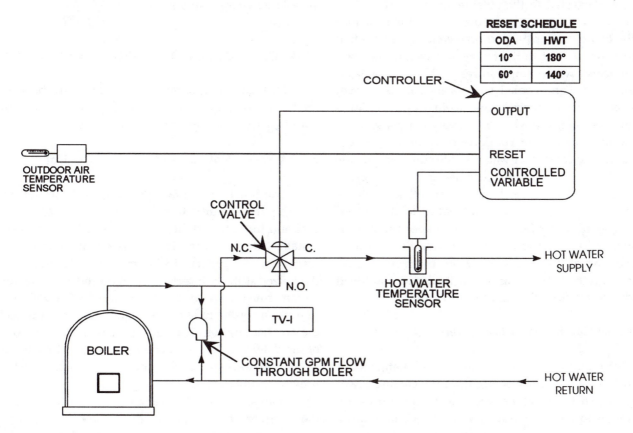

Figure 22.32 Outdoor reset of water.

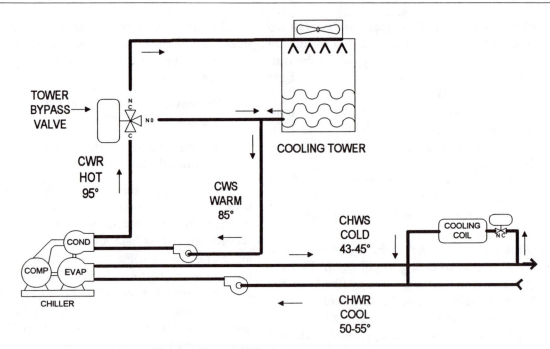

Figure 22.33 Chilled water system.

methods: Bypass valves, fan control, damper control or a combination of any or all. Condenser water from the chiller is piped to the tower so that evaporative cooling may take place. Through this evaporative cooling process, heat is rejected to the outside air.

A bypass valve located close to the chiller or perhaps the tower, under proportional control ensures that enough water is sent to the tower for evaporative cooling. Condenser water temperatures of 75° - 80° are desirable for most water-cooled chillers to achieve proper condensing temperatures. Given cold days, or "low-ambient" conditions, the bypass valve will bypass or re-route a portion of the condenser water to prevent overcooling the condenser water. In the summer, outside air as wet bulb temperatures increase and as condenser water temperatures exceed the set point, the bypass valve will modulate sending full water to the tower. Next, stages of tower fans or low-high speed motor arrangements are energized to provide additional evaporative cooling of the condenser water, by increasing air flow through the tower. Dampers may also be employed to modulate air flow through the tower.

22.11.4 Free Cooling Heat Exchangers

"Flat Plate" or "Plate to Plate" heat exchangers can be used to provide Free Cooling from the cooling tower on days when the outdoor wet bulb temperatures are low enough to lower the tower water to a temperature below 50 degrees by evaporation. Passing the cold tower

water through one side of the flat plate heat exchanger and building chilled water through the other side, you can cool the building by discharging the heat directly to the tower without running the chiller. This results in very large energy savings.

22.12 CONTROL OF DISTRIBUTION SYSTEM

Distribution systems supply and return heat transfer fluids, air and water. Regulating volume and pressure helps insure comfort for the occupants.

22.12.1 Fans - Volume and Pressure Control

The function of a fan is to move air through ducts at a required volume to deliver the quality of temperature and humidity to the zones. This is accomplished by installing ducts of various sizes, larger at the fan and coil discharge and progressively smaller to the end. Larger duct sizes at the coil section are required because the air stream must flow at a slower rate (500 - 700 ft. per minute) in order to allow heat transfer from the coil to the air (heating) or air to the coil (cooling). The volume of air delivered to each zone is determined by the heat gains and losses of each zone due to heat transfer as previously discussed.

If an air stream flows at a constant volume through the ductwork the temperature and humidity of that air varies as required by the zone sensors. This type of system is referred to as a *constant volume, variable temperature (CVVT) system.*

If air stream may be varied in volume as required by the zones, and is usually controlled at a relatively constant temperature (55°F) and moisture content. This type of system is referred to as a *variable air volume VAV, constant temperature (VVCT) system.*

Supply ducts which have variable flow rates have controls to ensure that the proper volume is delivered as required to ventilate and condition the zones. VAV supply fans are typically controlled by sensing static pressure. This control arrangement, *static pressure control*, is prevalent in variable air volume systems, as shown in Figure 22.34. Inlet vanes, discharge dampers, or variable speed drives modulate the fans volume output and hence control pressure within the duct.

The static pressure is the energy which pushes the volume of air through duct and the VAV box to meet the zones cooling requirement. As the temperature of the zone increases with a change in load, the zone's variable air volume box damper would be modulated open.

Pressures in the neighborhood of .75 to 1.5" W.G. are common. They are measured and controlled approximately 2/3 down stream of the longest duct run in the system. The reason for this location is to reduce the fluctuation of static pressure in the duct and also to make sure that the last variable air volume boxes have sufficient static pressure to operate.

Building pressurization control is required to maintain a slight positive pressure within the building. This is required to fend off unwanted infiltration into the building. Building pressure is sensed by an indoor sensor and often compared to outdoor atmospheric sensors. Pressures are typically maintained by modulating exhaust fans, return fans or relief dampers to control the buildings pressure at desirable levels.

22.12.2 Pumps - Volume and Pressure Control

Pumps are designed to achieve the same thing as fans—to move water at a volume required by the zones. The water must pass through boilers or heat exchangers,

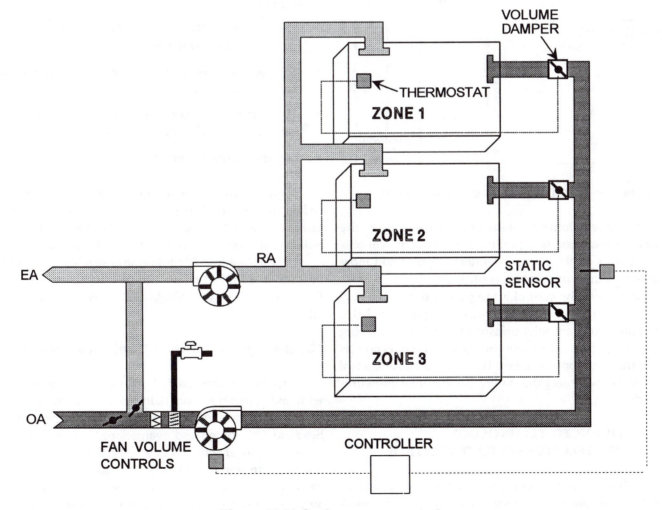

Figure 22.34 Static pressure control.

converters and chillers at a constant rate to allow heat transfer. Water piping then delivers the volume and temperature water for control by the thermostats in the zones as described in the previous section, Zone Control.

Much like air systems, pumping systems may be either constant volume or variable volume. Constant volume pumps are generally applied to primary loops through the primary equipment such as boilers and chillers. Flow in the secondary portion of the system, that which serves the terminal and air handling units, is variable in nature. Constant volume pumping arrangements may be used as long as provisions are made to compensate for the varying flow through the units.

Because the flow of water varies in the piping, and along with it the pressure, an appropriate control strategy is differential pressure control. The *differential pressure control* system senses differential pressure between the supply and return pipe lines and the resultant control signal controls a differential control bypass valve which relieves excess supply pressure and volume to the return. See Figure 22.35.

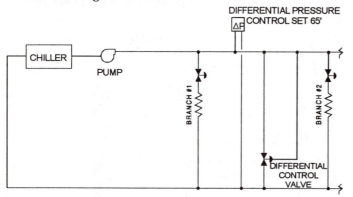

Figure 22.35 Differential pressure control.

Variable volume pumping arrangements are a wise choice for the secondary portion of the system. Variable speed drives on the pump is another popular and more efficient method of pressure control.

Whatever the control arrangement differential bypass or variable speed pumping the application helps maintain a constant inlet pressure to each valve. Doing this helps to prevent system pressures from overpowering of the zone control valves. The result is closer control of zone temperatures due to predictable inlet pressures at each valve.

22.13 ADVANCED TECHNOLOGY FOR EFFECTIVE FACILITY CONTROL

Advances in technology brought direct digital control, lighting control, fire management, security monitor-

ing, distributed networks, personal computers, and sophisticated graphics. Electronic chips replaced pneumatic controllers. Personal computers (PC's) replaced minicomputers. Software programs replaced hardwired logic.

Each new advancement in the electronics and communications industries was eagerly snapped up by Facilities Management System (FMS) designers. (Note FMS is also sometime referred to as EMS, but EMS are Energy management systems and FMS tend to be focused on other uses of the data beyond energy conservations such as computerized maintenance management.) Systems are now faster and more capable than ever before. Software programs, electronic components, sensors, actuators, hardware packaging, and communications networks are integrated, share information, and work together.

The overall purpose of a *Facilities Management System* is to make the job of facilities people easier, to make a facility more efficient, and to keep a facility's occupants comfortable and safe.

The FMS can save money for building owners in several ways:

*   By increasing the productivity from staff by doing mundane tasks for them.

*   By reducing energy consumption (energy management programs).

*   By identifying equipment needing maintenance, and even rotating the use of some equipment.

*   By managing information.

When considering the use of any FMS, you must define the desired functions, make a realistic financial analysis, and determine the amount of time available for building personnel to use and learn to use the system.

The following discussion investigates many of the options available throughout the industry, although there may not be any single FMS which includes them all.

22.13.1 Integrated Control - Distributed Networks

In older systems, a distinct headend communicates with and controls remote field gear. The field gear reads the signal from a controller or sensor and sends it to the headend for storage and evaluation. If the headend determines control action is necessary, for example, a high temperature signals that a cooling fan should be started, the headend computer sends a signal to the field gear associated with the fan telling it to start the fan.

All of the programming for storage, analysis, and necessary actions, is in the headend computer. The field devices, although possibly containing microprocessors largely for communications purposes, are primarily for converting and sending the signals from sensors, switches, and transducers so that the headend computer could monitor and control them.

Thousands of systems using the headend computer arrangement were installed in the 1970's and 1980's and are successfully in use today. Today's installations use more intelligent field panels. These field panels are typically direct digital control (DDC) panels compatible with most electronic and pneumatic sensors and actuators. They can control one large HVAC system or several smaller systems (including HVAC, boiler, chiller, and lighting systems). Many can interface to a headend computer for supervisory control and further data analysis or they can work independently if they lose communications with the headend.

Figures 22.36 a, 22.36 b, and 22.36 c compare how HVAC control has been traditionally done without a headend computer, how it can incorporate an FMS, and how it can be done by DDC field panels.

- All pneumatic closed loop control.
- Pneumatics control the setpoint.
- Pneumatic controller in command.
- FMS computer controls setpoint through an electric to pressure transducer or EPT.
- Global FMS control.
- DDC computer is the controller.
- Software control flexibility.
- Easy interface to the FMS network.

The DDC panels are often custom programmed or configured, individually, for complete freedom to suit a particular application.

The DDC panels might be lighting controls panels, fire panels, security panels, or even a separate minicomputer which can also interface to some of the same field devices for the purpose of Maintenance Management or Security Access Control, such as monitoring card readers and door access.

With intelligence in the DDC field panels, the line between the responsibilities of the headend and the field devices becomes less well-defined. Certain programs, such as energy management functions, might best be programmed into the field panels, while others, such as centralized alarm reporting, are best done by the headend computer.

However, the state-of-the-art avoids using the headend computer. The trend is toward a network of microprocessor based network control units or NCU's or

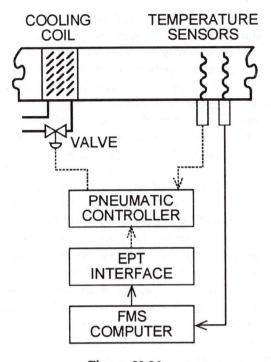

Figure 22.36 a

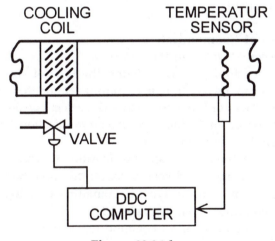

Figure 22.36 b

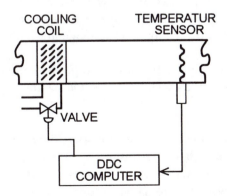

FMS Computer

Figure 22.36 c

distributed intelligent controllers, each monitoring and controlling designated equipment, and freely sharing data. Network expansion units or NEU's perform the localized control functions, such as starting stopping fans and closed loop control of valves and dampers. A personal computer is an equal partner in the network providing a sophisticated operator workstation, storing network data, and storing and executing sophisticated monitoring and controlling applications.

Figure 22.37 shows one possible configuration.

The distribution of intelligent network devices provides complete stand-alone control capability when needed, providing maximum reliability: no headend computer exists to be the primary center for energy management programs. The most sophisticated distributed network devices use state-of-the-art communication modes which can rapidly share data, giving building operators complete and consistent information about the facility.

Even with the high degree of sophistication in the technology available, costs of such systems have not increased significantly, packing in more value for each dollar spent.

**Types of Communication
between Devices on the Network**

In addition to considering the type of configuration, headend centralizing communication for remote field panels, or intelligent network devices consideration must be given to the type of communication link between devices in the system.

The means of communication determines the speed, distance, and cost of communication. Some systems can mix various types of communication to accommodate a more complex network.

Some of the most common types are:

- Coaxial cable and twisted-shielded pair may be least expensive and easiest to install, but are subject to electrical interference, such as lightning, which can cause component damage and loss of data.

Fiber optic cable offers a high degree of quality in transmission, as well as protection against electrical interference.

- Telephone lines can transport data long distances to include numerous remote buildings into a single FMS network. They are expensive to use and are subject to the same failure as normal telephone lines. To reduce costs, many FMS systems offer the ability for the operator to dial-up remote areas only

when needed, or the program can dial-up remote areas automatically when necessary to get data or control equipment.

- Some FMS systems can use existing building electrical or telephone wiring to transport FMS network data, reducing the cost of labor and materials for wiring to implement the FMS. Systems of this type have been named power line carrier systems.

Many networks can take advantage of existing communications links already installed in your building facility. If the communications scheme is an industry standard link such as ARCNET® that is used in thousands of office and industrial automation installations worldwide, installing and servicing the network should be easier and less expensive.

22.13.2 FMS Equipment Application

Small Building Systems

A smaller, less sophisticated (and less expensive) FMS might be appropriate:

- for a smaller building,
- where potential energy savings are too limited to yield a large dollar savings,
- where the operating staff has little time to devote to using, or learning to use, the system.

Many small systems are available with a limited number of functions, including Direct Digital Control, load control, time programming, and additional standard programs.

However, using the distributed network concept, the distinction between small building and large facility applications is not as clear. A smaller building simply uses fewer intelligent distributed devices on the appropriate communications network type.

22.13.3 Large Facility Systems

On a larger scale, an FMS can accommodate functions often associated with the needs of a larger facility:

- a greater number pieces of monitored and controlled equipment,
- greater distance requirements,
- flexible programming,
- sophisticated data management and reporting schemes.
- multiple operators can use the system simultaneously.

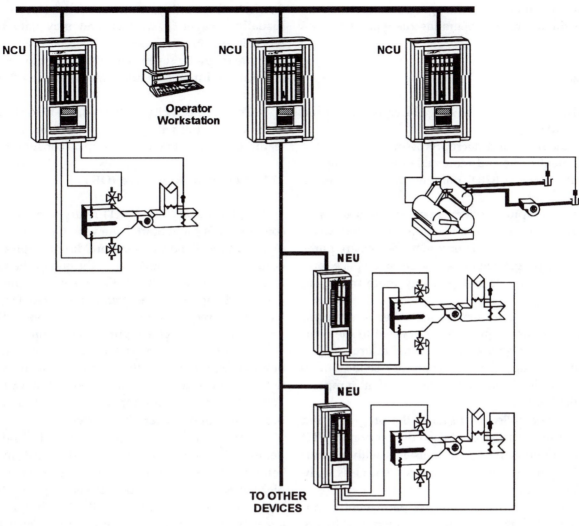

Figure 22.37

While a more sophisticated and complex system requires a greater initial investment, it can yield significantly better energy savings and increase productivity of personnel, improving the return on investment.

22.13.4 Computer Equipment

Computer Equipment is the heart of the FMS. Depending on the type and complexity of the system, the FMS may include any combination of computer types: microprocessors (least powerful), PCs, and minicomputers (most powerful).

When evaluating the computer equipment available with a system, focus on the actual work that the system can do rather than concentrating on the raw amount of memory or disk storage measured in bytes, or the speed of the processor, measured in msec.

Consider the tasks which are important to the efficient operation of your facility. For example, consider:

- the number of pieces of equipment it can accommodate,

- the amount of actual stored data like number of months of previous kilowatt hours (kWh) consumption it can store or the number of fans for which it can store a run time total,

- the ease of operation,

- the number and type of operator stations it can handle, such as video display terminals (VDT) screens, printers, hand-held modules.

If one or more PCs and/or printers will be part of the system, it may be advantageous if the equipment is the same as is already used in some other department of your business such as accounting, payroll and data processing. With similar computer equipment, you may be able to take advantage of existing agreements for purchasing, maintaining, and operating the equipment.

22.13.5 Hardware

Hardware is the actual tangible equipment used with a computer system, including:
* CPU, memory, and microprocessors.
* permanent disks and removable disks and tapes.
* operator devices (CRT, pointing devices, printer).

To use a computer system, a person needs an *operator device*. A system may have one or more such devices, in various combinations. Having more devices requires more complex programming, more storage space, and more processing power and probably more money to buy.

In general, operator devices are called *input/output devices*, or *I/O devices*. Input is any data sent to the computer, like the user typing in new temperature comfort limits. Output is any data the computer sends out to another device, like a printed summary of all temperature comfort limits.

Software is a collection of all of the programs and data the computer uses to do the job the programs tell it to do. The computer hardware is useless without software, just as your audio cassette player is useless without cassettes to play.

Firmware is a term for software permanently contained on usually one chip. Since the program software stored on the chip is entered into the computer by plugging in the chip hardware, firmware is not truly software or hardware, so is termed firmware.

22.14 FMS FEATURES

22.14.1 Features For Optimal Control

Automatic equipment controls are designed to improve building efficiency and maintain occupant comfort while saving as much energy as possible. These features often yield the most tangible and measurable energy and dollar savings for the building owner.

Overall, the features reduce the amount of electricity a facility uses. The electric bill of a commercial building complex is a large part of the building's operating costs. Lights, HVAC, and computers are a few of the major consumers of electricity in a commercial building.

The electric bill for a commercial building is largely based on the total amount of electricity used, measured in kWh (kilowatt hours). The charge for each kWh varies

from utility to utility, may vary from season to season (usually cheaper in winter), and may vary based on when it was used (usually cheapest at night). A rough estimate might be seven cents ($0.07) per kWh. The total amount used may also affect the rate, like a "bulk rate" discount.

The less electricity consumed, the lower that portion of the electric bill will be. This section discusses various features designed to reduce electrical consumption.

22.14.2 Optimal Run Time (ORT)

Optimal Run Time (ORT) refers to a single feature combining Optimal Start Time with Optimal Stop Time. Optimal Stop Time stops the building equipment before occupants leave the building at the end of the work day. Since stopping equipment also means closing the outdoor air dampers, this is often not allowed. (Many areas have codes requiring minimum ventilation.) Therefore, we will not investigate Optimal Stop Time.

Waiting until the last possible moment to start building equipment at the beginning of the work day will save on electrical consumption (kWh). *Optimal Start Time* (OST) delays morning start-up without sacrificing occupant comfort when they arrive.

Fans are normally turned off at night, and the temperature in the building is allowed to drift away from comfortable levels, the building may drop to 50°F at night in winter and may be allowed to get up to 90°F at night in summer. By the time the building is occupied again in the morning, the temperature must be up to about 68°F in winter and down to 78°F in summer. OST determines the latest time possible to start the fans in order to reach comfort levels by occupancy time. The amount of time necessary to reach comfort levels depends on many factors:
* the outdoor temperature,
* building insulation,
* the building's ability to gain and lose heat,
* how cold or warm the building got overnight,
* how warm or cool the building must be by occupancy to be considered comfortable.

OST uses these considerations and others to determine the latest possible equipment start time and still reach comfort at occupancy.

22.14.3 Load Rolling

Fans, pumps, and HVAC systems in a building are operated continuously during occupied periods to provide the heating, cooling, and ventilation for which they were designed. However, since the capacity of this equipment is large enough to maintain occupant comfort during the peak load conditions on the hottest and coldest days of the year, it is possible to turn off some of the equipment for short periods of time with no loss of occupant comfort.

The *Load Rolling* feature can significantly reduce overall kWh consumption by stopping (shedding) certain electrical loads (equipment) for predefined amounts of time periodically throughout the day.

Load Rolling is sometimes known as Load Cycling or Duty Cycling, among other names.

The program generally allows the user to define minimum On and minimum Off times to avoid short cycling which could cause more cost in equipment maintenance than is saved by shedding loads. Similarly, maximum Off times avoid discomfort caused by a single fan being off too long.

To maintain comfort, the program can automatically adjust the cycle times of loads to compensate for changes like space or outdoor air temperatures. In other words, before a load is shed, the controls can check a related temperature (such as space), and if too warm or too cool, the program can override (not shed) that load at that time, or it could just shed the load for a shorter time than usual.

For Load Rolling to reduce kWh consumption, it is important to shed only loads like constant volume fans and pumps which do not need to make up for lost time when they are started again. For example, if you stop a constant volume fan for 15 minutes, you do not have to run it faster later to make up for the time it was off. Such loads are considered to be expendable loads. The term "expendable" has nothing to do with relative importance.

Compare the idea of an expendable load like a constant volume fan or a pump, to the idea of a deferrable load like a chiller or a VAV fan. If you stop a VAV fan for 15 minutes, you would be saving kWh for that 15 minutes, but you would have to run the fan harder later to make up for the time it was off, thereby using the kWh avoided while the fan was off. A load which must "make up" for the time it is off is known as a *deferrable* load.

Since turning off a deferrable load for a short period of time does not really save kWh, you should not use deferrable loads with the Load Rolling feature. However, you will find that it is appropriate to shed deferrable loads with the Demand Limiting feature.

22.14.4 Demand Limiting

Most residential electric bills are based largely, or solely, on consumption. The electric bill of a commercial

building is based largely on the rate of consumption, not just the total amount of consumption. In other words, independent of the total consumption, the electric utility also continually monitors the rate of consumption (e.g., kWh per each 15-minute time period), known as the electrical demand. Demand is measured in kW (kilowatts). Each utility uses a slightly different method to calculate the highest demand it measures for the billing period. However, once it determines that peak demand, it adds an additional charge.

The period of time that the utility company routinely uses to measure demand is called the demand interval. The demand interval is determined by the utility and is commonly 15, 20, or 30 minutes during a specific time period each day. For example, if a utility uses a 15-minute demand interval, it measures the kWh consumption in each 15-minute period or demand interval of the billing period. It then calculates the average kW load "demand" for each interval. The highest 15-minute demand period for the corresponding billing period determines the demand portion of the bill.

If the use of electricity in a facility could be kept spread out to maintain a relatively constant level throughout each day, instead of using a lot of power in one short period of time, the resulting dollar charge would be smaller.

The Demand Limiting Feature keeps an eye on the rate of electrical consumption and starts shedding (turning off) loads when usage is exceeding a predefined demand limit demand target. Demand Limiting is sometimes also known as Peak Shaving or Load Shedding.

Figure 22.38 shows how turning off some expendable loads or during peak times, even shedding a deferrable load which would need to be run longer later, can save on the demand charge by flattening the demand curve.

22.14.5 Economizer Switchover

As a commercial building conditions the air it supplies to keep its occupants cool. It can supply air from one of two sources of air: it can take in outside air, or it can close the outdoor air to legal minimum limits and recondition air already in the building (the return air).

Various methods can be used to determine the air source to use: Simple Dry Bulb, Enthalpy Switchover, or Floating Switchover. Which method to use depends on the building and system.

For Simple Dry Bulb, the program compares outdoor air temperature with a predetermined switchover setpoint.

If the outdoor air temperature is lower than the

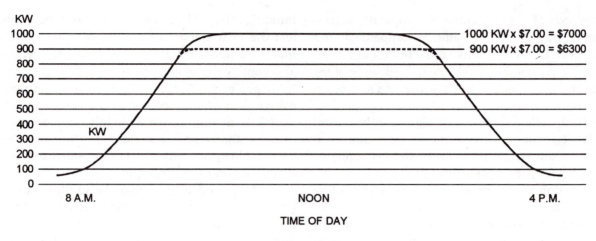

Figure 22.38

setpoint and cooling is needed, the dampers are allowed to modulate, providing free cooling. If the outdoor air temperature is greater than setpoint, the program holds the dampers at minimum position to recycle as much return air as possible, using mechanical cooling to recondition it.

Enthalpy Switchover is a more complex, but more accurate, method of determining economizer switchover. It is based on these facts: 1) at the same temperature, humid air contains more heat energy than less humid air, 2) you feel warmer when it is more humid, and 3) an HVAC system must work harder to cool that air, using more energy. The total heat content of air ("enthalpy") is calculated using dry bulb temperature and relative humidity (or dewpoint), among other values. Enthalpy is measured in Btus.

The program determines the enthalpy of the outdoor air and return air, and compares the results. If the outdoor air stream has less enthalpy than the return air stream, the dampers are allowed to modulate for free cooling. If the return air stream has lower enthalpy, the dampers are held at minimum position.

Floating Switchover has a high degree of accuracy without the need for humidity sensors. It uses two dry bulb sensors outdoor air and return air). Ideal Dry Bulb compares the outdoor air temperature with a variable or changing switchover temperature to determine when to allow the dampers to be commanded beyond minimum position. To use it, your specific geographical area must be analyzed and use some computer calculated values.

22.14.6 Supply Air Reset (SAR)

While in the occupied mode, a building has heating and cooling requirements throughout the day. Supply Air Reset is a strategy which monitors the heating and cooling loads in the building spaces and adjusts the discharge air temperature to the most efficient levels that satisfy the measured load.

Cooling discharge temperature is raised to the highest possible value which still cools and dehumidifies the warmest room served by the fan system. Heating discharge temperatures are reduced to the lowest possible levels which still heat the coolest room.

SAR works best with a constant volume system in which the amount of air being supplied to the zones is always the same.

The system really has two control loops, as illustrated in Fig. 22.39.

First, the room loop consists of the room temperature measuring element and the setpoint. As the room temperature varies around the setpoint, SAR calculates a new setpoint for the discharge loop.

The second loop uses the new discharge setpoint and measures the discharge air temperature. As the discharge temperature varies around the setpoint, the program sends a new command to the valve or dampers of the mechanical heating or cooling equipment.

Adjustment can be made of various values, such as setpoints, proportional bands, and deadbands. Some computers/controllers allow you to choose either Proportional Control or Proportional plus Integral Control (P.I.). P.I. Control is suitable for applications where the controlled variable must be right at setpoint, as in clean rooms or with static pressure control.

**22.14.7 Supply Water Reset
(Chilled Water or Hot Water)**

The Supply Water Reset feature automatically changes the setpoint of the water supplied to the cooling or heating loop to the highest (for chilled water) or lowest (for hot water) temperature possible, while satisfying the requirements of each zone it supplies.

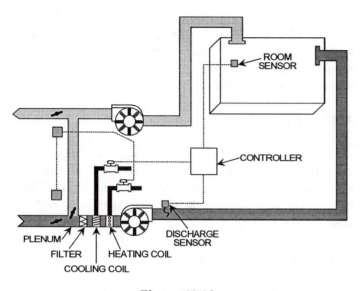

Figure 22.39

To cool a commercial building, water or coolant is cycled throughout the building zones. Water or coolant is chilled and then pumped to zones in the building. The zones air is cooled by transferring its heat to the water or coolant. The warmed water or coolant is passed back through a chiller to cool it down again. The water is then recycled back to the building zones for continued cooling.

If the chilled water setpoint is colder than necessary, the chiller wastes energy working to achieve the setpoint. Therefore, the *Chilled Water Reset* feature adjusts the chilled water setpoint as high as it can, while still satisfying the zones.

Valves vary the amount of available chilled water supplied to each zone. The position of each valve varies as more or less cooling is required. For example, a gym where a basketball game is taking place may require more cooling (requires more chilled water, so the valve opens farther) than an office where several people are quietly working at their desks.

The computer/controller checks the position of the valves for each zone. If none of the valves is fully open (each is bypassing), each room is cool enough and does not require all the chilled water available to it. The program determines that the temperature of the chilled water is colder than necessary to satisfy its current cooling requirements. The computer/controller adjusts the setpoint of the chilled water (supplying warmer chilled water). Warmer chilled water requires the chiller to use less energy, but will cause the chilled water valves to open farther to satisfy their zones. When One of the zone valves if fully open the controller will adjust the setpoint to a lower chill water temperature to meet the demand.

The Hot Water supply reset feature work similarly. As the zone temperatures are satisfied their zone valves begin to close and the hot water temperature is adjusted to meet the maximum need.

22.14.8 Condenser Water Reset

Chiller plants are usually sized to reject their rated capacity through cooling towers sized to operate at design outdoor air wet bulb conditions. This ensures that the plant will satisfy the design temperatures, but it is energy wasteful when conditions are not at the design wet bulb temperature.

As the chilled water removes heat from the building zones, it gets warmer. The chiller removes the heat from the chilled water. The condenser water removes the heat from the chiller itself. The chiller system pumps the condenser water to the cooling towers on top of the building to give off its collected heat to the outdoor air. The lower the condenser water temperature, the more heat it can remove from the chiller, reducing the energy necessary to cool the chilled water.

The *Condenser Water Reset* feature saves energy by lowering the condenser water to the lowest possible temperature setpoint based on the ambient wet bulb temperature and the actual load being handled by the chiller.

Cooling the condenser water to the lower setpoint requires the cooling tower fans to expend more energy. When properly implemented, lowering the condenser water by even 1°F, can save more energy at the chiller than is used by the tower fans to lower it. The system invests some energy to save more energy.

The program can automatically reset the setpoint of the condenser water as low as possible to save energy. It takes into account the wet bulb characteristics, and either uses more or less of the water cooled at the cooling tower to achieve the new setpoint.

22.14.9 Chiller Sequencing

When more than one chiller is required to cool the chilled water to the desired setpoint. On a very hot day, all of the available chillers may be required. On a cooler day, only one or two chillers be needed. The *Chiller Sequencing* feature determines the most efficient combination of chillers required to run. It allows each chiller to run only within its efficiency range (for example, between 40% and 90% of its design capacity), and automatically starts or stops another chiller to keep all operating chillers within their range. Optionally, the program also checks the run time of various chillers to determine which to turn on or off next as cooling requirements change.

For example, for a building with three chillers, assume that the DDC controller determines that one chiller running at 94% capacity could sufficiently chill enough water. Since 94% falls outside its efficiency range of 40-88%, the DDC controller would bring on another chiller. If both chillers on-line had equal capacity, each would run at 47% capacity to chill the water. Since 47% is in the most efficient operating range, using two chillers uses less energy than using one at 94% capacity. Whenever the DDC needs to decide which available chiller to bring on next, it can choose the one with less run time.

Other cases may have varying sizes of chillers and the DDC controller will determine which combination of chillers is the most economical to operate.

22.14.10 Information Management Features

The *Information Management* features are designed to help staff gather and analyze data to help them efficiently run the facility.

Once the FMS gathers and stores information about the facility and takes any actions as appropriate according to the program, that information is available to the user in various forms.

Many FMS vendors offer similar versions of each feature, but the features vary in the amount of detail they keep and in how easy they are to use.

The ability to export the information from the FMS system to other computerized programs such as spread sheets or computerized maintenance programs is another desirable feature.

22.14.11 Summaries

Summaries contain detailed information about specific aspects of the facility. For example, one summary might list all monitored and controlled equipment and variables with their current status. Another summary might just list the low/high temperature alarm limits. An FMS probably has many summaries, with at least one also associated with each information management feature such as Runtime Totalization or for energy management features like Load Rolling.

The data is probably stored in the headend computer of the system or, if the system has intelligent, independent network units, the data is stored in the network units and is also archived in a PC in the network. The summary data is gathered by the summary feature for output to the CRT, printer, or data file.

Depending on the system used, the names given to the controlled equipment and variables may simply be numbers, or they may be word-names assigned by the user for easy identification of the item.

In addition to what summaries are printed, most systems let you define when summaries should be automatically sent to a printer or data file. An example might be every weekday at 9 a.m.

22.14.12 Password

Password is a global security feature which prevents unauthorized people from using the FMS, and might even limit certain users to executing only certain FMS commands, and possibly to working with only certain areas of the facility.

22.14.13 Alarm Reporting

The FMS is usually programmed only to report abnormal alarm conditions. For example, a fan may normally cycle on and off during the day due to temperature changes or Load Rolling, and the FMS will not continually report each change though you can find out at any time what its current on/off status is. Should the fan fail to respond to the FMS commands, the FMS will issue an alarm message.

22.14.14 Time Scheduling

The operator can easily program events to occur at a certain time of certain days. For example, you could schedule fans and lights to turn on and off at certain times, schedule various summaries to print, and schedule different temperature alarm limits for occupied and unoccupied times. The types of events you can schedule is determined by the FMS programming. Some allow you to program holiday or special event programs.

22.14.15 Trending

Trends record the status of certain variables at various intervals such as every 30 minutes and stores that data for your analysis. Trending is useful for HVAC system troubleshooting, giving you data about your facility, without taxing your staff to routinely go, monitor, and record the data.

For example, you may want to see what happens to a room temperature associated with a fan; you could use Trending to sample the fan status and room temperature every 30 minutes and examine the Trend summary at the end of the day to see how the temperature is affected by the fan.

The capabilities of the FMS determine whether you can choose the time interval, how many samples the

FMS can store, how many different variables the FMS can trend at one time, and how the data appears on the Trend summary. Many systems can even output the Trend data in bar or line graph form.

22.14.16 Graphics

Graphics associated with the FMS can be a useful tool to help identify areas of concern, displaying in pictures, much of the same current information your staff could get from a typical summary of equipment status. However, a graphic has the advantage of visually associating one occurrence with another.

For example, to determine why a temperature is too high, a graphic could easily associate the temperature with whether a particular fan is on or with the temperature of the available chilled water or with the time of day when that area is occupied by 50 people. Without a graphic to visually pull all of the possible reasons together, the reason might never be realized as the explanation and possible solution.

22.14.17 Totalization

Totalization is a counting or summing process. It can often count:
the amount of time a particular status was in effect (for example, how long a fan was On or how long a temperature was high).

• the number of times an event occurred (how many times a motor started or a temperature was in alarm).

• the number of pulses recorded at a pulse sensor; for example, to determine the amount of chilled water used.

• the amount of a physical quantity used, calculated from the current rate of consumption such as, how many gallons of chilled water, pounds of steam, or Btus of cooling were used in a week, totalized from a different calculation determining the instantaneous rate of use.

Based on this easily accessible information, your staff can accurately schedule preventive maintenance, assess costs of running equipment, or even bill tenants.

22.15.1 Summary

In chapter 22 the main components of a control system from a functional block perspective were de-

scribed, to appreciate the wide and varied knowledge base of the control systems engineer.

Some of the main points covered were how the functional block approach to control systems simplifies the understanding of control systems. Sensors are available in many types depending on whether the control system is electromechanical (pneumatic or electric), electronic and the type of variable being sensed. Temperature, humidity and pressure are typical sensed variables. Controllers receive input from sensor and compare these inputs against a setpoint. The controller then outputs to the controlled device assembly. Controller outputs may be two position, proportional or proportional plus integral. Controlled devices are typically valves and dampers. Actuators may be either pneumatic or electric. Actuators may come in spring return or non spring return varieties. The fail-safe condition normally open or normally closed is a consideration for most control applications. HVAC processes, heating, cooling, humidification, dehumidification and pressurization are facilitated by their control agents.

The final condition under control eventually will produce a comfort level for the building occupants. A host of control strategies and control systems for boilers, chillers, pumps and distribution systems must play together to achieve occupant comfort. Feedback or system feedback is the communication of the final condition to the sensor and ultimately to the controller in control of the process.

Control strategies which when properly implemented provide the required occupancy comfort.

Zone control has the advantage of sensing the actual load in the space such as people, equipment and lights. A thermostat or humidistat in the zone senses zone conditions and depending on the deviation from set point, the control logic generates the appropriate response.

Air handling units can be controlled by any of four temperature control strategies; 1) Zone/room control, 2) Return air control, 3) Discharge air control and 4) Room reset of discharge. Two types of humidity strategies may be applied; 1) Room/return air control or 2) Dew point control of discharge air

Air Handling units of the mixed air variety have economizer cycles to choose whether the system should be using outdoor air for free cooling or return air. This is accomplished by measuring the outdoor air dry bulb temperature, the difference between outdoor air dry bulb temperature and return air, or the enthalpy of the outdoor versus the return air stream.

Control of primary equipment such as boilers, chillers and cooling towers is unique to each piece of equipment. Packaged control equipment generally control these pieces of equipment with classic control strategies.

Control of distribution systems regulate quantity and pressure of the heat transfer fluids, air and water.

The efficient operation on your facility ultimately depends on the efficient use of existing building equipment and the productivity of a building's people. The FMS is a tool to help a building's people make the most of their time on the job and to coordinate the equipment for optimal use.

When using an existing FMS, each feature should be used to maximize the benefit. Many existing FMSs have much power going unused, and the unused features could be implemented with a little investment of time and no investment of funds.

Remember, the more features an FMS has, the more time a staff may need to implement them.

Typical ways that an FMS aids a building's people from an energy management perspective are:

• delaying start-up or providing early shutdown of equipment (Optimal Run Time),

• turning off selected equipment for short times during the day (Load Rolling),

• limiting consumption is not concerned with the amount of consumption but more concerned about the rate (Demand Limiting)

• using the source of air that uses the least energy to cool (Economizer Switchover)

• conditioning the supply air only to the most efficient temperature (Supply Air Reset),

• conditioning the supply water to the most efficient temperature (Supply Water Reset),

• determining the most efficient combination of chillers to bring on-line (Chiller Sequencing).

Typical ways that an FMS aids the buildings people from a storing and manipulation of operational data (Information Management) perspective are:

• Summaries contain detailed information about specific aspects of the facility.

• Password global security feature which prevents unauthorized people from using the FMS.

• alarms report abnormal conditions

• Time Scheduling provides an easy method to program events.

• Trends record the status of measured variables at various intervals

• Graphics associated with the FMS display in pictures current information providing a visual summary of the data.

• Totalization provides a method to record consumption data such as energy or operating hours.

Chapter 22 is edited excerpts by James Smith from Johnsons Controls' Book, *Building Environments - HVAC Systems*, by Alan J. Zajac Global Learning Service copyright Johnson Controls Inc., 1996, all rights reserved.

James R. Smith CEM, is a Mechanical Engineer with 20 years field and engineering support experience in the Facility Performance Contracting area. He has extensive experience in the energy and Building Environments area. He currently is the Johnson Controls, Controls Group Intranet Web Administrator.

Alan Zajac is the Lead Instructional Technologist in the Global Learning Services organization at Johnson Controls, Inc., Milwaukee Wisconsin. He has extensive experience in developing learning solutions on HVAC Systems and Controls.

CHAPTER 23

ENERGY SECURITY AND RELIABILITY

BRADLEY L. BRACHER

Oklahoma City, OK.

23.1 INTRODUCTION

Reliable utility services are vital to all industrial, commercial and military installations. Loss of electricity, thermal fuels, water, environmental control, or communications systems can bring many operations to an immediate halt resulting in significant economic loss due to unscheduled downtime, loss of life, or threat to national security. These services are delivered by vast, complex networks with many components. A small number of damaged components is often sufficient to disable portions of these networks or halt operation of the entire system. These component failures can be caused by equipment failure, natural disaster, accidents, and sabotage.

Energy security is the process of assessing the risk of loss from unscheduled utility outages and developing cost effective solutions to mitigate or minimize that risk. This chapter will explore the vulnerable network nature of utility systems and the events or threats that disrupt utility services. Methods will be presented to assess these threats and identify those most likely to result in serious problems. Actions to counter these threats will be introduced along with a methodology to evaluate the cost effectiveness of proposed actions. Finally, the links between energy security and energy management will be discussed.

23.1.1 Utility Systems As Interdependent Networks

Modern industrial, institutional, and commercial facilities are dependent on many utility systems. The utility services supporting modern facilities include electricity, thermal fuels (natural gas, fuel oil, coal), water, steam, chilled water, compressed air, sanitary sewer, industrial waste, communication systems, and transportation systems. Some of these systems are restricted to the site. Others are a maze of distribution paths and processing stations connecting the site to distant generation or production facilities. The individual components of these systems form networks.

Any failed component in a network has the potential to disrupt or degrade the performance of the entire system. Some networks contain only a single path linking the site to the source of its utility supply. The loss of that single path results in total disruption at the site. Other networks have redundant paths. Alternate routes are available to serve the site at full or partial capacity should one of the routes be damaged. In a redundant system, loads normally carried by the damage portion of the system will be shifted to an operational part of the network. This places extra stress on the remaining portion of the system. Marginal components sometimes fail under this additional load, causing additional segments of the network to fail.

Utility networks do not operate in isolation; they are linked to each other in a web of dependence. Each provides a vital service to the other. The impact of one failed utility network is felt in many other systems. The effects spread like ripples on a pond. More systems will fail if redundant or back-up systems are not in place. Often, the easiest way for a saboteur to disable a utility system is to damage another system on which it depends.

Example. Figure 23.1 illustrates the mutual dependence utility systems share with each other at one facility. This facility generates some of its own electricity with natural gas engines, pumps a portion of its water from wells using electric pumps, utilizes steam absorption chillers, and produces compressed air from a mixture of electric and gas engine compressors. A disruption of natural gas will immediately halt the on-site generation of electricity and steam and curtail the production of compressed air since these systems depend directly on natural gas as their prime mover. Failure of one utility will cascade to other systems. The absorption chillers would also experience a forced outage since they are powered by steam which is experiencing a forced outage due to the lack of natural gas. All process equipment that requires any of these disrupted utilities will be idle until full utility service is restored.

23.1.2 Threats to Utility Systems

The individual components of a utility system can be damaged or disabled in many ways. The network

Figure 23.1 Interdependence of utility networks.

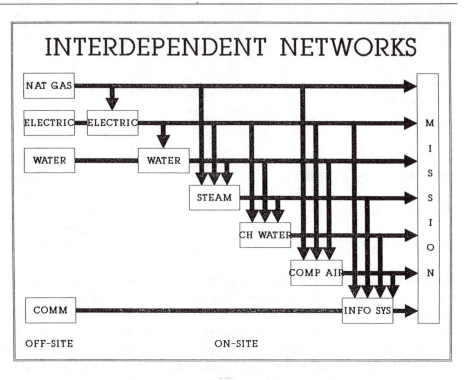

itself may be wholly or partially disrupted depending on the criticality of the component or components damaged. The threats to network components can be segregated into four categories: equipment failure, natural disaster, accidents, and sabotage.

Equipment failure is the normal loss of system components as they reach the end of their expected life. The reliability of most electrical and mechanical components follows a "bath-tub curve" illustrated in Figures 23.2 and 23.3. These curves begin with a high rate of failure early in the life cycle. This failure rate rapidly decreases until it reaches a minimum value that remains relatively stable throughout the normal operating period. The failure rate increases again once the component enters the wear-out phase.

Natural disasters are the most common cause of widespread network failure. They include earthquake, hurricane, tornado, wind, lightning, fire, flood, ice, and animal damage. Utility companies are well versed in dealing with these situations and generally have the means to effect repairs rapidly. However, widespread damage can leave some customers without service for days resulting in substantial economic loss.

Accidents are unintentional human actions such as traffic accidents, operator error, fires, and improper design or modification of the system. Operator error played a significant role in the failure of the Chernoble and Three Mile Island nuclear plants. It also contributed to both of the New York City blackouts. Once a failure sequence has begun, the system is in an abnormal state. Operators may lack the experience to know how the

system will respond to a given corrective action or might fail to respond quickly enough to a dynamic and rapidly changing situation.

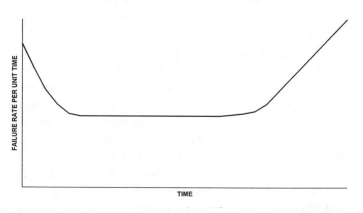

Figure 23.2 Typical "bath-tub curve" for electronic equipment showing failure rate per unit time.

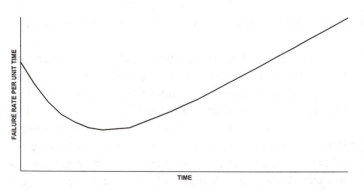

Figure 23.3 Typical "bath-tub curve" for mechanical equipment showing failure rate per unit time.

Sabotage covers the realm of intentional human action. It includes the entire spectrum from a single disgruntled employee to acts of terrorism to military action. Terrorism has been a fact of life in many countries for a long time. Its use in the United States has increased greatly in recent years. Most terrorist organizations have social or political motivations that could lead them target industrial or commercial operations. Military actions are beyond the scope of what most industrial or commercial facilities can handle. These types of threats are of interest only to the military itself.

23.2 RISK ANALYSIS METHODS

Risk analysis is a necessary first step in the process of minimizing the losses associated with unscheduled utility outages. The purpose of risk analysis is to understand the system under study and establish a knowledge base from which resources can be optimally allocated to counter known threats. This understanding comes from a systematic evaluation that collects and organizes information about the failure modes of a utility system. Once these failure modes are understood decisions can be made that reduce the likelihood of a failure, facilitate operation under duress conditions, and facilitate rapid restoration to normal operating conditions.

Two broad categories of risk analysis methods are available. The inductive methods make assumptions about the state of specific system components or some initiating event and then determine the impact on the entire system. They can be used to examine a system to any level of detail desired, but are generally only used to provide an overview. It is impractical and often unnecessary to examine every possible failure or combination of failures in a system. When the complexity or importance of a system merits more detailed analysis a deductive method is used.

Deductive analysis makes an assumption about the condition of the entire system and then determines the state of specific components that lead to the assumed condition. Fault tree analysis is the most common and most useful deductive technique. The deductive method is preferred because it imposes a framework of order and objectivity in place of what is often a subjective and haphazard process.

Probabilities can be incorporated into all of these methods to estimate the overall reliability of the system. Probabilistic techniques are best used when analyzing equipment failures but have also been used with some success in the evaluation of human error or accident. They are of less value when evaluating natural disasters and meaningless when applied to sabotage or other forms of organized hostility. The results would not only depend on the probability of a sabotage attempt occurring but also on the probability that all actions necessary to disable the system were successfully accomplished during the attack. This type of data is extremely difficult to obtain and is highly speculative. Therefore, the analysis is best conducted under the assumption that an adverse situation will definitely occur and proceed to determine what specific scenarios will result in system failure[1].

23.2.1 Inductive methods

A number of analysis tools are available to examine the effects of single component failures of a system. Three common and well developed techniques are Failure Mode and Effect Analysis (FMEA), Failure Mode Effect and Criticality Analysis (FMECA) and Fault Hazard Analysis (FHA). These methods are very similar and build upon the previous technique by gradually increasing in scope. Most utilize failure probabilities of individual components to estimate the reliability of the entire system. Preprinted forms are generally used to help collect and organize the information.

FMEA recognizes that components can fail in more than one way. All of the failure modes for each component are listed along with the probability of failure. These failure modes are then sorted into critical and non-critical failures. The non-critical failures are typically ignored in the name of economy. Any failure modes with unknown consequences should be considered critical. Figure 23.4 is a typical data sheet used in FMEA.

Failure Mode Effect and Criticality Analysis (FMECA) is very similar to FMEA except that the criticality of the failure is analyzed in greater detail and assurances and controls are described for limiting the likelihood of such failures[2]. FMECA has four steps. The first identifies the faulted conditions. The second step explores the potential effects of the fault. Next, existing corrective actions or countermeasures are listed that minimize the risk of failure or mitigate the impact of a failure. Finally, the situation is evaluated to determine if adequate precautions have been made and if not to identify additional action items. This technique is of particular value when working with the system operators. It can lead to an excellent understanding of how a system is actually operated as opposed to how the designer intended for it to be operated. Figure 23.5 is a data sheet used with FMECA.

Fault Hazard Analysis (FHA) is another permutation of Failure Mode and Effect Analysis (FMEA). Its

FAILURE MODE EFFECT AND CRITICALITY ANALYSIS

FAULT IDENTIFICATION	POTENTIAL EFFECTS OF FAULT	COMPENSATION OR CONTROL	ACTION REQUIRED

Figure 23.4 Data collection sheet for Failure Mode and Effect Analysis (FMEA).

FAILURE MODE AND EFFECTS ANALYSIS

COMPONENT	FAILURE PROBABILITY	FAILURE MODE	% FAILURES BY MODE	EFFECTS CRITICAL	EFFECTS NON-CRITICAL

Figure 23.5 Data collection sheet for Failure Mode Effect and Criticality Analysis (FMECA).

value lies in an ability to detect faults that cross organizational boundaries. Figure 23.6 is a data collection form for use with FHA. It is the same data form used with FMEA with three addition columns. Faults in column five are traced up to an organizational boundary. Column six is added to list upstream components that could cause the fault in question. Factors that cause secondary failures are listed in column seven. These are things like operational conditions or environmental variables known to affect the component. A remarks column is generally included to summarize the situation. FHA is an excellent starting point if more detailed examination is anticipated since the data is collected in a format that is readily used in Fault Tree Analysis.

These techniques concern themselves with the effects of single failures. Systems with single point failures tend to be highly vulnerable if special precautions are not taken and these methods will highlight those vital components. However, critical utility systems are normally designed to be redundant. The majority of forced outages in a redundant system will be caused by the simultaneous failure of multiple components. Techniques that only consider single point failures will significantly underestimate the vulnerability of a system. No matter how unlikely an event or combination of events may seem, experience has proven that improbable events do occur.

The Double Failure Matrix (DFM) examines the effects of two simultaneous failures. A square grid is laid out with every failure mode of every component listed along the columns and also along the rows. The intersection of any two failure modes in the matrix represents a double failure in the system. The criticality of that double failure is listed at the intersection. Critical and catastrophic failures are explored further to identify corrective actions or alternative designs. The diagonal along the grid is the intersection of a component failure with itself. This is the set of single mode failures. It can be used as a first cut analysis and later expanded to encompass the entire set of two component failures if desired. This method becomes quite cumbersome or prohibitively difficult with large or complex systems.

FAULT HAZARD ANALYSIS

COMPONENT	FAILURE PROBABILITY	FAILURE MODE	% FAILURES BY MODE	EFFECTS		UPSTREAM COMPONENTS THAT INITIATE FAULT	FACTORS THAT CAUSE SECONDARY FAILURES	COMMENTS
				CRITICAL	NON-CRITICAL			

Figure 23.6 Data collection sheet for Failure Hazard Analysis (FHA).

Event Tree Analysis is an exhaustive methodology that considers every possible combination of failed components. It is known as a tree because the pictorial illustration branches like a tree every time another component is included. The tree begins with a fully operational system that represents the trunk of the tree. The first component is added and the tree branches in two directions. One branch represents the normal operational state of the component and the other represents the failed state. The second component is considered next. A branch representing the operational and failed states is added to each of the existing branches, resulting in a total of four branches. Additional components are added in a like manner until the entire system has been included. The paths from the trunk to the tip of each branch are then evaluated to determine the state of the entire system for every combination of failed components. Complex systems can be in a state of total success, total failure, or some variant of partial success or failure.

Example. Figure 23.7 is a simplified one-line diagram of an electric distribution system for a facility with critical loads. The main switch gear at the facility is a double-ended substation fed from two different commercial power sources. All critical loads are isolated on a single buss which can receive power from either commercial source or an on-site emergency generator. The facility engineer conducted a Failure Mode Effect and Criticality Analysis. His findings are listed in Figure 23.8. Three single point failures were identified that result in a forced outage of the critical loads if any one of these components fails: automatic transfer switch, main breaker at the critical load switch gear, and the buss in the critical load switch gear. Action items were listed for all components to improve system reliability. Most of the breakers require on-site spares for maximum reliability, however, a number of these can be shared to minimize the expense. For example, the main breakers at switch gear "A" and "B" can share a spare since they are of the same size. The most important action items are those for the single point failures.

Figure 23.7 Simplified electric distribution system for example.

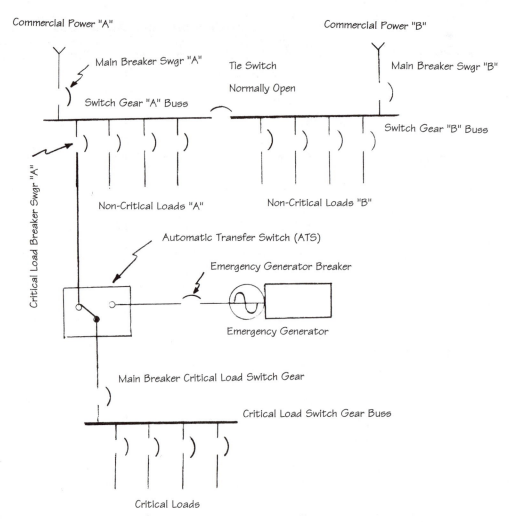

FMECA identifies only single point failures and can significantly underestimate the risk of a forced outage since improbable events like double failures do occur. A Double Failure Matrix (DFM) was constructed in Figure 23.9 to identify the combinations of two failed components that would deprive the critical loads of electricity. Rules were defined to gauge the criticality of each double failure. Loss of any one of the three sources of electricity is a level 2 failure with marginal consequences. Loss of any two of the three sources is a level 3 failure with critical consequences and loss of all three sources is a level 4 failure with catastrophic consequences since all critical loads are in an unscheduled forced outage. The upper portion of the matrix is left blank in this example since it is a mirror image of the lower portion.

The highlighted diagonal is the intersection of a component with itself and represents the single point failures. The results of the FMECA are verified since the three components identified as single point failures are now shown to be level 4 failures along the diagonal. Since these components are single point failures their

rows and columns are filled with 4's denoting a catastrophic loss of power when they and any other component are simultaneously failed. Six additional level 4 failures appear on the chart that represent true double component failures. All of these involve the inability to transfer commercial power to the critical load buss and a simultaneous failure of the emergency generator or its support equipment. This could lead the facility manager to investigate alternate systems configurations that have more than one way of transferring commercial power to the critical loads. Additionally, the threats that can reasonably be expected to damage the components involved in level 4 failures should be carefully explored and countermeasures implemented to reduce the likelihood of damage. The level 3 failures that occur along the diagonal also merit special attention.

23.2.2 Deductive method

Fault Tree Analysis is the most useful of the deductive methods and is preferred by the author above all the inductive methods. Benefits include an understanding of

FAILURE MODE EFFECT AND CRITICALITY ANALYSIS

FAULT IDENTIFICATION	POTENTIAL EFFECTS OF FAULT	COMPENSATION OR CONTROL	ACTION REQUIRED
Commercial Power "A"	No commercial power from source "A"	Close Tie and feed from Comm Power "B"	Investigate security & reliability of system supplying Comm Power "A"
Main Breaker Swgr "A"	No commercial power from source "A"	Close Tie and feed from Comm Power "B"	Conduct regular inspections & preventive maintenance & keep spare on-site
Buss Swgr "A"	No commercial power to critical loads or non-critical loads on Swgr "A"	ATS transfers critical loads to Emergency Generator Repair or replace buss	Conduct regular inspections & preventive maintenance Prevent animal & water damage
Commercial Power "B"	No commercial power from source "B"	Critical loads fed by preferred comm power source "A"	Investigate security & reliability of system supplying Comm Power "B"
Main Breaker Swgr "B"	No commercial power from source "B"	Critical loads fed by preferred comm power source "A"	Conduct regular inspections & preventive maintenance & keep spare on-site
Buss Swgr "B"	No commercial power from source "B" to crit loads & No power to non-crit loads "B"	Critical loads fed by preferred comm power source "A" Repair or replace Buss	Conduct regular inspections & preventive maintenance Prevent animal & water damage
Tie Between Swgr "A" and Swgr "B"	Reduced reliability Additional failures could lead to forced outage	Repair or replace tie switch	Conduct regular inspections & preventive maintenance
Critical Load Breaker Swgr "A"	No commercial power to critical loads	ATS transfers critical loads to Emergency Generator	Conduct regular inspections & preventive maintenance & keep spare on-site
Automatic Transfer Switch	Forced outage - No power to critical loads	Repair or replace ATS	Conduct regular inspections & preventive maintenance & test ATS monthly
Main Breaker Critical Load Swgr	Forced outage - No power to critical loads	Repair or replace breaker	Conduct regular inspections & preventive maintenance & keep spare on site
Buss Critical Load Swgr	Forced outage - No power to critical loads	Repair or replace buss	Conduct regular inspections & preventive maintenance Prevent animal & water damage
Emergency Generator	Reduced reliability - No on-site back-up to commercial power	Critical loads fed by preferred comm power source "A" - Repair gen.	Conduct regular inspections & preventive maintenance & test generator monthly
Breaker Emergency Generator	Reduced reliability - No on-site back-up to commercial power	Critical loads fed by preferred comm power source "A" Repair or replace breaker	Conduct regular inspections & preventive maintenance & keep spare on-site
Control System Emergency Generator	Reduced reliability - No on-site back-up to commercial power	Critical loads fed by preferred comm power source "A" - repair controls	Conduct regular inspections & preventive maintenance & test monthly

Figure 23.8 Failure Mode Effect and Criticality Analysis (FMECA) for example.

DOUBLE FAILURE MATRIX

KEY:
1 Negligible
2 Marginal
3 Critical
4 Catastrophic

	Commercial Power "A"	Main Breaker Swgr "A"	Buss Swgr "A"	Commercial Power "B"	Main Breaker Swgr "B"	Buss Swgr "B"	Tie Between Swgr "A" and Swgr "B"	Critical Load Breaker Swgr "A"	Automatic Transfer Switch	Main Breaker Critical Load Swgr	Buss Critical Load Swgr	Emergency Generator	Breaker Emergency Generator	Control System Emergency Generator
Commercial Power "A"	2													
Main Breaker Swgr "A"	2	2												
Buss Swgr "A"	3	3	3											
Commercial Power "B"	3	3	3	2										
Main Breaker Swgr "B"	3	3	3	2	2									
Buss Swgr "B"	3	3	3	2	2	2								
Tie Between Swgr "A" and Swgr "B"	3	3	3	2	2	2	2							
Critical Load Breaker Swgr "A"	3	3	3	3	3	3	3	3						
Automatic Transfer Switch	4	4	4	4	4	4	4	4	4					
Main Breaker Critical Load Swgr	4	4	4	4	4	4	4	4	4	4				
Buss Critical Load Swgr	4	4	4	4	4	4	4	4	4	4	4			
Emergency Generator	3	3	4	3	3	3	3	4	4	4	4	2		
Breaker Emergency Generator	3	3	4	3	3	3	3	4	4	4	4	2	2	
Control System Emergency Generator	3	3	4	3	3	3	3	4	4	4	4	2	2	2

Figure 23.9 Double Failure Matrix (DFM) for electric distribution system in example.

all system failure modes, identification of the most critical components in a complex network, and the ability to objectively compare alternate system configurations. Fault trees use a logic that is essentially the reverse of that used in event trees. In this method a particular failure condition is considered and a logic tree is constructed that identifies the various combinations and sequence of other failures that lead to the failure being considered. This method is frequently used as a qualitative evaluation method in order to assist the designer, planner or operator in deciding how a system may fail and what remedies may be used to overcome the cause of failure[3].

The fault tree is a graphical model of the various combinations of faults that will result in a predefined undesired condition. Examples of this undesired condition are:

a) Total loss of electricity to surgical suite.
b) Total loss of chilled water to computer facility.
c) Steam boiler unable to generate steam.
d) Loss of natural gas feedstock to fertilizer plant.
e) Water supply to major metropolitan area curtailed to half of minimum requirement.
f) Environmental conditioning of controlled experiment is interrupted for more than thirty minutes.

The faults can be initiated by sabotage actions, software failures, component hardware failures, human errors, or other pertinent events. The relationship between these events is depicted with logic gates.

The most important event and logic symbols are shown in Figure 23.10. A basic event is an initiating fault that requires no further development or explanation. The basic event is normally associated with a specific component or subsystem failure. The undeveloped event is a failure that is not considered in further detail because it is not significant or sufficient information is not available.

Logic gates are used to depict the relationship between two or more events and some higher level failure of the system. This higher level failure is known as the output of the gate. These higher level failures are combined using logic gates until they culminate in the top event of the tree, which is the previously defined undesired event. A simple fault tree is illustrated in Figure 23.11.

The "OR" gate shows that the higher level failure will occur if at least one of the input events occurs. An "OR" gate could be used to model two circuit breakers in series on a radial underground feeder. If either circuit breaker is opened the circuit path will be broken and all loads served by that feeder will be deprived of electricity.

The output event of an "AND" gate occurs only if all of the input events occur simultaneously. It would be used to model two redundant pumps in parallel. If one pump fails the other will continue to circulate fluid through the system and no higher order failure will occur. If both pumps fail at the same time for any reason the entire pumping subsystem fails.

The logic of the fault tree is analyzed using boolean algebra to identify the minimal cut sets. A minimal cut set is a collection of system components which, when

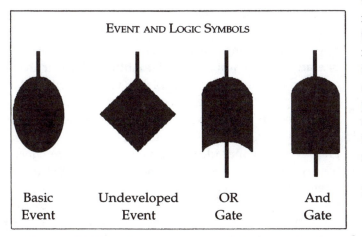

Figure 23.10 Fault tree event and logic symbols.

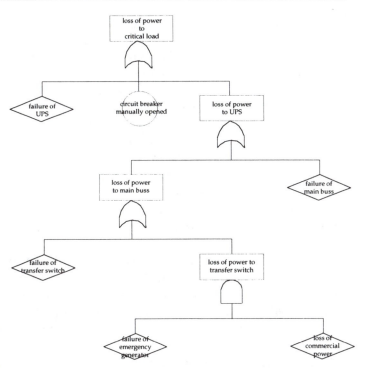

Figure 23.11 Sample fault tree for electric distribution system with uninterruptible power supply and emergency generator.

failed, cause the failure of the system. The system is not in a failed state if any one of the components in this set has not failed or is restored to operation. In fault tree terminology, a cut set is a combination of basic events that will result in the undesired top event of the tree. A computer is normally used to automate this tedious and error prone mathematical procedure.

Cut sets are utilized because they directly correspond to the modes of system failure. In a simple case, the cut sets do not provide any insights that are not already quite obvious. In more complex systems, where the system failure modes are not so obvious, the minimal cut set computation provides the analyst with a thorough and systematic method to identify the combinations of component failures which culminate in the top event. Once an exhaustive list of cut sets is assembled, they can be analyzed to determine which components occur in failure modes with the highest frequency. These, along with the single point failures, are the most critical components of the entire system and merit special attention to keep them out of harm's way.

23.3 COUNTERMEASURES

Corrective actions, or countermeasures, are implemented to reduce the risk of an unscheduled outage.

These measures should initially be focused on single component failures and those with the most catastrophic consequences when failed. The second priority are those components that occur in the largest number of cut sets. Countermeasures fall into three broad categories: protective measures, redundant systems, and rapid recovery.

Risk analysis contributes vital information to the countermeasure process by imparting an understanding of the system failure modes. By knowing how the system can fail and what components contribute the those failures, the facility manager can make informed decisions that allocate resources to those countermeasures that mitigate the most significant risks.

Facilities and the utility systems that support them are complex. No single countermeasure is sufficient to mitigate all risk. A multi-faceted approach is required. Just as a three-legged stool will not stand on one leg, neither will a facility be secure against disruption of utility support without a comprehensive approach.

23.3.1 Protective countermeasures

Protective countermeasures are actions taken before a crisis occurs to safeguard the components of a utility system against mechanical or electrical damage. Systems are normally constructed to withstand vandalism, severe weather, and other foreseeable events. Critical systems are normally designed with high reliability components.

Physical barriers are the first protective action that should be considered. They establish a physical and psychological deterrent to unauthorized access. Their purpose is to define boundaries for both security and safety, detect entry, and delay and impede unauthorized entry. Fences are the most common barrier. They are used to secure the perimeter of a site. In high value sites, fencing should be supplemented with barbed wire or razor wire. Walls are another excellent barrier. They are routinely used to isolate electrical vaults and mechanical rooms within a building for safety reasons. Proper access control makes these areas more secure as well as safe. It should be remembered that physical barriers are not sufficient to deter a determined adversary intent on causing damage. When the situation warrants, physical barriers must be supplemented with surveillance systems and guards. Area lighting is used in combination with physical barriers to further deter intrusion and aid security personnel in the detection of unauthorized entry.

Many components of utility systems are built in exposed locations that make them vulnerable to traffic accidents or tampering. Pipeline components such as valves and regulators can be buried in pits. Berms can protect many components like metering stations and regulators by sheltering them from vehicular traffic, deflecting blasts, and obscuring the line of sight necessary for fire arm damage. Bollards offer excellent protection against delivery vehicles and lawn mowers. Creek and river crossing should be designed to withstand the full force of flooding. This includes any large debris that might be carried by the current.

Site planning is an important aspect of energy security. Vital components should be located away from perimeter fences and shielded from view when possible. Dispersal of resources is another excellent strategy. When redundant system components like electrical transformers are co-located they can all be disabled by a single event. Hardening of components and the structures that house them is appropriate in some instances. Hardening makes a system tolerant to the blast, shock, and vibration caused by explosions or earthquakes.

Reliability improvement of individual components can significantly reduce outages related to equipment failures. This is especially applicable to single point failures. Equipment brands and models known to have high a mean time between failure (MTBF) should be specified. Newly constructed systems should be thoroughly tested at load to insure they are beyond the known infant mortality period. An ongoing preventive maintenance program should be implemented for existing facilities to keep them in a state of high reliability. Such a program should include through inspection, adjusting, lubrication, and replacement of failed redundant components. Aging components should be replaced before they enter the wear out region. Stand by systems, such as emergency electric generators, should be periodically tested at full load. These procedures require trained personnel, test equipment, and meticulous record keeping.

23.3.2 Redundant systems

Many actions can be taken to facilitate continued operation when a portion of the utility infrastructure is crippled or otherwise unavailable to the facility. The most important of these is redundancy. Critical facilities should be supported by multiple independent supply routes. Electrical feeders should follow different geographic routes and, ideally, come from different substations. Telephones lines can sometimes to routed to different switches. Back up generators and uninterruptible power supplies guard against disturbances and disruptions that occur off-site in the electric grid. Fuels subject to curtailment, such as natural gas, can be supplemented with alternate fuels such as fuel oil or propane-air mixing systems that can be stored on-site. Sites that burn coal should maintain a stockpile to guard against strikes or transportation uncertainties. Redundant systems should be sized with sufficient excess capacity to carry all critical loads and all support functions, such as lighting and environmental conditioning, that are required for continuous operation.

23.3.3 Rapid recovery

Once a crisis has developed, the overriding goal is to get the system operating in a normal state as rapidly as possible and resume full scale operations. Stockpiling critical or hard to get parts will reduce the recovery time more than any other action. These components should have been identified during the risk analysis. They must be stored separately from the components they are intended to replace to preclude the possibility of a single threat damaging both the primary component and the spare. Emergency response teams must be available to effect the repairs. If damage is expected to exceed the capabilities of in house personnel, additional parts suppliers and alternate repair crews should be identified and contracts in place to facilitate their rapid deployment. Simulation of recovery actions is an excellent training tool to prepare crews for operation under adverse conditions. Realistic simulation also helps identify unexpected obstacles like limited communication capability, electronic locks that won't open without electricity, and vehicles that can't be refueled.

Any number of things can and will go wrong once an emergency has been initiated. Operating with the lights off, both literally and figuratively, is a demanding task even for the prepared. Portable generators and lights are needed for twenty-four hour repair operations. Self contained battery-powered light carts should be available for use inside buildings. Clear lines of authority must be defined in advance. Crisis coordinators must be able to contact response teams even if the telephones don't work. Crew members need to know where to report and to whom. These things must be thought out in advance and documented in contingency plans.

23.3.4 Contingency planning

Contingency planning is a vital follow-up to the countermeasure process. Plans impart a much needed order to the chaos that surrounds a catastrophic event. A manager's thinking is not always clear in the fog of a crisis. Lines of communication break down. A plan provides the necessary framework of authority to imple-

ment restoration actions. The plan identifies the resources necessary to effect repairs and sets priorities to follow if the damage is extensive. Occasionally the planning process identifies critical components not previously identified or highlights bottlenecks such as communication systems that may become overloaded during the crisis.

The goal of contingency planning is rapid recovery to a normal operating state. Risk analysis plays an important role in planning. It identifies those high-risk components or sub-systems that are most vulnerable or result in catastrophic situations when failed. Contingency plans must be developed to complement the implementation of countermeasures that reduce the likelihood of damage occurring. Risk analysis also identifies those components that only yield moderate consequences when damaged. These components are frequently not vital enough to merit the expense of constructing physical barriers or redundant systems. A solid contingency plan identifying sources of parts and labor to effect repairs may be the least cost option.

The most important task performed in the plan is the designation of an emergency coordinator and delineation of authorities and responsibilities. A single coordinator is necessary to insure priorities are followed and to resolve conflicts that may develop. Each member of the response team should have a well defined role. Proper planning will insure no vital tasks "fall through the cracks" unnoticed. No plan can anticipate every contingency that may arise, but a well written plan provides a framework that can be adapted to any situation. An energy contingency plan should, as a minimum, contain the following items:

- Definition of specific authorities and responsibilities
- Priorities for plant functions and customers
- Priorities for protection and restoration of resources
- Curtailment actions
- Recall of personnel (in-house, contract, and mutual aid)
- Location of spares
- Location and contacts for repair equipment
- Equipment suppliers

Each critical component identified during risk analysis should be addressed in detail.

Contingency plans should be exercised on a regular basis. This familiarizes the staff with the specific roles they should assume during an emergency. Practicing these tasks allows them to become proficient and provides an opportunity to identify deficiencies in the plan.

Exercising also creates an awareness that a crisis can occur. This gets personnel thinking about actions that reduce risk in existing and new systems. Exercises can range from a paper game or role playing to a full scale simulation that actually interrupts the power to portions of the facility. While full scale simulation is expensive in terms of both labor and productivity, many valuable lessons can be learned that would not otherwise be apparent until an actual disruption occurred.

23.4 ECONOMICS OF ENERGY SECURITY AND RELIABILITY

The consequences of utility outages cover the spectrum from minor annoyances to catastrophic loss of life or revenue. Financial losses can be grouped in several convenient categories. Loss of work in progress covers many things. A few examples include parts damaged during machining operations or by being stranded in cleaning vats, spoiled meats or produces in cold storage, interruption of time sensitive chemical process, and lost or corrupted data. Lost business opportunities are common in retail outlets that depend on computerized registers. Businesses that require continuous contact with customers such as reservation centers, data processing facilities, and communications infrastructure also experience lost business opportunities when utilities are disrupted. Many data processing and reservation centers estimate the cost of interruptions in the six figure range per hour of downtime.

The cost of implementing countermeasures should be commensurate with the cost of an unscheduled outage. This requires a knowledge of the financial impact of an outage. If a probabilistic risk analysis technique was utilized, the frequency and duration of outages can be estimated using conventional reliability theory. Costs associated with recovery operations, damaged work in progress, and lost business opportunities are calculated for each type of outage under analysis. The expected annual loss can then be estimated for each scenario by multiplying the duration by the hourly cost and adding the cost of repairs and recovery operations. Countermeasure funding can be prioritized on the basis of expected financial loss, probability of occurrence, payback time, or other relevant measure.

For non-probabilistic techniques, a more subjective approach must be utilized. The duration of most outages can be estimated based on known repair times. The accuracy of these estimates becomes questionable when widespread damage requires repair crews to service multiple sites or stocks of spare parts are exhausted. The absence of hard data on the frequency of such outages

increases the subjective nature of the estimate. One technique is to simply assume that a crisis will occur and allocate resources to mitigate the risk on the basis of financial loss per occurrence or a cost/benefit ratio.

Payback period and cost/benefit ratio are the most common economic analysis tools used to rank countermeasures competing for budget dollars. They can also be used to determine if a particular measure should be funded at all. Payback is calculated by dividing the cost to implement a countermeasure by the expected annual loss derived from a probabilistic risk analysis. The payback period can be evaluated using normal corporate policy. A payback of two years or less is usually sufficient to justify the expenditure. The cost/benefit ratio is calculated by dividing the cost to implement a countermeasure by the losses associated with a single forced outage. If the ratio is greater than one the expenditure will be paid back after a single incident. If the ratio is between one-half and one, the expenditure will be paid back after two incidents.

Insurance should be considered to shift the risk of financial loss to another party. Business interruption policies are available to compensate firms for lost opportunities and help cover the cost of recovery operations. The cost of insurance policies should be treated as an alternate countermeasure and compared using payback or cost/benefit ratio.

23.5 LINKS TO ENERGY MANAGEMENT

Many energy security projects can piggy-back on energy conservation projects at a nominal incremental cost. Purchased utility costs can be reduced while reliability is improved. This can dramatically alter the economics of implementing certain countermeasures. Fuel switching strategies let installations take advantage of interruptible natural gas tariffs or transportation contracts. Peak shaving with existing or proposed generators will reduce electric demand charges. Time-of-use rates and real-time pricing make self-generation very attractive during certain seasons or at particular times of the day.

Many industrial facilities purchase utilities on interruptible supply contracts to reduce purchased utility costs. Curtailments are usually of short duration and are often contractually limited to a specified number of hours per year. Alternate fuel sources are normally installed prior to entering into an interruptible supply contract. An example would be to replace the natural gas supply to a steam plant with fuel oil or propane. This gives the facility the option of burning the least expensive fuel. Caution must be exercised when designing a dual-fuel system. Utility service is often curtailed at a time when it is most needed. Natural gas is typically curtailed during the coldest days of winter when demand is highest and supplies become limited due to freezing of wells. An extended curtailment could outlast the on-site supply of an alternate fuel. The same cold weather which caused the primary fuel to be curtailed may also result in shortages of the alternate fuel. An unscheduled plant shutdown could occur if the storage system is undersized or if proper alternate arrangement were not made in advance.

Thermal energy storage systems are installed to reduce on-peak electric demand charges. They can also play a key role in facility reliability. Chillers are frequently considered non-critical loads and are not connected to emergency generators. When a disruption occurs, critical equipment that requires cooling must be shut-down until electricity is restored and the chilled water system is returned to operation. The time required to restart the chilled water system and critical equipment can last many hours longer than the electrical disruption that initiated the event. When a storage system is in place, only the chilled water circulating pumps need to be treated as critical loads powered by the generator. Chilled water can be continuously circulated from the storage system to the equipment, thus precluding the need for a shut-down in all but the longest of outages. When the chillers are removed from the emergency generator additional process equipment can be connected in its place and kept operating during outages or a smaller generator can be installed at a lower first cost.

Energy conservation also has a direct impact on energy security. Conservation projects reduce the amount of energy required to perform at full capacity. By consuming less energy, an installation with a fixed quantity of alternate fuels stored on-site can remain in operation longer under adverse conditions. More efficient operations can also reduce the size of stand-by generators, uninterruptible power supplies, and similar systems. Smaller equipment usually means reduced construction costs and improved economics.

23.6 SUMMARY

All facilities require a continuous and adequate supply of utility support to function. Energy security is the process of evaluating utility systems and implementing actions that minimize the impact of unscheduled outages that prevent a facility from operating at full capacity. Utility systems are networks with many components. Loss of a few, or in some cases one, critical components is sufficient to disable a network or leave it operating at partial capacity. Additionally, utility networks are not independent. They support each other in a symbiotic manner. The collapse of one network can lead to a domino effect that causes other networks to fail.

The critical components of a utility network can be identified by using risk analysis techniques. The inductive methods make assumptions about the status of individual components and then determine what impact is felt on the entire system. Deductive methods, notably fault tree analysis, use an opposite approach. Fault tree analysis assumes the system is in some undesired condition and proceeds to determine what combinations of failed components will result in that condition. The combinations of failed components are called cut sets. The component failures can be caused by equipment failure, natural disaster, accident, or sabotage.

Countermeasures are actions that prevent or minimize the impact of utility disruptions. Three countermeasures that should not be neglected by any facility manager are physical protection, redundancy, and stockpiling of critical spare parts. Coupled with contingency planning, these counter measures will greatly enhance the energy security of any installation. Contingency plans should be exercised for maximum effectiveness.

Utility disruptions can cause a wide range of impacts on the affected facility. In most cases, this impact can be expressed as a dollar value. This financial impact is useful in evaluating the economics of implementing countermeasures. The most common economic analysis tools are payback and cost/benefit ratio. Many countermeasures can be made more cost effective by linking them to energy conservation projects.

References

1. Bracher, Bradley L., Utility Risk Analysis, Proceedings of the 8th World Energy Engineering Congress, 1989, p. 601.
2. U.S. Nuclear Regulatory Commission, *Fault Tree Handbook*, p. II-4.
3. Billinton and Allan, Reliability Evaluation of Engineering Systems: Concepts and Techniques, p. 113.

APPENDIX I
THERMAL SCIENCES REVIEW

L.C. WITTE

Professor of Mechanical Engineering
University of Houston
Houston, Texas

I.1. INTRODUCTION

Many technical aspects of energy management involve the relationships that result from the thermal sciences: classical thermodynamics, heat transfer, and fluid mechanics. For the convenience of the user of this handbook, brief reviews of some applicable topics in the thermal sciences are presented. Derivation of equations are omitted; for readers needing those details, references to readily available literature are given.

I.2. THERMODYNAMICS

Classical thermodynamics represents our understanding of the relationships of energy transport to properties and characteristics of various types of systems. This science allows us to describe the global behavior of energy-sensitive devices. The relationships that can be developed will find application in both fluid-flow and heat-transfer systems.

A thermodynamic system is a region in space that we select for analysis. The boundary of the system must be defined. Usually, the system boundary will coincide with the physical shell of a piece of hardware. A closed system is one where no mass may cross the boundary, whereas an open system, sometimes called a control volume, will generally have mass flowing through it.

We generally divide energy into two categories: stored and transient types of energy. The stored forms are potential, kinetic, internal, chemical, and nuclear. These terms are fairly self-descriptive in that they relate to ways in which the energy is stored. Chemical and nuclear energy represent the energy tied up in the structure of the molecular and atomic compounds themselves. These two types of stored energy presently form the prime energy sources for most industrial and utility applications and thus are of great importance to us.

Potential, kinetic, and internal energy forms generally are nonchemical and nonnuclear in nature. They relate to the position, velocity, and state of material in a thermodynamic system. More detailed representations of these energy forms will be shown later.

I.2.1 Properties and States

Thermodynamic systems are a practical necessity for the calculation of energy transformations. But to do this, certain characteristics of the system must be defined in a quantifiable way. These characteristics are usually called properties of the system. The properties form the basis of the state of the system. A state is the overall nature of the system defined by a unique relationship between properties.

Properties are described in terms of four fundamental quantities: length, mass, time, and temperature. Mass, length, and time are related to a force through Newton's second law.

In addition to the fundamental quantities of a system there are other properties of thermodynamic importance. They are pressure, volume, internal energy, enthalpy, and entropy—P, V, U, H, and S, respectively.

Equation of State. Returning to the concept of a state, we use an equation of state to relate the pertinent properties of a system. Generally, we use $P = P(m, V, T)$ as the functional equation of state. The most familiar form is the ideal gas equation of state written as

$$PV = mRT \tag{I.1}$$

or

$$PV = n\overline{R}T \tag{I.2}$$

Equation I.1 is based on the mass in a system, whereas equation I.2 is molal-based. R is called the gas constant and is unique to a particular gas. $\overline{R}$, on the other hand, is called the universal gas constant and retains the same value regardless of the gas (i.e., $\overline{R}$ = 1545 ft lb$_f$/lb$_m$ · mol · °R = 1.9865 Btu/lb$_m$ · mol · °R. It can be easily shown that R is simply $\overline{R}/M$, where M is the molecular weight of the gas.

The ideal gas equation is useful for superheated but not for saturated vapors. A vexing question that often arises is what is the range of application of the ideal gas equation. Figure I.1 shows the limits of applicability. The shaded areas demonstrate where the ideal gas equation applies to within 10% accuracy.

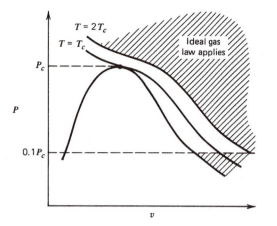

Fig. I.1 Applicability of ideal gas equation of state.

If high pressures or vapors near saturation are involved, other means of representing the equation of state are available. The compressibility factor Z for example $Z = PV/RT$, is a means of accounting for nonideal gas conditions. Several other techniques are available; see Ref. 1 for example.

Changes of state for materials that do not behave in an ideal way can be calculated by use of generalized charts for property changes. For example, changes in enthalpy and entropy can be presented in terms of reduced pressure and temperature, P_r and T_r. The reduced properties are the ratio of the actual to the critical properties. These charts (see Appendix II) can be used to calculate the property changes for any change in state for those substances whose thermophysical properties are well documented.

Ideal Gas Mixtures. Where several gases are mixed, a way to conveniently represent the properties of the mixture is needed. The simplest way is to treat the system as an ideal gas mixture. In combustion systems where fuel vapor and air are mixed, and in the atmosphere where oxygen nitrogen and water vapor form the essential elements of air, the concept of the ideal gas mixture is very useful.

There are two ways to represent gas mixtures. One is to base properties on mass, called the gravimetric approach. The second is based on the number of moles in a system, called a molal analysis. This leads to the definition of a mass fraction, $x_i = m_i/m_t$, where m_i is the mass of the ith component in the mixture and m_t is the total mass of the system, and the mole fraction, $y_i = n_i/n_t$, where n represents the number of moles.

Commonly, the equation of state for an ideal mixture involves the use of Dalton's law of additive pressures. This uses the concept that the volume of a system is occupied by all the components. Using a two-component system as an example, we write

and
$$P_1 V_1 = n_1 \overline{R} T_1$$
$$P_2 V_2 = n_2 \overline{R} T_2$$

Also
$$P_1 V_1 = n_1 \overline{R} T_2$$

Since V, $\overline{R}$, and T are the same for all three equations above, we see that

$$P_t = P_1 + P_2$$

and
$$(P_1 + P_2) V = (n_1 + n_2) \overline{R} T$$

for a mixture. Each gas in the mixture of ideal gases then behaves in all respects as though it exists alone at the volume and temperature of the total system.

I.2.2 Thermodynamic Processes

A transformation of a system from one state to another is called a process. A cycle is a set of processes by which a system is returned to its initial state. Thermodynamically, it is required that a process be quantifiable by relations between properties if an analysis is to be possible.

A process is said to be reversible if a system can be returned to its initial state along a reversed process line with no change in the surroundings of the system. In actual practice a reversible process is not possible. All processes contain effects that render them irreversible. For example, friction, nonelastic deformation, turbulence, mixing, and heat transfer are all effects that cause a process to be irreversible. The reversible process, although impossible, is valuable, because it serves as a reference value. That is, we know that the ideal process is a theoretical limit toward which we can strive by minimizing the irreversible effects listed above.

Many processes can be described by a phrase which indicates that one of its properties or characteristics remains constant during the process. Table I.1 shows the more common of these processes together with expressions for work, heat transfer, and entropy change for ideal gases.

I.2.3 Thermodynamic Laws

Thermodynamic laws are relationships between mass and energy quantities for both open and closed systems. In classical form they are based on the conservation of mass for a system with no relativistic effects. Table I.2 shows the conservation-of-mass relations for various types of systems. In energy conservation, the first and second laws of thermodynamics form the basis of most technical analysis.

Table I.1 Ideal Gas Processes[a]

Process	Describing Equations	$_1W_2$	$_1Q_2$	$S_2 - S_1$
Isometric or constant volume	$V = c,^{(1)} v = c, \dfrac{p}{T} = c$	0	$U_2 - U_1 = m(u_2 - u_1)$ $= mc_v(T_2 - T_1)^{(2)}$	$m\left(s_2^0 - s_1^0 - R\ \ln\dfrac{p_2}{p_1}\right)^{(3)}$ $= mc_v\ \ln\dfrac{T_2}{T_1}$
Isobaric, isopiestic, or constant pressure	$p = c, V/T = c, \dfrac{v}{T} = c$	$p(V_2 - V_1)$ $= mR(T_2 - T_1)$	$H_2 - H_1 = m(h_2 - h_1)$ $= mc_p(T_2 - T_1)$	$m\left(s_2^0 - s_1^0\right)$ $= mc_p\ \ln\dfrac{T_2}{T_1}$
Isothermal or constant temperature	$T = c, pV = pv = c$	$p_1V_1\ \ln\ r^{(4)} = p_2V_2\ \ln\ r$ $= mRT\ \ln\ r$	$_1W_2$	$mR\ \ln\ r$
Isentropic or reversible adiabatic	$s = c$ $pV^{k(5)} = c, TV^{k-1}$ $= c, p^{k-1}T^{-k} = c$	$U_1 - U_2 = m(u_1 - u_2)$ $\dfrac{P_1V_1 - P_2V_2}{k-1} = \dfrac{mR(T_1 - T_2)}{(k-1)}$	0	0
Polytropic	$pV^{n(6)} = c, TV^{n-1}$ $= c, p^{n-1}T^{-n} = c$	$\dfrac{P_1V_1\ P_2V_2}{n-1} = \dfrac{mR(T_1 - T_2)}{(n-1)}$	Use first law	$m\left(s_2^0 - s_1^0 - R\ \ln\dfrac{p_2}{p_1}\right)$ $= \dfrac{k-n}{k-1}\ \ln\ r$

[a](1) c stands for an unspecified constant; (2) the second line of each entry applies when c_p and c_v are independent of temperature; (3) $s_2 - s_1 = s^0(T_2) - s^0(T_1) - R\ \ln(p_2/p_1)$; (4) r = volume ratio or compression ratio = V_2/V_1; (5) $k = c_p/c_v > 1$; (6) n = polytropic exponent.

Table I.2 Law of Conservation of Mass

Closed system, any process: m = constant

Open system, SSSF: $\displaystyle\sum_{in} \dot{m} = \sum_{out} \dot{m}$

Open system, USUF: $(m_2 - m_1)_{c.v.} = \left[\displaystyle\sum_{in} m\ \sum_{out} m\right] \Delta t$

Open system, general case:

$\dfrac{d}{dt}\ m_{c.v.} = \displaystyle\sum_{in} m\ \sum_{out} m.$ or $\dfrac{d}{dt}\bigg|_V \int \rho\ dV = \int_A \rho\ V_{rn}\ dA$

For open systems two approaches to analysis can be taken, depending upon the nature of the process. For steady systems, the steady-state steady-flow assumption (SSSF) is adequate. This approach assumes that the state of the material is constant at any point in the system. For transient processes, the uniform flow uniform state (UFUS) assumption fits most situations. This involves the assumption that at points where mass crosses the system boundary, its state is constant with time. Also, the state of the mass in the system may vary with time but is uniform at any time.

Tables I.3 and I.4 give listings of the conservation-of-energy (first-law) and the second-law relations for various systems.

The first law simply gives a balance of energy during a process. The second law, however, extends the utility of thermodynamics to the prediction of both the possibility of a proposed process or the direction of a system change following a perturbation of the system. Although the first law is perhaps more directly valuable in energy conservation, the implications of the second law can be equally illuminating.

There are two statements of the second law. The two, although appearing to be different, actually can be shown to be equivalent. Therefore, we state only one of them, the Kelvin-Planck version:

Table I.3 First Law of Thermodynamics

Closed system, cyclic process:

$$\oint dQ = \oint dW$$

Closed system, state 1 to state 2:

$$_1Q_2 = E_2 - E_1 + {_1W_2}$$

E = internal energy + kinetic energy + potential energy = $m(u + V^2/2 + gz)$

Open system:

$$\dot{Q}_{c.v.} = \frac{d}{dt}\bigg|_v \rho\,(u + V^2/2 + gz)\,dV + \bigg|_A \rho\,(h + V^2/2 + gz)\,V_{rn}\,dA + W_{c.v.,}$$

where enthalpy per unit mass $h = u + pv$;

alternative form:

$$\dot{Q}_{c.v.} + \sum_{in} \dot{m}(h + V^2/2 + gz) = \dot{W}_{c.v.} + \sum_{out} \dot{m}(h + V^2/2 + gz) + \dot{E}_{c.v.,}$$

where

$$\dot{E}_{c.v.} = \frac{d}{dt}\bigg|_v \rho\,(u + V^2/2 + gz)\,dV$$

Open system, steady state steady flow (SSSF):

$$\dot{Q}_{c.v.} + \sum_{in} \dot{m}(h + V^2/2 + gz) = \dot{W}_{c.v.} + \sum_{out} \dot{m}(h + V^2/2 + gz)$$

Open system, uniform state uniform flow (USUF):

$$_1Q_{2c.v.} + \sum_{in} m(h + V^2/2 + gz) = {_1W_{2c.v.}} + \sum_{out} m(h + V^2/2 + gz) + [m(u + V^2/2 + gz)]_{1c.v.}^{2c.v.}$$

It is impossible for any device to operate in a cycle and produce work while exchanging heat only with bodies at a single fixed temperature.

The other statement is called the Clausius statement.

The implications of the second law are many. For example, it allows us to (1) determine the maximum possible efficiency of a heat engine, (2) determine the maximum coefficient of performance for a refrigerator, (3) determine the feasibility of a proposed process, (4) predict the direction of a chemical or other type of process, and (5) correlate physical properties. So we see that the second law is quite valuable.

I.2.4 Efficiency

Efficiency is a concept used to describe the effectiveness of energy conversion devices that operate in cycles as well as in individual system components that operate in processes. Thermodynamic efficiency, η, and coefficient of performance, COP, are used for devices that operate in cycles. The following definitions apply:

$$\eta = \frac{W_{net}}{Q_H} \qquad COP = \frac{Q_L}{W}$$

where Q_L and Q_H represent heat transferred from cold and hot regions, respectively, W_{net} is useful work produced in a heat engine, and W is the work required to drive the refrigerator. A heat engine produces useful work, while a

Table I.4 Second Law of Thermodynamics

Closed system, cyclic process:

$$\oint \frac{dQ}{T} \le 0$$

Closed system, state 1 to state 2:

$$\int_1^2 \frac{dQ}{T} \le S_2 - S_1 = m(S_2 - S_1)$$

Open system:

$$\frac{d}{dt}\Big|_V \int \rho s\, dV + \sum_{out} \dot{m}s \ge \sum_{in} \dot{m}s + \int_A \frac{Q}{T}\, dA$$

or

$$\frac{d}{dt}\Big|_V \int \rho s\, dV + \int_A \rho s\, V_{rn}\, dA \ge \int_{A} \frac{Q}{T} dA$$

Open system, SSSF:

$$\sum_{out} \dot{m}s \ge \sum_{in} \dot{m}s + \int_A \frac{Q}{T}\, dA$$

Open system, USUF:

$$(m_2 s_2 - m_1 s_1)_{c.v.} + \sum_{out} \dot{m}s \ge \sum_{in} \dot{m}s + \int_0^t \frac{Q_{c.v.}}{T}\, dt$$

refrigerator uses work to transfer heat from a cold to a hot region. There is an ideal cycle, called the Carnot cycle, which yields the maximum efficiency for heat engines and refrigerators. It is composed of four ideal reversible processes; the efficiency of this cycle is

$$\eta_c = 1 - \frac{T_L}{T_H}$$

and the COP is

$$[COP]_c = \frac{T_L/T_H}{1 - T_L/T_H}$$

These represent the best possible performance of cyclic energy conversion devices operating between tempera-

ture extremes, T_H and T_L. The thermodynamic efficiency should not be confused with efficiencies applied to devices that operate along a process line. This efficiency is defined as

$$\eta_{device} = \frac{actual\ energy\ transfer}{ideal\ energy\ transfer}$$

for a work-producing device and

$$\eta_{device} = \frac{ideal\ energy\ transfer}{actual\ energy\ transfer}$$

for a work-consuming device. Note these definitions are such that $\eta < 1$. These efficiencies are convenience factors in that the actual performance can be calculated from an ideal process line and the efficiency, which generally must be experimentally determined. Table I.5 shows the most commonly encountered versions of efficiencies.

Table I.5 Thermodynamic Efficiency

Heat engines and refrigerators:

Engine efficiency $\eta \equiv W/Q_H \le \eta_{Carnot} = (T_H - T_L)/T_H < 1$

Heat pump c.o.p. $\beta' \equiv Q_H/W \le \beta'_{Carnot} = T_H/(T_H - T_L) > 1$

Refrigerator c.o.p. $\beta \equiv Q_L/W \le \beta_{Carnot} = T_L/(T_H - T_L)$,

$0 < \beta < \infty, (Q_H/Q_L)_{Carnot} = T_H/T_L$

Process efficiencies:

$\eta_{ad,\ turbine} = w_{actual,\ adiabatic}/w_{isentropic}$

$\eta_{ad,\ compressor} = w_{isentropic}/w_{actual,\ adiabatic}$

$\eta_{ad,\ nozzle} = \Delta K.E._{actual,\ adiabatic}/\Delta K.E._{isentropic}$

$$\eta_{nozzle} = \frac{V_a^2/2g_c}{V_s^2/2g_c}$$

$\eta_{cooled\ compressor} = w_{isothermal,\ rev.}/w_{actual}$

I.2.5 Power and Refrigeration Cycles

Many cycles have been devised to convert heat into work, and vice versa. Several of these take advantage of the phase change of the working fluid: for example, the Rankine, the vapor compression, and the absorption cycles. Others involve approximations of thermodynamic processes to mechanical processes and are called air-standard cycles.

The Rankine cycle is probably the most frequently encountered cycle in thermodynamics. It is used in almost all large-scale electric generation plants, regardless of the energy source (gas, coal, oil, or nuclear). Many modern steam-electric power plants operate at supercritical pressures and temperatures during the boiler heat addition process. This leads to the necessity of reheating between high- and lower-pressure turbines to prevent excess moisture in the latter stages of turbine expansion (prevents blade erosion). Feedwater heating is also extensively used to increase the efficiency of the basic Rankine cycle (see Ref. 1 for details).

The vapor compression cycle is almost a reversed Rankine cycle. The major difference is that a simple expansion valve is used to reduce the pressure between the condensor and the evaporator rather than being a work-producing device. The reliability of operation of the expansion valve is a valuable trade-off compared to the small amount of work that could be reclaimed. The vapor compression cycle can be used for refrigeration or heating (heat pump).

In the energy conservation area, applications of the heat pump are taking on added emphasis. The device is useful for heating from an electrical source (compressor) in situations where direct combustion is not available. Additionally, the device can be used to upgrade the temperature level of waste heat recovered at a lower temperature.

Air-standard cycles, useful both for power generation and heating/cooling applications, are the thermodynamic approximations to the processes occurring in the actual devices. In the actual cases, a thermodynamic cycle is not completed, necessitating the approximations. Air-standard cycles are analyzed by using the following approximations:

1. Air is the working fluid and behaves as an ideal gas.
2. Combustion and exhaust processes are replaced by heat exchangers.

Other devices must be analyzed component by component using property data for the working fluids (see Appendix II). Figure I.2 gives a listing of various power systems with their corresponding thermodynamic cycle and other pertinent information.

I.2.6 Combustion Processes

The combustion process continues to be the most prevalent means of energy conversion. Natural and manufactured gases, coal, liquid fuel/air mixtures, even wood and peat are examples of energy sources requiring combustion.

There are two overriding principles of importance in analyzing combustion processes. They are the combustion equation and the first law for the combustion chamber. The combustion equation is simply a mass balance between reactants and products of the chemical reaction combustion process. The first law is the energy balance for the same process using the results of the combustion equation as input.

In practice, we can restrict our discussion to hydrocarbon fuels, meaning that the combustion equation (chemical balance) is written as

$$C_xH_y + a(O_2 + 3.76N_2) \rightarrow b\,CO_2 + c\,CO_2$$
$$+ e\,H_2O + d\,O_2 + 3.76a\,N_2$$

Table I.6 Characteristics of Some of the Hydrocarbon Families

Family	Formula	Structure	Saturated
Paraffin	C_nH_{2n+2}	Chain	Yes
Olefin	C_nH_{2n}	Chain	No
Diolefin	C_nH_{2n-2}	Chain	No
Naphthene	C_nH_{2n}	Ring	Yes
Aromatic Benzene	C_nH_{2n-6}	Ring	No
Aromatic Naphthalene	C_nH_{2n-12}	Ring	No

Molecular structure of some hydrocarbon fuels:

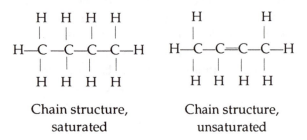

Chain structure,
saturated

Chain structure,
unsaturated

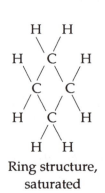

Ring structure,
saturated

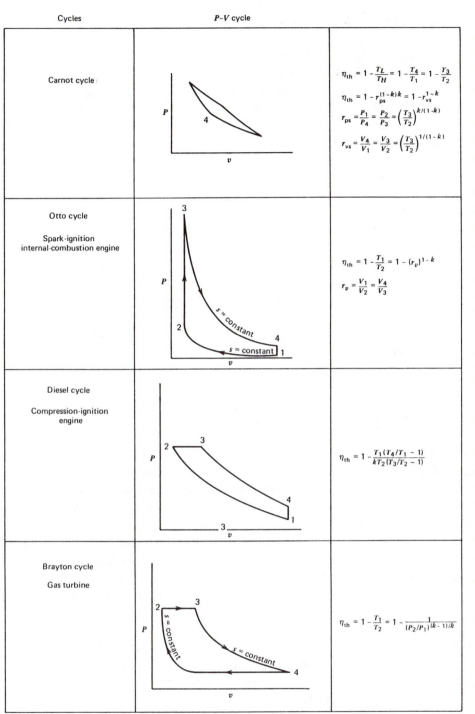

Fig. I.2 Air standard cycles.

Once a combustion process is decided upon (i.e., the fuel to be used and the heat transfer/combustion chamber are selected), the relative amount of fuel and air become of prime importance. This is because the air/fuel ratio (AF) controls the temperature of the combustion zone and the energy available to be transferred to a working fluid or converted to work. Stoichiometric air is that quantity of air required such that no oxygen would appear in the products. Excess air occurs when more than enough air is provided to the combustion process. Ideal combustion implies perfect mixing and complete reactions. In this case theoretical air (TA) would yield no free oxygen in the products. Excess air then is actual air less theoretical air.

Most industrial combustion processes conform closely to a steady-state, steady-flow case. The first law for an open control volume surrounding the combustion zone can then be written. If we assume that Q and W are zero and that ΔK.E. and ΔP.E. are negligible, then the following equation results:

$$\sum_{\text{products}} \left(H_e - H_{\text{ref}}\right) = \sum_{\text{reactants}} \left(H_i - H_{\text{ref}}\right) + \Delta H_{\text{comb}} \quad (\text{I.3})$$

Subscripts i and e refer to inlet and exit conditions, respectively. H_{ref} is the enthalpy of each component at some reference temperature. ΔHcomb represents the heat of combustion for the fuel and, in general, carries a negative value, meaning that heat would have to be transferred out of the system to maintain inlet and exit temperatures at the same level.

The adiabatic flame temperature occurs when the combustion zone is perfectly insulated. The solution of equation I.3 would give the adiabatic flame temperature for any particular case. The maximum adiabatic flame temperature would occur when complete combustion oc-

This equation neglects the minor components of air; that is, air is assumed to be 1 mol of O_2 mixed with 3.76 mol of N_2. The balance is based on 1 mol of fuel C_xH_y. The unknowns are determined for each particular application. Table I.6 gives the characteristics of some of the hydrocarbons. Table I.7 shows the volumetric analyses of several gaseous fuels.

Table I.7 Volumetric Analyses of Some Typical Gaseous Fuels

Constituent	Various Natural Gases				Producer Gas from Bituminous Coal	Carbureted Water Gas	Coke Oven Gas
	A	B	C	D			
Methane	93.9	60.1	67.4	54.3	3.0	10.2	32.1
Ethane	3.6	14.8	16.8	16.			
Propane	1.2	13.4	15.8	16.2			
Butanes plus[a]	1.3	4.2		7.4			
Ethene						6.1	3.5
Benzene						2.8	0.5
Hydrogen					14.0	40.5	46.5
Nitrogen		7.5		5.8	50.9	2.9	8.1
Oxygen					0.6	0.5	0.8
Carbon monoxide					27.0	34.0	6.3
Carbon dioxide					4.5	3.0	2.2

[a]This includes butane and all heavier hydrocarbons.

curs with a minimum of excess O_2 appearing in the products.

Appendix II gives tabulated values for the important thermophysical properties of substances important in combustion.

Gas Analysis. During combustion in heaters and boilers, the information required for control of the burner settings is the amount of excess air in the fuel gas. This percentage can be a direct reflection of the efficiency of combustion.

The most accurate technique for determining the volumetric makeup of combustion by-products is the Orsat analyzer. The Orsat analysis depends upon the fact that for hydrocarbon combustion the products may contain CO_2, O_2, CO, N_2, and water vapor. If enough excess air is used to obtain complete combustion, no CO will be present. Further, if the water vapor is removed, only CO_2, O_2, and N_2 remain.

The Orsat analyzer operates on the following principles. A sample of fuel gas is first passed over a desiccant to remove the moisture. (The amount of water vapor can be found later from the combustion equation.) Then the sample is exposed in turn to materials that absorb first the CO_2, then the O_2, and finally the CO (if present). After each absorption the volumetric change is carefully measured in a graduated pipette system. The remaining gas is assumed to be N_2. Of course, it could contain some trace of gases and pollutants.

I.2.7 Psychrometry

Psychrometry is the science of air/water vapor mixtures. Knowledge of the behavior of such systems is important, both in meteorology and industrial processes, especially heating and air conditioning. The concepts can be applied to other ideal gas/water vapor mixtures.

Air and water vapor mixed together at a total pressure of 1 atm is called atmospheric air. Usually, the amount of water vapor in atmospheric air is so minute that the vapor and air can be treated as an ideal gas. The air existing in the mixture is often called dry air, indicating that it is separate from the water vapor coexisting with it.

Two terms frequently encountered in psychrometry are relative humidity and humidity ratio. Relative humidity, ϕ, is defined as the ratio of the water vapor pressure to the saturated vapor pressure at the temperature of the mixture. Figure I.3 shows the relation between points on the T–s diagram that yield the relative humidity. Relative humidity cannot be greater than unity or 100%, as is normally stated.

The humidity ratio, ω, on the other hand, is defined as the ratio of the mass of water vapor to the mass of dry air in atmospheric air, $\omega = m_v/m_a$. This can be shown to be $\omega = v_a/v_v$, and a relationship between ω and ϕ exists, $\omega = (v_a/v_g)\phi$, where v_g refers to the specific volume of saturated water vapor at the temperature of the mixture.

A convenient way of describing the condition of atmospheric air is to define four temperatures: dry-bulb,

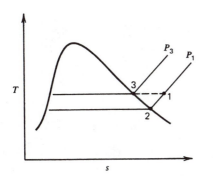

Fig. I.3 Behavior of water in air: $\phi = P_1/P_3$; T_2 = **dew point.**

wet-bulb, dew-point, and adiabatic saturation temperatures. The dry-bulb temperature is simply that temperature which would be measured by any of several types of ordinary thermometers placed in atmospheric air.

The dew-point temperature (point 2 on Figure I.3) is the saturation temperature of the water vapor at its existing partial pressure. In physical terms it is the mixture temperature where water vapor would begin to condense if cooled at constant pressure. If the relative humidity is 100% the dew-point and dry-bulb temperatures are identical.

In atmospheric air with relative humidity less than 100%, the water vapor exists at a pressure lower than saturation pressure. Therefore, if the air is placed in contact with liquid water, some of the water would be evaporated into the mixture and the vapor pressure would be increased. If this evaporation were done in an insulated container, the air temperature would decrease, since part of the energy to vaporize the water must come from the sensible energy in the air. If the air is brought to the saturated condition, it is at the adiabatic saturation temperature.

A psychrometric chart is a plot of the properties of atmospheric air at a fixed total pressure, usually 14.7 psia. The chart can be used to quickly determine the properties of atmospheric air in terms of two independent properties, for example, dry-bulb temperature and relative humidity. Also, certain types of processes can be described on the chart. Appendix II contains a psychrometric chart for 14.7-psia atmospheric air. Psychrometric charts can also be constructed for pressures other than 14.7 psia.

I.3 HEAT TRANSFER

Heat transfer is the branch of engineering science that deals with the prediction of energy transport caused by temperature differences. Generally, the field is broken down into three basic categories: conduction, convection, and radiation heat transfer.

Conduction is characterized by energy transfer by internal microscopic motion such as lattice vibration and electron movement. Conduction will occur in any region where mass is contained and across which a temperature difference exists.

Convection is characterized by motion of a fluid region. In general, the effect of the convective motion is to augment the conductive effect caused by the existing temperature difference.

Radiation is an electromagnetic wave transport phenomenon and requires no medium for transport. In fact, radiative transport is generally more effective in a vacuum, since there is attenuation in a medium.

I.3.1 Conduction Heat Transfer

The basic tenet of conduction is called Fourier's law,

$$\dot{Q} = -kA \frac{dT}{dx}$$

The heat flux is dependent upon the area across which energy flows and the temperature gradient at that plane. The coefficient of proportionality is a material property, called thermal conductivity k. This relationship always applies, both for steady and transient cases. If the gradient can be found at any point and time, the heat flux density, $\dot{Q}/A$, can be calculated.

Conduction Equation. The control volume approach from thermodynamics can be applied to give an energy balance which we call the conduction equation. For brevity we omit the details of this development; see Refs. 2 and 3 for these derivations. The result is

$$G + k\nabla^2 T = \rho C \frac{\partial T}{\partial \tau} \qquad (I.4)$$

This equation gives the temperature distribution in space and time, G is a heat-generation term, caused by chemical, electrical, or nuclear effects in the control volume. Equation I.4 can be written

$$\nabla^2 T + \frac{G}{k} = \frac{\rho C}{k} \frac{\partial T}{\partial \tau}$$

The ratio $k/\rho C$ is also a *material prop*erty called thermal diffusivity u. Appendix II gives thermophysical properties of many common engineering materials.

For steady, one-dimensional conduction with no heat generation,

$$\frac{d^2 T}{dx^2} = 0$$

This will give $T = ax + b$, a simple linear relationship between temperature and distance. Then the application of

Fourier's law gives

$$\dot{Q} = kA \frac{\Delta T}{\Delta x}$$

a simple expression for heat transfer across the Δx distance. If we apply this concept to insulation for example, we get the concept of the R value. R is just the resistance to conduction heat transfer per inch of insulation thickness (i.e., $R = 1/k$).

Multilayered, One-Dimensional Systems. In practical applications, there are many systems that can be treated as one-dimensional, but they are composed of layers of materials with different conductivities. For example, building walls and pipes with outer insulation fit this category. This leads to the concept of overall heat-transfer coeffficient, U. This concept is based on the definition of a convective heat-transfer coefficient,

$$\dot{Q} = hA \; \Delta T$$

This is a simplified way of handling convection at a boundary between solid and fluid regions. The heat-transfer coefficient h represents the influence of flow conditions, geometry, and thermophysical properties on the heat transfer at a solid-fluid boundary. Further discussion of the concept of the h factor will be presented later.

Figure I.4 represents a typical one-dimensional, multilayered application. We define an overall heat-transfer coefficient U as

$$Q = UA \left(T_i - T_o\right)$$

We find that the expression for U must be

$$U = \frac{1}{\dfrac{1}{h_i} + \dfrac{\Delta x_1}{k_1} + \dfrac{\Delta x_2}{k_2} + \dfrac{\Delta x_3}{k_3} + \dfrac{1}{h_o}}$$

This expression results from the application of the conduction equation across the wall components and the convection equation at the wall boundaries. Then, by noting that in steady state each expression for heat must be equal, we can write the expression for U, which contains both convection and conduction effects. The U factor is extremely useful to engineers and architects in a wide variety of applications.

The U factor for a multilayered tube with convection at the inside and outside surfaces can be developed in the same manner as for the plane wall. The result is

$$U_o = \frac{1}{\dfrac{1}{h_o} + \sum_j \dfrac{r_o \ln \left(r_j + 1/r_j\right)}{k_j} + \dfrac{1 \, r_o}{h_i \, r_i}}$$

where r_i and r_o are inside and outside radii.

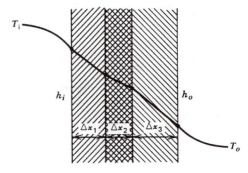

Fig. I.4 Multilayered wall with convection at the inner and outer surfaces.

Caution: The value of U depends upon which radius you choose (i.e., the inner or outer surface).
If the inner surface were chosen, we would get

$$U_i = \frac{1}{\dfrac{1}{h_o} \dfrac{r_i}{r_o} + \sum_j \dfrac{r_i \ln \left(r_j + 1/r_i\right)}{k_j} + \dfrac{1}{h_i}}$$

However, there is no difference in heat-transfer rate; that is,

$$Q_o = U_i A_i \Delta T_{\text{overall}} = U_o A_o \Delta T_{\text{overall}}$$

so it is apparent that

$$U_i A_i = U_o A_o$$

for cylindrical systems.

Finned Surfaces. Many heat-exchange surfaces experience inadequate heat transfer because of low heat-transfer coefficients between the surface and the adjacent fluid. A remedy for this is to add material to the surface. The added material in some cases resembles a fish "fin," thereby giving rise to the expression "a finned surface." The performance of fins and arrays of fins is an important item in the analysis of many heat-exchange devices. Figure I.5 shows some possible shapes for fins.

The analysis of fins is based on a simple energy balance between one-dimensional conduction down the length of the fin and the heat convected from the exposed surface to the surrounding fluid. The basic equation that applies to most fins is

$$\frac{d^2\theta}{dx^2} + \frac{1}{A} \frac{dA}{dx} \frac{d\theta}{dx} - \frac{h}{k} \frac{1}{A} \frac{dS}{dx} \theta = 0 \qquad (\text{I.5})$$

when θ is $(T - T_\infty)$, the temperature difference between fin and fluid at any point; A is the cross-sectional area of the fin; S is the exposed area; and x is the distance along the fin. Chapman[2] gives an excellent discussion of the develop-

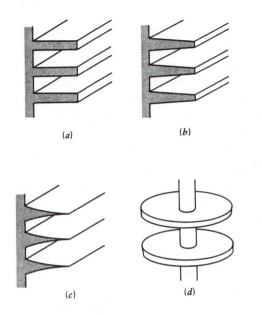

Fig. I.5 Fins of various shapes. (*a*) Rectangular, (*b*) Trapezoidal, (*c*) Arbitrary profile, (*d*) Circumferential.

ment of this equation.

The application of equation I.5 to the myriad of possible fin shapes could consume a volume in itself. Several shapes are relatively easy to analyze; for example, fins of uniform cross section and annular fins can be treated so that the temperature distribution in the fin and the heat rate from the fin can be written. Of more utility, especially for fin arrays, are the concepts of fin efficiency and fin surface effectiveness (see Holman[3]).

Fin efficiency η_f is defined as the ratio of actual heat loss from the fin to the ideal heat loss that would occur if the fin were isothermal at the base temperature. Using this concept, we could write

$$\dot{Q}_{\mathrm{fin}} = A_{\mathrm{fin}} h \left(T_b - T_\infty\right) \eta_f$$

η_f is the factor that is required for each case. Figure I.6 shows the fin efficiency for several cases.

Surface effectiveness K is defined as the actual heat transfer from a finned surface to that which would occur if the surface were isothermal at the base temperature. Taking advantage of fin efficiency, we can write

$$K = \frac{(A - A_f) h \, \theta_0 + \eta_f A_f \theta_0}{A h \, \theta_0} \qquad (I.6)$$

Equation I.6 reduces to

$$K = 1 \frac{A_f}{A} \left(1 - \eta_f\right)$$

which is a function only of geometry and single fin efficiency. To get the heat rate from a fin array, we write

$$\dot{Q}_{\mathrm{array}} = Kh \left(T_b - T_\infty\right) A$$

where A is the total area exposed.

Transient Conduction. Heating and cooling problems involve the solution of the time-dependent conduction equation. Most problems of industrial significance occur.when a body at a known initial temperature is suddenly exposed to a fluid at a different temperature. The temperature behavior for such unsteady problems can be characterized by two dimensionless quantities, the Biot number, $\mathrm{Bi} = hL/k$, and the Fourier modulus, $\mathrm{Fo} = \alpha\tau/L^2$. The Biot number is a measure of the effectiveness of conduction within the body. The Fourier modulus is simply a dimensionless time.

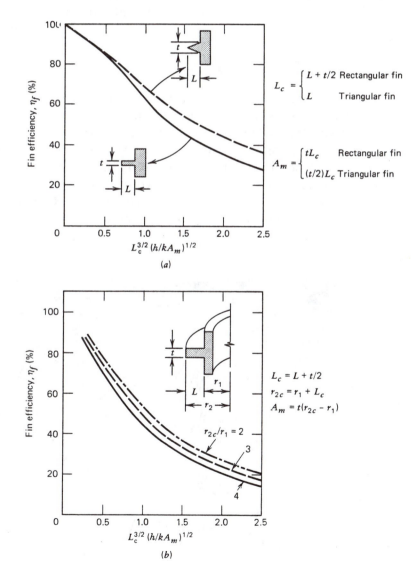

Fig. I.6 (*a*) Efficiencies of rectangular and triangular fins, (*b*) Efficiencies of circumferential fins of rectangular profile.

If Bi is a small, say Bi ≤ 0.1, the body undergoing the temperature change can be assumed to be at a uniform temperature at any time. For this case,

$$\frac{T - T_f}{T_i - T_f} = \exp\left[-\left(\frac{hA}{\rho CV}\right)\tau\right]$$

where T_f and T_i are the fluid temperature and initial body temperature, respectively. The term $(\rho CV/hA)$ takes on the characteristics of a time constant.

If Bi ≥ 0.1, the conduction equation must be solved in terms of position and time. Heisler[4] solved the equation for infinite slabs, infinite cylinders, and spheres. For convenience he plotted the results so that the temperature at any point within the body and the amount of heat transferred can be quickly found in terms of Bi and Fo. Figures I.7 to I.10 show the Heisler charts for slabs and cylinders. These can be used if h and the properties of the material are constant.

I.3.2 Convection Heat Transfer

Convective heat transfer is considerably more complicated than conduction because motion of the medium is involved. In contrast to conduction, where many geometrical configurations can be solved analytically, there are only limited cases where theory alone will give convective heat-transfer relationships. Consequently, convection is largely what we call a semi-empirical science. That is, actual equations for heat transfer are based strongly on the results of experimentation.

Convection Modes. Convection can be split into several subcategories. For example, forced convection refers to the case where the velocity of the fluid is completely independent of the temperature of the fluid. On the other hand, natural (or free) convection occurs when the temperature field actually causes the fluid motion through buoyancy effects.

We can further separate convection by geometry into external and internal flows. Internal refers to channel, duct, and pipe flow and external refers to unbounded fluid flow cases. There are other specialized forms of convection, for example the change-of-phase phenomena: boiling, condensation, melting, freezing, and so on. Change-of-phase heat transfer is difficult to predict analytically. Tongs gives many of the correlations for boiling and two-phase flow.

Dimensional Heat-Transfer Parameters. Because experimentation has been required to develop appropriate correlations for convective heat transfer, the use of generalized dimensionless quantities in these correlations is preferred. In this way, the applicability of experimental data covers a wider range of conditions and fluids. Some of these parameters, which we generally call "numbers," are given below:

Nusselt number: $\quad Nu = \dfrac{hL}{k}$

where k is the fluid conductivity and L is measured along the appropriate boundary between liquid and solid; the Nu is a nondimensional heat-transfer coefficient.

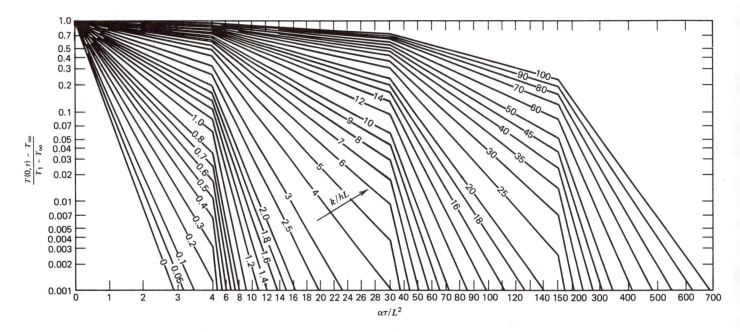

Fig. I.7 Midplane temperature for an infinite plate of thickness 2L. (From Ref. 4.)

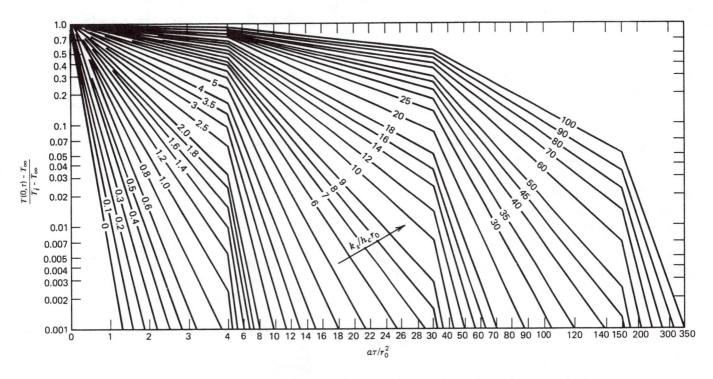

Fig. I.8 Axis temperature for an infinite cylinder of radius r_0. (From Ref. 4.)

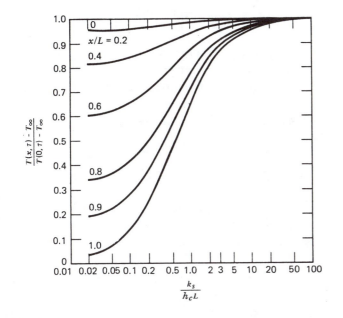

Fig. I.9 Temperature as a function of center temperature in an infinite plate of thickness 2L. (From Ref. 4.)

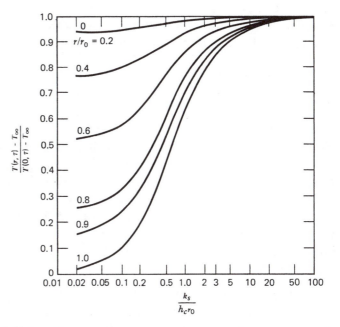
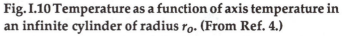

Fig. I.10 Temperature as a function of axis temperature in an infinite cylinder of radius r_0. (From Ref. 4.)

Reynolds number: $\text{Re} = \dfrac{Lu}{v}$

defined in Section I.4: it controls the character of the flow

Prandtl number: $\text{Pr} = \dfrac{C\mu}{k}$

ratio of momentum transport to heat-transport character-

istics for a fluid: it is important in all convective cases, and is a material property

Grashof number: $\text{Gr} = \dfrac{g\beta(T - T_\infty)L^3}{v^2}$

serves in natural convection the same role as Re in forced convection: that is, it controls the character of the flow

Stanton number: $\quad St = \dfrac{h}{\rho\, u C_p}$

also a nondimensional heat-transfer coefficient: it is very useful in pipe flow heat transfer

In general, we attempt to correlate data by using relationships between dimensionless numbers: for example, in many convection cases, we could write Nu = Nu(Re, Pr) as a functional relationship. Then it is possible either from analysis, experimentation, or both, to write an equation that can be used for design calculations. These are generally called working formulas.

Forced Convection Past Plane Surfaces. The average heat-transfer coefficient for a plate of length L may be calculated from

$$Nu_L = 0.664\,(Re_L)^{1/2}(Pr)^{1/3}$$

if the flow is laminar (i.e., if $Re_L \le 400{,}000$). For this case the fluid properties should be evaluated at the mean film temperature T_m, which is simply the arithmetic average of the fluid and the surface temperature.

For turbulent flow, there are several acceptable correlations. Perhaps the most useful includes both laminar leading edge effects and turbulent effects. It is

$$Nu = 0.0036\,(Pr)^{1/3}\big[(Re_L)^{0.8} - 18{,}700\big]$$

where the transition Re is 400,000.

Forced Convection Inside Cylindrical Pipes or Tubes. This particular type of convective heat transfer is of special engineering significance. Fluid flows through pipes, tubes, and ducts are very prevalent, both in laminar and turbulent flow situations. For example, most heat exchangers involve the cooling or heating of fluids in tubes. Single pipes and/or tubes are also used to transport hot or cold liquids in industrial processes. Most of the formulas listed here are for the $0.5 \le Pr \le 100$ range .

Laminar Flow. For the case where $Re_D < 2300$, Nusselt showed that $Nu_D = 3.66$ for long tubes at a constant tube-wall temperature. For forced convection cases (laminar and turbulent) the fluid properties are evaluated at the bulk temperature T_b. This temperature, also called the mixing-cup temperature, is defined by

$$T_b = \dfrac{\displaystyle\int_0^R uTr\,dr}{\displaystyle\int_0^R ur\,dr}$$

if the properties of the flow are constant.

Sieder and Tate developed the following more convenient empirical formula for short tubes:

$$Nu_D = 1.86\,(Re_D)^{1/3}(Pr)^{1/3}\left(\dfrac{D}{L}\right)^{1/3}\left(\dfrac{\mu}{\mu_s}\right)^{0.14}$$

The fluid properties are to be evaluated at T_b except for the quantity μ_s, which is the dynamic viscosity evaluated at the temperature of the wall.

Turbulent Flow. McAdams suggests the empirical relation

$$Nu_D = 0.023\,(Pr_D)^{0.8}(Pr)^n \qquad (I.7)$$

where $n = 0.4$ for heating and $n = 0.3$ for cooling. Equation I.7 applies as long as the difference between the pipe surface temperature and the bulk fluid temperature is not greater than 10°F for liquids or 100°F for gases.

For temperature differences greater then the limits specified for equation I.7 or for fluids more viscous than water, the following expression from Sieder and Tate will give better results:

$$Nu_D = 0.027\,(Pr_D)^{0.8}(Pr)^{1/3}\left(\dfrac{\mu}{\mu_s}\right)^{0.14}$$

Note that the McAdams equation requires only a knowledge of the bulk temperature, whereas the Sieder-Tate expression also requires the wall temperature. Many people prefer equation I.7 for that reason.

Nusselt found that short tubes could be represented by the expression

$$Nu_D = 0.0036\,(Pe_D)^{0.8}(Pr)^{1/3}\left(\dfrac{\mu}{\mu_s}\right)^{0.14}\left(\dfrac{D}{L}\right)^{1/18}$$

For noncircular ducts, the concept of equivalent diameter can be employed, so that all the correlations for circular systems can be used.

Forced Convection in Flow Normal to Single Tubes and Banks. This circumstance is encountered frequently, for example air flow over a tube or pipe carrying hot or cold fluid. Correlations of this phenomenon are called semi-empirical and take the form $Nu_D = C(Re_D)^m$. Hilpert, for example, recommends the values given in Table I.8. These values have been in use for many years and are considered accurate.

Flows across arrays of tubes (tube banks) may be even more prevalent than single tubes. Care must be exercised in selecting the appropriate expression for the tube bank. For example, a staggered array and an in-line array could have considerably different heat-transfer characteristics. Kays and London[6] have documented many of

Table I.8 Values of C and m for Hilpert's Equation

Range of N_{Re_D}	C	m
1-4	0.891	0.330
4-40	0.821	0.385
40-4000	0.615	0.466
4000-40,000	0.175	0.618
40,000-250,000	0.0239	0.805

these cases for heat-exchanger applications. For a general estimate of order-of-magnitude heat-transfer coefficients, Colburn's equation

$$Nu_D = 0.33 \, (Re_D)^{0.6} (Pr)^{1/3}$$

is acceptable.

Free Convection Around Plates and Cylinders. In free convection phenomena, the basic relationships take on the functional form $Nu = f(Gr, Pr)$. The Grashof number replaces the Reynolds number as the driving function for flow.

In all free convection correlations it is customary to evaluate the fluid properties at the mean film temperature T_m, except for the coefficient of volume expansion β, which is normally evaluated at the temperature of the undisturbed fluid far removed from the surface—namely, T_f. Unless otherwise noted, this convention should be used in the application of all relations quoted here.

Table I.9 gives the recommended constants and exponents for correlations of natural convection for vertical plates and horizontal cylinders of the form $Nu = C \cdot Ra^m$. The product Gr · Pr is called the Rayleigh number (Ra) and is clearly a dimensionless quantity associated with any specific free convective situation.

Table I.9 Constants and Exponents for Natural Convection Correlations

Ra	Vertical Plate[a]		Horizontal Cylinders[b]	
	c	m	c	m
$10^4 < Ra < 10^9$	0.59	1/4	0.525	1/4
$10^9 < Ra < 10^{12}$	0.129	1/3	0.129	1/3

[a]Nu and Ra based on vertical height L.
[b]Nu and Ra based on diameter D.

I.3.3 Radiation Heat Transfer

Radiation heat transfer is the most mathematically complicated type of heat transfer. This is caused primarily by the electromagnetic wave nature of thermal radiation. However, in certain applications, primarily high-temperature, radiation is the dominant mode of heat transfer. So it is imperative that a basic understanding of radiative heat transport be available. Heat transfer in boiler and fired-heater enclosures is highly dependent upon the radiative characteristics of the surface and the hot combustion gases. It is known that for a body radiating to its surroundings, the heat rate is

$$\dot{Q} = \varepsilon \sigma A \left(T^4 - T_s^4 \right)$$

where ε is the emissivity of the surface, σ is the Stefan-Boltzmann constant, $\sigma = 0.1713 \times 10^{-8}$ Btu/hr ft^2 · R^4. Temperature must be in absolute units, R or K. If $\varepsilon = 1$ for a surface, it is called a "blackbody," a perfect emitter of thermal energy. Radiative properties of various surfaces are given in Appendix II. In many cases, the heat exchange between bodies when all the radiation emitted by one does not strike the other is of interest. In this case we employ a shape factor F_{ij} to modify the basic transport equation. For two blackbodies we would write

$$\dot{Q}_{12} = F_{12} \, \sigma A_1 \left(T_1^4 - T_2^4 \right)$$

for the heat transport from body 1 to body 2. Figures I.11 to I.14 show the shape factors for some commonly encountered cases. Note that the shape factor is a function of geometry only.

Gaseous radiation that occurs in luminous combustion zones is difficult to treat theoretically. It is too complex to be treated here and the interested reader is referred to Siegel and Howell[7] for a detailed discussion.

I.4 FLUID MECHANICS

In industrial processes we deal with materials that can be made to flow in a conduit of some sort. The laws that govern the flow of materials form the science that is called fluid mechanics. The behavior of the flowing fluid controls pressure drop (pumping power), mixing efficiency, and in some cases the efficiency of heat transfer. So it is an integral portion of an energy conservation program.

I.4.1 Fluid Dynamics

When a fluid is caused to flow, certain governing laws must be used. For example, mass flows in and out of control volumes must always be balanced. In other words,

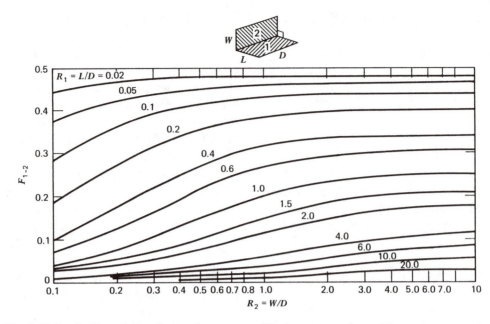

Fig. I.11 Radiation shape factor for perpendicular rectangles with a common edge.

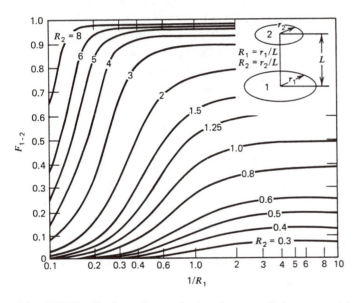

Fig. I.12 Radiation shape factor for parallel, concentric disks.

conservation of mass must be satisfied.

In its most basic form the continuity equation (conservation of mass) is

$$\iint_{c.s.} \rho(\overline{v} \cdot \overline{n}) \, dA + \frac{\partial}{\partial_t} \iiint_{c.v.} \rho \, dV = 0$$

In words, this is simply a balance between mass entering and leaving a control volume and the rate of mass storage. The $\rho(\overline{v} \cdot \overline{n})$ terms are integrated over the control surface, whereas the $\rho \, dV$ term is dependent upon an integration over the control volume.

For a steady flow in a constant-area duct, the continuity equation simplifies to

$$\dot{m} = \rho f A_c \overline{u} = \text{constant}$$

That is, the mass flow rate $\dot{m}$ is constant and is equal to the product of the fluid density ρf, the duct cross section A_c, and the average fluid velocity u.

If the fluid is compressible and the flow is steady, one gets

$$\frac{\dot{m}}{\rho f} = \text{constant} = \left(\overline{u}A_c\right)_1 = \left(\overline{u}A_c\right)_2$$

where 1 and 2 refer to different points in a variable area duct.

I.4.2 First Law—Fluid Dynamics

The first law of thermodynamics can be directly applied to fluid dynamical systems, such as duct flows. If there is no heat transfer or chemical reaction and if the internal energy of the fluid stream remains unchanged, the first law is

$$\frac{V_i^2 - V_e^2}{2g_c} + \frac{(z_i - z_e)}{g_c}g + \frac{p_i - p_e}{\rho} + (w_p - w_f) = 0 \qquad (\text{I.8})$$

where the subscripts i and e refer to inlet and exit conditions and w_p and w_f are pump work and work required to overcome friction in the duct. Figure I.15 shows schematically a system illustrating this equation.

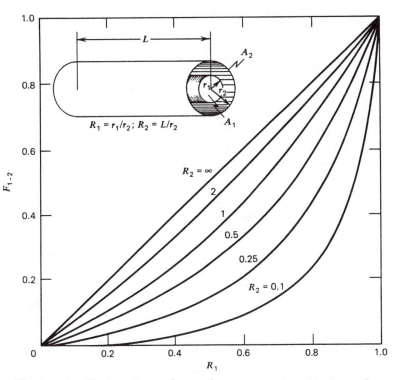

Fig. I.13 Radiation shape factor for concentric cylinders of finite length.

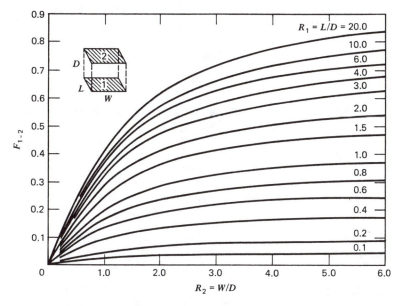

Fig. I.14 Radiation shape factor for parallel, directly opposed rectangles.

Any term in equation I.8 can be converted to a rate expression by simply multiplying by $\dot{m}$, the mass flow rate. Take, for example, the pump horsepower,

$$W \left(\frac{\text{energy}}{\text{time}}\right) = \dot{m} w_p \left(\frac{\text{mass}}{\text{time}}\right) \left(\frac{\text{energy}}{\text{mass}}\right)$$

In the English system, horsepower is

$$\text{hp} = \dot{m} \left(\frac{\text{lb}_m}{\text{sec}}\right) w_p \left(\frac{\text{ft} \cdot \text{lb}_f}{\text{lb}_m}\right) \times \left(\frac{1 \text{ hp-sec}}{500 \text{ ft-lb}_f}\right)$$

$$= \left(\frac{\dot{m} w_p}{550}\right)$$

Referring back to equation I.8, the most difficult term to determine is usually the frictional work term w_f. This is a term that depends upon the fluid viscosity, the flow conditions, and the duct geometry. For simplicity, w_f is generally represented as

$$w_f = \frac{\Delta p_f}{\rho}$$

when Δp_f is the frictional pressure drop in the duct. Further, we say that

$$\frac{\Delta p_f}{\rho} = \frac{2f \bar{u}^2 L}{g_c D}$$

in a duct of length L and diameter D. The friction factor f is a convenient way to represent the differing influence of laminar and turbulent flows on the friction pressure drop.

The character of the flow is determined through the Reynolds number, $\text{Re} = \rho u D / \mu$, where μ is the viscosity of the fluid. This nondimensional grouping represents the ratio of dynamic to viscous forces acting on the fluid.

Experiments have shown that if $\text{Re} \leq 2300$, the flow is laminar. For larger Re the flow is turbulent. Figure I.16 shows how the friction factor depends upon the Re of the flow. Note that for laminar flow the f vs. Re curve is single-valued and is simply equal to 16/Re. In the turbulent regime, the wall roughness e can affect the friction factor because of its effect on the velocity profile near the duct surface.

If a duct is not circular, the equivalent diameter D_e can be used so that all the relationships developed for circular systems can still be used. D_e is defined as

$$D_e = \frac{4A_c}{P}$$

P is the "wetted" perimeter, that part of the flow cross section that touches the duct surfaces. For a circular system $D_e = 4(\pi D^2 / 4\pi D) = D$, as it should. For an annular duct, we get

$$D_e = \frac{\left(\pi D_o^2 / 4 - \pi D_i^2 / 4\right) 4}{\pi D_o + \pi D_i} = \frac{\pi (D_o + D_i)(D_o - D_i)}{\pi (D_o + D_i)}$$

$$= D_o - D_i$$

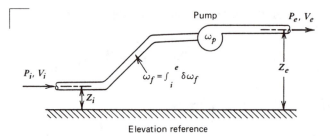

Fig. I.15 The first law applied to adiabatic flow system.

Pressure Drop in Ducts. In practical applications, the essential need is to predict pressure drops in piping and duct networks. The friction factor approach is adequate for straight runs of constant area ducts. But valves nozzles, elbows, and many other types of fittings are necessarily included in a network. This can be accounted for by defining an equivalent length L_e for the fitting. Table I.10 shows L_e/D values for many different fittings.

Pressure Drop across Tube Banks. Another commonly encountered application of fluid dynamics is the pressure drop caused by transverse flow across arrays of heat-transfer tubes. One technique to calculate this effect is to find the velocity head loss through the tube bank:

$$N_v = fNF_d$$

where f is the friction factor for the tubes (a function of the Re), N the number of tube rows crossed by the flow, and F_d is the "depth factor." Figures I.17 and I.18 show the f factor and F_d relationship that can be used in pressure-drop

Table I.10 L_e/D for Screwed Fittings, Turbulent Flow Only[a]

Fitting	L_e/D
45° elbow	15
90° elbow, standard radius	31
90° elbow, medium radius	26
90° elbow, long sweep	20
90° square elbow	65
180° close return bend	75
Swing check valve, open	77
Tee (as el, entering run)	65
Tee (as el, entering branch)	90
Couplings, unions	Negligible
Gate valve, open	7
Gate valve, 1/4 closed	40
Gate valve, 1/2 closed	190
Gate valve, 3/4 closed	840
Globe valve, open	340
Angle valve, open	170

[a]Calculated from Crane Co. Tech. Paper 409, May 1942.

calculations. If the fluid is air, the pressure drop can be calculated by the equation

$$\Delta p = N\left(\frac{30}{B}\right)\frac{T}{1.73\times10^5}\left(\frac{G}{10^3}\right)^2$$

where B is the atmospheric pressure (in. Hg), T is temperature (°R), and G is the mass velocity (lbm/ft^2 hr).

Bernoulli's Equation. There are some cases where the equation

$$\frac{p}{\rho} + \frac{u^2}{2} + gz = \text{constant}$$

which is called Bernoulli's equation, is useful. Strictly speaking, this equation applies for inviscid, incompressible, steady flow along a streamline. However, even in pipe flow where the flow is viscous, the equation can be applied because of the confined nature of the flow. That is, the flow is forced to behave in a streamlined manner. Note that the first law equation (I.8) yields Bernoulli's equation if the friction drop exactly equals the pump work.

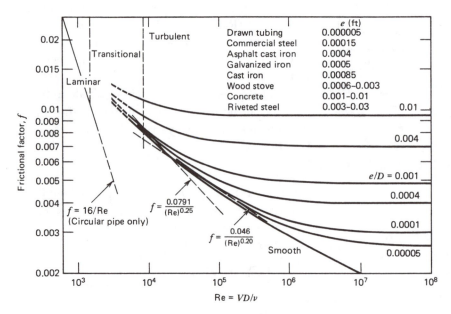

Fig. I.16 Friction factors for straight pipes.

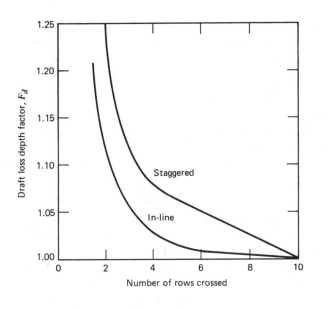

Fig. I.17 Depth factor for number of tube rows crossed in convection banks.

I.4.3 Fluid-Handling Equipment

For industrial processes, another prime application of fluid dynamics lies in fluid-handling equipment. Pumps, compressors, fans, and blowers are extensively used to move gases and liquids through the process network and over heat-exchanger surfaces. The general constraint in equipment selection is a matching of fluid handler capacity to pressure drop in the circuit connected to the fluid handler.

Pumps are used to transport liquids, whereas compressors, fans, and blowers apply to gases. There are features of performance common to all of them. For purposes of illustration, a centrifugal pump will be used to discuss performance characteristics.

Centrifugal Machines. Centrifugal machines operate on the principle of centrifugal acceleration of a fluid element in a rotating impeller/housing system to achieve a pressure gain and circulation.

The characteristics that are important are flow rate (capacity), head, efficiency, and durability. Q_f (capacity), h_p (head), and η_p (efficiency) are related quantities, dependent basically on the fluid behavior in the pump and the flow circuit. Durability is related to the wear, corrosion, and other factors that bear on a pump's reliability and lifetime.

Figure I.19 shows the relation between flow rate and related characteristics for a centrifugal pump at constant speed. Graphs of this type are called performance curves; f/hp and bhp are fluid and brake horsepower, respectively. The primary design constraint is a matching of flow rate to

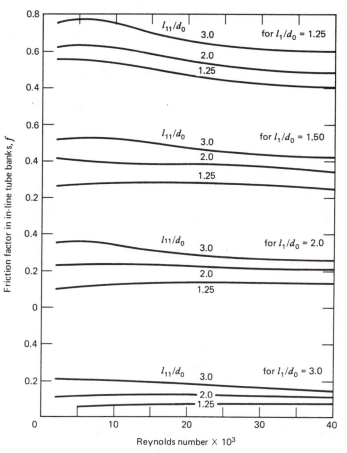

Fig. I.18 Friction factor f as affected by Reynolds number for various in-line tube patterns, crossflow gas or air, d_o, tube diameter; $l_\perp$, gap distance perpendicular to the flow; $l_\parallel$, gap distance parallel to the flow.

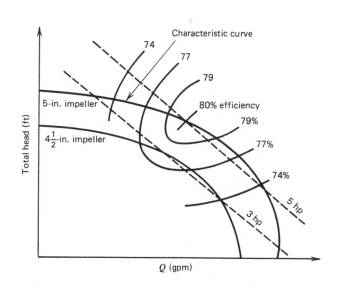

Fig. I.19 Performance curve for a centrifugal pump.

head. Note that as the flow-rate requirement is increased, the allowable head must be reduced if other pump parameters are unchanged.

Analysis and experience has shown that there are scaling laws for centrifugal pump performance that give the trends for a change in certain performance parameters. Basically, they are:

Efficiency:

$$\eta_p = f_1\left(\frac{Q_f}{D^3 n}\right)$$

Dimensionless head:

$$\frac{h_p g}{D^2 n^2} = f_2\left(\frac{Q_f}{D^3 n}\right)$$

Dimensionless brake horsepower:

$$\frac{bhp \cdot g}{\gamma D^2 n^3} = f_3\left(\frac{Q_f}{D^3 n}\right)$$

where D is the impeller diameter, n is the rotary impeller speed, g is gravity, and γ is the specific weight of fluid.

The basic relationships yield specific proportionalities such as $Q_f \propto n$ (rpm), $h_p \propto n^2$, $fhp \propto n^3$, $Q_f \propto D^3$, $h_p \propto D^2$, and $fhp \propto D^5$.

For pumps, density variations are generally negligible since liquids are incompressible. But for gas-handling equipment, density changes are very important. The scaling laws will give the following rules for changing density:

$$h_p \propto \rho \qquad (Q_f, n \text{ constant})$$
$$fhp \propto \rho$$
$$\left.\begin{array}{c} n \\ fhp \\ Q_f \end{array}\right\} \propto \rho^{-1/2} \qquad (h_p \text{ constant})$$
$$\left.\begin{array}{c} n \\ Q_f \\ h_p \end{array}\right\} \propto \frac{1}{\rho} \qquad (\dot{m} \text{ constant})$$
$$fhp \propto \frac{1}{\rho^2}$$

For centrifugal pumps, the following equations hold:

$$fhp = \frac{Q_f \rho g h_p}{550 g_c}$$

$$\eta_p p = \frac{Q_f \rho g h_p / 550 g_c}{bhp} = \frac{fhp}{bhp}$$

system efficiency $\eta_s = \eta_p \times \eta_m$ \qquad (motor efficiency)

It is important to select the motor and pump so that at nominal operating conditions, the pump and motor operate at near their maximum efficiency.

For systems where two or more pumps are present, the following rules are helpful. To analyze pumps in parallel, add capacities at the same head. For pumps in series, simply add heads at the same capacity.

There is one notable difference between blowers and pump performance. This is shown in Figure I.20. Note that the bhp continues to increase as permissible head goes to zero, in contrast to the pump curve when bhp approaches zero. This is because the kinetic energy imparted to the fluid at high flow rates is quite significant for blowers.

Manufactures of fluid-handling equipment provide excellent performance data for all types of equipment. Anyone considering replacement or a new installation should take full advantage of these data.

Fluid-handling equipment that operates on a principle other than centrifugal does not follow the centrifugal scaling laws. Evans[8] gives a thorough treatment of most types of equipment that would be encountered in industrial application.

REFERENCES

1. G.J. VAN WYLEN and R.E. SONNTAG, *Fundamentals of Classical Thermodynamics*, 2nd ed., Wiley, New York, 1973.
2. A.S. CHAPMAN, *Heat Transfer*, 3rd ed., Macmillan, New York, 1974.
3. J.P. HOLMAN, *Heat Transfer*, 4th ed., McGraw-Hill, New York, 1976.
4. M.P. HEISLER, *Trans. ASME, Vol. 69* (1947), p. 227.
5. L.S. TONG, *Boiling Heat Transfer and Two-Phase Flow*, Wiley, New York, 1965.
6. W.M. KAYS and A.L. LONDON, *Compact Heat Exchangers*, 2nd ed., McGraw-Hill, New York, 1963.
7. R. SIEGEL and J.R. HOWELL, *Thermal Radiation Heat Transfer*, McGraw-Hill, New York, 1972.
8. FRANK L. EVANS, JR., *Equipment Design Handbook for Refineries and Chemical Plants*, Vols. 1 and 2, Gulf Publishing, Houston, Tex., 1974.

SYMBOLS

Thermodynamics

AF	air/fuel ratio
C_p	constant-pressure specific heat
C_v	constant-volume specific heat
C_{p0}	zero-pressure constant-pressure specific heat
C_{v0}	zero-pressure constant-volume specufic heat
e, E	specific energy and total energy
g	acceleration due to gravity
g, G	specific Gibbs function and total Gibbs function
g_e	a constant that relates force, mass, length, and time
h, H	specific enthalpy and total enthalpy
k	specific heat ratio: C_p/C_v
K.E.	kinetic energy
lb_f	pound force
lb_m	pound mass

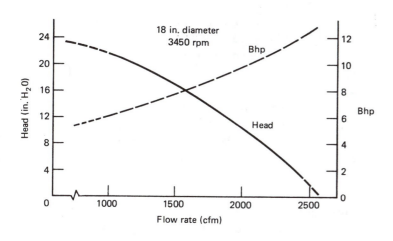

Fig. I.20 Variation of head and bhp with flow rate for a typical blower at constant speed.

lb mol	pound mole	ρ	density
m	mass	ϕ	relative humidity
$\dot{m}$	mass rate of flow	ω	humidity ratio or specific humidity
M	molecular weight		
n	number of moles		
n	polytropic exponent	**Subscripts**	
P	pressure	c	property at the critical point
P_i	partial pressure of component i in a mixture	c.v.	control volume
P.E.	potential energy	e	state of a substance leaving a control volume
P_r	relative pressure as used in gas tables	f	formation
q, Q	heat transfer per unit mass and total heat transfer	f	property of saturated liquid
		fg	difference in property for saturated vapor and saturated liquid
$\dot{Q}$	rate of heat transfer	g	property of saturated vapor
QH, QL	heat transfer from high- and low-temperature bodies	r	reduced property
R	gas constant	s	isentropic process
$\bar{R}$	universal gas constant		
s, S	specific entropy and total entropy	**Superscripts**	
t	time	$-$	bar over symbol denotes property on a molal basis (over V, H, S, U, A, G, the bar denotes partial molal property)
T	temperature		
u, U	specific internal energy and total internal energy	$\circ$	property at standard-state condition
v, V	specific volume and total volume	$*$	ideal gas
V	velocity	L	liquid phase
V_r	relative velocity	S	solid phase
w, W	work per unit mass and total work	V	vapor phase
W	rate of work, or power		
w_{rev}	reversible work between two states assuming heat transfer with surroundings	**Heat Transfer—Fluid Flow**	
x	mass fraction	A	surface area
Z	elevation	A_m	profile area for a fin
Z	compressibility factor	Bi	Biot number, (hL/k)
		c_p	specific heat at constant pressure
		c	specific heat
Greek Letters		D	diameter
β	coefficient of performance for a refrigerator	D_e	hydraulic diameter
β'	coefficient of performance for a heat pump	F_{i-j}	shape factor of area i with respect to area j
η	efficiency	f	friction factor

Gr   Grashof number, $g\,\beta\Delta T L_c^3/v^2$

g acceleration due to gravity

g_c gravitational constant

h convective heat-transfer coefficient

k thermal conductivity

m mass

m mass rate of flow

N number of rows

Nu   Nusselt number, hL/k

Pr   Prandtl number, $\mu C_p/k$

p pressure

Q rate of heat flow, volumetric flow rate

Ra   Rayleigh number, $g\,\beta\Delta T L_c^3/v\propto$

Re   Reynolds number, $\rho u_{av} L_c/\mu$

r radius

St   Stanton number, $h/C_p\,\rho u_\infty$

T temperature

U overall heat-transfer coefficient

u velocity

ux free-stream velocity

V volume

V velocity

W rate of work done

Greek Symbols

α thermal diffusivity

β coefficient of thermal expansion

Δ difference, change

ϵ surface emissivity

η_f fin effectiveness

μ viscosity

v kinematic viscosity

ρ density

σ Stefan-Boltzmann constant

τ time

Subscripts

b bulk conditions

cr   critical condition

c convection

cond   conduction

conv   convection

e entrance, effective

f fin, fluid

i inlet conditions

o exterior condition

0 centerline conditions in a tube at $r = 0$

o outlet condition

p pipe, pump

s surface condition

$\propto$ free-stream condition

CONVERSION FACTORS AND PROPERTY TABLES

Compiled by
L.C. WITTE

Professor of Mechanical Engineering
University of Houston
Houston, Texas

Table II.1 Conversion Factors

To Obtain:	Multiply:	By:
Acres	Sq miles	640.0
Atmospheres	Cm of Hg @ 0 deg C	0.013158
Atmospheres	Ft of H_2O @ 39.2 F.	0.029499
Atmospheres	Grams/sq cm	0.00096784
Atmospheres	In. Hg @ 32 F	0.033421
Atmospheres	In. H_2O @ 39.2 F	0.0024583
Atmospheres	Pounds/sq ft	0.00047254
Atmospheres	Pounds/sq in.	0.068046
Btu	Ft-lb	0.0012854
Btu	Hp-hr	2545.1
Btu	Kg-cal.	3.9685
Btu	Kw-hr	3413
Btu	Watt-hr	3.4130
Btu/(cu ft) (hr)	Kw/liter	96,650.6
Btu/hr	Mech. hp	2545.1
Btu/hr	Kw	3413
Btu/hr	Tons of refrigeration	12,000
Btu/hr	Watts	3.4127
Btu/kw hr	Kg cal/kw hr	3.9685
Btu/(hr) (ft) (deg F)	Cal/(sec) (cm) (deg C)	241.90
Btu/(hr) (ft) (deg F)	Joules/(sec) (cm) (deg C)	57.803
Btu/(hr) (ft) (deg F)	Watts/(cm) (deg C)	57.803
Btu/(hr) (sq ft)	Cal/(sec) (sq cm)	13,273.0
Btu/min	Ft-lb/min	0.0012854
Btu/min	Mech. hp	42.418
Btu/min	Kw	56.896
Btu/lb	Cal/gram	1.8
Btu/lb	Kg cal/kg	1.8
Btu/(lb) (deg F)	Cal/(gram) (deg C)	1.0
Btu/(lb) (deg F)	Joules/(gram) (deg C)	0.23889
Btu/sec	Mech. hp	0.70696
Btu/sec	Mech. hp (metric)	0.6971
Btu/sec	Kg-cal/hr	0.0011024
Btu/sec	Kw	0.94827
Btu/sq ft	Kg-cal/sq meter	0.36867

Table II.1 Continued

To Obtain:	Multiply:	By:
Calories	Ft-lb	0.32389
Calories	Joules	0.23889
Calories	Watt-hr	860.01
Cal/(cu cm) (sec)	Kw/liter	0.23888
Cal/gram	Btu/lb	0.55556
Cal/(gram) (deg C)	Btu/(lb) (deg F)	1.0
Cal/(sec) (cm) (deg C)	Btu/(hr) (ft) (deg F)	0.0041336
Cal/(sec) (sq cm)	Btu/(hr) (sq ft)	0.000075341
Cal/(sec) (sq cm) (deg C)	Btu/(hr) (sq ft) (deg F)	0.0001355
Centimeters	Inches	2.540
Centimeters	Microns	0.0001
Centimeters	Mils	0.002540
Cm of Hg @ 0 deg C	Atmospheres	76.0
Cm of Hg @ 0 deg C	Ft of H_2O @ 39.2 F	2.242
Cm of Hg @ 0 deg C	Grams/sq cm	0.07356
Cm of Hg @ 0 deg C	In. of H_2O @ 4 C	0.1868
Cm of Hg @ 0 deg C	Lb/sq in.	5.1715
Cm of Hg @ 0 deg C	Lb/sq ft	0.035913
Cm/deg C	In./deg F	4.5720
Cm/sec	Ft/min	0.508
Cm/sec	Ft/sec	30.48
Cm/(sec) (sec)	Gravity	980.665
Cm of H_2O @39.2 F	Atmospheres	1033.24
Cm of H_2O @39.2 F	Lb/sq in.	70.31
Centipoises	Centistokes	Density
Centistokes	Centipoises	l/density
Cu cm	Cu ft	28,317
Cu cm	Cu in.	16.387
Cu cm	Gal. (USA, liq.)	3785.43
Cu cm	Liters	1000 03
Cu cm	Ounces (USA, liq.)	29.573730
Cu cm	Quarts (USA, liq.)	946.358
Cu cm/sec	Cu ft/min	472.0
Cu ft	Cords (wood)	128.0
Cu ft	Cu meters	35.314
Cu ft	Cu yards	27.0
Cu ft	Gal. (USA, liq.)	0.13368
Cu ft	Liters	0.03532
Cu ft/min	Cu meters/sec	2118.9
Cu ft/min	Gal. (USA, liq./sec)	8.0192
Cu ft/lb	Cu meters/kg	16.02
Cu ft/lb	Liters/kg	0.01602
Cu ft/sec	Cu meters/min	0.5886
Cu ft/sec	Gal. (USA, liq.)/min	0.0022280
Cu ft/sec	Liters/min	0.0005886
Cu in.	Cu centimeters	0.061023
Cu in.	Gal. (USA, liq.)	231.0
Cu in.	Liters	61.03
Cu in.	Ounces (USA. liq.)	1.805

Table II.1 Continued

To Obtain:	Multiply:	By:
Cu meters	Cu ft	0.028317
Cu meters	Cu yards	0.7646
Cu meters	Gal. (USA. liq.)	0.0037854
Cu meters	Liters	0.001000028
Cu meters/hr	Gal./min	0.22712
Cu meters/kg	Cu ft/lb	0.062428
Cu meters/min	Cu ft/min	0.02832
Cu meters/min	Gal./sec	0.22712
Cu meters/sec	Gal./min	0.000063088
Cu yards	Cu meters	1.3079
Dynes	Grams	980.66
Dynes	Pounds (avoir.)	444820.0
Dyne-centimeters	Ft-lb	13,558,000
Dynes/sq cm	Lb/sq in.	68947
Ergs	Joules	10,000,000
Feet	Meters	3.281
Ft of H_2O @ 39.2 F	Atmospheres	33.899
Ft of H_2O @ 39.2 F	Cm of Hg @ 0 deg C	0.44604
Ft of H_2O @ 39.2 F	In. of Hg @ 32 deg F	1.1330
Ft of H_2O @ 39.2 F	Lb/sq ft	0.016018
Ft of H_2O @ 39.2 F	Lb/sq in.	2.3066
Ft/min	Cm/sec	1.9685
Ft/min	Miles (USA. statute)/hr	88.0
Ft/sec	Knots	1.6889
Ft/sec	Meters/sec	3.2808
Ft/sec	Miles (USA, statute)/hr	1.4667
Ft/(sec) (sec)	Gravity (sea level)	32.174
Ft/(sec) (sec)	Meters/(sec) (sec)	3.2808
Ft-lb	Btu	778.0
Ft-lb	Joules	0.73756
Ft-lb	Kg-calories	3087.4
Ft-lb	Kw-hr	2,655,200
Ft-lb	Mech. hp-hr	1,980,000
Ft-lb/min	Btu/min	778.0
Ft-lb/min	Kg cal/min	3087.4
Ft-lb/min	Kw	44,254.0
Ft-lb/min	Mech. hp	33,000
Ft-lb/sec	Btu/min	12.96
Ft-lb/sec	Kw	737.56
Ft-lb/sec	Mech. hp	550.0
Gal. (Imperial, liq.)	Gal. (USA. Liq.)	0.83268
Gal. (USA, liq.)	Barrels (petroleum, USA)	42
Gal. (USA. liq.)	Cu ft	7.4805
Gal. (USA. liq.)	Cu meters	264.173
Gal. (USA, liq.)	Cu yards	202.2
Gal. (USA. liq.)	Gal. (Imperial, liq.)	1.2010
Gal. (USA. liq.)	Liters	0.2642
Gal. (USA. liq.)/min	Cu ft/sec	448.83
Gal. (USA, liq.)/min	Cu meters/hr	4.4029

Table II.1 Continued

To Obtain:	Multiply:	By:
Gal. (USA. liq.)/sec	Cu ft/min	0.12468
Gal. (USA. liq.)/sec	Liters/min	0.0044028
Grains	Grams	15.432
Grains	Ounces (avoir.)	437.5
Grains	Pounds (avoir.)	7000
Grains/gal. (USA. liq.)	Parts/million	0.0584
Grams	Grains	0.0648
Grams	Ounces (avoir.)	28.350
Grams	Pounds (avoir.)	453.5924
Grams/cm	Pounds/in.	178.579
Grams/(cm) (sec)	Centipoises	0.01
Grams/cu cm	Lb/cu ft	0 .016018
Grams/cu cm	Lb/cu in.	27.680
Grams/cu cm	Lb/gal.	0.119826
Gravity (at sea level)	Ft/(sec) (sec)	0.03108
Inches	Centimeters	0.3937
Inches	Microns	0.00003937
Inches of Hg @ 32 F	Atmospheres	29.921
Inches of Hg @ 32 F	Ft of H_2O @ 39.2 F	0.88265
Inches of Hg @ 32 F	Lb/sq in.	2.0360
Inches of Hg @ 32 F	In. of H_2O @ 4 C	0.07355
Inches of H_2O@ 4 C	In. of Hg @ 32 F	13.60
Inches of H2O @ 39.2 F	Lb/sq in.	27.673
Inches/deg F	Cm/deg C	0.21872
Joules	Btu	1054.8
Joules	Calories	4.186
Joules	Ft-lb	1.35582
Joules	Kg-meters	9.807
Joules	Kw-hr	3,600,000
Joules	Mech. hp-hr	2,684,500
Kg	Pounds (avoir.)	0.45359
Kg-cal	Btu	0.2520
Kg-cal	Ft-lb	0.00032389
Kg-cal	Joules	0.0002389
Kg-cal	Kw-hr	860.01
Kg-cal	Mech. hp-hr	641.3
Kg-cal/kg	Btu/lb	0.5556
Kg-cal/kw hr	Btu/kw hr	0.2520
Kg-cal/min	Ft-lb/min	0.0003239
Kg-cal/min	Kw	14,33
Kg-cal/min	Mech. hp	10.70
Kg-cal/sq meter	Btu/sq ft	2.712
Kg/cu meter	Lb/cu ft	16.018
Kg/(hr) (meter)	Centipoises	3.60
Kg/liter	Lb/gal. (USA, liq.)	0.11983
Kg/meter	Lbm	1.488
Kg/sq cm	Atmospheres	1.0332
Kg sq cm	Lb/sq in .	0.0703
Kg/sq meter	Lb/sq ft	4.8824

Table II.1 Continued

To Obtain:	Multiply:	By:
Kg/sq meter	Lb/sq in.	703.07
Km	Miles (USA, statute)	1.6093
Kw	Btu/min	0.01758
Kw	Ft-lb/min	0.00002259
Kw	Ft-lb/sec	0.00135582
Kw	Kg-cal/hr	0.0011628
Kw	Kg-cal/min	0.069767
Kw	Mech. hp	0.7457
Kw-hr	Btu	0.000293
Kw-hr	Ft-lb	0.0000003766
Kw-hr	Kg-cal	0.0011628
Kw-hr	Mech. hp-hr	0.7457
Knots	Ft/sec	0.5921
Knots	Miles/hr	0.8684
Liters	Cu ft	28 . 316
Liters	Cu in.	0.01639
Liters	Cu centimeters	999.973
Liters	Gal. (Imperial. liq.)	4.546
Liters	Gal. (USA, liq.)	3.78533
Liters/kg	Cu ft/lb	62.42621
Liters/min	Cu ft/sec	1699.3
Liters/min	Gal. (USA. liq.)/min	3.785
Liters/sec	Cu ft/min	0.47193
Liters/sec	Gal./min	0.063088
Mech. hp	Btu/hr	0.0003929
Mech. hp	Btu/min	0.023575
Mech. hp	Ft-lb/sec	0.0018182
Mech. hp	Kg-cal/min	0.093557
Mech. hp	Kw	1.3410
Mech. hp-hr	Btu	0.00039292
Mech. hp-hr	Ft-lb	0.00000050505
Mech. hp-hr	Kg-calories	0.0015593
Mech. hp-hr	Kw-hr	1.3410
Meters	Feet	0.3048
Meters	Inches	0.0254
Meters	Miles (Int., nautical)	1852.0
Meters	Miles (USA, statute)	1609.344
Meters/min	Ft/min	0.3048
Meters/min	Miles (USA. statute)/hr	26.82
Meters/sec	Ft/sec	0.3048
Meters/sec	Km/hr	0.2778
Meters/sec	Knots	0.5148
Meters/sec	Miles (USA, statute)/hr	0.44704
Meters/(sec) (sec)	Ft/(sec) (sec)	0.3048
Microns	Inches	25,400
Microns	Mils	25.4
Miles (Int., nautical)	Km	0.54
Miles (Int., nautical)	Miles (USA, statute)	0.8690
Miles (Int., nautical)/hr	Knots	1.0

Table II.1 Continued

To Obtain:	Multiply:	By:
Miles (USA, statute)	Km	0.6214
Miles (USA, statute)	Meters	0.0006214
Miles (USA, statute)	Miles (Int., nautical)	1.151
Miles (USA, statute)/hr	Knots	1.151
Miles (USA, statute)/hr	Ft/min	0.011364
Miles (USA, statute)/hr	Ft/sec	0.68182
Miles (USA, statute)/hr	Meters/min	0.03728
Miles (USA, statute)/hr	Meters/sec	2.2369
Milliliters/gram	Cu ft/lb	62.42621
Millimeters	Microns	0.001
Mils	Centimeters	393.7
Mils	Inches	1000
Mils	Microns	0.03937
Minutes	Radians	3437.75
Ounces (avoir.)	Grains (avoir.)	0.0022857
Ounces (avoir.)	Grams	0.035274
Ounces (USA, liq.)	Gal. (USA, liq.)	128.0
Parts/million	Gr/gal. (USA, liq.)	17.118
Percent grade	Ft/100 ft	1.0
Pounds (avoir.)	Grains	0.0001429
Pounds (avoir.)	Grams	0.0022046
Pounds (avoir.)	Kg	2.2046
Pounds (avoir.)	Tons, long	2240
Pounds (avoir.)	Tons, metric	2204.6
Pounds (avoir.)	Tons, short	2000
Pounds/cu ft	Grams/cu cm	62.428
Pounds/cu ft	Kg/cu meter	0.062428
Pounds/cu ft	Pounds/gal.	7.48
Pounds/cu in .	Grams/cu cm	0.036127
Pounds/ft	Kg/meter	0.67197
Pounds/hr	Kg/min	132.28
Pounds/(hr) (ft)	Centipoises	2.42
Pounds/inch	Grams/cm	0.0056
Pounds/(sec) (ft)	Centipoises	0.000672
Pounds/sq inch	Atmospheres	14.696
Pounds/sq inch	Cm of Hg @ 0 deg C	0.19337
Pounds/sq inch	Ft of H_2O @ 39.2 F	0.43352
Pounds/sq inch	In. Hg @ l 32 F	0.491
Pounds/sq inch	In. H_2O @ 39.2 F	0.0361
Pounds/sq inch	Kg/sq cm	14 . 223
Pounds/sq inch	Kg/sq meter	0.0014223
Pounds/gal. (USA, liq.)	Kg/liter	8.3452
Pounds/gal. (USA, liq.)	Pounds/cu ft	0.1337
Pounds/gal. (USA, liq.)	Pounds/cu inch	231
Quarts (USA, liq.)	Cu cm	0.0010567
Quarts (USA, liq.)	Cu in.	0.01732
Quarts (USA, liq.)	Liters	1.057
Sq centimeters	Sq ft	929.0
Sq centimeters	Sq inches	6.4516

Table II.1 Continued

To Obtain:	Multiply:	By:
Sq ft	Acres	43,560
Sq ft	Sq meters	10.764
Sq inches	Sq centimeters	0.155
Sq meters	Acres	4046.9
Sq meters	Sq ft	0.0929
Sq mlles (USA. statute)	Acres	0.001562
Sq mils	Sq cm	155.000
Sq mils	Sq inches	1,000.000
Tons (metric)	Tons (short)	0.9072
Tons (short)	Tons (metric)	1.1023
Watts	Btu/sec	1054.8
Yards	Meters	1.0936

Table II.2-1 Saturated Steam

Temp. t (°F)	Abs. Press. p (psi)	Specific Volume			Enthalpy			Entropy			Temp. t (°F)
		Sat. Liquid v_f	Evap v_{fg}	Sat. Vapor v_g	Sat. Liquid h_f	Evap h_{fg}	Sat. Vapor h_g	Sat. Liquid s_f	Evap s_{fg}	Sat. Vapor s_g	
					Temperature Table						
32.0"	0.08859	0.016022	3304.7	3304.7	−0.0179	1075.5	1075.5	0.0000	2.1873	2.1873	32.0"
34.0	0.09600	0.016021	3061.9	3061.9	1.996	1074.4	1076.4	0.0041	2.1762	2.1802	34.0
36.0	0.10395	0.016020	2839.0	2839.0	4.008	1073.2	1077.2	0.0081	2.1651	2.1732	36.0
38.0	0.11249	0.016019	2634.1	2634.2	6.018	1072.1	1078.1	0.0122	2.1541	2.1663	38.0
40.0	0.12163	0.016019	2445.8	2445.8	8.027	1071.0	1079.0	0.0162	2.1432	2.1594	40.0
42.0	0.13143	0.016019	2272.4	2272.4	10.035	1069.8	1079.9	0.0202	2.1325	2.1527	42.0
44.0	0.14192	0.016019	2112.8	2112.8	12.041	1068.7	1080.7	0.0242	2.1217	2.1459	44.0
46.0	0.15314	0.016020	1965.7	1965.7	14.047	1067.6	1081.6	0.0282	2.1111	2.1393	46.0
48.0	0.16514	0.016021	1830.0	1830.0	16.051	1066.4	1082.5	0.0321	2.1006	2.1327	48.0
50.0	0.17796	0.016023	1704.8	1704.8	18.054	1065.3	1083.4	0.0361	2.0901	2.1262	50.0
52.0	0.19165	0.016024	1589.2	1589.2	20.057	1064.2	1084.2	0.0400	2.0798	2.1197	52.0
54.0	0.20625	0.016026	1482.4	1482.4	22.058	1063.1	1085.1	0.0439	2.0695	2.1134	54.0
56.0	0.22183	0.016028	1383.6	1383.6	24.059	1061.9	1086.0	0.0478	2.0593	2.1070	56.0
58.0	0.23843	0.016031	1292.2	1292.2	26.060	1060.8	1086.9	0.0516	2.0491	2.1008	58.0
60.0	0.25611	0.016033	1207.6	1207.6	28.060	1059.7	1087.7	0.0555	2.0391	2.0946	60.0
62.0	0.27494	0.016036	1129.2	1129.2	30.059	1058.5	1088.6	0.0593	2.0291	2.0885	62.0
64.0	0.29497	0.016039	1056.5	1056.5	32.058	1057.4	1089.5	0.0632	2.0192	2.0824	64.0
66.0	0.31626	0.016043	989.0	989.1	34.056	1056.3	1090.4	0.0670	2.0094	2.0764	66.0
68.0	0.33889	0.016046	926.5	926.5	36.054	1055.2	1091.2	0.0708	1.9996	2.0704	68.0
70.0	0.36292	0.016050	868.3	868.4	38.052	1054.0	1092.1	0.0745	1.9900	2.0645	70.0
72.0	0.38844	0.016054	814.3	814.3	40.049	1052.9	1093.0	0.0783	1.9804	2.0587	72.0
74.0	0.41550	0.016058	764.1	764.1	42.046	1051.8	1093.8	0.0821	1.9708	2.0529	74.0
76.0	0.44420	0.016063	717.4	717.4	44.043	1050.7	1094.7	0.0858	1.9614	2.0472	76.0
78.0	0.47461	0.016067	673.8	673.9	46.040	1049.5	1095.6	0.0895	1.9520	2.0415	78.0
80.0	0.50683	0.016072	633.3	633.3	48.037	1048.4	1096.4	0.0932	1.9426	2.0359	80.0
82.0	0.54093	0.016077	595.5	595.5	50.029	1047.3	1097.3	0.0969	1.9334	2.0303	82.0
84.0	0.57702	0.016082	560.3	560.3	52.029	1046.1	1098.2	0.1006	1.9242	2.0248	84.0
86.0	0.61518	0.016087	227.5	527.5	54.026	1045.0	1099.0	0.1043	1.9151	2.0193	86.0
88.0	0.65551	0.016093	496.8	496.8	56.022	1043.9	1099.9	0.1079	1.9060	2.0139	88.0

Table II.2-1   Continued

Temp. t (°F)	Abs. Press. p (psi)	Specific Volume			Enthalpy			Entropy			Temp. t (°F)
		Sat. Liquid v_f	Evap v_{fg}	Sat. Vapor v_g	Sat. Liquid h_f	Evap h_{fg}	Sat. Vapor h_g	Sat. Liquid s_f	Evap s_{fg}	Sat. Vapor s_g	
90.0	0.69813	0.016099	468.1	468.1	58.018	1042.7	1100.8	0.1115	1.8970	2.0086	90.0
92.0	0.74313	0.016105	441.3	441.3	60.014	1041.6	1101.6	0.1152	1.8881	2.0033	92.0
94.0	0.79062	0.016111	416.3	416.3	62.010	1040.5	1102.5	0.1188	1.8792	1.9980	94.0
96.0	0.84072	0.016117	392.8	392.9	64.006	1039.3	1103.3	0.1224	1.8704	1.9928	96.0
98.0	0.89356	0.016123	370.9	370.9	66.003	1038.2	1104.2	0.1260	1.8617	1.9876	98.0
100.0	0.94924	0.016130	350.4	350.4	67.999	1037.1	1105.1	0.1295	1.8530	1.9825	100.0
102.0	1.00789	0.016137	331.1	331.1	69.995	1035.9	1105.9	0.1331	1.8444	1.9775	102.0
104.0	1.06965	0.016144	313.1	313.1	71.992	1034.8	1106.8	0.1366	1.8358	1.9725	104.0
106.0	1.1347	0.016151	296.16	296.18	73.99	1033.6	1107.6	0.1402	1.8273	1.9675	106.0
108.0	1.2030	0.016158	280.28	280.30	75.98	1032.5	1108.5	0.1437	1.8188	1.9626	108.0
110.0	1.2750	0.016165	265.37	265.39	77.98	1031.4	1109.3	0.1472	1.8105	1.9577	110.0
112.0	1.3505	0.016173	251.37	251.38	79.98	1030.2	1110.2	0.1507	1.8021	1.9528	112.0
114.0	1.4299	0.016180	238.21	238.22	81.97	1029.1	1111.0	0.1542	1.7938	1.9480	114.0
116.0	1.5133	0.016188	225.84	225.85	83.97	1027.9	1111.9	0.1577	1.7856	1.9433	116.0
118.0	1.6009	0.016196	214.20	214.21	85.97	1026.8	1112.7	0.1611	1.7774	1.9386	118.0
120.0	1.6927	0.016204	203.25	203.26	87.97	1025.6	1113.6	0.1646	1.7693	1.9339	120.0
122.0	1.7891	0.016213	192.94	192.95	89.96	1024.5	1114.4	0.1680	1.7613	1.9293	122.0
124.0	1.8901	0.016221	183.23	183.24	91.96	1023.3	1115.3	0.1715	1.7533	1.9247	124.0
126.0	1.9959	0.016229	174.08	174.09	93.96	1022.2	1116.1	0.1749	1.7453	1.9202	126.0
128.0	2.1068	0.016238	165.45	165.47	95.96	1021.0	1117.0	0.1783	1.7374	1.9157	128.0
130.0	2.2230	0.016247	157.32	157.33	97.96	1019.8	1117.8	0.1817	1.7295	1.9112	130.0
132.0	2.3445	0.016256	149.64	149.66	99.95	1018.7	1118.6	0.1851	1.7217	1.9068	132.0
134.0	2.4717	0.016265	142.40	142.41	101.95	1017.5	1119.5	0.1884	1.7140	1.9024	134.0
136.0	2.6047	0.016274	135.55	135.57	103.95	1016.4	1120.3	0.1918	1.7063	1.8980	136.0
138.0	2.7438	0.016284	129.09	129.11	105.95	1015.2	1121.1	0.1951	1.6986	1.8937	138.0
140.0	2.8892	0.016293	122.98	123.00	107.95	1014.0	1122.0	0.1985	1.6910	1.8895	140.0
142.0	3.0411	0.016303	117.21	117.22	109.95	1012.9	1122.8	0.2018	1.6534	1.8852	142.0
144.0	3.1997	0.016312	111.74	111.76	111.95	1011.7	1123.6	0.2051	1.6759	1.8810	144.0
146.0	3.3653	0.016322	106.58	106.59	113.95	1010.5	1124.5	0.2084	1.6684	1.8769	146.0
148.0	3.5381	0.016332	101.68	101.70	115.95	1009.3	1125.3	0.2117	1.6610	1.8727	148.0
150.0	3.7184	0.016343	97.05	97.07	117.95	1008.2	1126.1	0.2150	1.6536	1.8686	150.0
152.0	3.9065	0.016353	92.66	92.68	119.95	1007.0	1126.9	0.2183	1.6463	1.8646	152.0
154.0	4.1025	0.016363	88.50	88.52	121.95	1005.8	1127.7	0.2216	1.6390	1.8606	154.0
156.0	4.3068	0.016374	84.56	84.57	123.95	1004.6	1128.6	0.2248	1.6318	1.8566	156.0
158.0	4.5197	0.016384	80.82	80.83	125.96	1003.4	1129.4	0.2281	1.6245	1.8526	158.0
160.0	4.7414	0.016395	77.27	77.29	127.96	1002.2	1130.2	0.2313	1.6174	1.8487	160.0
162.0	4.9722	0.016406	73.90	73.92	129.96	1001.0	1131.0	0.2345	1.6103	1.8448	162.0
164.0	5.2124	0.016417	70.70	70.72	131.96	999.8	1131.8	0.2377	1.6032	1.8409	164.0
166.0	5.4623	0.016428	67.67	67.68	133.97	998.6	1132.6	0.2409	1.5961	1.8371	166.0
168.0	5.7223	0.016440	64.78	64.80	135.97	997.4	1133.4	0.2441	1.5892	1.8333	168.0
170.0	5.9926	0.016451	62.04	62.06	137.97	996.2	1134.2	0.2473	1.5822	1.8295	170.0
172.0	6.2736	0.016463	59.43	59.45	139.98	995.0	1135.0	0.2505	1.5753	1.8258	172.0
174.0	6.5656	0.016474	56.95	56.97	141.98	993.8	1135.8	0.2537	1.5684	1.8221	174.0
176.0	6.8690	0.016486	54.59	54.61	143.99	992.6	1136.6	0.2568	1.5616	1.8184	176.0
178.0	7.1840	0.016498	52.35	52.36	145.99	991.4	1137.4	0.2600	1.5548	1.8147	178.0
180.0	7.5110	0.016510	50.21	50.22	148.00	990.2	1138.2	0.2631	1.5480	1.8111	180.0
182.0	7.850	0.016522	48.172	18.189	150.01	989.0	1139.0	0.2662	1.5413	1.8075	182.0
184.0	8.203	0.016534	46.232	46.249	152.01	987.8	1139.8	0.2694	1.5346	1.8040	184.0
186.0	8.568	0.016547	44.383	44.400	154.02	986.5	1140.5	0.2725	1.5279	1.8004	186.0
188.0	8.947	0.016559	42.621	42.638	156.03	985.3	1141.3	0.2756	1.5213	1.7969	188.0
190.0	9.340	0.016572	40.941	40.957	158.04	984.1	1142.1	0.2787	1.5148	1.7934	190.0
192.0	9.747	0.016585	39.337	39.354	160.05	982.8	1142.9	0.2818	1.5082	1.7900	192.0
194.0	10.168	0.016598	37.808	37.824	162.05	981.6	1143.7	0.2848	1.5017	1.7865	194.0
196.0	10.605	0.016611	36.348	36.364	164.06	980.4	1144.4	0.2879	1.4952	1.7831	196.0
198.0	11.058	0.016624	34.954	34.970	166.08	979.1	1145.2	0.2910	1.4888	1.7798	198.0
200.0	11.526	0.016637	33.622	33.639	168.09	977.9	1146.0	0.2940	1.4824	1.7764	200.0
204.0	12.512	0.016664	31.135	31.151	172.11	975.4	1147.5	0.3001	1.4697	1.7698	204.0
208.0	13.568	0.016691	28.862	28.878	176.14	972.8	1149.0	0.3061	1.4571	1.7632	208.0
212.0	14.696	0.016719	26.782	26.799	180.17	970.3	1150.5	0.3121	1.4447	1.7568	212.0
216.0	15.901	0.016747	24.878	24.894	184.20	967.8	1152.0	0.3181	1.4323	1.7505	216.0
220.0	17.186	0.016775	23.131	23.148	188.23	965.2	1153.4	0.3241	1.4201	1.7442	220.0
224.0	18.556	0.016805	21.529	21.545	192.27	962.6	1154.9	0.3300	1.4081	1.7380	224.0
228.0	20.015	0.016834	20.056	20.073	196.31	960.0	1156.3	0.3359	1.3961	1.7320	228.0
232.0	21.567	0.016864	18.701	18.718	200.35	957.4	1157.8	0.3417	1.3842	1.7260	232.0
236.0	23.216	0.016895	17.454	17.471	204.40	954.8	1159.2	0.3476	1.3725	1.7201	236.0
240.0	24.968	0.016926	16.304	16.321	208.45	952.1	1160.6	0.3533	1.3609	1.7142	240.0
244.0	26.826	0.016958	15.243	15.260	212.50	949.5	1162.0	0.3591	1.3494	1.7085	244.0

Table II.2-1 Continued

Temp. t (°F)	Abs. Press. p (psi)	Specific Volume			Enthalpy			Entropy			Temp. t (°F)
		Sat. Liquid v_f	Evap v_{fg}	Sat. Vapor v_g	Sat. Liquid h_f	Evap h_{fg}	Sat. Vapor h_g	Sat. Liquid s_f	Evap s_{fg}	Sat. Vapor s_g	
248.0	28.796	0.016990	14.264	14.281	216.56	946.8	1163.4	0.3649	1.3379	1.7028	248.0
252.0	30.883	0.017022	13.358	13.375	220.62	944.1	1164.7	0.3706	1.3266	1.6972	252.0
256.0	33.091	0.017055	12.520	12.538	224.69	941.4	1166.1	0.3763	1.3154	1.6917	256.0
260.0	35.427	0.017089	11.745	11.762	228.76	938.6	1167.4	0.3819	1.3043	1.6862	260.0
264.0	37.894	0.017123	11.025	11.042	232.83	935.9	1168.7	0.3876	1.2933	1.6808	264.0
268.0	40.500	0.017157	10.358	10.375	236.91	933.1	1170.0	0.3932	1.2823	1.6755	268.0
272.0	43.249	0.017193	9.738	9.755	240.99	930.3	1171.3	0.3987	1.2715	1.6702	272.0
276.0	46.147	0.017228	9.162	9.180	245.08	927.5	1172.5	0.4043	1.2607	1.6650	276.0
280.0	49.200	0.017264	8.627	8.644	249.17	924.6	1173.8	0.4098	1.2501	1.6599	280.0
284.0	52.414	0.01730	8.1280	8.1453	253.3	921.7	1175.0	0.4154	1.2395	1.6548	284.0
288.0	55.795	0.01734	7.6634	7.6807	257.4	918.8	1176.2	0.4208	1.2290	1.6498	288.0
292.0	59.350	0.01738	7.2301	7.2475	261.5	915.9	1177.4	0.4263	1.2186	1.6449	292.0
296.0	63.084	0.01741	6.8259	6.8433	265.6	913.0	1178.6	0.4317	1.2082	1.6400	296.0
300.0	67.005	0.01745	6.4483	6.4658	269.7	910.0	1179.7	0.4372	1.1979	1.6351	300.0
304.0	71.119	0.01749	6.0955	6.1130	273.8	907.0	1180.9	0.4426	1.1877	1.6303	304.0
308.0	75.433	0.01753	5.7655	5.7830	278.0	904.0	1182.0	0.4479	1.1776	1.6256	308.0
312.0	79.953	0.01757	5.4566	5.4742	282.1	901.0	1183.1	0.4533	1.1676	1.6209	312.0
316.0	84.688	0.01761	5.1673	5.1849	286.3	897.9	1184.1	0.4586	1.1576	1.6162	316.0
320.0	89.643	0.01766	4.8961	4.9138	290.4	894.8	1185.2	0.4640	1.1477	1.6116	320.0
324.0	94.826	0.01770	4.6418	4.6595	294.6	891.6	1186.2	0.4692	1.1378	1.6071	324.0
328.0	100.245	0.01774	4.4030	4.4208	298.7	888.5	1187.2	0.4745	1.1280	1.6025	328.0
332.0	105.907	0.01779	4.1788	4.1966	302.9	885.3	1188.2	0.4798	1.1183	1.5981	332.0
336.0	111.820	0.01783	3.9681	3.9859	307.1	882.1	1189.1	0.4850	1.1086	1.5936	336.0
340.0	117.992	0.01787	3.7699	3.7878	311.3	878.8	1190.1	0.4902	1.0990	1.5892	340.0
344.0	124.430	0.01792	3.5834	3.6013	315.5	875.5	1191.0	0.4954	1.0894	1.5849	344.0
348.0	131.142	0.01797	3.4078	3.4258	319.7	872.2	1191.1	0.5006	1.0799	1.5806	348.0
352.0	138.138	0.01801	3.2423	3.2603	323.9	868.9	1192.7	0.5058	1.0705	1.5763	352.0
356.0	145.424	0.01806	3.0863	3.1044	328.1	865.5	1193.6	0.5110	1.0611	1.5721	356.0
360.0	153.010	0.01811	2.9392	2.9573	332.3	862.1	1194.4	0.5161	1.0517	1.5678	360.0
364.0	160.903	0.01816	2.8002	2.8184	336.5	858.6	1195.2	0.5212	1.0424	1.5637	364.0
368.0	169.113	0.01821	2.6691	2.6873	340.8	855.1	1195.9	0.5263	1.0332	1.5595	368.0
372.0	177.648	0.01826	2.5451	2.5633	345.0	851.6	1196.7	0.5314	1.0240	1.5554	372.0
376.0	186.517	0.01831	2.4279	2.4462	349.3	848.1	1197.4	0.5365	1.0148	1.5513	376.0
380.0	195.729	0.01836	2.3170	2.3353	353.6	844.5	1198.0	0.5416	1.0057	1.5473	380.0
384.0	205.294	0.01842	2.2120	2.2304	357.9	840.8	1198.7	0.5466	0.9966	1.5432	384.0
388.0	215.220	0.01847	2.1126	2.1311	362.2	837.2	1199.3	0.5516	0.9876	1.5392	388.0
392.0	225.516	0.01853	2.0184	2.0369	366.5	833.4	1199.9	0.5567	0.9786	1.5352	392.0
396.0	236.193	0.01858	1.9291	1.9477	370.8	829.7	1200.4	0.5617	0.9696	1.5313	396.0
400.0	247.259	0.01864	1.8444	1.8630	375.1	825.9	1201.0	0.5667	0.9607	1.5274	400.0
404.0	258.725	0.01870	1.7640	1.7827	379.4	822.0	1201.5	0.5717	0.9518	1.5234	404.0
408.0	270.600	0.01875	1.6877	1.7064	383.8	818.2	1201.9	0.5766	0.9429	1.5195	408.0
412.0	282.894	0.01881	1.6152	1.6340	388.1	814.2	1202.4	0.5816	0.9341	1.5157	412.0
416.0	295.617	0.01887	1.5463	1.5651	392.5	810.2	1202.8	0.5866	0.9253	1.5118	416.0
420.0	308.780	0.01894	1.4808	1.4997	396.9	806.2	1203.1	0.5915	0.9165	1.5080	420.0
424.0	322.391	0.01900	1.4184	1.4374	401.3	802.2	1203.5	0.5964	0.9077	1.5042	424.0
428.0	336.463	0.01906	1.3591	1.3782	405.7	798.0	1203.7	0.6014	0.8990	1.5004	428.0
432.0	351.00	0.01913	1.30266	1.32179	410.1	793.9	1204.0	0.6063	0.8903	1.4966	432.0
436.0	366.03	0.01919	1.24887	1.26806	414.6	789.7	1204.2	0.6112	0.8816	1.4928	436.0
440.0	381.54	0.01926	1.19761	1.21687	419.0	785.4	1204.4	0.6161	0.8729	1.4890	440.0
444.0	397.56	0.01933	1.14874	1.16806	423.5	781.1	1204.6	0.6210	0.8643	1.4853	444.0
448.0	414.09	0.01940	1.10212	1.12152	428.0	776.7	1204.7	0.6259	0.8557	1.4815	448.0
452.0	431.14	0.01947	1.05764	1.07711	432.5	772.3	1204.8	0.6308	0.8471	1.4778	452.0
456.0	448.73	0.01954	1.01518	1.03472	437.0	767.8	1204.8	0.6356	0.8385	1.4741	456.0
460.0	466.87	0.01961	0.97463	0.99424	441.5	763.2	1204.8	0.6405	0.8299	1.4704	460.0
464.0	485.56	0.01969	0.93588	0.95557	446.1	758.6	1204.7	0.6454	0.8213	1.4667	464.0
468.0	504.83	0.01976	0.89885	0.91862	450.7	754.0	1204.6	0.6502	0.8127	1.4629	468.0
472.0	524.67	0.01984	0.86345	0.88329	455.2	749.3	1204.5	0.6551	0.8042	1.4592	472.0
476.0	545.11	0.01992	0.82958	0.84950	459.9	744.5	1204.3	0.6599	0.7956	1.4555	476.0
480.0	566.15	0.02000	0.79716	0.81717	464.5	739.6	1204.1	0.6648	0.7871	1.4518	480.0
484.0	587.81	0.02009	0.76613	0.78622	469.1	734.7	1203.8	0.6696	0.7785	1.4481	484.0
488.0	610.10	0.02017	0.73641	0.75658	473.8	729.7	1203.5	0.6745	0.7700	1.4444	488.0
492.0	633.03	0.02026	0.70794	0.72820	478.5	724.6	1203.1	0.6793	0.7614	1.4407	492.0
496.0	656.61	0.02034	0.68065	0.70100	483.2	719.5	1202.7	0.6842	0.7528	1.4370	496.0
500.0	680.86	0.02043	0.65448	0.67492	487.9	714.3	1202.2	0.6890	0.7443	1.4333	500.0
504.0	705.78	0.02053	0.62938	0.64991	492.7	709.0	1201.7	0.6939	0.7357	1.4296	504.0

Table II.2-1   Continued

Temp. t (°F)	Abs. Press. p (psi)	Specific Volume Sat. Liquid v_f	Evap v_{fg}	Sat. Vapor v_g	Enthalpy Sat. Liquid h_f	Evap h_{fg}	Sat. Vapor h_g	Entropy Sat. Liquid s_f	Evap s_{fg}	Sat. Vapor s_g	Temp. t (°F)
508.0	731.40	0.02062	0.60530	0.62592	497.5	703.7	1201.1	0.6987	0.7271	1.4258	508.0
512.0	757.72	0.02072	0.58218	0.60289	502.3	698.2	1200.5	0.7036	0.7185	1.4221	512.0
516.0	784.76	0.02081	0.55997	0.58079	507.1	692.7	1199.8	0.7085	0.7099	1.4183	516.0
520.0	812.53	0.02091	0.53864	0.55956	512.0	687.0	1199.0	0.7133	0.7013	1.4146	520.0
524.0	841.04	0.02102	0.51814	0.53916	516.9	681.3	1198.2	0.7182	0.6926	1.4108	524.0
528.0	870.31	0.02112	0.49843	0.51955	521.8	675.5	1197.3	0.7231	0.6839	1.4070	528.0
532.0	900.34	0.02123	0.47947	0.50070	526.8	669.6	1196.4	0.7280	0.6752	1.4032	532.0
536.0	931.17	0.02134	0.46123	0.48257	531.7	663.6	1195.4	0.7329	0.6665	1.3993	536.0
540.0	962.79	0.02146	0.44367	0.46513	536.8	657.5	1194.3	0.7378	0.6577	1.3954	540.0
544.0	995.22	0.02157	0.42677	0.44834	541.8	651.3	1193.1	0.7427	0.6489	1.3915	544.0
548.0	1028.49	0.02169	0.41048	0.43217	546.9	645.0	1191.9	0.7476	0.6400	1.3876	548.0
552.0	1062.59	0.02182	0.39479	0.41660	552.0	638.5	1190.6	0.7525	0.6311	1.3837	552.0
556.0	1097.55	0.02194	0.37966	0.40160	557.2	632.0	1189.2	0.7575	0.6222	1.3797	556.0
560.0	1133.38	0.02207	0.36507	0.38714	562.4	625.3	1187.7	0.7625	0.6132	1.3757	560.0
564.0	1170.10	0.02221	0.35099	0.37320	567.6	618.5	1186.1	0.7674	0.6041	1.3716	564.0
568.0	1207.72	0.02235	0.33741	0.35975	572.9	611.5	1184.5	0.7725	0.5950	1.3675	568.0
572.0	1246.26	0.02249	0.32429	0.34678	578.3	604.5	1182.7	0.7775	0.5859	1.3634	572.0
576.0	1285.74	0.02264	0.31162	0.33426	583.7	597.2	1180.9	0.7825	0.5766	1.3592	576.0
580.0	1326.17	0.02279	0.29937	0.32216	589.1	589.9	1179.0	0.7876	0.5673	1.3550	580.0
584.0	1367.7	0.02295	0.28753	0.31048	594.6	582.4	1176.9	0.7927	0.5580	1.3507	584.0
588.0	1410.0	0.02311	0.27608	0.29919	600.1	574.7	1174.8	0.7978	0.5485	1.3464	588.0
592.0	1453.3	0.02328	0.26499	0.28827	605.7	566.8	1172.6	0.8030	0.5390	1.3420	592.0
596.0	1497.8	0.02345	0.25425	0.27770	611.4	558.8	1170.2	0.8082	0.5293	1.3375	596.0
600.0	1543.2	0.02364	0.24384	0.26747	617.1	550.6	1167.7	0.8134	0.5196	1.3330	600.0
604.0	1589.7	0.02382	0.23374	0.25757	622.9	542.2	1165.1	0.8187	0.5097	1.3284	604.0
608.0	1637.3	0.02402	0.22394	0.24796	628.8	533.6	1162.4	0.8240	0.4997	1.3238	608.0
612.0	1686.1	0.02422	0.21442	0.23865	634.8	524.7	1159.5	0.8294	0.4896	1.3190	612.0
616.0	1735.9	0.02444	0.20516	0.22960	640.8	515.6	1156.4	0.8348	0.4794	1.3141	616.0
620.0	1786.9	0.02466	0.19615	0.22081	646.9	506.3	1153.2	0.8403	0.4689	1.3092	620.0
624.0	1839.0	0.02489	0.18737	0.21226	653.1	496.6	1149.8	0.8458	0.4583	1.3041	624.0
628.0	1892.4	0.02514	0.17880	0.20394	659.5	486.7	1146.1	0.8514	0.4474	1.2988	628.0
632.0	1947.0	0.02539	0.17044	0.19583	665.9	476.4	1142.2	0.8571	0.4364	1.2934	632.0
636.0	2002.8	0.02566	0.16226	0.18792	672.4	465.7	1138.1	0.8628	0.4251	1.2879	636.0
640.0	2059.9	0.02595	0.15427	0.18021	679.1	454.6	1133.7	0.8686	0.4134	1.2821	640.0
644.0	2118.3	0.02625	0.14644	0.17269	685.9	443.1	1129.0	0.8746	0.4015	1.2761	644.0
648.0	2178.1	0.02657	0.13876	0.16534	692.9	431.1	1124.0	0.8806	0.3893	1.2699	648.0
652.0	2239.2	0.02691	0.13124	0.15816	700.0	418.7	1118.7	0.8868	0.3767	1.2634	652.0
656.0	2301.7	0.02728	0.12387	0.15115	707.4	405.7	1113.1	0.8931	0.3637	1.2567	656.0
660.0	2365.7	0.02768	0.11663	0.14431	714.9	392.1	1107.0	0.8995	0.3502	1.2498	660.0
664.0	2431.1	0.02811	0.10947	0.13757	722.9	377.7	1100.6	0.9064	0.3361	1.2425	664.0
668.0	2498.1	0.02858	0.10229	0.13087	731.5	362.1	1093.5	0.9137	0.3210	1.2347	668.0
672.0	2566.6	0.02911	0.09514	0.12424	740.2	345.7	1085.9	0.9212	0.3054	1.2266	672.0
676.0	2636.8	0.02970	0.08799	0.11769	749.2	328.5	1077.6	0.9287	0.2892	1.2179	676.0
680.0	2708.6	0.03037	0.08080	0.11117	758.5	310.1	1068.5	0.9365	0.2720	1.2086	680.0
684.0	2782.1	0.03114	0.07349	0.10463	768.2	290.2	1058.4	0.9447	0.2537	1.1984	684.0
688.0	2857.4	0.03204	0.06595	0.09799	778.8	268.2	1047.0	0.9535	0.2337	1.1872	688.0
692.0	2934.5	0.03313	0.05797	0.09110	790.5	243.1	1033.6	0.9634	0.2110	1.1744	692.0
696.0	3013.4	0.03455	0.04916	0.08371	804.4	212.8	1017.2	0.9749	0.1841	1.1591	696.0
700.0	3094.3	0.03662	0.03857	0.07519	822.4	172.7	995.2	0.9901	0.1490	1.1390	700.0
702.0	3135.5	0.03824	0.03173	0.06997	835.0	144.7	979.7	1.0006	0.1246	1.1252	702.0
704.0	3177.2	0.04108	0.02192	0.06300	854.2	102.0	956.2	1.0169	0.0876	1.1046	704.0
705.0	3198.3	0.04427	0.01304	0.05730	873.0	61.4	934.4	1.0329	0.0527	1.0856	705.0
705.47[b]	3208.2	0.05078	0.00000	0.05078	906.0	0.0	906.0	1.0612	0.0000	1.0612	705.47[b]

Pressure Table

0.08865	32.018	0.016022	3302.4	3302.4	0.0003	1075.5	1075.5	0.0000	2.1872	2.1872	0.08865
0.25	59.323	0.016032	1235.5	1235.5	27.382	1060.1	1087.4	0.0542	2.0425	2.0967	0.25
0.50	79.586	0.016071	641.5	641.5	47.623	1048.6	1096.3	0.0925	1.9446	2.0370	0.50
1.0	101.74	0.016136	333.59	333.60	69.73	1036.1	1105.8	0.1326	1.8455	1.9781	1.0
5.0	162.24	0.016407	73.515	73.532	130.20	1000.9	1131.1	0.2349	1.6094	1.8443	5.0
10.0	193.21	0.016592	38.404	38.420	161.26	982.1	1143.3	0.2836	1.5043	1.7879	10.0
14.696	212.00	0.016719	26.782	26.799	180.17	970.3	1150.5	0.3121	1.4447	1.7568	14.696
15.0	213.03	0.016726	26.274	26.290	181.21	969.7	1150.9	0.3137	1.4415	1.7552	15.0
20.0	227.96	0.016834	20.070	20.087	196.27	960.1	1156.3	0.3358	1.3962	1.7320	20.0
30.0	250.34	0.017009	13.7266	13.7436	218.9	945.2	1164.1	0.3682	1.3313	1.6995	30.0

Table II.2-1 Continued

Temp. t (°F)	Abs. Press. p (psi)	Specific Volume			Enthalpy			Entropy			Temp. t (°F)
		Sat. Liquid v_f	Evap v_{fg}	Sat. Vapor v_g	Sat. Liquid h_f	Evap h_{fg}	Sat. Vapor h_g	Sat. Liquid s_f	Evap s_{fg}	Sat. Vapor s_g	
40.0	267.25	0.017151	10.4794	10.4965	236.1	933.6	1169.8	0.3921	1.2844	1.6765	40.0
50.0	281.02	0.017274	8.4967	8.5140	250.2	923.9	1174.1	0.4112	1.2474	1.6586	50.0
60.0	292.71	0.017383	7.1562	7.1736	262.2	915.4	1177.6	0.4273	1.2167	1.6440	60.0
70.0	302.93	0.017482	6.1875	6.2050	272.7	907.8	1180.6	0.4411	1.1905	1.6316	70.0
80.0	312.04	0.017573	5.4536	5.4711	282.1	900.9	1183.1	0.4534	1.1675	1.6208	80.0
90.0	320.28	0.017659	4.8779	4.8953	290.7	894.6	1185.3	0.4643	1.1470	1.6113	90.0
100.0	327.82	0.017740	4.4133	4.4310	298.5	888.6	1187.2	0.4743	1.1284	1.6027	100.0
110.0	334.79	0.01782	4.0306	4.0484	305.8	883.1	1188.9	0.4834	1.1115	1.5950	110.0
120.0	341.27	0.01789	3.7097	3.7275	312.6	877.8	1190.4	0.4919	1.0960	1.5879	120.0
130.0	347.33	0.01796	3.4364	3.4544	319.0	872.8	1191.7	0.4998	1.0815	1.5813	130.0
140.0	353.04	0.01803	3.2010	3.2190	325.0	868.0	1193.0	0.5071	1.0681	1.5752	140.0
150.0	358.43	0.01809	2.9958	3.0139	330.6	863.4	1194.1	0.5141	1.0554	1.5695	150.0
160.0	363.55	0.01815	2.8155	2.8336	336.1	859.0	1195.1	0.5206	1.0435	1.5641	160.0
170.0	368.42	0.01821	2.6556	2.6738	341.2	854.8	1196.0	0.5269	1.0322	1.5591	170.0
180.0	373.08	0.01827	2.5129	2.5312	346.2	850.7	1196.9	0.5328	1.0215	1.5543	180.0
190.0	377.53	0.01833	2.3847	2.4030	350.9	846.7	1197.6	0.5384	1.0113	1.5498	190.0
200.0	381.80	0.01839	2.2689	2.2873	355.5	842.8	1198.3	0.5438	1.0016	1.5454	200.0
210.0	385.91	0.01844	2.16373	2.18217	359.9	839.1	1199.0	0.5490	0.9923	1.5413	210.0
220.0	389.88	0.01850	2.06779	2.08629	364.2	835.4	1199.6	0.5540	0.9834	1.5374	220.0
230.0	393.70	0.01855	1.97991	1.99846	368.3	831.8	1200.1	0.5588	0.9748	1.5336	230.0
240.0	397.39	0.01860	1.89909	1.91769	372.3	828.4	1200.6	0.5634	0.9665	1.5299	240.0
250.0	400.97	0.01865	1.82452	1.84317	376.1	825.0	1201.1	0.5679	0.9585	1.5264	250.0
260.0	404.44	0.01870	1.75548	1.77418	379.9	821.6	1201.5	0.5722	0.9508	1.5230	260.0
270.0	407.80	0.01875	1.69137	1.71013	383.6	818.3	1201.9	0.5764	0.9433	1.5197	270.0
280.0	411.07	0.01880	1.63169	1.65049	387.1	815.1	1202.3	0.5805	0.9361	1.5166	280.0
290.0	414.25	0.01885	1.57597	1.59482	390.6	812.0	1202.6	0.5844	0.9291	1.5135	290.0
300.0	417.35	0.01889	1.52384	1.54274	394.0	808.9	1202.9	0.5882	0.9223	1.5105	300.0
350.0	431.73	0.01912	1.30642	1.32554	409.8	794.2	1204.0	0.6059	0.8909	1.4968	350.0
400.0	444.60	0.01934	1.14162	1.16095	424.2	780.4	1204.6	0.6217	0.8630	1.4847	400.0
450.0	456.28	0.01954	1.01224	1.03179	437.3	767.5	1204.8	0.6360	0.8378	1.4738	450.0
500.0	467.01	0.01975	0.90787	0.92762	449.5	755.1	1204.7	0.6490	0.8148	1.4639	500.0
550.0	476.94	0.01994	0.82183	0.84177	460.9	743.3	1204.3	0.6611	0.7936	1.4547	550.0
600.0	486.20	0.02013	0.74962	0.76975	471.7	732.0	1203.7	0.6723	0.7738	1.4461	600.0
650.0	494.89	0.02032	0.68811	0.70843	481.9	720.9	1202.8	0.6828	0.7552	1.4381	650.0
700.0	503.08	0.02050	0.63505	0.65556	491.6	710.2	1201.8	0.6928	0.7377	1.4304	700.0
750.0	510.84	0.02069	0.58880	0.60949	500.9	699.8	1200.7	0.7022	0.7210	1.4232	750.0
800.0	518.21	0.02087	0.54809	0.56896	509.8	689.6	1199.4	0.7111	0.7051	1.4163	800.0
850.0	525.24	0.02105	0.51197	0.53302	518.4	679.5	1198.0	0.7197	0.6899	1.4096	850.0
900.0	531.95	0.02123	0.47968	0.50091	526.7	669.7	1196.4	0.7279	0.6753	1.4032	900.0
950.0	538.39	0.02141	0.45064	0.47205	534.7	660.0	1194.7	0.7358	0.6612	1.3970	950.0
1000.0	544.58	0.02159	0.42436	0.44596	542.6	650.4	1192.9	0.7434	0.6476	1.3910	1000.0
1050.0	550.53	0.02177	0.40047	0.42224	550.1	640.9	1191.0	0.7507	0.6344	1.3851	1050.0
1100.0	556.28	0.02195	0.37863	0.40058	557.5	631.5	1189.1	0.7578	0.6216	1.3794	1100.0
1150.0	561.82	0.02214	0.35859	0.38073	564.8	622.2	1187.0	0.7647	0.6091	1.3738	1150.0
1200.0	567.19	0.02232	0.34013	0.36245	571.9	613.0	1184.8	0.7714	0.5969	1.3683	1200.0
1250.0	572.38	0.02250	0.32306	0.34556	578.8	603.8	1182.6	0.7780	0.5850	1.3630	1250.0
1300.0	577.42	0.02269	0.30722	0.32991	585.6	594.6	1180.2	0.7843	0.5733	1.3577	1300.0
1350.0	582.32	0.02288	0.29250	0.31537	592.3	585.4	1177.8	0.7906	0.5620	1.3525	1350.0
1400.0	587.07	0.02307	0.27871	0.30178	598.8	576.5	1175.3	0.7966	0.5507	1.3474	1400.0
1450.0	591.70	0.02327	0.26584	0.28911	605.3	567.4	1172.8	0.8026	0.5397	1.3423	1450.0
1500.0	596.20	0.02346	0.25372	0.27719	611.7	558.4	1170.1	0.8085	0.5288	1.3373	1500.0
1550.0	600.59	0.02366	0.24235	0.26601	618.0	549.4	1167.4	0.8142	0.5182	1.3324	1550.0
1600.0	604.87	0.02387	0.23159	0.25545	624.2	540.3	1164.5	0.8199	0.5076	1.3274	1600.0
1650.0	609.05	0.02407	0.22143	0.24551	630.4	531.3	1161.6	0.8254	0.4971	1.3225	1650.0
1700.0	613.13	0.02428	0.21178	0.23607	636.5	522.2	1158.6	0.8309	0.4867	1.3176	1700.0
1750.0	617.12	0.02450	0.20263	0.22713	642.5	513.1	1155.6	0.8363	0.4765	1.3128	1750.0
1800.0	621.02	0.02472	0.19390	0.21861	648.5	503.8	1152.3	0.8417	0.4662	1.3079	1800.0
1850.0	624.83	0.02495	0.18558	0.21052	654.5	494.6	1149.0	0.8470	0.4561	1.3030	1850.0
1900.0	628.56	0.02517	0.17761	0.20278	660.4	485.2	1145.6	0.8522	0.4459	1.2981	1900.0
1950.0	632.22	0.02541	0.16999	0.19540	666.3	475.8	1142.0	0.8574	0.4358	1.2931	1950.0
2000.0	635.80	0.02565	0.16266	0.18831	672.1	466.2	1138.3	0.8625	0.4256	1.2881	2000.0
2100.0	642.76	0.02615	0.14885	0.17501	683.8	446.7	1130.5	0.8727	0.4053	1.2780	2100.0
2200.0	649.45	0.02669	0.13603	0.16272	695.5	426.7	1122.2	0.8828	0.3848	1.2676	2200.0
2300.0	655.89	0.02727	0.12406	0.15133	707.2	406.0	1113.2	0.8929	0.3640	1.2569	2300.0
2400.0	662.11	0.02790	0.11287	0.14076	719.0	384.8	1103.7	0.9031	0.3430	1.2460	2400.0
2500.0	668.11	0.02859	0.10209	0.13068	731.7	361.6	1093.3	0.9139	0.3206	1.2345	2500.0
2600.0	673.91	0.02938	0.09172	0.12110	744.5	337.6	1082.0	0.9247	0.2977	1.2225	2600.0

Table II.2-1 Continued

Abs. Temp. t (°F)	Press. p (psi)	Specific Volume			Enthalpy			Entropy			Temp. t (°F)
		Sat. Liquid v_f	Evap v_{fg}	Sat. Vapor v_g	Sat. Liquid h_f	Evap h_{fg}	Sat. Vapor h_g	Sat. Liquid s_f	Evap s_{fg}	Sat. Vapor s_g	
2700.0	679.53	0.03029	0.08165	0.11194	757.3	312.3	1069.7	0.9356	0.2741	1.2097	2700.0
2800.0	684.96	0.03134	0.07171	0.10305	770.7	285.1	1055.8	0.9468	0.2491	1.1958	2800.0
2900.0	690.22	0.03262	0.06158	0.09420	785.1	254.7	1039.8	0.9588	0.2215	1.1803	2900.0
3000.0	695.33	0.03428	0.05073	0.08500	801.8	218.4	1020.3	0.9728	0.1891	1.1619	3000.0
3100.0	700.28	0.03681	0.03771	0.07452	824.0	169.3	993.3	0.9914	0.1460	1.1373	3100.0
3200.0	705.08	0.04472	0.01191	0.05663	875.5	56.1	931.6	1.0351	0.0482	1.0832	3200.0
3208.2^c	705.47	0.05078	0.00000	0.05078	906.0	0.0	906.0	1.0612	0.0000	1.0612	3208.2^c

^{a} The states shown are metastable.
^{b} Critical temperature.
^{c} Critical pressure.
Source: Copyright 1967 ASME (Abridged); reprinted by permission.

Table II.2-2 Superheated Steamᵃ

Abs. Press. (psi) (Sat. Temp.)		Sat. Water	Sat. Steam	Temperature (°F)												
				200	250	300	350	400	450	500	600	700	800	900	1000	1100	1200
1 (101.74)	Sh			98.26	148.26	198.26	248.26	298.26	348.26	398.26	498.26	598.26	698.26	798.26	898.26	998.26	1098.26
	v	0.01614	333.6	392.5	422.4	452.3	482.1	511.9	541.7	571.5	631.1	690.7	750.3	809.8	869.4	929.0	988.6
	h	69.73	1105.8	1150.2	1172.9	1195.7	1218.7	1241.8	1265.1	1288.6	1336.1	1384.5	1433.7	1483.8	1534.9	1586.8	1639.7
	s	0.1326	1.9781	2.0509	2.0841	2.1152	2.1445	2.1722	2.1985	2.2237	2.2708	2.3144	2.3551	2.3934	2.4296	2.4640	2.4969
5 (162.24)	Sh			37.76	87.76	137.76	187.76	237.76	287.76	337.76	437.76	537.76	637.76	737.76	837.76	937.76	1037.76
	v	0.01641	73.53	78.14	84.21	90.24	96.25	102.24	108.23	114.21	126.15	138.08	150.01	161.94	173.86	185.78	197.70
	h	130.20	1131.1	1148.6	1171.7	1194.8	1218.0	1241.3	1264.7	1288.2	1335.9	1384.3	1433.6	1483.7	1534.7·	1586.7	1639.6
	s	0.2349	1.8443	1.8716	1.9054	1.9369	1.9664	1.9943	2.0208	2.0460	2.0932	2.1369	2.1776	2.2159	2.2521	2.2866	2.3194
10 (193.21)	Sh			6.79	56.79	106.79	156.79	206.79	256.79	306.79	406.79	506.79	606.79	706.79	806.79	906.79	1006.79
	v	0.01659	38.42	38.84	41.93	44.98	48.02	51.03	54.04	57.04	63.03	69.00	74.98	80.94	86.91	92.87	98.84
	h	161.26	1143.3	1146.6	1170.2	1193.7	1217.1	1240.6	1264.1	1287.8	1335.5	1384.0	1433.4	1483.5	1534.6	1586.6	1639.5
	s	0.2836	1.7879	1.7928	1.8273	1.8593	1.8892	1.9173	1.9439	1.9692	2.0166	2.0603	2.1011	2.1394	2.1757	2.2101	2.2430
14.696 (212.00)	Sh				38.00	88.00	138.00	188.00	238.00	288.00	388.00	488.00	588.00	688.00	788.00	888.00	988.00
	v	.0167	26.799		28.42	30.52	32.60	34.67	36.72	38.77	42.86	46.93	51.00	55.06	59.13	63.19	67.25
	h	180.17	1150.5		1168.8	1192.6	1216.3	1239.9	1263.6	1287.4	1335.2	1383.8	1433.2	1483.4	1534.5	1586.5	1639.4
	s	.3121	1.7568		1.7833	1.8158	1.8459	1.8743	1.9010	1.9265	1.9739	2.0177	2.0585	2.0969	2.1332	2.1676	2.2005
15 (213.03)	Sh				36.97	86.97	136.97	186.97	236.97	286.97	386.97	486.97	586.97	686.97	786.97	886.97	986.97
	v	0.01673	26.290		27.837	29.899	31.939	33.963	35.977	37.985	41.986	45.978	49.964	53.946	57.926	61.905	65.882
	h	181.21	1150.9		1168.7	1192.5	1216.2	1239.9	1263.6	1287.3	1335.2	1383.8	1433.2	1483.4	1534.5	1586.5	1639.4
	s	0.3137	1.7552		1.7809	1.8134	1.8437	1.8720	1.8988	1.9242	1.9717	2.0155	2.0563	2.0946	2.1309	2.1653	2.1982
20 (227.96)	Sh			22.04	72.04	122.04	172.04	222.04	272.04	372.04	472.04	572.04	672.04	772.04	872.04	972.04	
	v	0.01683	20.087	20.788	22.356	23.900	25.428	26.946	28.457	31.466	34.465	37.458	40.447	43.435	46.420	49.405	
	h	196.27	1156.3	1167.1	1191.4	1215.4	1239.2	1263.0	1286.9	1334.9	1383.5	1432.9	1483.2	1534.3	1586.3	1639.3	
	s	0.3358	1.7320	1.7475	1.7805	1.8111	1.8397	1.8666	1.8921	1.9397	1.9836	2.0244	2.0628	2.0991	2.1336	2.1982	
25 (240.07)	Sh			9.93	59.93	109.93	159.93	209.93	259.93	359.93	459.93	559.93	659.93	759.93	859.93	959.93	
	v	0.01693	16.301	16.558	17.829	19.076	20.307	21.527	22.740	25.153	27.557	29.954	32.348	34.740	37.130	39.518	
	h	208.52	1160.6	1165.6	1190.2	1214.5	1238.5	1262.5	1286.4	1334.6	1383.3	1432.7	1483.0	1534.2	1586.2	1639.2	
	s	0.3535	1.7141	1.7212	1.7547	1.7856	1.8145	1.8415	1.8672	1.9149	1.9588	1.9997	2.0381	2.0744	2.1089	2.1418	
30 (250.34)	Sh				49.66	99.66	149.66	199.66	249.66	349.66	449.66	549.66	649.66	749.66	849.66	949.66	
	v	0.01701	13.744		14.810	15.859	16.892	17.914	18.929	20.945	22.951	24.952	26.949	28.943	30.936	32.927	
	h	218.93	1164.1		1189.0	1213.6	1237.8	1261.9	1286.0	1334.2	1383.0	1432.5	1482.8	1534.0	1586.1	1639.0	
	s	0.3682	1.6995		1.7334	1.7647	1.7937	1.8210	1.8467	1.8946	1.9386	1.9795	2.0179	2.0543	2.0888	2.1217	
35 (259.29)	Sh				40.71	90.71	140.71	190.71	240.71	340.71	440.71	540.71	640.71	740.71	840.71	940.71	
	v	0.01708	11.896		12.654	13.562	14.453	15.334	16.207	17.939	19.662	21.379	23.092	24.803	26.512	28.220	
	h	228.03	1167.1		1187.8	1212.7	1237.1	1261.3	1285.5	1333.9	1382.8	1432.3	1482.7	1533.9	1586.0	1638.9	
	s	0.3809	1.6872		1.7152	1.7468	1.7761	1.8035	1.8294	1.8774	1.9214	1.9624	2.0009	2.0372	2.0717	2.1046	
40 (267.25)	Sh				32.75	82.75	132.75	182.75	232.75	332.75	432.75	532.75	632.75	732.75	832.75	932.75	
	v	0.01715	10.497		11.036	11.838	12.624	13.398	14.165	15.685	17.195	18.699	20.199	21.697	23.194	24.689	
	h	236.14	1169.8		1186.6	1211.7	1236.4	1260.8	1285.0	1333.6	1382.5	1432.1	1482.5	1533.7	1585.8	1638.8	
	s	0.3921	1.6765		1.6992	1.7312	1.7608	1.7883	1.8143	1.8624	1.9065	1.9476	1.9860	2.0224	2.0569	2.0899	
45 (274.44)	Sh				25.56	75.56	125.56	175.56	225.56	325.56	425.56	525.56	625.56	725.56	825.56	925.56	
	v	0.01721	9.399		9.777	10.497	11.201	11.892	12.577	13.932	15.276	16.614	17.950	19.282	20.613	21.943	
	h	243.49	1172.1		1185.4	1210.4	1235.7	1260.2	1284.6	1333.3	1382.3	1431.9	1482.3	1533.6	1585.7	1638.7	
	s	0.4021	1.6671		1.6849	1.7173	1.7471	1.7748	1.8010	1.8492	1.8934	1.9345	1.9730	2.0093	2.0439	2.0768	
50 (281.02)	Sh				18.98	68.98	118.98	168.98	218.98	318.98	418.98	518.98	618.98	718.98	818.98	918.98	
	v	0.01727	8.514		8.769	9.424	10.062	10.688	11.306	12.529	13.741	14.947	16.150	17.350	18.549	19.746	
	h	250.21	1174.1		1184.1	1209.9	1234.9	1259.6	1284.1	1332.9	1382.0	1431.7	1482.2	1533.4	1585.6	1638.6	
	s	0.4112	1.6586		1.6720	1.7048	1.7349	1.7628	1.7890	1.8374	1.8816	1.9227	1.9613	1.9977	2.0322	2.0652	
55 (287.07)	Sh				12.93	62.93	112.93	162.93	212.93	312.93	412.93	512.93	612.93	712.93	812.93	912.93	
	v	0.01733			7.945	8.546	9.130	9.702	10.267	11.381	12.485	13.583	14.677	15.769	16.859	17.948	
	h	256.43			1182.9	1208.9	1234.2	1259.1	1283.6	1332.6	1381.8	1431.5	1482.0	1533.3	1585.5	1638.5	
	s	0.4196			1.6601	1.6933	1.7237	1.7518	1.7781	1.8266	1.8710	1.9121	1.9507	1.987	2.022	2.055	
60 (292.71)	Sh				7.29	57.29	107.29	157.29	207.29	307.29	407.29	507.29	607.29	707.29	807.29	907.29	
	v	0.01738	7.174		7.257	7.815	8.354	8.881	9.400	10.425	11.438	12.446	13.450	14.452	15.452	16.450	
	h	262.21	1177.6		1181.6	1208.0	1233.5	1258.5	1283.2	1332.3	1381.5	1431.3·	1481.8	1533.2	1585.3	1638.4	
	s	0.4273	1.6440		1.6492	1.6934	1.7134	1.7417	1.7681	1.8168	1.8612	1.9024	1.9410	1.9774	2.0120	2.0450	
65 (297.98)	Sh				2.02	52.02	102.02	152.02	202.02	302.02	402.02	502.02	602.02	702.02	802.02	902.02	
	v	0.01743	6.653		6.675	7.195	7.697	8.186	8.667	9.615	10.552	11.484	12.412	13.337	14.261	15.183	
	h	267.63	1179.1		1180.3	1207.0	1232.7	1257.9	1282.7	1331.9	1381.3	1431.1	1481.6	1533.0	1585.2	1638.3	
	s	0.4344	1.6375		1.6390	1.6731	1.7040	1.7324	1.7590	1.8077	1.8522	1.8935	1.9321	1.9685	2.0031	2.0361	
70 (302.93)	Sh					47.07	97.07	147.07	197.07	297.07	397.07	497.07	597.07	697.07	797.07	897.07	
	v	0.01748	6.205			6.664	7.133	7.590	8.039	8.922	9.793	10.659	11.522	12.382	13.240	14.097	
	h	272.74	1180.6			1206.0	1232.0	1257.3	1282.2	1331.6	1381.0	1430.9	1481.5	1532.9	1585.1	1638.2	
	s	0.4411	1.6316			1.6640	1.6951	1.7237	1.7504	1.7993	1.8439	1.8852	1.9238	1.9603	1.9949	2.0279	
75 (307.61)	Sh					42.39	92.39	142.39	192.39	292.39	392.39	492.39	592.39	692.39	792.39	892.39	
	v	0.01753	5.814			6.204	6.645	7.074	7.494	8.320	9.135	9.945	10.750	11.553	12.355	13.155	
	h	277.56	1181.9			1205.0	1231.2	1256.7	1281.7	1331.3	1380.7	1430.7	1481.3	1532.7	1585.0	1638.1	
	s	0.4474	1.6260			1.6554	1.6868	1.7156	1.7424	1.7915	1.8361	1.8774	1.9161	1.9526	1.9872	2.0202	

Table II.2-2   Continued

Abs. Press. (psi) (Sat. Temp.)		Sat. Water	Sat. Steam	Temperature (°F)													
				350	400	450	500	550	600	700	800	900	1000	1100	1200	1300	1400
80 (312.04)	Sh			37.96	87.96	137.96	187.96	237.96	287.96	387.96	487.96	587.96	687.96	787.96	887.96	987.96	1087.96
	v	0.01757	5.471	5.801	6:218	6.622	7.018	7.408	7.794	8.560	9.319	10.075	10.829	11.581	12.331	13.081	13.829
	h	282.15	1183.1	1204.0	1230.5	1256.1	1281.3	1306.2	1330.9	1380.5	1430.5	1481.i	1532.6	1584.9	1638.0	1692.0	1746.8
	s	0.4534	1.6208	1.6473	1.6790	1.7080	1.7349	1.7602	1.7842	1.8289	1.8702	1.9089	1.9454	1.9800	2.0131	2.0446	2.0750
85 (316.26)	Sh			33.74	83.74	133.74	183.74	233.74	283.74	383.74	483.74	583.74	683.74	783.74	883.74	983.74	1083.74
	v	0.01762	5.167	5.445	5.840	6.223	6.597	6.966	7.330	8.052	8.768	9.480	10.190	10.898	11.604	12.310	13.014
	h	286.52	1184.2	1203.0	1229.7	1255.5	1280.8	1305.8	1330.6	1380.2	1430.3	1481.0	1532.4	1584.7	1637.9	1691.9	1746.8
	s	0.4590	1.6159	1.6396	1.6716	1.7008	1.7279	1.7532	1.7772	1.8220	1.8634	1.9021	1.9386	1.9733	2.0063	2.0379	2.0682
90 (320.28)	Sh			29.72	79.72	129.72	179.72	229.72	279.72	379.72	479.72	579.72	679.72	779.72	879.72	979.72	1079.72
	v	0.01766	4.895	5.128	5.505	5.869	6.223	6.572	6.917	7.600	8.277	8.950	9.621	10.290	10.958	11.625	12.290
	h	290.69	1185.3	1202.0	1228.9	1254.9	1280.3	1305.4	1330.2	1380.0	1430.1	1480.8	1532.3	1584.6	1637.8	1691.8	1746.7
	s	0.4643	1.6113	1.6323	1.6646	1.6940	1.7212	1.7467	1.7707	1.8156	1.8570	1.8957	1.9323	1.9669	2.0000	2.0316	2.0619
95 (324.13)	Sh			25.87	75.87	125.87	175.87	225.87	275.87	375.87	475.87	575.87	675.87	775.87	875.87	975.87	1075.87
	v	0.01770	4.651	4.845	5.205	5.551	5.889	6.221	6.548	7.196	7.838	8.477	9.113	9.747	10.380	11.012	11.643
	h	294.70	1186.2	1200.9	1228.1	1254.3	1279.8	1305.0	1329.9	1379.7	1429.9	1480.6	1532.1	1584.5	1637.7	1691.7	1746.6
	s	0.4694	1.6069	1.6253	1.6580	1.6876	1.7149	1.7404	1.7645	1.8094	1.8509	1.8897	1.9262	1.9609	1.9940	2.0256	2.0559
100 (327.82)	Sh			22.18	72.18	122.18	172.18	222.18	272.18	372.18	472.18	572.18	672.18	772.18	872.18	972.18	1072.18
	v	0.01774	4.431	4.590	4.935	5.266	5.588	5.904	6.216	6.833	7.443	8.050	8.655	9.258	9.860	10.460	11.060
	h	298.54	1187.2	1199.9	1227.4	1253.7	1279.3	1304.6	1329.6	1379.5	1429.7	1480.4	1532.0	1584.4	1637.6	1691.6	1746.5
	s	0.4743	1.6027	1.6187	1.6516	1.6814	1.7088	1.7344	1.7586	1.8036	1.8451	1.8839	1.9205	1.9552	1.9883	2.0199	2.0502
105 (331.37)	Sh			18.63	68.63	118.63	168.63	218.63	268.63	368.63	468.63	568.63	668.63	768.63	868.63	968.63	1068.63
	v	0.01778	4.231	4.359	4.690	5.007	5.315	5.617	5.915	6.504	7.086	7.665	8.241	8.816	9.389	9.961	10.532
	h	302.24	1188.0	1198.8	1226.6	1253.1	1278.8	1304.2	1329.2	1379.2	1429.4	1480.3	1531.8	1584.2	1637.5	1691.5	1746.4
	s	0.4790	1.5988	1.6122	1.6455	1.6755	1.7031	1.7288	1.7530	1.7981	1.8396	1.8785	1.9151	1.9498	1.9828	2.0145	2.0448
110 (334.79)	Sh			15.21	65.21	115.21	165.21	215.21	265.21	365.21	465.21	565.21	665.21	765.21	865.21	965.21	1065.21
	v	0.01782	4.048	4.149	4.468	4.772	5.068	5.357	5.642	6.205	6.761	7.314	7.865	8.413	8.961	9.507	10.053
	h	305.80	1188.9	1197.7	1225.8	1252.5	1278.3	1303.8	1328.9	1379.0	1429.2	1480.1	1531.7	1584.1	1637.4	1691.4	1746.4
	s	0.4834	1.5950	1.6061	1.6396	1.6698	1.6975	1.7233	1.7476	1.7928	1.8344	1.8732	1.9099	1.9446	1.9777	2.0093	2.0397
115 (338.08)	Sh			11.92	61.92	111.92	161.92	211.92	261.92	361.92	461.92	561.92	661.92	761.92	861.92	961.92	1061.92
	v	0.01785	3.881	3.957	4.265	4.558	4.841	5.119	5.392	5.932	6.465	6.994	7.521	8.046	8.570	9.093	9.615
	h	309.25	1189.6	1196.7	1225.0	1251.8	1277.9	1303.3	1328.6	1378.7	1429.0	1479.9	1531.6	1584.0	1637.2	1691.4	1746.3
	s	0.4877	1.5913	1.6001	1.6340	1.6644	1.6922	1.7181	1.7425	1.7877	1.8294	1.8682	1.9049	1.9396	1.9727	2.0044	2.0347
120 (341.27)	Sh			8.73	58.73	108.73	158.73	208.73	258.73	358.73	458.73	558.73	658.73	758.73	858.73	958.73	1058.73
	v	0.01789	3.7275	3.7815	4.0786	4.3610	4.6341	4.9009	5.1637	5.6813	6.1928	6.7006	7.2060	7.7096	8.2119	8.7130	9.2134
	h	312.58	1190.4	1195.6	1224.1	1251.2	1277.4	1302.9	1328.2	1378.4	1428.8	1479.8	1531.4	1583.9	1637.1	1691.3	1746.2
	s	0.4919	1.5879	1.5943	1.6286	1.6592	1.6872	1.7132	1.7376	1.7829	1.8246	1.8635	1.9001	1.9349	1.9680	1.9996	2.0300
130 (347.33)	Sh			2.67	52.67	102.67	152.67	202.67	252.67	352.67	452.67	552.67	652.67	752.67	852.67	952.67	1052.67
	v	0.01796	3.4544	3.4699	3.7489	4.0129	4.2672	4.5151	4.7589	5.2384	5.7118	6.1814	6.6486	7.1140	7.5781	8.0411	8.5033
	h	318.95	1191.7	1193.4	1222.5	1249.9	1276.4	1302.1	1327.5	1377.9	1428.4	1479.4	1531.1	1583.6	1636.9	1691.1	1746.1
	s	0.4998	1.5813	1.5833	1.6182	1.6493	1.6775	1.7037	1.7283	1.7737	1.8155	1.8545	1.8911	1.9259	1.9591	1.9907	2.0211
140 (353.04)	Sh				46.96	96.96	146.96	196.96	246.96	346.96	446.96	546.96	646.96	746.96	846.96	946.96	1046.96
	v	0.01803	3.2190		3.4661	3.7143	3.9526	4.1844	4.4119	4.8588	5.2995	5.7364	6.1709	6.6036	7.0349	7.4652	7.8946
	h	324.96	1193.0		1220.8	1248.7	1275.3	1301.3	1326.8	1377.4	1428.0	1479.1	1530.8	1583.4	1636.7	1690.9	1745.9
	s	0.5071	1.5752		1.6085	1.6400	1.6686	1.6949	1.7196	1.7652	1.8071	1.8461	1.8828	1.9176	1.9508	1.9825	2.0129
150 (358.43)	Sh				41.57	91.57	141.57	191.57	241.57	341.57	441.57	541.57	641.57	741.57	841.57	941.57	1041.57
	v	0.01809	3.0139		3.2208	3.4555	3.6799	3.8978	4.1112	4.5298	4.9421	5.3507	5.7568	6.1612	6.5642	6.9661	7.3671
	h	330.65	1194.1		1219.1	1247.4	1274.3	1300.5	1326.1	1376.9	1427.6	1478.7	1530.5	1583.1	1636.5	1690.7	1745.7
	s	0.5141	1.5695		1.5993	1.6313	1.6602	1.6867	1.7115	1.7573	1.7992	1.8383	1.8751	1.9099	1.9431	1.9748	2.0052
160 (363.55)	Sh				36.45	86.45	136.45	186.45	236.45	336.45	436.45	536.45	636.45	736.45	836.45	936.45	1036.45
	v	0.01815	2.8336		3.0060	3.2288	3.4413	3.6469	3.8480	4.2420	4.6295	5.0132	5.3945	5.7741	6.1522	6.5293	6.9055
	h	336.07	1195.1		1217.4	1246.0	1273.3	1299.6	1325.4	1376.4	1427.2	1478.4	1530.3	1582.9	1636.3	1690.5	1745.6
	s	0.5206	1.5641		1.5906	1.6231	1.6522	1.6790	1.7039	1.7499	1.7919	1.8310	1.8678	1.9027	1.9359	1.9676	1.9980
170 (368.42)	Sh				31.58	81.58	131.58	181.58	231.58	331.58	431.58	531.58	631.58	731.58	831.58	931.58	1031.58
	v	0.01821	2.6738		2.8162	3.0288	3.2306	3.4255	3.6158	3.9879	4.3536	4.7155	5.0749	5.4325	5.7888	6.1440	6.4983
	h	341.24	1196.0		1215.6	1244.7	1272.2	1298.8	1324.7	1375.8	1426.8	1478.0	1530.0	1582.6	1636.1	1690.4	1745.4
	s	0.5269	1.5591		1.5823	.1.6152	1.6447	1.6717	1.6968	1.7428	1.7850	1.8241	1.8610	1.8959	1.9291	1.9608	1.9913
180 (373.08)	Sh				26.92	76.92	126.92	176.92	226.92	326.92	426.92	526.92	626.92	726.92	826.92	926.92	1026.92
	v	0.01827	2.5312		2.6474	2.8508	3.0433	3.2286	3.4093	3.7621	4.1084	4.4508	4.7907	5.1289	5.4657	5.8014	6.1363
	h	346.19	1196.9		1213.8	1243.4	1271.2	1297.9	1324.0	1375.3	1426.3	1477.7	1529.7	1582.4	1635.9	1690.2	1745.3
	s	0.5328	1.5543		1.5743	1.6078	1.6376	1.6647	1.6900	1.7362	1.7784	1.8176	1.8545	1.8894	1.9227	1.9545	1.9849
190 (377.53)	Sh				22.47	72.47	122.47	172.47	222.47	322.47	422.47	522.47	622.47	722.47	822.47	922.47	1022.47
	v	0.01833	2.4030		2.4961	2.6915	2.8756	3.0525	3.2246	3.5601	3.8889	4.2140	4.5365	4.8572	5.1766	5.4949	5.8124
	h	350.94	1197.6		1212.0	1242.0	1270.1	1297.1	1323.3	1374.8	1425.9	1477.4	1529.4	1582.1	1635.7	1690.0	1745.1
	s	0.5384	1.5498		1.5667	1.6006	1.6307	1.6581	1.6835	1.7299	1.7722	1.8115	1.8484	1.8834	1.9166	1.9484	1.9789
200 (381.80)	Sh				18.20	68.20	118.20	168.20	218.20	318.20	418.20	518.20	618.20	718.20	818.20	918.20	1018.20
	v	0.01839	2.2873		2.3598	2.5480	2.7247	2.8939	3.0583	3.3783	3.6915	4.0008	4.3077	4.6158	4.9165	5.2191	5.5209
	h	355.51	1198.3		1210.1	1240.6	1269.0	1296.2	1322.6	1374.3	1425.5	1477.0	1529.1	1581.9	1635.4	1689.8	1745.0
	s	0.5438	1.5454		1.5593	1.5938	1.6242	1.6518	1.6773	1.7239	1.7663	1.8057	1.8426	1.8776	1.9109	1.9427	1.9732

Table II.2-2 Continued

Abs. Press. (psi) (Sat. Temp.)		Sat. Water	Sat. Steam	Temperature (°F)													
				400	450	500	550	600	700	800	900	1000	1100	1200	1300	1400	1500
210 (385.91)	Sh			14.09	64.09	114.09	164.09	214.09	314.09	414.09	514.09	614.09	714.09	814.09	914.09	1014.09	1114.09
	v	0.01844	2.1822	2.2364	2.4181	2.5880	2.7504	2.9078	3.2137	3.5128	3.8080	4.1007	4.3915	4.6811	4.9695	5.2571	5.5440
	h	359.91	1199.0	1208.02	1239.2	1268.0	1295.3	1321.9	1373.7	1425.1	1476.7	1528.8	1581.6	1635.2	1689.6	1744.8	1800.8
	s	0.5490	1.5413	1.5522	1.5872	1.6180	1.6458	1.6715	1.7182	1.7607	1.8001	1.8371	1.8721	1.9054	1.9372	1.9677	1.9970
220 (389.88)	Sh			10.12	60.12	110.12	160.12	210.12	310.12	410.12	510.12	610.12	710.12	810.12	910.12	1010.12	1110.12
	v	0.01850	2.0863	2.1240	2.2999	2.4638	2.6199	2.7710	3.0642	3.3504	3.6327	3.9125	4.1905	4.4671	4.7426	5.0173	5.2913
	h	364.17	1199.6	1206.3	1237.8	1266.9	1294.5	1321.2	1373.2	1424.7	1476.3	1528.5	1581.4	1635.0	1689.4	1744.7	1800.6
	s	0.5540	1.5374	1.5453	1.5808	1.6120	1.6400	1.6658	1.7128	1.7553	1.7948	1.8318	1.8668	1.9002	1.9320	1.9625	1.9919
230 (393.70)	Sh			6.30	56.30	106.30	156.30	206.30	306.30	406.30	506.30	606.30	706.30	806.30	906.30	1006.30	1106.30
	v	0.01855	1.9985	2.0212	2.1919	2.3503	2.5008	2.6461	2.9276	3.2020	3.4726	3.7406	4.0068	4.2717	4.5355	4.7984	5.0606
	h	368.28	1200.1	1204.4	1236.3	1265.7	1293.6	1320.4	1372.7	1424.2	1476.0	1528.2	1581.1	1634.8	1689.3	1744.5	1800.5
	s	0.5588	1.5336	1.5385	1.5747	1.6062	1.6344	1.6604	1.7075	1.7502	1.7897	1.8268	1.8618	1.8952	1.9270	1.9576	1.9869
240 (397.39)	Sh			2.61	52.61	102.61	152.61	202.61	302.61	402.61	502.61	602.61	702.61	802.61	902.61	1002.61	1102.61
	v	0.01860	1.9177	1.9268	2.0928	2.2462	2.3915	2.5316	2.8024	3.0661	3.3259	3.5831	3.8385	4.0926	4.3456	4.5977	4.8492
	h	372.27	1200.6	1202.4	1234.9	1264.6	1292.7	1319.7	1372.1	1423.8	1475.6	1527.9	1580.9	1634.6	1689.1	1744.3	1800.4
	s	0.5634	1.5299	1.5320	1.5687	1.6006	1.6291	1.6552	1.7025	1.7452	1.7848	1.8219	1.8570	1.8904	1.9223	1.9528	1.9822
250 (400.97)	Sh				49.03	99.03	149.03	199.03	299.03	399.03	499.03	599.03	699.03	799.03	899.03	999.03	1099.03
	v	0.01865	1.8432		2.0016	2.1504	2.2909	2.4262	2.6872	2.9410	3.1909	3.4382	3.6837	3.9278	4.1709	4.4131	4.6546
	h	376.14	1201.1		1233.4	1263.5	1291.8	1319.0	1371.6	1423.4	1475.3	1527.6	1580.6	1634.4	1688.9	1744.2	1800.2
	s	0.5679	1.5264		1.5629	1.5951	1.6239	1.6502	1.6976	1.7405	1.7801	1.8173	1.8524	1.8858	1.9177	1.9482	1.9776
260 (404.44)	Sh				45.56	95.56	145.56	195.56	295.56	395.56	495.56	595.56	695.56	795.56	895.56	995.56	1095.56
	v	0.01870	1.7742		1.9173	2.0619	2.1981	2.3289	2.5808	2.8256	3.0663	3.3044	3.5408	3.7758	4.0097	4.2427	4.4750
	h	379.90	1201.5		1231.9	1262.4	1290.9	1318.2	1371.1	1423.0	1474.9	1527.3	1580.4	1634.2	1688.7	1744.0	1800.1
	s	0.5722	1.5230		1.5573	1.5899	1.6189	1.6453	1.6930	1.7359	1.7756	1.8128	1.8480	1.8814	1.9133	1.9439	1.9732
270 (407.80)	Sh				42.20	92.20	142.20	192.20	292.20	392.20	492.20	592.20	692.20	792.20	892.20	992.20	1092.20
	v	0.01875	1.7101		1.8391	1.9799	2.1121	2.2388	2.4824	2.7186	2.9509	3.1806	3.4084	3.6349	3.8603	4.0849	4.3087
	h	383.56	1201.9		1230.4	1261.2	1290.0	1317.5	1370.5	1422.6	1474.6	1527.1	1580.1	1634.0	1688.5	1743.9	1800.0
	s	0.5764	1.5197		1.5518	1.5848	1.6140	1.6406	1.6885	1.7315	1.7713	1.8085	1.8437	1.8771	1.9090	1.9396	1.9690
280 (411.07)	Sh				38.93	88.93	138.93	188.93	288.93	388.93	488.93	588.93	688.93	788.93	888.93	988.93	1088.93
	v	0.01880	1.6505		1.7665	1.9037	2.0322	2.1551	2.3909	2.6194	2.8437	3.0655	3.2855	3.5042	3.7217	3.9384	4.1543
	h	387.12	1202.3		1228.8	1260.0	1289.1	1316.8	1370.0	1422.1	1474.2	1526.8	1579.9	1633.8	1688.4	1743.7	1799.8
	s	0.5805	1.5166		1.5464	1.5798	1.6093	1.6361	1.6841	1.7273	1.7671	1.8043	1.8395	1.8730	1.9050	1.9356	1.9649
290 (414.25)	Sh				35.75	85.75	135.75	185.75	285.75	385.75	485.75	585.75	685.75	785.75	885.75	985.75	1085.75
	v	0.01885	1.5948		1.6988	1.8327	1.9578	2.0772	2.3058	2.5269	2.7440	2.9585	3.1711	3.3824	3.5926	3.8019	4.0106
	h	390.60	1202.6		1227.3	1258.9	1288.1	1316.0	1369.5	1421.7	1473.9	1526.5	1579.6	1633.5	1688.2	1743.6	1799.7
	s	0.5844	1.5135		1.5412	1.5750	1.6048	1.6317	1.6799	1.7232	1.7630	1.8003	1.8356	1.8690	1.9010	1.9316	1.9610
300 (417.35)	Sh				32.65	82.65	132.65	182.65	282.65	382.65	482.65	582.65	682.65	782.65	882.65	982.65	1082.65
	v	0.01889	1.5427		1.6356	1.7665	1.8883	2.0044	2.2263	2.4407	2.6509	2.8585	3.0643	3.2688	3.4721	3.6746	3.8764
	h	393.99	1202.9		1225.7	1257.7	1287.2	1315.2	1368.9	1421.3	1473.6	1526.2	1579.4	1633.3	1688.0	1743.4	1799.6
	s	0.5882	1.5105		1.5361	1.5703	1.6003	1.6274	1.6758	1.7192	1.7591	1.7964	1.8317	1.8652	1.8972	1.9278	1.9572
310 (420.36)	Sh				29.64	79.64	129.64	179.64	279.64	379.64	479.64	579.64	679.64	779.64	879.64	979.64	1079.64
	v	0.01894	1.4939		1.5763	1.7044	1.8233	1.9363	2.1520	2.3600	2.5638	2.7650	2.9644	3.1625	3.3594	3.5555	3.7509
	h	397.30	1203.2		1224.1	1256.5	1286.3	1314.5	1368.4	1420.9	1473.2	1525.9	1579.2	1633.1	1687.8	1743.3	1799.4
	s	0.5920	1.5076		1.5311	1.5657	1.5960	1.6233	1.6719	1.7153	1.7553	1.7927	1.8280	1.8615	1.8935	1.9241	1.9536
320 (423.31)	Sh				26.69	76.69	126.69	176.69	276.69	376.69	476.69	576.69	676.69	776.69	876.69	976.69	1076.69
	v	0.01899	1.4480		1.5207	1.6462	1.7623	1.8725	2.0823	2.2843	2.4821	2.6774	2.8708	3.0628	3.2538	3.4438	3.6332
	h	400.53	1203.4		1222.5	1255.2	1285.3	1313.7	1367.8	1420.5	1472.9	1525.6	1578.9	1632.9	1687.6	1743.1	1799.3
	s	0.5956	1.5048		1.5261	1.5612	1.5918	1.6192	1.6680	1.7116	1.7516	1.7890	1.8243	1.8579	1.8899	1.9206	1.9500
330 (426.18)	Sh				23.82	73.82	123.82	173.82	273.82	373.82	473.82	573.82	673.82	773.82	873.82	973.82	1073.82
	v	0.01903	1.4048		1.4684	1.5915	1.7050	1.8125	2.0168	2.2132	2.4054	2.5950	2.7828	2.9692	3.1545	3.3389	3.5227
	h	403.70	1203.6		1220.9	1254.0	1284.4	1313.0	1367.3	1420.0	1472.5	1525.3	1578.7	1632.7	1687.5	1742.9	1799.2
	s	0.5991	1.5021		1.5213	1.5568	1.5876	1.6153	1.6643	1.7079	1.7480	1.7855	1.8208	1.8544	1.8864	1.9171	1.9466
340 (428.99)	Sh				21.01	71.01	121.01	171.01	271.01	371.01	471.01	571.01	671.01	771.01	871.01	971.01	1071.01
	v	0.01908	1.3640		1.4191	1.5399	1.6511	1.7561	1.9552	2.1463	2.3333	2.5175	2.7000	2.8811	3.0611	3.2402	3.4186
	h	406.80	1203.8		1219.2	1252.8	1283.4	1312.2	1366.7	1419.6	1472.2	1525.0	1578.4	1632.5	1687.3	1742.8	1799.0
	s	0.6026	1.4994		1.5165	1.5525	1.5836	1.6114	1.6606	1.7044	1.7445	1.7820	1.8174	1.8510	1.8831	1.9138	1.9432
350 (431.73)	Sh				18.27	68.27	118.27	168.27	268.27	368.27	468.27	568.27	668.27	768.27	868.27	968.27	1068.27
	v	0.01912	1.3255		1.3725	1.4913	1.6002	1.7028	1.8970	2.0832	2.2652	2.4445	2.6219	2.7980	2.9730	3.1471	3.3205
	h	409.83	1204.0		1217.5	1251.5	1282.4	1311.4	1366.2	1419.2	1471.8	1524.7	1578.2	1632.3	1687.1	1742.6	1798.9
	s	0.6059	1.4968		1.5119	1.5483	1.5797	1.6077	1.6571	1.7009	1.7411	1.7787	1.8141	1.8477	1.8798	1.9105	1.9400
360 (434.41)	Sh				15.59	65.59	115.59	165.59	265.59	365.59	465.59	565.59	665.59	765.59	865.59	965.59	1065.59
	v	0.01917	1.2891		1.3285	1.4454	1.5521	1.6525	1.8421	2.0237	2.2009	2.3755	2.5482	2.7196	2.8898	3.0592	3.2279
	h	412.81	1204.1		1215.8	1250.3	1281.5	1310.6	1365.6	1418.7	1471.5	1524.4	1577.9	1632.1	1686.9	1742.5	1798.8
	s	0.6092	1.4943		1.5073	1.5441	1.5758	1.6040	1.6536	1.6976	1.7379	1.7754	1.8109	1.8445	1.8766	1.9073	1.9368
380 (439.61)	Sh				10.39	60.39	110.39	160.39	260.39	360.39	460.39	560.39	660.39	760.39	860.39	960.39	1060.39
	v	0.01925	1.2218		1.2472	1.3606	1.4635	1.5598	1.7410	1.9139	2.0825	2.2484	2.4124	2.5750	2.7366	2.8973	3.0572
	h	418.59	1204.4		1212.4	1247.7	1279.5	1309.0	1364.5	1417.9	1470.8	1523.8	1577.4	1631.6	1686.5	1742.2	1798.5
	s	0.6156	1.4894		1.4982	1.5360	1.5683	1.5969	1.6470	1.6911	1.7315	1.7692	1.8047	1.8384	1.8705	1.9012	1.9307

Table II.2-2 Continued

Abs. Press. (psi) (Sat. Temp.)		Sat. Water	Sat. Steam	450	500	550	600	650	700	800	900	1000	1100	1200	1300	1400	1500
400 (444.60)	Sh			5.40	55.40	105.40	155.40	205.40	255.40	355.40	455.40	555.40	655.40	755.40	855.40	955.40	1055.40
	v	0.01934	1.1610	1.1738	1.2841	1.3836	1.4763	1.5646	1.6499	1.8151	1.9759	2.1339	2.2901	2.4450	2.5987	2.7515	2.9037
	h	424.17	1204.6	1208.8	1245.1	1277.5	1307.4	1335.9	1363.4	1417.0	1470.1	1523.3	1576.9	1631.2	1686.2	1741.9	1798.2
	s	0.6217	1.4847	1.4894	1.5282	1.5611	1.5901	1.6163	1.6406	1.6850	1.7255	1.7632	1.7988	1.8325	1.8647	1.8955	1.9250
420 (449.40)	Sh			.60	50.60	100.60	150.60	200.60	250.60	350.60	450.60	550.60	650.60	750.60	850.60	950.60	1050.60
	v	0.01942	1.1057	1.1071	1.2148	1.3113	1.4007	1.4856	1.5676	1.7258	1.8795	2.0304	2.1795	2.3273	2.4739	2.6196	2.7647
	h	429.56	1204.7	1205.2	1242.4	1275.4	1305.8	1334.5	1362.3	1416.2	1469.4	1522.7	1576.4	1630.8	1685.8	1741.6	1798.0
	s	0.6276	1.4802	1.4808	1.5206	1.5542	1.5835	1.6100	1.6345	1.6791	1.7197	1.7575	1.7932	1.8269	1.8591	1.8899	1.9195
440 (454.03)	Sh				45.97	95.97	145.97	195.97	245.97	345.97	445.97	545.97	645.97	745.97	845.97	945.97	1045.97
	v	0.01950	1.0554		1.1517	1.2454	1.3319	1.4138	1.4926	1.6445	1.7918	1.9363	2.0790	2.2203	2.3605	2.4998	2.6384
	h	434.77	1204.8		1239.7	1273.4	1304.2	1333.2	1361.1	1415.3	1468.7	1522.1	1575.9	1630.4	1685.5	1741.2	1797.7
	s	0.6332	1.4759		1.5132	1.5474	1.5772	1.6040	1.6286	1.6734	1.7142	1.7521	1.7878	1.8216	1.8538	1.8847	1.9143
460 (458.50)	Sh				41.50	91.50	141.50	191.50	241.50	341.50	441.50	541.50	641.50	741.50	841.50	941.50	1041.50
	v	0.01959	1.0092		1.0939	1.1852	1.2691	1.3482	1.4242	1.5703	1.7117	1.8504	1.9872	2.1226	2.2569	2.3903	2.5230
	h	439.83	1204.8		1236.9	1271.3	1302.5	1331.8	1360.0	1414.4	1468.0	1521.5	1575.4	1629.9	1685.1	1740.9	1797.4
	s	0.6387	1.4718		1.5060	1.5409	1.5711	1.5982	1.6230	1.6680	1.7089	1.7469	1.7826	1.8165	1.8488	1.8797	1.9093
480 (462.82)	Sh				37.18	87.18	137.18	187.18	237.18	337.18	437.18	537.18	637.18	737.18	837.18	937.18	1037.18
	v	0.01967	0.9668		1.0409	1.1300	1.2115	1.2881	1.3615	1.5023	1.6384	1.7716	1.9030	2.0330	2.1619	2.2900	2.4173
	h	444.75	1204.8		1234.1	1269.1	1300.8	1330.5	1358.8	1413.6	1467.3	1520.9	1574.9	1629.5	1684.7	1740.6	1797.2
	s	0.6439	1.4677		1.4990	1.5346	1.5652	1.5925	1.6176	1.6628	1.7038	1.7419	1.7777	1.8116	1.8439	1.8748	1.9045
500 (467.01)	Sh				32.99	82.99	132.99	182.99	232.99	332.99	432.99	532.99	632.99	732.99	832.99	932.99	1032.99
	v	0.01975	0.9276		0.9919	1.0791	1.1584	1.2327	1.3037	1.4397	1.5708	1.6992	1.8256	1.9507	2.0746	2.1977	2.3200
	h	449.52	1204.7		1231.2	1267.0	1299.1	1329.1	1357.7	1412.7	1466.6	1520.3	1574.4	1629.1	1684.4	1740.3	1796.9
	s	0.6490	1.4639		1.4921	1.5284	1.5595	1.5871	1.6123	1.6578	1.6990	1.7371	1.7730	1.8069	1.8393	1.8702	1.8998
520 (471.07)	Sh				28.93	78.93	128.93	178.93	228.93	328.93	428.93	528.93	628.93	728.93	828.93	928.93	1028.93
	v	0.01982	0.8914		0.9466	1.0321	1.1094	1.1816	1.2504	1.3819	1.5085	1.6323	1.7542	1.8746	1.9940	2.1125	2.2302
	h	454.18	1204.5		1228.3	1264.8	1297.4	1327.7	1356.5	1411.8	1465.9	1519.7	1573.9	1628.7	1684.0	1740.0	1796.7
	s	0.6540	1.4601		1.4853	1.5223	1.5539	1.5818	1.6072	1.6530	1.6943	1.7325	1.7684	1.8024	1.8348	1.8657	1.8954
540 (475.01)	Sh				24.99	74.99	124.99	174.99	224.99	324.99	424.99	524.99	624.99	724.99	824.99	924.99	1024.99
	v	0.01990	0.8577		0.9045	0.9884	1.0640	1.1342	1.2010	1.3284	1.4508	1.5704	1.6880	1.8042	1.9193	2.0336	2.1471
	h	458.71	1204.4		1225.3	1262.5	1295.7	1326.3	1355.3	1410.9	1465.1	1519.1	1573.4	1628.2	1683.6	1739.7	1796.4
	s	0.6587	1.4565		1.4786	1.5164	1.5485	1.5767	1.6023	1.6483	1.6897	1.7280	1.7640	1.7981	1.8305	1.8615	1.8911
560 (478.84)	Sh				21.16	71.16	121.16	171.16	221.16	321.16	421.16	521.16	621.16	721.16	821.16	921.16	1021.16
	v	0.01998	0.8264		0.8653	0.9479	1.0217	1.0902	1.1552	1.2787	1.3972	1.5129	1.6266	1.7388	1.8500	1.9603	2.0699
	h	463.14	1204.2		1222.2	1260.3	1293.9	1324.9	1354.2	1410.0	1464.4	1518.6	1572.9	1627.8	1683.3	1739.4	1796.1
	s	0.6634	1.4529		1.4720	1.5106	1.5431	1.5717	1.5975	1.6438	1.6853	1.7237	1.7598	1.7939	1.8263	1.8573	1.8870
580 (482.57)	Sh				17.43	67.43	117.43	167.43	217.43	317.43	417.43	517.43	617.43	717.43	817.43	917.43	1017.43
	v	0.02006	0.7971		0.8287	0.9100	0.9824	1.0492	1.1125	1.2324	1.3473	1.4593	1.5693	1.6780	1.7855	1.8921	1.9980
	h	467.47	1203.9		1219.1	1258.0	1292.1	1323.4	1353.0	1409.2	1463.7	1518.0	1572.4	1627.4	1682.9	1739.1	1795.9
	s	0.6679	1.4495		1.4654	1.5049	1.5380	1.5668	1.5929	1.6394	1.6811	1.7196	1.7556	1.7898	1.8223	1.8533	1.8831
600 (486.20)	Sh				13.80	63.80	113.80	163.80	213.80	313.80	413.80	513.80	613.80	713.80	813.80	913.80	1013.80
	v	0.02013	0.7697		0.7944	0.8746	0.9456	1.0109	1.0726	1.1892	1.3008	1.4093	1.5160	1.6211	1.7252	1.8284	1.9309
	h	471.70	1203.7		1215.9	1255.6	1290.3	1322.0	1351.8	1408.3	1463.0	1517.4	1571.9	1627.0	1682.6	1738.8	1795.6
	s	0.6723	1.4461		1.4590	1.4993	1.5329	1.5621	1.5884	1.6351	1.6769	1.7155	1.7517	1.7859	1.8184	1.8494	1.8792
650 (494.89)	Sh				5.11	55.11	105.11	155.11	205.11	305.11	405.11	505.11	605.11	705.11	805.11	905.11	1005.11
	v	0.02032	0.7084		0.7173	0.7954	0.8634	0.9254	0.9835	1.0929	1.1969	1.2979	1.3969	1.4944	1.5909	1.6864	1.7813
	h	481.89	1202.8		1207.6	1249.6	1285.7	1318.3	1348.7	1406.0	1461.2	1515.9	1570.7	1625.9	1681.6	1738.0	1794.9
	s	1.6828	1.4381		1.4430	1.4858	1.5207	1.5507	1.5775	1.6249	1.6671	1.7059	1.7422	1.7765	1.8092	1.8403	1.8701
700 (503.08)	Sh					46.92	96.92	146.92	196.92	296.92	396.92	496.92	596.92	696.92	796.92	896.92	996.92
	v	0.02050	0.6556			0.7271	0.7928	0.8520	0.9072	1.0102	1.1078	1.2023	1.2948	1.3858	1.4757	1.5647	1.6530
	h	491.60	1201.8			1243.4	1281.0	1314.6	1345.6	1403.7	1459.4	1514.4	1569.4	1624.8	1680.7	1737.2	1794.3
	s	1.6928	1.4304			1.4726	1.5090	1.5399	1.5673	1.6154	1.6580	1.6970	1.7335	1.7679	1.8006	1.8318	1.8617
750 (510.84)	Sh					39.16	89.16	139.16	189.16	289.16	389.16	489.16	589.16	689.16	789.16	889.16	989.16
	v	0.02069	0.6095			0.6676	0.7313	0.7882	0.8409	0.9386	1.0306	1.1195	1.2063	1.2916	1.3759	1.4592	1.5419
	h	500.89	1200.7			1236.9	1276.1	1310.7	1342.5	1401.5	1457.6	1512.9	1568.2	1623.8	1679.8	1736.4	1793.6
	s	0.7022	1.4232			1.4598	1.4977	1.5296	1.5577	1.6065	1.6494	1.6886	1.7252	1.7598	1.7926	1.8239	1.8538
800 (518.21)	Sh					31.79	81.79	131.79	181.79	281.79	381.79	481.79	581.79	681.79	781.79	881.79	981.79
	v	0.02087	0.5690			0.6151	0.6774	0.7323	0.7828	0.8759	0.9631	1.0470	1.1289	1.2093	1.2885	1.3669	1.4446
	h	509.81	1199.4			1230.1	1271.1	1306.8	1339.3	1399.1	1455.8	1511.4	1566.9	1622.7	1678.9	1735.7	1792.9
	s	0.7111	1.4163			1.4472	1.4869	1.5198	1.5484	1.5980	1.6413	1.6807	1.7175	1.7522	1.7851	1.8164	1.8464
850 (525.24)	Sh					24.76	74.76	124.76	174.76	274.76	374.76	474.76	574.76	674.76	774.76	874.76	974.76
	v	0.02105	0.5330			0.5683	0.6296	0.6829	0.7315	0.8205	0.9034	0.9830	1.0606	1.1366	1.2115	1.2855	1.3588
	h	518.40	1198.0			1223.0	1265.9	1302.8	1336.0	1396.8	1454.0	1510.0	1565.7	1621.6	1678.0	1734.9	1792.3
	s	0.7197	1.4096			1.4347	1.4763	1.5102	1.5396	1.5899	1.6336	1.6733	1.7102	1.7450	1.7780	1.8094	1.8395
900 (531.95)	Sh					18.05	68.05	118.05	168.05	268.05	368.05	468.05	568.05	668.05	768.05	868.05	968.05
	v	0.02123	0.5009			0.5263	0.5869	0.6388	0.6858	0.7713	0.8504	0.9262	0.9998	1.0720	1.1430	1.2131	1.2825
	h	526.70	1196.4			1215.5	1260.6	1298.6	1332.7	1394.4	1452.2	1508.5	1564.4	1620.6	1677.1	1734.1	1791.6
	s	0.7279	1.4032			1.4223	1.4659	1.5010	1.5311	1.5822	1.6263	1.6662	1.7033	1.7382	1.7713	1.8028	1.8329

Table II.2-2   Continued

Abs. Press. (psi) (Sat. Temp.)		Sat. Water	Sat. Steam	550	600	650	700	750	800	850	900	1000	1100	1200	1300	1400	1500
950 (538.39)	Sh			11.61	61.61	111.61	161.61	211.61	261.61	311.61	361.61	461.61	561.61	661.61	761.61	861.61	961.61
	v	0.02141	0.4721	0.4883	0.5485	0.5993	0.6449	0.6871	0.7272	0.7656	0.8030	0.8753	0.9455	1.0142	1.0817	1.1484	1.2143
	h	534.74	1194.7	1207.6	1255.1	1294.4	1329.3	1361.5	1392.0	1421.5	1450.3	1507.0	1563.2	1619.5	1676.2	1733.3	1791.0
	s	0.7358	1.3970	1.4098	1.4557	1.4921	1.5228	1.5500	1.5748	1.5977	1.6193	1.6595	1.6967	1.7317	1.7649	1.7965	1.8267
1000 (544.58)	Sh			5.42	55.42	105.42	155.42	205.42	255.42	305.42	355.42	455.42	555.42	655.42	755.42	855.42	955.42
	v	0.02159	0.4460	0.4535	0.5137	0.5636	0.6080	0.6489	0.6875	0.7245	0.7603	0.8295	0.8966	0.9622	1.0266	1.0901	1.1529
	h	542.55	1192.9	1199.3	1249.3	1290.1	1325.9	1358.7	1389.6	1419.4	1448.5	1505.4	1561.9	1618.4	1675.3	1732.5	1790.3
	s	0.7434	1.3910	1.3973	1.4457	1.4833	1.5149	1.5426	1.5677	1.5908	1.6126	1.6530	1.6905	1.7256	1.7589	1.7905	1.8207
1050 (550.53)	Sh				49.47	99.47	149.47	199.47	249.47	299.47	349.47	449.47	549.47	649.47	749.47	849.47	949.47
	v	0.02177	0.4222		0.4821	0.5312	0.5745	0.6142	0.6515	0.6872	0.7216	0.7881	0.8524	0.9151	0.9767	1.0373	1.0973
	h	550.15	1191.0		1243.4	1285.7	1322.4	1355.8	1387.2	1417.3	1446.6	1503.9	1560.7	1617.4	1674.4	1731.8	1789.6
	s	0.7507	1.3851		1.4358	1.4748	1.5072	1.5354	1.5608	1.5842	1.6062	1.6469	1.6845	1.7197	1.7531	1.7848	1.8151
1100 (556.28)	Sh				43.72	93.72	143.72	193.72	243.72	293.72	343.72	443.72	543.72	643.72	743.72	843.72	943.72
	v	0.02195	0.4006		0.4531	0.5017	0.5440	0.5826	0.6188	0.6533	0.6865	0.7505	0.8121	0.8723	0.9313	0.9894	1.0468
	h	557.55	1189.1		1237.3	1281.2	1318.8	1352.9	1384.7	1415.2	1444.7	1502.4	1559.4	1616.3	1673.5	1731.0	1789.0
	s	0.7578	1.3794		1.4259	1.4664	1.4996	1.5284	1.5542	1.5779	1.6000	1.6410	1.6787	1.7141	1.7475	1.7793	1.8097
1150 (561.82)	Sh				39.18	89.18	139.18	189.18	239.18	289.18	339.18	439.18	539.18	639.18	739.18	839.18	939.18
	v	0.02214	0.3807		0.4263	0.4746	0.5162	0.5538	0.5889	0.6223	0.6544	0.7161	0.7754	0.8332	0.8899	0.9456	1.0007
	h	564.78	1187.0		1230.9	1276.6	1315.2	1349.9	1382.2	1413.0	1442.8	1500.9	1558.1	1615.2	1672.6	1730.2	1788.3
	s	0.7647	1.3738		1.4160	1.4582	1.4923	1.5216	1.5478	1.5717	1.5941	1.6353	1.6732	1.7087	1.7422	1.7741	1.8045
1200 (567.19)	Sh				32.81	82.81	132.81	182.81	232.81	282.81	332.81	432.81	532.81	632.81	732.81	832.81	932.81
	v	0.02232	0.3624		0.4016	0.4497	0.4905	0.5273	0.5615	0.5939	0.6250	0.6845	0.7418	0.7974	0.8519	0.9055	0.9584
	h	571.85	1184.8		1224.2	1271.8	1311.5	1346.9	1379.7	1410.8	1440.9	1499.4	1556.9	1614.2	1671.6	1729.4	1787.6
	s	0.7714	1.3683		1.4061	1.4501	1.4851	1.5150	1.5415	1.5658	1.5883	1.6298	1.6679	1.7035	1.7371	1.7691	1.7996
1300 (577.42)	Sh				22.58	72.58	122.58	172.58	222.58	272.58	322.58	422.58	522.58	622.58	722.58	822.58	922.58
	v	0.02269	0.3299		0.3570	0.4052	0.4451	0.4804	0.5129	0.5436	0.5729	0.6287	0.6822	0.7341	0.7847	0.8345	0.8836
	h	585.58	1180.2		1209.9	1261.9	1303.9	1340.8	1374.6	1406.4	1437.1	1496.3	1554.3	1612.0	1669.8	1727.9	1786.3
	s	0.7843	1.3577		1.3860	1.4340	1.4711	1.5022	1.5296	1.5544	1.5773	1.6194	1.6578	1.6937	1.7275	1.7596	1.7902
1400 (587.07)	Sh				12.93	62.93	112.93	162.93	212.93	262.93	312.93	412.93	512.93	612.93	712.93	812.93	912.93
	v	0.02307	0.3018		0.3176	0.3667	0.4059	0.4400	0.4712	0.5004	0.5282	0.5809	0.6311	0.6798	0.7272	0.7737	0.8195
	h	598.83	1175.3		1194.1	1251.4	1296.1	1334.5	1369.3	1402.0	1433.2	1493.2	1551.8	1609.9	1668.0	1726.3	1785.0
	s	0.7966	1.3474		1.3652	1.4181	1.4575	1.4900	1.5182	1.5436	1.5670	1.6096	1.6484	1.6845	1.7185	1.7508	1.7815
1500 (596.20)	Sh				3.80	53.80	103.80	153.80	203.80	253.80	303.80	403.80	503.80	603.80	703.80	803.80	903.80
	v	0.02346	0.2772		0.2820	0.3328	0.3717	0.4049	0.4350	0.4629	0.4894	0.5394	0.5869	0.6327	0.6773	0.7210	0.7639
	h	611.68	1170.1		1176.3	1240.2	1287.9	1328.0	1364.0	1397.4	1429.2	1490.1	1549.2	1607.7	1666.2	1724.8	1783.7
	s	0.8085	1.3373		1.3431	1.4022	1.4443	1.4782	1.5073	1.5333	1.5572	1.6004	1.6395	1.6759	1.7101	1.7425	1.7734
1600 (604.87)	Sh					45.13	95.13	145.13	195.13	245.13	295.13	395.13	495.13	595.13	695.13	795.13	895.13
	v	0.02387	0.2555			0.3026	0.3415	0.3741	0.4032	0.4301	0.4555	0.5031	0.5482	0.5915	0.6336	0.6748	0.7153
	h	624.20	1164.5			1228.3	1279.4	1321.4	1358.5	1392.8	1425.2	1486.9	1546.6	1605.6	1664.3	1723.2	1782.3
	s	0.8199	1.3274			1.3861	1.4312	1.4667	1.4968	1.5235	1.5478	1.5916	1.6312	1.6678	1.7022	1.7347	1.7657
1700 (613.13)	Sh					36.87	86.87	136.87	186.87	236.87	286.87	386.87	486.87	586.87	686.87	786.87	886.87
	v	0.02428	0.2361			0.2754	0.3147	0.3468	0.3751	0.4011	0.4255	0.4711	0.5140	0.5552	0.5951	0.6341	0.6724
	h	636.45	1158.6			1215.3	1270.5	1314.5	1352.9	1388.1	1421.2	1483.8	1544.0	1603.4	1662.5	1721.7	1781.0
	s	0.8309	1.3176			1.3697	1.4183	1.4555	1.4867	1.5140	1.5388	1.5833	1.6232	1.6601	1.6947	1.7274	1.7585
1800 (621.02)	Sh					28.98	78.98	128.98	178.98	228.98	278.98	378.98	478.98	578.98	678.98	778.98	878.98
	v	0.02472	0.2186			0.2505	0.2906	0.3223	0.3500	0.3752	0.3988	0.4426	0.4836	0.5229	0.5609	0.5980	0.6343
	h	648.49	1152.3			1201.2	1261.1	1307.4	1347.2	1383.3	1417.1	1480.6	1541.4	1601.2	1660.7	1720.1	1779.7
	s	0.8417	1.3079			1.3526	1.4054	1.4446	1.4768	1.5049	1.5302	1.5753	1.6156	1.6528	1.6876	1.7204	1.7516
1900 (628.56)	Sh					21.44	71.44	121.44	171.44	221.44	271.44	371.44	471.44	571.44	671.44	771.44	871.44
	v	0.02517	0.2028			0.2274	0.2687	0.3004	0.3275	0.3521	0.3749	0.4171	0.4565	0.4940	0.5303	0.5656	0.6002
	h	660.36	1145.6			1185.7	1251.3	1300.2	1341.4	1378.4	1412.9	1477.4	1538.8	1599.1	1658.8	1718.6	1778.4
	s	0.8522	1.2981			1.3346	1.3925	1.4338	1.4672	1.4960	1.5219	1.5677	1.6084	1.6458	1.6808	1.7138	1.7451
2000 (635.80)	Sh					14.20	64.20	114.20	164.20	214.20	264.20	364.20	464.20	564.20	664.20	764.20	864.20
	v	0.02565	0.1883			0.2056	0.2488	0.2805	0.3072	0.3312	0.3534	0.3942	0.4320	0.4680	0.5027	0.5365	0.5695
	h	672.11	1138.3			1168.3	1240.9	1292.6	1335.4	1373.5	1408.7	1474.1	1536.2	1596.9	1657.0	1717.0	1771.1
	s	0.8625	1.2881			1.3154	1.3794	1.4231	1.4578	1.4874	1.5138	1.5603	1.6014	1.6391	1.6743	1.7075	1.7389
2100 (642.76)	Sh					7.24	57.24	107.24	157.24	207.24	257.24	357.24	457.24	557.24	657.24	757.24	857.24
	v	0.02615	0.1750			0.1847	0.2304	0.2624	0.2888	0.3123	0.3339	0.3734	0.4099	0.4445	0.4778	0.5101	0.5418
	h	683.79	1130.5			1148.5	1229.8	1284.9	1329.3	1368.4	1404.4	1470.9	1533.6	1594.7	1655.2	1715.4	1775.7
	s	0.8727	1.2780			1.2942	1.3661	1.4125	1.4486	1.4790	1.5060	1.5532	1.5948	1.6327	1.6681	1.7014	1.7330
2200 (649.45)	Sh					.55	50.55	100.55	150.55	200.55	250.55	350.55	450.55	550.55	650.55	750.55	850.55
	v	0.02669	0.1627			0.1636	0.2134	0.2458	0.2720	0.2950	0.3161	0.3545	0.3897	0.4231	0.4551	0.4862	0.5165
	h	695.46	1122.2			1123.9	1218.0	1276.8	1323.1	1363.3	1400.0	1467.6	1530.9	1592.5	1653.3	1713.9	1774.4
	s	0.8828	1.2676			1.2691	1.3523	1.4020	1.4395	1.4708	1.4984	1.5463	1.5883	1.6266	1.6622	1.6956	1.7273
2300 (655.89)	Sh						44.11	94.11	144.11	194.11	244.11	344.11	444.11	544.11	644.11	744.11	844.11
	v	0.02727	0.1513				0.1975	0.2305	0.2566	0.2793	0.2999	0.3372	0.3714	0.4035	0.4344	0.4643	0.4935
	h	707.18	1113.2				1205.3	1268.4	1316.7	1358.1	1395.7	1464.2	1528.3	1590.3	1651.5	1712.3	1773.1
	s	0.8929	1.2569				1.3381	1.3914	1.4305	1.4628	1.4910	1.5397	1.5821	1.6207	1.6565	1.6901	1.7219

Table II.2-2 Continued

Abs. Press. (psi) (Sat. Temp.)		Sat. Water	Sat. Steam	700	750	800	850	900	950	1000	1050	1100	1150	1200	1300	1400	1500
2400 (662.11)	Sh			37.89	87.89	137.89	187.89	237.89	287.89	337.89	387.89	437.89	487.89	537.89	637.89	737.89	837.89
	v	0.02790	0.1408	0.1824	0.2164	0.2424	0.2648	0.2850	0.3037	0.3214	0.3382	0.3545	0.3703	0.3856	0.4155	0.4443	0.4724
	h	718.95	1103.7	1191.6	1259.7	1310.1	1352.8	1391.2	1426.9	1460.9	1493.7	1525.6	1557.0	1588.1	1649.6	1710.8	1771.8
	s	0.9031	1.2460	1.3232	1.3808	1.4217	1.4549	1.4837	1.5095	1.5332	1.5553	1.5761	1.5959	1.6149	1.6509	1.6847	1.7167
2500 (668.11)	Sh			31.89	81.89	131.89	181.89	231.89	281.89	331.89	381.89	431.89	481.89	531.89	631.89	731.89	831.89
	v	0.02859	0.1307	0.1681	0.2032	0.2293	0.2514	0.2712	0.2896	0.3068	0.3232	0.3390	0.3543	0.3692	0.3980	0.4259	0.4529
	h	731.71	1093.3	1176.7	1250.6	1303.4	1347.4	1386.7	1423.1	1457.5	1490.7	1522.9	1554.6	1585.9	1647.8	1709.2	1770.4
	s	0.9139	1.2345	1.3076	1.3701	1.4129	1.4472	1.4766	1.5029	1.5269	1.5492	1.5703	1.5903	1.6094	1.6456	1.6796	1.7116
2600 (673.91)	Sh			26.09	76.09	126.09	176.09	226.09	276.09	326.09	376.09	426.09	476.09	526.09	626.09	726.09	826.09
	v	0.02938	0.1211	0.1544	0.1909	0.2171	0.2390	0.2585	0.2765	0.2933	0.3093	0.3247	0.3395	0.3540	0.3819	0.4088	0.4350
	h	744.47	1082.0	1160.2	1241.1	1296.5	1341.9	1382.1	1419.2	1454.1	1487.7	1520.2	1552.2	1583.7	1646.0	1707.7	1769.1
	s	0.9247	1.2225	1.2908	1.3592	1.4042	1.4395	1.4696	1.4964	1.5208	1.5434	1.5646	1.5848	1.6040	1.6405	1.6746	1.7068
2700 (679.53)	Sh			20.47	70.47	120.47	170.47	220.47	270.47	320.47	370.47	420.47	470.47	520.47	620.47	720.47	820.47
	v	0.03029	0.1119	0.1411	0.1794	0.2058	0.2275	0.2468	0.2644	0.2809	0.2965	0.3114	0.3259	0.3399	0.3670	0.3931	0.4184
	h	757.34	1069.7	1142.0	1231.1	1289.5	1336.3	1377.5	1415.2	1450.7	1484.6	1517.5	1549.8	1581.5	1644.1	1706.1	1767.8
	s	0.9356	1.2097	1.2727	1.3481	1.3954	1.4319	1.4628	1.4900	1.5148	1.5376	1.5591	1.5794	1.5988	1.6355	1.6697	1.7021
2800 (684.96)	Sh			15.04	65.04	115.04	165.04	215.04	265.04	315.04	365.04	415.04	465.04	515.04	615.04	715.04	815.04
	v	0.03134	0.1030	0.1278	0.1685	0.1952	0.2168	0.2358	0.2531	0.2693	0.2845	0.2991	0.3132	0.3268	0.3532	0.3785	0.4030
	h	770.69	1055.8	1121.2	1220.6	1282.2	1330.7	1372.8	1411.2	1447.2	1481.6	1514.8	1547.3	1579.3	1642.2	1704.5	1766.5
	s	0.9468	1.1958	1.2527	1.3368	1.3867	1.4245	1.4561	1.4838	1.5089	1.5321	1.5537	1.5742	1.5938	1.6306	1.6651	1.6975
2900 (690.22)	Sh			9.78	59.78	109.78	159.78	209.78	259.78	309.78	359.78	409.78	459.78	509.78	609.78	709.78	809.78
	v	0.03262	0.0942	0.1138	0.1581	0.1853	0.2068	0.2256	0.2427	0.2585	0.2734	0.2877	0.3014	0.3147	0.3403	0.3649	0.3887
	h	785.13	1039.8	1095.3	1209.6	1274.7	1324.9	1368.0	1407.2	1443.7	1478.5	1512.1	1544.9	1577.0	1640.4	1703.0	1765.2
	s	0.9588	1.1803	1.2283	1.3251	1.3780	1.4171	1.4494	1.4777	1.5032	1.5266	1.5485	1.5692	1.5889	1.6259	1.6605	1.6931
3000 (695.33)	Sh			4.67	54.67	104.67	154.67	204.67	254.67	304.67	354.67	404.67	454.67	504.67	604.67	704.67	804.67
	v	0.03428	0.0850	0.0982	0.1483	0.1759	0.1975	0.2161	0.2329	0.2484	0.2630	0.2770	0.2904	0.3033	0.3282	0.3522	0.3753
	h	801.84	1020.3	1060.5	1197.9	1267.0	1319.0	1363.2	1403.1	1440.2	1475.4	1509.4	1542.4	1574.8	1638.5	1701.4	1763.8
	s	0.9728	1.1619	1.1966	1.3131	1.3692	1.4097	1.4429	1.4717	1.4976	1.5213	1.5434	1.5642	1.5841	1.6214	1.6561	1.6888
3100 (700.28)	Sh				49.72	99.72	149.72	199.72	249.72	299.72	349.72	399.72	449.72	499.72	599.72	699.72	799.72
	v	0.03681	0.0745		0.1389	0.1671	0.1887	0.2071	0.2237	0.2390	0.2533	0.2670	0.2800	0.2927	0.3170	0.3403	0.3628
	h	823.97	993.3		1185.4	1259.1	1313.0	1358.4	1399.0	1436.7	1472.3	1506.6	1539.9	1572.6	1636.7	1699.8	1762.5
	s	0.9914	1.1373		1.3007	1.3604	1.4024	1.4364	1.4658	1.4920	1.5161	1.5384	1.5594	1.5794	1.6169	1.6518	1.6847
3200 (705.08)	Sh				44.92	94.92	144.92	194.92	244.92	294.92	344.92	394.92	444.92	494.92	594.92	694.92	794.92
	v	0.04472	0.0566		0.1300	0.1588	0.1804	0.1987	0.2151	0.2301	0.2442	0.2576	0.2704	0.2827	0.3065	0.3291	0.3510
	h	875.54	931.6		1172.3	1250.9	1306.9	1353.4	1394.9	1433.1	1469.2	1503.8	1537.4	1570.3	1634.8	1698.3	1761.2
	s	1.0351	1.0832		1.2877	1.3515	1.3951	1.4300	1.4600	1.4866	1.5110	1.5335	1.5547	1.5749	1.6126	1.6477	1.6806
3300	Sh																
	v				0.1213	0.1510	0.1727	0.1908	0.2070	0.2218	0.2357	0.2488	0.2613	0.2734	0.2966	0.3187	0.3400
	h				1158.2	1242.5	1300.7	1348.4	1390.7	1429.5	1466.1	1501.0	1534.9	1568.1	1632.9	1696.7	1759.9
	s				1.2742	1.3425	1.3879	1.4237	1.4542	1.4813	1.5059	1.5287	1.5501	1.5704	1.6084	1.6436	1.6767
3400	Sh																
	v				0.1129	0.1435	0.1653	0.1834	0.1994	0.2140	0.2276	0.2405	0.2528	0.2646	0.2872	0.3088	0.3296
	h				1143.2	1233.7	1294.3	1343.4	1386.4	1425.9	1462.9	1498.3	1532.4	1565.8	1631.1	1695.1	1758.2
	s				1.2600	1.3334	1.3807	1.4174	1.4486	1.4761	1.5010	1.5240	1.5456	1.5660	1.6042	1.6396	1.6728
3500	Sh																
	v				0.1048	0.1364	0.1583	0.1764	0.1922	0.2066	0.2200	0.2326	0.2447	0.2563	0.2784	0.2995	0.3198
	h				1127.1	1224.6	1287.8	1338.2	1382.2	1422.2	1459.7	1495.5	1529.9	1563.6	1629.2	1693.6	1757.2
	s				1.2450	1.3242	1.3734	1.4112	1.4430	1.4709	1.4962	1.5194	1.5412	1.5618	1.6002	1.6358	1.6691
3600	Sh																
	v				0.0966	0.1296	0.1517	0.1697	0.1854	0.1996	0.2128	0.2252	0.2371	0.2485	0.2702	0.2908	0.3106
	h				1108.6	1215.3	1281.2	1333.0	1377.9	1418.6	1456.5	1492.6	1527.4	1561.3	1627.3	1692.0	1755.9
	s				1.2281	1.3148	1.3662	1.4050	1.4374	1.4658	1.4914	1.5149	1.5369	1.5576	1.5962	1.6320	1.6654
3800	Sh																
	v				0.0799	0.1169	0.1395	0.1574	0.1729	0.1868	0.1996	0.2116	0.2231	0.2340	0.2549	0.2746	0.2936
	h				1064.2	1195.5	1267.6	1322.4	1369.1	1411.2	1450.1	1487.0	1522.4	1556.8	1623.6	1688.9	1753.0
	s				1.1888	1.2955	1.3517	1.3928	1.4265	1.4558	1.4821	1.5061	1.5284	1.5495	1.5886	1.6247	1.6584
4000	Sh																
	v				0.0631	0.1052	0.1284	0.1463	0.1616	0.1752	0.1877	0.1994	0.2105	0.2210	0.2411	0.2601	0.2783
	h				1007.4	1174.3	1253.4	1311.6	1360.2	1403.6	1443.6	1481.3	1517.3	1552.2	1619.8	1685.7	1750.6
	s				1.1396	1.2754	1.3371	1.3807	1.4158	1.4461	1.4730	1.4976	1.5203	1.5417	1.5812	1.6177	1.6516
4200	Sh																
	v				0.0498	0.0945	0.1183	0.1362	0.1513	0.1647	0.1769	0.1883	0.1991	0.2093	0.2287	0.2470	0.2645
	h				950.1	1151.6	1238.6	1300.4	1351.2	1396.0	1437.1	1475.5	1512.2	1547.6	1616.1	1682.6	1748.0
	s				1.0905	1.2544	1.3223	1.3686	1.4053	1.4366	1.4642	1.4893	1.5124	1.5341	1.5742	1.6109	1.6452
4400	Sh																
	v				0.0421	0.0846	0.1090	0.1270	0.1420	0.1552	0.1671	0.1782	0.1887	0.1986	0.2174	0.2351	0.2519
	h				909.5	1127.3	1223.3	1289.0	1342.0	1388.3	1430.4	1469.7	1507.1	1543.0	1612.3	1679.4	1745.3
	s				1.0556	1.2325	1.3073	1.3566	1.3949	1.4272	1.4556	1.4812	1.5048	1.5268	1.5673	1.6044	1.6389

Table II.2-2 Continued

Abs. Press. (psi) (Sat. Temp.)		Sat. Water	Sat. Steam	Temperature (°F)													
				750	800	850	900	950	1000	1050	1100	1150	1200	1250	1300	1400	1500
4600	Sh																
	v			0.0380	0.0751	0.1005	0.1186	0.1335	0.1465	0.1582	0.1691	0.1792	0.1889	0.1982	0.2071	0.2242	0.2404
	h			883.8	1100.0	1207.3	1277.2	1332.6	1380.5	1423.7	1463.9	1501.9	1538.4	1573.8	1608.5	1676.3	1742.7
	s			1.0331	1.2084	1.2922	1.3446	1.3847	1.4181	1.4472	1.4734	1.4974	1.5197	1.5407	1.5607	1.5982	1.6330
4800	Sh																
	v			0.0355	0.0665	0.0927	0.1109	0.1257	0.1385	0.1500	0.1606	0.1706	0.1800	0.1890	0.1977	0.2142	0.2299
	h			866.9	1071.2	1190.7	1265.2	1323.1	1372.6	1417.0	1458.0	1496.7	1533.8	1569.7	1604.7	1673.1	1740.0
	s			1.0180	1.1835	1.2768	1.3327	1.3745	1.4090	1.4390	1.4657	1.4901	1.5128	1.5341	1.5543	1.5921	1.6272
5000	Sh																
	v			0.0338	0.591	0.0855	0.1038	0.1185	0.1312	0.1425	0.1529	0.1626	0.1718	0.1806	0.1890	0.2050	0.2203
	h			854.9	1042.9	1173.6	1252.9	1313.5	1364.6	1410.2	1452.1	1491.5	1529.1	1565.5	1600.9	1670.0	1737.4
	s			1.0070	1.1593	1.2612	1.3207	1.3645	1.4001	1.4309	1.4582	1.4831	1.5061	1.5277	1.5481	1.5863	1.6216
5200	Sh																
	v			0.0326	0.0531	0.0789	0.0973	0.1119	0.1244	0.1356	0.1458	0.1553	0.1642	0.1728	0.1810	0.1966	0.2114
	h			845.8	1016.9	1156.0	1240.4	1303.7	1356.6	1403.4	1446.2	1486.3	1524.5	1561.3	1597.2	1666.8	1734.7
	s			0.9985	1.1370	1.2455	1.3088	1.3545	1.3914	1.4229	1.4509	1.4762	1.4995	1.5214	1.5420	1.5806	1.6161
5400	Sh																
	v			0.0317	0.0483	0.0728	0.0912	0.1058	0.1182	0.1292	0.1392	0.1485	0.1572	0.1656	0.1736	0.1888	0.2031
	h			838.5	994.3	1138.1	1227.7	1293.7	1348.4	1396.5	1440.3	1481.1	1519.8	1557.1	1593.4	1663.7	1732.1
	s			0.9915	1.1175	1.2296	1.2969	1.3446	1.3827	1.4151	1.4437	1.4694	1.4931	1.5153	1.5362	1.5750	1.6109
5600	Sh																
	v			0.0309	0.0447	0.0672	0.0856	0.1001	0.1124	0.1232	0.1331	0.1422	0.1508	0.1589	0.1667	0.1815	0.1954
	h			832.4	975.0	1119.9	1214.8	1283.7	1340.2	1389.6	1434.3	1475.9	1515.2	1552.9	1589.6	1660.5	1729.5
	s			0.9855	1.1008	1.2137	1.2850	1.3348	1.3742	1.4075	1.4366	1.4628	1.4869	1.5093	1.5304	1.5697	1.6058
5800	Sh																
	v			0.0303	0.0419	0.0622	0.0805	0.0949	0.1070	0.1177	0.1274	0.1363	0.1447	0.1527	0.1603	0.1747	0.1883
	h			827.3	958.8	1101.8	1201.8	1273.6	1332.0	1382.6	1428.3	1470.6	1510.5	1548.7	1585.8	1657.4	1726.8
	s			0.9803	1.0867	1.1981	1.2732	1.3250	1.3658	1.3999	1.4297	1.4564	1.4808	1.5035	1.5248	1.5644	1.6008
6000	Sh																
	v			0.0298	0.0397	0.0579	0.0757	0.0900	0.1020	0.1126	0.1221	0.1309	0.1391	0.1469	0.1544	0.1684	0.1817
	h			822.9	945.1	1084.6	1188.8	1263.4	1323.6	1375.7	1422.3	1465.4	1505.9	1544.6	1582.0	1654.2	1724.2
	s			0.9758	1.0746	1.1833	1.2615	1.3154	1.3574	1.3925	1.4229	1.4500	1.4748	1.4978	1.5194	1.5593	1.5960
6500	Sh																
	v			0.0287	0.0358	0.0495	0.0655	0.0793	0.0909	0.1012	0.1104	0.1188	0.1266	0.1340	0.1411	0.1544	0.1669
	h			813.9	919.5	1046.7	1156.3	1237.8	1302.7	1358.1	1407.3	1452.2	1494.2	1534.1	1572.5	1646.4	1717.6
	s			0.9661	1.0515	1.1506	1.2328	1.2917	1.3370	1.3743	1.4064	1.4347	1.4604	1.4841	1.5062	1.5471	1.5844
7000	Sh																
	v			0.0279	0.0334	0.0438	0.0573	0.0704	0.0816	0.0915	0.1004	0.1085	0.1160	0.1231	0.1298	0.1424	0.1542
	h			806.9	901.8	1016.5	1124.9	1212.6	1281.7	1340.5	1392.2	1439.1	1482.6	1523.7	1563.1	1638.6	1711.1
	s			0.9582	1.0350	1.1243	1.2055	1.2689	1.3171	1.3567	1.3904	1.4200	1.4466	1.4710	1.4938	1.5355	1.5735
7500	Sh																
	v			0.0272	0.0318	0.0399	0.0512	0.0631	0.0737	0.0833	0.0918	0.0996	0.1068	0.1136	0.1200	0.1321	0.1433
	h			801.3	889.0	992.9	1097.7	1188.3	1261.0	1322.9	1377.2	1426.0	1471.0	1513.3	1553.7	1630.8	1704.6
	s			0.9514	1.0224	1.1033	1.1818	1.2473	1.2980	1.3397	1.3751	1.4059	1.4335	1.4586	1.4819	1.5245	1.5632
8000	Sh																
	v			0.0267	0.0306	0.0371	0.0465	0.0571	0.0671	0.0762	0.0845	0.0920	0.0989	0.1054	0.1115	0.1230	0.1338
	h			796.6	879.1	974.4	1074.3	1165.4	1241.0	1305.5	1362.2	1413.0	1459.6	1503.1	1544.5	1623.1	1698.1
	s			0.9455	1.0122	1.0864	1.1613	1.2271	1.2798	1.3233	1.3603	1.3924	1.4208	1.4467	1.4705	1.5140	1.5533
8500	Sh																
	v			0.0262	0.0296	0.0350	0.0429	0.0522	0.0615	0.0701	0.0780	0.0853	0.0919	0.0982	0.1041	0.1151	0.1254
	h			792.7	871.2	959.8	1054.5	1144.0	1221.9	1288.5	1347.5	1400.2	1448.2	1492.9	1535.3	1615.4	1691.7
	s			0.9402	1.0037	1.0727	1.1437	1.2084	1.2627	1.3076	1.3460	1.3793	1.4087	1.4352	1.4597	1.5040	1.5439
9000	Sh																
	v			0.0258	0.0288	0.0335	0.0402	0.0483	0.0568	0.0649	0.0724	0.0794	0.0858	0.0918	0.0975	0.1081	0.1179
	h			789.3	864.7	948.0	1037.6	1125.4	1204.1	1272.1	1333.0	1387.5	1437.1	1482.9	1526.3	1607.9	1685.3
	s			0.9354	0.9964	1.0613	1.1285	1.1918	1.2468	1.2926	1.3323	1.3667	1.3970	1.4243	1.4492	1.4944	1.5349
9500	Sh																
	v			0.0254	0.0282	0.0322	0.0380	0.0451	0.0528	0.0603	0.0675	0.0742	0.0804	0.0862	0.0917	0.1019	0.1113
	h			786.4	859.2	938.3	1023.4	1108.9	1187.7	1256.6	1318.9	1375.1	1426.1	1473.1	1517.3	1600.4	1679.0
	s			0.9310	0.9900	1.0516	1.1153	1.1771	1.2320	1.2785	1.3191	1.3546	1.3858	1.4137	1.4392	1.4851	1.5263
10000	Sh																
	v			0.0251	0.0276	0.0312	0.0362	0.0425	0.0495	0.0565	0.0633	0.0697	0.0757	0.0812	0.0865	0.0963	0.1054
	h			783.8	854.5	930.2	1011.3	1094.2	1172.6	1242.0	1305.3	1362.9	1415.3	1463.4	1508.6	1593.1	1672.8
	s			0.9270	0.9842	1.0432	1.1039	1.1638	1.2185	1.2652	1.3065	1.3429	1.3749	1.4035	1.4295	1.4763	1.5180
10500	Sh																
	v			0.0248	0.0271	0.0303	0.0347	0.0404	0.0467	0.0532	0.0595	0.0656	0.0714	0.0768	0.0818	0.0913	0.1001
	h			781.5	850.5	923.4	1001.0	1081.3	1158.9	1228.4	1292.4	1351.1	1404.7	1453.9	1500.0	1585.8	1666.7
	s			0.9232	0.9790	1.0358	1.0939	1.1519	1.2060	1.2529	1.2946	1.3371	1.3644	1.3937	1.4202	1.4677	1.5100

Table II.2-2   Continued

Abs. Press. (psi) (Sat. Temp.)		Sat. Water	Sat. Steam	\multicolumn{14}{c}{Temperature (°F)}													
				750	800	850	900	950	1000	1050	1100	1150	1200	1250	1300	1400	1500
11000	v			0.0245	0.0267	0.0296	0.0335	0.0386	0.0443	0.0503	0.0562	0.0620	0.0676	0.0727	0.0776	0.0868	0.0952
	h			779.5	846.9	917.5	992.1	1069.9	1146.3	1215.9	1280.2	1339.7	1394.4	1444.6	1491.5	1578.7	1660.6
	s			0.9196	0.9742	1.0292	1.0851	1.1412	1.1945	1.2414	1.2833	1.3209	1.3544	1.3842	1.4112	1.4595	1.5023
11500	v			0.0243	0.0263	0.0290	0.0325	0.0370	0.0423	0.0478	0.0534	0.0588	0.0641	0.0691	0.0739	0.0827	0.0909
	h			777.7	843.8	912.4	984.5	1059.8	1134.9	1204.3	1268.7	1328.8	1384.4	1435.5	1483.2	1571.8	1654.7
	s			0.9163	0.9698	1.0232	1.0772	1.1316	1.1840	1.2308	1.2727	1.3107	1.3446	1.3750	1.4025	1.4515	1.4949
12000	v			0.0241	0.0260	0.0284	0.0317	0.0357	0.0405	0.0456	0.0508	0.0560	0.0610	0.0659	0.0704	0.0790	0.0869
	h			776.1	841.0	907.9	977.8	1050.9	1124.5	1193.7	1258.0	1318.5	1374.7	1426.6	1475.1	1564.9	1648.8
	s			0.9131	0.9657	1.0177	1.0701	1.1229	1.1742	1.2209	1.2627	1.3010	1.3353	1.3662	1.3941	1.4438	1.4877
12500	v			0.0238	0.0256	0.0279	0.0309	0.0346	0.0390	0.0437	0.0486	0.0535	0.0583	0.0629	0.0673	0.0756	0.0832
	h			774.7	838.6	903.9	971.9	1043.1	1115.2	1184.1	1247.9	1308.8	1365.4	1418.0	1467.2	1558.2	1643.1
	s			0.9101	0.9618	1.0127	1.0637	1.1151	1.1653	1.2117	1.2534	1.2918	1.3264	1.3576	1.3860	1.4363	1.4808
13000	v			0.0236	0.0253	0.0275	0.0302	0.0336	0.0376	0.0420	0.0466	0.0512	0.0558	0.0602	0.0645	0.0725	0.0799
	h			773.5	836.3	900.4	966.8	1036.2	1106.7	1174.8	1238.5	1299.6	1356.5	1409.6	1459.4	1551.6	1637.4
	s			0.9073	0.9582	1.0080	1.0578	1.1079	1.1571	1.2030	1.2445	1.2831	1.3179	1.3494	1.3781	1.4291	1.4741
13500	v			0.0235	0.0251	0.0271	0.0297	0.0328	0.0364	0.0405	0.0448	0.0492	0.0535	0.0577	0.0619	0.0696	0.0768
	h			772.3	834.4	897.2	962.2	1030.6	1099.1	1166.5	1229.7	1291.0	1348.1	1401.5	1451.8	1545.2	1631.9
	s			0.9045	0.9548	1.0037	1.0524	1.1014	1.1495	1.1948	1.2361	1.2749	1.3098	1.3415	1.3705	1.4221	1.4675
14000	v			0.0233	0.0248	0.0267	0.0291	0.0320	0.0354	0.0392	0.0432	0.0474	0.0515	0.0555	0.0595	0.0670	0.0740
	h			771.3	832.6	894.3	958.0	1024.5	1092.3	1158.5	1221.4	1283.0	1340.2	1393.8	1444.4	1538.8	1626.5
	s			0.9019	0.9515	0.9996	1.0473	1.0953	1.1426	1.1872	1.2282	1.2671	1.3021	1.3339	1.3631	1.4153	1.4612
14500	v			0.0231	0.0246	0.0264	0.0287	0.0314	0.0345	0.0380	0.0418	0.0458	0.0496	0.0534	0.0573	0.0646	0.0714
	h			770.4	831.0	891.7	954.3	1019.6	1086.2	1151.4	1213.8	1275.4	1332.9	1386.4	1437.3	1532.6	1621.1
	s			0.8994	0.9484	0.9957	1.0426	1.0897	1.1362	1.1801	1.2208	1.2597	1.2949	1.3266	1.3560	1.4087	1.4551
15000	v			0.0230	0.0244	0.0261	0.0282	0.0308	0.0337	0.0369	0.0405	0.0443	0.0479	0.0516	0.0552	0.0624	0.0690
	h			769.6	829.5	889.3	950.9	1015.1	1080.6	1144.9	1206.8	1268.1	1326.0	1379.4	1430.3	1526.4	1615.9
	s			0.8970	0.9455	0.9920	1.0382	1.0846	1.1302	1.1735	1.2139	1.2525	1.2880	1.3197	1.3491	1.4022	1.4491
15500	v			0.0228	0.0242	0.0258	0.0278	0.0302	0.0329	0.0360	0.0393	0.0429	0.0464	0.0499	0.0534	0.0603	0.0668
	h			768.9	828.2	887.2	947.8	1011.1	1075.7	1139.0	1200.3	1261.1	1319.6	1372.8	1423.6	1520.4	1610.8
	s			0.8946	0.9427	0.9886	1.0340	1.0797	1.1247	1.1674	1.2073	1.2457	1.2815	1.3131	1.3424	1.3959	1.4433

^{a} Sh-superheat, °F; v specific volume, ft³/lb; h enthalpy, Btu/lb; s entropy, Btu/°F · lb.

Source: Copyright 1967 ASME (Abridged); reprinted by permission.

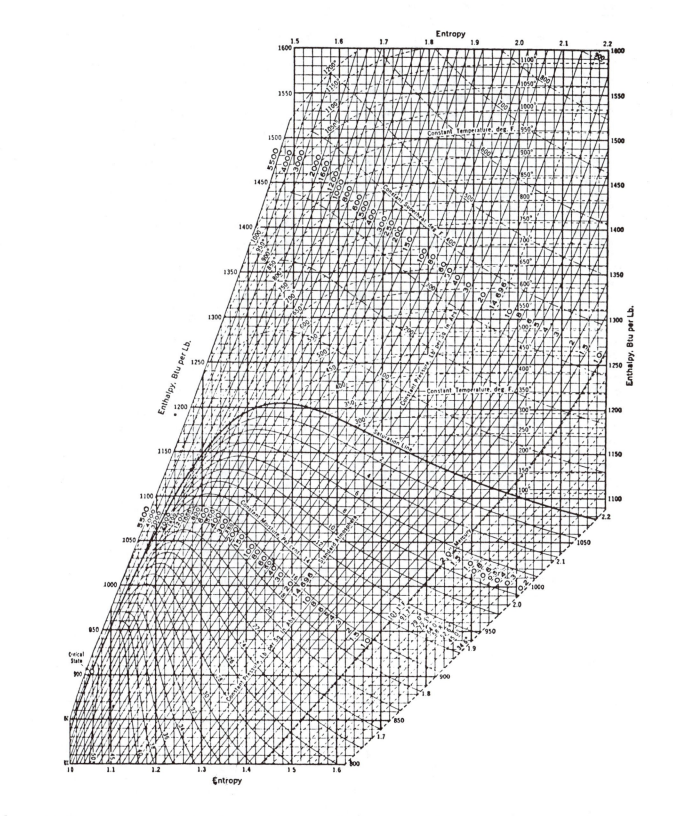

Source: Modified and greatly reduced from J.H. Keenan and F.G. Keyes, *Thermodynamic Properties of Steam*, John Wiley & Sons Inc., New York, 1936; reproduced by permission of the publishers.

Table II.2-4 Thermodynamic Properties of Saturated Freon-12

Temp. T (°F)	Abs. Press. P (lb$_f$/in.^{2})	Specific Volume (ft^3/lb$_m$) Sat. Liquid v_f	Evap. v_{fg}	Sat. Vapor v_g	Enthalpy (Btu/lb$_m$) Sat. Liquid h_f	Evap. h_{fg}	Sat. Vapor h_g	Entropy (Btu/lb$_m$ · °R) Sat. Liquid s_f	Evap. s_{fg}	Sat. Vapor s_g
−130	0.41224	0.009736	70.7203	70.730	−18.609	81.577	62.968	−0.04983	0.24743	0.19760
−120	0.64190	0.009816	46.7312	46.741	−16.565	80.617	64.052	−0.04372	0.23731	0.19359
−110	0.97034	0.009899	31.7671	31.777	−14.518	79.663	65.145	−0.03779	0.22780	0.19002
−100	1.4280	0.009985	21.1541	22.164	−12.466	78.714	66.248	−0.03200	0.21883	0.18683
−90	2.0509	0.010073	15.8109	15.821	−10.409	77.764	67.355	−0.02637	0.21034	0.18398
−80	2.8807	0.010164	11.5228	11.533	−8.3451	76.812	68.467	−0.02086	0.20229	0.18143
−70	3.9651	0.010259	8.5584	8.5687	−6.2730	75.853	69.580	−0.01548	0.19464	0.17916
−60	5.3575	0.010357	6.4670	6.4774	−4.1919	74.885	70.693	−0.01021	0.18716	0.17714
−50	7.1168	0.010459	4.9637	4.9742	−2.1011	73.906	71.805	−0.00506	0.18038	0.17533
−40	9.3076	0.010564	3.8644	3.8750	0	72.913	72.913	0	0.17373	0.17373
−30	11.999	0.010674	3.0478	3.0585	2.1120	71.903	74.015	0.00496	0.16733	0.17229
−20	15.267	0.010788	2.4321	2.4429	4.2357	70.874	75.110	0.00983	0.16119	0.17102
−10	19.189	0.010906	1.9628	1.9727	6.3716	69.824	76.196	0.01462	0.15527	0.16989
0	23.849	0.011030	1.5979	1.6089	8.5207	68.750	77.271	0.01932	0.14956	0.16888
10	29.335	0.011160	1.3129	1.3241	10.684	67.651	78.335	0.02395	0.14403	0.16798
20	35.736	0.011296	1.0875	1.0988	12.863	66.522	79.385	0.02852	0.13867	0.16719
30	43.148	0.011438	0.90736	0.91880	15.058	65.361	80.419	0.03301	0.13347	0.16648
40	51.667	0.011588	0.76198	0.77357	17.273	64.163	81.436	0.03745	0.12841	0.16586
50	61.394	0.011746	0.64362	0.65537	19.507	62.926	82.433	0.04184	0.12346	0.16530
60	72.433	0.011913	0.54648	0.55839	21.766	61.643	83.409	0.04618	0.11861	0.16479
70	84.888	0.012089	0.46609	0.47818	24.050	60.309	84.359	0.05048	0.11386	0.16434
80	98.870	0.012277	0.39907	0.41135	26.365	58.917	85.282	0.05475	0.10917	0.16392
90	114.49	0.012478	0.34281	0.35529	28.713	57.461	86.174	0.05900	0.10453	0.16353
100	131.86	0.012693	0.29525	0.30794	31.100	55.929	87.029	0.06323	0.09992	0.16315
110	151.11	0.012924	0.25577	0.26769	33.531	54.313	87.844	0.06745	0.09534	0.16279
120	172.35	0.013174	0.22019	0.23326	36.013	52.597	88.610	0.07168	0.09073	0.16241
130	195.71	0.013447	0.19019	0.20364	38.553	50.768	89.321	0.07583	0.08609	0.16202
140	221.32	0.013746	0.16424	0.17799	41.162	48.805	89.967	0.08021	0.08138	0.16159
150	249.31	0.014078	0.14156	0.15564	43.850	46.684	90.534	0.08453	0.07657	0.16110
160	279.82	0.014449	0.12159	0.13604	46.633	44.373	91.006	0.08893	0.07260	0.16053
170	313.00	0.014871	0.10386	0.11873	49.529	41.830	91.359	0.09342	0.06643	0.15985
180	349.00	0.015360	0.08794	0.10330	52.562	38.999	91.561	0.09804	0.06096	0.15900
190	387.98	0.015942	0.073476	0.089418	55.769	35.792	91.561	0.10284	0.05511	0.15793
200	430.09	0.016659	0.060069	0.076728	59.203	32.075	91.278	0.10789	0.04862	0.15651
210	475.52	0.017601	0.047242	0.064843	62.959	27.599	90.558	0.11332	0.03921	0.15453
220	524.43	0.018986	0.035154	0.053140	67.246	21.790	89.036	0.11943	0.03206	0.15149
230	577.03	0.021854	0.017581	0.039435	72.893	12.229	85.122	0.12739	0.01773	0.14512
233.6 (critical)	596.9	0.02870	0	0.02870	78.86	0	78.86	0.1359	0	0.1359

Source: Copyright 1955 and 1956 E. I. du Pont de Nemours & Company, Inc. reprinted by permission.

Table II.2-5 Superheated Freon-12

Temp. (°F)	v (5 lb$_f$/in.^{2})	h	s	v (10 lb$_f$/in.^{2})	h	s	v (15 lb$_f$/in.^{2})	h	s
0	8.0611	78.852	0.19663	3.9809	78.246	0.18471	2.6201	77.902	0.17751
20	8.4265	81.309	0.20244	4.1691	81.014	0.19061	2.7494	80.712	0.18349
40	8.7903	84.090	0.20812	4.3556	83.828	0.19635	2.8770	83.561	0.18931
60	9.1528	86.922	0.21367	4.5408	86.689	0.20197	3.0031	86.451	0.19498
80	9.5142	89.806	0.21912	4.7248	89.596	0.20746	3.1281	89.383	0.20051
100	9.8747	92.738	0.22445	4.9079	92.548	0.21283	3.2521	92.357	0.20593
120	10.234	95.717	0.22968	5.0903	95.546	0.21809	3.3754	95.373	0.21122
140	10.594	98.743	0.23481	5.2720	98.586	0.22325	3.4981	98.429	0.21640
160	10.952	101.812	0.23985	5.4533	101.669	0.22830	3.6202	101.525	0.22148
180	11.311	104.925	0.24479	5.6341	104.793	0.23326	3.7419	104.661	0.22646
200	11.668	108.079	0.24964	5.8145	107.957	0.23813	3.8632	107.835	0.23135
220	12.026	111.272	0.25441	5.9946	111.159	0.24291	3.9841	111.046	0.23614

Table II.2-5 Continued

Temp. (°F)	v	h	s	v	h	s	v	h	s
	20 lb$_f$/in.^{2}			25 lb$_f$/in.^{2}			30 lb$_f$/in.^{2}		
20	2.0391	80.403	0.17829	1.6125	80.088	0.17414	1.3278	79.765	0.17065
40	2.1373	83.289	0.18419	1.6932	83.012	0.18012	1.3969	82.730	0.17671
60	2.2340	86.210	0.18992	1.7723	85.965	0.18591	1.4644	85.716	0.18257
80	2.3295	89.168	0.19550	1.8502	88.950	0.19155	1.5306	88.729	0.18826
100	2.4241	92.164	0.20095	1.9271	91.968	0.19704	1.5957	91.770	0.19379
120	2.5179	95.198	0.20628	2.0032	95.021	0.20240	1.6600	94.843	0.19918
140	2.6110	98.270	0.21149	2.0786	98.110	0.20763	1.7237	97.948	0.20445
160	2.7036	101.380	0.21659	2.1535	101.234	0.21276	1.7868	101.086	0.20960
180	2.7957	104.528	0.22159	2.2279	104.393	0.21778	1.8494	104.258	0.21463
200	2.8874	107.712	0.22649	2.3019	107.588	0.22269	1.9116	107.464	0.21957
220	2.9789	110.932	0.23130	2.3756	110.817	0.22752	1.9735	110.702	0.22440
240	3.0700	114.186	0.23602	2.4491	114.080	0.23225	2.0351	113.973	0.22915
	35 lb$_f$/in.^{2}			40 lb$_f$/in.^{2}			50 lb$_f$/in.^{2}		
40	1.1850	82.442	0.17375	1.0258	82.148	0.17112	0.80248	81.540	0.16655
60	1.2442	85.463	0.17968	1.0789	85.206	0.17712	0.84713	84.676	0.17271
80	1.3021	88.504	0.18542	1.1306	88.277	0.18292	0.89025	87.811	0.17862
100	1.3589	91.570	0.19100	1.1812	91.367	0.18854	0.93216	90.953	0.18434
120	1.4148	94.663	0.19643	1.2309	94.480	0.19401	0.97313	94.110	0.18988
140	1.4701	97.785	0.20172	1.2798	97.620	0.19933	1.0133	97.286	0.19527
160	1.5248	100.938	0.20689	1.3282	100.788	0.20453	1.0529	100.485	0.20051
180	1.5789	104.122	0.21195	1.3761	103.985	0.20961	1.0920	103.708	0.20563
200	1.6327	107.338	0.21690	1.4236	107.212	0.21457	1.1307	106.958	0.21064
220	1.6862	110.586	0.22175	1.4707	110.469	0.21944	1.1690	110.235	0.21553
240	1.7394	113.865	0.22651	1.5176	113.757	0.22420	1.2070	113.539	0.22032
260	1.7923	117.175	0.23117	1.5642	117.074	0.22888	1.2447	116.871	0.22502
	60 lb$_f$/in.^{2}			70 lb$_f$/in.^{2}			80 lb$_f$/in.^{2}		
60	0.69210	84.126	0.16892	0.58088	83.552	0.16556	...	...	...
80	0.72964	87.330	0.17497	0.61458	86.832	0.17175	0.52795	86.316	0.16885
100	0.76588	90.528	0.18079	0.64685	90.091	0.17768	0.55734	89.640	0.17489
120	0.80110	93.731	0.18641	0.67803	93.343	0.18339	0.58556	92.945	0.18070
140	0.83551	96.945	0.19186	0.70836	96.597	0.18891	0.61286	96.242	0.18629
160	0.86928	100.776	0.19716	0.73800	99.862	0.19427	0.63943	99.542	0.19170
180	0.90252	103.427	0.20233	0.76708	103.141	0.19948	0.66543	102.851	0.19696
200	0.93531	106.700	0.20736	0.79571	106.439	0.20455	0.69095	106.174	0.20207
220	0.96775	109.997	0.21229	0.82397	109.756	0.20951	0.71609	109.513	0.20706
240	0.99988	113.319	0.21710	0.85191	113.096	0.21435	0.74090	112.872	0.21193
260	1.0318	116.666	0.22182	0.87959	116.459	0.21909	0.76544	116.251	0.21669
280	1.0634	120.039	0.22644	0.90705	119.846	0.22373	0.78975	119.652	0.22135
	90 lb$_f$/in.^{2}			100 lb$_f$/in.^{2}			125 lb$_f$/in.^{2}		
100	0.48749	89.175	0.17234	0.43138	88.694	0.16996	0.32943	87.407	0.16455
120	0.51346	92.536	0.17824	0.45562	92.116	0.17597	0.35086	91.008	0.17087
140	0.53845	95.879	0.18391	0.47881	95.507	0.18172	0.37098	94.537	0.17686
160	0.56268	99.216	0.18938	0.50118	98.884	0.18726	0.39015	98.023	0.18258
180	0.58629	102.557	0.19469	0.52291	102.257	0.19262	0.40857	101.484	0.18807
200	0.60941	105.905	0.19984	0.54413	105.633	0.19782	0.42642	104.934	0.19338
220	0.63213	109.267	0.20486	0.56492	109.018	0.20287	0.44380	108.380	0.19853
240	0.65451	112.644	0.20976	0.58538	112.415	0.20780	0.46081	111.829	0.20353
260	0.67662	116.040	0.21455	0.60554	115.828	0.21261	0.47750	115.287	0.20840
280	0.69849	119.456	0.21923	0.62546	119.258	0.21731	0.49394	118.756	0.21316
300	0.72016	122.892	0.22381	0.64518	122.707	0.22191	0.51016	122.238	0.21780
320	0.74166	126.349	0.22830	0.66472	126.176	0.22641	0.52619	125.737	0.22235
	150 lb$_f$/in.^{2}			175 lb$_f$/in.^{2}			200 lb$_f$/in.^{2}		
120	0.28007	89.800	0.16629	...	...	...	...	...	...
140	0.29845	93.498	0.17256	0.24595	92.373	0.16859	0.20579	91.137	0.16480
160	0.31566	97.112	0.17849	0.26198	96.142	0.17478	0.22121	95.100	0.17130
180	0.33200	100.675	0.18415	0.27697	99.823	0.18062	0.23535	98.921	0.17737
200	0.34769	104.206	0.18958	0.29120	103.447	0.18620	0.24860	102.652	0.18311
220	0.36285	107.720	0.19483	0.30485	107.036	0.19156	0.26117	106.325	0.18860
240	0.37761	111.226	0.19992	0.31804	110.605	0.19674	0.27323	109.962	0.19387
260	0.39203	114.732	0.20485	0.33087	114.162	0.20175	0.28489	113.576	0.19896
280	0.40617	118.242	0.20967	0.34339	117.717	0.20662	0.29623	117.178	0.20390
300	0.42008	121.76!	0.21436	0.35567	121.273	0.21137	0.30730	120.775	0.20870
320	0.43379	125.290	0.21894	0.36773	124.835	0.21599	0.31815	124.373	0.21337
340	0.44733	128.833	0.22343	0.37963	128.407	0.22052	0.32881	127.974	0.21793

Table II.2-5   Continued

Temp. (°F)	v	h	s	v	h	s	v	h	s
	250 lb$_f$/in.^{2}			300 lb$_f$/in.^{2}			400 lb$_f$/in.^{2}		
160	0.16249	92.717	0.16462	...	...	...	...	...	...
180	0.17605	96.925	0.17130	0.13482	94.556	0.16537	...	...	...
200	0.18824	100.930	0.17747	0.14697	98.975	0.17217	0.091005	93.718	0.16092
220	0.19952	104.809	0.18326	0.15774	103.136	0.17838	0.10316	99.046	0.16888
240	0.21014	108.607	0.18877	0.16761	107.140	0.18419	0.11300	103.735	0.17568
260	0.22027	112.351	0.19404	0.17685	111.043	0.18969	0.12163	108.105	0.18183
280	0.23001	116.060	0.19913	0.18562	114.879	0.19495	0.12949	112.286	0.18756
300	0.23944	119.747	0.20405	0.19402	118.670	0.20000	0.13680	116.343	0.19298
320	0.24862	123.420	0.20882	0.20214	122.430	0.20489	0.14372	120.318	0.19814
340	0.25759	127.088	0.21346	0.21002	126.171	0.20963	0.15032	124.235	0.20310
360	0.26639	130.754	0.21799	0.21770	129.900	0.21423	0.15668	128.112	0.20789
380	0.27504	134.423	0.22241	0.22522	133.624	0.21872	0.16285	131.961	0.21258
	500 lb$_f$/in.^{2}			600 lb$_f$/in.^{2}					
220	0.064207	92.397	0.15683	...	...	...			
240	0.077620	99.218	0.16672	0.047488	91.024	0.15335			
260	0.087054	104.526	0.17421	0.061922	99.741	0.16566			
280	0.094923	109.277	0.18072	0.070859	105.637	0.17374			
300	0.10190	113.729	0.18666	0.078059	110.729	0.18053			
320	0.10829	117.997	0.19221	0.084333	115.420	0.18663			
340	0.11426	122.143	0.19746	0.90017	119.871	0.19227			
360	0.11992	126.205	0.20247	0.95289	124.167	0.19757			
380	0.12533	130.207	0.20730	0.10025	128.355	0.20262			
400	0.13054	134.166	0.21196	0.10498	132.466	0.20746			
420	0.13559	138.096	0.21648	0.10952	136.523	0.21213			
440	0.14051	142.004	0.22087	0.11391	140.539	0.21664			

Source: Copyright 1955 and 1956 E. I. du Pont de Nemours & Company, Inc.; reprinted by permission.

Table II.2-6   Thermodynamic Properties of Saturated Ammonia

Temp. (°F)	Abs. Press. P (lb$_f$/in.^{2})	Specific Volume (ft^3/lb$_m$)			Enthalpy (Btu/lb$_m$)			Entropy (Btu/lb$_m$ · °R)		
		Sat. Liquid v_f	Evap. v_{fg}	Sat. Vapor v_g	Sat. Liquid h_f	Evap. h_{fg}	Sat. Vapor h_g	Sat. Liquid s_f	Evap. s_{fg}	Sat. Vapor s_g
−60	5.55	0.0228	44.707	44.73	−21.2	610.8	589.6	−0.0517	1.5286	1.4769
−55	6.54	0.0229	38.357	38.38	−15.9	607.5	591.6	−0.0386	1.5017	1.4631
−50	7.67	0.0230	33.057	33.08	−10.6	604.3	593.7	−0.0256	1.4753	1.4497
−45	8.95	0.0231	28.597	28.62	−5.3	600.9	595.6	−0.0127	1.4495	1.4368
−40	10.41	0.02322	24.837	24.86	0	597.6	597.6	0.000	1.4242	1.4242
−35	12.05	0.02333	21.657	21.68	5.3	594.2	599.5	0.0126	1.3994	1.4120
−30	13.90	0.0235	18.947	18.97	10.7	590.7	601.4	0.0250	1.3751	1.4001
−25	15.98	0.0236	16.636	16.66	16.0	587.2	603.2	0.0374	1.3512	1.3886
−20	18.30	0.0237	14.656	14.68	21.4	583.6	605.0	0.0497	1.3277	1.3774
−15	20.88	0.02381	12.946	12.97	26.7	580.0	606.7	0.0618	1.3044	1.3664
−10	23.74	0.02393	11.476	11.50	32.1	576.4	608.5	0.0738	1.2820	1.3558
−5	26.92	0.02406	10.206	10.23	37.5	572.6	610.1	0.0857	1.2597	1.3454
0	30.42	0.02419	9.092	9.116	42.9	568.9	611.8	0.0975	1.2377	1.3352
5	34.27	0.02432	8.1257	8.150	48.3	565.0	613.3	0.1092	1.2161	1.3253
10	38.51	0.02446	7.2795	7.304	53.8	561.1	614.9	0.1208	1.1949	1.3157
15	43.14	0.02460	6.5374	6.562	59.2	557.1	616.3	0.1323	1.1739	1.3062
20	48.21	0.02474	5.8853	5.910	64.7	553.1	617.8	0.1437	1.1532	1.2969
25	53.73	0.02488	5.3091	5.334	70.2	548.9	619.1	0.1551	1.1328	1.2879
30	59.74	0.02503	4.8000	4.825	75.7	544.8	620.5	0.1663	1.1127	1.2790
35	66.26	0.02518	4.3478	4.373	81.2	540.5	621.7	0.1775	1.0929	1.2704
40	73.32	0.02533	3.9457	3.971	86.8	536.2	623.0	0.1885	1.0733	1.2618
45	80.96	0.02548	3.5885	3.614	92.3	531.8	624.1	0.1996	1.0539	1.2535
50	89.19	0.02564	3.2684	3.294	97.9	527.3	625.2	0.2105	1.0348	1.2453
55	98.06	0.02581	2.9822	3.008	103.5	522.8	626.3	0.2214	1.0159	1.2373
60	107.6	0.02597	2.7250	2.751	109.2	518.1	627.3	0.2322	0.9972	1.2294
65	117.8	0.02614	2.4939	2.520	114.8	513.4	628.2	0.2430	0.9786	1.2216
70	128.8	0.02632	2.2857	2.312	120.5	508.6	629.1	0.2537	0.9603	1.2140
75	140.5	0.02650	2.0985	2.125	126.2	503.7	629.9	0.2643	0.9422	1.2065
80	153.0	0.02668	1.9283	1.955	132.0	498.7	630.7	0.2749	0.9242	1.1991
85	166.4	0.02687	1.7741	1.801	137.8	493.6	631.4	0.2854	0.9064	1.1918
90	180.6	0.02707	1.6339	1.661	143.5	488.5	632.0	0.2958	0.8888	1.1846
95	195.8	0.02727	1.5067	1.534	149.4	483.2	632.6	0.3062	0.8713	1.1775
100	211.9	0.02747	1.3915	1.419	155.2	477.8	633.0	0.3166	0.8539	1.1705
105	228.9	0.02769	1.2853	1.313	161.1	472.3	633.4	0.3269	0.8366	1.1635
110	247.0	0.02790	1.1891	1.217	167.0	466.7	633.7	0.3372	0.8194	1.1566
115	266.2	0.02813	1.0999	1.128	173.0	460.9	633.9	0.3474	0.8023	1.1497
120	286.4	0.02836	1.0186	1.047	179.0	455.0	634.0	0.3576	0.7851	1.1427
125	307.8	0.02860	0.9444	0.973	185.1	448.9	634.0	0.3679	0.7679	1.1358

Source: National Bureau of Standards Circular No. 142, *Tables of Thermodynamic Properties of Ammonia*.

Table II.2-7 Thermodynamic Properties of Superheated Ammonia

Abs. Press. (Sat. Temp.) (lbf/in.²)		0	20	40	60	80	100	120	140	160	180	200	220
10 (−41.34)	v	28.58	29.90	31.20	32.49	33.78	35.07	36.35	37.62	38.90	40.17	41.45	
	h	618.9	629.1	639.3	649.5	659.7	670.0	680.3	690.6	701.1	711.6	722.2	
	s	1.477	1.499	1.520	1.540	1.559	1.578	1.596	1.614	1.631	1.647	1.664	
15 (−27.29)	v	18.92	19.82	20.70	21.58	22.44	23.31	24.17	25.03	25.88	26.74	27.59	
	h	617.2	627.8	638.2	648.5	658.9	669.2	679.6	690.0	700.5	711.1	721.7	
	s	1.427	1.450	1.471	1.491	1.511	1.529	1.548	1.566	1.583	1.599	1.616	
20 (−16.64)	v	14.09	14.78	15.45	16.12	16.78	17.43	18.08	18.73	19.37	20.02	20.66	21.3
	h	615.5	626.4	637.0	647.5	658.0	668.5	678.9	689.4	700.0	710.6	721.2	732.0
	s	1.391	1.414	1.436	1.456	1.476	1.495	1.513	1.531	1.549	1.565	1.582	1.598
25 (−7.96)	v	11.19	11.75	12.30	12.84	13.37	13.90	14.43	14.95	15.47	15.99	16.50	17.02
	h	613.8	625.0	635.8	646.5	657.1	667.7	678.2	688.8	699.4	710.1	720.8	731.6
	s	1.362	1.386	1.408	1.429	1.449	1.468	1.486	1.504	1.522	1.539	1.555	1.571
30 (−.57)	v	9.25	9.731	10.20	10.65	11.10	11.55	11.99	12.43	12.87	13.30	13.73	14.16
	h	611.9	623.5	634.6	645.5	656.2	666.9	677.5	688.2	698.8	709.6	720.3	731.1
	s	1.337	1.362	1.385	1.406	1.426	1.446	1.464	1.482	1.500	1.517	1.533	1.550
35 (5.89)	v		8.287	8.695	9.093	9.484	9.869	10.25	10.63	11.00	11.38	11.75	12.12
	h		622.0	633.4	644.4	655.3	666.1	676.8	687.6	698.3	709.1	719.9	730.7
	s		1.341	1.365	1.386	1.407	1.427	1.445	1.464	1.481	1.498	1.515	1.531
40 (11.66)	v		7.203	7.568	7.922	8.268	8.609	8.945	9.278	9.609	9.938	10.27	10.59
	h		620.4	632.1	643.4	654.4	665.3	676.1	686.9	697.7	708.5	719.4	730.3
	s		1.323	1.347	1.369	1.390	1.410	1.429	1.447	1.465	1.482	1.499	1.515
45 (16.87)	v		6.363	6.694	7.014	7.326	7.632	7.934	8.232	8.528	8.822	9.115	9.406
	h		618.8	630.8	642.3	653.5	664.6	675.5	686.3	697.2	708.0	718.9	729.9
	s		1.307	1.331	1.354	1.375	1.395	1.414	1.433	1.450	1.468	1.485	1.501
50 (21.67)	v			5.988	6.280	6.564	6.843	7.117	7.387	7.655	7.921	8.185	8.448
	h			629.5	641.2	652.6	663.7	674.7	685.7	696.6	707.5	718.5	729.4
	s			1.317	1.340	1.361	1.382	1.401	1.420	1.437	1.455	1.472	1.488
60 (30.21)	v			4.933	5.184	5.428	5.665	5.897	6.126	6.352	6.576	6.798	7.019
	h			626.8	639.0	650.7	662.1	673.3	684.4	695.5	706.5	717.5	728.6
	s			1.291	1.315	1.337	1.358	1.378	1.397	1.415	1.432	1.449	1.466
70 (37.7)	v	4.401	4.615	4.822	5.025	5.224	5.420	5.615	5.807	6.187	6.563		
	h	636.6	648.7	660.4	671.8	683.1	694.3	705.5	716.6	738.9	761.4		
	s	1.294	1.317	1.338	1.358	1.377	1.395	1.413	1.430	1.463	1.494		
80 (44.4)	v	3.812	4.005	4.190	4.371	4.548	4.722	4.893	5.063	5.398	5.73		
	h	634.3	646.7	658.7	670.4	681.8	693.2	704.4	715.6	738.1	760.7		
	s	1.275	1.298	1.320	1.340	1.360	1.378	1.396	1.414	1.447	1.478		
90 (50.47)	v	3.353	3.529	3.698	3.862	4.021	4.178	4.332	4.484	4.785	5.081		
	h	631.8	644.7	657.0	668.9	680.5	692.0	703.4	714.7	737.3	760.0		
	s	1.257	1.281	1.304	1.325	1.344	1.363	1.381	1.400	1.432	1.464		
100 (56.05)	v	2.985	3.149	3.304	3.454	3.600	3.743	3.883	4.021	4.294	4.562		
	h	629.3	642.6	655.2	667.3	679.2	690.8	702.3	713.7	736.5	759.4		
	s	1.241	1.266	1.289	1.310	1.331	1.349	1.368	1.385	1.419	1.451		
140 (74.79)	v		2.166	2.288	2.404	2.515	2.622	2.727	2.830	3.030	3.227	3.420	
	h		633.8	647.8	661.1	673.7	686.0	698.0	709.9	733.3	756.7	780.0	
	s		1.214	1.240	1.263	1.284	1.305	1.324	1.342	1.376	1.409	1.440	
180 (89.78)	v			1.720	1.818	1.910	1.999	2.084	2.167	2.328	2.484	2.637	
	h			639.9	654.4	668.0	681.0	693.6	705.9	730.1	753.9	777.7	
	s			1.199	1.225	1.248	1.269	1.289	1.308	1.344	1.377	1.408	
220 (102.42)	v				1.443	1.525	1.601	1.675	1.745	1.881	2.012	2.140	2.265
	h				647.3	662.0	675.8	689.1	701.9	726.8	751.1	775.3	799.5
	s				1.192	1.217	1.239	1.260	1.280	1.317	1.351	1.383	1.413
240 (108.09)	v				1.302	1.380	1.452	1.521	1.587	1.714	1.835	1.954	2.069
	h				643.5	658.8	673.1	686.7	699.8	725.1	749.8	774.1	798.4
	s				1.176	1.203	1.226	1.248	1.268	1.305	1.339	1.371	1.402
260 (113.42)	v				1.182	1.257	1.326	1.391	1.453	1.572	1.686	1.796	1.904
	h				639.5	655.6	670.4	684.4	697.7	723.4	748.4	772.9	797.4
	s				1.162	1.189	1.213	1.235	1.256	1.294	1.329	1.361	1.391
280 (118.45)	v				1.078	1.151	1.217	1.279	1.339	1.451	1.558	1.661	1.762
	h				635.4	652.2	667.6	681.9	695.6	721.8	747.0	771.7	796.3
	s				1.147	1.176	1.201	1.224	1.245	1.283	1.318	1.351	1.382

Source: National Bureau of Standards Circular No. 142, *Tables of Thermodynamic Properties of Ammonia*.

Table II.2-8 Thermodynamic Properties of Saturated Nitrogen

Temp. (°R)	Abs. Press. P (lb$_f$/in.^{2})	Specific Volume (ft^3/lb$_m$) Sat. Liquid v_f	Evap. v_{fg}	Sat. Vapor v_g	Enthalpy (Btu lb$_m$) Sat. Liquid h_f	Evap. h_{fg}	Sat. Vapor h_g	Entropy (Btu/lb$_m$ · °R) Sat. Liquid s_f	Evap. s_{fg}	Sat. Vapor s_g
113.670	1.813	0.01845	23.793	23.812	0.000	92.891	92.891	0.00000	0.81720	0.81720
120.000	3.337	0.01875	13.570	13.589	3.113	91.224	94.337	0.02661	0.76020	0.78681
130.000	7.654	0.01929	6.3208	6.3401	8.062	88.432	96.494	0.06610	0.68025	0.74634
139.255	14.696	0.01984	3.4592	3.4791	12.639	85.668	98.306	0.09992	0.61518	0.71510
140.000	15.425	0.01989	3.3072	3.3271	13.006	85.436	98.443	0.10253	0.61026	0.71279
150.000	28.120	0.02056	1.8865	1.9071	17.945	82.179	100.124	0.13628	0.54786	0.68414
160.000	47.383	0.02132	1.1469	1.1682	22.928	78.458	101.476	0.16795	0.49093	0.65888
170.000	74.991	0.02219	0.7299	0.7521	28.045	74.383	102.427	0.19829	0.43754	0.63584
180.000	112.808	0.02323	0.4789	0.5021	33.411	69.478	102.889	0.22805	0.38599	0.61404
190.000	162.761	0.02449	0.3190	0.3435	39.153	63.582	102.735	0.25789	0.33464	0.59254
200.000	226.853	0.02613	0.2119	0.2380	45.283	56.474	101.757	0.28780	0.28237	0.57017
210.000	307.276	0.02845	0.1354	0.1639	52.061	47.474	99.536	0.31894	0.22607	0.54501
220.000	406.739	0.03249	0.0750	0.1075	60.336	34.536	94.872	0.35494	0.15698	0.51192
226.000	477.104	0.03806	0.0374	0.0755	68.123	20.423	88.546	0.38789	0.09037	0.47826

Source: Abstracted from National Bureau of Standards Technical Note 129A, *The Thermodynamic Properties of Nitrogen from 114 to 540 R between 1.0 and 3000 psia, Supplement A (British Units)*, by Thomas R. Strobridge.

Table II.2-9 Thermodynamic Properties of Superheated Nitrogen

Temp. (°R)	v (ft^3/lb$_m$)	h (Btu/lb$_m$)	s (Btu/lb$_m$ · °R)	v (ft^3/lb$_m$)	h (Btu/lb$_m$)	s (Btu/lb$_m$ · °R)	v (ft^3/lb$_m$)	h (Btu/lb$_m$)	s (Btu/lb$_m$ · °R)
	14.7 lb$_f$/in.^{2}			20 lb$_f$/in.^{2}			50 lb$_f$/in.^{2}		
150	3.7782	101.086	0.7343	2.7395	100.715	0.7109			
200	5.1366	113.849	0.8078	3.7538	113.625	0.7852	1.4534	112.315	0.715
250	6.4680	126.443	0.8640	4.7397	126.293	0.8418	1.8663	125.432	0.774
300	7.7876	138.958	0.9096	5.7138	138.850	0.8875	2.2662	138.239	0.821
350	9.1015	151.432	0.9481	6.6820	151.351	0.9261	2.6599	150.896	0.860
400	10.412	163.882	0.9814	7.6469	163.821	0.9594	3.0502	163.471	0.893
450	11.721	176.319	1.0107	8.6098	176.271	0.9887	3.4385	175.997	0.923
500	13.028	188.748	1.0368	9.5714	188.710	1.0149	3.8255	188.492	0.949
540	14.073	198.690	1.0560	10.340	198.657	1.0341	4.1344	198.474	0.968
	100 lb$_f$/in.^{2}			200 lb$_f$/in.^{2}			500 lb$_f$/in.^{2}		
200	0.6834	109.931	0.6585	0.2884	103.911	0.5875			
250	0.9078	123.948	0.7212	0.4272	120.763	0.6631	0.1321	108.378	0.560
300	1.1169	137.205	0.7696	0.5420	135.076	0.7153	0.1966	128.168	0.633
350	1.3192	150.133	0.8094	0.6490	148.589	0.7570	0.2473	143.838	0.681
400	1.5181	162.888	0.8435	0.7522	161.718	0.7921	0.2932	158.205	0.720
450	1.7149	175.540	0.8733	0.8532	174.630	0.8225	0.3368	171.933	0.752
500	1.9103	188.129	0.8998	0.9529	187.408	0.8494	0.3790	185.292	0.780
540	2.0660	198.170	0.9192	1.0319	197.567	0.8690	0.4120	195.807	0.801
	1000 lb$_f$/in.^{2}			2000 lb$_f$/in.^{2}			3000 lb$_f$/in.^{2}		
250	0.0384	78.126	0.4145	0.0286	70.290	0.3596	0.0261	69.719	0.337
300	0.0828	115.224	0.5514	0.0398	97.820	0.4599	0.0321	93.216	0.422
350	0.1150	135.789	0.6150	0.0552	122.614	0.5366	0.0403	116.066	0.493
400	0.1417	152.487	0.6597	0.0699	142.869	0.5908	0.0493	136.883	0.549
450	0.1659	167.637	0.6954	0.0833	160.406	0.6321	0.0582	155.522	0.593
500	0.1887	181.969	0.7256	0.0958	176.411	0.6659	0.0667	172.551	0.628
540	0.2063	193.069	0.7470	0.1053	188.526	0.6892	0.0732	185.361	0.653

Source: Abstracted from National Bureau of Standards Technical Note 129A, *The Thermodynamic Properties of Nitrogen from 114 to 540 R between 1.0 and 3000 psia, Supplement A (British Units)*, by Thomas R. Strobridge.

Table II.2-10   Pressure/Enthalpy Diagram, FREON-22

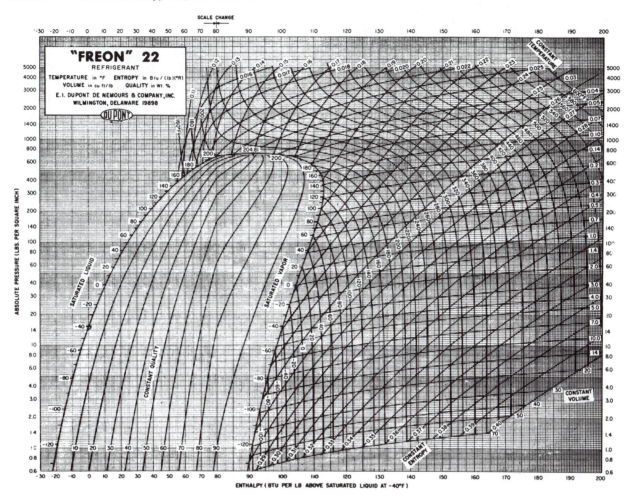

Table II.2-11   Pressure/Enthalpy Diagram, FREON-11

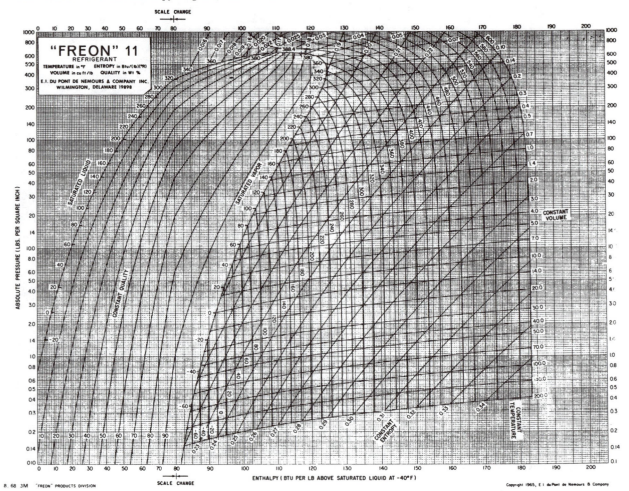

SCALE CHANGE   ENTHALPY (BTU PER LB ABOVE SATURATED LIQUID AT -40°F)   Copyright 1965, E I duPont de Nemours & Company

Table II.2-12 Thermodynamic Properties of Air at Low Pressure

T (°R)	h (Btu/lb$_m$)	u (Btu/lb$_m$)	$s°$ (Btu/lb$_m$ · °R)	T (°R)	h (Btu/lb$_m$)	u (Btu/lb$_m$)	$s°$ (Btu/lb$_m$ · °R)
200	47.67	33.96	0.36303	1200	291.30	209.05	0.79628
220	52.46	37.38	0.38584	1220	296.41	212.78	0.80050
240	57.25	40.80	0.40666	1240	301.52	216.53	0.80466
260	62.03	44.21	0.42582	1260	306.65	220.28	0.80876
280	66.82	47.63	0.44356	1280	311.79	224.05	0.81280
300	71.61	51.04	0.46007	1300	316.94	227.83	0.81680
320	76.40	54.46	0.47550	1320	322.11	231.63	0.82075
340	81.18	57.87	0.49002	1340	327.29	235.43	0.82464
360	85.97	61.29	0.50369	1360	332.48	239.25	0.82848
380	90.75	64.70	0.51663	1380	337.68	243.08	0.83229
400	95.53	68.11	0.52890	1400	342.90	246.93	0.83604
420	100.32	71.52	0.54058	1420	348.14	250.79	0.83975
440	105.11	74.93	0.55172	1440	353.37	254.66	0.84341
460	109.90	78.36	0.56235	1460	358.63	258.54	0.84704
480	114.69	81.77	0.57255	1480	363.89	262.44	0.85062
500	119.48	85.20	0.58233	1500	369.17	266.34	0.85416
520	124.27	88.62	0.59173	1520	374.47	270.26	0.85767
540	129.06	92.04	0.60078	1540	379.77	274.20	0.86113
560	133.86	95.47	0.60950	1560	385.08	278.13	0.86456
580	138.66	98.90	0.61793	1580	390.40	282.09	0.86794
600	143.47	102.34	0.62607	1600	395.74	286.06	0.87130
620	148.28	105.78	0.63395	1620	401.09	290.04	0.87462
640	153.09	109.21	0.64159	1640	406.45	294.03	0.87791
660	157.92	112.67	0.64902	1660	411.82	298.02	0.88116
680	162.73	116.12	0.65621	1680	417.20	302.04	0.88439
700	167.56	119.58	0.66321	1700	422.59	306.06	0.88758
720	172.39	123.04	0.67002	1720	428.00	310.09	0.89074
740	177.23	126.51	0.67665	1740	433.41	314.13	0.89387
760	182.08	129.99	0.68312	1760	438.83	318.18	0.89697
780	186.94	133.47	0.68942	1780	444.26	322.24	0.90003
800	191.81	136.97	0.69558	1800	449.71	326.32	0.90308
820	196.69	140.47	0.70160	1820	455.17	330.40	0.90609
840	201.56	143.98	0.70747	1840	460.63	334.50	0.90908
860	206.46	147.50	0.71323	1860	466.12	338.61	0.91203
880	211.35	151.02	0.71886	1880	471.60	342.73	0.91497
900	216.26	154.57	0.72438	1900	477.09	346.85	0.91788
920	221.18	158.12	0.72979	1920	482.60	350.98	0.92076
940	226.11	161.68	0.73509	1940	488.12	355.12	0.92362
960	231.06	165.26	0.74030	1960	493.64	359.28	0.92645
980	236.02	168.83	0.74540	1980	499.17	363.43	0.92926
1000	240.98	172.43	0.75042	2000	504.71	367.61	0.93205
1020	245.97	176.04	0.75536	2020	510.26	371.79	0.93481
1040	250.95	179.66	0.76019	2040	515.82	375.98	0.93756
1060	255.96	183.29	0.76496	2060	521.39	380.18	0.94026
1080	260.97	186.93	0.76964	2080	526.97	384.39	0.94296
1100	265.99	190.58	0.77426	2100	532.55	388.60	0.94564
1120	271.03	194.25	0.77880	2120	538.15	392.83	0.94829
1140	276.08	197.94	0.78326	2140	543.74	397.05	0.95092
1160	281.14	201.63	0.78767	2160	549.35	401.29	0.95352
1180	286.21	205.33	0.79201	2200	560.59	409.78	0.95868
				2300	588.82	431.16	0.97123
				2400	617.22	452.70	0.98331

Source: Abridged from Table 1 in Joseph H. Keenan and Joseph Kaye, *Gas Tables*, John Wiley & Sons, Inc., New York; copyright 1948.

Table II.2-14   Generalized Entropy Correction Chart

Table II.2-13.   Generalized Enthalpy Correction Chart

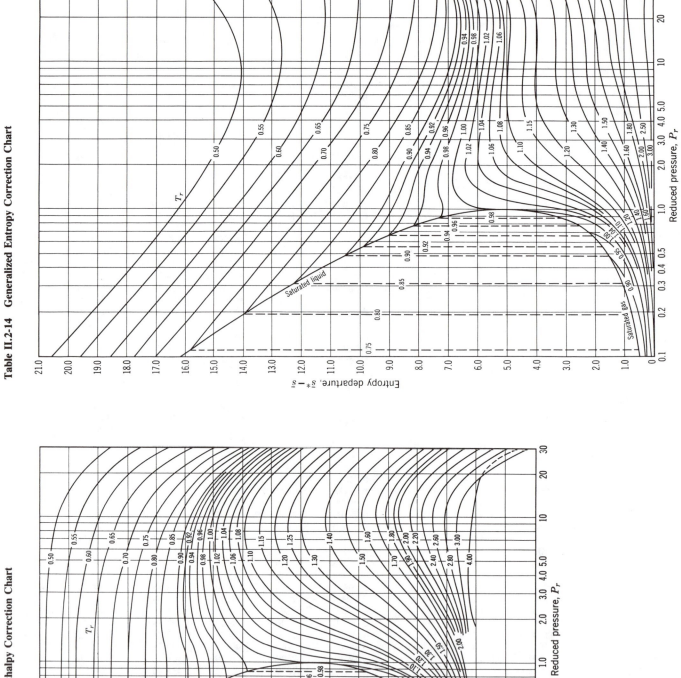

Table II.3 Critical Constants

Substance	Formula	Molecular Weight	Temperature °K	Temperature °R	Pressure atm	Pressure $lb_f/in.^2$	Volume (ft³/lb mol)
Ammonia	NH_3	17.03	405.5	729.8	111.3	1636	1.16
Argon	Ar	39.944	151	272	48.0	705	1.20
Bromine	Br_2	159.832	584	1052	102	1500	2.17
Carbon dioxide	CO_2	44.01	304.2	547.5	72.9	1071	1.51
Carbon monoxide	CO	28.01	133	240	34.5	507	1.49
Chlorine	Cl_2	70.914	417	751	76.1	1120	1.99
Deuterium (normal)	D_2	4.00	38.4	69.1	16.4	241	—
Helium	He	4.003	5.3	9.5	2.26	33.2	0.926
Helium	He	3.00	3.34	6.01	1.15	16.9	—
Hydrogen (normal)	H_2	2.016	33.3	59.9	12.8	188.1	1.04
Krypton	Kr	83.7	209.4	376.9	54.3	798	1.48
Neon	Ne	20.183	44.5	80.1	26.9	395	0.668
Nitrogen	N_2	28.016	126.2	227.1	33.5	492	1.44
Nitrous oxide	N_2O	44.02	309.7	557.1	71.7	1054	1.54
Oxygen	O_2	32.00	154.8	278.6	50.4	736	1.25
Sulfur dioxide	So_2	64.06	430.7	775.2	77.8	1143	1.95
Water	H_2O	18.016	647.4	1165.3	218.3	3204	0.90
Xenon	Xe	131.3	289.75	521.55	58.0	852	1.90
Benzene	C_6H_6	78.11	562	1012	48.6	714	4.17
n-Butane	C_4H_{10}	58.120	425.2	765.2	37.5	551	4.08
Carbon tetrachloride	CCl_4	153.81	556.4	1001.5	45.0	661	4.42
Chloroform	$CHCl_3$	119.39	536.6	965.8	54.0	794	3.85
Dichlorodifluoromethane	CCl_2F_2	120.92	384.7	692.4	39.6	582	3.49
Dichlorofluoromethane	$CHCl_2F$	102.93	451.7	813.0	51.0	749	3.16
Ethane	C_2H_6	30.068	305.5	549.8	48.2	708	2.37
Ethyl alcohol	C_2H_5OH	46.07	516.0	929.0	63.0	926	2.68
Ethylene	C_2H_4	28.052	282.4	508.3	50.5	742	1.99
n-Hexane	C_6H_{14}	86.172	507.9	914.2	29.9	439	5.89
Methane	CH_4	16.012	191.1	343.9	45.8	673	1.59
Methyl alcohol	CH_3OH	32.04	513.2	923.7	78.5	1154	1.89
Methyl chloride	CH_3Cl	50.49	416.3	749.3	65.9	968	2.29
Propane	C_3H_8	44.094	370.0	665.9	42.0	617	3.20
Propene	C_3H_6	42.078	365.0	656.9	45.6	670	2.90
Propyne	C_3H_4	40.062	401	722	52.8	776	—
Trichlorofluoromethane	CCl_3F	137.38	471.2	848.1	43.2	635	3.97

Source: K. A. Kobe and R. E. Lynn, Jr., *Chemical Reviews*, Vol. 52 (1953), pp. 117–236.

Table II.4-1 Enthalpy of Formation, Gibbs Function of Formation, and Absolute Entropy of Various Substances at 77 °F (25°C) and 1 Atm Pressure

Substance	Formula	M	State	$\bar{h}_f^\circ$ cal/g mol	$\bar{h}_f^\circ$ Btu/lb mol	$\bar{g}_f^\circ$ cal/g mol	$\bar{g}_f^\circ$ Btu/lb mol	$\bar{s}^\circ$ cal/g mol °K, Btu/lb mol °R
Carbon monoxide[a]	CO	28.011	Gas	−26,417	−47,551	−32,783	−59,009	47.214
Carbon dioxide[a]	CO_2	44.011	Gas	−94,054	−169,297	−94,265	−169,677	51.072
Water[a,b]	H_2O	18.016	Gas	−57,798	−104,036	−54,636	−98,345	45.106
Water[b]	H_2O	18.016	Liquid	−68,317	−122,971	−56,690	−102,042	16.716
Methane[a]	CH_4	16.043	Gas	−17,895	−32,211	−12,145	−21,861	44.490
Acetylene[a]	C_2H_2	26.038	Gas	54,190	97,542	49,993	89,987	48.004
Ethene[a]	C_2H_4	28.054	Gas	12,496	22,493	16,281	29,306	52.447
Ethane[c]	C_2H_6	30.070	Gas	−20,236	−36,425	−7,860	−14,148	54.85
Propane[c]	C_3H_8	44.097	Gas	−24,820	−44,676	−5,614	−10,105	64.51
Butane[c]	C_4H_{10}	58.124	Gas	−30,150	−54,270	−4,100	−7,380	74.12
Octane[c]	C_8H_{18}	114.23	Gas	−49,820	−89,680	3,950	7,110	111.55
Octane[c]	C_8H_{18}	114.23	Liquid	−59,740	−107,532	1,580	2,844	86.23
Carbon[a] (graphite)	C	12.011	Solid	0	0	0	0	1.359

[a] From JANAF *Thermochemical Tables*, Thermal Research Laboratory, The Dow Chemical Company, Midland, Mich.
[b] From *Circular 500*, National Bureau of Standards, Washington D.C.
[c] From F. D. Rossini et al., *API Research Project 44.*

Table II.4-2   Enthalpy of Combustion of Some Hydrocarbons at 25°C (77°F)

Hydrocarbon	Formula	Liquid H₂O in Products (Negative of Higher Heating Value)		Vapor H₂O in Products (Negative of Lower Heating Value)	
		Liquid Hydrocarbon (Btu/lb$_m$ fuel)	Gaseous Hydrocarbon (Btu/lb$_m$ fuel)	Liquid Hydrocarbon (Btu/lb$_m$ fuel)	Gaseous Hydrocarbon (Btu/lb$_m$ fuel)
Parafin family					
Methane	CH_4		−23.861		−21.502
Ethane	C_2H_6		−22.304		−20.416
Propane	C_3H_8	−21.490	−21.649	−19.773	−19.929
Butane	C_4H_{10}	−21.134	−21.293	−19.506	−19.665
Pentane	C_5H_{12}	−20.914	−21.072	−19.340	−19.499
Hexane	C_6H_{14}	−20.772	−20.930	−19.233	−19.391
Heptane	C_7H_{16}	−20.668	−20.825	−19.157	−19.314
Octane	C_8H_{18}	−20.591	−20.747	−19.100	−19.256
Decane	$C_{10}H_{22}$	−20.484	−20.638	−19.020	−19.175
Dodecane	$C_{12}H_{26}$	−20.410	−20.564	−18.964	−19.118
Olefin family					
Ethene	C_2H_4		−21.626		−20.276
Propene	C_3H_6		−21.033		−19.683
Butene	C_4H_8		−20.833		−19.483
Pentene	C_5H_{10}		−20.696		−19.346
Hexene	C_6H_{12}		−20.612		−19.262
Heptene	C_7H_{14}		−20.552		−19.202
Octene	C_8H_{16}		−20.507		−19.157
Nonene	C_9H_{18}		−20.472		−19.122
Decene	$C_{10}H_{20}$		−20.444		−19.094
Alkylbenzene family					
Benzene	C_6H_6	−17.985	−18.172	−17.259	−17.446
Methylbenzene	C_7H_8	−18.247	−18.423	−17.424	−17.601
Ethylbenzene	C_8H_{10}	−18.488	−18.659	−17.596	−17.767
Propylbenzene	C_9H_{12}	−18.667	−18.832	−17.722	−17.887
Butylbenzene	$C_{10}H_{14}$	−18.809	−18.970	−17.823	−17.984

Table II.4-3   Enthalpy of Formation at 25°C, Ideal Gas Enthalpy, and Absolute Entropy at 1 Atm Pressure

Temperature		Nitrogen, Diatomic (N₂) (Sept. 30, 1965) $(\bar{h}_f^\circ)_{298} = 0$ cal/g mol = 0 Btu/lb mol, $M = 28.016$			Nitrogen Monatomic (N) (Mar. 31, 1961) $(\bar{h}_f^\circ)_{298} = 112.965$ cal/g mol = 203.337 Btu/lb mol, $M = 14.008$		
°K	°R	$\bar{h}^\circ - \bar{h}_{298}^\circ$ (cal/g mol)	$\bar{s}^\circ$ (cal/g mol · °K, Btu/lb mol · °R)	$\bar{h}^\circ - \bar{h}_{537}^\circ$ (Btu/lb mol)	$\bar{h}^\circ - \bar{h}_{298}^\circ$ (cal/g mol)	$\bar{s}^\circ$ (cal/g mol · °K, Btu/lb mol · °R)	$\bar{h}^\circ - \bar{h}_{537}^\circ$ (Btu/lb mol)
0	0	−2,072	0	−3,730	−1,481	0	−2,666
100	180	−1,379	38.170	−2,483	−984	31.187	−1,771
200	360	−683	42.992	−1,229	−488	34.631	−878
298	537	0	45.770	0	0	36.614	0
300	540	13	45.813	23	9	36.645	16
400	720	710	47.818	1,278	506	38.074	911
500	900	1,413	49.386	2,543	1,003	39.183	1,805
600	1,080	2,125	50.685	3,825	1,500	40.089	2,700
700	1,260	2,853	51.806	5,135	1,996	40.855	3,593
800	1,440	3,596	52.798	6,473	2,493	41.518	4,487
900	1,620	4,335	53.692	7,839	2,990	42.103	5,382
1000	1,800	5,129	54.507	9,232	3,487	42.627	6,277
1100	1,980	5,917	55.258	10,651	3,984	43.100	7,171
1200	2,160	6,718	55.955	12,092	4,481	43.532	8,066
1300	2,340	7,529	56.604	13,552	4,977	43.930	8,959
1400	2,520	8,350	57.212	15,030	5,474	44.298	9,853
1500	2,700	9,179	57.784	16,522	5,971	44.641	10,748
1600	2,880	10,015	58.324	18,027	6,468	44.962	11,642
1700	3,060	10,858	58.835	19,544	6,965	45.263	12,537

Table II.4-3 Continued

Temperature		Oxygen, Diatomic (O_2) (Sept. 30, 1965) $(\bar{h}_f^\circ)_{298} = 0$ cal/g mol = 0 Btu/lb mol, $M = 32.00$			Oxygen, Monatomic (O) (June 30, 1962) $(\bar{h}_f^\circ)_{298} = 59{,}559$ cal/g mol = 107,206 Btu/lb mol, $M = 16.00$		
°K	°R	$\bar{h}^\circ - \bar{h}_{298}^\circ$ (cal/g mol)	$\bar{s}^\circ$ (cal/g mol · °K, Btu/lb mol · °R)	$\bar{h}^\circ - \bar{h}_{537}^\circ$ (Btu/lb mol)	$\bar{h}^\circ - \bar{h}_{298}^\circ$ (cal/g mol)	$\bar{s}^\circ$ (cal/g mol · °K, Btu/lb mol · °R)	$\bar{h}^\circ - \bar{h}_{537}^\circ$ (Btu/lb mol)
1800	3,240	11,707	59.320	21,073	7,461	45.547	13,430
1900	3,420	12,560	59.782	22,608	7,958	45.815	14,324
2000	3,600	13,418	60.222	24,152	8,455	46.070	15,219
2100	3,780	14,280	60.642	25,704	8,952	46.313	16,114
2200	3,960	15,146	61.045	27,263	9,449	46.544	17,903
2300	4,140	16,015	61.431	28,827	9,946	46.765	17,008
2400	4,320	16,886	61.802	30,395	10,444	46.977	18,799
2500	4,500	17,761	62.159	31,970	10,941	47.180	19,694
2600	4,680	18,638	62.503	33,548	11,439	47.375	20,590
2700	4,860	19,517	62.835	35,131	11,938	47.563	21,488
2800	5,040	20,398	63.155	36,716	12,437	47.745	22,387
2900	5,220	21,280	63.465	38,304	12,936	47.920	23,285
3000	5,400	22,165	63.765	39,897	13,437	48.090	24,187
3200	5,760	23,939	64.337	43,090	14,441	48.414	25,994
3400	6,120	25,719	64.877	46,294	15,451	48.720	27,812
3600	6,480	27,505	65.387	49,509	16,469	49.011	29,644
3800	6,840	29,295	65.871	52,731	17,495	49.288	31,491
4000	7,200	31,089	66.331	55,960	18,531	49.554	33,356
4200	7,560	32,888	66.770	59,198	19,580	49.810	35,244
4400	7,920	34,690	67.189	62,442	20,643	50.057	37,157
4600	8,280	36,496	67.591	65,693	21,721	50.297	39,098
4800	8,640	38,306	67.976	68,951	22,816	50.530	41,069
5000	9,000	40,119	68.346	72,214	23,928	50.757	43,070
5200	9,360	41,935	68.702	75,483	25,059	50.978	45,106
5400	9,720	43,755	69.045	78,759	26,210	51.195	47,178
5600	10,180	45,579	69.377	82,042	27,380	51.408	49,284
5800	10,540	47,406	69.698	85,331	28,570	51.617	51,426
6000	10,800	49,237	70.008	88,627	29,780	51.822	53,604
0	0	−2,075	0	−3,735	−1,608	0	−2,894
100	180	−1,381	41.395	−2,486	−1,080	32.466	−1,944
200	360	−685	46.218	−1,233	−523	36.340	−941
298	537	0	49.004	0	0	38.468	0
300	540	13	49.047	23	10	38.501	18
400	720	724	51.091	1,303	528	39.991	950
500	900	1,455	52.722	2,619	1,038	41.131	1,868
600	1,080	2,210	54.098	3,978	1,544	42.054	2,779
700	1,260	2,988	55.297	5,378	2,048	42.831	3,686
800	1,440	3,786	56.361	6,815	2,550	43.501	4,590
900	1,620	4,600	57.320	8,280	3,052	44.092	5,494
1000	1,800	5,427	58.192	9,769	3,552	44.619	6,394
1100	1,980	6,266	58.991	11,279	4,051	45.095	7,292
1200	2,160	7,114	59.729	12,805	4,551	45.529	8,192
1300	2,340	7,971	60.415	14,348	5,049	45.928	9,088
1400	2,520	8,835	61.055	15,903	5,548	46.298	9,986
1500	2,700	9,706	61.656	17,471	6,046	46.642	10,883
1600	2,880	10,583	62.222	19,049	6,544	46.963	11,779
1700	3,060	11,465	62.757	20,637	7,042	47.265	12,676
1800	3,240	12,354	63.265	22,237	7,540	47.550	13,572
1900	3,420	13,249	63.749	23,848	8,038	47.819	14,468
2000	3,600	14,149	64.210	25,468	8,536	48.074	15,365
2100	3,780	15,054	64.652	27,097	9,034	48.317	16,261
2200	3,960	15,966	65.076	28,739	9,532	48.549	17,158
2300	4,140	16,882	65.483	30,388	10,029	48.770	18,052
2400	4,320	17,804	65.876	32,047	10,527	48.982	18,949
2500	4,500	18,732	66.254	33,718	11,026	49.185	19,847
2600	4,680	19,664	66.620	35,395	11,524	49.381	20,743
2700	4,860	20,602	66.974	37,084	12,023	49.569	21,641
2800	5,040	21,545	67.317	38,781	12,522	49.751	22,540
2900	5,220	22,493	67.650	40,487	13,022	49.926	23,440
3000	5,400	23,446	67.973	42,203	13,522	50.096	24,340
3200	5,760	25,365	68.592	45,657	14,524	50.419	26,143
3400	6,120	27,302	69.179	49,144	15,529	50.724	27,952
3600	6,480	29,254	69.737	52,657	16,537	51.012	29,767
3800	6,840	31,221	70.269	56,198	17,549	51.285	31,588

Table II.4-3   Continued

Temperature		Carbon Dioxide (CO_2) (Sept. 30, 1965) $(\bar{h}_f^\circ)_{298} = -94,054$ cal/g mol $= -169,297$ Btu/lb mol, $M = 44.011$			Carbon Monoxide (CO) (Sept 30, 1965) $(\bar{h}_f^\circ)_{298} = -26,417$ cal/g mol $= -47,551$ Btu/lb mol, $M = 28.011$		
°K	°R	$\bar{h}^\circ - \bar{h}_{298}^\circ$ (cal/g mol)	$\bar{s}^\circ$ (cal/g mol · °K, Btu/lb mol · °R)	$\bar{h}^\circ - \bar{h}_{537}^\circ$ (Btu/lb mol)	$\bar{h}^\circ - \bar{h}_{298}^\circ$ (cal/g mol)	$\bar{s}^\circ$ (cal/g mol · °K, Btu/lb mol · °R)	$\bar{h}^\circ - \bar{h}_{537}^\circ$ (Btu/lb mol)
4000	7,200	33,201	70.776	59,762	18,565	51.546	33,417
4200	7,560	35,193	71.262	63,347	19,586	51.795	35,255
4400	7,920	37,196	71.728	66,953	20,611	52.033	37,100
4600	8,280	39,208	72.176	70,574	21,641	52.262	38,954
4800	8,640	41,229	72.606	74,212	22,676	52.482	40,817
5000	9,000	43,257	73.019	77,863	23,715	52.695	42,687
5200	9,360	45,292	73.418	81,526	24,760	52.899	44,568
5400	9,720	47,332	73.803	85,198	25,809	53.097	46,456
5600	10,180	49,377	74.175	88,879	26,863	53.289	48,353
5800	10,540	51,426	74.535	92,567	27,921	53.475	50,258
6000	10,800	53,479	74.883	96,262	28,984	53.655	25,171
0	0	−2,238	0	−4,028	−2,072	0	−3,730
100	180	−1,543	42.758	−2,777	−1,379	39.613	−2,483
200	360	−816	47.769	−1,469	−683	44.435	−1,229
298	537	0	51.072	0	0	47.214	0
300	540	16	51.127	29	13	47.257	23
400	720	958	53.830	1,724	711	49.265	1,280
500	900	1,987	56.122	3,577	1,417	50.841	2,551
600	1,080	3,087	58.126	5,557	2,137	52.152	3,847
700	1,260	4,245	59.910	7,641	2,873	53.287	5,171
800	1,440	5,453	61.522	9,815	3,627	54.293	6,529
900	1,620	6,702	62.992	12,064	4,397	55.200	7,915
1000	1,800	7,984	64.344	14,371	5,183	56.028	9,329
1100	1,980	9,296	65.594	16,733	5,983	56.790	10,769
1200	2,160	10,632	66.756	19,138	6,794	57.496	12,229
1300	2,340	11,988	67.841	21,578	7,616	58.154	13,709
1400	2,520	13,362	68.859	24,052	8,446	58.769	15,203
1500	2,700	14,750	69.817	26,550	9,285	59.348	16,713
1600	2,880	16,152	70.722	29,074	10,130	59.893	18,234
1700	3,060	17,565	71.578	31,617	10,980	60.409	19,764
1800	3,240	18,987	72.391	34,177	11,836	60.898	21,305
1900	3,420	20,418	73.165	36,752	12,697	61.363	22,855
2000	3,600	21,857	73.903	39,343	13,561	61.807	24,410
2100	3,780	23,303	74.608	41,945	14,430	62.230	25,974
2200	3,960	24,755	75.284	44,559	15,301	62.635	27,542
2300	4,140	26,212	75.931	47,182	16,175	63.024	29,115
2400	4,320	27,674	76.554	49,813	17,052	63.397	30,694
2500	4,500	29,141	77.153	52,454	17,931	63.756	32,276
2600	4,680	30,613	77.730	55,103	18,813	64.102	33,863
2700	4,860	32,088	78.286	57,758	19,696	64.435	35,453
2800	5,040	33,567	78.824	60,421	20,582	64.757	37,048
2900	5,220	35,049	79.344	63,088	21,469	65.069	38,644
3000	5,400	36,535	79.848	65,763	22,357	65.370	40,243
3200	5,760	39,515	80.810	71,127	24,139	65.945	43,450
3400	6,120	42,507	81.717	76,513	25,927	66.487	46,669
3600	6,480	45,508	82.574	81,914	27,719	66.999	49,894
3800	6,840	48,518	83.388	87,332	29,516	67.485	53,129
4000	7,200	51,538	84.162	92,768	31,316	67.946	56,369
4200	7,560	54,566	84.901	98,219	33,121	68.387	59,618
4400	7,920	57,601	85.607	103,682	34,930	68.807	62,874
4600	8,280	60,644	86.284	109,159	36,741	69.210	66,134
4800	8,640	63,695	86.933	114,651	38,557	69.596	69,403
5000	9,000	66,753	87.557	120,155	40,375	69.967	72,675
5200	9,360	69,819	88.158	125,674	42,196	70.325	75,953
5400	9,720	72,893	88.738	131,207	44,021	70.669	79,238
5600	10,180	75,976	89.299	136,757	45,849	71.001	82,528
5800	10,540	79,068	89.841	142,322	47,679	71.332	85,822
6000	10,800	82,168	90.367	147,902	49,513	71.663	89,123

Table II.4-3   Continued

Temperature		Water (H₂O) (Mar. 31, 1961)			Hydroxyl (OH) (Mar. 31, 1966)		
		$(\bar{h}_f^\circ)_{298} = -57,798$ cal/g mol $= -104,036$ Btu/lb mol, $M = 18.016$			$(\bar{h}_f^\circ)_{298} = 9432$ cal/g mol $= 16,978$ Btu/lb mol, $M = 17.008$		
°K	°R	$\bar{h}^\circ - \bar{h}_{298}^\circ$ (cal/g mol)	$\bar{s}^\circ$ (cal/g mol·°K, Btu/lb mol·°R)	$\bar{h}^\circ - \bar{h}_{537}^\circ$ (Btu/lb mol)	$\bar{h}^\circ - \bar{h}_{298}^\circ$ (cal/g mol)	$\bar{s}^\circ$ (cal/g mol·°K, Btu/lb mol·°R)	$\bar{h}^\circ - \bar{h}_{537}^\circ$ (Btu/lb mol)
0	0	−2,367	0	−4,261	−2,192	0	−3,946
100	180	−1,581	36.396	−2,846	−1,467	35.726	−2,641
200	360	−784	41.916	−1,411	−711	40.985	−1,280
298	537	0	45.106	0	0	43.880	0
300	540	15	45.155	27	13	43.925	23
400	720	825	47.484	1,485	725	45.974	1,305
500	900	1,654	49.334	2,977	1,432	47.551	2,578
600	1,080	2,509	50.891	4,516	2,137	48.837	3,847
700	1,260	3,390	52.249	6,102	2,845	49.927	5,121
800	1,440	4,300	53.464	7,740	3,556	50.877	6,401
900	1,620	5,240	54.570	9,432	4,275	51.724	7,695
1000	1,800	6,209	55.592	11,176	5,003	52.491	9,005
1100	1,980	7,210	56.545	12,978	5,742	53.195	10,336
1200	2,160	8,240	57.441	14,832	6,491	53.847	11,684
1300	2,340	9,298	58.288	16,736	7,252	54.455	13,054
1400	2,520	10,384	59.092	18,691	8,023	55.027	14,441
1500	2,700	11,495	59.859	20,691	8,805	55.566	15,849
1600	2,880	12,630	60.591	22,734	9,596	56.077	17,273
1700	3,060	13,787	61.293	24,817	10,397	56.563	18,715
1800	3,240	14,964	61.965	26,935	11,207	57.025	20,173
1900	3,420	16,160	62.612	29,088	12,024	57.467	21,643
2000	3,600	17,373	63.234	31,271	12,849	57.891	23,128
2100	3,780	18,602	63.834	33,484	13,681	58.296	24,626
2200	3,960	19,846	64.412	35,723	14,520	58.686	26,136
2300	4,140	21,103	64.971	37,985	15,364	59.062	27,655
2400	4,320	22,372	65.511	40,270	16,214	59.424	29,185
2500	4,500	23,653	66.034	42,575	17,069	59.773	30,724
2600	4,680	24,945	66.541	44,901	17,929	60.110	32,272
2700	4,860	26,246	67.032	47,243	18,794	60.436	33,829
2800	5,040	27,556	67.508	49,601	19,662	60.752	35,392
2900	5,220	28,875	67.971	51,975	20,535	61.058	36,963
3000	5,400	30,201	68.421	54,362	21,411	61.355	38,540
3200	5,760	32,876	69.284	59,177	23,174	61.924	41,713
3400	6,120	35,577	70.102	64,039	24,949	62.462	44,908
3600	6,480	38,300	70.881	68,940	26,735	62.973	48,123
3800	6,840	41,043	71.622	73,877	28,532	63.458	51,358
4000	7,200	43,805	72.331	78,849	30,338	63.922	54,608
4200	7,560	46,583	73.008	83,849	32,153	64.364	57,875
4400	7,920	49,375	73.658	88,875	33,976	64.788	61,157
4600	8,280	52,181	74.281	93,926	35,807	65.195	64,453
4800	8,640	55,000	74.881	99,000	37,644	65.586	67,759
5000	9,000	57,829	75.459	104,092	39,489	65.963	71,080
5200	9,360	60,669	76.016	109,204	41,340	66.326	74,412
5400	9,720	63,520	76.553	114,336	43,197	66.676	77,755
5600	10,180	66,381	77.074	119,486	45,060	67.015	81,108
5800	10,540	69,251	77.577	124,652	46,929	67.343	84,472
6000	10,800	72,131	78.065	129,836	48,803	67.661	87,845

Table II.4-3   Continued

Temperature		Hydrogen, Diatomic (H₂) (Mar. 31, 1961) $(\bar{h}_f^\circ)_{298} = 0$ cal/g mol $= 0$ Btu/lb mol, $M = 2.016$			Hydrogen, Monatomic (H) (Sept. 30, 1965) $(\bar{h}_f^\circ) = 52,100$ cal/g mol $= 93,780$ Btu/lb mol, $M = 1.008$		
°K	°R	$\bar{h}^\circ - \bar{h}^\circ_{298}$ (cal/g mol)	$\bar{s}^\circ$ (cal/g mol · °K, Btu/lb mol · °R)	$\bar{h}^\circ - \bar{h}^\circ_{537}$ (Btu/lb mol)	$\bar{h}^\circ - \bar{h}^\circ_{298}$ (cal/g mol)	$\bar{s}^\circ$ (cal/g mol · °K, Btu/lb mol · °R)	$\bar{h}^\circ - \bar{h}^\circ_{537}$ (Btu/lb mol)
0	0	−2,024	0	−3,643	−1,481	0	−2,666
100	180	−1,265	24.387	−2,277	−984	21.965	−1,771
200	360	−662	28.520	−1,192	−488	25.408	−878
298	537	0	31.208	0	0	27.392	0
300	540	13	31.251	23	9	27.423	16
400	720	707	33.247	1,273	506	28.852	911
500	900	1,406	34.806	2,531	1,003	29.961	1,805
600	1,080	2,106	36.082	3,791	1,500	30.867	2,700
700	1,260	2,808	37.165	5,054	1,996	31.632	3,593
800	1,440	3,514	38.107	6,325	2,493	32.296	4,487
900	1,620	4,226	38.946	7,607	2,990	32.881	5,382
1000	1,800	4,944	39.702	8,899	3,487	33.404	6,277
1100	1,980	5,670	40.394	10,206	3,984	33.878	7,171
1200	2,160	6,404	41.033	11,527	4,481	34.310	8,066
1300	2,340	7,148	41.628	12,866	4,977	34.708	8,959
1400	2,520	7,902	42.187	14,224	5,474	35.076	9,853
1500	2,700	8,668	42.716	15,602	5,971	35.419	10,748
1600	2,880	9,446	43.217	17,003	6,468	35.739	11,642
1700	3,060	10,233	43.695	18,419	6,965	36.041	12,537
1800	3,240	11,030	44.150	19,854	7,461	36.325	13,430
1900	3,420	11,836	44.586	21,305	7,958	36.593	14,324
2000	3,600	12,651	45.004	22,772	8,455	36.848	15,219
2100	3,780	13,475	45.406	24,255	8,952	37.090	16,114
2200	3,960	14,307	45.793	25,753	9,449	37.322	17,088
2300	4,140	15,146	46.166	27,263	9,945	37.542	17,901
2400	4,320	15,993	46.527	28,787	10,442	37.754	18,796
2500	4,500	16,848	46.875	30,326	10,939	37.957	19,690
2600	4,680	17,708	47.213	31,874	11,436	38.152	20,585
2700	4,860	18,575	47.540	33,435	11,933	38.339	21,479
2800	5,040	19,448	47.857	35,006	12,430	38.520	22,374
2900	5,220	20,326	48.166	36,587	12,926	38.694	23,267
3000	5,400	21,210	48.465	38,178	13,423	38.862	24,161
3200	5,760	22,992	49.040	41,386	14,417	39.183	25,951
3400	6,120	24,794	49.586	44,629	15,410	39.484	27,738
3600	6,400	26,616	50.107	47,909	16,404	39.768	29,527
3800	6,840	28,457	50.605	51,223	17,398	40.037	31,316
4000	7,200	30,317	51.082	54,571	18,391	40.292	33,104
4200	7,560	32,194	51.540	57,949	19,385	40.534	34,893
4400	7,920	34,088	51.980	61,358	20,379	40.765	36,682
4600	8,280	35,999	52.405	64,798	21,372	40.986	38,470
4800	8,640	37,926	52.815	68,267	22,366	41.198	40,259
5000	9,000	39,868	53.211	71,762	23,359	41.400	42,046
5200	9,360	41,825	53.595	75,285	24,353	41.595	43,835
5400	9,720	43,797	53.967	78,835	25,347	41.783	45,625
5600	10,180	45,783	54.328	82,409	26,340	41.963	47,412
5800	10,540	47,783	54.679	86,009	27,334	42.138	49,201
6000	10,800	49,796	55.020	89,633	28,328	42.306	50,990

Source: Thermochemical data are from the JANAF *Thermochemical Tables*, Thermal Research Laboratory, The Dow Chemical Company, Midland Mich. The date each table was issued is indicated.

Table II.5-1 Thermal Properties of Metals

Metal	Properties at 68°F				k(Btu/hr·ft·°F)									
	ρ (lbm/ft³)	C_p (Btu/lbm·°F)	k (Btu/hr·ft·°F)	α (ft²/hr)	−148°F −100°C	32°F 0°C	212°F 100°C	392°F 200°C	572°F 300°C	752°F 400°C	1112°F 600°C	1472°F 800°C	1832°F 1000°C	2192°F 1200°C
Aluminum														
Pure	169	0.214	132	3.665	134	132	132	132	132					
Al-Cu (Duralumin): 94–96 Al, 3–5 Cu, trace Mg	174	0.211	95	2.580	73	92	105	112						
Al-Mg (Hydronalium): 91–95 Al, 5–9 Mg	163	0.216	65	1.860	54	63	73	82						
Al-Si (Silumin): 87 Al, 13 Si	166	0.208	95	2.773	86	94	101	107						
Al-Si (Silumin, copper bearing): 86.5 Al, 12.5 Si, 1 Cu	166	0.207	79	2.311	69	79	83	88	93					
Al-Si (Alusil): 78–80 Al, 20–22 Si	164	0.204	93	2.762	83	91	97	101	103					
Al-Mg-Si: 97 Al, 1 Mg, 1 Si, 1 Mn	169	0.213	102	2.859	—	101	109	118						
Lead	710	0.031	20	0.924	21.3	20.3	19.3	18.2	17.2					
Iron														
Pure	493	0.108	42	0.785	50	42	39	36	32	28	23	21	20	21
Wrought iron (C < 0.50%)	490	0.11	34	0.634	—	34	33	30	28	26	21	19	19	19
Cast iron (C ≈ 4%)	454	0.10	30	0.666										
Steel (C_{max} ≈ 1.5%)														
Carbon steel (C ≈ 0.5%)	489	0.111	31	0.570	—	32	30	28	26	24	20	17	17	18
1.0%	487	0.113	25	0.452	—	25	25	24	23	21	19	17	16	17
1.5%	484	0.116	21	0.376	—	21	21	21	20	19	18	16	16	17
Nickel steel (Ni ≈ 0%)	493	0.108	42	0.785										
10%	496	0.11	15	0.279										
20%	499	0.11	11	0.204										
30%	504	0.11	7	0.118										
40%	510	0.11	6	0.108										
50%	516	0.11	8	0.140										
60%	523	0.11	11	0.182										
70%	531	0.11	15	0.258										
80%	538	0.11	20	0.344										
90%	547	0.11	27	0.452										
100%	556	0.106	52	0.892										
Invar (Ni ≈ 36%)	508	0.11	6.2	0.108										
Chrome steel (Cr = 0%)	493	0.108	42	0.785	50	42	39	36	32	28	23	21	20	21
1%	491	0.11	35	0.645	—	36	32	30	27	24	21	19	19	
2%	491	0.11	30	0.559	—	31	28	26	24	22	19	18	18	
5%	489	0.11	23	0.430	—	23	22	21	21	19	17	17	17	17
10%	486	0.11	18	0.344	—	18	18	18	17	17	16	16	17	
20%	480	0.11	13	0.258	—	13	13	13	13	14	14	15	17	
30%	476	0.11	11	0.204										
Cr-Ni (chrome-nickel)														
15 Cr, 10 Ni	491	0.11	11	0.204										
18 Cr, 8 Ni (V2A)	488	0.11	9.4	0.172	—	9.4	10	10	11	11	13	15	18	
20 Cr, 15 Ni	489	0.11	8.7	0.161										
25 Cr, 20 Ni	491	0.11	7.4	0.140										
Ni-Cr (nickel-chrome)														
80 Ni, 15 Cr	532	0.11	10	0.172										
60 Ni, 15 Cr	516	0.11	7.4	0.129										
40 Ni, 15 Cr	504	0.11	6.7	0.118										
20 Ni, 15 Cr	491	0.11	8.1	0.151	—	8.1	8.7	8.7	9.4	10	11	13		
Cr-Ni-Al: 6 Cr, 1.5 Al, 0.5 Si (Sicromal 8)	482	0.117	13	0.237										
24 Cr, 2.5 Al, 0.5 Si (Sicromal 12)	479	0.118	11	0.194										
Manganese steel (Ma = 0%)	493	0.118	42	0.784										
1%	491	0.11	29	0.538										
2%	491	0.11	22	0.376	—	22	21	21	21	20	19			
5%	490	0.11	13	0.247										
10%	487	0.11	10	0.194										
Tungsten steel (W = 0%)	493	0.108	42	0.785										
1%	494	0.107	38	0.720										
2%	497	0.106	36	0.677	—	36	34	31	28	26	21			
5%	504	0.104	31	0.591										
10%	519	0.100	28	0.527										
20%	551	0.093	25	0.484										
Silicon steel (Si = 0%)	493	0.108	42	0.785										
1%	485	0.11	24	0.451										
2%	479	0.11	18	0.344										
5%	463	0.11	11	0.215										

Table II.5-1   Continued

Metal	Properties at 68°F ρ (lbm/ft³)	Cp (Btu/lbm·°F)	k (Btu/hr·ft·°F)	α (ft²/hr)	k(Btu/hr·ft·°F) -148°F -100°C	32°F 0°C	212°F 100°C	392°F 200°C	572°F 300°C	752°F 400°C	1112°F 600°C	1472°F 800°C	1832°F 1000°C	2192°F 1200°C
Copper														
Pure	559	0.0915	223	4.353	235	223	219	216	—	210	204			
Aluminum bronze: 95 Cu, 5 Al	541	0.098	48	0.903										
Bronze: 75 Cu, 25 Sn	541	0.082	15	0.333										
Red brass: 85 Cu, 9 Sn, 6 Zn	544	0.092	35	0.699	—	34	41							
Brass: 70 Cu, 30 Zn	532	0.092	64	1.322	51	—	74	83	85	85				
German silver: 62 Cu, 15 Ni, 22 Zn	538	0.094	14.4	0.290	11.1	—	18	23	26	28				
Constantan: 60 Cu, 40 Ni	557	0.098	13.1	0.237	12	—	12.8	15						
Magnesium														
Pure	109	0.242	99	3.762	103	99	97	94	91					
Mg-Al (electrolytic) 6–8% Al, 1–2% Zn	113	0.24	38	1.397	—	30	36	43	48					
Mg-Mn: 2% Mn	111	0.24	66	2.473	54	64	72	75						
Molybdenum	638	0.060	79	2.074	80	79	79							
Nickel														
Pure (99.9%)	556	0.1065	52	0.882	60	54	48	42	37	34				
Impure (99.2%)	556	0.106	40	0.677	—	40	37	34	32	30	32	36	39	40
Ni-Cr: 90 Ni, 10 Cr	541	0.106	10	0.172	—	9.9	10.9	12.1	13.2	14.2				
80 Ni, 20 Cr	519	0.106	7.3	0.129	—	7.1	8.0	9.0	9.9	10.9	13.0			
Silver														
Purest	657	0.0559	242	6.601	242	241	240	238						
Pure (99.9%)	657	0.0559	235	6.418	242	237	240	216	209	208				
Tungsten	1208	0.0321	94	2.430	—	96	87	82	77	73	65	44		
Zinc, pure	446	0.0918	64.8	1.591	66	65	63	61	58	54				
Tin, pure	456	0.0541	37	1.505	43	38.1	34	33						

Source: From E. R. G. Eckert and R. M. Drake, *Heat and Mass Transfer*; copyright 1959 McGraw-Hill; used with the permission of McGraw-Hill Book Company.

Table II.5-2   Thermal Properties of Some Nonmetals

Substance	Cp (Btu/lbm·°F)	ρ (lbm/ft³)	t (°F)	k (Btu/hr·ft²·°F)	α (ft²/hr)
Structural					
Asphalt			68	0.43 a	
Bakelite	0.38 b	79.5 b	68	0.134 b	0.0044
Bricks					
Common	0.20 d	100 d	68	0.40 a	0.02
Face		128 d	68	0.76 a	
Carborundum brick			{ 1110	10.7 a	
			2550	6.4 a	
Chrome brick	0.20 d	188 d	{ 392	1.34 a	0.036
			1022	1.43 a	0.038
			1652	1.15 a	0.031
Diatomaceous earth (fired)			{ 400	0.14 a	
			1600	0.18 a	
Fire clay brick (burnt 2426 F)	0.23 d	128 d	{ 932	0.60 a	0.020
			1472	0.62 a	0.021
			2012	0.63 a	0.021
Fire clay brick (burnt 2642 F)	0.23 d	145 d	{ 932	0.74 a	0.022
			1472	0.79 a	0.024
			2012	0.81 a	0.024
Fire clay brick (Missouri)	0.23 d	165 f	{ 392	0.58 a	0.015
			1112	0.85 a	0.022
			2552	1.02 a	0.027
Magnesite	0.27 d		{ 400	2.2 a	
			1200	1.6 a	
			2200	1.1 a	
Cement, portland		94		0.17 a	
Cement, mortar			75	0.67 a	
Concrete	0.21 b	119–144 b	68	0.47–0.81 b	0.019–0.027
Concrete, cinder			75	0.44 a	

Table II.5-2   Continued

Substance	C_p (Btu/lb$_m$·°F)		ρ (lb$_m$/ft^3)		t (°F)	k (Btu/hr·ft^2·°F)		α (ft^2/hr)
Glass, plate	0.2	b	169	b	68	0.44	b	0.013
Glass, borosilicate			139	b	86	0.63	b	
Plaster, gypsum	0.2	d	90	d	70	0.28	a	0.016
Plaster, metal lath					70	0.27	a	
Plaster, wood lath					70	0.16	a	
Stone								
Granite	0.195	d	165	d		1.0–2.3	a	0.031–0.071
Limestone	0.217	d	155	d	210–570	0.73–0.77	a	0.022–0.023
Marble	0.193	b	156–169	b	68	1.6	b	0.054
Sandstone	0.17	b	135–144	b	68	0.94–1.2	b	0.041–0.049
Wood, cross grain								
Balsa			8.8	a	86	0.032	a	
Cypress			29	d	86	0.056	a	
Fir	0.65	d	26.0	b	75	0.063	a	0.0037
Oak	0.57	d	38–30	b	86	0.096	a	0.0049
Yellow pine	0.67	d	40	d	75	0.085	a	0.0032
White pine			27	d	86	0.065	a	
Wood, radial								
Oak	0.57	b	38–30	b	68	0.10–0.12	b	$\begin{cases} 0.0043- \\ 0.0047 \end{cases}$
Fir	0.65	b	26.0–26.3	b	68	0.08	b	0.0048
Insulating								
Asbestos			29.3	b	$\begin{cases} -328 \\ 32 \end{cases}$	0.043 0.090	b b	
			36.0	b	$\begin{cases} 32 \\ 212 \\ 392 \\ 752 \end{cases}$	0.087 0.111 0.120 0.129	b b b b	
			43.5	b	$\begin{cases} -328 \\ 32 \end{cases}$	0.09 0.135	b b	
Asbestos cement						1.2	a	
Asbestos cement board					68	0.43	a	
Asbestos sheet					124	0.096	a	
Asbestos felt (40 laminations per inch)					$\begin{cases} 100 \\ 300 \\ 500 \end{cases}$	0.033 0.040 0.048	a a a	
Asbestos felt (20 laminations per inch)					$\begin{cases} 100 \\ 300 \\ 500 \end{cases}$	0.045 0.055 0.065	a a a	
Asbestos, corrugated (4 plies per inch)					$\begin{cases} 100 \\ 200 \\ 300 \end{cases}$	0.05 0.058 0.069	a a a	
Balsam wool			2.2	a	90	0.023	a	
Cardboard, corrugated						0.037	a	
Celotex					90	0.028	a	
Corkboard			10	b	86	0.025	b	
Cork, expanded scrap	0.45	b	2.8–7.4	b	68	0.021	b	0.006–0.017
Cork, ground			9.4	b	86	0.025	b	
Insulating								
Diatomaceous earth (powdered)			10	e	$\begin{cases} 200 \\ 400 \\ 600 \end{cases}$	0.029 0.038 0.048	e e e	
Diatomaceous earth (powdered)			14	e	$\begin{cases} 200 \\ 400 \\ 600 \end{cases}$	0.033 0.039 0.046	e e e	
Diatomaceous earth (powdered)			18	e	$\begin{cases} 200 \\ 400 \\ 600 \end{cases}$	0.040 0.045 0.049	e e e	
Felt, hair			8.2	c	$\begin{cases} 20 \\ 100 \\ 200 \end{cases}$	0.0237 0.0269 0.0310	c c c	

Table II.5-2   Continued

Substance	C_p (Btu/lb$_m$·°F)		ρ (lb$_m$/ft^3)		t (°F)	k (Btu/hr·ft^2·°F)		α (ft^2/hr)
Felt, hair			11.4	c	20	0.0212	c	
					100	0.0254	c	
					200	0.0299	c	
Felt, hair			12.8	c	20	0.0233	c	
					100	0.0262	c	
					200	0.0295	c	
Fiber insulating board			14.8	b	70	0.028	b	
Glass wool			1.5	c	20	0.0217	c	
					100	0.0313	c	
					200	0.0435	c	
Glass wool			4.0	c	20	0.0179	c	
					100	0.0239	c	
					200	0.0317	c	
Glass wool			6.0	c	20	0.0163	c	
					100	0.0218	c	
					200	0.0288	c	
Kapok					86	0.020	a	
Magnesia, 85%			16.9	c	100	0.039	a	
					200	0.041	a	
					300	0.043	a	
					400	0.046	a	
Rock wool			4.0	c	100	0.0224	c	
					200	0.0317	c	
Rock wool			8.0	c	20	0.0171	c	
					100	0.0228	c	
					200	0.0299	c	
Rock wool			12.0	c	20	0.0183	c	
					100	0.0226	c	
					200	0.0281	c	
Miscellaneous								
Aerogel, silica			8.5	b	248	0.013	b	
Clay	0.21	b	91.0	b	68	0.739	b	0.039
Coal, anthracite	0.30	b	75–94	b	68	0.15	b	0.005–0.006
Coal, powdered	0.31	b	46	b	86	0.067	b	0.005
Cotton	0.31	b	5	b	68	0.034	b	0.075
Earth, coarse	0.44	b	128	b	68	0.30	b	0.0054
Ice	0.46	b	57	b	32	1.28	b	0.048
Rubber, hard			74.8	b	32	0.087	b	
Sawdust					75	0.034	a	
Silk	0.33	b	3.6	b	68	0.021	b	0.017

Source: Adapted from (a) A. I. Brown and S. M. Marco, *Introduction to Heat Transfer*, 3rd ed., McGraw-Hill, New York, 1958; (b) E. R. G. Eckert, *Introduction to the Transfer of Heat and Mass*, McGraw-Hill, New York, 1950; (c) R. H. Heilman, *Industrial and Engineering Chemistry*, Vol. 28 (1936), p. 782; (d) L. S. Marks, *Mechanical Engineers' Handbook*, 6th ed., McGraw-Hill, New York, 1958; (e) R. Calvert, *Diatomaceous Earth*, Chemical Catalog Company, Inc., 1930; (f) H. F. Norton, *Journal of the American Ceramic Society*, Vol. 10 (1957), p. 30.

Table II.5-3 Property Values of Fluids in Saturated State

t (°F)	ρ (lb/ft^3)	C_p (Btu/ lb·°F)	ν (ft^2/sec)	k (Btu/ hr·ft·°F)	α (ft^2/hr)	Pr	β (1/R)
			Water (H$_2$O)				
32	62.57	1.0074	1.925×10^{-5}	0.319	5.07×10^{-3}	13.6	
68	62.46	0.9988	1.083	0.345	5.54	7.02	0.10×10^{-3}
104	62.09	0.9980	0.708	0.363	5.86	4.34	
140	61.52	0.9994	0.514	0.376	6.02	3.02	
176	60.81	1.0023	0.392	0.386	6.34	2.22	
212	59.97	1.0070	0.316	0.393	6.51	1.74	
248	59.01	1.015	0.266	0.396	6.62	1.446	
284	57.95	1.023	0.230	0.395	6.68	1.241	
320	56.79	1.037	0.204	0.393	6.70	1.099	
356	55.50	1.055	0.186	0.390	6.68	1.004	
392	54.11	1.076	0.172	0.384	6.61	0.937	
428	52.59	1.101	0.161	0.377	6.51	0.891	
464	50.92	1.136	0.154	0.367	6.35	0.871	
500	49.06	1.182	0.148	0.353	6.11	0.874	
537	46.98	1.244	0.145	0.335	5.74	0.910	
572	44.59	1.368	0.145	0.312	5.13	1.019	
			Ammonia (NH$_3$)				
-58	43.93	1.066	0.468×10^{-5}	0.316	6.75×10^{-3}	2.60	
-40	43.18	1.067	0.437	0.316	6.88	2.28	
-22	42.41	1.069	0.417	0.317	6.98	2.15	
-4	41.62	1.077	0.410	0.316	7.05	2.09	
14	40.80	1.090	0.407	0.314	7.07	2.07	
32	39.96	1.107	0.402	0.312	7.05	2.05	
50	39.09	1.126	0.396	0.307	6.98	2.04	
68	38.19	1.146	0.386	0.301	6.88	2.02	1.36×10^{-3}
86	37.23	1.168	0.376	0.293	6.75	2.01	
104	36.27	1.194	0.366	0.285	6.59	2.00	
122	35.23	1.222	0.355	0.275	6.41	1.99	
			Carbon Dioxide (CO$_2$)				
-58	72.19	0.44	0.128×10^{-5}	0.0494	1.558×10^{-3}	2.96	
-40	69.78	0.45	0.127	0.0584	1.864	2.46	
-22	67.22	0.47	0.126	0.0645	2.043	2.22	
-4	64.45	0.49	0.124	0.0665	2.110	2.12	
14	61.39	0.52	0.122	0.0635	1.989	2.20	
			Eutectic Calcium Chloride Solution (29.9% CaCl$_2$)				
-58	82.39	0.623	39.13×10^{-5}	0.232	4.52×10^{-3}	312	
-40	82.09	0.6295	26.88	0.240	4.65	208	
-22	81.79	0.6356	18.49	0.248	4.78	139	
-4	81.50	0.642	11.88	0.257	4.91	87.1	
14	81.20	0.648	7.49	0.265	5.04	53.6	
32	80.91	0.654	4.73	0.273	5.16	33.0	
50	80.62	0.660	3.61	0.280	5.28	24.6	
68	80.32	0.666	2.93	0.288	5.40	19.6	
86	80.03	0.672	2.44	0.295	5.50	16.0	
104	79.73	0.678	2.07	0.302	5.60	13.3	
122	79.44	0.685	1.78	0.309	5.69	11.3	
			Glycerin [C$_3$H$_5$(OH)$_3$]				
32	79.66	0.540	0.0895	0.163	3.81×10^{-3}	84.7×10^3	
50	79.29	0.554	0.0323	0.164	3.74	31.0	
68	78.91	0.570	0.0127	0.165	3.67	12.5	0.28×10^{-3}
86	78.54	0.584	0.0054	0.165	3.60	5.38	
104	78.16	0.600	0.0024	0.165	3.54	2.45	
122	77.72	0.617	0.0016	0.166	3.46	1.63	

Table II.5-3   Continued

t (°F)	ρ (lb/ft³)	C_p (Btu/ lb·°F)	ν (ft²/sec)	k (Btu/ hr·ft·°F)	α (ft²/hr)	Pr	β (1/R)
\multicolumn{8}{c}{*Ethylene Glycol [C₂H₄(OH₂)]*}							
32	70.59	0.548	61.92×10^{-5}	0.140	3.62×10^{-3}	615	
68	69.71	0.569	20.64	0.144	3.64	204	0.36×10^{-3}
104	68.76	0.591	9.35	0.148	3.64	93	
140	67.90	0.612	5.11	0.150	3.61	51	
176	67.27	0.633	3.21	0.151	3.57	32.4	
212	66.08	0.655	2.18	0.152	3.52	22.4	
\multicolumn{8}{c}{*Engine Oil (Unused)*}							
32	56.13	0.429	0.0461	0.085	3.53×10^{-3}	47100	
68	55.45	0.449	0.0097	0.084	3.38	10400	0.39×10^{-3}
104	54.69	0.469	0.0026	0.083	3.23	2870	
140	53.94	0.489	0.903×10^{-3}	0.081	3.10	1050	
176	53.19	0.509	0.404	0.080	2.98	490	
212	52.44	0.530	0.219	0.079	2.86	276	
248	51.75	0.551	0.133	0.078	2.75	175	
284	51.00	0.572	0.086	0.077	2.66	116	
320	50.31	0.593	0.060	0.076	2.57	84	
\multicolumn{8}{c}{*Mercury (Hg)*}							
32	850.78	0.0335	0.133×10^{-5}	4.74	166.6×10^{-3}	0.0288	
68	847.71	0.0333	0.123	5.02	178.5	0.0249	1.01×10^{-4}
122	843.14	0.0331	0.112	5.43	194.6	0.0207	
212	835.57	0.0328	0.0999	6.07	221.5	0.0162	
302	828.06	0.0326	0.0918	6.64	246.2	0.0134	
392	820.61	0.0375	0.0863	7.13	267.7	0.0116	
482	813.16	0.0324	0.0823	7.55	287.0	0.0103	
600	802	0.032	0.0724	8.10	316	0.0083	
\multicolumn{8}{c}{*Carbon Dioxide (CO₂)*}							
32	57.87	0.59	0.117	0.0604	1.774	2.38	
50	53.69	0.75	0.109	0.0561	1.398	2.80	
68	48.23	1.2	0.098	0.0504	0.860	4.10	7.78×10^{-3}
86	37.32	8.7	0.086	0.0406	0.108	28.7	
\multicolumn{8}{c}{*Sulfur Dioxide (SO₂)*}							
−58	97.44	0.3247	0.521×10^{5}	0.140	4.42×10^{-3}	4.24	
−40	95.94	0.3250	0.456	0.136	4.38	3.74	
−22	94.43	0.3252	0.399	0.133	4.33	3.31	
−4	92.93	0.3254	0.349	0.130	4.29	2.93	
14	91.37	0.3255	0.310	0.126	4.25	2.62	
32	89.80	0.3257	0.277	0.122	4.19	2.38	
50	88.18	0.3259	0.250	0.118	4.13	2.18	
68	86.55	0.3261	0.226	0.115	4.07	2.00	1.08×10^{-3}
86	84.86	0.3263	0.204	0.111	4.01	1.83	
104	82.98	0.3266	0.186	0.107	3.95	1.70	
122	81.10	0.3268	0.174	0.102	3.87	1.61	
\multicolumn{8}{c}{*Methyl Chloride (CH₃Cl)*}							
−58	65.71	0.3525	0.344×10^{5}	0.124	5.38×10^{-3}	2.31	
−40	64.51	0.3541	0.342	0.121	5.30	2.32	
−22	63.46	0.3561	0.338	0.117	5.18	2.35	
−4	62.39	0.3593	0.333	0.113	5.04	2.38	
14	61.27	0.3629	0.329	0.108	4.87	2.43	
32	60.08	0.3673	0.325	0.103	4.70	2.49	
50	58.83	0.3726	0.320	0.099	4.52	2.55	
68	57.64	0.3788	0.315	0.094	4.31	2.63	
86	56.38	0.3860	0.310	0.089	4.10	2.72	
104	55.13	0.3942	0.303	0.083	3.86	2.83	
122	53.76	0.4034	0.295	0.077	3.57	2.97	

Source: From E. R. G. Eckert and R. M. Drake, *Heat and Mass Transfer;* copyright 1959 McGraw-Hill; Used with the permission of McGraw-Hill Book Company.

Table II.5-4 Property Values of Gases at Atmospheric Pressure

T (°F)	ρ (lb/ft^3)	C_p (Btu/lb·°F)	μ (lb/sec·ft)	ν (ft^2/sec)	k (Btu/hr·ft·°F)	α (ft^2/hr)	Pr
Air							
−280	0.2248	0.2452	0.4653×10^{-5}	2.070×10^{-5}	0.005342	0.09691	0.770
−190	0.1478	0.2412	0.6910	4.675	0.007936	0.2226	0.753
−100	0.1104	0.2403	0.8930	8.062	0.01045	0.3939	0.739
−10	0.0882	0.2401	1.074	10.22	0.01287	0.5100	0.722
80	0.0735	0.2402	1.241	16.88	0.01516	0.8587	0.708
170	0.0623	0.2410	1.394	22.38	0.01735	1.156	0.697
260	0.0551	0.2422	1.536	27.88	0.01944	1.457	0.689
350	0.0489	0.2438	1.669	31.06	0.02142	1.636	0.683
440	0.0440	0.2459	1.795	40.80	0.02333	2.156	0.680
530	0.0401	0.2482	1.914	47.73	0.02519	2.531	0.680
620	0.0367	0.2520	2.028	55.26	0.02692	2.911	0.680
710	0.0339	0.2540	2.135	62.98	0.02862	3.324	0.682
800	0.0314	0.2568	2.239	71.31	0.03022	3.748	0.684
890	0.0294	0.2593	2.339	79.56	0.03183	4.175	0.686
980	0.0275	0.2622	2.436	88.58	0.03339	4.631	0.689
1070	0.0259	0.2650	2.530	97.68	0.03483	5.075	0.692
1160	0.0245	0.2678	2.620	106.9	0.03628	5.530	0.696
1250	0.0232	0.2704	2.703	116.5	0.03770	6.010	0.699
1340	0.0220	0.2727	2.790	126.8	0.03901	6.502	0.702
1520	0.0200	0.2772	2.955	147.8	0.04178	7.536	0.706
1700	0.0184	0.2815	3.109	169.0	0.04410	8.514	0.714
1880	0.0169	0.2860	3.258	192.8	0.04641	9.602	0.722
2060	0.0157	0.2900	3.398	216.4	0.04880	10.72	0.726
2240	0.0147	0.2939	3.533	240.3	0.05098	11.80	0.734
2420	0.0138	0.2982	3.668	265.8	0.05348	12.88	0.741
2600	0.0130	0.3028	3.792	291.7	0.05550	14.00	0.749
2780	0.0123	0.3075	3.915	318.3	0.05750	15.09	0.759
2960	0.0116	0.3128	4.029	347.1	0.0591	16.40	0.767
3140	0.0110	0.3196	4.168	378.8	0.0612	17.41	0.783
3320	0.0105	0.3278	4.301	409.9	0.0632	18.36	0.803
3500	0.0100	0.3390	4.398	439.8	0.0646	19.05	0.831
3680	0.0096	0.3541	4.513	470.1	0.0663	19.61	0.863
3860	0.0091	0.3759	4.611	506.9	0.0681	19.92	0.916
4160	0.0087	0.4031	4.750	546.0	0.0709	20.21	0.972
Helium							
−456		1.242	5.66×10^{-7}		0.0061		
−400	0.0915	1.242	33.7	3.68×10^{-5}	0.0204	0.1792	0.74
−200	0.211	1.242	84.3	39.95	0.0536	2.044	0.70
−100	0.0152	1.242	105.2	69.30	0.0680	3.599	0.694
0	0.0119	1.242	122.1	102.8	0.0784	5.299	0.70
200	0.00829	1.242	154.9	186.9	0.0977	9.490	0.71
400	0.00637	1.242	184.8	289.9	0.114	14.40	0.72
600	0.00517	1.242	209.2	404.5	0.130	20.21	0.72
800	0.00439	1.242	233.5	531.9	0.145	25.81	0.72
1000	0.00376	1.242	256.5	682.5	0.159	34.00	0.72
1200	0.00330	1.242	277.9	841.0	0.172	41.98	0.72
Hydrogen							
−406	0.05289	2.589	1.079×10^{-6}	2.040×10^{-5}	0.0132	0.0966	0.759
−370	0.03181	2.508	1.691	5.253	0.0209	0.262	0.721
−280	0.01534	2.682	2.830	18.45	0.0384	0.933	0.712
−190	0.01022	3.010	3.760	36.79	0.0567	1.84	0.718
−100	0.00766	3.234	4.578	59.77	0.0741	2.99	0.719
−10	0.00613	3.358	5.321	86.80	0.0902	4.38	0.713
80	0.00511	3.419	6.023	117.9	0.105	6.02	0.706
170	0.00438	3.448	6.689	152.7	0.119	7.87	0.697
260	0.00383	3.461	7.300	190.6	0.132	9.95	0.690
350	0.00341	3.463	7.915	232.1	0.145	12.26	0.682
440	0.00307	3.465	8.491	276.6	0.157	14.79	0.675
530	0.00279	3.471	9.055	324.6	0.169	17.50	0.668
620	0.00255	3.472	9.599	376.4	0.182	20.56	0.664
800	0.00218	3.481	10.68	489.9	0.203	26.75	0.659
980	0.00191	3.505	11.69	612	0.222	33.18	0.664
1160	0.00170	3.540	12.62	743	0.238	39.59	0.676
1340	0.00153	3.575	13.55	885	0.254	46.49	0.686
1520	0.00139	3.622	14.42	1039	0.268	53.19	0.703
1700	0.00128	3.670	15.29	1192	0.282	60.00	0.715
1880	0.00118	3.720	16.18	1370	0.296	67.40	0.733
1940	0.00115	3.735	16.42	1429	0.300	69.80	0.736

Table II.5-4   Continued

T (°F)	ρ (lb/ft³)	C_p (Btu/ lb·°F)	μ (lb/sec·ft)	ν (ft²/sec)	k (Btu/ hr·ft·°F)	α (ft²/hr)	Pr
			Oxygen				
−280	0.2492	0.2264	5.220×10^{-6}	2.095×10^{-5}	0.00522	0.09252	0.815
−190	0.1635	0.2192	7.721	4.722	0.00790	0.2204	0.773
−100	0.1221	0.2181	9.979	8.173	0.01054	0.3958	0.745
−10	0.0975	0.2187	12.01	12.32	0.01305	0.6120	0.725
80	0.0812	0.2198	13.86	17.07	0.01546	0.8662	0.709
170	0.0695	0.2219	15.56	22.39	0.01774	1.150	0.702
260	0.0609	0.2250	17.16	28.18	0.02000	1.460	0.695
350	0.0542	0.2285	18.66	34.43	0.02212	1.786	0.694
440	0.0487	0.2322	20.10	41.27	0.02411	2.132	0.697
530	0.0443	0.2360	21.48	48.49	0.02610	2.496	0.700
620	0.0406	0.2399	22.79	56.13	0.02792	2.867	0.704
			Nitrogen				
−280	0.2173	0.2561	4.611×10^{-6}	2.122×10^{-5}	0.005460	0.09811	0.786
−100	0.1068	0.2491	8.700	8.146	0.01054	0.3962	0.747
80	0.0713	0.2486	11.99	16.82	0.01514	0.8542	0.713
260	0.0533	0.2498	14.77	27.71	0.01927	1.447	0.691
440	0.0426	0.2521	17.27	40.54	0.02302	2.143	0.684
620	0.0355	0.2569	19.56	55.10	0.02646	2.901	0.686
800	0.0308	0.2620	21.59	70.10	0.02960	3.668	0.691
980	0.:267	0.2681	23.41	87.68	0.03241	4.528	0.700
1160	0.0237	0.2738	25.19	98.02	0.03507	5.404	0.711
1340	0.0213	0.2789	26.88·	126.2	0.03741	6.297	0.724
1520	0.0194	0.2832	28.41	146.4	0.03958	7.204	0.736
1700	0.0178	0.2875	29.90	168.0	0.04151	8.111	0.748
			Carbon Dioxide				
−64	0.1544	0.187	7.462×10^{-6}	4.833×10^{-5}	0.006243	0.2294	0.818
−10	0.1352	0.192	8.460	6.257	0.007444	0.2868	0.793
80	0.1122	0.208	10.051	8.957	0.009575	0.4103	0.770
170	0.0959	0.215	11.561	12.05	0.01183	0.5738	0.755
260	0.0838	0.225	12.98	15.49	0.01422	0.7542	0.738
350	0.0744	0.234	14.34	19.27	0.01674	0.9615	0.721
440	0.0670	0.242	15.63	23.33	0.01937	1.195	0.702
530	0.0608	0.250	16.85	27.71	0.02208	1.453	0.685
620	0.0558	0.257	18.03	32.31	0.02491	1.737	0.668
			Carbon Monoxide				
−64	0.09699	0.2491	9.295×10^{-6}	9.583×10^{-5}	0.01101	0.4557	0.758
−10	0.0525	0.2490	10.35	12.14	0.01239	0.5837	0.750
80	0.07109	0.2489	11.990	16.87	0.01459	0.8246	0.737
170	0.06082	0.2492	13.50	22.20	0.01666	1.099	0.728
260	0.05329	0.2504	14.91	27.98	0.01864	1.397	0.722
350	0.04735	0.2520	16.25	34.32	0.0252	1.720	0.718
440	0.04259	0.2540	17.51	41.11	0.02232	2.063	0.718
530	0.03872	0.2569	18.74	48.40	0.02405	2.418	0.721
620	0.03549	0.2598	19.89	56.04	0.02569	2.786	0.724
			Ammonia (NH₃)				
−58	0.0239	0.525	4.875×10^{-6}	2.04×10^{-4}	0.0099	0.796	0.93
32	0.0495	0.520	6.285	1.27	0.0127	0.507	0.90
122	0.0405	0.520	7.415	1.83	0.0156	0.744	0.88
212	0.0349	0.534	8.659	2.48	0.0189	1.015	0.87
302	0.0308	0.553	9.859	3.20	0.0226	1.330	0.87
392	0.0275	0.572	11.08	4.03	0.0270	1.713	0.84
			Steam (H₂O Vapor)				
224	0.0366	0.492	8.54×10^{-6}	2.33×10^{-4}	0.0142	0.789	1.060
260	0.0346	0.481	9.03	2.61	0.0151	0.906	1.040
350	0.0306	0.473	10.25	3.35	0.0173	1.19	1.010
440	0.0275	0.474	11.45	4.16	0.0196	1.50	0.996
530	0.0250	0.477	12.66	5.06	0.0219	1.84	0.991
620	0.0228	0.484	13.89	6.09	0.0244	2.22	0.986
710	0.0211	0.491	15.10	7.15	0.0268	2.58	0.995
800	0.0196	0.498	16.30	8.31	0.0292	2.99	1.000
890	0.0183	0.506	17.50	9.56	0.0317	3.42	1.005
980	0.0171	0.514	18.72	10.98	0.0342	3.88	1.010
1070	0.0161	0.522	19.95	12.40	0.0368	4.38	1.019

Table II.5-5 Zero-Pressure Properties of Gases

Gas	Chemical Formula	Molecular Weight	R (ft $lb_f/lb_m \cdot °R$)	C_{p0} (Btu/$lb_m \cdot °R$)[a]	C_{v0} (Btu/$lb_m \cdot °R$)[a]	k[a]
Air	—	28.97	53.34	0.240	0.171	1.400
Argon	Ar	39.94	38.66	0.1253	0.0756	1.667
Carbon dioxide	CO_2	44.01	35.10	0.203	0.158	1.285
Carbon monoxide	CO	28.01	55.16	0.249	0.178	1.399
Helium	He	4.003	386.0	1.25	0.753	1.667
Hydrogen	H_2	2.016	766.4	3.43	2.44	1.404
Methane	CH_4	16.04	96.35	0.532	0.403	1.32
Nitrogen	N_2	28.016	55.15	0.248	0.177	1.400
Oxygen	O_2	32.000	48.28	0.219	0.157	1.395
Steam	H_2O	18.016	85.76	0.445	0.335	1.329

[a] C_{p0}, C_{v0}, and k are at 80°F.

Table II.5-6 Constant-Pressure Specific Heats of Various Substances at Zero Pressure

Gas or Vapor	Equation: $\tilde{C}_{p0}$ in Btu/lb mol $\cdot °R$, T in °R	Range (°R)	Max. Error (%)
O_2	$\tilde{C}_{p0} = 11.515 - \dfrac{172}{\sqrt{T}} + \dfrac{1530}{T}$	540–5000	1.1
	$= 11.515 - \dfrac{172}{\sqrt{T}} + \dfrac{1530}{T}$ $+ \dfrac{0.05}{1000}(T - 4000)$	5000–9000	0.3
N_2	$\tilde{C}_{p0} = 9.47 - \dfrac{3.47 \times 10^3}{T} + \dfrac{1.16 \times 10^6}{T^2}$	540–9000	1.7
CO	$\tilde{C}_{p0} = 9.46 - \dfrac{3.29 \times 10^3}{T} + \dfrac{1.07 \times 10^6}{T^2}$	540–9000	1.1
H_2	$\tilde{C}_{p0} = 5.76 + \dfrac{0.578}{1000}T + \dfrac{20}{\sqrt{T}}$	540–4000	0.8
	$= 5.76 + \dfrac{0.578}{1000}T + \dfrac{20}{\sqrt{T}}$ $- \dfrac{0.33}{1000}(T - 4000)$	4000–9000	1.4
H_2O	$\tilde{C}_{p0} = 19.86 \dfrac{597}{\sqrt{T}} + \dfrac{7500}{T}$	540–5400	1.8
CO_2	$\tilde{C}_{p0} = 16.2 - \dfrac{6.53 \times 10^3}{T} + \dfrac{1.41 \times 10^6}{T^2}$	540–6300	0.8
CH_4	$\tilde{C}_{p0} = 4.52 + 0.00737T$	540–1500	1.2
C_2H_4	$\tilde{C}_{p0} = 4.23 + 0.01177T$	350–1100	1.5
C_2H_6	$\tilde{C}_{p0} = 4.01 + 0.01636T$	400–1100	1.5
C_3H_8	$\tilde{C}_{p0} = 2.258 + 0.0320T - 5.43 \times 10^{-6}T^2$	415–2700	1.8
C_4H_{10}	$\tilde{C}_{p0} = 4.36 + 0.0403T - 6.83 \times 10^{-6}T^2$	540–2700	1.7
C_8H_{18}	$\tilde{C}_{p0} = 7.92 + 0.0601T$	400–1100	est. 4
$C_{12}H_{26}$	$\tilde{C}_{p0} = 8.68 + 0.0889T$	400–1100	est. 4

Source: From R. L. Sweigert and M. W. Beardsley, *Bulletin No. 2*, Georgia School of Technology, 1938, except C_3H_8 and C_4H_{10}, which are from H. M. Spencer, *Journal of the American Chemical Society*, Vol. 67 (1945), p. 1859.

Table II.6 Psychrometric Chart for Air/Steam Mixtures

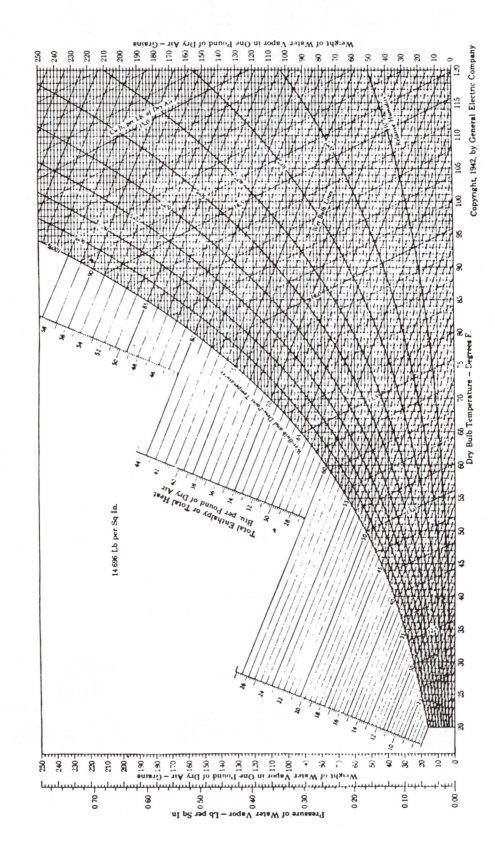

Table II.7 Total Emissivity Data

Surface	°C	°F	ϵ
	Metals		
Aluminum			
Polished, 98% pure	200–600	400–1100	0.04–0.06
Commercial sheet	100	200	0.09
Rough plate	40	100	0.07
Heavily oxidized	100–550	200–1000	0.20–0.33
Antimony			
Polished	40–250	100–500	0.28–0.31
Bismuth			
Bright	100	200	0.34
Brass			
Highly polished	250	500	0.03
Polished	40	100	0.07
Dull plate	40–250	100–500	0.22
Oxidized	40–250	100–500	0.46–0.56
Chromium			
Polished sheet	40–550	100–1000	0.08–0.27
Cobalt			
Unoxidized	250–550	500–1000	0.13–0.23
Copper			
Highly polished electrolytic	100	200	0.02
Polished	40	100	0.04
Slightly polished	40	100	0.12
Polished, lightly tarnished	40	100	0.05
Dull	40	100	0.15
Black oxidized	40	100	0.76
Gold			
Pure, highly polished	100–600	200–1100	0.02–0.035
Inconel			
X, stably oxidized	230–900	450–1600	0.55–0.78
B, stably oxidized	230–1000	450–1750	0.32–0.55
X and B, polished	150–300	300–600	0.20
Iron and Steel			
Mild steel, polished	150–500	300–900	0.14–0.32
Steel, polished	40–250	100–500	0.07–0.10
Sheet steel, ground	1000	1700	0.55
Sheet steel, rolled	40	100	0.66
Sheet steel, strong rough oxide	40	100	0.80
Steel, oxidized at 1100°F	250	500	0.79
Cast iron, with skin	40	100	0.70–0.80
Cast iron, newly turned	40	100	0.44
Cast iron, polished	200	400	0.21
Cast iron, oxidized	40–250	100–500	0.57–0.66
Iron, red rusted	40	100	0.61
Iron, heavily rusted	40	100	0.85
Wrought iron, smooth	40	100	0.35
Wrought iron, dull oxidized	20–360	70–680	0.94
Stainless, polished	40	100	0.07–0.17
Stainless, after repeated heating and cooling	230–930	450–1650	0.50–0.70
Lead			
Polished	40–250	100–500	0.05–0.08
Gray, oxidized	40	100	0.28
Oxidized at 390°F	200	400	0.63
Oxidized at 1100°F	40	100	0.63
Magnesium			
Polished	40–250	100–500	0.07–0.13
Manganin			
Bright rolled	100	200	0.05
Mercury			
Pure, clean	40–100	100–200	0.10–0.12

Table II.7 Continued

Surface	°C	°F	ϵ
Molybdenum			
Polished	40–250	100–500	0.06–0.08
Polished	550–1100	1000–2000	0.11–0.18
Filament	550–2800	1000–5000	0.08–0.29
Monel			
After repeated heating and cooling	230–930	450–1650	0.45–0.70
Oxidized at 1100°F	200–600	400–1100	0.41–0.46
Polished	40	100	0.17
Nickel			
Polished	40–250	100–500	0.05–0.07
Oxidized	40–250	100–500	0.35–0.49
Wire	250–1100	500–2000	0.10–0.19
Platinum			
Pure, polished plate	200–600	400–1100	0.05–0.10
Oxidized at 1100°F	250–550	500–1000	0.07–0.11
Electrolytic	250–550	500–1000	0.06–0.10
Strip	550–1100	1000–2000	0.12–0.14
Filament	40–1100	100–2000	0.04–0.19
Wire	200–1370	400–2500	0.07–0.18
Silver			
Polished or deposited	40–550	100–1000	0.01–0.03
Oxidized	40–550	100–1000	0.02–0.04
German silver," polished	250–550	500–1000	0.07–0.09
Tin			
Bright tinned iron	40	100	0.04–0.06
Bright	40	100	0.06
Polished sheet	100	200	0.05
Tungsten			
Filament	550–1100	1000–2000	0.11–0.16
Filament	2800	5000	0.39
Filament, aged	40–3300	100–6000	0.03–0.35
Polished	40–550	100–1000	0.04–0.08
Zinc			
Pure polished	40–250	100–500	0.02–0.03
Oxidized at 750°F	400	750	0.11
Galvanized, gray	40	100	0.28
Galvanized, fairly bright	40	100	0.23
Dull	40–250	100–500	0.21
Nonmetals			
Asbestos			
Board	40	100	0.96
Cement	40	100	0.96
Paper	40	100	0.93–0.95
Slate	40	100	0.97
Brick			
Red, rough	40	100	0.93
Silica	1000	1800	0.80–0.85
Fireclay	1000	1800	0.75
Ordinary refractory	1100	2000	0.59
Magnesite refractory	1000	1800	0.38
White refractory	1100	2000	0.29
Gray, glazed	1100	2000	0.75
Carbon			
Filament	1050–1420	1900–2600	0.53
Lampsoot	40	100	0.95
Clay			
Fired	100	200	0.91
Concrete			
Rough	40	100	0.94

Table II.7 Continued

Surface	°C	°F	ϵ
Corundum			
Emery rough	100	200	0.86
Glass			
Smooth	40	100	0.94
Quartz glass (2 mm)	250–550	500–1000	0.96–0.66
Pyrex	250–550	500–1000	0.94–0.75
Gypsum	40	100	0.80–0.90
Ice			
Smooth	0	32	0.97
Rough crystals	0	32	0.99
Hoarfrost	−18	0	0.99
Limestone	40–250	100–500	0.95–0.83
Marble			
Light gray, polished	40	100	0.93
White	40	100	0.95
Mica	40	100	0.75
Paints			
Aluminum, various ages and compositions	100	200	0.27–0.62
Black gloss	40	100	0.90
Black lacquer	40	100	0.80–0.93
White paint	40	100	0.89–0.97
White lacquer	40	100	0.80–0.95
Various oil paints	40	100	0.92–0.96
Red lead	100	200	0.93
Paper			
White	40	100	0.95
Writing paper	40	100	0.98
Any color	40	100	0.92–0.94
Roofing	40	100	0.91
Plaster			
Lime, rough	40–250	100–500	0.92
Porcelain			
Glazed	40	100	0.93
Quartz	40–550	100–1000	0.89–0.58
Rubber			
Hard	40	100	0.94
Soft, gray rough	40	100	0.86
Sandstone	40–250	100–500	0.83–0.90
Snow	(−12) − (−6)	10–20	0.82
Water			
0.1 mm or more thick	40	100	0.96
Wood			
Oak, planed	40	100	0.90
Walnut, sanded	40	100	0.83
Spruce, sanded	40	100	0.82
Beech	40	100	0.94
Planed	40	100	0.78
Various	40	100	0.80–0.90
Sawdust	40	100	0.75

Source: From E. M. Sparrow and R. D. Cess, *Radiation Heat Transfer*, rev. ed.; copyright ©
1970 by Wadsworth Publishing Company, Inc.; reprinted by permission of the publisher, Brooks/
Cole Publishing Company, Monterey, Calif.

[a] German silver is actually an alloy of copper, nickel, and zinc.

APPENDIX III
REVIEW OF ELECTRICAL SCIENCE

RUSSELL L. HEISERMAN, Ed.D.

School of Technology
Oklahoma State University
Stillwater, Oklahoma

III. 1 INTRODUCTION

This brief review of electrical science is intended for those readers who may use electrical principles only on occasion and is intended to be supportive of the material found in those chapters of the handbook based on electrical science. The review consists of selected topics in basic ac circuit theory presented at a nominal analytical level. Much of the material deals with power in ac circuits and principles of power-factor improvement.

III.2 REVIEW OF VECTOR ALGEBRA

Vector algebra is the mathematics most appropriate for ac circuit problems. Most often electric quantities, voltage and current, are not in phase in ac circuits, so that phase relationships as well as magnitude have to be considered. This brief review will cover the basic idea of a vector quantity and then refresh the process of adding, subtracting, multiplying, and dividing vectors.

III.2.1 Review

A vector is a quantity having both direction and magnitude. Familiar vector quantities are velocity and force. Other familiar quantities, such as speed, volume, area, and mass, have magnitude only.

A vector quantity is expressed as having both magnitude and direction, such as

$$Ae^{\pm j\theta}$$

where A is the magnitude and $e^{\pm j\theta}$ expresses the direction in the complex plane (Figure III.1).

The important feature of this vector notation is to note that the angle of displacement is in fact an exponent. This feature is significant, since it will allow the use of the law of exponents when multiplying, dividing, or raising to a power.

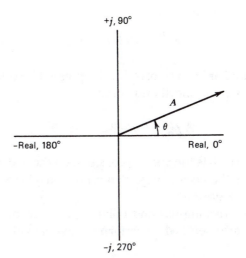

Fig. III.1 The generalized vector $Ae^{j\theta}$ shown in the complex plane. If j is positive, it is referenced to the positive real axis with a counterclockwise displacement. If j is negative, it is referenced to the positive real axis with a clockwise displacement.

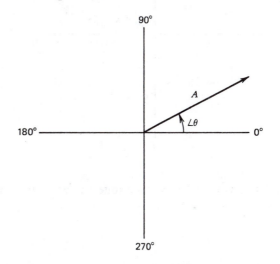

Fig. III.2 The vector A $\underline{/\theta}$ shown in the polar coordinate system. A vector expressed as A $\underline{/\theta}$ is said to be in polar form.

Common practice has created a shorthand for expressing vectors. This method is quicker to write and for many, more clearly expresses the idea of a vector:

$$A \underline{/\theta}$$

This shorthand is read as a vector magnitude A operating or pointing in the direction θ. It is termed the polar representation of a vector as shown in Figure III.2.

Now the function $e^{j\theta}$ may be expressed or resolved into its horizontal and vertical components in the complex plane:

$$e^{j\theta} = \cos\theta + j\sin\theta$$

The vector has been resolved and expressed in rectangular form. Using the shorthand notation

$$A\,\underline{/\theta} = A\cos\theta + jA\sin\theta$$

where $A\cos\theta$ is the vector projection on the real axis and $jA\sin\theta$ is the vector projection on the imaginary axis, as shown in Figure III.3.

Both rectangular and polar expressions of a vector quantity are useful when performing mathematical operations.

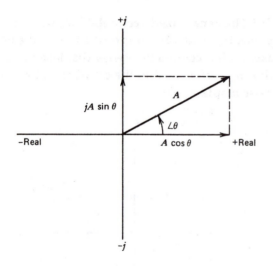

Fig. III.3 The vector A $\underline{/\theta}$ shown together with its rectangular components.

III.2.2 Addition and Subtraction of Vectors

When adding or subtracting vectors, it is most convenient to use the rectangular form. This is best demonstrated through an example. Suppose that we have two vectors, $20\underline{/30°}$ and $25\underline{/-45°}$, and these vectors are to be added. The quickest way to accomplish this is to resolve each vector into its rectangular components, add the real components, then add the imaginary components, and, if needed, express the results in polar form:

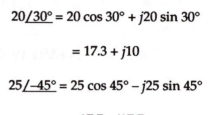

$$20\underline{/30°} = 20\cos 30° + j20\sin 30°$$

$$= 17.3 + j10$$

$$25\underline{/-45°} = 25\cos 45° - j25\sin 45°$$

$$= 17.7 - j17.7$$

A calculator is a handy tool for resolving vectors. Many calculators have automatic programs for converting vectors from one form to another.

Now adding we obtain

$$
\begin{array}{r}
17.3 + j10 \\
(+)\quad \underline{17.7 - j17.7} \\
35.0 - j7.7
\end{array}
$$

By inspection, this vector is seen to be slightly greater in magnitude than 35.0 and at a small angle below the positive real axis. Again using a calculator to express the vector in polar form: $35.8\underline{/-12.4°}$, an answer in agreement with what was anticipated. Figure III.4 shows roughly the same result using a graphical technique. Subtraction is accomplished in much the same way. Suppose that the vector $25\underline{/-45°}$ is to be subtracted from the vector $20\underline{/30°}$.

$$20\underline{/30°} = 17.3 + j10$$

$$25\,\underline{/-45°} = 17.7 - j17.7$$

To subtract

$$
\begin{array}{l}
17.3 + j10 \\
\underline{17.7 - j17.7}
\end{array}
$$

first change sign of the subtrahend and then add:

$$
\begin{array}{r}
17.3 + j10 \\
\underline{-17.7 + j17.7} \\
-0.4 + j27.7
\end{array}
$$

The effect of changing the sign of the subtrahend is to push the vector back through the origin. as shown in Figure III.5.

The resulting vector appears to be about 28 units long and barely in the second quadrant. The calculator gives $27.7\underline{/90.8°}$.

III.2.3 Multiplication and Division of Vectors

Vectors are expressed in polar form for multiplication and division. The magnitudes are multiplied or divided and the angles follow the rules governing exponents, added

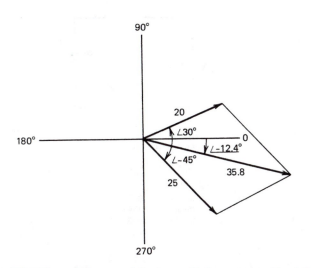

Fig. III.4 Use of the graphical parallelogram method for adding two vectors. The result or sum is the diagonal originating at the origin of the coordinate system.

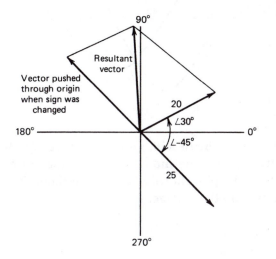

Fig. III.5 Graphical solution to subtraction of vectors.

when multiplying, subtracted when dividing. Consider

$$20\underline{/30°} \times 25\underline{/-45°} = 500\underline{/-15°}$$

The magnitudes are multiplied and the angles are added. Consider

$$\frac{20\underline{/30°}}{25\underline{/-45°}} = 0.8 \ \underline{/75°}$$

The magnitudes are divided and the angle of the divisor is subtracted from the angle of dividend.

Raising to powers is a special case of multiplication. The magnitude is raised to the power and the angle is multiplied by the power. Consider

$$(20\underline{/30°})^3 = 8000\underline{/90°}$$

or consider

$$(20\underline{/30°})^{1/2} = 4.47\underline{/15°}$$

III.2.4 Summary

Vector manipulation is straightforward and easy to do. This presentation is intended to refresh those techniques most commonly used by those working at a practical level with ac electrical circuits. It has been the author's intent to exclude material on dot and cross products in favor of techniques that tend to allow the user more of a feeling for what is going on.

III.3 RESISTANCE, INDUCTANCE, AND CAPACITANCE

The three types of electric circuit elements having distinct characteristics are resistance, inductance, and capacitance. This brief review will focus on the characteristics of these circuit elements in ac circuits to support later discussions on circuit impedance and power-factor-improvement principles.

III.3. 1 Resistance

Resistance R in an ac circuit is the name given to circuit elements that consume real power in the form of heat, light. mechanical work, and so on. Resistance is a physical property of the wire used in a distribution system that results in power loss commonly called I^2R loss. Resistance can be thought of as a name given that portion of a circuit load that performs real work, that is, the portion of the power fed to a motor that results in measurable mechanical work being accomplished.

If resistance is the only circuit element in an ac circuit, the physical properties of that circuit are easily summarized, as shown in Figure III.6. The important property is that the voltage and current are in phase.

Since the current and voltage are in phase and the ac source is a sine wave, the power used by the resistor is easily computed from root-mean-square (rms) (effective) voltage and current readings taken with a typical multimeter. The power is computed by taking the product of the measured voltage in volts or kilovolts and the measured current in amperes:

$$P(\text{watts}) = V(\text{volts}) \times I \ (\text{amperes})$$

where V is the voltage measured in volts and I is the current measured in amps. Both quantities are measured with an rms reading meter.

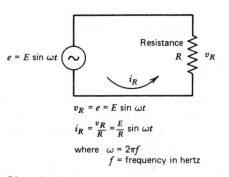

$$v_R = e = E \sin \omega t$$

$$i_R = \frac{v_R}{R} = \frac{E}{R} \sin \omega t$$

where $\omega = 2\pi f$
       f = frequency in hertz

Fig. III.6 Circuit showing an ac source with radian frequency ω. The current through the resistor is in phase with the voltage across the resistor.

In many industrial settings the voltage may be measured in kilovolts and current in amperes. The power is computed as the product of current and voltage and expressed as kilowatts:

$$P \text{ (kilowatts)} = V \text{ (kilovolts)} \times I \text{ (amps)}$$

If it is unhandy to measure both voltage and current, one can compute power using only voltage or current if the resistance R is known:

$$P \text{ (watts)} = I^2 \text{(amps)} \times R \text{ (ohms)}$$

or

$$P \text{ (watts)} = \frac{V^2 \text{ (volts)}}{R \text{ (ohms)}}$$

III.3.2 Inductance

Inductance L in an ac circuit is usually formed as coils of wire, such as those found in motor windings, solenoids, or inductors. In a real circuit it is impossible to have only pure inductance, but for purposes of establishing background we will take the theoretical case of a pure inductance so that its circuit properties can be isolated and presented.

An inductor is a circuit element that uses no real power; it simply stores energy in the form of a magnetic field and will give up this stored energy, alternately storing energy and giving it up every half-cycle. The result of this storing and giving up energy when an inductor is driven by a sine-wave source is to put the measured magnetizing current (i_c.) 90° out of phase with the driving voltage. The magnetizing current lags behind the driving voltage by 90°. If pure inductance were the load of a sine-wave generator, we could summarize its characteristics as in Figure III.7.

An inductor limits the current flowing through it by reacting with the voltage change across it. This property is

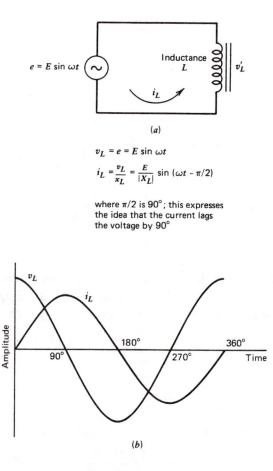

$$v_L = e = E \sin \omega t$$

$$i_L = \frac{v_L}{x_L} = \frac{E}{|X_L|} \sin (\omega t - \pi/2)$$

where π/2 is 90°; this expresses the idea that the current lags the voltage by 90°

Fig. III.7 (a) Ac circuit with pure inductance. (b) Plot of the voltage across the inductor v_L, and the current i_L through it. The plot shows a 90° displacement between the current and the voltage.

called inductive reactance X_L. The inductive reactance of a coil whose inductance is known in henrys (H) may be computed using the expression

$$X_L = 2\pi f L$$

where f is the frequency in hertz and L is the coil's inductance in henrys.

III.3.3 Capacitance

Capacitance C, like inductance, only stores and gives up energy. However, the voltage and current phasing is exactly opposite that of a inductor in an ac circuit. The current in an ac circuit containing only capacitance leads the voltage by $\underline{/90°}$. Figure III.8 summarizes the characteristics of an ac circuit with a pure capacity load. A capacitor also reacts to changes. This property is called capacitive reactance X_c. The capacity reactance may be computed by using the expression

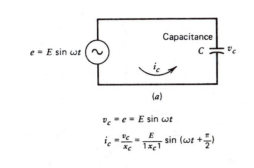

$$v_c = e = E \sin \omega t$$

$$i_c = \frac{v_c}{x_c} = \frac{E}{1 x_c 1} \sin \left(\omega t + \frac{\pi}{2} \right)$$

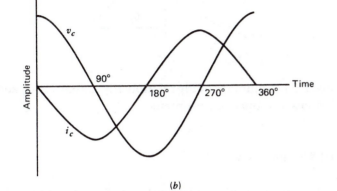

Fig. III.8 (*a*) Ac circuit with pure capacitance. (*b*) Plot of the voltage across the capacitor v_c, and the current i_c through it. The plot shows a 90° displacement between the current and the voltage.

$$X_c = \frac{1}{2\pi f C}$$

where f is the frequency in hertz and C is the capacity in farrads.

III.3.4 Summary

Circuit elements are resistance that consumes real power and two reactive elements that only store and give up energy. These two reactive elements, capacitors and inductors, have opposite effects on the phase displacement between the current and voltage in ac circuits. These opposite effects are the key to adding capacitors in an otherwise inductive circuit for purposes of reducing the current-voltage phase displacement. Reducing the phase displacement improves the power factor of the circuit. (Power factor is defined and discussed later.)

III.4 IMPEDANCE

In the preceding section it was mentioned that pure inductance does not occur in a real-world circuit. This is because the wire that is used to form the most carefully made coil

still has resistance. This section considers circuits containing resistance and inductive reactance and circuits containing resistance and capacity reactance. Attention will be given to the notation used to describe such circuits since vector algebra must be used exclusively.

III.4.1 Circuits with Resistance and Inductive Reactance

Figure III.9 shows a circuit that has both resistive and inductive elements. Such a circuit might represent a real inductor with the resistance representing the wire resistance, or such a circuit might be a simple model of a motor, with the inductance reflecting the inductive characteristics of the motor's windings and the resistance representing both the wire resistance and the real power consumed and converted to mechanical work performed by the motor.

In Figure III.9 the current is common to both circuit elements. Recall that the voltage across the resistor is in phase with this current while the voltage across the inductor leads the current. This idea is shown by plotting these quantities in the complex plane. Since i is the reference, it is plotted on the positive real axis as shown in Figure III.10.

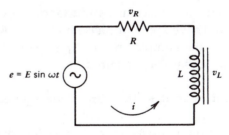

Fig. III.9 Circuit with both resistance and inductance. The circuit current i is common to both elements.

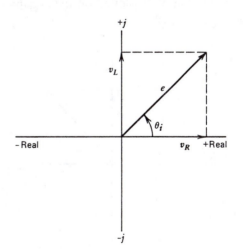

Fig. III.10 Circuit voltages and current plotted in the complex plane. Both i and V_R are on the positive real axis since they are in phase, V_L is on the positive j axis since it leads the current by 90°.

The voltage across the resistor is in phase with the current, so it is also on the positive real axis. whereas the voltage across the inductor is on the positive j axis since it leads the current by 90°. However, the sum of the voltages must be the source voltage e. Figure III.10 shows that the two voltages must be added as vectors:

$$e = v_r + jv_L$$

$$e = iR + jiX_L$$

If we call the ratio of voltage to current the circuit impedance, then

$$Z = \frac{e}{i} = R + jX_L$$

Z, the circuit impedance, is a complex quantity and may be expressed in either polar or rectangular form:

$$Z = R + jX_L$$

or

$$Z = |Z| \underline{/\theta}$$

In circuits with resistance and inductance the complex impedance will have a positive phase angle and if R and X_L are plotted in the complex plane, X_L is plotted on the positive j axis, as shown in Figure III.11.

III.4.2 Circuits with Resistance and Capacity Reactance

Circuits containing resistance and capacitance are approached about the same way. Going through a similar analysis and looking at the relationship among R, X_c, and Z would show that X_c is plotted on the negative j axis, as shown in Figure III.12.

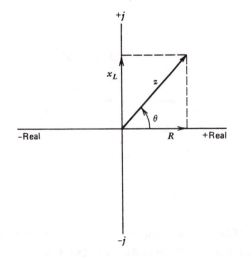

Fig. III.11 Plot in the complex plane showing the complex relationship of R, X_L, and Z.

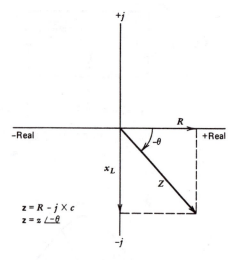

Fig. III.12 Summary of the relationship among R, X_c, and Z shown in the complex plane.

III.4.3 Summary

In circuits containing both resistive and reactive elements, the resistance is plotted on the positive real axis while the reactances are plotted on the imaginary axis. The fact that inductive and capacitive reactance causes opposite phase displacements (has opposite effects in ac circuits) is further emphasized by plotting their reactance effects in opposite directions on the imaginary axis of the complex plane. The case is building for why capacitors might be used in an ac circuit with inductive loading to improve the circuit's power factor.

III.5 POWER IN AC CIRCUITS

This section considers three aspects of power in ac circuits. First, the case of a circuit containing resistance and inductance is discussed, followed by the introduction of the power triangle for circuits containing resistance and inductance. Finally, power-factor improvement by the use of capacitors is presented.

III.5.1 Power in a Circuit Containing Both Resistance and Inductance

Figure III.13 reviews this situation through a circuit drawing and the voltages and currents shown in the complex plane. Meters are in place that read the effective or rms voltage V across the complex load and the effective or rms line current I.

Power is usually thought of as the product of voltage and the current in a circuit. The question is: The current I times *which* voltage will yield the correct or true power?

This is an important question, since Figure III.13b shows three voltages in the complex plane.

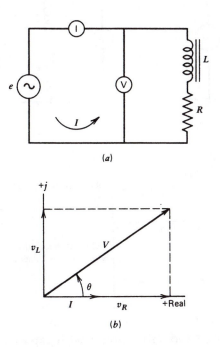

(a)

(b)

Fig. III.13 (*a*) **Circuit having resistance and inductance; meters are in place to measure the line current *I* and the voltage V. (*b*) Relationship between the various voltages and the line current for this circuit.**

Each of the three products may be taken, and each has a name and a meaning. Taking the ammeter reading *I* times the voltmeter reading yields the *apparent power*. The apparent power is the load current-load voltage product without regard to the phase relationship of the current and voltage. This figure by itself is meaningless:

$$P(\text{apparent}) = IV$$

If the voltmeter could be connected across the resistor only, to measure v_R, then the line current-voltage product would yield the *true power*, since the current and voltage are in phase.

$$P(\text{true}) = Iv_R$$

Usually, this connection cannot be made, so the true power of a load is measured with a special meter called a wattmeter that automatically performs the following calculation.

$$P(\text{true}) = IV \cos \theta$$

Note that in Figure III.13b, the circuit voltage V and the resistance voltage V_R are related through the cosine of θ. The third product that could be taken is called *imaginary power*

or VAR, the voltampere reactive product.

$$P(\text{imaginary}) = Iv_L$$

This is the power that is alternately stored and given up by the inductor to maintain its magnetic field. None of this reactive power is actually used.

If the voltages in the foregoing examples were measured in kilovolts, the three values computed would be the more familiar:

$$P(\text{apparent}) = kVA$$

$$P(\text{real}) = kW$$

$$P(\text{imaginary}) = kVAR$$

This discussion, together with Figure III.13b, leads to the power triangle.

III.5.2 The Power Triangle

The power triangle consists of three values, kVA, kW, and kVAR, arranged in a right triangle. The angle between the line current and voltage, H, becomes an important factor in this triangle. Figure III.14 shows the power triangle.

To emphasize the relationship between these three quantities, an example may be helpful. Suppose that we have a circuit with inductive characteristics and using a voltmeter, ammeter, and wattmeter the following values are measured:

$$\text{watts} = 1.5 \text{ kW}$$

$$\text{line current} = 10 \text{ A}$$

$$\text{line voltage} = 240 \text{ V}$$

From this information we should be able to determine the kVA, H, and the kVAR.

The kVA can be computed directly from the voltmeter and ammeter readings:

$$kVA = (10 \text{ A})(0.24) \text{ kV} = 2.4 \text{ kVA}$$

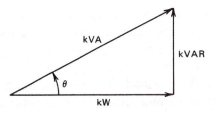

Fig. III.14 Power triangle for an inductive load. The angle θ is the angle of displacement between the line voltage and the line current.

Looking at the triangle in Figure III.14 and recalling some basic trigonometry, we have

$$\cos \theta = \frac{kW}{kVA} = \frac{1.5}{2.4} = 0.625$$

and θ is the angle whose cosine equals 0.625. This can be looked up in a table or calculated using a hand calculator that computes trig functions:

$$\theta = \cos^{-1} 0.625 = 51.3°$$

Again referring to the power triangle and a little trig, we see that

$$kVAR = kVA \sin \theta$$

$$= 2.4 \ kVA \sin 51.3°$$

$$= 1.87 \ kVAR$$

Figure III.15 puts all these measured and calculated data together in a power triangle.

Of particular interest is the ratio kW/kVA. This ratio is called the *power factor* (PF) of the circuit. So the power factor is the ratio of true power to apparent power in a circuit. This is also the cosine of the angle θ, the angle of displacement between the line voltage and the line current. To improve the power factor, the angle θ must be reduced. This could be accomplished by reducing the kVAR side of the triangle.

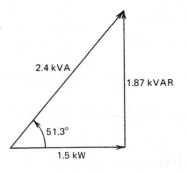

Fig. III.15 Organization of the measured and computed data of the example into a power triangle.

III.5.3 Power-Factor Improvement

Recall that inductive reactance and capacity reactance are plotted in opposite directions on the imaginary axis, *j*. Thus it should be no surprise to consider that kVAR produced by a capacity load behave in an opposite way to kVAR produced by inductive loads. This is the case and is the reason capacitors are commonly added to circuits having inductive loads to improve power factor (reduce the angle θ).

Suppose in the example being considered that enough capacity is added across the load to offset the effects of 90%

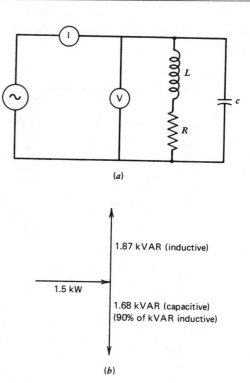

(a)

(b)

Fig. III.16 (a) Inductive circuit with capacity added to correct power factor. (b) Power vectors showing the relationship among kW, kVAR inductive, and kVAR capacitive.

of the inductive load. That is, we will try to improve the power factor by better than 90%. Figure III.16 shows the circuit arrangement with the kW and kVAR vectors drawn to show their relationship.

Following the example through, consider Figure III.17, where 90% of the kVAR inductive load has been neutralized by adding the capacitor.

Working with the modified triangle in Figure III.17, we can compute the new θ, call it θ_2.

$$\theta_2 = \tan^{-1} \frac{0.19}{1.5}$$

$$= 7.2°$$

Again, a calculator comes in handy.

Since the new power factor is the cosine of θ_2, we compute

$$PF \ new = \cosine \ 7.2 = 0.99$$

certainly an improvement.

Recall that the power factor can be expressed as a ratio of kW to kVA. From this idea we can compute a new kVA value:

$$PF = 0.99 = \frac{1.5 \ kW}{kVA \ new}$$

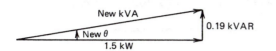

Fig. III.17 Resulting net power triangle when the capacitor is added. A new kVA can be calculated as well as a new θ.

or

$$kVA \text{ new} = \frac{1.5 \text{ kW}}{0.99}$$

$$= 1.52 \text{ kVA}$$

The line voltage did not change, so the line current must be lower.

$$1.52 \text{ kVA new} = 0.24 \text{ kV} \times I \text{ new}$$

$$I \text{ new} = \frac{1.52 \text{ kVA}}{0.24} = 6.3 \text{ A}$$

Comparing the original circuit to the circuit after adding capacity, we have:

	Inductive Circuit	Improved Circuit
Line voltage	240 V	240 V
Line current	10 A	6.3 A
PF	62.5%	99%
kVA	2.4 kVA	1.52 kVA
kW	1.5 kW	1.5 kW
kVAR	1.87 kVAR	0.19 kVAR

The big improvement noted is the reduction of line current by 37% with no decrease in real power, kW, used by the load. Also note the big change in kVA; less generating capacity is used to meet the same real power demand.

III.5.4 Summary

Through an example it has been demonstrated how the addition of a capacitor across an inductive load can improve power factor, reduce line current, and reduce the amount of generating capacity required to supply the load. The way this comes about is by having the capacitor supply the inductive magnetizing current locally. Since inductive and capacitive elements store and release power at different times in each cycle, this reactive current simply flows back and forth between the capacitor and inductor of the load. This idea is reinforced by Figure III.18. Adding capacitors to inductive loads can free generating capacity, reduce line loss, improve power factor, and in general be cost effective in controlling energy bills.

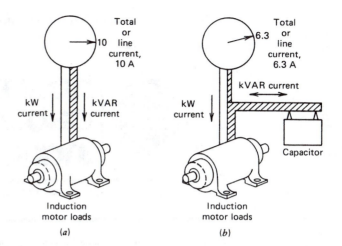

Fig. III.18 (a) Pictorial showing the inductive load of the example in this section. (b) The load with a capacitor added. With the exchange of the kVAR current between the capacitor and inductive load, very little kVAR current is supplied by the generator.

III.6 THREE-PHASE POWER

Three-phase power is the form of power most often distributed to industrial users. This form of transmission has three advantages over single-phase systems: (1) less copper is required to supply a given power at given voltage; (2) if the load of each phase of the three-phase source is identical, the instantaneous output of the alternator is constant; and (3) a three-phase system produces a magnetic field of constant density that rotates at the line frequency—this greatly reduces the complexity of motor construction.

The author realizes that both delta systems and wye systems exist. but will concentrate on four-wire wye systems as being representative of internal distribution systems. This type of internal distribution system allows the customer both single-phase and three-phase service. Our focus will be on measuring power and determining power factor in four-wire three-phase wye-connected systems.

III.6.1 The Four-Wire Wye-Connected System

Figure III.19 shows a generalized four-wire wye-connected system. The coils represent the secondary windings of the transformers at the site substation while the generalized loads represent phase loads that are the sum loads on each phase. These loads may be composites of single-phase services and three-phase motors being fed by the distribution system. N is the neutral or return.

To determine the power and power factor of any phase A, B, or C, consider that phase as if it were a single-

Table III.1 How to Select Capacitor Ratings for Induction Motors/*Source: 1.*

how to select capacitor ratings for induction motors

Reference No. 1: For motor designs <u>pre-dating</u> TRI-CLAD 700® and CUSTOM 8000® Motors
(See GED-6063-02, Reference No. 2, for TRI-CLAD and CUSTOM 8000 motor designs)

Now it's easy to choose the right capacitors for your induction motors. Just refer to the following tables to find the kvar required by your particular motors. Locate motors by horsepower, rpm, and number of poles. All ratings are based on General Electric motor designs.

Tables I and V are also applicable to standard wound-rotor, open-type, three-phase, 60-cycle motors, provided the kvar values in the table are multiplied by a factor of 1.1, and the reduction in line current is increased by multiplying the values in the table by 1.05.

When selecting and installing capacitors, keep in mind the following: A capacitor located at the motor releases the maximum system capacity and is most effective in reducing system losses. Also, for a motor that runs continuously, or nearly so, it is usually most economical to locate the capacitor right at the motor terminals and switch it with the motor.

TABLE I—220-, 440-, AND 550-VOLT MOTORS, ENCLOSURE OPEN—INCLUDING DRIPPROOF AND SPLASHPROOF, GENERAL ELECTRIC TYPE K (NEMA DESIGN "B"), NORMAL STARTING TORQUE AND CURRENT

Induction Motor Horsepower Rating	3600 / 2		1800 / 4		1200 / 6		900 / 8		720 / 10		600 / 12	
	Kvar	% AR	Kvar	% AR	Kvar	% AR	Kvar	% AR	Kvar	% AR	Kvar	% AR
2	1	16	1	20	1	22	1	24	...	..	...	..
3	1	10	1	16	2	21	2	24	...	..	...	..
5	1	9	2	16	2	21	2	21	...	..	...	..
7½	1	8	2	13	2	15	4	21	5	29	...	..
10	2	8	2	13	4	15	5	21	5	25	7.5	34
15	4	8	4	13	4	13	5	15	7.5	23	7.5	25
20	4	7	5	9	5	12	7.5	15	10	23	10	24
25	5	7	5	9	7.5	11	7.5	12	10	23	10	23
30	5	7	7.5	9	5	11	10	12	10	15	10	19
40	5	5	7.5	9	10	11	10	12	10	15	15	19
50	7.5	5	10	7	10	9	15	12	25	15	25	19
60	7.5	5	10	7	10	9	15	11	20	15	30	19
75	10	5	10	7	15	9	15	10	30	15	40	19
100	15	5	20	7	25	9	30	10	40	15	45	17
125	15	5	20	7	30	9	35	10	45	15	50	17
150	15	5	25	6	30	9	40	9	50	13	60	17
200	40	5	40	6	45	8	50	9	70	13	75	17
250	45	5	50	6	50	8	70	9	75	12	90	17
300	50	5	50	6	70	8	75	9	105	11	105	17
350	50	5	50	6	75	8	80	9	105	11	105	17
400	60	5	60	6	75	8	100	9	100	11	110	17
450	60	5	75	5	75	6	100	9	100	11	110	17
500	70	5	90	5	90	6	110	9	120	11	120	17

TABLE II—220-, 440-, 550-VOLT MOTOR, TOTALLY-ENCLOSED, FAN-COOLED, GENERAL ELECTRIC TYPE K (NEMA DESIGN "B"), NORMAL STARTING TORQUE, NORMAL STARTING CURRENT

Induction Motor Horsepower Rating	3600 / 2		1800 / 4		1200 / 6		900 / 8		720 / 10		600 / 12	
	Kvar	% AR	Kvar	% AR	Kvar	% AR	Kvar	% AR	Kvar	% AR	Kvar	% AR
2	1	17	1	20	1	23	1	24	...	..	...	..
3	1	11	1	16	1	19	2	24	...	..	...	..
5	1	9	2	15	2	19	2	20	...	..	...	..
7½	1	6	2	13	2	19	4	20	...	..	...	..
10	2	6	2	11	4	16	5	15	7.5	20	7.5	27
15	4	6	4	11	4	13	5	15	7.5	20	7.5	24
20	4	6	5	11	5	13	7.5	15	10	20	10	24
25	5	6	5	8	5	9	7.5	15	10	17	10	18
30	5	6	7.5	8	7.5	9	10	15	10	15	10	18
40	7.5	6	10	8	10	9	10	15	10	15	15	17
50	7.5	6	10	8	10	9	15	12	15	12	20	17
60	10	6	10	8	10	8	15	12	20	12	25	17
75	15	6	15	8	15	9	20	11	25	12	35	17
100	15	6	20	8	25	9	25	11	40	12	45	17
125	20	6	25	8	30	9	30	11	45	12	45	15
150	25	6	30	7	30	9	40	11	45	12	50	15
200	35	6	40	7	40	9	60	11	55	11	60	13
250	40	5	50	6	60	9	80	11	60	11	100	13
300	50	5	45	6	80	8	80	10	80	10	125	13
350	60	5	70	6	80	8	80	9	...	..	...	..
400	60	5	80	6	80	6	160		...	..	...	..
450	70	5	100	6	...	..	...	..	...	..	...	..
500	70	.5	...	..	...	..	...	..	...	..	...	..

TABLE III—220-, 440-, 550-VOLT MOTORS, ENCLOSURE OPEN—INCLUDING DRIPPROOF AND SPLASHPROOF, GENERAL ELECTRIC TYPE KG (NEMA DESIGN "C"), HIGH STARTING TORQUE AND NORMAL STARTING CURRENT

Induction Motor Horsepower Rating	1800 / 4		1200 / 6		900 / 8		720 / 10	
	Kvar	% AR	Kvar	% AR	Kvar	% AR	Kvar	% AR
3	...	..	2	19	2	30	...	...
5	1	15	2	19	4	26	...	...
7½	2	15	2	15	4	22	...	...
10	2	12	4	12	4	17	...	...
15	4	12	4	12	5	17	...	...
20	4	12	5	12	7.5	17	...	...
25	7.5	9	7.5	11	10	17	...	...
30	7.5	9	7.5	11	10	15	...	...
40	10	9	10	11	10	13	...	...
50	10	9	10	9	15	13	...	...
60	10	8	15	9	20	13	20	18
75	15	8	15	9	20	12	30	18
100	20	8	25	9	30	12	40	18
125	20	7	30	8	30	12	50	18
150	35	7	35	8	45	12	...	...
200	40	7	50	8	70	12	...	...
250	50	7	50	8	80	12	...	...
300	60	7	80	8	80	10	...	...
350	60		80	8	80	9	...	...

NOTE: A capacitor located on the motor side of the overload relay reduces current through the relay, and therefore, a smaller relay may be necessary. The motor-overload relay should be selected on the basis of the motor full-load nameplate current reduced by the percent reduction in line current (% AR) due to capacitors.

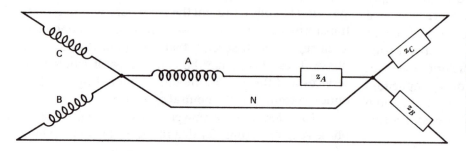

Fig. III.19 Generalized four-wire wye-connected system. The coils A, B, and C represent the three transformer secondaries at the site substation; while Z_A, Z_B and Z_C are the generalized loads seen by each phase.

TABLE IV—220-, 440-, 550-VOLT MOTORS, TOTALLY-ENCLOSED, FAN-COOLED, GENERAL ELECTRIC TYPE KG (NEMA DESIGN "C"), HIGH STARTING TORQUE, NORMAL STARTING CURRENT

Induction Motor Horsepower Rating	1800 / 4		1200 / 6		900 / 8		720 / 10		600 / 12	
	Kvar	%AR	Kvar	%AR	Kvar	%AR	Kvar	%AR	Kvar	%AR
3	...	...	2	20	2	27	...	...	...	...
5	2	10	2	19	2	21	...	...	...	...
7½	2	10	4	17	4	21				
10	2	9	4	17	5	21				
15	4	9	5	14	5	16				
20	5	9	5	11	7.5	16				
25	7.5	9	5	10	10	15	10	15	10	18
30	7.5	9	7.5	10	10	15	10	15	10	18
40	10	9	10	9	10	12	15	15	15	18
50	10	9	10	9	15	12	15	15	25	18
60	15	9	15	9	20	12	30	15	30	18
75	15	7	15	8	20	11	30	15	40	16
100	20	7	25	8	25	11	40	15	45	16
125	25	7	35	8	35	11	50	15	50	15
150	35	7	35	7	50	11	60	14	50	13
200	40	7	45	7	50	10	70	14	70	13

TABLE V—2300- AND 4000-VOLT MOTORS, ENCLOSURE OPEN—INCLUDING DRIPPROOF AND SPLASHPROOF, GENERAL ELECTRIC TYPE K (NEMA DESIGN "B"), NORMAL STARTING TORQUE AND CURRENT

Induction Motor Horsepower Rating	3600 / 2		1800 / 4		1200 / 6		900 / 8		720 / 10		600 / 12	
	Kvar	%AR	Kvar	%AR	Kvar	%AR	Kvar	%AR	Kvar	%AR	Kvar	%AR
100	...	...	25	10	25	9	25	11	25	11	25	11
125	25	7	25	8	25	9	25	9	25	10	50	15
150	25	6	25	7	25	7	25	9	50	11	50	14
200	25	6	25	6	50	8	50	9	50	11	75	14
250	25	7	50	6	50	8	50	9	75	11	100	14
300	50	6	50	6	75	8	75	9	75	9	100	13
350	50	5	50	5	75	8	75	9	100	9	100	12
400	50	5	50	5	75	6	100	9	100	9	125	12
450	75	5	75	5	75	5	100	8	100	8	100	8
500	75	6	75	5	100	6	125	8	125	8	125	8
600	75	5	100	5	125	6	125	7	125	8	150	8
700	100	5	100	5	100	5	125	7	150	8	150	8
800	100	5	125	5	125	5	150	7	150	8	150	8

TABLE VI—2300- AND 4000-VOLT MOTORS, TOTALLY-ENCLOSED, FAN-COOLED, GENERAL ELECTRIC TYPE K (NEMA DESIGN "B"), NORMAL STARTING TORQUE, NORMAL STARTING CURRENT

Induction Motor Horsepower Rating	3600 / 2		1800 / 4		1200 / 6		900 / 8		720 / 10		600 / 12	
	Kvar	%AR	Kvar	%AR	Kvar	%AR	Kvar	%AR	Kvar	%AR	Kvar	%AR
100	...	...	25	8	25	8	25	9	25	10	25	12
125	25	5	25	7	25	8	25	9	25	9	50	14
150	25	6	25	6	25	8	50	10	50	10	50	14
200	25	6	25	6	50	9	50	11	50	10	75	14
250	25	4	50	7	50	8	75	10	75	12	75	13
300	50	5	50	7	50	7	75	10	100	12	100	14
350	50	5	50	6	75	9	75	9	100	11	125	15
400	75	5	75	7	125	9	100	9	100	9	125	14
450	50	4	100	8	100	8	100	8	100	10	125	13
500	75	5	125	8	125	8	125	8	125	10	150	13

TABLE VII—2300- AND 4000-VOLT MOTORS, ENCLOSURE OPEN—INCLUDING DRIPPROOF AND SPLASHPROOF, GENERAL ELECTRIC TYPE KG (NEMA DESIGN "C"), HIGH STARTING TORQUE AND NORMAL STARTING CURRENT

Induction Motor Horsepower Rating	1800 / 4		1200 / 6		900 / 8		720 / 10	
	Kvar	%AR	Kvar	%AR	Kvar	%AR	Kvar	%AR
100	25	7	25	10	25	10	25	10
125	25	7	25	9	25	9	25	10
150	25	8	25	9	25	9		
200	50	9	50	9	50	10		
250	50	8	50	8	50	9		
300	50	7	75	9	75	10		
350	50	6	75	8	75	9		

TABLE VIII—2300- AND 4000-VOLT MOTORS, TOTALLY-ENCLOSED, FAN-COOLED, GENERAL ELECTRIC TYPE KG (NEMA DESIGN "C"), HIGH-STARTING TORQUE, NORMAL STARTING CURRENT

Induction Motor Horsepower Rating	1200 / 6		900 / 8		720 / 10		600 / 12	
	Kvar	%AR	Kvar	%AR	Kvar	%AR	Kvar	%AR
75					25	16	25	12
100	25	9	25	9	25	12	50	12
125	25	9	25	8	50	18	50	17
150	25	8	50	14	50	15	50	15
200	25	7	50	10	75	15	74	14

TABLE IX—440-VOLT OPEN, DRIPPROOF, OIL-FIELD MOTORS, 1200-RPM, GENERAL ELECTRIC TYPE KG, KR AND KOF

Induction Motor Horsepower Rating	High Starting Torque—Low Starting Current					
	Type KG Motor		Type KR Motor		Type KOF Motor	
	Kvar	% AR	Kvar	% AR	Kvar	% AR
5	3	25½	3	14½	4	24
7½	5	25½	4	20½	6	26
10	5	13½	4	12½	6	19
15	6	17½	6	14	6	13
20	8	17	10	20½	10	16
25	10	15	10	14½	10	15
30	10	16	10	14½	10	13
40	10	13½	10	9	10	10
50	10	10½	10	7½	15	10
60	10	9	15	9½	15	9
75	15	10½	15	8½	...	...
100	30	13	...	...	...	...

CAPACITORS BENEFIT YOUR DISTRIBUTION SYSTEM BY

- Reducing power costs
- Releasing system capacity
- Improving voltage levels
- Reducing system losses

INDUSTRIAL AND POWER CAPACITOR PRODUCTS DEPT. ● GENERAL ELECTRIC CO. ● HUDSON FALLS, N. Y.

GENERAL ⊛ ELECTRIC

GED-6063.01A
12-72 (10M) 6200

phase system. Measure the real power, kW, delivered by the phase by use of a wattmeter and measure and compute the volt-ampere product, apparent power, kVA, using a voltmeter and ammeter.

The power factor of the phase can then be determined and corrected as needed. Each phase can be treated independently in turn. The only caution to note is to make the measurements during nominal load periods, this will allow power-factor correction for the most common loading.

If heavy motors are subject to intermittent duty, additional power and power-factor information can be gathered while they are operating. Capacitors used to correct power factor for these intermittent loads should be connected to relays so that they are across the motors and on phase only when the motor is on; otherwise, overcorrection can occur.

In the special case of a four-wire wye-connected system with balanced loading. two wattmeters may be used to monitor the power consumed on the service and also allow computation of the power factor from the two wattmeter readings.

Table III.2 How to Select Capacitor Ratings for Induction Motors/*Source: 2.*

how to select capacitor ratings for induction motors

Reference No. 2: For TRI-CLAD 700® and CUSTOM 8000® motors only.
(See GED-6063-01, Reference No. 1, for motor designs pre-dating TRI-CLAD 700 and CUSTOM 8000 Line)

Now it's easy to choose the right capacitors for your induction motors. Just refer to the following tables to find the kvar required by your particular motors. Locate motors by horsepower, rpm, and number of poles. All ratings are based on General Electric motor designs.

Tables I and V are also applicable to standard, wound-rotor, open-type, three-phase, 60-cycle motors, provided the kvar values in the table are multiplied by a factor of 1.1, and the reduction in line current is increased by multiplying the values in the table by 1.05.

When selecting and installing capacitors, keep in mind the following: A capacitor located at the motor releases the maximum system capacity and is most effective in reducing system losses. Also, for a motor that runs continuously, or nearly so, it is usually most economical to locate the capacitor right at the motor terminals and switch it with the motor.

TABLE II—230-, 460-, 575-VOLT MOTOR, TOTALLY ENCLOSED, FAN-COOLED GENERAL ELECTRIC TYPE K (NEMA DESIGN "B"), NORMAL STARTING TORQUE, NORMAL STARTING CURRENT

Induction Motor Horse-power Rating	Nominal Motor Speed in Rpm and Number of Poles											
	3600		1800		1200		900		720		600	
	Kvar	%AR	Kvar	%AR	Kvar	%AR	Kvar	%AR	Kvar	%AR	Kvar	%AR
2	1	15	2	24	1	28	2	42	...	...	3	50
3	1	15	2	24	2	26	4	29	4	35	4	49
5	2	15	3	20	4	22	4	29	4	35	5	49
7½	3	15	4	20	4	22	5	29	7.5	35	10	49
10	3	11	4	17	5	22	5	29	7.5	35	10	41
15	4	10	5	17	5	22	10	29	7.5	24	10	34
20	5	10	5	17	10	21	10	23	10	23	15	34
25	5	10	7.5	17	10	21	10	22	10	22	20	31
30	5	10	7.5	17	10	21	15	22	15	22	20	31
40	10	10	10	12	15	21	20	22	20	22	30	31
50	10	10	15	12	25	21	23	21	25	22	35	31
60	12	9	15	12	25	21	25	19	30	22	40	30
75	15	9	20	12	25	15	30	19	35	21	40	30
100	20	9	30	12	25	13	40	19	40	12	50	30
125	20	9	35	12	30	13	45	18	50	12	50	30
150	25	9	40	8	25	13	55	18	50	12	70	30
200	30	9	40	8	60	13	60	18	70	12	75	30
250	60	9	50	8	60	13	115	18	100	12	125	30
300	65	7	50	8	60	13	140	18	125	12	150	30
350	70	7	55	8	80	13	160	18	150	12	150	30
400	70	7	60	8	130	13	160	17	175	12	175	30
450	90	7	95	8	145	13	160	17	175	12	200	30
500	100	7	110	7	170	13	210	17	...	...	...	...

TABLE I—230-, 460-, AND 575-VOLT MOTORS, ENCLOSURE OPEN—INCLUDING DRIPPROOF GENERAL ELECTRIC TYPE K (NEMA DESIGN "B"), NORMAL STARTING TORQUE AND CURRENT

Induction Motor Horse-power Rating	Nominal Motor Speed in Rpm and Number of Poles											
	3600 2		1800 4		1200 6		900 8		720 10		600 12	
	Kvar	%AR	Kvar	%AR	Kvar	%AR	Kvar	%AR	Kvar	%AR	Kvar	%AR
2	1	14	1	24	1	28	2	42	...	...	3	50
3	1	14	2	24	2	28	4	42	4	40	4	49
5	2	14	2	21	3	26	4	31	4	40	5	49
7½	2	14	4	21	4	26	5	26	7.5	36	10	49
10	4	14	4	17	5	21	7.5	26	7.5	31	10	41
15	5	12	5	17	5	19	10	26	7.5	31	10	34
20	5	11	7.5	17	7.5	19	10	23	10	29	15	34
25	7.5	11	7.5	17	7.5	19	10	23	10	24	20	34
30	7.5	10	7.5	17	10	19	15	23	15	24	20	32
40	7.5	10	15	17	15	19	20	23	20	24	30	32
50	10	10	20	17	20	19	25	23	20	24	35	32
60	10	10	20	17	20	19	30	23	30	23	45	32
75	15	10	25	14	30	16	30	17	35	21	40	19
100	15	10	30	14	30	12	35	16	40	15	45	17
125	30	10	35	12	30	12	50	16	50	15	50	17
150	30	10	35	11	35	12	50	14	50	13	60	17
200	35	10	50	11	55	12	70	14	70	13	90	17
250	35	10	55	9	70	12	85	14	90	13	100	17
300	35	10	65	9	75	12	95	14	100	13	110	17
350	40	10	80	9	85	12	125	14	120	13	150	17
400	100	10	80	8	100	12	140	14	150	13	150	17
450	100	10	90	8	140	12	150	12	150	13	175	17
500	100	8	115	8	150	12	150	12	175	13	175	17

TABLE III—230-, 460-, 575-VOLT MOTORS, ENCLOSURE OPEN—INCLUDING DRIPPROOF, GENERAL ELECTRIC TYPE KG (NEMA DESIGN "C") HIGH STARTING TORQUE AND NORMAL STARTING CURRENT

Induction Motor Horse-power Rating	Nominal Motor Speed in Rpm and Number of Poles									
	1800 4		1200 6		900 8		720 10		600 12	
	Kvar	%AR	Kvar	%AR	Kvar	%AR	Kvar	%AR	Kvar	%AR
3	2	21	2	28	4	42	...	...	...	...
5	4	21	4	26	4	32	...	...	...	...
7.5	4	17	5	22	4	29	...	...	...	...
10	5	17	5	22	10	29	...	...	20	40
15	5	17	7.5	21	10	25	...	...	...	...
20	5	17	7.5	21	10	23	...	...	30	40
25	7.5	17	10	21	10	23	...	...	35	39
30	7.5	17	10	21	15	23	20	28	45	39
40	10	17	15	21	20	23	30	28		
50	20	17	15	21	23	23	30	28		
60	20	15	30	21	30	23	35	28		
75	25	14	30	17	40	23	45	28		
100	30	13	30	14	50	23	40	15	45	17
125	35	12	40	14	50	16	45	15	50	17
150	35	10	45	13	50	14	50	13	60	17
200	50	10	55	11	70	14	70	13	90	17
250	55	9	70	11	85	14	90	13	100	17
300	65	9	75	10	95	14	100	13	110	17
350	80	8	85	9	125	14	120	13	150	17

NOTE: A capacitor located on the motor side of the overload relay reduces current through the relay, and therefore, a smaller relay may be necessary. The motor-overload relay should be selected on the basis of the motor full-load nameplate current reduced by the percent reduction in line current (% AR) due to capacitors.

Balanced Four-Wire Wye-Connected System

Figure III.20 shows a balanced system containing two wattmeters. The sum of these two wattmeter readings are the total real power being used by the service:

$$P_T = P_1 + P_2$$

Further, the angle of displacement between each line current and voltage can be computed from P_1 and P_2:

$$\theta = \tan^{-1} \sqrt{3}\, \frac{P_2 - P_1}{P_2 + P_1}$$

and the power factor PF = cos θ.

This quick method for monitoring power and power factor is useful in determining both fixed capacitors to be tied across each phase for the nominal load, and the capacitors that are switched in only when intermittent loads come on-line.

TABLE IV—230-, 460-, 575-VOLT MOTORS, TOTALLY-ENCLOSED, FAN-COOLED, GENERAL ELECTRIC TYPE KG (NEMA DESIGN "C"), HIGH STARTING TORQUE, NORMAL STARTING CURRENT

Induction Motor Horse-power Rating	Nominal Motor Speed in Rpm and Number of Poles									
	1800 / 4		1200 / 6		900 / 8		720 / 10		600 / 12	
	Kvar	%AR	Kvar	%AR	Kvar	%AR	Kvar	%AR	Kvar	%AR
3	...	...	2	28	4	42	...	...	...	...
5	2	20	3	26	4	32	...	...	...	...
7.5	4	19	4	22	4	30	...	...	...	...
10	4	19	5	22	5	30	...	...	...	...
15	4	19	7.5	22	10	30	...	...	...	...
20	7.5	19	10	21	10	29	...	...	20	40
25	7.5	19	10	19	10	29	...	...	...	...
30	7.5	19	10	19	20	29	25	31	30	40
40	15	19	15	19	19	24	...	...	35	38
50	15	19	20	20	20	22	30	28	40	37
60	25	19	25	18	25	22	35	28	40	37
75	25	13	25	15	35	22	45	28	40	37
100	30	13	25	12	45	20	40	28	50	37
125	35	12	30	12	45	20	50	28	50	37
150	40	11	40	12	50	20	50	28	70	37
200	40	8	60	12	60	20	70	28	75	38

TABLE V—2400- AND 4160-VOLT MOTORS, ENCLOSURE OPEN—INCLUDING DRIPPROOF AND SPLASHPROOF, GENERAL ELECTRIC TYPE K (NEMA DESIGN "B") NORMAL STARTING TORQUE AND CURRENT

Induction Motor Horse-power Rating	Nominal Motor Speed in Rpm and Number of Poles											
	3600 / 2		1800 / 4		1200 / 6		900 / 8		720 / 10		600 / 12	
	Kvar	%AR	Kvar	%AR	Kvar	%AR	Kvar	%AR	Kvar	%AR	Kvar	%AR
100	...	..	25	11	25	12	50	24	25	14	25	20
125	...	..	25	9	25	12	25	13	25	14	50	20
150	25	9	25	9	50	13	50	14	25	14	75	20
200	25	9	50	9	50	12	50	13	75	14	100	20
250	50	9	50	8	75	12	75	13	75	14	100	20
300	50	9	50	8	75	12	100	13	100	14	125	20
350	75	9	50	8	75	12	100	12	100	14	125	19
400	75	9	50	8	100	12	100	12	100	14	150	19
450	100	9	75	8	100	12	125	11	125	14	150	19
500	100	8	100	8	125	12	125	11	150	14	200	19
600	125	8	125	8	175	12	150	11	150	14	200	17
700	150	8	150	8	200	11	150	10	200	14	200	15
800	175	8	150	7	175	10	175	10	225	13	250	15

TABLE VI—2400- AND 4160-VOLT MOTORS, TOTALLY-ENCLOSED, FAN-COOLED, GENERAL ELECTRIC TYPE K (NEMA DESIGN "B"), NORMAL STARTING TORQUE, NORMAL STARTING CURRENT

Induction Motor Horse-power Rating	Nominal Motor Speed in Rpm and Number of Poles											
	3600 / 2		1800 / 4		1200 / 6		900 / 8		720 / 10		600 / 12	
	Kvar	%AR	Kvar	%AR	Kvar	%AR	Kvar	%AR	Kvar	%AR	Kvar	%AR
100	...	...	25	17	...	...	50	22	25	12	50	15
125	...	...	50	17	25	15	50	17	25	12	50	15
150	25	6	25	12	30	15	50	17	50	12	75	15
200	25	6	50	12	75	15	50	17	50	12	100	15
250	25	6	50	11	75	15	75	17	75	12	100	15
300	50	6	50	11	75	13	125	17	100	12	125	15
350	50	6	50	11	75	13	125	17	125	12	150	15
400	75	6	125	11	125	13	150	17	150	12	200	15
450	75	6	125	10	150	13	175	17	200	12	225	15
500	75	6	125	8	175	13	225	17	225	12	225	15

TABLE VII—2400- AND 4160-VOLT MOTORS, ENCLOSURE OPEN—INCLUDING DRIPPROOF AND SPLASHPROOF, GENERAL ELECTRIC TYPE KG (NEMA DESIGN "C"), HIGH STARTING TORQUE AND NORMAL STARTING CURRENT

Induction Motor Horse-power Rating	Nominal Motor Speed in Rpm and Number of Poles							
	1800 / 4		1200 / 6		900 / 8		720 / 10	
	Kvar	%AR	Kvar	%AR	Kvar	%AR	Kvar	%AR
100	25	10	25	11	25	13	25	14
125	25	8	25	9	25	13	25	14
150	25	7	25	9	50	13	50	14
200	25	8	50	12	75	13	75	14
250	50	8	50	12	75	13	75	14
300	50	8	75	12	100	13	100	14
350	50	8	75	12	100	12	100	14

TABLE VIII—2400- AND 4160-VOLT MOTORS, TOTALLY-ENCLOSED, FAN-COOLED, GENERAL ELECTRIC TYPE KG (NEMA DESIGN "C"), HIGH STARTING TORQUE, NORMAL STARTING CURRENT

Induction Motor Horse-power Rating	Nominal Motor Speed in Rpm and Number of Poles							
	1200 / 6		900 / 8		720 / 10		600 / 12	
	Kvar	%AR	Kvar	%AR	Kvar	%AR	Kvar	%AR
75	...	...	...	...	25	12	50	15
100	25	10	50	17	25	12	50	15
125	25	..	50	17	50	12	75	15
150	...	...	50	17	50	12	75	15
200	75	15	50	17	50	12	100	15

CAPACITORS BENEFIT YOUR DISTRIBUTION SYSTEM BY

- Reducing power costs
- Releasing system capacity
- Improving voltage levels
- Reducing system losses

INDUSTRIAL AND POWER CAPACITOR PRODUCTS DEPT. ● GENERAL ELECTRIC CO. ● HUDSON FALLS, N.Y.

GENERAL ⊕ ELECTRIC

GED-6063.02A
12-72 (10M) 6200

The two-wattmeter method is useful for determining real power consumed in either wye- or delta-connected systems with or without balanced loads:

$$P_T = P_1 + P_2$$

However, the use of these readings for determining phase power factor as well is restricted to the case of balanced loads.

III.6.2 Summary

This brief coverage of power and power-factor determination in three-phase systems covers only the very basic ideas in this important area. It is the aim of this brief coverage to recall or refresh ideas once learned but seldom used.

Tables III.1 and III.2 were supplied by General Electric. who gave permission for the reproduction of their materials in this handbook.

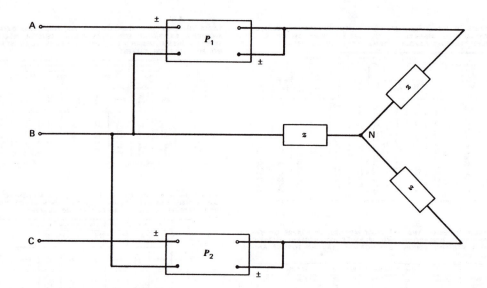

Fig. III.20 Four-wire wye-connected system with wattmeter connections detailed. Solid circle voltage connections to wattmeter; open circle. current connections to wattmeter.

INDEX